KU-385-630

FISHING BOATS OF THE WORLD 1

FISHING BOATS OF THE WORLD 1

Edited by

JAN-OLOF TRAUNG
Naval Architect, Fisheries Division, Food and Agriculture Organization of the United Nations, Rome, Italy

Published by

FISHING NEWS BOOKS LIMITED
FARNHAM, SURREY
ENGLAND

First published 1955
1st Reprint 1958
2nd Reprint 1966
3rd Reprint 1969
4th Reprint 1975
5th Reprint 1978

British Library CIP Data

International Fishing Boat Congress, Paris, etc. 1953

Fishing boats of the world.
1. Fishing boats – Congresses
I. Title II. Traung, Jan-Olof III. Food and Agriculture Organization
623.82′8 VM431

ISBN 0-85238-073-9

The views expressed in the papers and discussions are those of the contributors and not necessarily those of the Food and Agriculture Organization of the United Nations.

PRINTED IN GREAT BRITAIN BY THE WHITEFRIARS PRESS LTD
LONDON AND TONBRIDGE

CONTENTS

CONTENTS

CONTENTS

LIST OF CONTRIBUTORS

Page No.

Aguinaldo, James C. 363

c/o Kennard, Secretary of the Society of Naval Architects and Marine Engineers, Markey Machinery Company Inc., 85 Horton Street, Seattle, 4, Wash., U.S.A.

Akasaka, M. 424

Akasaka Iron Works Co. Ltd., No. 3, 6-Chome, Ginza, Chuo-ku, Tokyo, Japan.

Allan, J. F., D.Sc., M.I.N.A. . . 354, 369, 473

Superintendent, Ship Division, National Physical Laboratory, Teddington, Middlesex, England.

Allan, Robert F. 92, 360, 377

Naval Architect, 2026 West 63rd Avenue, Vancouver 14, B.C., Canada.

Andersen, Bent G. 469, 470, 473

Managing Director, Tuxham A/S, 122 Trekronergade, Copenhagen Valby, Denmark.

Andersen, R. G. 402

Chairman, Tuxham A/S, 122 Trekronergade, Copenhagen Valby, Denmark.

Astrup, N. Chr. 355

Scientific Officer, Skipsmodelltanken, Trondheim, Norway.

Bain, P. L. 233, 498, 512

Directeur Technique de la Société MacGregor-Comarain, Neuilly-sur-Seine, France.

Beever, C. 234, 267

Fisheries Economist, Fisheries Division, FAO, Rome, Italy.

Bergius, E. George 457

The Bergius Company Ltd., 254 Dobbies Loan, Glasgow C.4, Scotland.

Bertram, Captain H. 255, 379

Bundesministerium für Ernährung, Landwirtschaft und Forsten, Neuer Wall 72, Hamburg 36, Germany.

Bindloss, John B. 267

Atlantic Marine Fisheries, Stonington, Conn., U.S.A.

Boavida, Joaquim Gormicho 73

Fishery Research Investigator, Gabinete de Estudos de Pesca, Av. da Liberdade 211, 4°, D, Lisbon, Portugal.

Bochet, Pierre 463

Président Directeur Général, Moteurs Badouin, 37 rue Galilee, Paris 16, France.

Bordoli, Dr. Ing. Gian Guido . . 380, 385, 461

President, Rome Section of A.T.E.N.A., Ministry of Merchant Marine, Rome, Italy.

Page No.

Bradbury, C. H. 459, 463

Chief Engineer, Petters Ltd., Causeway Works, Staines, Middlesex, England.

Brard, Professor R. 353, 361

Ingénieur en Chef du Génie Maritime; Professeur à l'Ecole Polytechnique et à l'Ecole Nationale Supérieure du Génie Maritime; Directeur du Bassin d'Essai des Carenes de la Marine, 34 rue Rayouard, Paris 16e, France.

Bromfield, I. 374
464, 466, 471, 472, 508, 509, 510, 511, 562, 566, 568

President, Bromfield Manufacturing Co. Inc., 246-288 Border Street, East Boston 28, Mass., U.S.A.

Brownlow, D. L. 258, 409, 465

Director and Chief Engineer, Mirrlees, Bickerton and Day Ltd., Hazel Grove, Stockport, England

Bruce, Georg 460, 469

Engineer-in-Chief, Seffle Motorverkstad, Seffle, Sweden.

Carlson, Carl B. 476, 494

Chief, Gear Development and Research Program, Branch of Commercial Fisheries, U.S. Fish and Wildlife Service, c/o University of Miami, Marine Laboratory, Coral Gables, Flo., U.S.A.

Chapelle, Howard I. . 1, 34, 243, 247, 251, 257, 260, 262, 263, 266, 358, 360, 373, 378, 383, 387, 458, 467, 507, 511

Lt.-Col., U.S.A.R.C., Naval Architect, Cambridge, Maryland, U.S.A.

Chardome, Alexandre 449

Director of Beliard Crighton and Co., Chantier Naval, 27 rue Gérard David, Bruges, Belgium.

Christensen, Anders N. 238

Managing Director, Ancas Traeskibsbyggeri, Prinsensgate 2, Oslo, Norway.

Christensen, Poul A. 384, 464, 473

Mechanical Engineer, President, Burmeister and Wain's American Corporation, 17 Battery Place, New York 4, N.Y., U.S.A.

Conhagen, A. 509

President, Alfred Conhagen Inc., 429 West 17th Street, New York, N.Y., U.S.A.

Costantini, Ing. Mario 502, 545

Direttore del Cantiere San Marco, Trieste, Italy.

Coste, D. 461

Société Industrielle Générale de Mécanique Appliqués SIGMA, 61 Avenue Franklin D. Roosevelt, Paris 8e, France.

LIST OF CONTRIBUTORS

Page No.

Huet, G. O. . 260, 262, 267, 359, 383, 385, 386, 388, 507

Naval Architect and Engineer, Higgins Inc., P.O. Box 8001, New Orleans, La., U.S.A.

Huse *see* Vestre Huse

Hylton, P. H., Assoc. I.N.A. 462, 471

Marine Sales Manager, Ruston and Hornsby Ltd., Lincoln, England.

Jaeger, Prof. Ir. H. E., M.I.N.A. . 151, 246, 264, 367, 369, 472, 543, 563

Director of Research, Shipbuilding Department, Research Centre for Shipbuilding and Navigation; Department of Naval Architecture, Technical University, Delft, Netherlands.

Jul, Mogens 264, 265, 511, 551, 564, 568

Chief, Technology Branch, Fisheries Division, FAO, Rome, Italy; *now:* Director, Slagteriernes Laboratorium, Sdr. Ringvej 16, Roskilde, Denmark.

Kannt, Hans S. 166, 246

Dipl. Ing., Director, Werft Seebeck, Bremerhaven-G, Bachstrasse 6, Germany.

Kendall, Ronald, M.I.N.A. 568

Burness, Kendall and Partners Ltd., 15 St. Helen's Place, London, E.C.3, England.

Kesteven, G. L., D.Sc. 213

Chief, Biology Branch, Fisheries Division, FAO, Rome, Italy.

Kloess, H. K., M.I.N.A. 357, 375

Naval Architect, Maierform, Lothringerstrasse 65, Bremen, Germany.

Kolbeck, Dipl. Ing. Robert 444

Maschinenfabrik Augsburg Nürnberg A.G., Augsburg, Germany.

Kristjonsson, Hilmar 505, 507, 566

Fisheries Engineer, Technology Branch, Fisheries Division, FAO, Rome, Italy.

Kühl, Dr. H. 227

Bundesforschungsanstalt für Fischerei, Institut für Küsten-u. Binnenfischerei, Laboratorium Cuxhafen, Bei der Alten Lieve, 1, Cuxhaven (24a), Germany.

Kullenberg, Dr. B. 259

Oceanographic Institute, Göteborg V, Sweden.

Larsson, Karl Hugo 470

Naval Architect, Stadsgarden 10, Stockholm 10, Sweden.

Leach, T. S., M.C. 255, 457

Chief Inspector of Fisheries, Ministry of Agriculture and Fisheries, 3 Whitehall Place, London, S.W.1, England.

Lembke, Paul A. H. (VDI) 250

Ing.-Büro Schiffbau, Hamburg 19, Tornquiststrasse 1, Germany.

Lindblom, Jarl, A.M.I.N.A. . 248, 262, 263, 355

Naval Architect, Abo (Turku), Finland.

Lochridge, W. 565

Chr. Salvesen & Co., 29 Bernard Street, Leith, England.

MacCallum, W. A., B.E.(Mech.), M.Sc., M.F.I.C. 230, 263, 264, 265

Development Engineer, Fisheries Research Board of Canada, Atlantic Fisheries Experimental Station, Halifax, Nova Scotia, Canada.

McInnis, Walter J. 323

Naval Architect, Eldredge-McInnis Inc., 131 State Street Boston 9 Mass., U.S.A.

Magnusson, Henry 25

Naval Architect, FAO technical assistance programme expert, *now:* Prästvägen 15, Göteborg, Sweden.

Mardesic, Pavac 256

Naval Architect, Shipyard V, Kratulovic, Split, Yugoslavia.

Messiez-Poche, J. 456, 464, 467, 470

Ingénieur A et M, 15 Avenue Henry-Marcel, Maisons-Lafitte, (S-et-O.), France.

Miller, William C. . . 337, 381, 383, 384, 385, 386, 388, 510, 511

Marine and Aviation Surveyor, Marine Engineer, 577 Spreckels Building, San Diego 1, California, U.S.A.

Miller, William P. 251, 260, 266, 457

Chairman, Fishing Boat Builders' Association, Aberdeen; Jas. N. Miller and Sons Ltd., East Shore, St. Monance, Fife, Scotland.

Minot, Francis . 216, 257, 258, 259, 353, 369, 382

Naval Architect, Marine and Fisheries Research Institute Inc., Woods Hole, Mass., U.S.A.

Möckel, Captain Walter . . 326, 370, 371, 376

Naval Architect, Hamburgische Schiffbau Versuchsanstalt, Bramfelderstrasse 164, (24) Hamburg 33, Germany.

Mollekleiv, Olai 469, 507

A/S De Forenede Motorfabriker, Bergen, Norway.

Nickum, George C. . 254, 258, 263, 265, 266, 320, 357, 366, 368, 369, 371, 383, 384, 385, 388, 458, 506, 509, 513, 561, 566

Naval Architect, W. C. Nickum and Sons, 300 Polson Building, Seattle, Wash., U.S.A.

Normand, A. Augustin, Jnr., M.I.N.A. . . 428

Director, Les Chantiers et Ateliers Augustin Normand, Le Havre, France.

Oldershaw, C. G. P. 535

Mechanical Engineer, Fishery Technological Laboratory, Branch of Commercial Fisheries, Fish and Wildlife Service, U.S. Department of the Interior, Boston, Mass., U.S.A.

O'Meallain, Seamus 265

Civil Engineer, Chairman, An Bord Iascaigh Mhara, 67 Lower Mount Street, Dublin, Ireland.

Parkes, Basil 122

Managing Director, St. Andrew's Steam Fishing Company, Hull, England.

Parkes, Fred 245, 357, 370, 457

Chairman, The Boston Deep Sea Fishing and Ice Co. Ltd., Fleetwood, England.

LIST OF CONTRIBUTORS

Page No.

VANAARSEN, Walter H. 466

c/o General Motors Continental, Noorderlaan, Antwerp, Belgium; *now:* General Motors Overseas Operations, 1775 Broadway at 57th, New York 19, N.Y., U.S.A.

VAN AKEN, J. A. 438

Head of Propeller Design Department, Lips Propeller Works, Drunen, Netherlands.

VAN LAMMEREN, Prof. Dr. Ing., W. A. P., M.I.N.A. 269, 353, 371

Superintendent, Wageningen Model Basin, Wageningen, Netherlands.

VARRIALE, Ing. Leone 461, 464

Ansaldo, Genova-Sampierdarena, Italy.

VESTRE HUSE, Hans 488, 507, 509

Technical Director, Hydraulic A/S, Brattvaag, Norway.

VIBRANS, F. C., Jnr. 469, 472

Marine Engineer, Diesel Engineering Department, Fairbanks, Morse and Co., Beloit, Wis., U.S.A.

VON BRANDT, Oberregierungsrat Dr. A. . 474, 505

Direktor des Instituts für Netz- und Materialforschung, Bundesforschungsanstalt für Fischerei, Hamburg 36, Neuer Wall 72, Germany.

Page No.

WEGMANN, A. J. 382

President, Bagille's Seafood Co., New Orleans, La., U.S.A.

WHITAKER, James, B.Sc.Eng.(Hons.), A.M.I.Mech.E., A.M.I.Mar.E., A.M.I.P.E. . . . 433, 469

H. Widdop and Co. Ltd., Keighley, England.

WHITELEATHER, R. T. . 257, 258, 259, 372, 383, 385, 508, 509, 512

Assistant Chief, Branch of Commercial Fisheries, Fish and Wildlife Service, U.S. Department of the Interior, Washington 25, D.C., U.S.A.

YAGI, K. 503

Izui Iron Works Co. Ltd., Muroto-machi, Aki-gum, Kochi-ken, Japan.

ZIENER, Paul 393

Naval Architect; Lecturer FAO Latin American Fisheries Training Center, Santiago, Chile; *now* FAO technical assistance programme expert, c/o Fisheries Division, FAO, Rome, Italy.

ZIMMER, Hans K. 51, 253, 255, 470

Naval Architect; FAO Consultant; *now:* Jaegermyren 13, Bergen, Norway.

ZWOLSMAN, W. 154, 261, 460, 462

Naval Architect, Hollandse Motorboot N.V., 2 Ketelstraat, Amsterdam, Netherlands.

ABBREVIATIONS

Am	=	Area of midship section.
B	=	Beam in water line.
BHP	=	Brake horse power.
C_1	=	Admiralty constant.
D	=	Draft, also displacement.
EHP	=	Effective (tow rope) horse power.
G	=	Centre of gravity.
GM	=	Metacentric height.
GZ	=	Stability lever (arm).
g	=	Acceleration due to gravity.
IHP	=	Indicated horse power.
K	=	Keel.
KB	=	Distance from keel to centre of buoyancy.
KG	=	Distance from keel to centre of gravity.
KM	=	Distance from keel to metacentre.
L	=	Length in water line.
L.B.P.	=	Length between perpendiculars.
LCB	=	Location of the centre of buoyancy.
LCG	=	Location of the centre of gravity.
L.O.A.	=	Length over-all.
L.p.p.	=	Length between perpendiculars.
L.W.L.	=	Length in water line.
M	=	Metacentre.
n	=	Revolutions per minute.
P_e	=	Effective (tow rope) horse power.
r.p.m.	=	Revolutions per minute.
RT	=	Tons of refrigeration.
SHP	=	Shaft horse power.
T	=	Draft.
t	=	Period of roll, waves, etc.
V	=	Speed in knots.
v	=	Speed in m./sec.
α	=	Water plane coefficient.
$\frac{1}{2}\alpha E$	=	Half angle of entrance.
∇	=	Displacement in cu. m.
∇_2	=	Displacement in cu. ft.
Δ	=	Displacement, salt water, in metric tons.
Δ_2	=	Displacement, salt water, in long tons.
δ	=	Block coefficient.
φ	=	Prismatic coefficient.
⊙	=	Location of the centre of buoyancy.
⌧	=	Midship section.

Note

Distinction should be made between short ton (2,000 lb., 907.2 kg.), long ton (2,240 lb., 1,016 kg.) and metric ton (2,204.6 lb., 1,000 kg.). When not otherwise indicated, conversions have been made to metric ton. For practical reasons they can be regarded as long tons.

PREFACE

INTERNATIONAL organizations were created to perform tasks which can be accomplished better by joint action amongst nations than by any one of them alone. Among other things, FAO was created to bring action to bear upon the problems of achieving an equilibrium between an expanding world population on the one hand and, on the other, food production, adequate distribution and consumption.

The sea contains a renewable food resource which, though known and used by man from earliest times, has until recently been comparatively neglected and one, moreover, which holds the possibility of greater development. Because of this, it became the duty of FAO to pay attention to the fishing boat, the instrument concerned in primary exploitation of food resources of the sea.

The present book—an assembly of papers and discussions presented at FAO's International Fishing Boat Congress 1953 in Paris, France, and Miami, U.S.A.—furnishes an example of one of the functions of an international organization.

It might have been possible for any one of the member-nations of FAO to have arranged for and carried through this Congress but it is unlikely that any one would have done so. The benefit accrues to a whole community of nations and their nationals and it would not have been fair to impose the task on any one nation or to expect it to furnish the driving force. The existence of FAO made this unnecessary.

The basis for the Congress lay in the heretofore inarticulate feeling on the part of many naval architects, engineers and fishermen that there should be improvement in the design and the engineering of fishing vessels. Much effort had been devoted to design and experiment on large vessels but not much had been done for the smaller fishing vessels—except by a very few individuals, acting alone, who were scattered in various parts of the world. Yet more than fifty per cent. of the investment of the fishing industry is in its floating equipment.

This book is based upon the experience of workers in many parts of the world, on critical discussion, and the exchange of ideas. It is hoped that it will be useful not only to those who are in the forefront of practical achievement but also to the many educational institutions where young designers and engineers are preparing to assume responsible roles in the development of the expanding fishing industry.

The authors of the papers gave freely of their time. Their only compensation is that of knowing they have made a significant contribution to development. I feel sure that this will bring satisfaction more enduring than mere tangible reward.

The thanks of FAO are also due to the many people who worked for the success of this venture. It is possible to mention only a few of them. One to be named must be Commander A. C. Hardy, who first thought of organizing the Congress, who worked so hard to make it a success and who acted as president of the Paris session. Another is Mr. H. C. Hanson, president of the Miami session, who presented four papers, attended both sessions and was a most enthusiastic and helpful supporter.

Lastly, our thanks must be given to the Government of France (Ministère de la Marine Marchande) and the Government of the United States of America (Fish and Wildlife Service). By their generous invitations, they made possible the meetings in Paris and Miami and their officials did much to help organize and run the Congress successfully.

D. B. FINN

Rome, Italy, April, 1955

INTRODUCTION

THE fishing boats of the world of all types, shapes and sizes, represent the largest single collective investment in the world's fishing industry. Upon their efficiency of design and operation depends the economy of millions of people and the lives of an untold number of fishermen. This book contains the experience of specialists from most parts of the world in which fishing is carried out, and it is a symposium of to-day's knowledge of fishing boat design and construction.

In every part of the world designers attempt to produce the ideal fishing boat; opinions as to what constitutes the ideal, however, vary considerably, in technical as in other matters. In this book opinions are presented as they exist, and just as often as they occur. No attempt is made to talk down to the reader or to tell him what he ought to do.

One of the most important aspects of fishing boat design is the study of hull shape, and there is no doubt that many designs can be materially improved. For example, by a change in the distribution of displacement, the same " cubic " or cargo capacity will be retained and the ships will not be more expensive to build or operate. The result might be a better and faster boat in both calm and rough water. It will be noticed that the FAO experts stress the importance of the correct selection of prismatic coefficient; they are of the opinion that much remains to be done before there is general appreciation of the importance of a sharp bow and full midship-section, for it is really contrary to commonsense that a sharp forebody should produce a better sea-boat. But . . . the laws of naval architecture frequently seem to fly in the face of commonsense. The fact, for example, that less ballast produces a more sea-kindly hull, even if the stability proper decreases, is a case in point. One of the papers in this book describes an incident when a large wave pooped a trawler and flooded the foredeck. Had speed not been reduced on that occasion, but increased as commonsense would appear to indicate, the investigator would have been drowned and this valuable contribution to FISHING BOATS OF THE WORLD would not have been available.

Study of the chapters of the book make it very difficult to emphasize any particular contribution. Those from Japan are valuable by virtue of their detailed nature. For financial reasons it was impossible for a Japanese delegation to attend the meetings and to contribute to the discussion. Japan is one of the few countries in which the government has thoroughly appreciated the importance of fishing boat design. There is a large technical staff working on problems connected with it, and there is even a special tank for testing fishing vessel models. Fishing, of course, is a life blood industry for Japan.

The American contributions reveal a wealth of detail which is generally unknown and often unappreciated in Europe. They show a brutal honesty typical of the American " open door " policy which never fails to attract admiration and friendship among technicians in other parts of the world.

It is to the discussions generally that I would direct the attention of readers. They occupy a large part of the book and are really full of meat. They prove how much benefit can arise from an international meeting at its best. They prove how worth while it is to have sessions of a Congress in two parts of the world; a third and fourth session in the Far East and South America respectively would have yielded even better results.

It was said many times that the Congress marked a big step forward in fishing boat design, and this book which records it, is but a milestone on the way to further progress. Much remains to be done before we can say categorically that fishing boat design as a whole, embracing all types and sizes, has reached the same technical level as that for other types of ships. Much more experience remains to be correlated and interpreted and much more to be discovered. One paper clearly shows how much new knowledge can be obtained from actual measurements of the behaviour of trawlers at sea; such activities could be well extended to boats of other nations. And it was gratifying to note that several trawler owners were ready to put units from their fleets at the disposal of anyone interested in research. Who can pay for such activities? It does not seem possible that FAO can afford it, yet recent losses among the North European fishing fleet speak eloquently of the need of such research, and the discussion about American boats in the Safety at Sea chapter confirms it.

The speed of technical development which has so affected ordinary ships, whether passenger or cargo liners, or tramps, has not passed by the fishing vessel, whether she be a large Grand Banks schooner or a small crude in-shore vessel operating, say, from relatively isolated coastal spots, or from the shores of one of the under-developed countries which it is part of FAO's task to assist.

While it is true that increasing mechanization means increasing first cost, yet mechanization must take place if efficient fishing is to be carried out, and this can only be made effective on a rational basis if there is a full and free interchange of information about the direction in which design is proceeding, and potential processes for hull construction. For example, it is possible that, within the very near future, a complete range of standard, quantity-produced, plastic hulls will be available, suitable for some of the smaller classes of fishing vessels. A drifter trawler with this kind of construction is already well past the design stage, and its logical propulsive unit is a standardized high-speed oil engine. Study of the engineering section will indicate that, in general, fishermen in European countries are in favour of the slow, direct-coupled, direct-reversing oil engine; whilst in the United States there is a ready acceptance of reduction or reduction-reverse gears for ships of even quite respectable size. It is true that in the continental countries of Europe a great deal has been done towards the development of trawler drive with gear boxes and also with the father-and-son method. But it is evident that even in such a common fishing type as the trawler there are considerable variations of opinion. A bridge could and should be thrown between the different techniques, which are merely opposite ends of an opinion about the same subject.

This book is very modern in the studies which it presents on fishing boat propulsion. There is even a paper on the free piston gas generator, gasifier and turbine, a type of drive which has already been applied with success in small warships and in two coasters. This may be a pointer to the future. We do not take you in this book as far as atomic propulsion, nor indeed would we dare to do so because most people would agree that, already, the personnel problems in fishing vessels are a sufficient headache. Nevertheless, on a marine engineering note, it is essential to recall that the first atomic driven vessel—a submarine—has already made its trials, and we do not know where atomic propulsion will take us within the next quarter of a century. If atomic power can be harnessed to some kind of jet propulsion then it might solve many problems as far as large trawlers are concerned, and it might also meet the increased demands for electricity. The use of electric fields to attract fish is already being investigated but practical solutions have not yet been worked out to overcome the high power requirements in salt water. Atomic power might provide part of the answer to this problem.

Great credit is due to FAO for having arranged the Congress and for the large amount of work which members of the Fisheries Division have put into the production of this book. Let us hope that FAO will realize that what has been done so far is no more than a beginning and that the next step is to arrange a second Congress. Thus we shall reach the second milestone. After all, world fishing is world feeding.

A. C. Hardy

London, March 1955

SOME AMERICAN FISHING LAUNCHES

by

HOWARD I. CHAPELLE

SMALL craft, under 45 ft. (13.7 m.) in length, represent the largest single investment and the largest part of the fishing fleets of Canada and the United States but there has been little technical information published concerning their hull design, construction and powering. They provide, however, an interesting study of the effects of economic factors on commercial boat design.

The launches have been developed by trial and error to meet the demands of local conditions which are so rigid that boats must be efficient to survive. As a type, they generally represent a very high level of design, but they can be usefully improved provided that any improvement is based on a working knowledge of the local economic and physical factors. For example, introduction of a larger, faster and more highly mechanized boat will not constitute improvement unless it can produce greater annual income and more financial security to the owner, or increase his personal safety. Too often, attempts to improve a local fishing boat have resulted only in an uneconomic increase in the costs of building, equipping, operating and maintaining it.

Small fishing launches do not lend themselves to the usual mode of academic analysis by mathematical comparisons because there is a great range in the hull-forms, proportions and, in fact, all elements, and of course the lack of accurate plans creates an almost insurmountable difficulty. The standards used in large vessel design do not apply to North American small power craft because the prismatic coefficient is commonly well below the lower limits used in large vessels. Model-testing (and the accumulation of the necessary data) is far too costly for any individual and there is, at present, no organization or governmental body in North America with the necessary funds, or the necessary information on the models that ought to be tested, to undertake investigations on an effective scale. The small naval craft that have been tested in model basins are not of the hullforms nor of the weight-power-speed ratios applicable to small fishing and commercial craft. Presentation of the results of model-testing, or of modes of calculation based upon scientific investigation, must be in compact and simple form to be of real value because the boat-owner cannot afford to pay the small boat designer for extensive calculations or research.

As the result of such limitations there is much use of simple approximations and " short-cuts ". The margin between successful design and failure is apparently much narrower in small boats than in large vessels, so that slight departures in weight and trim become of great importance. The recommended range in a coefficient, for example, may not apply very well to the design of small launches.

There is very little steel construction of small fishing launches in North America and little likelihood that there will be in the near future because the available facilities for building them are not suitable. But there are plenty of boat-yards to construct and maintain small wooden vessels and, where the fisheries will not support a local yard, fishermen build their own boats.

North American fishing launches have developed almost entirely from two sources: (1) the old local sailing boat which has been adapted to motor propulsion. Some of these are now being replaced because they are not fast enough for fishermen today; (2) the high-speed pleasure launch. Fishing launches developed by such craft are popular—and gaining in numbers—because they are very fast, but some are unnecessarily expensive and even unsafe in heavy weather.

THE GASPÉ BOAT

An example of a powered fishing boat developed entirely by slight modifications is an older sailing model, the Gaspé boat, fig. 1. This craft is used on the northern tip of the Gaspé Peninsula, Quebec, Canada, known as the English shore. The popular size is about 35 ft. (10.7 m.) in length and between 9 and 10 ft. (2.75 to 3.05 m.) in beam, with a draft of about 3½ ft. (1.07 m.). The hulls are roughly built, without any attempt at a smooth finish, but are strong and lasting. The planking is either lapstrake (clench) or caravel; the former seems to be the most popular. The boats carry a simple schooner rig —jib, loose-footed foresail and boomed mainsail of the gaff type—and are powered with a one or two cylinder heavy duty gasoline engine manufactured in Nova

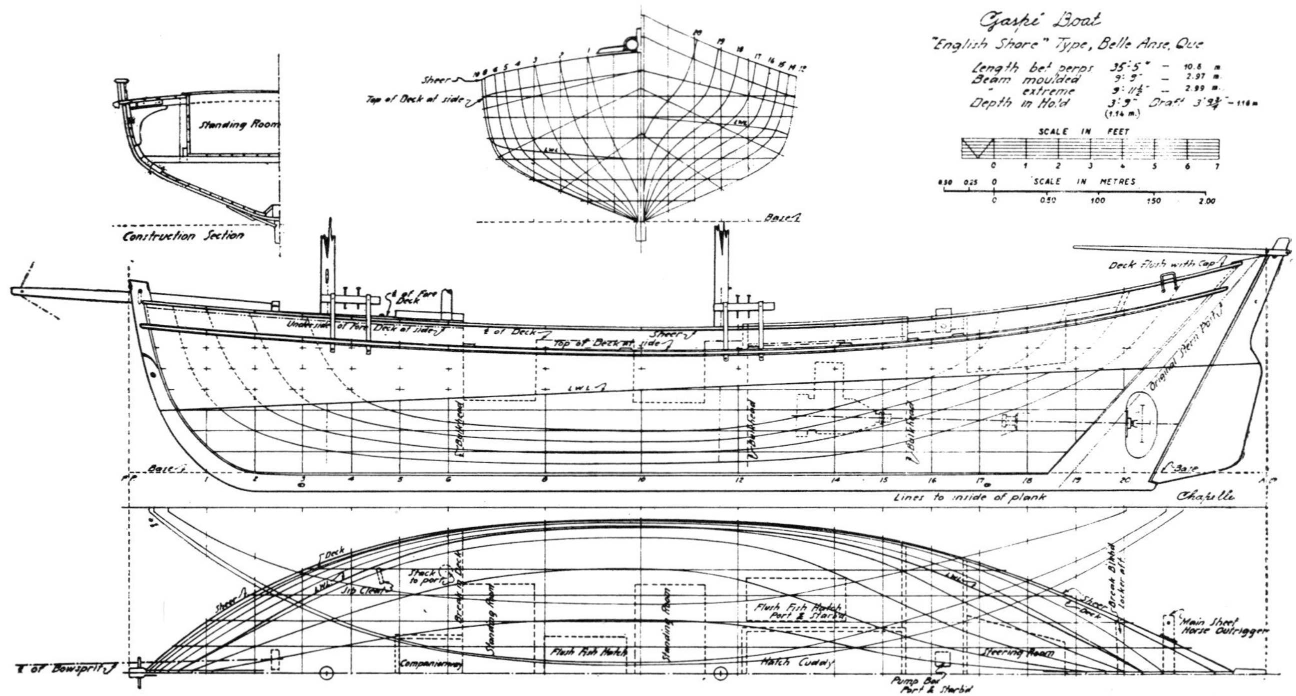

Fig. 1

Scotia. Most have make-and-break ignition, are manually started, and give a speed of 5 to 7 knots. The boats usually stay at sea only for two or three days and have a record of safety in this area so subject to severe gales. As they commonly work out of semi-exposed coves or small harbours they are well designed for beaching. The usual process is to discharge the stone ballast and haul the boat up on a grid made of two long spruce logs laid at right angles to the shore, over which spruce poles closely spaced are spiked, or pinned, with wooden tree-nails. An iron bolt is passed through the hole, bow or stern, in the keel and a hauling line is secured to the ends of the bolt or pin. A truck is often used to do the hauling but in some places the old capstan or crab is still in use. No cradle or other support is used and the operation is rapid. The spruce-hole grids are still used in some ports where the Government has furnished breakwaters.

The similar boats used by the French Canadian fishermen on the Bay Chaleur side of the Gaspé Peninsula have more rake to the ends and their midsections show some hollow at the garboards, combined with a rather low and hard bilge. The topsides flare a good deal. They range from 32 to 45 ft. (9.75 to 13.7 m.) in length, and work from harbours fitted with breakwaters and are not often hauled out on grids. They are lap-strake planked, have engines similar to those used in the boats on the English Shore, are worked in the same manner and do long-line and some net fishing.

Nearly all are designed as sailing craft but have been converted to power by adding a wide sternpost to the original and cutting an aperture as shown in fig. 1. Power boats have been built on this general model although they retain a small schooner rig. All the boats are inexpensively built. The steam-bent frames are often saplings run twice through the saw so that they have flitch-edges, and the hulls, built of locally grown wood (spruce, birch, larch), are iron-fastened. No floor timbers are used, the heels of the frames being nailed to the top of the keel where they butt at the centreline of the hull. There are usually two berths and a small stove in the forecastle. The foresail has no boom, lug-fashion, and the rigging is of the simplest nature but the boats are most seaworthy and safe.

Gaspé boats are economical to operate, and adequate for their work. An improvement in design would be to lengthen the run to obtain greater speed under power. A lighter engine of greater power would also be an advantage. Retention of the sailing rig, for the present, seems desirable as the cost of fuel is high.

The Gaspé boat, with these modifications, would be well suited for fisheries in primitive areas, such as the coast of northern Labrador, in which case the possible installation of a lightweight air-cooled manual-starting diesel of, say, 15 h.p., might be examined. The engine used now is simple and reliable but is heavy and has high fuel consumption. The installation is often poorly done and there is sometimes danger of gasoline explosion.

Under the auspices of the Canadian Government, through a loan system, some Gaspé fishermen are obtaining diesel-powered long-liners 55 ft. (16.8 m.) long (built on the Nova Scotian Cape Island model), having a speed of between 10 and 12 knots. The boats are capable of making more trips during the season and fishing at greater distances from the home port. But there are scanty facilities for hauling and repairing craft of the long-liner size. The larger capital investment required, and the greater cost of operation and maintenance may make such craft unprofitable, especially if the price of fish drops.

The Provincial Government has built some fish freezing and storing plants on the Gaspé Peninsula, a most important first step in the improvement of undeveloped or primitive areas. The improvement of transportation methods is next and then should come the efforts to improve the existing type of fishing boat. Caution should be exercised in introducing new and larger or more highly mechanized craft, for their capital and running costs might cancel all advantages and a fall in fish prices might be dangerous to the economy of the area.

THE CAPE ISLAND BOAT

The Cape Island boat is the most popular fishing launch in Nova Scotia and New Brunswick, Canada, and in eastern New England, U.S.A., where it is called the Jonesport Boat. The date and place of origin of the type are uncertain but it was clearly copied from an early form of high-speed motorboat. It has a very long, fine entrance and an equally long and flat run, with the greatest beam of the loadwaterline well abaft midships. On the north coast of New Brunswick, it has a canoe or cruiser stern and is very flat underneath. Elsewhere, it is square-sterned and the wide, flat transom is either plumb or rakes forward slightly at the top. The sternpost is well under the boat and is planked up, schooner-fashion, though the New England boats often have skegs, without much departure otherwise in the lines. In Nova Scotia the boats have their forefoot cut away, fig. 2, to allow them to run upon a steep shore, end on, and unload where no wharf exists and there is no surge on the beach. The basic lines of the Cape Island model are shown in fig. 2. They are commonly between 35 and 45 ft. (10.7 to 13.7 m.) long, 10 to 12 ft. (3.05 to 3.66 m.) beam, and draw light 2¼ to 4 ft. (0.69 to 1.22 m.). Some heavily powered boats are reported to have speeds up to 17 to 18 knots per hour in smooth water.

The boats are light, the frames being of ash or oak, wide and thin in cross-section and very closely spaced. The planking is white pine or white cedar and is rarely more than 1¼ in. (32 mm.) thick before finishing. The keels are rectangular in cross-section and rabbeted. At the sternpost there is often weakness; the turn of the tuck becomes so quick that a frame cannot be bent sharply enough, so shaped chocks are set up on the keel and the frames let into their tops. The horn timber is often merely bolted on top of the wide, short sternpost. In some boats, to strengthen the stern, there is a sort of A-frame laid flat on the frames, with its apex kneed to the transom. The heels of the arms run well forward of the sternpost, on the flat of the run outboard, and are fastened to each frame, with a cross-piece laid over the arms and drifted to them and to the top of the horn timber, directly over the short sternpost. Another construction employs a sternpost with deadwood inboard of it, to which the horn timber is drifted in the same manner as the horn timber of a wooden fishing schooner is fastened.

Forward is a low trunk, or a raised fore deck with a trunk mounted on it. Here is the cuddy which contains two to four berths, a galley, a provision locker and, in some large boats, a toilet. Abaft the cuddy is the steersman's position, sometimes semi-enclosed. Many boats have engines farther forward than in the example. They usually have a light, high-speed car motor, but some have marine engines. The motor is in an engine box in the

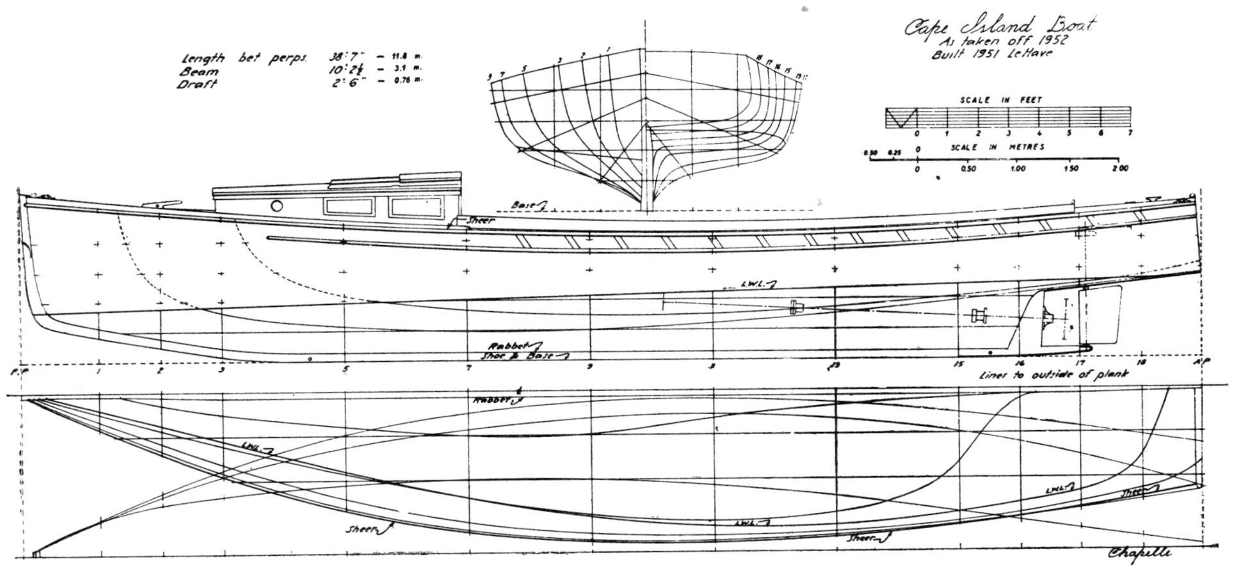

Fig. 2

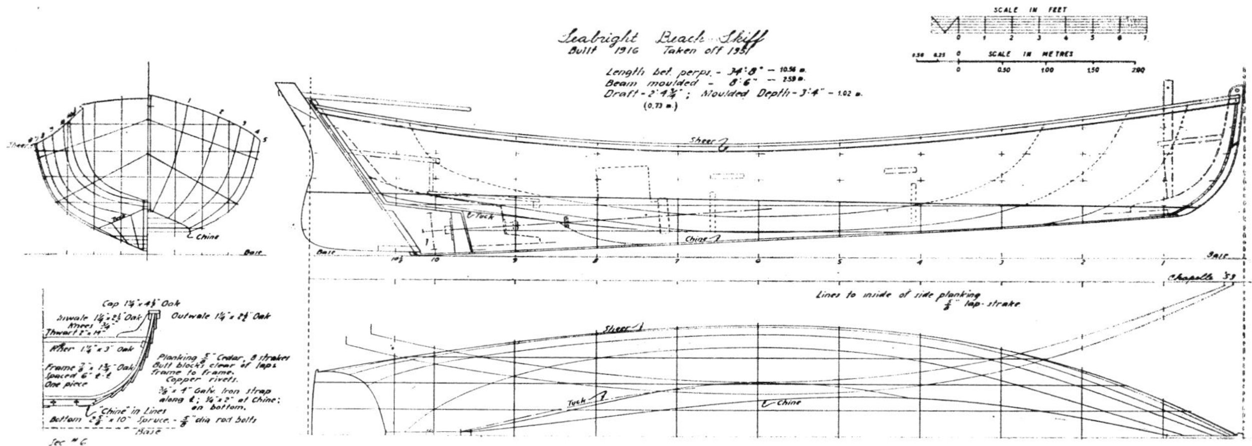

Fig. 3

cockpit with the battery. Fuel tanks are placed at either end of the hull. The cockpit often extends to the extreme stern, or there is a short stern deck as in the example, which is unusual in having a cockpit coaming, whereas most boats have only a narrow covering-board along the sides. The cockpit sole and ceiling are often caulked, and the sides are usually strengthened by a cross-timber laid athwartship over the keel, and knees are secured to the ends of this against the sides. In fig. 2 the knees were located near Station 13.

The Cape Island boat and its counterpart in New England, the Jonesport boat, have replaced the older fishing launches which were usually modifications of old sailing boats, slow under power but safe and seaworthy. The high speed of the Cape Island model is a great attraction to fishermen and has led to the introduction of a 55 ft. (16.8 m.) standard design of long-liner which can go 60 and 70 miles to sea. Boats of this model have been built to 57 ft. (17.4 m.) length.

Hardly a season passes in which there are not fatalities in this class of fishing launch although the men using them are among the most skilled North American boatmen. Inquiry has led to the conclusion that accidents are due to the hull design, the large cockpit and, in a few instances, to structural failure. The faults are obvious. The long sharp bow, combined with a broad, flat stern and shallow body, produces a hull that will broach on a heavy following sea. There is no bearing forward and, if the wide stern is thrown out of water, capsizing results. For instance, on the Nova Scotian South Shore two years ago a boat was capsized in tide rip, and another boat, returning from a sport fishing trip, approached a pier during a heavy rain and suddenly capsized when the passengers crowded forward.

Knowing the weaknesses of the model, the fishermen take care to load the boat aft, but this may produce a danger due to the cockpit. Low freeboard being a desirable feature in a fishing launch, the boats, when loaded, are very low aft and in a following sea they are easily swamped. In a gale the boats must heave-to under motor and, perhaps, a small riding sail set on a short mast placed well aft. If this is not done soon enough, disaster may occur.

The long stern abaft the sternpost is not very strong so no load is placed there. In any case it is liable to pound under certain conditions and is a source of weakness. The builders and fishermen recognize the faults of the boat far more than do some who have praised it in print, but there has been no effort made to produce a safer launch for open sea fishing. The present model would have to be altered radically to produce a really safe launch and fishermen would object to any loss of speed. The cruiser stern used in New Brunswick is an improvement over the square-sterned model, but it is also too fine forward, and increasing the bearing forward would result in a slower hull.

A completely new form of launch, fast yet seaworthy, is wanted, and the only one of this description now in use is

THE SEABRIGHT SKIFF

This is a development of an old sailing beach boat which is now in use off the New Jersey, U.S.A., coast. The skiffs range from 20 to 40 ft. (6.1 to 12.2 m.) in length and, when heavily powered, can run at speeds up to 21 knots. The boats working off the beaches were usually completely open, with their engine housed in a box aft of amidships. The boat shown in fig. 3 was built for beach work and had a four-cylinder automobile engine of an obsolete make that gave her a running speed of about 11 knots when light.

Skiffs are lap strake planked, with steam-bent frames. The bottom is flat and narrow, made of three to five plank of spruce or yellow pine cleated together between the bent frames with oak battens. In the example these were $1\frac{1}{4} \times 3$ in. (32×76 mm.) oak, not shown in the plan. They are lightly but strongly built. The peculiar form of the stern, with its reverse chine in the tuck, was used in the early sailing model and has been retained because it produces a steady-steering boat. Another advantage is that the boat can be pumped by the helmsman, the skiffs being often operated by one man.

Older power skiffs had the short sternpost at the transom, as in the sailing model, but the sternpost has been moved forward to produce greater speed. However, the projection of the transom abaft the post is not great and the horn timber and transom knee are one in most boats. The post is often a massive cedar knee. The boat shown in fig. 3 has some rocker in the bottom, but in many skiffs there is none. Large skiffs are often decked and have trunk cabins. During the prohibition period in the United States, the skiff became very popular with liquor smugglers and many very large, fast and capacious ones were built.

The Seabright skiff is less expensive to build than the Cape Island model, is just as fast or faster for a given power, and is far safer. Proportions of beam and depth to length need not be those shown in the example; they are often proportionately wider and deeper and the boats are very lively in a sea, the price paid for their otherwise fine qualities.

THE SHARPIE LAUNCH

In the United States there are a number of flat-bottom launches used in shoal and relatively protected waters, such as rivers, lakes and bays. The Sharpie is one, and is used on Long Island Sound and the rivers emptying into it, the Mississippi river, and on the Chesapeake, particularly in the crab fisheries. There is much variety of the model but, in the main, construction is standard. Fig. 4 represents a small Chesapeake Sharpie launch not much different from that used on Long Island Sound. It is like a flat-bottom rowing skiff except that the transom is slightly immersed and the fore-and-aft rocker of the bottom is slight. The beam is usually small on the bottom, the flare giving the appearance of greater width than is actually the case. It is one of the least expensive launches and is suitable for fishing shallow, semi-protected waters. Experience has shown that a wide beam is undesirable if speed is wanted and, as the speed is increased, the amount of rocker must be decreased to obtain the best results. It also appears important to make the chine profile straight, toward the ends of the hull, as in the example shown here. The models in use are up to about 50 ft. (15.3 m.) in length. They venture to sea in good weather but they are not good boats in open water as they pound heavily in a rough sea. But in sheltered waters they are superior boats and most economical to build.

Sharpies are usually powered with one- or two-cylinder marine engines, 3 to 9 h.p., which give them speeds ranging from 5 to 7 knots. With an automobile engine, large skiffs will run at speeds in excess of 17 knots in smooth water.

The usual construction is that employed in a rowing skiff. The sides are bent around moulds placed bottom up. A bent plank-keelson is laid down and the bottom plank is laid on athwartship. The stems are usually in two parts, an inner piece, or liner, to which the sides are nailed and a cutwater added afterwards, to cover the ends of the side planks. The side frames are no more than cleats set on edge, wide at the head to support the side deck, no deckbeams being required along the run of the cockpit coaming. Stiffness in the open hull is obtained by the use of bulkheads, one at bow and one at the stern, with a deep floor formed of a wide plank on edge amidships or, as an alternative, one or more thwarts. Large boats have a cross floor at each end of the engine bearers, with the ends kneed to the sides. The construction is plainly shown in fig. 4.

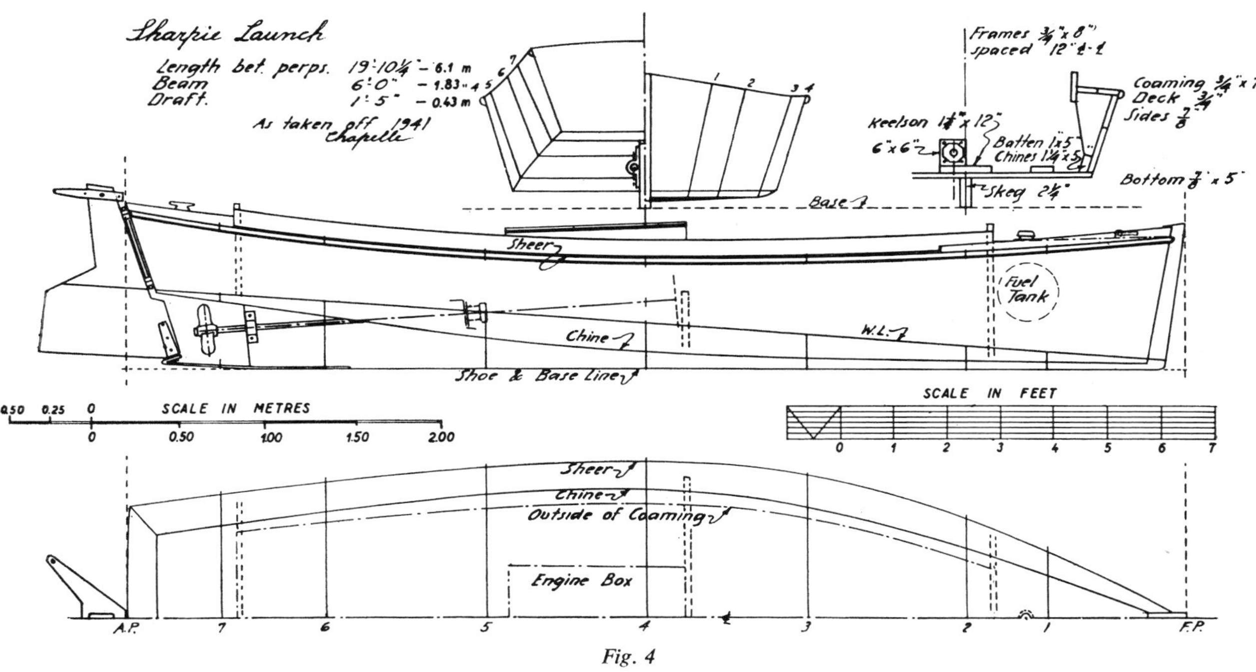

Fig. 4

As building costs must be kept down it is common to have the propeller shaft alongside the skeg rather than passed through it. The side position not only avoids the labour of boring the shaft hole but also simplifies the aligning of the shaft and its installation. The stern bearing is through-bolted to the side of the skeg and usually has a rubber bushing, water-lubricated. The stuffing box is inside the hull, mounted on a block. This places the shaft and engine slightly off the centre line of the hull and it is usual to put the shaft on the side opposite to which the propeller turns. The effect of the off-centre position of the engine on transverse trim cannot be observed, the weight of the battery probably being used to counteract such influence in the narrow skiffs.

engines of the industrial type and these are now being used in launches under 25 ft. (7.6 m.) in length.

THE SCOW OR GARVEY

Another variety of flat-bottom fishing launch is the Scow or Garvey which is extensively used in United States, in southern New Jersey, along the Atlantic coast of Maryland and Virginia and, to a slight extent, in the lower end of the Chesapeake Bay. The Garvey was long popular in southern New Jersey as a sailing fishing boat for use on bays and streams. When engines of low cost became available, the sailing Garvey was converted to a motor boat by deepening the stern to give a flatter run. The result was an inexpensive and useful launch

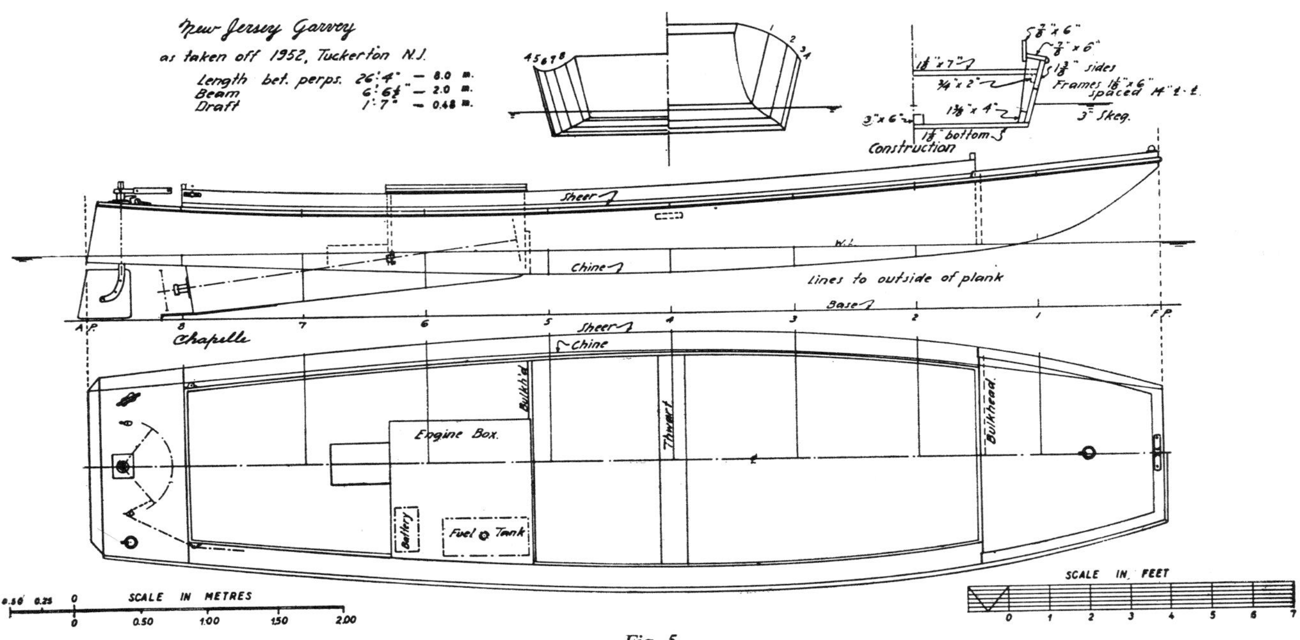

Fig. 5

Some launches work in very grassy waters and, to prevent weed from fouling the propeller and shaft, are fitted with "weedless" propellers, the shaft alongside the skeg, from bottom to stern bearing and the propeller hub, being enclosed in loosely-fitting steam hose. The hose revolves with the shaft until it gathers weed, then it will stop and the weed will be torn away by the wake. During this operation the hose revolves in the opposite direction of the shaft. The launches used in the crab fisheries usually have the skeg carried forward as an outside keel to the stem. This enables the launch to run unattended in a straight line at low speed, in spite of a moderate side wind, which is necessary in trot-line fishing for crabs, as practiced on the Chesapeake. The example shown has an unusually shallow keel; in most launches the keel is nearly 6 in. (153 mm.) deep at its shallowest part.

Skiffs can be driven by small air-cooled gasoline

of greater carrying capacity than the Sharpie and, perhaps, less costly to build. Fig. 5 shows an example of a fishing Garvey from New Jersey. The construction is the same as in a Sharpie, except for the stem. The boats use a variety of engines, ranging from one-cylinder marine motors to high-speed automobile engines and the Garvey, with a proper chine profile and enough power, is a very fast launch in smooth water. Although not suited for use in open water, it is not uncommon for them to be used alongshore outside the Maryland beaches. When designed for the purpose they can be beached in moderate weather.

The skeg is often not bored for the propeller shaft and tunnel sterns are much used. In a fast Garvey the chine profile is nearly straight from the end of the fore rake to the transom and, when the boat is at rest, it is nearly parallel to the waterline. The boat shown in fig. 5 had an old four-cylinder automobile engine and ran about

9 knots with the throttle about two-thirds open. Even the larger boats, which are about 35 ft. (10.7 m.) in length rarely draw more than 2 ft. (0.6 m.) at the skeg. The bottom is commonly planked athwartship, but a few boats are framed out and have the bottom plank run lengthwise. As far as can be determined, the cross-planking does not affect the speed and is the strongest and by far the most inexpensive mode of construction. Large Garveys are fitted with cabin trunks and steering shelters. In New Jersey and Maryland the launches are used in oyster and clam fisheries, as offal boats and for tending the fish traps.

BEACH LANDING SCOW OR GARVEY FOR TRAPFISHING

Fig. 6 shows a motor Scow or Garvey used for beaching, a standard boat at the lower end of the Chesapeake, near Hampton Roads. It is used to tend fish traps and does not run great distances. Trap boats must be able to land in quite a heavy surf as the beaches from which they operate are exposed to the full sweep of the wind for almost the entire length of the bay so they are very heavily built with much flare and sheer. The most curious feature is the crude retractable propeller shaft. The lift rod of the shaft serves as the strut and has a pin through it—above the pipe flange in the keelson—which not only serves as a stop but also ensures realignment when the shaft is lowered into operating position. Fig. 6 should be sufficiently complete to show how it is arranged and operated. Obviously the arrangement would be unsatisfactory at great speed but the boats have small motors and run between 5 and 6 knots. Single-cylinder marine engines of 3 to 5 h.p. seem to be common. The centre " dividing board " has extensions that can be fitted to form almost a complete bulkhead to the height of the deck at sides. The boats seem to be almost standard in dimensions and fittings, although the position of the bulkheads varies with the make and size of engine installed. Some boats have been fitted with a thrust bearing between the stuffing box and the engine to allow a fabric type flexible coupling to be used, presumably to make a smoother running boat. No stern bearing was visible on any boat inspected, though one owner stated he had a flax packing inside the plate on the skeg. The bearing on the lift rod was bronze-bushed.

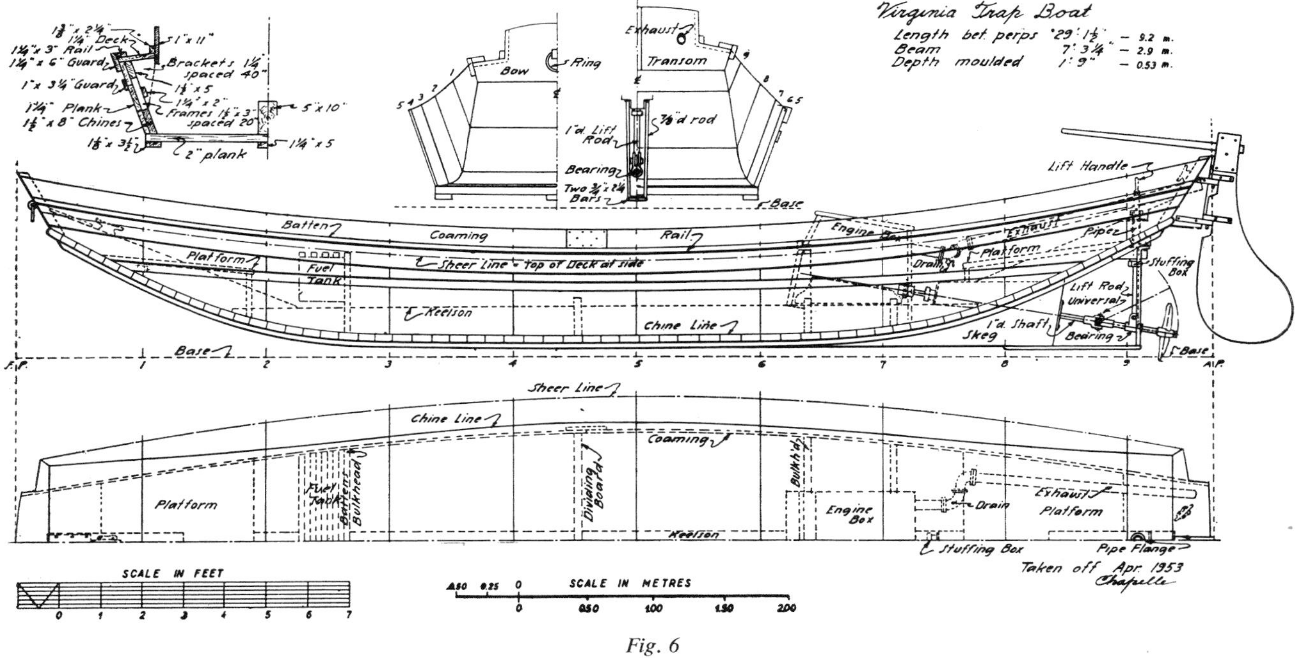

Fig. 6

USE OF V-BOTTOMS

The power Garvey has spread rapidly along the coast of Maryland and has reached into Virginia at Chincoteague Island where the boats are sometimes built with a V-bottom of very moderate deadrise, but the most common design has a slight V, formed only in the fore rake near the waterline. The bottom is covered with thick plank for a couple of feet each way from the fore-end of the waterline, then a shallow V is dubbed into the plank. A popular size is between 28 and 30 ft. (8.55 to 9.15 m.) length with a beam of 6 to 7½ ft. (1.83 to 2.3 m.). Tunnel sterns are becoming common. The usual arrangement is to have a steering shelter aft, with the engine box just forward, and an open cockpit from engine box to the fore rake, where a short fore deck is placed. The launches are cheaper and easier to build than the V-bottom types now in use in that area.

One of the reasons for the adoption of the V-bottom

along the Atlantic coast is its use in pleasure boats, which brings it to the attention of fishermen in areas where, otherwise, it would remain unknown.

There is probably no model in which the economic factor is more generally misunderstood than that of the V-bottom. It is usual to assume that the use of straight lines in the body plan of a V-bottom wooden boat indicates an inexpensive hull to build. This is incorrect. The V-bottom is economical to build only when the bottom plank is laid at an angle to the keel, as in the Chesapeake Bay Sharpie. In the V-bottom, the frame must be made up of five pieces, the floor timber, two-bottom futtocks and the topside timbers, and these must usually be kneed or bracketed at the chines, so they require more labour to form than does a steam-bent frame of three members. A framed V-bottom is about as costly to form as a sawn-frame round-bottom boat, in material and labour.

It is sometimes stated that "cross" planking adds to the resistance in driving the hull, which may be true in theory but not in practice. It is so insignificant that it can be overcome with slight variation in engine revolutions as experiments in two launches showed.

Another advantage of the Chesapeake Bay mode of construction is that it gives greater freedom in the shape of the chines. The problems of design in the use of the V-bottom at moderate speed-lengths ratios have not been explored in model-testing, but in some tests the chines, in profile, have been carried high forward as in the high-speed planing model of V-bottom. There are practical reasons for suspecting that a different chine profile forward is required in the lower speed-length ratios, and that the high chines usually cause full bow lines and, therefore, too great an angle of attack forward. This is particularly the case when the cross-sections forward are made up of straight, or nearly straight, lines from rabbet to chines. It seems reasonable to hope that model testing of a variety of these hull-forms might show that a V-bottom could be designed to drive as easily and efficiently at moderate speeds as a round bottom, and a number of experiments with full-size launches of comparable designs have been undertaken which seem to indicate this to be so. It is obvious, of course, that minute differences indicated in model tests would be of no practical importance in full-sized craft, where slight departures in lines or in the efficiency of the engines, or propellers, might make a far greater difference in observable performance.

One reason why V-bottom boats intended for relative low speed-length ratios have had high chines forward is that it is easier to plank when the bottom is laid lengthwise the hull. With a low chine forward and lengthwise planking there is often an excessive twist in the fore ends of the lower plank and the upper planking of the bottom runs off to feather-edges on the chines.

There are many variations of the model in use on the Chesapeake and to the southward, but the basic problems of V-bottom design can be shown by the use of three examples. Fig. 7 shows the lines of a V-bottom fishing launch from lower Chesapeake Bay. This boat is of moderate speed and must be seaworthy enough to withstand the vicious, short steep sea of the lower Bay. It is built almost entirely of southern pine, oak being used only for a rubbing shoe on the keel and for the inner stem, outer outwater, sternpost and shoe on the keel, and for the inner stem, outer outwater, sternpost and the topside frames. The boat has very low chines forward so that, when loaded, the chines show only at the extreme stern. The sides increase in flare sharply at the bow to their maximum and this continues with no real change to the stern. The rise of bottom is moderate and carried well aft by means of a reverse curve in the rabbet as the stern is approached, becoming flat over the propeller, across from chine to chine, which is thought to prevent squatting at full speed. The favoured stern is round, but boats are built with square transoms. The arrangement shown in fig. 7 is for a combination launch that can be used for crabbing, oystering and fishing and for light freighting as well.

Almost the same in form, but with greater rise to the bottom, are the noted Hatteras boats used for fishing the notoriously rough water area. Examination of the boats show that they are usually sharper forward, having a cross-section a few feet abaft the bow that is a straight-sided V from rabbet to sheer, and the chine is faired into this. The stern is formed much like that of the boat shown in fig. 7 but with a slight reverse in the rabbet. The chines are submerged and the forebody, viewed from ahead, appears noticeably hollow at the load waterline, due to the V-section just mentioned. The boats are usually planked lengthwise on the bottom and, powered by car engines, have a top speed of about 10½ knots.

Boats of the model shown in fig. 7 run very cleanly at 7 to 9 knots but begin to settle aft if driven faster. It is not uncommon to find these boats much overpowered and driven too fast, and to overcome the resulting "squatting" or settling aft, "squatboards" have been added. These are flat wooden fins or planes placed at and nearly parallel to the waterline under and abaft the stern, to hold the stern up when the boat is driven hard. The fins are usually braced by a pair of iron rods from the quarters, or may be chocked to the bottom, with their fore edges faired into the bottom of the boat over the propeller.

HOOPER ISLAND LAUNCHES

Fig. 8 shows the lines and much construction details of a popular style of crabbing and oyster-tonging launch found in the middle portion of Chesapeake Bay. It is an "old style" launch which is no longer being built but is still quite prevalent. It first appeared about 1905 and was a copy of a local racing motorboat, the half-model of which is now in the National Museum at Washington, D.C. This racer, in turn, was apparently inspired by a design for the Dolphin, published in the American yachting magazine *Rudder* about 1903. The

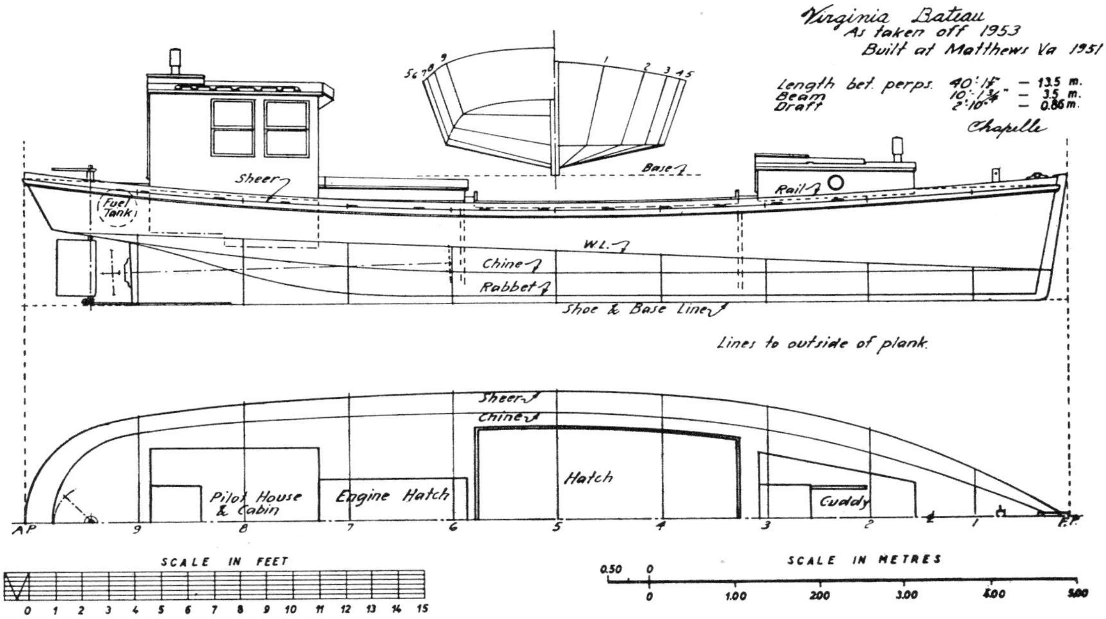

Fig. 7

Dolphin and the Chesapeake Bay racer, as well as the subsequent fishing launches, were modelled with dead straight lines for the chines in profile and it was intended that the chines and the line of flotation should coincide. The hull-forms were basically the same, the " double-wedge ", in which the greatest draft at the rabbet was at or very near the heel of the stem, from which the rabbet ran up fair and finally straight to the waterline at the transom. In plan view, the boat was all bow, the greatest beam being at the transom or nearly that far aft. From this extreme form, the model used was gradually changed until it developed into the form illustrated in fig. 8. This style was usually narrow, with a beam of about one-fifth the overall length, or even a little less. The greatest depth of the rabbet was finally about one-fifth the overall length from the stem and the greatest beam about seven-twelfths the length from the stem at the chine but slightly forward of amidships at sheer. These boats, curiously enough, were found very satisfactory in the short steep sea of the Bay and were also useful in cabbing with a trot-line, as they ran straight unattended. The peculiar stern was favoured for many years by oyster tongers, as a man could work over the stern with safety without the quarters of the boat rocking the tongs.

The narrow beam remained popular as long as the

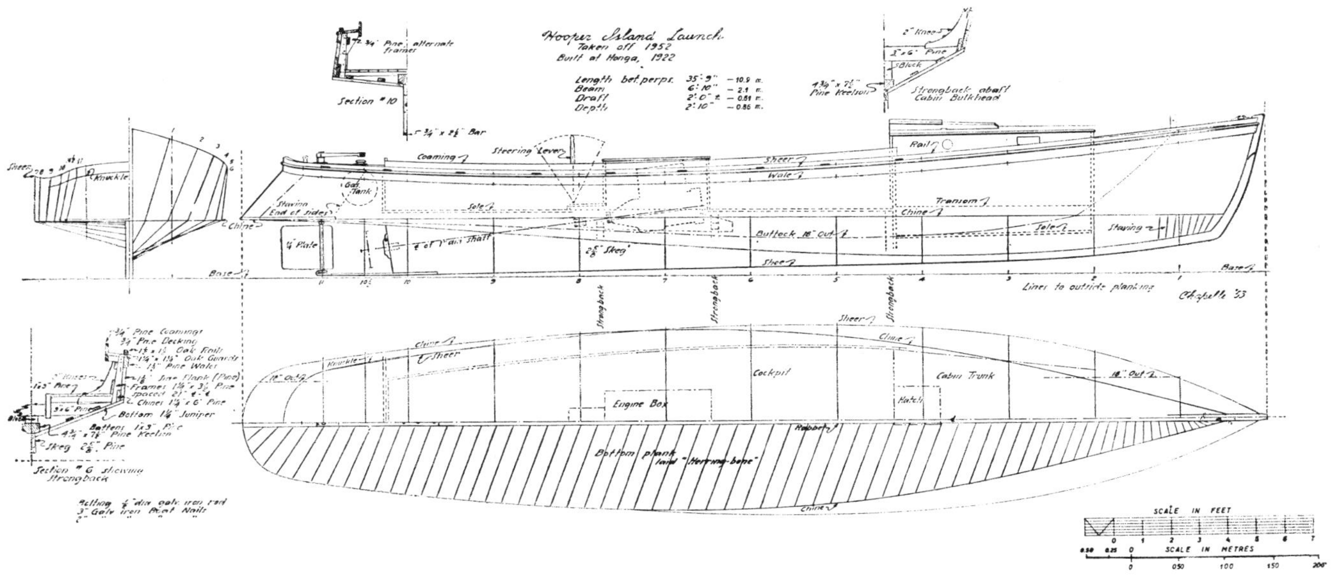

Fig. 8

boats used only small marine engines of low power but when automobile engines became common, the beam was increased. With small power, the narrow boats ran fast; one of 25 ft. (7.6 m.) length, and 4 ft. 2 in. (1.27 m.) beam, made 13 knots with a single cylinder engine rated 7 h.p.

The effect of the straight chine on the lines can be seen in the example of the type. A buttock line 18 in. (0.456 m.) out has been projected to show the low angle of attack of the bottom along the chine, in spite of the straight-line sections. It is very noticeable that these old launches run very cleanly and show a fine turn of speed with moderate power. In the boat shown, capable of a speed of about 12.2 knots with her six-cylinder automobile engine, she lifts forward at between 8.7 and 9.6 knots per hour. At full speed she squats slightly. The owner uses her in the open Bay for winter tonging and considers her a far better sea boat than his larger and more modern launch.

The drawing shows the appearance of the bottom planked. The bow is staved up with thick plank bevelled on the inside, where each piece bears on chine and rabbet. After all is in place, it is dubbed off smooth with adze and planed. The angle at which the bottom plank stands to the keel is determined by trial, so that the slight twist necessary can be worked in; in some instances it can almost be eliminated. With any form of round stern, it is usual to fan the plank at the stern.

It is apparent that the bottom, laid in this manner, furnishes no support to the long keel, so strongbacks are worked into the hull, consisting of a heavy timber running square across the boat, inside, from chine to chine. Due to the amount of deadrise, the timber can be secured to the keel only through heavy chocks resting on top of the keel member. The ends of the cross-timber or strongback are kneed to each side of the hull. There are usually at least two of these in a boat, one at the after bulkhead of the cuddy and one at the fore end of the engine, and many boats have two at the engine (as in the example) with only the fore one kneed to the sides, the one aft being a mere floor-timber used to support the fore ends of the engine beds or bearers. The ends of this floor-timber are merely pinned to the chine logs. Some boats have a similar floor-timber at the after end of the skeg.

The bottom planking is not caulked but is fitted with the inner edge of the seams tight and payed with seam-compound, before the hull is painted. A common feature is the stern-bearing, made of an oak cleat, which contains a rubber bushing made of a length of steam-hose. This sockets an inch or so into the shaft hole of the skeg, the shaft hole having been reamed out for the purpose. The after end of the hose is split, and the tabs formed bent down and tacked to the after face of the oak cleat with copper tacks. A hole is bored athwart the shaft hole in the skeg, just forward of the rubber hose, and the result is a very inexpensive water-lubricated rubber stern bearing, which lasts for years.

The round stern is built with an upper and lower frame and vertical staving. The top frame is sometimes padded up to the crown of the deck and faced off with a steam-bent moulding, as in the example, otherwise the deck comes down flat around the stern. The stem is made of an inner member and a cutwater, sharpie fashion. The keel member is really a keelson, being wholly inside the boat, and is hewn in a single length from a curved tree to profile. The skeg and the keel are then bolted to the keelson, so that no rabbet for the bottom plank is required. In some boats the keelson is brought to the outwater and a rabbet cut across it for the bow staving, then the projecting end is dubbed off with the staving which allows a curved rabbet to be formed in appearance between the chines and the keel. The wale is of a thicker plank than the sides but around the stern it is made of very short vertical staving, or blocks, nailed to the stern frame at deck level. Raised fore decks are very rare in Chesapeake launches, the trunk cabin being preferred.

From the drawing and description, it will be seen that the Chesapeake manner of building avoids a complete framing system and spiling of the bottom plank is not required. The round stern was estimated to add £110 (U.S.$300) to the cost of a launch in 1949.

The availability of powerful light gasoline engines at low cost caused changes in this style of launch, the beam being increased and the depth of hull, or amount of deadrise in the bottom, lessened. The builders also made the topsides curved in frame and the curved, vertical transom of the yacht also became popular. Fig. 9 shows an example of a modern version of the Hooper Island launch of the smaller size, which are built in lengths from 20 to about 55 ft. (6.1 to 16.8 m.). The boat shown was fitted with a low-cost automobile engine which drove the launch at 21.8 knots. At this speed she was running with her stem raised and without much settling aft.

PILOT HOUSES AND STEERING SHELTERS

In recent years, in the United States, most types of fishing launches have had pilot houses or semi-enclosed steering positions fitted. In the larger boats, the wholly enclosed pilot-house can be justified but in the smaller launches it is very doubtful if these serve a useful purpose. Their windows soon salt up in a fresh breeze and must be lowered so the helmsman has little real protection. But the real objection to half-houses or semi-enclosed steering shelters is the false sense of security they apparently give, which leads to large cabin companion-ways without proper coamings. Hence, if the sea breaks into the cockpit, it is not uncommon for a launch to swamp by the filling of the cuddy or cabin. Some launches, due to large semi-enclosed shelters, are top heavy and can be blown over. Accidents of this kind are on record. Shelter for the helmsman is admittedly desirable in boats working during cold weather and might be obtained, in small launches with cuddys, in the manner used in many Hooper Island launches and in some New England boats. It consists of a trunk over the companionway high enough to permit the helmsman to stand in the hatch, or to sit there if the boat is small, and steer. Windows, to open outward, are fitted to the trunk. With a wheel or steering lever, or yoke lines, steering in the companionway is readily possible and it is made comfortable if the cuddy has a heater in it. The small trunk steering shelter produces no dangerous windage, leaves the cockpit clear and does not naturally lead to a lack of a coaming in the companionway door.

SAFETY PROBLEMS

One of the gravest dangers in most fishing launches in North America is gasoline explosions, caused by leaks in the fuel tank or piping. This has been a very common cause of loss of both boats and crews and the carelessness with which fuel tanks, piping and electrical wiring are installed is amazing. More accidents would occur were that most engines and tanks are not enclosed in unventilated spaces. A common cause of fuel leakage is the absence of coils or other flexible units in the piping, which allows vibration to fracture the fuel lines. The widespread use of automobile engines has led to carelessness about carburretor leakage and the ignition wiring on the engines is often improper for use in a boat; grounding, shorting and sparking may cause trouble. Many batteries are badly secured and protected and cables are exposed to blows or fouling with fishing gear. It is comparatively rare to find a fishing launch fitted with bilge-blowers even when the engine is enclosed under deck. Many have no fire extinguishing equipment, lifebelts or flares aboard. With the increase in mechanization, it will be necessary to educate fishermen in safety requirements.

ENGINES

The powering of fishing launches is an almost unexplored field. Among the launches of Canada and the United States, only a small proportion have engines built for marine use, the rest having car engines.

An automobile engine is not designed for constant load, so lubrication and cooling troubles often occur, and while they are inefficient when propelling a boat at low speed (and also wiring and ignition are not waterproof under marine conditions), fishermen will use them because they are cheap to buy in comparison with the marine engine of approximately the same weight and rated power. The motor-car engine is often " converted " by adding a marine water pump and installing the

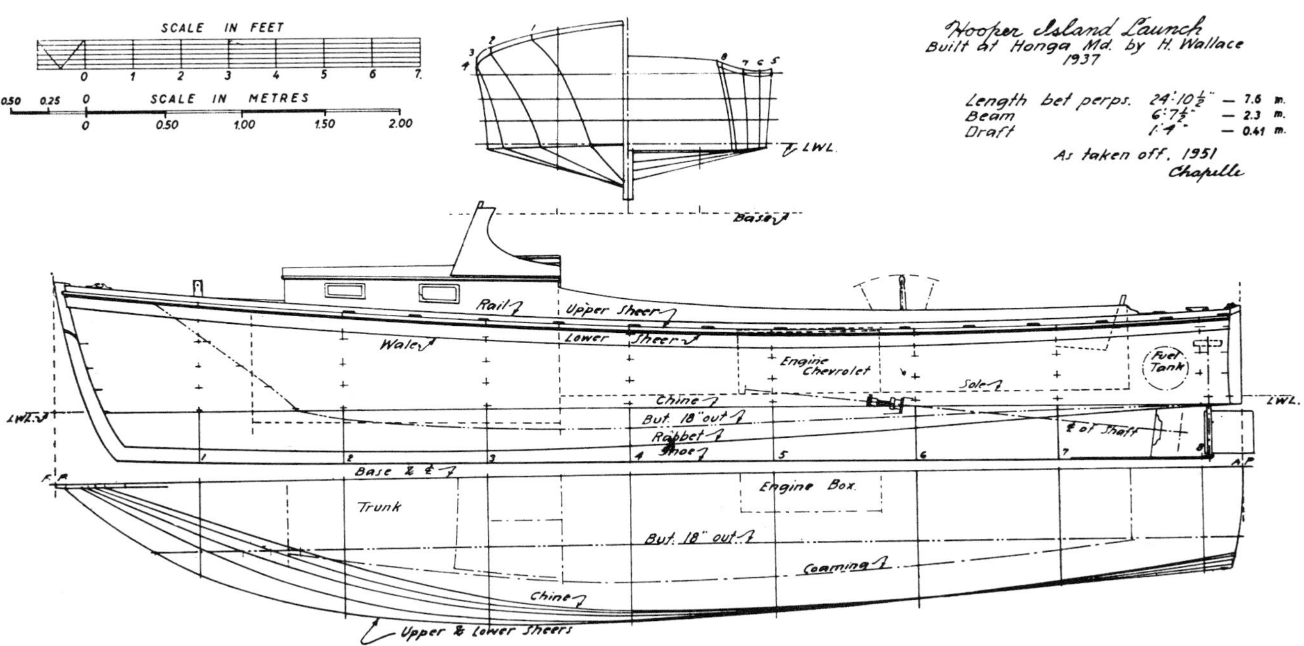

Fig. 9

engine in a boat with a fresh-water cooling system consisting of an expansion tank and a cooling grid, or pipe, outboard along the keel rabbet. The engine having a fresh-water cooling system lasts well if operation is fairly constant with no prolonged lay-ups. Tongers and crab-fishermen often work close to home and run slowly in the morning, while they get their gear ready, and then run slowly home while they clean up. About the only time the boat is operated at top speed is on some pleasure trip. Automobile engines are attractive because they are readily maintained, parts and maintenance skill being available in any village and town and even if the engine lasts only three seasons, it is considered inexpensive power.

There is little likelihood of many small diesels being used in the United States fishing launches as the cost is prohibitive compared with the automobile motor. The diesel would be highly satisfactory in the large launches working well offshore but, in all small craft, the diesels suitable would be those that could be manually started, should the electrical system fail.

Boats, such as the Dory and Sharpie models, are using outboard motors. The boats are often fitted with wells inboard, in which a large outboard engine of 5 to 15 h.p. is mounted. Some boats have the wells forward, so are propelled in the " tractor " manner. The outboard motors now on the market are not designed for long, steady operation, nor for use in heavy boats. Moreover, the casings of the shaft gearing are aluminium alloy, subject to corrosion when left exposed to salt water, and it is reasonable to expect that the use of outboard motors in commercial fishing boats will not be widespread until they are designed for long operation in heavy-duty craft.

CONCLUSIONS

It is very dangerous to expect perfection in a small fishing boat, and it is often noticeable that naval architects when discussing small craft, make much of some minor variation in engineering efficiency which cannot be significant in the finished boats. Engineering must be judged by what is needed and what can be afforded. It is easy and popular to overdo mechanization and produce too complicated a boat for the work of fishing. This has become apparent in some small draggers, for instance, whose operations are reduced by the need of laying up while some equipment or fitting is repaired by factory representatives.

Sail should not be ejected from fishing boats by fiat of the naval architect and engineer. There are still many areas in the world—in North America for instance—where fuel for gasoline or diesel engines is very expensive, or where the skill and facilities to maintain engines of any kind, are lacking, and where the climate or working conditions are very hard on machinery. It might be more progressive, in such areas, to improve the sailing craft than to motorize, for the latter may merely increase the cost of fishing without producing one cent more income to the fishermen.

The average fishing launch costs too much in proportion to the income she produces over a period of years. A cause is building with unnecessarily expensive materials and construction methods based on a false opinion of what " good construction " is in fishing boats. It has led to launches being built more to yacht standard than to commercial requirements. An example is to insist upon letting the heels of steam-bent frames into the keel, when floor-timbers are called for in the plans. Excessive fuel cost is common through all fishing fleets because there is too much emphasis on high speed which leads to over-powering. The theory that the increased number of fishing trips possible with high-speed boats makes for greater income for the fisherman ignores practical considerations. It supposes that the speed is actually used to and from fishing and assumes that it is possible to catch more fish annually, a not infallible rule. The record shows that the excessive number of breakdowns, after a short period of intensive operation, has usually prevented the fast fishing boat from performing as expected. High speed adds to the cost of boat, operation and maintenance, and profitable working of the boat depends largely on a high price for fish. Once the price breaks below an accepted normal, the fast boats quickly become uneconomic to operate.

For fishing launches the most useful range of speed is between 9 and 11 knots which will meet the practical requirements for almost any fishery. In fact, the launches that reach 8 knots are generally very satisfactory. Fishermen need to be convinced that their demands for speeds of 13 to 18 knots are, practically, a waste of money.

Cost of fittings aboard launches at present is very small, although some larger boats have ship-to-shore telephone and other devices installed. It would be a good rule to scrutinize such devices to be certain that they will actually aid in finding and catching fish, or in meeting market fluctuations. Widespread complaints of the alleged misuse of communication systems raises the question of how much real use they are in fishing.

Finally, problems surrounding the improvement of small fishing craft are not simply those of technical improvement and are rarely to be settled on engineering and design functions, except as far as safety at sea is concerned. The work of a naval architect goes far beyond the boat alone, for the result of his labours must lead to more profitable operation by fishermen.

PACIFIC GILLNETTERS

by

H. C. HANSON

GILLNET fishing takes place on the Pacific Coast of North America from central California to Oregon, Washington, British Columbia and Alaska. Gillnetters are particularly numerous in the many inlets in British Columbia where they catch a substantial amount of fish.

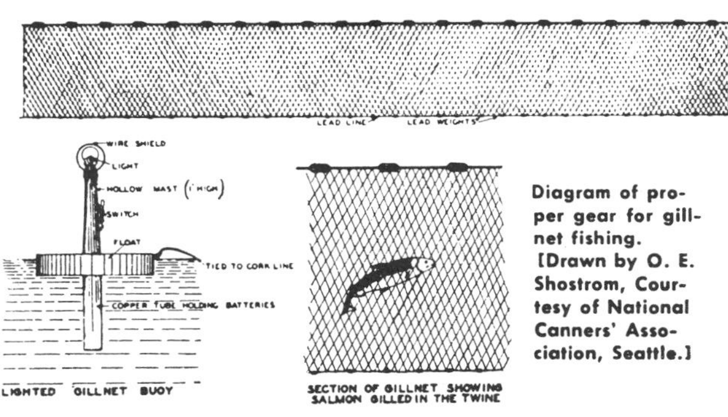

Diagram of proper gear for gillnet fishing. [Drawn by O. E. Shostrom, Courtesy of National Canners' Association, Seattle.]

Fig. 10

Fig. 12

Fig. 11

Most gillnets are fished from boats drifting with the tide and the nets may be either on the surface or along the bottom. They may also be fished from the shore or between posts or buoys. The nets, fig. 10, are made of fine linen thread twine, of 5 to 10 in. (12.5 to 25 cm.) mesh for salmon. They are often up to 900 ft. (275 m.) long, or even 1,500 ft. (460 m.), and are from 14 to 20 ft. (4.3 to 6.1 m.) deep. Fine threads and muddy waters improve the efficiency and where the water is clear, fishing is usually done at night.

Gillnetters developed from small handlining craft of 28 to 32 ft. (8.5 to 9.8 m.) in length, operated first by oars and sail, then by gasoline engines. This development was recently repeated in Alaska, where fishing with motors had been prohibited. As soon as permission was given, in 1951, the Bristol Bay fishery began to adopt motors and completed the change-over in two years.

Fig. 11 shows the original rowing craft formerly used on the Columbia river and other areas on the Pacific Coast and fig. 12 shows a similar vessel, with sail added. They were double-ended and had a fair deadrise. The original vessels used in the Bristol Bay fisheries, fig. 13, had a comparatively flat deadrise because of the shallower water, but otherwise they were almost identical to the Columbia River type shown in fig. 14. The boats with sail had large and cumbersome centreboards.

The Columbia river boats were equipped with 3 to 5 h.p. engines in the early 1900s, and they made 6 to 7 knots. The motorization is indicated in fig. 14 and 15. When larger engines became available, the speed of the double enders increased to 9½ to 10 knots, which is the maximum possible with that design. Then transom sterns were introduced and still larger-engines employed, so that in the 1920s up to 25 knots were obtained. Such a vessel, with a 140 h.p. engine, is illustrated in fig. 16.

The gillnetters were built with different general

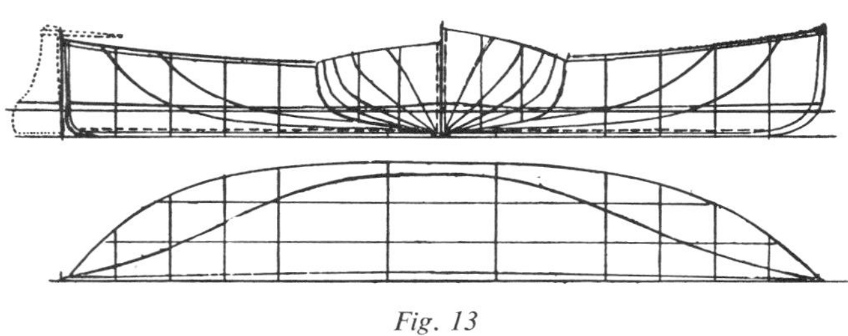

Fig. 13

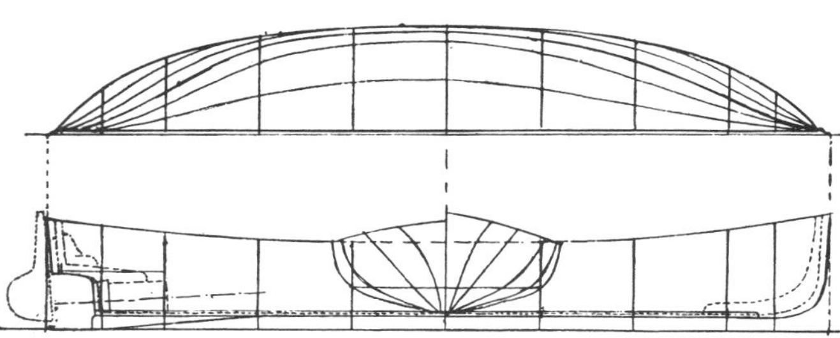

Fig. 14

Fig. 15

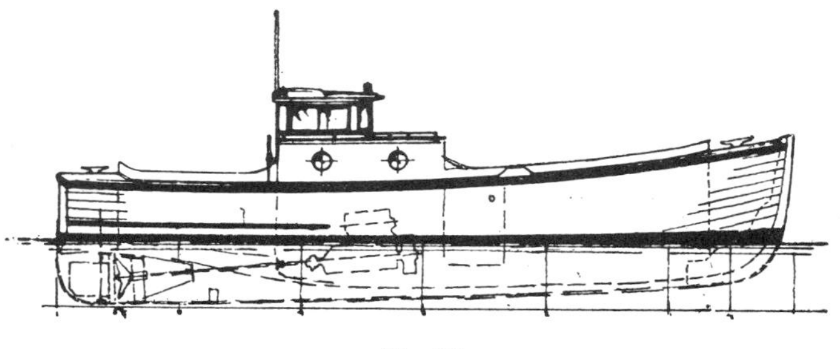

Fig. 16

Fig. 17

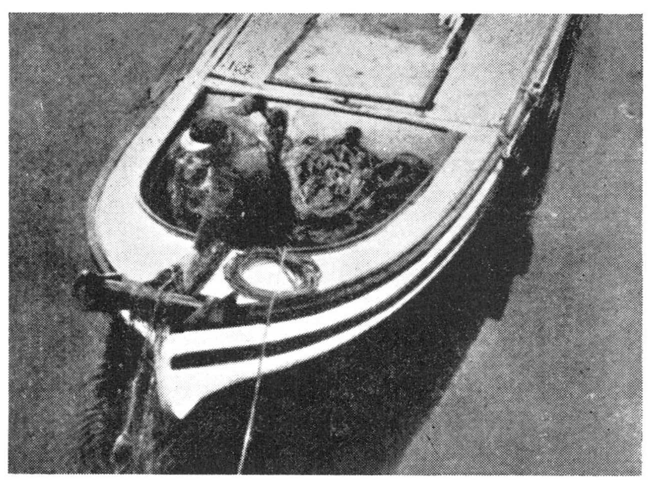

Fig. 18

Fig. 19

Fig. 20

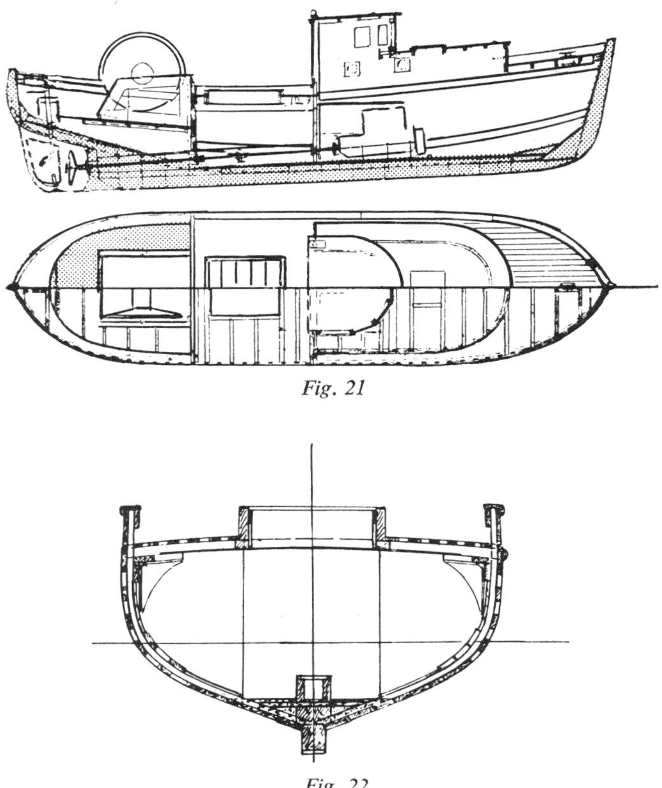

Fig. 21

Fig. 22

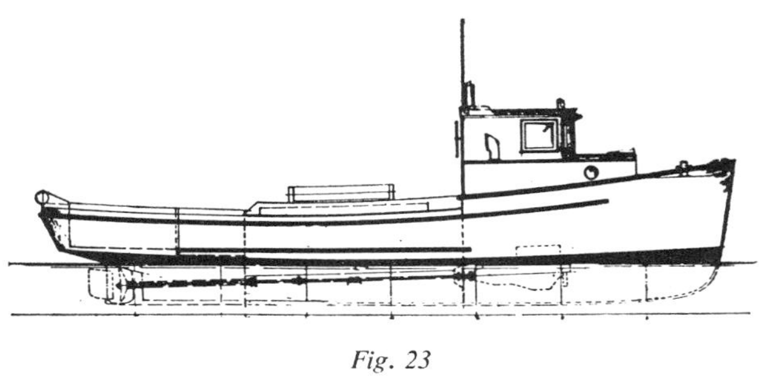

Fig. 23

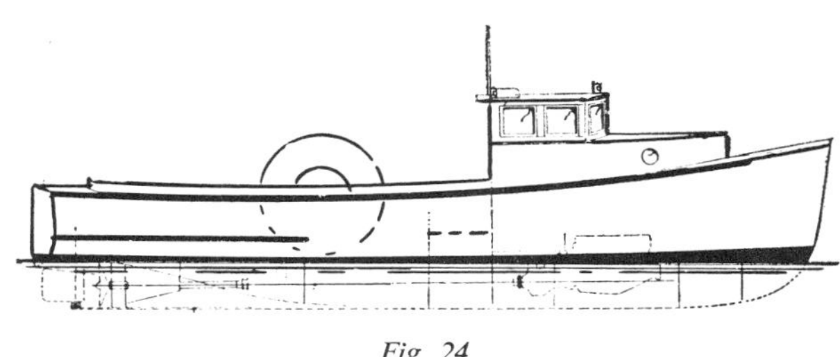

Fig. 24

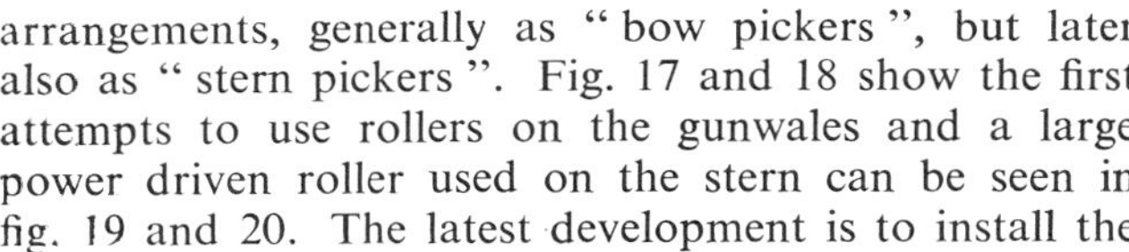

arrangements, generally as " bow pickers ", but later also as " stern pickers ". Fig. 17 and 18 show the first attempts to use rollers on the gunwales and a large power driven roller used on the stern can be seen in fig. 19 and 20. The latest development is to install the roller so that it swivels on a base to facilitate the hauling of the net from any angle. Most rollers are mechanically driven, but hydraulic drive is used to some extent. The rudder and propeller have steel guards to prevent the net from becoming entangled.

The boats are from 26 to 32 ft. (7.9 to 9.8 m.) long, and from 9 ft. to 10 ft. 6 in. (2.74 to 3.2 m.) wide. They are designed to be as shallow as possible, but for work in deeper waters, the deadrise is increased to get better seaworthiness. The " stern picker " is well exemplified by the common British Columbia type, fig. 21. The crew consists of one or two men. The midship section, fig. 22,

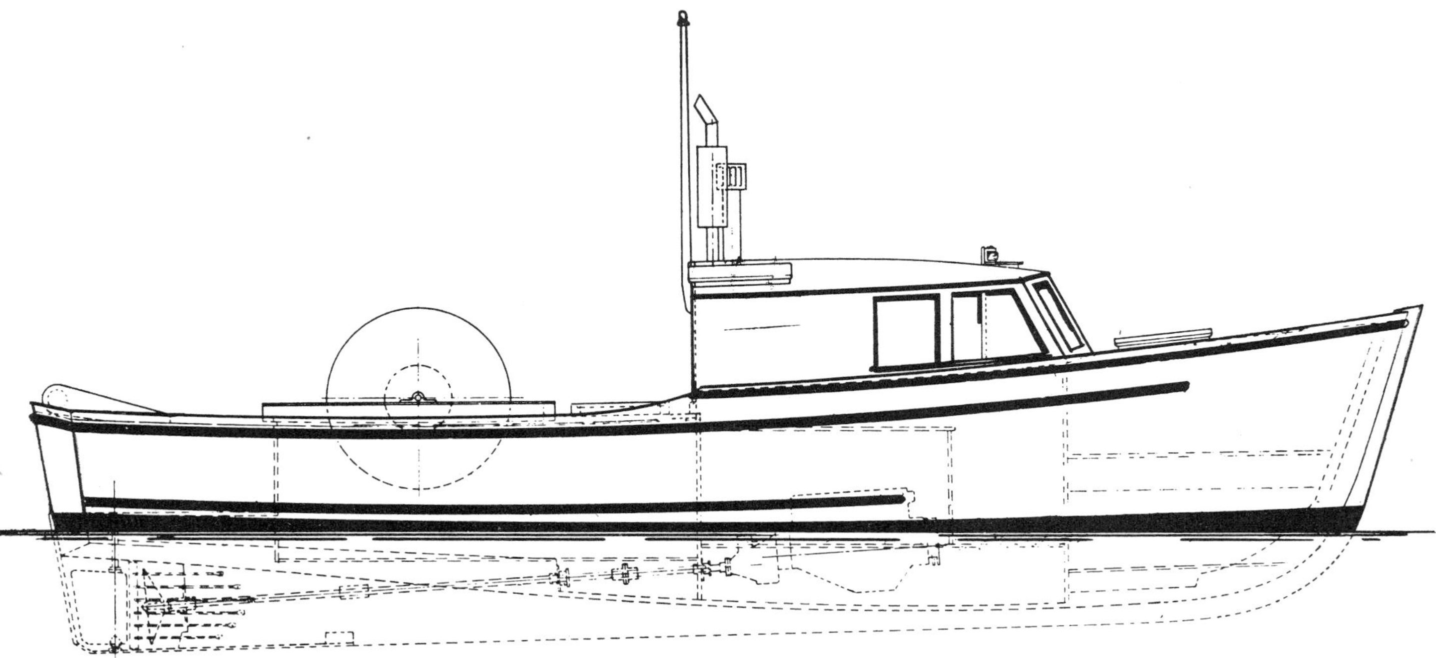

Fig. 25

shows a fair deadrise with the bilge carried well out. The scantlings and fastenings are:

Keel, 4 × 5½ in. (10.2 × 14 cm.), ½ in. (13 mm.) bolts;
Keelson, 5½ × 6 to 8 in. (14 × 15.2 to 20.3 cm.), ½ in. (13 mm.) bolts;
Floors, 2 × 7½ in. (5.1 × 19.1 cm.), ½ in. (13 mm.) bolts;
Frames, 1½ × 2 in. (3.8 × 5.1 cm.) on 9 to 10 in. (23 to 25.4 cm.) centres;
Garboard, 1½ in. (3.8 cm.), 2¼ to 2½ in. (57 to 64 mm.) boat nails;
Planking, 1 in. (2.5 cm.), 2 in. (51 mm.) boat nails;
Decking, 1¼ × 2 in. (3.2 × 5.1 cm.), 2½ in. (64 mm.) boat nails;
Shelves, 1¾ × 5½ in. (4.5 × 13.4 cm.), ⅜ in. (9.5 mm.) carriage bolts;
Clamps, 1¾ × 5½ in. (4.5 × 13.4 cm.), 5/16 in. (8 mm.) carriage bolts and 2½ in. (64 mm.) boat nails;
Bilge Stringers, four separate or three together, 1¾ × 5½ in. (4.5 × 13.4 cm.), 5/16 in. (8 mm.) carriage bolts;
Ceiling, 1 in. (2.5 cm.), 2 in. (51 mm.) boat nails.

The capacity is 1,800 to 2,000 salmon and, in some of the latest designs, as much as 2,500, such cargo having a weight of 5 to 7 tons.

A few steel vessels, fig. 23 and 24, have been built in the last years and their greater strength should result in longer life and less upkeep. They are less liable to sink, but, unlike wooden craft, they require insulation.

In future there will be a demand for more speed. At present the author is designing a V-bottom boat, planked with plywood, which will result in a light and inexpensive construction. Fig. 25 shows a 30 × 10 ft. (9.1 × 3.1 m.) gillnetter in wooden V-bottom design, which will be powered with a 120 h.p. gasoline engine.

BOMBAY FISHING BOATS

by

S. B. SETNA

THERE are more than 6,000 fishing vessels of 3 to 10 tons along the coast of Bombay, India. They are built by local carpenters and vary in size and design in different parts of the coast, each being made to meet the preferences of local fishermen and their methods of fishing.

Relatively few of these boats are used to catch fish with a floating or bottom drift net. Most of them work in stake net fisheries (bag nets), and merely transport fish trapped in the nets which are fixed to stakes in the sea. A few of the larger type are used for the "rampan" (shore seine) fishery, the net being hauled from the shore and the vessel being used only to set it around the shoal of fish.

DESCRIPTION OF FISHING CRAFT

The majority of the fishing craft are small, being about 40 ft. (12.2 m.) overall by 9 ft. (2.74 m.) beam with a draft of $2\frac{1}{2}$ ft. (7.6 m.) light to $3\frac{1}{2}$ ft. (1.07 m.) loaded. The largest may be 45 ft. (13.7 m.), including the overhanging bow. As a rule, the size of boat is determined by the type of fishing and the amount of money available for her construction.

The most noticeable features of the craft are the long fine entry bow and the rather abrupt and rounded stern. The bow shape, in profile, varies from place to place: those characteristic of Bassein are long and straight, while those from Satpati are curved and shorter. This overhanging bow with its V-sections is selected to give to these undecked boats a reserve of buoyancy and spray deflection in rough water.

The broad, rounded and rather low sterns, which are decked with loose boards, provide living space for the crew and the helmsman. The vessels are sturdy, although the method of construction and fastening is primitive.

One characteristic of the underwater hull in places such as Varsova is the arched keel with its deep forefoot and less deep heel, on which is hung the rudder. This characteristic is said to give the boats their ability to sail close to and tack against the wind.

Indian teak is used largely in construction with other timbers for crooks and spars. Boats are built mainly at Satpati, Dahanu, Bulsar, Billimora, Navasari, Surat and Broach in the north and Ratnagiri in the south. At all these sites boatbuilding yards are improvised, the local fishermen themselves being capable shipwrights.

Many of the craft are being mechanized, but still carry sails for use with favourable winds and, although the craft are not ideal, the problems of engine and propeller installation have been overcome. One result of mechanization has been to encourage the formation of co-operative societies to construct vessels to meet the demand for mechanized boats from fishermen in surrounding villages.

MECHANICAL PROBLEMS

Technical difficulties encountered in the installation of marine engines are surprisingly few in spite of the fact that the vessels were not built for mechanical propulsion. The main difficulty is in boring the stern post for the stern tube. In building, it is the usual practice to drive innumerable iron nails into the stern post at right angles to the direction in which the stern tube is fitted. Before boring, therefore, it is necessary to remove all nails and, as the majority of the craft have been in use for years, the nails are rusty and difficult to remove. If a nail should break then it is almost impossible to dislodge the submerged fragments.

In fastening the planks, the usual practice of using brass and copper is not always followed, and iron nails are used without any regard to their size or thickness. The nails are driven into the planks and turned over the timber. Contact between the nails and other metallic material in the vessel sets up galvanic action. The nails corrode and the original firmness of the joined planks is weakened and they are further loosened by the vibrations of engines not properly installed.

To stop corrosion and the entry of seawater through the nail holes, the Indian boat builder resorts to an ingenious process. Cotton waste, soaked in resin and mustard oil, is wound around the nail, under the head, before it is driven into the plank. The effect of this packing is to make the groove water-tight. A depression is formed in the plank by the head of the nail being driven into it and is filled up with a mixture of resin and oil and covered with coal tar.

Efforts are being made by the Government to impress

Fig. 26. A small dinghy for use in creeks from Versova into which a $1\frac{1}{2}$ h.p. stationary engine is installed

to allow for the propeller, which results in poor manoeuvrability. Experience shows that the best way to fix the propeller is to fit a dummy stern post to the existing stern post and to hand the rudder in the usual manner which gives the vessel good manoeuvring qualities. Fairing of the dummy stern post improves performance, as it allows a smooth flow of the water to the propeller. The stern posts in some vessels are not wide enough to allow for the stern tube and it is necessary to add pieces of wood on either side of the dead wood.

The size of the propeller is determined by the basic design of the boat itself, its displacement, resistance factor, draft, etc. The standard propellers supplied by the manufacturers are, generally, not ideal. Indian fishing craft are constructed without plans and it is difficult for a manufacturer abroad to select the ideal propeller not knowing the type of hull. It has been suggested that extra propellers of different dimensions should be ordered, that the best one could be found out by trials.

Fishing vessels around Bombay are generally suitable for obtaining sufficient submergence of the propeller without inclining the engine too much. Fishermen like the engine as far aft as possible. When it is impossible to get correct submergence of the propeller without surpassing the maximum angle of the rake, it is necessary to move the engine forward, which generally arouses considerable opposition from the boat owner as it makes stepping the mast difficult.

A set of instructions for the construction of improved types of hulls, conforming to the type of the existing sailing craft, has been drawn up by the Mercantile Marine Department. The instructions are:

" The keelson and keel should be bolted together with $\frac{1}{2}$ in. (12.7 mm.) dia. through-bolts, spaced about 3 ft. (0.9 m.) apart and driven from the keel ; the point of the bolt to be clenched over a nut and washer on the keelson. Iron bolts should be used for this purpose, galvanized if possible.

" All the frames should be of well selected timber, free from knots and shakes and grown to the form of the boat. The timbers should be continuous across the keel and in one length from gunwale, if possible. If the frames have to be worked in shorter lengths, care should be

upon the fishermen that fastenings of the component parts of the hull should be copper, clenched up on rooves or washers, and used to secure the outside planking to the framing of the hull. Where through-fastenings cannot be worked, brass screws of proper gauge should be used. The difference between the cost of iron fastened craft and the craft fastened with copper is approximately Rs.2,000 (£150, U.S.$420), a considerable extra expenditure as the cost of an iron fastened boat is only Rs.6,000 (£450, U.S.$1,260).

It is good practice in marine installations to have the engine bearers as long as the vessel will permit, which is not appreciated by the average fishermen, who insist on the engine bearers being just the length of the engine. This places unnecessary stress on the hull in the area where the engine is installed.

Most craft around Bombay have the rudder attached to the stern post by means of a rope. On installation of an engine, it is not uncommon to cut space in the rudder taken that the butts in adjacent frames are separated by a distance of about 2 to 2½ ft. (0.61 to 0.76 m.). Joints of frames should be well scraped. Joints in the frames should be avoided in the round part of the bilge.

" The spacing of the frames should not exceed 12 in. (0.305 m.) centre to centre.

" The bulkheads at the fore end of the engine room and at the forward store locker should be of strong construction, say 1 in. (25 mm.) tongue and groove with 3 × 3 in. (76 × 76 mm.) vertical (or horizontal) stiffeners spaced about 18 in. (0.46 m.) apart and well secured at top and bottom.

" Substantial deck beams about 2½ × 3 in. (63 × 76 mm.) should be fitted at each bulkhead, well secured to the frames and gunwale by stout bracket knees, iron for preference. Additional beams 2 × 2½ in. (51 × 63 mm.) on alternate frames, should be fitted between the engine room bulkhead and the stern post. All the beams should be dovetailed into the gunwale (covering board) and bracketed to the frames.

" A skylight would be required over the engine space and the coamings should be fitted at time of building. The opening could be closed by a hatch if necessary until such time as the engine is installed when the skylight could be completed."

New craft, being built specially for power propulsion, are beamier and have modified sterns while retaining the general characteristic of the sailing boats. They have greater capacity for nets, fish and ice, and have far greater range of operation. At the same time, the saddle-like or arched keel of vessels is disappearing, the fishermen realizing that this design has no merit.

Sooner or later it will be necessary to re-design the boats completely. This does not, however, mean blind imitation of the craft of other countries but the design will have to be adapted to meet local requirements.

Apart from the deep sea cutters and trawlers with refrigerated holds, capable of ranging further afield, two types of inshore or short-range craft might be developed. The first is a vessel of 35 ft. (10.7 m.) overall length, with a shallow draft and equipped with a 24 h.p. engine, for use in the many creeks and inlets which dry up at low water. It is, therefore, necessary to have a craft which will sit upright on the beach or on the mud. Such craft will not normally remain at sea over 24 hours and will meet almost all the various needs of fishermen in under-developed areas.

The second is a boat of about 45 ft. (13.7 m.) overall length with moderately deep draft and a large insulated hold. This craft would carry ice, remain at sea up to three to four days, and operate from deep-water bases unaffected by the tide.

Both types must be varied in detail and equipment according to the fishing to be done—drift nets, long-line, etc.—and must carry sails of small area for steadying purposes and for use in favourable winds.

The desire for mechanization has reached such proportions that fishermen show no concern about the seaworthiness of their vessels so long as they have engines. Accordingly, the Government has laid down the principle that every vessel, prior to installation of an engine, must be certified by the Mercantile Marine Department. To ensure compliance with the Government's directive, an officer of the Mercantile Marine Department, accompanied by the Superintendent of Fisheries (Marine), visits fishing centres to examine vessels before engines are installed. This officer suggests the modifications necessary for installation and to keep the boats seaworthy.

DIFFICULTIES SURMOUNTED

A large part of the credit for the great strides made by the fishing industry is due to Messrs. Burmah-Shell Oil

Fig. 27. Three-cylinder 30 h.p. engine installed in one approx. 45 ft. (13.7 m.) fishing boat in Satpati, India

Storage and Distributing Co. of India Ltd. and Messrs. Greaves Cotton and Co. Were it not for their foresight and encouragement, mechanization of Bombay fishing craft would not have progressed as much as it has.

Messrs. Greaves Cotton and Co. mechanized, at their own cost, a decked hull and gave it to the fishermen to use. The results were unsatisfactory. The fishermen seemed to think that their position on deck did not give them that control over the manipulation of their nets which they have in an ordinary undecked sailing craft. The fishermen have a rooted prejudice against fishing by standing on deck, and the firm abandoned the experiment.

It was realized that fishermen felt more at home on the existing type of vessels and two mechanized vessels, corresponding in essential features to the present undecked sailing craft, were built. The vessels are operated at key centres on the coast; and the results have convinced the fishermen of their utility and remunerativeness. They feel that such boats are the answer to the urgent problem of mechanization which is so essential for the expansion of the industry.

The success which has attended such small attempts at mechanization has roused the fishermen into action and over 200 engines have already been installed. Of this number, 46 were acquired with the aid of a subsidy from the Government, 57 on loans and 97 and more purchased by fishermen with their own private funds. The expectations are that no difficulty will be experienced in the absorption of a thousand engines in the fishing villages north of Bombay alone.

It is generally agreed that for the type of vessels along Bombay's coastline an engine of between 20 and 40 h.p. is necessary. It has been established by experience that for vessels up to 42 ft. (12.8 m.) overall length, a 22 to 30 h.p. engine at 1,200 r.p.m. with a 2:1 reducing gear is best. It will give a loaded boat approximately 6 knots of speed. For vessels up to 55 ft. (16.8 m.) in length, 35 to 40 h.p. engines at 1,200 r.p.m. are required, giving a speed of approximately 7 knots. There is, however, a tendency among some fishermen to install highly powered engines in order to get more speed. In Varsova, near Bombay, for example, a vessel with an overall length of 34 ft. 2 in. (10.4 m.), breadth 7 ft. (2.14 m.), depth 2 ft. 3 in. (0.69 m.) and 3.93 net register tons, was equipped with a 65 h.p., 3-cylinder diesel, much against the advice of the Fisheries Department.

The bulk of the engines installed are of British manufacture, as spare parts are readily available. The types are: Gardners, Ruston Hornsby, Thornycroft, Kelvin, Lister, Central (Japanese), Greymarine (American), Samofa (Dutch).

BIGGER CATCHES

Rapidity of movement is not the only advantage accruing from mechanization of existing fishing craft. For instance the catching power of sailing craft is greatly increased when they are equipped with engines. A good example of the range and the catching potentiality of powered craft is found in fishing for " dara " (*Polynemus indicus*), popularly known as the giant threadfin, and " ghol " (*Sciaena diacanthus*), popularly known as the Jew fish. " Dara " is, as is well known, a bottom fish, looming large in hauls made by trawlers. Sailing vessels, in order to capture this fish, have to employ bottom drift nets. Though both powered and sailing craft use the same type of nets, the catches made by powered vessels are substantially greater, because a sailing boat employs 60 to 75 nets and a powered vessel employs 100 to 110. Sailing

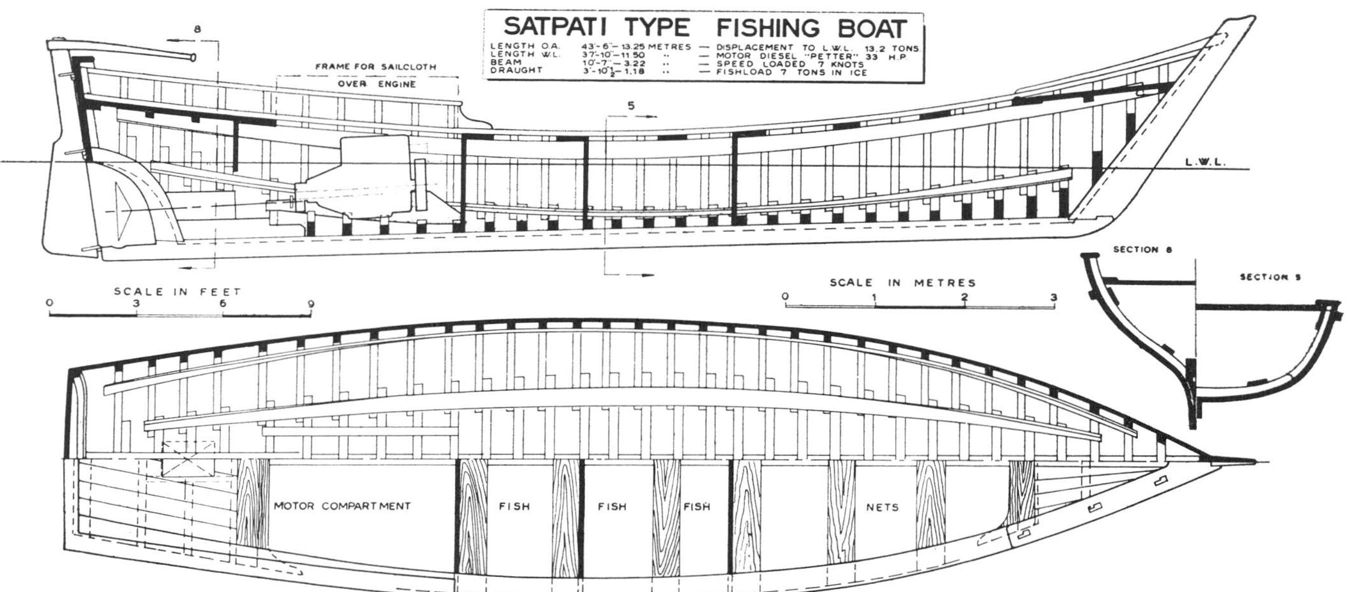

Fig. 28. Drawing of 43.5 ft. (13.25 m.) Satpati fishing boat, one original measured by Paul W. Ziener, FAO Naval Architect

craft caught on an average 485 of these large fish in a season, lasting 5 months, whereas powered craft landed, on an average, 2,000 fish (*see* Table I). The fish were caught in the same area. The additional catch made by powered vessels represents a substantial benefit to fishermen, and they also caught other fish which are not included in the Table. The fishermen were astounded by the magnitude of the catch ; sailing craft without engines have never landed as much fish.

The overhead cost of operation of these vessels is small as the entire crew is drawn from the fishermen themselves, and the maintenance cost of the vessels and the engines is very little.

Fig. 29. Vessel built at Satpati by local carpenters at a cost of Rs.9,000 (£680; U.S.$1,900). It is equipped with a 24 h.p. engine and is used for both transport and capture of fish

LABOUR-SAVING EQUIPMENT

The performance of the mechanized boats would have been still more remunerative if they had possessed small winches. It is realized that mechanization will not be fully complete without the use of mechanical equipment aboard. At present, power is only used for propulsion; it can be profitably employed to perform work which engages the tedious and exhausting labour of a number of hands. For instance, power vessels using drift nets carry seven to eight men for setting and hauling the nets. Similar work is, in other countries, done by four men through the use of winches driven by the engines. This economy can be achieved in India also and the problem of installing winches in mechanized craft is now before the Department of Fisheries. This will lend to an increase in the number of boats as crews will be smaller and there will be more fishermen available to man new boats and, consequently, increased supplies of fish made available.

TABLE I

COMPARISON OF CATCHES OF SAILING CRAFT AND POWERED VESSELS—BOMBAY, INDIA
(from January to end of May, 1953)

Sailing Craft (used drift nets and bag nets)

Name of owner	*Dara* (number of fish)*	*Ghol† (number of fish)*	*Total (number of fish)*
1. Sukarya Hira	539	26	565
2. Rajaram Damodar Meher	180	16	196
3. Kashinath Keshav	521	28	549
4. Krishna Shravan Chowdhary	587	15	602
5. Harischandra Keshav	483	32	515
		Average	485

Powered Craft (used drift nets only)

Name of vessel	*Dara* (number of fish)*	*Ghol† (number of fish)*	*Total (number of fish)*
1. Bharatmata	2047	544	2591
2. Pramilaprasad	1981	116	2097
3. Padmanowka	1793	64	1857
4. Bharatprasad	1709	101	1810
5. Jalazad	2091	76	2167
6. Dhanasagar	2155	77	2232
7. Lilawatiprasad	2061	163	2224
8. Padmaprasad	1692	78	1770
9. Jayashreeprasad	1649	116	1765
10. Jivannowka	1514	170	1684
11. Pratibha nowka	2273	114	2387
12. Jalaprasad	1833	135	1968
		Average	2046

* Common vernacular name—*Dara*. Scientific name—*Polynemus indicus*. Popular English name—*Giant Threadfin*.

† Common vernacular name—*Ghol*. Scientific name—*Sciaena diacanthus*. Popular English name—*Jewfish*.

MAINTENANCE OF ENGINES

Maintenance of the engines is a big problem, the importance of which has not yet been realized sufficiently by the fishermen. They expect that once an engine has been installed it will operate without any trouble, and they are so busy during the season that they do not carry out ordinary repairs. Minor troubles develop into major defects, with the result that the cost of repairs mount and the fishermen cannot make the maximum number of voyages. They have yet to learn the value of the advice in the old adage: " A stitch in time saves nine ". Their indifference is such that they do not overhaul the engine, even when the vessel is drawn ashore for the monsoon. They do not realize that the life of an engine can be prolonged greatly if, for example, the water cooling passages are flooded with fresh water to remove the salt deposits made by cooling with sea water. This simple action would greatly reduce corrosion during the three or four months the boat is laid up.

At present the fishermen greatly benefit from the free services rendered by firms supplying engines. Once they

Fig. 30. A small about 35 ft. (10.7 m.) Satpati boat for drift-net fishing

Fig. 31. A mechanized fishing craft from Bassein fishing village, near Bombay

withdraw help, the fishermen will be on their own resources and they should learn to prepare for that day. Now is the time to convince them that periodic surveys of their vessels and engines are beneficial and that an annual overhaul is most important. Technical staff ought to be provided at fishing centres, where vessels have been mechanized on a large scale, to instruct fishermen in these matters.

TRAINING OF CREW

Unless fishermen themselves operate the engines, and have part in the mechanization programme, prospects of the expansion of power fishing are remote. These objectives underlie the apprenticeship scheme—drawn up by the Department and initiated in 1933—to train apprentices, so that they qualify, in due course, as engine drivers and navigators.

The position to-day is that mechanization has been so rapid that trained engine drivers are not available in sufficient numbers. This cannot be viewed with indifference for, if it were to continue, absence of trained personnel is bound to tell on the efficiency of the engines. If a large number of vessels were tied up on account of engine trouble, it would be worse than if the vessels were not mechanized at all, because they would not be able to operate even as sailing craft on account of the weight of the engines.

Fig. 32. One improved type of fishing craft built by the Bombay Department of Fisheries. The vessel is equipped with a 21½ h.p. marine diesel engine, mast and sail. The boat is 45 ft. (13.7 m.) in length, 11.5 ft. (3.5 m.) broad and 5 ft. (1.5 m.) deep. The picture does not show the mast and sail

Opportunities are being offered to the young fishermen in Bombay State to visit centres where powered vessels are operating and to learn to fish with such boats. On return to their villages they will then be able to import to others the benefit of mechanized fishing. So far, 70 youths have been trained as mechanic-drivers. Their number is very small considering that about 1,000 engines are required to mechanize the existing fishing fleet in the north of Bombay.

FINANCIAL AID

The scheme for the grant of loans and subsidies for mechanization of fishing craft was amended by the Government during 1953. Up till then, loans and subsidies for purchase of engines were granted only to co-operative societies. The engines were, in turn, allotted by the societies to their members who operated the vessels individually or

Fig. 33. The rich fishing fields of the Baluchistan coast have been brought within easy reach of Bombay by launches. Chandrika *is operated from Ormara, along the Baluchistan coast, 700 miles from Bombay. It was built by a shipyard in Bombay*

collectively. The Government has now ruled that subsidies for engines will not be granted to individuals, but will be made available to groups, irrespective of their numerical composition, the determining factor being the type of fisheries and the usual number of crew on the vessel. Share of profits need not necessarily be on uniform basis. For instance, the more skilled may draw a higher proportion of the profits.

If the group is in a position to pay the loan portion in cash, the Government will have no objection to such a course, but will insist on an agreement from the group not to transfer the engine to another group for a period of five years. Direct assistance to individuals will be available in the form of loans for engines. These loans will be given without the intervention of societies.

Each group invariably has a good helmsman, an expert fisherman and a person with some mechanical skill to attend to minor engine troubles. Experience shows that these groups work successfully and their formation is, accordingly, encouraged. One of the advantages resulting from the institution of such groups has been to promote healthy rivalry, as they vie each other to catch more fish. It is an arrangement to recommend to countries where the introduction of power fishing is contemplated.

Experiments in which crews were not drawn from among the fishermen and in which they were paid regular monthly wages were unsatisfactory. They were only content to receive their wages at the end of each month and had no inducement to try their best to catch more fish. But when the fishermen are actively associated with the experiments in a way to make them feel that they have a stake in the industry, they work with enthusiasm and success.

The total amount of loans to fishermen and their co-operative societies, under the State Aid to Industries Rules, exceeds Rs.2,600,000 (£195,000; U.S.$550,000). A sum of Rs.400,000 (£30,000; U.S.$84,500) provided in the budget estimates for 1954-55 for loans to fishermen proved insufficient due to increasing demand for engines. A further sum of Rs.400,000 (£30,000; U.S. $84,500) for loans and Rs.130,000 (£9,800; U.S. $27,500) for subsidies was, therefore, provided.

NAVIGATION PERMITS

The following conditions govern the issue of permits to fishing vessels:

1. Vessels should be in sound and seaworthy condition, and efficient battening down arrangements, satisfactory to a Government Surveyor of the Mercantile Marine Department, are to be provided for hatches and other openings, to prevent admission of water in adverse weather conditions.
2. All vessels are to carry lights, shapes (two cane baskets joined together), and sound signals in conformity with the International Regulations for preventing collision.
3. In vessels under 60 ft. (18.3 m.) registered length, an approved buoyant apparatus (or more than one, if one is not sufficient) will be accepted, in lieu of a lifeboat, to support all persons on board. This buoyant apparatus (or apparatuses) is to be stowed to the satisfaction of the Surveyor. Two efficient paddles, suitably lashed, are to be provided with each buoyant apparatus.
4. The following equipment is to be carried and maintained in an efficient condition:
   - (*a*) Two approved lifebuoys, fitted with self-igniting lights;
   - (*b*) One approved lifejacket for each person on board;
   - (*c*) Two one-gallon (4.5 l.) watertight tins, containing fresh water, lashed to the buoyant apparatus;
   - (*d*) Two fire buckets with lines;
   - (*e*) One two-gallon (9 l.) foam fire extinguisher;
   - (*f*) One fireman's hatchet;
   - (*g*) One box of sand with a scoop;
   - (*h*) Six red flares *or* six distress rockets.

New vessels should be provided with substantial bulwarks and hatch coamings. In existing vessels, each case is considered on its merits (*see* 1 above).

Carriage of all this equipment is not actually necessary. Provision of a lifejacket for each member of the crew is essential. Similarly, three lifebuoys, as suggested by the Mercantile Marine Department, are understandable, but insistence on the provision of two buoyant apparatuses seems superfluous. It is not possible for small vessels to carry even one buoyant apparatus.

The various proposals suggested by the Mercantile Marine Department, are an intolerable burden as they will not only raise the cost of operation but also diminish the limited space available in the launches. The main aim underlying encouragement of the use of power vessels is to induce fishermen to carry bigger quantities and improved types of nets. Attainment of this objective will be handicapped if the limited space on the vessels is cluttered up with life saving equipment.

The Fisheries Department fully realizes the value of safety measures and endeavours to insist that all essential life-saving appliances are carried, so that crews are not exposed to unnecessary risks, but it is felt that the Act is being enforced with undue severity, which is a discouraging influence on the enterprise of the fishermen. There will always be hazards at sea, but reasonable precautions are enough.

WEST PAKISTAN FISHING CRAFT

by

M. RAHIMULLAH QURESHI, HENRY MAGNUSSON and JAN-OLOF TRAUNG

WEST PAKISTAN fisheries are located in the Arabian Sea and on the Pakistan coast bordering it, a total length of about 500 miles (800 km.). This includes the Makran coast (350 miles/560 km.), made up of windswept, surf-ridden large bays, and the Sind coast (about 150 miles/240 km.), which has a network of creeks and includes the delta of the River Indus. A boat can pass from the fishing village Ibrahim Hyderi, situated just south of Karachi, to the border of India without once entering the open sea.

The fish yield is estimated at some 37,000 tons per annum and is delivered, to a large extent, to Karachi for marketing. It is landed mostly by the fishing boats, but some is transported by special carrier vessels. The efficiency of both production and distribution is greatly handicapped by the lack of berthing, landing and handling accommodation, and of organized marketing facilities.

There are reported to be 1,103 boats within the Karachi Administration area (not including the boats on the Makran and Sind coasts) and their distribution, by types, is as follows:

Tony	234
Ekdar	131
Dhatti Hora	232
Hora or Gharat Hora	276
Bedi boats	230

To obtain records of the present types of boats (as a basis for improved designs), measurement drawings were made of ten typical types, in January to March 1953 (fig. 34 to 58). The measurement technique is described in the Appendix A. Two representative boats have been investigated in the Swedish Shipbuilding Experimental Tank, and work is going on at FAO, Rome, Italy, on the problems of engine installations and drawings of proposed new types.

DESCRIPTION OF PAKISTAN FISHING BOATS

Tony

Small dug-outs manufactured in Calicut, Southwest India. They are used mostly in creeks and are propelled by oars and, occasionally, a small sail is used. The Calicut dug-outs are sold all around the Arabian Sea. Similar small boats, planked and framed, are being built in Pakistan because importation from India has been limited. Cast nets and hand lines are used in this one-man boat.

Typical example (fig. 34): length, 20½ ft. (6.25 m.); cost, Rs. 300/- (£32; U.S. $90); wood, amb (*Hopea parviflora*).

Dhatti Hora

Narrow double-ended boats built mostly on the plan of the hora (gharat hora). There are two types, one made of imported dug-outs with the upper part planked, and provided with an outrigger (*dhatti*) and a balancing plank, the other constructed in Pakistan entirely by planking. The length varies between 30 to 40 ft. (9.1 to 12.2 m.); registered tonnage, 2.0 to 2.5. Only one sail is carried and oars are used for short trips. Most of the fishing is done by a crew of four in creeks and inshore waters, operations lasting six to eight hours. Equipment used includes the *bunn* (small gill net), *dora* (beach seine), and *kundi* lines with 100 to 150 hooks.

Typical example (fig. 35, 36): length, 33½ ft. (10.2 m.); registered tonnage, 2.0. The hull (*khall*) consists of the dug-out made of amb (*Hopea parviflora*), and the top planking of teakwood. The outrigger (*dhatti*) consists of poles of phoona (*Callophyllum inophyllum*) and the outrigger weight of dayal (*Cedrus deodara*).

Ekdar

Double-ended, comparatively narrow boats built on the principle of a hora (gharat hora), but heavily constructed with rounded, strong stem and stern timbers. The registered tonnage varies from 4.5 to 6.0. A balancing plank is used and only one sail is carried.

The boats were formerly imported, but are now being built at Karachi and at Buleji fishing village (Hawke's Bay). Towards Lasbella and Makran, the boats are used on the open surf-ridden coast and, as their construction shows, the rounded ends help in pulling them ashore. The name " ekdar " means that the keel is made

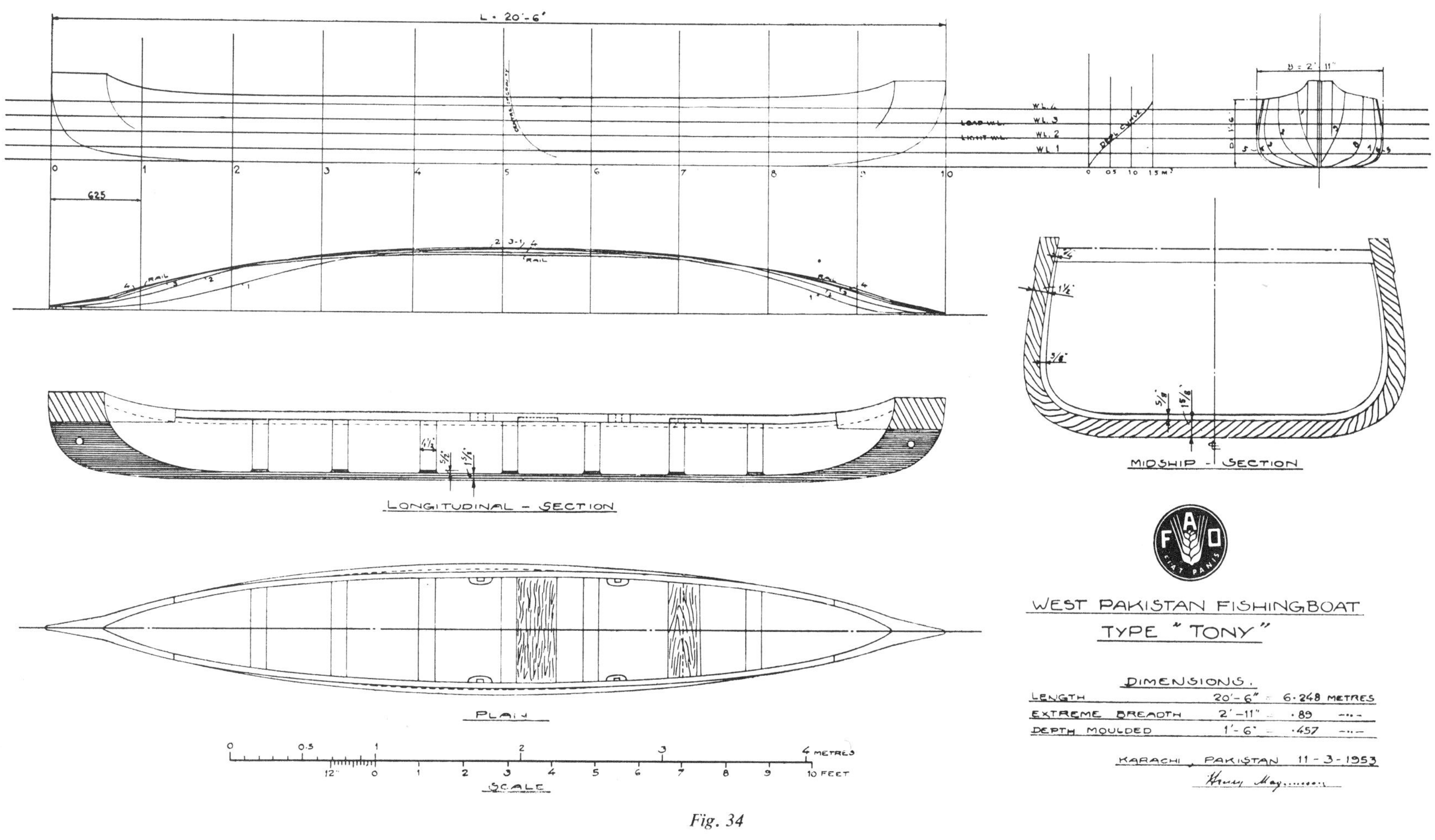
L = 20'-6'
WL 4
WL 3
WL 2
WL 1
LOAD W.L.
625
RAIL
LONGITUDINAL - SECTION
PLAN
0
0.5
1
2
3
4 METRES
12"
10 FEET
SCALE
MIDSHIP - SECTION
WEST PAKISTAN FISHINGBOAT
TYPE "TONY"
DIMENSIONS.
LENGTH 20'-6" 6.248 METRES
EXTREME BREADTH 2'-11" .89
DEPTH MOULDED 1'-6' .457
KARACHI, PAKISTAN 11-3-1953

Fig. 34

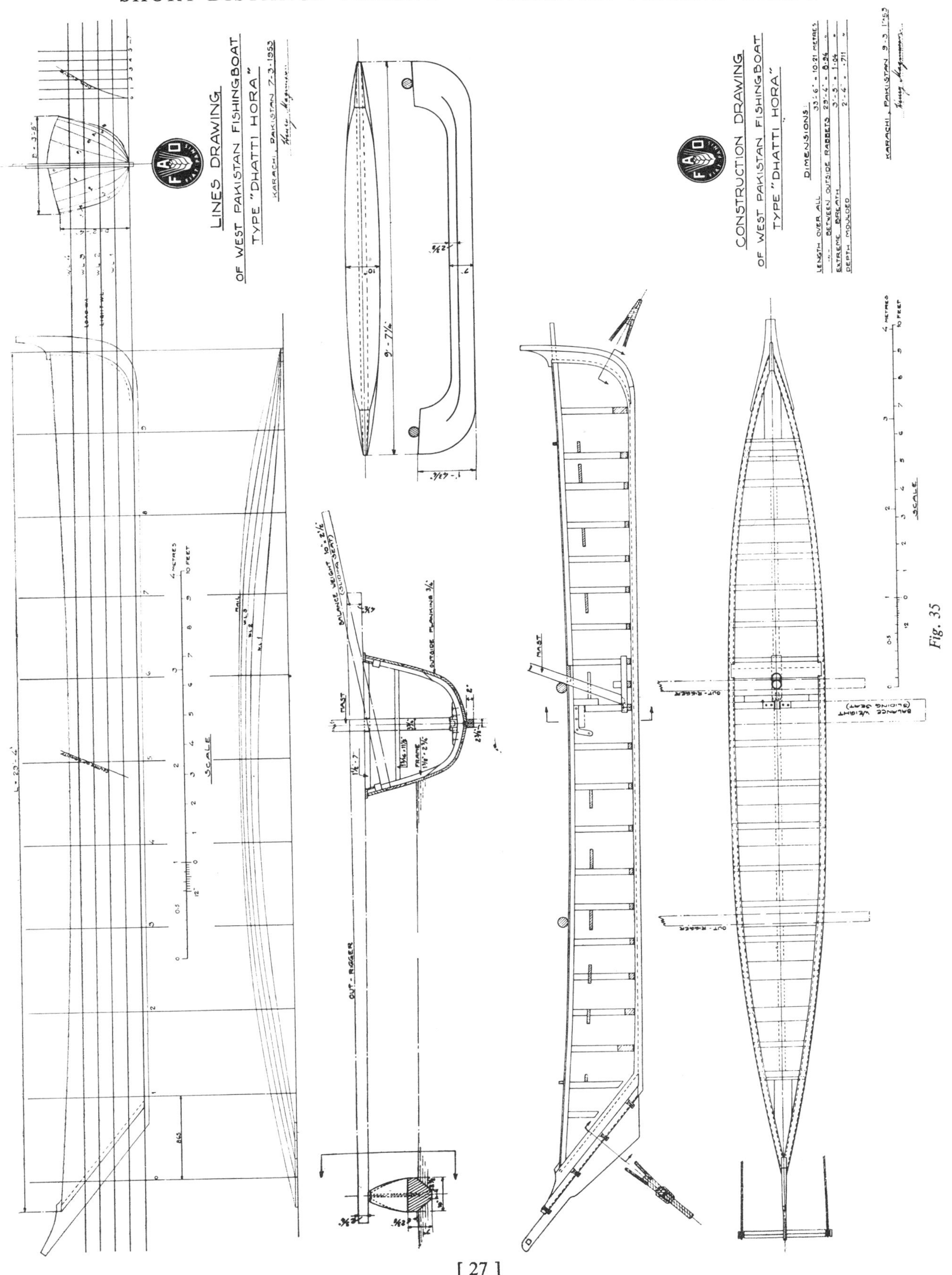
LINES DRAWING
OF WEST PAKISTAN FISHINGBOAT
TYPE "DHATTI HORA"
KARACHI, PAKISTAN 7-3-1953
CONSTRUCTION DRAWING
OF WEST PAKISTAN FISHINGBOAT
TYPE "DHATTI HORA"
DIMENSIONS
LENGTH OVER ALL 33'-6" = 10.21 METRES
BETWEEN OUTSIDE RABBETS 29'-4" = 8.94
EXTREME BREADTH 3'-5" = 1.04
DEPTH MOULDED 2'-4" = .711
SCALE
OUT-RIGGER
MAST
BALANCE WEIGHT (SLIDING SEAT)
FRAME
OUTSIDE PLANKING

Fig. 35

up of only one piece of wood. Actually, the lower part of the hull is usually a dug-out. The boats, with a crew of four, remain out only for the day.

Typical example (fig. 37 and 38): Length, 46 ft. 3 in. (14.1 m.); registered tonnage 4.9; hull—upper part of teak, lower made of phoona; keel and ribs, babul (*Acacia scorpiodes*); mast, phoona (*Callophyllum inophyllum*); nets used, *khori jal* (an encircling net), *rebi, duck* (gill nets), *kundi* (long lines), etc. Owner's name: Piri s/o Jamal. Name of boat: *Jung Bahadur*. Year of construction, 1942. Cost, Rs. 2,500 (£265; U.S. $750).

Hora (Gharat Hora)

Long and relatively narrow boats with a round forefoot, a vertical stem and a long overhanging stern. The sheerline is almost straight and the hull full at midship-section and having sharp ends. The length is between 35 and 68 ft. (10.7 to 20.7 m.) and registered tonnage, 20 to 35. The sail plan is similar to the bedi boats. The large boats fish in the open sea, the smaller ones near the shore and in the creeks. The capacity and arrangement are much the same as in the bedi boats, and they are built in Karachi with wood imported from Burma and Malaya. Nets used are the *bunn* (a gill net of small meshes) and *dora* (beach seine). Fish caught are sardines, etc., and prawns (shrimps).

Typical example (fig. 44, 45 and 46): Length, 67 ft. (20.4 m.); registered tonnage 30.24. It is called *Safina-e-Pakistan* and the owner is Hajee Kassim of Ibrahim Hyderi fishing village. Year of construction: 1952. Cost: Rs. 10,000 to 12,000 (£1,100 to 1,300; U.S. $3,000 to 3,600). Fig. 39 and 40 and fig. 41, 42 and 43 show a 48 ft. 2½ in. (14.7 m.) hora and a 54 ft. 5 in. (16.6 m.) hora, respectively.

Bedi Boats

Large boats with relatively good beam and an overhanging stem and stern timber. The hull has a full midship-section with a hard bilge, sharp entrance and a transom stern. The sheerline is almost straight and the length is from 35 to 70 ft. (10.7 to 21.3 m.). Registered tonnage is 20 to 45, the sail area 350 to 1,500 sq. ft. (32.5 to 140 sq. m.), as many as four sails being occasionally used on the bigger boats, but generally only one is hoisted. The boats have relatively many inboard fittings: loose decks to carry one large wooden box for fresh fish and ice, a small hold for cured fish, freshwater tank and space for cured fish, nets and cooking utensils. The deck is open and ballast is carried in sandbags on the deck level, always placed on the windward side when tacking.

These seaworthy boats fish in deep waters and are all built in Karachi by local craftsmen who use no drawing or plans.

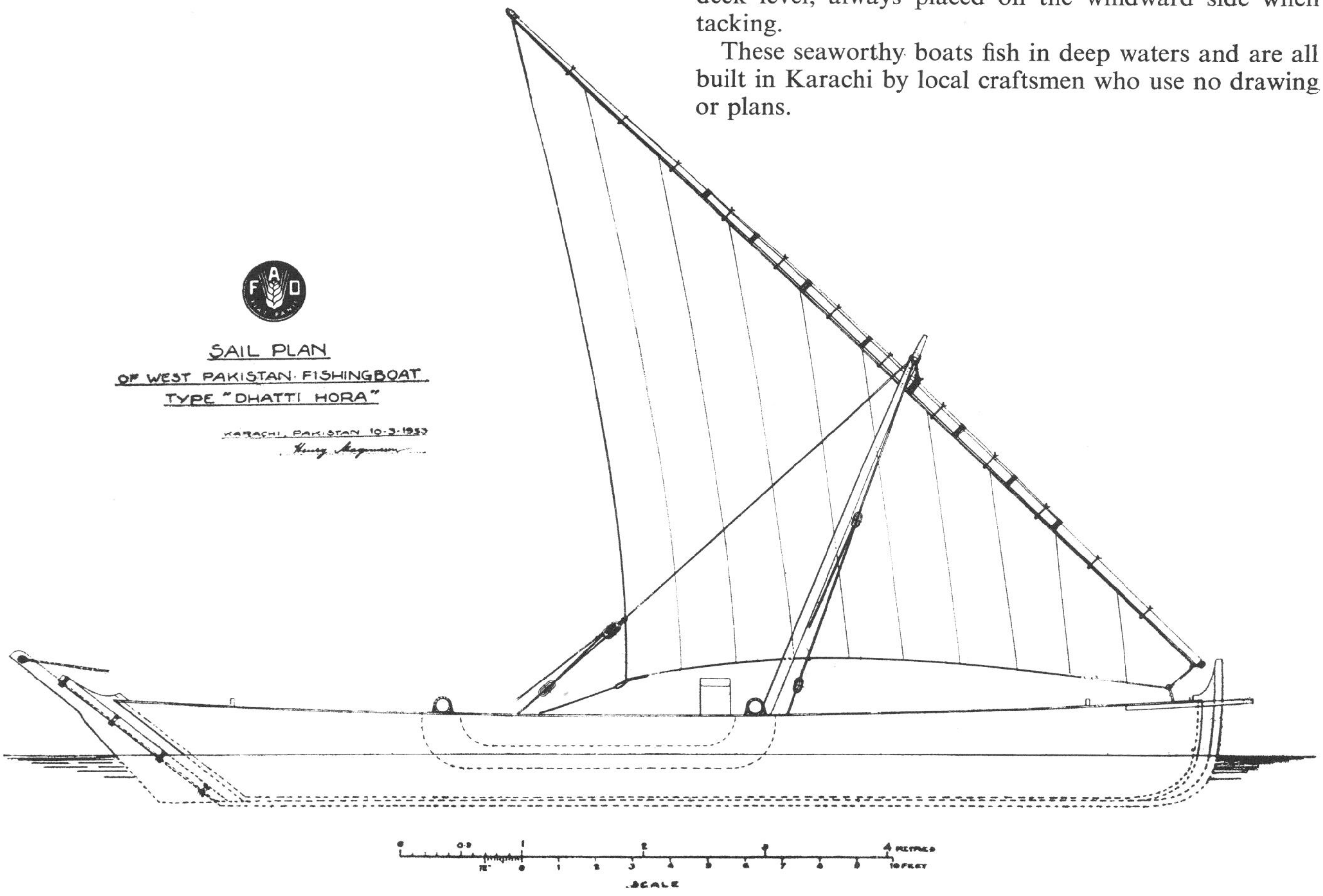

Fig. 36

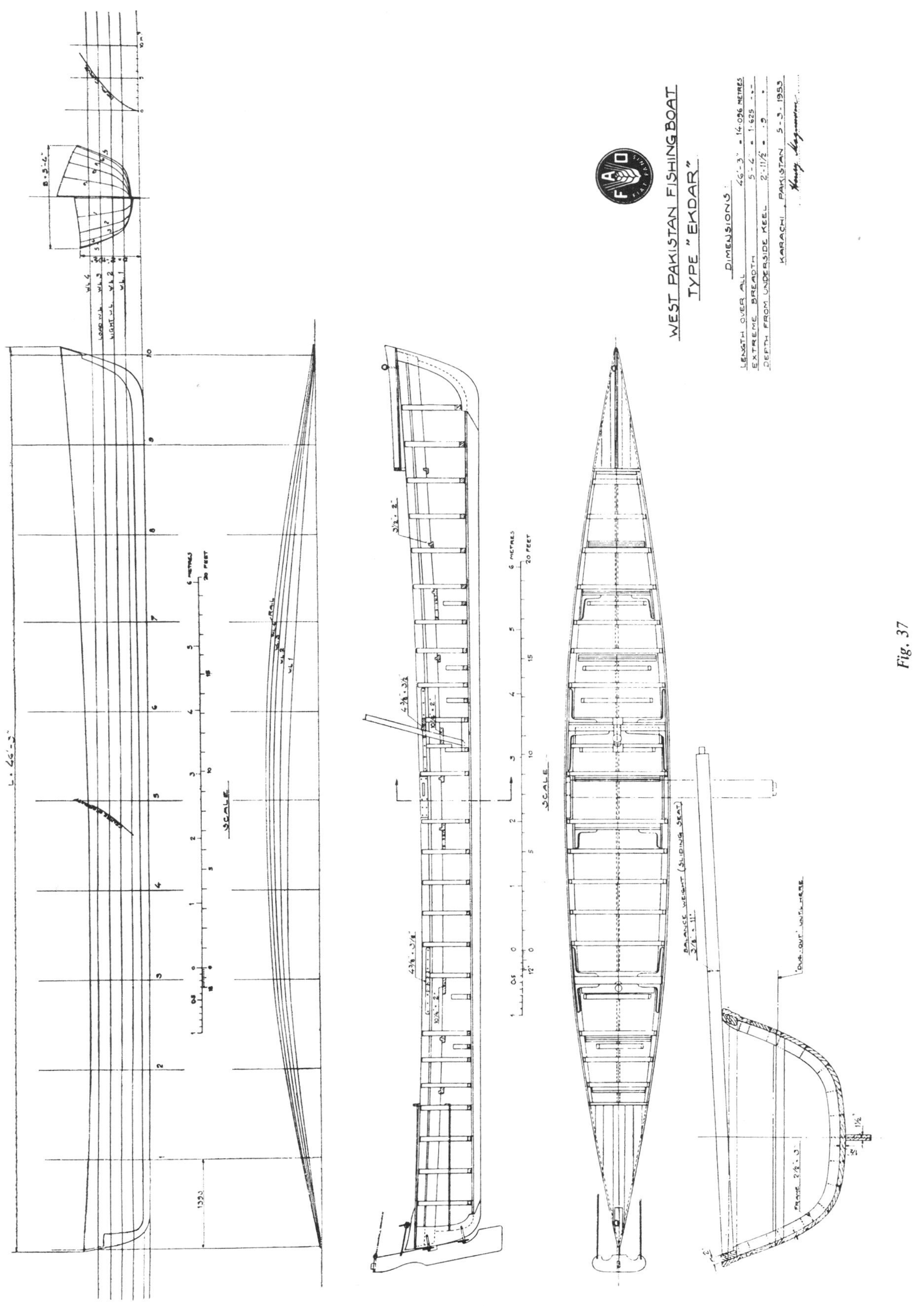
FAO
FIAT PANIS
WEST PAKISTAN FISHING BOAT
TYPE "EKDAR"
DIMENSIONS:
LENGTH OVER ALL 46'-3" = 14.096 METRES
EXTREME BREADTH 5'-4" = 1.625
DEPTH FROM UNDERSIDE KEEL 2'-11½" = .9
KARACHI, PAKISTAN 5-3-1953
L = 46'-3"
SCALE
SCALE
6 METRES
20 FEET
BALANCE WEIGHT (SLIDING SEAT)
'DUG-OUT' UNTIL HERE
1393

Fig. 37

Gill nets are used (*duck* is the local name), and are now prepared from parachute nylon cord consisting of 20 pieces 75 ft. (22.9 m.) long and 3 to 5 fathoms (5.5 to 9.1 m.) deep. The meshes are 5 to 6½ in. (12.7 to 16.5 cm.) and the nets weigh between 12 to 15 maunds (984 to 1,230 lb.; 446 to 538 kg.). Coir ropes are used at the top and bottom of the net; cylindrical floats of a light wood, called gugar (*Compiphora mukul*) and belapat (*Hibiscus tiliaceus*), 1½ ft. (46 cm.) long, about 4 in. (10 cm.) in diameter, are attached at intervals of one fathom (1.83 m.). The sinkers—about the same in number as the floats—are made of lead. The cost of a net varies from Rs. 12,000 to 15,000 (£1,300 to 1,600; U.S. $3,600 to 4,500). There is usually a 12-man crew, one working as skipper (*sarang* or *nakhuda*).

The owner gets 50 per cent. of the catch and 50 per cent. goes to the crew, the share of the *sarang* being double that of any one fisherman. If the boat is owned by the *sarang* 50 per cent. goes to him and 50 per cent. to the crew.

These boats catch only big fish such as thread-fins, croakers, etc., fishing up to 10 miles at sea. They have no navigational instruments and have to keep land in sight but, parallel to the coast, they sometimes travel 150 miles in search of fish.

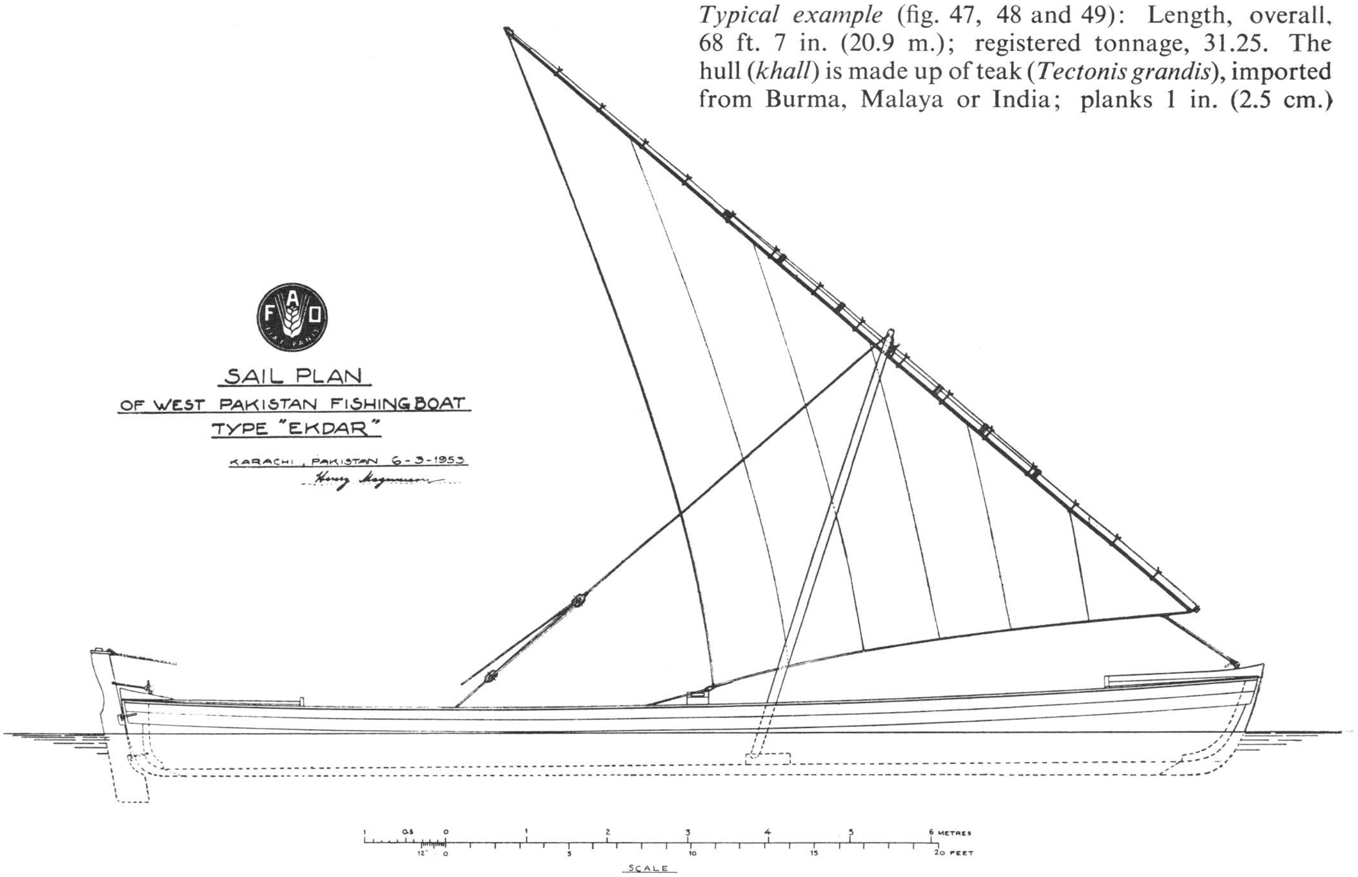

If the boats go out for three to five days, ice is taken to keep the fish fresh in a wooden box. If they go out for longer trips, salt is carried to cure the fish. On an ordinary three to five days trip, each boat brings back 500 to 600 big fish, weighing between 20 to 25 lb. (9 to 11.3 kg.) each.

Crude fish oil, prepared from the livers of elasmobranch fish, is used for the preservation of the boats against the action of sea water and the attack of borer worm, which is very common in these waters.

Bedi Boat I

Typical example (fig. 47, 48 and 49): Length, overall, 68 ft. 7 in. (20.9 m.); registered tonnage, 31.25. The hull (*khall*) is made up of teak (*Tectonis grandis*), imported from Burma, Malaya or India; planks 1 in. (2.5 cm.) thick. The total quantity of wood used is 1,400 cu. ft. (39.6 cu. m.) at Rs. 32/- per cu. ft. Wooden strips 1½ in. (3.8 cm.) thick (*takoon*) are used for ribband and stringers. The keel (*tar*) is of teak, about 1½ cu. ft. (0.042 cu. m.). Iron nails (*kil*), locally made, totalling 8 maunds (656 lb./298 kg.) and priced at Rs. 70/- per md. ($0.27 per lb.; $0.59 per kg.) are used for the boat. Ribs (*a*) straight of tali (*Dalbergia sissoo*), babul (*Acacia arabica*) and neem (*Azaderachta indica*), 18 pieces; (*b*) angle ribs (*ada*) of tali lohra (*Tacoma undulata*), 100 pieces at Rs. 10/- (£1; U.S. $3) each. Mast (*khuwa*) of phoona wood (*Callophyllum inophyllum*) from Calicut (India). Price Rs. 500/- to 600/- (£55 to 65; U.S. $150 to 180).

Fig. 38

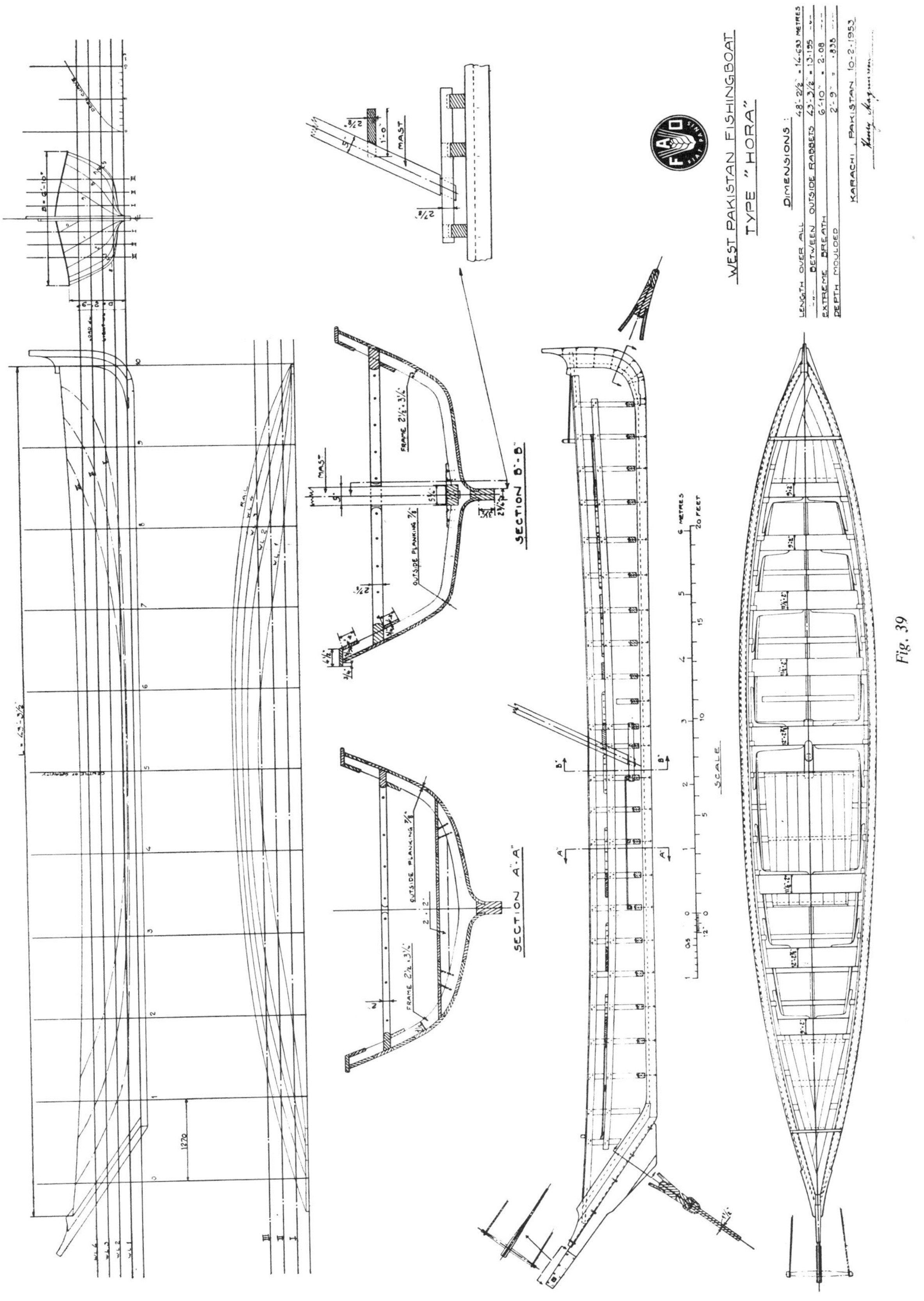
WEST PAKISTAN FISHINGBOAT
TYPE "HORA"
DIMENSIONS:
SECTION B'-B'
SECTION A'-A'
SCALE

Fig. 39

Poles for stretching the sails: (*a*) for two larger ones of phoona and (*b*) for the smaller ones, bamboo (*Bambusa arundinacea*). Sails (*sir*): two high ones made up of handwoven cloth, 2 maunds (184 lb./83.5 kg.) at Rs. 280/- per maund (7s.; U.S. $1.03 per lb.; 16s.; U.S. $2.37 per kg.), and two smaller sails of drill, 200 yards at Rs. 2/4/- (4s.; U.S. $0.68) per yard. Ropes for the mast (*kabar*): coir ropes, 6 maunds (492 lb./223 kg.), at Rs. 35/- per maund (10d.; U.S. $0.13 per lb.; $0.28 per kg.), and cotton rope (*katan*). Pulleys (*charakh*): 6 made locally of wood, Rs. 15/- to 20/- (£1 12s. to £2 5s.; U.S. $4.50 to 6) each. Anchor: 2, made locally, Rs. 50/- (£5 7s.; U.S. $15) each, 2 maunds (164 lb./74 kg.) coir rope required for both the anchors. Rudder (*sokhan*): Rs. 15/- to 20/- (£1 10s.; U.S. $4.50 to 6). Owner's name: Ali Mohammad s/o Ibrahim, Ibrahim Hyderi fishing village. Name of the boat: *Cairim Shahi*. Year of construction: 1926. Cost: Rs. 12,000 (£1,300; U.S. $3,600).

Bedi Boat II

(Fig. 50, 51 and 52.) The details are much the same as in I, with some variations, due to the smaller size. Length: 55 ft. 3 in. (16.84 m.); registered tonnage, 25.87 Owner's name: Jamot Wali Mohammed, Ibrahim Hyderi fishing village. Name of boat: *Pakistan*. Year of construction: 1948. Cost: Rs. 10,000/- (£1,100; U.S. $3,000).

Bedi Boat III

(Fig. 50, 51 and 52.) Length: 50 ft. 6½ in. (15.4 m.); registered tonnage 9.84. Owner's name: Ismail s/o Ishaq. Name of boat: *Maula Madad*. Year of construction: 1950. Cost: Rs. 5,000/- (£540; U.S. $1,500).

Bedi Boat IV

(Fig. 56, 57 and 58.) Length: 43 ft. 9 in. (13.3 m.); registered tonnage 7.12. Owner's name: Mohammad s/o Ishaq. Name of boat: *Husaini*. Year of construction: 1942. Cost: Rs. 3,500/- (£375; U.S. $1,050).

The details are much the same, except: No more than two sails are used instead of four. The net is called *rebi* (gill net), which is either made of nylon or cotton twine and the mesh size is smaller. The catch is medium-size fish, such as snappers, breams, etc., and, with small meshed nets, small threadfins and shrimps (prawns). Net costs run from Rs. 900/- to 1,000/- (£100 to 110, U.S. $270 to 300). The crew does not exceed seven and the range is not more than four days.

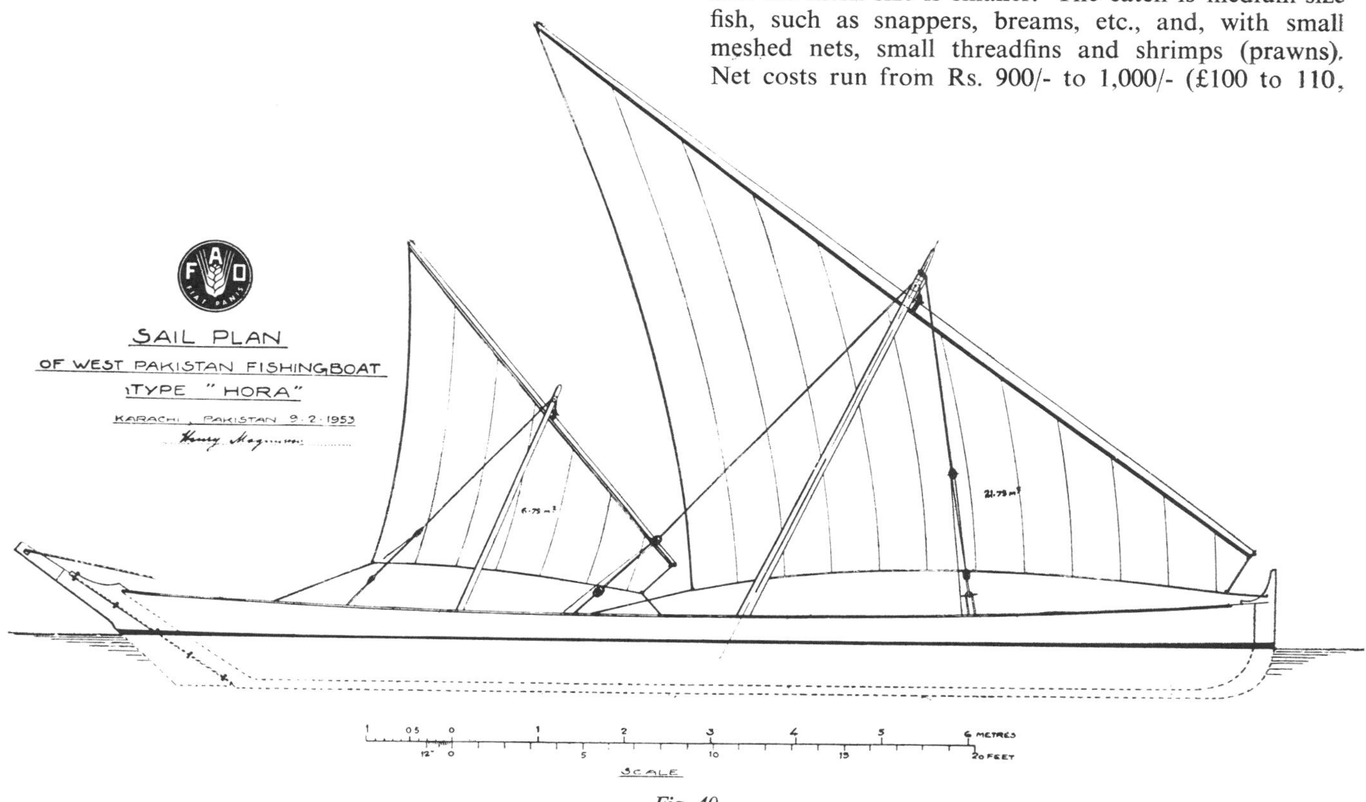

Fig. 40

NAVAL ARCHITECTURE

When studying drawings, fig. 34 to 58, it will be found that bedi boats are most suitable for mechanization and development. They have a sharp entrance, full midships section and a transom stern, giving straight buttocks and a flat run. The displacement, in relation to the length, is lower than normal for European fishing boats, due to the fact that the boats have no engines and are not decked and are, therefore, light. Two independent naval architects having good knowledge of small fishing boat

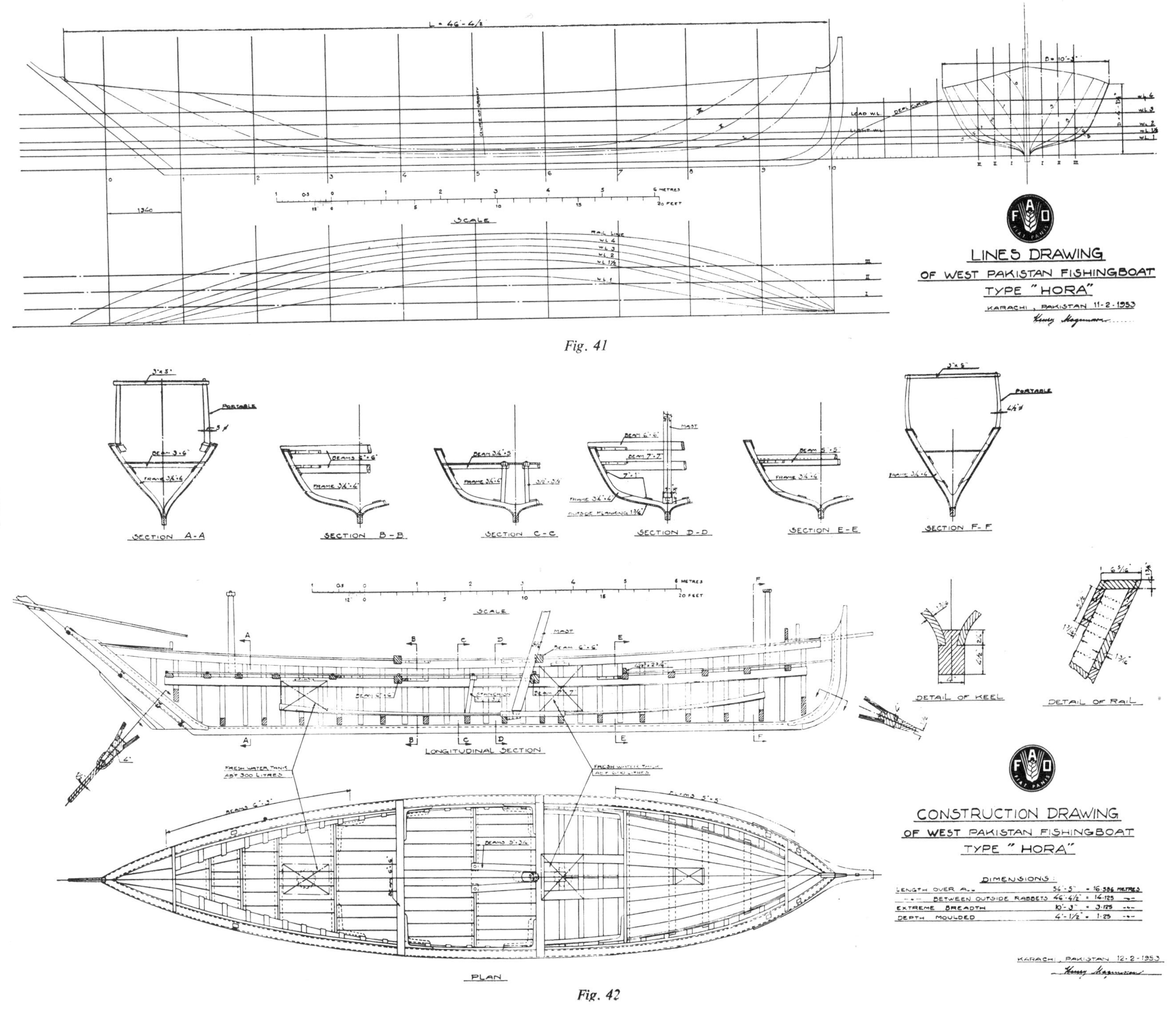
LINES DRAWING
OF WEST PAKISTAN FISHINGBOAT
TYPE "HORA"
KARACHI, PAKISTAN 11-2-1953
L = 46'-4½"
SCALE
SECTION A-A
SECTION B-B
SECTION C-C
SECTION D-D
SECTION E-E
SECTION F-F
LONGITUDINAL SECTION
PLAN
DETAIL OF KEEL
DETAIL OF RAIL
CONSTRUCTION DRAWING
OF WEST PAKISTAN FISHINGBOAT
TYPE "HORA"
DIMENSIONS
LENGTH OVER A. 54'-5" = 16.586 METRES
BETWEEN OUTSIDE RABBETS 46'-4½" = 14.125
EXTREME BREADTH 10'-3" = 3.125
DEPTH MOULDED 4'-1½" = 1.25
KARACHI, PAKISTAN 12-2-1953

Fig. 41

Fig. 42

design have studied the designs, and their remarks are as follows:

Howard I. Chapelle, Cambridge, Md., U.S.A. on the 7th July, 1953:

" Here is a sample of a local, native type that can be 'modernized' merely by adding small engines, and I dare say there are many others . . .

" I can see no practical difficulty in motorizing the bedi and hora types. Were I altering such craft, I would remove the after portion of their sternposts outside the rabbet, bore the remaining post for the propeller shaft, and then build up the outside deadwood to form an aperture and rehang the rudder. This would cause the rudder to stand at less rake than formerly, so the whole rudder would have to be shortened. I would like to use an air-cooled diesel. None of the boats will require large power, and any simple low-powered engine will serve. I would place the engine in the bedi type at the break of the quarterdeck, so as not to lose much cargo space. I think the bedi and hora types can be powered with engines in the 10–35 h.p. medium duty range or light duty. I do not think the largest bedi or hora will require more than 30 h.p. for effective fishing operations, unless light high-speed engines are used.

" I am rather astonished at the very complete framing of the boats and the evidence of great strength. This is not true of many Eastern craft—at least some Malaya types I have seen appear more fragile and less completely framed.

" I agree that these Pakistan craft should be powered without attempting to introduce a new form of craft.

" I shall regret that testing of the two small bedi types is not possible for I think their lines indicate better form for auxiliary power than the large examples. The flat run and wide stern of the smallest example seems very promising and is a boat I would like to power.

" The hora type undoubtedly can be effectively motorized if low power is employed. The slack quarters will prevent great power being utilized, I suspect, as squatting will probably develop above a moderate speed-length ratio.

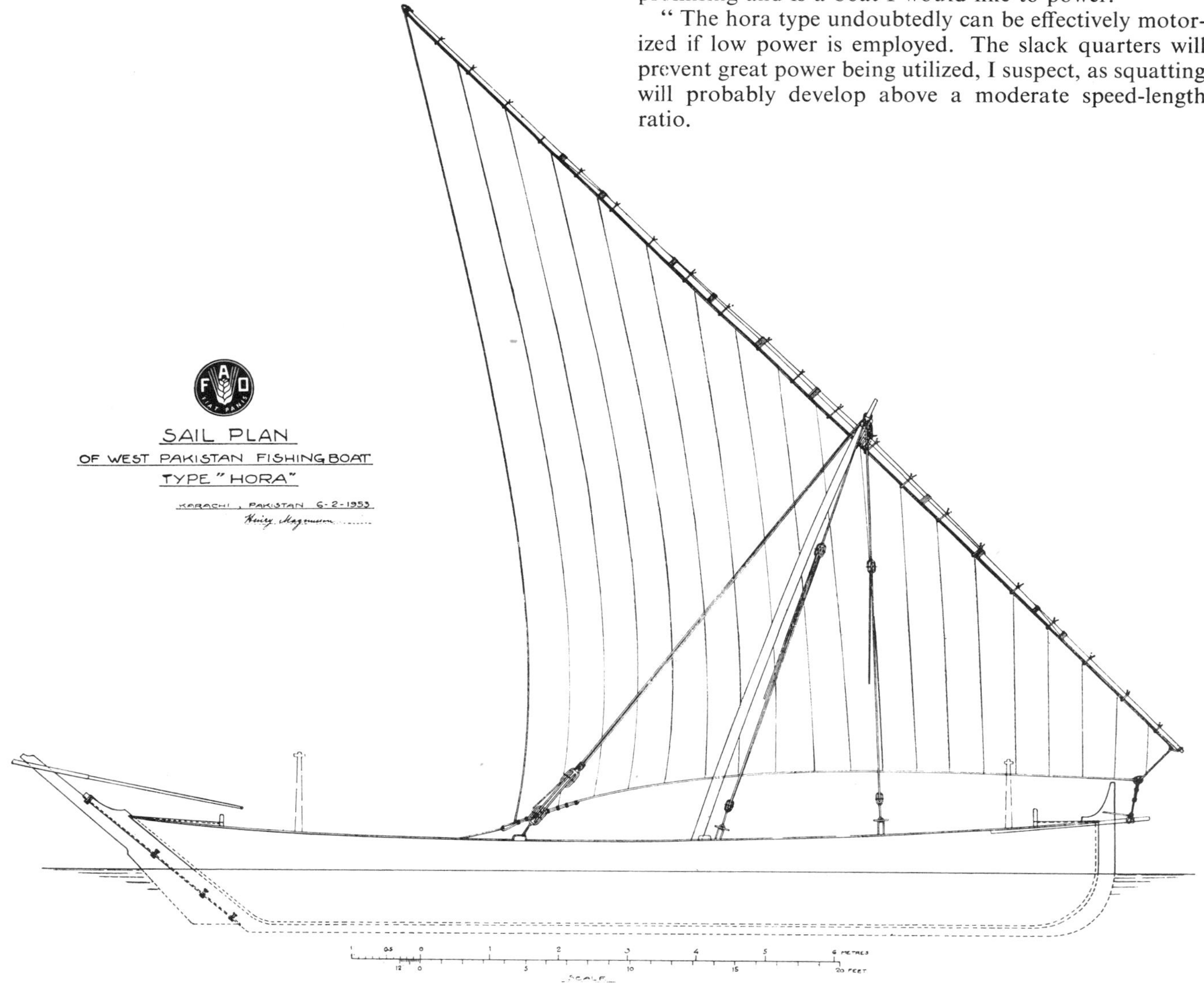

Fig. 43

" This whole class of boats is an illustration of my personal idea that it would be very wise indeed to motorize and otherwise improve the small primitive types of fishing boats than to introduce European and American launches, draggers, seiners and trawlers.

" I do not suppose it will be practical to motorize the ekdar boats, as they can be capsized too readily, but the outrigged dhatta hora could be. A friend of mine used a double outrigged Luzon Island boat for survey work in which he has installed a 7 h.p. aircooled stationary or industrial type of gasoline engine as an auxiliary motor. It drove the boat about 7 knots in smooth water. One of the more powerful outboard engines used in this country, say in the 12–20 h.p. range, could be readily employed if maintenance facilities were made available. I am certain that any planning for mechanization should begin with the establishment of proper maintenance and repair facilities in a primitive area, and this visualizes physical plant, not just educational facilities."

Arthur M. Swinfield, A.M.I.N.A., Sydney, N.S.W., Australia, on the 24th August, 1953:

" No doubt the builders have very good reasons for doing a job one way rather than another. These may be practical or economical reasons, and any comments I make are so done in ignorance of local conditions.

" The most obvious weakness in construction seems to be at the joint of the stem to the keel, and the stern post of the keel, the length of the scarp being inadequate in both cases. I presume, of course, that stopwaters are used in the normal way, and at the same time wonder if undue leaking occurs at these points as the vessel begins to age. I suggest that the ' anchor stock ' method would no doubt strengthen these connections, and such a joint is inexpensive and easy to make.

" The builders have evolved the ideal shape for performance and easy construction. This is apparent when one considers the ' angle ' of the stem and stern post in relation to the keel. The plank edges should show comparatively straight edges with very little moulding.

" It would be most interesting to know why every keel is straight and level with waterline. Although bilge water would be satisfactorily contained between the upright garboards, a slight ' drop ' aft would perhaps assist overboard discharge of bilge from a common pumping point.

" The ' hog piece ' method of forming the keel is not universal and one wonders why not. The bedi boats of 55 ft. 3 in. and 68 ft. 7 in. have this desirable feature. I suggest that ' treenails ' would be of great help where this method of construction is adopted, as the ' through fastenings ' in use appear to be clenched (no size given)."

The large hora boat (fig. 44, 45 and 46) and the largest bedi boat (fig. 47, 48 and 49) were tested in the Swedish State Shipbuilding Experimental Tank on a heavily loaded displacement of 20 cu. m. (21.6 tons) and with three different trims, the trim water lines being indicated on fig. 44 and 47.

Data about the main dimensions of the models, during these tests, are listed in Table II and III and effective (tow-rope) power and C_1 values are given in fig. 59 and 60 for the hora and bedi boats, respectively. Table II, hora boat, indicates that, because of the overhanging stern, the water line lengths vary to some extent. The prismatic coefficient, therefore, varies between 0.57 and 0.61. The diagram, fig. 59, shows that stern trim at lower speeds gives less resistance. The optimum location of an

TABLE II

Hora SSPA 587-A

Trim		*Level Keel*	*By Head*	*By Stern*
Series		I	II	III
L	ft.	55.1	53.8	56.1
	m.	16.80	16.42	17.12
B	ft.	12	12	12
	m.	3.65	3.65	3.65
T	ft.	3.15	3.21	3.11
	m.	0.96	0.98	0.95
∇	cu. ft.	707	707	707
	cu. m.	20	20	20
Am	sq. ft.	21.5	21.7	22.0
	sq. m.	2.0	2.02	2.05
δ		.340	.340	.338
φ		.596	.605	.570
$\odot$		—2.27	—2.34	—5.97
$\frac{1}{2}\alpha_E$		14°	19°	9°
$L/\nabla^{\frac{1}{3}}$		6.15	6.05	6.31

engine should therefore be far aft, producing an aft position of the L.C.B.

The main dimensions of the bedi boat appear in Table III, which indicate a very low prismatic coefficient of about 0.51. Test of the bedi boat (fig. 60) shows that the original trim parallel to the keel gives the best results. Trimming down the bow gives somewhat better value at very high speeds, stern trim gives worse values, indicating that the transom is too large.

A study of the power curves shows that a speed of 7 knots could be reached in a hora boat with an engine of about 20 h.p. and in a bedi boat with an engine of about 18 h.p. This is for the loaded condition; the speed would be somewhat higher in the light condition.

The best C_1 curves of the hora and bedi boats have been assembled and drawn with heavy lines on fig. 61. On this diagram, the boats have been compared with some typical fishing boats from European countries, the main dimensions of which have been listed in Table IV. The Denny-0 model (Allan 1950) is a Scotch ringnetter; SMT 25A and 28A are Norwegian 40 ft. (12.2 m.) and 65 ft. (19.8 m.) boats respectively; SSPA 17 is a Swedish west-coast fishing boat and VNR (Roma) C479 is a postwar Italian trawler. It is apparent from the diagram that

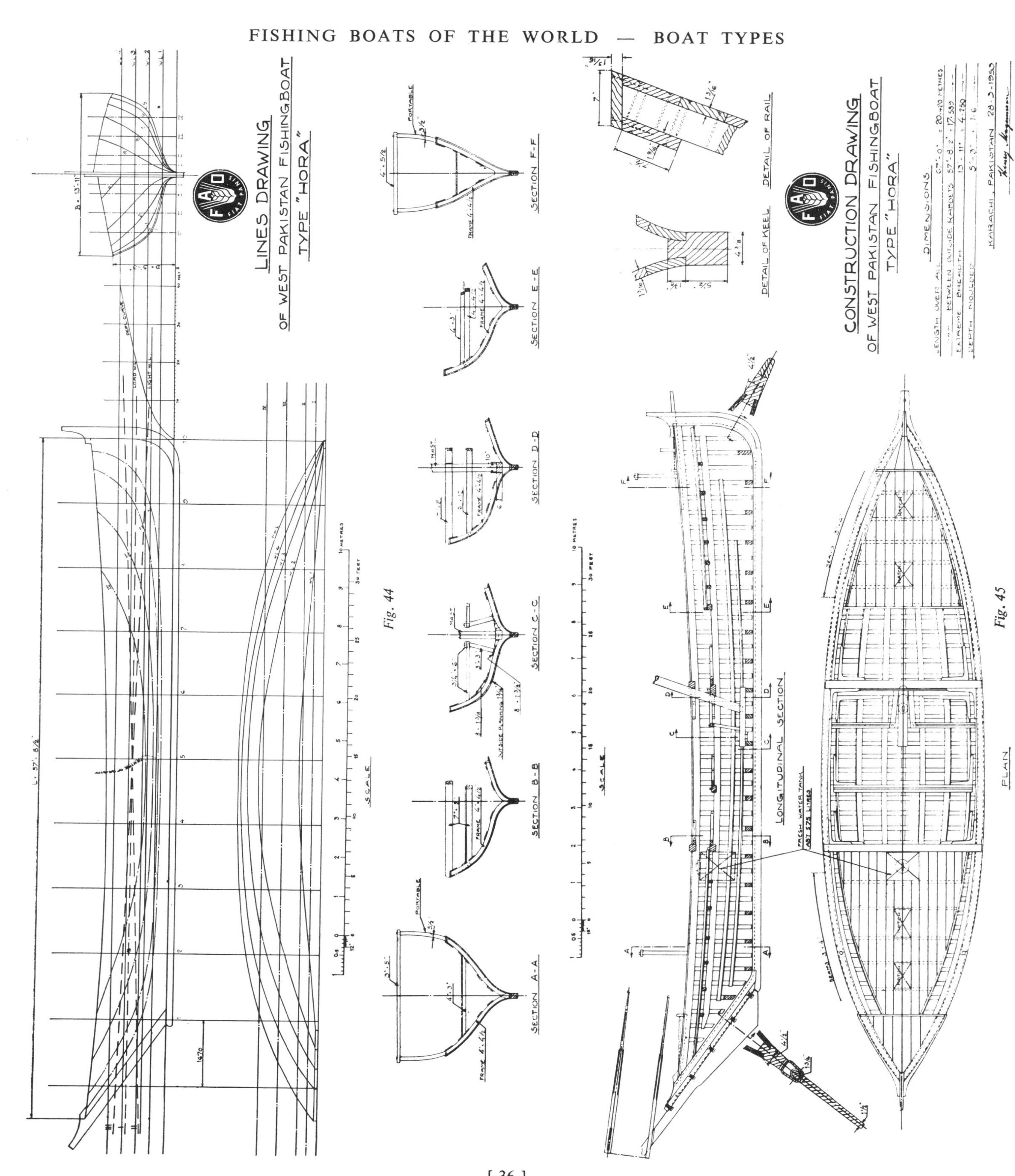
LINES DRAWING
OF WEST PAKISTAN FISHINGBOAT
TYPE "HORA"
SECTION A-A
SECTION B-B
SECTION C-C
SECTION D-D
SECTION E-E
SECTION F-F
SCALE
CONSTRUCTION DRAWING
OF WEST PAKISTAN FISHINGBOAT
TYPE "HORA"
DETAIL OF KEEL
DETAIL OF RAIL
DIMENSIONS
LONGITUDINAL SECTION
PLAN

Fig. 44

Fig. 45

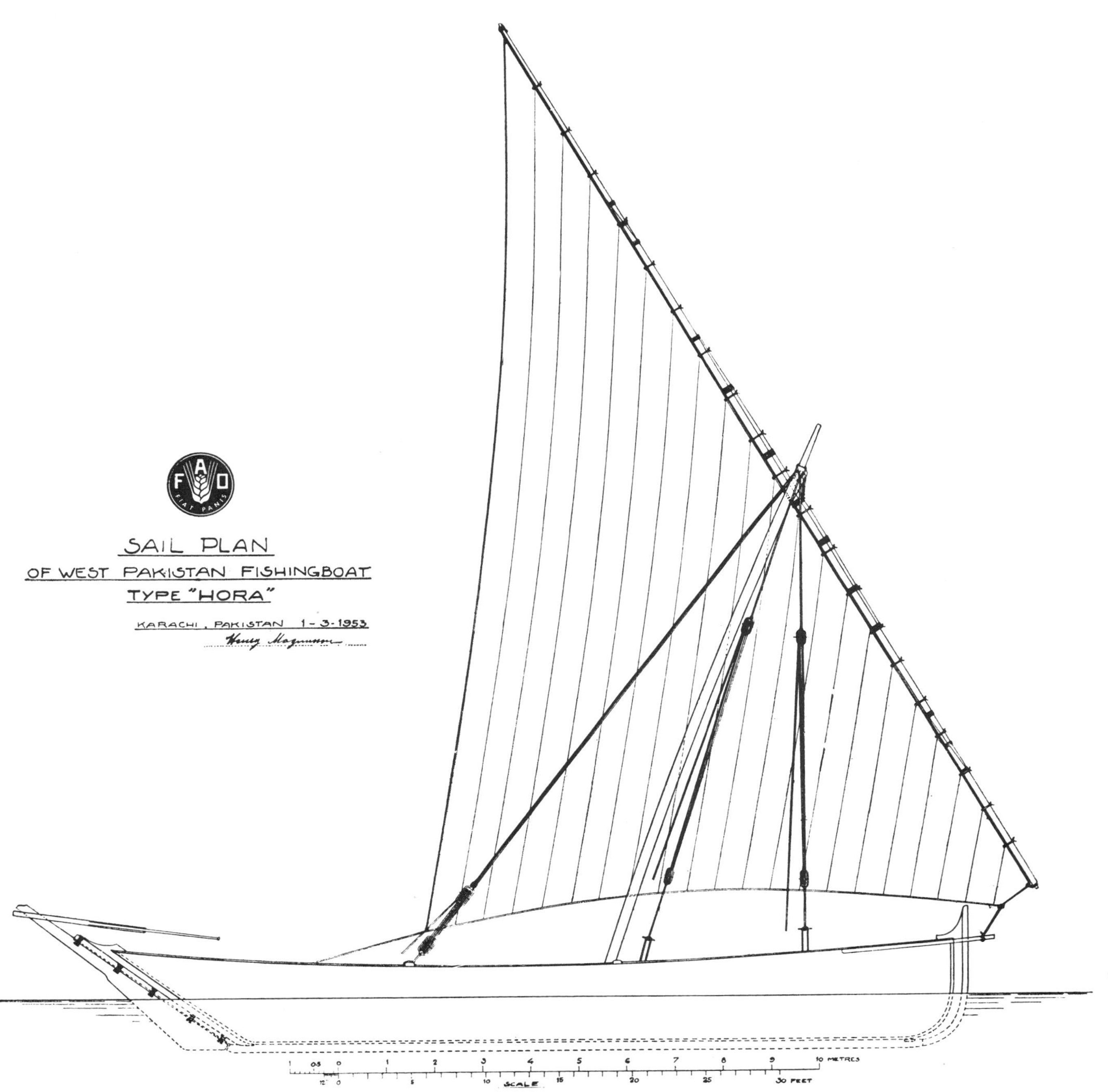
FAO
FIAT PANIS
SAIL PLAN
OF WEST PAKISTAN FISHINGBOAT
TYPE "HORA"
KARACHI, PAKISTAN 1-3-1953
10 METRES
SCALE
30 FEET

Fig. 46

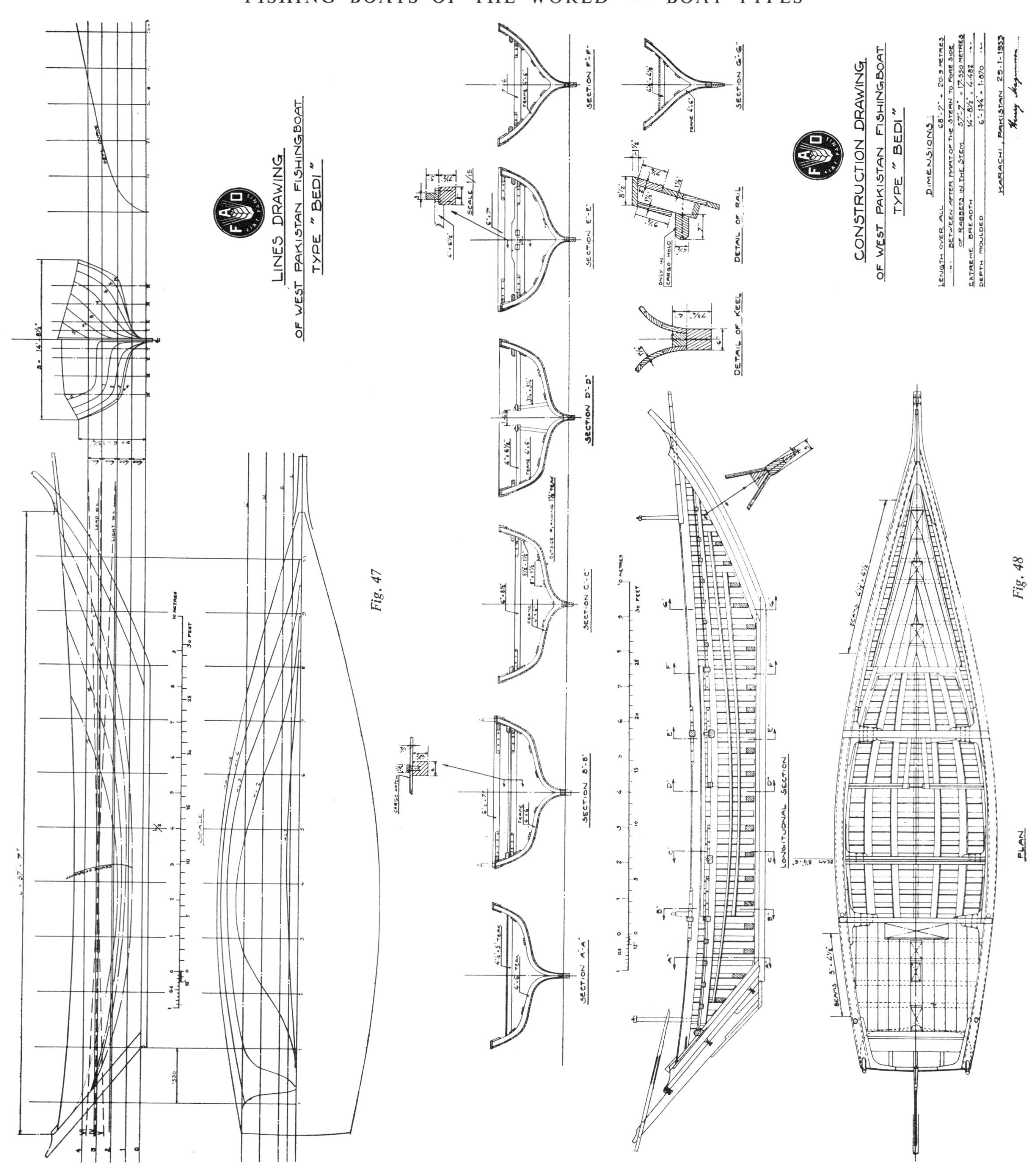
LINES DRAWING
OF WEST PAKISTAN FISHINGBOAT
TYPE "BEDI"
SCALE
SECTION A'-A'
SECTION B'-B'
SECTION C'-C'
SECTION D'-D'
SECTION E'-E'
SECTION F'-F'
SECTION G'-G'
DETAIL OF KEEL
DETAIL OF RAIL
LONGITUDINAL SECTION
PLAN
CONSTRUCTION DRAWING
OF WEST PAKISTAN FISHINGBOAT
TYPE "BEDI"
DIMENSIONS
LENGTH OVER ALL 68'-7" = 20.9 METRES
" BETWEEN AFTER PART OF THE STERN TO FORE SIDE OF RABBETS IN THE STEM 57'-7" = 17.550 METRES
EXTREME BREADTH 14'-8½" = 4.482 "
DEPTH MOULDED 6'-1¾" = 1.870 "
KARACHI, PAKISTAN 23-1-1953

Fig. 47

Fig. 48

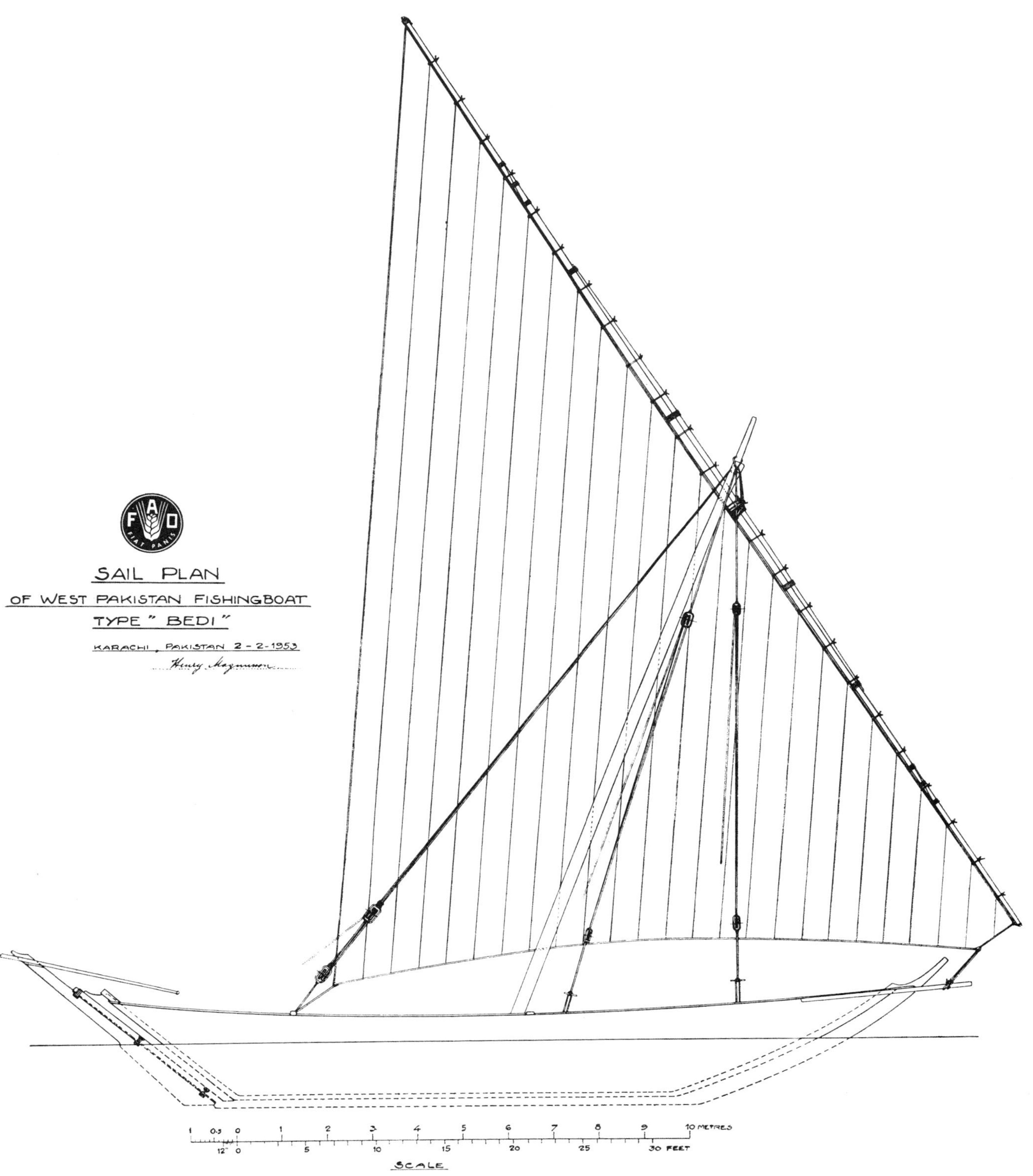
FAO
FIAT PANIS
SAIL PLAN
OF WEST PAKISTAN FISHINGBOAT
TYPE " BEDI "
KARACHI, PAKISTAN 2 - 2 - 1953
Henry Magnusson
1 0.5 0 1 2 3 4 5 6 7 8 9 10 METRES
12" 0 5 10 15 20 25 30 FEET
SCALE

Fig. 49

TABLE III

Bedi SSPA 588-A

Trim		*Level Keel*	*By Head*	*By Stern*
Series		IV	V	VI
L	ft.	48.6	49.6	47.6
	m.	14.85	15.15	14.54
B	ft.	13	13	13
	m.	3.97	3.97	3.97
T	ft.	4.0	4.2	3.9
	m.	1.22	1.28	1.19
∇	cu. ft.	707	707	707
	cu. m.	20	20	20
Am	sq. ft.	28.5	27.2	29.4
	sq. m.	2.65	2.53	2.74
δ		.280	.260	.292
φ		.510	.522	.502
$\odot$		−9.4	−7.45	−11.8
$\frac{1}{2}\alpha_E$		8°	11°	6°
$L/\nabla^{\frac{1}{3}}$		5.48	5.59	5.36

the Pakistan boats are equally good, if not better, than typical fishing boats in use in Europe to-day. Because of their higher length-displacement ratio, they show low C_1 values at lower speeds, but the required power is so low anyway that it does not matter.

The aft body of the bedi boat model was changed in order to study the effect of different sizes of transoms and a canoe stern. Three modifications were tested. Full beam was assumed to represent 100 per cent. transom, thus giving the bedi model SSPA 588A a transom width of 68 per cent. and the models B, C, and D, 44, 22 and 0 per cent. respectively.

The main dimensions of the three models during the tests are listed in Table V. The three transoms and the canoe stern are shown in fig. 62 while fig. 63 gives C_1 values at three Froude's numbers with trim conditions, and fig. 64 gives P_e (effective tow-rope HP) values for the models run at level keel. The tests show that the present transom on the bedi boat is somewhat too large. Furthermore the 44 per cent. transom B does not change the C_1 values appreciably with different trims and gives the highest average C_1 value for all speeds, thus indicating the best size.

The curves show that by reducing the transom too much the resistance increases considerably at higher speeds. Trim by the head gives, in such cases, better results. This somewhat startling fact may perhaps be explained by the fact that, with the original transom at higher speeds, a head trim gave better results (see fig. 60). On the other hand stern trim with the smaller transoms is slightly better than level keel trim but not as good as head trim; (stern trim with the original transom was much worse than the level keel trim). The level keel tests indicate that a 44 per cent. transom is better than one of 22 per cent. But when the 22 per cent. transom is trimmed by the stern, it increases in size towards 44 per cent. and by this means the loss sustained in reducing the transom to 22 per cent. is offset.

Diagrams, fig. 65 and 66, show power estimates for the different types of hora and bedi boats, although only the largest of each type was tested in the tank. The curves are, therefore, assumptions, but they promise to be fairly accurate because the boats can be regarded as geometrically similar. The power curves for the two smaller hora and the three smaller bedi boats were obtained by using the C_1 curves of the tested boats. Power values were then calculated for the appropriate Froude's numbers and plotted on a base of knots. The diagrams show effective (tow rope) horse power. If the

TABLE IV

Data of Fishing Boats compared in Fig. 61

		L	*B*	*T*	∇ *cu. m.*	$L/\nabla^{\frac{1}{3}}$	δ	φ	$\odot$	$\frac{1}{2}\alpha_E$
DENNY 0	ft.	62	17.83	6.00	71	4.58	.383	.645	−1.25	26.5°
	m.	18.9	5.44	1.83						
SMT 25A	ft.	64.5	18	7.64						
	m.	19.66	5.49	2.33	119.2	4.00	.476	.648	−0.18	34°
SMT 28A	ft.	39.4	13	5.54						
	m.	12.00	3.97	1.69	35	3.67	.435	.634	−0.5	34°
SSPA 17	ft.	62	19.8	8.2						
	m.	18.9	6.04	2.5	112	3.92	.393	.628	0	33°
VNR C479	ft.	48	13	5.47						
(ROMA)	m.	14.64	3.95	1.67	51.75	3.92	.534			
SSPA 587A III	ft.	56.1	12	3.11						
	m.	17.12	3.65	0.95	20	6.31	.338	.558	−5.97	9
588A IV	ft.	48.6	13	4						
	m.	14.85	3.97	1.22	20	5.48	.280	.510	−9.40	8°

TABLE V

Bedi SSPA 588B, C, D

Trim			*Level Keel*	*By Head*	*By Stern*
L		ft. m.	48.6 14.85	49.6 15.15	47.6 14.54
B		ft. m.	12.0 3.97	12.0 3.97	12.0 3.97
T		ft. m.	4.0 1.22	4.2 1.28	3.9 1.19
	588B	cu. ft. cu. m.	698 19.75	700 19.8	696 19.7
∇	588C	cu. ft. cu. m.	689 19.5	700 19.8	686 19.4
	588D	cu. ft. cu. m.	680 19.25	695 19.65	674 19.06
Am		sq. ft. sq. m.	28.5 2.65	27.2 2.53	29.4 2.74
	588B		.275	.257	.287
δ	588C		.271	.257	.282
	588D		.267	.255	.277
	588B		.502	.516	.495
φ	588C		.496	.516	.486
	588D		.489	.512	.478
	588B		−7.4	−3.75	−11.5
$\odot$	588C		−7.16	−3.24	−11.2
	588D		−6.7	−3.07	−10.6
$\frac{1}{2}\alpha E$			8°	11°	6°
	588B		5.5	5.6	5.3
$L/\nabla^{\frac{1}{3}}$	588C		5.51	5.6	5.42
	588D		5.54	5.61	5.45

efficiency of propulsion is 50 per cent., the required engine power will be twice the effective power shown. Tables VI and VII contain the main dimensions corresponding to the curves for the hora and bedi boats. The displacement corresponds to the loaded waterline.

Fig. 65—hora boats—shows that for a speed of 7 to 8 knots the power requirements vary considerably in the different boats. This is due to the fact that all the boats are relatively long but there are big differences in the displacements of the three types.

Fig. 66—bedi boats—shows that all types at 7.5 knots require about 10 effective (tow rope) h.p. or about 20 h.p. engines. Over 7.6 knots the smaller boats require more power than the larger ones. This is because the length decreases much more than in the case of the hora boats, therefore the boats will run with higher Froude's numbers and with less C_1 values, which offset the reduction of power made possible by smaller displacement.

The bedi boats, because of their transom and larger deck area, seem to be more suitable for mechanization. It might then be practical to standardize a 20 h.p. engine for all boats, giving them the same speed and making it possible to import engines of the same size. This would probably lessen the problem of keeping spare parts in stock.

TABLE VI

Hora Boats

Dimensions corresponding to loaded waterline and power estimate in fig. 65

Type *Model*	*Fig.*	*44, 45, 46,* *587A*	*41, 42, 43*	*39, 40*
L	ft. m.	55.1 16.8	44.6 13.6	42.6 13.0
B	ft. m.	12.0 3.65	8.73 2.66	5.90 1.80
T	ft. m.	3.15 0.96	2.46 0.75	1.7 0.52
∇	cu. ft. cu. m.	707 20	395 11.2	176.5 5.0
δ		0.340	0.413	0.410
φ		0.596	0.566	0.557
$L/\nabla^{\frac{1}{3}}$		6.15	6.08	7.6

TABLE VII

Bedi Boats

Dimensions corresponding to loaded waterline and power estimate in fig. 66

Type *Model*	*Fig.*	*47, 48, 49* *588A*	*50, 51, 52*	*53, 54, 55*	*56, 57, 58*
L	ft. m.	48.6 14.85	38.20 11.65	34.40 10.48	30.45 9.28
B	ft. m.	13.00 3.97	11.45 3.5	9.67 2.95	8.59 2.62
T	ft. m.	4.0 1.22	3.71 1.13	3.45 1.05	2.95 0.9
∇	cu. ft. cu. m.	707 20	459 13	353.5 10	229.5 6.5
δ		0.280	0.282	0.309	0.297
φ		0.510	0.557	0.561	0.540
$L/\nabla^{\frac{1}{3}}$		5.48	4.96	4.86	4.98

APPENDIX A

DESCRIPTION OF THE RADIAL MEASUREMENT METHOD (See fig. 67 on page 48)

The following method enables one man only to take off the lines of a boat.

The boat is first divided in its length into several stations, normally eight, which do not need to have equal distances

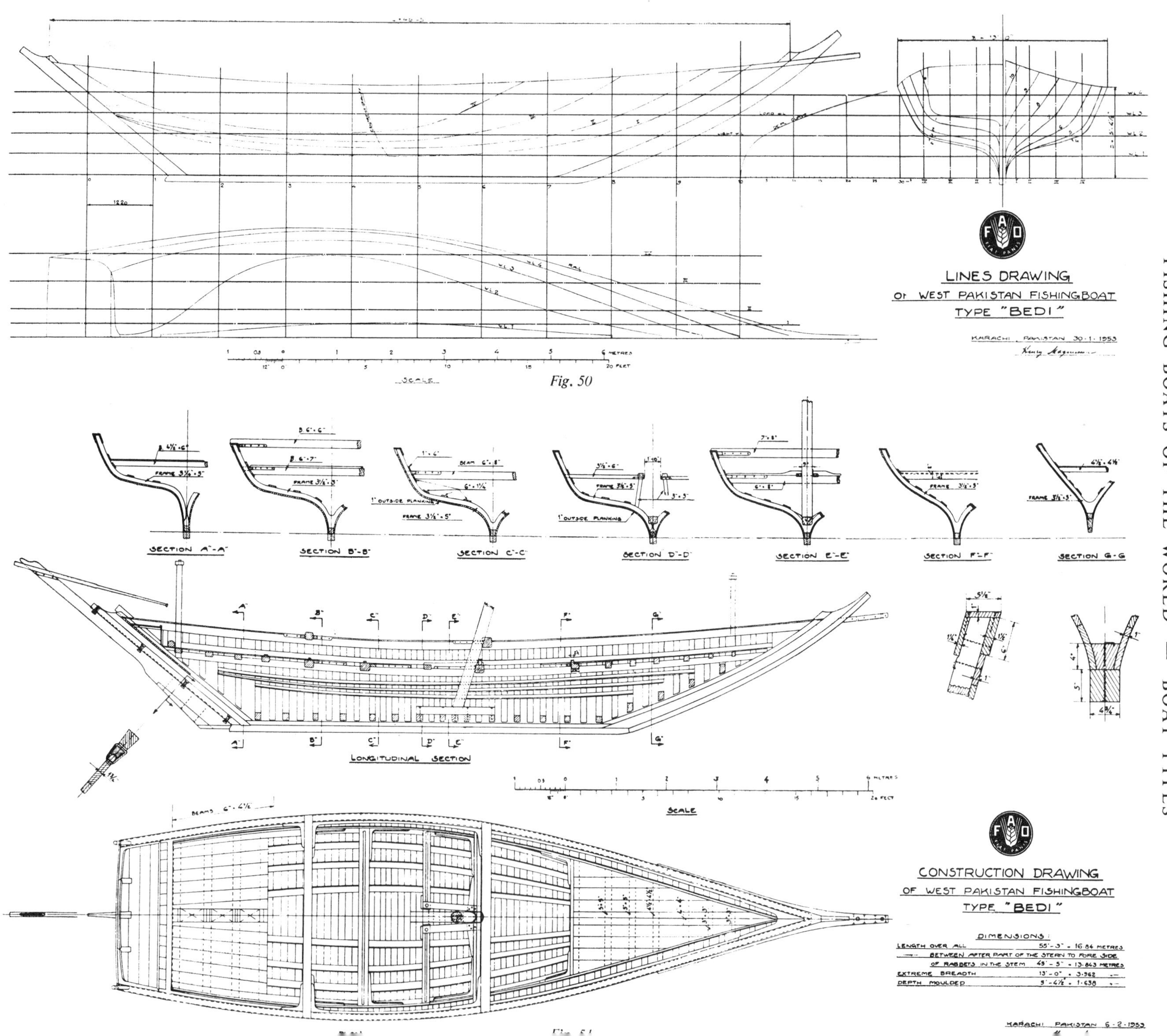
LINES DRAWING
OF WEST PAKISTAN FISHINGBOAT
TYPE "BEDI"
KARACHI PAKISTAN 30-1-1953
SCALE
SECTION A'-A'
SECTION B'-B'
SECTION C'-C'
SECTION D'-D'
SECTION E'-E'
SECTION F'-F'
SECTION G-G
LONGITUDINAL SECTION
SCALE
CONSTRUCTION DRAWING
OF WEST PAKISTAN FISHINGBOAT
TYPE "BEDI"
DIMENSIONS:
LENGTH OVER ALL 55'-3" = 16.84 METRES
BETWEEN AFTER PART OF THE STERN TO FORE SIDE OF RABBETS IN THE STEM 45'-5" = 13.843 METRES
EXTREME BREADTH 13'-0" = 3.962
DEPTH MOULDED 5'-4½" = 1.638
KARACHI PAKISTAN 6-2-1953
Fig. 50

Fig. 51

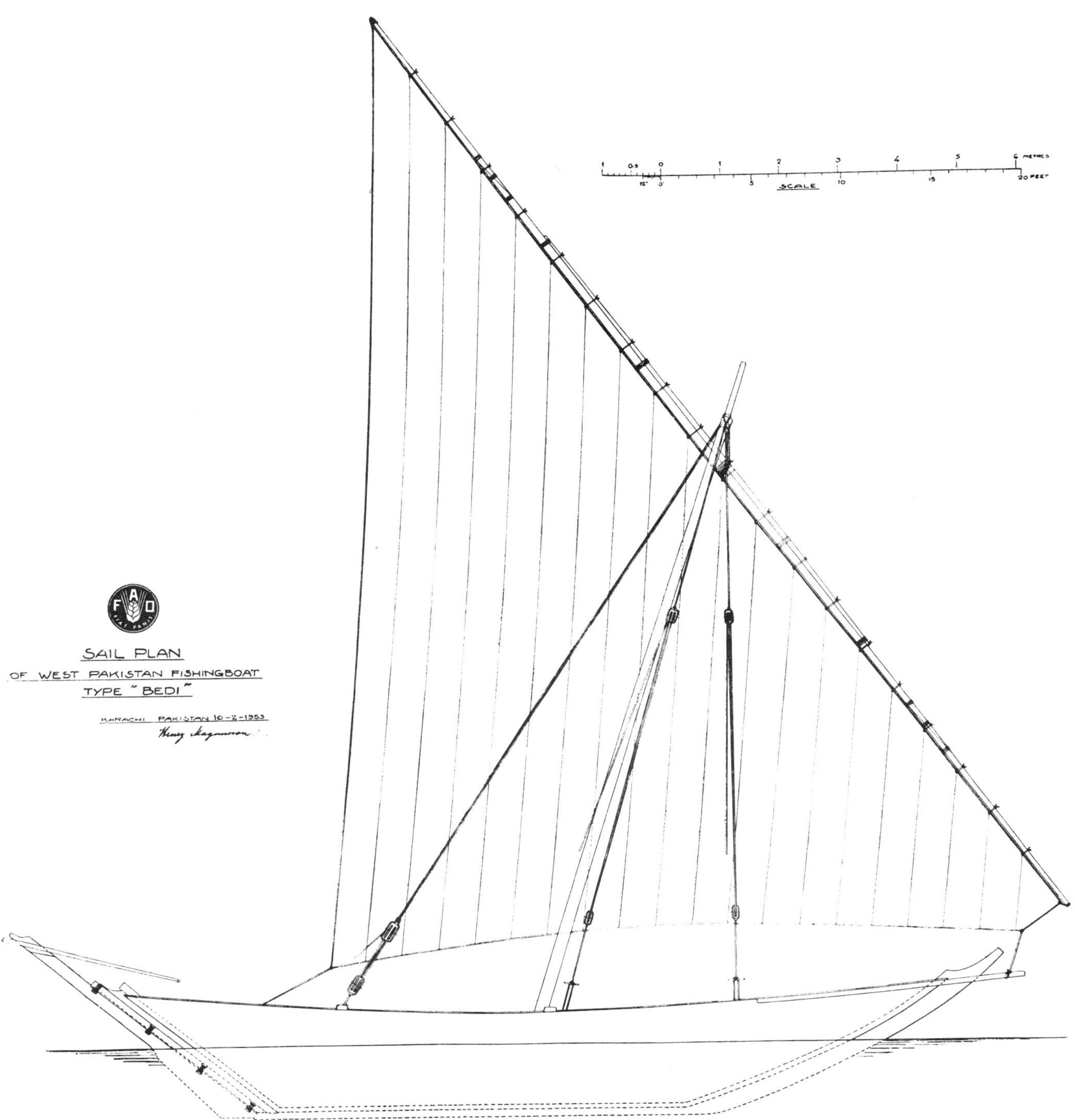
1 0.5 0 1 2 3 4 5 6 METRES
12" 0' 5 10 15 20 FEET
SCALE
FAO
FIAT PANIS
SAIL PLAN
OF WEST PAKISTAN FISHINGBOAT
TYPE "BEDI"
KARACHI PAKISTAN 10-2-1953
Henry Magnusson

Fig. 52

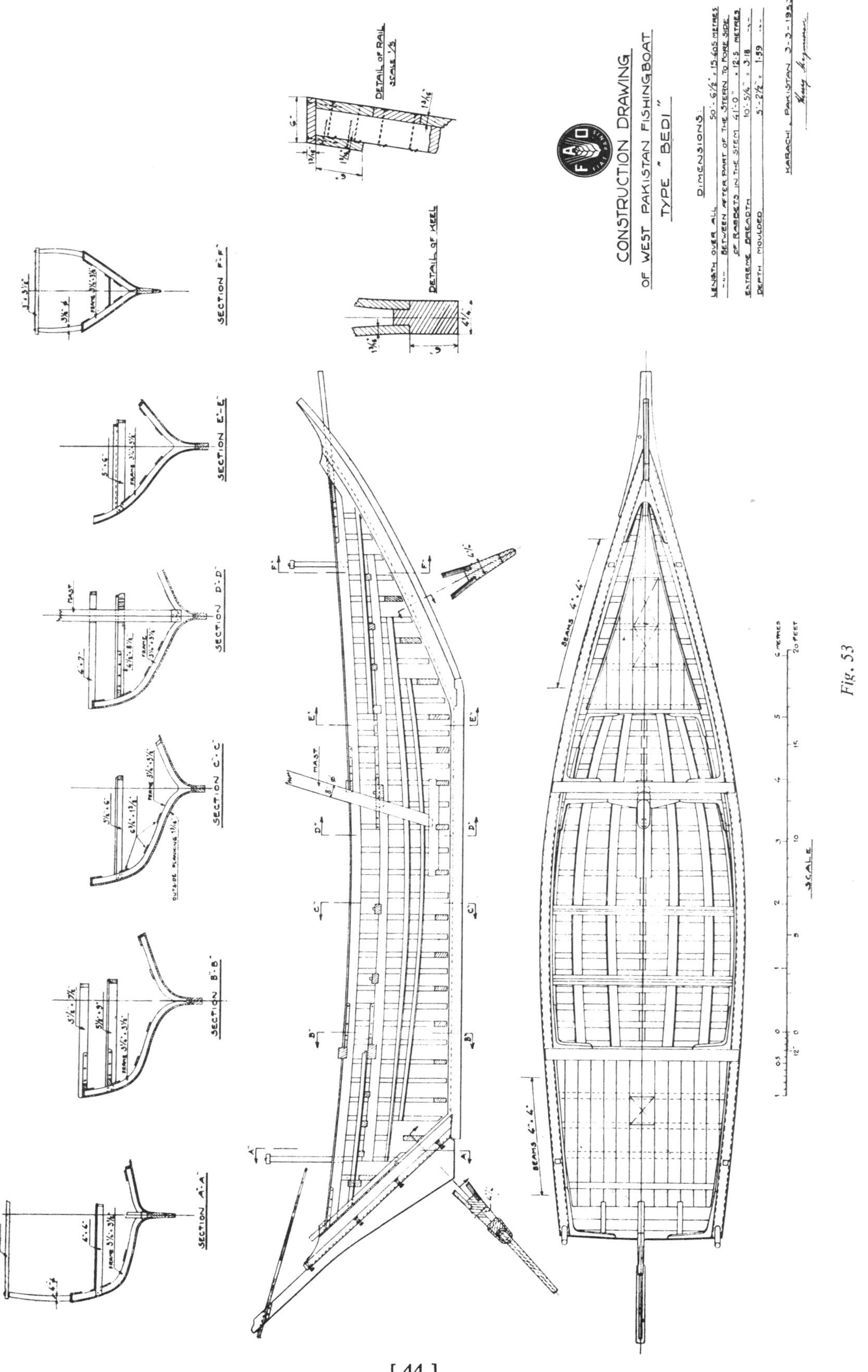
FAO
FIAT PANIS
CONSTRUCTION DRAWING
OF WEST PAKISTAN FISHING BOAT
TYPE " BEDI "
DIMENSIONS:
LENGTH OVER ALL 50'-6½" = 15·405 METRES
-"- BETWEEN AFTER PART OF THE STERN TO FORE SIDE OF RABBETS IN THE STEM 41'-0" = 12·5 METRES
EXTREME BREADTH 10'-5¼" = 3·18 -"-
DEPTH MOULDED 5'-2½" = 1·59 -"-
KARACHI , PAKISTAN 3-3-1953.
DETAIL OF RAIL SCALE 1/5
DETAIL OF KEEL
SECTION A'-A'
SECTION B'-B'
SECTION C'-C'
SECTION D'-D'
SECTION E'-E'
SECTION F'-F'
MAST
SCALE

Fig. 53

between them. They can be arranged so that they do not coincide with places where the boat is supported by stanchions, etc. The stations fore and aft should be somewhat closer than those of the middle to get fuller details of the curves. All the stations must be chalked on the hull, and the distances between them measured.

The rake and bend on the stem or stern posts are measured from a station vertical to the keel, or they can be measured according to the radial method described below. When measuring the sheer line, one places a plank horizontally at right angles to the keel, checking it with an ordinary bubble-level. The vertical distance from the plan to the gunwale is then measured at the different stations. If the keel of the boat is not horizontal, one measures the deviation at the stem and stern post to get a correct datum line for determining the outboard profile of the boat. The breadth of the hull is measured on deck at the different stations, or wherever possible, so that one can arrive at a correct deck line.

After having drawn the profile and the contour of the deck, one starts with the sections. On each station one marks a number of places which can be called, for example, Nos. 1 to 9 (fig. 67). The number will depend on the size and the shape of the section. Each section has to be at a right angle to the keel line (the centre line of the boat). Then one has to take two fixed points A and B in the same plane as the section as the general points for the radial measurements. One point can be selected on the ground, by placing a plank there and the other can be chosen on a ladder or a plank placed between the gunwale and the ground, as can be seen from fig. 67. In the two foci, A and B, one fixes the measurement tape with a nail and the radii A–1, A–2, A–3, etc., and B–1, B–2, B–3 are measured. The same procedure is repeated for each section.

In working out a lines plan one first draws the profile and the deck contour. These determine the position of the keel and gunwale in the body-plan. With these points determined it is easy to circle-in the focal points A and B using the radii A–1 and A–9 and B–1 and B–9 respectively. After A and B have been found, it is easy to determine the section by drawing a line through the bisection points of the arcs of radii A_2, B_2; A_3, B_3, etc., swung from A and B. The resulting body-plan will have the sections at different distances from each other. From this, longitudinals, water-lines, buttocks and diagonals must be drawn next. It will then be easy to draw a new body-plan with equal distances between the sections.

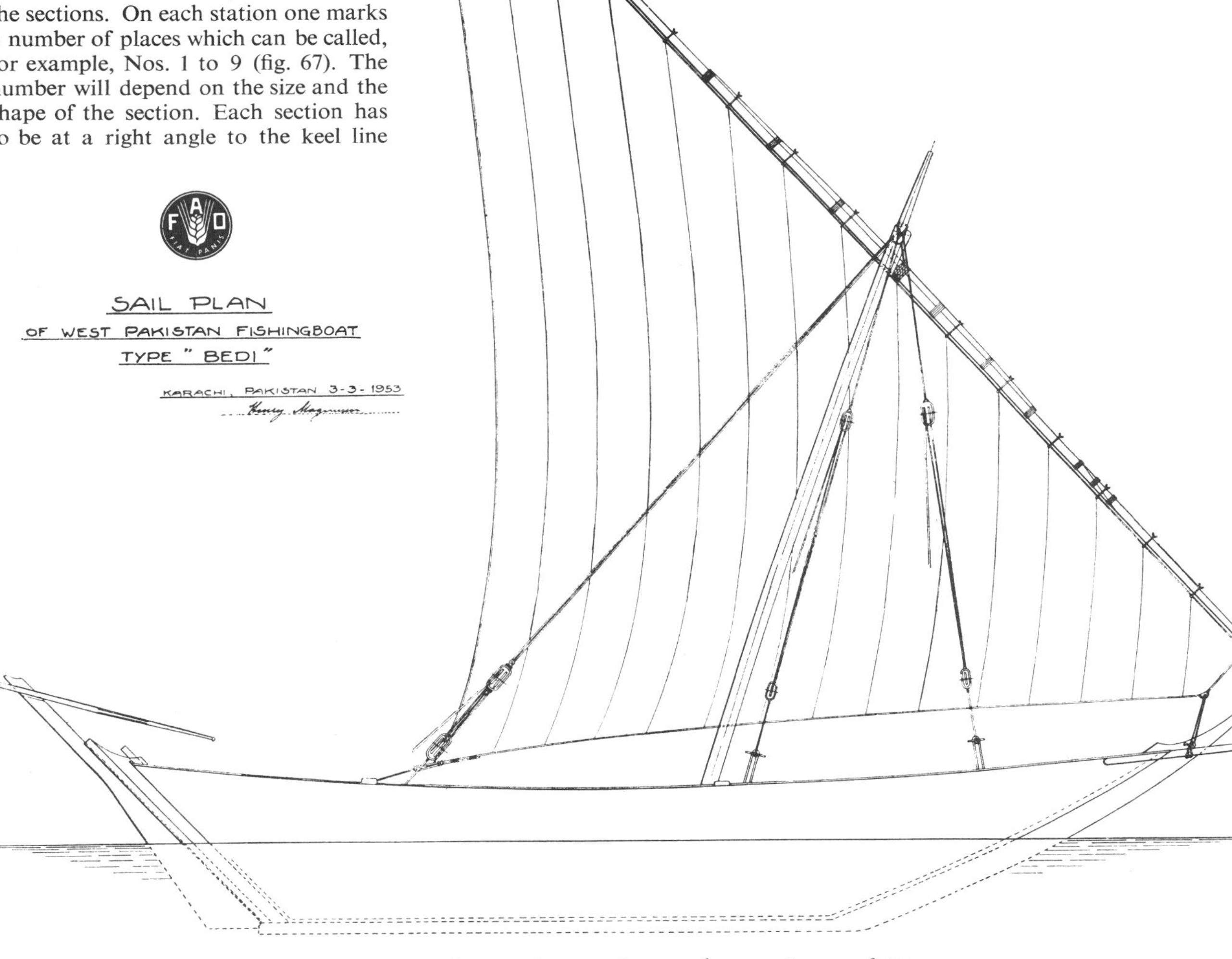

Fig. 54

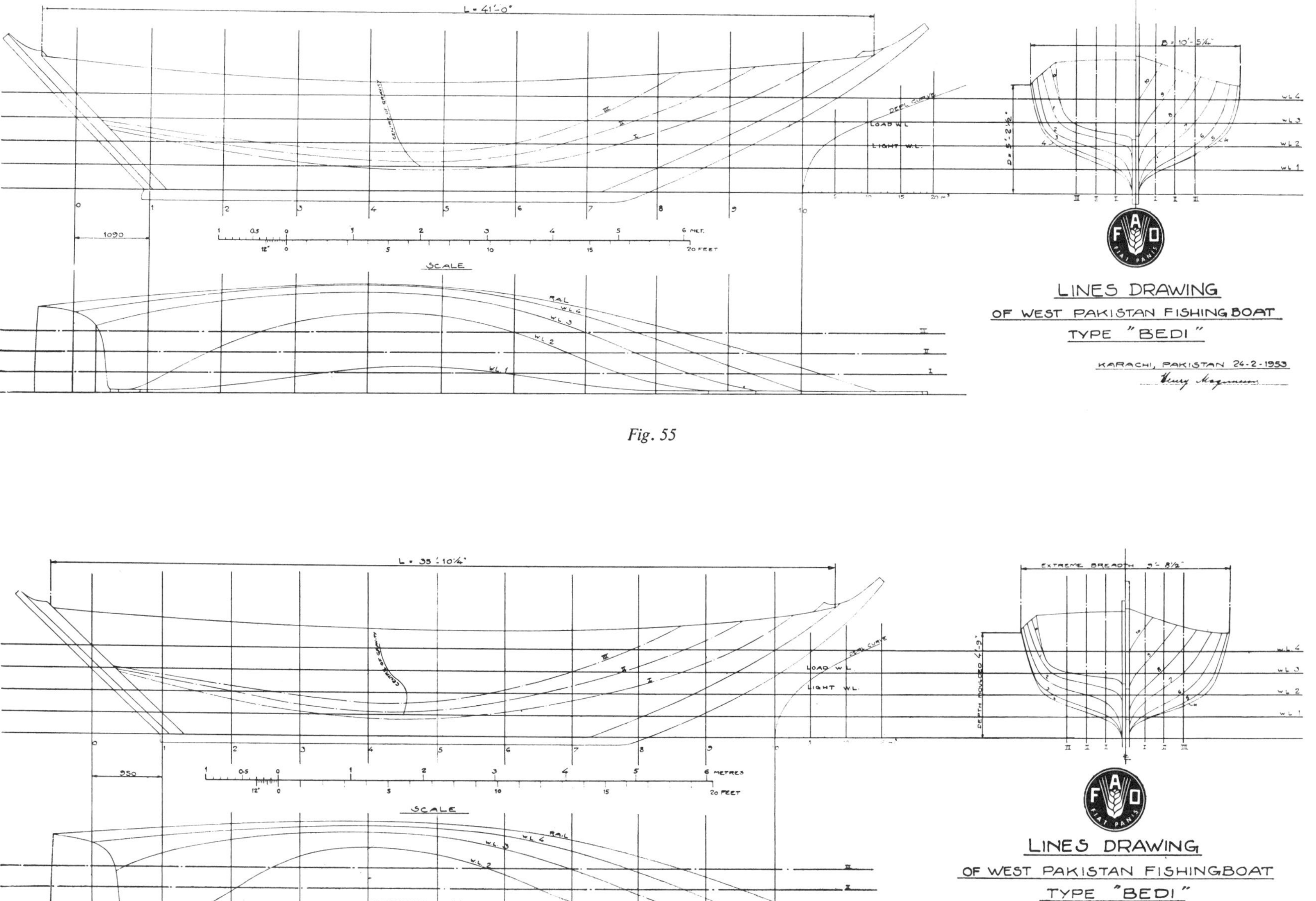
L = 41'-0"
B = 10'-5¼"
D = 5'-2½"
LOAD W.L.
LIGHT W.L.
1090
SCALE
LINES DRAWING
OF WEST PAKISTAN FISHING BOAT
TYPE "BEDI"
KARACHI, PAKISTAN 24-2-1953
L = 35'-10¼"
EXTREME BREADTH
LOAD W.L.
LIGHT W.L.
550
SCALE
LINES DRAWING
OF WEST PAKISTAN FISHINGBOAT
TYPE "BEDI"
KARACHI, PAKISTAN 18-2-1953

Fig. 55

Fig. 56

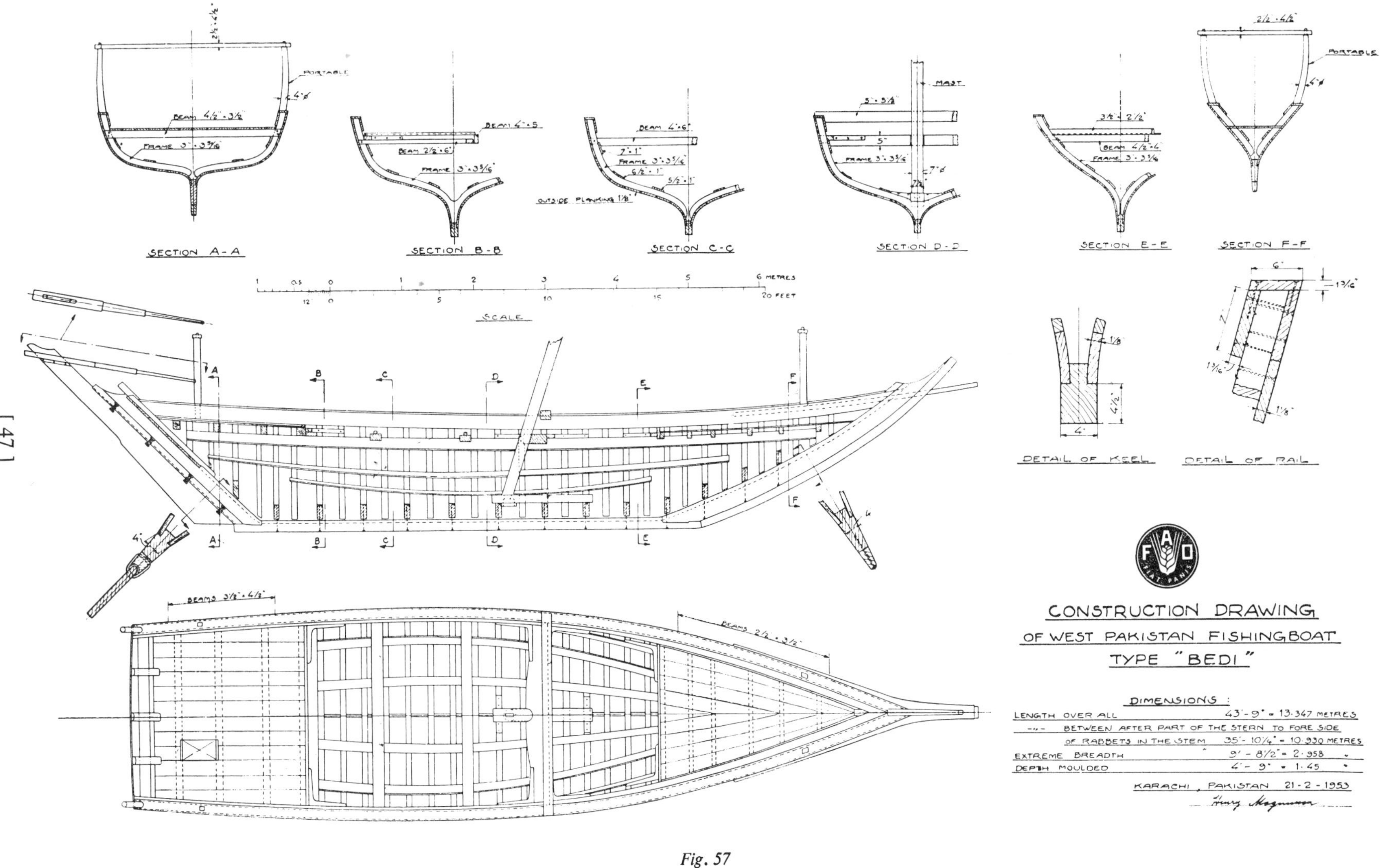

SECTION A-A
SECTION B-B
SECTION C-C
SECTION D-D
SECTION E-E
SECTION F-F
PORTABLE
MAST
OUTSIDE PLANKING 1⅛"
SCALE
6 METRES
20 FEET
DETAIL OF KEEL
DETAIL OF RAIL
CONSTRUCTION DRAWING
OF WEST PAKISTAN FISHINGBOAT
TYPE "BEDI"
DIMENSIONS :
LENGTH OVER ALL 43'-9" = 13·347 METRES
-"- BETWEEN AFTER PART OF THE STERN TO FORE SIDE OF RABBETS IN THE STEM 35'-10¼" = 10·930 METRES
EXTREME BREADTH 9'-8½" = 2·958 "
DEPTH MOULDED 4'-9" = 1·45 "
KARACHI, PAKISTAN 21-2-1953

Fig. 57

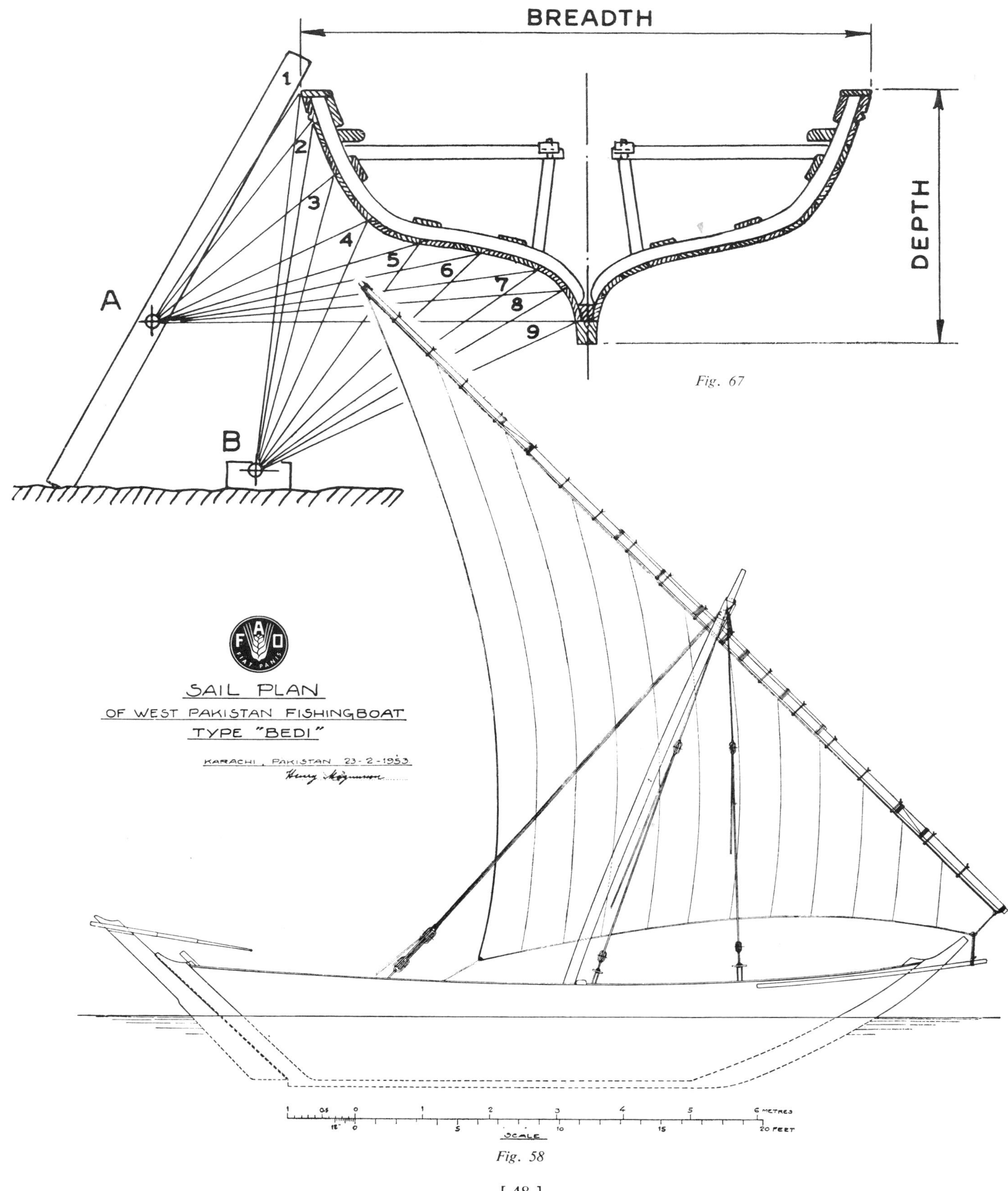
BREADTH
DEPTH
1
2
3
4
5
6
7
8
9
A
B
FAO
FIAT PANIS
SAIL PLAN
OF WEST PAKISTAN FISHINGBOAT
TYPE "BEDI"
KARACHI, PAKISTAN 23-2-1953
1 0.5 0 1 2 3 4 5 6 METRES
12" 0 5 10 15 20 FEET
SCALE

Fig. 67

Fig. 58

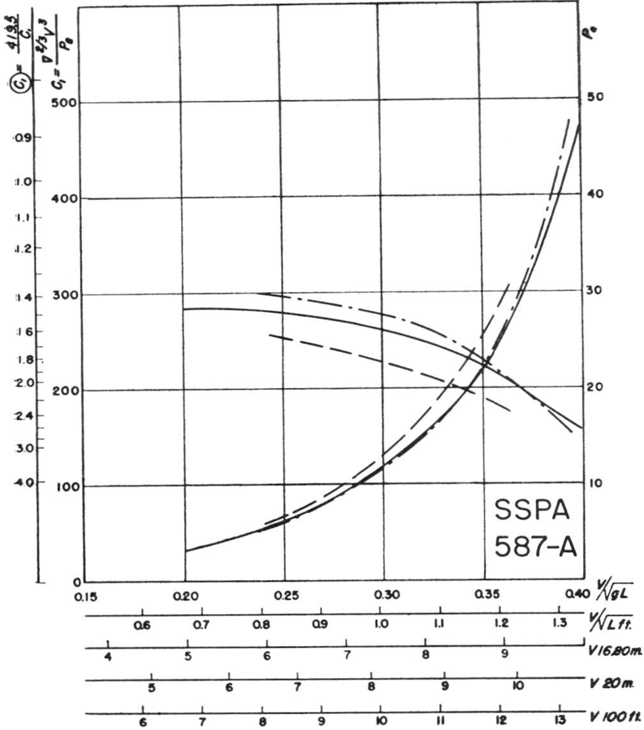

Fig. 59. C_1 and P_e (tow-rope h.p.) curves for hora model 587-A according to fig. 44 and Table II. Trim level keel, series I (———). Trim by head, series II (— — —). Trim by stern, series III (— · — · —). Stern trim is somewhat better, but head trim worse

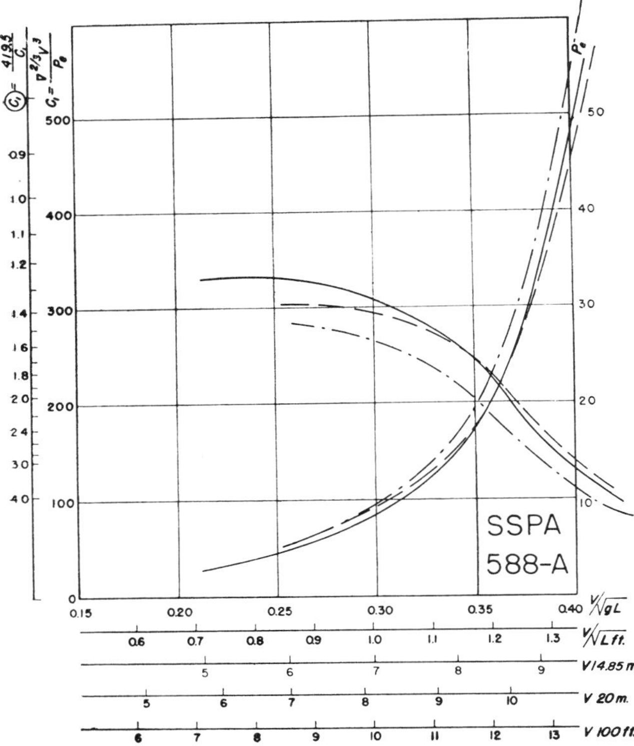

Fig. 60. C_1 and P_e (tow-rope h.p.) curves for bedi model 588-A according to fig. 47 and Table III. Trim level keel, series IV (———). Trim by head, series V (— — — —). Trim by stern, series VI (— · — · — · —). Head trim is slightly better at very high speed, but stern trim much worse at all speeds

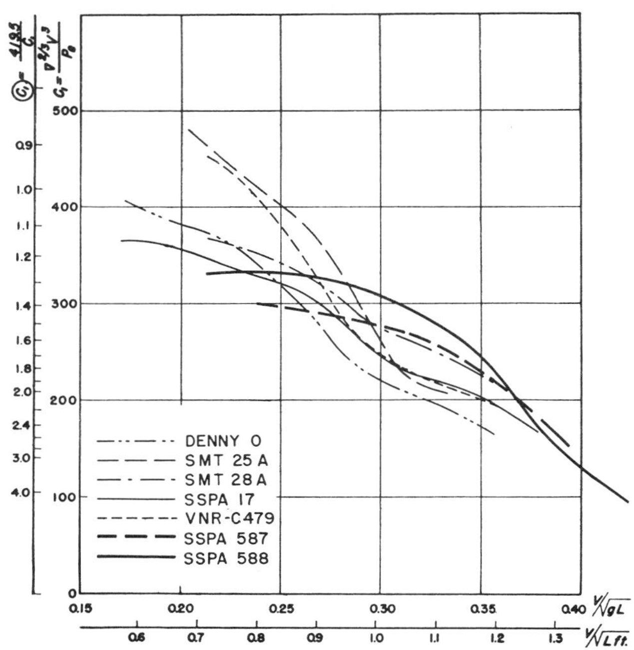

Fig. 61. Hora boat, series III (stern trim), and bedi boat, series VI (level keel trim), compared with different European fishing boats of original design, e.g. before they were by aid of tank tests improved

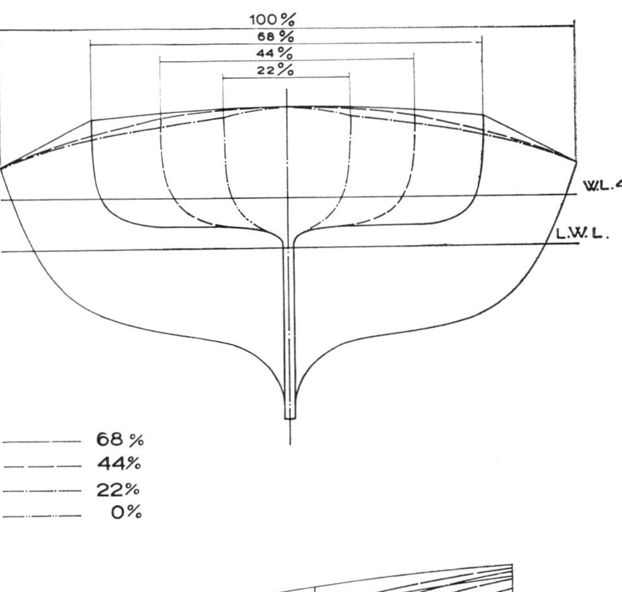

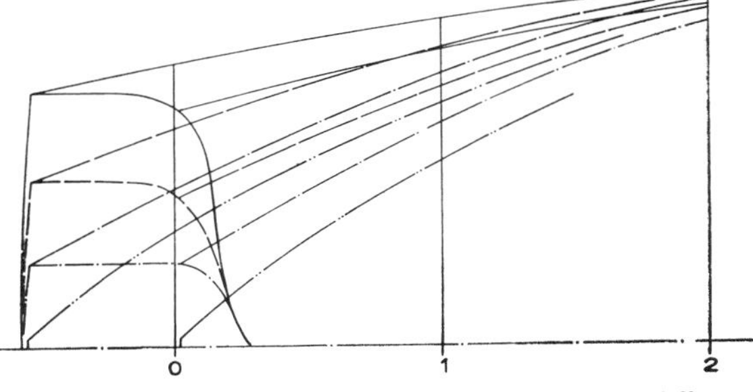

Fig. 62. The bedi model 588-A, was tried with different sizes of transom sterns and a canoe stern. If the width of the midship section is taken as 100 per cent., then the original transom A is 68 per cent. (———), and the three alternatives are 44 per cent. (B — — — —), 22 per cent. (C — · — · —), and 0 per cent. (D — ·· — ·· —) alternatively. For dimensions see Table V and test results fig. 63 and 64

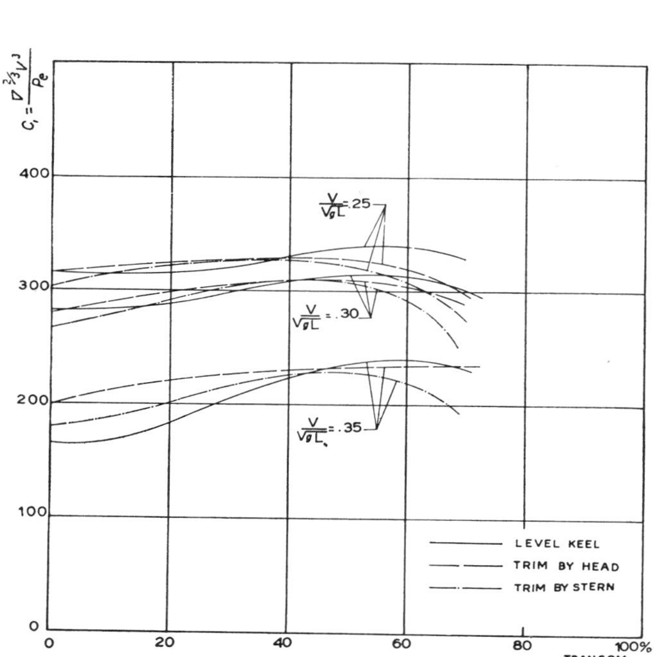

Fig. 63. C_1 values at different Froude's numbers for the different transoms and canoe stern (0 per cent.) tried out on the bedi model 588. Level trim is better with the large transom but head trim better with the smaller one. The best transom is somewhat smaller than the original one

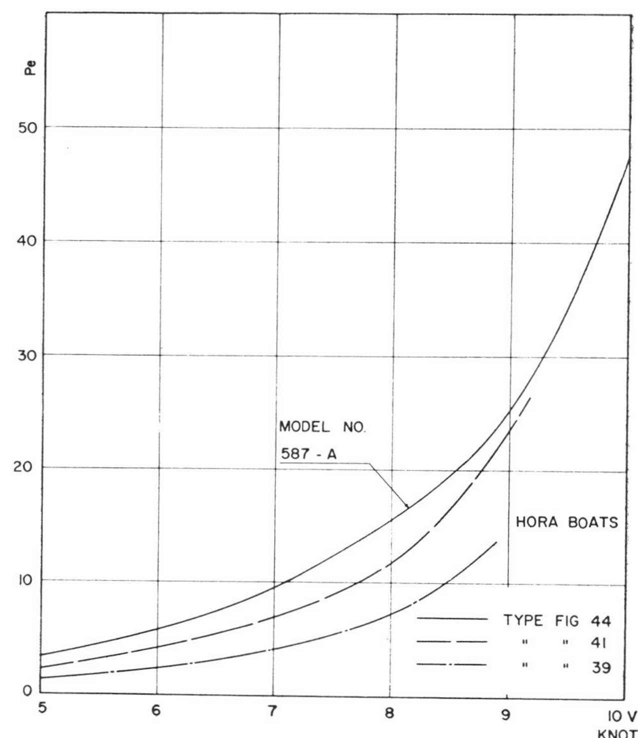

Fig. 65. P_e (tow-rope h.p.) curves for the hora boats according to fig. 39, 41 and 44 and listed in Table VI. The curves have been calculated on the results of the tank test with model 587-A. The necessary power of engines to propel the boats is about twice the P_e

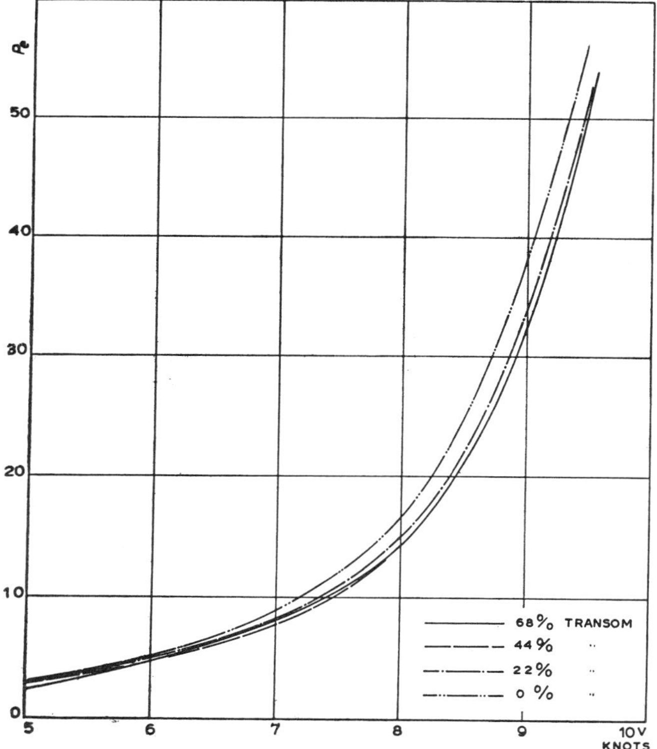

Fig. 64. P_e (tow-rope h.p.) curves for the transom tests shown in fig. 62 and 63

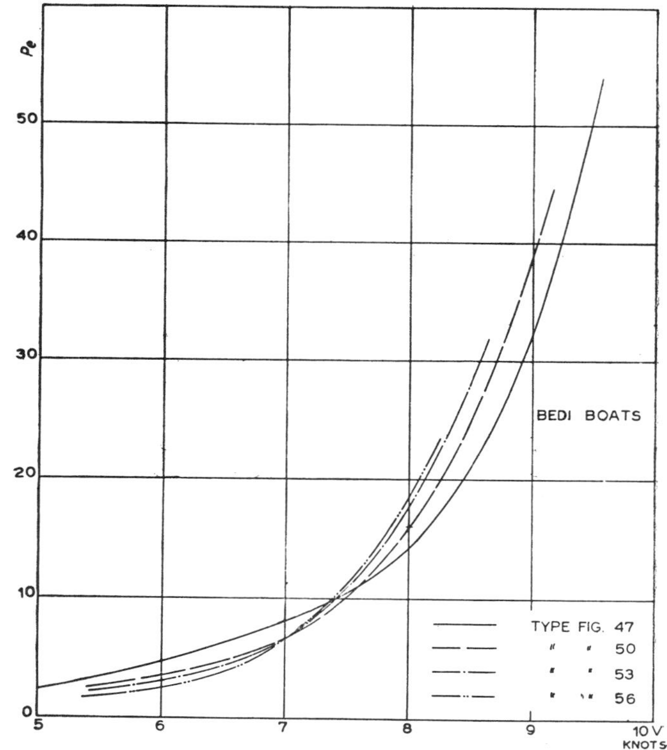

Fig. 66. P_e (tow-rope h.p.) curves for the bedi boats according to fig. 47, 50, 53 and 56 and listed in Table VII. The curves have been calculated on the results of the tank test with Model 588-A. The necessary power of engines to propel the boats is about twice the P

EUROPEAN BEACH LANDING CRAFT

by

HANS K. ZIMMER

THERE are many places where natural harbours do not exist, where the construction of harbours is too expensive or where water and current conditions make such construction virtually impossible. Fishing has therefore been carried out from open beaches in many places, quite large and efficient boats having been designed for this purpose, and such fisheries have often assumed great economic importance.

Many countries, now in the process of developing fisheries, would benefit from studying the designs of existing beach landing craft in Europe and the techniques of operation which have been successfully developed by generations of fishermen.

DENMARK

Beach Fishing Places

The main Danish beach fishing places are on the north-east coast of Jutland, between the natural harbour of Tyborøn, in the south, and the artificial harbour of Hirtshals, in the north. The weather is rough and there are not more than about 120 fishing days a year. The beaches are fairly steep and sandreefs are seldom built up outside the landing places. The unprotected parts of the beach are not very stable and most of the landings have natural rocks or artificial stabilizers. This is necessary, as the boats are beached and launched by powerful winches installed in permanent buildings on the beach. To save transport, buildings for net stores, fish handling, freezing, etc., are also placed near the beach.

The boats are of strong, clinker-built wood construction. The hulls are relatively flat-bottomed and they are fuller than the usual Scandinavian fishing boats, especially in the midship section and the part of the stern that is above water. The boats range up to 40 ft. (12.19 m.) length over-all, and 12 gross registered tons (Danish measurement). The larger boats are propelled by hot-bulb internal combustion engines (semi-diesels) of 35 h.p. giving a speed of 7 to 8 knots.

Launching and Beaching

In the last century the fishing boats were smaller and were launched by man-power. With the introduction of the motor, boats became heavier and launching and beaching procedure consequently grew more difficult. A simple hand-winch was introduced to make the work easier, the bow of the boats being supplied with a sturdy steel eye-bolt to take a hook of the winch wire. Later, portable rollers on wooden flats were put under the keel instead of the usual plain wooden sleepers, fig. 68. The rollers are made of cast-iron in wooden bearings. A more modern version is made in Thisted (Krogh and Christensen) which rolls on a fixed steel shaft on a greased " needle " bearing. The rollers have two ropes attached to aid transport along the beach. While they greatly reduce friction they also elevate the boat, causing difficulty in keeping it transversely balanced. Fig. 69 shows a typical beaching at night. The searchlight is on and the shore crew is directing the boat to the rollers with the help of long sticks. Portable tripping stanchions are used to avoid the boat tipping over once on shore, fig. 70.

Crews from three boats usually work together as a team to launch the boats. The motor is started when the boat is on the beach, a special cooling water tank making this possible. With the exception of the helmsman on board, the combined crews push the boat out into the sea. The clutch is engaged as soon as the boat is in the water and, with the controllable pitch propeller in astern position, it is seaborne. All engine controls are in handy positions in the wheel-house. In the same manner, the second, third and last boats are launched but after the second launching the crews are reduced by the helmsmen of the first and second boats. As each boat has usually a crew of four, it means there are only nine men left to launch the last boat. When the last boat reaches the water, the launching crews jump on board with the help of a ladder or by steps on the stem itself, fig. 70, and are transferred to the other boats.

This procedure has the following disadvantages:

(*a*) beaching is still heavy work and, when using rollers, boats often are supported by one roller only which puts a great strain on the hull and there is a tendency to hog;

(*b*) launching is heavy manual work and can only be carried out by co-operation of the crews of a number of boats;

Fig. 68

Fig. 69

(*c*) a temptation to put the propeller in action before it is totally submerged, to aid in the last stages of the launching, causes undue wear on the stern-gear by the sandy water of the breakers;

(*d*) the boats have to come alongside each other in open sea to distribute the crews after launching so that damages to bulwarks and topsides are frequent.

flexible steel wire. The two driving handles fit both the primary and secondary shafts. The wooden construction is made of $6\frac{1}{2}\times 3$ in. (16.5×7.6 cm.) planks with stiffeners and a ground-frame of $5\frac{1}{2}\times 2\frac{1}{2}$ in. (13.9×6.3 cm.) planks. The connections are dove-tailed and stiffened with steel straps. The winches are permanently anchored to the beach and require very little maintenance.

At a number of landing places such as Stenbjerg, Vorupør, Klitmøller and Lild Strand, power-driven winches have been installed with a pulley block permanently anchored in the sea, fig. 71. Anchors and the submerged pulley block, which is about 400 ft. (120 m.) from the shore, are indicated at the top of the diagram. The pulley is about 20 in. ($\frac{1}{2}$ m.) in diameter and of the usual cargo block construction. The anchor wire is connected to a hoop made of $1\frac{1}{2}$ in. (38 mm.) diameter steel; a 16.4 ft. (5 m.) beam of $3.2\times 3.2\times \frac{3}{8}$ in. ($80\times 80\times 10$ mm.) steel angle is screwed to the block to prevent tripping. A spare block with a tripping beam is usually kept in the winch-house. The main wires are led from the anchored block to the two drums of the

Fig. 70

Handwinches were introduced, and are still used, to reduce labour, fig. 70. They are usually made locally from cast-iron gearwheels and bearings, steel shafts and wood. The single type has a reduction on the gearwheels of about $6\frac{1}{2}:1$ and a drum diameter of 4 in. (100 mm.) minimum for 1 in. (25.4 mm.) flexible steel wire. Some winches have foot- or hand-brakes and a stopper working on the larger gearwheel. The double type has a 1-in. diameter primary shaft with a reduction of about $3:1$ to the $1\frac{1}{8}$ in. (29 mm.) diameter second shaft, which in turn has about 6:1 reduction to the $1\frac{1}{4}$ in. (32 mm.) diameter drum shaft. The drum has a diameter of 5 in. (12.5 cm.) minimum for $1\frac{1}{4}$ to $1\frac{1}{2}$ in. (32 to 38 mm.)

winch. They usually consist of 5 parts with a $1\frac{1}{4}$ in. (32 mm.) link and two 1 in. (25 mm.) shackles between. The length of the outer, or sea wire, " A " of fig. 71, is about 720 ft. (220 m.) $3\frac{3}{4}$ in. (95 mm.) flexible steel wire. This wire is shackled to the intermediate wires " B ", which are each about 185 ft. (56 m.) long. The intermediate wires are shackled to the drum wires, each about 400 ft. (120 m.) long. The winch has two drums which are

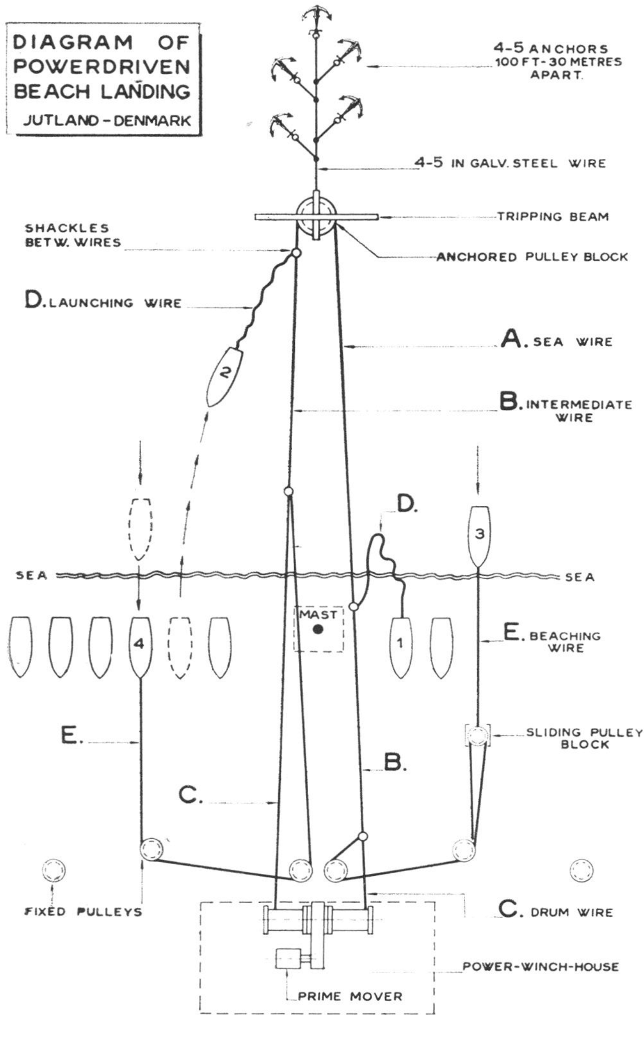

Fig. 71

connected by clutches to the prime mover, an electric motor of about 30 h.p., or an internal combustion engine of 30 to 60 h.p. Both drums have ample brakes. Some of the winches are converted steam-trawler-winches, others are especially manufactured at a price of about Danish kr.90,000 (£4,650; U.S.$13,000). The converted winches are cheaper in first cost.

The launching wire is connected to a special fitting at the after end of the sternpost, fig. 72. The eye of the launching wire is put between steel bands secured to the

Fig. 72

sternpost and the after end of the keel, and is locked in position by a bolt, the upper end of which is carried through a steel tube in the stern to the after deck. There is an aperture between the steel bands on either side, to enable the boats to be launched from either side of the main wires. The other end of the launching wire is shackled to the link between the sea-wire and the intermediate wire. In fig. 71 the boat marked " 1 " on the right-hand side is connected for launching. The far side drum is connected to the prime mover by means of the clutch, the other drum is let loose, and the launching

Fig. 73

wire and the boat are carried out to sea. When the boat is seaborne, the skipper raises the locking bolt and the launching wire drops off. To aid this, a rope attached to the boat end of the launching wire may be pulled by one of the shore gang. Once away from the launching wire, the boat is free to move by engaging the propeller clutch. The motor is always started when on the beach. On the left-hand side of fig. 71, the boat " 2 " has just been launched. The time taken for one launching is about four minutes.

The beaching wire " E ", fig. 71, 2½ to 2¾ in. (64 to 70 mm.) galvanized flexible steel wire, is connected to the stem of the incoming boats, " 3 " and " 4 ". The beaching wire may be attached to a pulley block which slides over the beach on a wooden cradle, or the wire may be led directly to the main pulleys fixed on concrete blocks in the beach, the sliding block being used to ease the strain on the winch when beaching larger boats.

The other end of the beaching or sliding-block wire is attached to the link between the intermediate wire and the drum wire. Fig. 71 shows boat " 3 ", right-hand side, using a sliding block. Boat " 3 " will be beached by the same haul that launches boat " 1 ". On the left-hand side of fig. 71, boat " 4 " has just been beached. The beach fairlead mast indicated as " Mast " in fig. 71, *see also* fig. 73, is sometimes necessary to break loose the anchored block when it has been buried in sand by bad weather. In front of this mast is a fixed pulley that may be used for beaching purposes instead of the anchored block. The intermediate wires are then disconnected from the drum wires and these are shackled to a wire running over the mast pulley (fig. 73, but not indicated in fig. 71).

In heavy weather it may not be possible to beach the boats but they are strong and sea-kindly and can stand heavy storms. A mast is erected on a nearby hill, from which special signals in local code are sent to the boats in case of difficult or impossible beaching. Most boats have wireless receivers and listen to weather reports. Fishing is considered safe work and there have been no casualties since World War II.

The first cost of a power-driven beach landing installation is estimated to be about Danish kr.110,000 (£5,700; U.S.$15,900) and the upkeep, inclusive of the salary of the winch operator, is about Danish kr.25,000 to 30,000 (£1,300 to 1,550; U.S.$3,600 to 4,350) a year. One power winch will serve 12 to 18 boats of about 10 gross registered tons. At Vorupør, there are two power-winches serving about 25 large boats.

The fishing boats pay a fixed fee and, in addition, 1 to 2 per cent. of their gross income. To this " harbour due " may be added the excessive wear on the bottom of the boats which slide on the beach itself, without any sleepers or rollers, but such expenditures are negligible compared to the cost of a permanent artificial harbour. At Hanstholm a permanent harbour is under construction but the work has almost stopped, as it is estimated conservatively that it will cost more than Danish kr. 70,000,000 (£3,500,000; U.S.$10,000,000) to finish the job (Hanstholm-Udvalget 1946).

The Boats

The boats are built locally based on tradition, experience and imagination. The builders do not work to drawings nor do they take their measurements off models. The materials are bought and the boat is built according to the fisherman's order and to the mental picture this order creates in the boatbuilder's mind. The smaller boats are almost invariably of the transom stern type and all are clinker built.

Rowboats

A rowboat of 16½ ft. (5 m.) length and 5½ ft. (1.7 m.) beam, will have the following scantlings: keel, 2½ in. (6.4 cm.), oak; gunwale 2½ × ¾ in. (6.4 × 1.9 cm.), oak or fir; nine frames with floor timbers 2 × 1½ in. (5.1 × 3.8 cm.), oak or fir; two floor timbers, 8 in. (20.3 cm.) high to form a fish hold; skin planking, seven planks each side, ⅝ in. (1.6 cm.), beech or fir, with butts having inside straps, riveted by galvanized steel or sometimes copper nails ; stern thwart, about 3½ ft. (1 m.) long; three fixed thwarts with pillars to the keel; thwartstringers, 1¾ × ⅝ in. (4.4 × 1.6 cm.). The oars have metal-lined holes to fit single pintle rowlocks. Boats have up to three men crews who fish near the beach with simple gear.

Small Motorboats

A small open motorboat, 20 × 7½ ft. (6.1 × 2.3 m.) of the " pram " type may have the following scantlings: keel, 2½ × 1 in. (6.4 × 2.5 cm.), oak; hog, 4 × 2½ in. (10.2 × 6.4 cm.); floor timbers, 3 × 2 in. (7.6 × 5.1 cm.) about 16 in. (40 cm.) apart; frames, 2½ × 1½ in. (6.4 × 3.8 cm.); planking, 10 planks each side, ⅝ in. (1.6 cm.); garboard strake, ¾ in. (1.9 cm.); gunwale, 2½ × 1½ in. (6.4 × 3.8 cm.) with 1½ in. (3.8 cm.) halfround; 4 thwarts. A 5 h.p. slow, single cylinder gasoline engine is fitted aft of amidship, the propeller being the usual reversible type located under the stern and protected by the keel aft. " Pram " boats use a fin, or sailing bilge keel, under the forefoot, to get better course-keeping.

Large Motorboats

Larger boats have a closed-in engine room aft and a cabin forward. When longer than about 30 ft. (9.1 m.), they are decked and have a fish-hold amidship. The bulwarks must be low and there is a danger of the crew falling overboard when fishing eagerly. To prevent this, the vessels have small square hatches to accommodate the fishermen, two on either side of the fish-hold hatch. The hatches fit watertight in oak coamings, flush with the deck. The fish-hold hatch, about 6 × 6 ft. (1.8 × 1.8 m.) has steel or oak coamings with wooden covers which are sometimes hinged. The hold itself is divided by portable wooden boards.

The wheel house at the after end of the wooden engine casing usually has its floorboards below the level of the deck. From the front of the wheelhouse there is access to the engine-room, but all engine and propeller-controls are located in the wheelhouse. Rudders are mostly fitted on steel posts and are raised above the keels by upturned skegs to avoid contact with the beach or they may be elevated by raising the rudder-stock and tiller from the deck. The more advanced boats have hydraulic steering arrangements.

The engine room has this arrangement: a single cylinder hot-bulb (semi-diesel) engine, driving (*a*) the hand-operated controllable pitch propeller, usually two-bladed, connected through a clutch at the after end; (*b*) the electric dynamo, about 300 w., 12 v. D.C., through a belt drive from the flywheel; (*c*) the fishing winch at the forward end of the engine casing, by a belt or chaindrive from the forward end of the engine. A bilge pump of the same size as the cooling water pump is usually built into the engine. Cooling water may be drawn from a 11 imp. gal. (13 gal., 50 l.) steel tank in the engine room " hopper cooling system " or from the sea.

Fuel oil is carried in steel tanks containing up to 130 imp. gal. (160 gal., 600 l.). As these tanks are placed low down, it is usually necessary to transfer the oil by handpump to a daily service tank, about 9 imp. gal. (11 gal., 40 l.), fitted in the engine casing. Extra fuel for long trips is carried in 9 imp. gal. (11 gal., 40 l.) drums. Lubricating oil, about one-tenth of the fuel, is in a separate tank or in drums.

A tank in the engine room delivers compressed exhaust gas, for starting, and to the blow lamp. Charging is done by a pipeline from the engine cylinder. Exhaust gases are discharged from the engine by a pipe leading through the casing, ending above the wheel house. The dynamo charges a storage battery which supplies current to the signal, deck and cabin lights, and to a search or floodlight. Sails are carried for an emergency and for steadying the boat in a seaway.

Drinking water is stored in a glass or wooden cask and a paraffin stove is used on long trips for cooking. Usually the boats leave the beach before dawn and return in the evening but some trips may take about four days.

Description of a typical boat

To give a better impression of these outstanding beach landing craft, the *Jylland* is described in detail. (*See* linesplan fig. 74 and sailplan fig. 75.) The lines, fig. 74, are drawn to the frames and not to the outside of the skin planking as is usual for wooden hulls. Fig. 76 shows the boat on the beach.

Built 1950 by Thisted Skibvaerft; 10.16 gross and 3.0 net registered tons (Danish measurement); length over-all, 36.8 ft. (11.25 m.); length between perpendiculars 32 ft. (9.8 m.); beam 14 ft. (4.20 m.); depth, 4 ft. (1.25 m.); displacement to load water line, inclusive of keel and planking, about 15 tons. Hot-bulb engine 35 h.p. at 450 r.p.m., with compressed air starting, driving a controllable pitch propeller through a clutch; belt drives for electric dynamo and fishing winch. Two fuel oil storage tanks, each 66 imp. gal. (79 gal., 300 l.) and one 9 imp. gal. (11 gal., 40 l.) daily service tank, hand-fuel-transfer pump; lubricating oil tank 13 imp. gal. (16 gal., 60 l.); cooling water tank for beach-running

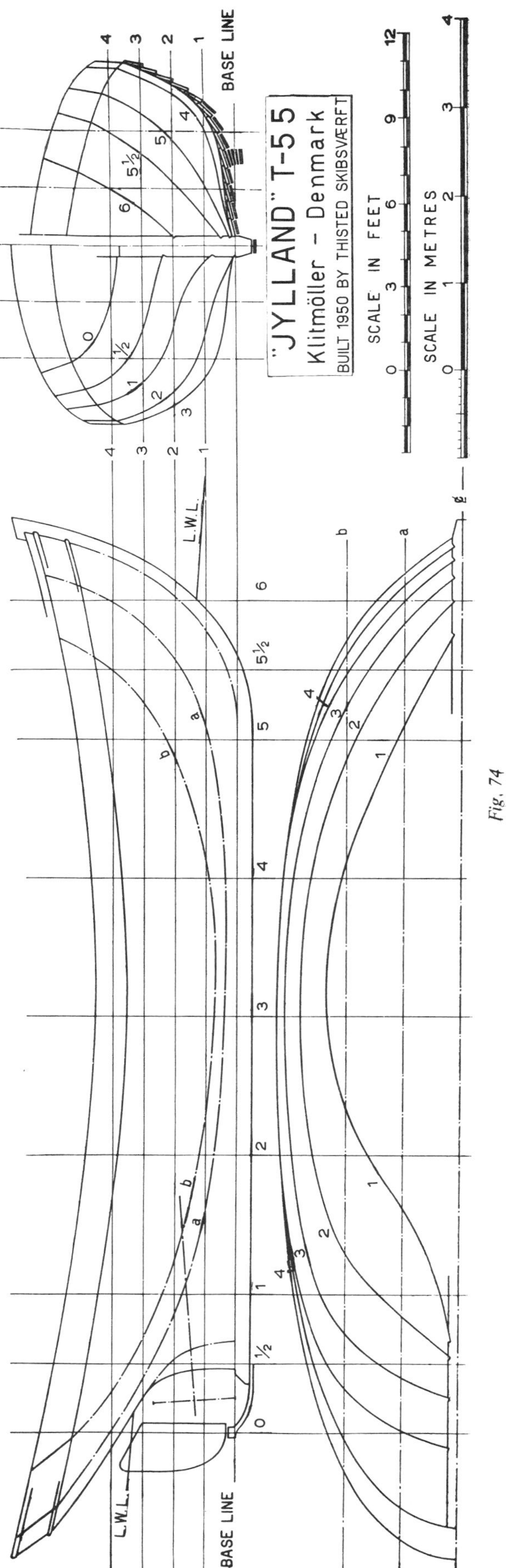

Fig. 74

about 11 imp. gal. (13 gal., 50 l.); speed, 7-8 knots. Scantlings: keel, 10×9 in. (25.4×22.9 cm.), Danish oak, with 3½×1 in. (89×25 mm.) steel bar; hog, 6×6 in. (15.2×15×2 cm.) in the midship (fish-hold) only; floor timbers, 4×3 in. (10.2×7.6 cm.) pine, 13 in. (33 cm.) apart, ⅝ in. (16 mm.) bolts; floor timbers, 3 in. (7.6 cm.) thick, with grown knees at sides to the frames; in engine room 5 in. (12.7 cm.) thick; in stern, steel floors, 1¾× 1½ in. (44×38 mm.); frames, 3 in. (7.6 cm.) sided×5 in. (12.7 cm.) moulded, 6½ in. (16.5cm.) moulded at the bilge; shelf, 8×2¼ in. (20.3×5.7 cm.), pine, tapered towards the ends; deck beams, 3½×3 in. (8.9×7.6 cm.); deck clamps, 5×2 in. (12.7×5.1 cm.), from the engine casing to cabin; in engine room, 7×3½ in. (17.8×8.9 cm.), tapered to the stern; planking, 13 strakes, 1 in. (2.5cm.), American oak: the bottom planking is reinforced by wear planks, 1 in. beech or oak, renewed about every third year, at a cost of Danish kr.1,000 (£52; U.S.$145); bilge keels, four layers of 1 in. beech or oak (mainly to steady the boat on the beach); deck, ¾×4 in. (1.9×10.2 cm.), tongue-and-groove fir; gunwale slotted at the frames; bulwark rail, 6×1½ in. (15.2×3.8 cm.), oak, at the stern 10½× 1½ in. (26.7×3.8 cm.); height of bulwark about 6½ in. (16.5 cm.) above deck; bulkheads, ¾ in. (1.9 cm.), tongue-and-groove pine; ceiling in hold and floorboards, 1 in. thick.

The stern of *Jylland* is a departure from the usual, which is a full elliptical stern with a pronounced knuckle just above the L.W.L., fig. 77.

ENGLAND

English beach fishing places are on the east coast of Yorkshire and Northumberland and on the Channel and south-east coasts. The boats do not go far out as there is competition from larger vessels operating from harbours.

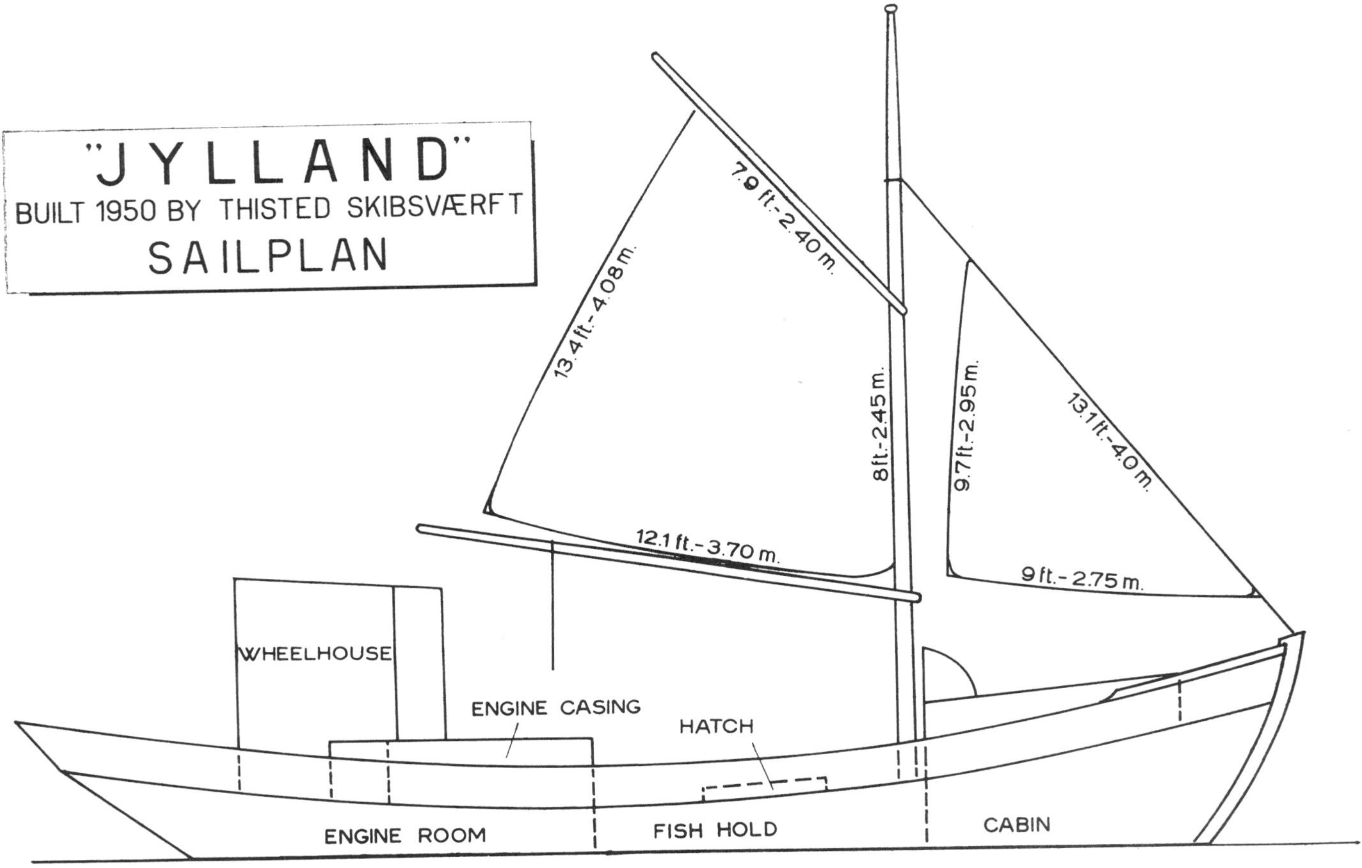

Fig. 75

The Yorkshire Coble

One of the most distinctive of English beach landing boats is the Yorkshire coble, still operating off the East coast at such places as Flamborough and Filey. (Carr 1934, Change in Stern Design, 1948, White 1950).

The primary purpose of the design of this open boat was to enable it to be launched from beaches against heavy breakers and to reach the open sea, requirements for which the coble has proved itself to be so admirably

Fig. 76

Fig. 77

suited, in addition to being a magnificent sea-boat. Great skill, however, is required in handling these craft and most of the fishermen have only acquired the art by long and habitual usage. But in unskilled hands or through lack of experience these craft are liable to involve disaster, as so truthfully expressed by a competent yachtsman who, in 1869, so aptly described the coble as " boat and harbour in one, and each of the very worst and most dangerous order".

In the landing operation the coble is manoeuvred to get the bow to seaward and, after the rudder has been unshipped and the mast lowered, the boat is put on the crest of a wave and beached. This practice of beaching stern first is reponsible for a distinctive structural feature, the use of two side keels instead of the usual central keel, extending from stem to stern. They are attached to the after part of the underwater hull, which has an almost flat form, rising gradually to a transom stern. With the side keels resting firmly on the sand or gravel beach, the boat can be hauled out of the water without risk of overturning.

The coble is unique in form, possessing remarkable forward sheer which gives a high bow with a deep well-rounded forefoot, while the stern is low and terminates in a flat sharply-raking transom. As a consequence, the craft have a pronounced curved gunwale and are most graceful in appearance. In the above-water form of its midship section there is an evident likeness to Norwegian herring boats of recent times. The broad planking, arranged clinker fashion, the sharp " tumble home " of the sheer strake causing the maximum beam to occur at its lower edge, and the high hollow bow with its powerful shoulder, are all features of marked similarity.

The coble is not built with a continuous projecting keel but its central member, known as the " ram plank ", is fashioned from a substantial piece of timber. The foremost section is narrow and deep but it is gradually widened and reduced in depth as the after-end is approached, finally assuming a plank-like form with an upturned end, so important when landing stern first on shelving gravel beaches. The frames, inserted after the boat is planked, are of oak, carefully fitted to the angular assembly of the strakes, the result of the use of unusually wide planks. Owing to the absence of a continuous central keel, two beeching keels, termed " skorvels ", are fitted to the after part of the hull to provide the necessary protection when beaching. The planking is prepared from locally grown larch or oak and the sheer-strake is 18 in. (46 cm.) wide in a large coble given a noticeably sharp " tumble home ".

One peculiarity of the coble is the narrow and very long rudder attached to the sharply raking transom. It extends a considerable distance below the hull, to serve as a deep keel or centre board. The tiller is of remarkable length and is so shaped to be readily accessible to the helmsman. The normal rig of cobles consists of one tall mast, with a well-defined rake aft, carrying a single large dipping lug-sail which is provided with bowlines to enable the boat to sail unusually close to the wind. Some of the larger cobles, however, are fitted with a mizen-mast to carry a small standing lug-sail.

Cobles do no not conform to a standard pattern, local variations having been built. For example, the ports of Scarborough, Filey and Whitby introduced a double-ended type in which a sharp raking stern was substituted for the flat transom. Furthermore, a normal projecting keel was adopted but it curved slightly upwards aft as does the typical coble. This hybrid type, combining features of the coble and yawl, was appropriately termed a " mule ". The rigging resembled that of the normal coble but often smaller lug-sail was employed to avoid reefing down the larger lug-sail.

Nearly all remaining cobles are fitted with petrol engines so that, notwithstanding the fact that their characteristic form has been specially developed for the purpose of drawing them well up on beaches, there is now a tendency for these craft to lie at moorings whenever possible.

It is interesting to note that cobles of the transom stern type are rowed, towed and beached stern first, with the rudder unshipped. A transom sterned coble will draw 2 to 2½ ft. (0.61 to 0.76 m.) forward and practically nothing aft. With the wind abeam it makes little head-way and in running before heavy seas it is apt to broach, and there is the risk of the long rudder breaking or being unshipped.

Launching and Beaching

The boats are transported by placing them, amidships, on a pair of old lorry or bus wheels. The boats seem to withstand this unusual strain very well.

At Filey there is a flat smooth beach, sheltered on both sides. When launching, the boats are pushed on the wheels into the sea, bow first, The two propeller blades are kept horizontal to avoid damage from the beach, the propeller shaft being marked inside the boat so that the blades can be set in the horizontal position at any time.

When beaching, the engine is stopped and the boat led stern first to the beach, where the shore gang is waiting with the wheels and the tractor. The gang consists of 9 men, inclusive of the tractor driver. The boat's crew is usually two, but when longlining three. The tractor is of a usual agriculture type and seems to work well with the axles awash. The boat is lifted on to the wheels and is pulled out of the sea by the tractor. The weight of the bigger cobles is about 3 tons, inclusive of engine and gear, and the catch may be up to 300 stones, or about 2 tons. The average catch is 80 to 100 stones (1,120 to 1,400 lb., 510 to 640 kg.) a day.

At Flamborough Head there are two landing places, north and south of the Head, the north being the main one. It has a narrow beach in a bay surrounded by rocks and, on the hill overlooking the bay, is a winch house containing a 45 h.p. gasoline engine, with a friction clutch and belt drive to the winch, and carrying a 2½ in. (64 mm.) flexible steel wire on its drum. When a boat is being beached the wire is led over fixed pulleys and

Fig. 78. Transom sterned Yorkshire coble

*Fig. 79. **Stern** view of Yorkshire coble on its beach cradle*

Fig. 80. Bow view of Yorkshire coble

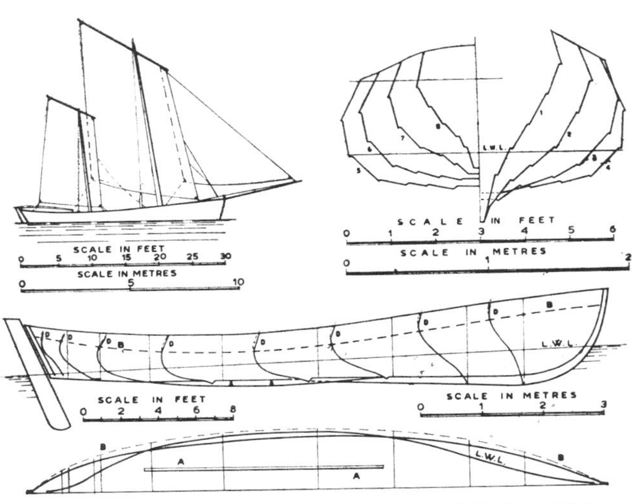

Fig. 81. Typical lines and sail plans of a 31 ft. 8 in. (9.65 m.) Yorkshire coble of the " mule " type

connected to two substantial bollards at the stern of the coble, one on each side. Greased wooden sleepers, or skids, are placed under the boat when it is hauled along the beach. The winch owner makes a fixed weekly charge for beaching. In launching, the boats are lowered down the hill without using the winch and are pushed into the water.

Description of boat

Coble engines are mostly of the gasoline or gasoline-paraffin type, with reverse gear, from 10 to 35 h.p., giving speeds from 6 to 9 knots. Small, high-speed diesels are also used, sometimes with reduction gear, but the high initial cost is usually out of the fishermen's reach. Power is taken off the forward end through a dog clutch and chain drive to a shaft along the boat, beneath the floorboards. It is carried to the fishing, or lobsterpot, capstan aft, which may be reversible and will always have an emergency cut out. The 1 in. (25 mm.) hand-driven bilge pump is amidship.

The coble is usually 25 to 30 ft. (7.6 to 9.1 m.) long, the pointed stern or " mule " type being longer than the transom sterner. Both types are clinker-built with nine strakes each side, excluding the rising plank forward. The propellers work in shallow tunnels formed by a lifted central keel and deep sidekeels on each side. The maximum depth of tunnel is about 18 in. (46 cm.). The " mules " have their propellers about one quarter of the length from the stern.

Scantlings: keel, $4\frac{1}{2}\times6$ in. (11.4×15.2 cm.), oak; hog, $4\frac{1}{2}\times2\frac{1}{2}$ in. (11.4×6.4 cm.); planking, 8 strakes, $\frac{5}{8}$ in. (1.6 cm.), larch; sheer-strake, 1 in. (2.5 cm.), oak; floor timbers, 4×2in. (10.2×5cm.), oak, about 10 in. (25.4cm.) apart amidship; frames, $1\frac{1}{2}\times1\frac{1}{2}$ in. (3.8×3.8 cm.), oak, scarped to the floor timbers; rubbing piece under the sheerstrake, $1\frac{3}{4}$ in. (4.4 cm.), oak; gunwale, $3\frac{1}{2}\times2$ in. (8.9×5.1 cm.), oak or ash; thwart stringer, $2\frac{1}{2}\times1\frac{1}{4}$ in. (6.4×3.2 cm.); thwarts, 5 to 6, $11\times1\frac{1}{4}$ in. (27.9×3.2 cm.); middle thwart aft of engine, fastened with double steel knees at sides. The mast is placed aft of the bow thwart.

The cost (1952) of a 26 ft. (7.9 m.) coble was approximately:

Hull and rigging . . .	£650 (U.S.$1,820)
Engine and capstan (15 h.p., petrol)	£350 (U.S.$ 980)
	£1,000 (U.S.$2,800)

The Hastings Lugger

Launching and beaching

Hastings, on the south coast, has a steep pebble beach and the boats must be brought over shelves of pebbles to a safe position above high water level. Power-driven winches are used in conjunction with sliding blocks and strong tallow lubricated sleepers. Most of the winches are driven by second hand lorry gasoline engines developing about 35 h.p. although some are electrically driven, but they are smaller and only about 5 h.p. The smaller boats are hauled by steel hand-winches. As at Flamborough the winches are used for beaching only. The $2\frac{1}{2}$ in. (64 mm.) circumference winch wire is led through a pulley block sliding on the beach and back to a fixed position near the winch. The hook of the sliding block takes a $\frac{5}{8}$ in. (16 mm.) chain that is led through a hole in the forefoot. The sleepers (" throws " at Hastings) are substantially built of oak, $30\times24\times4$ in. ($76\times61\times10$ cm.) with two through-bolts on each side, to prevent splitting. It is not unusual for the beach to be faired to make beaching and launching possible. Normally the beach is steep enough for launching by gravity. To control the speed one fluke of an ordinary stock anchor is made into a handle and the other fluke dropped into the beach to act as a brake.

Description of boat

To withstand these labourious operations over such a rugged surface, the Hastings luggers, fig. 82, must be robust and have considerable breadth with little rise of floor. They are a particular type developed on the south coast. During the first half of the nineteenth century the sterns of all Hastings boats were of the typical square transom form, but about 1880 the " lute " stern, or a " beaching counter ", became universal on the larger boats. The present boats have a more rounded form of stern, fig. 83.

A typical Hastings lugger, fig. 82 and 83, is the *Edward and Mary*, R.X.74, 22 gross and 7.56 net registered tons. Her registered dimensions are: 27 ft. 9 in. × 11 ft. × 3 ft. 7 in. (8.5 × 3.4 × 1.1 m.). The sails consist of jib, foresail and mizzen, and fishing is done by nets, lines and small trawls. There are, of course, several smaller types operating, a typical example being the *Rose Mary Ann*, about 28 ft. (8.5 m.). Two of the many

Fig. 82

Fig. 83

Fig. 84

different stern forms common on the Hastings beach are shown in fig. 84.

Engine arrangements in the boats are unusual. The boats are equipped with up to three propellers usually driven by different type of engines with differing horse power.

NETHERLANDS

The beach fishing fleet at Scheveningen, about 150 of the " Bomschuit " type, was badly hit by a great storm on 22 December, 1894, when more than 25 boats were lost and the others damaged. Since then, boats fishing off the open beach have given way in Holland to other types operating from sheltered harbours. The " Bom ", as the Dutch beach landing craft was called, was a beamy, full-lined, decked craft, clinker built and of heavy construction. It was hauled along the beach on long wooden rollers operating on wooden boards, or rails.

The life-saving institutions of the Netherlands do still launch some of their boats off the beaches, and the Koninglijke Noord-en Zuid-Hollandsche Redding-Maatschappij have some advanced designs of boats.

PORTUGAL

Portugal has many different types of beach landing fishing craft, both oar, sail and engine-propelled, all carvel built, a contrast to the clinker built hulls of the northern European countries. Fishing from the beaches takes place all along the Portuguese coast.

Rowing and sailing boats

Small flat-bottomed boats, called *Chata*, are used off Caparica, south of Lisbon. They are about 16 ft. (4.9 m.) long, and have planking of ½ in. (1.3 cm.), supported by floor timbers, 2×1½ in. (5.1×3.8 cm.) and frames which are kneed to the floors, about 11 in. (28 cm.) apart. There is a watertight floor extending to the thwart amidship, and floorboards are in the after part only.

Fig. 85

The boat weighs approximately 1,800 lb. (800 kg.) and costs about Escudos 2,000 (£25; U.S.$70). (The sardine net itself has a value of about Esc.8,000 (£100; U.S.$280). The boats are difficult to mechanize, but at Cascais one boat has been fitted with a well to take an outboard engine.

Another type at Caparica is called *Saveiro*, used for sardine beach seining, fig. 85. Such a boat has a crew of 12, consisting of master, chief fisherman, poleman, eight oarsmen and one brailing man. The bottom planking is ⅞ in. (2.2 cm.) and the side-planking ¾ in. (1.9 cm.), supported by floors and frames, 2 in. (5.1 cm.) thick and spaced about 16 in. (41 cm.) apart. The *Saveiro* is also carried to the water by man-power, working on timbers lashed transversely over the gunwales. The form of the flat bottom makes it possible to move the boat along

Fig. 86

the beach by rocking it longitudinally and swinging it when supported at the ends.

In the north of Portugal, at Póvoa de Varzim and Vila do Conde, the small type of beach landing craft, the *Catraia*, is double-ended and almost without sheer. A typical Póvoa de Varzim type, fig. 86, has the dimensions: 14 ft. 5 in. × 5 ft. 10 in. × 1 ft. 10 in. (4.4 × 1.8 × 0.55 m.) and is about 1.2 registered tons.

Because the boats ship water over the ends when going through the breakers, they are provided with waterways and scuppers all round. Although there is a small artificial harbour at Póvoa—built before the war, at a cost of more than Esc.30,000,000 (£370,000; U.S.$1,040,000)—it is too small to give any protection in bad weather. In the district there are some 1,000 fishermen operating about 325 boats, a number of which are kept in nearby harbours during the off season. There is an average of about 100 fishing days a year.

Nearly the same conditions prevail at Vila do Conde, but the waterways and scuppers on the boats are omitted, otherwise the *Catraias* are of the same form.

A third form of small rowing boat is the *Xavega*, which is found on the Nazaré beach. The *Xavega* and its sister, the *Candil*, which is without the pointed stern, are used as auxiliary craft in beach seining. Their planking of $\frac{1}{2}$ to $\frac{5}{8}$ in. (1.3 to 1.6 cm.) is supported by $2\frac{1}{2}$ in. (6.4 cm.) frames, not more than 8 in. (20 cm.) apart.

At Nazaré there is an average of 200 fishing days a year and about 1,500 fishermen operate about 150 sailing craft and 50 motorboats. The landing is protected from the prevailing north-west winds by a rocky head, and as the fishing village is built close to the beach, the boats must be " parked " in the streets and squares during the

Fig. 87

winter storms. Besides the boats already described there are larger types called *Barcos*, small powered transom sterners, and the *Peniche* types.

The *Barcos* are double-ended rowboats with a full midship section and are used mostly for sardine transport from the fishing grounds to the beach and for beach seining. An average *Barco*, fig. 87, is 34×11×3.6 ft. (10.5×3.4×1.1 m.). The transom sterners are about

to the frames. The longitudinal and transverse timbers have beautifully scarped connections, reinforced by galvanized steel straps. The dimensions of such a boat are: 21.6 (19.1) × 8.1 × 3.4 ft. (6.57 (5.81) × 2.46 × 1.05 m.). The cost, with a 10 h.p. single cylinder hot-bulb engine, mast rigging and sail, is about Esc.55,000 (£680; U.S.$1,900), the engine accounting for more than half of it.

Fig. 88

25 ft. (7.6 m.) long and are easier to build than those with a rounded stern.

Larger boats

Of the larger Portuguese beach boats the *Peniche* type is the most popular, especially on the coast between Peniche and Aveiro. On the beach of Nazaré the boats reach 17 gross registered tons and have engines up to 75 h.p.

A typical small *Peniche* type has substantial keel, hog and engine bearers to floors and frames, riveted with galvanized steel bolts, but the planking is simply nailed

The *Linda Pastora* ex *Digna Flor* of Nazaré is a " Peniche " type with the dimensions: 32.8 (28.1)×10.4 ×3.4 ft. (10 (8.57)× 3.18×1.04 m.) and 5.5 gross registered tons. It is powered by a 25 h.p. hot-bulb engine with reverse gear, giving a speed of 6 knots. The cost of the *Linda Pastora* is estimated at Esc.100,000 (£1,240; U.S.$3,460) and she carries a crew of 18.

The *Estrelinha*, fig. 88, is 11 gross registered tons and is powered by a 50 h.p. hot bulb engine. Main dimensions: length, over-all, 38.5 ft. (11.7 m.); length, pp., 33.7 ft. (10.25 m.); breadth 10.5 ft. (3.5 m.); depth 3.6 ft. (1.1 m.). The general lay-out is shown in the lines plan, midship sections and longitudinal plan of figs. 89, 90, 91.

The *Gloria a Deus* is one of the largest boats at Nazaré, 16.7 gross registered tons. Dimensions: 41.6 (36.4)× 13.2×4.1 ft. (12.68 (11.05)×4.02×1.24 m.). She has a 75 h.p., three-cylinder hot-bulb engine, giving a speed of 9 knots in light condition. Connection between the wheel house and the engine room is by means of a string-operated bell and a speaking tube. As in Denmark, a hopper cooling tank is fitted for running the engine on the beach, and there is a mechanical drive to the winch on the foredeck. *Gloria a Deus* carries a fishing crew of 20 but she cannot accommodate them all below deck so her trips are not very long.

All boats have fixed blade propellers with reverse gears, or directly reversible engines, controllable pitch propellers are not used.

The boats of Nazaré are beached and hauled by manpower or by bulls. Up to 20 bulls are required for beaching big boats. Launching, bow first, is aided by the crew pulling the cable of previously laid anchors. In the calm season the larger boats are kept at mooring buoys some distance from the beach.

The Portuguese *Direçao da Marinha Mercante* (Ministry of Transport) keeps a strict eye on all boatbuilding activities. Even for small craft, drawings must be submitted for approval and, for vessels above 25 gross registered tons, hydrostatic curves and cross curves of stability are required. The *Direçao* also assures that the boats are built according to the plans. The boatbuilders seldom make drawings ; they usually work on measurements taken off a wooden scale model, so they send their models to a naval architect or draughtsman to have the required plans made. These plans, although adding to the cost of the boats, are valuable when investigating the reasons for disasters, failures and defects of design. In the long run, a secondary advantage will be a safer development of new forms and constructional details.

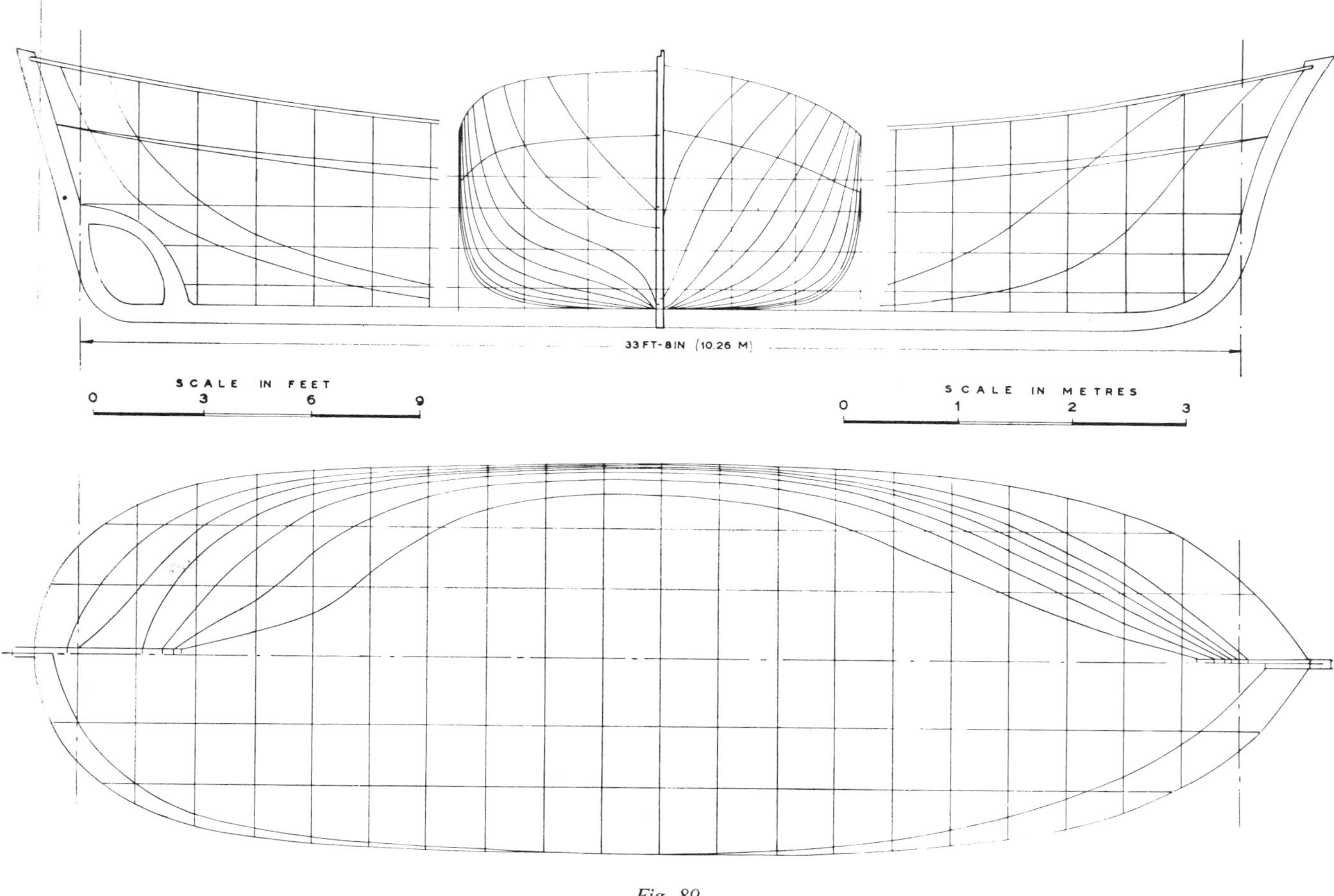

Fig. 89

SPAIN

Fishing from open beaches is done along the Mediterranean coast by boats of oar, sail and engine propulsion. At places near Barcelona—Badalona, Vilasar de Mar and Vilanueva—there are permanent electrical winches for beaching and sometimes also for launching, fig. 92. When launching, a block is fixed on an anchored buoy, some distance from the beach, and the winch wire is

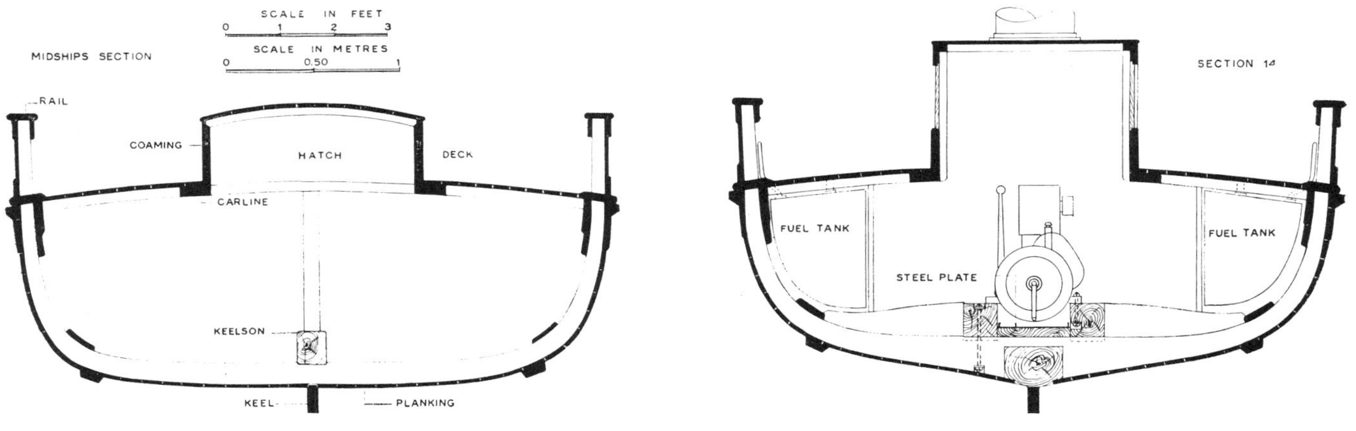

Fig. 90

led over fairleads on the beach to the buoy and over the block back to the boat. The buoy is removed in winter. Steel hand-winches are also used and, at Vilanueva and Sitges, a particular kind of capstan made of wood reinforced by steel bands, is used. Wooden bars are inserted in it and up to 12 men work them. The beaching wire is led through one or more sliding blocks before it is turned round the capstan barrel.

Even the small double-ended sailing boats are decked, fig. 93. The dimensions are generally: $19 \times 7.25 \times 1.6$ ft. ($5.7 \times 2.2 \times 0.5$ m.). There is a 1 ft. (0.3 m.) bulwark all around, with free openings at the sides of the heavily cambered deck. Two side keels are set at an angle, over the midship part, down to the level of the keel, to act as stabilizers when on the beach, fig. 92 and 94. The scantlings are: keel. $5 \times 1\frac{1}{2}$ in. (12.7×3.8 cm.); sidekeels, $1\frac{1}{2}$ in. (3.8 cm.); hog, $7 \times 2\frac{1}{4}$ in. (17.8×5.7 cm.) midship to $3\frac{1}{2} \times 2\frac{1}{4}$ in. (8.9×5.7 cm.) at ends; floor timbers, $2\frac{1}{4} \times 1\frac{1}{2}$ in. (5.7×3.8 cm.) double, about 10 in. (25 cm.) apart; planking, $\frac{3}{4}$ to $\frac{7}{8}$ in. (1.9 to 2.2 cm.);

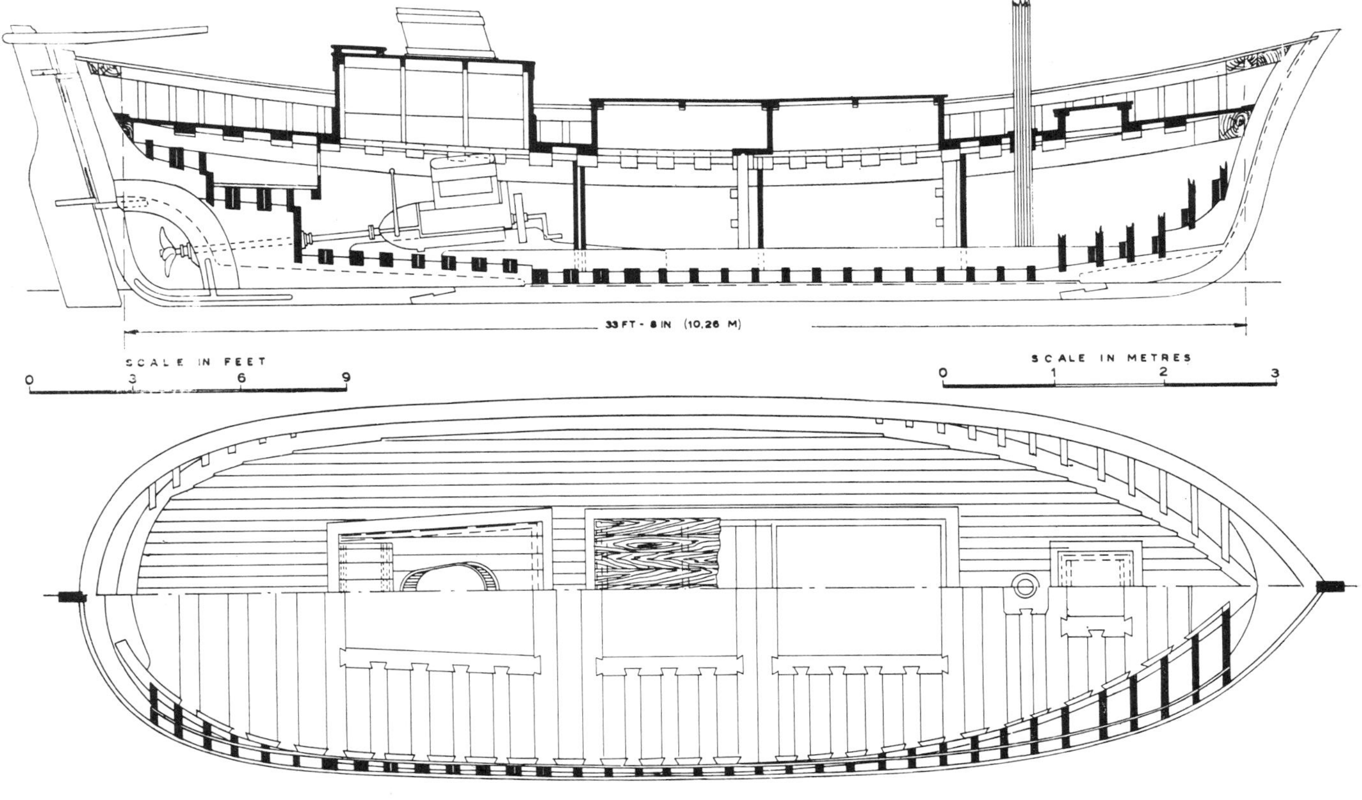

Fig. 91

deck, $\frac{3}{4}$ in. (1.9 cm.); covering board, 1 in. (2.5 cm.); beams, $2 \times 1\frac{1}{2}$ in. (5.1×3.8 cm.), 10 in. (25 cm.) apart; shelf, 5×1 in. (12.7×2.5 cm.); rail, 3×2 in. and $3 \times 1\frac{1}{4}$ in. (7.6×5.1 and 7.6×3.2 cm.); and three hatches, 20×20 in., 41×27 in. and 30×27 in. (51×51 cm., 104×69 cm. and 76×69 cm.).

The boats have a crew of three to four men and the average value of the catch per boat per day is about 300 pesetas (£3; U.S.$8). Near Valencia the beach boats are laid up during the winter, and the fishermen join the bigger vessels which need more men in winter than in summer. A Masnou motorboat with sail and petrol engine is shown in fig. 93.

Fig. 92

The Spanish way of securing the beaching wire is somewhat unusual. There is a transverse, substantial wooden member through the bow, just above the forefoot. A hole in this member takes the wire, which is also slung over the top of the stem, fig. 94.

SPECIFIC BEACHING PROBLEMS

Solutions by the Lifeboat Institutions

The beach lifeboats are faced with much the same problems as the beach fishing boats but lifeboats must be prepared to go out under much harder conditions of weather and the boats and gear must be the very best obtainable. Much ingenuity and scientific research have been spent on developing the beach lifeboat hullform, propulsion, constructional details, choice of materials and launching gear, etc. (Report of the International Lifeboat Conference 1936, 1947, 1951; Svetsaren 1943; A stainless steel boat 1938; A new type of lifeboat 1946; De Reddingboot 1952).

Lifeboats are launched by the fishermen's winches on the Jutland coast, but have usually their own beaching winch. Lifeboat launching gear is in general too elaborate and expensive to suit commercial fishing. The lifeboats in the Netherlands and England are very often launched from the beach under the lee of the vessel in distress and it calls for rapid movements along the beach from the station to the wreck. The boats are placed on a specially-constructed carriage and wheeled along the beach and launched by the aid of a team of horses or by a water-proofed tractor.

The lifeboat's hullform is developed from Scandinavian practice. The first lifeboats were double-enders, but in the Netherlands a modified " pram " type hull has been very popular. The present tendency seems to indicate fuller sterns, keeping the reserve buoyancy with low free-board at the stern. The bow is made finer with more freeboard. Buttock lines are kept flat, particularly in the after body. Lifeboats can be, and are mostly, self-righting, a quality desirable also for a fishing boat, but the necessary tanks at the ends would greatly affect its efficiency. In fact, the self-righting tanks are also rather a nuisance on the lifeboats and many stations prefer to do without them—which indicates that fishing boats should not be made self-righting.

Lifeboats have their propellers working in tunnels to protect them from damage when grounding and from being entangled in ropes and wreckage. This feature is, to a certain extent, used in fishing boats such as the Yorkshire coble, but most fishing boats are able to control the beaching better than a lifeboat. But even a fishing boat's propeller would be safer if protected in a tunnel.

Propellers are not the sole means of propulsion. There is jet propulsion, several forms of which have the advantage of being protected inside the hull. Such inventions have been, and will be, tried by the lifeboat institutions and may save a lot of trouble for the fishermen. Choice of engines is easier for lifeboats. They have expert engineers and can spend more on maintenance, but fishermen require cheap engines that will run economically with little expert attendance and maintenance. Speed is important in both life-saving and fishing operations. The speed of the lifeboats, however, is mostly higher than is economical for fishing boats. The maximum speed, in knots, of a seagoing fishing boat in light condition should be $1.3 \times \sqrt{\text{length of waterline}}$ in feet. While this speed is already the maximum for fishing boats, the life-boat institutions have increased speed and are trying to develop boats of the planing type still retaining the necessary seaworthiness for operation in bad weather.

Smaller fishing boats in Europe are usually made of wood, with the exception of the Netherlands ; so are many lifeboats. It may be possible that wood is the better material, but lifeboat institutions are also experimenting in this field. While fishing boat builders argue about the advantages of carvel or clinker construction, lifeboats are being made of double diagonal wood, laminated mouldings, plywood, steel, stainless steel, aluminium and plastics. These experiments should be watched with interest by the fishermen.

Solutions by the Navy

Military invasions from the sea take the easiest path, which very often leads over the beaches. The main object of the naval landing craft is to unload the men and war material; what becomes of the ship is of secondary importance. The craft has to be a shallow draft vessel that can be beached bow first with the bow as near dry land as possible, features which are not easily incorporated in a seagoing ship.

One solution to this military problem is the amphibium, a vessel that can sail from a transport ship to the beach, proceed under its own power up the beach, unload, and go back again (Lifeboat Conference 1947, Jackson 1952). Such amphibious abilities would be an asset to a beach fishing boat but in its present form the naval amphibium is too expensive and elaborate to interest fishermen. The amphibium, however, is there to be simplified and adapted to peace-time work.

General Problems

Mechanical propulsion of beach fishing boats

The easiest engine to instal in beach fishing boats is undoubtedly the outboard. Such engines are commercially available in a range from $\frac{1}{2}$ to more than 25 h.p., weighing from 10 to 110 lb. (4.5 to 50 kg.). Their approximate weight in kg. is $7\times \text{h.p.}^{\frac{2}{3}}$ A high-speed, 4,000 r.p.m. gasoline engine, drives the vertical shaft of the outboard and the built-in tank for the gasoline and oil mixture has usually a capacity for one hour's running at full speed. Refuelling at sea is possible by a special arrangement, including an air pump on the storage tank and a flexible pipeline to the engine tank. Outboard engines are not considered to be very reliable, but they

Fig. 93

are cheap in first cost and handy in the smaller sizes. For fishing near the shore with oars and sail, outboards may be worth a trial as an auxiliary means of propulsion, particularly where existing boats are of the transom stern type as the installation requires little or no alterations to the hull. For such installations, however, extra long shafts are recommended.

If inboard engines are to be installed the hull should be designed as a motor-boat, as conversions such as the Hastings lugger, fig. 83, are not likely to be very efficient. This means the introduction of new types of boats, which is always a difficult matter. Handling the boats in surf and at sea requires a skill developed locally through generations and any change in hull form will require adjustments in handling because the movements and behaviour of the boat will be different. The change may be to the better, as judged by an unbiased naval architect, but the fisherman may not agree. The smaller and more primitive the boats, the more difficult will be the change to the more advanced mechanized types. Development must, on account of the human element, be gradual, and the main problem is the persuasion and education of the fishermen.

Fig. 94

The fact that the fishermen really have to know their engines—not merely press a starter button like a car driver but to be able to do emergency repairs at sea—calls for the simplest and most robust type of engine. Apart from England, where gasoline, gasoline-paraffin and diesel engines are used, the most popular beach fishing boat engine is the hot-bulb or semi-diesel. This engine has the drawback of consuming more fuel, particularly lubricating oil, than a full diesel. It is also heavy and slow running, but robust and reliable, able to stand up to a lot of abuse, and the heavy construction gives greater wear tolerances. It is a crankcase scavenged, 2-cycle engine that requires a minimum of routine maintenance and repairs can be done locally. The gasoline- and gasoline-paraffin engine has a vulnerable electrical ignition system and uses a highly taxed and inflammable fuel, but despite these drawbacks, the 1-cyl. variety is a very simple engine to run.

From a technical point of view, the best and most efficient engine is the compression-ignition engine, the diesel. Suitable types are commercially available from about 5 h.p. upwards. Gasoline and diesel engines are also available with air-cooling, instead of the usual sea-water-cooling. Where there is frost or much dirt in the water, air-cooling may be of special advantage, and the elimination of seawater-cooling would be an asset as water near a beach is usually contaminated with sand. The installation of a cooling water tank in the boat is one solution to avoid the sand, but a fresh-water cooling system might work better. In frosty places, a suitable anti-freeze cooling medium could be applied and fresh water cooling would permit engines to be run at a higher and more profitable temperature, and corrosion of the cylinder block would be diminished.

If the engine is driving a fixed propeller, there will have to be a reverse gear between the engine and the propeller shaft. A controllable pitch propeller will require a clutch so that the propeller can be stopped when the engine is running. On the decked boats, one crew member can be spared by extending the controls of the engine and propeller to a suitable position near the helm or the steering wheel. With such controls the engineer does not need to be permanently in the engine room and can join the deckhands during fishing operations.

Jet-propulsion, inclusive of the Hotchkiss and Gill propellers, has so far proved to be inferior to ordinary propellers. The new Kermath Hydrojet has still to be tested in commercial beach fishing operations, and is unfortunately only available with one rather large-size gasoline engine. Inventions, such as the Voith-Schneider propeller, could be installed with advantage, but the added mechanical complications and the cost are big drawbacks, and as a Voith-Schneider propeller also acts as a rudder, it must have unrestricted water in all directions. The same applies to the Kitchen rudder, and these devices are difficult to install in conjunction with sidekeels in the afterbody. The Kort nozzle would improve the thrust when dragging nets, setting lines, and against adverse weather, but would also impair the manoeuvrability.

The solution of the problem is an ordinary propeller with fixed blades or, preferably, a controllable pitch propeller, protected in a tunnel formed by deep side-keels or by having a substantial skeg. In Japan, and also at the island of St. Pierre (La pêche aux îles 1953), the propeller can be lifted during the launching and beaching to keep it in a safe, elevated position, fig. 95. This vertical movement is made possible by fitting a universal joint on the propeller-shaft, just outside the stern-tube. At the front of the propeller, aft of the universal joint, is a bearing supported by a screw (tubes or rods are also used), extending to convenient height in the boat or at the stern. The position of the propeller may then be controlled by turning the supporting screw or by lifting the tube or rod. Apart from the wear on the universal joint and the outer propeller bearing, the device seems to be a practical one, particularly for conversions of many native types of hulls to engine propulsion. The boats of St. Pierre Island are of the flat bottom dory type.

The beaching and launching gear on shore will depend on the local conditions and the size of the boats. The mechanical system used in Jutland is the most advanced. In some places it may be advantageous to have the anchored block floating, preferably between two buoys, to counteract tripping of the block. The wear on the anchored block would be diminished with no chance of it being buried in sand, but the strain on the anchors, particularly during storms, might prohibit such arrangement. In warm climates, hand-winches and the Filey wheels may be the simplest way of beaching, launching and transport.

Building Materials

Notes on scantlings and modes of construction

Beach boats must be built stronger than common boats so that the hulls can stand the longitudinal strain of one-point support and the bottoms withstand bumping on the beach and over reefs.

European beach fishing boats are normally built of wood. The advantages of wood are its low cost, favourable strength-weight ratio, elasticity, thermal conductivity and the ease with which it can be formed and assembled by simple and familiar operations. The drawbacks are its hygroscopic, heterogeneous structure and its vulnerability to rot and decay, features that make strength calculations difficult and uncertain (Henderson 1948, Wood Handbook 1940, Bjursten 1947, Norén 1948, Holst 1952).

To overcome the first of these drawbacks, double- or triple-diagonal construction of panels like skin and bulkheads is one solution, but such panels are difficult to repair. A number of beach lifeboats and naval craft are built according to this method, which gives light and strong construction with good results. Another solution, that of laminated wood and plywood panels, is coming into use with the aid of modern marine glues (Stevens 1948, Hearmon 1948). Boat forms, such as the fishing dory and hard chine transom sterners, lend themselves favourably to plywood construction, particularly for smaller boats, when plywood panels are long enough to omit butt joints of the planking. But it is not feasible for the larger boats of, say, Denmark and Portugal, except perhaps for the construction of bulkheads, casings, deckhouses and other smaller items.

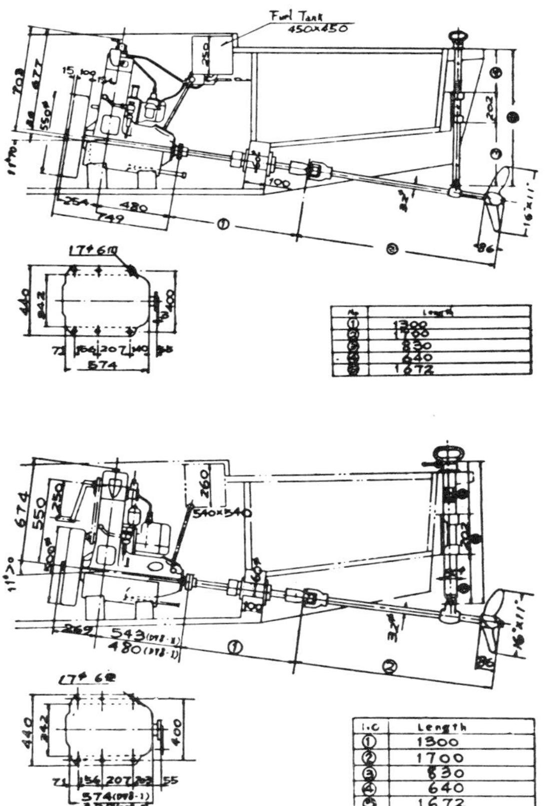

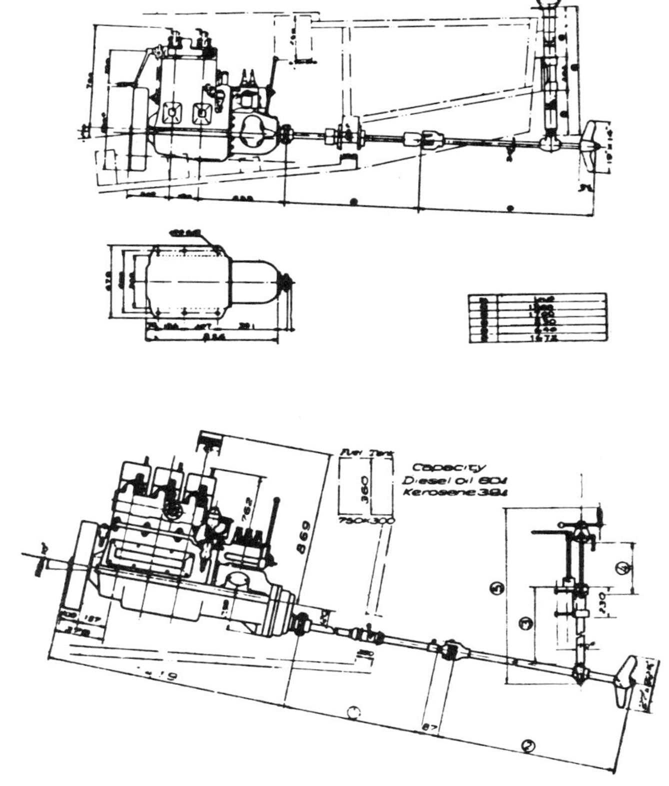

Fig. 95

In the Netherlands, steel is a popular building material, even for small lifeboats. The main drawbacks are corrosion and high thermal conductivity. For the small boats the plating must be very thin, a matter of a sixteenth of an inch or a few millimetres and such thin shells are soon rusted through and are vulnerable to local deformations, so that extra thickness of the bottoms seems to be necessary. Composite construction of steel and wood may be feasible if the steel is made to carry all the main strains on the hull and the wood is used for planking and deck, i.e. only to keep out the water.

Aluminium is now being used extensively in the building of boats and ships. It is lighter, antimagnetic and, in certain alloys, less corrosive than steel. The welding of aluminium is, however, not so easy as the welding of steel and most aluminium structures are riveted. Aluminium beach boats would be more expensive than the present wooden ones, but lighter and stronger, except for local deformations. Items such as tanks, including ballast tanks, may be built into the hull structure. In Jutland, where the boats are beached on sand, aluminium hulls should be well worth a trial. Composite aluminium and wood construction is being used in British boatbuilding for Royal Naval Motor Minesweepers, about 150 ft. (46 m.) in length. This mode of construction may be a practical solution for future large beach boats.

In addition to metal hulls, much research and investigation is being done, particularly by the U.S. Navy and Coast Guard, on plastic hulls (Lifeboat Conference 1951, Jackson 1952, Plastics in boat building, 1951, 1953). The moulds in which these boats are made are expensive and mass production is, therefore, of primary importance.

General Beach Problems

Effect on hull form

In order to maintain an upright position when on the beach, the bottom of a boat should be flat or steadying sidekeels should be applied, preferably near the end that is beached first and launched last (e.g. Yorkshire coble) as the draught at the beached end is then at a minimum, which saves the crew from going too far into the water. In launching, the real trouble is the passing of the breaking waves. Once outside the surf, the beach problems are left behind, and the boats should preferably behave as do other good seakindly vessels.

The simplified form of ocean waves and their mathematical and geometrical analysis may be known from text-books (Taylor 1933). The effect of such waves coming across shallow water is a retardation of the speed of the wave ((Taylor fig. 11 and 12, show the relation between length of wave and speed of advance in various depths of water). In shallow water of uniform depth, the orbits which the water particles are describing, and which are circular in deep water, tend to flatten owing to interference with the bottom. Near the surface, the orbit becomes an ellipse with the major axis horizontal, and as the bottom is approached, the vertical movement is reduced until the particles only move to and fro at the bottom. This becomes effective when the water depth is less than half the wavelength.

Waves are not reflected from a beach as from a steep rock, but the energy is broken down gradually. When uniform ocean waves approach the coast and the water becomes shallow, the velocity decreases and successive waves tend to close up. The effect of this closing up is to make the shape of the waves conform to that of the coastline so that a long line of a wave may break almost simultaneously against a flat beach. The waves themselves become sharp at the crests as they are slowed down and, because of this, the crests become unstable, break down and finally become waves of translation rushing up the beach.

The dangerous zone for the beach boat is where the wavecrests are starting to break. Water is likely to come over the end launched first and wash the decks clean or fill an open boat, and there is always the danger of being thrown broadside on by a wave and being capsized by the next. The speed of the crests, according to Taylor (1933), exceeds 10 knots at 10 ft. (3.05 m.) still water depth. Conditions of the surf will vary with the size of the waves and the slope of the beach. At a certain beach there will be a size of wave that prohibits any beaching attempts but, then, the weather will most likely be too rough for fishing.

A hull form like the coble seems to be ingeniously designed for passing the surf with the bow facing the direction of the waves, but it is not a boat easily handled, particularly in the open sea. Hull forms of the Jutland and the Portuguese types are a more judicious compromise between a seagoing and a beach boat. They are able to tackle the surf with either the stern or the bow and in cases of emergency such as sudden storms, there is no need to perform the risky turning of the boat outside the beach surf. Because of their broad gunwales, or decks, particularly at the ends, the *Peniche* type of Portugal must be lively in a seaway, but as they usually work in a regular swell or in calm, liveliness makes little difference. Actually, the difference in a seagoing and a beach boat of the *Peniche* type is the flatter bottom and buttock lines of the beach boat, the deck line being about the same in both forms. The Jutland boats have largely retained their original seagoing qualities and, as they are mechanically beached and launched, their transverse stability on the beach is sufficient.

SMALL FISHING BOATS IN PORTUGAL

by

JOAQUIM GORMICHO BOAVIDA

THE Portuguese motorized fishing boats are almost entirely developed from older local types of sailing boats. They usually stay at sea for only one or two days. They are suited to local conditions and are reasonably safe, but some are not as fast or economical as fishermen would like them to be.

Powering small boats presents different problems in the north, central and south coastal fishing centres of Portugal, but in all these places the fishermen are anxious to install motors in their boats, particularly in coastal long-liners and lobster catchers.

This is a direct result of the improved standard of life of Portuguese fishermen and of the interest free loans granted them by their " Casas dos Pescadores " (" Fishermen's Houses "). The primary object of the credit programme is to increase production and to lower the price of fish by the use of more efficient boats and gear. To achieve this, it is necessary to remove some technical and economical factors, mainly in the north and south coast fishing centres. It is felt that the engines sometimes are larger than necessary, and that in some instances, they are not suitable for marine operation. Overpowering makes operation more expensive. The engines weigh down the hull and increase fuel consumption without contributing much to higher speed.

Investigations are being made by the Gabinete de Estudos das Pescas (Fisheries Research Office) with the object of selecting standard designs of powered fishing boats, and types of sail and oar fishing craft suitable for motorizing. The selected boats will be those whose crew and owners confirm their seaworthiness and economical operation as being superior to all other types. Later, the investigators hope to improve the seaworthiness and economical operation of the boats by tank tests of models. The plans resulting from this study will be at the disposal of the fishermen, free of charge, so that they can build the boats through the credit programme.

It is intended to avoid, as much as possible, the building of larger, faster and more complicated boats in the

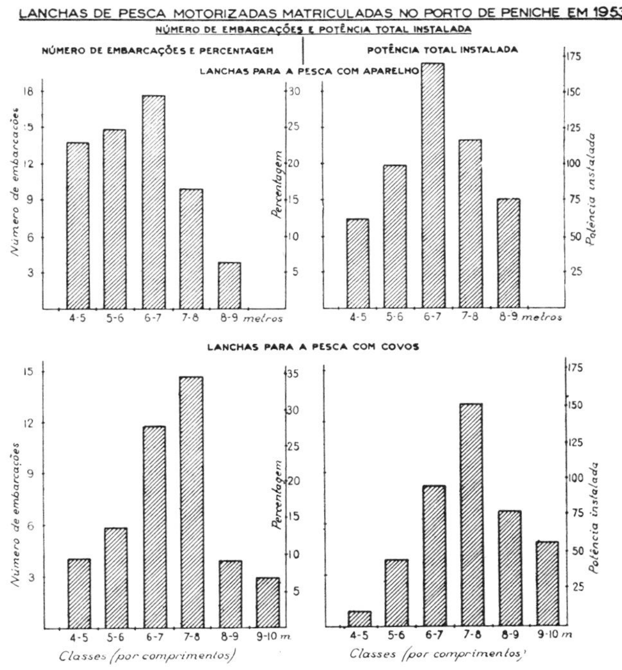

Fig. 97. Peniche's engine propelled fishing craft, according to length groups: 13 to 16 ft. (4 to 5 m.); 16 to 20 ft. (5 to 6 m.); 20 to 23 ft. (6 to 7 m.); 23 to 26 ft. (7 to 8 m.); 26 to 30 ft. (8 to 9 m.); 30 to 33 ft. (9 to 10 m.)

Fig. 96. Motorization of the fishing craft at Ericeira

TABLE VIII

Lobster Catcher of Peniche

Main dimensions:

	ft.	*m.*
Length, over-all	21.7	6.62
Length, between perpendiculars	18.5	5.63
Length, registration	21.0	6.40
Breadth, moulded	7.4	2.26
Breadth, registration	7.4	2.26
Draught, forward	1.2	0.38
Draught, amidships	1.9	0.59
Draught, aft	2.7	0.81
Depth, amidships	3.4	1.05
Depth, registration	3.4	1.05
Depth, for tonnage calculations	2.6	0.80
Gross tonnage	2.66	
Net tonnage	1.57	

GENERAL ARRANGEMENT—*fig. 101*

Scantlings:

	in.	*mm.*
Keel	4.9×2.2	125×55
Floor timbers	3.9×1.8	100×45
Middle futtocks	2.8×0.6	70×15
Shelf	5.3×0.8	135×20
Stringers	5.3×0.8	135×20
Bilge-keels	4.7×1.0	120×25
Waterways	8.3×1.2	210×30
Rail	4.7×1.0	120×25

TABLE IX

Lobster catcher of Peniche

Main dimensions:

	ft.	*m.*
Length, over-all	27.6	8.40
Length, registration	23.6	7.20
Length, for tonnage calculations	24.9	7.60
Breadth, over-all	9.2	2.80
Breadth, registration	9.0	2.75
Breadth, for tonnage calculations	8.7	2.65
Depth, forward	5.2	1.57
Depth, amidships	3.8	1.15
Depth, aft	5.5	1.68
Depth, registration	3.8	1.15
Depth, for tonnage calculations	3.0	0.90
Gross tonnage	4.53	
Net tonnage	2.3	

GENERAL ARRANGEMENT—*fig. 102*

Scantlings:

	in.	*mm.*	*Materials*
Keel	5.1×2.4	130×60	Oak
Stems	5.1×2.4	130×60	,,
Aprons	5.1×5.1	130×130	Pinaster
Middle futtocks	2.4×2.0	60×50	Stonepine
Floor timbers	2.8×2.0	70×50	,,
Keelson	5.5×5.1	140×130	Pinaster
Stringers	5.5×1.1	140×27	,,
Shelf	5.1×1.0	130×25	,,
Beams	6.3×2.0	160×50	,,
Garboards	5.1×0.8	130×20	,,
Outside planking	0.8	20	,,
Bilge keels	5.5×1.6	140×40	,,
Sheerstrake	5.9×1.2	150×30	,,
Waterways	8.7×1.2	220×30	,,
Wales	3.1×2.4	80×60	,,
Engine bearers	4.7	120	,,

TABLE X

Long-liner and lobster catcher of Ericeira

Main dimensions:

	ft.	*m.*
Length, over-all	20.3	6.20
Length, between perpendiculars	17.1	5.20
Breadth, extreme	7.1	2.15
Depth, forward	3.8	1.15
Depth, amidships	2.7	0.82
Depth, aft	4.0	1.23
Moulded draught	1.6	0.50

GENERAL ARRANGEMENT—*fig. 106*

Scantlings:

	in.	*mm.*
Keel	5.5×2.0	140×50
Shelf	3.1×0.8	80×20
Stringers	3.1×0.8	80×20
Waterways	10.4×1.8	265×45
Wale	1.6×1.6	40×40
Sheerstrake	4.7×1.0	120×25

TABLE XI

Long-liner of Sesimbra

Main dimensions:

	ft.	*m.*
Length, over-all	21.3	6.50
Moulded, breadth	8.2	2.50
Moulded, depth	2.8	0.85

GENERAL ARRANGEMENT—*fig. 108*

Scantlings:

	in.	*mm.*
Keel	7.1×2.0	180×50
Floor timbers	2.8×1.4	70×35
Middle futtocks	2.8×1.4	70×35
Keelson	4.3×3.9	110×100
Shelf	3.9×0.7	100×18
Stringers	3.1×0.8	80×20
Bilge keels	2.4×2.0	60×50
Beams	7.1×1.6	180×40
Coamings	8.7×1.6	220×40
Waterways	7.9×1.6	200×30
Deck	7.9×0.8	200×20
Rail	3.7×1.0	95×25
Wale	3.9×2.0	100×50
Outside planking	7.9×0.7	200×18
Sheerstrake	3.1×1.6	80×40
Topside—planking	3.9×0.8	100×20

TABLE XII

Engine-propelled long-liner of Sesimbra

Main dimensions:

	ft.	*m.*
Length, over-all	29.5	9.00
Length, between perpendiculars	28.8	8.56
Breadth, extreme	9.8	3.00
Depth, forward	5.4	1.66
Depth, amidships	4.1	1.25
Depth, aft	5.4	1.65
Depth, moulded	3.0	0.90

GENERAL ARRANGEMENT—*fig. 109*

Scantlings:

	in.	*mm.*
Stem	9.1×2.8	230×70
Stern-post	6.7×2.8	170×70
Keel	5.9×3.5	150×90
Keelson	5.5×5.1	140×130
Deadwood, forward	7.9	200
Deadwood, aft	11.8	300
Apron	7.9×6.3	200×160
Inner-post	7.9×5.9	200×150
Floor timbers	2.0	50
Middle futtocks	2.0	50
Deck planking	1.0	25
Waterways	7.9×1.6	200×40
Rail	5.5×1.4	140×35
Coamings	9.8×1.6	250×40
Beams	7.1×2.0	180×50
Shelf	5.9×1.2	150×30
Stringers	5.9×1.2	150×30
Hatchway-carling	4.7×2.0	120×50
Wale	3.9×2.8	100×70
Topside-planking	7.1×1.0	180×25
Outside-planking	1.0	25
Bilge keels	3.1×2.4	80×60

TABLE XIII

Sardine seiner of Peniche

Main dimensions:

	ft.	*m.*
Length, over-all	53.8	16.40
Length, between perpendiculars	39.4	12.00
Breadth, extreme	14.7	4. 48
Depth, moulded	5.6	1.72
Mean draught	4.1	1.25
Displacement to 10 WL	39.06 cu. m.	
Block coefficient	0.57	

GENERAL ARRANGEMENT—*fig. 112*

Scantlings:

	in.	*mm.*
Keel	9.1×5.1	230×130
Keelson	8.6×8.6	220×220
Shelf	7.9×2.4	200× 60
Stringers	5.9×2.0	150× 50
Bilge keels	6.3×2.4	160× 60
Beams	7.1×3.5	180× 90
Coamings	15.7×3.0	400× 75
Deck planking	5.5×1.5	140× 38
Rail	7.1×2.0	180× 50
Wale	6.3×0.8	160× 20
Outside planking	5.9×1.5	150× 38
Sheerstrake	6.3×2.4	160× 60
Topside-planking	6.7×2.0	170× 50

TABLE XIV

Sardine seiner of Peniche

Main dimensions:

	ft.	*m.*
Length, over-all	42.6	13.00
Length, between perpendiculars	39.7	12.10
Breadth, moulded	13.1	4.00
Breadth, extreme	13.5	4.10
Depth, moulded	4.2	1.28
Distance from underside of keel to topside of beam	3.3	1.02
Draught, forward	3.1	0.94
Draught, aft	4.5	1.38

GENERAL ARRANGEMENT—*fig. 113*

Scantlings:

	in.	*mm.*
Keel	6.7×3.9	170×100
Keelson	9.1×7.9	230×200
Shelf	6.7×2.0	170× 50
Stringers	7.1×2.0	180× 50
Bilge keels	7.1×2.0	180× 50
Beams	8.3×2.8	210× 70
Hatchway-carling	7.5×2.8	190× 70
Coamings	15.7×2.4	400× 60
Rail	7.1×1.2	180× 30
Wale	5.5×4.7	140×120
Topside-planking	7.1×2.0	180× 50

TABLE XV

Sardine seiner of Sesimbra

Main dimensions:

	ft.	*m.*
Length, over-all	45.3	13.80
Length, between perpendiculars	36.4	11.10
Breadth, extreme	13.3	4.06
Depth, forward	8.1	2.47
Depth, amidships	5.9	1.80
Depth, aft	7.7	2.34
Depth, moulded	4.3	1.30
Depth, registration	4.1	1.25
Draught, moulded	3.6	1.10
Volume of displacement, in cu. m.		26.90
Displacement, in tons	27.59	
Block coefficient	0.54	
Water plane coefficient	0.76	
Midship coefficient	0.80	
Prismatic coefficient	0.675	

GENERAL ARRANGEMENT—*fig. 114*

Scantlings:

	in.	*mm.*
Keel	7×4.3	180×110
Stem	,,	,,
Stern-post	,,	,,
Apron	,,	,,
Deadwood	,,	,,
Floor timbers	5.5×2.8	140× 70
Middle futtocks	4.7×2.8	120× 70
Keelson	8.7×7.9	220×200
Shelf	6.7×1.6	170× 40
Stringers: central	6.3×2.0	160× 50
lateral	4.7×1.2	120× 30
Bilge keels	6.3×2.0	160× 50
Beams	7.9×2.8	200× 70
Hatchway-carling	4.7×2.8	120× 70
Coamings	11.8×2.8	300× 70
Waterways	9.4×2.0	240× 50
Deck planking	4.7×1.18	120× 30
Rail	7.1×2.0	180× 50
Wale	3.9×1.6	100× 40
Outside planking	3.9×1.6	100× 40
Sheerstrake	3.9×1.6	100× 40
Topside-planking	7.9×1.6	200× 40
Bulwark	3.9×1.4	100× 35
Garboards	6.3×1.6	160× 40

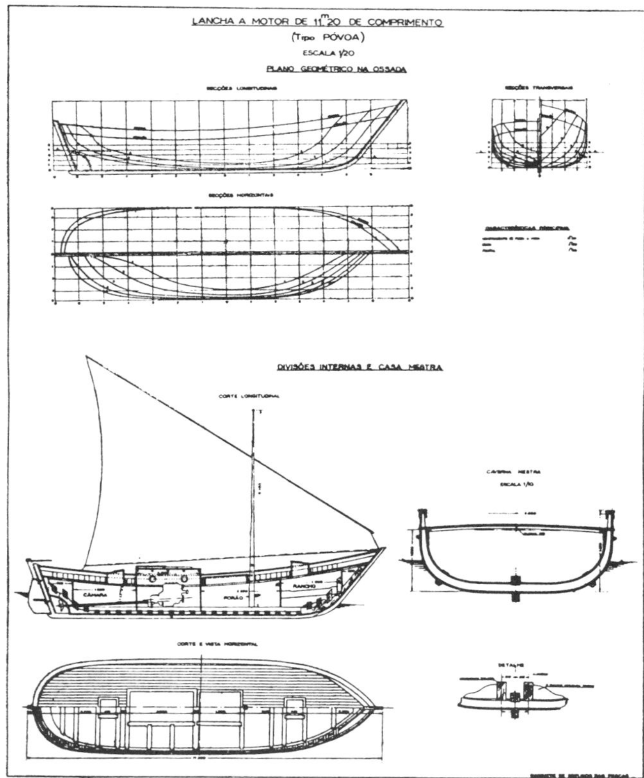

Fig. 98. Long-liner of Póvoa do Varzim and Vila do Conde. Length, over-all, 36.8 ft. (11.20 m.); breadth, 9.8 ft. (3.00 m.); depth, 3.3 ft. (1.00 m.)

coastal fisheries. Experience shows that such boats are not more seaworthy nor do they produce a greater annual income for the owner. The " improved models " should not result in an increase in the cost of building, operating, and maintaining, but in higher productivity and profit.

At Ericeira, a small fishing centre in the central coast of Portugal, all fishing boats were sail and oar propelled a few years ago, but today 30 of the 151 fishing boats in the centre are engine propelled. Ericeira has a small beach without breakwater or natural protection from the strong winds blowing from south-west to north-west. It is backed by high cliffs so there is no space for stowing the boats which, in winter, must be taken over the cliff by way of a very steep ramp. Fig. 96 shows the development of motorization in Ericeira and gives a fair idea of the actual rate of installations of engines in fishing boats in Portugal.

At Nazaré there are some 200 small fishing boats, 50 of which have engines, while at Peniche and Sesimbra the motorization of small craft is nearly complete.

Fig. 97 gives a picture of mechanization of small fishing boats at Peniche. An analysis of the data shows there are too many sizes and types of boats in use.

Fig. 98 to 111 show some selected small boats at northern and central coast harbours. These are the most efficient owner-operated boats.

Fig. 100. Hand- and long-liner (Leixões harbour). Main dimensions: length, over-all, 27.3 ft. (8.31 m.); breadth, 8.8 ft. (2.69 m.); depth, 2.8 ft. (0.86 m.); powered by a 10 h.p. engine

Fig. 99. Hand and long-line fishing is carried out aboard these types of small motor boats (Póvoa de Varzim harbour). Main dimensions: length, over-all, 36.8 ft. (11.20 m.); breadth, 9.8 ft. (3.00 m.); depth, 3.3 ft. (1.00 m.)

The first three illustrations show long-liners which operate from the northern harbours (Póvoa de Varzim, Vila do Conde and Leixões). Fig. 101 to 104 and Tables VIII and IX refer to engine-propelled fishing craft at Peniche, in the northern part of the central coast. Fig. 105 to 107 and Table X present some information regarding Ericeira's small motorized boats and finally, fig. 108 to 111 and Tables XI and XII refer to Sesimbra's long-line and lobster boats.

Fig. 112 to 115 show some of the principal types of small sardine seiners, operating from the central coast harbours which are engaged in the industrial fisheries. The sardine fleet, including 381 " traineiras " and seiners, produced, in 1953, 140,587 metric tons of sardine and similar species, to a value of 300 million escudos ($10.725 million—£3.8 million). Fig. 112 and 113 and Table XIII and XIV give information on Peniche's types while fig. 114 and 115 and Table XV present data on the Sesimbra's type.

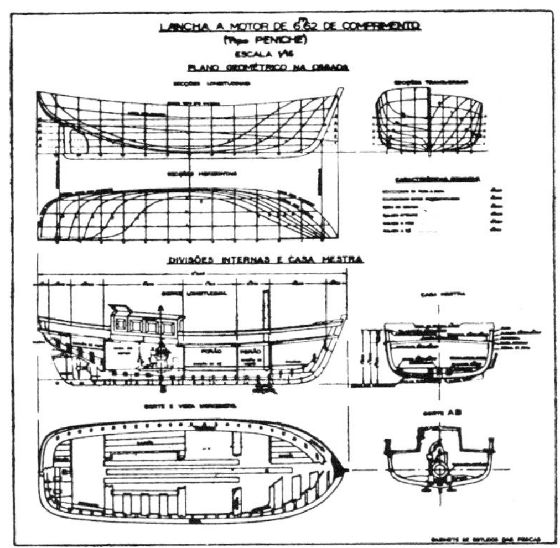

Fig. 101. Lobster catcher of Peniche. Length, over-all, 21.7 ft. (6.62 m.). See Table VIII

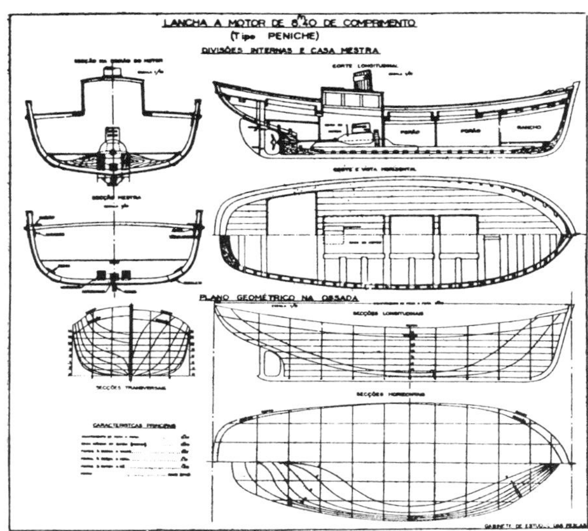

Fig. 102. Lobster catcher of Peniche. Length, over-all, 27.6 ft. (8.40 m.). See Table IX

Fig. 103. Small lobster catcher of Peniche. Main dimensions: length, over-all, 24 ft. (7.30 m.); breadth, 8.5 ft. (2.55 m.); depth, 3.3 ft. (1.03 m.); powered by a 8 h.p. engine

Fig. 104. Small long-liner from Peniche's harbour. Main dimensions: length, over-all, 24.7 ft. (7.55 m.); breadth, 8.5 ft. (2.59 m.); depth, 3.9 ft. (1.20 m.); powered by a 9 h.p. engine

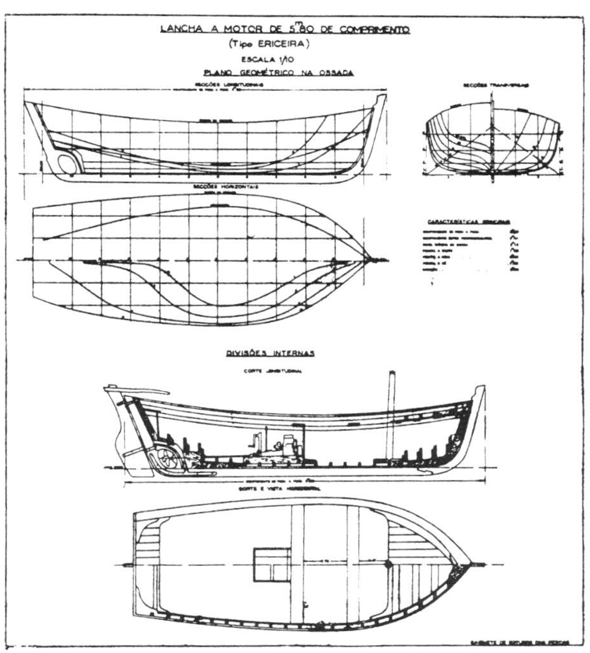

Fig. 105. Lobster catcher from Ericeira; also used as long-liner. Length, over-all, 19 ft. (5.80 m.); length between perpendiculars, 19 ft. (5.80 m.); breadth, extreme, 7.1 ft. (2.16 m.); mean draught 1.6 ft. (0.49 m.)

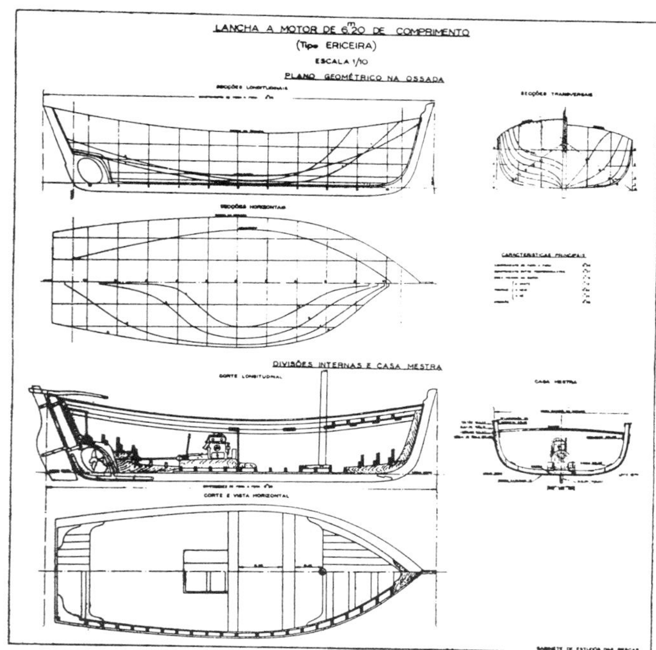

Fig. 106. Long-liner and lobster catcher from Ericeira. Length, over-all, 20.3 ft. (6.20 m.). See Table X

Fig. 107. Motor fishing boats on the beach (Ericeira). Main dimensions: length, over-all, 20.3 ft. (6.20 m.); breadth, 7 ft. (2.15 m.); depth, 2.7 ft. (0.82 m.)

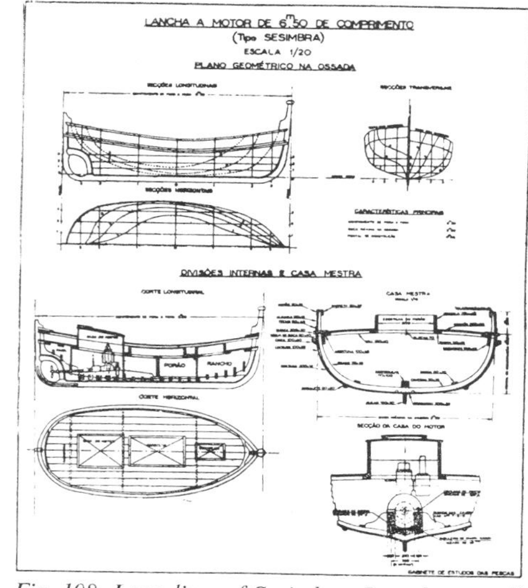

Fig. 108. Long-liner of Sesimbra. Length, over-all, 21.3 ft. (6.50 m.) See Table XI

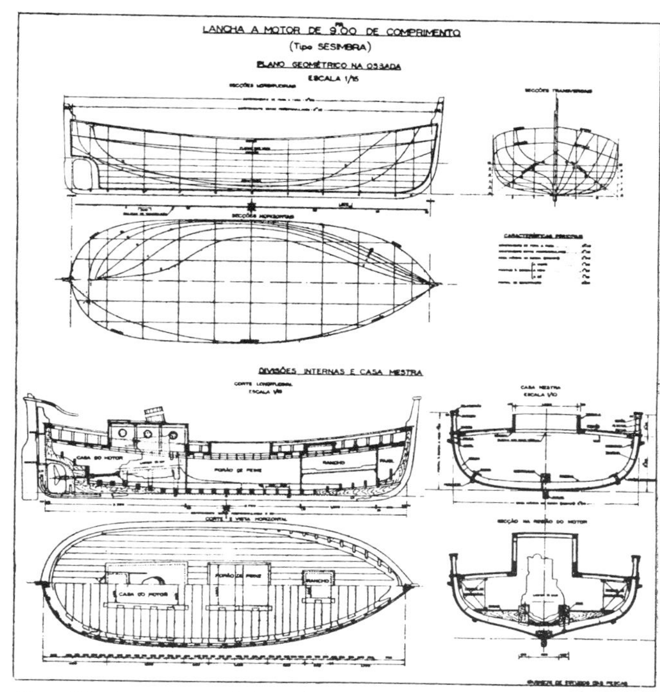

Fig. 109. Long-liners of Sesimbra. Length, over-all, 29.5 ft. (9.0 m.) See Table XII

Fig. 110. Small long-liner of Sesimbra. Main dimensions: length, over-all, 21.3 ft. (6.50 m.); breadth, 8.2 ft. (2.50 m.); depth, 2.8 ft. (0.85 m.)

Fig. 111. Small long-liner of Sesimbra. Main dimensions: length, over-all, 29.5 ft. (9.0 m.); breadth, 9.8 ft. (3.0 m.); depth, 4.1 ft. (1.25 m.)

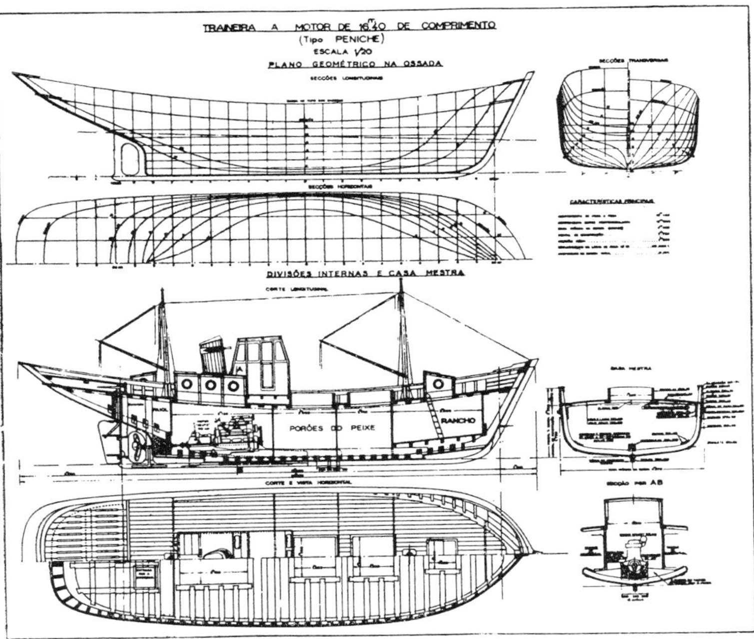

Fig. 112. Sardine seiner of Peniche. Length, over-all, 53.8 ft. (16.40 m.).
See Table XIII

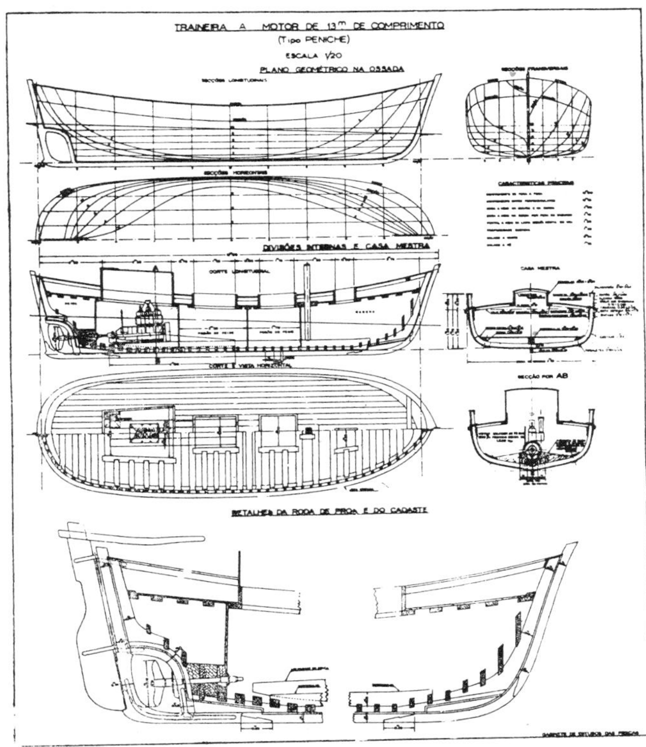

Fig. 113. Sardine seiner of Peniche. Length, over-all 42.6 ft. (13.0 m.)
See Table XIV

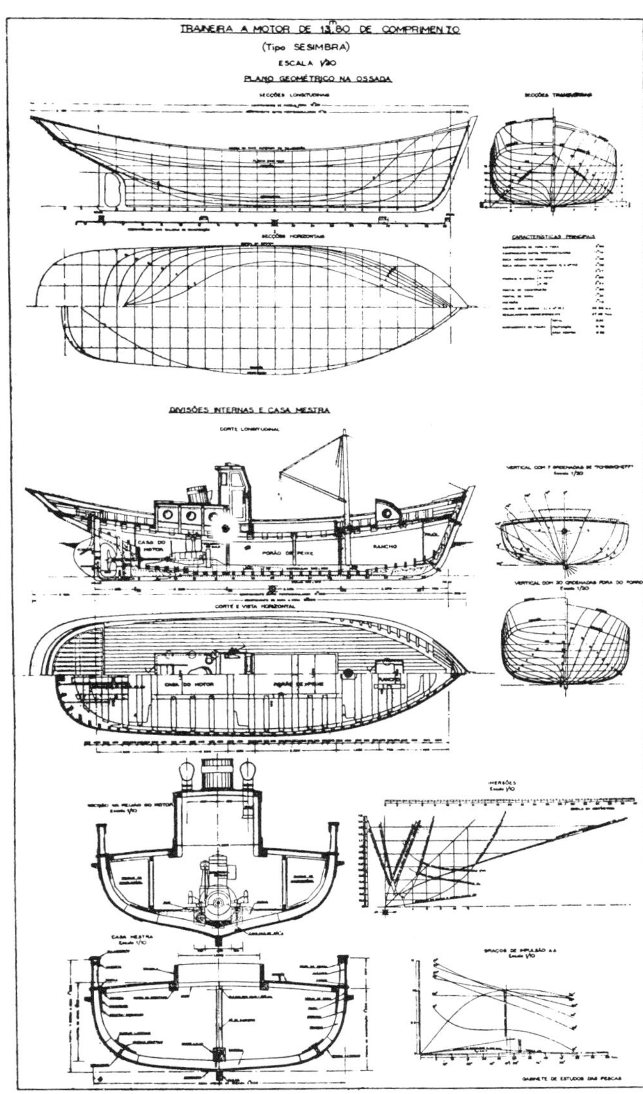

Fig. 114. Sardine seiner of Sesimbra. Length, over-all, 45.3 ft. (13.80 m.) See Table XV

Fig. 115. Small Sesimbra seiner (sardine catcher)

JAPANESE PELAGIC FISHING BOATS

by

ATSUSHI TAKAGI

TO feed the Japanese population adequately some 6,350,000 metric tons of sea food are required but in 1952 production was only 4,300,000 tons. The fishing fleet is the most obvious means for producing the food although during World War II the Japanese fishing fleet suffered great damage and decreased from 350,000 to 280,000 boats. An additional handicap has been the poor quality, because of lack of repairs, of the vessels which escaped war damage.

Immediately after the war reconstruction began and in 1952 the Japanese fishing fleet consisted of 450,000 vessels totalling about 1,200,000 gross tons and 2,350,000 h.p. For marine fisheries there were 129,048 powered boats (860,644 gross tons, 2,334,818 h.p.) of which 101,647 were less than 5 gross tons, 20,824 of 5 to 20 tons, 6,097 of 20 to 100 gross tons and 480 boats larger than 100 gross tons. The 20 to 100 tons group makes up more than 30 per cent. of the total tonnage and horse-power while the group larger than 100 tons makes up 20 per cent. of the total gross tonnage and 10 per cent. of the total horse-power. Table XVI groups the power boats used for marine fisheries according to the different fishing methods.

The Japanese fishing fleet uses almost every known fishing method. A few of the present types of boat in the fleet are explained in the following pages and in the paper about drag net boats on page 137.

TABLE XVI

Japanese powered marine fishing boats in 1952 according to fishing method

	No.	*Gross tons*	*Horse-power*
Fixed net	3,753	14,627	45,059
Pole and line (excluding tuna and skipjack) . . .	32,791	81,185	281,659
Gill net (including drift net) .	7,816	36,611	106,041
Purse seine	8,024	91,002	292,455
Square net . . .	4,114	20,037	65,634
Tuna and skipjack (pole and line, long line) . . .	1,590	108,319	246,106
Whaling	131	59,063	98,245
Long line (exluding tuna, swordfish and shark long line)	15,739	52,757	163,093
Total—pelagic fishing .	73,958	463,601	1,298,392
Small trawler . . .	23,888	140,573	414,727
Medium trawler and otter trawler (west of 130°E) .	729	69,979	151,130
Miscellaneous drag net (including beach seines) .	6,942	25,828	89,179
Total—bottom fishing .	31,559	236,380	655,036
Government vessels . .	323	11,919	32,405
Fish carriers . . .	5,192	104,106	209,228
Miscellaneous . . .	18,016	44,639	139,857
GRAND TOTAL . .	129,048	860,644	2,334,818

MACKEREL POLE AND LINE FISHING BOATS

Mackerel pole and line fishing formerly was carried on at night by small boats of less than 20 gross tons, but since 1950, with abundant catches of mackerel, the size of boats has increased up to 70 gross tons and the fishing method has been much improved. The fish are caught by combined use of bait and lights. The crew is quite large as in skipjack pole and line fishing, and the bigger boats now being built are of the skipjack type. *Minobu Maru* is a typical modern mackerel pole and line fishing boat and has the following principal characteristics:

TABLE XVII

Draft and Stability

	Load			
	Light		*Full*	
Displacement .	60 tons	60.919 ton	88 tons	89.200 ton
Draft, fore .	2.1 ft.	0.640 m.	4.05 ft.	1.235 m.
„ aft . .	6.6 ft.	2.020 m.	7.15 ft.	2.187 m.
„ mean .	4.35 ft.	1.330 m.	5.6 ft.	1.711 m.
Freeboard . .	3.4 ft.	1.035 m.	2.1 ft.	0.654 m.
KM . . .	6.8 ft.	2.070 m.	6.25 ft.	1.910 m.
GM . . .	1.5 ft.	0.451 m.	1.24 ft.	0.380 m.
G from midship .	5.3 ft.	1.617 m. aft	3.1 ft.	0.933 m. aft
Co-efficients:				
Block . .		0.590		0.655
Prismatic .		0.638		0.685
Midship . .		0.926		0.947
Waterplane .		0.774		0.837

Length, between perpendiculars, 66.5 ft. (20.30 m.); beam 14 ft. (4.25 m.); depth, 6.5 ft. (2.00 m.); 46.20 gross tons, 15.78 net tons; fish holds 520 cu. ft. (14.76 cu. m.); ice holds 510 cu. ft. (14.36 cu. m.); fresh water tanks 860 imp. gal. (1,030 gal., 3.90 cu. m.); fuel oil tanks, 1,530 imp. gal. (1,830 gal., 6.95 cu. m.); officers and crew, 35; main engine: single-acting, four-cycle diesel, four cylinders with 9.85 in. (250 mm.) diameter and 15 in. (380 mm.) stroke, 160 b.h.p. at 380 r.p.m.; three-bladed propeller with 53 in. (1,346 mm.) diameter and 33 in. (838 mm.) pitch; 5 h.p. auxiliary hot bulb engine driving a 3 kw., D.C., generator, and 10 kw. generator driven by main engine; 50 w. and 30 w. wireless.

Fig. 116 shows the lines indicating the V-bottom design; fig. 117 is the general arrangement drawing;

TABLE XVIII

Speed Trial

Date June 5, 1952	*Load*	*Speed (in knots)*	*r.p.m.*	*Slip (in %)*	*b.h.p.*
Draft:					
Fore, 2.5 ft. (0.75 m.)	over	9.224	404	15.9	192
Aft, 7.4 ft. (2.25 m.)	4/4	8.829	380	14.4	160
Mean, 4.9 ft. (1.50 m.)	3/4	8.381	340	9.3	120
	1/2	7.814	302	4.7	80
Displacement:	1/4	6.940	240	6.4	40
71.5 tons (72.5 ton)					

fig. 118 the midship section, indicating the sawn frame construction, and fig. 119 is a photograph of *Minobu Maru*.

Table XVII gives particulars of the drafts and stability and Table XVIII gives results from the speed trial.

SKIPJACK POLE AND LINE FISHING BOAT

The boats engaged in skipjack, pole, and line fishing from spring to autumn, and in tuna long line fishing from fall to spring, vary in size from quite small to large—in wood up to 200 gross tons and in steel up to 250 gross tons. The number of crew for pole and line fishing is twice that needed for tuna long line fishing.

For pole and line fishing a platform is installed outside the bulwark for fishermen to stand on while handling the gear. As live fish is used for bait, the centre compartment between the two longitudinal bulkheads is used as a bait tank, with sea cocks on the bottom shell plates.

The sea water in the bait tanks circulates naturally by the pitching, rolling, dipping and heaving of the boat. The tanks, when emptied and with the sea cocks closed, are used as fish holds.

For tuna long lining, the arrangements for skipjack pole and line fishing are removed and a line hauler on board is operated.

The first wooden boat of this type of more than 100 ft. (30 m.) length was built in 1952 and is named *Kashio Maru*. It can operate in rough seas and has the following principal data: Length, between perpendiculars, 105 ft.

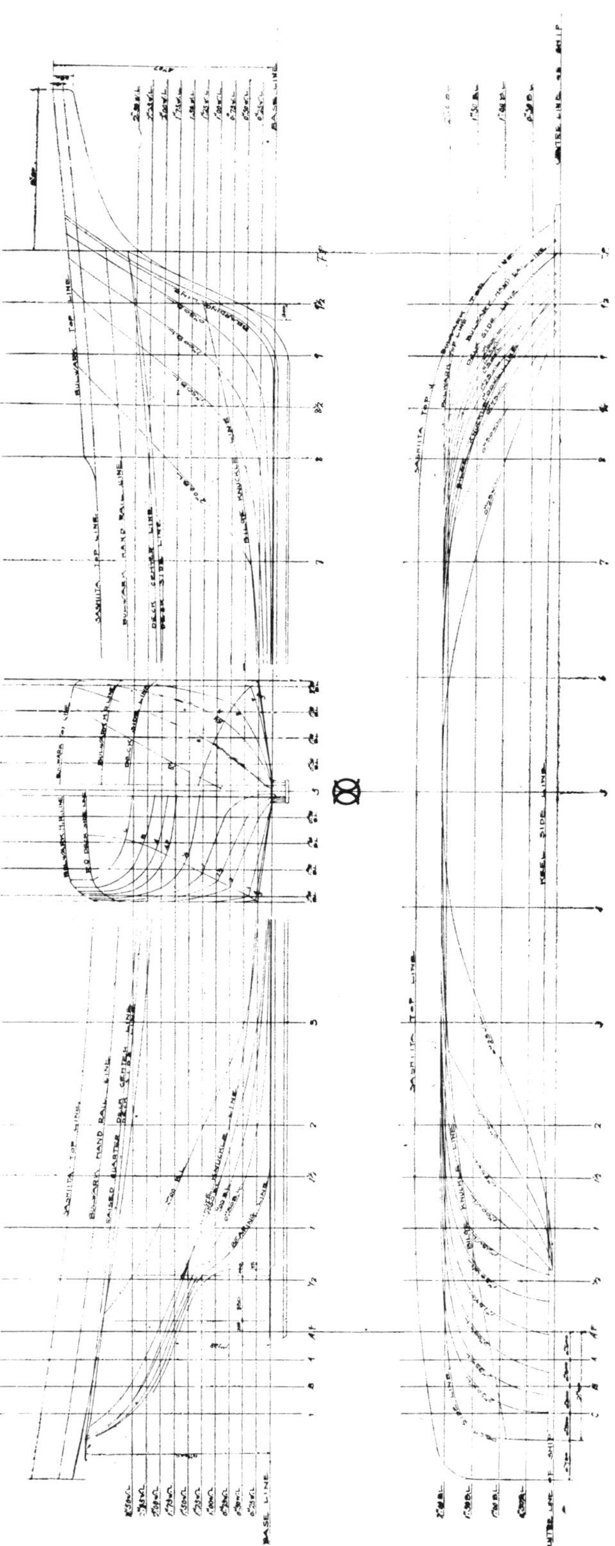

Fig. 116. Lines of the 66.5 ft. (20.3 m.) mackerel pole and line fishing boat Minobu Maru

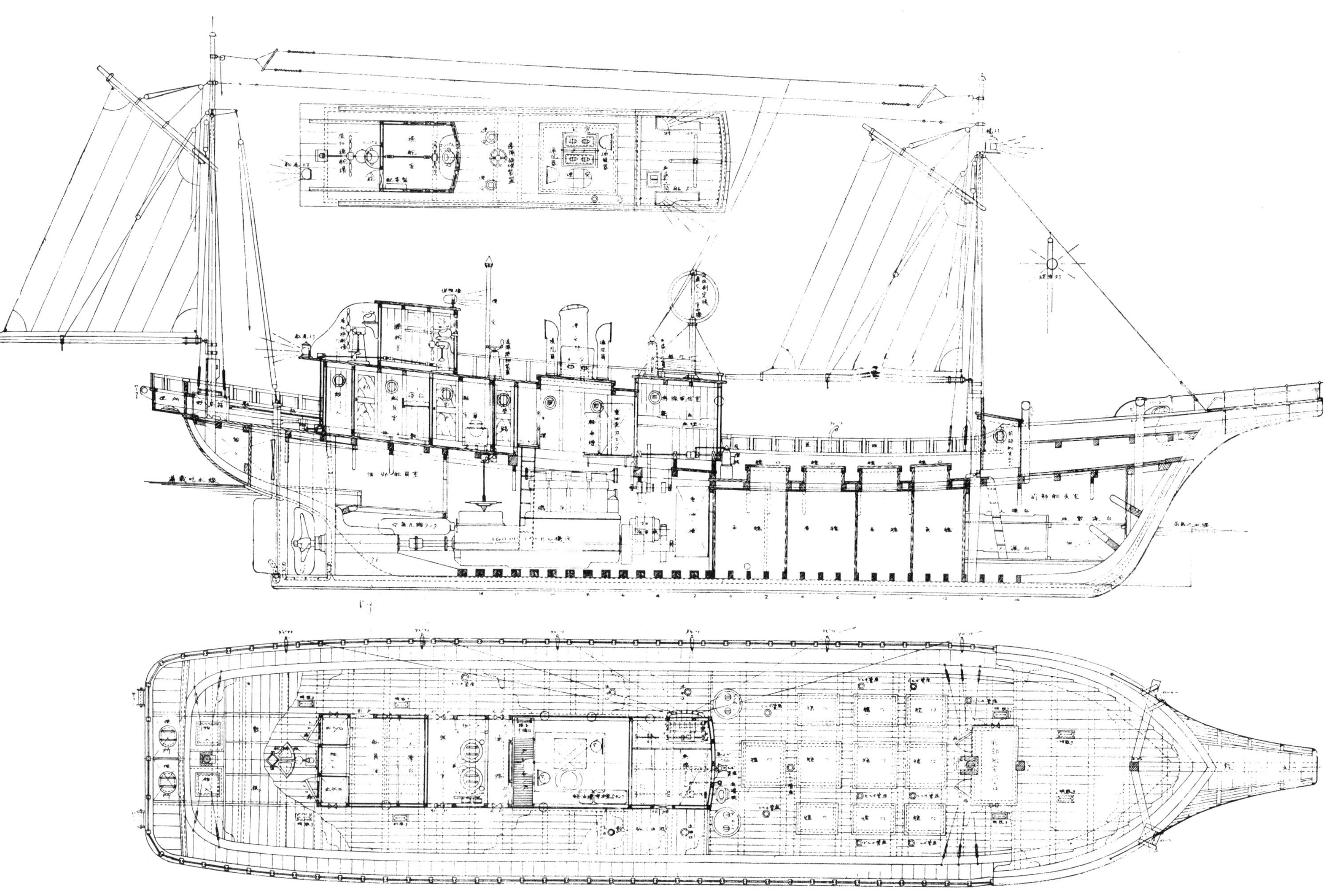

Fig. 117. General arrangement of the 66.5 ft. (20.3 m.) mackerel pole and line fishing boat Minobu Maru

(32.00 m.); beam, 21.7 ft. (6.60 m.); depth, 11 ft. (3.40 m.); 197.36 gross tons; 108.82 net tons; bait tanks, 3.270 imp. gal. (3,920 gal., 14.85 cu. m.); ice holds, 2,920 cu. ft. (82.58 cu. m.); fuel oil tanks, 14,700 imp. gal. (17,600 gal., 66.57 cu. m.); fresh water tanks, 3.270 imp. gal.

TABLE XIX

Draft and stability

	Light		Full	
Displacement	251 tons	255.320 ton	462 tons	470.0 ton
Draft, fore	4 ft.	1.22 m.	9.3 ft.	2.83 m.
„ aft	11 ft.	3.37 m.	13.5 ft.	4.10 m.
„ mean	7.5 ft.	2.295 m.	11.4 ft.	3.465 m.
Freeboard	5.25 ft.	1.605 m.	1.43 ft.	0.437 m.
KM	10.65 ft.	3.250 m.	10.1 ft.	3.080 m.
KG	8.25 ft.	2.520 m.	8 ft.	2.440 m.
GM	2.4 ft.	0.73 m.	2.1 ft.	0.64 m.
G from midship	6 ft.	1.84 m. aft	5.2 ft.	1.58 m. aft
Coefficients:				
Block	0.60		0.69	
Prismatic	0.65		0.73	
Midship	0.92		0.95	
Waterplane	0.77		0.88	

(3,920 gal., 14.85 cu. m.); officers and crew, 70; main engine, single acting four-cycle diesel, six cylinders with 15 in. (380 mm.) diameter and 20½ in. (520 mm.) stroke, 550 b.h.p. at 300 r.p.m.; four-bladed propeller with 79 in. (2,000 mm.) diameter and 45.2 in. (1,150 mm.) pitch; 35 h.p. auxiliary diesel driving two 5 kw., D.C. generators, two 6-in. 150 mm.) centrifugal water spray pumps; 150 w. and 25 w. wireless.

Careful consideration has been given to the refrigeration system, at some sacrifice of speed and fish-hold capacity.

Fig. 120 shows the V-bottom lines, fig. 121 the general arrangement with the fishing platform all along the side,

TABLE XX

Speed trial

Date June 7, 1952	Load	Speed (in knots)	r.p.m.	Slip (in %)	b.h.p.
Draft:					
Fore, 4.75 ft. (1.45 m.)	Over	10.937	320	7.9	660
Aft, 10.85 ft. (3.31 m.)	4/4	10.562	300	5.5	550
Mean, 7.8 ft. (2.38 m.)	3/4	10.050	273	1.0	412
	1/2	9.441	238	− 6.3	275
Displacement:	1/4	8.112	189	−15.2	137
266 ton (270.3 ton)					

fig. 122 the sawn frame construction of the midship section and fig. 123 the *Kashio Maru* under speed.

Table XIX gives particulars of draft and stability, and Table XX results from the speed trials.

TUNA LONG LINERS

Boats for the exclusive use of tuna long line have been operated for the past 20 years, and their size is increasing continuously. The average size up to 1950 was about 200 gross tons but now it exceeds 500. Only a line hauler is used and the boat can be operated by a small crew. Large fish holds allows the installation of refrigerating coils within them. *Kaiko Maru No. 12* is a typical steel

TABLE XXI

Draft and stability

	Light		Full	
Displacement	310 tons	315.43 ton	550 tons	559.56 ton
Draft, fore	3.5 ft.	1.056 m.	7.1 ft.	2.165
„ aft	8.9 ft.	2.711 m.	12.3 ft.	3.736
„ mean	6.1 ft.	1.884 m.	9.7 ft.	2.950
Freeboard	5.9 ft.	1.801 m.	2.4 ft.	0.735
KM	11.25 ft.	3.430 m.	10.4 ft.	3.180
KG	9.55 ft.	2.910 m.	8.6 ft.	2.635
GM	1.7 ft.	0.520 m.	1.8 ft.	0.545
G from midship	3.3 ft.	1.014 m. aft	3.3 ft.	1.014 m. aft.
Coefficients:				
Block	0.608		0.676	
Prismatic	0.652		0.711	
Midship	0.952		0.952	
Waterplane	0.820		0.821	

tuna long line fishing boat, and has the following principal data: Length, over-all, 142 ft. (43.30 m.); length, between perpendiculars, 127 ft. (38.67 m.); beam, 23.5 ft. (7.20 m.); depth, 11.8 ft. (3.60 m.); 300.56 gross tons, 153.02 net tons; fish holds, 10,000 cu. ft. (287.60 cu. m.); pre-cooling tanks, 1,500 imp. gal. (1,800 gal., 6.76 cu. m.); fuel oil tanks, 28,000 imp. gal. (33,000 gal., 126.22 cu. m.); fresh water tank, 3,700 imp. gal. (4,500 gal., 17.00 cu. m.); officers and crew, 33; main engine, single-acting four-cycle diesel, six cylinders with 15 in. (370 mm.) diameter and 20½ in. (520 mm.) stroke and 600 b.h.p. at 300 r.p.m.; four-bladed propeller with 75 in. (1,900 mm.) diameter

TABLE XXII

Speed trial

Date December 30, 1951	Load	Speed (in knots)	r.p.m.	Slip) (in %	b.h.p.
Draft:					
Fore, 3.5 ft. (1.08 m.)	Over	11.331	307	16.8	660
Aft, 9.1 ft. (2.76 m.)	4/4	11.292	300	15.8	600
Mean, 6.45 ft. (1.97 m.)	3/4	10.680	270	11.0	450
	1/2	10.076	236	4.5	300
Displacement:	1/4	8.797	190	4.1	150
329 tons (334 ton)					

and 54 in. (1,370 mm.) pitch; 75 and 40 b.h.p. auxiliary diesels driving 55 kw., A.C. and 30 kw., A.C. generators; 5 kw. A.C. generator driven by main engine; line hauler and windlass driven by 225 v., A.C. 15 h.p. motor; 3 h.p. capstan; 200 w. and 50 w. wireless; NH_3 refrigerating compressor, two cylinder 20 h.p., capacity 9.6 RT (29,000 kcal.).

Fig. 124 shows the general arrangement and fig. 125

the *Kaiko Maru No. 12* under way. Table XXI gives particulars of drafts and stability and Table XXII results from the speed trial.

WHALE CATCHERS

Since the first participation of the Japanese whaling vessels in the Antarctic in 1934, the fleet has been steadily improved and the good catches made reflect this progress.

The first trial of a diesel in a whale catcher was made by Japan in 1937. Whales were thought to be sensitive to sound and diesels had for that reason been avoided because of the noise they made—reciprocating engines were used instead. But a test was made in 1936 by a submarine chaser to determine the maximum speed of whales, which was found to be 14 knots. By this test, it was discovered that the noise of the diesel driven submarine chaser did not matter if the speed was high. In consequence a diesel was installed in the *Seki Maru*, built by the Hayashikane shipbuilding yard, and having the following data: length × beam × depth, 129 × 24 × 13.7 ft. (39.24 × 7.35 × 4.19 m.); gross tonnage, 297.80; engine output, 900 h.p.

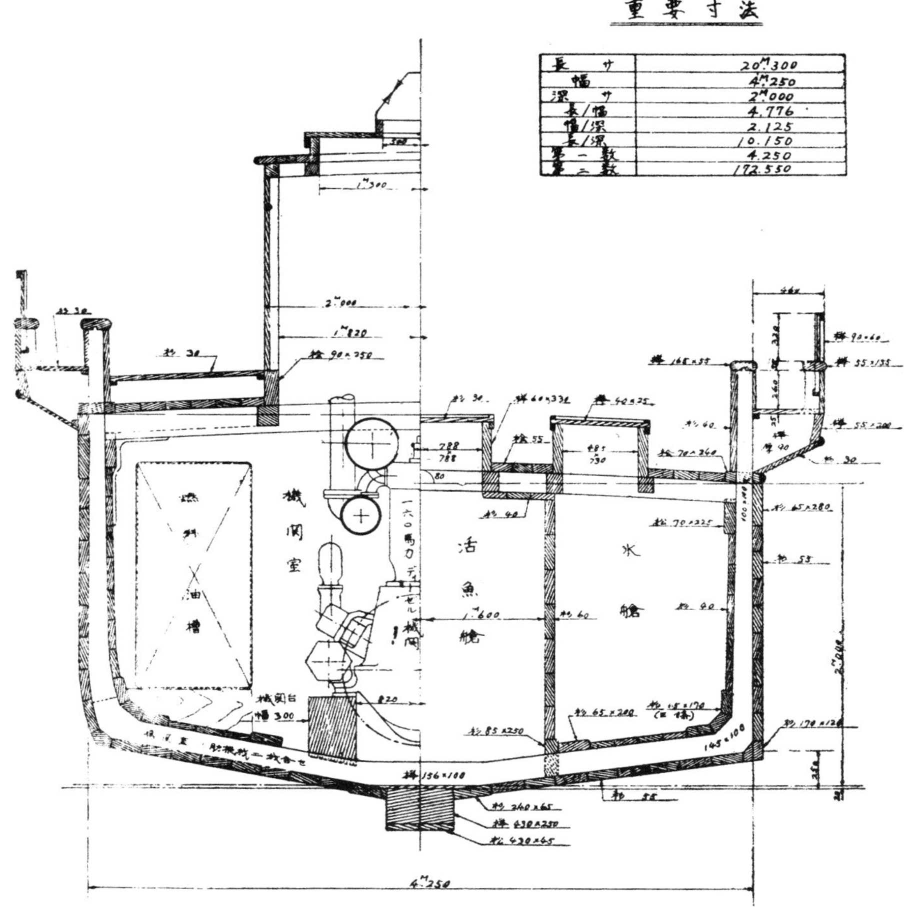

Fig. 118. Midship section of the 66.5 ft. (20.3 m.) mackerel pole and line fishing-boat Minobu Maru

Because of the low fuel oil consumption of diesels, the cruising radius is large and a frequent supply of oil *en route* to the Antarctic Ocean is therefore not needed. The disadvantage of diesels in whale catchers is that a low speed can hardly be obtained and manoeuvring is difficult when the boat approaches a whale. It is also difficult to tow the whale when caught.

As a result of the experiment of the *Seki Maru*, all vessels built later have been equipped with diesels and recently electricity has

Fig. 119. Minobu Maru *66.5 ft. (20.3 m.) mackerel pole and line fishing boat*

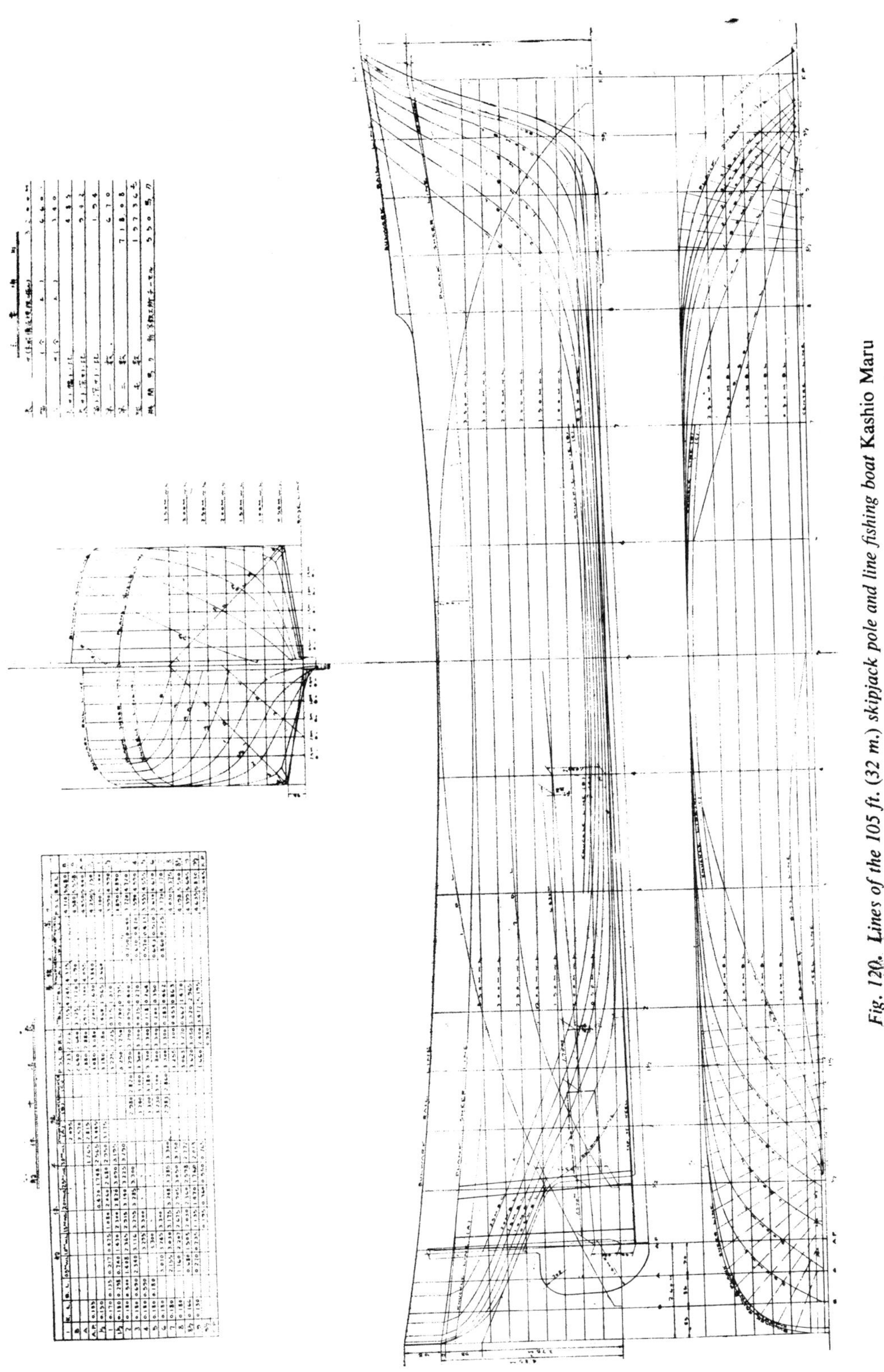

Fig. 120. Lines of the 105 ft. (32 m.) skipjack pole and line fishing boat Kashio Maru

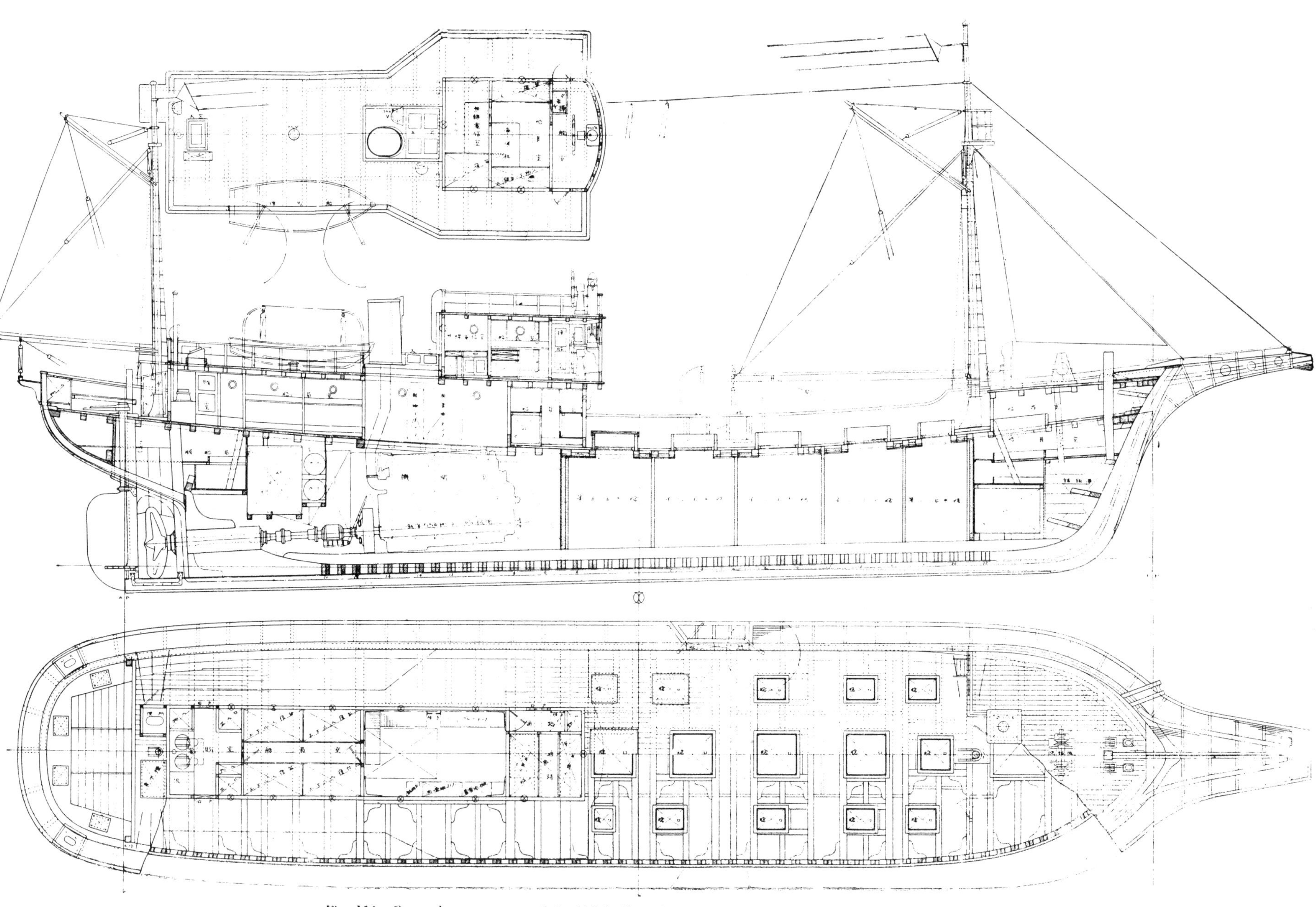

Fig. 121. General arrangement of the 105 ft. (32 m.) skipjack pole and line fishing boat Kashio Maru

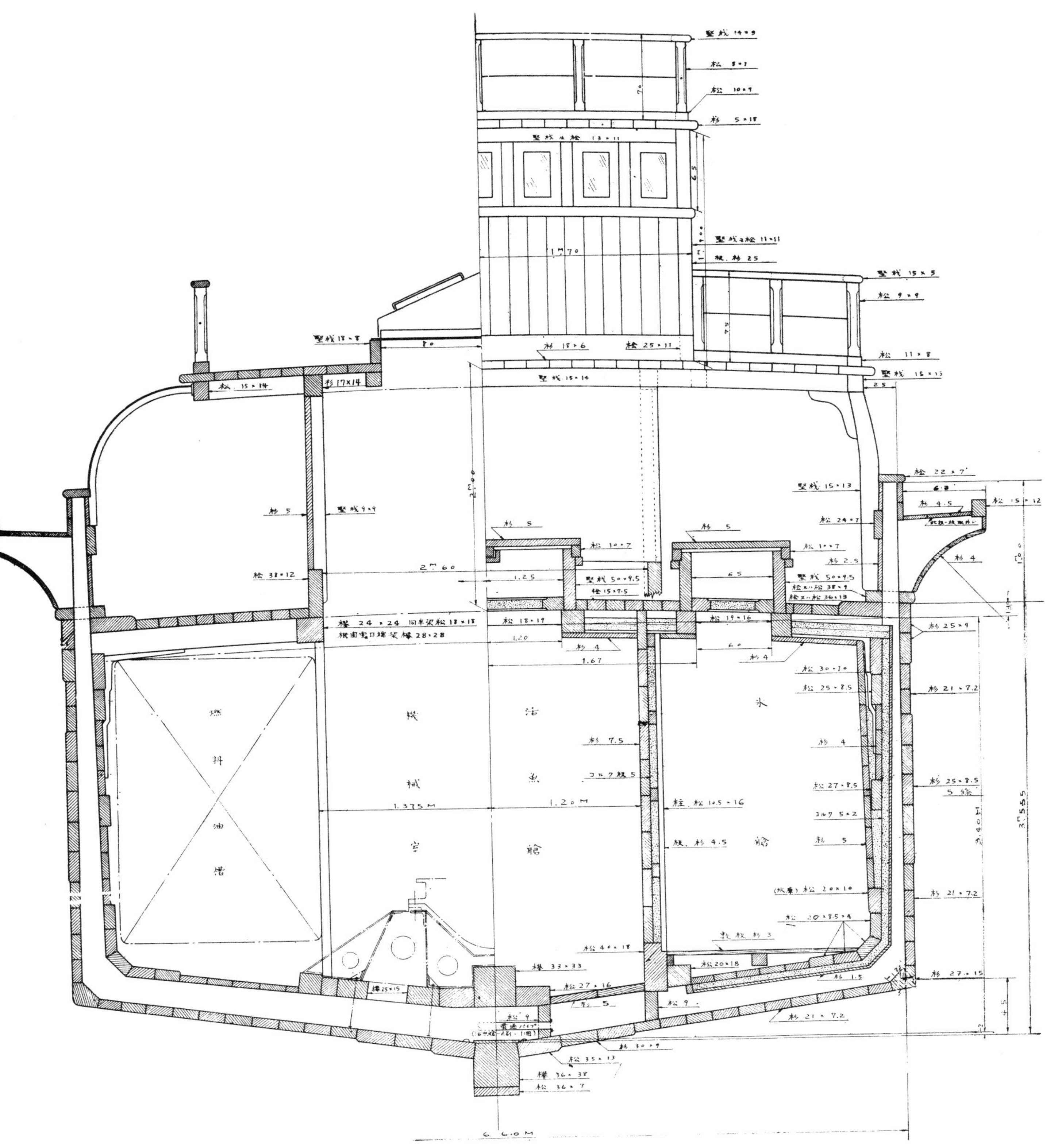

Fig. 122. Midship section of the 105 ft. (32 m.) skipjack pole and line fishing boat Kashio Maru

Fig. 123. Kashio Maru, *105 ft. (32 m.) skipjack pole and line fishing boat*

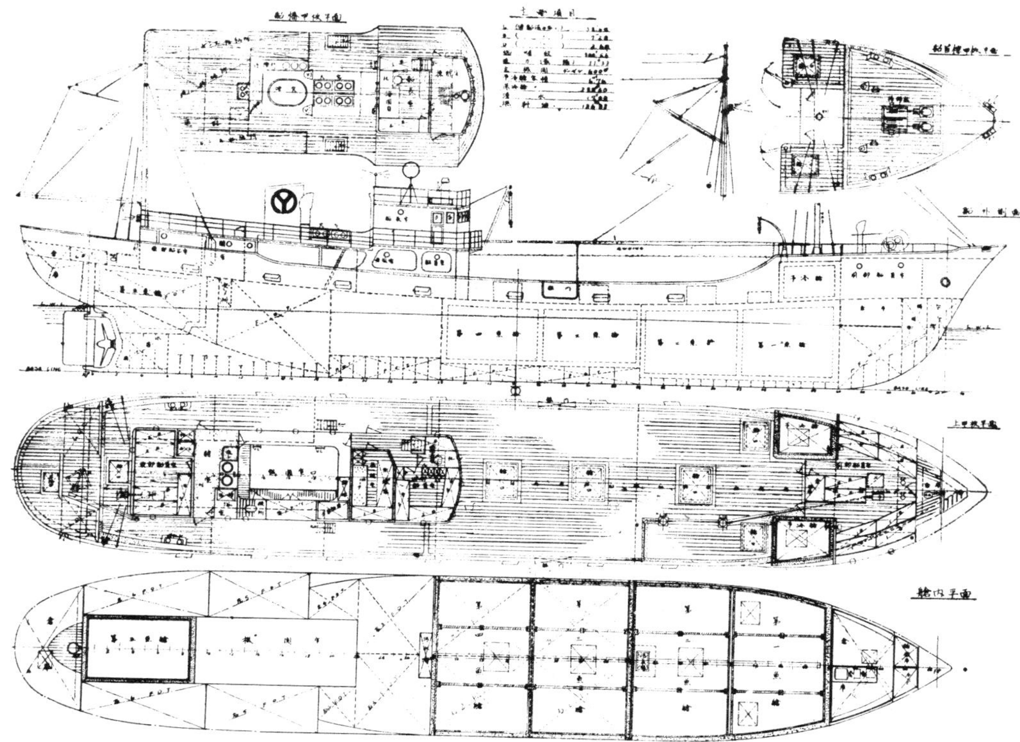

Fig. 124. General arrangement of Kaiko Maru No. 12, *142 ft. (43.3 m.) tuna long liner*

Fig. 125. Kaiko Maru No. 12, *142 ft. (43.3 m.) tuna long liner*

been used in place of steam to drive the donkey boiler.

The 300 to 350 gross tons type was considered to be the most profitable size, but the success of the larger foreign boats has persuaded Japanese owners to build bigger vessels. But the process has been gradual because of the high cost of shipbuilding, only five or six larger boats being constructed every year. It is expected that whale catchers of over 600 gross tons will eventually be built for Antarctic whaling.

Kyo Maru No. 7 is a typical Japanese steel whale catcher and has the following principal data: Length, over-all, 175 ft. (53.250 m.); length between perpendiculars, 157 ft. (48.00 m.); beam 27.5 ft. (8.400 m.); depth 14.9 ft. (4.546 m.); 399.43 gross tons, 123.67 net tons; fuel oil tanks, 42,000 imp. gal. (50,000 gal., 189.85 cu. m.); feed water tanks, 6,600 imp. gal. (7,900 gal., 29.87 cu. m.); fresh water tanks, 4,000 imp. gal. (4,750 gal., 17.98 cu. m.); lubricating oil tanks, 3,300 imp. gal. (4,000 gal,, 15.12 cu. m.); ballast tanks, 10,500 imp. gal. (12,600 gal., 47.89 cu. m.); 11,000 nautical miles cruising radius; 12 knots speed; officers and crew, 26; main engine, two-cycle single acting diesel, eight cylinders with 19 in. (480 mm.) diameter and 30 in. (760 mm.) stroke, 2,300 b.h.p., at 200 r.p.m., fuel oil consumption, 0.38 lb. (173 gr.)/b.h.p./hr.; four-bladed propeller with 122 in. (3,100 mm.) diameter and 105 in. (2,670 mm.) pitch; two 100 h.p. auxiliary diesels driving 60 kw., 230 v., D.C., generators; 20 kw., 230 v., D.C. generator driven by main engine; donkey boiler with 9 ft. (2.750 mm.) length and 8.5 ft. (2,600 mm.) diameter; 230 lb./sq. in. (16 kg./sq. cm.) steam pressure; 5-ton/day vertical evaporator; steam horizontal reciprocating whale winch, 590 ft. ton/min. (180 m. ton/min.); 20 h.p. capstan; two oil pressure pump steering engines, 75 h.p.; 250 w. and 50 w. wireless, 3.5 in. (90 mm.) harpoon gun.

Electric welding was used to decrease the weight of the boat and a sharp ship form was applied. Special care was taken to get a hull with a low centre of gravity and to reduce the surface area of the boat above the water line so as to minimize wind pressure. Bilge keels were fitted and a flat plate keel was used instead of the bar keel for damping the rolling.

Fig. 126 shows the lines, fig. 127 the general arrangement and fig. 128 shows *Kyo Maru No.* 7 under speed. Table XXIII gives particulars of drafts and stability and Table XXIV results from speed trials.

TABLE XXIII

Drafts and stability

	Load			
	Light load		*Whaling load*	
Displacement .	585 tons	595.04 ton	720 tons	731.08 ton
Draft, fore .	6.25 ft.	1.900 m.	9.15 ft.	2.790 m.
„ aft .	14 ft.	4.260 m.	14 ft.	4.290 m.
„ mean .	10.1 ft.	3.080 m.	11.6 ft.	3.540 m.
Freeboard .	4.8 ft.	1.464 m.	3.5 ft.	1.025 m.
KM .	13.6 ft.	4.150 m.	13.7 ft.	4.170 m.
KB .	6.15 ft.	1.880 m.	7 ft.	2.150 m.
KG .	11.6 ft.	3.540 m.	11.1 ft.	3.380 m.
GM .	2 ft.	0.610 m.	2.6 ft.	0.790 m.
G from midship .	6.1 ft.	1.870 m. aft	3.3 ft.	1.010 m. aft
Coefficients:				
Block	0.465		0.500	
Prismatic .	0.570		0.596	
Midship .	0.814		0.840	
Waterplane .	0.712		0.764	
GZmax .	1.1 ft.	0.334 m.	1.08 ft.	0.330 m.
GZmax at .	32.5°		39.0°	
Range .	64.3°		66.4°	

TABLE XXIV

Speed trial

Date October 21, 1952	*Load*	*Speed (in knots)*	*r.p.m.*	*Slip (in %)*	*i.h.p.*	*b.h.p.*
Draft:						
Fore, 6.6 ft. (2.02 m.)	Over	16.36	212.3	10.6	3,150	2,610
Aft, 14.1 ft. (4.31 m.)	4/4	15.96	201.0	7.8	2,610	2,140
Mean, 10.4ft. (3.17 m.)	3/4	15.18	183.0	4.3	1,950	1,540
	1/2	13.10	151.3	— 0.7	1,195	860
Displacement:	1/4	11.24	123.9	— 6.4	710	472
617 tons (627.7 ton)						

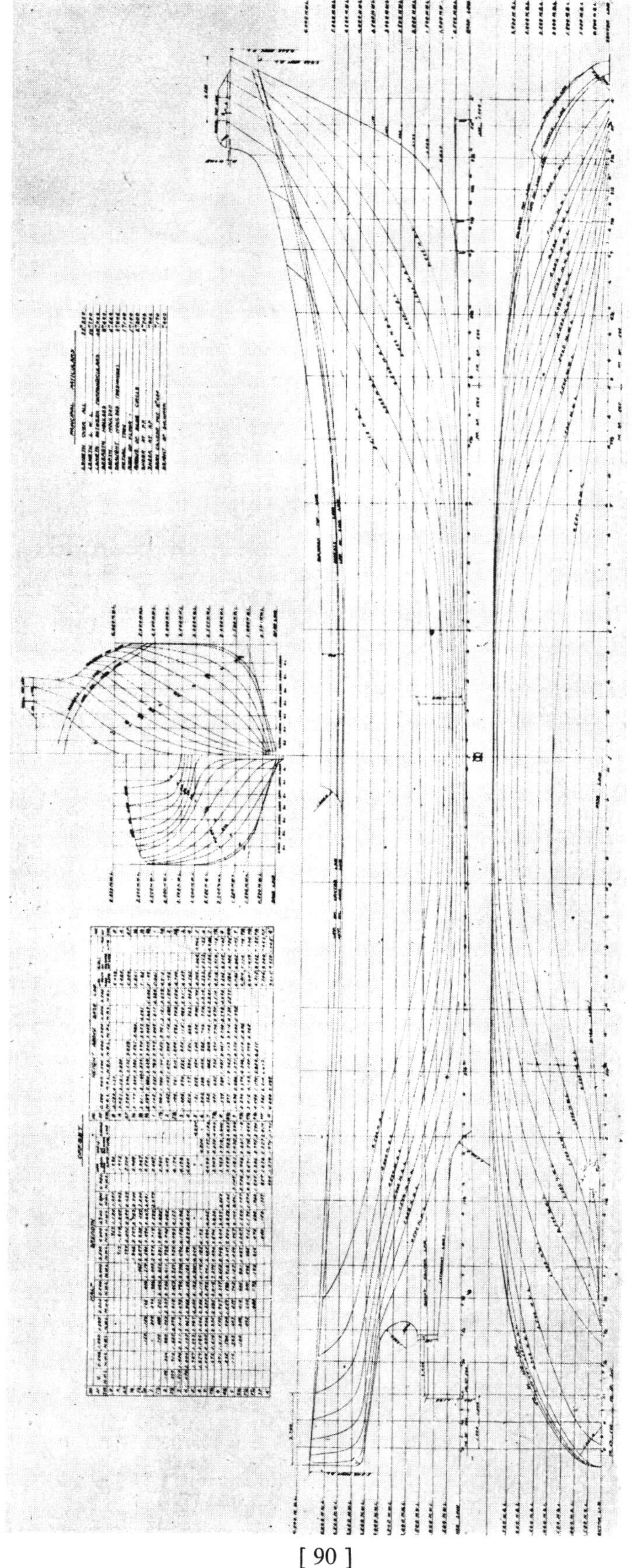

Fig. 126. Lines of Kyo Maru No. 7, *175 ft. (53.25 m.) diesel whale catcher*

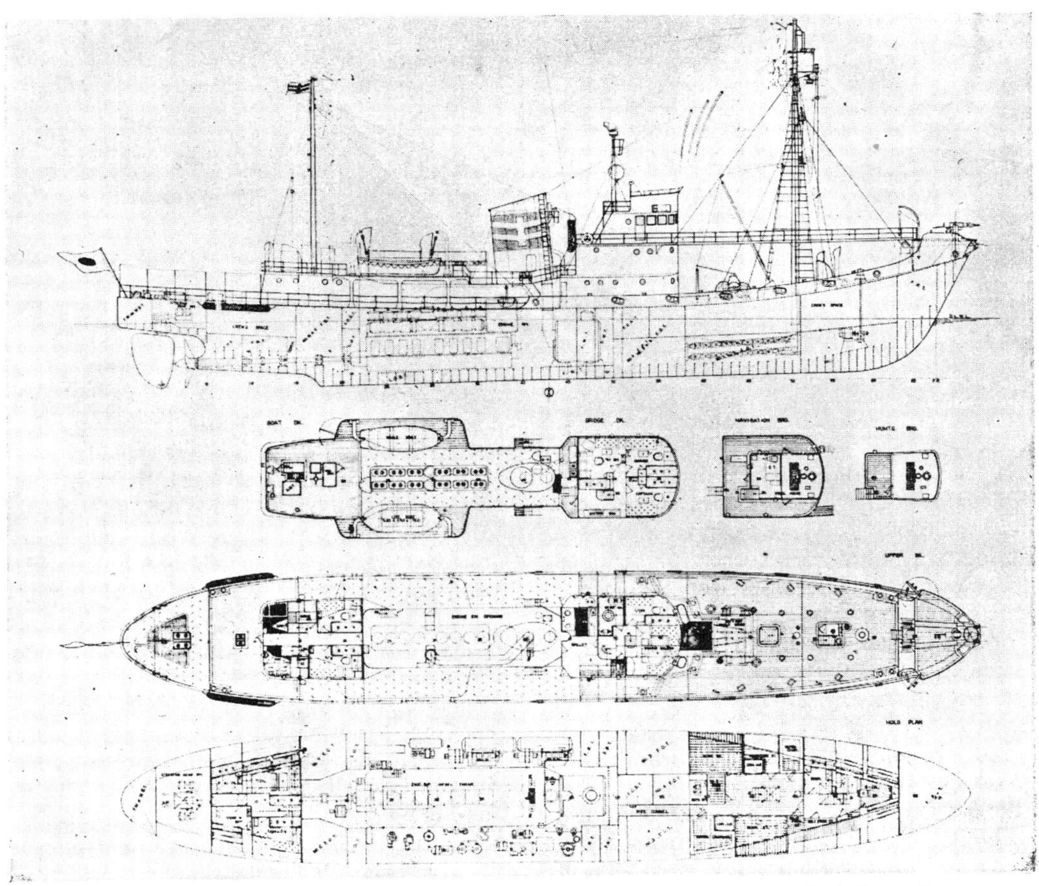

Fig. 127. General arrangement of Kyo Maru No. 7, *175 ft. (53.25 m.) diesel whale catcher*

Fig. 128. Kyo Maru No. 7, *175 ft. (53.25 m.) diesel whale catcher*

BRITISH COLUMBIA (CANADA) FISHING VESSELS

by

ROBERT F. ALLAN

FISHING is British Columbia's third industry, exceeded in value only by forest products and minerals, and main part of the harvest from the sea is, of course, salmon, which has been successfully canned and exported for more than 60 years. Second in value is the halibut which, in frozen and fresh form, is absorbed mainly in the domestic market. Third is the herring of which approximately 100,000 tons are caught in one short season and are almost entirely converted into fish meal for export. A minor part of Canadian fishing is for the local fresh sea-food market and there is a small crab fishing industry in the Queen Charlotte Islands.

Fisheries on the British Columbia coast are, of course, shared with the United States fishing fleets from Puget Sound and Alaska and the development of methods of fishing and types of boats and details of equipment in British Columbia waters has to a large extent followed their lead. But there are numerous differences in detail, particularly in the case of gillnetters, which easily identify the British Columbia craft to the expert eye.

The fishing vessels may be divided into two general classes: (*a*) those under 15 tons gross which are largely owned or operated by individuals or partners; (*b*) those over 15 tons gross, most of which are owned by companies or individuals employing crews on a share basis.

GOVERNMENT REGULATIONS

The Canadian Steamship Inspection service has a definite influence on the design and construction of modern fishing craft, which further emphasizes the difference between Canadian and American vessels, as the latter are subject only to the "Motorboat Regulations", administered by the Coast Guard. These require nothing more than compliance with certain regulations concerning navigation lights, fire extinguishers and life-jackets, and apply to all motor vessels under 150 tons gross which are not licensed to carry passengers.

In Canada, since 1942, all new vessels over 15 tons gross have been subject to inspection under the Canada Shipping Act and, since 1951, all existing fishing vessels over 15 tons have come within the scope of the Act, which requires that vessels over 15 tons must be built under inspection and must be re-inspected each four years. Before construction, plans have to be submitted showing: General arrangement, details of hull scantlings and fastenings, details of bulkheads, details of rudder and rudder stock, diagram of bilge pumping arrangements and detail of propeller shafting and fuel tanks. If the vessel is under 60 ft. (18.3 m.) registered length, these plans are passed by the regional office of the Inspection Service—if over 60 ft., they must be examined by the "Board of Steamship Inspection" (Ottawa, Ont., Canada), which is empowered to make regulations concerning all matters relating to the safety of vessel and crew and to pass judgment on "unusual methods of construction."

In effect, these regulations are concerned mainly with minimum standards for lifesaving equipment, fire extinguishing equipment, bilge pumping, ventilation to crew and machinery spaces, ground tackle, hatch covers and coamings, height of door sills, handrails and bulwarks and deadlights for ports and windows.

For hull construction, the standard is that set by previous submission. Although this may appear to be a haphazard method, it is on the whole satisfactory, allowing, as it does, a different standard for the East coast, including Nova Scotia and Newfoundland where the methods of construction and materials are very different from those employed on the West coast.

Although there has been little evidence of structural failures, even in the older boats, to justify much concern over hull scantlings, Government regulation has put a stop to unfair competition, between builders, by way of cheap or skimpy construction. And it is vastly superior to the method of definite rules, similar to Classification Society Rules, such as those in force in Denmark or Japan which, apparently, require scantlings and fastenings far in excess of actual need as dictated by experience. This places an economic handicap on the boat owner and is a barrier to progress in design.

MATERIALS AND CONSTRUCTION METHODS

The materials commonly used in construction of Pacific coast vessels are Douglas fir, red cedar, yellow cedar,

Australian gumwood or ironbark and white oak from Georgia or Indiana (U.S.A.). Larger vessels, built with sawn frames, are almost entirely constructed of fir, with some yellow cedar used in locations most subject to rot, such as bulwark stanchions and rim timbers at the stern. Bent frame vessels may be planked with fir or yellow cedar over white oak ribs. Stems and propeller posts are almost universally of gum, a hard wood resistant to abrasion and having enormous nail holding power. Small vessels up to 40 ft. (12.2 m.) in length are generally planked with edge grain red cedar, which is a light, clear wood easily worked when steamed.

The properties of Douglas fir are well known, but some special mention should be made of the unique properties of yellow cedar, sometimes referred to as Alaska cedar. Closely resembling red cedar, the tree grows in higher altitudes in the coast range. The wood is a light canary yellow, is pleasantly aromatic, has a high natural oil content and is a pleasure to work in any direction with or across grain. Its special value in boatbuilding lies in its low density (28 to 32 lb./cu. ft.; 448 to 512 kg./cu. m.) ease of working and handling and its excellent resistance to rot. Being lower in density and strength than Douglas fir, it is more commonly used in lower-stressed structure such as beams, house coamings and framing, where it helps to lower the centre of gravity. It is always used in stern timbers of seine boats and, frequently, for hold ceiling, bilge stringers, beam shelves, sheer planks behind guards, covering boards and hatch coamings. Its resistance to rot is remarkable. Where the life of fir is commonly only 5 to 10 years, yellow cedar has been observed in good condition after 30 years.

The quality of all local woods has fallen off as the larger trees have been felled and the better grades have been exported. The boat building industry does not use enough wood to merit special consideration from the mills.

Fir plywood is widely used for sheathing of deckhouses inside and out and for tops of houses to be canvas covered. It is also widely used over deckhouse beams and under light caulked decks where it helps maintain watertightness and forms an attractive deckhead finish.

The majority of British Columbia fishing vessels are built on the bent frame system, although three larger seiners and one 60 ft. (18.3 m.) crab boat have been constructed on the sawn frame system. In contrast to British or European practice, in British Columbia the latter involves a double frame built up of straight grained pieces sawn to shape and connected together with staggered butts. This type of frame is obviously not as strong as a single or double frame utilizing grown crooks, so more emphasis is placed on keelsons and bilge planking or bilge stringers as they are generally known in the West. An example of such construction is shown in fig. 129.

All fastenings are of galvanized iron, planking being secured by square sectioned boat or bridge spikes. No through fastenings are used which means, in the event of damage to planking, it is not necessary to have access to

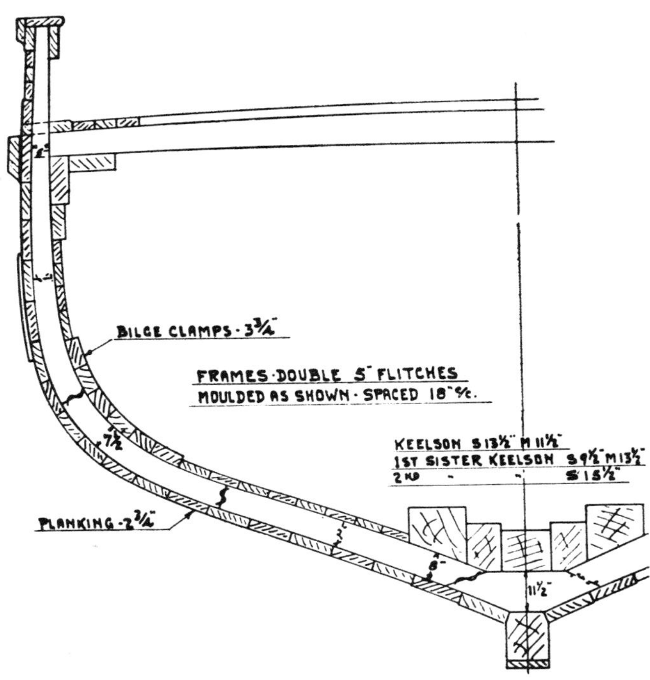

Fig. 129. Sawn frame construction as used in British Columbia in a 86×22½ ft. (26.2×6.9 m.) boat

nuts or bolts or treenail wedges behind tanks, machinery or accommodation lining. The contention of all local builders is that there is no advantage to through-fastening, at any rate in vessels up to 80 ft. (24.4 m.), with which they are familiar. They also contend that a galvanized spike has better holding power in fir and oak than has a treenail because the zinc undergoes a chemical reaction which tends to unite it with the wood to such an extent that, on being withdrawn, wood splinters are seen to be adhering to the fastenings. In larger craft where both the inner and outer planking are united in forming the planking envelope, the virtues of through-fastenings are recognized.

The bent frame system of construction is used with success for all types of fishing boats up to the largest seine boats 80 ft. (24.4 m.) long and 21½ ft. (6.55 m.) beam and has been used for vessels as long as 130 ft. (39.6 m.). Sawn frame construction requires every frame to be drawn on the loft floor and all the bevels and patterns for each flitch or frame member must be prepared by the loftsmen. The bent frame boat requires only 8, 10 or 12 moulds corresponding to the designer's displacement sections. When the lines plan has been prepared on a large scale such as ¾ in. or 1 in. to the foot (1:16 or 1:12), the only loft work consists of laying down the sections and end profiles as necessary for moulds and patterns for end timbers.

The main strength of any small boat hull, wood or metal, is in its skin or planking and if the shape is kept by transverse framing and bulkheads, the structure will maintain true alignment. The British or European fishing boat, with its rigid natural crook frame, represents the

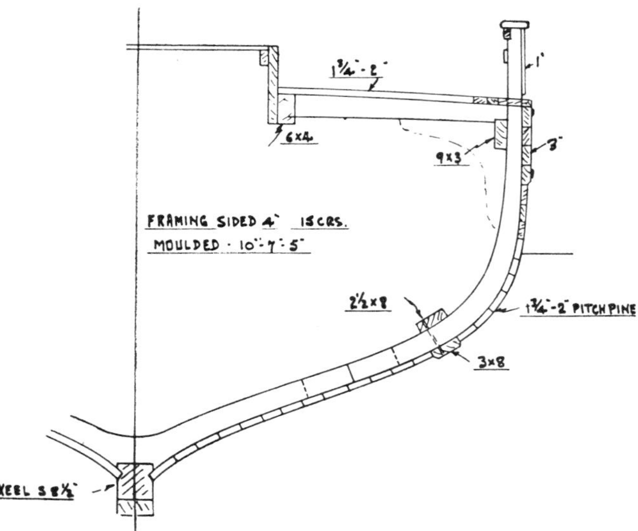

Fig. 130. Sawn frame construction of a British 62 ft. (18.9 m.) motor-fishing vessel (M.F.V.)

most economical way of maintaining the transverse shape, as illustrated in fig. 130, but it is dependent on a supply of suitable crooks.

The better class of East coast boat generally has a bent oak frame of almost square section and if laminated frames are used they are generally greater in moulding than in siding. On the West coast, and Nova Scotia (East coast), however, there is a tendency to use flat frames closer spaced. A typical section is 2×4 in. (5.1×10.2 cm.) and in order to maintain transverse strength with this comparatively limber member, the frames are reinforced by longitudinal members in the form of bilge stringers and keelsons. Compared to the British

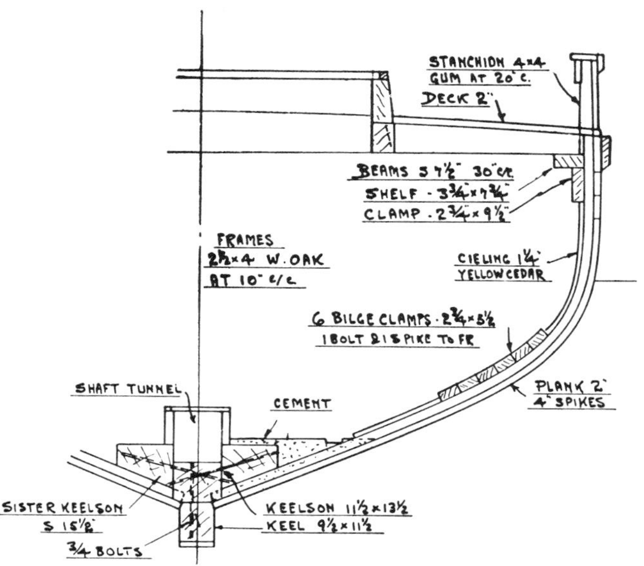

Fig. 131. Bent frame construction used in British Columbia in a 68×18 ft. (20.7×5.5 m.) seine boat

type of construction, it is wasteful of fastenings and labour but it is by no means inferior.

Eastern boats generally have floor timbers across the keel, alongside each frame, but this system has been superseded in the West by a keelson assembly as shown in the section of a typical 68 ft. (20.7 m.) seine boat, fig. 131. This method is very practical in fishing boats because it provides a centre duct for bilge water and cargo drainage as well as a place for suction pipes or shafting and, at the same time, creates a sanitary floor in the fish hold without need for cement between each floor or a caulked floor over the bilge. It has also been proved that shaft alignment is more effectively maintained with this " keelson construction ".

The keelson type of construction is used in small and large craft, but in boats under 50 ft. (15.2 m.) in length, the sister keelsons become redundant as one wide timber can be used in lieu of three smaller ones. In such cases a duct is formed by 2 in. or 3 in. (5.1 or 7.6 cm.) boards, on edge, forming a shaft tunnel. For boats close to the 15 tons gross measure, floors are often used in order to reduce the tonnage, which is measured from the top of the floors.

To connect all main structural members, such as keel and keelson, stem, shaft log, beam ends, etc., galvanized screw bolts are used. In certain cases where through-bolts would be too long, drift or blind bolts are used, and are driven into $\frac{1}{16}$ in. (1.6 mm.) undersize holes. Shorter connections, such as clamps to frame, are made with galvanized carriage bolts which have an oval head with a short length of square section immediately under, to prevent bolt turning when the nut is tightened.

TYPES

The fishing craft under 15 tons gross consist mainly of: gillnetters, trollers, long-liners, small table or drum seiners. The larger class are: halibut long-liners, salmon seiners, herring seiners, packers, draggers. As there is a tendency to combine types, all large seiners are at least a combination salmon and herring seiner-packer, whilst several of the latest boats have been intended for all five purposes. In the same manner, gillnetters are often used for long-lining and less frequently for trolling.

GILLNETTERS

British Columbia gillnetters are different from the American Columbia river or Bristol bay boats, as they handle the net by power drum over the stern, whereas the Columbia river boats pull theirs in over the forward quarter, by hand, with the aid of a powered roller. The modern Bristol bay power boats now favour the roller at the stern or aft quarter.

The original gillnetter, as used on the Fraser river estuary, developed along the lines of the Columbia river type, being a double-ended open boat with oars and sail, later adapted to power. This type was replaced by a double-ended model developed by Japanese fishermen, having a type of cruiser stern, commonly known as a Jap stern. An average boat of the type was 30 ft. (9.1 m.) long with a beam of 7 ft. 6 in. (2.3 m.) and was fairly low in profile. It was usually powered with a heavy duty gasoline engine of local manufacture, the power being 8 to 18 h.p. at 600 to 900 r.p.m. Such boats were economical to build and operate and were most suitable for estuary and river fishing. But since World War II the average fisherman cannot earn enough in one short season on the Fraser or Skeena and he now goes north for the opening runs on Rivers Inlet and others which precede the Fraser river runs. To do this and be able to venture further in the Gulf, has called for larger and faster boats with improved accommodation and seaworthiness. At the same time, the car type engine with reduction gear has displaced, almost entirely, the heavy duty type. A car engine of about 250 cu. in. (4,097 cu. cm.) displacement is rated at about 90 h.p., yet it sells for approximately the same price as the 18 h.p. heavy duty model. In practice, the engines are not called upon to deliver over 30 h.p. for general cruising and most boats are unable to use more, but there is a trend towards boat models which can.

Fig. 132 shows a modern design, 33 ft. $\times 9\frac{1}{2}$ ft. (10 $\times$ 2.9 m.), having a built up stern—a transom with all the corners washed off to ease recovery of the net. Other points are the extreme forward position of propeller post and rudder to keep them clear of the net, immersion of transom to minimise slap and spray when pulling the net, and the forefoot cut away in an easy sweep in order more readily to cross over the nets. Not shown is the basket type propeller guard which surrounds the propeller. Such a boat displaces approximately 7 tons and is capable of 9 knots when equipped with a gasoline engine with $2\frac{1}{2}:1$ reduction, turning a 26 in. (660 mm.) diameter propeller. Many fishermen want higher speed, but when it is realised that 9 knots represent a $\frac{V}{\sqrt{L}}$ of 1.6 $\left(\frac{v}{\sqrt{gL}}=0.48\right)$ it is evident that higher speeds require a semi-planing craft of lighter displacement. Although it is possible to design a gillnetter of minimum size and displacement, such a large proportion of the weight reduction has to be taken from hull proper that there is a sacrifice of durability, load carrying ability and seaworthiness.

In hull form, the modern gillnetters approach pleasure cruisers, having a fairly flat mid-section and flat straight buttocks in conjunction with a prismatic coefficient of about .63. To combat the tendency to broach or sheer, the waterline of a square stern boat should be fairly full forward and the forward sections should have a good slope for quick increase in buoyancy. In order to balance the weight of the net drum, drum rollers, and drum drive in the aft cockpit the centre of buoyancy must be at least 55 per cent. of the fore perpendicular. Although no measurements of GM have been taken, it has been noted that modern "stiff" boats are much favoured over older cranky types.

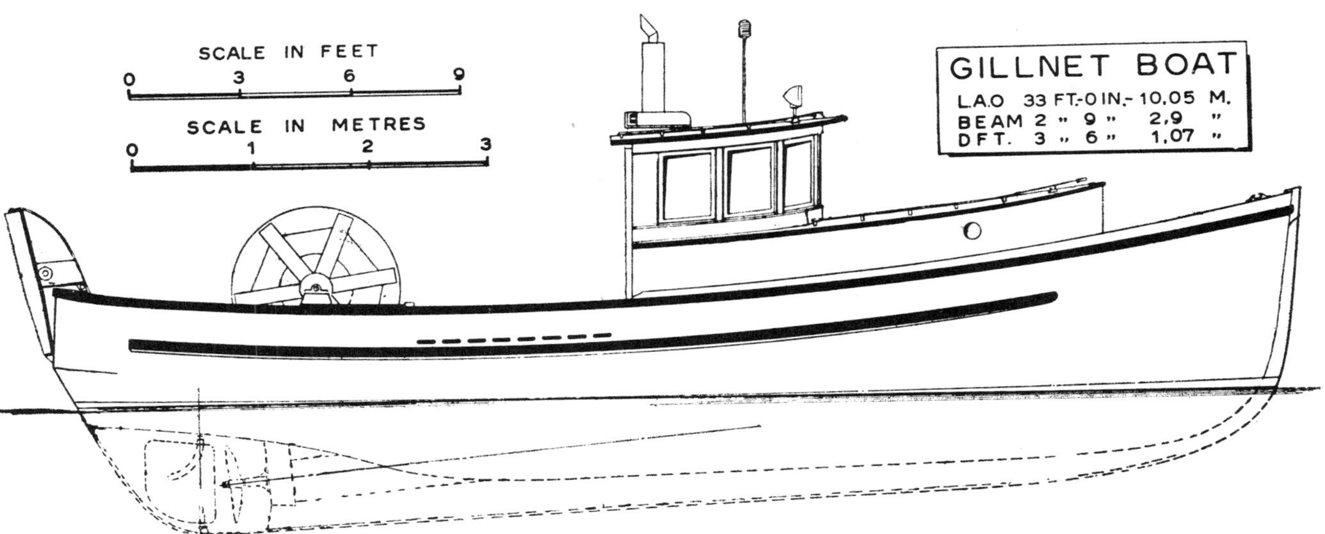

Fig. 132. Modern gillnetter 33 $\times 9\frac{1}{2} \times 3\frac{1}{2}$ ft. (10.1 $\times$ 2.9 $\times$ 1.1 m.) designed by the author

Construction of gillnetters has been more or less standardized by experience and cost. There are for instance, minor variations between builders in methods of keelson construction but most gillnetters have planking of $1\frac{1}{8}$ in. (2.9 cm.) red cedar and ribs of 1×2 in. (2.5 $\times$ 5.1 cm.) white oak, fig. 133.

The arrangement and detail of trunk cabin and wheelhouse is also largely standardized and, although so-called streamlined cabins have been introduced, the cheapness and efficiency of the standard arrangement is still favoured by a majority of fishermen and builders. Points to be noted are: reduction of the width of the wheelhouse on the trunk to open the passage forward on deck and at the same time, bring the windows close to the helmsman's position near the clutch lever so that he may lean out in fog or at night.

Trollers

Trollers are generally required to travel farther afield than gillnetters so usually they have deeper draft and displacement, otherwise the older boats resemble gillnetters in general appearance. Particulars about some trollers are given in Table XXV.

A typical older troller would be 36 × 9 ft. (11 × 2.7 m.) with a draft of 5 ft. (1.52 m.). Later models have tended towards larger sizes and greater beam up to the limit of 15 tons gross and have been arranged for long-lining or seining as well as trolling. Most trollers have a full type of canoe stern, fig. 134. The midship section of a typical 42 ft. (12.8 m.) troller is shown in fig. 135. But with a trend towards combination boats, a seine stern has found favour as it allows more room on deck for long line or seine operation. The drum seiner, fig. 138 and 139, is a typical example of this type. A Norwegian counter stern is also popular, particularly for boats designed as long liners, with trolling a secondary consideration.

Speed is secondary to seaworthiness for trolling, but a good form is necessary in order to obtain a $\frac{V}{\sqrt{L}}$ of 1.25 to 1.35 $\left(\frac{v}{\sqrt{gL}} = 0.37 \text{ to } 0.40\right)$ with fairly heavy displacement length ratios. The $\frac{\nabla}{(.01L)^3}$ varies from 250 ($L/\nabla^{\frac{1}{3}}$ = 4.8) in a short range gasoline-powered boat to 390 ($L/\nabla^{\frac{1}{3}}$ = 4.1) for a 42 ft. (12.8 m.) diesel vessel intended for long-lining.

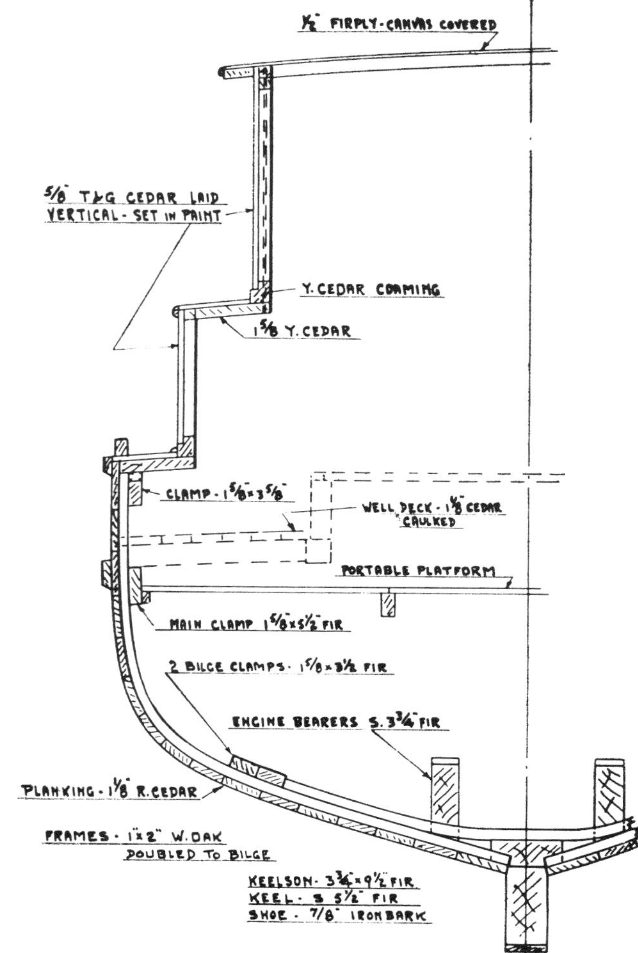

Fig. 133. Midship section to 33 × 9½ ft. (10.1 × 2.9 m.) gillnetter

TABLE XXV

PARTICULARS OF TROLLERS

No.		*L.O.A.*	*L.W.L.*	*Beam*	*Draft, aft*	$\frac{\Delta}{(.01L)^3}$	$L/\nabla^{\frac{1}{3}}$	*Max. h.p.*	*Machinery type*	*Remarks*
1	ft. m.	34 ft. 0 in. 10.38	32 ft. 0 in. 9.75	9 ft. 6 in. 2.9	3 ft. 8 in. 1.13	268	4.7	70	R/G gasoline	Double ender
2	ft. m.	36 ft. 0 in. 10.98	34 ft. 0 in. 10.38	10 ft. 6 in. 3.2	4 ft. 9 in. 1.45	300	4.5	70	,, ,,	,, ,,
3	ft. m.	38 ft. 0 in. 11.56	36 ft. 0 in. 10.98	11 ft. 6 in. 3.51	4 ft. 7 in. 1.40	285	4.6	85	,, ,,	,, ,,
4	ft. m.	42 ft. 0 in. 12.80	40 ft. 0 in. 12.2	12 ft. 0 in. 3.66	5 ft. 0 in. 1.525	295	4.6	85	,, ,,	,, ,,
5	ft. m.	45 ft. 6 in. 13.90	40 ft. 0 in. 12.2	12 ft. 0 in. 3.66	5 ft. 0 in. 1.525	315	4.4	120	R/G diesel	Seine boat stern
6	ft. m.	46 ft. 0 in. 14.03	43 ft. 0 in. 13.1	14 ft. 0 in. 4.27	6 ft. 0 in. 1.83	380	4.2	80	,, ,,	,, ,, ,,

Long-line and troller fishermen have favoured a fairly narrow boat with deep draft and great deadrise so that when not loaded with ice, it is necessary to carry rock ballast to maintain sufficient stability for safety. This small beam was wanted in order to obtain a long rolling period for comfort when trolling at low speeds or lying to at the halibut banks. However, as so often happens, the final stability has turned out less than anticipated and the boats have been very cranky. To combat this, it is desirable to design a midsection form with reverse curve, as is popular with most British fishing boats, thereby securing adequate draft with good stability. Another development leading to beamier boats is the invention of stabilizers in the form of small paravanes suspended from the outrigger poles. These damp the rolling motion to such a degree that it is possible to use flat-bottomed pleasure boat types for trolling.

Long-Liners (under 15 tons)

These are invariably built to the maximum dimensions within the tonnage limit, usually about 42 ft. (12.8 m.) long and 12 ft. (3.66 m.) beam. Fig. 136 shows a typical example. Larger boats, up to 50 ft. (15.24 m.) would, in many cases, be more desirable, but the additional expense involved in a vessel under inspection, and the difficulty of finding space for required equipment, particularly a lifeboat, have prevented their construction.

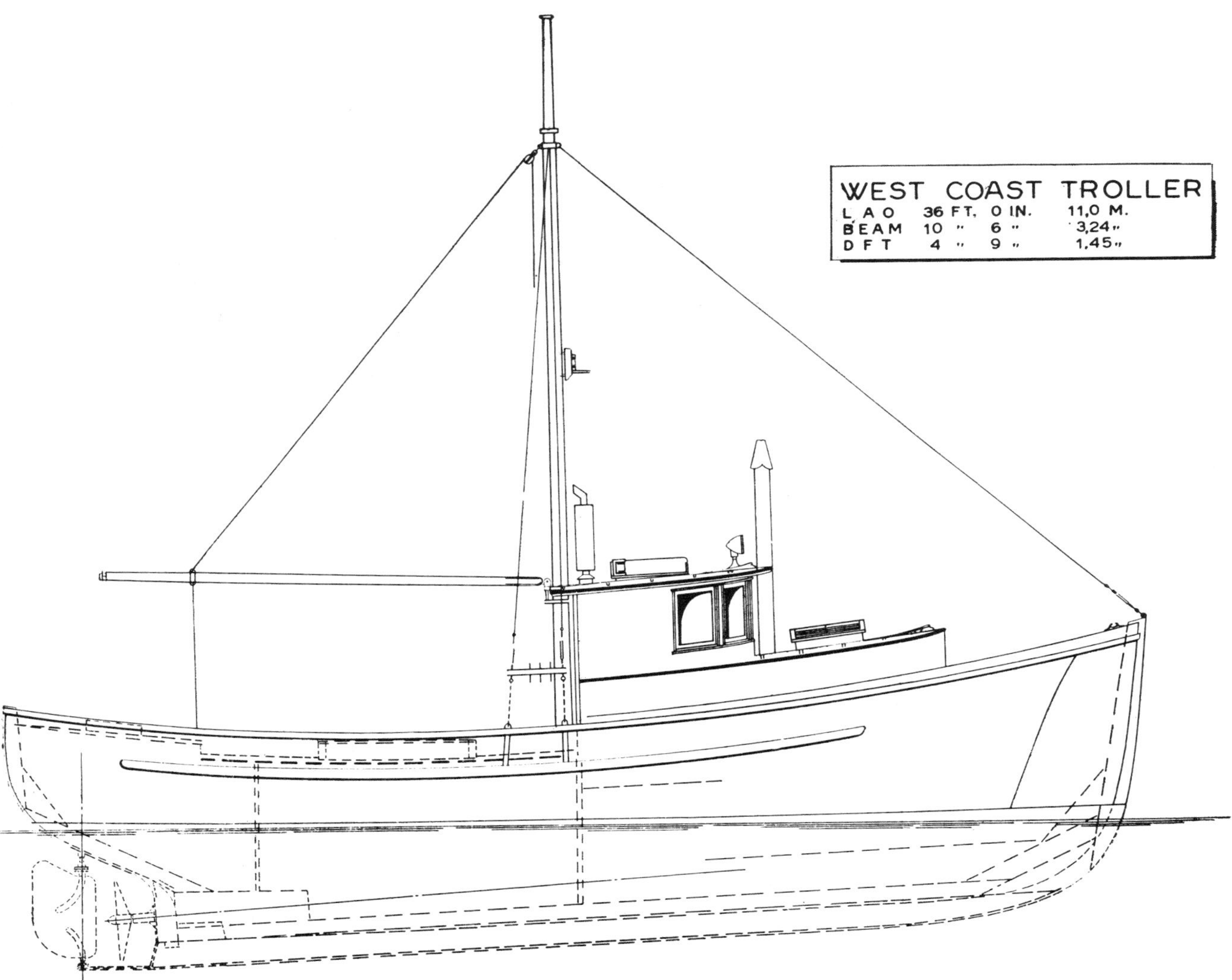

Fig. 134. 36 × 10½ × 4¾ ft. (11.32 × 1.45 m.) troller with canoe stern designed by the author

The halibut fishing season is hard on men and boats. Although it lasts only about a month, it is in winter and involves a long passage to the fishing grounds. Fishing goes on 20 hours a day, an arduous type of life, which appeals mainly to the hardy Norseman, and the typical halibut boat shows strong Norse influence, particularly at the stern, which is generally a Norwegian counter type modified to give a fuller deck line. Unlike European

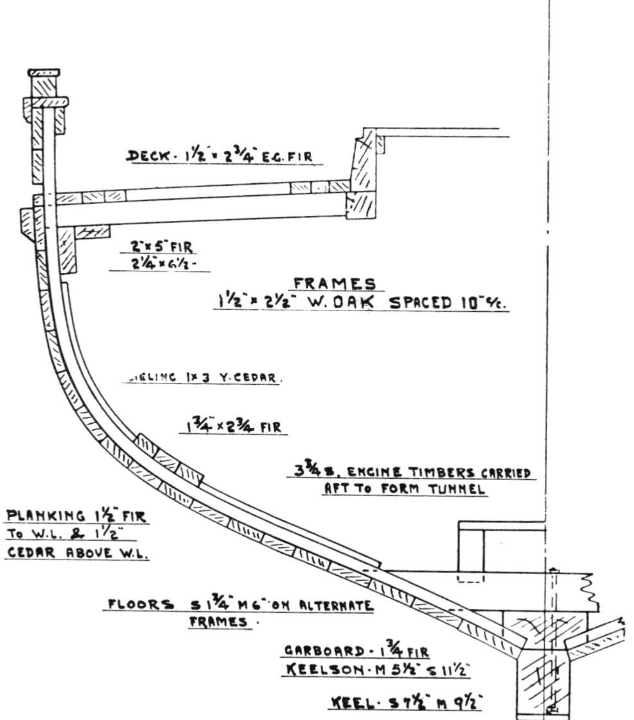

Fig. 135. Midship section of a typical 42 × 12 ft. (12.8 × 3.7 m.) troller

Fig. 136. Halibut long-liner under 15 gross tons, 42 ft. 4 in. × 11 ft. 8 in. × 7½ ft. (12.9 × 3.55 × 2.3 m.); displacement 20 tons, speed 8 knots with 82-h.p. diesel

craft, which have engines astern and little top hamper, the local vessels have been influenced by the seine boat types and have the engine, fuel and fairly large deckhouse all forward, with considerable increase in height of the centre of gravity.

When in service, a halibut boat is always loaded, with ice and stores outward bound and with fish and ice on the return trips. Under these conditions, they are generally trimmed well and have adequate stability, but should occasion arise that a light ship voyage is necessary, it is usually advisable to carry ballast to ensure safe stability and to counteract excessive trim by the head.

To ensure better stability under all conditions more beam is wanted, but cannot be allowed in the 15-ton limit, another example of the effect of imposing arbitrary restrictions.

Small long liners are short in waterline length and heavy in displacement, consequently the $\frac{\Delta}{(.01L)^3}$ is usually over 300 ($L/\nabla^{\frac{1}{3}}=4.5$). Because of the long range, and the power required to handle the boat in rough weather, diesels of about 80 h.p. are fitted, usually high speed,

TABLE XXVI

PARTICULARS OF LONG-LINERS

No.		*L.O.A.*	*L.W.L.*	*Beam*	*Draft, aft*	$\frac{\Delta}{(.01L)^3}$	$L/\nabla^{\frac{1}{3}}$	*Max h.p.*	*Machinery type*	*Remarks*
1	ft. m.	40 ft. 4 in. 12.3	37 ft. 0 in. 11.27	11 ft. 8 in. 3.56	5 ft. 3 in. 1.60	390	4.1	80	R/G diesel	Counter stern
2	ft. m.	56 ft. 0 in. 17.05	50 ft. 0 in. 15.25	16 ft. 0 in. 4.88	5 ft. 9 in. 1.75	380	4.2	125	,, ,,	,, ,,
3	ft. m.	57 ft. 0 in. 17.38	50 ft. 0 in. 15.25	15 ft. 0 in. 4.57	6 ft. 6 in. 1.98	338	4.4	114	,, ,,	,, ,,
4	ft. m.	72 ft. 0 in. 21.9	66 ft. 8 in. 20.3	19 ft. 6 in. 5.95	8 ft. 0 in. 2.44	315	4.4	150	D/D diesel	Combination packer
5	ft. m.	73 ft. 0 in. 22.25	67 ft. 0 in. 20.4	19 ft. 6 in. 5.95	8 ft. 6 in. 2.59	310	4.5	300	R/G diesel	Combination seiner packer

1,200 to 1,600 r.p.m. with 3:1 reduction gears. The speed of a 42-ft. (12.8 m.) boat with such power is about $7\frac{3}{4}$ knots, a $\frac{V}{\sqrt{L}}$ of 1.25 $\left(\frac{V}{\sqrt{gL}}=0.37\right)$ on a W.L. length of 37 ft. (12.28 m.) and $\frac{\triangle}{(.01L)^3}=390$ $(L/\nabla^{\frac{1}{3}}=4.1)$.

Better results can be obtained with cruiser stern types as No. 4 in Table XXV. The corresponding speed for $\frac{V}{\sqrt{L}}$ of 1.30 is 8.2 knots.

Long-Liners (over 15 tons)

In the Vancouver area only three boats have been built since 1944 for long-lining—one 72 ft. (22 m.), one 56 ft.

Fig. 137. Halibut long-liner over 15 gross tons, 57 ft. 4 in. × 16 ft. × 8 in. (17.5 × 4.9 × 2.45 m.) displacement 48 tons, speed $9\frac{1}{2}$ knots with 125 h.p. diesel

(17 m.) and one 57 ft. (17.4 m.). In design they do not differ radically from seine boats, except that they carry more fuel and are designed to a partly-loaded condition, not to a light waterline, because of the ice carried at all times in service. Particulars of the three vessels are given in Table XXVI and a typical one is illustrated by fig. 137.

Seiners (under 15 tons)

A few small seine boats up to 50 ft. (15.24 m.), but mainly below 15 tons, have been built in recent years for salmon fishing, but in the last 2 or 3 years there has been a good deal of interest in smaller seiners using a drum to handle the seine similar to a gillnetter. The object of this is to cut the number of hands from 6 to 4 or even 3.

Most of the boats have been just below the 15-ton limit—about $42\frac{1}{2}$ ft. × 12 ft. (13 × 3.7 m.), fig. 138 and 139. The design requires careful consideration of trim and stability as the net and turntable or drum are mounted high and far aft and the combined weight is about $2\frac{1}{2}$ tons. By limiting the draft of these boats, it is possible to obtain an apparently adequate stability. But, although the GM may be no less than that of the average fishing boat, the centre of gravity is too high for adequate range. One such craft which, by superficial examination appeared to have excellent initial stability, lay over on its side in a tide rip off Cape Mudge and would not right itself.

As in the case of trollers and long-liners, drum seiners in small sizes would be beamier and often longer, and consequently more seaworthy, were it not for the 15-ton limit set paradoxically to increase safety. A few drum seiners have been built up to 50 ft. (15.24 m.) but in most cases their design has followed the pattern of existing seiners with turntable. When building is resumed, a trend to beamier flatter boats with transom or deep rim sterns and, consequently, great initial stability, is expected. It is hoped there will be less tendency to keep below the 15-ton limit and a step to help such a trend has been taken by the Steamship Inspection Department. In 1951, a regulation was adopted authorizing small fishing boats to carry a flat-bottomed skiff of simple construction, but equipped with buoyancy tanks, instead of a standard lifeboat. The skiffs may be stowed on deck wherever convenient and need not have the chocks and lifting arrangements required for the standard lifeboat. The standard seine skiff has always been accepted as lifesaving equipment for seine boats.

Seiners (over 15 tons)

Salmon seiners

Purse seine boats of less than 65 ft. (19.8 m.) in length are too small for herring fishing or packing (carrying), and few have been built in recent years. In the future, all small salmon seiners will probably be of the drum type.

Salmon-herring seiners

Purse seiners of 65 to 80 ft. (19.8 to 24.4 m.) length are used for salmon seining, herring seining, trawling and transportation. The larger boats necessitate a large investment but as long as they can make capacity catches they pay greater dividends, because expenses for fuel and crew are only little more than those for smaller boats.

This class is probably the most interesting to the designer, because of the problems of design and competition between owners to have the fastest boat. This has led to absurd over-powering, particularly in smaller boats. In Europe there has been careful work to improve hull form to get greater speed with the same power or the same speed with less power; in British Columbia the drive has been for more speed regardless of power. But it is reasonably certain that worthwhile increase in speed can only be got by reducing displacement and not by varying the hull form. The opportunity to reduce displacement is limited, as machinery, fuel and fishing equipment are fixed by owners' requirements. This confines economies to the hull and superstructure where the major economy is obtained by the use of yellow cedar wherever possible. An illustration of the reduction of displacement for a stated increase of speed may be seen on fig. 140, a profile and general arrangement of a modern 68 ft. (20.7 m.) seine boat in fig. 141 and 142, the corresponding midship section in fig. 131, a photo in fig. 143, and a summary of particulars of a few representative vessels in Table XXVII. Model test data and form particulars of a successful 65 ft. seiner designed by the author may be found on SNAME data sheet 127, reprinted on FAO No. 22.

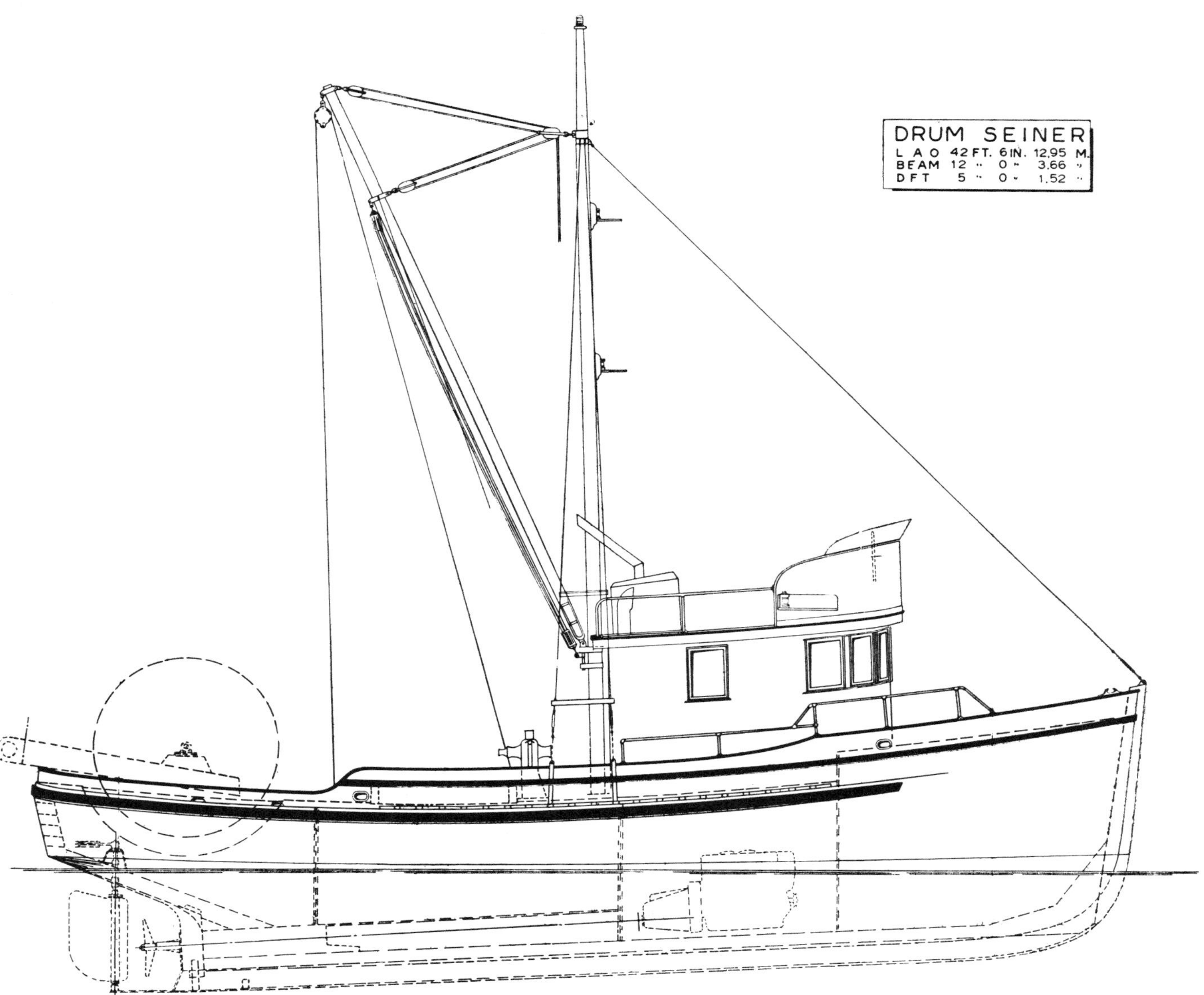

Fig. 138. Profile of a 42½ × 12 × 5 ft. (13 × 3.7 × 1.5 m.) seiner designed by the author

Packers

The larger seine boats are often used as carriers or transport vessels and are then called packers. This happens especially when another vessel makes a set containing more fish than can be handled in her own hold as, for example, a set of 800 tons of herring, made in a recent season. Large seiners are, therefore, designed with the largest possible hold space and sufficient reserve buoyancy to ensure freeboard when the hold is filled to capacity.

110 ft. (33.5 m.) sub-chasers, 138 ft. (42.1 m.) Y.M.S., 106 ft. (32.3 m.) A.P.C., 96 ft. (29.3 m.) A.Mc. and 107 ft.

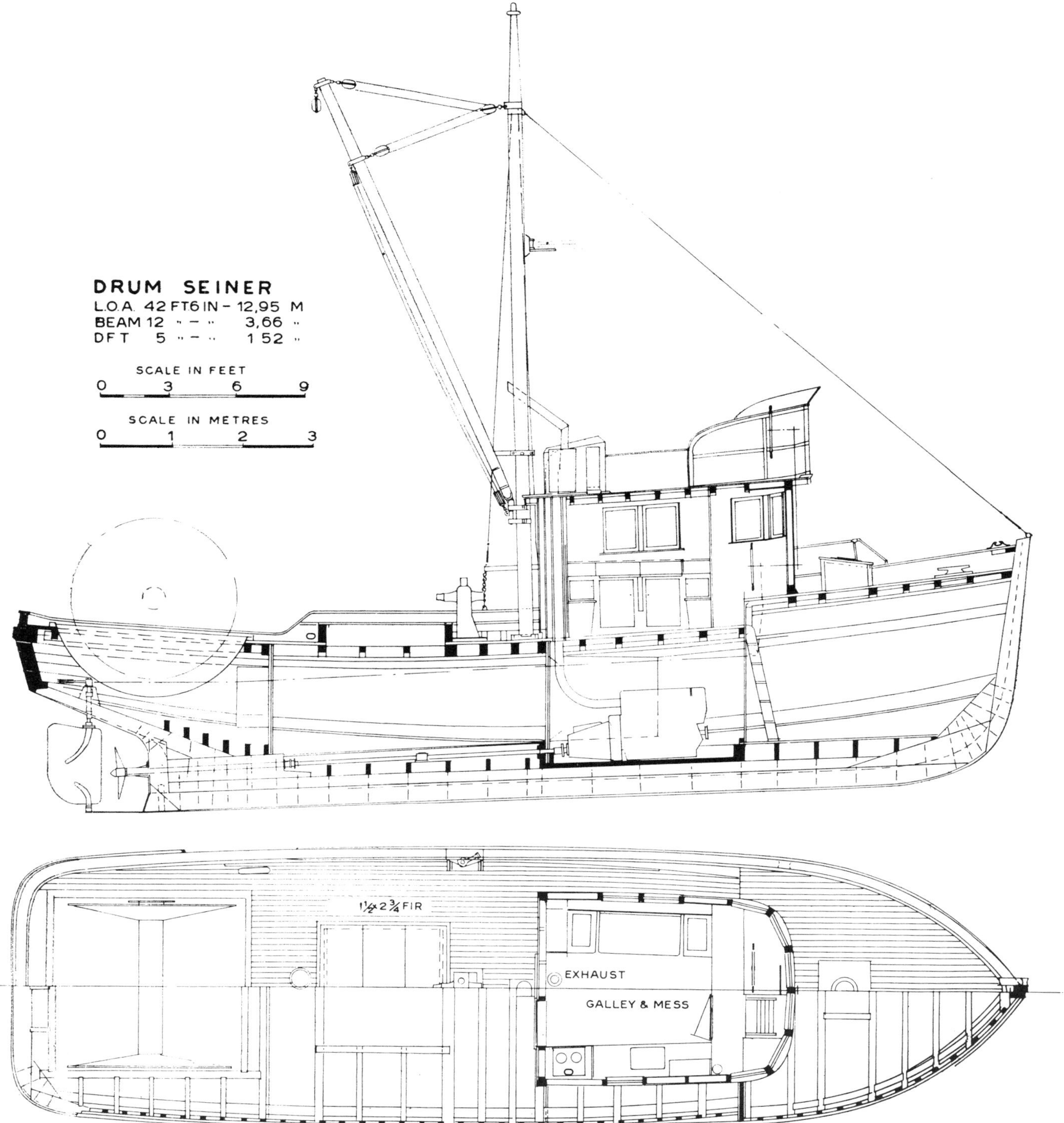

Fig. 139. General arrangement of a 42½ × 12 × 5 ft. (13 × 3.7 × 1.5 m.) seiner designed by the author

In addition to seiners, the packer fleet consists of miscellaneous craft, mostly conversions. The most recent have been conversions of war surplus craft such as (32.6 m.) Canadian minesweepers. The policy has been to concentrate on fewer but faster vessels, capable of more trips per season.

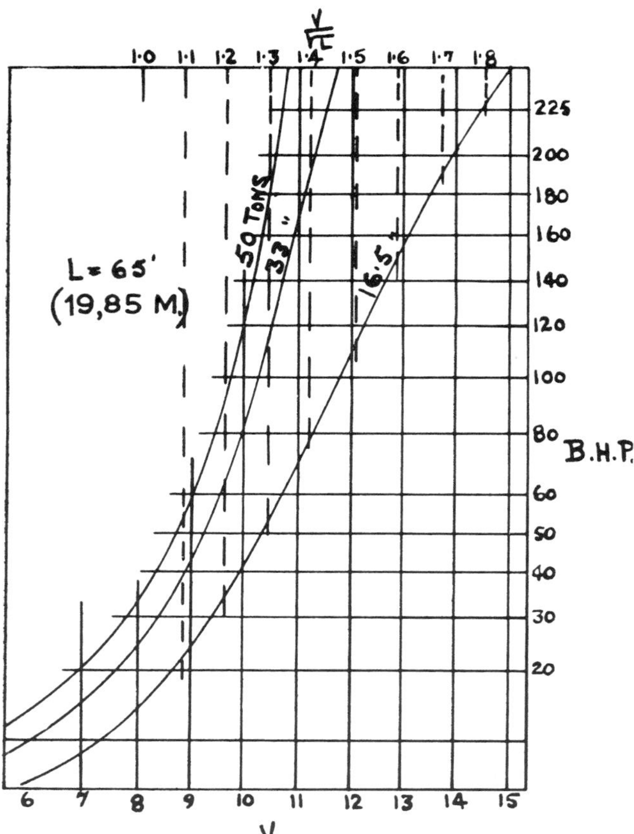

Fig. 140. Increase of speed of a 65 ft. (19.8 m.) seiner if displacement could be reduced

Trawlers

A small number of vessels trawl exclusively in British Columbia waters and are similar to halibut long-liners in all general characteristics. Recently, however, many seine boats of larger type have been fitted with winches, portable davits and other gear necessary for trawling thereby extending their yearly operating time.

DESIGN

General

It is not possible to foretell the displacement of a wooden vessel with the same accuracy as a steel vessel for several reasons: the weight of materials is variable; different builders vary in choice of materials and attitude to weight; owners are never very specific in choice of equipment, fuel and water capacity or even main machinery at early stages of design; and the amount of cement in bilges varies.

There are many dangers to under-estimating displacement. For instance, the metacentre is lowered, the trim of a broad sterned vessel is increased by the head and the carrying capacity is reduced.

There has been a fondness for extreme deadrise as great as 30°, resulting in, perhaps, a midsection coefficient of .60 and, consequently, increasing the overall prismatic coefficient to about .66, instead of a more reasonable .72 midsection and .62 prismatic. If such extreme deadrise sections were constructed so that the equivalent area was obtained whilst maintaining the waterline breadth, the resulting section would have a metacentre located at the same level relative to the waterline, as the fuller section, but the keel would be lower and possibly some weights, such as engine and shafting, could be lower, thereby lowering the centre of gravity and increasing GM. But usually the section was fined, with consequent loss of displacement and increased draft, which meant a dangerous drop in metacentre resulting in insufficient GM.

The reaction from the deadrise section can be equally dangerous. A very full midsection gives a lower position of the metacentre than the deadrise form for equivalent area on the same beam. It is essential, therefore, to choose proportions of beam and draft which maintain a proper height of metacentre and, at the same time, the displacement must not be under-estimated so that these proportions are upset.

A limited deadrise and reasonable midsection coefficient in order to maintain a correct prismatic for the $\frac{V}{\sqrt{L}}$ at which the boat will be generally cruising is required. If it is necessary to obtain draft without endangering stability, the reverse curve type of section, as used in British or East Canada boats, is favoured.

Most west coast commercial vessels, with exception of tugs in which other factors govern, are in the terminology of resistance data, medium to high-speed craft, and the form of the curve representing the longitudinal distribution of displacement should be chosen accordingly. In this respect the British Columbia troller is no different from one of Her Majesty's cruisers and, in fact, a form derived from a Royal Navy cruiser has been most successful. It is, of course, necessary to modify such curves to adjust for trim and it is preferable to confine such alterations to the aft end.

In this connection, it is interesting to note that seine boats, in particular, impose several limitations in form selection. This cannot be stated in too general terms, for the location of the longitudinal centre of gravity gradually tends to move aft as the length of the vessel decreases, mainly because the length required for crew and machinery space does not decrease in proportion to overall length. Another limiting factor is the full after deck line which imposes a full waterline which, in turn, imposes full after areas in order to avoid curtailing the volume at the aft end of the fish-hold.

The afterbody form is not too critical, but these limitations compel an afterbody prismatic approaching .72, suggested as the limit of good practice. At the same time, the forebody prismatic, usually about .56, is about ideal for the speed. It could be increased without adverse effect at higher speeds but this is limited, particularly in smaller craft, by consideration of trim.

The form of the waterline has received a good deal of

attention from researchers and it has been demonstrated that a sharper angle of entrance has beneficial effects on resistance, particularly at $\frac{V}{\sqrt{L}}$ up to 1.1 $\left(\frac{v}{\sqrt{gL}}=0.33\right)$. It is interesting to note, however, that at higher speeds, a fuller waterline, with about 21°—23° half angle of entrance, gives best results. It is advisable to maintain a fairly full waterline in broad-sterned craft to help minimize their tendency to broach in a following sea.

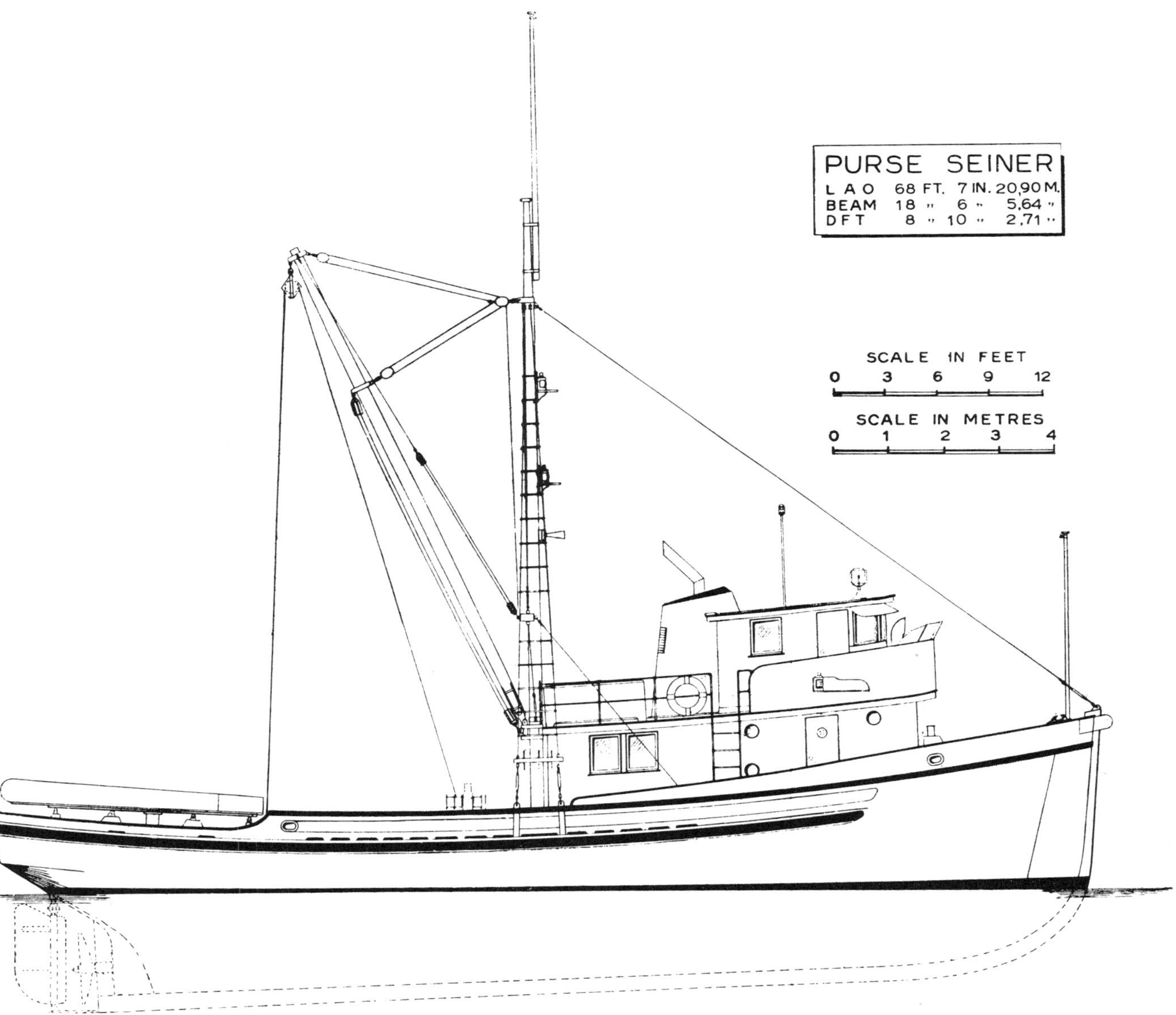

Fig. 141. Profile of a 68 ft. 7 in. × 18½ ft. × 8 ft. 10 in. (20.9 × 5.6 × 2.7 m.) seiner designed by the author

Vertical Centre of Gravity

Compared to European or East Canada vessels, most West coast types have a higher centre of gravity in relation to the waterline, or the centre of buoyancy, for several reasons: engines are forward and higher due to keel rake; accommodation is more elaborate and includes large deckhouses forward at the highest part of the ship; fishing gear, nets, seine table, winches, etc., are all located on deck.

General Seaworthiness

Except for double-ended craft, such as trollers, most West coast fishing vessels are inferior to corresponding European or East Canada vessels for these reasons: centre of gravity is high; unsymmetrical waterline in square sterned types has the centre of flotation far aft, thereby aggravating the tendency to broach; in light condition, most types have the centre of gravity forward of the centre of the lateral plane, thereby inviting broaching.

To minimize these faults it is first necessary to ensure adequate GM. No limiting value can be given, as it should increase with the height of the centre of gravity and, inversely, with the size of the boat, but it can be said that a value of 21 in. (53 cm.) or less is suspect. Secondly, a vessel should have the tankage so distributed that the designed trim can be maintained in the light condition. Thirdly, adequate freeboard should be maintained in all conditions. To ensure this, the hold capacity should not be so large that a capacity load will reduce the freeboard below a reasonable minimum.

Research on Hull Design

In recent years there have been published several papers describing the results of tank and full-scale tests of fishing vessel designs. Allan (1950) describes work carried out at the Experiment Tank of William Denny and Bros. for the Herring Board of Great Britain. Four models were tested, the first model O being representative of the best current models. Model A represented a moderate departure from O with finer entrance and

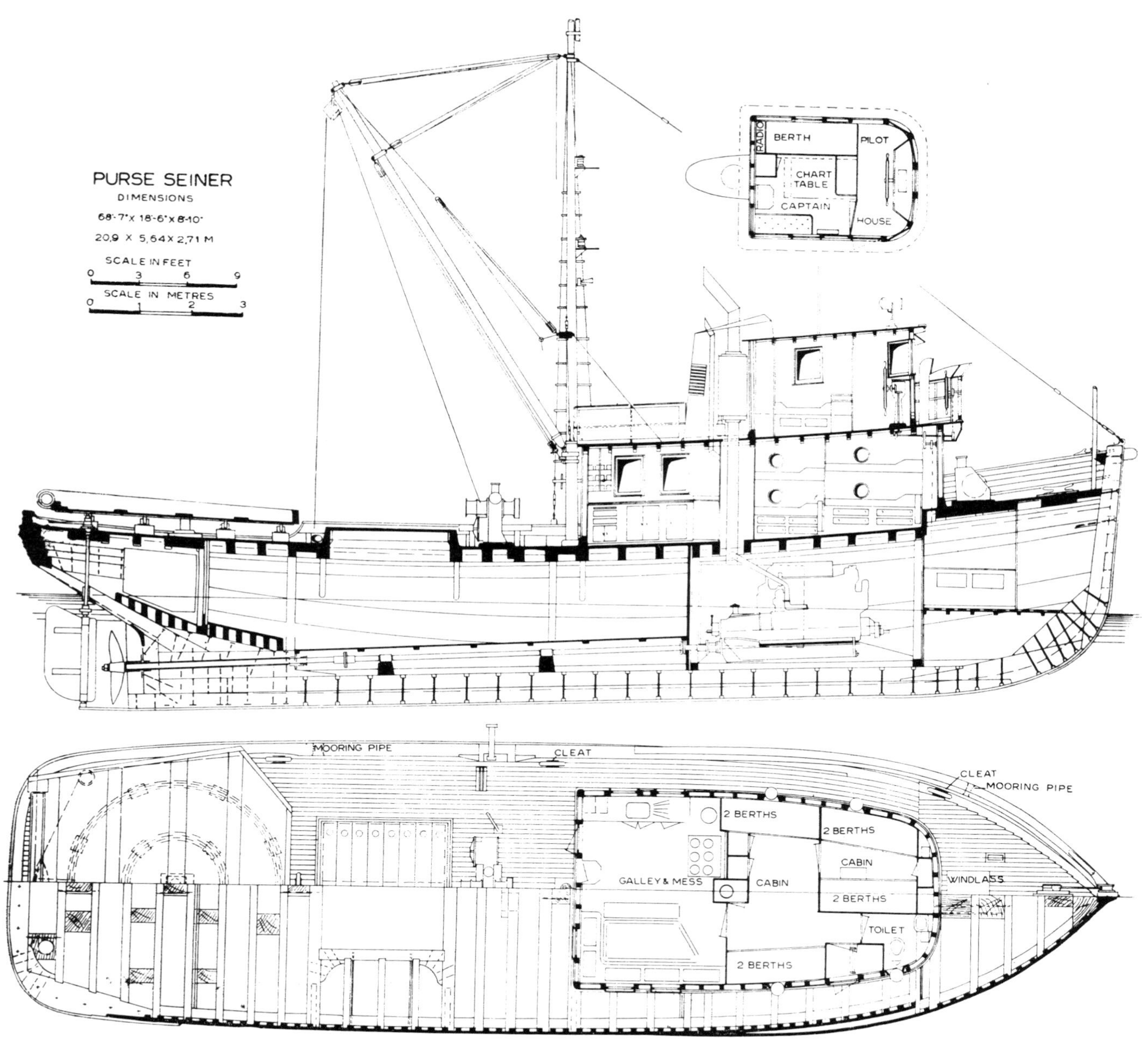

Fig. 142. General arrangement of a 68 ft. 7 in. × 18½ ft. × 8 ft. 10 in. (20.9 × 5.6 × 2.7 m.) seiner designed by the author

fuller midsection. Model B was freely altered to give a good speed performance. As this model lacked stability, a model C was constructed to give adequate stability with the low resistance characteristics of model B. Results of bare hull and self-propulsion tests showed a marked reduction in h.p. for model B and C at $\frac{V}{\sqrt{L}}=1.0 \left(\frac{v}{\sqrt{gL}}=0.3,\right)$ with diminishing reduction at higher speeds until approximate equality with model A was reached at $\frac{V}{\sqrt{L}}=1.3 \left(\frac{v}{\sqrt{gL}}=0.38.\right)$

Two drifters were afterwards constructed, one to design A and the other to design C. In spite of radical differences in waterline fineness represented by 9° half angle of entrance in C, and 21° in model A, there was no appreciable difference in seagoing behaviour except that, contrary to expectations, the finer boat with its flaring bow was drier forward than the other. The speed of the finer hull was, of course, considerably better than the other.

As the displacement and general dimensions of these drifters corresponded closely to a typical 65 ft. (19.8 m.) seiner, of which several were built in British Columbia in recent years, the models have been examined with care to determine if they could be improved. The conclusion is that the seiner agrees with the model A type and, being powered for 10 knots speed or $\frac{V}{\sqrt{L}}=1.3 \left(\frac{v}{\sqrt{gL}}=0.38\right)$ nothing would be gained by adopting a finer form, but there would be a loss of space in the fore end which is required for accommodation and machinery. Fig. 144 shows a comparison between the form of model C and a typical seiner. Fig. 145 shows the remarkable similarity between model A and a successful 78 ft. (23.8 m.) seiner.

Another valuable work on fishing vessel design is that of Traung (1951), in which he describes various researches on fishing vessel design, and gives results from tank tests

TABLE XXVII

PARTICULARS OF SEINE BOATS

No.		*L.O.A.*	*L.W.L.*	*Beam*	*Draft Aft*	$\frac{\Delta}{(.01L)^3}$	$\frac{L}{\nabla^{\frac{1}{3}}}$	*B.H.P. max.*	*R/G or D/D*	*Prop. R.P.M.*	*Trial speed knots*	*Year built*	*Remarks*
1	ft.	42 ft. 6 in.	40 ft. 0 in.	12 ft. 0 in.	5 ft. 0 in.	380	4.2	80	3/1	533	—	1953	Drum seine
	m.	12.96	12.2	3.66	1.525								
2	ft.	50 ft. 0 in.	46 ft. 3 in.	15 ft. 0 in.	6 ft. 3 in.	360	4.2	95	4.5/1	400	—	53	Table seine
	m.	15.25	14.12	4.58	1.91								
3	ft.	57 ft. 8 in.	51 ft. 8 in.	15 ft. 6 in.	6 ft. 3 in.	310	4.5	115	2/1	450	9.7	45	„ „
	m.	17.6	15.75	4.73	1.91								
4	ft.	60 ft. 6 in.	50 ft. 5 in.	17 ft. 0 in.	5 ft. 9 in.	320	4.4	150	3/1	400	—	47	„ „
	m.	18.45	15.4	5.18	1.755								
5	ft.	65 ft. 0 in.	59 ft. 0 in.	17 ft. 6 in.	5 ft. 9 in.	275	4.7	150	3/1	400	10.0	50	„ „
	m.	19.82	18.0	5.34	1.755								
6	ft.	68 ft. 0 in.	64 ft. 9 in.	17 ft. 6 in.	7 ft. 3 in.	350	4.3	125	D.D.	350	—	43	„ „
	m.	20.75	15.77	5.34	2.21								
7	ft.	68 ft. 0 in.	63 ft. 9 in.	18 ft. 6 in.	7 ft. 6 in.	290	4.6	270	3/1	400	10.8	51	„ „
	m.	20.75	19.47	5.65	2.28								
8	ft.	70 ft. 3 in.	65 ft. 0 in.	19 ft. 3 in.	8 ft. 0 in.	350	4.3	270	3/1	370	10.7	52	Seine Hal Pack
	m.	21.45	19.82	5.87	2.44								
9	ft.	73 ft. 2 in.	66 ft. 8 in.	19 ft. 6 in.	8 ft. 3 in.	350	4.3	300	2/1	375	—	52	„ „ „
	m.	22.31	20.32	5.95	2.52								
10	ft.	75 ft. 0 in.	70 ft. 0 in.	20 ft. 0 in.	8 ft. 6 in.	280	4.6	200	D.D.	325	10.6	44	Seine
	m.	22.9	21.35	6.1	2.59								
11	ft.	78 ft. 0 in.	73 ft. 0 in.	20 ft. 6 in.	8 ft. 9 in.	328	4.4	200	D.D.	325	—	44	„
	m.	23.8	22.26	6.26	2.67								
12	ft.	78 ft. 0 in.	73 ft. 0 in.	21 ft. 0 in.	9 ft. 0 in.	338	4.3	200	D.D.	325	10.3	45	„
	m.	23.8	22.26	6.41	2.745								
13	ft.	78 ft. 0 in.	73 ft. 0 in.	21 ft. 0 in.	8 ft. 6 in.	320	4.4	300	2.5/1	440	—	50	War surplus machinery
	m.	23.8	22.26	6.41	2.59								
14	ft.	80 ft. 0 in.	75 ft. 0 in.	21 ft. 6 in.	9 ft. 0 in.	335	4.4	260	D.D.	390	—	52	Combination
	m.	24.4	22.9	6.56	2.745								

Fig. 143. Purse-seiner, 68 ft. 7 in. × 18½ ft. × 8 ft. 10 in. (20.9 × 5.6 × 2.7 m.), displacement 75 tons, 11 knots with 270 h.p. diesel

carried out by himself in Sweden, and a valuable section on the general design of fishing boats. Traung's own tests are similar in conclusion to Allan's, in that a worthwhile reduction in power for a $\frac{V}{\sqrt{L}}$ of about 1.0 to 1.2 $\left(\frac{V}{\sqrt{gL}}=0.30 \text{ to } 0.36\right)$ may be obtained by an increase of the midsection and reduction of the entrance angle. The type of boats tested were of very high displacement length ratio.

The general conclusion from these two works do not suggest any radical alteration in present British Columbia designs but that does not rule out improvement. It would, in fact, be interesting to have made a series of tank tests on West coast types under the control of someone familiar with the limitations imposed. Such tests should investigate particularly the effect of afterbody fullness on the necessary h.p. and also a comparison between a usual seine boat stern and a modified cruiser stern.

CONCLUSION

Post-war years were a boom period for the British Columbia fishing industry and many new boats were built, but failure to market the 1951 and 1952 salmon catch has resulted in extreme retrenchment and almost no new craft have been ordered or built in 1953. When a revival occurs, the trend is expected to be mainly towards a revision of the fishing technique in order to economize on man-power, rather than changes in hull form. But economy will also require more moderate powering and less elaborate outfitting in larger boats and, for gillnetters, smaller boats with a minimum of trimmings.

Fig. 144. Comparison of waterlines and section area curves of a Pacific seiner and a British drifter (Denny "C" model)

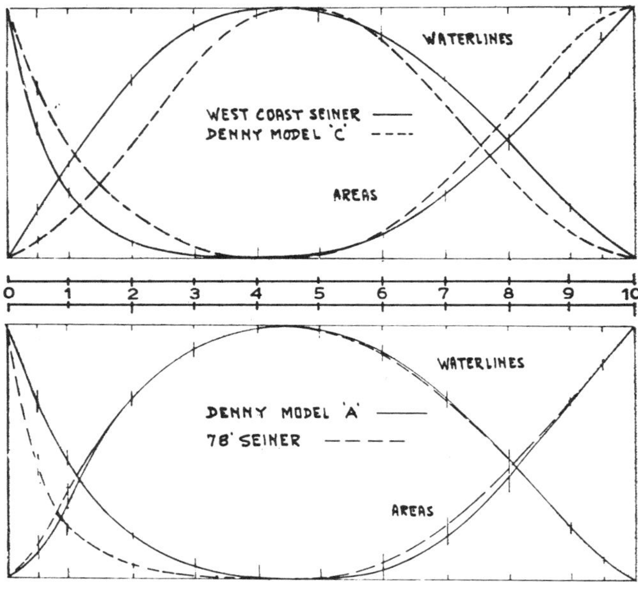

Fig. 145. Comparison of waterlines and sectional area curves of a 78 ft. (23.8) Pacific seiner and a typical British drifter (Denny "A" model)

PACIFIC TROLLERS

by

H. C. HANSON

TROLLERS may be up to 60 ft. (18.3 m.) in length and are the most numerous fishing vessels on the Pacific coast of North America. The first were converted gillnetters and fishing was done with cotton lines

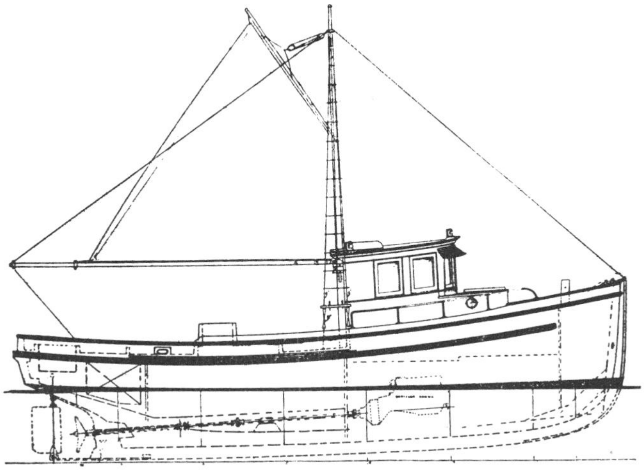

Fig. 146. 37 × 10 ft. × 5 ft. 3 in. (11.3 × 3.05 × 1.6 m.) wooden troller

hauled in by hand. Later, simple spools or drums were introduced, and when the vessels went further from their ports and their power and size was increased, metal

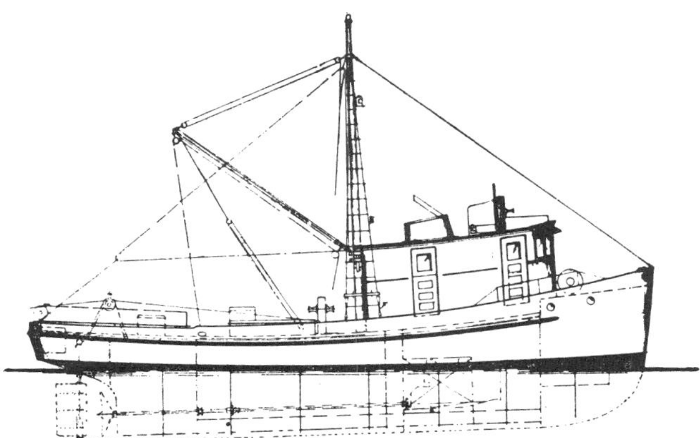

Fig. 147. 45 × 13 ft. 4 in. (13.7 × 4.08 m.) steel combination boat designed for trolling, seining and trawling

spools called trolling gurdies were installed to handle wire fishing lines. The first trolling wires were of copper but today they are of stainless steel.

Fig. 146 to 148 show some popular types of trollers.

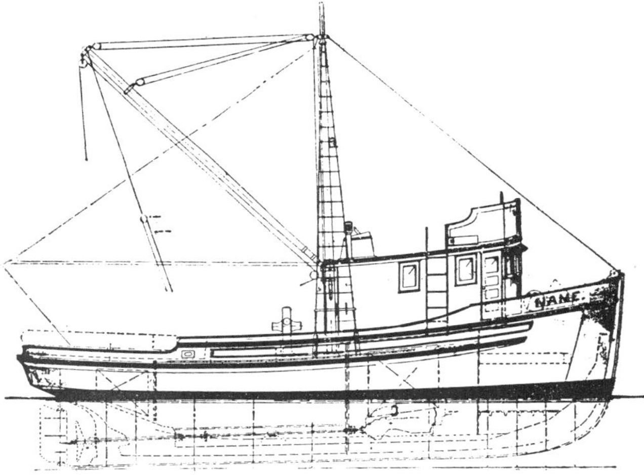

Fig. 148. 46 ft. 6 in. × 14 ft. × 6 ft. 6 in. (14.2 × 4.27 × 1.98 m.) wooden troller

The first trollers were double-enders but the stern has been changed to the seine or square stern suitable also for other fishing methods.

The present boats are as much as 25 per cent. wider and deeper than they were 30 years ago, a development brought about by increased engine power and the knowledge that greater speed is attained by greater waterline width. Formerly the main area of the midship section was lower where the resistance was greatest. Greater stability and speed are obtained with the displacement well out, as in fig. 149. A few of the trolling boats have a V-bottom; they are less expensive to build, attain higher speeds with

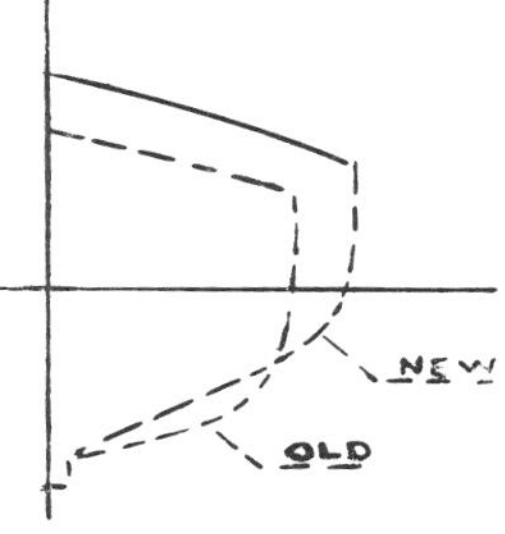

Fig. 149. The new type of midship section gives better stability and speed

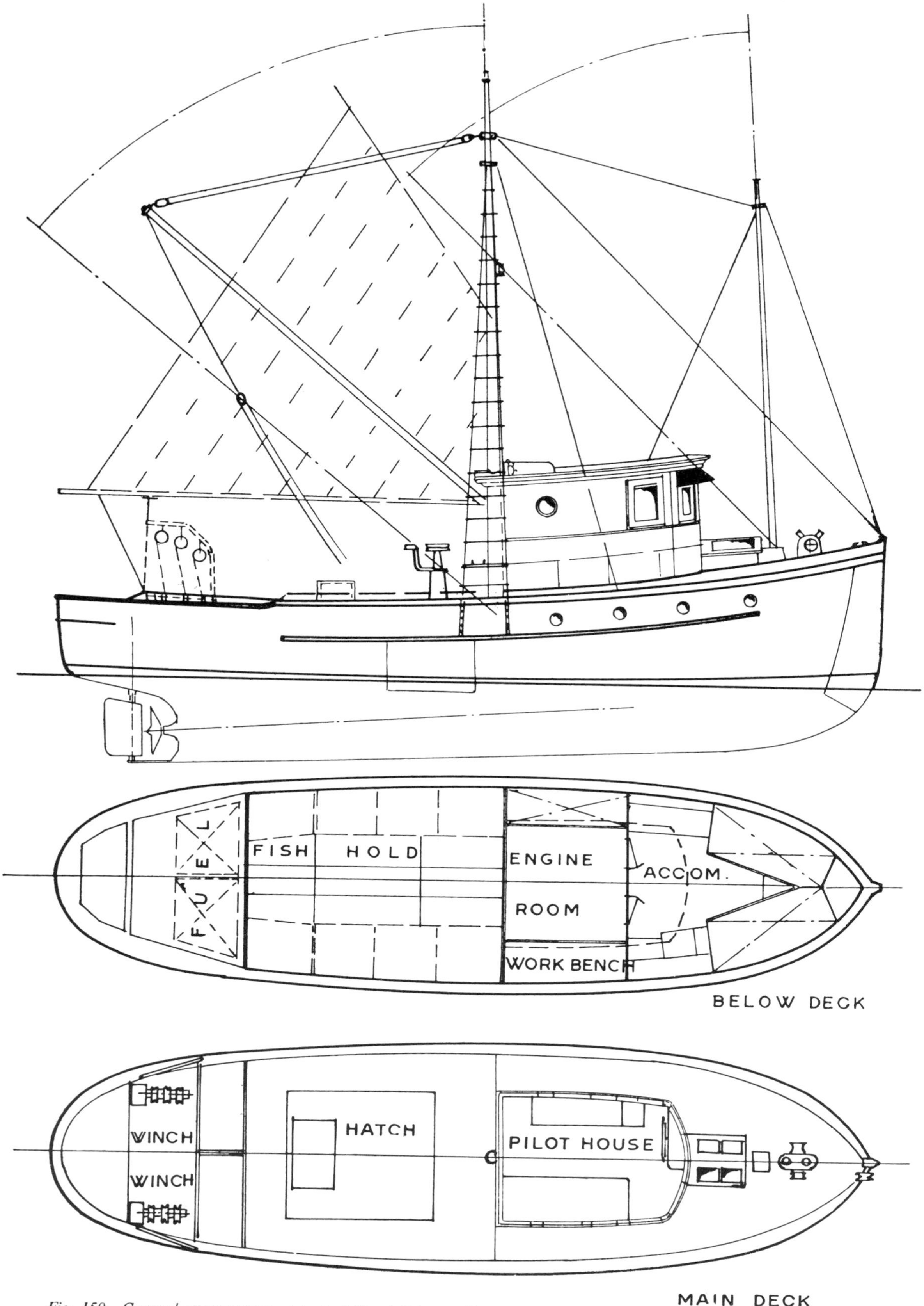

Fig. 150. General arrangement of typical North West troller 42 × 12 ft. × 4 ft. 6 in. (12.8 × 3.66 × 1.37 m.)

the same power and are as seaworthy as round-bottom boats. They should be more popular.

Fig. 150 shows the general arrangement of a typical North-west troller. The boat is 42×12×4 ft. 6 in. (12.8×3.66×1.37 m.), fine-lined for speed and therefore having a cruiser stern. The high sheer makes a dry boat

frame type construction, some with floor timbers but most with the west coast keelson type of construction—less expensive and more space-saving and stronger.

The troller in fig. 151 has an 80 h.p. diesel, with a 36-in. (915 mm.) diameter, 500 r.p.m., propeller and is capable of 10 knots in light condition.

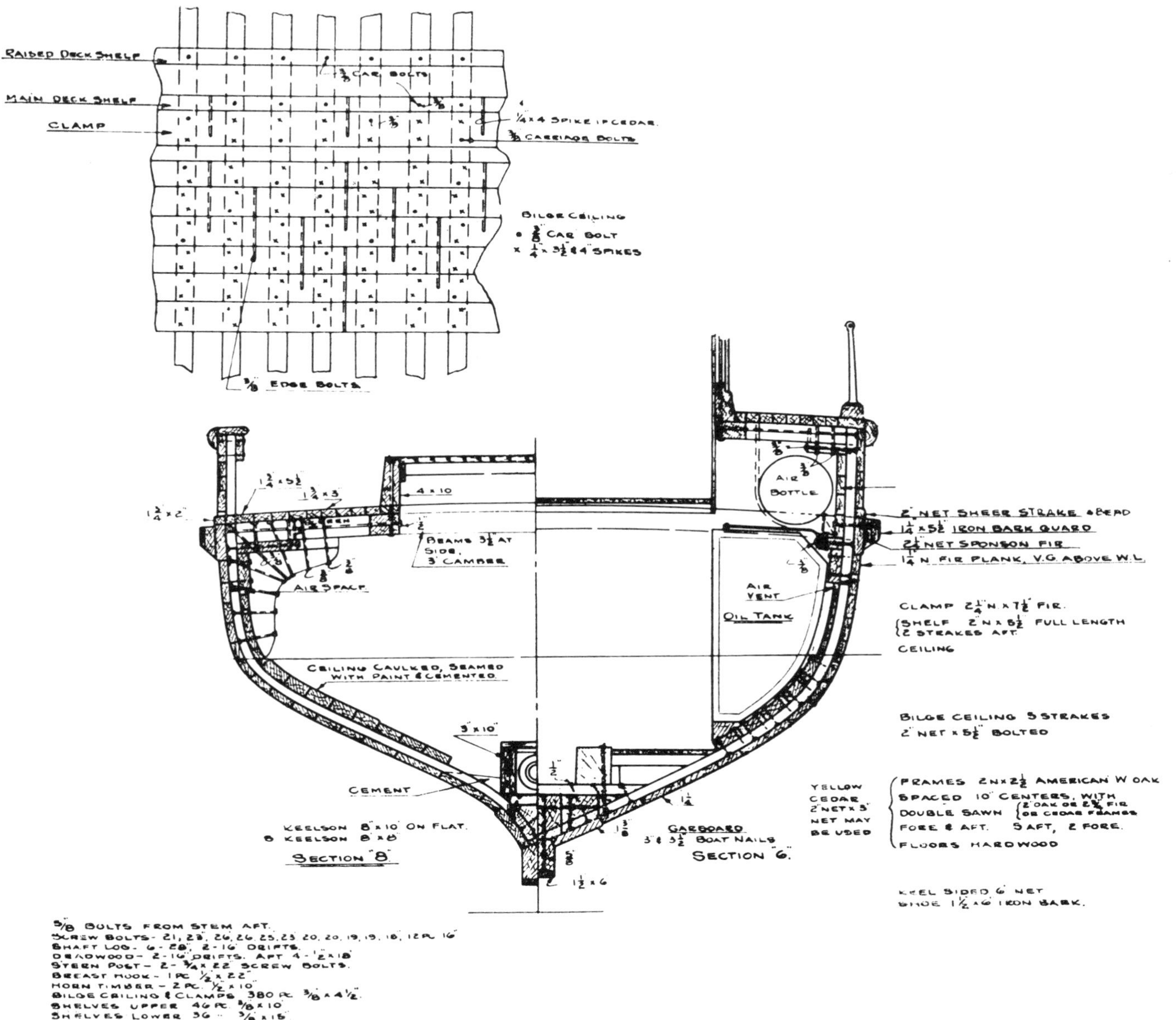

Fig. 151. Typical midship section and table of scantlings and fastenings

and the deck is also raised to give more room in the forecastle. The step in the deck prevents loading by the bow and provides a stopper for the fish which can accumulate on deck until there is time for stowing. The vessel will load on an even keel at all times—a very important feature in fishing of this nature. Fig. 151 shows a typical midship section, as well as a table of scantlings and fastenings. Most craft are of bent oak

Some vessels sail 100 miles off shore and up to 1,000 miles from home port. The fuel capacity is here 670 imp. gal. (800 gal., 3.04 cu. m.) and the water capacity 250 imp. gal. (300 gal., 1.14 cu. m.). The galley and living quarters are in the forecastle but some boats are built with a larger deck house containing the galley and the berths, which allows easier handling if only one man is aboard.

The cargo hold is 14 ft. (4.26 m.) long, with 12 tons of payload capacity. It is divided into bins, with portable bin boards. The boat carries normally 3 tons of chipped ice. Few trollers have mechanical refrigeration because of the cool temperatures in the salmon fishing regions and the relatively short trips.

Fig. 152. Set of 3-spool trolling gurdies with reverse gear boxes and hydraulic drive unit

The cockpit is 2½ ft. (0.76 m.) fore and aft, and is built as wide as is possible whilst retaining a depth of at least 2 ft. (0.61 m.) to give safe leg room for work in a seaway. The cockpit is best placed near the stern to keep the trolling lines clear and to reduce the height the fish have to be lifted. Checkers hold the fish until they can be dressed and stowed in ice.

The power gurdies, fig. 152, are placed near the cockpit. Though more poles can be used, the craft would normally use four for trolling: two main ones aft of the house and two at the bow to hinge back and rest in stowed position, as in fig. 153. Some, however, prefer a mast forward of the house to stow the poles upright, fig. 150 and 154.

The trolling gurdy sets, one on each side of the cockpit, are operated hydraulically, or mechanically with shaft, mitre gear and chain drives. The clutch is set near the cockpit for ease of control. The wire leads to the trolling poles through pulleys (fig. 150) hung on the pulley davit abreast of the gurdy spools. Some owners

Fig. 153. Troller with forward poles hinged back to rest on the mast

Fig. 154. Troller with a forward mast for stowing of poles

Fig. 155. Right: *bronze bow shock absorber with 6 in. (150 mm.) block*
Left: *davit block clamp securing 4 in. (100 mm.) block to davit*

Fig. 157. Cross arm at masthead for stowing trolling poles

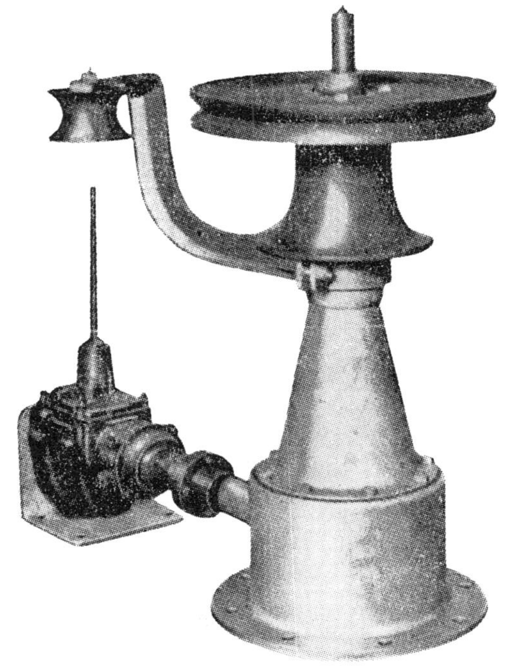

Fig. 156. Longline gurdy

Fig. 158. Bracket for trolling poles and socket for allowing poles swivelling and hinging outboards

prefer to place the gurdy spools on the centre-line of the vessel, fore and aft, which dispenses with the davits, and the pulley leads are then hung from a spar beam secured horizontally to the main boom. This method requires more lead blocks. The stainless wire, which is $\frac{3}{64}$ to $\frac{5}{64}$ in. (1.2 to 2 mm.) in diameter, and up to 100 fm. (183 m.) long, is led to the outer end of each trolling pole where it goes through a pulley hung on spring shock absorbers (fig. 155). A longline gurdy, such as that illustrated in fig. 156 is sometimes used on this type of vessel and is placed between hatch and mast.

The mast is installed aft of the house and has a cross arm at the masthead for stowing the trolling poles upright, fig. 157, If a foremast is used, a cross arm is fitted on it, as in fig. 154. The trolling poles—up to 56 ft. (17.1 m.) in length and holding 50 lures—are set in sockets, fig. 158, to allow swivelling and hinging outboard. The guy wires for the poles are generally $\frac{3}{16}$ in. (4.8 mm.) diameter and are attached to a stainless steel thimble or socket.

As the vessels are often in heavy weather they have steadying sails, fig. 150, and some have stabilizers—heavy galvanized plates hung from the main poles with a three-point suspension. As the craft rises and falls, these plates offer resistance to the roll.

If an anchor windlass is used, the capacity is about 100 fm. (183 m.) of wire and 10 fm. (18.3 m.) galvanized chain. Otherwise the wire and chain comes in over a sheave fitted on top of the bow chock and the wire goes back to the niggerhead or reel on the gurdy via lead blocks set on the forecastle deck.

Some of the larger craft are fitted with echo sounders and direction finders and most are equipped with radio transmitters and receiving sets.

In recent years a few welded steel trollers have been built and drawings of two types are given here in fig. 147 and 159, the latter being a 40 ft. (12.2 m.) V-bottom troller. Whilst these have proved satisfactory, many fishermen still prefer the warmth of older type wooden vessels.

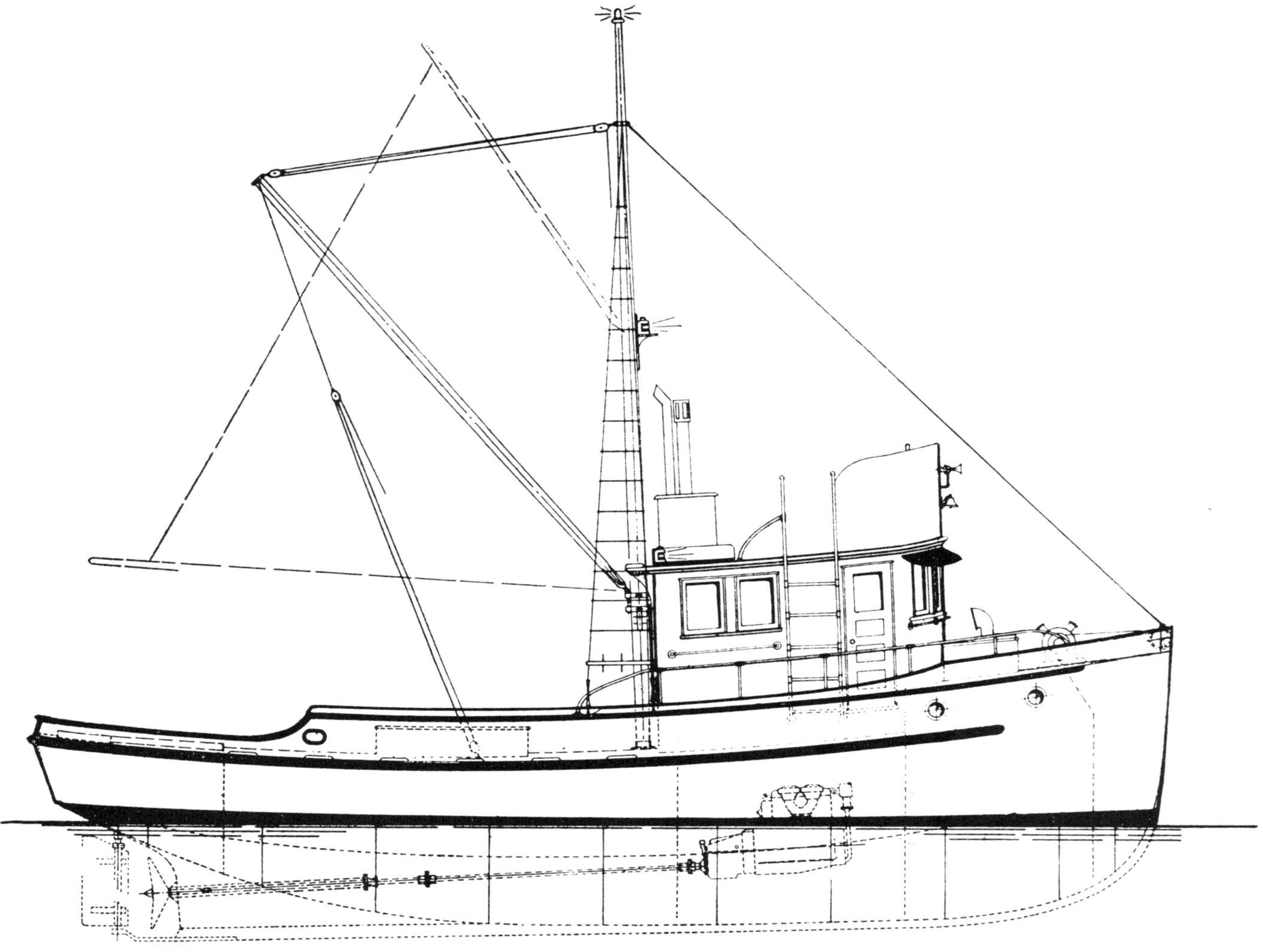

Fig. 159. 40 ft. (12.2 m.) V-bottom steel troller

OPERATIONAL INFLUENCES ON TUNA CLIPPERS

by

JAMES F. PETRICH

THE tuna clipper, one of the largest, most complex and romantic of the fishing vessels of the world, has been developed around a simple method of fishing—that of using live bait, hook and line.

In the early days of tuna fishing off southern California before the 1920's, fishermen learned to attract the schools of tuna to their boats by throwing out live bait, small sardines or anchovies, found in abundance along the coast. The tuna would churn the water in their strikes for the bait and in this frenzy would also strike at the feathered hooks held by the fishermen on a short line at the end of a bamboo pole. The hooks were made to resemble squids when played in the water but sometimes even this ruse was unnecessary as the tuna would strike at a bare hook. The hooks were left barbless so that they could be easily dislodged from the mouth of the fish.

When the tuna were too big for one man to lift safely over the rail and into the boat, the fishermen attached two poles on a bridle to a single hook and two fishermen worked as a team. For even bigger fish, three-pole or four-pole teams were used and soon the fishermen acquired great skill in fishing all sizes of tuna by this method.

Since those early days the tuna fishery was expanded to include the waters off the west coast of Mexico and Central America and the northwest coast of South America, as far as 3,000 miles from the home ports of San Diego and San Pedro, California. The fishing boat has developed from the small flush-deck launch, without living accommodation and able only to make daily trips out of port, to the modern high seas clipper that can travel 10,000 to 12,000 miles. Clippers now range from 90 to 169 ft. (27.4 to 51.5 m.) long, are able to carry huge quantities of live bait as well as fuel oil and provisions, and can bring home 130 to 630 short tons (110 to 570 long or metric tons) of frozen tuna. The crews, though mainly of Portuguese extraction, also include other nationalities such as Japanese, Italians, Slavonians and Scandinavians, and the clippers have been influenced by these peoples.

The modern clipper is a good looking, yacht-like vessel, with a raised deck forward reaching back past amidships to the deck boxes built on the main deck aft. The stern is low in the water, the main deck aft being practically awash at all times. Around the stern are the steel fishing racks hung outside the main deck rail and extending on the port or fishing side up to the start of the raised deck. There are usually three deck boxes on the stern, in line fore and aft, and set inboard of the main deck rail to allow for a 4 or 5 ft. (1.2 or 1.5 m.) passageway at the sides and around the stern. Over the deck boxes is built a canopy strong enough to withstand the pistol shot impact of an errant flying hook, and to protect the "chummer" at his station atop the boxes. Forward of the deck boxes, about amidships, is the single mast and boom used to lift the power tender and fishing skiffs carried on the raised deck. Atop the mast is the covered crow's nest large enough to seat two men, while ahead of the mast is the deck house with accommodation for 10 to 20 men in rooms that extend the full width of the house to allow free passage of cooling sea air through open doors on each side. Above the deck house is the pilot house and chart room and, behind them, the stack.

The hull and houses are painted white, the deck boxes and working spaces aft are usually grey, and the guards and rails stained their natural finish to accentuate the long lines of the vessel.

Below decks is the machinery to operate and propel the clipper and also the hold, which is divided into many water-tight compartments. The engine room below the main deck is well forward, just aft of a large fuel oil tank in the bow. From the engine room aft leads a full height shaft alley flanked on either side by the 8 to 16 watertight wells used to carry fuel oil and live bait on the outbound trip and frozen tuna on the homeward voyage. Aft of the wells, in the very stern, are tanks for carrying fresh water and more fuel oil.

In the engine room, besides the single main diesel propulsion engine, there are installed two large diesel-driven generating sets as well as two large bait water pumps, bilge and fire pumps, fuel oil transfer pumps, lubricating oil tanks and other auxiliary machinery and equipment. The shaft alley contains, besides the propeller shaft, the brine circulating pumps and piping, and the complex ammonia refrigeration lines for each well and deck box.

Above the engine room, in the 'tween deck space, is the upper engine room containing the main switchboard, the ammonia refrigeration compressors, air compressors and air bottles, the engineer's work bench, CO_2 fire-fighting equipment, etc. The forecastle space is for ship storage while aft are two watertight doors (one port and one starboard) that lead to the open 'tween deck space which contains the hatches to the wells below deck through which the fish are loaded.

In the centre of this space, aft of the upper engine room bulkhead, is the spacious galley cabin. The open 'tween deck space can be enclosed with storm doors at the sides, and storm doors aft on either side of the forward edge of the deck box give protection from severe weather.

The modern tuna clipper is built either of steel or wood, the wooden vessel being preferred by the majority of fishermen. Steel construction, however, has gained popularity since the end of World War II because it allows more space in the vessel, and because building costs are competitive with the rising costs of wooden construction. But, recently, there seems to be a trend back to wooden vessels. The corrosion problems of the steel vessel in the warm tropical waters have been a headache to the fishermen, and rust streaks, appearing through the paint when only two weeks out of port, have been a blow to the pride of an owner-captain.

The very first tuna clippers had simple design requirements. Almost any kind of boat could take fishermen out a few miles to where the tuna were running and return in the afternoon with the catch on deck. A small wooden box, through which sea water was pumped, was installed on deck to carry live bait for the day's fishing. The weight of the deck box conveniently lowered the stern so that the fishermen, standing close to water level, could easily fish with pole, line and hook.

The distinguishing mark of these first tuna clippers was a low freeboard at the stern, and in all development since that time this mark has remained. It is a vitally necessary condition for efficient operation.

Fishing Trim and Stability

To fish effectively, to brace himself against the terrific impact of a tuna striking at his hook, and to have enough leverage to lift the heavy fish out of the water past his shoulder and into the boat, the fisherman must be standing only a few inches above the water. He stands in the steel racks, hung outside the rail, to trail his hook as far out from the hull as possible, for the tuna is timid and will not come too close to the boat. Such an arrangement leaves the deck space behind him free to receive the tuna. When twelve men are fishing in a good school of tuna, they can toss fish onto the deck faster than a man can count and fill the deck up to the rail height with 15 to 20 tons inside of an hour. But everything must be just right for the fisherman. He must not be too high off the water as the tuna may easily pull him off balance; the rail behind him must not be too high or he will tire lifting the tuna into the boat; nor must it be too low otherwise it will not give him support as he leans back

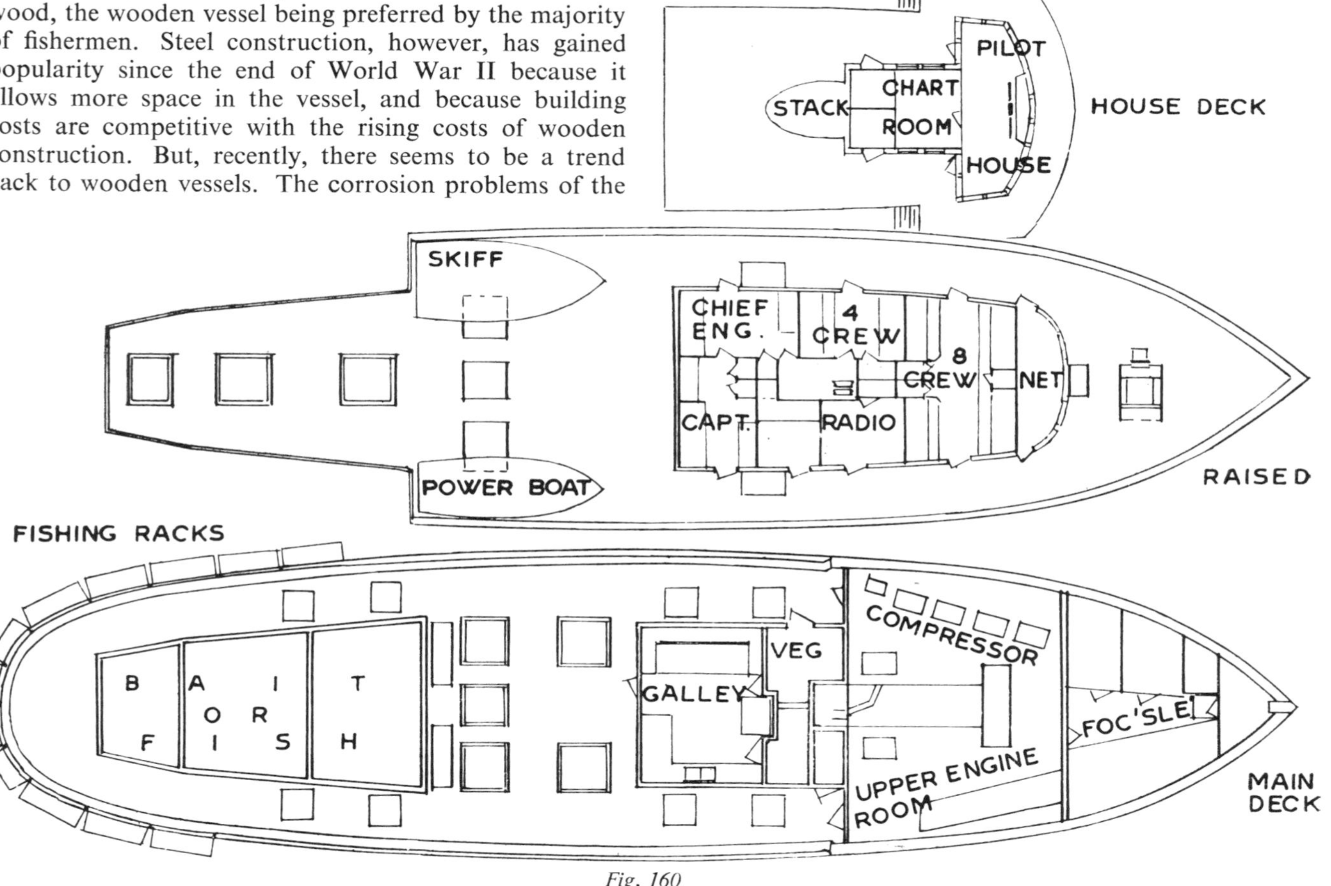

Fig. 160

his full weight to heave in the fish. Again, he does not want the deck rail to be too low because, instead of 15 tons, he may only have space for 10 tons out of each school before he must stop and pass the fish on deck into the storage wells. And, once he has stopped fishing a school, he may lose it. (See fig. 169 on page 121).

It has been found best to have the fish racks at the same level as the deck, and the rail behind the fisherman to reach to the height of his buttocks; and then to trim the ship with just a few inches of freeboard at the stern, enough to be safe and seaworthy in the rough waters where they fish and yet hold the fishing platform close to the water level. The designer, then, must calculate closely his displacements and weights and tank capacities to obtain this condition of trim all the time the vessel is on the fishing grounds.

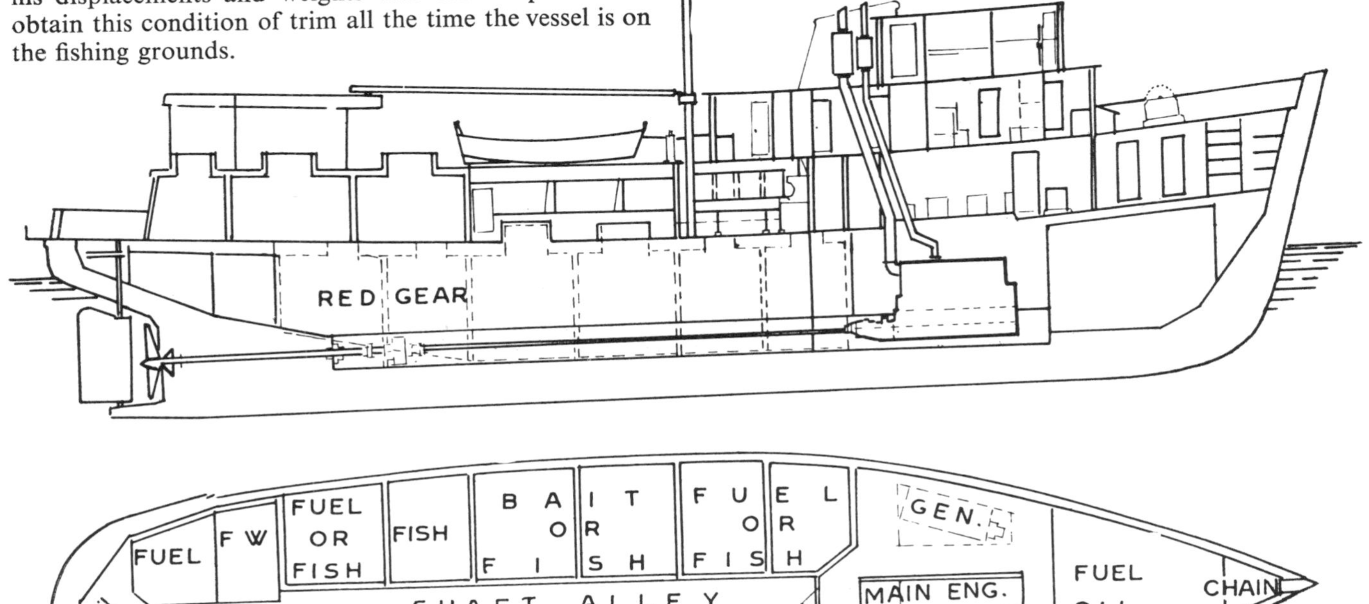

Fig. 161

"Safe and seaworthy" may appear to some incongruous with only a few inches of freeboard but, in a real sense, the low freeboard design is no more dangerous than the low freeboard oil tanker. Many a storm has seen a tuna clipper riding it out more easily than a large cargo ship.

There have been times, however, when some clippers were very unstable and unsafe, and the fundamental features of stability were forsaken.

In the period from 1926 to 1930 there was a building boom of large long-range clippers up to 130 ft. (39.6 m.), designed to sail farther south to fish the larger schools of tuna. These boats had their engines installed as far forward as possible to make room for the hold aft to carry more fish; and they carried crushed ice to preserve the catch. They were flush-deck vessels with a deck house for crew accommodation, a single mast and boom amidships and, on the after-deck, a large bait box divided into two compartments, one forward of the other.

The need for more bait increased proportionately with the size of the boat but, rather than put all this bait carrying capacity in boxes on deck and make the vessel top-heavy and unsafe, the designers built two additional water-tight "wells" in part of the hold. This was inconvenient for the fishermen-owners because they had to transfer bait as it was being used from the wells to the boxes. They also had to move the ice about more often in the smaller hold, and then move it from the hold to the wells as the hold was filled with fish. To overcome these inconveniences, many owners eliminated the bait wells in the hold and installed a third large bait box on deck. The result was a very tender and top-heavy boat. In fact, two of these vessels capsized and a third was lost with all hands. The rest were kept afloat and fishing mainly, perhaps, through the constant vigilance and ability of their skippers and crews. These disasters put a stop to such changes, induced the insurance companies finally to require stability tests on the vessels, and resulted in a much improved consciousness and treatment of the stability problems of the tuna clipper. All this helped the designer who, though always aware that the stability problems were especially critical because of the clipper's low freeboard, now had less trouble convincing builders and owners of their importance.

Since that time the need for bait-carrying capacity has increased three-fold, and this has been installed without sacrifice to stability or safety.

Carrying Live Bait

Because live bait is essential and is becoming scarce, keeping it healthy and alive is vital. The chummer, who looks after the bait and tosses it into the sea, to " chum up " the schools of tuna, is considered next in importance to the skipper and chief engineer. He has given much information to the designer so that now specifications for all bait wells and boxes call for white-painted interiors, and water-tight glass lenses in the tops and sides through which electric lamps shine to light up the interiors and keep the bait from being frightened.

A large bait pump is installed to pump enough sea water through all bait wells and boxes to change the water in them every 12 minutes. To guard against a breakdown of this pump, a stand-by pump is installed of the same capacity because if the water supply were to stop for 15 minutes, the bait may die. The pumps are the large propeller-type, 10 in. (25.4 cm.) size, with 10 h.p. motors, in the smaller tuna clipper, and up to 14 in. (35.6 cm.) size, with 50 h.p. motors, in the large clippers. They are usually placed on each side of the main engine in the aft end of the engine room and directly over their own sea chest valves. They discharge into a common header pipe, which passes through the length of the shaft alley and feeds through individual valves into the bottom of each bait well and deck box. The water flow through each tank is dispersed by a screen at the inlet, rises up through the tank, passes through an outlet screen and then goes overboard through an overflow pipe. The level of water in the tanks is kept well up in the hatch, which is built about 2 ft. (61 cm.) above the deck. This is done by regulating the amount of overflow through the overflow pipes. These hatches are 4 to 5 ft. (1.2 to 1.5 m.) square, large enough to provide easy access to the bait and small enough to minimize the effect on the vessel's stability of the free surface in the tank.

Refrigeration of the Catch

Another great problem of the tuna clipper is preserving its catch from the time when the first tuna is caught to the time when, after the long trip home, the fish is discharged at the cannery.

The early long-range clipper carried crushed ice in its hold to preserve the fish and used ammonia refrigeration coils, banked overhead and on the sides and bulkheads of the hold, to supplement the ice and overcome the heat absorbed through the hull. Usually the forward deck box was refrigerated, too, and used to carry iced tuna after the hold was full, the bait having been used up by this time or transferred to another deck box.

Some fishermen tried a method of circulating chilled sea water to preserve the tuna in the forward box, and this proved so satisfactory that soon all clippers were converting their water-tight wells and forward boxes to this type of freezing. A brine cooler was installed on deck through which the ammonia refrigerant was used to cool the sea water that was circulated constantly between it and the wells or boxes that were being chilled.

In the late 1930's, some " all brine " boats were built, having the hold divided into water-tight compartments so that the entire catch could be frozen " in brine." This added more complicated piping equipment and demanded a much greater investment by the owner. In fact, it resulted in a boat with less fish-carrying capacity than a comparable size " ice boat ". A good deal of space was taken up by a full height shaft alley extending the complete length of the hold between the wells, necessary to accommodate the extra piping and machinery and provide buoyancy to compensate for the heavier loaded " brine " wells.

However, the popularity of the " all brine " clipper was immediately established with the fishermen. After a hard day's fishing, those on the " all brine " boats did not have to change into warm, clean clothes and work most of the night stowing their catch in ice in the cold fish-hold. They merely had to pass the fish into one of the wells or boxes. already partially filled with chilled sea water, top off the tank, batten down the hatch, and then they were free to go to their bunk for a full night's sleep and be ready for another hard day's fishing the next morning. Though they might not be able to bring in as large a load as a comparable " ice boat ", these boats soon had all the best fishermen, and from that time newly built clippers were " all brine ".

The boats had other advantages, too. They could use some of their wells for carrying bait, thereby decreasing the size of the deck boxes and cutting down some of the top-heaviness of the clipper. They lined the forward wells with steel, making them suitable for carrying fuel oil for use on the outbound voyage. When empty, the wells could be cleaned for carrying fish.

A little later, the story goes, a skipper whose vessel was a bit " too heavy " (that is, has a little more tank capacity than she could safely carry) caught a full load of fish, but feared of the hazards of the long trip home. Reluctant to throw any away to lighten up the vessel, he decided to pump out the chilled sea water from some of the wells that had been under refrigeration for a considerable time, which meant returning with the fish concerned in a " dry " condition, kept cold only by the ammonia coils used to supplement refrigeration. His much-lightened vessel made better speed than usual though the weather was bad, and when the fish were unloaded from the " dry wells " they were in excellent condition, perhaps even better than those carried all the way in brine because they experienced less salt penetration.

The skipper tried the same procedure again and again, each time with very satisfactory results and soon the whole fleet was bringing in its load " dry frozen " after the initial brine freezing.

Some fish preserved in this manner were spoiled

because of a " warm spot " in a tank that had otherwise kept cool when the cold sea water was circulating it, and sometimes, too, the engineer would pump out the sea water too soon—before all the latent heat had been removed from the fish. In such cases the boat would come home with the tank full of spoiled fish. But, with more and more operating experience, these troubles were mostly overcome.

Other refinements followed to improve and simplify brine freezing. After the initial freezing of the fish with chilled sea water at 28° to 30°F. (—2° to —1° C.), salt was added to the sea water to make heavy brine and the temperature of the circulating brine and fish further reduced to about 20° to 25° F. (—7° to —4° C.). The brine was not pumped out until this temperature had been maintained in the tank for two days, by which time all the fish in the well were assured of being frozen. The fish were then kept " dry " at this temperature, or even colder, for the remainder of the trip.

Then it was found that the refrigeration could be done by the ammonia coils alone, and the coils are now specified at 6 or 8 in. (15.2 or 20.3 cm.) spacings instead of the 10 or 12 in. (25.4 or 30.5 cm.) used with brine coolers. A small 2 or 3 h.p. brine circulating pump, mounted in the shaft alley, one for each well and box, circulates the brine from bottom to top of the tank. The coils draw heat from the brine and fish by means of the direct expansion of ammonia, supplied from a large reservoir located in the aft of the shaft alley. The supply is through automatic expansion valves at each well. The ammonia gas is drawn from the coils in any one of three suction headers, maintained at three different suction pressures by back pressure valves corresponding to the temperature required: (1) to chill sea water and perform the initial chilling of the tuna, (2) to freeze the tuna in heavy brine at from 20° to 25° F. (—7° to —4°C.), and (3) to maintain the tuna "dry frozen" at about 10°F. (—12°C.). The 3 suction headers are piped through manifold valves to motor-driven reciprocating type of ammonia compressors, which are located in the upper engine room. The number of compressors may vary from 3 to 6 and their total refrigeration capacity from 35 to 200 tons (100,000 to 600,000 kcal./h.). Hot gas discharge from the ammonia compressors is piped to two large condensers (usually of a vertical tube flooded type especially developed for the clipper) located on the main deck ahead of the deck boxes. The condensers are cooled by circulating sea water which may have a temperature of 85° to 90° F. (29° to 32° C.). The condensed ammonia then flows by gravity to the ammonia receiver in the shaft alley. This method of refrigeration is simple, effective, and especially suitable to the freezing of tuna and to the specialized clipper operation.

A Touch of its Spirit

The early long-range tuna clippers used only the slow heavy-duty diesel engines and machinery. Now, as then, reliability of machinery is the measure of its popularity as there are no service ships or spare parts available when they are thousands of miles from home and days away from any port. On the continuous operation of the machinery depends, perhaps, a $100,000 (£36,000) catch, a $300,000 or $400,000 (£110,000 to £140,000) vessel, and lives of the fishermen.

Tropical waters are particularly corrosive and monel fastenings are used on under-water fittings, copper-nickel heat exchangers are installed, monel pump shafts and propeller tail shafts are used (the largest monel tail shaft in the world is installed in a clipper). Tropical airs do little to cool the engine room so oversized electric motors are placed on important pieces of machinery to prevent over-heating.

Fig. 162. Modern tuna clipper designed by the author

Cannerymen long ago learned that, with few notable exceptions, the cannery-owned boat was never successful. It was the boat owned by the fishermen themselves that consistently brought back the tuna and the canneries wisely encouraged good fishermen to own their own boats, helped to finance them and co-operated with them in every way. These fishermen were quick to use their earnings to increase the efficiency of the clipper operation. They went on long voyages over unknown seas to find new fishing grounds, and they charted the ocean depths where they fished. They added the raised deck forward to make their vessel more seaworthy; they studied, with a critical eye, the plans of a new vessel, to make the lines more pleasing and the accommodation better. They installed a small chapel aboard and their wives furnished it with the finest laces and appointments. It is such fishermen, proud, colourful, jealous, courageous and bold, and at the same time fearful, born to the sea and with a deep religious culture, who have put the deepest mark on the modern tuna clipper.

PACIFIC TUNA CLIPPERS

by

H. C. HANSON

THE first raised deck tuna clipper, *Northwestern*, fig. 163, was built in 1928, and was 125 ft. (38.1 m.) long, 27 ft. (8.23 m.) wide, and 12 ft. 8 in. (3.9 m.) deep. It had a 500 h.p. heavy-duty diesel engine and a three compartment wooden bait tank, built aft. It was of sawn frame construction and was insulated between the frames but so that an air space was also left between the ceiling and the planking to aid ventilation and prevent decay. Previously the insulation had been placed inside the ceiling. The new insulation method was successful, and after 23 years of operation, when the vessel was modernized in 1951, there was no trouble because of wood decay. Fig. 164 shows the clipper after conversion.

Until this vessel was built it was only possible for bait to be carried alive for a few days. A new method of piping was installed in the large bait tank allowing more water to be circulated and this was so successful that after the first trip of *Northwestern*, of more than four weeks, there was still bait alive.

One of the largest clippers, built expressly for tuna fishing, is the 130×30×17 ft. (39.6×9.1×5.2 m.) welded steel craft, *Sun Dial*, which can carry about 400 tons (360 ton). She is a full-bodied vessel with a good beam, is full aft and carries her load without squatting. At full speed—10.1 knots—and loaded, the propeller does not pull her stern down. She has a supercharged 850 h.p. diesel—8 cylinders with 12 in. (305 mm.) bore and 15 in. (380 mm.) stroke—turning a 3-bladed propeller 72 in. (1,830 mm.) in diameter, with 49 in. (1,245 mm.) pitch at speeds up to 380 r.p.m. The auxiliaries consist of three 120 h.p. diesels each driving a 90 k.w. generator. Two of them are in the lower engine room, the other being in the upper engine room where there are four $6\frac{1}{2}\times6\frac{1}{2}$ in. (165×165 mm.) twin cylinder refrigerating compressors driven by 40 h.p. electric motors equipped with manual starters. All engines are fresh water cooled.

Fig. 163. 125/117×27×12 ft. 8 in. (38.1/35.7×8.23×3.9 m.) tuna clipper Northwestern

Fig. 164. Northwestern *after conversion in 1951*

There is permanent fuel bunkerage for 26,700 imp. gal. (32,000 gal., 121 cu. m.) and another 27,600 imp. gal. (33,000 gal., 125 cu. m.) can be carried in cargo tanks on the outgoing trip. The lubricating oil tank capacity is 2,000 imp. gal. (2,400 gal., 9.1 cu. m.) and that for fresh water, 4,200 imp. gal. (5,000 gal., 18.9 cu. m.)

The refrigeration receiver and brine coolers are located in the shaft alley forward upper end, while the refrigerating condensers are on deck at the forward end of the bait tanks. The 2 in. (51 mm.) circulating pumps for the 16 cargo tanks, two $7\frac{1}{2}$ h.p. brine transfer pumps, and a 15 h.p. pump for the condenser, are in the shaft alley. There are also three 30 h.p. 14×12 in. (356×305 mm.) vertical bait pumps, two 10 h.p. vertical bilge pumps, one 15 h.p. high pressure fire pump and some smaller pumps for general service. There are 19 magnetic starters for the larger pumps and 16 push button starters for the brine well circulation pumps.

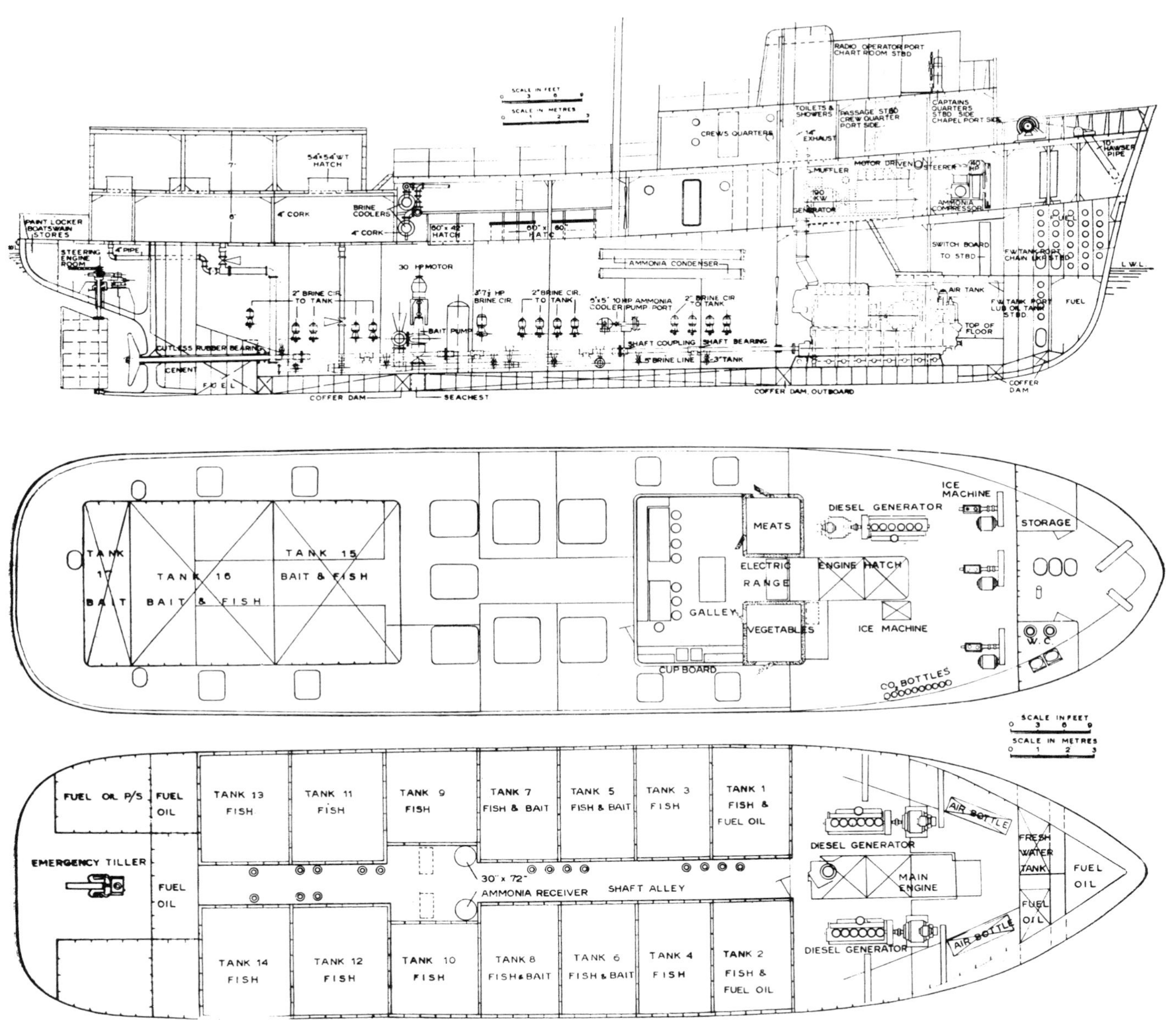

Fig. 165. Sun Dial—*inboard profile and deck plans*

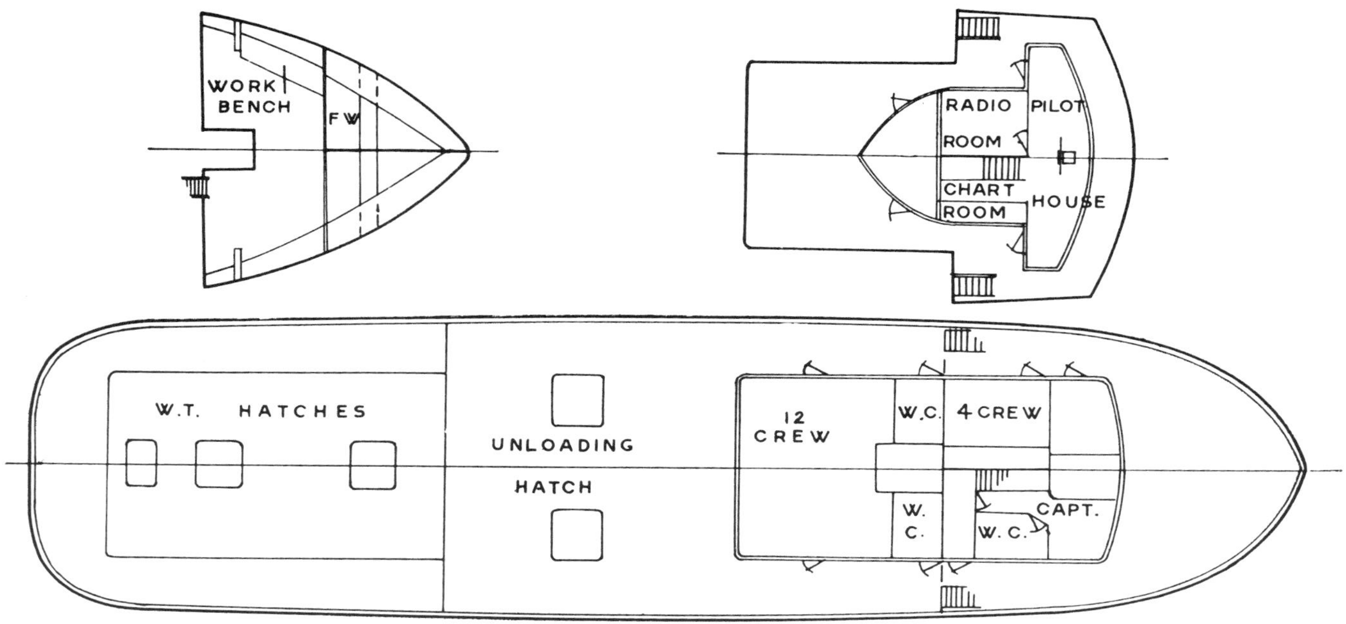

Fig. 166. Sun Dial—*deck plans*

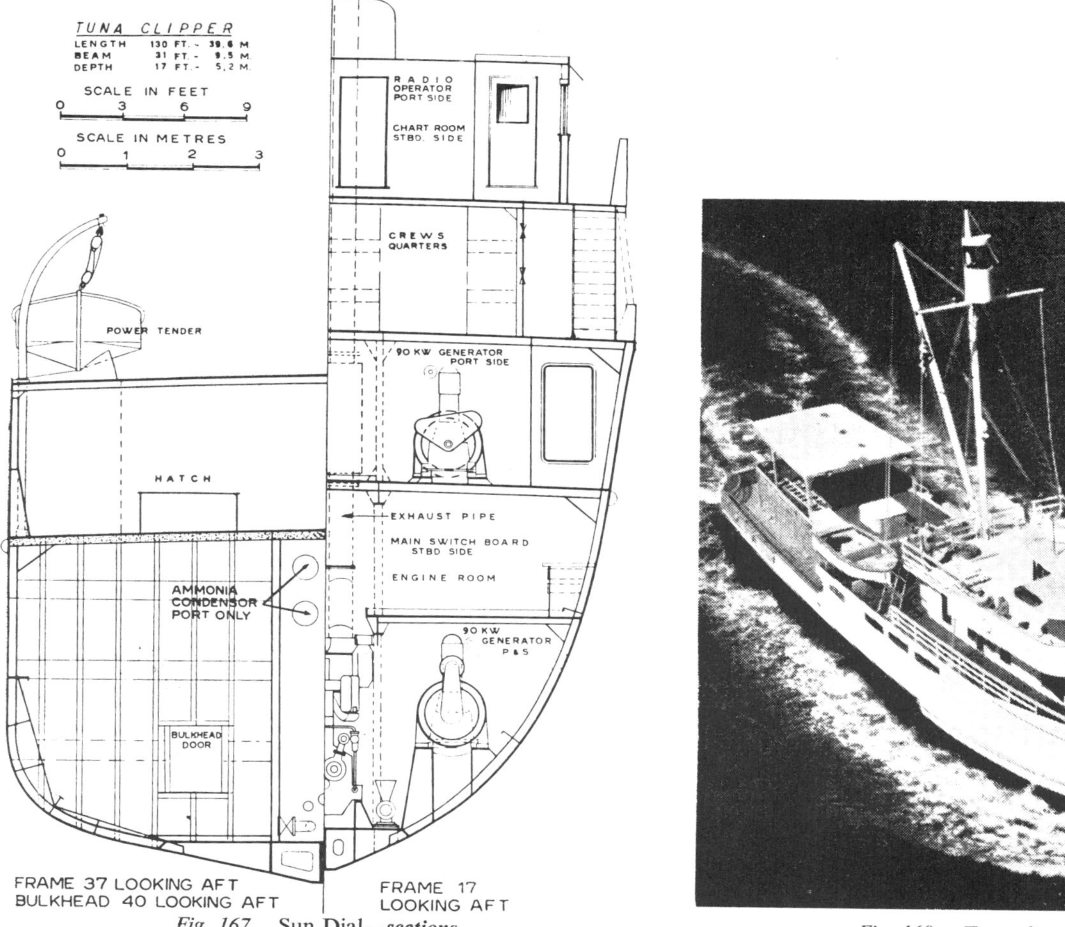

Fig. 167. Sun Dial—*sections*

Fig. 168. Tuna clipper Sun King

The welded bait tank on the aft deck has the two forward compartments refrigerated for stowing the last of the catch. In all there are 14 refrigerated cargo holds as well as the two in the bait tank on deck, and they are equipped with 17,000 ft. (5,200 m.) of $1\frac{1}{4}$ in. (32 mm.) ammonia piping.

Navigational instruments include a gyro compass, a two-unit power-driven gyro pilot, a steering repeater compass, a bearing repeater on the bridge deck, an echo-sounder, a 65 w. radiophone, an eye-level radio direction finder, a 10-station teletalk intercommunication system, pyrometers for the diesel engine and mercury thermometers for each brine circulating pipe-line to give the temperature in all cargo wells. Steering, whether manual or automatic, is through a chain-type steering engine in the forepeak of the upper engine room.

The anchor windlass carries 100 fm. (183 m.) of wire rope, $\frac{7}{8}$ in. (22 mm.) diameter, with 15 fm. (27 m.) of 1 in. (25 mm.) stud link chain at the outer end. It is driven with a 20 h.p. motor.

There are quarters, and a chapel, for 17 men in staterooms on the deck. The boom is controlled by an electric driven winch, and an electric gear driven winch is installed on the boom itself for handling cargo. Two boats are carried: a fast 18 ft. (5.49 m.) power tender and a bait net skiff fitted with buoyancy-tanks. Plans of the general arrangements are shown in fig. 165 and 166, and typical sections in fig. 167. The plating is $\frac{5}{16}$ in. (7.8 mm.), garboards $\frac{3}{8}$ in. (9.5 mm.) and the transverse framing $5 \times 3\frac{1}{2}$ in. (127×89 mm.) on 24 in. (610 mm.) centres. The deck is $\frac{3}{16}$ in. (4.8 mm.), complying with standard construction rules. A bird's eye view of one of the tuna clippers of the *Sun* class is shown in fig. 168. The difficult task of tuna catching by hook and line is indicated in fig. 169.

Fig. 170 is a photograph of the first steel tuna boat built and used in Hawaiian waters. Vessels of this kind are used as day boats. The fish tanks are arranged with openings to the sea, to allow water circulation by the surging of the vessel. Live bait is carried successfully and when the stocks are depleted the tanks are used to store the catch. The water openings may be opened or shut as required for wet or dry storage and water may be pumped out of the tanks.

Fig. 169

Fig. 170. First steel tuna clipper built in Hawaii. 75×16×7 ft. 2in. (22.9×4.9×2.2 m.)

THE OWNER'S VIEWPOINT

by

BASIL PARKES

BRITISH post-war trawlers are of three types: deep-sea, middle water and near water.

DEEP SEA TRAWLERS

The deep sea trawler must be a seaworthy vessel that can battle against North Atlantic gales throughout the year and stand up to the cold, hard weather experienced in the Arctic Circle during the winter months. There are two main reasons why the majority of British owners and skippers still prefer steamships: (1) reliability; (2) because fishing is often carried on in forty and fifty degrees of frost (–20 to –30° C.) and a liberal supply of hot water and steam must be available to prevent ice forming in dangerous quantities on the superstructure, fishing gear, etc. But a number of diesel vessels have been built and are fitted with an auxiliary steam boiler. Altogether, more than one hundred deep sea trawlers have been built in Great Britain since the World War II.

The vessels operate on grounds stretching from the Davis Straits, Grand Banks of Newfoundland and Greenland in the west, Iceland, Bear Island, Seahorse Island in the north and Kolveu Island and Nova Sembla in the east. The average distance to and from the grounds is 1,600 sea miles and fishing is continued throughout the year.

Dimensions of the vessels vary from 160 to 190 ft. (48.8 to 57.9 m.) between perpendiculars, 29 to 32 ft. (8.8 to 9.75 m.) beam and 14 to 17 ft. (4.3 to 5.2 m.) moulded depth. All owners would like a speed much in excess of that economically obtainable at present, but speed is somewhat dependent on the length of the vessel. All would like a speed of at least 12½ knots and most of the post-war deep-sea vessels do attain loaded speeds of 12½ to 13½ knots. Engine power, according to the size and displacement of ship, varies from 900 to 1,400 continuous b.h.p.

The British owner requires a winch hauling speed in excess of 250 ft. (76 m.) per minute and many of the post-war ships are heaving at more than this speed. It means installing a very powerful winch, developing from 200/350 h.p. at the winch shaft, depending on size of ship and trawling conditions.

Many types of fishrooms have been installed in the post-war ships but the majority are still built of wood. Very few owners are satisfied with wood and are looking for something better. It is essential to have a fully insulated fishroom and the fishroom ceilings and bulkheads should be constructed of a non-absorbent material which can be washed down and kept clean with the least effort. Interior partitions should be of the kind that can be totally dismantled to facilitate stowing, cleaning and discharging fish, and fishroom capacity varies from 12,000 to 20,000 cu. ft. (340 to 565 cu. m.).

The most important factors governing crew accommodation are: (1) to ensure the maximum safety for the crew and (2) to provide comfort. To accomplish this it has become customary to accommodate all the crew aft and amidships, with internal access to and from the wheel-house, engine room, messrooms, bathrooms, drying rooms, etc. Once the vessel leaves the fishing grounds she can be battened down and, particularly in bad weather, the crew have no need to go out on the exposed decks. It is also customary to instal inter-communication so that the skipper can speak to any part of the crew's quarters, engine room, etc. It also enables the wireless operator to switch on news and music when the crew are off duty, which gives a comforting link with home.

Navigational and fishing equipment requirements for the average deep sea trawler are: Two independent recording or indicating echo sounders, a long range wireless transmitter and receiver capable of transmitting messages round the world; a smaller wireless telephony transmitter and receiver for inter-ship communication and ship to shore telephone communication at short range; direction finders and radar.

Most British deep sea trawlers have two direction finders, one for navigation and one for locating other ships, and nearly all of them have radar, with ranges varying from 25 to 40 miles. This aid is invaluable when fishing in foggy weather amongst ice. It also assists the skippers with navigational and fishing activities, particularly in fog and snow, and in the winter months in the northern hemisphere when there is over 20 hours darkness in midwinter. Some vessels are now fitted with the Fischlupe instrument for detecting shoals of fish and some have electric logs and an electrical instrument to tell the skipper how much trawl warp he has got out when paying away, fishing or hauling in.

MIDDLE WATER TRAWLERS

The requirements are almost as varied in the middle water as the deep sea vessel, and they must be stalwart, seaworthy vessels built to 100 A.1 Lloyds Class. Middle water trawlers fish off the West coast of Scotland and Ireland, Faroe Isles, Norwegian Deeps and as far north as Iceland, and it is necessary for the vessels to have a sea range of at least 25 days.

Dimensions range from 115 to 130 ft. (35 to 39.6 m.) between perpendiculars, 25 to 27 ft. (7.6 to 8.2 m.) beam, and 12 ft. to 13 ft. 6 in. (3.66 to 4.12 m.) moulded

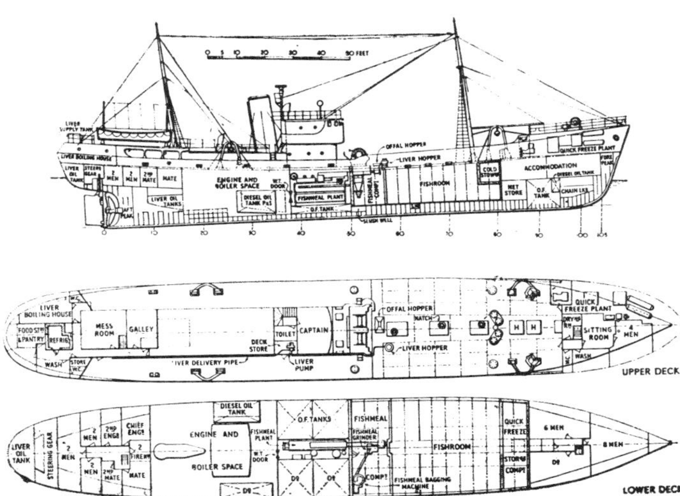

Fig. 171, 172, 173. Profile and plans of the deep sea trawler Olafur Johannesson, *built in 1950. Particulars: 198/183.5 × 30 × 17 ft. (60.4/55.9 × 9.1 × 5.2 m.); 1,200 h.p. triple expansion steam engine; 38 men crew*

depth. The speed required is not less than 10 knots, more if economically possible. If a diesel is installed it should be from 600 to 750 b.h.p. Most owners would prefer direct driven, direct reversing, slow-running diesel engines of 200 to 250 r.p.m., similar to those fitted in the post-war French fleet.

An electrically-driven winch is usually installed and should heave at 250 ft. (76 m.) per minute—up to 300 ft. (91 m.) if possible—and have a capacity of not less than 900 fm. (1,600 m.) on each drum.

Fishrooms should be as large as possible, well insulated and arranged so that a big proportion of the catch can be laid one fish deep on each shelf. All the interior of the fishroom should be removable so that it can be cleaned with the maximum efficiency, and all battens should be detachable, made from galvanized steel or aluminium. The most suitable material for the ceilings and bulkheads is a debatable point and, as in the case of the deep-sea trawler owners are looking for something better than wood.

Most British owners now consider it just as important to have all the crew accommodation aft and amidships as in the deep sea trawlers. This is considered a great step forward for the safety and comfort of crews. It is also necessary to have a messroom separate from the sleeping quarters (so that the men can use it for their recreation), hot and cold water, bathrooms, wash-houses and toilets all built aft and accessible without going on the exposed deck.

Regarding navigational equipment, the owners requirements are much the same as for the deep-sea vessel. The essential equipment includes wireless transmitter and receiver, two independent echo sounders, a recording type being preferable, and a direction finder. Radar is rapidly becoming standard equipment and most owners want inter-communication throughout the crew accommodation and on the main deck.

Unfortunately Great Britain has not built a large number of middle water trawlers since World War II, and the fleet is much depleted. For this reason, fewer improvements have been made as compared with those carried out in the deep water vessel, but the aid which the British Government is now giving for building near and middle-water vessels should stimulate construction. The majority of British owners now realize they must turn to diesel power both for economy and range. Owners want a vessel which can fish economically on the various near water grounds at different seasons and in depths varying from 200 to 300 fm. (366 to 550 m.); at other seasons they want a vessel to go to the Faroe Isles and Iceland fishing grounds. The diesel-powered vessel can do this so there is no need to have a large, uneconomical trawler.

NEAR WATER TRAWLERS

British near water vessels are of two types: the near water trawler and the near-water drifter trawler. Dealing with the former vessel first, it is essential to have a seaworthy, robust, 100 per cent. A.1 Lloyds Class craft and one that will attract a good class skipper, mate and engineer, and also encourage young men to become fishermen. She must also be so economical to operate that she can make a living on the depleted fishing grounds around the British Isles.

The fishing grounds comprise mainly of the North Sea, to a lesser degree the English Channel, the Bristol Channel, the Irish Sea and Morecambe Bay, and the shallow water around the coast of Scotland. Trips usually last 8 to 12 days but the vessels generally have a seagoing range of three to four weeks.

Dimensions range from 100 to 115 ft. (30.5 to 35 m.) length between perpendiculars, 23.5 to 22 ft. (7.2 to 6.7 m.) beam and 12 to 10 ft. (3.7 to 3.1 m.) moulded depth. The average speed is between 9 and 10 knots.

The power of the engine installed depends largely on whether it is direct driven or driven through a gear box.

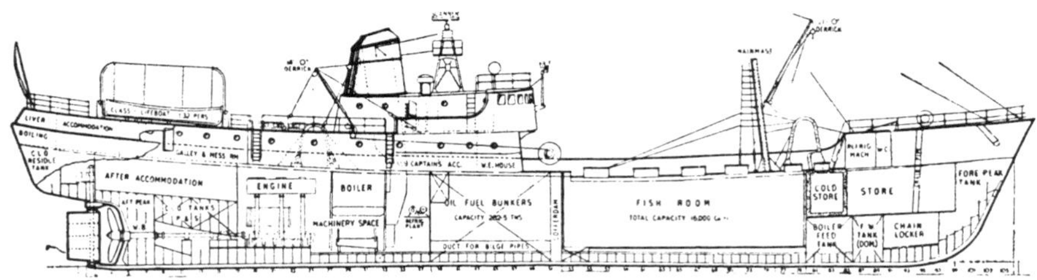

Fig. 174. Profile of the deep sea trawler Prince Charles, *built in 1950. Particulars: 178 × 30.5 × 16 ft. (54.3 × 9.3 × 4.9 m.); 712 gross tons; 16,000 cu. ft. (450 cu. m.) fish-room with chilling plant; experimental filleting-freezing plant with 800 cu. ft. (23 cu. m.) frozen storage; 32 men crew*

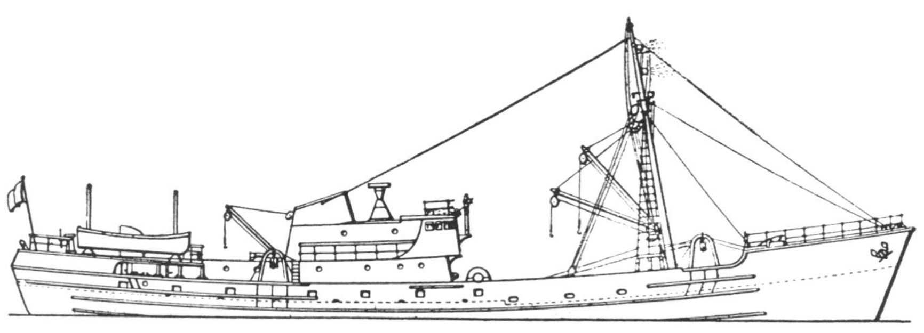

Fig. 175. Profile of the deep sea trawler Princess Elizabeth, *built in 1952. Particulars: 160.5 × 29 × 14.5 ft. (48.9 × 8.8 × 4.4 m.); 41 ft. (12.5 m.) fish-room with 12,725 cu. ft. (360 cu. m.) capacity; 1,200 h.p., 250 r.p.m., diesel; 300 h.p., 500 r.p.m., auxiliary diesel driving a 165 kw., 175/220 v., 940 amp. generator delivering power to the 200 h.p., 650/950 r.p.m., electric motor for the trawlwinch with 1,200 fm. (2,200 m.) capacity; 60 h.p. auxiliary diesel set; oil-fired boiler for liver processing and heating; 23 men crew*

Fig. 176. Stern view of the deep sea trawler Princess Elizabeth

Fig. 177. Middle water trawler St. Leonard, *built in 1953. Particulars: 115 × 25 × 11 ft. (35 × 7.6 × 3.35 m.); 275 tons gross; 600 h.p. diesel; 10.5 knots*

Direct driven it is 350/550 b.h.p. with 230 to 300 r.p.m. If a reduction gear is used the power varies from 300 b.h.p. to 450 b.h.p. and the main engine speed from 400 to 650 r.p.m.

The majority of winches are belt or chain driven from the main engine, but a few of the larger vessels have electric or hydraulic drive. The winches have a capacity of 300 to 600 fm. (550 to 1,100 m.) on each drum.

Fish rooms are equipped with shelf cod pounds and are fitted up in the same way as those of the middle-water trawlers.

The majority of vessels accommodate their crew aft

Fig. 178. Near water trawler Boston Victor, *built in 1953. Particulars: 102 × 22.1 × 9.8 ft. (31 × 6.7 × 3 m.); 188 gross tons; 435 h.p. diesel; 10 knots; 12 men crew*

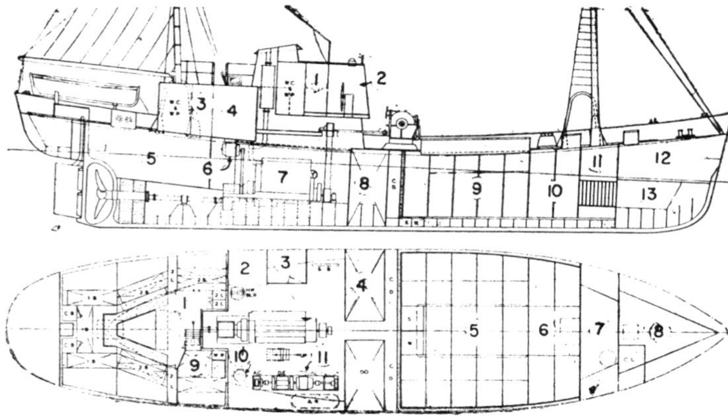

Fig. 179, 180. Profile and plan of drifter-trawler Underley Queen, *built in 1953. Particulars: 94.25/84 × 21.5 × 9.75 ft. (28.7/25.6 × 6.6 × 2.97 m.); 114 gross tons; 3,500 cu. ft. (100 cu. m.) fish-room; 9.8 knots.*

Profile: *1, radio room; 2, wheel house; 3, food store; 4, galley; 5, accommodation; 6, 3 kw. generator; 7, 230 h.p., 475 r.p.m., $2\frac{1}{2}$:1 reduction gear, diesel; 8, fuel; 9, fish; 10, nets; 11, ropes; 12, stores; 13, water ballast.*

Plan: *1, crew; 2, fresh water; 3, stores; 4, fuel; 5, fish; 6, nets; 7, ropes; 8, stores; 9, captain; 10, lubricating oil; 11, auxiliary diesel set.*

and amidships and new vessels are being built with mess rooms adjacent to the galley on the main deck. Hot and cold water shower baths, drying rooms, and other amenities are the same as in the larger vessels.

Navigational equipment includes a radio telephony transmitter and receiver, direction finder, and one echo sounder (usually the recording type). Many ships are now fitting a Decca-navigator, which is of great assistance in keeping their position on the fishing grounds and avoiding wrecks, etc., and, of course, in navigating in fog and thick weather.

Building of this class of vessel has been very slow in Great Britain, the strength of the fleet being about half of the pre-war figure. Intense overfishing of the near water grounds by all countries bordering the North Sea, high construction costs and Government control of fish prices have discouraged British owners from rebuilding the fleet.

The near water drifter trawler is for trawling in the North Sea and other near water grounds, and for drifting during the herring season, a requirement that limits the size of the vessel.

Dimensions are from 75 to 85 ft. (23 to 26 m.) length between perpendiculars, 20 to 22 ft. (6.1 to 6.7 m.) beam and 9.5 to 10.5 ft. (2.9 to 3.2 m.) moulded depth. A speed of $9\frac{1}{2}$ to 10 knots is required so that herring can be landed fresh.

If the vessels were built for drifting only, the h.p. could probably be reduced, but as they are built for two types of fishing, it has to be kept reasonably high, from 200 to 300 b.h.p.

The winch is driven from the main engine in the same manner as the 100 ft. (30.5 m.) trawler, and fishrooms are fitted with a large type hatch suitable for drifting. They are constructed in such a way that a large proportion of the hatch can be battened down permanently when trawling. Some vessels that box herrings are also fitted with a large after well for packing herrings caught overnight, the main fishroom being used for the catch to be landed the same day it is caught.

Crew accommodation and navigational equipment are the same as in the near-water trawlers.

THE DEVELOPMENT OF THE NEW ENGLAND TRAWLER

by

DWIGHT S. SIMPSON

THE yawl *Resolute* was the first American beam trawler, built in 1891 by Arthur D. Story, of Essex, Mass., from designs by the U.S. Fish Commission and she was unmistakably English in her origin. Her dimensions were: 85 ft. (25.9 m.) length, 22 ft. (6.7 m.) beam, and 9.6 ft. (2.93 m.) depth, with a fish-hold capacity of about 2,500 cu. ft. (71 cu. m.). She was operated by an English crew.

Many other attempts at trawling in New England waters were made in the following years—with steam and sail, with and without English crews—but it was 1904 before the present industry started. That year (Symonds and Trowbridge 1947) Admiral Bowles got the support of a group of Boston men, imported a set of plans from England and built the steam trawler *Spray*, fig. 181. The vessel was so successful that others were quickly built and soon a fleet was operating out of Boston. World War I depleted the fleet but it was rapidly rebuilt and there were 43 vessels of the *Spray* type on the register in 1926. Although similar to the *Spray*, the vessels were larger, ranging from 135 to 140 ft. (41 to 43 m.) in length and 24 to 25 ft. (7.3 to 7.6 m.) in beam. They were fitted with Scotch boilers and triple expansion steam engines of 550 to 600 i.h.p.

At the same time, sailing vessels which had taken to auxiliary power—first the petrol engine, then the diesel—also took up trawling and by 1920 a fleet of converted schooners with top masts and bowsprits removed, and carrying only a riding sail, were in operation. The next development was to build vessels such as the *Pioneer* (1918), 129.1 ft. (39.4 m.) registered length fitted with a Fairbanks Morse 450 h.p. engine, and the *Mariner* (1919) which was fitted with a diesel—electric drive. No such unit has ever again been installed in a New England vessel. H. Munro Smith (1924) has described this trawler as follows:

" L.W.L. 140 ft. (42.7 m.); beam 24 ft. 3 in. (7.40 m.); mean draft 11 ft. 9 in. (3.57 m.), displacement, 500 tons; cruising radius, 6,000 miles, at 10 knots. The propelling machinery consists of two eight-cylinder four-cycle Nelseco diesel engines, rated at 240 b.h.p. at 350 r.p.m. direct coupled to two generators, each being rated 165 kw. at 125 volts. For propulsion, these generators are run in series and supply power to the main motor,

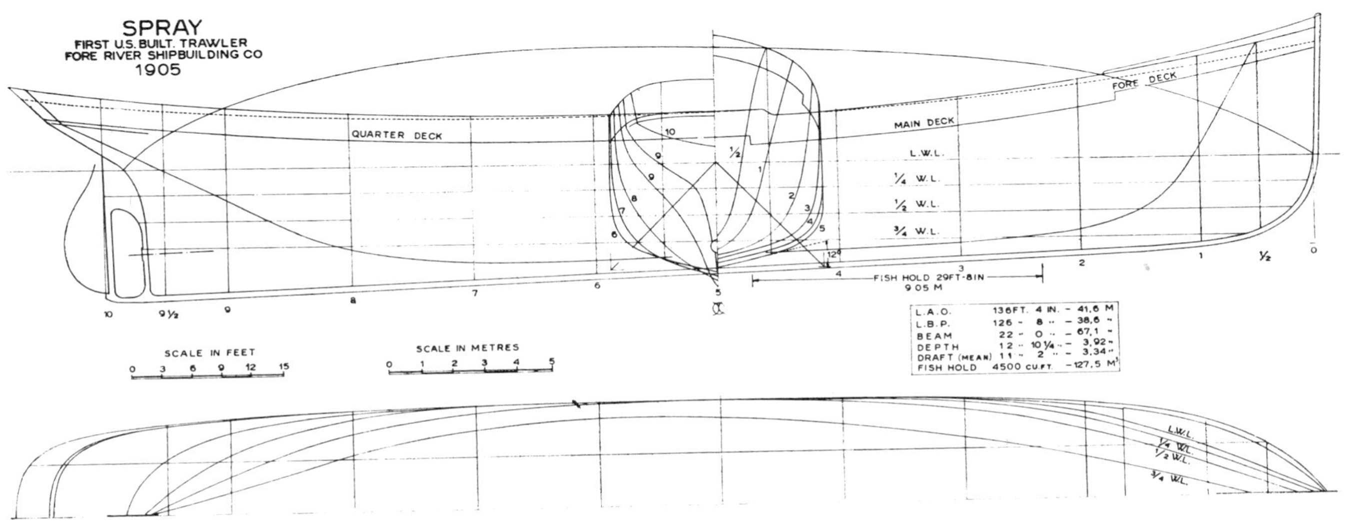

Fig. 181

which is rated at 400 b.h.p. at 200 r.p.m. with voltage at 250. Connections and means of adjustment are provided so that any range of voltage and current can be supplied to the motor and the load be divided between the generating sets as desired, or one generator can be shut down. The winch motor is 125 volts as is that of all the auxiliary and lighting equipment and the power can be taken from either main generator. Fuel consumption was just over 30 gal. (25 imp. gal., 114 l.) per hour and full speed just over 10 knots. Full ahead to full astern in 15 seconds ".

Another trawler, called *Fabia*, was also built in 1919. She was 131.6 ft. (40 m.) in length, 25 ft. (7.62 m.) in beam and 13.4 ft. (4.8 m.) in depth with a capacity of about 6,000 cu. ft. (170 cu. m.). Her main engine was a Nelseco, six-cylinder, 13 in. (330 mm.) diameter by 18 in. (457 mm.) stroke, single-acting, four cycle, air injection, developing 360 h.p. at 240 r.p.m. weighing 57,000 lb. (25,800 kg.), and giving a speed of 9 knots in a smooth sea. Later, a 520 h.p. engine with mechanical, solid injection, was put into her and, although free running speed was increased by only half a knot the trawling speed is said to have improved considerably. Her winch engine was a 100 h.p. diesel with two cylinders 14×17 in. (356×432 mm.), weighing 24,000 lb. (10,900 kg.) and her auxiliary was a 15 h.p. horizontal single-cylinder, double flywheel engine weighing 4,000 lb. (1,814 kg.). These rugged, heavy engines were still in service when the trawler was converted to a freighter a few years ago. It is interesting to note that the original engine installed in the *Fabia*, and the auxiliaries, developed a total of 475 h.p. with weight of 85,000 lb. (38,550 kg.) whereas a new trawler, the *Judith Lee Rose*, fig. 190, will carry machinery developing 795 h.p. for a weight of only 46,000 lb. (22,860 kg.), while another vessel now (1953) in the design stage will have an even better ratio: 1,190 h.p. for a total weight of 49,000 lb. (22,230 kg.) of machinery.

This development of diesel trawlers brought about the eclipse of the steam vessel and by 1930 there were only 30 in operation, while by 1939 the figure had dropped to 23 in the register compared with 88 diesel trawlers of 100 tons or more.

While the World War II again decimated the fleet, the U.S. Navy built about three hundred 100-ft. (30.5 m.) vessels with trawler hulls and some of them have now been converted to fishing. Many of the larger trawlers were bought by the U.S. Army for use in Germany immediately after the war, a fact which added to a boom in construction from 1945 to 1947, but since then no new large trawler, and very few small ones, have been built because the price of fish fell, the wages for crews increased, cost of building vessels soared and the nearer fishing grounds were depleted. According to the 1952 register there are no steam and only 120 diesel trawlers of over 100 gross tons. These vessels, together with a multitude of small craft, make up the present new fleet.

Although the schooners have disappeared, many small trawlers have been built of the " schooner type ", with two masts and a riding sail. This is a hangover which the author deems useless.

The earlier diesel vessels were of two distinct types. The steel hulls, designed and built primarily by Fore River and the Bath Iron Works of Bath, Maine, and the generally smaller, wooden vessels modelled and built by James, Story, Rice and the Morses (three brothers, three different yards). The steel vessels differed little from the previous steam trawlers, as can be seen by the lines of *Spray* (1905, fig. 181) and *Notre Dame* (1929, fig. 182).

When the building of smaller diesel vessels started the boatyards had no precedents for them but they

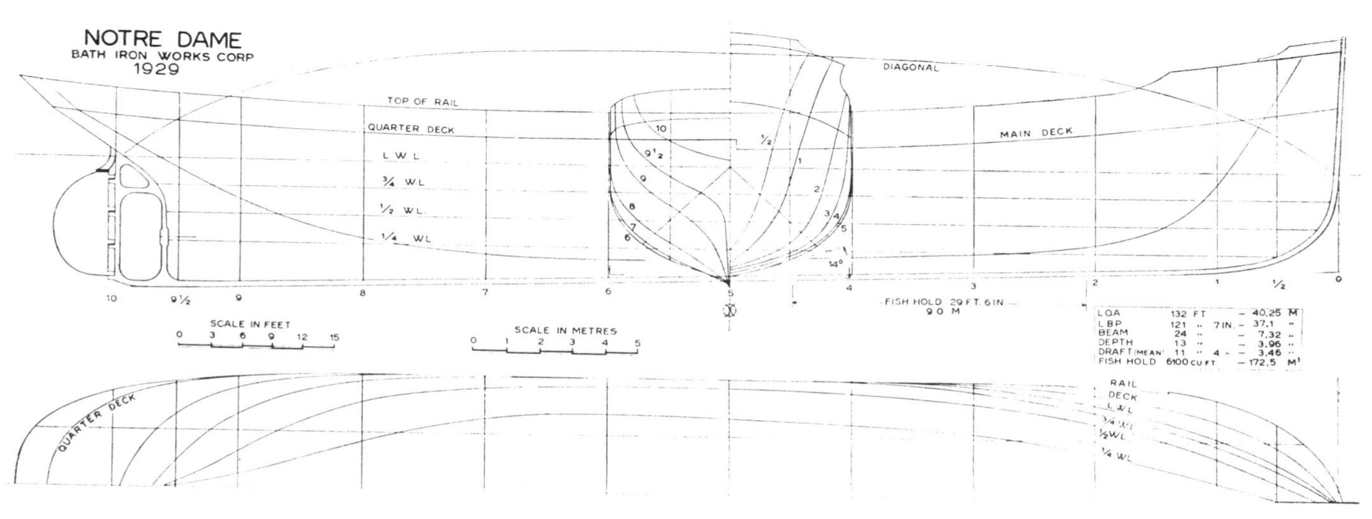

Fig. 182

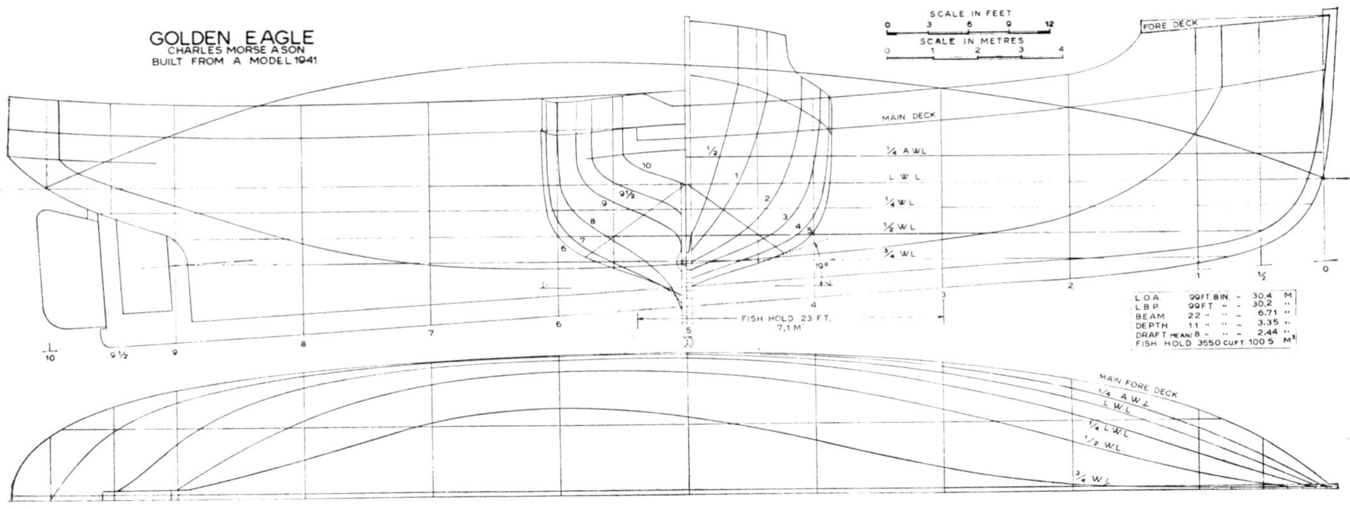

Fig. 183

evolved reasonably efficient hulls, often constructing several vessels from the same model with a change here and there. The *Golden Eagle*, fig. 183, was one of the last and best, although she shows the heavy after quarters and fine fore end that brought the vessels by the head and made them hard to manage in a loaded condition.

About 1935 attempts were made to avoid extreme change of trim; accomplished partly by a change in the hull form and partly by redistribution of weights, notably the shifting of 20/25 per cent. of the fuel load forward of the fish-hold. In recent years the use of light weight short engines has allowed the fish-hold to be placed nearer the centre of the vessels. An interesting comparison between the old and the new is that of the *Golden Eagle* and the *Judith Lee Rose*. The *Golden Eagle* has a change of trim of more than 30 in. (76 cm.) between going out and coming in loaded, with 25 per cent. of the fuel left. In exactly similar circumstances the *Judith Lee Rose* has a calculated change of only 12 in. (30.5 cm.).

Conditions in 1945 called for a different type of boat to that built before the war. Building and running costs had risen and the price of fish, which had been high, was falling off but they were plentiful and nearby. To save construction costs smaller vessels were needed and as capacity had to be maintained the trend was to fuller hulls. At the same time skippers had found that the faster they dragged their nets, the more fish they caught, so power plants were considerably increased. A very good example of a vessel built to meet these requirements is the Alden-designed *Bay*, fig. 184 and 185, which is a highly successful trawler in the Boston fleet.

When the nearer fishing grounds became depleted trawlers had to make voyages up to 1,200 miles from port instead of 300, and as the majority of the fleet can

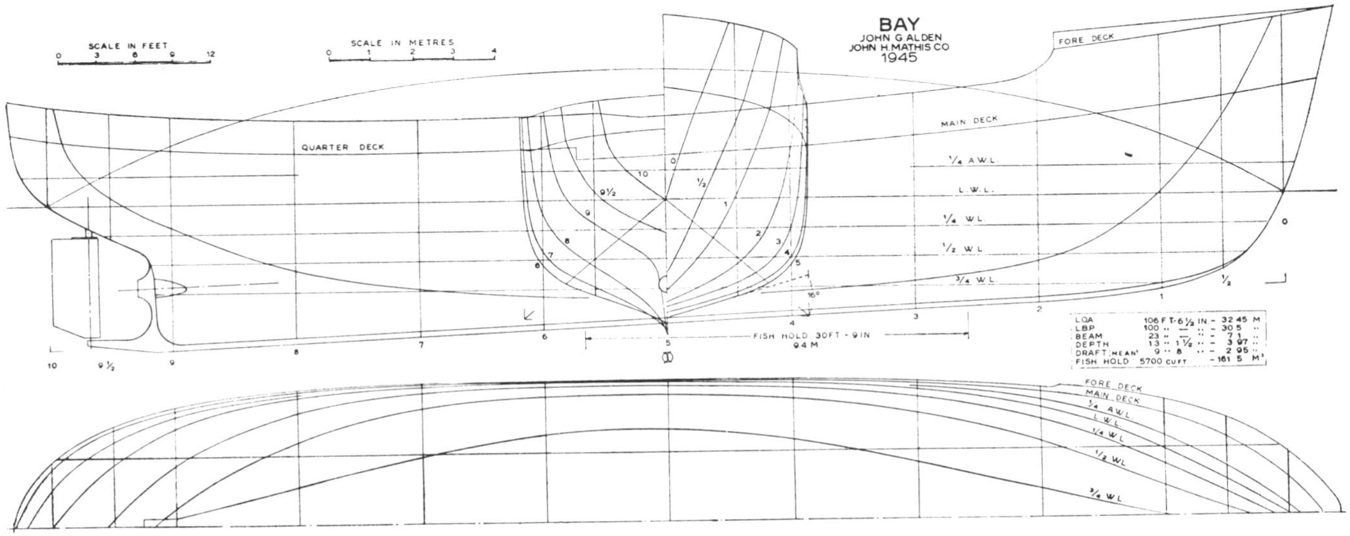

Fig.184

Fig. 185

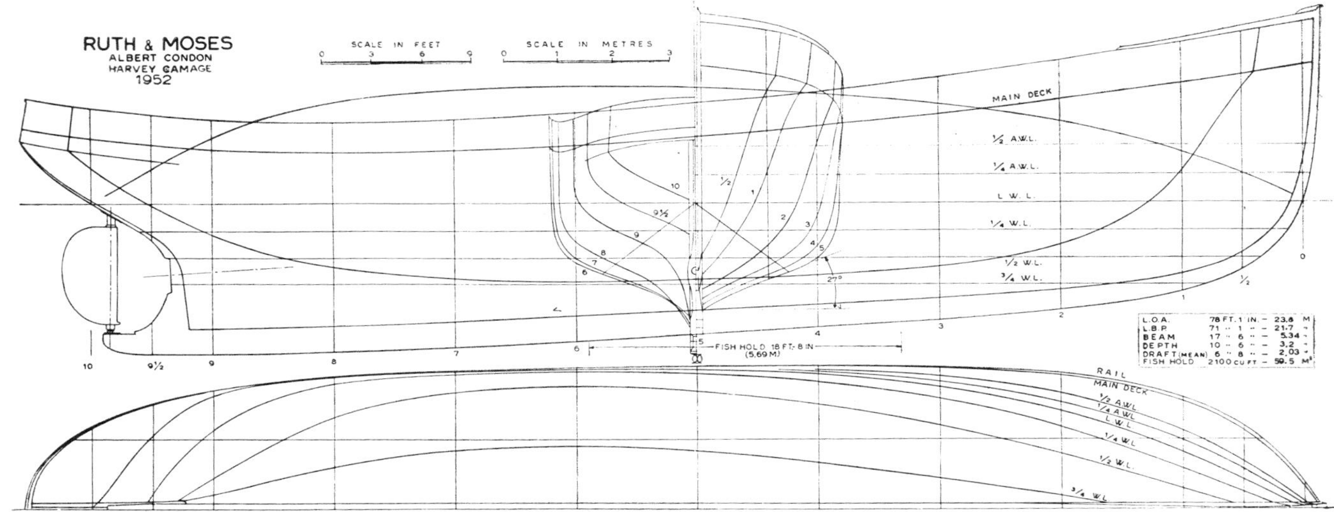

Fig. 186

Fig. 187

only make 9 to 10 knots, the period of actual fishing time on each trip has been drastically curtailed. In addition, the Labour Union enforced short-time trips and three day lay-ups between trips so that where, a few years ago, a vessel might make 30 trips a year, to-day 22 trips is a record and 18 or 20 about the average. Present New England trawlers spend three times as many days free-running as they do fishing, therefore speed is in demand and the aim of designers and owners is now to produce faster vessels which are also able to fish in winter and carry large catches of redfish.

In an attempt to meet the changed conditions many of the older trawlers have been re-engined to give them greater sailing and trawling speed, but while some have shown considerable improvement, others have been disappointing. In this respect, the same is true of the new vessels that have been built and which seem greatly over-powered. Yet these unsatisfactory results might have been anticipated by an analysis of the hull forms, as a study of fig. 183, 184 and 186 will show. All have high displacement-length ratio, high prismatic co-efficients and the speed-length ratio is more than 1.0, i.e. on the hump of the speed-power curve. The hulls are unsuitable for high speed as the high horse power per ton ratio and speed-length ratio of *Ruth and Moses*, fig. 186 and 187 indicate. But the type of shape is forced on them by need to carry the heavy slow speed diesel as far aft as possible, and the understandable desire to limit the length of the vessel. The small midship section is an accompanying evil if the displacement is to be in reason.

In putting new engines into old ships it has often been necessary to use super-charging, reduction gears, or perhaps both, because of space limitations. But now when these new engines have been accepted there should be a change in hull form to one more suitable for the speeds required. This indicates a trend towards larger and even faster vessels and the development of 140 ft. (42.7 m.) 15 knot trawlers is already under way, and a further change in the form—possibly knuckled after-sections and transoms—will be seen. Experiment will be needed to see what happens to sea-comfort in such new vessels, which will be powered by the lighter and faster geared engines.

Another possible development is the use of the controllable pitch propellers and there are now several manufacturers in the United States working on the problem, in an effort to reduce the initial cost, whereas, for many years, only one manufacturer had been engaged in the work.

It is interesting to note that wood is being used for all the new fishing boats being built or contemplated. The reason for this is that the cost of the wooden vessel is about 75 per cent. of that of steel, and in any case, a properly designed and built wooden vessel is every bit as good as one of steel.

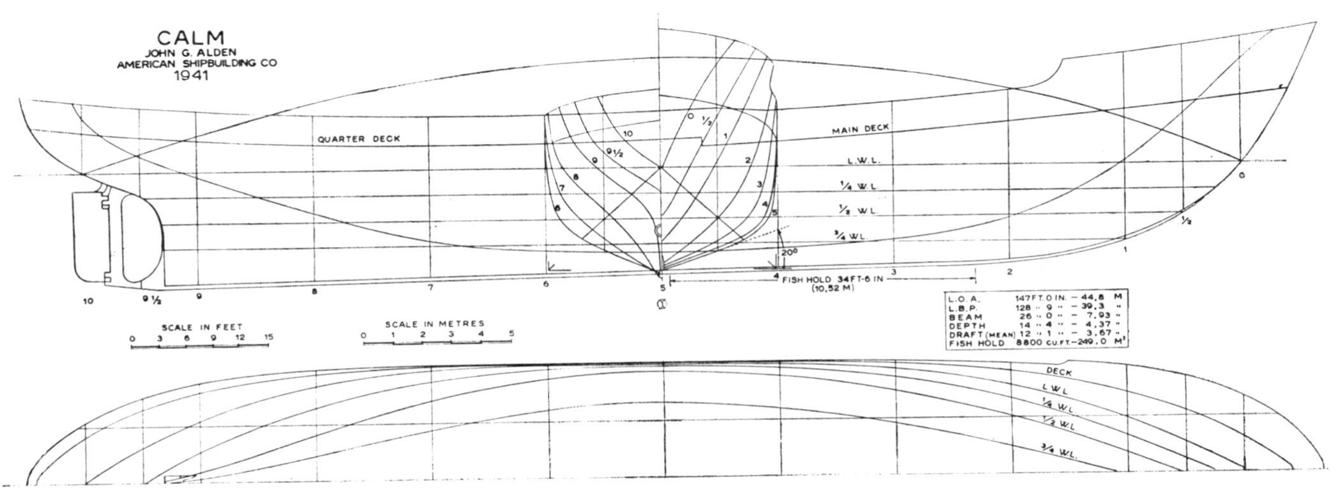

Fig. 188

The Plans

The vessels shown in the plans are in the " ready for sea " condition—fuel, water, ice, stores, equipment and crew aboard. This is the only operating condition that can be approximately calculated and checked on the vessel, although in winter and summer the condition varies. In summer a vessel may carry from 20 to 50 tons of ice and only 15 to 35 tons in winter. The plans deal with summer loading. Vessels returning from the fishing grounds are often close to the " ready for sea " condition because consumption of fuel, water, ice and stores, tends to balance the weight of fish caught. This might therefore be called an average condition. Waterlines are spaced one quarter of the draft to the rabbet in wooden vessels and to the intersection of shell and keel in steel vessels. The buttock is drawn at one-quarter beam of the waterline at Station 5 and the diagonal is that of the rectangle of the draft and one-half the water-line beam at Station 5.

Spray, fig. 181, was the first American steam trawler

and was built by the Fore River Shipbuilding Co., Quincy, Mass., from English plans. The lines, however, are unquestionably drawn at Fore River. It is a close copy of English vessels of the period. Her lines are typical and could almost serve for all trawlers built during the following 25 years.

The dotted sheer line is reproduced from the original which bears this note in script. " The dotted line shows the usual sheer of a trawler. The full line shows the actual but excessive sheer required by the owners of this particular trawler." This " excessive " sheer is retained in most trawlers to-day, so that the development of the New England trawler started with the first one, although of English origin.

The trawler remained in service for 35 years and had the distinction of being the first steam trawler to be converted to diesel power. She now forms a bulkhead at the head of a slip of the United Shipbuilding Corporation, East Boston, Mass.

Notre Dame, fig. 182, was designed and built by the Bath Iron Works of Bath, Maine, in 1929. The profile is almost identical with that of the *Spray*, except for a slight increase in the L.W.L. and the section areas at Stations 7, 8 and 9, fig. 191 and 192. The fullness in the after quarters seems due to the attempt to get the diesel as near the stern as possible. It is evident in nearly all trawlers and especially in smaller vessels built without plans. They need a lot of ballast under the fish-hold to get them into reasonable " ready for sea " trim, and, when loaded, they naturally go heavily by the head. This, in turn, accounts for the excessive drag given to the keel. The *Notre Dame* had rather full waterlines forward but not so extreme as some of the later steam and contemporary diesel trawlers. Indeed, some of the steamers were so easily stopped in a sea, making no headway, that they had difficulty in getting crews. The *Notre Dame* was powered with a 500 h.p. Cooper Bessemer engine, 230 r.p.m., giving her a speed of 10.1 knots which, considering the horse power—displacement ratio and the high displacement-length ratio, is excellent.

Fig. 189

The *Golden Eagle*, fig. 183, is one of the last and best of the whittled boats and was built in 1941 by Charles Morse and Son, Thomaston, Maine. She has a 425 h.p. Cooper Bessemer engine which gives her a speed of about 10 knots. Originally she carried about 25,000 lb. (11,340 kg.) cement ballast, but this has been replaced by light weight concrete, an estimated saving of nearly 10,000 lb. (4,540 kg.). While she has considerable change in trim, the heavy drag of the keel makes her handle well under all conditions and, according to her captain, her sea performance has greatly improved since the new ballast was put in.

The builders who used models instead of drawings usually built several boats from the same model, making changes as experience would indicate, lifting bow or stern, adding a few frames, etc., until the actual boat was so far from the model that it was necessary to make a new model and start over again.

During the 1930's, consulting naval architects began to come into the field and even the small boat yards bought from them designs at reasonable prices. However boats of various sizes are usually built from the same set of plans. A 116-footer (35.4 m.) has just been launched, built from a naval architect's design drawn for a 98-foot (29.9 m.) vessel. This practice accounts for a large number of fairly narrow vessels in the about 100 ft. (30.5 m.) size.

Calm, fig. 188, was designed in 1935 by John G. Alden, Boston, Mass. under the supervision of the author and in collaboration with Maierform of America Inc., and she is another example from this period. The type was one of the first of the few trawlers to be tank tested before construction started, and three vessels were completed in 1936 and two in 1937, by the Fore River plant of the Bethlehem Shipbuilding Corporation. Although considered generally satisfactory, the original vessels were too lean aft and were uncomfortable in a following sea. In 1938 four more vessels of the *Calm*-class were built with about 20 tons displacement added in the last quarter length, and they were lighter owing to the use of welded construction throughout, see fig. 189 the *Annapolis*. The trim change of this class was calculated to about 12 in. (0.3 m.) but was never checked in practice.

Bay, fig. 184 and 185, was designed in 1945 by John G. Alden under the supervision of the author and is an example of the attempt to meet prevailing conditions when owners demanded a larger carrying capacity in a small vessel. As she was built to fish nearby grounds, speed was not too important but, due to the demand for trawling speed she is somewhat overpowered for the hull form. She has a 650 b.h.p. Atlas supercharged

TABLE XXVIII

PRINCIPAL DATA FOR TRAWLERS ILLUSTRATED

	Name		*Spray*	*Notre Dame*	*Golden Eagle*	*Calm*	*Bay*	*Judith Lee Rose*	*Ruth and Moses*
	Year Built		1905	1929	1941	1938	1945	1953	1952
	L.W.L. actual	ft.	126.0	122.5	96.5	131.0	99.0	109.75	72.8
		m.	38.45	37.35	29.45	39.95	30.2	33.5	22.2
Calculated for 100 ft. (30.5 m.) L.W.L.	Beam, L.W.L.	ft.	17.45	19.50	22.58	19.80	22.75	21.30	23.81
		m.	5.33	5.95	6.88	6.04	6.94	6.5	7.26
	Draft to rabbet	ft.	8.85	9.10	8.18	9.18	9.70	8.66	9.86
		m.	2.7	2.775	2.495	2.8	2.96	2.64	3.01
	△ tons, S.W.		263.5	291.5	247.2	248.0	320.5	261.0	294.0
	L.C.B. per cent. L.W.L.		.517	.532	.526	.495	.5095	.5225	.539
	Block coefficient		.596*	.569*	.479*	.477	.5075	.508	.428*
	Prismatic coefficient		.683*	.694*	.588*	.605	.647	.610	.645*
	Prismatic fwd. coefficient		.625	.641	.498	.612	.620	.546	.546
	Prismatic aft coefficient		.741	.747	.676	.599	.674	.674	.741
	Area L.W.L.	sq. ft.	1416	1643	1677	1472	1813	1634	1855
		sq. m.	131.3	152.6	155.8	136.9	168.4	152.0	172.4
	Water plane C.G., per cent. L.W.L.		.523	.531	.534	.509	.518	.525	.540
	Water plane coefficient		.811	.839	.746	.764	.787	.767	.778
	Water plane fwd. coefficient		.763	.776	.676	.746	.745	.692	.693
	Water plane aft coefficient		.859	.903	.817	.788	.829	.842	.862
	Area ⊕	sq. ft.	135.4*	147.2*	144.2*	139.6*	173.5	153.5	158.0*
		sq. m.	12.6	13.68	13.4	12.95	16.12	14.26	14.68
	Midship coefficient		.842	.800	.745	.788	.784	.834	.672
	L/B		5.72	5.10	4.43	5.04	4.37	4.69	4.20
	L/D		11.30	10.94	12.21	10.92	10.30	11.55	10.1
	B/D		1.974	2.145	2.760	2.180	2.355	2.220	2.416
	Drag/L.W.L.		.035	.025	.064	.021	.044	.049	.053
	Half angle entrance		19°	30°	23°	26°	31°	23°	26°
	Angle Deadrise		13°	16°	19°	20°	16°	13°	22°
	Rudder/Lateral plate		.029	.035	.0535	.0303†	.0345†	.0347	.035
	Rudder/⊕		.192*	.222*	.330	.177†	.167†	.206	.212*
	$V/\sqrt{L}$		.94	.912	1.02	1.065	1.045	1.12	1.085
	$V/\sqrt{gL}$		.279	.271	.303	.3165	.311	.333	.323
	B.H.P./Tons △		.855	.933	1.32	1.17	1.77	1.78	2.13
	Volume fish-hold/△		.244	.327	.468	.451	.523	.589	. 33

* Based on station No. 6 † Double plate rudders

Tons of 2,240 lb. = 1.016 metric ton

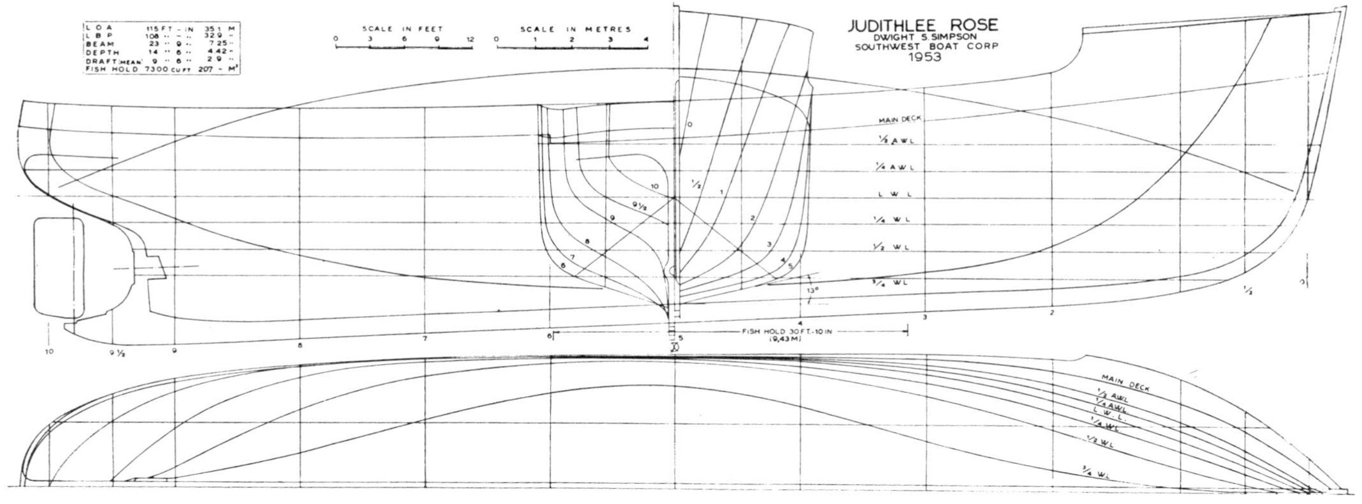

Fig. 190

diesel and her displacement-length and h.p.-ton ratios are high, while the speed-length ratio of 1.045 is probably as good as can be expected. Her ratio of fish-hold to displacement is the highest attained at that time, Table XXVIII. Four vessels of this type were built by John H. Mathis and Co., of Camden, New Jersey and, later, two others were constructed with 10 ft. (3 m.) middle-body added. In these, 700 h.p. Fairbanks Morse engines were installed but speed was very little increased. All six vessels have a reputation for seakindliness and comfort.

Judith Lee Rose, fig. 190, designed by the author is under construction (1953) by the Southwest Boat Corporation, Southwest Harbour, Maine, and is the first vessel designed expressly to meet current conditions, which are:

(*a*) Round trips to and from the best fishing grounds are from 2,000 to 2,500 miles free " steaming ", indicating a demand for speed;

(*b*) Redfish are, and apparently will be for some time, the principal source of income. They seem to school in large quantities and many vessels bring home full cargoes, trip after trip. Hence large capacity is called for ;

(*c*) To be profitable a vessel must fish all the year round and get in as many trips as possible. As winter fishing yields the best profit, a vessel should

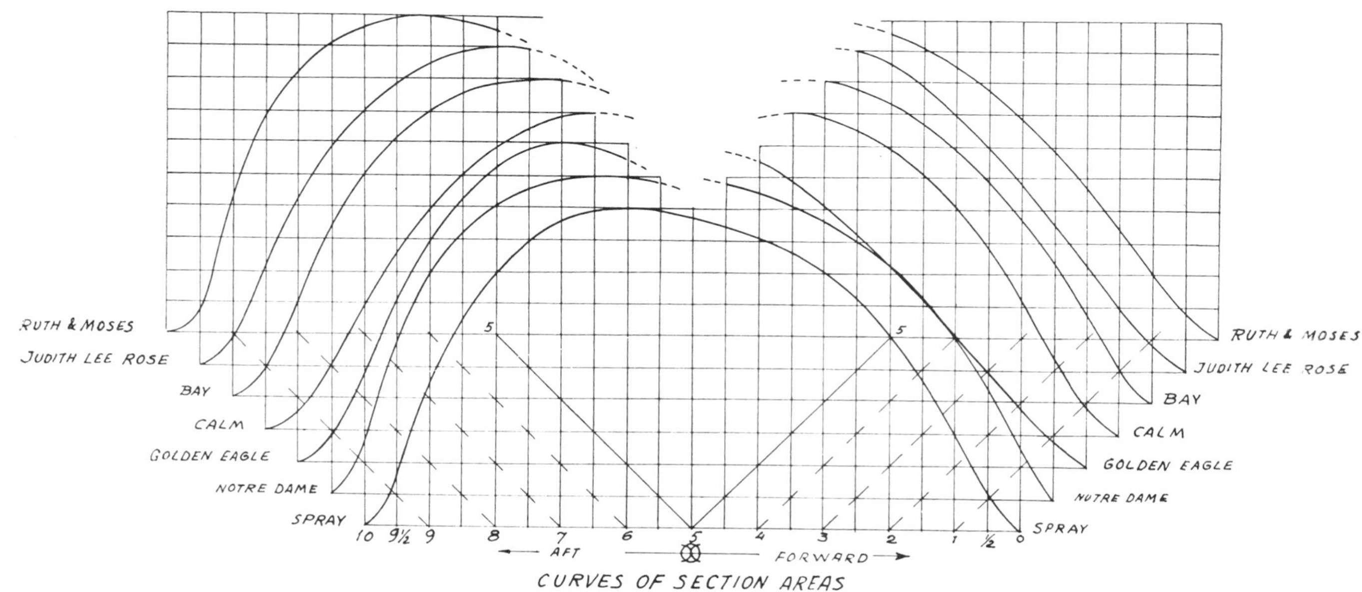

Fig. 191

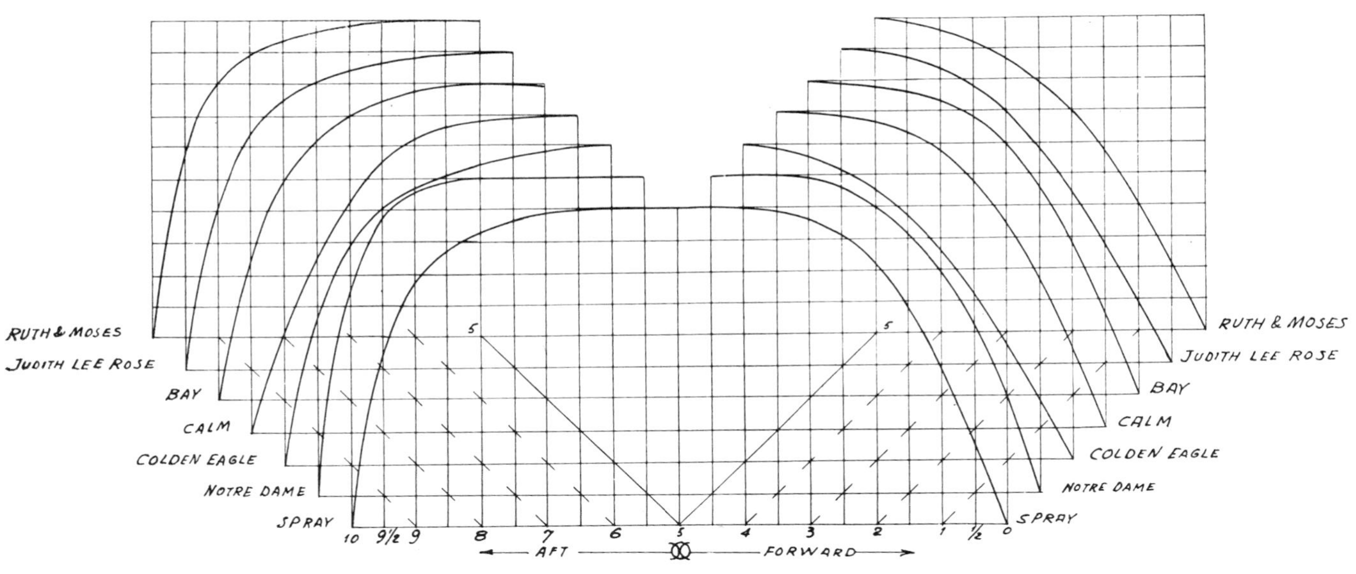

LOAD WATER LINES

Fig. 192

be designed to go to the banks in any weather, to stay there through winter storms and to fish in as rough water as the gear will stand. A theoretical investigation of conditions at sea and on the banks, together with the experience of many captains, indicate that such a vessel cannot be less than about 110 ft. (33.5 m.) long but 130 or 140 ft. (39.6 or 42.7 m.) would be better;

(*d*) Hull and equipment should be of the lowest cost without sacrifice of quality, ability and endurance.

The wooden vessel, whose qualities have already been discussed, is definitely indicated.

While these considerations were kept in mind in designing the *Judith Lee Rose*, the length was fixed, the hull contract let, and the engine selected, before the designer was called in, therefore the vessel does not exactly meet all requirements. The length is about right and the engine is close to the desired weight and minimum length, but to meet the demand for speed the

Table XXIX

OFFSETS OF SECTION AREAS, L.W.L. AND DIAGONALS

Stations	Spray			Notre Dame			Golden Eagle			Calm			Bay			Judith L. Rose			Ruth and Moses			Stations
	Sec. A.	*L.W.L.*	*Diag.*	*Sec. A.*	*L.W.L.*	*Diag.*	*Sec. A.*	*L.W.L.*	*Diag.*	*Sec. A.*	*L.W.L.*	*Diag.*	*Sec. A.*	*L.W.L.*	*Diag.*	*Sec. A.*	*L.W.L.*	*Diag.*	*Sec. A.*	*L.W.L.*	*Diag.*	
0	0	0	0	0	0	0	0	0	0	0	0	0	0	0	0	0	0	0	0	0	0	0
½	.120	.243	.268	.161	.309	.344	.074	.183	.203	.084	.254	.234	.103	.255	.258	.075	.197	.199	.090	.214	.250	½
1	.302	.477	.476	.344	.530	.555	.177	.368	.364	.230	.462	.414	.274	.483	.460	.197	.381	.372	.216	.408	.424	1
2	.615	.828	.758	.653	.822	.817	.405	.668	.606	.582	.784	.693	.589	.799	.730	.461	.686	.623	.482	.696	.684	2
3	.810	.964	.892	.800	.967	.928	.625	.855	.764	.813	.935	.870	.794	.926	.873	.695	.886	.805	.701	.868	.834	3
4	.908	1.00	.966	.904	1.00	.971	.810	.964	.885	.948	.993	.969	.933	.978	.963	.885	.974	.930	.850	.960	.921	4
5	.968	1.00	1.00	.946	1.00	.998	.941	1.00	.970	1.00	1.00	1.00	1.00	.992	1.00	1.00	1.00	1.00	.951	.992	.984	5
6	1.00	1.00	1.00	1.00	1.00	1.00	1.00	.983	1.00	.948	.993	.941	.988	1.00	.986	.988	.944	.969	1.00	1.00	1.00	6
7	.966	.982	.955	.955	1.00	.972	.916	.946	.928	.807	.966	.819	.895	.972	.906	.884	.976	.875	.956	.982	.972	7
8	.801	.925	.805	.822	.993	.861	.696	.863	.750	.543	.850	.620	.686	.895	.751	.685	.914	.712	.838	.942	.855	8
9	.449	.768	.532	.481	.872	.584	.277	.684	.441	.204	.550	.351	.303	.693	.460	.322	.731	.428	.461	.808	.563	9
9½	.170	.532	.287	.173	.634	.332	.084	.473	.238	.052	.319	.183	.084	.473	.258	.081	.505	.229	.107	.512	.272	9½
10	0	0	0	0	0	0	0	0	0	0	0	0	0	0	0	0	0	0	0	0	0	10

midship section has been made larger than customary, the midship section coefficient .834 being higher than that of any typical vessel. This permits finer ends without losing the desired displacement, which is rather high, and leaves the prismatic coefficient close to good practice. In spite of the very high capacity ratio, the fish-hold is nearer the centre of the ship than in any of the others which, with the spreading of tanks forward and aft, means little change in trim. The lines are shown by fig. 190. The speed/length ratio given in Table XXVIII is developed from the tank tests conducted under the auspices of the F.A.O. Fisheries Division, using a 60 per cent. propulsive efficiency.

Ruth and Moses, fig. 186 and 187 was designed by Albert Condon of Thomaston, Maine, one of the most prolific designers of vessels between 60 and 100 ft. (18.3 and 30.5 m.). She was built by Harvey Gamage of South Bristol, Maine, in 1952. This latest of Condon's designs still shows the influence of the whittled model and she is heavily over-powered, having more than 2 h.p. per ton of displacement. It seems that 9 knots is the maximum speed to be obtained with this hull, and on the 100 ft. standard she is the widest and deepest vessel shown and has a small midship section, which probably accounts for her resistance.

It is pointed out that proportional reduction of vessels discussed to any standard length is useful for comparison but is deceiving. The figures should not be used. for the design of a vessel very different in length from the original, the reason being that plotting beam, draft, etc., against the waterline length, never gives a mean curve that reaches zero, that is to say the curve is not in the form of $X=aY$ but in the form of $X=aY \pm C$, and this constant variation should be considered when using the figures in Table XXVIII, which are proportional only.

The diagrams, fig. 191 and 192, are drawn to distorted scales in order to give a ready and easy visual comparison of the shape of load water lines and section areas. Table XXIX gives the offsets for L.W.L., section areas and bilge diagonal, based on the greatest dimension as unity.

JAPANESE DRAG NET BOATS

by

ATSUSHI TAKAGI

LARGE TRAWLERS

THE first Japanese trawler was built around 1900. In 1929, the first diesel trawler, the *Kushiro Maru*, was built and it was quickly recognized as a very economical boat. Motors for driving the trawl winches and a mechanical refrigeration of the catch were also introduced at the same time. Thus, Japanese trawlers were very efficient before the war and some vessels exceeded 1,000 gross tons.

Since the war, with limited grounds for fishing, most new trawlers are only about 300 gross tons. However, as a result of the peace treaty, the restrictions on fishing have been abolished and larger trawlers will be built in the future. Actually three large trawlers of 750 gross tons and 1,200 h.p. were in 1953 under construction.

Yamato Maru is a typical steel diesel otter trawler. It has the following principal data: Length, over-all, 148 ft. (45.20 m.); length between perpendiculars, 134 ft. (41.00 m.); beam, 23.5 ft. (7.20 m.); depth 13.1 ft. (4.00 m.); 291.42 gross tons, 127.43 net tons; freezing rooms, 830 cu. ft. (23.6 cu. m.); fish holds, 5,000 cu. ft. (141.34 cu. m.); fresh water tanks, 10,000 imp. gal. (12,000 gal., 45.71 cu. m.); fuel oil tanks, 14,300 imp. gal. (17,200 gal., 65.08 cu. m.); officers and crew, 26; main engine, four-cycle single-acting diesel, six cylinders with 13.8 in. (350 mm.) diameter and 20.5 in. (520 mm.) stroke; 550 b.h.p. at 300 r.p.m.; four-bladed propeller with 77 in. (1,950 mm.) diameter and 51 in. (1,290 mm.) pitch; 120 b.h.p. auxiliary diesels driving two 70 kw. 225 v., D.C. generators; 5 kw. generator driven by main engine; 70 h.p. trawl winch driven by 220 v. motor; 15 h.p. windlass, 220 v. motor; NH_3 refrigerating compressor, two cylinder, 400 r.p.m., driven by a 230 h.p. motor and having a capacity of 15.5 RT (47,000 kcal./hr.); 250 w. and 50 w. wireless.

In the design of this vessel special consideration has been given to the installation of an air freezing system and there are also freezing shelves of ammonia coils.

Fig. 193 shows the lines, fig. 194 the general arrangement, fig. 195 the installation of the fishing gear and fig. 196 the *Yamato Maru* under speed. Table XXX gives particulars of drafts and stability; Table XXXI results from speed trials.

TABLE XXX

Draft and Stability

	Load			
	Light		*Full*	
Displacement	431 tons	438.26 ton	596 tons	605.64 ton
Draft, fore	3.8 ft.	1.15 m.	7.9 ft.	2.40
„ aft	14.8 ft.	4.52 m.	15.4 ft.	4.68
„ mean	9.3 ft.	2.835 m.	11.6 ft.	3.54
Freeboard	4.7 ft.	1.42 m.	2.3 ft.	0.71
KM	11.4 ft.	3.46 m.	11.3 ft.	3.45
KG	10.4 ft.	3.16 m.	9.3 ft.	2.84
GM	1.0 ft.	0.30 m.	2.0 ft.	0.61
G from midship	8.8 ft.	2.67 m. aft	4.4 ft.	1.34 m. aft
Coefficients:				
Block		0.535		0.585
Prismatic		0.619		0.655
Midship		0.864		0.893
GZmax	0.42 ft.	0.128 m.	0.71 ft.	0.216 m.
GZmax at		28.4°		27.2°
Range J		51.0°		66.8°

TABLE XXXI

Speed Trial

Date October 12th, 1950	*Load*	*Speed (in knots)*	*r.p.m.*	*Slip (in %)*	*i.h.p.*	*b.h.p.*
Draft:						
Fore, 4.5 ft. (1.36 m.)	4/4	12.50	304.5	2.11	732	**603**
Aft, 14.5 ft. (4.41 m.)	3/4	11.77	274.5	−2.17	548	**434**
Mean, 9.5 ft. (2.885 m.)	1/2	10.93	245.6	−6.12	408	**299**
Displacement:	1/4	8.54	187.0	−8.93	232	**132**
443 tons (450 ton)						

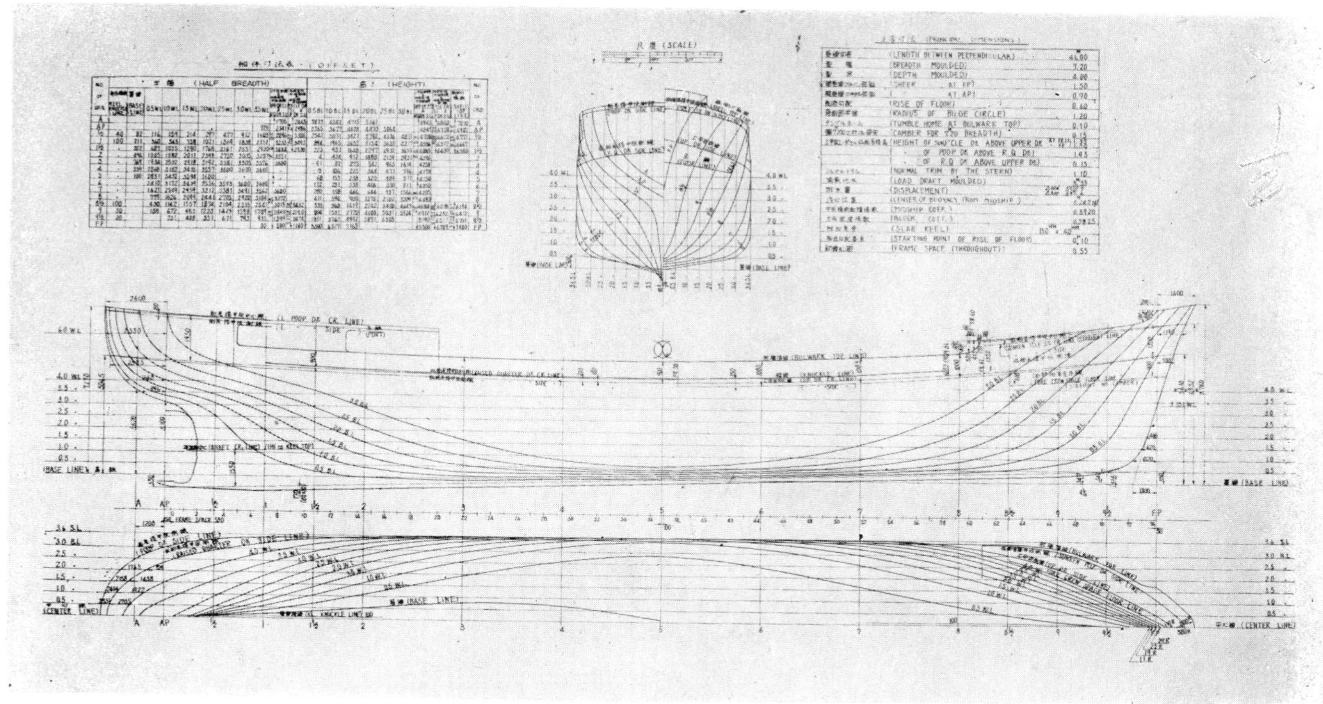

Fig. 193. Lines of 148 ft. (45.2 m.) otter trawler Yamato Maru

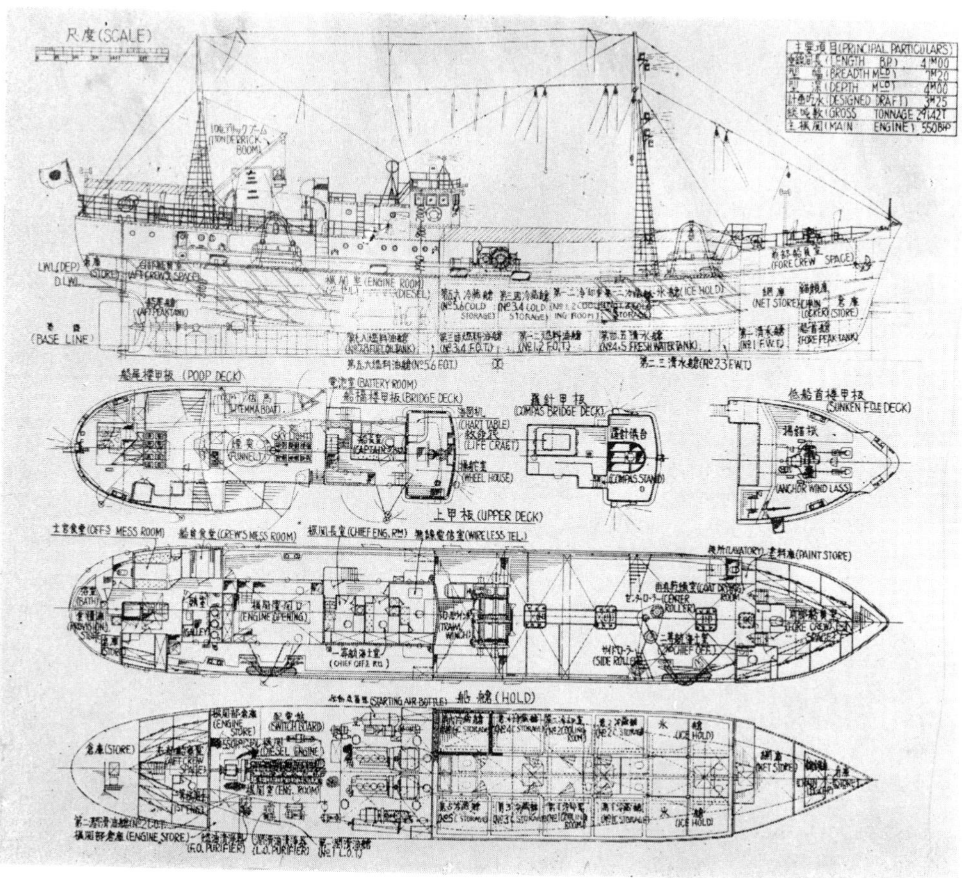

Fig. 194. General arrangement of 148 ft. (45.2 m.) otter trawler Yamato Maru

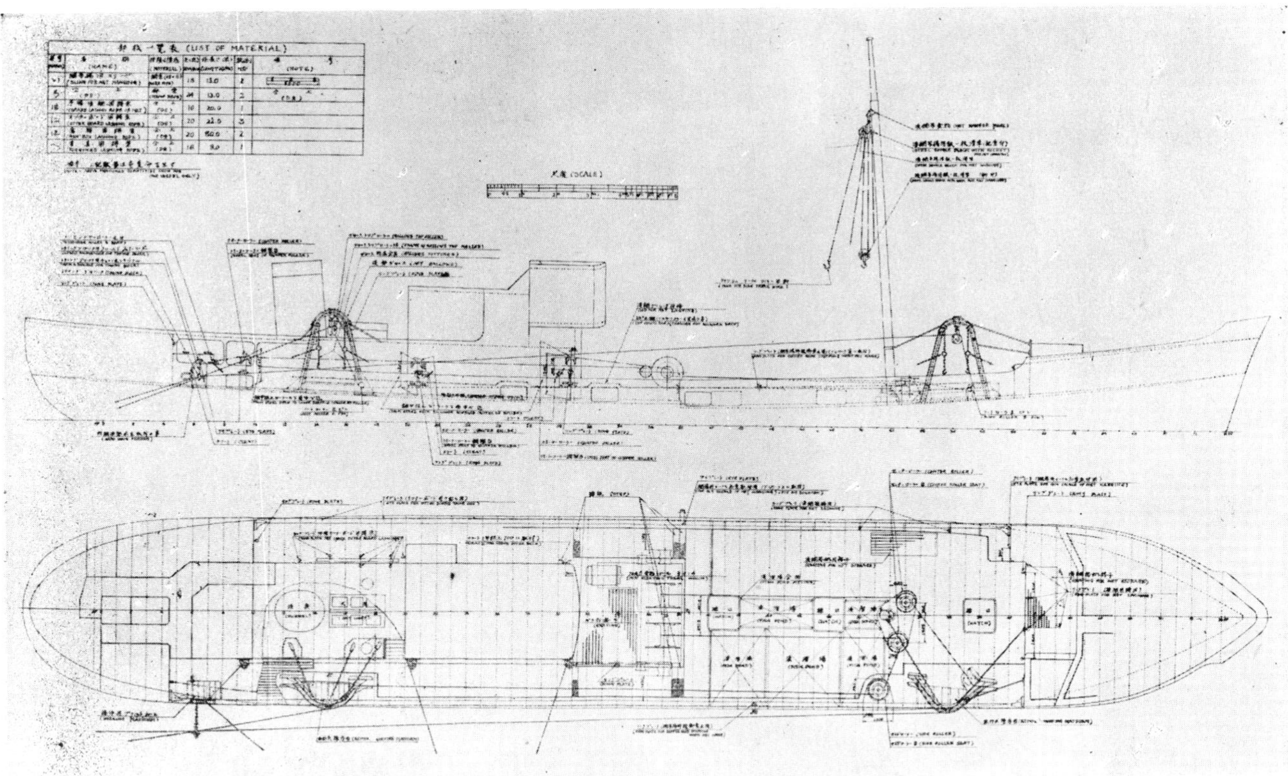

Fig. 195. Arrangement of gear installations on the 148 ft. (45.2 m.) otter trawler Yamato Maru

Fig. 196. Yamato Maru, *148 ft. (45.2 m.) otter trawler*

Fig. 197. Tomi Maru No. 11, *85 ft. (25.8 m.) pair trawler*

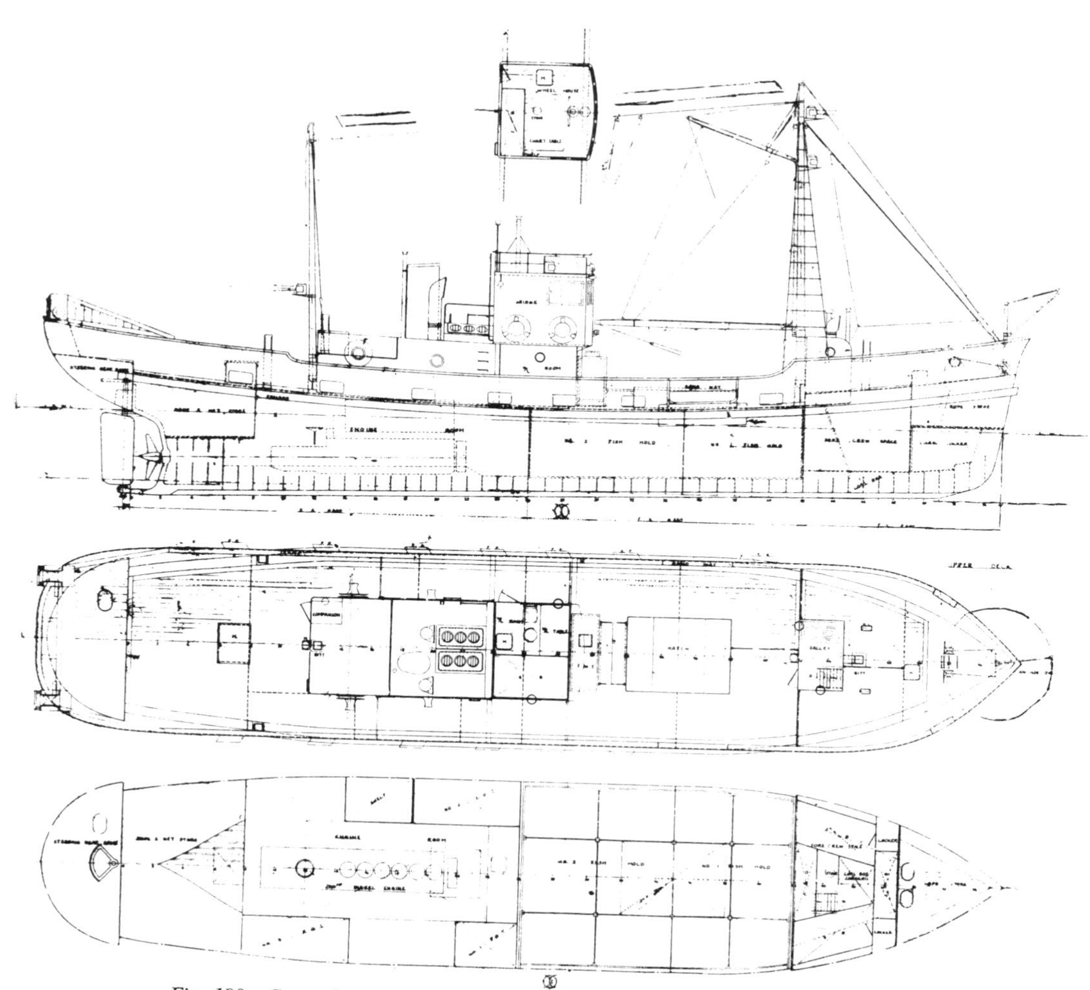

Fig. 198. General arrangement of Tomi Maru No. 11, *85 ft. (25.8 m.) pair trawler*

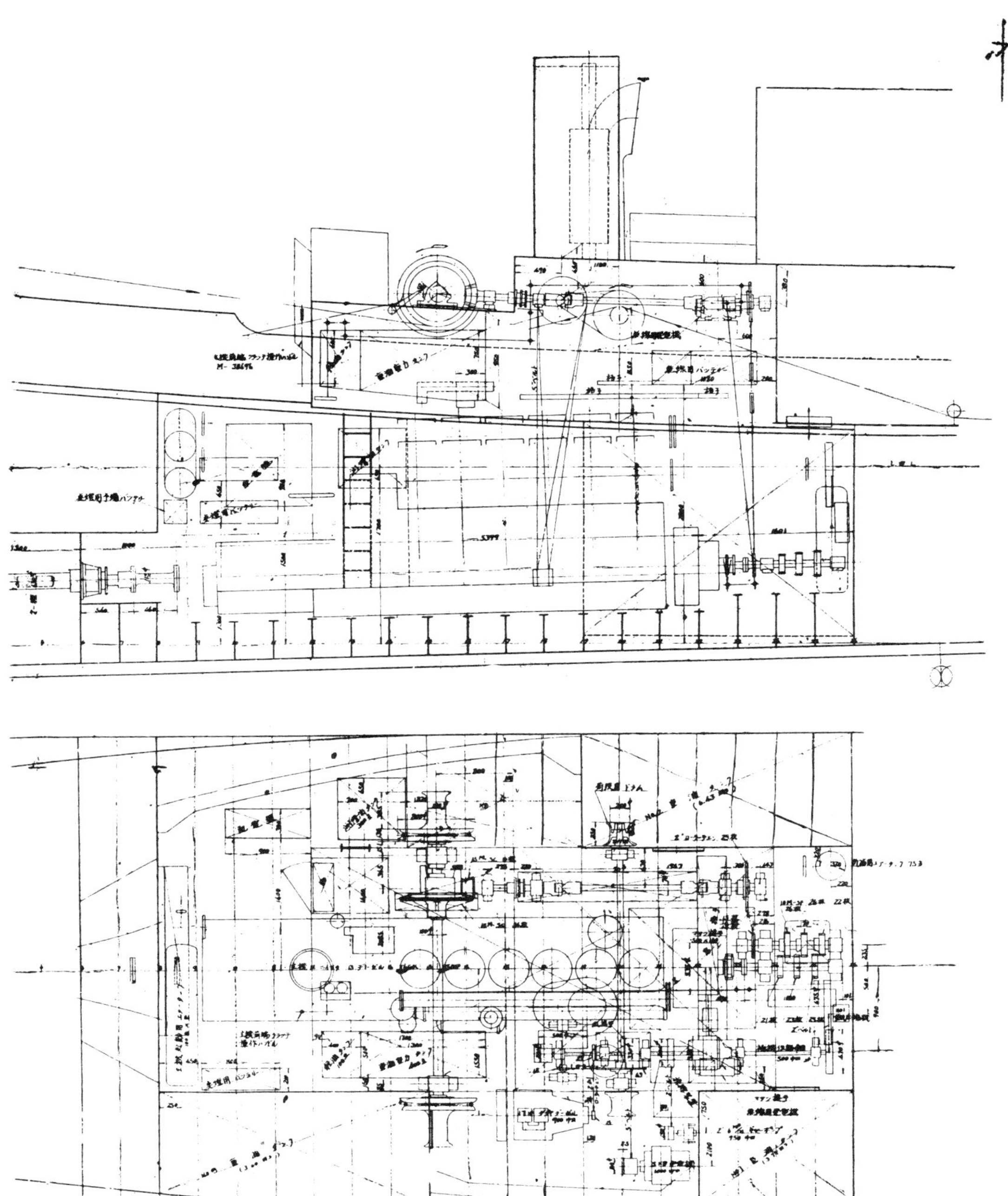

Fig. 199. Installation of trawl winch in Tomi Maru No. 11, *85 ft. (25.8 m.) pair trawler*

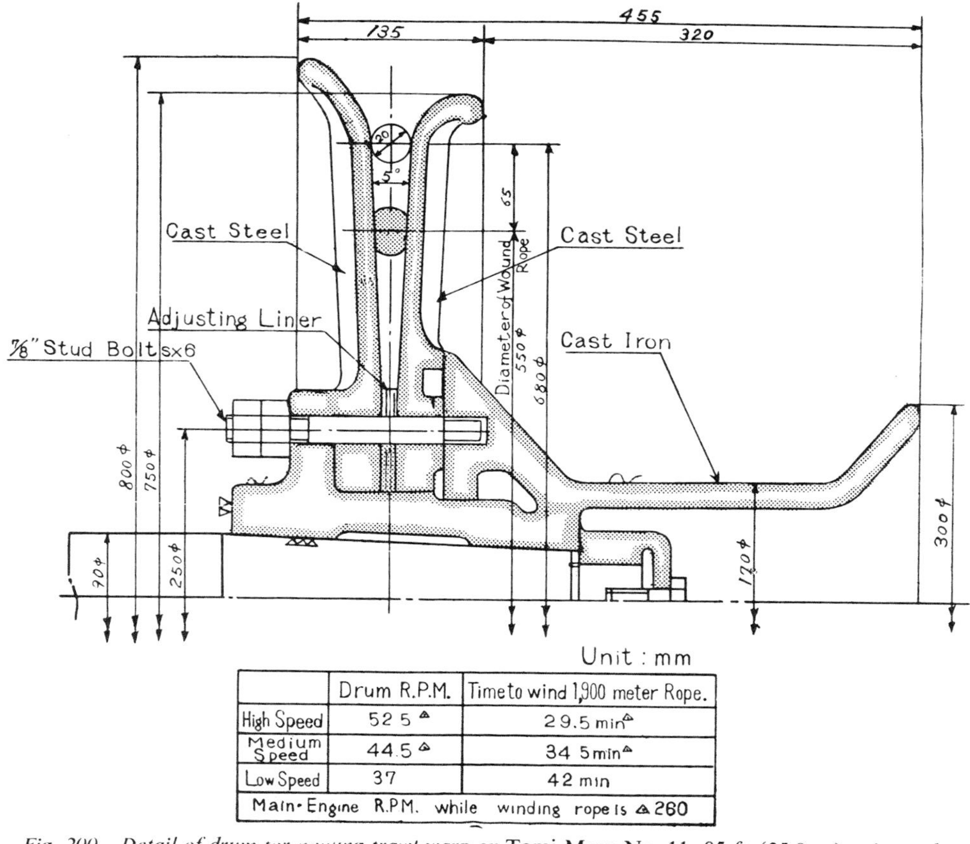

	Drum R.P.M.	Time to wind 1,900 meter Rope.
High Speed	52.5 ▲	29.5 min▲
Medium Speed	44.5 ▲	34.5 min▲
Low Speed	37	42 min
Main-Engine R.P.M. while winding rope is ▲260		

Fig. 200. Detail of drum for hauling trawl warp on Tomi Maru No. 11, *85 ft. (25.8 m.) pair trawler*

SMALLER TRAWLERS AND DANISH SEINERS

Two-boat or pair trawling is operated in the Yellow Sea and the China Sea, and Danish seining on the banks in the Japan Sea. These fishing methods are quite effective, but the operators are trying to find more distant fishing grounds. The boats are built for both methods and are mostly wooden. They are called medium or smaller trawlers and a typical example of a steel one, the *Tomi Maru No. 11*, operating around Hokkaido, is given below. This boat is also used as a salmon and trout drifter in summer to prevent overcatching by trawling, so the stern has been widened to set drift nets from the deck.

The *Tomi Maru No. 11*, has the following: Length, overall, 85 ft. (25.80 m.); length between perpendiculars, 75.5 ft. (23.00 m.); beam, 16.4 ft. (5.00 m.); depth, 7.4 ft. (2.25 m.); 59.12 gross tons, 25.83 net tons; fish holds, 2,100 cu. ft. (59.40 cu. m.); fuel oil tanks, 3,260 imp. gal. (3,900 gal., 14.83 cu. m.); fresh water tanks, 330 imp. gal. (395 gal., 1.50 cu. m.); officers and crew, 12; main engine, four-cycle single-acting diesel, six cylinders with 10-in. (250 mm.) diameter and 15-in. (380 mm.) stroke; 250 b.h.p. at 380 r.p.m.; four-bladed propeller with 59 in. (1,500 mm.) diameter and 34.2 in. (870 mm.) pitch; 17 h.p. auxiliary diesel driving a 5 kw., 110 v., D.C. generator; 15 w. wireless telephone.

Fig. 197 shows the *Tomi Maru No. 11* under speed, fig. 198 the general arrangement, fig. 199 the installation of the trawl winch, fig. 200 a detail of the drums for hauling the trawl warps. Table XXXII gives particulars of drafts and stability and Table XXXIII results from speed trials.

TABLE XXXII
Draft and stability

	Load			
	Light		*Full*	
Displacement	93.5 tons	94.92 ton	115.5 tons	117.44 ton
Draft, fore	2.5 ft.	0.77 m.	2.7 ft.	0.82 m.
,, aft	8.0 ft.	2.42 m.	9.4 ft.	2.88 m.
,, mean	5.25 ft.	1.595 m.	6.1 ft	1.85 m.
Freeboard	2.9 ft.	0.875 m.	2.0 ft.	0.62 m.
KM	7.7 ft.	2.33 m.	7.4 ft.	2.26 m.
KG	5.9 ft.	1.79 m.	5.9 ft.	1.80 m.
GM	1.8 ft.	0.54 m.	1.5 ft.	0.46 m.
G from midship	3.7 ft.	1.12 m. aft	4.7 ft.	1.44 m. aft

TABLE XXXIII
Speed trial

Date March 6, 1951	*Load*	*Speed (in knots)*	*r.p.m.*	*Slip (in %)*	*b.h.p.*
Draft:					
Fore, 2.7 ft. (0.82 m.)	Over	10.64	403	6.7	262
Aft, 8.3 ft. (2.54 m.)	4/4	9.99	372	5.0	213
Mean, 5.5 ft. (1.68 m.)	3/4	9.07	329	2.3	139
Displacement:	1/2	8.62	303	0.7	105
100.5 tons (102 ton)	1/4	7.24	246	3.9	60

FRENCH MOTOR TRAWLERS

by

E. R. GUEROULT

THE French fleet of trawlers, almost completely destroyed during the war, has been rebuilt on new lines. A number of standard designs were made of 28, 32, 38, 42 and 68 m. (92, 105, 125, 138 and 223 ft.) lengths and immediately after the end of hostilities, contracts were placed in France, Great Britain, Canada and U.S.A. In all, 158 vessels (round 60,000 tons gross) of moderate or large size, steam and motor, have been put in commission between 1946 and 1952. The owners' experience prior to 1940 was pooled and the new fleet is the result of their common effort.

The distance between port and the fishing grounds is the important factor influencing the choice of the main particulars of the trawlers. From the small coastal types, the size is constantly increasing with the need to fish further away and to maintain a normal proportion between the time spent under way and that of actually fishing. Each fishing ground calls for a certain tonnage and type and the financial results depend for a great part on the correct choice of tonnage. This relation between tonnage and distance of the fishing grounds has been the outcome of years of experience, but propulsion by diesels and trawling on more distant grounds has made it necessary to revise the characteristics of the new vessels.

A standardized fleet, designed for the lowest operating cost, calls for particulars somewhat different from those typical of the many individual contracts between owners and builders, based mainly on capital cost. The naval architect, free of price competition problems, can design the most economical vessel for its size which is, in the end, profitable to the owner. In the past, when trawlers were built to a length specified by the owner, other dimensions were very often reduced to keep down the size, and the steam trawlers were generally narrow boats with a little midship section.

The landings of the trawlers show that they are normally seldom loaded to more than 65 per cent. of their capacity and the naval architect must decide whether the boats should be designed for a maximum dead weight or a mean carrying capacity, corresponding to the average landings. On the other hand, motor trawlers, leaving port fully loaded, can easily be overloaded in case of a heavy catch on nearby grounds. Tank tests, particularly the recent work at the Teddington Tank (Allan 1953), have shown the heavy penalty exacted by too great a fullness of form. It is, therefore, for a mean condition of loading that the proportions and lines should be based upon, with a good reserve of freeboard in case of overloading.

The density of the load can be taken at 0.45 to 0.50, whether the fish are stored in boxes or on shelves, and the last figure should be used for the fresh fish vessels. The density of salt cod cargo should be taken at 0.85.

To classify the trawlers described in this paper, the length between perpendiculars in metres has been kept as a convenient reference dimension, but the fish-hold capacity was the dominant standardization factor.

Large trawlers are normally catching greater quantities of fish than the smaller ones. The fishing gears are not very different and the power required for trawling is not proportional to the tonnage. The greater part of the superiority of large trawlers comes from their ability to work in bad weather. It seems that in the North Sea, trawlers of a displacement of around 800 tons can still work during a wind force of up to 8 Beaufort.

There are no draft limitations because the harbours and the maximum draft of the trawlers depend only on the propeller's diameter and immersion. The normal r.p.m. in 1944 of diesels from 400 to 1,200 b.h.p. was between 300 and 175 and the propeller diameter resulted from these figures, which are reasonably close to the optimum r.p.m. for speeds of 10 to 12 knots. The direct drive was retained for all the classes, except the smallest one, the 28 m. (92 ft.) class for which a reduction gear was provided. With a propeller immersion of half the propeller diameter and a deep cruiser stern, the danger of propeller racing in rough weather is greatly reduced.

The owners always insist on the highest possible speed and the trawlers are running at high Froude's numbers, particularly the smaller types, which are often overdriven. From tank tests and experience in service, the economical speed has been ascertained and the results for the new French trawlers are given in fig. 215.

An easy motion and small rolling amplitude are necessary to facilitate the men's work on deck. It is difficult to reach a compromise between the forms which give good initial stability and seakindliness, and those which give low resistance.

PARTICULARS OF THE VARIOUS TYPES

The main characteristics of the standardized motor trawlers are given in Table XXXIV. The boats are shown in longitudinal section on fig. 201 to 205, and the body plans, fig. 206 to 210. The various proportions and coefficients are listed in Table XXXV.

A good balance of volumes and lines is obtained by a proper inclination of the keel. The lines have been drawn with a trim equal to two-fifths of the mean draft.

The underwater form and the position of the centre of buoyancy lead to volumes above the water which are not balanced. A well raked and rounded stem will add a volume forward, much needed for seaworthiness. The 42 m. (138 ft.) type could, for example, with advantage have the stem more raked.

Table XXXIV

Main Particulars

Length, between perpendiculars, L	28 m. (92 ft.)		32 m. (105 ft.)		38 m. (125 ft.)		42 m. (138 ft.)		68 m. (223 ft.)	
	ft.	*m.*	*ft.*	*m.*	*ft.*	*m.*	*ft.*	*m.*	*ft.*	*m.*
Length, over all	103	31.45	117	35.85	138	41.90	152	46.45	239	73.00
Length, waterline	96.5	29.50	110	33.50	129	39.35	144	44.00	228	69.35
Breadth, moulded, B	21.3	6.50	225	6.85	24.6	7.50	27	8.25	38.5	11.75
Depth	12	3.65	12.8	3.90	13.6	4.25	15.4	4.70	20.6	6.30
Draft, moulded, T	9.7	2.95	10.7	3.25	11.5	3.50	12.6	3.85	17.2	5.25
	cu. ft.	*cu. m.*	*cu. ft.*	*cu. m.*	*cu. ft.*	*cu. m.*	*cu. ft.*	*cu. m.*	*cu. ft.*	*cu. m.*
Fish room volume	35,300	100	42,300	120	75,500	214	113,000	320	530,000	1,500
Bunker capacity	22 tons		34 tons		74 tons		120 tons		500 tons	
Displacement, light	200 tons		280 tons		334 tons		505 tons		1,280 tons	
Displacement, loaded	261.5 tons		356 tons		484 tons		718 tons		2,713 tons	
Installed power	375 h.p.		450 h.p.		600 h.p.		750 h.p.		1,100 h.p.	

FORMS AND SEAKINDLINESS

The distribution of weights gives a centre of gravity aft of midships which facilitates a desirable fine entrance. The forebody is therefore relatively easy to shape, with straight lines near the waterline and slightly hollow near the keel, the shoulder being carried well aft. The aft body is necessarily rather full, with fine sections, however, at the foot, near the propeller, to ensure good flow and wake conditions to the screw at the low trawling speeds.

The optimum ratio of the waterline and block coefficients (with these lines) lies in the region of 1.5 for the 28/42 m. (92/138 ft.) types, fig. 216. It appeared quite early that the forms above the waterline required the same attention as the underwater body and that the fining down of this latter called for a compensation in the upper part.

The right selection of sheer line is important and it can be determined as follows: the vessel is placed on a series

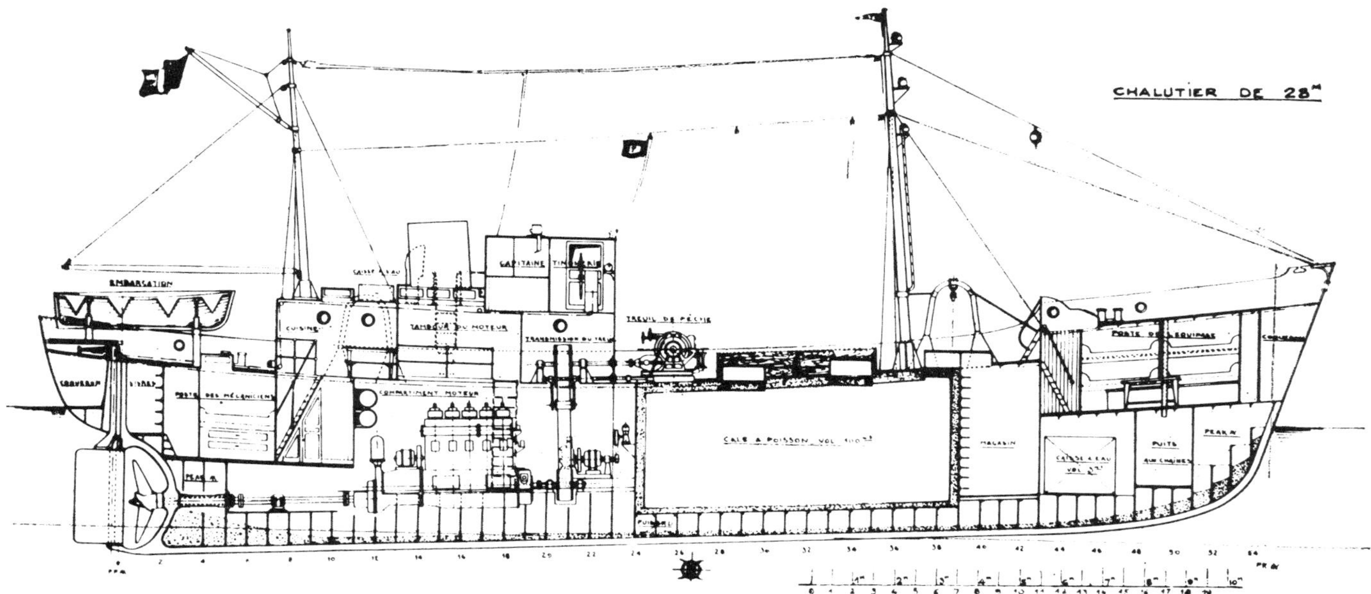

Fig. 201

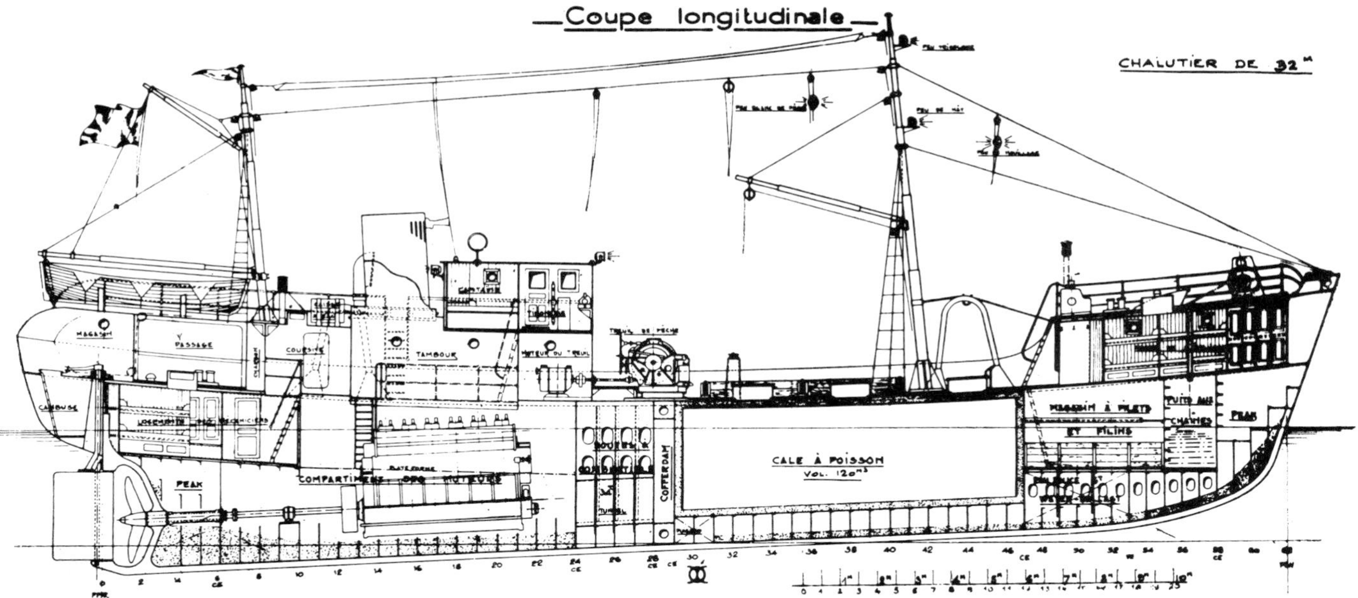

Fig. 202

of trochoidal waves of the same length as the vessel, so that the displacement and centre of buoyancy are correct. The wave crests give a wetting line of the hull and the envelope to the crests supplies a sheer line in function of the volumes of that particular ship. Fig. 211 illustrates an application to a small trawler, which has been tested on waves at the Paris tank. The model was actually shipping water by the stern, as a consequence of too little sheer aft. The standardized trawlers have accordingly been given a greater sheer aft than the boats built before 1940 and this proved well justified in service.

Seakindliness for a trawler depends for a great part on the correct choice of the breadth and the ratio between beam and draft. When hauling in the net, the trawler is put across the wind and sea, and rolling in this condition must be such that easy operation on deck is still possible. For the safety of the vessel the righting arm must be increasing up to an angle well above the immersion of the bulwarks amidships. This and the above requirement are conflicting. Tank tests (Graff and Heckscher 1941) have shown that, from the point of view of the resistance, the breadth can be increased

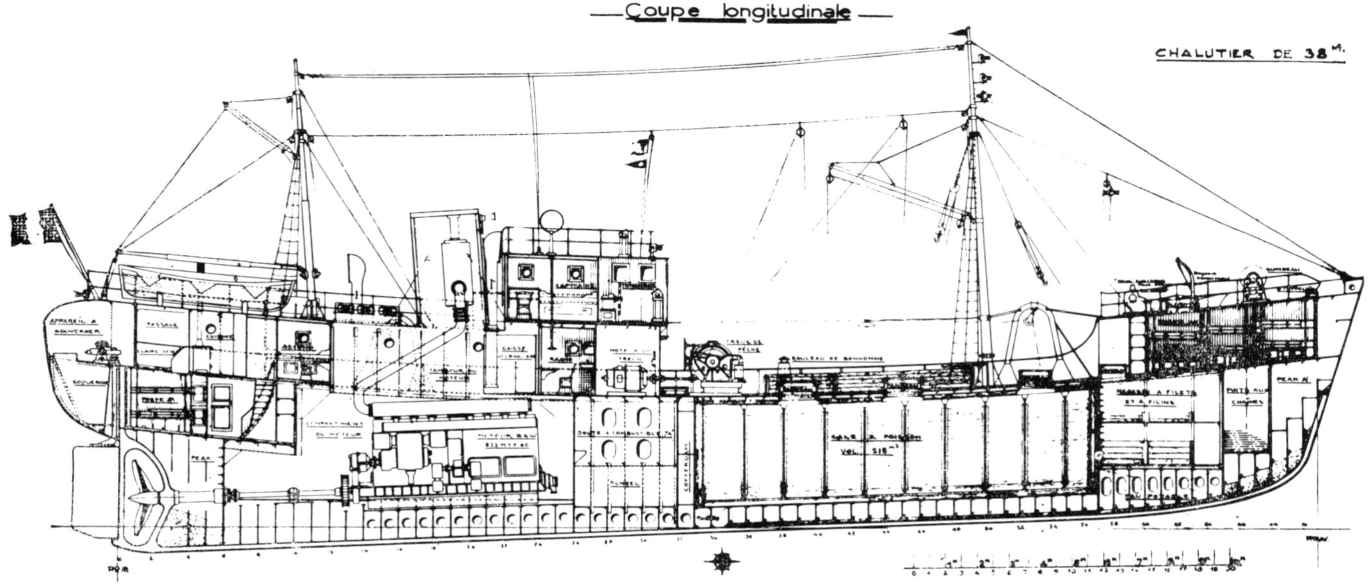

Fig. 203

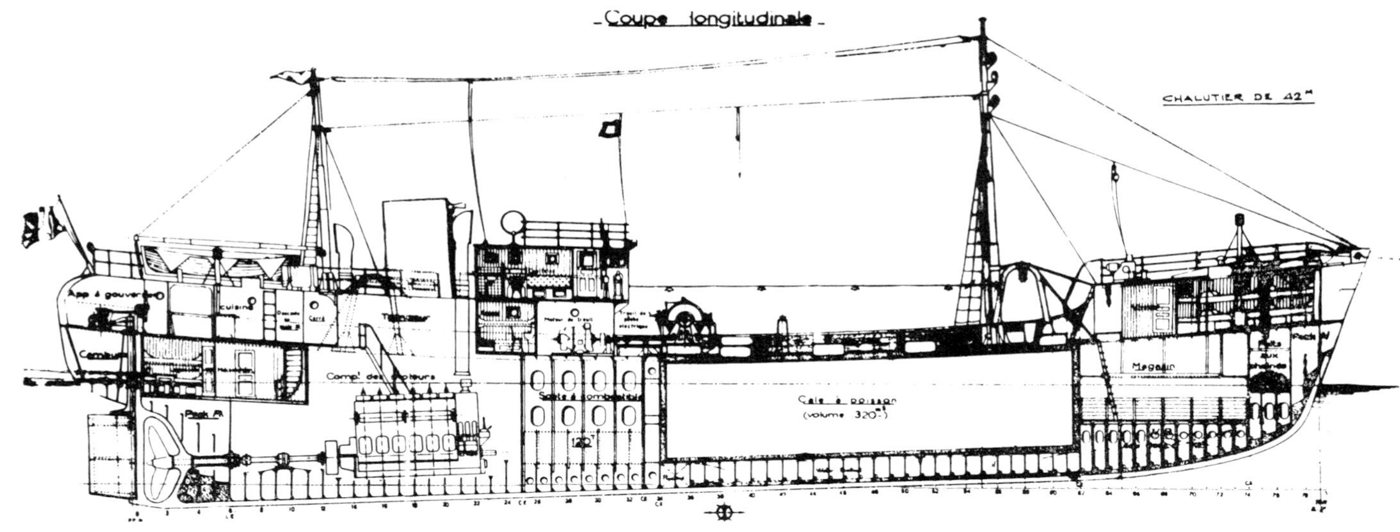

Fig. 204

TABLE XXXV

Ratios and Coefficients

Length, between perpendiculars	28 m. (92 ft.)	32 m. (105 ft.)	38 m. (125 ft.)	42 m. (138 ft.)	68 m. (223 ft.)
L/B	4.31	4.67	5.07	5.09	5.79
B/T	2.2	2.11	2.14	2.14	2.24
B^2/T	14.3	14.4	16.0	17.7	26.4
$L/\nabla^{\frac{1}{3}}$	4.38	4.50	4.84	4.69	4.88
Coefficients, block	0.475	0.49	0.479	0.524	0.651
midship section	0.788	0.81	0.820	0.820	0.971
prismatic	0.604	0.605	0.586	0.64	0.67
waterplane	0.745	0.755	0.750	0.775	0.797

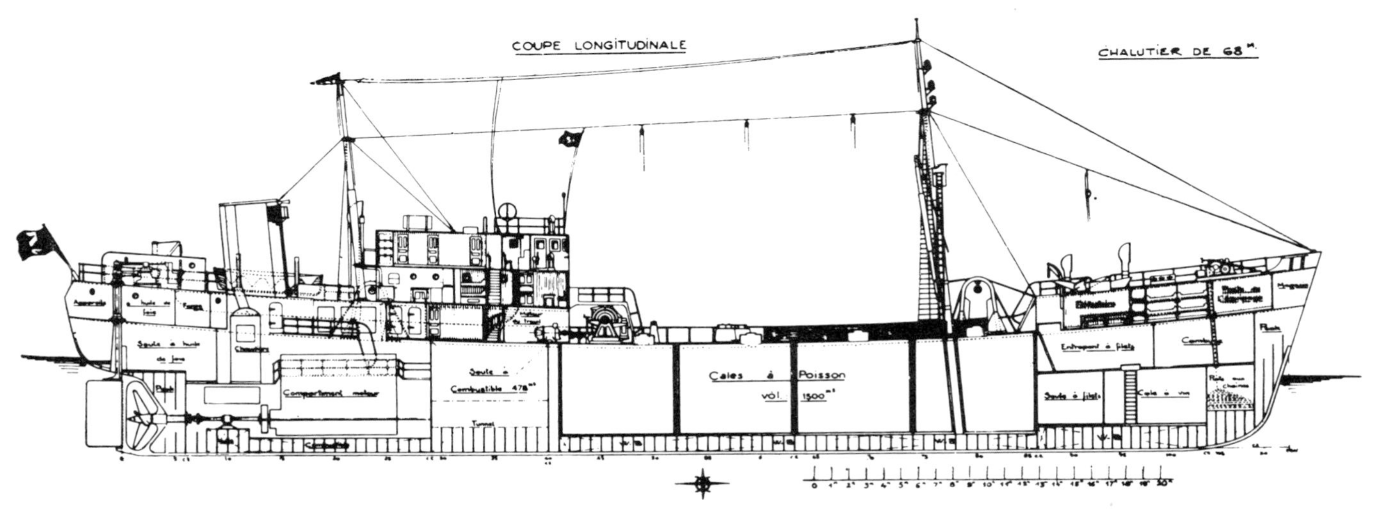

Fig. 205

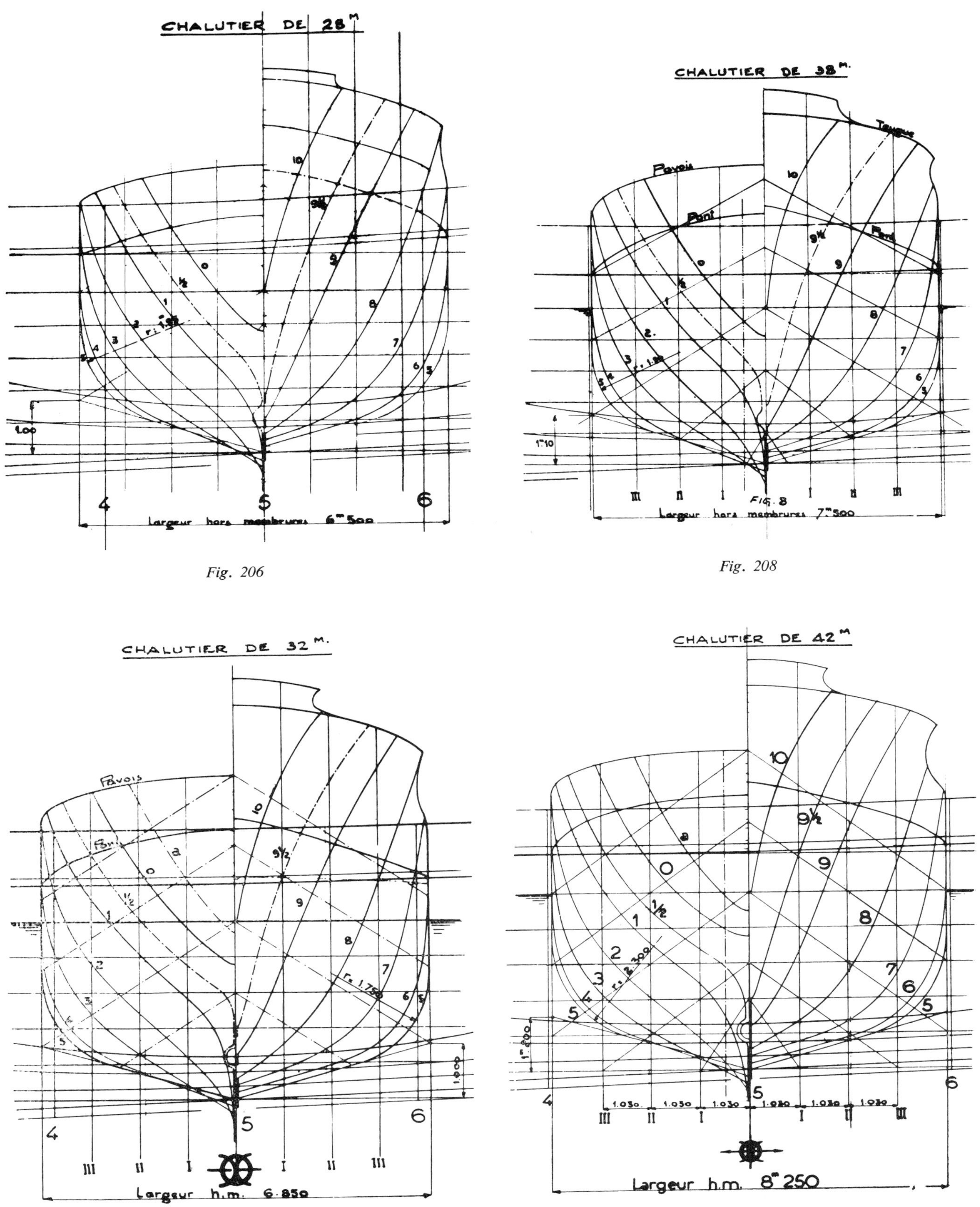

Fig. 206

Fig. 208

Fig. 207

Fig. 209

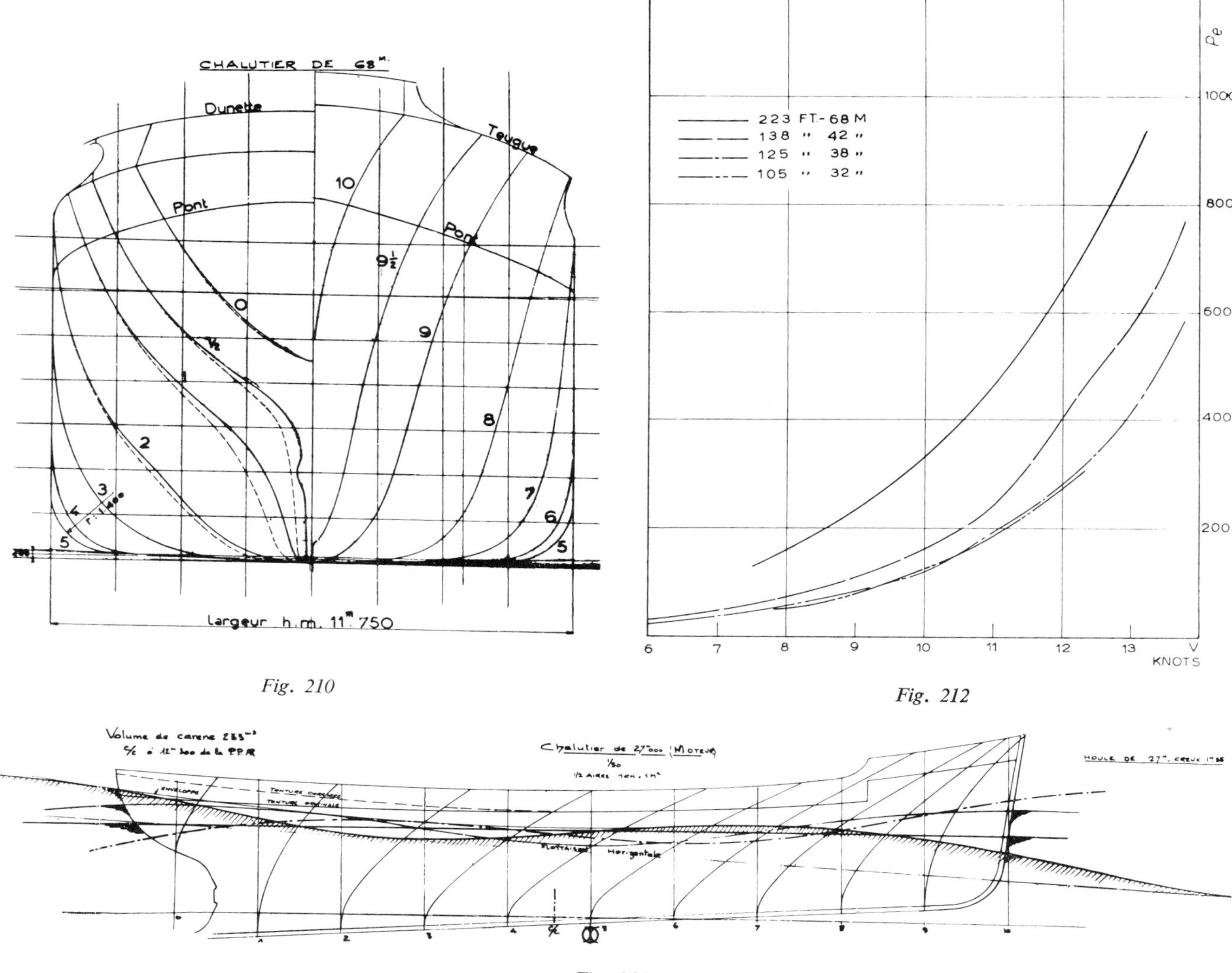

Fig. 210

Fig. 212

Fig. 211

considerably with the corresponding fining down of the ends, without penalty, but that initial stability and freeboard had to be increased for the broader models. An investigation on the effect of breadth for the 42 m. (138 ft.) type with three different values of B: 26.2, 27 and 27.9 ft. (8.00, 8.25 and 8.50 m.), led to the same conclusion and finally the mean value 27 ft. (8.25 m.) was retained.

With the extended superstructures of the present-day trawler, some permanent ballast is needed, up to about ½ ton per metre (3.3 ft.) in length, to avoid the necessity of giving the vessel a high form stability producing quick motions when the vessel is put across the seas. The skippers very often alter the vertical distribution of weights to decrease GM. The fisherman's gear brings always an underestimated increase in the height of the vertical centre of gravity. There is a difference of around 8 in. (20 cm.) on the vertical centre of gravity between the light ship, as delivered by the shipyard, and what the owner calls the light ship, i.e. without bunkers, ice or water, but with fishing gear, provisions and crew. This difference, which has been ascertained after quite a number of inclining experiments on trawlers in service, must be taken into account during the early stage of the design. With time, an augmentation of the weight and the height of the centre of gravity must be reckoned with.

Rolling periods have been measured at sea and, for nearly all the series was found to be in the region of 1 second per metre (3.3 ft.) in breadth in normal loaded conditions with three-quarters of the bunkers and water in the double bottom. This period is practically the same throughout the voyage. It seems to suit the requirements of the crew for work and comfort.

RESISTANCE AND PROPULSION

Results from tank tests are given in fig. 212 and Table XXXVI for the 32 to 68 m. (105 to 223 ft.) classes. For various types having forms of similar character the

TABLE XXXVI

Tank Test—Resistance in e.h.p.

Speed in knots	*Trawlers* 32 m. (105 ft.)	38 m. (125 ft.)	42 m. (138 ft.)	68 m. (223 ft.)
5	11.4		16	40
6	19.6		27	69
7	31.4		45	107
8	49	54	69	157
9	75.6	82	107	230
10	120.6	117	162	329
11	186.9	193	240	466
12	271.8	282	408	651
13		407	581	887
13.5		520		

resistance curves are also very similar. The 38 m. (125 ft.) type, designed when the other models' results were known, compares favourably with the 32 and 42 m. (138 ft.) types. The 42 m. type could be improved by straightening the waterline forward for the 12 knot speed. The trials at sea were run at other displacements than the tank tests. Comparisons are therefore not possible and the correlation between tank tests and service at sea could not be established. The propulsion data are given in Table XXXVII.

TABLE XXXVII

Tank Tests

Propulsion Tests

Length, between perpendiculars	32 m. (105 ft.)	42 m. (138 ft.)	68 m. (223 ft.)
Displacement in tons	358	718	2,810
Wake in %	18.5	15	26
Thrust in %	23.5	23.3	20
Total propulsive efficiency	0.571	0.65	0.63
Speed in knots	12	12	12.5
Shaft horse power	476	642	1,125
r.p.m.	227	139*	185

* The normal r.p.m. is 200.

PROPELLERS

All the propellers have been designed for the normal torque of the engines and a speed of 3 knots. The propellers have a low value of pitch: diameter, consistent with good efficiency when trawling and at reduced speed in bad weather, which are the prevailing conditions of a trawler in service. Since the vessels have been put in commission, no undue wear of the motors has been noted and such engine troubles, which were frequent before 1940, have been practically absent. However, it is almost certain that the normal torque of a motor at reduced r.p.m. is different from the torque at the full number of revolutions. If engine manufacturers would supply the naval architects with the real value, then great progress regarding wear and upkeep would follow. The conditions of a propeller working at 3 knots are, at present, very little known; they have not been investigated in shipbuilding experimental tanks in the same manner as for the free steaming. The values of wake and thrust deduction are uncertain and there is no data about the beginning of cavitation behind the hull, at 3 knots. It is highly desirable to have these problems investigated, to obtain an economy in the upkeep and a greater duration of the propellers.

AUXILIARIES

All the auxiliaries in both engine room and on deck have been standardized and also the deck fitting, such as gallows and rollers. The electric trawl winch, in particular, has been thoroughly studied and the results in service are highly satisfactory. Ward-Leonard couplings and slow-running machines have been adopted exclusively.

ENGINE ROOM INSTALLATION

Special care has been taken with the engine room arrangement in bringing all the controls to the manoeuvring platform, so that only one engineer is required on each watch on the smaller types up to the 42 m. (138 ft.) boats. Three men are necessary on the 68 m. (223 ft.) boats. All the auxiliaries are placed on a common foundation which provides great strengthening to the bottom structure and, with the elimination of the individual seatings, the vessels are reasonably free from vibrations.

FISH HOLD—INSULATION

Very few boats have been equipped with a refrigerating plant. Experience in service has proved that the insulation was sufficient to keep a temperature close to freezing point and that the melting of the ice was slow. Cement and light alloy have been used at the sides and wood for the partitions and pond boards only. The bottom is cemented. The suction well has been fitted with a syphon, isolating the fish hold from the bilge suction line.

FUTURE DESIGNS

The operation of the rebuilt French fishing fleet has given experiences which will be useful for future design work.

Fig. 213 shows weights and proportions. A beam/draft ratio of about 2.0 with a superstructure of about half the ship's breadth, is safe for transverse stability. For the cod vessels the beam/draft will depend on the amount of superstructure carried. The weights are for a construction with a fair proportion of welding, slow-running diesel motors and steel superstructure. The weights could be thus reduced, but the values given are safe for an early design stage.

The resistance in effective horsepower (EHP) per ton of displacement is shown in fig. 214, and with the data of Table XXXVII, it will be possible to calculate the necessary power for a given speed. With the diagram in fig. 215, a quick estimate of the maximum economical speed can be made, taking into consideration not only the resistance of the hull but also the requirements of the owner and the possibility of installing a maximum power with the present-day motors.

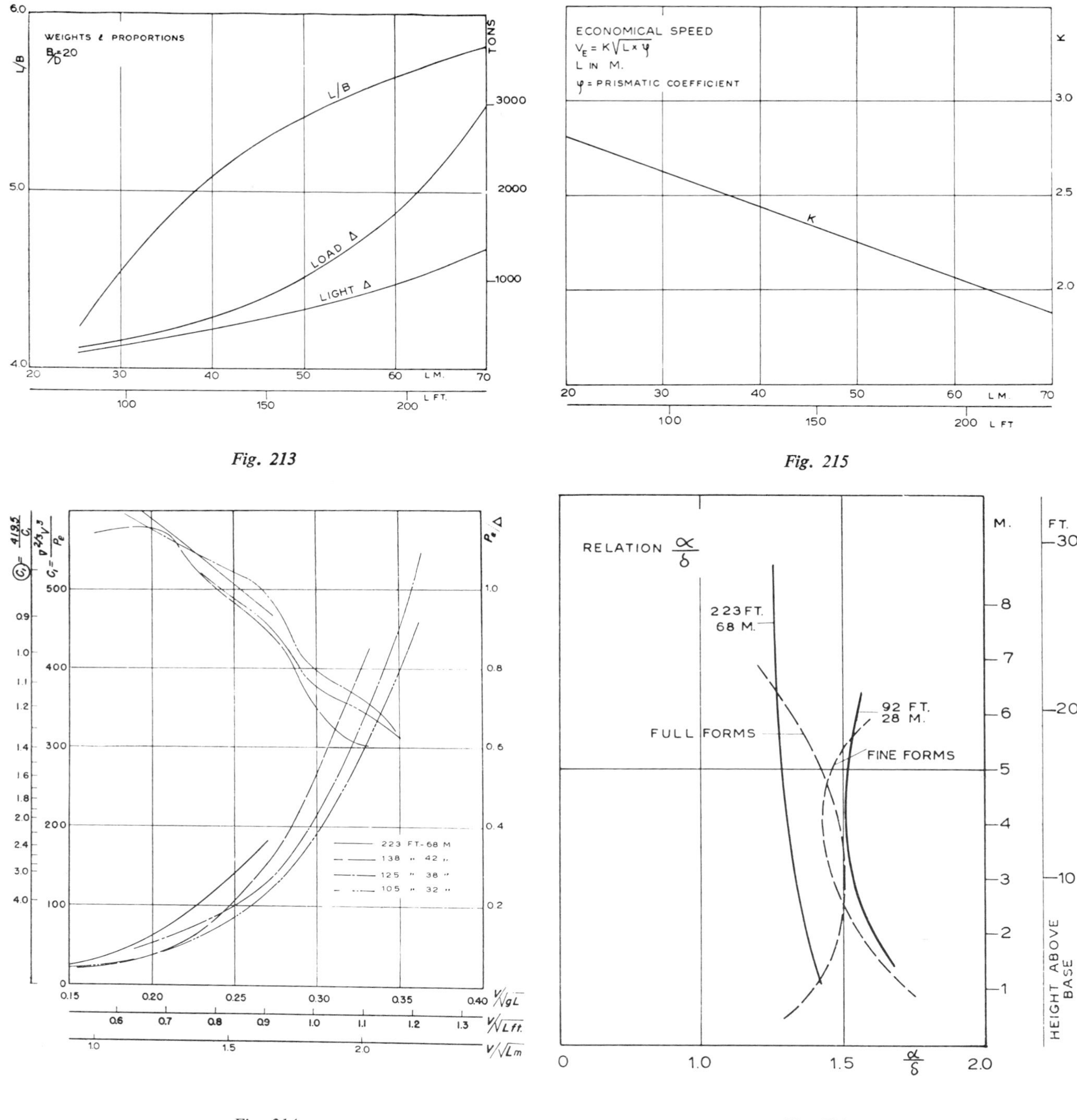

Fig. 213

Fig. 215

Fig. 214

Fig. 216

Attention is drawn to the fact that none of the trawlers are safe in case of damage to the ship's side in one of the main compartments. Additional bulkheads in the fish hold do not meet the approval of the operating crew, but a reduction of the floodable lengths seems unavoidable in future designs.

Since a few years ago the hold of the 68 m. (223 ft.) cod trawlers is divided into two compartments by a watertight bulkhead. There are some watertight doors for the transfer of salt or fish from one hold to the other. This causes extra work for the crew but from the safety point of view it is necessary.

In the engine room there is the same danger and it is very difficult to find a solution in view of the present engines and the present regulations with regard to gross tonnage.

SOME NOTES ON LARGE TRAWLERS

by

H. E. JAEGER

EVEN a large trawler is a small vessel and must therefore have good sea-keeping qualities. The trim and stability conditions are quite different from those of ships carrying a constant cargo. The freeboard must be low for easy handling of the trawl and the deck area must be large for easy handling of the gear and sorting of the catch. Pitching and rolling must be smooth to enable the fishing operations in winds of 6 to 7 Beaufort and a good lateral plane is needed for keeping a straight course while fishing and when hauling in the nets.

PRINCIPAL DIMENSIONS

The largest fresh fish trawlers for Iceland and Newfoundland fishing built in recent years are 165 to 180 ft. (50 to 55 m.) long and have a displacement of 1,000 to 1,300 tons. Owing to the proportion between fish-hold and bunker capacity, this size will probably be retained for some time in the future.

Because of the increasing distance between port and the fishing grounds due to lack of fish nearby, the speed has been increased to 13 knots and more, and the length is also increased to 180 ft. (55 m.) to keep within an acceptable value of Froude's number. A radius of action of 2,000 miles, three to four weeks at sea for about ten days fishing, a power of 850 to 1,200 b.h.p. with a corresponding bunker capacity, are factors dominating the design. Therefore the steam engine as prime mover becomes more and more scarce. Except for a number of British and German boats, nearly all the modern trawlers are motorships. The maximum displacement in that case occurs when leaving the fishing grounds.

Some modern trawlers have a raised quarter deck which has the following advantages:

(1) More room in accommodation aft and more space on deck for handling fish.
(2) Water and fish refuse cannot flow aft.
(3) With the trawl winch on the raised quarter deck, the trawl warps are clear of obstacles on deck.
(4) The wheelhouse is higher.

The raised quarter deck gives a greater freeboard from the stability point of view and not too much for fishing. This freeboard of trawlers is always larger than requested by the international rules. The minimum freeboard recommended by Schleufe (1948) is 1/75 of the length between perpendiculars. Spanner (1946) advises also to increase the normal freeboard by about 1 ft. (0.3 m.) for ships of 165 to 180 ft. (50 to 55 m.) in length, in addition to the load-line committee's rules. The figures given by Schleufe should be considered as a minimum for sufficient stability. The trim according to Spanner must not exceed 5 ft. (1.5 m.) and Cunningham (1949) suggests 4 ft. (1.2 m.) when leaving the fishing grounds. He also suggests that the maximum draught aft should not exceed 10 per cent. of the length of the ship. With a propeller immersion of $4\frac{1}{2}$ ft. (1.35 m.) as suggested by Spanner the propeller diameter is fixed and this fixes again the base for propeller and machinery design.

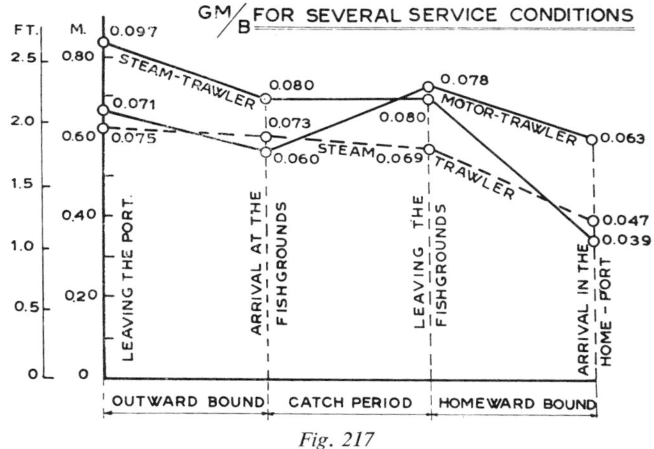

Fig. 217

STABILITY

Both Spanner and Cunningham give the same value of the metacentric height, GM=2 ft. (0.6 m.) for the ship leaving the fishing grounds. Roorda (1944) gives a value of GM in proportion to the breadth of the ship leaving port, GM/B=0.10 and returning to port GM/B=0.06. The stability criterions by Rahola (1939) do not apply to trawlers. The question is: what kind of standard can we use? Fig. 217 gives a variation of GM/B for two

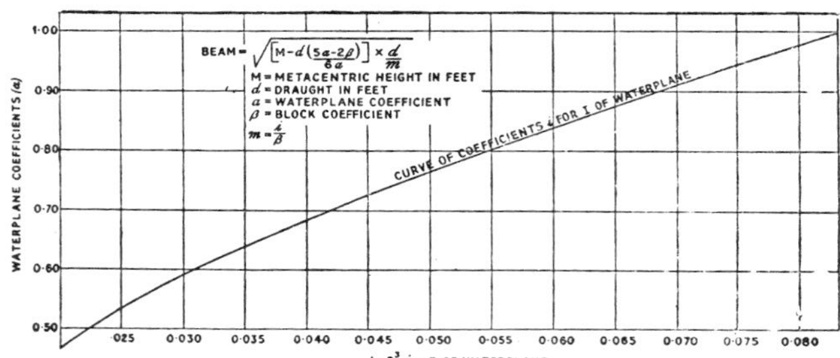

Fig. 218

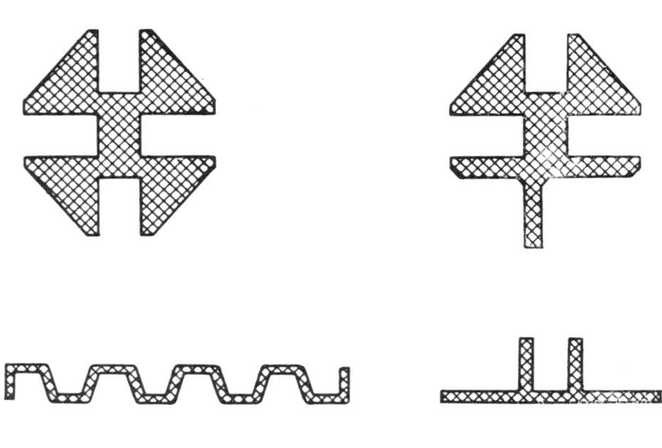

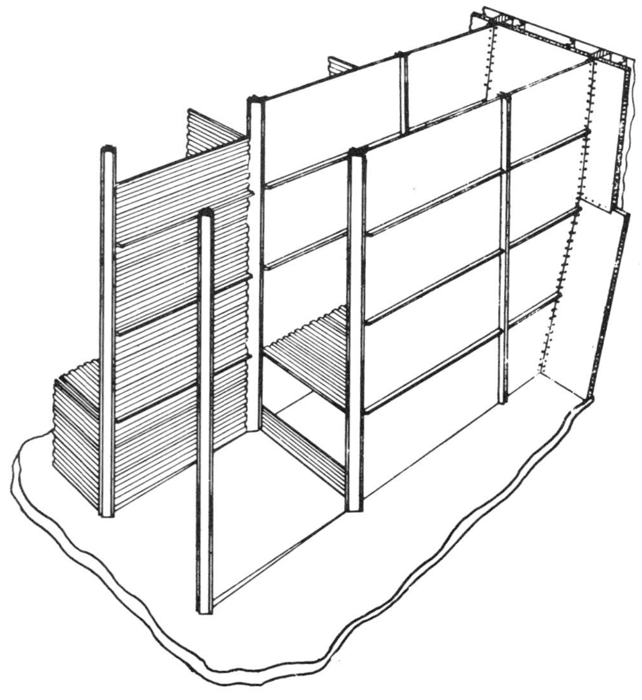

Fig. 219

large steam trawlers and one motorship. This variation can be considered as normal for this type of vessel, and these trawlers never had stability trouble. It is interesting to note the difference between motorships and steamships. The GM value of motorships increases during the fishing, due to the greater proportion of fishing load to the much smaller quantity of fuel and fresh water as compared with displacement. Motor trawlers can have a smaller value of GM than steamships which must have enough initial stability when leaving port to maintain sufficient stability till the end of the voyage.

A severe condition for a steam trawler is when, at the beginning of the fishing period, 30 to 40 tons of fish are loaded on deck and usually the last 3 tons are still in the cod end hanging in from the tackle. Sometimes this is aggravated by ice on deck and rigging, and in this connection Cunningham states that the value of GM might decrease by 8 in. (0.2 m.)

For large trawlers the most critical condition takes place at the arrival in port. The ratio of Roorda, GM/B =.06, is not always possible but is an absolute minimum for all large motor trawlers. The stability range for big trawlers would be in the region of 70 deg. when leaving port, and about 60 deg. when reaching port. If the stability range is inferior to the above values, the initial GM should be increased. The statical stability righting arm will reach a maximum at an angle between 25 deg. and 30 deg. in all stability conditions. The superstructure should be taken into account, in particular the forecastle, and the angle at which the deck is immersed must be not less than 7 deg. to 9 deg. in fully loaded condition. If the angle is smaller, the freeboard should be increased.

Eichler (1942) has attempted to give stability standards which can be guaranteed by the shipbuilder. These standards are generally in agreement with the above indications and suggestions but are too severe to be adopted. For instance to attain the maximum stability arm only at angles above 30 deg. is practically impossible.

A stability criterion is given by the Simpson formula quoted by Taylor (1943) and reproduced in fig. 218.

This formula is based on the Morrish stability formula and is only an approximation but it is a valuable control of the minimum initial metacentric height.

PROPULSION

Because of the distance to the fishing grounds, the time spent fishing is shorter than the time spent on the way. Normally tank tests for the design of propellers are based on the steaming condition. In Holland the propellers are calculated with 5 per cent. more r.p.m. and for 10 per cent. less power than the maximum power given by the engine builders. The propeller has to tow the trawl at 3½ to 4 knots and it absorbes then less power than when the trawler is steaming.

The controllable pitch propeller, has the following advantages when used in trawlers:

(1) Better efficiency and adaptation to all conditions of load.

(2) Bridge control.
(3) Immediate manoeuvring possibilities.
(4) Smaller speed loss against the wind and gain with the wind.
(5) Engine starting under lighter load.
(6) Non-reversible type of engine.
(7) Full efficiency of the motor when trawling and under way.
(8) Higher trawling speed for better fishing.
(9) Greater speed under way which is important when bringing back the catch.
(10) Progressive pull on the nets.

The controllable pitch propeller of the system de Schelde, with four blades in two pairs at 90 deg. to each other, is very strong, and its optimum efficiency is only 4 per cent. smaller than that of a fixed-blade propeller.

FUEL CONSUMPTION

The total fuel consumption is difficult to estimate on trawlers. The consumption under way is easy to calculate but not when fishing. An analysis shows that the consumption when trawling is 80 per cent. of that when under way for motor trawlers and 85 per cent. for steam trawlers.

WEIGHT ESTIMATION

The weight of the hull in tons can be estimated by the formula $P_s = \text{Length} \times \text{breadth} \times \text{depth} \times \text{coefficient}\ C_s$. C_s is the weight coefficient, which varies between .225 and .230 for modern trawlers, the dimensions being in meters. For the weight of the whole engine room installation one can take approximately twice the weight of the main motor the reduction gear included. Weight of fuel, lubricating oil, fresh water and provisions can be easily estimated and, assuming a number of days underway and fishing, the displacement can be ascertained at any time during the voyage.

FISH ROOM CONSIDERATIONS

The density of .7 for fish and ice gives the capacity of the fish-hold. It is generally refrigerated by ice, but small refrigerating plants are sometimes used to save ice from melting during outbound voyages and to take care of transmission losses. Cunningham gives for a fish-hold capacity of 17,700 cu. ft. (500 cu. m.) a weight of 80 tons of ice. Such a fish-hold will be filled to the extent of 70 per cent. in 10 or 11 days at a rate of 25 to 30 tons a day. A period of about 10 days may be considered as a maximum in order to arrive at the home port with the catch still fresh. For the inside installation of the fish-hold light alloy sheathing, fig. 219 has been recommended by the Aluminium Development Association. In this construction one should notice that the vertical profiles of aluminium must not be used as pillars. They must be separate and independent of the fish-hold partitions. Insulating materials, such as granulated cork or isoflex, are usually employed and, whenever possible, it is advisable to place the insulating panels on the frame to reduce the heat transmission surfaces. The light alloy sheets can be glued direct to the cork or isoflex panels. The author has recommended (1946 and 1949) that in order to prevent contamination of the fish by germs, all shelves and partitions should be made in light alloy. No wood should be used in the fish-hold.

DUTCH COASTAL FISHING BOATS

by

W. ZWOLSMAN

THE Dutch fishing fleet once mainly consisted of flat-bottomed wooden vessels, but a start was made, during 1914–16, in converting the fleet to power, and a powered steel vessel—also of the flat-bottomed type—was built. The first engines had an output of 14 to 18 b.h.p. and were single-cylinder semi-diesels. Flat-bottomed craft were still used in the coastal fishery between 1914 and 1930, but nearly all new vessels were built of steel. Meanwhile the output of the engines had increased to 80 or 100 h.p.

A fisherman and a boatyard together designed a vessel with finer lines, similar to the herring luggers. The result was a transitional type, the first two Netherlands steel fishing cutters, *H.D. 12* and *T.X. 4*. In the 1930s there was some construction of vessels of a sharper design but, because of the world economic crisis, the fleet was not rebuilt on a large scale.

After the war it was decided to construct only steel vessels. The difference in the cost of repair between a wooden vessel and a steel one is substantial and profitable operation along the Dutch coast is vitally affected by the cost of maintenance. An investigation covering five post-war years of operation of both steel and wooden vessels, fitted with 30, 60 and 200 h.p. engines, revealed that in all three cases the cost of upkeep of the steel vessels was about 40 per cent. of that of the wooden ones. It was found that the cost of maintenance of steel fishing cutters, which were put into service in 1930, was confined so far as the hull was concerned, to cleaning and painting.

During the war, three types of vessels were designed by the author to serve as a pattern for the reconstruction of the fleet. They have the following dimensions:

	No. 1		*No. 2*		*No. 3*	
	ft.	*m.*	*ft.*	*m.*	*ft.*	*m.*
Length, overall	54	16.45	59	18.00	65.5	20.00
Breadth	15.4	4.70	16.7	5.10	17.7	5.40
Depth	7	2.15	7.5	2.30	8.2	2.50

TABLE XXXVIII

PARTICULARS OF MOTOR FISHING CUTTERS

	MAIN DIMENSIONS					COEFFICIENTS			GROSS REGISTER TONS	VOLUME OF FISH HOLD	ENGINE		SPEED IN KNOTS	REMARKS
Fig. No.	*L.O.A.*	*L.p.p.*	*Beam*	*Depth*	*Draft aft*	*Block*	*Midship*	*Prismatic*			*b.h.p.*	*r.p.m.*		
221	49.2 ft. (15.00 m.)	43.3 ft. (13.20 m.)	13.4 ft. (4.10 m.)	6.4 ft. (1.95 m.)	5.7 ft. (1.75 m.)	0.485	0.789	.615	21.40	10 cu. m. 350 cu. ft.	50	320	8	With fishwells
222	54 ft. (16.45 m.)	47.5 ft. (14.55 m.)	15.4 ft. (4.70 m.)	7.1 ft. (2.15 m.)	5.9 ft. (1.80 m.)	0.427	0.744	.572	32.50	28 cu. m. 1,000 cu. ft.	60/80	1,000/ 1,300 Red. 2.5: 1	8	With Kort nozzle rudder
223	54 ft. (16.45 m.)	47.5 ft. (14.55 m.)	15.4 ft. (4.70 m.)	7.1 ft. (2.15 m.)	5.9 ft. (1.80 m.)	0.427	0.744	.572	32.50	21 cu. m. 750 cu. ft.	80	350	8	
224	59 ft. (18.00 m.)	52.5 ft. (16.00 m.)	16.7 ft. (5.10 m.)	7.5 ft. (2.30 m.)	6.1 ft. (2.00 m.)	0.455	0.710	.641	42.38	36 cu. m. 1,300 cu. ft.	100	320	8½	
225	62.2 ft. (19.00 m.)	55.5 ft. (16.97 m.)	17.4 ft. (5.30 m.)	8 ft. (2.45 m.)	7.1 ft. (2.15 m.)	0.434	0.718	.605	46.34	31 cu. m. 1,100 cu. ft.	120	350	9	
—	65.5 ft. (20.00 m.)	58 ft. (17.70 m.)	17.7 ft. (5.40 m.)	8.2 ft. (2.50 m.)	7.2 ft. (2.20 m.)	0.437	0.712	.615	52.48	49 cu. m. 17,000 cu. ft.	120	350	8¾	With Kort nozzle
226/7	69 ft. (21.00 m.)	62 ft. (18.90 m.)	18.4 ft. (5.60 m.)	9 ft. (2.75 m.)	8.2 ft. (2.50 m.)	0.448	0.726	.616	63.59	46 cu. m. 1,600 cu. ft.	150	320	9¼	With Kort nozzle
228	74 ft. (22.50 m.)	63.5 ft. (19.35 m.)	19 ft. (5.80 m.)	9 ft. (2.75 m.)	8.2 ft. (2.50 m.)	0.460	0.746	.616	67.14	53 cu. m. 1,900 cu. ft.	200	320	9¾	
229	79 ft. (24.00 m.)	71 ft. (21.70 m.)	19.7 ft. (6.00 m.)	9.8 ft. (3.00 m.)	8.5 ft. (2.60 m.)	0.437	0.740	.590	80.89	65 cu. m. 2,300 cu. ft.	200	320	9½	With Kort nozzle

Fig. 220

Type No. 3 was fitted with a three-cylinder two-cycle marine diesel with an output of 120 h.p. at 350 r.p.m., the 59-ft. (18 m.) boat with a two-cylinder two-cycle marine diesel with 80 h.p. at 350 r.p.m., and the 54 ft. (16.45 m.) boat with a six cylinder high-speed marine diesel with an output of 60/80 h.p. at 1,000 to 1,300 r.p.m.

The difficulty of getting raw materials prevented immediate construction of the vessels, but in January 1946 the first were ready to be put into commission. To reduce the cost as much as possible and to save materials, the boats were welded electrically and since then almost all the new vessels have been built in this way. The present design conforms, without any major alteration, to that of the first three vessels planned in 1942.

There was some difference of opinion in 1942 about the shape the vessels should have. The design to give the highest speed did not produce the best sea boats, and the fishermen, when asked which they would prefer—a smoothly going vessel or a vessel making a quarter of a knot more, but less smoothly—chose the former.

Many shipyards have built these boats to the same drawings but they have not confined themselves only to the types developed during the war. In Table XXXVIII are particulars of various types of vessels. Table XXXIX shows the scantlings. Fig. 220 illustrates a fishing cutter of average dimensions and fig. 226 shows the typical lines of the standard form. Fig. 221 to 229 give plans of the general arrangement of the boats; fig. 230 shows the midship section of the welded construction and fig. 231 shows a similar drawing of the riveted design.

An interesting fact is that the 60 to 80 h.p. engine, running at 1,000 to 1,300 r.p.m., has been replaced in some cutters by a two-cylinder two-cycle marine diesel with 80 h.p. at 350 r.p.m. The Dutch fishermen have adapted themselves to the use of slow-speed engines, maintenance of which is much less costly than that of high-speed engines.

Three vessels, one of each type developed and built during the war, are fitted with a Kort nozzle rudder. Greater tractive power is obtained and it has not been necessary to renew the propeller of any post-war vessel using the Kort nozzle. Some vessels, built after the war, are, however, fitted with a fixed nozzle because of the great stress on the points of suspension of the rudder of larger vessels. This may entail costly maintenance. One vessel has a 3.6 ft. (1,100 mm.) diameter propeller and a Kort nozzle rudder, which is about the maximum size. If this size is exceeded the stress, in a high sea, becomes so great that repeated repairs are necessary. An advantage of the Kort system is that it prevents nets and lines getting entangled in the propeller and beating against the trawl boards.

Steel vessels are much more economical than wooden ones, as the investigation showed, and they are rapidly gaining in popularity. Last year (1952), the Netherlands shipyards built a few of the new type vessels for Belgian fishermen and there are now several others under construction. Some Belgian yards, too, are preparing to build boats from the same drawings, while several vessels are now in service in Israel, New Guinea and Curacao, having made the voyage to these countries under their own power. Their seaworthiness was praised by their crews.

All vessels are built from drawings approved and under the supervision of the Dutch Navigation Inspection, a service which took over the supervision of the coastal fishing fleet when the boats started to have more powerful

TABLE XXXIX

SCANTLINGS OF FISHING CUTTERS

Fig. No.	*Stem Plate*	*Stern Post*	*Rudder Shaft*	*Keel Plate*	*Sheer Strake*	*Other Shell Plates*	*Deck*	*Frames*	*Floors*	*Deck beams*	*Bar keel*	*Bilge keels*	*Bulk-heads*
221	.275 in. (7 mm.)	3 × 1 in. (75 × 25 mm.)	2.55 in. (65 mm.)	.275 in. (7 mm.)	.235 in. (6 mm.)	.235 in. (6 mm.)	.2 in. (5 mm.)	2 × 2 × .235 in. (50 × 50 × 6 mm.)	12 × .200 in. (300 × 5 mm.)	2½ × 2 × .235 in. (65 × 50 × 6 mm.)	6 × .315 in. (150 × 8 mm.)	5½ × .275 in. (140 × 7 mm.)	.235 in. (6 mm.)
222	.315 in. (8 mm.)	4 × 1¾ in (100 × 45 mm.)	3.15 in. (80 mm.)	,,	.275 in. (7 mm.)	.275 in. (7 mm.)	.235 in. (6 mm.)	2½ × 2 × .235 in. (65 × 50 × 6 mm.)	,,	4 × 2 × .235 in. (100 × 50 × 6 mm.)	5½ × .315 in. (140 × 8 mm.)	6 × .275 in. (150 × 7 mm.)	
223	,,	,,	,,	,,	,,	,,	,,	,,	,,	,,	,,	,,	,,
224	,,	4 × 2 in. (100 × 50 mm.)	3⅜ in. (85 mm.)	,,	,,	,,	,,	,,	12 × .235 in. (300 × 6 mm.)	4 × 2½ × .275 in. (100 × 65 × 7 mm.)	7 × ⅜ in. (180 × 10 mm.)	6 × .315 in. (150 × 8 mm.)	.235 to .275 in. (6 to 7 mm.)
225	,,	,,	3½ in. (90 mm.)	.315 in. (8 mm.)	.315 in. (8 mm.)	,,	,,	2½ × 2 × .275 in. (65 × 50 × 7 mm.)	14 × .235 in. (350 × 6 mm.)	,,	6 × 1 in. (150 × 25 mm.)	7 × .315 in. (180 × 8 mm.)	
—	,,	,,	,,	,,	,,	,,	,,	,,	,,	,,	7 × ⅜ in. (180 × 10 mm.)	,,	
226/7	.355 in. (9 mm.)	5 × 2 in. (125 × 50 mm.)	,,	,,	,,	,,	,,	(3 × 2½ × .235 in. (75 × 65 × 6 mm.)	,,	4 × ⅜ in. (100 × 10 mm.)	8 × ⅜ in. (200 × 10 mm.)	8 × .315 in. (200 × 8 mm.)	
228	,,	,,	,,	,,	,,	.315 in. (8 mm.)	,,	3 × 2½ × .275 in. (75 × 65 × 7 mm.)	14 × .275 in. (350 × 7 mm.)	4 × 2½ × .275 in. (100 × 65 × 7 mm.)	4¾ × ¾ in. (120 × 20 mm.)	,,	
229	,,	,,	3¾ in. (95mm.)	.355 in. (9mm.)	,,	,,	,,	4 × .275 in. (100 × 7 mm.)	,,	4¾ × .315 in. (120 × 8 mm.)	8 × ½ in. (200 × 12 mm.)	,,	

engines fitted in them. The standards required are about the same as those demanded by Lloyds Register of Shipping and the Bureau Veritas.

The reconstruction of the Dutch fishing fleet has been effected without any aid from the Government. Those who lost their vessels because of the war were, of course, compensated to some extent but not sufficiently to cover the price of a new vessel. It is therefore true to say that the Dutch fishermen have rebuilt the coastal fishing fleet from their own resources, and several fishermen who started with a small wooden fishing vessel are now owners of fishing cutters from 50 to 80 tons, costing from 120,000 to 200,000 Dutch guilders (£11,000 to £19,000; U.S. $32,000 to $53,000) each.

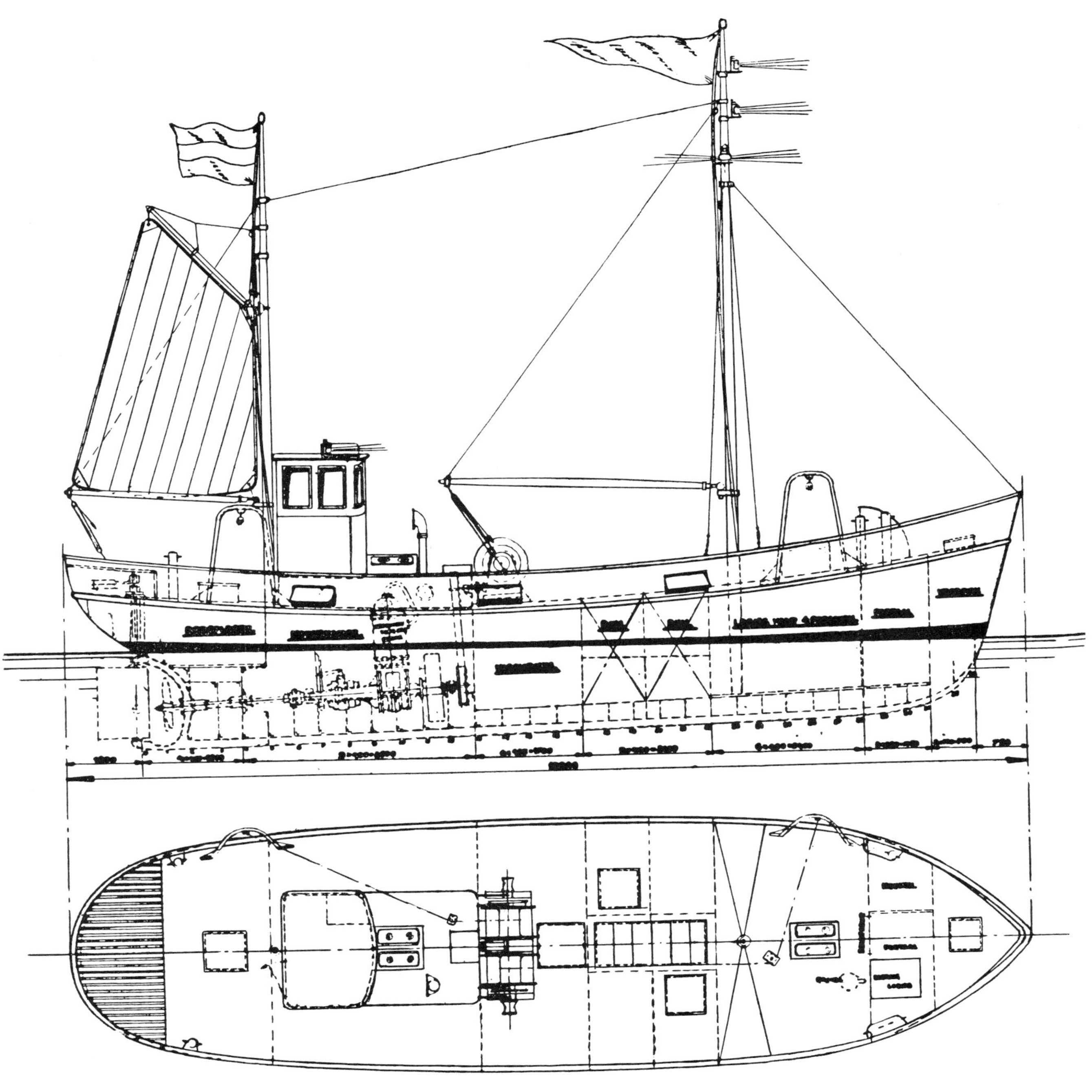

Fig. 221

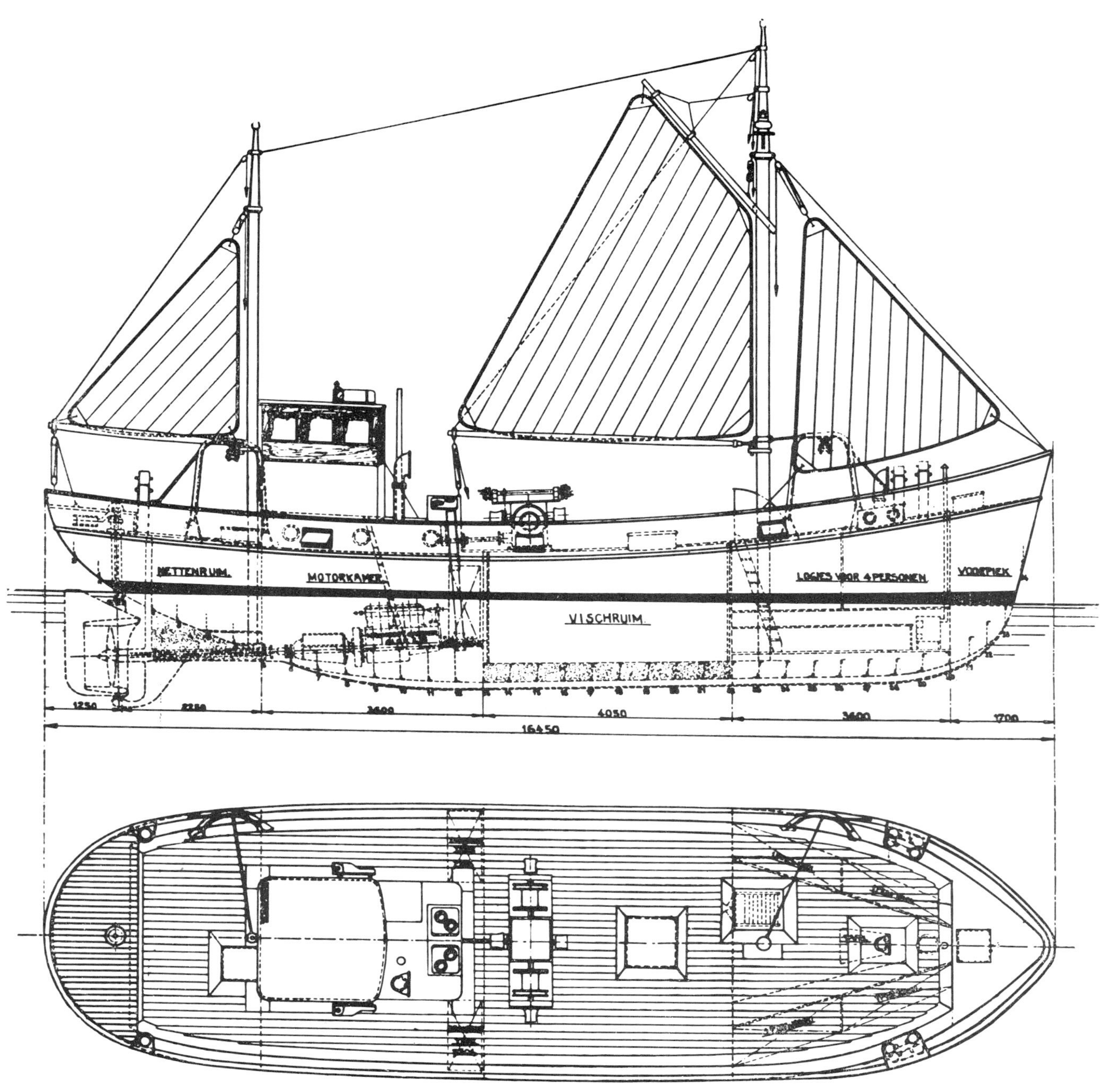
NETTENRUIM.
MOTORKAMER
VISCHRUIM.
LOGIES VOOR 4 PERSONEN
VOORPIEK
1250
2250
3600
4050
3600
1700
16450

Fig. 222

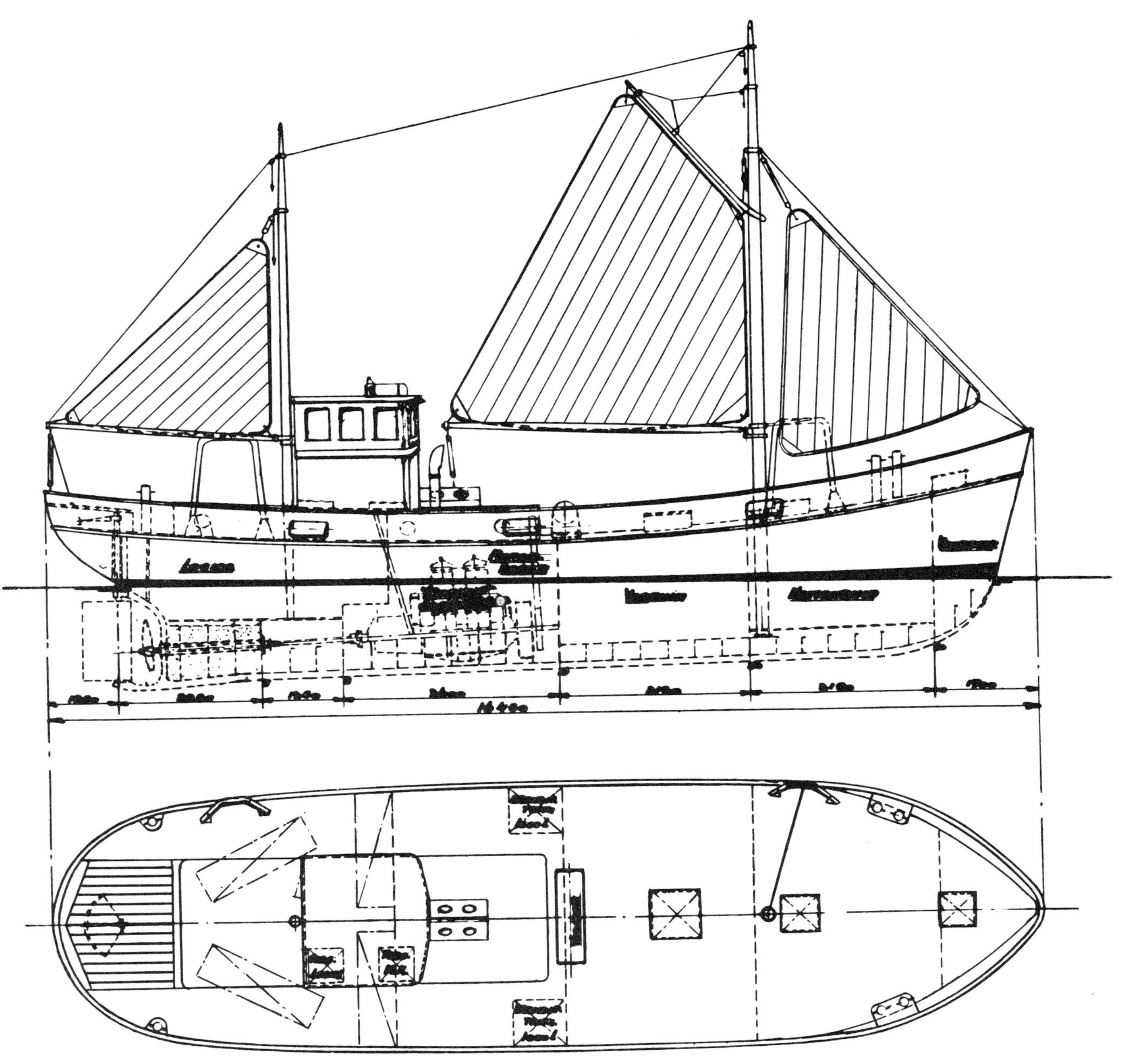

Fig. 223

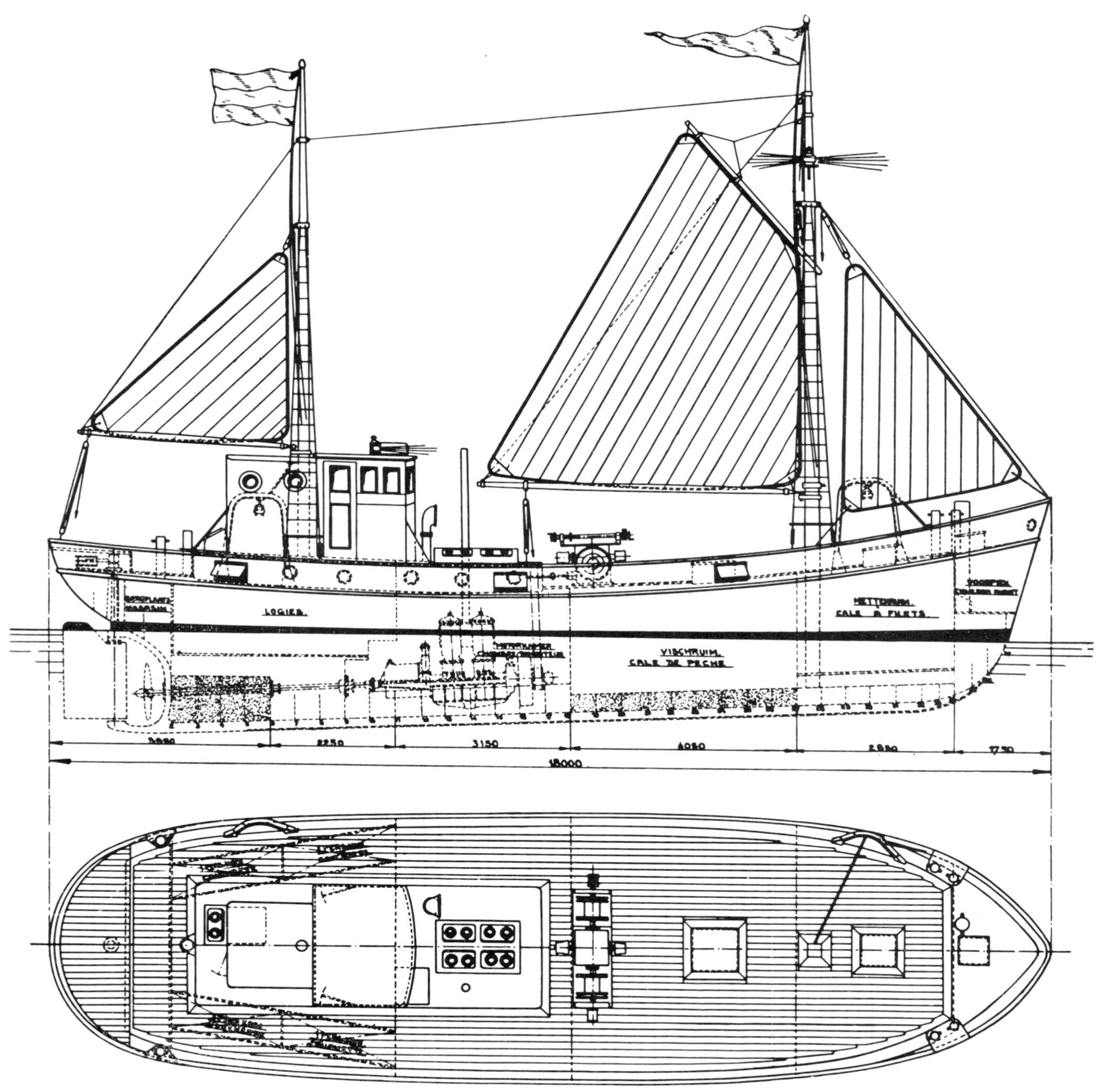

Fig. 224

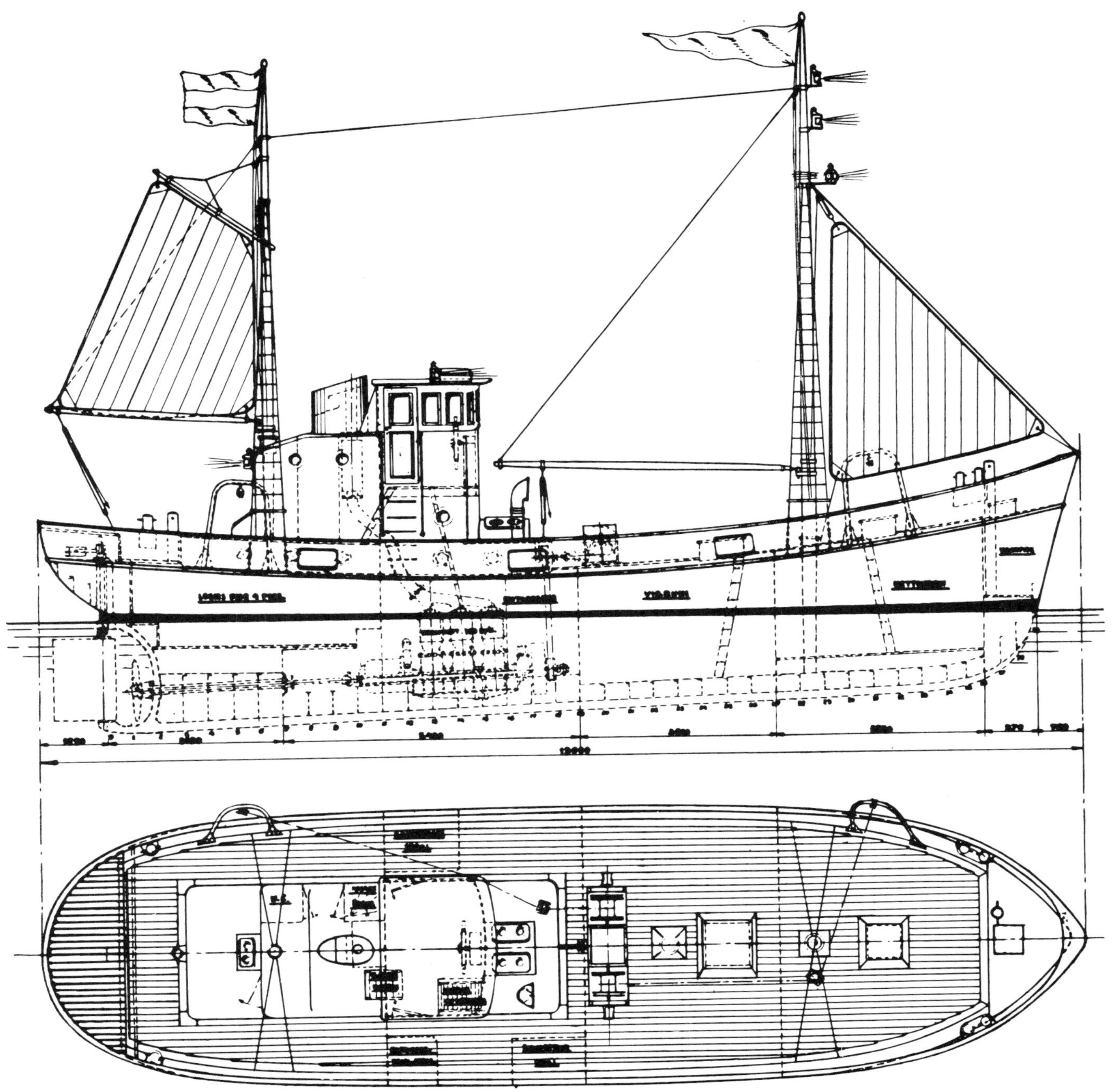

Fig. 225

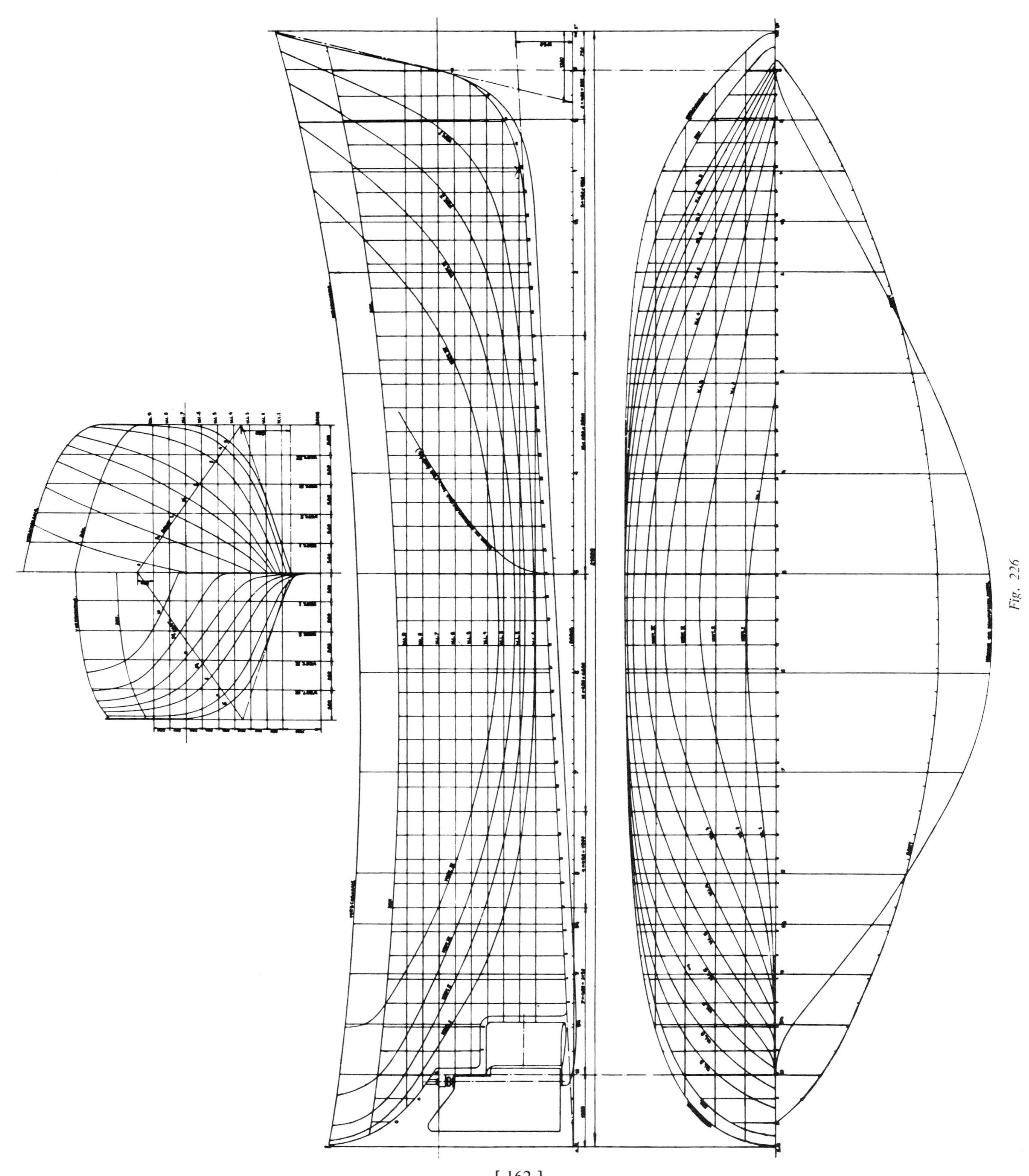

Fig. 226

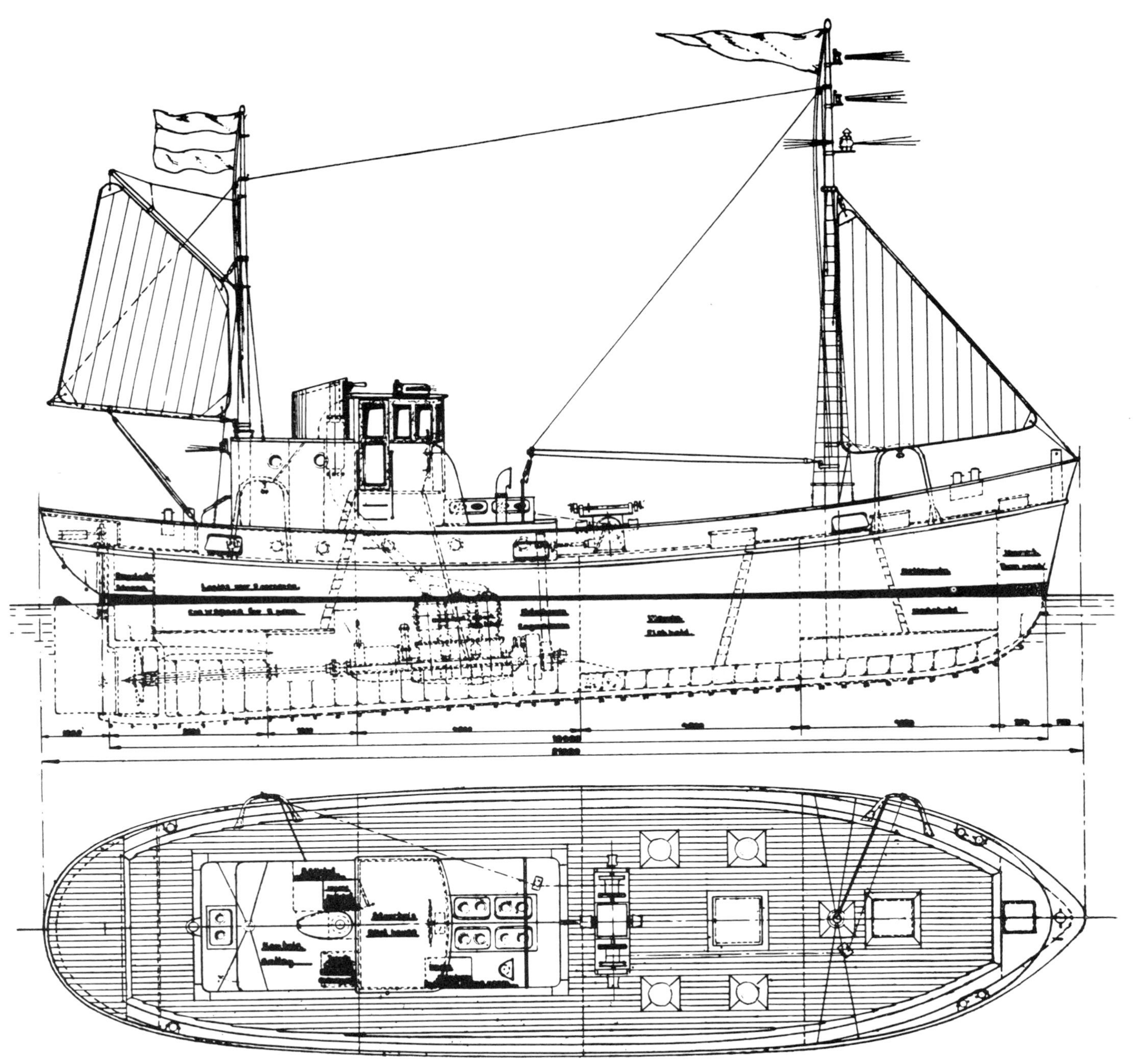

Fig. 227

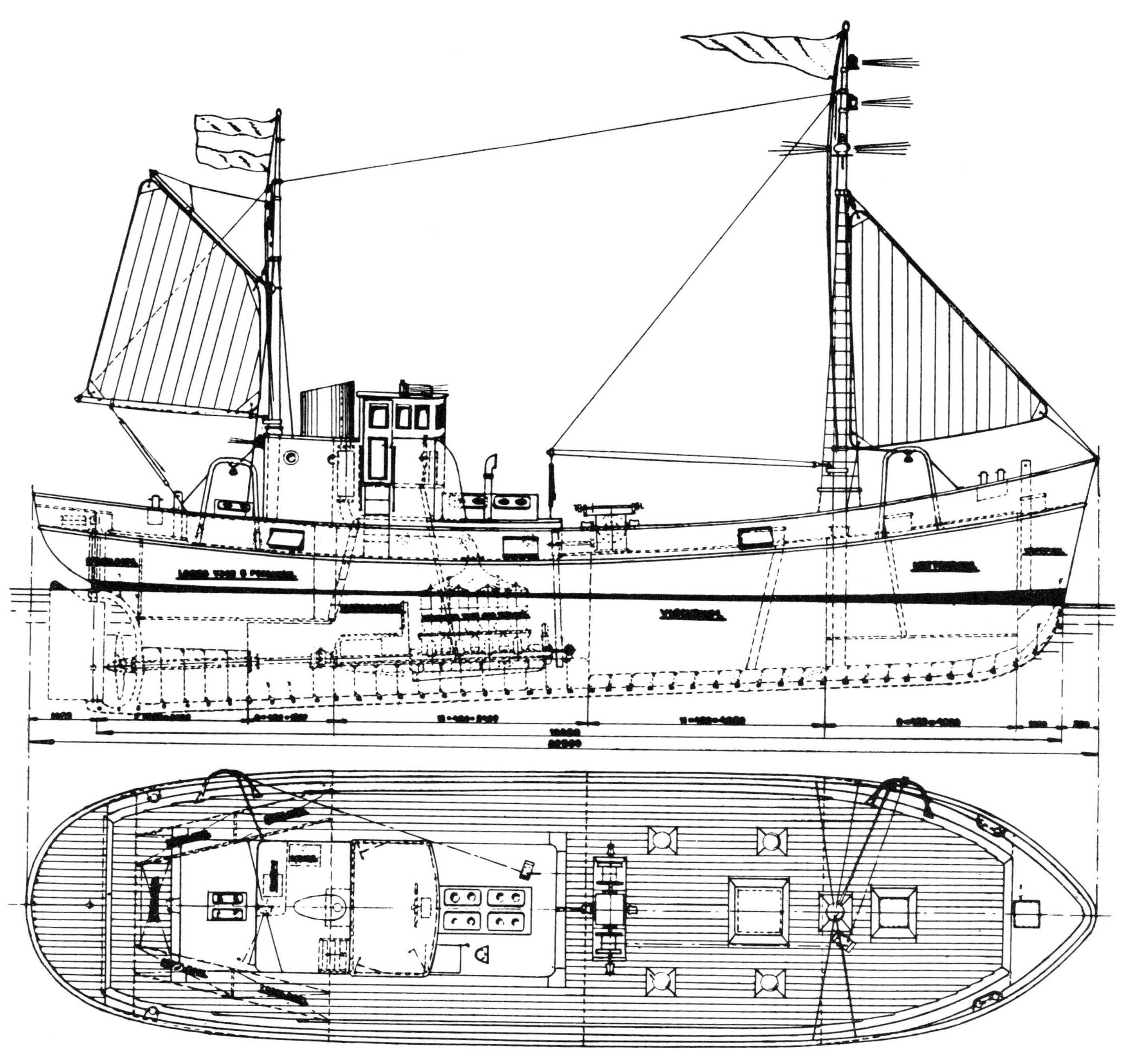

Fig. 228

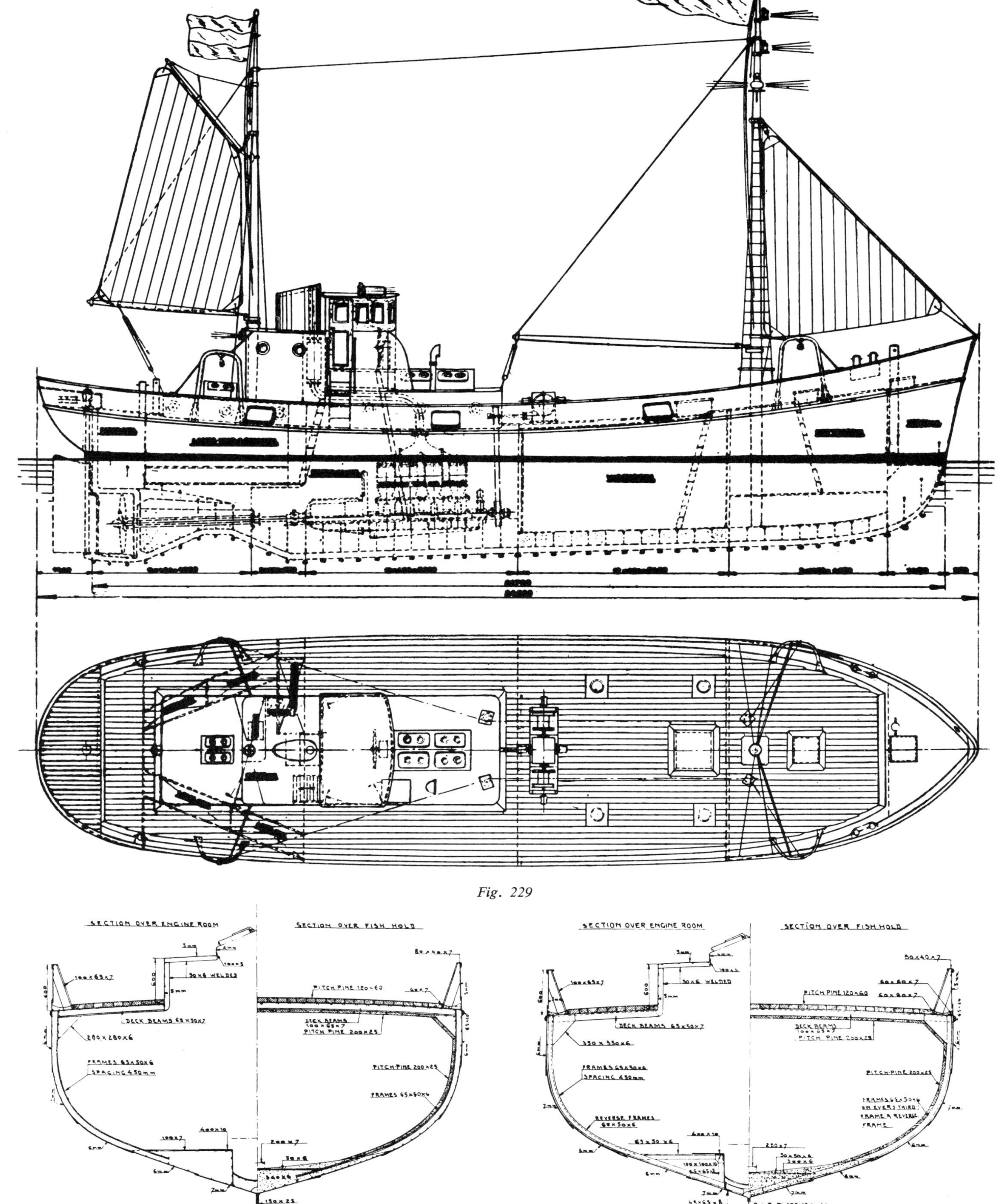

Fig. 229

Fig. 230

Fig. 231

GERMAN FISHING VESSELS

by

H. KANNT

A SURVEY of the last ten years of German trawler construction shows that the problem of economy has been particularly considered in relation to existing resources, fishing areas, methods of processing, consumer demand and the question of distribution. The modern German trawlers are the result.

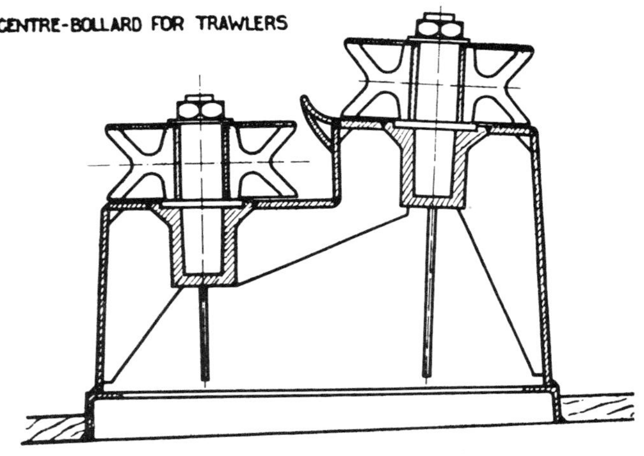

Fig. 232

After the World War II the German market for fresh fish declined and the problem was to find a satisfactory relationship between operating costs and revenue. In this respect it was considered necessary to keep both size and propulsion power of trawlers within limits not exceeding 600 gross tons and 1,200 i.h.p. This means

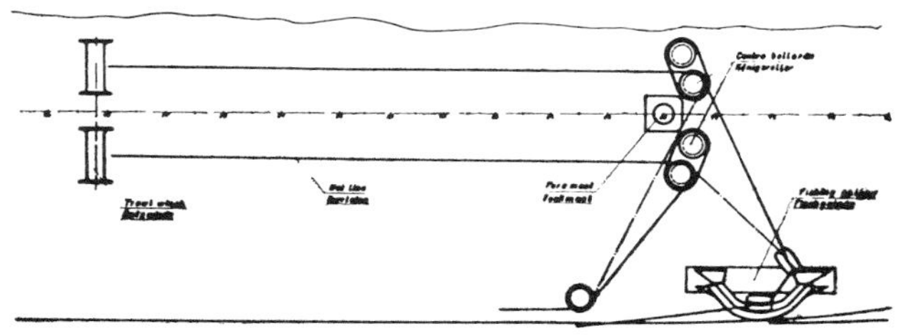

Fig. 233

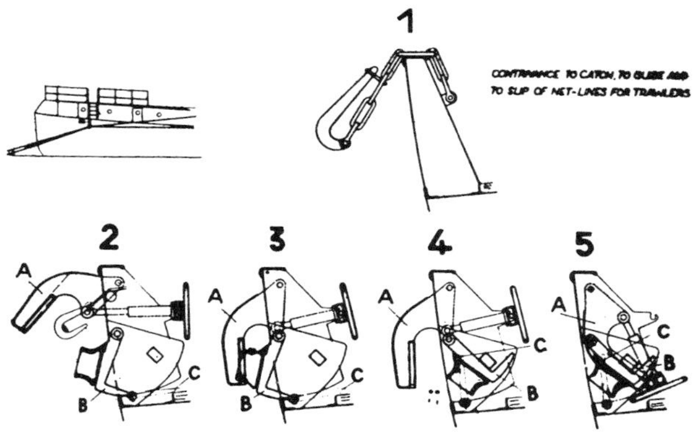

Fig. 234

that cargo space for fish has never exceeded 18,000 cu. ft. (500 cu. m.) in any German post-war trawler.

There is not a " freezing chain " in Germany to guarantee fish reaching the consumer in first-class condition. For this reason, two experiments with " deep

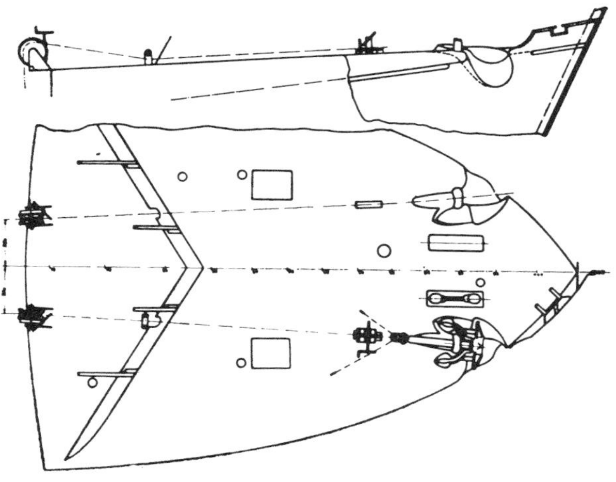

Fig. 235

freeze " vessels have not proved a success and both were reconverted into normal trawlers.

In the new German trawlers the fish-holds have the usual wooden linings, but recently " Iporka " and " Alfol ", both weight-saving materials, have been used as substitutes for cork. Aluminium linings have not found much favour because of their high cost. A saving of 20 per cent. in ice consumption has been made by fitting refrigerating coils in the ceilings of the fish-holds.

Although there has been no change in the catching methods, the conventional catching gear has been improved and strengthened. The fairlead system has been standardized and the centre fairleads arranged to enable the adjustment to equal length of the trawl warps, fig. 232 and 233. A mechanical towing block, fig. 234, has been developed. This gives better handling of the trawl warps and also helps to prevent accidents likely to happen through the gripping and slipping of the warps.

More powerful trawl winches have been installed and their driving unit shifted from the open deck into the superstructure. Because present-day captains prefer to operate over one side only, the gallows have been arranged on the starboard side. This provides a larger working area.

An innovation is the " Heuer " anchor hawse, fig. 235. This enables the anchor to be heaved from its exposed position into a recess on the forecastle ensuring that it always moves into correct position, an arrangement that has been especially appreciated in narrow ports and at discharging points.

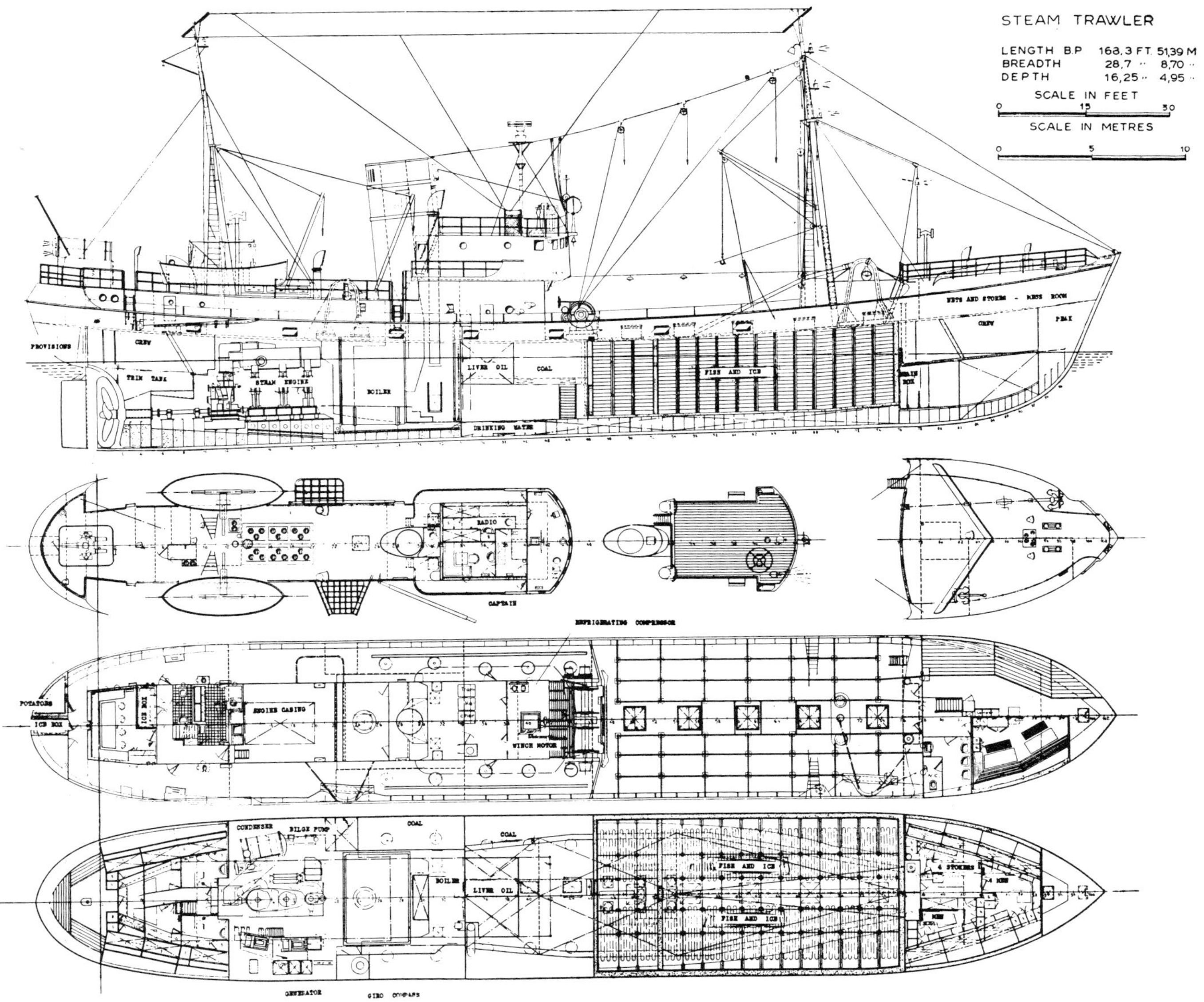

Fig. 236

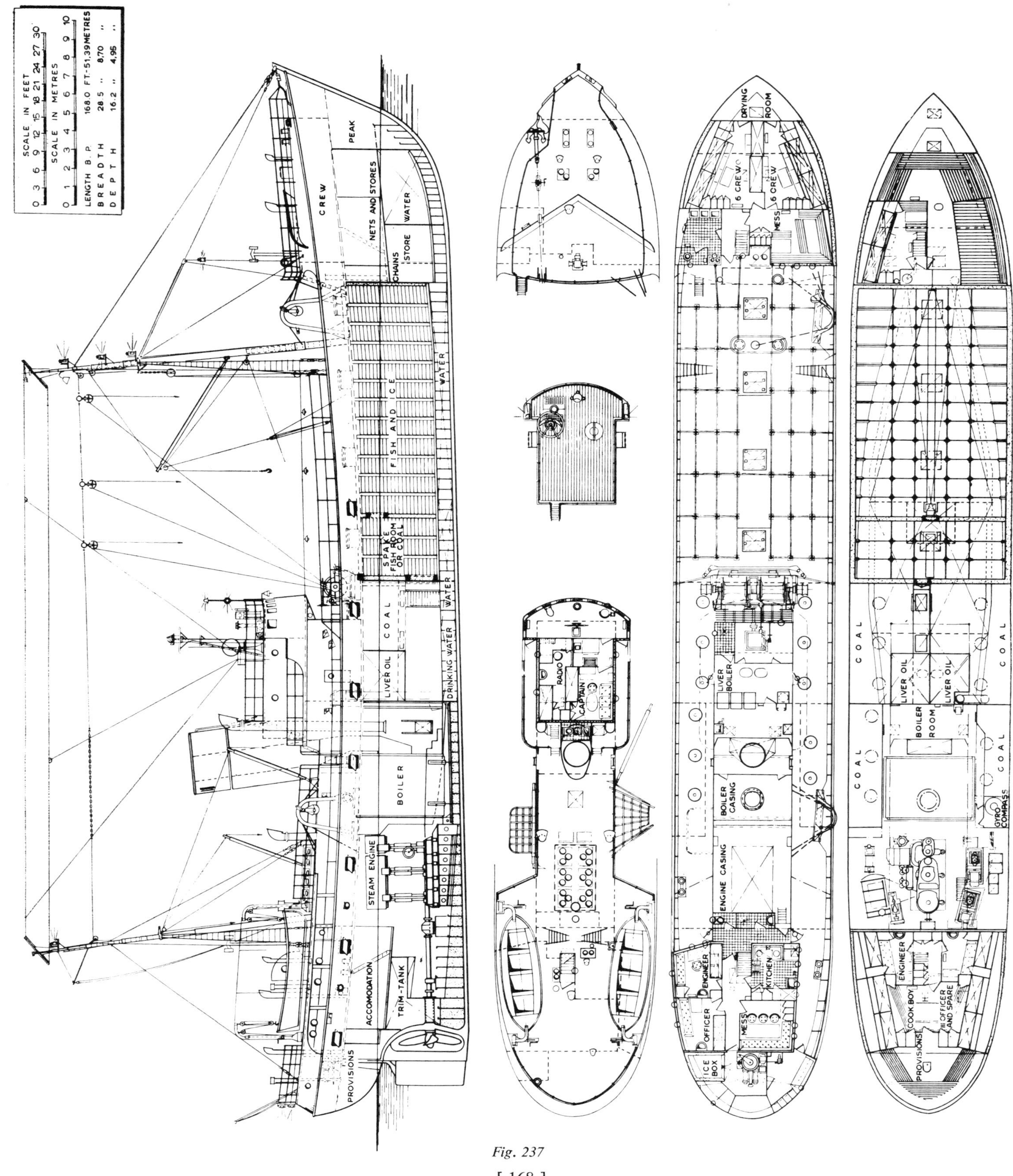

Fig. 237

There are no special legal provisions governing the accommodation of crews on German trawlers. Fig. 236 gives an example of the arrangement on board *Helgoland* but other owners sometimes have different ideas regarding use of available space, such as fig. 237, the trawler *Wartburg*.

Foremost attention has been paid to the problem of finding the most economical propulsion power for a trawler. In steam-driven vessels the Bauer-Wach exhaust vessel by using oil-burning machinery is only a theoretical one because fuel oil tanks can never be used for the storage of fish or fish products, but the spare cargo space for fish, on a coal burner, can always take coal. Water tube boilers are not favoured by the owners. It is considered that such boilers may be too difficult to handle and not able to cover adequately the high peaks of steam required during fishing operations. Recently, however, a combination of Scotch-type and water-tube

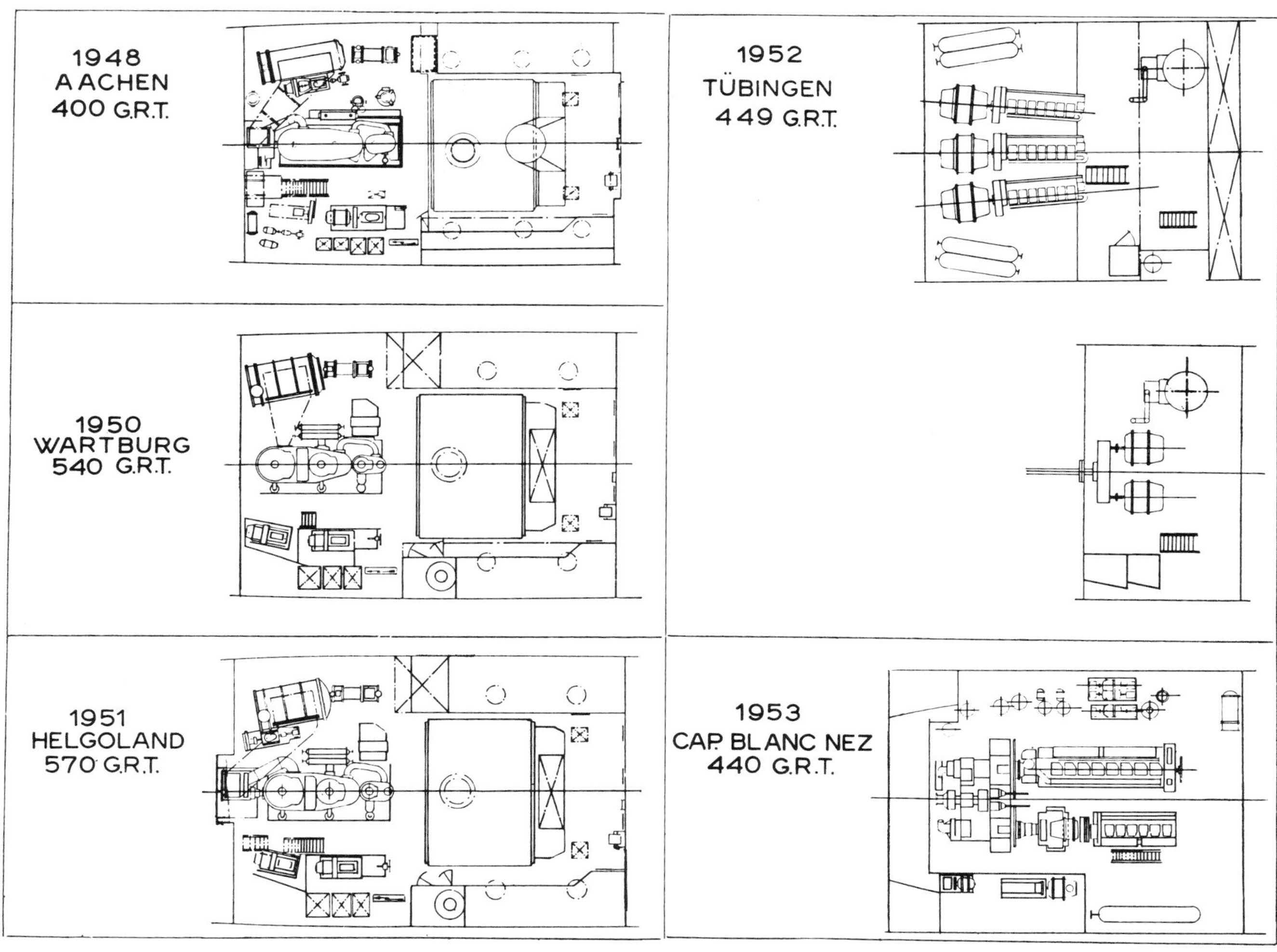

Fig. 238

turbine, which makes a 22 per cent. saving of steam or gives a 30 per cent. increase of power, is now predominant in Germany. A few steam trawlers which, for financial reasons, could not be fitted with exhaust turbines straight away, have been designed to permit the addition of such turbines at a later date.

The triple expansion engine, using coal burning Scotch-type boilers, is generally preferred. Oil burning did not lead to economical results, due to the high price of oil as compared with that of coal in Germany. Moreover, the possibility of increasing the operation range of a boiler, the so-called " Capus " boiler, has attracted attention because of its lower weight and better water circulation. Typical examples of modern trawlers, with and without an exhaust turbine, are *Wartburg* and *Helgoland.* Fig. 238 illustrates a number of different engine-room arrangements.

The development of the more distant fishing grounds has raised the demand to extend the radius of action without increasing the size of the trawlers. This has meant the design of a propulsion plant having a smaller fuel consumption per power unit. The influence of such a plant on the general arrangement may be seen in the trawlers *Tübingen* and *Cap Blanc Nez*. The *Cap Blanc Nez* has a propulsion plant consisting of two diesel engines of different power (" Father and Son "), coupled and geared to the shafting by a " Vulcan " gear. The total output of both engines may be transmitted to the propeller shaft for cruising, while under trawling operations the larger engine propels the boat while the smaller engine generates power for the electric trawl winch. Both engines may be used for trawling, the smaller engine being switched over to winch operation immediately it is required. This combination avoids the provision of an additional powerful generating set which previously was required to operate the trawl winch and was never an economical proposition. The diesel-electric plant in the trawler *Tübingen* has been installed in a similar way, an arrangement which provides for engine control from the bridge.

The trend of trawler development, in view of existing problems, may point towards the " factory " ship rather than to a revolutionary type of catching vessel. How far the development of the factory ship will go—a ship which processes its catch to ready packed fillets, marinated products, fish oils, and fish meals—will eventually be decided by a detailed calculation of operating cost and revenue. Such processing at sea will no doubt result in an improvement of the quality of fish products, but production costs will be increased by factory hands claiming the same level of wages and conveniences as those of the sea-faring personnel. Costs will also be affected by the fact that it will never be possible to make full use of the floating factories.

Prevailing conditions suggest that more use should be made of the offal and trash fish catch, rather than relying on the conventional landing of ice-cooled fresh fish to make the operation economical. There is plenty of steam available on board, which could be used to press offal and trash fish at sea. Such a valuable concentrate could be carried in special refrigerated spaces to be delivered to shore factories for processing.

SPANISH FISHING VESSELS

by

JOSE Ma. GONZALEZ-LLANOS Y CARUNCHO

ABOUT 2,000 boats are engaged in surface fishing off the Spanish coast. In addition, about 10,000 smaller, less profitable ones fish along the coast and employ a total of some 80,000 men. Trawling, mainly for hake, is done by dragging a net between two boats, known as pair trawling (" pareja ") and for other species than hake, such as large horse mackerel, by " bou " or otter trawling.

Boats that trawl in pairs near the coast are usually made of wood and the biggest is 75.4 ft. (23 m.) in length between perpendiculars. Main dimensions for this boat are given in Table XL, column 1, and the general arrangement drawings in fig. 239. Such boats do not go out for more than seven days at a time. The fishing grounds are depleted and landings small. The rising cost of fuel and supplies has worsened the situation and many of the vessels are fitted with uneconomical, old engines. Otherwise they are strong and seaworthy; the keel, stem, stern post and frames are oak, the keelson of eucalyptus, and the remainder of the hull, pine.

As the fishing grounds within easy reach of the Spanish coast become exhausted it was necessary to go further afield. It was clear that these boats were not the most suitable for arduous and distant trips. They were slow and had small capacity, but, even so, they have been taken to the distant fishing grounds, sometimes with 20 tons of coal stacked on deck to supplement their bunker capacity. Often this extra coal has been washed overboard.

With the development of distant fishing grounds came a trend towards building bigger boats, highly-powered and of better sea-worthiness. For the main dimensions of a typical boat of this class see Table XL, column 2, and for the general arrangement drawings, fig. 240.

MODERN " PAREJA " BOATS

Since 1940 diesels have been mostly used in new boats and the modern Spanish trawler fleet have virtually abandoned the steam engine. Furthermore, the rigid structure required by the diesel and the larger size boat needed for fishing the distant grounds has led to the widespread use of steel hulls. Such boats are built on more technically advanced lines, their shape being determined by the results of tests in shipbuilding experimental tanks in Madrid and elsewhere.

The freeboard of Spanish " parejas " is perhaps higher than in trawlers of other countries, but it is based on experience and these boats can continue to fish when bigger boats have to heave-to or drift with the wind. Further, in spite of the height of the freeboard, there is no difficulty in lifting the net on board.

Welding is being used now in the construction of the superstructures, bulkheads—with the exception of their connection to the hull—and decks, except for the beams and deck stringers. It is also being used in the double bottoms and tank tops, stem and stern posts, engine foundation, and the whole of the structure except the outer plating where the longitudinal seams and the connection with the frames are riveted, but the vertical seams are welded. The fish-room is usually insulated with compressed cork and wood linings, and wood is used for pound boards and stanchions where the catch is stowed in layers of ice after gutting and heading. Because of the long trips that have to be made, it would be better if aluminium alloy materials were used as in other countries.

A few boats now have refrigerating machinery to reduce the amount of ice they otherwise have to carry.

Diesels are used without exception in these boats. They range in b.h.p. from 350 to 600, usually direct-coupled, two- or four-cycle, solid injection and direct reversing. In some cases change of direction is through a reverse gear, and sometimes engines are fitted with reduction gears. A good many of the engines are still imported but firms in Spain are now building engines in considerable numbers.

The trawl winch is usually operated by an auxiliary 75 b.h.p. diesel with belt drive and this is found to be most satisfactory. A 6-kw. engine dynamo compressor set is usually installed along with an auxiliary dynamo which can be coupled to the trawl winch motor by belt. The steering gear is usually of the tiller-chain type and hand operated, although hydraulic transmission has recently been introduced. For main dimensions of a typical boat of this class, see Table XL, column 3.

TABLE XL

	Column 1	*Column 2*	*Column 3*	*Column 4*	*Column 5*	*Column 6*
	75.4 ft. (23 m.) Pareja trawler Fig. 239	*91.2 ft. (27.8 m.) Pareja trawler Fig. 240*	*88.9 ft. (27.1 m.) Pareja trawler Fig. 241/2*	*88.9 ft. (27.1 m.) Pareja trawler Fig. 246*	*209 ft. (63.75 m.) Otter trawler Fig. 253/4*	*108 ft. (32.8 m.) Pareja trawler Fig. 258*
Length, over-all	82 ft. (25.00 m.)	100 ft. (30.33 m.)	105 ft. (31.97 m.)	101 ft. (30.80 m.)	235 ft. (71.75 m.)	116 ft. (35.49 m.)
Length, waterline	77 ft. (23.58 m.)	94 ft. (28.30 m.)	95 ft. (29 m.)	95 ft. (29 m.)	218 ft. (66.60 m.)	108 ft. (32.80 m.)
Length, between perpendiculars	75.4 ft. (23 m.)	91.2 ft. (27.86 m.)	88.9 ft. (27.13 m.)	88.9 ft. (27.13 m.)	209 ft. (63.75 m.)	98.5 ft. (30.00 m.)
Breadth, moulded	18.2 ft. (5.55 m.)	21.6 ft. (6.60 m.)	20.5 ft. (6.25 m.)	20.5 ft. (6.25 m.)	35.2 ft. (10.75 m.)	22.5 ft. (6.85 m.)
Depth	12.0 ft. (3.66 m.)	13.1 ft. (4.0 m.)	12.3 ft. (3.75 m.)	11.6 ft. (3.45 m.)	19.3 ft. (5.90 m.)	12.8 ft. (3.90 m.)
Mean draft, loaded	9.9 ft. (3.01 m.)	11.1 ft. (3.37 m.)	10.8 ft. (3.28 m.)	10.3 ft. (3.13 m.)	17.0 ft. (5.20 m.)	10.7 ft. (3.25 m.)
Dead weight	65 tons	114 tons	117 tons	99 tons	1,300 tons	149 tons
Effective load	40 tons (catch and ice)	90 tons (catch and ice)	90 tons (catch and ice)	75 tons (catch and ice)	1,000 tons	110.0 tons (catch and ice)
Fish-room capacity	2,720 cu. ft. (77 cu. m.)	5,440 cu. ft. (154 cu. m.)	5,370 cu. ft. (152.25 cu. m.)	4,450 cu. ft. (126 cu. m.)	46,000 cu. ft. (1,300 cu. m.)	8,200 cu. ft. (232.5 cu. m.)
Fuel	45 tons coal	53 tons oil	46 tons oil	44 tons oil	480 tons oil	70 tons oil
Water tanks	19 tons	20 tons	14 tons	4 tons	89 tons	20 tons
Propulsion machinery	Triple expansion 120 i.h.p. engine and Scotch boiler 180 lb. per sq. in. (12.2 kg./sq. cm.)	Triple expansion 180 i.h.p. engine and Scotch boiler 200 lb. per sq. in. (13.6 kg./sq. cm.)	430 b.h.p. diesel	315 b.h.p. diesel	1,200 b.h.p. diesel	450 b.h.p. diesel
Speed, service	8.5 knots	9 knots	10.5 knots	10.5 knots	10.5 knots	10.5 knots
Speed, trial	8.9 knots	9.5 knots	12 knots	12 knots	13 knots	12 knots
Radius under full power	15 days	22 days	24 days	26 days	30,000 miles	9,250 miles
Displacement, loaded	237 tons	351 tons	310 tons	256 tons	2,300 tons	346 tons
Gross register tonnage	124.51 tons	187 tons	202.27 tons	154 tons	1,360 tons	224.14 tons
Crew	13	13	13	13	53	18
Block coefficient	0.610	0.575	0.519	0.470	0.672	0.484
Prismatic coefficient	0.680	0.640	0.643	0.598	0.706	0.621
Midship section coefficient	0.898	0.900	0.798	0.784	0.956	0.780
Waterplane coefficient	0.820	0.780	0.792	0.720	0.860	0.752
Longitudinal location of centre of buoyancy	1.28 ft. forward (0.390 m.)	1.08 ft. forward (0.33 m.)	1.05 ft. aft (0.32 m.)	1.44 ft. aft (0.44 m.)	1.28 ft. aft (0.39 m.)	0.23 ft. aft (0.07 m.)
Half-entrance angle at waterline	36°	48°	33°	23°	31°	24°
Half-run angle, tangent at waterline in plane of screw	37°	41°	35°	22°	26°	26°
Metacentric height, unloaded	1.31 ft. (0.4 m.)	1.84 ft. (0.56 m.)	1.18 ft. (0.36 m.)	1.69 ft. (0.515 m.)	1.84 ft. (0.56 m.)	1.87 ft. (0.57 m.)
Metacentric height on leaving port	1.25 ft. (0.38 m.)	1.84 ft. (0.56 m.)	1.5 ft. (0.46 m.)	2.16 ft. (0.659 m.)	2.7 ft. (0.825 m.)	1.81 ft. (0.55 m.)
Metacentric height, on reaching fishing ground	1.34 ft. (0.41 m.)	1.74 ft. (0.53 m.)	1.6 ft. (0.94 m.)	2.13 ft. (0.650 m.)	2.55 ft. (0.775 m.)	1.71 ft. (0.52 m.)
Metacentric height, at beginning of return journey	1.48 ft. (0.45 m.)	2.00 ft. (0.61 m.)	1.87 ft. (0.57 m.)	1.97 ft. (0.600 m.)	2.29 ft. (0.698 m.)	2.06 ft. (0.63 m.)
Metacentric height, on reaching port	1.34 ft. (0.41 m.)	1.77 ft. (0.54 m.)	1.67 ft. (0.51 m.)	1.84 ft. (0.560 m.)	2.1 ft. (0.67 m.)	1.97 ft. (0.60 m.)

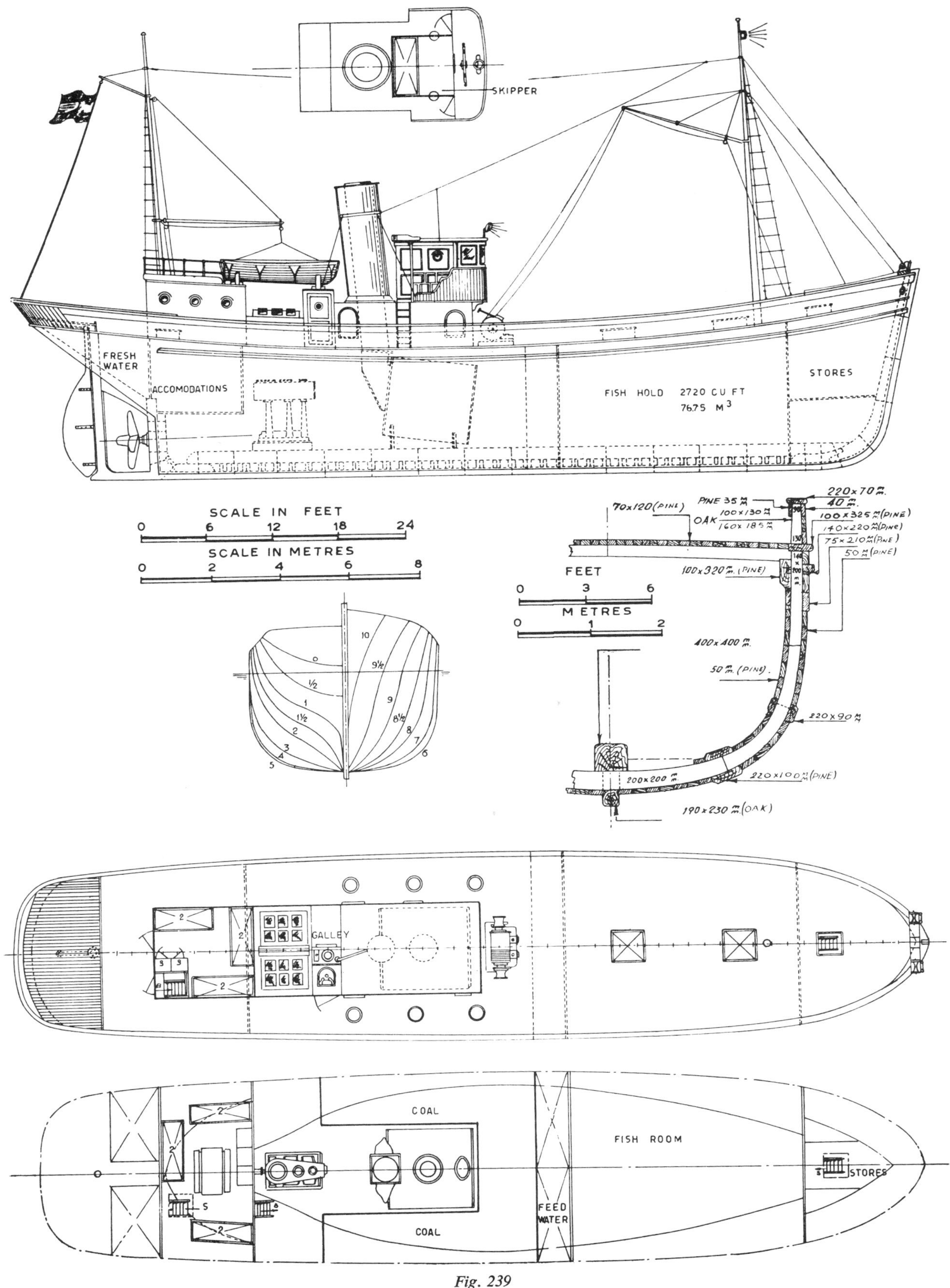
SKIPPER
FRESH WATER
ACCOMODATIONS
FISH HOLD 2720 CU FT
76.75 M3
STORES
SCALE IN FEET
0 6 12 18 24
SCALE IN METRES
0 2 4 6 8
FEET
0 3 6
METRES
0 1 2
70x120 (PINE)
PINE 35 m/m
OAK 100x130 m/m
160x185 m/m
220x70 m/m
40 m/m
100x325 m/m (PINE)
140x220 m/m (PINE)
75x210 m/m (PINE)
50 m/m (PINE)
100x320 m/m (PINE)
400x400 m/m
50 m/m (PINE)
220x90 m/m
200x200 m/m
220x100 m/m (PINE)
190x230 m/m (OAK)
GALLEY
COAL
FISH ROOM
FEED WATER
STORES

Fig. 239

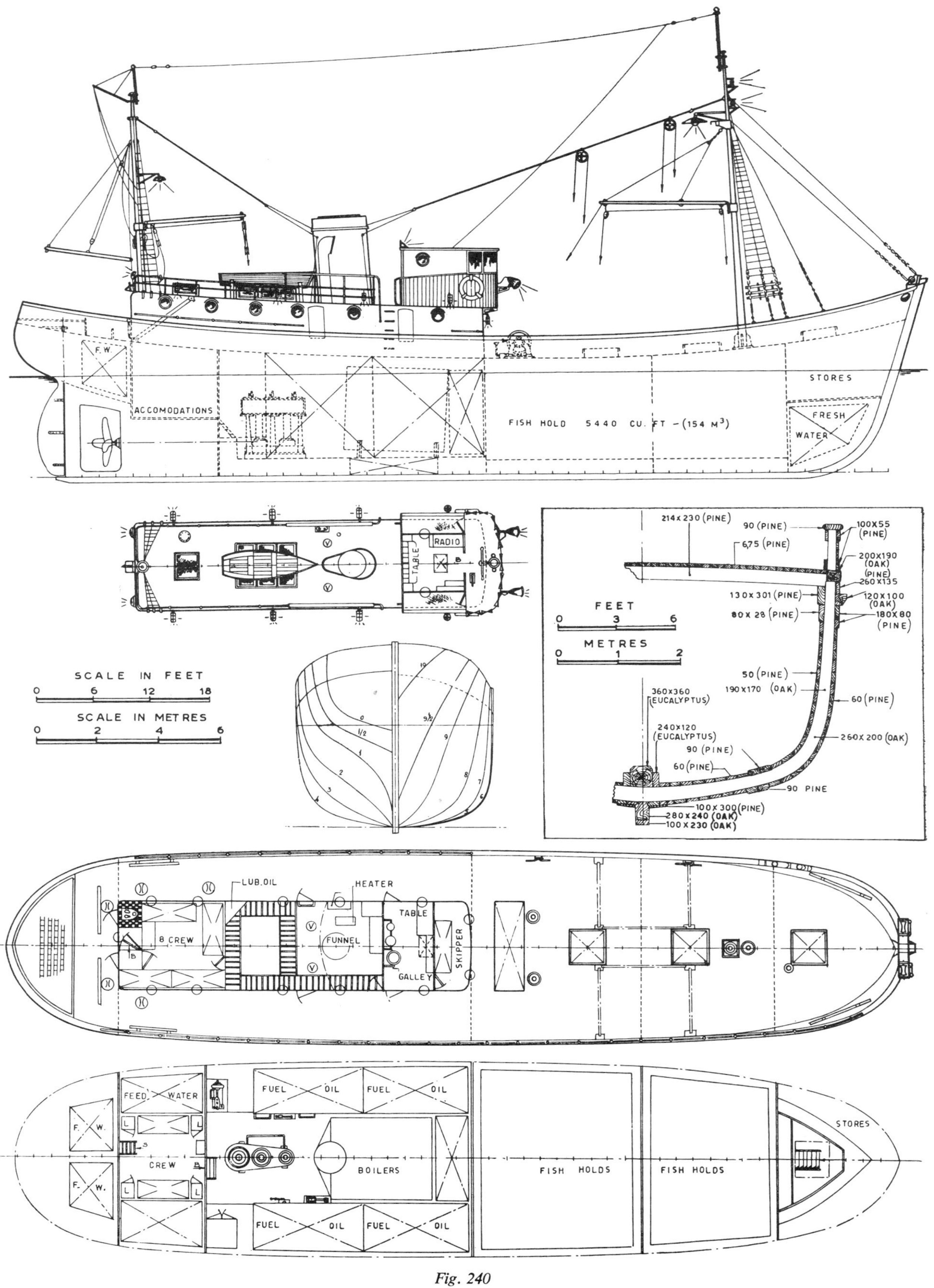
F.W.
ACCOMODATIONS
FISH HOLD 5440 CU. FT – (154 M³)
STORES
FRESH
WATER
RADIO
TABLE
SCALE IN FEET
0 6 12 18
SCALE IN METRES
0 2 4 6
FEET
0 3 6
METRES
0 1 2
214x230 (PINE)
90 (PINE)
100X55 (PINE)
6,75 (PINE)
200x190 (OAK)
(PINE)
260X135
130x301 (PINE)
120X100 (OAK)
80X 28 (PINE)
180X80 (PINE)
50 (PINE)
190x170 (OAK)
60 (PINE)
360x360 (EUCALYPTUS)
260X200 (OAK)
240x120 (EUCALYPTUS)
90 (PINE)
60 (PINE)
90 PINE
100X300 (PINE)
280x240 (OAK)
100X230 (OAK)
LUB. OIL
HEATER
TABLE
8 CREW
FUNNEL
SKIPPER
GALLEY
FEED WATER
FUEL OIL
FUEL OIL
F. W.
F. W.
CREW
BOILERS
FISH HOLDS
FISH HOLDS
STORES
FUEL OIL
FUEL OIL

Fig. 240

Table XLIA

Condition	Leaving port						Arrival at fishing ground						Leaving fishing ground						Return to port					
	Weight in ton	Centre of gravity in m.					Weight in ton	Centre of gravity in m.					Weight in ton	Centre of gravity in m.					Weight in ton	Centre of gravity in m.				
		Vertical from base		Horizontal				Vertical from base		Horizontal				Vertical from base		Horizontal				Vertical from base		Horizontal		
				Lever	Moment					Lever	Moment					Lever	Moment					Lever	Moment	
		Height	Moment		Forward	Aft		Height	Moment		Forward	Aft		Height	Moment		Forward	Aft		Height	Moment		Forward	Aft
Hull and equipment	160.51		471.04			70.21																		
Machinery	26.49		61.64			96.93																		
Displacement, unloaded	187.00	2.85	532.68	+ 0.893		167.14	187.00	2.85	532.68	+ 0.893		167.14	187.00	2.85	532.68	+ 0.893		167.14	187.00	2.85	532.68	+ 0.893		167.14
Additional items:																								
Officers and effects																								
Crew and effects	1.40	4.80	6.72	+ 7.00		+9.80	1.40	4.80	6.72	+ 7.00		9.80	1.40	4.80	6.72	+ 7.00		9.80	1.40	4.80	6.72	+ 7.00		9.80
Provisions	0.70	5.00	3.50	—11.50	8.05		0.60	5.00	3.00	—11.50	6.90		0.55	5.00	2.75	—11.50	6.33		0.20	5.00	1.00	—11.50	2.30	
Chandlery (soap, lamp-oil)	0.50	2.30	1.15	+ 10.00		5.00	0.40	2.30	0.92	+ 10.00		4.00	0.30	2.30	0.69	+ 10.00		3.00	0.20	2,30	0.46	+ 10.00		2.00
Fresh water	13.74	3.21	44.11	+ 12.04		165.43	11.00	3.00	33.00	+ 12.04		132.44	6.00	2.80	16.80	+ 12.04		72.24	3.00	2.60	7.80	+ 12.04		36.12
Salt water																								
Hold luggage																								
Mail																								
Cargo—fish													60.00	2.40	144.0	— 6.0	360.0		60.0	2.40	144.0	— 6.0	360.0	
,, ice	30.0	1.8	54.0	— 5.0	150.0		28.0	1.8	50.4	— 5.0	140.0		27.0	1.8	48.0	— 5.0	135.0		20.0	1.8	36.0	— 5.0	100.0	
Rigging	3.5	4.55	15.93	+ 13.38		46.83	3.50	4.55	15.93	+ 13.38		46.83	3.50	4.55	15.93	+ 13.38		46.83	3.5	4.55	15.93	+ 13.38		46.83
Feed water																								
,, ,, in reserve																								
Fuel—coal	2.00	4.1	8.2	+ 10.25		20.50	1.5	4.1	6.15	+ 10.25		15.35	1.0	4.10	4.1	+ 10.25		10.25	0.5	4.1	2.05	+ 10.25		5.12
,, diesel oil	45.72	2.43	111.10	+ 1.12		51.21	34.0	2.0	68.0	+ 1.8		61.2	22.0	1.8	35.6	+ 3.4		74.8	10.0	1.5	15.0	+ 4.3		43.0
,, boiler oil																								
,, gasoline																								
Lubricating oil	3.5	1.0	3.50	+ 7.10		24.85	3.00	1.00	3.00	+ 7.10		21.3	1.0	1.0	1.0	+ 7.0		7.0	0.5	1.0	0.5	+ 7.0		3.55
Liver oil																								
Trim water ballast, forward	3.0	1.78	5.35	—11.95	35.85		3.00	1.78	5.35	—11.95	35.85													
Displacement	291.06	2.70	786.24	+ 1.02	193.9	490.76 296.86	273.40	2.66	725.15	+ 1.01	182.75	458.06 275.31	309.75	2.62	812.27	—0.374	501.33 110.27	391.06	286.3	2.67	762.14	— 0.52	462.3 148.74	313.56

Table XLIB

Condition	Displacement in salt water	Mean draft		Static stability															Trim			
				KM_T		KG		GM_T		Free surface		GM_T corrected		GZ maximum			Range of stability		KM_L		KG corrected	
	ton	ft.	m.	ft.	m.	ft.	m.	ft.	m.	in.	m.	ft.	m.	ft.	m.	degrees	degrees		ft.	m.	ft.	m.
Unloaded	187.0	7.92	2.42	10.5	3.21	9.35	2.85	1.18	0.36			1.18	0.36	0.8	0.244	35	74.3		107	32.75	9.35	2.85
Leaving port	291.06	10.4	3.16	10.6	3.22	8.85	2.70	1.7	0.52	2.4	0.06	1.5	0.46	0.52	0.158	30	80.0		91.5	27.90	9	2.76
Arrival to fishing ground	273.4	10	3.04	10.5	3.20	8.7	2.66	1.78	0.54	2	0.05	1.6	0.49	0.59	0.180	33	> 90.0		94	28.60	8.9	2.71
Leaving ,, ,,	309.75	10.7	3.28	10.6	3.23	8.6	2.62	2	0.61	1.6	0.04	1.87	0.57	0.69	0.210	33	> 90.0		90	27.50	8.7	2.66
Return to port	286.3	10.8	3.13	10.5	3.21	8.75	2.67	1.78	0.54	1.2	0.03	1.67	0.51	0.64	0.196	33	> 90.0		92	28.10	8.85	2.70

Condition	Trim																							
	GM_L		L.C.B.		L.C.G.		Difference between B and G		Change of trim		Centre of water plane buoyancy		Corrected trim				Resulting draft				Displacement
													Forward		Aft		Forward		Aft		
	ft.	m.	ft.	m.	ft.	m.	ft.	m.	ft.	m.	ft.	m.	ft.	m.	ft.	m.	ft.	m.	ft.	m.	ton
Unloaded	96	29.9	+ .95	+ 0.290	+ 2.67	+ 0.813	+ 1.72	+ 0.603	+ 1.92	+ 0.584	+ 1.56	+ 0.475	1	0.304	.92	0.280	5.2	1.552	10.7	3.264	187.0
Leaving port	82	25.14	+ .985	+ 0.300	+ 3.35	+ 1.02	+ 2.36	+ 0.720	+ 2.72	+ 0.830	+ 2.95	+ 0.900	1.45	0.441	1.28	0.389	7.2	2.155	13.5	4.113	291.06
Arrival to fishing ground	85	25.89	+ .92	+ 0.280	+ 3.31	+ 1.01	+ 2.39	+ 0.730	+ 2.67	+ 0.818	+ 2.79	+ 0.850	1.46	0.433	1.26	0.385	6.7	2.043	13.1	3.989	273.4
Leaving ,, ,,	81.5	24.84	+ 1.05	+ 0.320	— 1.22	—0.374	— 2.27	—0.694	— 2.65	—0.809	+ 3	+ 0.910	1.31	0.429	1.25	0.380	10.3	3.145	11.4	3.464	309.75
Return to port	83.5	25.40	+ .975	+ 0.297	— 1.71	—0.520	—2.69	—0.817	— 3.06	—0.934	+ 2.89	+ 0.880	1.62	0.495	1.44	0.439	10.0	3.061	10.7	3.255	286.3

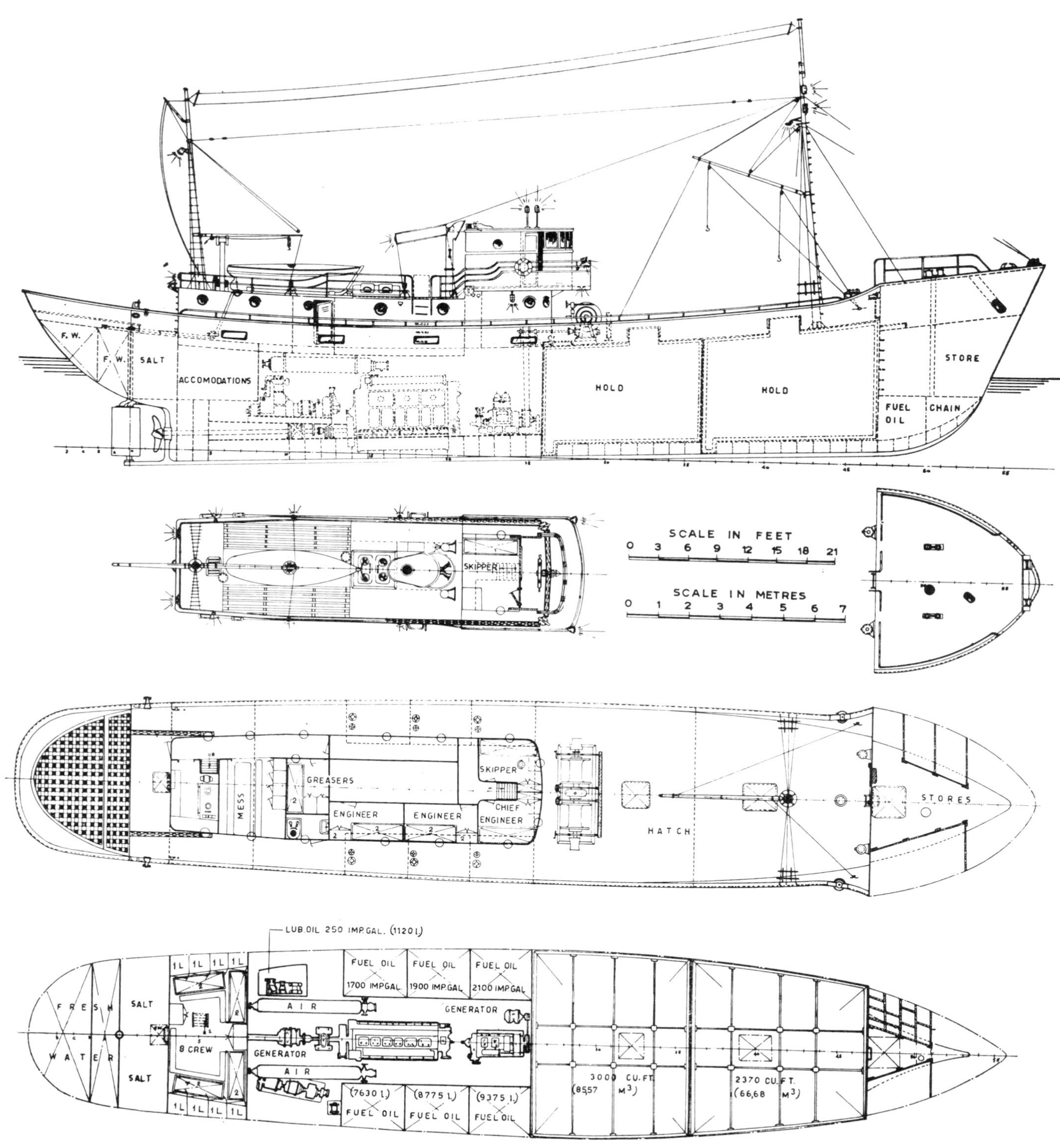
F. W.
F. W.
SALT
ACCOMODATIONS
HOLD
HOLD
STORE
FUEL OIL
CHAIN
SKIPPER
SCALE IN FEET
0 3 6 9 12 15 18 21
SCALE IN METRES
0 1 2 3 4 5 6 7
GREASERS
MESS
SKIPPER
ENGINEER
ENGINEER
CHIEF ENGINEER
HATCH
STORES
LUB.OIL 250 IMP.GAL. (1120 l.)
FUEL OIL 1700 IMP.GAL.
FUEL OIL 1900 IMP.GAL.
FUEL OIL 2100 IMP.GAL.
FRESH WATER
SALT
SALT
AIR
AIR
GENERATOR
GENERATOR
8 CREW
(7630 l.) FUEL OIL
(8775 l.) FUEL OIL
(9375 l.) FUEL OIL
3000 CU.FT. (85,57 M3)
2370 CU.FT. (66,68 M3)

Fig. 241

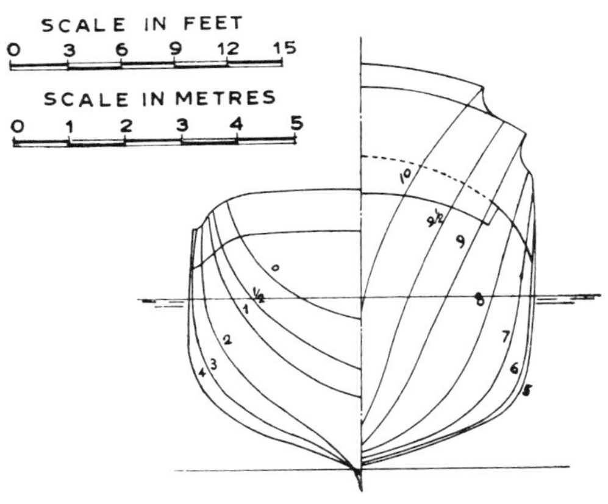

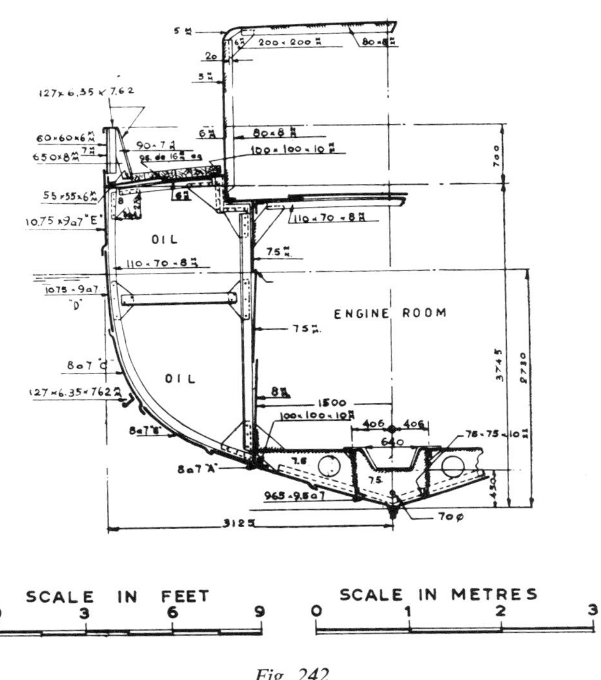

Fig. 242

Table XLI gives distribution of weights, displacements and metacentric heights under different service conditions. Fig. 241 and 242 give the general arrangement drawings; fig. 243 a photograph; fig. 244 the curves of stability at inclinations; fig. 245 the power-speed graph.

Dimensions of another design of this type of vessel, with the same beam and length but with less depth and displacement, is given in Table XL, column 4. Fig. 246 shows the general arrangement drawings, and fig. 247 a photograph.

The greater cargo capacity of the former boat, and its sea-worthiness, make it the most popular type. The fact that it can yield greater profit is most important in view of the very steep rise in the cost of building. In 1949 that cost was 3,500,000 pesetas (£38,000; U.S.$89,000). To-day the figure is 5,500,000 pesetas (£50,000; U.S.$140,000). This fact, and the lower catches, the increase in fuel and outfitting costs, and the higher fishing port fees, has led to fewer of this class of boats being built because the principal payments and insurance premiums on such high costs make it difficult to operate a boat at a profit.

The towing of the pair trawl requires much skill and experience to maintain the proper vertical and horizontal opening during trawling. The operation is conducted much more slowly than with the otter trawl. The speed is usually 2 to 3 knots and the power required is 25 to 35 per cent. less than under ordinary sailing conditions.

Towing of the net is different from that of the otter trawl, in that one warp is attached to each boat, as can be seen in fig. 248. The net has no doors for opening the mouth horizontally. It is set out from one boat over the stern as shown in fig. 249. Once it is in the water, the second boat approaches and throws a line to pick up the other warp which has to be attached to it.

The trawling gear can be seen in fig. 250. The length of the warps naturally depends upon the depth of the water. Once the necessary length has been released, it is fixed to both boats by towing hooks.

The method of hauling in the net is shown in fig. 251. Only 110 fm. (200 m.) of the warp is left in the water; then the boats come together. One of them, with a throw-line, hauls in the warp from the other boat and reeves it, as is shown in fig. 252, to bring the net alongside for hoisting aboard by the derrick, leaving the cod end in the water until it has been sufficiently emptied by dip-net to allow it to be hauled on board.

The duration of each haul is considerably longer than in the case of otter trawling, averaging from 5 to 7 hours, so that in winter there is only one a day, and in summer up to three. At night the boats are moored to buoys with engines stopped, unless they use the time to change fishing grounds.

Often three boats are operated together, so that while two are fishing, the third, with the previous catch on board, returns to base and comes back to take over from the one in which the next catch has been stowed. The advantage is that fishing can be continuous and the time lost in the first part of the normal "pareja" trip is avoided. Depending on the distance from base to fishing grounds, the abundance of fish, the season, etc., sometimes four or five boats operate together in this way as a team.

There are roughly 300 modern "pareja" boats fishing distant grounds.

FAR-DISTANT FISHING

The first Spanish codfishing company was founded in the early twenties, and large boats were equipped with otter trawls to catch fish on the Newfoundland, Greenland, and Labrador banks. Two trips were made a year, each lasting about five months, and the cod and other

Fig. 243

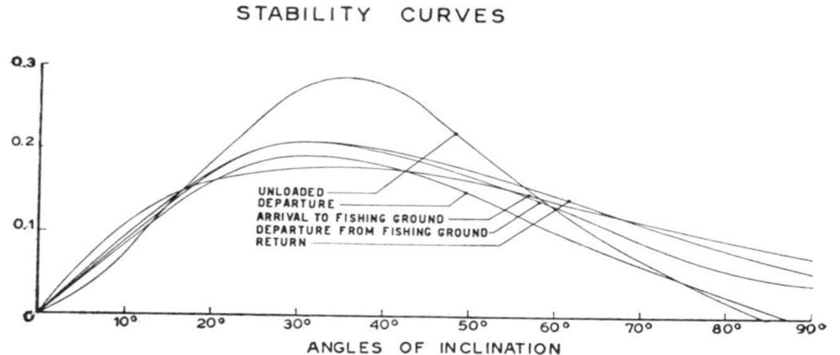

Fig. 244

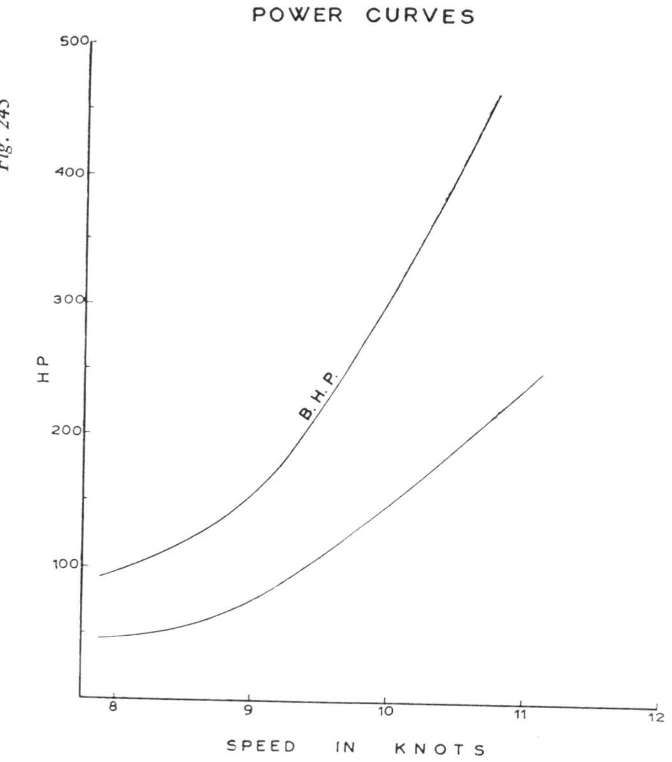

Fig. 245. Curves of necessary engine power in BHP (top curve) and effective (tow rope) power (lower curve) of 88.9 ft. (27.1 m.) "Pareja" trawler

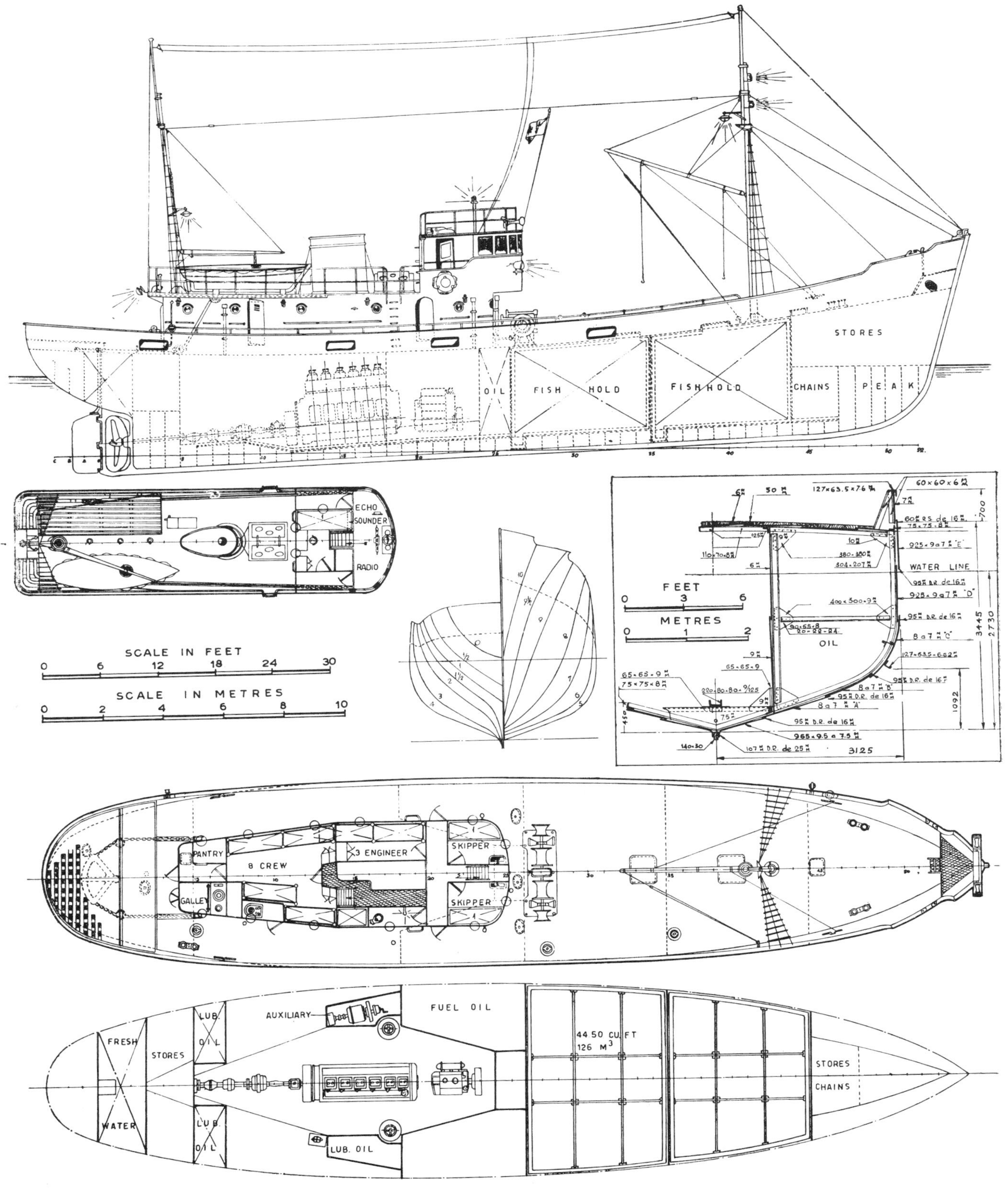
STORES
OIL
FISH HOLD
FISH HOLD
CHAINS
PEAK
ECHO SOUNDER
RADIO
SCALE IN FEET
0 6 12 18 24 30
SCALE IN METRES
0 2 4 6 8 10
FEET
0 3 6
METRES
0 1 2
WATER LINE
OIL
3445
2730
1092
3125
PANTRY
8 CREW
3 ENGINEER
SKIPPER
GALLEY
SKIPPER
AUXILIARY
FUEL OIL
FRESH
WATER
STORES
LUB. OIL
LUB. OIL
LUB. OIL
4450 CU. FT
126 M³
STORES
CHAINS

Fig. 246

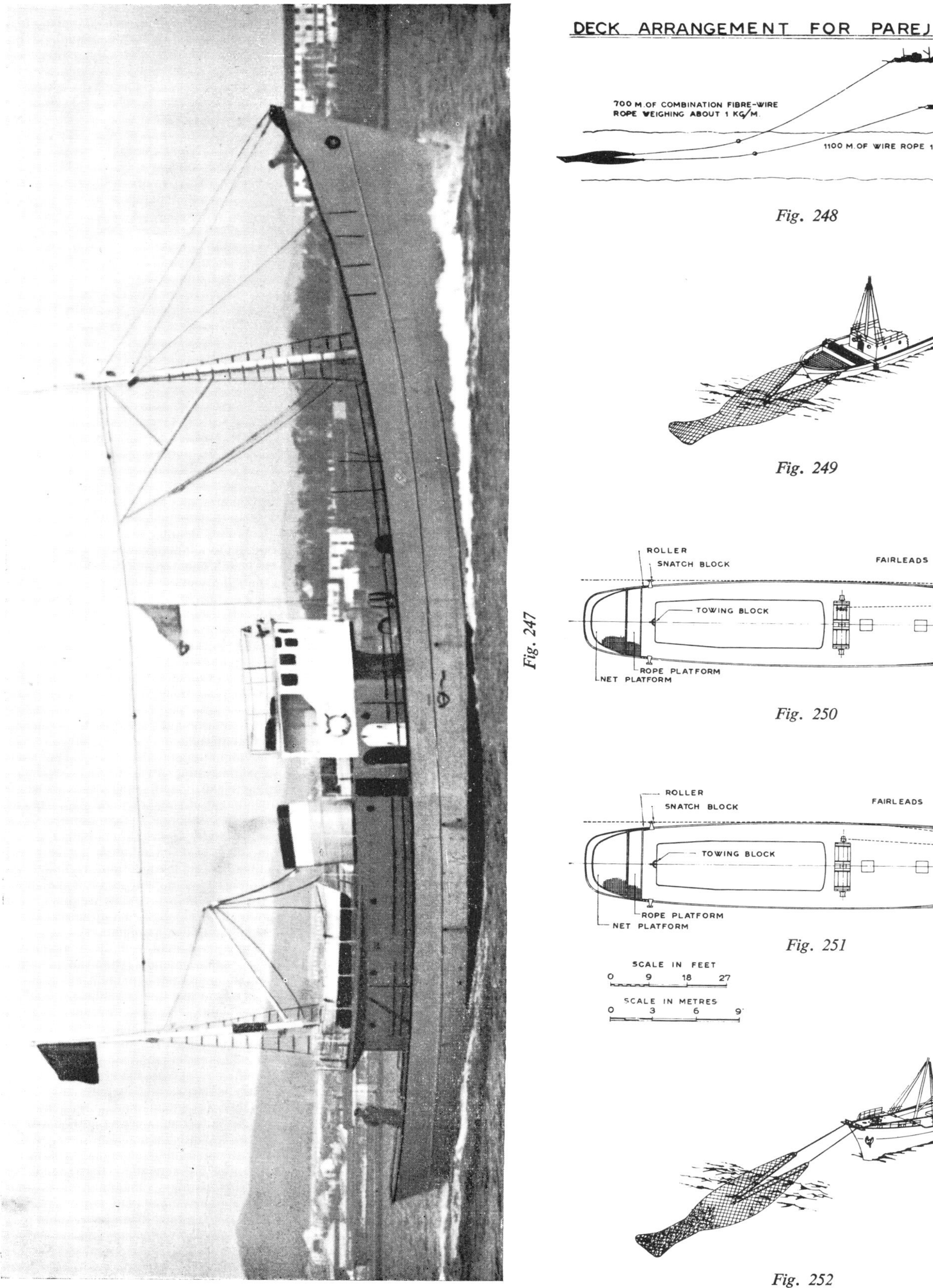

Fig. 247

Fig. 248

Fig. 249

Fig. 250

Fig. 251

Fig. 252

Table XLIIA

Condition	Leaving port: Weight in ton	Vertical from base: Height	Vertical from base: Moment	Horizontal: Lever	Horizontal Moment: Forward	Horizontal Moment: Aft	Arrival at fishing ground: Weight in ton	Vertical from base: Height	Vertical from base: Moment	Horizontal: Lever	Horizontal Moment: Forward	Horizontal Moment: Aft
ater in donkey boiler	3.00	11.00	33.00	+ 14.80		44.40						
ull and equipment	1010.74	5.00	5061.64	+ 2.17		2195.94						
achinery	142.01	3.97	563.46	+ 16.90		2397.11						
isplacement, unloaded	1155.75	4.90	5658.1	+ 3.96		4565.74	1155.75		5658.10			4565.74
dditional items:												
Officers and effects	1.20	8.80	10.56	+ 15.00		18.00	1.20	8.80	10.56	+ 15.00		18.00
Crew and effects	6.10	8.15	49.75	— 16.00	97.50		6.10	8.15	49.75	— 16.00	97.50	
Provisions: misc. and wine	16.50	5.50	90.75	+ 28.20		465.00	15.20	5.50	83.60	+ 28.20		428.00
Potatoes	5.00	5.30	26.50	— 21.00	105.00		4.50	5.30	23.85	— 21.00	94.50	
Chandlery (soap, lamp-oil, etc.)	3.00	1.50	4.50	+ 17.50		52.50	2.70	1.50	4.05	+ 17.50		47.20
Fresh water	89.00	5.72	508.90	— 18.50	1644.00		80.00	5.65	452.00	— 18.20	1455.00	
Salt water	2.40	11.50	27.60	— 10.00	24.00		2.40	11.50	27.60	— 10.00	24.00	
Hold luggage	2.25	11.20	25.20	— 2.09	4.70		2.25	11.20	25.20	— 2.09	4.70	
Mail	7.20	7.63	54.90	+ 11.60		76.40	7.20	7.63	54.90	+ 11.60		76.40
Cargo, salt	600.00	2.20	1420.00	— 8.00	4800.00		570.00	2.20	1420.00	— 8.00	4560.00	
,, fish												
Rigging	27.00	6.06	164.00	— 20.40	550.60		27.00	6.06	164.00	— 20.40	550.60	
Feed water												
,, ,, in reserve	2.00	4.30	8.60	+ 15.00		30.00	2.00	4.30	8.60	+ 15.00		30.00
Fuel, coal												
,, diesel oil	375.00	3.31	1242.00	+ 1.97		738.00	330.00	3.25	1072.00	+ 1.99		666.00
,, boiler oil	3.20	7.10	22.70	+ 15.10		48.30	3.20	7.10	22.70	+ 15.10		48.30
,, gasoline												
Lubricating oil	8.85	0.70	6.20	+ 15.70		139.00	8.00	0.65	5.20	+ 15.70		135.60
Liver oil												
Ballast												
SPLACEMENT	2294.45	4.06	9320.26	— 0.475	7221.80 1088.86	6132.94	2217.50	4.10	9082.11	— 0.348	6786.30 771.06	6015.24

Condition	Leaving fishing ground: Weight in ton	Vertical from base: Height	Vertical from base: Moment	Horizontal: Lever	Horizontal Moment: Forward	Horizontal Moment: Aft	Return to port: Weight in ton	Vertical from base: Height	Vertical from base: Moment	Horizontal: Lever	Horizontal Moment: Forward	Horizontal Moment: Aft
ater in donkey boiler												
ull and equipment												
achinery												
isplacement, unloaded	1156.75		5658.10			4565.74	1155.75		5658.10			4565.74
dditional items:												
Officers and effects	1.20	8.80	10.56	+ 15.00		18.00	1.20	8.80	10.56	+ 15.00		18.00
Crew and effects	6.10	8.15	49.75	— 16.00	97.50		6.10	8.15	49.75	— 16.00	97.50	
Provisions: misc. and wine	2.00	5.50	11.00	+ 28.20		56.40	0.70	5.50	3.85	+ 28.00		19.74
Potatoes	0.70	5.30	3.71	— 21.00	14.70		0.20	5.30	1.06	— 21.00	4.20	
Chandlery (soap, lamp-oil, etc.)	0.60	1.30		+ 17.50			0.30	1.30	0.39	+ 17.50		5.25
Fresh water	15.00	2.80	42.00	+ 28.20		424.00	5.00	2.50	12.50	+ 28.20		141.00
Salt water	2.40	11.50	27.60	— 10.00	24.00		2.40	11.50	27.60	— 10.00	24.00	
Hold luggage	2.25	11.20	25.20	— 2.09	4.70		2.25	11.20	25.20	— 2.09	4.70	
Mail	7.20	7.63	54.90	+ 11.60		76.40	7.20	7.63	54.90	+ 11.60		76.40
Cargo, salt												
,, fish	1000.00	3.50	5500.00	— 4.00	4000.00		1000.00	3.50	3500.00	— 4.00	4000.00	
Rigging	27.00	6.06	164.00	— 20.40	350.60		27.00	5.05	164.00	— 20.40	550.60	
Feed water												
,, ,, in reserve	2.00	4.30	8.60	+ 15.00		30.00	2.00	4.30	8.60	+ 15.00		30.00
Fuel, coal												
,, diesel oil	80.00	1.67	133.70	+ 15.37		1232.00	40.00	1.50	60.00	+ 15.37		615.00
,, boiler oil	3.20	7.10	22.70	+ 15.10		48.30	3.20	7.10	22.70	+ 15.10		48.30
,, gasoline												
Lubricating oil	1.35	0.45	0.61	+ 15.70		21.20	0.50	0.30	0.15	+ 15.70		8.85
Liver oil	20.00	8.00	160.00	+ 24.00		480.00	20.00	8.00	160.00	+ 24.00		480.00
Ballast												
SPLACEMENT	2326.75	4.25	9872.43	+ 0.974	4691.50	6952.04 2260.54	2273.80	4.30	9759.36	+ 0.584	4691.00	6008.28 1327.28

Table XLIIB

Condition	Displacement in salt water	Mean draft		Static stability: KM_T		KG		GM_T		Free surface		GM_T corrected		GZ maximum			Range of stability	Trim: KM_L		KG corrected	
	ton	ft.	m.	ft.	m.	ft.	m.	ft.	m.	ft.	m.	ft.	m.	ft.	m.	degrees	degrees	ft.	m.	ft.	m.
nloaded	1155.75	9.55	2.910	17.95	5.46	16.1	4.90	1.84	0.560			1.84	0.560	1.64	0.499	40	>90	302	92.0	16.1	4.900
eaving port	2294.45	16.35	4.980	16.35	4.98	13.3	4.06	3.02	0.920	3.75	0.095	2.70	0.825	1.66	0.504	54	>90	223	68.0	13.6	4.155
rrival at fishing ground	2217.50	15.9	4.850	16.3	4.97	13.45	4.10	2.85	0.870	3.75	0.095	2.55	0.775	1.7	0.518	52	>90	226	69.0	13.7	4.195
eaving ,, ,,	2326.75	16.55	5.030	16.35	4.98	13.95	4.25	2.35	0.730	1.26	0.032	2.29	0.698	1.63	0.496	55	>90	229	68.0	14	4.282
eturn to port	2273.80	16.20	4.940	16.30	4.97	14.1	4.30	2.2	0.670	1.18	0.030	2.1	0.640	1.68	0.512	54	>90	225	68.5	14.2	4.330

Condition	Trim: GM_L		L.C.B.		L.C.G.		Difference between B and G		Change of trim		Centre of water plane buoyancy		Corrected trim: Forward		Corrected trim: Aft		Resulting draft: Forward		Resulting draft: Aft		Displacement
	ft.	m.	ft.	m.	ft.	m.	ft.	m.	ft.	m.	ft.	m.	ft.	m.	ft.	m.	ft.	m.	ft.	m.	ton
nloaded	286	87.100	+ .75	+ 0.23	+ 13.00	+ 3.96	+ 12.25	+ 3.73	+ 8.95	+ 2.73	— 0.62	— 0.19	4.42	1.348	4.55	1.382	4.15	1.262	15	4.592	1155.75
eaving port	209	63.845	+ .85	+ 0.26	— 1.56	— 0.475	— 2.41	— 0.735	— 2.71	— 0.735	+ 2.72	+ 0.83	1.23	0.375	1.18	0.360	16.55	5.055	16.1	4.920	2294.45
rrival to fishing ground	212	64.805	+ .79	+ 0.24	— 1.14	— 0.348	— 1.93	— 0.538	— 1.9	— 0.579	+ 2.46	+ 0.75	0.97	0.295	0.96	0.284	15.8	4.825	15.95	4.866	2217.50
eaving ,, ,,	209	63.718	+ .89	+ 0.27	+ 3.2	+ 0.974	+ 2.31	+ 0.704	+ 2.31	+ 0.704	+ 2.82	+ 0.86	1.18	0.359	1.13	0.345	14.35	4.371	18.6	5.675	2326.75
eturn to port	211	64.170	+ .82	+ 0.25	+ 1.92	+ 0.584	+ 1.1	+ 0.334	+ 1.09	+ 0.332	+ 2.65	+ 0.81	0.55	0.169	0.53	0.163	14.6	4.471	17.7	5.403	2273.80

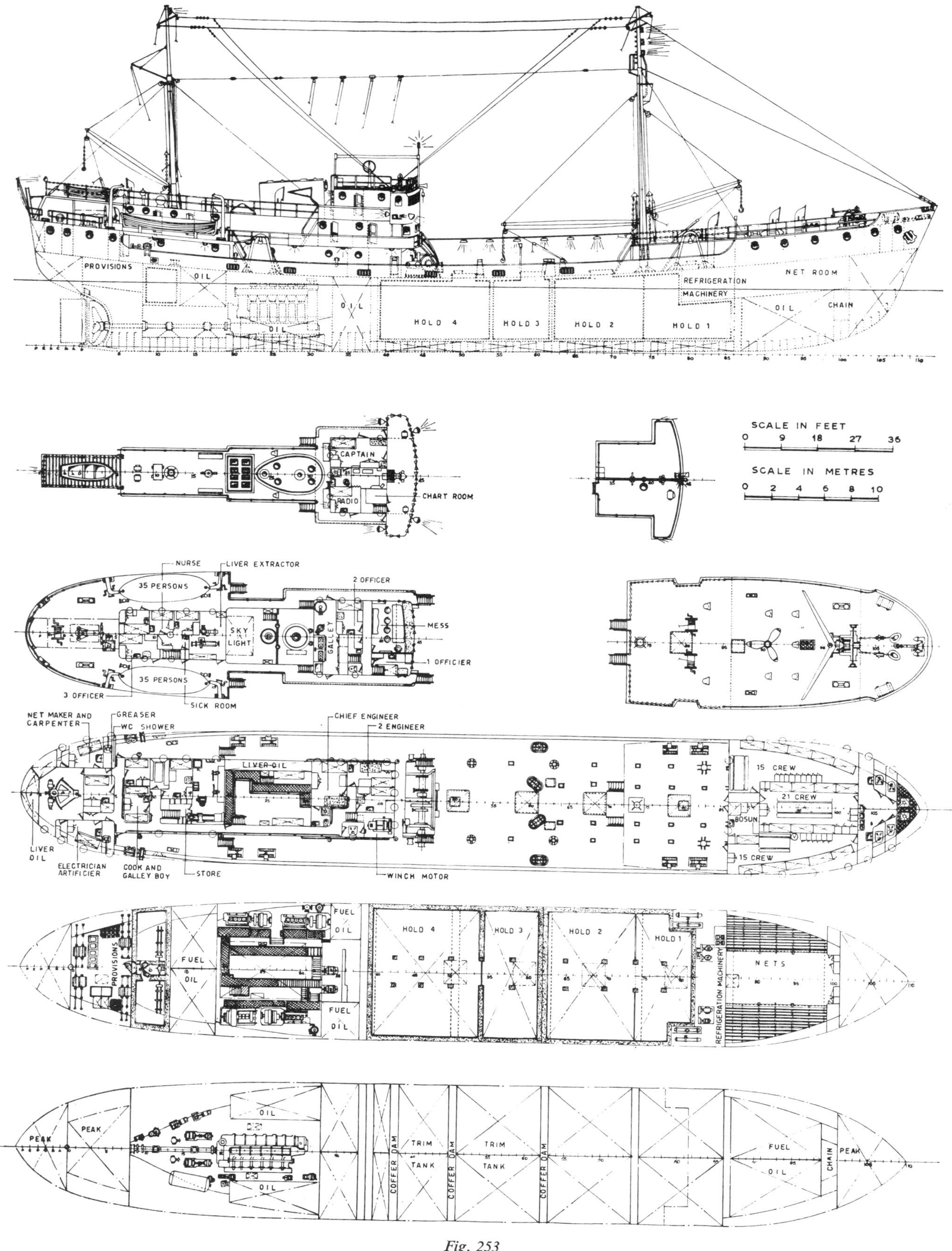
PROVISIONS
OIL
OIL
OIL
HOLD 4
HOLD 3
HOLD 2
HOLD 1
REFRIGERATION
MACHINERY
NET ROOM
OIL
CHAIN
CAPTAIN
RADIO
CHART ROOM
SCALE IN FEET
0 9 18 27 36
SCALE IN METRES
0 2 4 6 8 10
NURSE
LIVER EXTRACTOR
35 PERSONS
2 OFFICER
SKY LIGHT
GALLEY
MESS
1 OFFICIER
35 PERSONS
3 OFFICER
SICK ROOM
NET MAKER AND CARPENTER
GREASER
WC SHOWER
CHIEF ENGINEER
2 ENGINEER
LIVER OIL
15 CREW
21 CREW
BOSUN
15 CREW
LIVER OIL
ELECTRICIAN ARTIFICIER
COOK AND GALLEY BOY
STORE
WINCH MOTOR
PROVISIONS
FUEL OIL
FUEL OIL
FUEL OIL
HOLD 4
HOLD 3
HOLD 2
HOLD 1
REFRIGERATION MACHINERY
NETS
PEAK
PEAK
OIL
OIL
COFFER DAM
TRIM TANK
COFFER DAM
TRIM TANK
COFFER DAM
FUEL OIL
CHAIN
PEAK

Fig. 253

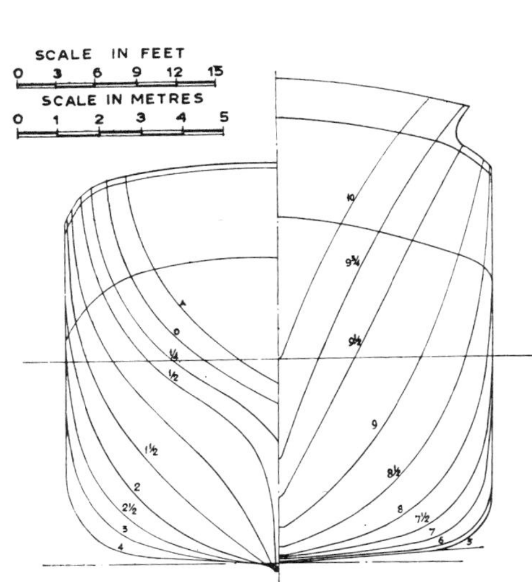

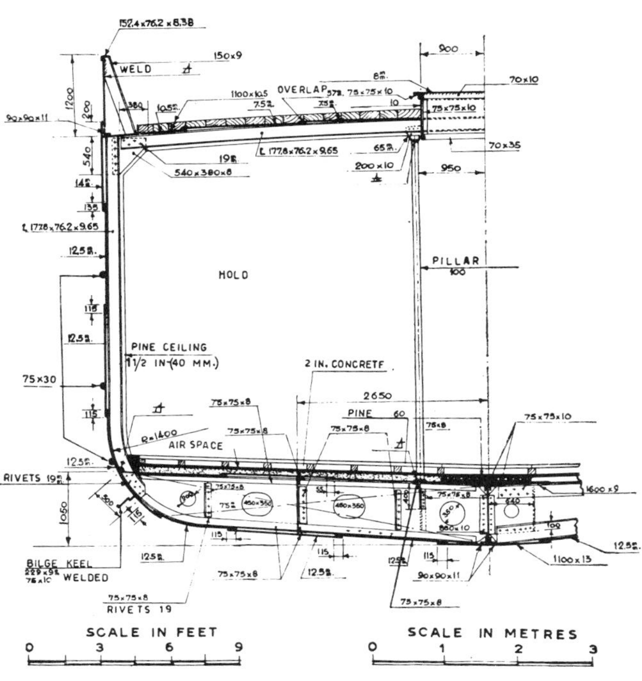

Fig. 254

Fig. 255

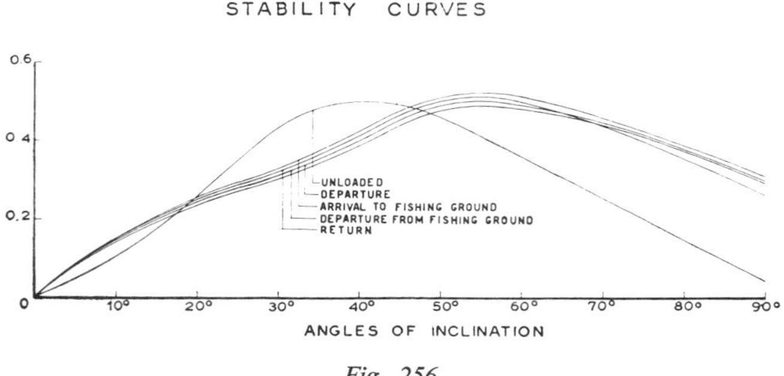

Fig. 256

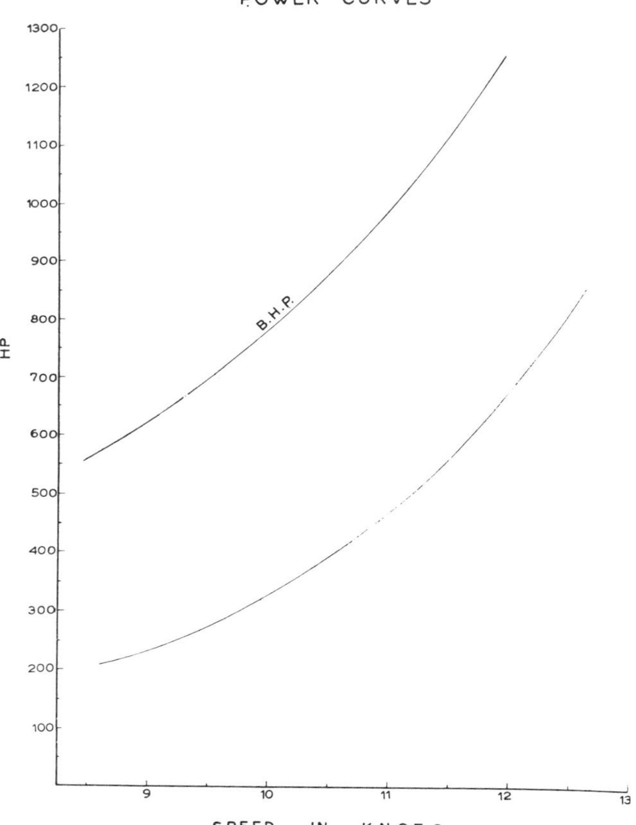

Fig. 257. Curves of necessary engine power in BHP (top curve) and effective (tow rope) power (lower curve) of 209 ft. (63.75 m.) otter trawler

fish was salted for delivery to the curing plants. The Spanish market consumes about 70,000 tons of salted and dried fish a year, and there are now 22 Spanish codfishers that supply between 35-40,000 tons a year. Another 10 codfishers are under construction. For main specifications of one of the more modern boats of the fleet see Table XL, column 5; Table XLII gives the distribution of weights, displacements and metacentric heights under various conditions of service; fig. 253 and 254 show the general arrangement, body plan and midship section of a boat of this type, and fig. 255 is a photograph of one. Fig. 256 gives the stability curves and fig. 257 the speed-power diagram.

This type of boat is built for sailing through ice. The auxiliary deck machinery, windlass, trawl winch and steering servo-motor are electrically driven, and so are the fish washing, bilge, fire and ballast pumps. There is a 4-h.p. ammonia refrigerator for the crew's food, and two 14-h.p. motors for fish-room refrigeration, and machinery for the extraction of cod liver oil. The radio, chart and navigation equipment is up to date and complete.

The main propelling engine is four-cycle, single-acting super-charged direct-coupled and direct-reversible, six cylinders, developing 1,200 b.h.p. at 205 r.p.m. In addition to the bilge and ventilation pumps incorporated in the main engine, there is the following equipment in the engine room: three diesel-generator of 100 kw. at 220 volts with Ward-Leonard couplings, one 60-kw. diesel-generator, one 20-kw. diesel-generator, two electric and one hand air-compressor for starting, two centrifugal electric pumps for water-cooling of engine, one centrifugal electric pump for water-cooling of auxiliary machinery, two electric pumps for lubrication, one Duplex electric ballast and fire pump, one Duplex electric bilge pump, one centrifugal electric pump for washing fish, one Duplex electric fuel pump, one electric fuel pump for daily use, one centrifugal lubricating oil filter, one centrifugal fuel filter.

The steam used for extracting liver oil, for the drinking water evaporator and for the heating of living quarters, is provided by a small boiler heated by exhaust gases and fuel-oil burners. This vessel costs (1953) some 30 million pesetas (£270,000; U.S.$760,000).

When the first Spanish " parejas " fished on the Newfoundland banks in 1949, they brought back good catches of prime cod, but, sold as fresh fish, it was not popular and fetched poor prices. As a result of this, the character of the " pareja " fishing on the Newfoundland banks changed. Instead of the boat returning with fresh fish from 20 to 30 day trips they landed salted fish after 60 to 90 days. More hands were taken on board and a small hold was built aft to increase capacity and improve trim. Bunker capacity was also increased and the nets made larger. For main dimensions of one of these boats see table XL, column 6, and for general arrangement, body plan and midship section, see fig. 258, and for a photograph see fig. 259. It costs about one million pesetas (£9,200; U.S.$25,700) to fit out such a " pareja " for each trip, but reasonable results have been obtained.

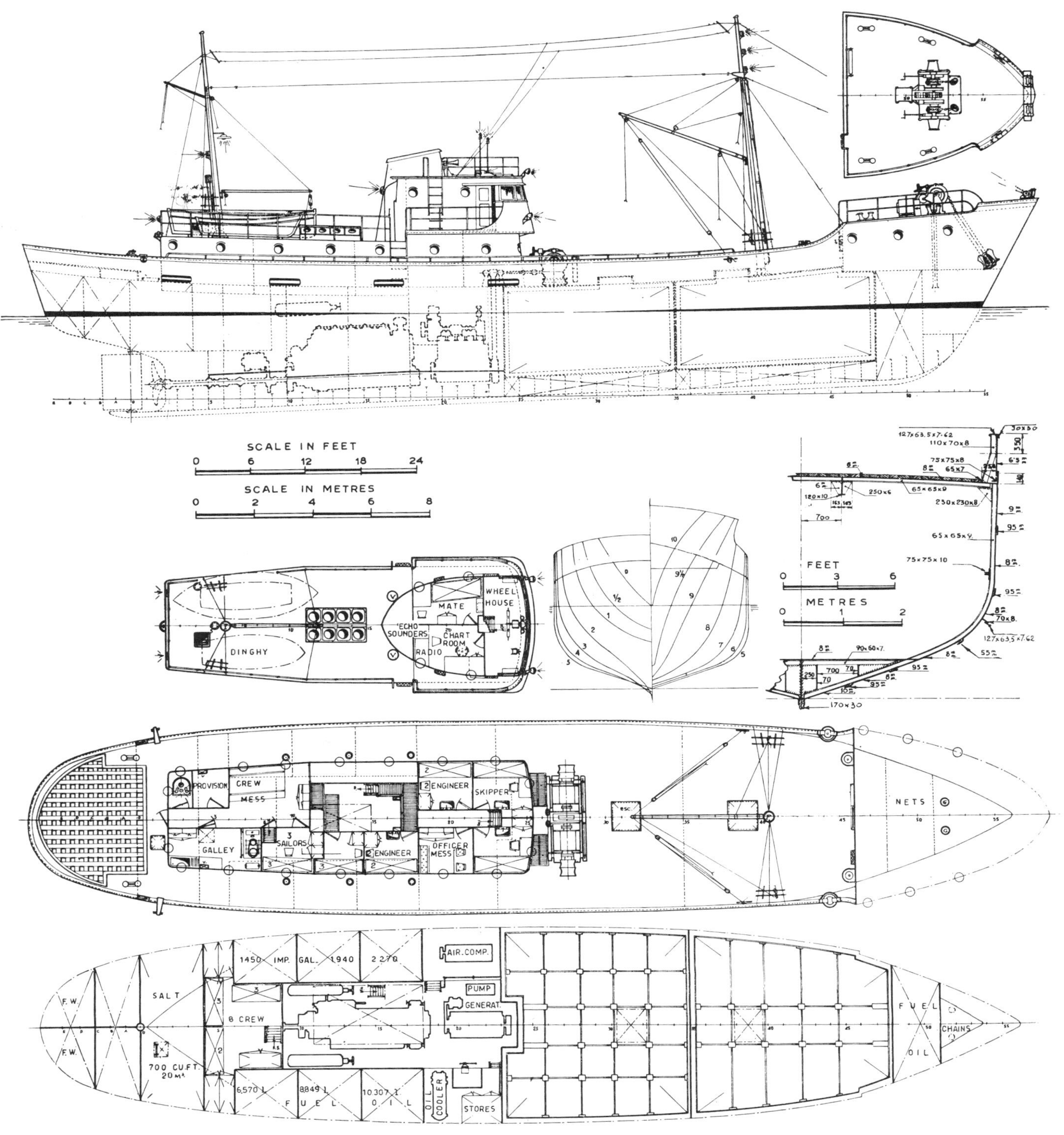
SCALE IN FEET
0 6 12 18 24
SCALE IN METRES
0 2 4 6 8
DINGHY
ECHO SOUNDERS
RADIO
MATE
CHART ROOM
WHEEL HOUSE
FEET
0 3 6
METRES
0 1 2
PROVISION
CREW MESS
GALLEY
3 SAILORS
ENGINEER
SKIPPER
OFFICER MESS
NETS
F.W.
SALT
8 CREW
700 CU.FT. 20 m³
1450 IMP. GAL. 1940 2270
6,570 L 8849 L 10307 L
FUEL OIL
OIL COOLER
STORES
AIR COMP.
PUMP
GENERAT.
FUEL OIL
CHAINS

Fig. 258

Fig. 259

PACIFIC COMBINATION FISHING VESSELS

by

H. C. HANSON

PACIFIC combination fishing vessels are designed to purse seine for salmon, herring, etc., trawl for bottom fish, long line for halibut, etc., fish with live bait for tuna and troll for salmon. One purpose craft are wasteful to the extreme. Some types, such as the large tuna clippers, can perhaps be justified, but the author hopes that their design can soon be modified so that fishing gear, such as long lines, can be used.

Combination craft can be employed practically all the year round in the Pacific area. Because of their versatility, utility and limited use of scarce and expensive construction materials, the author is of the opinion that they are also best for fisheries in other parts of the world.

A good sheer forward and a good flare of the bow help to make the boats dry. It is particularly important to have a high chock rail to protect the deckhouse. The stern sheer must be designed to be adaptable to the different fishing methods; it has to be low for trolling, trawling and long lining, but high enough to carry the heavy purse seine. Therefore the design must be a compromise. The long overhang stern, characteristic of the author's designs, is to keep the nets clear of the propeller. The bilge has to be carried well out, as shown by the midship section, fig. 261. If such vessels are designed with a full bilge, ballast is necessary to keep the vessel down. The indicated midship section makes for long, easy lines and is the best for seaworthiness and speed, giving the propeller and the boat the necessary grip on the water in heavy weather. The fishermen like vessels with such lines, whereas the rolling characteristic of full midship sections bothers them. The angle of entrance of the waterline should not be too full, because the fishermen drive the wooden vessels too hard in heavy seas.

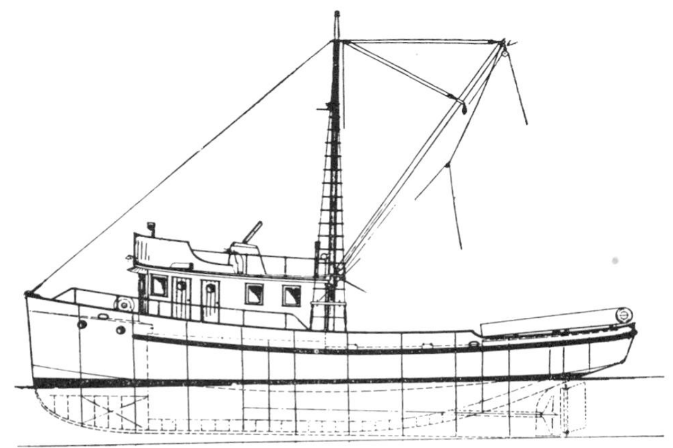

Fig. 262. 57 × 16 × 6 ft. (17.4 × 4.9 × 1.8 m.) steel combination boat in V-bottom or round-bottom design

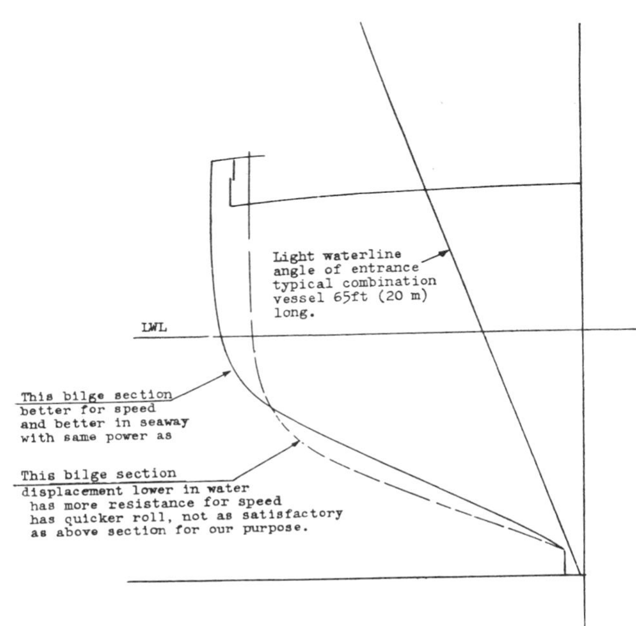

Fig.261. Influence of midship section upon speed and sea behaviour

Combination boats can be of steel or wood, the latter being preferred for sizes up to 80 ft. (24.4 m.). Construction, as shown in the drawings of this paper, will produce strong and seaworthy vessels, good for twenty or more years if proper materials are used.

Bent oak frames in one piece have been used in vessels up to 90 ft. (27.4 m.) long and, by using two pieces one on top of the other, up to 135 ft. (41.1 m.). It now seems

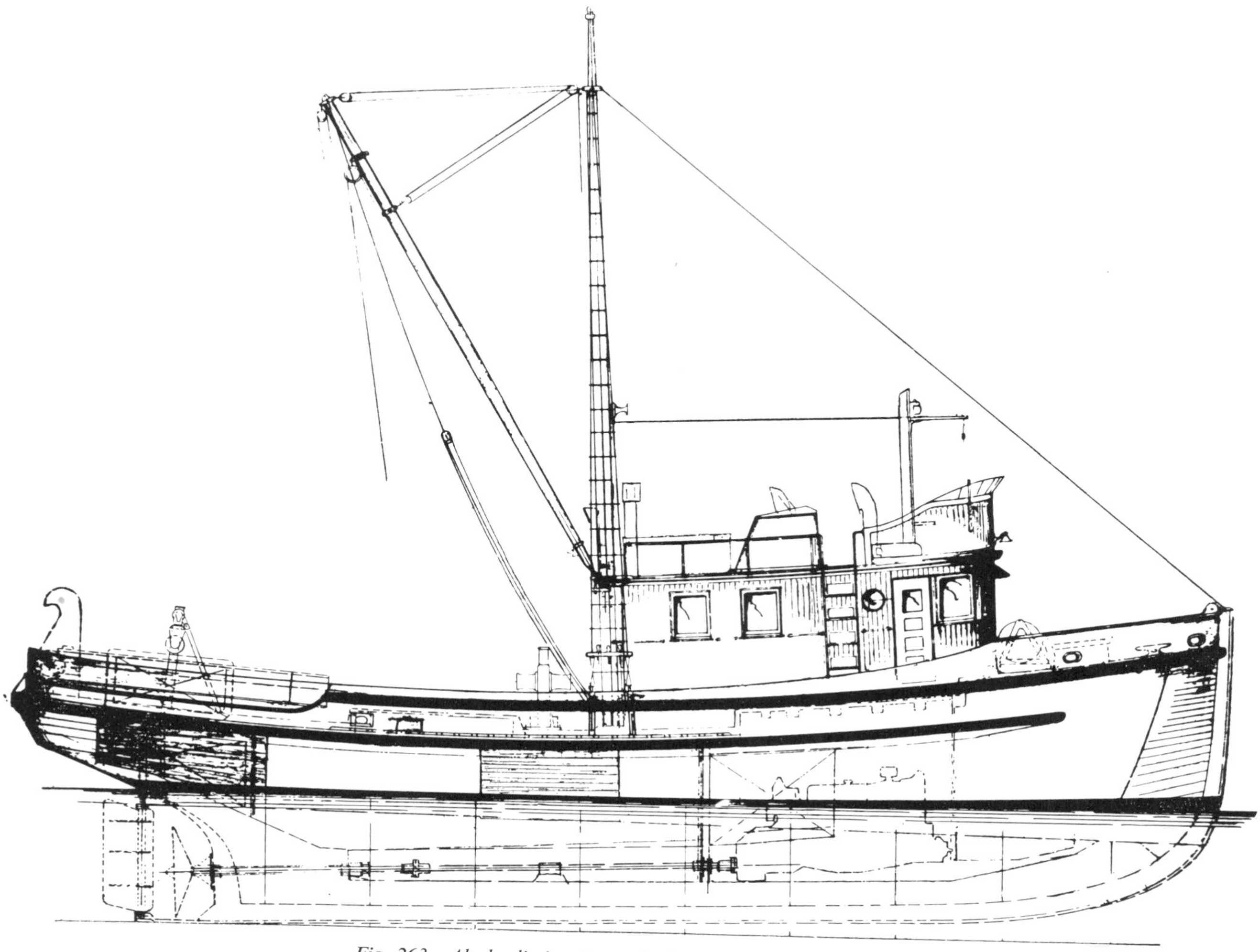

Fig. 263. *Alaska limit seiner. Outboard profile*

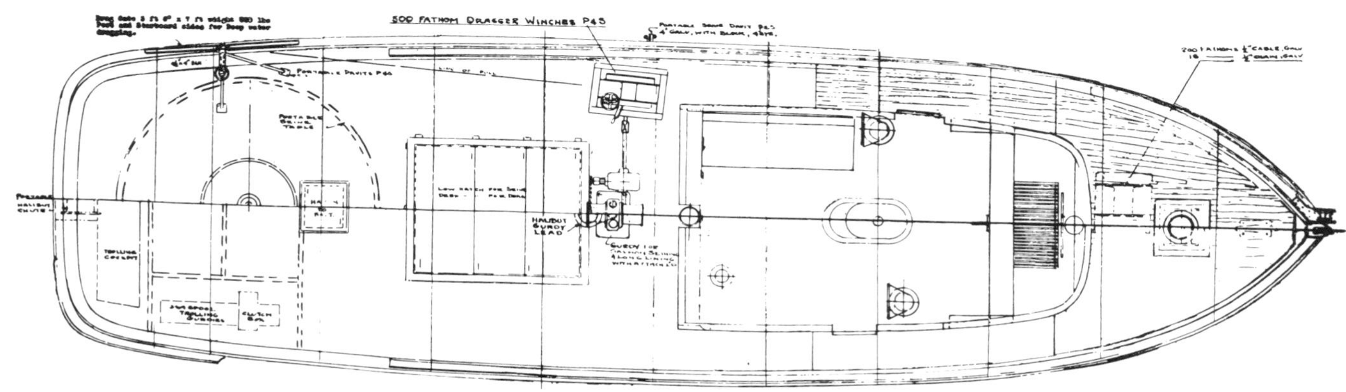

Fig. 264. *Alaska limit seiner. Deck arrangement A, with side winches for trawling with 500 fm. (915 m.) wire. The plan indicates also arrangement of trolling gurdies and the longline halibut chute*

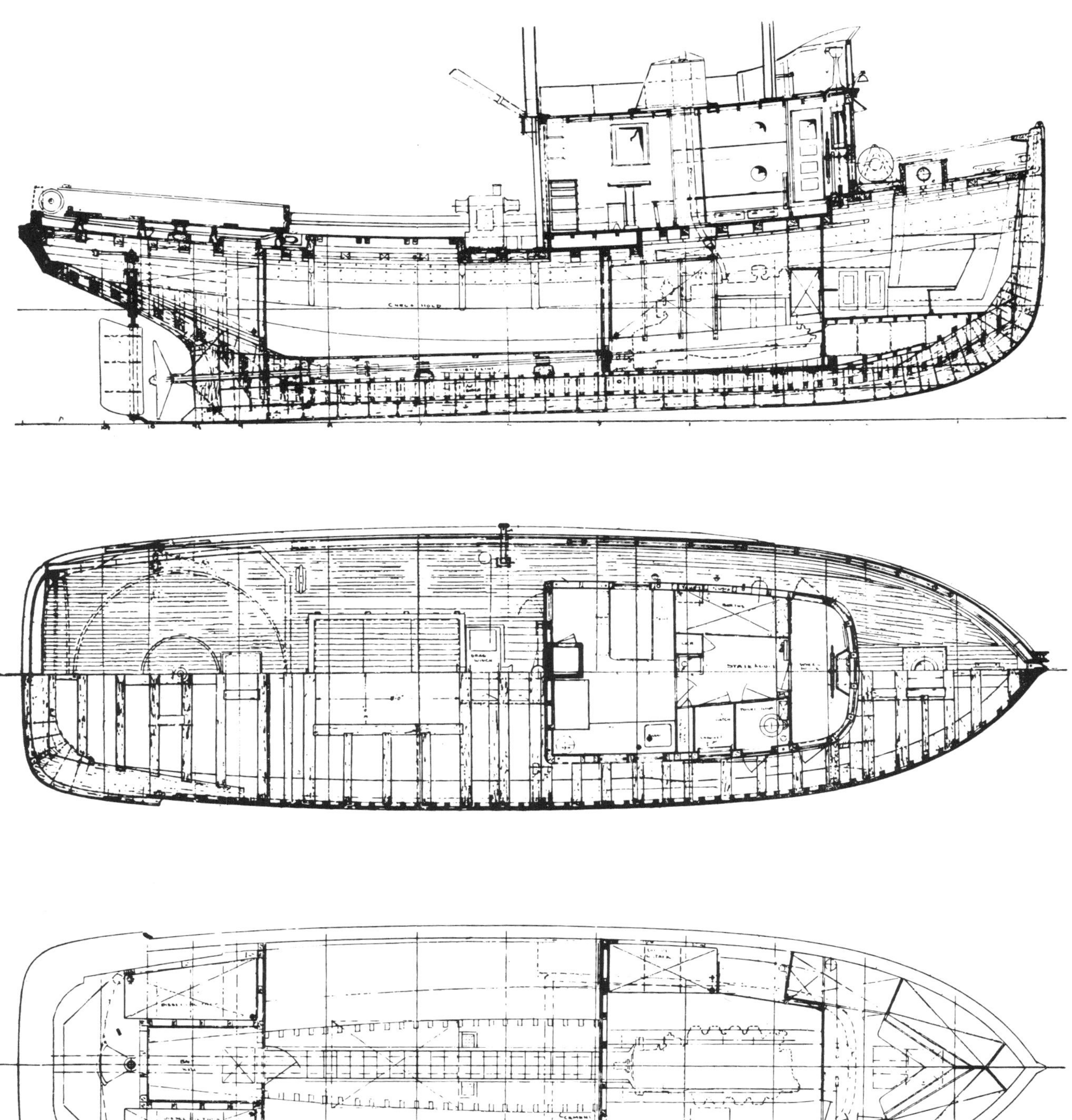

Fig. 265, 266, 267. Alaska limit seiner. Deck arrangement B with central single trawl winch for shallow water trawling, each drum holding 300 fm. (500 m.) wire. The plan shows also turntable and seine davit

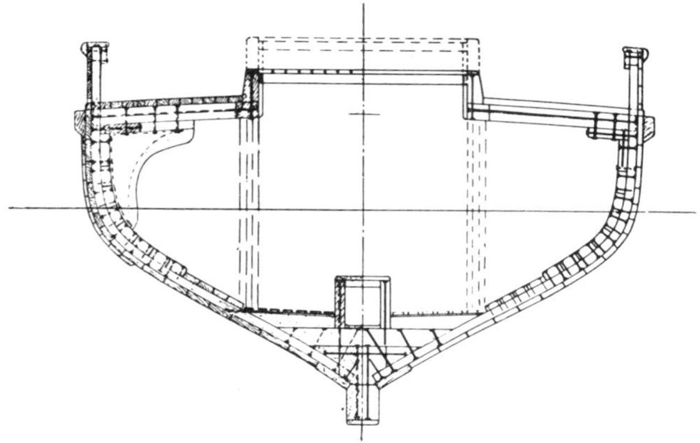

Fig. 268. Alaska limit seiner. Midship section

to be impossible to get suitable oak longer than 20 ft. (6.1 m.) for the frames and therefore a boat length of 80 ft. (24.4 m.) must be regarded as the maximum. Laminated frames can be recommended for vessels larger than 80 ft. (24.4 m.), provided the oak is not dried out so much so that it is weakened. If green oak is used moisture accumulates and decay is accelerated and for these reasons some laminated frames, made of oak strips artificially dried and glued together, are unsatisfactory. At present several vessels are being built in the U.S.A., using 20 per cent. sapwood for the laminated frames. In the author's experience, rotting will take place in a few years. Much of this trouble could be avoided, and to the advantage of the industry, if a satisfactory method could be developed for glueing oak with a moisture content as high as 18 to 20 per cent.

For lengths up to 60 ft. (18.3 m.), steel vessels cost approximately the same as wooden ones; beyond this they are cheaper and also upkeep costs are less, but for comfort they do have to be insulated. The fishermen, however, have not yet overcome their preference for wood. Their current objection is partly because of the lack of insulation in present steel vessels, but the time will soon come when steel will be more favoured. A fine example of a steel vessel is shown in fig. 262.

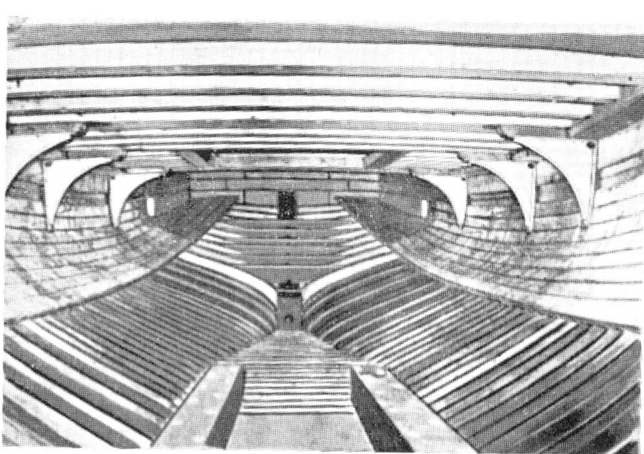

Fig. 269. Alaska limit seiner. Details from construction

SMALL COMBINATION BOATS

The Alaska Limit Seiner is a good example of an efficient combination boat.

The name limit seiner derives from a law in Alaska restricting fishing boats to a maximum registered length, from the rabbet at stem to foreside of the rudder stock, measured at deck level, of 50 ft. (15.2 m.). Within this regulation it is possible to design a length over-all of 57 ft. (17.4 m.). Fig. 263 to 268 show the drawings for the general arrangement and midship section of one design. Fig. 269 gives some details of the actual construction and fig. 270 the bolting of the ceiling. Design particulars and scantlings for this type, as well as for 65 ft. (19.8 m.) and 74 ft. (22.6 m.) combination boats are given in Table XLIII.

The illustrated boat is equipped with a 150 h.p., 1,000 r.p.m., diesel engine driving, through a 3 : 1 reduction gear, a three-bladed bronze propeller with a diameter of 52 in. (1,320 mm.), a pitch of 46 in. (1,168 mm.), a developed blade area of 920 sq. in. (5,934 sq. cm.), a mean width ratio of .272 and an

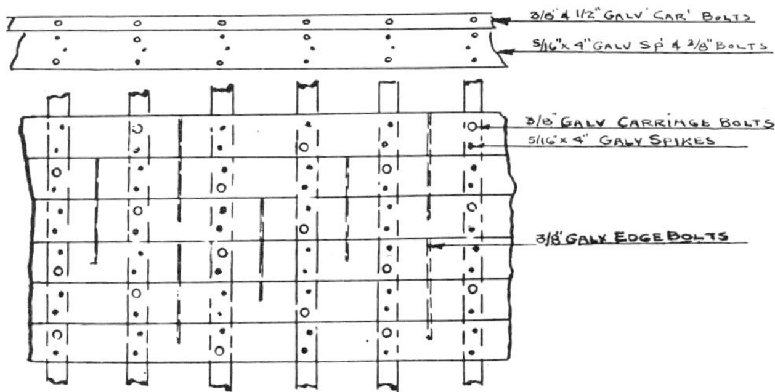

Fig. 270. Alaska limit seiner. Bolting of ceiling

Fig. 271. The combination boats are designed to maintain an even trim

TABLE XLIII

Design particulars for 57 ft. (Alaska limit seiner), 65 ft. and 74 ft. (17.4, 19.8, 22.6 m.) combination vessels

Length, over all	57 ft.	17.4 m.	65 ft.	19.8 m.	74 ft.	22.6 m.
Length, loaded waterline	50 ft.	15.2 m.	58 ft.	17.7 m.	67 ft. 6 in.	20.6 m.
Breadth, over all	16 ft.	4.88 m.	17 ft. 3 in.	5.26 m.	20 ft.	6.1 m.
Breadth, loaded waterline	15 ft. 7 in.	4.75 m.	17 ft.	5.18 m.	19 ft. 8 in.	5.99 m.
Block coefficient	.383		.418		.368	
Midship coefficient	.607		.674		.567	
Prismatic coefficient	.630		.620		.647	
Waterplane coefficient	.710		.740		.768	
LCB	−3.9 per cent.		−1.0 per cent.		−2 per cent.	
Displacement lwl	44.05 long tons	44.8 cu. m.	59 long tons	60 cu. m.	98 long tons	99.5 cu. m.
Cargo capacity	35.0 short tons	31.7 tons	55 short tons	50 tons	90 short tons	81.5 tons
Cruising speed	9.75 knots		9.8 knots		10.5 knots	
Maximum speed	10.5 knots		10.5 knots		11.4 knots	
Tonnage, gross register	42.7 tons		58.5 tons		103.0 tons	
Tonnage, net register	29.0 tons		40.0 tons		70.0 tons	
Main engine	150 h.p.		210 h.p.		275 h.p.	
Revolutions	1,000 r.p.m.		1,000 r.p.m.		1,000 r.p.m.	
Reduction gear	3 : 1		3 : 1		3 : 1	
Diesel oil tanks	1,600 gal.	6.1 cu. m.	1,200 gal.	4.5 cu. m.	4,000 gal.	15.2 cu. m.
Water	350 gal.	1.3 cu. m.	600 gal.	2.3 cu. m.	1,300 gal.	4.9 cu. m.
Propeller 3 bl.	52 × 46 in.	1,320 × 1,168 mm.	52 × 44 in.	1,320 × 1,117 mm.	60 × 41–45 in.	1,524 × 1,041–1,143 mm.
Scantlings	*in.*	*cm.*	*in.*	*cm.*	*in.*	*cm.*
Keel	9½ × 9½	24.1 × 24.1	11½ × 11½	29.2 × 29.2	11½ × 11½	29.2 × 29.2
Keelson	13½ × 13½	34.3 × 34.3	13½ × 13½	34.3 × 34.3	13½ × 17½	34.3 × 44.5
Shoe	1¾	4.5	1¾	4.5	1¾	4.5
Stern post	9½ × 13½	24.1 × 34.3	11½ × 15½	29.2 × 39.4	13½ × 17½	34.3 × 44.5
Sister keelson	7½ × 9½	19.1 × 24.1	7½ × 7½	19.1 × 19.1	9½ × 9½	19.1 × 19.1
Frames, oak	2¼ × 4	5.7 × 10.2	3 × 4	7.6 × 10.2	3 × 4	7.6 × 10.2
Spaced	10	25.4	12	30.5	12	30.5
Beams	5 × 7½	12.7 × 19.1	5¼ × 7¾	13.4 × 19.7	5½ × 8	14 × 20.3
Clamps	2¾ × 10	7 × 25.4	2¾ × 11½	7 × 29.2	3½ × 11½	8.9 × 29.2
Shelf	2¾ × 11½	7 × 29.2	2¾ × 11½	7 × 29.2	2¾ × 11½	7 × 29.2
Garboard	2 × 11½	5.1 × 29.2	2 × 11½	5.1 × 29.2	2½ × 11½–14	6.4 × 29.2–35.6
Planking	1¾	4.5	1¾	4.5	2 net	5.1
Bilge ceiling	6–2¾ × 5½	6–7 × 14	5–3½ × 6	5–8.9 × 15.2	7–5½ × 7½	7–14 × 19.1
Decking	2 × 3½	5.1 × 8.9	2 × 3½	5.1 × 8.9	2¼ × 3½	5.7 × 8.9

8 deg. rake aft—this latter being quite important on a wooden vessel. The vessel makes 10½ knots in light condition and can carry 35 short tons of fish, equivalent to 25,000 salmon.

The deck is raised to make room in the forecastle for the large crew necessary for handling the seines and so that the break provides a stopper for the fish. Trim can be maintained even with a full load (*fig. 271*).

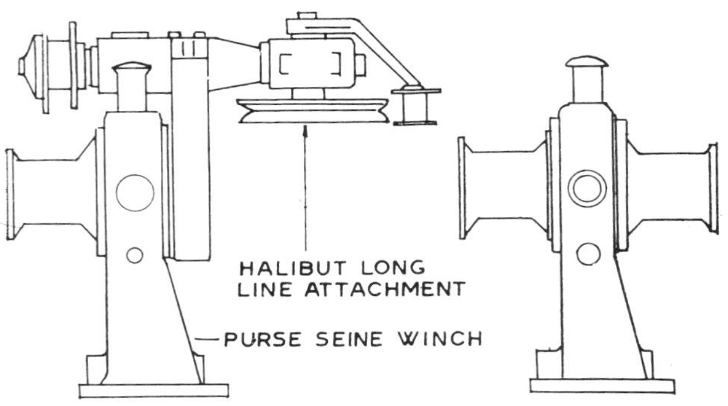

Fig. 272. Gurdy winch for purse seining and long lining

The main engine has a power take-off for driving the anchor windlass, and the drive shaft is carried aft to the purse seine and trawl winches. Sometimes it is used for the drive to the turntable roller, otherwise another take-off operates a 2 in. (51 mm.) water pump used to turn the turntable roller hydraulically. This pump can also be used for washing, for fire-fighting and as a bait pump. A 3 kw. generator is driven by power take-off on the main engine also.

An auxiliary engine of 5 h.p. installed in the side of the engine room drives a 1½ kw. auxiliary generator, and an air compressor, as well as a small fuel and lubricating oil pump. A small refrigerating compressor for the

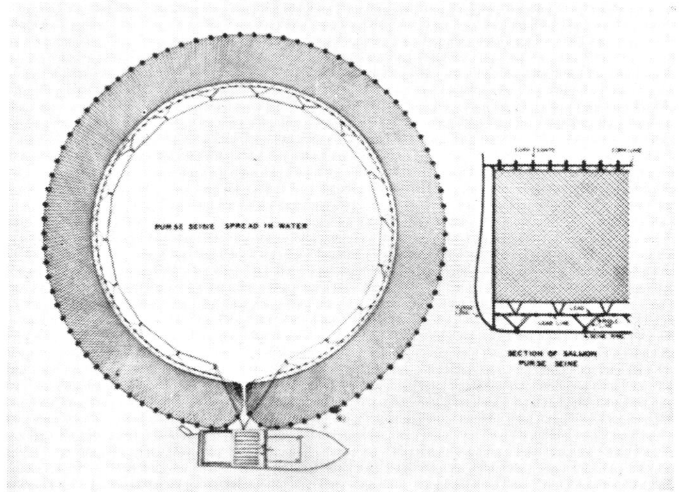

Fig. 273. Diagram of purse seine

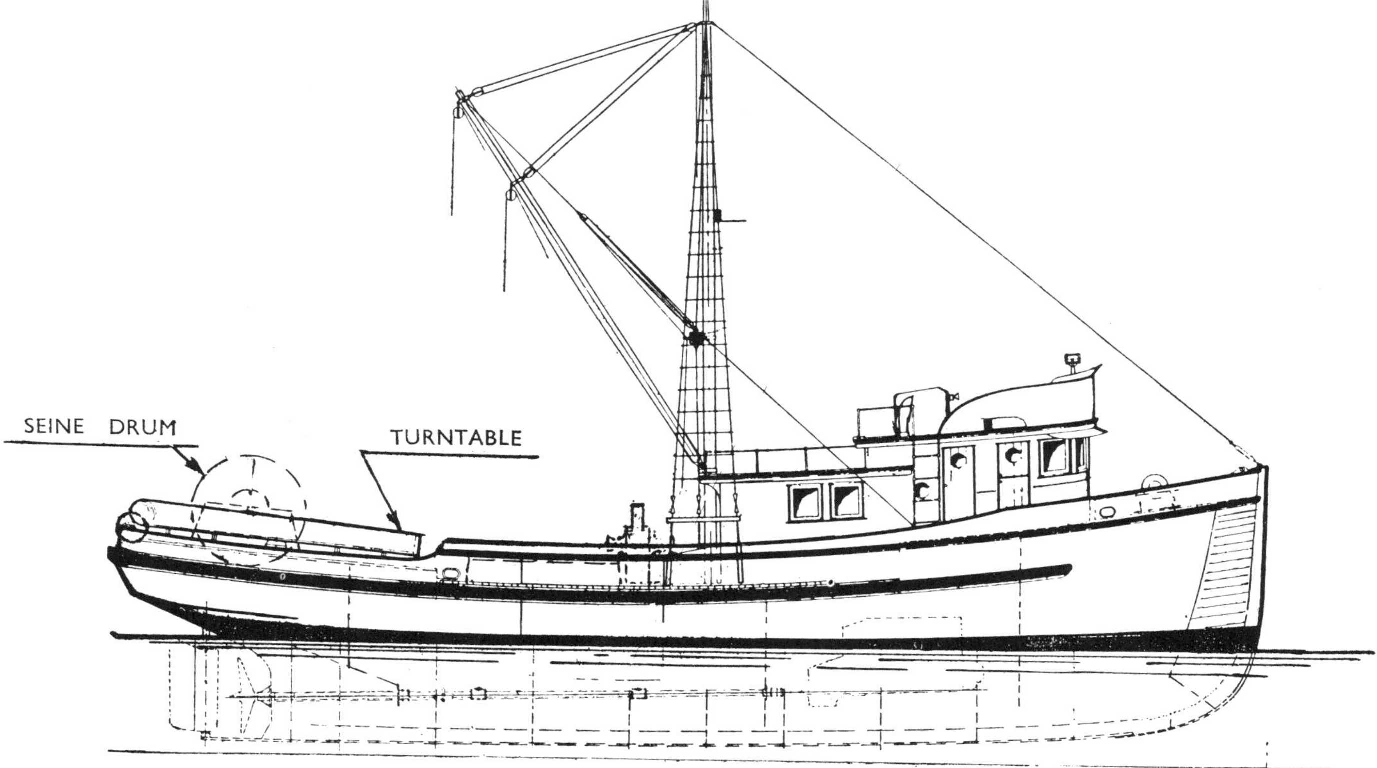

Fig. 274. 65 ft. × 17 ft. 4 in. × 8 ft. (19.8 × 5.28 × 2.44 m.). Combination seiner and trawler with purse seine drum compared with turntable

galley cold box is located over the port tank. The tanks have a total capacity of 1,330 imp. gal. (1,600 gal., 6 cu. m.) of diesel oil, 58 imp. gal. (70 gal., 265 l.) of lubricating oil and 250 imp. gal. (300 gal., 1.1 cu. m.) of water. The bulkhead aft of the engine room has a portable section so that the engine can be removed.

Aft of the hold is a bait well for circulating sea water

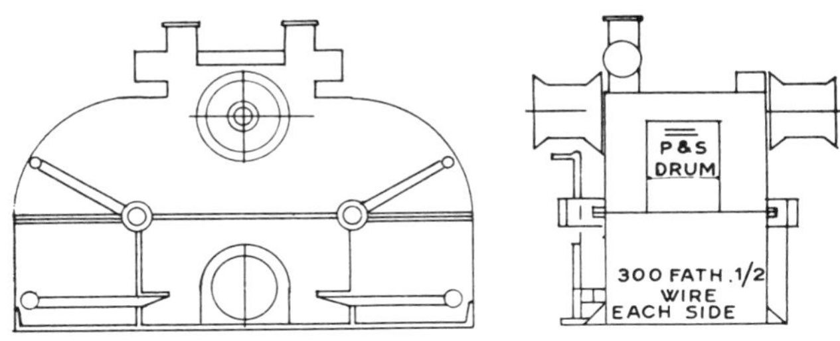

Fig. 275. Combined trawl and purse seine winch for shallow water trawling

(capacity 5 tons in this vessel) or dry storage. When live bait is carried in the tank, properly screened intakes and outlets, as well as lights, are arranged, but some owners prefer a portable tank on deck. When fishing for tuna, the owner may also use the trolling cockpit on the aft deck for the catch.

The turntable is installed on a portable pivot and

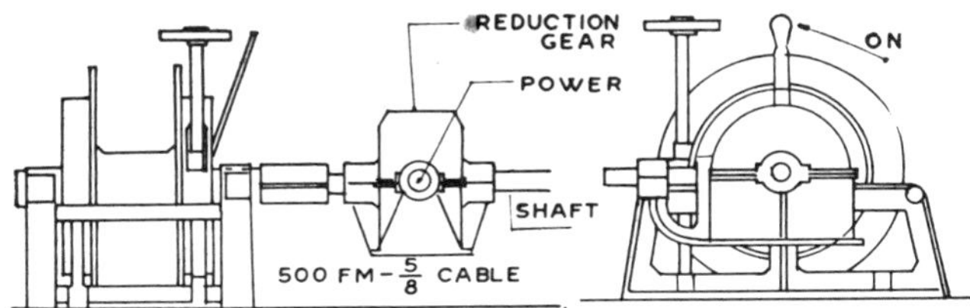

Fig. 276. Deep-water trawl winch

rollers on the deck, and the bulwarks are cut down. It is quite heavy—about 3½ tons including the 1.6 h.p. electric drive and the portable base—and the seine net itself weighs about 5 tons. A seine davit is installed on each side in the bulwarks abreast of the purse seine winch (*fig. 272*), between hatch and mast.

Fig. 273 shows a purse seine for a 65 ft. (19.8 m). boat. It is about 250 fm. (457 m.) long with 4¼ in. (108 mm.) mesh, 12 and 15 thread material, 450 meshes (4½ strips) deep. The lead line of the seine is 75 fm. (137 m.).

A recent trend in Canada is to use, instead of a turntable, a drum reel similar to those used on the gillnetters. It is said that two to four men less are needed and, if this proves to be so, the fleet will be converted rapidly. The reels might be set on a swivel base as on modern gillnetters. Fig. 274 shows the profile of a 65 ft. (19.8 m.) seiner in which such a reel is compared with a turntable.

When trawling in shallow water, the purse seine winch is replaced by a combined trawl and seine winch, fig. 275.

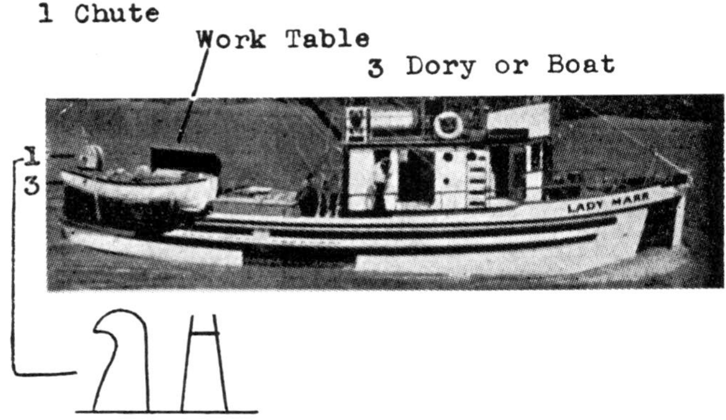

Fig. 277. Halibut longlining arrangement

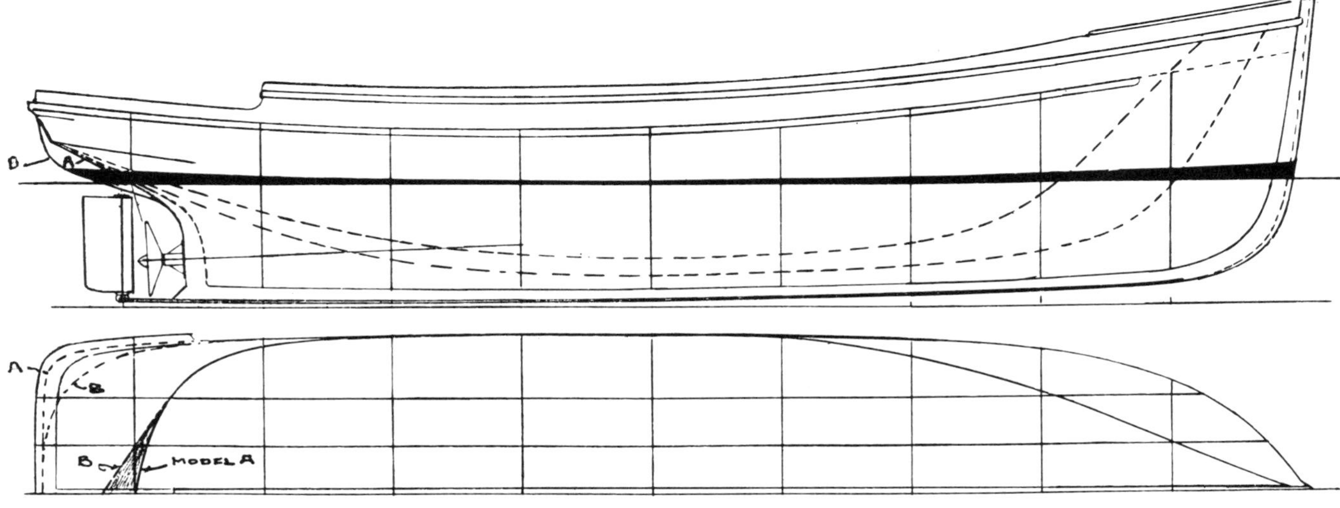

Fig. 278. 89 × 24 ft. (27.1 × 7.3 m.). Combination boat lines

This has niggerheads and double drums for 300 fm. (550 m.) wire. The wire leads through blocks at the side and aft to blocks on the trawl davits at the quarters. Steel bulwarks are installed when trawling and the davits for handling the 4×7 ft. (1.22×2.1 m.) 600 lb. (270 kg.) trawl boards are very substantially made.

For deep trawling separate winch drums are installed at the sides, fig. 276. Each drum carries 500 fm. (915 m.) of wire and is mechanically driven. A small purse seine winch is then installed amidships near the hatch.

convertor is used to transfer from D.C. to A.C. when required. There are power and steering controls both on top of the wheel house and inside it.

MEDIUM-SIZED COMBINATION BOATS

Models of the described 89×24 ft. (27.1×7.3 m.) wooden tuna seiner, which is convertible to bait fishing, trawling and long lining, were tested at the University of Michigan, under Professor L. A. Baier, with both a typical seine stern and a cruiser stern (Hanson 1952).

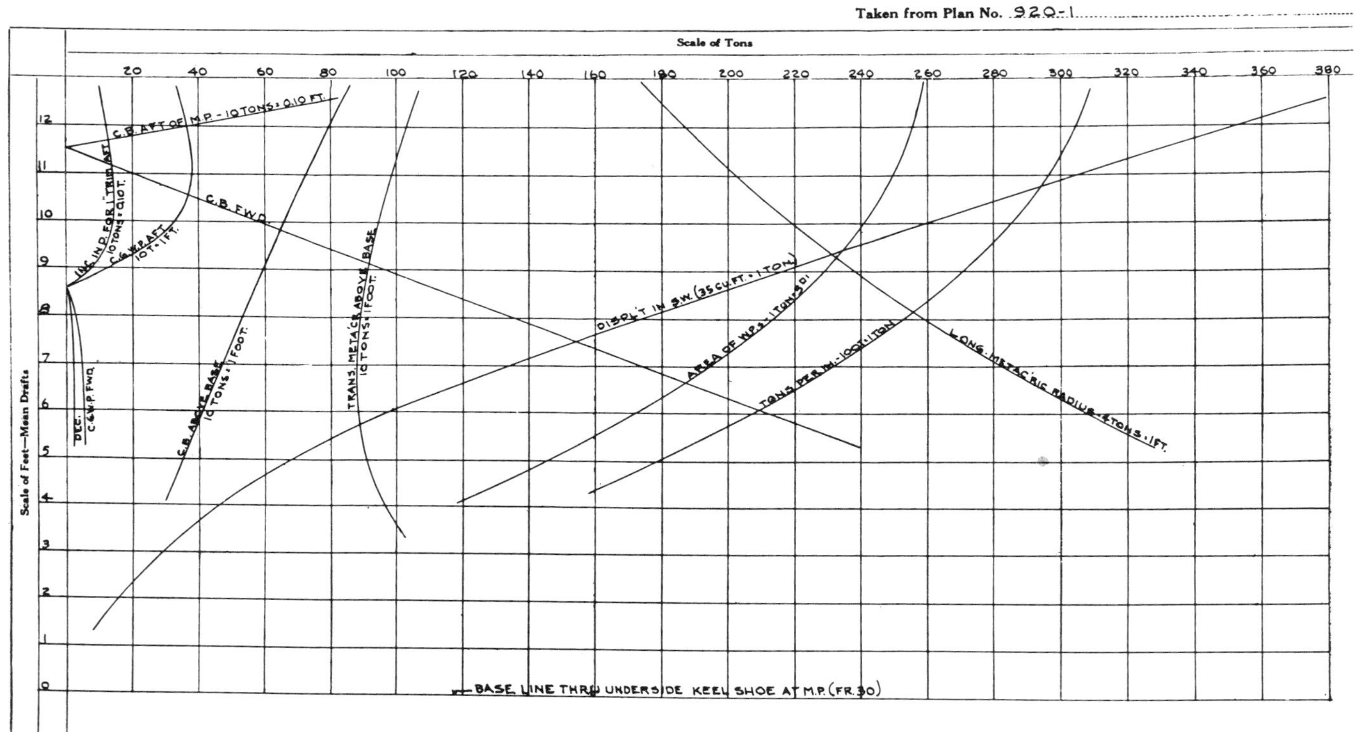

Fig. 279. 89 × 24 ft. (27.1 × 7.3 m.). Combination boat. Curves of form

For halibut long lining an attachment is fitted on the purse seine winch. The galvanised iron long line skate chute is midships aft and the dressing table is on the port quarter (fig. 277).

When salmon trolling, gurdies, fig. 152 (page 110) are installed on both sides and ahead of the cockpit and spring block davits are installed in bulwarks abreast of the gurdies. All of these are portable. Two poles are installed abreast of the mast and two on the deck forward of the house. They are stowed as shown in fig. 153/4/7.

The galley has an oil-burning stove, a refrigerator and working and messing tables. In the deck house there is a stateroom for two, a toilet, and the wheel house containing echo sounder, direction finder, automatic pilot and radio transmitter. The electrical wiring for 32 volts, D.C. is very complete and includes a shore plug. A

Fig. 278 shows a hull with long easy lines, giving good speed and the best trim. An alternative cruiser stern, Model B, is indicated. The midship section shows a bilge carried well out for seaworthiness. The curves of form are given in fig. 279, and Table XLIV gives the main dimensions of the tested models.

The data from the towing tank tests in fig. 280 are well worth close scrutiny. The different sterns require quite different power, especially in the load condition, since the drag of the seiner stern is eliminated.

The tuna clipper, fig. 281, with a raised instead of a flush deck, has the same underbody lines as the tested model and shows the versatility of this type. The raised deck design has produced successful vessels for years.

Fig. 282 and 283 show the general arrangements of the tuna seiner. The midship section, fig. 284, and fig. 285, indicate the assembly of the sawn frames by the

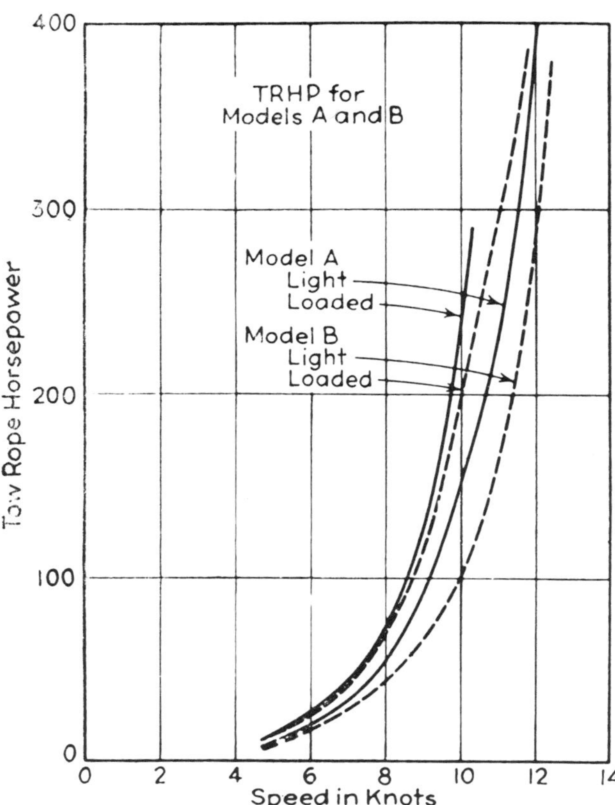

Fig. 280. 89 × 24 ft. (27.1 × 7.3 m.). Combination boat. Results from tank tests

Fig. 281. 89 × 24 ft. (27.1 × 7.3 m.). Combination boat. Tuna clipper version

double flitch method. The rules of the American Construction Bureau do not cover this size of vessel, and the material and fastenings given in fig. 286 might form a standard of construction. This table also covers a 125 ft. (38.1 m.) wooden tuna clipper.

The dimensions quoted are all for soft woods. If oak is used the thickness can be reduced by approximately 20 per cent. The timbers should be free of sap or wane. Small hard knots are permitted in frames, ceiling, beams, etc. Lumber exposed to sun and air should be vertically

TABLE XLIV

	MODEL A				*MODEL B*			
Type . . .	Light		Loaded		Light		Loaded	
Stern . . .	Fantail				Cruiser			
L.O.A. . .	89.00 ft.	27.1 m.	89.00 ft.	27.1 m.	95.00 ft.	28.96 m.	95.00 ft.	28.96 m.
L. . . .	80.42 ft.	24.5 m.	86.58 ft.	26.39 m.	90.00 ft.	27.40 m.	94.00 ft.	28.65 m.
T fwd. . .	Even keel		9.74 ft.	2.97 m.	6.75 ft.	2.06 m.	7.75 ft.	2.36 m.
T aft. . .	Even keel		12.45 ft.	3.8 m.	8.75 ft.	2.67 m.	12.5 ft.	3.82 m.
T. . . .	8.70 ft.	2.65 m.	11.10 ft.	3.38 m.	7.75 ft.	2.36 m.	10.12 ft.	3.08 m.
B.O.A. . .	23.42 ft.	7.14 m.	23.42 ft.	7.14 m.	24.50 ft.	7.47 m.	24.50 ft.	7.47 m.
B. . . .	23.42 ft.	7.14 m.	23.42 ft.	7.14 m.	23.50 ft.	7.17 m.	24.20 ft.	7.38 m.
Δ . . .	192.7 lg. tons	195 cu. m.	308.1 lg. tons	312 cu. m.	194.6 lg. tons	198 cu. m.	320.3 lg. tons	325 cu. m.
S total . .	2.202 sq. ft.	205 sq. m.	2,701 sq. ft.	251 sq. m.	2,290 sq. ft.	213 sq. m.	2,780 sq. ft.	258 sq. m.
S rudder one side	23.8 sq. ft.	2.2 sq. m.	23.8 sq. ft.	2.2 sq. m.	20 sq. ft.	1.9 sq. m.	20 sq. ft.	1.9 sq. m.
Coefficients:								
Block . .	0.465		0.526		0.415		0.487	
Midship section	0.725		0.796		0.722		0.766	
Prismatic .	0.641		0.661		0.575		0.635	
Waterplane .	0.776		0.826		0.726		0.767	
$\Delta/0.01L^3$ . .	368		476		267		385	
$L/\nabla^{\frac{1}{3}}$ . .	4.2		3.9		4.7		4.2	
L/B . . .	3.432		3.695		3.675		3.83	
B/T . . .	2.7		2.12		3.17		2.42	
$\odot\% = \pm\frac{1}{2}L$.	−3.4		−2.5		−2.5		−7.0	
Entrance . .	55.2%		51.6%		52.5%		56.0%	
Run . . .	44.8%		48.4%		47.5%		44.0%	

Coefficients—ex. appendages, and based on draft to rabbet line

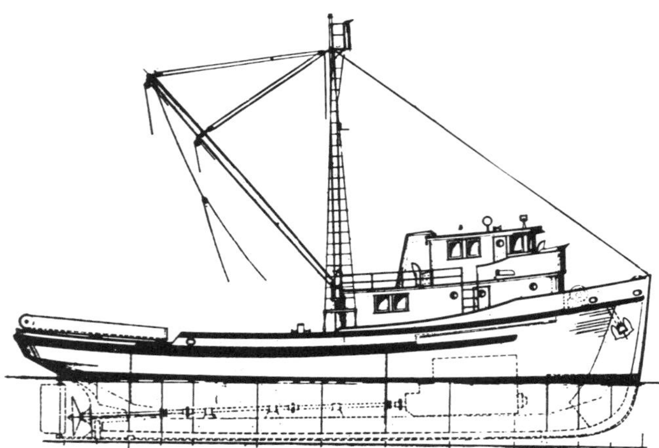

Fig. 282. 89 × 24 ft. (27.1 × 7.3 m.). Combination boat. Outside profile

grained and free of knots. Slash grain lumber is most satisfactory underwater and it should be free of knots; close-grained material gives longest life. The ceiling is the main strength member. It should be heavy, well fayed to the frames and each strake well fayed to the other. Thorough edge-bolting from the bottom to the top line of the ceiling on every third bay, sometimes extending through the deck shelves, gives the best results. The engine room contains a 400 h.p. heavy-duty diesel, auxiliaries and tanks. The fish-hold is 35 ft. (10.7 m.) long, fitted with fish bins, and has a capacity of up to 110 tons (100 ton) of fish and approximately 20 tons (18 ton) of chipped ice.

When used as a tuna clipper with brine and bait tanks, the vessel carries approximately 25 tons (22.5 cu.m.) circulating water besides the weight of the bait tank itself.

This wooden type is comparatively rigid against hogging. The author usually specifies this length of keel to be laid with a reverse camber of approximately 2½ in. (64 mm.). Where bent oak frames are used, the camber would be approximately 3 in. (76 mm.) depending upon the type and size of timber and fastenings. There is no hogging problem in small steel vessels.

All fishing gear would be similar to that on the larger 100 ft. (30.5 m.) steel combination vessel described next. A centreline double drum winch is used instead of the wing winches carried by the smaller combination vessels described in the preceding chapter.

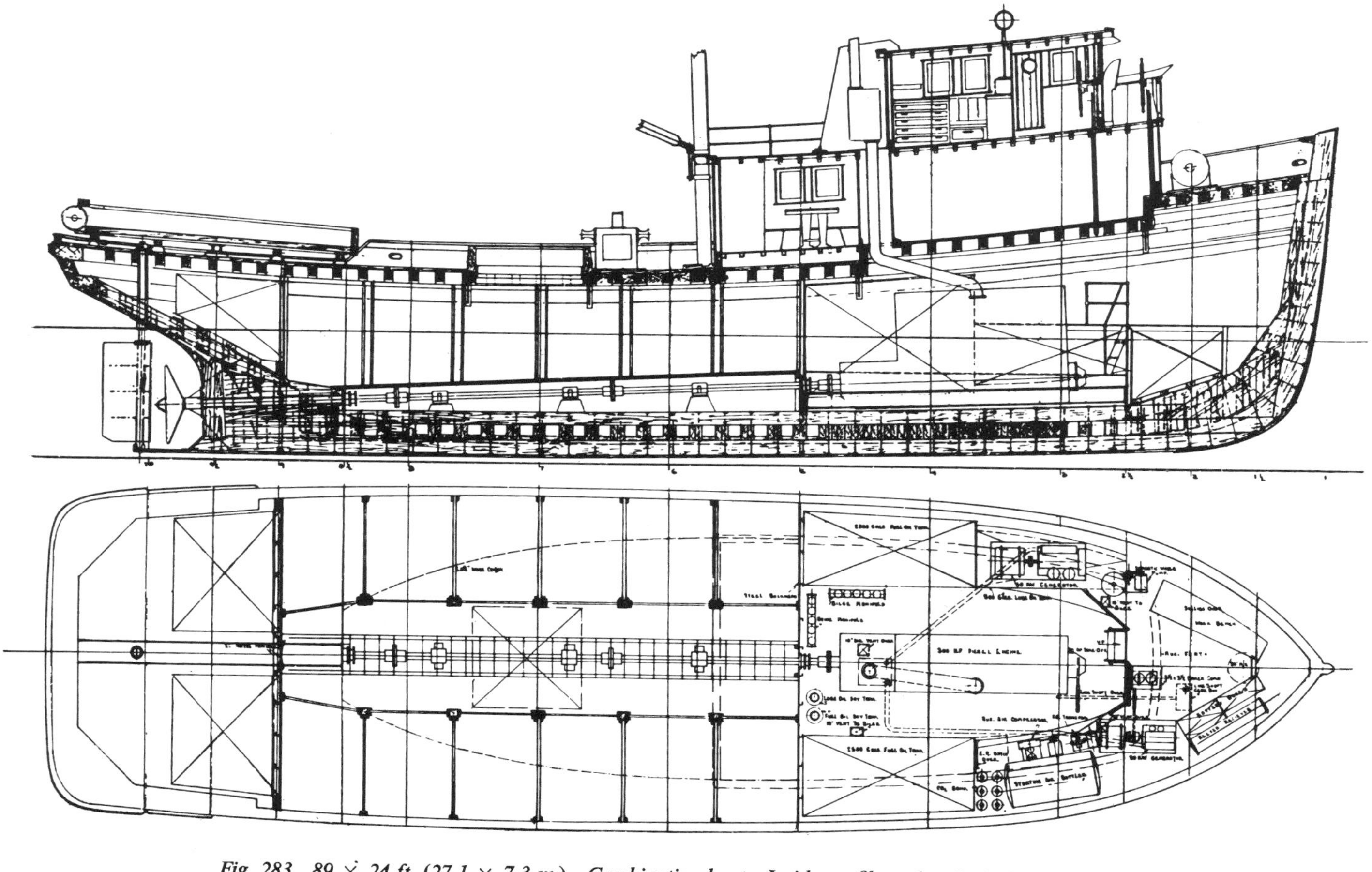

Fig. 283. 89 × 24 ft. (27.1 × 7.3 m.). Combination boat. Inside profile and underdeck arrangement

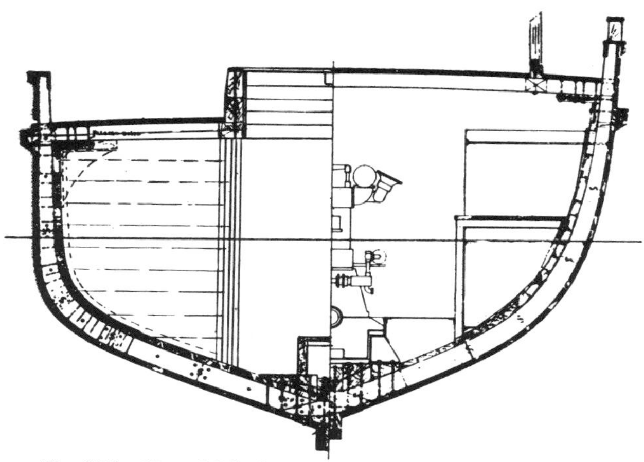

Fig. 284. 89 × 24 ft. (27.1 × 7.3 m.). Combination boat. Midship section

LARGE COMBINATION VESSELS

A typical large combination vessel is 100 × 26 × 13 ft. 9 in. (30.5 × 7.9 × 4.19 m.), all welded steel, for trawling, tuna bait fishing and purse seining.

Fig. 287 shows one as a trawler and fig. 288 as a tuna vessel. During tuna fishing the stern is trimmed down so that the deckline is almost flush with the water. Bait is taken from the deck bait tanks. Tuna, with a body temperature up to 80 deg. F. (27 deg. C.), is brine refrigerated and kept for an extended period before delivery to a mother ship or home port. As trawlers, these vessels catch bottom fish and crabs. They can also be arranged for long lining. Trawling and long lining put a severe test on any craft in the North Pacific and Bering Sea, probably as severe at times as anywhere in the world. Therefore seaworthiness and sea-keeping ability are a prime consideration and must be the best possible. The working deck is best aft, and for trawling and long lining it must be clear.

The lines have a decided sheer forward for heavy weather. The stem rake helps to give a good hull flare which has made these vessels very dry and the deck is the fullest possible.

Fig. 285. Assembling of frames by the double flitch method

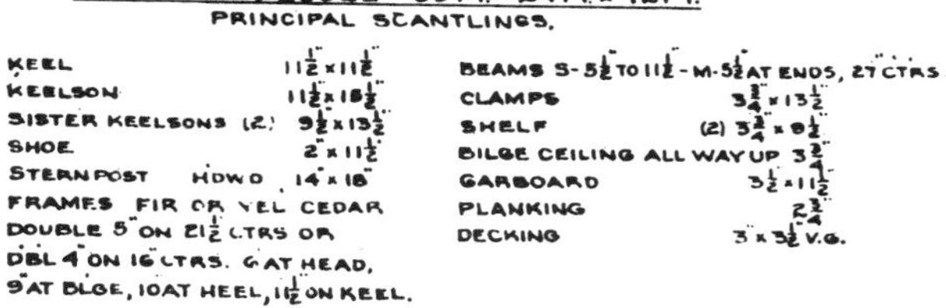

VESSEL 89 FT. × 24 FT. × 12 FT.
PRINCIPAL SCANTLINGS.

KEEL	$11\frac{1}{2}'' \times 11\frac{1}{2}''$	BEAMS S-$5\frac{1}{2}''$ TO $11\frac{1}{2}''$-M-$5\frac{1}{2}''$ AT ENDS, 27" CTRS	
KEELSON	$11\frac{1}{2}'' \times 15\frac{1}{2}''$	CLAMPS	$3\frac{3}{4}'' \times 13\frac{1}{2}''$
SISTER KEELSONS (2)	$9\frac{1}{2}'' \times 13\frac{1}{2}''$	SHELF	(2) $3\frac{3}{4}'' \times 8\frac{1}{2}''$
SHOE	$2'' \times 11\frac{1}{2}''$	BILGE CEILING ALL WAY UP	$3\frac{3}{4}''$
STERNPOST HDWD	$14'' \times 18''$	GARBOARD	$3\frac{1}{2}'' \times 11\frac{1}{2}''$
FRAMES FIR OR YEL CEDAR DOUBLE 5" ON $21\frac{1}{2}''$ CTRS OR DBL 4" ON 16" CTRS. 6" AT HEAD, 9" AT BLGE, 10" AT HEEL, $11\frac{1}{2}''$ ON KEEL.		PLANKING	$2\frac{3}{4}''$
		DECKING	$3'' \times 3\frac{1}{2}''$ V.G.

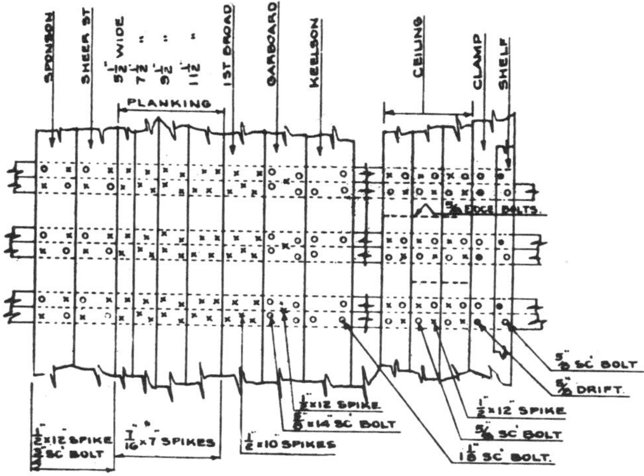

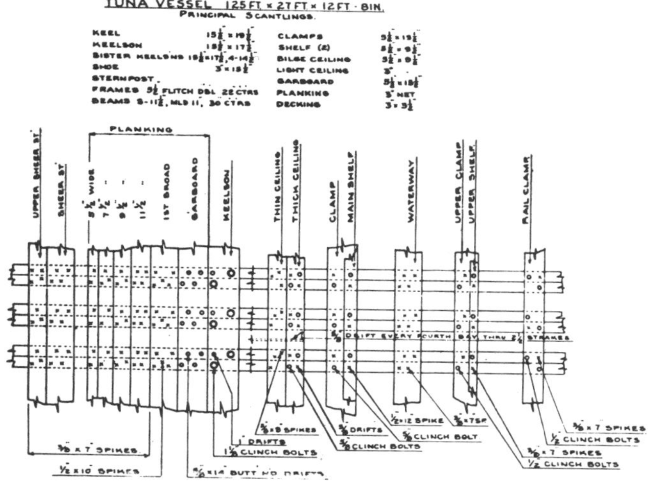

TUNA VESSEL 125 FT. × 27 FT. × 12 FT. · 8 IN.
PRINCIPAL SCANTLINGS.

KEEL	$15\frac{1}{2}'' \times 19\frac{1}{2}''$	CLAMPS	$5\frac{1}{2}'' \times 13\frac{1}{2}''$
KEELSON	$15\frac{1}{2}'' \times 17\frac{1}{2}''$	SHELF (2)	$5\frac{1}{2}'' \times 9\frac{1}{2}''$
SISTER KEELSONS	$15\frac{1}{2}'' \times 17\frac{1}{2}''$, $4 - 14\frac{1}{2}''$	BILGE CEILING	$5\frac{1}{2}'' \times 9\frac{1}{2}''$
SHOE	$3'' \times 15\frac{1}{2}''$	LIGHT CEILING	3"
STERNPOST		GARBOARD	$5\frac{1}{2}'' \times 15\frac{1}{2}''$
FRAMES $5\frac{1}{2}''$ FLITCH DBL 22" CTRS		PLANKING	3" NET
BEAMS S-$11\frac{1}{2}''$, MLD 11", 30" CTRS		DECKING	$3'' \times 5\frac{1}{2}''$

Fig. 286. Scantlings and fastenings for 89 ft (27.1 m.) combination boat and 125 ft. (38.1 m.) tuna clipper

The general arrangements are shown in fig. 289 to 291 and the midship section in fig. 292. The latter has a full moulded shape, giving a carrying capacity of 220 short tons (200 ton) which is exceptional for a 100 ft. (30.5 m.) craft. The author believes that transverse framing makes better use of the inside space of the hull than longitudinals. The scantlings are ample and in accord with existing rules and regulations.

The frames are $4\frac{1}{2} \times 3 \times \frac{5}{16}$ in. (114 × 76 × 8 mm.) on 22 in. (560 mm.) centres. Plating is $\frac{5}{16}$ in. (8 mm.) with

Fig. 287. 100 × 26 × 13.75 ft. (30.5 × 7.9 × 4.19 m.). Combination vessels—as trawler

$\frac{3}{8}$ in. (9.5 mm.) garboards; deck is $\frac{1}{4}$ in. (6.35 mm.), angle stringers intercostal in hold, $4 \times 3 \times \frac{5}{16}$ in. ($102 \times 76 \times 8$ mm.), lining $\frac{3}{16}$ in. (4.8 mm.), plate keel, $1\frac{3}{8} \times 6$ in. (35×152 mm.), floors, $\frac{1}{4}$ in. (6.35 mm.) and centre vertical keelson, $\frac{5}{16}$ in. (8 mm.). The hold and bulkheads are insulated with 4 in. (102 mm.) cork.

The main deck has two breaks. The house is placed far forward down into the hull and protected by the sheer. Unlike present-day clippers, it has no overhang and, in fact, reverts to the earlier style. This design eliminates top weight and improves the seaworthiness.

The bait tanks, fishing racks, etc., are portable to facilitate conversion to trawling. The bait tanks are smaller than the usual tuna clipper's but, as compensation, two large brine tanks are arranged below deck to carry bait on the outboard voyage.

The portable bait and brine tank on deck has three compartments and is of steel with plywood sheathing. The sides are $\frac{1}{4}$ in. (6.35 mm.) and the bulkheads $\frac{3}{16}$ in. (4.8 mm.) plate with $4 \times 3 \times \frac{1}{4}$ in. ($102 \times 76 \times 6.35$ mm.) angle stiffeners. The 4 in. (102 mm.) cork insulation all around and between the tanks is covered with $\frac{1}{2}$ in. (13 mm.) plywood over the exterior. This plywood serves two purposes—heat reduction, and prevention against damage to the hooks when pole fishing. The hatches have 24 in. (0.61 m.) high coamings and 1 in. (25 mm.) plywood covers on the top with a 3 in. (76 mm.) cork insulated cover below. The bait and cargo tanks have supply lines led into the top. In each intake a segment with a perforated screen, arranged to create a circular water motion, is welded in full depth of tank to break down the water stream so it will not damage the bait. The overflow duct is opposite and of much larger area. It is divided in half by weir boards right up to the coamings. The water flows over the weir boards down to deck level and overboard through a discharge pipe. The tanks are painted with white plastic paint and watertight flood lights are fitted inside each bait well.

The main engine is a 600 h.p. turbo supercharged fresh-water cooled diesel, turning a 68 in. (1,727 mm.) diameter, 38 to 41 in. (965 to 1,041 mm.) pitch, three-bladed propeller 400 r.p.m. through a 6 in. (152 mm.) tail shaft in rubber bearings.

The fuel capacity is 30,700 imp. gal. (36,800 gal., 140 cu. m.) bunkered in four wing tanks and the double bottom. Lubricating oil storage is in the lower engine room wings to capacity of 625 imp. gal. (750 gal., 2.85 cu. m.). Water capacity is 1,670 imp. gal. (2,000 gal., 7.6 cu. m.), carried in the lower forward engine room.

There are two 120 h.p. auxiliary diesels each driving a 50 kw. generator and each equipped with a power take-off. They drive a central overhead shaft through clutches by means of sprockets and roller chains. The shaft drives the windlass which handles 100 fm. (183 m.) of wire and a 15 fm. (27 m.) chain to a 750 lb. (340 kg.) anchor. The shaft extends to the trawl winch with two main drums, each for 400 fm. (730 m.) of $\frac{3}{4}$ in. (19 mm.) wire plus the dandelion gear, and two cargo drums with 150 ft. (46 m.) capacity each. It has niggerheads, bitts and a gurdy head suitable for long lining and net hauling.

The vessel has hand and power steering, echosounder, direction finder, radio telephone, a five-bottle carbon dioxide fire extinguishing system in the engine room, a hot water heating plant with oil burner, oil purifier, fire, bilge, fresh water and sanitary pumps, and a 10 in. (305 mm.) bait pump. There are two $5\frac{1}{2} \times 5\frac{1}{2}$ in. (140×140 mm.) twin refrigerating compressors at the forward end of engine room.

Aft of the engine room, and on either side of the shaft alley, are two large brine tanks having 3,242 ft. (1,000 m.) of $1\frac{1}{4}$ in. (32 mm.) ammonia refrigeration coils. Fish are loaded through the deck hatches and unloaded through the bolted manhole openings into the main hold after freezing. The wells are fitted with bronze overflow valves and four watertight lights for use when bait is carried.

The main hold is divided by a steel bulkhead and is refrigerated, one compartment by 2,212 ft. (675 m.) and the other by 1,892 (580 m.) coils. Fish is kept either in ice, aided by mechanical refrigeration, or it is frozen, since the hold refrigeration is designed to maintain the frozen fish.

The stern rail has spray attachments and a series of individual bait cans arranged inside the rail so that they can easily be removed and cleaned. They are connected with hose for water supply to carry a few live bait for special chumming.

Heavy gallows for trawling are installed at the quarters, and heavy block leads allow the wire to be carried to the trawl winch.

The bulwarks aft are removable so that a turntable can be used.

A 16 ft. (4.9 m.) high-speed tender is carried athwartship on brackets extending from the top of the bait tank. Full headroom under the brackets allows the use of the deck underneath.

Fig. 288. 100 × 26 × 13.75 ft. (30.5 × 7.9 × 4.19 m.). Combination vessels—as tuna vessel

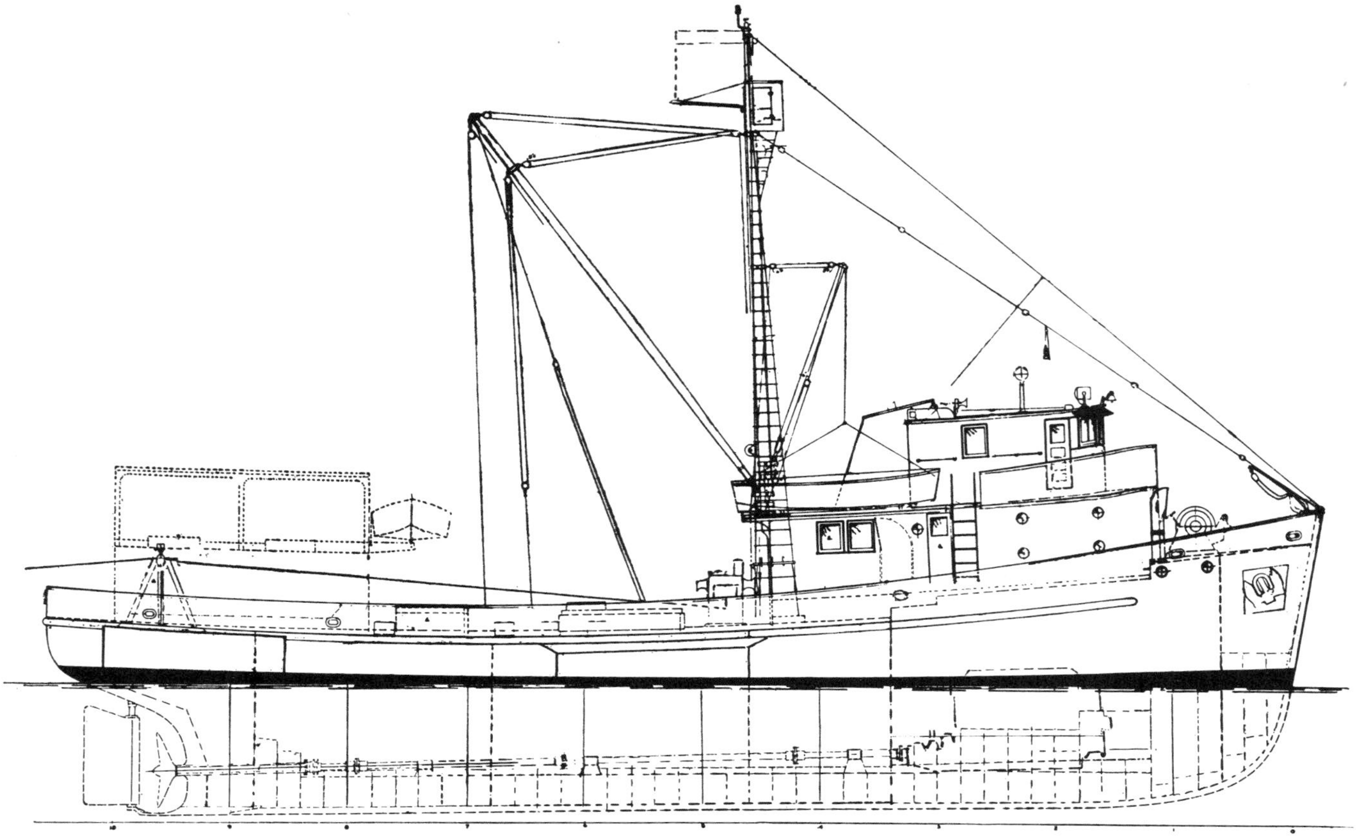

Fig. 289. 100 × 26 × 13.75 ft. (30.5 × 7.9 × 4.19 m.). Combination vessel—outside profile

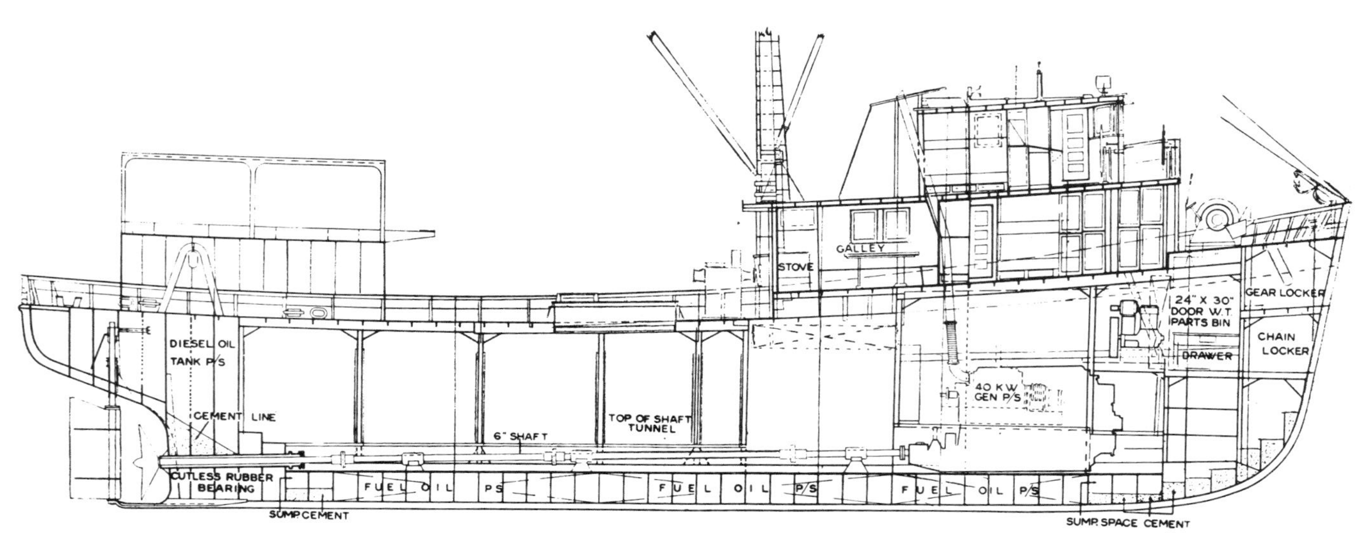

Fig. 290. 100 × 26 × 13.75 ft. (30.5 × 7.9 × 4.19 m.). Combination vessel—inside profile

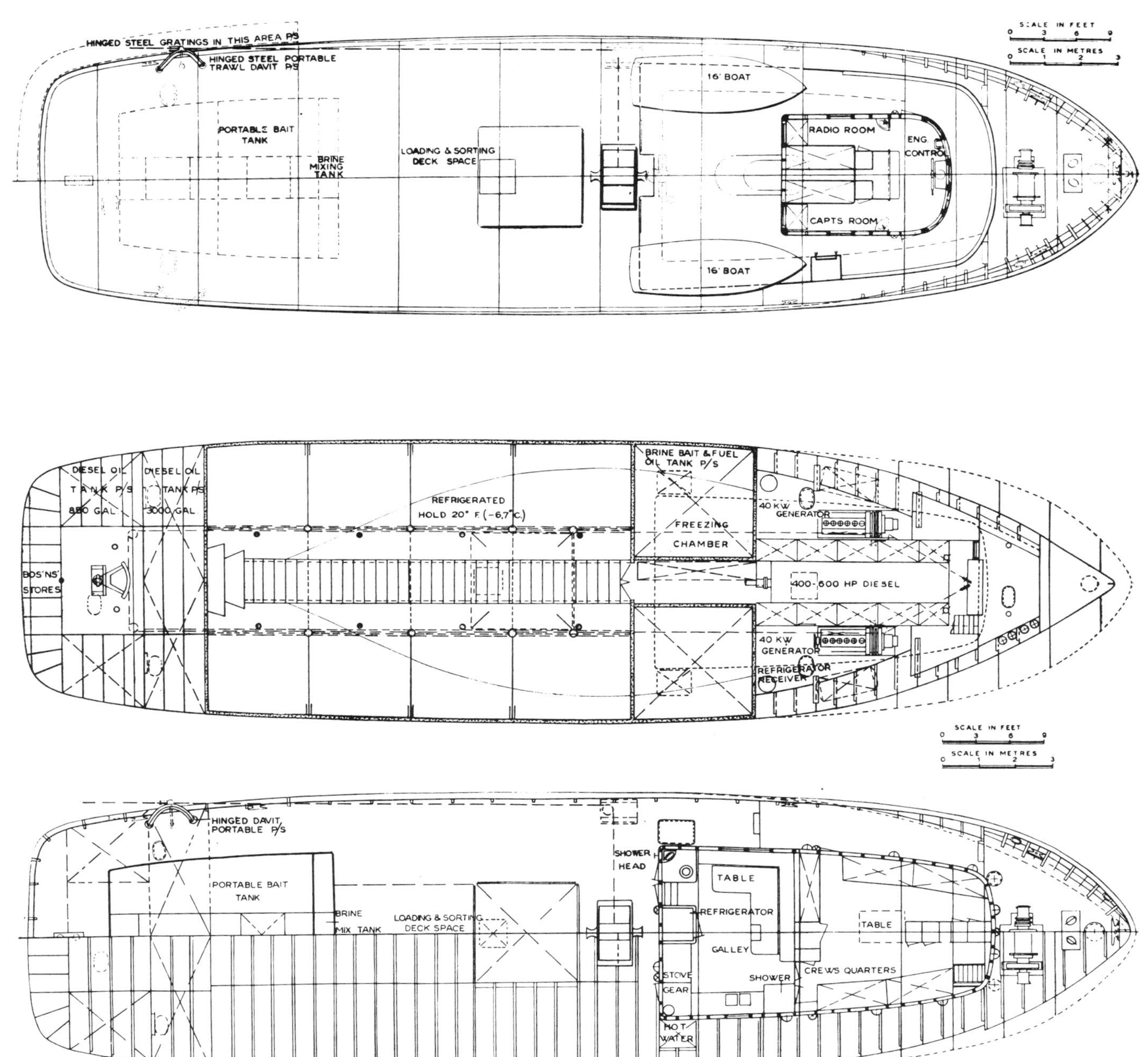

Fig. 291. 100 × 26 × 13.75 ft. (30.5 × 7.9 × 4.19 m.). Combination vessel—deck arrangements

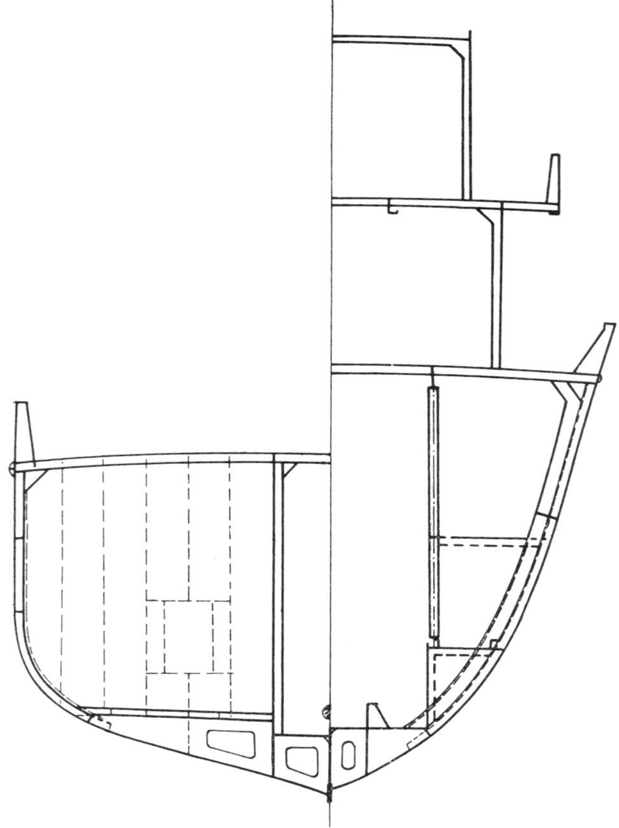

Fig. 292. 100 × 26 × 13.75 ft. (30.5 × 7.9 × 4.19 m.). Combination vessel—sections

Quarters are provided for 10 men and three officers.

Awning stanchions are fitted into sockets on deck. The steel deck is covered with wood grating.

It is believed that, with prospects of tuna long lining in deeper water as well as near the surface, this type of vessel will be used in the future for extended fishing off shore, for which purpose it readily lends itself.

APPENDIX

MACHINERY FOR COMBINATION 100 ft. (30.5 m.) TRAWLER AND TUNA CLIPPER

1 600 h.p. 6 cylinder reversible diesel.
2 120 h.p. auxiliary diesel engines, direct connected to 50 k.W. 220 volt, 3 phase 60 cycle A.C. generators on common base.
1 10 in. (254 mm.) 15 h.p., 2-speed propeller pump.
1 2 × 3 in. (51 × 76 mm.) 5 h.p. centrifugal pump (ammonia cooler).
2 1½ in. (38.1 mm.) 1 h.p. centrifugal pump (brine agitator).
1 3 × 3 in. (76 × 76 mm.) centrifugal pump—general service —driven off auxiliary engine.
1 3 × 3 in. (76 × 76 mm.) centrifugal pump—fire and bilge— driven off propeller shaft.
1 ¾ in. (19 mm.) 1 h.p., 166 imp. gal. (200 gal., 760 litres) per hour sanitary pump.
1 ¾ in. (19 mm.) 1 h.p., 166 imp. (200 gal., 760 litres) per hour fresh water pump.
1 2 in. (51 mm.) 2 h.p. centrifugal pump (fresh water cooling).
1 2 in. (51 mm.) 2 h.p. centrifugal pump (salt water cooling).
1 1¼ in. (32 mm.) 1 h.p. rotary pump (lubricating oil).
1 1¼ in. (32 mm.) 1 h.p. rotary pump (fuel oil transfer).
2 5½ × 5½ in. (140 × 140 mm.) twin cylinder single acting ammonia compressors, direct V belt drive with 20 h.p. motors.
1 ⅓ h.p. water cooled freon refrigeration unit.
1 23 cu. ft. (0.65 cu. m.) air compressor driven off auxiliary engine.
1 ⅓ h.p. motor for the water heater.
1 anchor windlass, driven off auxiliary engine take-off.
1 30 h.p. trawl winch, driven off aux. engine take-off.
1 oil burning galley range with ⅓ h.p. motor.
1 echosounder.
1 radio direction finder.
1 radio, radio telephone.
3 flood lights.

IRISH FISHING BOATS

by

JOHN TYRRELL

THE design and the construction of sailing fishing boats underwent very considerable development in Ireland after 1850 when, because of expanding export markets for herring and mackerel, it became necessary to have large, seaworthy boats, to operate up to 80 miles off the South and West coasts where there were large quantities of fish. The vessel developed was a powerful and fast sailing craft, generally 55 to 65 ft. (16.8 to 19.8 m.) in overall length, 14 to 15½ ft. (4.3 to 4.7 m.) beam and 8 to 9 ft. (2.4 to 2.7 m.) draft, ketch rigged, with long bowsprit and mizzen boom. Such vessels fished from ports all around the coast of Ireland. The largest fleet was based on Arklow where, in 1890, there were about 90 large boats locally built.

With the development of steam drifters and trawlers, the boats were gradually laid up but, after the introduction of oil engines, some of the better vessels were converted to power and a few of them remained in operation until recently.

In 1905 Mr. Michael Tyrrell, of John Tyrrell and Sons, Arklow, designed and built a model for a motor fishing boat which, he claimed, could be worked profitably in face of steam drifter competition. This vessel (fig. 293) was 50 ft. (15.2 m.) overall length, 14 ft. (4.3 m.) beam and 6 ft. (1.8 m.) draft. She was to be propelled by a 25 h.p. Danish " Dan " engine of hot-bulb type, having two cylinders and a controllable pitch propeller. The winch was belt-driven by the main engine. A version of

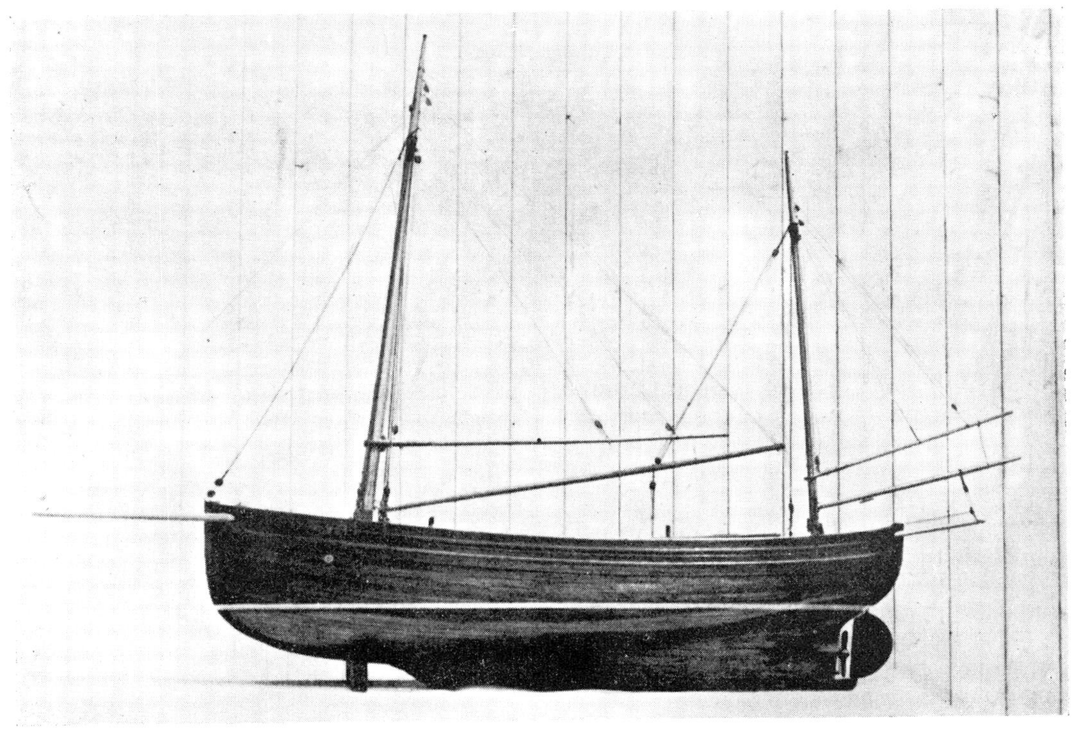

Fig. 293

the cruiser stern was adopted and the hull was a considerable departure from the sailing type, being much more suitable for power propulsion. In 1907 the Department of Agriculture had an experimental boat built to this design and she was named *Ovoca*. The vessel was an immediate success, and is believed to be the first specially designed motor fishing boat in the British Isles. She was intended mainly for drift-net fishing and to do some seine and long-line fishing. After her first year it was found that the large sail area was not required so both masts were shortened and the bowsprit and mizzen boom removed. Subsequent boats developed along this general type but, as fishermen disliked the short cruiser stern, most later vessels had a longer canoe-type stern. The boats also became larger, up to about 70 ft. (21.3 m.) long by 18 ft. (5.5 m.) beam, with engines of 75 h.p., chiefly hot-bulb type.

Fishing suffered a serious decline after 1918 from which it began to recover with the establishment of the Government-sponsored Sea Fisheries Association in 1932. By then many of the earlier motor boats were obsolete, so it may be said that the modern types of Irish fishing boats have been developed within the past 25 years. It is interesting to record, however, that the *Ovoca*, with a modern engine, winch, wheelhouse, etc., is still operating successfully, and shows little sign of her age.

Most of the small building yards on the Irish coasts ceased to exist when the sailing boat declined, and in 1932 the only commercial yards still in business were John Tyrrell and Sons of Arklow, and Wm. Skinner and Sons of Baltimore, both old-established family concerns.

STANDARD 50 ft. (15 m.) BOAT

Experience shows that the most popular boat with Irish fishermen is one of 50 ft. (15.24 m.) length, suitable for drift-net, seine-net, long-line and trawl fishing. Working in collaboration with the Irish Sea Fisheries Association, it has been possible largely to standardize the hull form, variation being allowed in lay-out to suit the needs of different localities. Fig. 294 shows the hull shape of the latest type of 50-ft. boat, the main dimensions of which are:

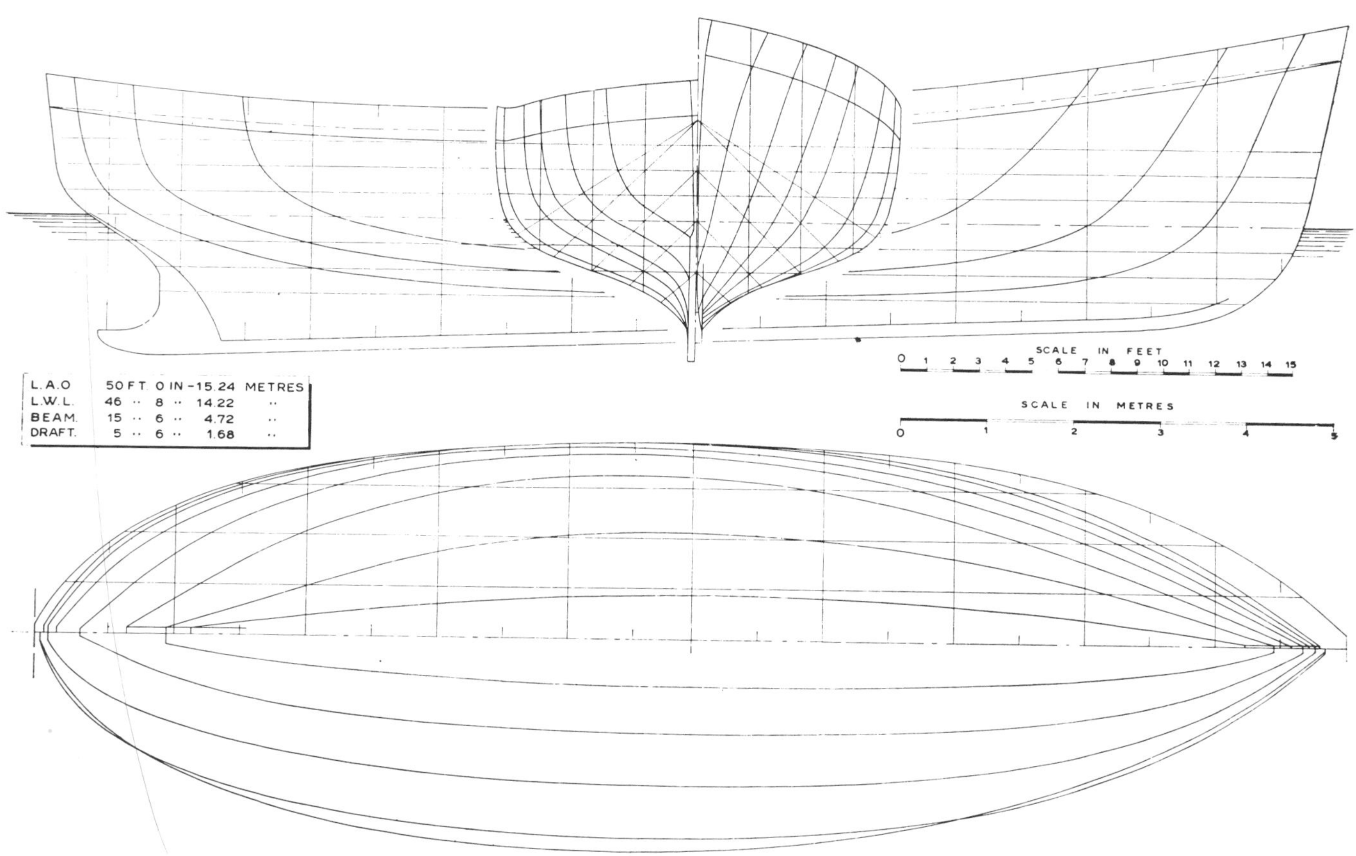

Fig. 294

Length, overall	50 ft.	(15.24 m.)
„ water-line	46 ft. 8 in.	(14.2 m.)
Beam, outside planking	15 ft. 6 in.	(4.7 m.)
Draft, forward	3 ft. 9 in.	(1.15 m.)
„ aft	5 ft. 6 in.	(1.67 m.)
Displacement	28.43 tons	(28.9 ton)
Midship-section area	32.72 sq. ft.	(3.04 sq.m.)
Prismatic coefficient	.652	

The design has proved a satisfactory compromise to embody as many as possible of the requirements at different ports. Limited draft, particularly on the east coast, is essential because of shallow harbours. It is not, however, good practice to design a boat to suit a harbour and, with the promised deepening of several of the smaller harbours, it is hoped that future boats will have greater draft. A big deck area is required in all cases, and full ends are preferred to prevent large changes in trim due to variation in loading.

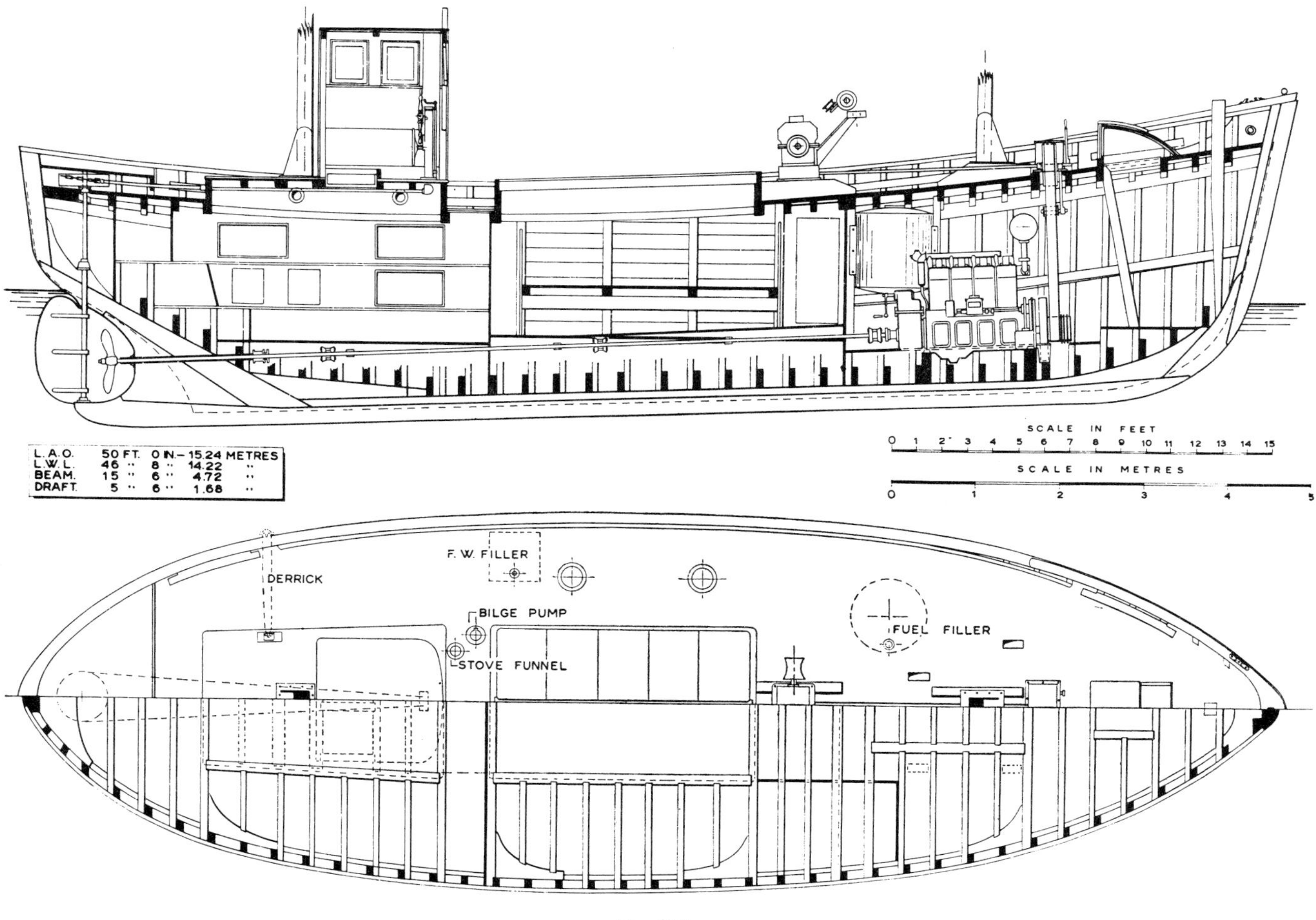

Fig. 295

The propeller position, well aft, has been found satisfactory, and appears to allow more level running than when it is in a more forward position, but a certain amount of roughening of the blades occurs. This may be due to the propeller position near the surface, which causes cavitation. The rudder area of 7½ sq. ft. (0.696 sq. m.) is, in practice, ample, and no steering troubles have occurred. In a few cases a balance of about 4 in. (100 mm.) forward of the stock has been fitted, but this is not considered to be of great benefit.

The main propelling engine is, in most cases, an 88 h.p., four-cylinder, four-cycle diesel at 750 r.p.m., driving a three-bladed propeller 32 in. (813 mm.) diameter by 19 in. (483 mm.) pitch. A full speed of 9 knots is attained with the boat trimmed to designed waterline, carrying 3 tons of fixed ballast and ½ ton fuel, consumption being 4 imp. gal. (4.8 gal., 18.2 l.) per hour.

It is doubtful whether any considerable improvement will be possible in hull shape to reduce resistance (without reducing the value of other aspects of the design) until it is possible to adopt a greater draft.

Fig. 295 and 296 show the internal arrangement most favoured when the vessels are intended for multi-purpose fishing and long periods are spent away from the home port. In Irish boats the engine is installed forward, an unusual position, while the fish-hold is amidships and the living accommodation aft. The chief advantages of this arrangement are:

1. Trim is unaffected by variation in loading, due to the hold being situated about the longitudinal centre of buoyancy.
2. Full use is made of the space in the stern for cabin accommodation, the aft bunks extending to the rudder.

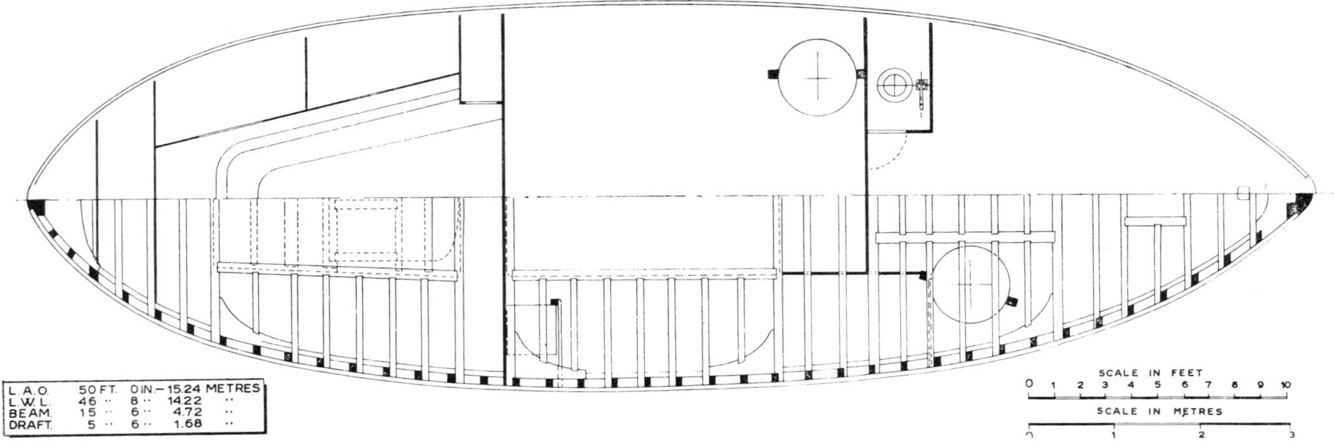

Fig. 296

3. Forward installation of the engine permits a convenient drive to the winch which, in multi-purpose vessels, is located in the best compromise position on the forward deck.
4. The engine is completely accessible, with good floor space clear below the crank-case doors, and is at all times clear of bilge water.

The chief disadvantages are:

1. Much increased length of propeller shaft with its support bearings. Usually, about 12 to 18 months after launching, a general re-alignment of shafting and engine is required, after which no further alignment trouble is experienced.

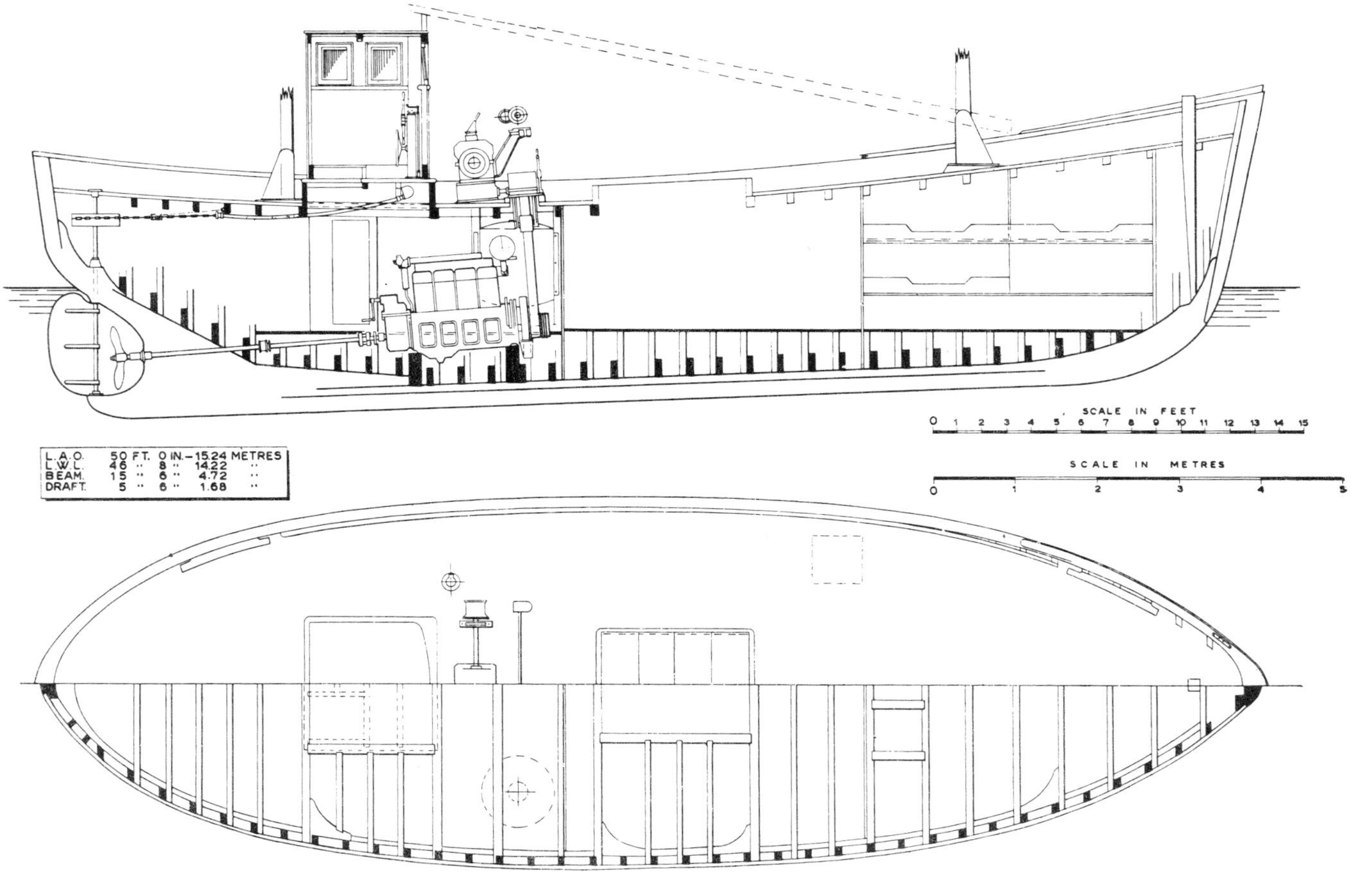

Fig. 297

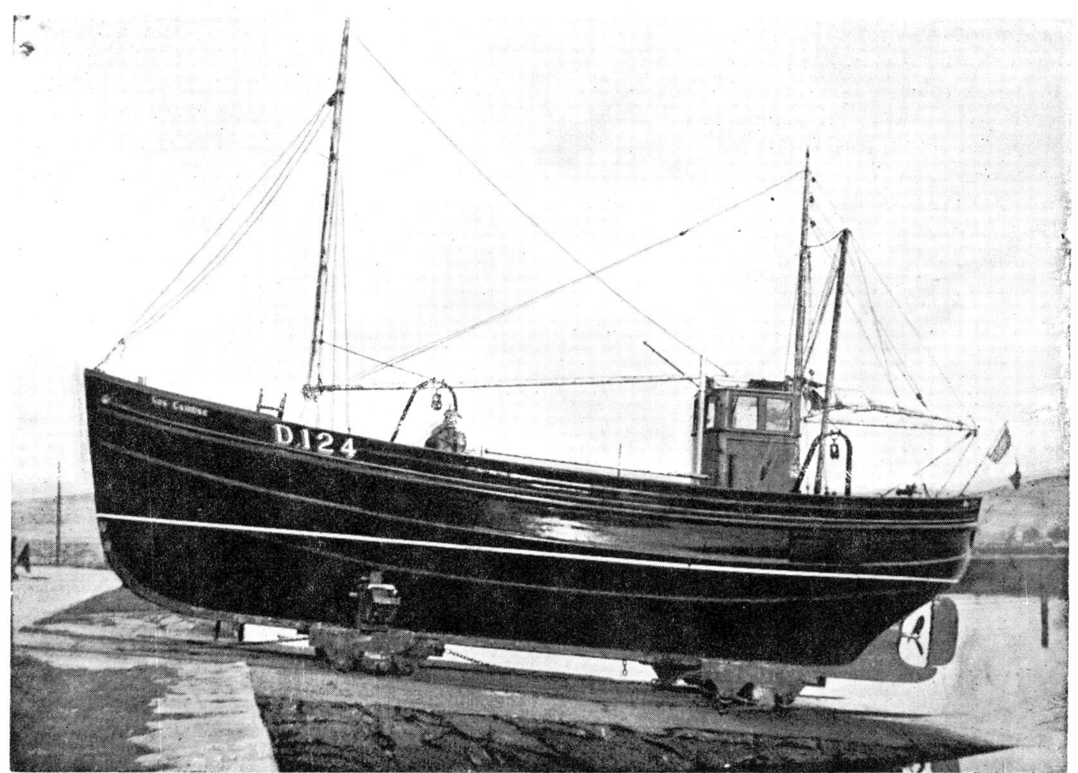

D124

Fig. 298

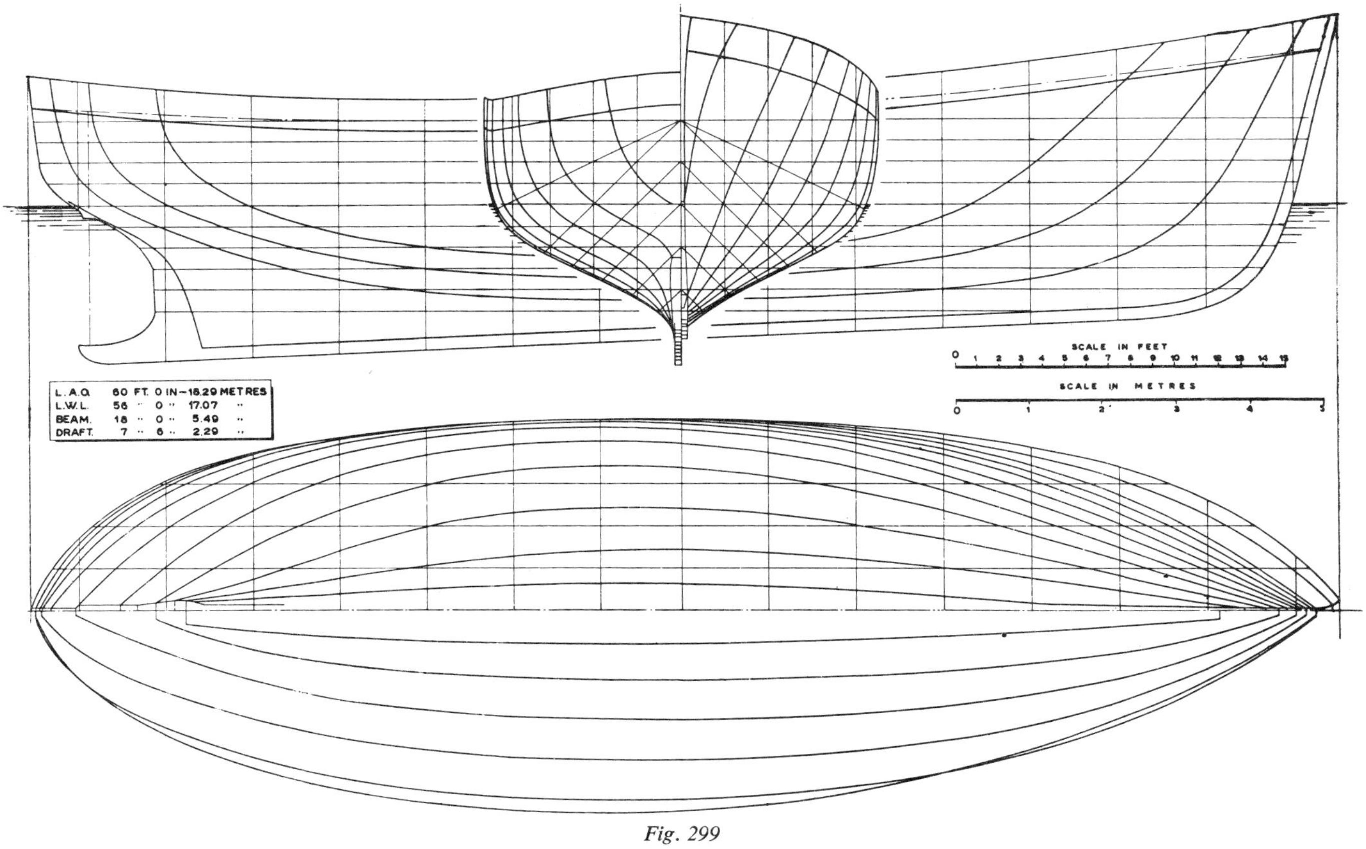
SCALE IN FEET
0 1 2 3 4 5 6 7 8 9 10 11 12 13 14 15
SCALE IN METRES
0 1 2 3 4 5
L.A.O. 60 FT. 0 IN—18.29 METRES
L.W.L. 56 " 0 " 17.07 "
BEAM. 18 " 0 " 5.49 "
DRAFT. 7 " 6 " 2.29 "

Fig. 299

2. Complication of mechanical controls of reverse and governor over the considerable distance from engine to wheelhouse. The reverse controls are usually carried by means of chains running over pulleys from the clutch to the steering column, on which the reverse control wheel is mounted. This is simple, if cumbersome, but when properly installed, with correct alignment and well-lubricated, it works satisfactorily. In engines provided with hydraulic clutch controls no difficulties are experienced; it is simply a matter of running two oil pipes by the most convenient route from the gearbox to the operating unit in the wheelhouse.
3. The relatively large weight located so far forward has a tendency to produce heavier pitching into head seas. While this tendency is always present its effects are minimized by the design of the fore-body, which has considerable flare in the forward sections, the waterlines not being excessively sharp. The vessels behave well and are dry forward; in bad weather and head seas, most spray is shipped over the weather side, amidships.

The boats are fitted with electric lighting throughout from 24 volt batteries of about 120 amp. hours capacity, charged by a generator belt-driven from the main engine. Echo sounders are fitted and take current from the main batteries. A single oscillator tank is fitted in the vessel's bottom, about midships, to avoid turbulence from the propeller. It is as close as possible to the keel, and stands vertical, being welded to a square plate outside, shaped to the hull curvature and bolted to the frames and planking. In the centre of the plate a thin stainless steel diaphragm. of the internal diameter of the tank, is welded to minimize interference with the impulses of the sounder. The shaping and fitting of the outer plate requires careful work to make a watertight and eddy-free job. All boats built in the last years are fitted with a radio telephone, usually of the coastal radio "Seafarer" type. This instrument is installed in the wheelhouse and has proved to be of considerable value.

Fig. 297 shows the internal arrangement adopted occasionally in the 50 ft. standard hull when required for seine fishing daily from the home port. The winch is placed forward of the wheelhouse, which gives all

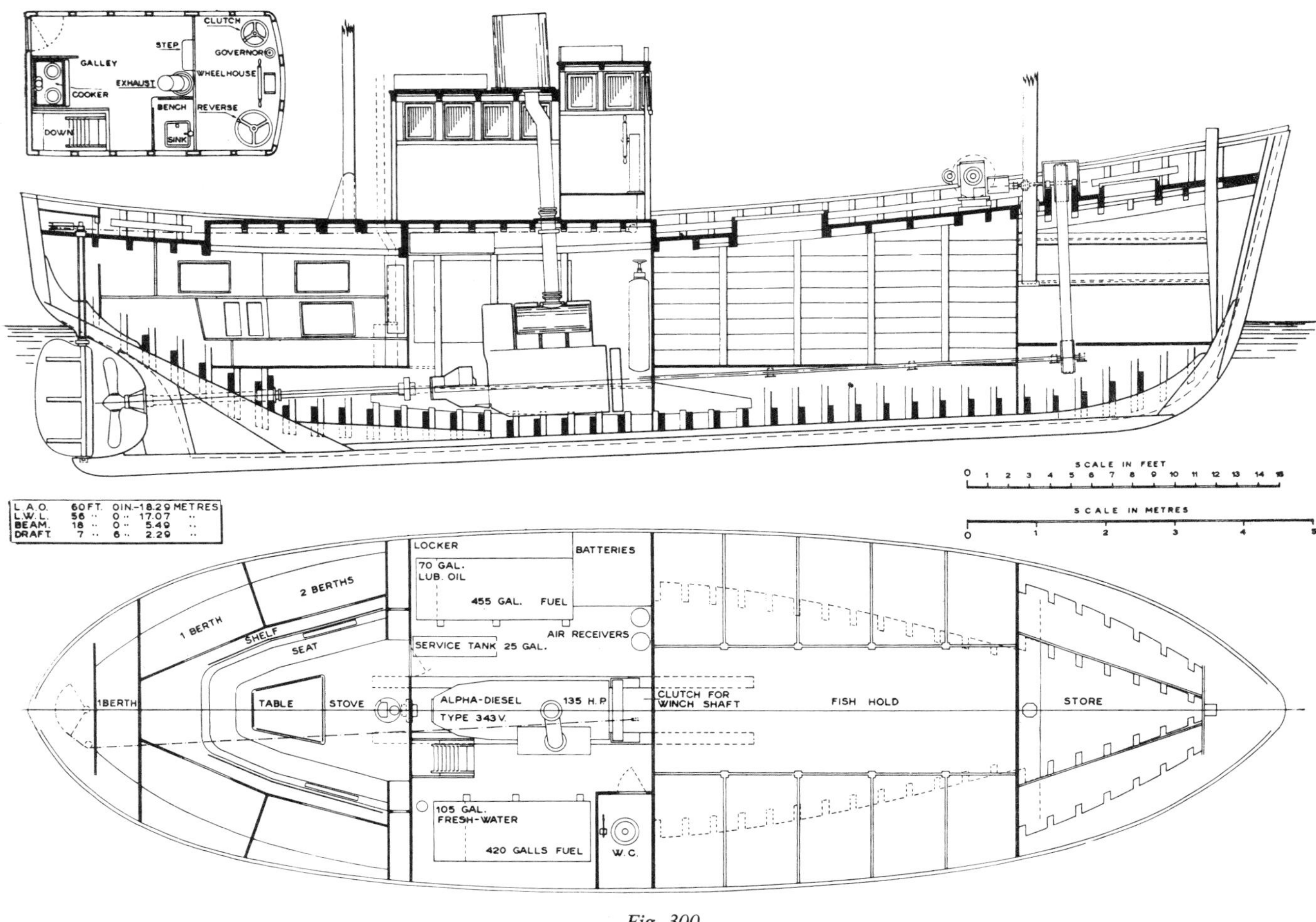

Fig. 300

possible deckroom forward for the seine ropes, and a convenient winch drive from the engine, while difficulties with the long propeller shaft and engine controls are eliminated. The aft bulkhead of the fish-hold is approximately 3 ft. (0.9 m.) further forward than with the arrangement shown in fig. 295. The problems of loading, however, do not arise in this case, since the fish are boxed on deck and properly stowed in the hold, nor do catches approach in weight and bulk those when drift-net fishing for herring. Fig. 298 shows one of the 50-footers on the slipway.

These boats will operate at greater distances than the 50-footers, will not always land their catches daily, and, when working the Irish Sea grounds, will use chiefly deep water harbours for landing. The larger draft will make the boats superior in performance compared with the shallower types. The deeper section will give better motion; resistance will be improved by the finer entrance; and the heavier displacement will provide a steadier towing pull so that larger nets will be used in deeper water.

Both the crew accommodation and engine-room are located aft of the fish-hold (see fig. 300 and 301 for interior arrangement and deck lay-out). The problem of trim will not arise because the vessels will not be dealing with large catches of herring but with fish boxed on deck and stowed in the hold.

The vessel shown in fig. 300 will be fitted with a 135 h.p., three-cylinder, two-cycle, 450 r.p.m. engine, driving a 53 in. (1,346 mm.) diameter, two-bladed controllable pitch propeller. The winch will be driven by a shafting through a clutch on the forward end of the engine. Controls of propeller, governor, engine and winch clutches will be mounted in the wheelhouse. An echo sounder and radio telephone will be installed. Trawl gallows

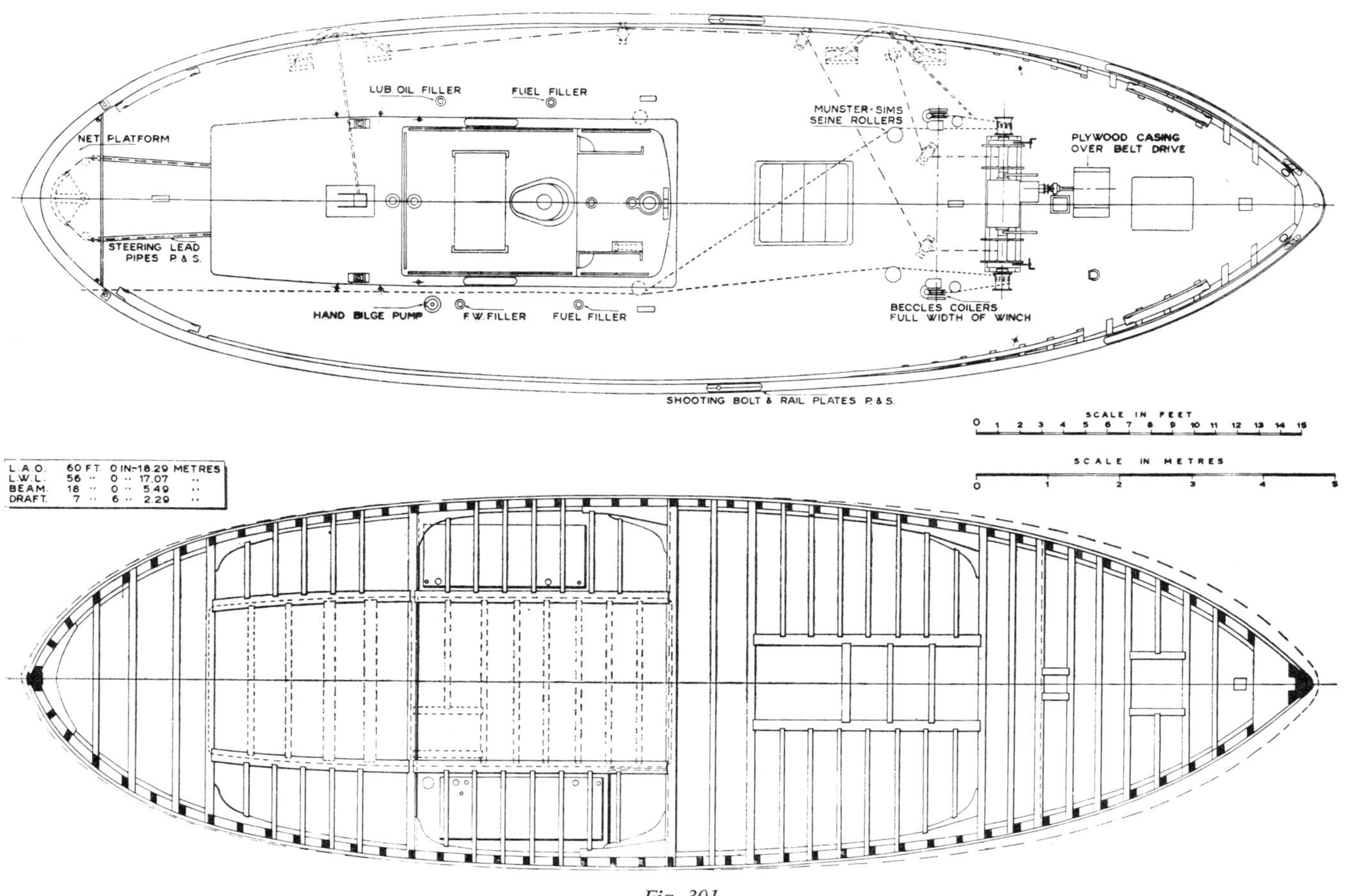

Fig. 301

STANDARD 60 ft. (18.3 m.) BOAT

Fig. 299 shows the lines of a new larger type of boat now being built for seine-net fishing and trawling only. The main dimensions are:

Length, overall . .	60 ft.	(18.3 m.)
,, waterline .	56 ft.	(17.1 m.)
Beam, outside planking	18 ft.	(5.5 m.)
Draft, forward . .	5 ft. 3 in.	(1.6 m.)
,, aft . . .	7 ft. 6 in.	(2.28 m.)
Displacement . .	56.96 tons	(57.8 tons)
Midship-section area .	55.84 sq. ft.	(5.187 sq. m.)
Prismatic coefficient .	.637	

and leads will be arranged so that they can be readily removed when seine fishing. The galley, fitted abaft the wheelhouse, will have the usual cooking arrangements, and a heating stove will be fitted in the cabin.

A second boat of the same hull design and general arrangement will have a 114 h.p., six-cylinder, four-cycle, 900 r.p.m. diesel, driving a 48 in. (1,220 mm.) three-bladed fixed-blade propeller through a 3 : 1 reduction gear. A reduction gear of 2 : 1 will be fitted on the forward end of the engine for the winch drive.

CONSTRUCTION

All these vessels are constructed of grown oak single frames with a separate floor of the same siding fitted to each frame. The planking is of good-quality Irish larch in lengths up to 40 ft. (12 m.). Workmanship, is now of a very high standard and good fitting of all faying surfaces is regarded as essential. Particular attention is given to accurate frame bevelling. By careful loft work and good bandsawing, satisfactory fitting of planks to frames is achieved; almost no hand trimming of frames is required after erection. Bilge strakes and stringers of good size are fitted, through bolted in every frame, and similar through fastening is used for top belting and deck stringer. Decks are of oregon pine, usually 3 in. (7.6 cm.) wide planks. Closely spaced and fairly light deck beams provide a strong and watertight deck, while deck erections and interior correspond to good yacht practice.

ICELANDIC FISHING BOATS

by

BARDUR G. TOMASSON

THERE are 516 Icelandic fishing boats of 100 gross tons and less, the total tonnage being 16,442, making an average of about 32 gross tons. The boats are used for fishing with purse seine, drift net, cod net, longline and handline.

Boats from 10 to 30 gross tons are intended for fishing with longline and cod net. They are strongly built of carvel oak planking on sawn frames of oak. Crew space for sleeping and cooking is arranged forward.

Boats from 30 to 50 gross tons are arranged for fishing with longline and cod net, and with drift net for herring. Hull construction is much the same as for the 10 to 30 gross tons. There is better crew space with a cabin aft of the engine room, and a cabin in the deckhouse. Most of the fishing boats under 50 gross tons are built with an

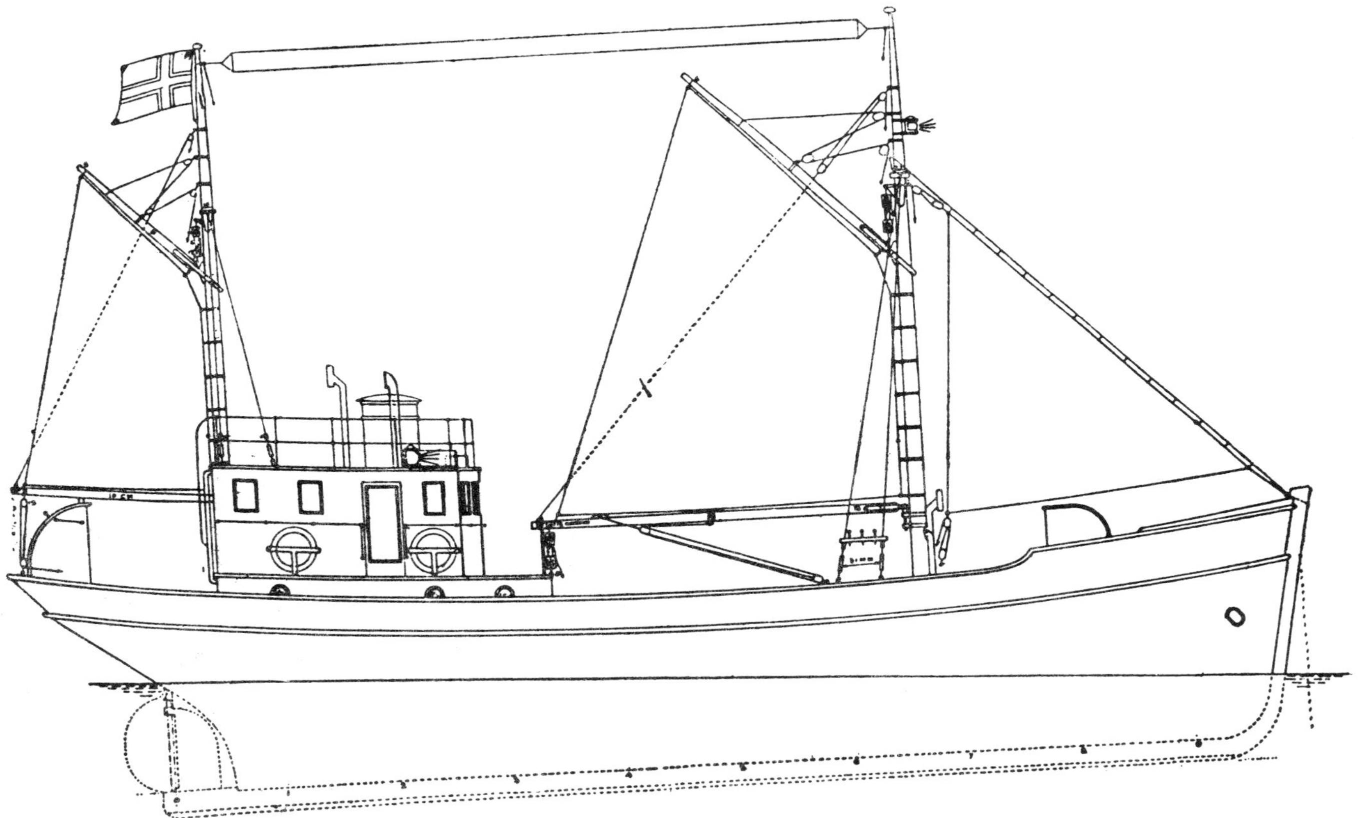

Fig. 302. Icelandic fishing boat

Boats below 10 gross tons are mostly used for fishing with longline and handline. They are of very light construction, clinker built with $\frac{3}{4}$ to 1 in. (20 to 25 mm.) boards, originally on sawn frames but now on bent frames, which makes a lighter construction.

elliptical stern to make more space for the fishing gear on the aft deck.

Boats from 50 to 100 gross tons are intended for fishing with longline, cod net, drift net and purse seine. Hull construction is of the same type as for 30 to 50 gross tons boats and the boats have mostly a cruiser stern. In the stern is a cabin, usually for four men, then the engine room, cargo room, and forecastle. In the deck-house is the mess and galley; above it is the wheel house and aft of it is the skipper's room, fitted with such navigation instruments as echo sounder, direction-finder, radio-telephone, radio, etc.

About half of all boats up to 100 gross tons have diesels and the other half have hot bulb engines. The speed varies from 7 to 9 knots. It is thought that the engines, in most cases, are too far aft, placed there because owners want the fish-hold to be as large as possible. In the author's view, such a full stern makes for great resistance at high speeds, and the thick propeller post—which is necessary to accommodate the stern tube—is a drawback to efficient propulsion. In fitting a large diameter propeller in a small boat with a cruiser stern, it is very difficult to get sufficient distance between the stern timber and the tips of the propeller blades. But fishermen want more speed, which is expensive, and it seems that shipbuilders should consult with propeller makers regarding both propeller design and location. *Resistance, Propulsion and Steering of Ships*, by Van Lammeren, Troost and Koning (1948) is a very useful reference work when planning the propulsion of small craft. It is considered that a reasonable speed for a fishing boat is very near to the square root of the waterline length in feet of the vessel. In accordance with this, a reasonable speed for a 64 ft. (19.5 m.) long boat would be 8 knots.

The stability of a fishing vessel is different during various loading conditions and is a difficult problem to solve. In the new Icelandic boats, in light condition with fuel oil, the GM is about 18 to 22 in. (0.45 to 0.56 m.) with the beam/draft ratio varying from 1.86 to 2.08. When a boat has reached 80 gross tons, the fishermen want a whaleback to give shelter and less shipment of water. Having built up the foreship by 6 ft. (1.8 m.), a " lock out " problem arises and it is necessary to lift the wheel house sufficiently to place its floor 6 to 7 ft. (1.8 to 2.1 m.) above the deck. The next step is a boat deck, heavy davits and boats slung outside port and starboard, with the purse seine gear inboard. All this lifts the centre of gravity, which has to be considered when determining dimensions and designing the lines.

In good herring fishing, the hold is filled and the hatch is supposed to be closed, but some fishermen prefer to keep it open as long as possible to make a " herring feeder " similar to the " grain feeder " of a grain ship. And when the herring catch is also loaded on the deck to a height of 4 to 5 ft. (1.2 to 1.5 m.) the sea level is about 1 ft. (0.3 m.) above the deck at midships but the boats are found to be remarkably seaworthy.

According to the author's experience, this should affect stability, but in fact does not, probably because the bulk of herring on the deck is a homogeneous load which the sea cannot penetrate, thus making it a part of the ship. When the purse seine boats and fishing gear are hoisted in the davits with about half the fish load on deck, the centre of gravity is lifted. It seems to the author that the increased radius of gyration and the diminution of the metacentric heights results, within a certain limit, in an increased seaworthiness of the boats in this condition of loading.

A valuable paper is Spanner's *Some notes on the design of trawlers and drifters, with particular reference to seaworthiness and stability* (1947). Regarding the stability of small craft, it is thought that a combined form and weight-stability will give the best results for fishing boats, taking into consideration the great variety of loading conditions involved.

FISHERY RESEARCH AND EXPERIMENTAL VESSELS

by

G. L. KESTEVEN

THE present decade is seeing a considerable growth of interest in the possibilities of marine fisheries and, in fact, in the entire hydrosphere. This interest springs from certain significant improvements in technique and equipment for investigating and exploiting the seas, and from very important advances which have been made in theories of marine productivity and of fishing.

The need for greater production from fisheries finds immediate realistic expression in the drive by the industry to improve, extend and increase operations and gain better returns. As a result there are urgent demands for bigger and better equipment and for more extensive and more precise knowledge of the seas and, as a consequence, the industry is achieving an almost startling evolution of its craft and equipment. At the same time, Governments are spending more money on the craft and equipment required for that part of the work which is, and must remain, their responsibility. Each year vessels are being planned and commissioned for various types of government fishery work and this will lead to wider and deeper knowledge.

These developments cannot be regarded entirely without misgiving for there is evidence that some of the planning and building is not well advised and is, in fact, based upon misconceptions as to the nature of the work to be done by research and experimental vessels. A brief review of the types of work undertaken under governmental programme and consideration of the designing, equipping, manning and operating of the vessels, may make clear the differences between the types of work.

GOVERNMENTAL FISHERY PROGRAMMES

Governmental patrol vessels, although they may make observations of weather, the sea or other elements and police fisheries, are not in the research and experimental class. Their design and operation are very different.

There are four principal categories of research and experimental work in sea fisheries, namely:

1. Fundamental (oceanographic) research;
2. Applied exploratory surveys;
3. Experimental fishing;
4. Continuing observation of fishery resources.

In fundamental research, observations are made on the water masses and their biota, on the sea floor and on the atmosphere. They are made according to established scientific systems and irrespective of the commercial value of any element under observation.

In applied exploratory surveys, selected types of fishing gear are used to determine the presence of commercially exploitable stocks within an unexplored area, or to determine the limits of distribution of a resource whose presence is already established.

Experimental fishing is the trial of various types of fishing gear (including, in some cases, novel equipment) in an area where improvements may be effected in the methods of fishing.

In continuing observation of fishery resources, normal fishing gear, or especially designed sampling gear, is used to obtain samples of a stock which is being fished.

These different types of work require different sets of equipment and plans of operation and, when coupled with differences in circumstances of operation, they also call for different designs of craft. A research vessel should be designed, equipped, manned and operated for the type of work which she is intended to do, and, if this principle is substantially departed from, there is bound to be serious loss of time, efficiency and money. For instance, it may not matter much if, in attempting to transport a great iron casting with a wheel barrow, the barrow is crushed so long as the casting is not dropped. Nor does the wastage much matter in carrying a crate of eggs with a 10-ton truck. But the cost of building, equipping and operating a sea-going vessel is a very different matter. And, with important issues at stake, inappropriate equipment cannot be regarded with equal complacency. The task then is to consider the requirements, for each type of work, in respect of general design, size, speed, equipment, accommodation of various kinds, crew, scientific personnel, and other features.

PROGRAMME CHARACTERISTICS

In this analysis the programme characteristics may be considered under two principal headings: (1) the investigations in respect of their nature, methods, patterns of work (number and frequency of station, etc.), and the observations or work at each station; (2) the vessel movements *per se* in respect of frequency of cruise, the duration of cruise, various types of travelling and working time, and the conditions under which the vessel sails. Apart from placing these headings in their present order, it would be difficult to suggest any other order of priority for the various matters to be considered, and it is, perhaps, questionable whether such an order exists. The following analysis, therefore, must not be taken to imply any views in this sense.

1. *Nature of investigations*:

(*a*) General objective:
fundamental research;
applied exploratory survey;
experimental fishing;
continuing observation of fishery resources.

(*b*) Objects of investigation: within each of the divisions of (*a*) there is a range of possibilities. For example, in fundamental research the work may relate to:

meteorological elements;
hydrographical ,,
hydrological ,,
geological ,,
biological ,,

Within each of these there is, of course, a great range for investigation. In each of the other three the primary object is fishery resources, but there must always be parallel observations of other elements, chiefly of the environment of the resource. The resources may be classified for these purposes, in two ways: neritic or oceanic; and surface, middle or bottom water.

(*c*) Methods of investigations: it serves no purpose here to attempt any listing of fundamental research methods, but it may be stressed that the complexity of the equipment, its delicacy or otherwise, and its bulk, as required by the selected method, have considerable bearing on what must be required of the vessel.

(*d*) Scope of investigation: under this heading must be considered the area which is to be investigated, the range of elements to be observed, and the duration of the project.

(*e*) Plan of investigation: because all these programmes are, in essence, sampling programmes, their plan should conform essentially to the principles of sampling. The methods of observations (including considerations of sample size), and the pattern of observation in time and space, should be planned in accordance with a logical analysis of available information on the behaviour of the element under observation. A primary distinction must be made between those objects which can be regarded as virtually stationary and those which move; thus, the method to be used in searching for tuna cannot be the same as that used in measuring the abundance of a more or less stationary demersal stock and the frequency of observation will be much less across a very slight gradient of change (say of water temperature), than across a steep gradient.

2. *Vessel movements*

(*a*) duration of cruise;
(*b*) frequency of cruises each year; a function of (*a*) with duration of working period of the year;
(*c*) conditions of operation, chiefly climatic and sea-surface;
(*d*) distance to investigation area;
(*e*) movement within the investigation area.

VESSEL REQUIREMENTS

Theoretically the decision on requirements of a vessel may be taken in three stages: (1) the decision on the nature of the investigations; (2) the decision on the vessel movements which will be required to implement the first decision, and on the equipment which must be carried; (3) the decision on design itself. With all the subdivisions given in the analysis of the previous section, there would be an almost unlimited number of combinations, but, in practice, these could be placed in a number of main groups. When vessels are being designed or purchased for these purposes, the whole range of considerations listed here should be taken in view. And while no rules are proposed for fitting vessel design to programme characteristics, it is possible to indicate some of the errors which can be avoided.

From time to time governments have sought to economize by combining fundamental research programmes with programmes of other types and even, in the extreme, with commercial fishing operations. In such circumstances, the loss of efficiency is considerable and, at times, the usefulness of both lines of work is greatly impaired or even entirely lost. This does not mean that the one vessel cannot at different times do different jobs. Subject to limitations of design and equipment, it is possible to make an arrangement of this kind. But it is to be insisted that the operational requirements of one type of programme are often inconsistent with those of another.

Generally a vessel designed for fundamental research is characterized by making the fullest possible provision for research equipment and personnel, both above and below deck. To reduce these requirements to accommodate fishing equipment lowers the efficiency of the research programme without gaining much advantage from the fishing. In the other three types of programme, the emphasis is on the fishing equipment and the research equipment is a minor complement. It will, however, command more attention in the case of the applied exploratory survey than in either of the other two.

In exploratory work there should be research equipment and personnel of quality and kind sufficient to observe various environmental elements and thus furnish data which should serve to guide the conduct of the operations. Such operations (exploratory fishing) imply the use of a systematic logic of searching along determined co-ordinates: the use of geographic co-ordinates alone is, in the light of our current knowledge of fish distribution and behaviour, not a great deal better than merely haphazard searching. This means that hydrographic and biological observations should be regarded as essential to the programme as the fishing operation. This statement is made in full knowledge of the reluctance of the " fisherman " to accommodate such work, in either physical or programme sense.

In respect of the last two types of programme (experimental and continued fishing observation), there should be certain minimal routine observations to permit the fishing results of such work to be fitted to the earlier fundamental work.

In many cases the planning of fishery work and the commissioning of a vessel are made under conditions of compromise; the research programme is presented with a vessel and obliged to convert it to the programme requirements. It is contended that the principles discussed can still be applied and that the result should be the selection of the limited range permitted by the vessel in terms of these principles. It should not be taken to mean an abandonment of the principles.

RESEARCH VESSEL DESIGN ASPECTS

by

FRANCIS MINOT

ONLY through the collaboration of science, engineering, economics, sociology and business management—together with forward-looking support from government and private enterprise—can world fish production be expanded on a scale and within a time which can aid in meeting ever more pressing food problems.

As evidence of the growing realization of the importance of oceanography, the writer was recently privileged to make a comprehensive study of the requirements of oceanographic research vessels, which was directly concerned, of course, only with the needs of oceanography in the United States and was made for the Woods Hole Oceanographic Institution and the Office of Naval Research. The writer was fortunate in having associated with him in this study, Dr. Colombus O'D. Iselin, of the Woods Hole Oceanographic Institution, Professor Emeritus Herbert L. Seward, of Yale University (a vice-president of the Society of Naval Architects and Marine Engineers), Dr. Kenneth S. M. Davidson, of the Experimental Towing Tank at the Stevens Institute of Technology, and Mr. Mandell Rosenblatt of New York, a designer of small craft and vessels for specialized services.

The following is an outline of this study and some additional thoughts with regard to the more important characteristics of fisheries research vessels.

OCEANOGRAPHIC VESSELS

Seakeeping Ability

Oceanographic research vessels must have sufficient metacentric height and adequate range of stability in all operating conditions, ample draft, freeboard and sheer, and protected working area. In addition, they must embody characteristics and dimensions which result in minimum motion in a seaway, so that work may be continued in relatively severe weather, and so that they will fulfil Kent's (1950) description of a seakindly ship as one which " rides the sea in rough weather with decks free from seawater; that is, green seas are not shipped and little spray comes inboard. No matter in which direction the wind and waves meet the ship; she will stay on her course with only an occasional use of helm, she will respond quickly to small rudder angles and maintain a fair speed without slamming, abnormal fluctuations in shaft torque, or periodical racing of her engines. Open decks will be easy to traverse in all weathers, without danger or discomfort to her passengers or crew, and her behaviour in a seaway, i.e., her rolling, pitching, yawing, heaving, surging, and leeward drift, will be free and smooth from baulks or shocks ".

The most seaworthy of current research ships cannot work effectively in stronger winds than 5 Beaufort scale (17 to 21 knots or 9 to 11 m./sec.), and ability to do so would add materially to the number of working days at sea. Seakindliness has direct bearing also on " habitability ", a most important quality for research vessels. No matter how well arranged the living spaces may be, they will be comfortable only to the degree that the ship is seakindly. It is, therefore, important that the ship shall maintain displacement, trim and GM as constant as possible at all times. A series of double bottom and trim tanks, properly subdivided, will permit replacement of fuel by salt water ballast.

In the opinion of an experienced master of oceanographic research vessels, Captain Adrian Lane, of the Woods Hole Oceanographic Institution, who has commanded *Atlantis* and other research ships, heavy equipment can be handled on vessels of the Atlantis type and size only in winds not exceeding 4 Beaufort scale (11 to 16 knots or 6 to 8 m./sec.) and only medium and light gear can be handled in stronger winds.

Dr. C. O'D. Iselin, of the Woods Hole Oceanographic Institution, has summed up the matter as follows: " There are certain aspects of oceanography that are essentially independent of time. For example, most problems in submarine geology can be equally well studied at any time of the year or can be taken up next year or the year after. But often the problem involves seasonal variations and, less frequently, it is a question of comparing data gained at the same season and place one year after the other.

It is usually unprofitable to send a ship north in the winter months because there may be so few calm days and, beyond the limits of Loran reception, even navigation may be uncertain in stormy weather. It becomes impossible to carry out effective field observations if the wind increases beyond about 5 Beaufort scale (17 to 21 knots or 9 to 11 m./sec.). Even in low latitudes, tropical storms have resulted in big gaps in our knowledge concerning seasonal changes. What can be done by dodging bad weather has been done. A major need in the marine sciencies today is to round out both the picture of seasonal changes and to secure observations—for example, acoustical observations—during the progress of a storm.

The interest of the oceanographer corresponds exactly to that of the fisherman. In both cases, the maximum number of working days in a selected area counts, rather than a round voyage at minimum expense, which is the objective of merchant ships. The merchant ship can lengthen the course in order to avoid storms, but the oceanographer does not want to have to give way to the weather. That he has had to do so in the past is very evident. For example, almost nothing is known about the Gulf Stream between the months of November and April. In winter, the oceanographic vessel usually heads south because only in that way can the number of working days per voyage be kept at a favourable level.

Some seaworthiness can, of course, be gained by increasing the size of the vessel but even if expense were no consideration, the oceanographer prefers a not too large ship. As in fishing, the height of the working deck above the water should be kept as small as is consistent with reasonably dry decks because the safe handling of instruments and nets in and out of the water is a prime consideration. Further, both operations are carried out while stopped and with the sea abeam, during which time it becomes most desirable to reduce leeway. A deep draft and a well-balanced abovewater profile are, therefore, important design features."

The research ship is also required to be manoeuvrable under severe conditions. She must be brought on an exact heading and held there, often at zero speed, while observations or measurements are made, a difficult manoeuvre in rough water. The required manoeuvrability will be attained through good underwater hull design and through the installation of twin screws and twin rudders. Consideration has to be given to the installation of some mechanism such as, an "active rudder", as a positive aid to precise manoeuvring. While it is hoped to avoid such features, it is strongly felt that all possible means should be thoroughly investigated and appraised during the design stage. Obviously, rudders should be of design and size to assure rapid steering effects at moderate helm angles.

Manoeuvrability can readily conflict with directional stability, which becomes of equal importance in holding the headings and in proceeding from station to station. Directional stability will also be attained through good underwater hull design, which will take note of course-keeping influences, and will consider the possible advantages of fitting twin skegs as well as twin screws and double rudders (Saunders, 1947).

The research ship must be able to heave-to and to maintain her position during heavy weather because it affords rest and comfort to the personnel during storms and gales and it saves the time otherwise required to regain position. The design must adopt some characteristics of the sailing vessel, particularly the fishing schooner and pilot cutter, whose ability to heave-to in comfort and maintain position is superlative. Ample draft and lateral plane and the judicious placing of masts and riding sails will assist in attaining the desired ability.

Since an important requirement is that the research ship shall provide a steady platform, a moderate, although sufficient GM will be maintained closely, and a long range of stability provided, through good sheer, adequate freeboard and ample reserve buoyancy. It is planned that working decks will be protected from sea and wind from ahead by the superstructure, which houses the main laboratory, the navigational and operational headquarters.

Possibly the best assurance that research ships of the proposed type will give superior performance in heavy weather is that seakeeping qualities can be given prime consideration. All favourable factors will be incorporated during the design stage and refined through comprehensive experimental tank tests.

Reliability

The oceanographic research ship is often remote from any port for long periods of time, sails the loneliest and least travelled ocean reaches, and is likely to do so more frequently in the future. Where her work is done there is no repair yard, no machine shop, no marine surveyor, no aid of any sort, other than the experience and ingenuity of the personnel.

Reliability is of the utmost importance and it is obvious that if the propelling equipment is divided into two independent parts, the chance of complete breakdown is reduced, and the ship's manoeuvrability and steering are improved. The conviction is that twin screw installation is essential on the basis of added reliability alone. The choice of engines is unusually complicated by the need for effective noise control because the ship must be suitable for use in research on underwater sound and in the development of underwater acoustical instruments and techniques. Other requirements to be met in the choice of the machinery are:

(1) simplicity of operation and maintenance;
(2) low fuel consumption;
(3) flexibility, covering the ability to operate steadily at very low speeds, and the ability to develop close to maximum torque and power when towing heavy gear or trawls at great depths;
(4) complete independence of starboard and port main engines and essential auxiliaries;
(5) ample reserve power;
(6) auxiliary machinery suited to the unusual loads and demands of a deep sea winch, deck machinery and laboratory requirements;

(7) well immersed, well protected propellers, completely free from cavitation, and turning at perhaps 200 to 450 r.p.m.; the lower the r.p.m. the better the vibration characteristics;
(8) least objectionable properties from the acoustical viewpoint, and capable of maximum noise control.

With regard to the choice of propulsion systems, the following have been considered as being possibly applicable:

(*a*) reciprocating steam engines with oil-fired boilers;
(*b*) uniflow steam engines with oil-fired boilers;
(*c*) geared steam turbines with oil-fired boilers;
(*d*) direct-connected diesels;
(*e*) direct-connected diesels with controllable pitch propellers;
(*f*) diesels with reduction gears;
(*g*) diesel-electric drive with multiple engine-generator units.

In view of the very long cruising range which is required of offshore research ships, regardless of size, and the resultant disproportionate weight of fuel oil, all types of steam propulsion have been rejected as impracticable even though they satisfy many of the necessary conditions, since their fuel consumption is so much greater than the comparative diesel. Except in case of vessels engaged in underwater sound investigations—and perhaps, therefore, on relatively short voyages—it is unlikely that any type of steam reciprocating machinery will be considered for future research ships.

All forms of geared drive, turbo or diesel alike, are unsuitable for installation in the research ship because only some very experimental and expensive form of flexible gear mounting could possibly subdue the noise. Acoustical advisors have discouraged any consideration of propulsion systems which involve reduction gears.

Finally, while direct-connected diesels satisfy the cruising range requirements through low fuel consumption and minimum weight of fuel, they do not meet other important performance standards. They do not run steadily at the low revolutions required for very slow speeds. It is doubted whether any heavy duty diesel would be guaranteed to perform continuously at less than 25 to 30 per cent. full power revolutions. They do not develop sufficient power and torque at lower r.p.m.s, unless fitted with superchargers. Direct-connected diesels do not possess the necessary manoeuvrability as their direction of rotation must be reversed with compressed air and they require powerful compressors and large capacity compressed air tanks.

Among the propulsion systems obtainable, there appear to be only two choices:

(*a*) direct-connected diesels with controllable pitch propellers;
(*b*) diesel-electric drive with multiple engine-generator units with fixed blade propellers.

Decision must be made on the basis of individual designs—the size of the proposed ship, the relative importance of reliability and simplicity or acoustical qualities, and the overall cost of machinery and installation. Both systems have the advantage of using undirectional engines at constant speed under most operating conditions, but, in general, better noise control may be expected from the diesel-electric system because noises from low propeller revolutions (outside) and high speed engines (inside) are more readily dealt with acoustically. This system is particularly effective at very slow ship speeds with very low propeller revolutions, conditions under which much underwater sound work will be conducted.

In order to reach equally slow ship speeds, the direct-connected diesel with controllable pitch propeller may have to be reduced to its slowest steady revolutions (about 30 per cent. of the full revolutions) and the propeller pitch reduced as well, thus resulting in lower engine and higher propeller revolutions than in diesel generators in similar circumstances, a less favourable condition on each end of the propeller shaft. The direct-connected diesel will also have to be fitted with some sort of fluid coupling in order to be acoustically acceptable at all.

By and large, the advantages of reliability, simplicity and fuel economy (it costs about 15 per cent. to make up electrical losses) and first cost lie with the direct-connected diesel with controllable pitch propellers, although, in a twin screw installation, this reliability is to a very considerable degree offset by the multiple units and cross wiring of electrical generators and motors.

It may be argued that controllable pitch propellers are much more damageable than fixed blade propellers but experience does not seem to bear out this contention (Rupp, 1948). The original controllable pitch propeller installed in the research vessel *Atlantis* performed with entire satisfaction and with almost no attention for more than 15 years.

Possibly—and this view is ventured without exhaustive study—the direct-connected diesel with controllable pitch propeller may be more suitable for the smaller offshore research vessels where weight, space, first cost and simplicity are of great importance.

Likewise, diesel-electric drive may more closely approach the ideal installation for deep sea research ships in which acoustical investigations are a major undertaking and cost is of less moment.

Finally, while no serious consideration has been given to the gas turbine for obvious reasons, it is felt that its development should be carefully followed, as it may prove particularly appropriate to meet the requirements of the research ship.

Range and Speed

Oceanographic research ships may be called upon to engage in worldwide expeditions which require complete independence from any shore establishment for a period of not less than three months, or they may be away from port for only thirty days. Such a range might run from as much as 25,000 to as little as 5,000 nautical miles. But whatever her size and cruising range, the capacity of any research vessel will on occasion be taxed to the limit of fuel and stores.

Speed, to satisfy the requirements for range, must be moderate for cruising, yet extra speed is of considerable importance, because physical oceanography is entering the synoptic stage, and there is an increasing need to secure a network of observations from a limited area in the shortest possible time, so that a current pattern can be mapped before it changes. The synoptic type of survey does not require great endurance. If some reserve in power is available, the scientist can be given some choice between speed and mileage in any given investigation. He can thus spend the available fuel quickly if it will contribute most to the scientific results, or he can elect to stretch the fuel out over a great number of days when endurance seems more important. It is certain, however, that for an all-purpose research ship, speed must give way to range, endurance and seaworthiness. A cruising speed corresponding to a speed length ratio of ·85 should in most cases be adequate.

Precise control over the speed from stand-still to maximum is essential, and the propulsion systems recommended have been selected, in part, because of their flexibility of operation and control. Evaporators and a reserve of potable water, generous allowance for consumable stores and scientific supplies, a sizable refrigerated hold space, etc., will, of course, match the intended range and endurance.

Arrangement of Main Laboratory and Working Deck

The main laboratory must be on the same level as the exposed working deck and there must be direct and easy access from one to the other. Most important, these two areas must be as large as possible. Furthermore, the main laboratory and the working deck must be located as close as possible to the point of minimum pitching. The superstructure, which houses the main laboratory, must be forward of the working deck to protect it and shelter the laboratory entrance from wind, sea and spray from ahead.

The main laboratory and working deck should be adjacent to the control centre of the ship, having ready access to the bridge and to operations headquarters. And they should be conveniently located with regard to the engine room, but this is not so essential. It is helpful to relate the open working deck space and the main laboratory on a horizontal axis and the lower laboratory, main deck laboratory (again), the operations headquarters and the pilot house, on a vertical axis. The main deck laboratory is thus located at the origin of the axial arrangement.

The reasons for this arrangement are clear. On the horizontal plane there must be a free flow of communication, personnel and apparatus between the open working deck space and the main deck laboratory. In the vertical plane, the nearness of the main deck laboratory to the lower deck laboratory is necessary to obtain a free flow of equipment and specimens from the main deck for secondary study, and to clear the main deck laboratory for other work quickly and easily. Above is the operations office with the necessary chart tables, tracking and navigating instruments. Physical communication from the operations headquarters to the pilot house above and to the main deck laboratory below should be easy, if not direct, so that during operations the scientists may have ready access to the navigational and control centres.

Generally speaking, the midship position of the working deck has been the one adopted on oceanographic vessels used in Atlantic expeditions, and the stern working deck position has been generally used in Pacific service. There are good reasons for each arrangement both from the weather and the operational point of view.

Obviously, the midship position is close to the most stable point on the ship. The open deck is protected fore and aft by superstructure and at the sides by bulwarks. As the deck laboratory entrance faces aft, there is little danger of a sea breaking into the laboratory when the doors are open.

In the stern position, the open working deck space is certainly not located near the point of least motion of the ship and, in heavy weather, the after deck is subject to flooding unless the freeboard is so great that much of the advantage of this location is sacrificed.

Oceanographers agree that some operations are best performed from the midship position, and others from the stern position, and they would like to see working deck areas provided in both these positions. The position of the working deck, however, is determined by the size of the ship and the dual position might only be arranged in ships of perhaps 170 ft. (52 m.) and upwards in length.

With regard to freeboard, the wishes of the scientific personnel and navigating personnel are diametrically opposed. The master of the ship wants as much freeboard and as dry a deck as is compatible with stability and good handling; the oceanographer wants to work as close to the water as he can. The obvious answer is to provide the least freeboard possible at the primary position of overside work, but sufficient to maintain seaworthiness and reasonably dry decks.

Centre Well

Much thought has been given to provision of some internal access to the sea, a " centre well ", through which instruments may be suspended, or even towed at very moderate speeds. It may well take the form of a rectangular well, or trunk, which extends from the working deck down through the bottom of the ship. The minimum useful dimensions of such a well might be 8 ft. (2.44 m.) in length and 5 ft. (1.52 m.) in breadth at the deck, increasing in length at the bottom so that towing wires might avoid contact when the vessel is under way.

The purpose of the well is twofold: to be able to suspend certain instruments from a point close to the centre of least motion in such a manner that they will have no contact except at the point of suspension, and to be able to carry on certain work through the well under weather conditions which would make it impossible to do so over the side or stern. Much more consideration of this proposed feature will be required during the design stage.

Laboratories

The oceanographic vessel is required to carry equipment which is becoming increasingly delicate and complicated. Much of it consists of hastily-wired electronic components which must be protected from moisture and vibrations if they are to function. As the objectives are constantly changing and the laboratory space is limited, much of the old equipment must be dismantled between each cruise to make room for the new. This is very hard on the ship's permanent wiring, on the bulkheads, on the work benches and even on the decks. At least three continuing classes of work have acquired some more or less standard equipment and, if possible, the laboratory space should be divided into three parts: (1) hydrographic; (2) biological and chemical; (3) electronics, geophysical and acoustical.

In the hydrographic laboratory standard equipment includes reversing thermometers, water bottles, echo-ranging instrument, the recording echo-sounder, some form of surface temperature and salinity recorder, racks for spare bathythermographs, salinity sample bottles, oxygen sample bottles, etc. This section requires the use of a large, permanent table, and should have ready access to the ship's navigational centre and to the deck. The personnel will be standing regular watches so that not more than three persons are likely to be using the laboratory at any one time.

The biological and chemical laboratory requires sinks, running salt water, racks for glassware, chemical reagents, etc. Except for buckets, bottles and glass jars of all kinds, the marine biologists have few bulky permanent instruments. As they are acquiring increasing interest in underwater sound, their space might well be near the ship's permanent echo-ranging gear. A space for four or five men should be provided.

The electronics, geophysical and acoustical laboratory needs the largest space and it must be dry. The connection to the deck can be mainly through wires. Most of the equipment can be mounted in standard electrical racks, bolted to the deck. The laboratory also makes use of much standard asdic equipment.

Hydrography, biology and chemistry can, if necessary, be combined into one " wet " laboratory, especially if a small office space is provided for paper work. But electronics do not combine well with these activities. The joint facility needed by all is a well-equipped photographic dark room, preferably with some supply of fresh air.

The laboratory spaces should not be used for recreation spaces for the scientific party or the crew, nor should they be passageways for the ship's company. The scientific party needs a small ward-room where they can read and relax. The alternative is to give each man his own cabin, which is wasteful of space. To combine dining-room and ward-room is, if possible, to be avoided.

Winches and Gear-handling Equipment

The winch is the most important piece of permanent equipment on board the research ship and at least two winches for different purposes, and of quite different design and capacity, are usually carried.

The main (deep sea) winch must be effective for obtaining long cores of bottom sediment and for deep anchoring. For both purposes it must have great cable capacity and power. It must run equally well throughout the whole speed range and must be readily reversible. Reserve power and a sensitive dynamometer are important. In so specialized and expensive a piece of equipment, the user should be consulted before design characteristics are decided, because probably each user will have slightly different requirements depending upon his special field of interest.

There is at least one deep sea winch in use in the Atlantic (and one in the Pacific) which functions reasonably well. The one on *Atlantis* of the Woods Hole Oceanographic Institution, which was designed for that vessel, has worked for some 22 years and was largely rebuilt only five years ago (1948). It is harder than desirable on the cable and has less power than is now needed, but on the whole it has given good service.

A new deep sea winch of advanced design has just completed two years' service on the Danish research ship, *Galathea*, and the reports of its performance are awaited with interest. Likewise, the new winch just developed by the Scripps Institute of Oceanography and now on trial will certainly contribute much to deep sea winch design.

Anchoring in deep water is not likely to become an important requirement. Long cores of bottom sediments are already being accumulated more rapidly than they can be fully studied so it is difficult to argue that a new vessel needs particularly elaborate deep sea winches until it is known that real use can be made of such costly equipment. Space will, of course, be provided for the installation of a very large winch, but it is felt that one of about the capacity of that on *Atlantis* will probably suffice.

Greater flexibility of operation might result if new winches were provided with two drums for different kinds of cable. It will probably soon be practical, at some sacrifice in strength, to obtain a deep sea cable with electrical leads as the core, which would open the way for many new kinds of observations. Probably a tapered cable would be best or perhaps one of greater diameter than those adequate for dredging, grappling or towing nets off bottom. For such purposes one drum might be fitted with a spooling device to handle ½ in. (12.7 mm.) diameter cable and the other take cable of varying diameter. For handling otter trawls, the two drums could be used simultaneously, as on fishing vessels. Shorter lengths of extra heavy cable could go on the drum with the flexible spooling device. Both drums should be of such a size as to take 5,000 fm. (9,150 m.) ½-in. diameter wire, and all lead blocks should be large enough to handle 1-in. (25.4 mm.) diameter wire.

The requirements for a hydrographic winch have become more standardized and are much easier to meet. It should be simple and rugged so that it can operate for hours without attention. The new hydraulic hydrographic winches provide nicely for flexibility in speed, but as yet they have only been used for a few years. It is thought that this design also should be deferred as long as possible so that experience can be gained.

High latitude voyages call for extra heating capacity and for extensive use of steam on deck for freeing deck machinery from ice. The relative advantages of steam as well as electric and hydraulic deck machinery should be thoroughly considered.

Past experience indicates that the scientists will want to lower any sort of equipment that can be assembled on the available deck space. If it is too large to assemble in one piece, they will want to place half of it over the side in the water and then attach the other half. Weights will often be as great as booms and winches can stand. Actually, oceanographic vessels must combine many of the qualities of the fishing vessel, the cable ship, the salvage ship, the buoy setter, and the lighter.

An oceanographic vessel may have an anchor windlass, a steering engine, a deep sea winch and a hydrographic winch, together with extensive engine room auxiliaries for varied ship's and scientific purposes, which, altogether, make up a formidable array of machinery. Much of the ability of the ship to function smoothly and to meet unusual requirements and emergencies depends upon the skilful choice and installation of this equipment.

Explosives Magazine

Research ships must carry explosives, and the location, accessibility and construction of the explosives magazine is important. The entire load of explosives should be easy to dump overboard. The regulations of the U.S. Coast Guard concerning the " Storage and Handling of Explosives and other Dangerous Articles " require that the magazine must be adequately ventilated and wood lined, with no protruding fastenings of any sort. It may not be located adjacent to or in line with any berthing spaces, collision bulkhead or machinery bulkhead.

Noise Control

The development, testing and use of underwater acoustical instruments has become a most important function of the oceanographic research ship and it seems certain that acoustical instruments will be increasingly used in biological oceanography. Generally, acoustical work requires a ship which does not radiate noise in excess of certain prescribed values. It may be shown that the amount of noise control required to produce an acoustically ideal vessel is in the range of 40 to 60 decibels. No measures for noise suppression or control were incorporated in the original design or conversion plan of present day research ships and an improvement on today's standard of 20 decibels is required as a minimum.

Habitability

It is important that every provision is made which is conducive to good health, good spirits and good work. Instead of routine temperature and salinity surveys, the trend is more towards actual research and analysis at sea, rather than proceeding with a predetermined plan of observation. The scientific party should be able to work in the same sort of atmosphere that prevails in the laboratory ashore. The senior scientist should be encouraged to go to sea rather than to leave the field work to technicians and students. Where possible, separate facilities should be provided for the scientific party, whose members are to a degree transient, and for the ship's complement, which is permanent. The needs and privacy of both groups must be provided for, each in a somewhat different fashion.

Proposed Type

For purpose of comparison, a parent form was evolved which incorporated the basic arrangements as outlined above, and expanded systematically from 125 ft. (38.2 m.) length between perpendiculars, and 400 tons loaded displacement to 250 ft. (76.4 m.) length and 3,000 tons loaded displacement.

Then, costs of construction and operation were estimated from known data, and these costs, together with range and endurance, and the areas of laboratories and deck spaces, were set up on a comparative basis.

Finally, a programme was outlined for further development and design procedure.

FISHERIES RESEARCH VESSELS

The principal requirements for vessels for fisheries investigations are similar to those of ships for general oceanographic research. These are seakeeping ability, reliability, a considerable cruising range, appropriate facilities for science, effective noise control, habitability and versatility. The difference is one of degree rather than one of kind. For instance, there is less science and more commerce on board a fisheries research vessel.

Maximum seakeeping ability is of paramount importance and an appropriate balance between seaworthiness and seakindliness is again called for if the largest dividends in terms of practical fisheries knowledge is desired.

The importance of reliability too must be be given

great weight but the more specialized duties suggest that it may be evaluated and provided for in a different manner for two reasons: (1) because the cruising range, the distance from ports, and the need for total self-reliance are less than those required for the oceanographic ship, and (2) because the handling of commercial fishing equipment, in both conventional and advanced style is essential.

The required degree of reliability need not, and in all likelihood cannot, be attained through twin-screw installation of propelling machinery. Because the fisheries vessel must have maximum towing power available at reduced ship's speed when trawling, and because she must, like the trawler, keep clear of the trawl warps, it is felt that the required degree of reliability must be attained with a single-screw installation. In addition, in trawling operations, the fisheries research vessel requires the propulsive characteristics associated with a large diameter single screw and low r.p.m.

Both diesel-electric drive with multiple generator units, and the direct drive diesel with controllable pitch propeller, satisfy the particular requirements of the fisheries research vessel. And, in view of the less severe requirements for long range, it would appear that the steam reciprocating engine of modern type and oil-fired Scotch boilers could be considered suitable, particularly for fisheries investigations in cold climates where an ample supply of steam is a necessity both below and above decks.

Steam, it may be added, is without doubt the simplest, the least troublesome, and the most flexible source of power for the trawl winch, a piece of equipment of first importance to the fisheries research vessel. Steam, again, through the triple expansion engine, will approach the necessary flexibility in power and speed so essential to trawler performance.

The scientific facilities required are more limited. They must include the deck arrangement, the equipment and the gear necessary for all major fishing operations, such as trawling, purse seining, longlining, etc. The vessel must be prepared to fish at depths and under conditions once considered impractical. It is through such experimentation that new fisheries, or new sources of old fisheries, may be found to be commercially profitable. There should be more deck space around the stern than is common to the trawler, so that large deep-water seines may be effectively handled. There must be provision for the storage of a large seine boat of perhaps 35 ft. (10.7 m.) length, and there must be facilities for the stowing and handling of dories and for their ready launching and recovery at all times.

For experimental purposes, the vessel must be prepared to test and develop improved preservation of the catch, such as the freezing of fish in the round, the filleting and packaging of fish at sea, and explore the use of all parts of fish for the manufacture of fish meals, liver oils, etc. for commercial processes.

And, finally, there must be ample laboratory and other facilities for a wide range of biological study, and the necessary arrangements and handling equipment on deck for the proper management of scientific instruments. Obviously, practical fisheries biology and development of advanced fishing tools and techniques are the responsibility of the fisheries research vessel.

In the matter of acoustical requirements, the vessel must be silenced to whatever degree is necessary for the development, the testing and use of underwater sound equipment for the detection and identification of schools of fish, and she must be equipped to use the latest instruments in echo sounding and echo ranging. The detection and identification of schools of fish may well be the most revolutionary fishing development since the earliest times.

The fisheries research vessel is the key tool to fishing development, and it merits all the scientific and engineering skills which can be enlisted in its design.

NEW MATERIALS FOR CONSTRUCTION AND OPERATION

by

E. C. GOLDSWORTHY

MANY new materials are used in fishing vessel construction, fishing equipment, interior linings to fish-holds, insulation and for deck coverings and upholstery. They may be classified: (1) structural and semi-structural applications; (2) insulation; (3) miscellaneous equipment.

The demand for higher speeds and increased efficiency in finding, catching and returning fish to port, has resulted in changes in form and propulsion of fishing boats. The hulls are tank tested to obtain the highest speed consistent with carrying capacity and great care is taken to ensure that a vessel be seakindly both when fishing and in bad weather.

The problem of stability has become more acute because of the demands made by the modern vessel at certain stages of the fishing trip, particularly if she is diesel driven. And the difficulties have been increased by the improved standards of living in accommodation above deck. It is not by chance, therefore, that designers turned to new materials.

Aluminium alloys, specially developed to resist corrosion from sea water and marine atmospheres, were introduced in 1930, but the world-wide depression in shipbuilding restricted their use until 1933. From then until 1939 they were used in yachts and other small craft, and for lifeboats and scuttles in a number of big passenger ships. They were, however, mostly ignored in fishing boat construction although their use for bridge and upper deck structures, including the funnel casing, would reduce top weight and improve stability, and make it possible to place extra accommodation above deck without reducing the metacentric height. And then there is freedom from corrosion, cutting the cost of upkeep and lengthening the life of the structure. Venus (1951) gives the comparative costs for steel and aluminium funnels obtained from a firm of trawler owners, as shown in table opposite.

Although these specific figures are given for funnels, there will be a corresponding saving in maintenance and painting costs over the whole of the deck structures.

Aluminium alloys began to be used generally in the structures of the fishing vessels in 1945. Their availability immediately after the war, when steel and wood were in short supply, led to a number of small deck houses being erected, particularly in converting minesweepers to fishing craft. The importance of these pioneer installations was appreciated by the aluminium producers, who gave every assistance through the Industry's Development Association and their own technical staffs. Today many of the largest deep-sea trawlers and whale catchers have their deck houses and deck ladders, funnel casings, radar structures and lifeboats made of aluminium. It is not uncommon for 20 tons of aluminium alloy to be used above the main deck in a modern trawler; the same structures in an all-steel vessel would weigh at least 40 tons. And this is in a position where saving in weight is of the greatest benefit to stability.

	Steel funnel average life 10 years		*Aluminium funnel estimated life (after 5 years experience) 20 years*	
Initial cost	£50	$140	£126	$353
Cost of painting at £4 (steel monthly; aluminium six-monthly)	£480	$1,344	£160	$448
Repairs over life	£50	$140	Nil	
	£580	$1,624	£286	$801
Total cost over 20 years .	£1,160	$3,248	£286	$801

The corrosion resisting alloy used in shipbuilding today contains a percentage of magnesium and the strength of the alloy varies with the amount of magnesium. For internal use, such as trunking or linings which are not subject to heavy blows, the alloy with 2 per cent. magnesium will be strong enough. For lifeboats, radar masts, accommodation ladders, funnel casings and the like, 3 per cent. magnesium alloy will be most suitable, and for lifeboat davits and some of the more highly-stressed bulwarks and bulkheads, 5 per cent. alloy would be used.

Riveting is the normal method of joining and rivets can be driven cold up to ⅝ in. (16 mm.) diameter without any difficulty. Above that size some form of heating is advisable. Welding with the Argon arc is fast becoming

Fig. 303. Aluminium superstructure of Red Hackle *in course of construction*

Fig. 304. Insulated fish-hold of Cape Scatari *in course of construction*

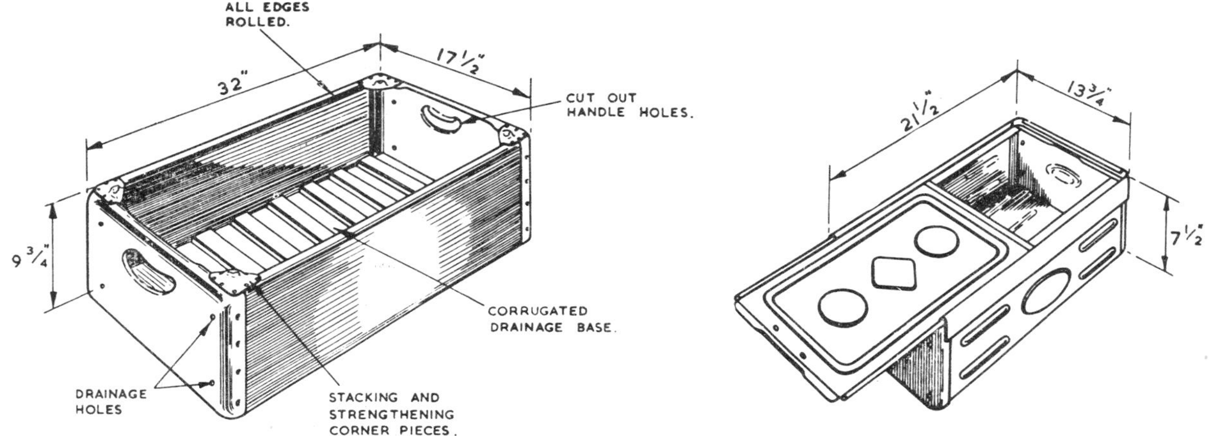

Fig. 306. Typical aluminium fish boxes

Fig. 305. Quayside scene at an English fishing port showing aluminium landing trunks and the trawler Jörundur *with aluminium superstructure and fish-hold lining*

Fig. 307. Aluminium fish kits of the type seen here at Hull are rapidly replacing the traditional wooden barrel

standard practice for work in closed spaces and it is anticipated that, before long, the difficulties of using it in the open air will be overcome. This will be important because the Argon arc, requiring no flux, is inert to any corrosive influence. It is already used in the construction of lifeboat davits, and many small structures, which can be welded under cover for subsequent erection on board.

The technique of using aluminium alloys presents no difficulties to the fishing vessel builder, no matter how small he may be. He can obtain all the advice and guidance necessary from the aluminium industry and virtually no additional equipment is required. Certain elementary precautions have to be taken to avoid galvanic action where aluminium is joined to dissimilar materials such as steel or wood and, in particular, copper-bearing alloys. Many small wood boat builders have stated that, after very little practice, they have found work with aluminium quicker and easier than with wood.

A few typical examples of the modern fishing vessel with aluminium superstructures, funnel casings, fish-holds, radar stands, compass platforms, etc. are the *Princess Elizabeth*, built by Messrs. Cook, Welton and Gemmell Ltd., Beverley, the *Jörunder*, by Brooke Marine Ltd., the *Red Hackle*, by John Lewis and Sons Ltd., the Canadian dragger *Cape Scatari*, built by Smith and Ruland, Lunenberg, Nova Scotia, and whalers such as *Enern*, built by Moss Vaerft og Dokk, Norway, and refrigerating fish carriers, a typical example being *Bemar*, built by Haugesund Mek. Verksted, Norway.

New materials are also necessary in the fish-hold linings to satisfy the demand for better quality fish which had to be caught at far greater distances than in the past. Wood was, for a long time, considered the most satisfactory material for lining a fish-hold and is still used. Provided it is well varnished and kept scrupulously clean, it serves its purpose. But when the varnish wears off, fish slime accumulates in the crevices and bacterial action takes place to the detriment of the catch. Some owners have lined the wood with zinc but this is heavy whilst others used a thin gauge aluminium sheet, with or without some insulating material between the linings and the hull.

Aluminium proved an ideal material for this purpose, being non-absorbent, easy to clean and non-odorous, and much thought has been given to find a method of using it in conjunction with insulating materials. In some cases aluminium lining, sufficiently thick to withstand the stresses and blows in service without wood backing, has been installed. Another method is to make the hold lining from a number of prefabricated panels taken off templates or cut to shape on the ship. The advantages claimed for this method are that the panels, faced on both sides with aluminium alloy and having a core of 2 in. (5 cm.) Onazote insulating material, are rigid and can withstand the heaviest blows from the moving mass of ice or fish and of local blows from ice axes. There are various methods of joining the panels by aluminium screws or properly galvanized or cadmium-plated screws. The panels can be removed for inspection purposes without disturbing the others and without removing loose insulation. Observations over the past seven years show that a working surface of aluminium is able to withstand rigorous service conditions and remain immune from corrosion. Soap and water can remove any film of grey substance that might appear on idle surfaces such as deckheads.

INSULATING MATERIALS

The search for light weight non-hydroscopic non-inflammable insulating material has never ceased. Cork was virtually supreme in the past, but, because of its tendency to absorb moisture, disintegrate and cease to have insulating properties, it is being superseded by new materials derived from glass, cellulose acetate or rubber, to mention three of those most commonly used.

A typical example is fibre glass which consists of a mass of very fine drawn glass wire. It is light, does not absorb moisture and is impervious to attacks by insects and bacteria. The air trapped between the wires gives high resistance to leakage of heat. Its weight, when packed between the frames and fish-hold insulation, is 5 lb./cu. ft. (80 kg./cu. m.) with a conductivity of about 0.24 B.Th.U./sq. ft. per hour per °F for 1 in. thickness (0.14 kcal./sq.m., hr., °C for 10 mm.).

The synthetic material, " Isoflex ", is made from non-inflammable cellulose acetate film and is supplied in sheets, manufactured as a series of alternating corrugated layers, which allows free drainage for condensation moisture. It is stated to be proof against vermin and bacteria and will resist atmospheric conditions. The sheets are placed between the frames and cut to the desired dimensions, the weight being 12 oz./cu. ft. (12 kg./cu. m.) with conductivity of 0.32 B.Th.U. (0.19 kcal.).

The expanded rubber material, known as " Onazote ", with a conductivity of 0.20 B.Th.U. (0.12 kcal.) is unique in that during manufacture completely sealed minute gas cells are formed, making it an ideal light weight insulating material. The density most commonly used for fishing vessel insulation is 4 lb./cu. ft. (64 kg./cu. m.) with a tensile strength of 80 lb./sq. in. (5.6 kg./sq. cm.) and compression strength of 40 lb./sq. in. (2.8 kg./sq. cm.). It is manufactured in slabs which can be cut to suit dimensions. It can be put between the frames (as other normal insulating materials) but it has also been used extensively in panel form, faced by an aluminium alloy sheet. The panels form a strong and rigid lining to the hold.

MISCELLANEOUS EQUIPMENT

Many fish-holds are equipped throughout with pound boards, either of a corrugated section or of hollow floatable boxes with Argon arc welded ends. These boards are easy to clean and, although expensive in the first instance, are proving themselves economical by their longer life. Some vessels use the standard fish-hold pound board for the deck pounds, and even the storage bins and deck tanks for washing are made of aluminium.

It is also used in making the para-otters for stern trawling, marker buoys with radio transmitters for whaling vessels, and trawl net and purse net floats. Aluminium floats have been in use in North America since the beginning of the century and, indeed, some of the original models are still in service. They are usually formed from two die-cast hemispheres joined by welding. Oval purse net floats are a comparatively recent development. Aluminium deck plating is also increasingly used in the form of tread plates which help to prevent slipping by fishermen in rubber boots.

Nylon cordage, the strength of which is approximately twice that of the best manilla, is also being used in some fishing vessels. Other advantages claimed are that it remains pliable under all conditions and handles better than other cordage. It is rot proof and, although more expensive, is economical because it lasts much longer.

ACCOMMODATION

The improved standards of accommodation have led to the use of many new materials in the living quarters and for furniture and upholstery. They should be capable of standing up to hard service conditions and the marine atmosphere, be fireproof or nearly so, and resistant to attack by vermin.

Rubber, as a flooring material, is extensively used in sheet form or in a composition laid in a semi-plastic state. More use is being made of expanded or sponge rubber for mattresses and seats, and cabin bulkheads and furniture are being made from synthetic materials derived from a mixture of wood fibres. There are also synthetic resin products, some of hollow form. These materials are said to be resistant to water, oil and acids and have good fire-resisting properties. They are also impervious to attack by vermin. They can be supplied in different colours. They are easy to cut and form and, in the case of the hollow panels, give good sound and heat insulation.

SHORE HANDLING

In all modern British fishing ports, wood for crates and boxes, kits and other handling receptacles, is being superseded by aluminium. The fish is landed in specially-constructed boxes or kits and transported in lidded boxes or liner crates to the wholesale market and retailer. The boxes and kits are entirely of aluminium whilst the liner crates consist of a wood slatted crate in which is a detachable pressed lining of aluminium, the lid being sheathed with the same material.

Another development is the use of very thin aluminium sheet consumer packs for the direct supply of fish in packages weighing from 1 to 14 lb. (0.45 to 6.3 kg.). Aluminium cans have long been a favourite because they are about one-third the weight of tin plate, are non-toxic, maintain their brightness and are easily opened.

In the processing of fish, the smoke house racks or horses and the screens are of aluminium, so also are the filleting tables, fillet trays, quick freeze trays, and, indeed, most of the equipment now used in the modern fish factory.

The new insulating materials are being used to a considerable extent in transportation of fish by road and rail. Road containers are often made up of aluminium alloy sheet and extrusions, with insulating materials between the inner and outer walls. The interior is partitioned by shelves on which the fish boxes rest. Railway engineers are also experimenting with new materials and, recently, in constructing 400 fish vans, the British Railways used aluminium extrusions and inner sheet with " Onazote " insulation.

TEREDOS AND FOULING

by

H. KÜHL

THE Federal Fisheries Research Institute (Inshore and Freshwater) Hamburg, working with a committee of representatives of all the port authorities and industrial organizations interested in controlling marine pests, has found an increase in the activities of the teredo all along the German coast and in other European regions.

Damage to wooden vessels and port structures by teredo worms was a particularly important problem when fishing cutters, built of pine—which is particularly susceptible to attack—and used, during the war, as patrol boats (kriegsfischkutter—KFKs) were returned to the fishermen. The damage done in the Elbe estuary, near Cuxhaven, was particularly great, infestation doubled between 1936 (when the last survey was made) and 1949–51.

The Prussian Institute of Sea Bottom and Air Hygiene in 1925 suggested the following grading for teredo infestation:

Degree 1—very few borers.
,, 2—only local infestation.
,, 3—necessity to replace a pine pole, 10 in. (25 cm.) in diameter after 10 to 20 years.
,, 4—necessity to replace a pine pole, 10 in. (25 cm.) in diameter after 5 to 10 years.
,, 5—necessity to replace a pine pole, 10 in. (25 cm.) in diameter after 2 to 5 years.
,, 6—necessity to replace a pine pole, 10 in. (25 cm.) in diameter after every year.

For example, whereas Windolf (1936) estimated Cuxhaven as being infested to degree 3 in 1936, the author had to raise this classification to degree 5 in 1949/50. Poles as much as 16 in. (40 cm.) in diameter, were completely destroyed within two years. Further examples can be seen in Table XLV.

Investigations were confined to harbour structures as they are the source of infestation of wooden vessels. Experiments were also carried out on small wooden poles and it was found that in four months the teredos had bored $1\frac{5}{8}$ to 2 in. (4 to 5 cm.) long channels in those poles which were insufficiently or not at all protected.

Three or four species of *teredo* are found in the Adriatic and investigations of these were carried out along the Italian coast in co-operation with the research station at Venice.

Timber may be protected from the teredo by: (1) preventing the larvae from entering the wood, (2) changing the condition of the timber so that the pest is killed because it cannot digest it, (3) killing the teredo by changing the condition of the water.

Painting will prevent the larvae from entering the timber—a method as old as seafaring. In the case of carvel built or any smooth bottom vessel, copper or galvanized sheathing of the bottom will prevent growth there of marine pests. Anti-fouling paint, too, can be used effectively as the larvae of the teredo are killed by the poison in the paint. Investigations showed that highly poisonous, first-class paint is needed to give good protection and it should be applied at regular intervals to maintain that protection. It is especially necessary to apply a protective coat of paint before the time when spawning begins and larvae are produced. The ordinary tar or bitumen coats so often used by fishermen do not afford sure protection but some special tar products can be used with success.

A complete soaking of wood with creosote under pressure gives long and safe protection. An example of its effectiveness can be observed near the engine of any boat where fuel and lubricating oils have flowed on the planking and soaked the timber. Infestation rarely occurs in such wood. Frequent coating of the planking with tar oil products such as " Xylamon ", gives suitable protection. Another process is to soak the wood with chemicals, such as fluorine, chromium and phenol compounds, with poisons added, but as the experiments with these are not finished, the effectiveness of this method is not yet known. Different kinds of wood react in varying ways when soaked. For example, oily substances sink deep into pine but spruce and fir generally resist such protection.

Rust hampers the growth of the teredo, but the use of iron nails is hardly practical as some 400 would have to be driven into each square foot (4,000 into each sq. m.) of wood. Again, burning off the surface of timber is only effective in killing the teredo already in the wood—and

TABLE XLV

Locality	*Structure*	*Age in years*	*Depth of infested layer*	*Timber species*	*Remarks*
Lentzkai	exterior pole	7	sound		The poles mentioned were made of pine wood with an average diameter of 18 in. (46 cm.)
Steubenhoft	exterior pole	8	1⅛ in. (29 mm.)		
,,	construction pole 1.R	23	4⅛ in. (105 mm.)		
,,	rear-pole of dolphin	23	sound		
Fishery harbour east-side	exterior pole	12	2–2⅜ in. (51–60 mm.)		
,, ,, west-side	exterior pole	14	totally destroyed		
Landing bridge	construction pole	31	3⅛ in. (79 mm.)		
Old harbour west-side	exterior pole	13	2⅜–2¾ in. (60–70 mm.)		

Summary of Findings, 1948-1950 (Jahn, Kühl, Osterndorf 1951)

Locality	*Structure*	*Age in years*	*Depth of infested layer*	*Timber species*	*Remarks*
Outside edge Steubenhöft	30 construction poles	36	8 in. (203 cm.)	pine	all poles uniformly infested
,, ,, ,,	5 ,, ,, 2.R	36	completely destroyed	,,	
,, ,, ,,	4 ,, ,, 3.R	36	,, ,,	,,	
,, ,, ,,	4 fender dolphins	8	,, ,,	fir	
Inside edge Steubenhöft	20 construction poles	36	6⅜ in. (162 mm.)	pine	
,, ,, ,,	8 ,, ,,	36	completely destroyed	,,	
Outside edge Steubenhöft	squared timber 10/12 in.	36	heavily destroyed	,,	impregnated after the Rüping process
,, ,, ,,	,, ,, 10/12 in.	36	slightly damaged	,,	
,, ,, ,,	gang ways 2 in.	3	heavily destroyed	,,	
Osterhöft	250 construction poles	36	6 in. (152 mm.)	,,	
,,	40 ,, ,,	36	completely destroyed	,,	
Amerikahafen	9 poles dolphin	2½	5¼ in. (134 mm.)	fir	
Landing bridge outside	construction poles	34	completely destroyed	pine	
,, ,, inside	,, ,,	15	slightly infested	,,	in the vicinity of the sewer
Alte Liebe	,, ,,	17	6 in. (152 mm.)	,,	
,, ,,	,, ,,	17	7¼ in. (184 mm.)	,,	
,, ,,	squared timber 10/12 in.	17	completely destroyed	,,	
Old docks	exterior pole	2½	,, ,,	fir	
Schleusenpriel	Slipholms 11/12 in.	24	5¼ in. (134 mm.)	pine	
Osterhöft landing dolphins	4 sets of 9 pole dolphins	9	4⅛ in. (105 mm.)	Pommeranian pine wood rammed in unprepared original condition	
,, ,, ,,	frame timber 10/12 in.	9	heavily destroyed		

even that is not sure. It will certainly not prevent re-infestation.

The teredo cannot live in water containing less than 0.9 per cent. of salt and there is a possibility of clearing a vessel of the pest by leaving it at anchor in fresh water. Three weeks is the minimum time needed to ensure killing the teredo because the pest protects itself by closing its hole with lime covers. Underwater explosions will also kill the teredo but there are no details available of experiments in this method.

To sum up, the best—but an expensive—method for protection against teredo infestation is to soak the timber in an oily product.

Fouling increases resistance and fuel consumption and affects manoeuvrability. All the processes using mechanical, electrical and other principles to prevent fouling have failed, with the exception of that employing poisonous paints, particularly copper and mercury compounds. But even the length of their effectiveness is limited, being determined by the rate at which they are released by the sea water. A costly way to increase the period is to add more poison but that cannot be done indefinitely. Experiments have shown that there is still a lot of poison left in an anti-fouling coat when it has become ineffective. The binding agent containing the poison is of great importance. If a very active poison is used with an unsuitable agent—or vice versa—it will not be effective. One experiment showed that a protective coat containing 60 per cent. of effective poison did not prevent fouling because the binding agent had, through action of the sea water, become too hard to release the poison—a mercuric Oxide HgO and copper Oxide CO_2O mixture. The binding agent and the poisonous substance should have about equal solubility, which will ensure the gradual release of the poison, so prolonging the period of effectiveness.

Corrosion and fouling of iron ships go hand in hand and protection against corrosion is a necessary prerequisite to the prevention of fouling. Attempts to cope with the two problems with one paint have failed because paints, to repel corrosion, must be hard, and those to kill marine pests and growths must be soluble in sea-water. Further, tests have shown that coats of anti-corrosion paints sometimes have a detrimental effect on the anti-fouling paint.

A great variety of plants and pests are found on the bottom of ships (Marine . . . 1952) and it is an advantage if only a few non-specific poisons can be used to kill them. Heavy metal compositions, especially copper and mercury, are poisons of this type, affecting all organisms that attach themselves to the bottom of a ship. A greatly enhanced effect is gained when two or more poisons are added together to the paint, being much deadlier in this form than when they are added separately.

It is however necessary to protect the iron hull against poisoned paints. For example, if copper and mercury combinations are used as poison there is likely to be an electro-chemical reaction so that the poison will not dissolve (Kühl, 1950), and the corrosion of the iron is increased. The anti-corrosive and other appropriate insulating coats are applied first. There are good insulating paints but it is difficult to apply anti-fouling paint containing copper powder to them when they are on an iron ground. " Copper-bronze " paints are, however, excellent for use on wood. For the same reason, experiments in sheathing iron bottoms in copper have failed because it was impossible to obtain adequate insulation between the iron and the copper.

This danger is acute if the metal ground is of aluminium because of the position of this metal in the electrochemical series. Experiments were made by using non-metallic poisons but it was found that purely organic poisons only affected certain pests to a limited extent (Kühl 1948). For example, combinations of the DDT type had a poisonous effect on balanids, a most important fouling growth, but no effect at all on serpulids, bryozoa and other organisms. While a DDT coat remained free of balanids, great numbers of bryozoa (membranipora) developed and covered the paint, and then balanids grew on top of the lot. As the specific poisonous effect of pure organic preparations is known they should be added to the antifouling paint. They will not detrimentally affect each other in the water. Other features of anti-fouling paints, such as surface condition, shade of colour, and so on, play some part but, generally speaking, they are of little practical importance. Fouling control is an intricate problem involving a great number of questions which have yet to be answered and, so far, the best results are obtained by using poisons in paint.

JACKETED, REFRIGERATED FISH HOLDS

by

W. A. MacCALLUM

IF watertight jacketed metal linings are fitted in fish holds, mechanical refrigeration can be used as a buffer against the flow of heat which tends to melt ice lying against pen peripheries: the bottoms, bulkheads, ceilings and tops of holds. Wood and cement linings, not being watertight, cannot be used successfully in bottom and ceiling jackets. Mechanical refrigeration, therefore, in a wood or cement-lined fish hold would influence the temperature in only two pen areas, the pen top and the pen front.

In 1951 the fish holds of the two 115 ft. (35 m.) wooden trawlers, the *Cape Fourchu* and the *Cape Scatari*, were fitted with jacketed linings of welded aluminium alloy, believed to be the first installed aboard ship in this manner.

The linings were the result of seven years' investigation at the Atlantic Fisheries Experimental Station of the Fisheries Research Board of Canada, in co-operation with National Sea Products Ltd., Halifax, Nova Scotia. A paper on the subject has been published (MacCallum, 1953), with illustrations and constructional details. An account of the experience with the construction and use of these, as well as conventional fish holds, is available in a Bulletin of the Fisheries Research Board of Canada.

REFRIGERATION REQUIREMENTS EXPRESSED IN TONS OF ICE

Based on summer temperatures of surface sea water and air in Nova Scotia and in Newfoundland fishing areas, the calculated requirements of ice over a six-day period

TABLE XLVI

Calculated ice consumption aboard medium-sized wooden Canadian Atlantic Coast fishing craft for six-day period

Source of heat flowing into hold	*Non-insulated craft*		*Insulated craft with 4 in. (10 cm.) of cork equivalent on underside of deck and on end bulkheads of hold; 2 in. (5 cm.) of cork equivalent on ceilings*		*Coldwall or jacketed and insulated craft. (Insulation and cold wall neutralizing all conduction heat gain into holds).*	
	Short Tons	*Metric Tons*	*Short Tons*	*Metric Tons*	*Short Tons*	*Metric Tons*
Heat transfer through deck	3.6	3.3	1.1	1	—	—
Heat transfer through hull	5.0	4.5	2.5	2.3	—	—
Heat transfer through bulkheads	2.8	2.5	0.8	0.7	—	—
Warm air infiltration, heat from lights and workers in hold	2.9	2.6	2.9	2.6	—*	—
Initial cooling of hull	1.0	0.9	1.0	0.9	—*	—
Theoretical ice requirements to cool cargo of 200,000 lb. (90,000 kg.) round fish from 60 deg. F. (16 deg. C.) to 32 deg. F. (0 deg. C.) plus 25 per cent. surplus†	20.9	19.8	20.9	19.8	20.9	19.8
TOTALS	36.2	33.6	29.2	27.3	20.9	19.8

* Cold walls, floors, and decks possess sufficient refrigeration capacity to take care of these heat loads.

† An estimate of the extra ice used for " insurance " that some shall remain at time of discharge.

for a typical 6,000 cu. ft. (170 cu. m.) fish hold are found in Table XLVI for a wooden craft where the hold is:

(1) uninsulated, and not artificially refrigerated;
(2) insulated only;
(3) insulated and cold wall jacketed.

The table does not show the quantities of ice required to shield fish from contaminated wood of pen or division boards, shelf boards, and on pen partitions and ceilings of many wood fish rooms. The need for this extra icing in wood constructed fish holds tips the balance even more strongly in favour of the metal pens (MacCallum, 1954).

The calculated thicknesses of crushed or fine ice to combat leakage of heat into insulated and uninsulated fish holds over a six-day period are given in Table XLVII. The peripheral icing specified is difficult to achieve in practice. In both uninsulated and insulated non-refrigerated holds, some fish from these peripheral areas were found to be in poor condition when landed. As even a small percentage of poor quality fish can cause loss of consumer confidence, it is felt that all fish should be efficiently refrigerated.

DESCRIPTION OF JACKETED INSTALLATIONS

In the fish holds of the *Cape Fourchu* and the *Cape Scatari* the ceilings, bottoms, bulkheads, and tops were insulated and metal-lined. The lining is supported clear of the surfaces of the ship to provide a channel through which refrigerated air may be circulated. Aluminium alloy sheets and extrusions were combined to provide an uninterrupted metal interior lining in the fish room. All joints and seams were argon-arc welded, with the exception of those at the deck where mechanically-constructed joints were used because there is no water pressure. All pen partitions or " wings " are of heavy gauge aluminium alloy sheet. The shelf board supporting angles and pen shelf and division boards and the hold stanchions are all of aluminium alloy extrusions.

A Freon-12 marine condensing unit and all motors required for the refrigeration system were installed in the engine room together with an A.C. supply for the motors. The blowers were located opposite the blower motors but on the fish room side of the engine room bulkhead.

As shown in fig. 308, air is forced by two blowers, mounted at F, through the finned evaporator coils A, where it is cooled to a temperature of about 30.5 deg. F. (—0.8 deg. C.). The circuit includes the cold air supply ducts extending on both sides of the keel, as shown at B. Mechanical and thermal pressure differences in the

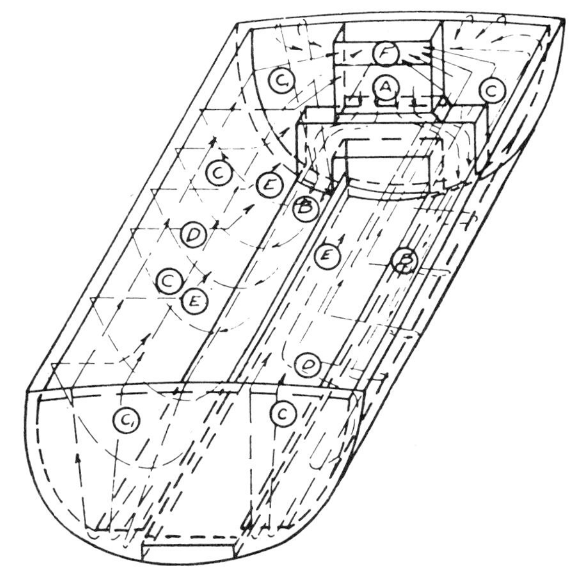

Diagrammatic sketch of cold air circuit within jacket in the fish hold of the trawler "Cape Fourchu".

Fig. 308

TABLE XLVII

Calculated thickness of crushed ice required to combat leakage heat into fish holds over a period of six days

Locality of ice	*Thickness of ice melted Wooden vessel* — *Uninsulated*	*Insulated**
Above top layer of fish in pounds	4 in. (10 cm.)	1½ in. (3.8 cm.)
Against ceilings	3 in. (7.5 cm.)	1½ in. (3.8 cm.)
Against engine room bulkhead	5 in. (12.5 cm.)	2 in. (5 cm.)
Against forward bulkhead	4 in. (10 cm.)	1¾ in. (4.5 cm.)

* The following thicknesses of corkboard were assumed in the calculations:

Between deck beams	4 in. (10 cm.)
Against ceilings	3 in. (7.5 cm.)
Against bulkheads	4 in. (10 cm.)

Included in each of the above totals is ¾ in. of ice which would be required to contribute to the cooling of the fish found immediately adjacent. In all calculations, thicknesses are based on the density of fine ice, not block ice.

enclosed circuit provide for the passage of air up the ceiling at C and also vertically in the hollow construction of end bulkheads, as at C_1. Other parallel circuits provide for refrigeration of the hollow stanchions and ventilation of deckbeams and ships frames. D and E show the return, within the deck jacket, of all air from the parallel circuits.

OPERATING CHARACTERISTICS OF FISH HOLD INSTALLATIONS

Refrigeration Effect and Use of Ice

A back pressure control which works a magnetic starting switch, is used to control the compressor. Blower motors, manually switched, are operated continuously to maintain constant air temperature in the jacket.

When the compressor operating time amounts to about 50 per cent. of the period at sea, temperatures in the jacket average 30.5 deg. F. (—0.8 deg. C.). This is sufficiently low to prevent the ice against jacket walls from melting once the fish have been chilled, and ice is preserved for the whole of a trip. Air temperature in the hold proper is maintained at 33.5 deg. F. (+0.8 deg. C.). This may be compared with temperatures of 40 to 42 deg. F., (5 to 6 deg. C.), observed in insulated and non-refrigerated holds and in those insulated and refrigerated by conventional grids installed below the deck. The ineffectiveness of grid refrigeration is due to frost build-up on the coils at an early stage of the fishing trip. There is no problem in defrosting the coils of the *Cape Fourchu* and *Cape Scatari* because the circuit of the refrigerated air is closed.

Consumption of ice in jacketed holds is about half that used to obtain equivalent refrigeration in an uninsulated non-refrigerated hold. There is less cost in ice handling and an assurance of an adequate supply of ice throughout a fishing trip.

Hold Linings

Inspections of seams are made periodically. The argon-arc welds are of high quality.

Electric Power Requirements

The provision of electric energy for the refrigeration equipment is not difficult but it is an expensive item. It can be provided most economically by installing tail-shaft and auxiliary generators, each capable of handling all lighting, pumping, air compressor and refrigeration loads. Another possibility is the use of an hydraulic drive.

Future Pen Construction

" Unit " fish pens of shell-like, heavy gauge metal construction fabricated ashore in standard size and shape for assembly aboard ship, should be considered. Construction and installation costs are reduced, there is greater resistance to seam and sheet failure, and the cost of refrigerating the hold is lower. When repairs are necessary the pens can be removed without delay or damage to pen or hold. Also it would be worthwhile to consider changes in ship deck construction and in hatch layouts to provide for quick removal of the pens, loaded with fish and ice, at each landing.

CONCLUSIONS

Ice is the logical choice on Canada's East Coast to take care of the product heat-load when large quantities of fresh fish are stowed in a short time. However, the leakage heat-load cannot be offset effectively by the use of ice alone, nor by using conventional grid refrigeration in conjunction with ice. It has been demonstrated that an insulated watertight and cold wall jacketed metal hold provides refrigeration most efficiently These holds make proper handling of fish easier and lessen the effect of human error.

PROTECTION OF INSULATION AGAINST HUMIDITY

by

P. BAIN

WATER vapour is continually moving from the outside towards the inside of refrigerated holds through the insulating material. When it reaches a zone where the saturation pressure is less than the vapour pressure, condensation occurs. Later on this condensed water and the water already contained in the insulant goes from zones with high to zones with low water content due to the capillary action.

A Swedish engineer, Mr. Munther, investigating this process, has found that the heat transfer coefficient of an insulant can be increased by 30 per cent. for each volume per cent. of water in it and in expanded cork of first quality, after several years in use, the water content has been as high as 18 volume per cent. The heat losses caused by humidity in the insulant may easily be twice those of the dry material. Water in insulation material also shortens its life, and where serious ice formation occurs the material is destroyed.

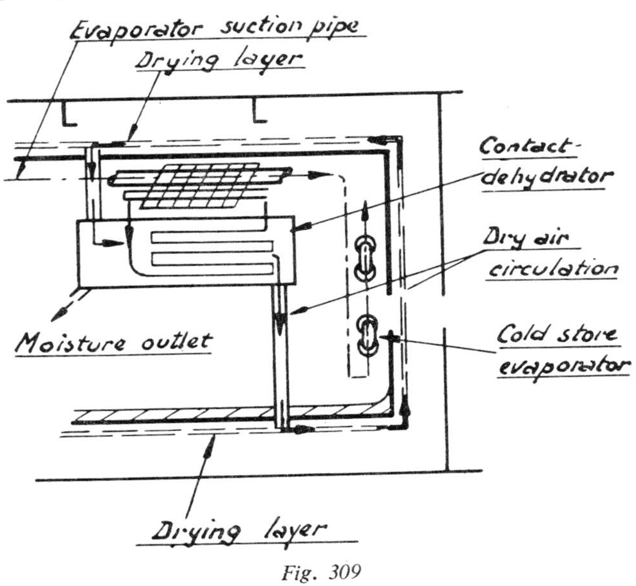

Fig. 309

The Minikay system of dehydration is designed to prevent water accumulating inside the insulation of the refrigerated holds. It consists of a drying layer placed between the insulant and the inner lining of the cold chamber. This layer is an insulating material in which dry air is circulated through ducts. Any vapour of water which penetrates from the insulant to the layer is brought with the air to a dehydrator where it is condensed by a rapid fall in temperature. The dry air is then recirculated through the ducts of the drying layer. The dehydrator is periodically heated to remove the condensed water.

The whole installation consists of: (*a*) the drying layer; (*b*) the ventilator; (*c*) dehydrator; (*d*) pipes between the drying layer and the dehydrator; and (*e*) an automatic control device.

The advantages claimed are:

1. The heat transfer coefficient is kept at an almost constant value which increases the efficiency of the refrigerating plant and saves power.
2. The thickness of the insulation can be reduced, giving more space in the fish hold.
3. The insulation does not suffer damage from humidity and the maintenance cost is, therefore, lower.

The cost of installing the system in small is comparison with the saving on maintenance. Fig. 309 shows the principle of such an installation.

ECONOMIC INFLUENCE ON DESIGN OF FISHING CRAFT

by

C. BEEVER

AMONG the developed fishery industries of the world it is clearly recognized that the size and design of fishing craft are governed by many factors, including variation in catch and markets available, which differ greatly in their importance according to local conditions. They are familiar to owners, fishermen and designers and they explain both the wide variations in fishing craft design, and the orthodoxy observed in individual areas.

Theoretically, it might be possible to take a particular fish stock on a particular fishing ground, and to calculate what range, seaworthiness, stability, capacity and speed, etc., would provide the most economic means of taking catches. But the practicability of introducing such an ideal vessel would be determined by the demand for fish products, by the availability of trained fishermen, by pilotage, berthing and maintenance facilities, by the level of capital reserves or the availability of credit and by numerous other factors which, in varying degrees, are reflected in the profits and losses of day-to-day fishing.

This question of design can never be dissociated from the number of the craft engaged in a particular fishery and, in theory, the best design has to be related to a maximum fishing effort which is consistent with sustained yields and the capacity of the market. In other words, commercial efficiency dictates that there should be an ideal number, as well as an ideal design, of fishing craft in any fishery. A precarious equilibrium is often established by the force of competition, although many fishermen show extraordinary tenacity before they relinquish the struggle to wrest a living from depleted or unprofitable fisheries.

In more recent years attempts have been made by governments, or by mutual agreement amongst the owners, to regulate the fishing effort in the interests of the resources or the markets, or of both. Today the extension of the fishing operations over wide ocean areas often requires the sanction of international agreement, which largely rests on an integration of national economies beyond the control of individuals.

There are many examples of the problem, but the case of the North Sea especially is of historic and topical interest. Here the conservation of young stocks is hoped to restore the level of the sustained yield. The question of design, and especially of size, will then turn, *inter alia*, on the level of catches, i.e. what share of the total yield can each craft hope to take. This will obviously depend on the number of vessels engaged and the methods employed. Owners are now critically examining the problem of whether to build at all and, if so, what size of craft and to what extent the North Sea should be exploited as a seasonal fishery.

On the more distant grounds of the Arctic and North Atlantic, the question of size is largely determined by the nature of the seagoing conditions, but beyond that it is really a question of what level of investment can be supported by markets in which a preponderance of cod may create uncertain and unfavourable price structures. Similarly, the future of the magnificently equipped but costly tuna fleets of California, built to range over 3,000 miles or more, hangs on the assurance with which the elusive fish can be located, and on the stability of a market, which has recently shown signs of disturbance.

If it is true that the picture of the world's fishing fleets today reflects the wide range of achievements in the development of hull design, propulsion units, constructional materials and techniques, fishing equipment and so on, it is also true that it reflects, in the very age and composition of the fleets, the variegated economic pattern of the fisheries. Historically, owners in the more advanced fishing industries have been able to draw upon years of experience in determining the number and design of their craft. Their intimate experience of fishing conditions, the accumulated knowledge of generations about the habits and behaviour of the stocks, and their ability to judge the vagaries of the most uncertain markets in the food trade, have all contributed to that judgment on which the decision to build is based.

Most owners will agree that there are many limiting factors to their judgment. Fishing is still a gamble which deters all but the most knowledgeable and courageous investors, and there is perhaps no other trade where financial stability depends so much on the reciprocity between unpredictable fluctuations in both supply and demand. Slowly, developments in research are making possible a more scientific analysis of the technical and economic factors which govern the level of investment in fisheries and, thereby, the size and design of the fleets.

There is a division between those industries which have on call the highly developed skills and techniques of modern shipbuilding practice—and which are clearly aware of the economic factors determining the design of their craft—and those in the underdeveloped economies of the world where there is virtually no experience, training, equipment or financial resources to draw upon, other than the energy and will of indigenous fishermen engaged in specialized but highly localized operations.

The fundamental problem is familiar: the nutritional standards of large populations are seriously undermined by acute shortages of animal protein which there is little prospect of relieving from agriculture alone. The existence of unexploited or partially exploited fish resources has inevitably directed the attention of governments to the great social, economic and strategic value of increased fish supplies. Many of the problems concerned are rooted in low purchasing power, consumer prejudice or ignorance and the illiteracy and poverty of the fishermen in undeveloped countries.

Efforts to improve or replace local fishing craft vary considerably in character, and reflect estimates of the economic or commercial justification for investing in certain types of craft. In many countries of the Far East, Africa and Latin America, the first steps have been taken in the mechanization of local craft and the introduction of new designs adapted to local experience and training. There is already a well-established demand for marine engines in several areas of rapid development, i.e. Hong Kong, Singapore and Bombay, which might extend rapidly with the provision of technical training, financial assistance and adequate servicing facilities. From the economic standpoint the characteristic feature of these developments is that they preserve, in essentials, the traditional patterns of production, i.e. short trips, small catches, close-range operations.

In some areas, however, developments of a more dramatic character have been promoted and have been strongly influenced, firstly by the belief in the existence of substantial deep sea resources, and secondly by the highly commercialized undertakings of north west Europe and North America. Deep water trawling and seining, from vessels of European design, have been introduced in several places in Asia, notably Bombay, Calcutta and Colombo. The remarkable feature of these ventures is that they represent a revolutionary departure from traditional practice and are based on an urgent desire to bridge, as quickly as possible, the gap between indigenous fisheries and the highly organized enterprises needed to bring resources into full productivity.

DEMAND

Development of the major fisheries of Europe and North America has been inextricably linked with the general heightening of economic activity brought about by industrialization. The fishing effort expanded and intensified spontaneously to meet demand for fish products among growing populations with rising levels of purchasing power. It is of profound significance for the underdeveloped fisheries that the same modernized industries which they seek to emulate are now facing critical marketing problems. In many of the areas where the need for fishery development is so acute, the initial demand is often lacking and is frustrated by ignorance, prejudice and, above all, by poverty. Even where it exists, it has often been conditioned by old habits related to familiar species deriving from traditional fishing operations and processed by traditional, but often primitive, techniques.

In many places consumer demand has been built up in coastal and large urban areas only, and the habit of fish consumption among rural inland populations is often weak and restricted to certain forms of preserved products, e.g. sun-dried, pastes, sauces, etc. Commercially, there is little prospect of success if unfamiliar products or, for that matter, unexpectedly large quantities of a familiar product, are introduced into the traditional markets. And the necessary purchasing power must become available, which, in turn, will depend upon overall improvements in the economy.

The capacity and design of fishing vessels, with which must be associated also the nature and location of the operations, must keep in step with demand if fishing ventures are to be commercially viable in these areas. Efforts to improve and extend the fishing operations must be accompanied by efforts to improve and extend consumption. It is worth noting in passing that if the nutritional needs of these great populations ever become expressed, as anticipated, in an active demand for fish products at a price commensurate with costs of production, then the combined activities of all the fishing fleets of the world would be hard pressed to fill the gap.

INCENTIVES

The improvement of traditional fishing craft designs and equipment, especially mechanization, does not always have the desired effect because of the absence of incentives for the fishermen to catch more. There are some situations where mechanization of the fleets has proceeded smoothly but the effect has been to enable the fishermen to meet his simple needs for food and shelter more quickly and with less effort. This reduces the financial increment and impedes further development. How far the fisherman is willing and wants to go in improving his professional and social status, has an important bearing on the nature and cost of investment in superior craft and better techniques. A similar lack of incentive manifests itself among the small merchants and dealers who dispose of the catches, and here again the reluctance to increase turnover or to change traditional methods of handling and distribution imposes its own limitations on the fishing effort. Development has to be founded first of all on the desire for, and then on the belief in, higher standards of social and economic well-being.

RESOURCES

The location, composition and behaviour of the fish stocks naturally govern the type of craft to be employed. It is only comparatively recently that this problem has received serious attention, and a number of countries with important fisheries are conducting investigations to provide the fishing industries with more reliable information on which to plan their operations. In the under-developed areas, however, very little has yet been done to make even a preliminary appraisal of the resources which might be brought into profitable exploitation. Even in the traditionally exploited fisheries, few reliable estimates have been made of the total fishing effort which can be indefinitely sustained. Clearly, until more is reliably known about both the onshore and offshore resources, the introduction of new or better craft will need to be cautious and, to some extent, exploratory. In the case of the smaller units especially, the seasonal variations of the resources and, possibly, the variety and scattered distribution of the stocks, may require a large measure of flexibility in the operations, for which the designer must help to provide.

FINANCE

The problem of finding capital for essential equipment is particularly acute, and the special risks attached to fluctuations of supply and demand offer a serious deterrent to the ordinary investor. Development, therefore, depends either on courageous private speculation or on planned financial assistance in the form of cheap credit facilities, direct or indirect subsidies, tax concessions, and other incentives. Even in the most highly-organized modern fisheries cheap credit is proving to be a corner-stone of development, but in the underdeveloped countries it is virtually indispensable. The widespread indebtedness and poverty of the fishermen militate heavily against rapid development. Their meagre incomes are often insufficient to finance even the replacement of essential gear without the help of private loans, which serve only to perpetuate and increase the burden of debt.

Increased production based on improved fishing craft and techniques is unlikely to occur on any extensive scale until cheap long-term credit facilities are made available, and commercial organization permits the accumulation, collectively or individually, of capital reserves. In such circumstances there is a need to consider very carefully how the sums invested in fishing craft, for example, can be used to the best advantage. The danger of having all one's eggs in one basket is particularly real in the more primitive fisheries where there is so little past experience to assist in measuring the financial risk. Heavy concentrations of capital as, for example, in costly deep-sea vessels, may not only prove commercially unjustifiable, but may divert capital from more essential, more immediate uses—such as the purchase of small engines or better fishing equipment.

The need for capital in the fishing fleets is matched by the need for a fairly heavy investment in harbour, market, transport, processing and ancillary undertakings. Where this is not forthcoming, the types and sizes of fishing craft need to be adjusted accordingly. There may be very sound financial, as well as geographical or navigational reasons, for concentrating on types of craft which can be beached. It is also important to consider the costs of labour and the bearing which they may have on the design of craft. Where labour is plentiful and cheap, the use of expensive labour-saving devices, either ashore or afloat, may prove commercially unjustifiable.

EDUCATION AND TRAINING

The lack of education, which in so many fisheries amounts to illiteracy, is reflected in the general standards of commercial organization and practice. The very remarkable skills which have been developed over the centuries in many of the more primitive fisheries, have been taught by demonstration and imitation. New techniques or the use of superior equipment must be taught in a similar fashion and must be adapted to the capacity of the fishermen to absorb instruction. Revolutionary changes of craft and techniques may alarm and repel, while too many changes all at once usually confuse the fisherman. It is usually more profitable to introduce changes one at a time so that the characteristic features are retained. A fisherman who successfully tries out a new craft or a new engine on his familiar grounds with his own methods is more likely to co-operate in further improvements than one who is introduced simultaneously to new grounds, different working hours, new equipment, new craft and so on. That is one of the more important lessons learned from past efforts at development.

At a different level, the lack of training is also reflected in the commercial management of production and distribution. Extended operations and heavier catches need a corresponding improvement in business practice. Commercial acumen is a factor of obvious importance and there is little doubt that some of the underdeveloped fisheries could be made more profitable if the executive personnel could perceive and grasp trading opportunities.

In the technical field there is a similar need for training to operate and maintain hulls, machinery and engines.

The problems can perhaps be illustrated by a typical case where, in the same area, a small craft mechanization programme has been developing simultaneously with pilot deep-sea trawling operations.

The highest catch recorded by the trawler is reported to be 80 tons for a voyage of ten days. Assuming such catches could be regularly obtained, the monthly catch on the basis of two and a half trips per month would be 200 tons (in point of fact, the average monthly catch has been considerably lower than this). The trawler carries a crew of twenty, and therefore the catch per man-month would be 10 tons. In the case of the small mechanized craft, the record of landings of 46 craft during the last quarter of 1952 showed an average catch per man-month of 1.5 tons (taking into account breakdowns, bad fishing, etc.). On average, these small craft carry a crew of ten. In other words, the quantity landed by the one trawler could be landed in the same time by 13 to 14 mechanized local craft manned by 135 fishermen.

In this comparison, the important point is that the operations of the small craft involve so many more local fishermen who are gaining experience in improving their own conditions and are getting a chance to accumulate capital with which to build the larger craft they require to extend their operations. Moreover, the regular and frequent small catches, of good quality and composed usually of more popular varieties, which the local mechanized craft can land, are more consistent with, and more profitable under present marketing arrangements. In fact, the landing of even a moderate catch by a large trawler is apt to over-supply the local market and cause resentment amongst the other fishermen. In this particular area there are some 15,000 local sailing craft, of which it might be possible perhaps to mechanize one-tenth. The theoretical catching power of these 1,500 boats, when mechanized, would require the equivalent of 110 deep-sea trawlers of the type now engaged. The total cost of mechanization of this number of local craft would be approximately £1,100,000 ($3,000,000), whereas the cost of a fleet of trawlers with the same catching capacity would be £8,500,000 to £11,000,000 ($23,000,000 to $30,000,000).

Such a case emphasizes the wisdom of concentrating primarily on the needs and prospects of the domestic fishing industry. It may be desirable to experiment with larger unfamiliar craft as a demonstration project, and to test hitherto inaccessible resources, but it should never be at the expense of developments based on existing industry.

From the point of view of the naval architect, the questions which naturally arise turn on the adaptation of local designs, local building experience and local materials, etc., to the need for more efficient operations. In fact, the ideal naval architect for the underdeveloped fisheries would also be a mixture of an economist, psychologist, biologist and teacher. Unlike his more fortunate colleagues in the advanced fishing industries, he cannot simply follow the judgment and wishes of owners and fishermen. His designs will invariably reflect something of the traditional economic pattern of the fisheries and some element of speculation as to the future economic trends.

HOW COSTINGS RATIONALIZE CONSTRUCTION

by

ANDERS N. CHRISTENSEN

THE humanistic ideas behind FAO's world-wide activities will result in new and difficult problems for boat builders. On the whole it can probably be said that boat builders have not kept pace with the technical development of the times, partly because it is difficult to break away from thousand-year old traditions, but chiefly because fishing boat construction has not been sufficiently profitable in most countries to allow investment in plants and machines and there is a lack of people with the necessary theoretical and technical experience.

Our generation of boat builders lives in a transition period where machines must replace manual work. New means must be found and it is of vital importance that old-fashioned accounting should be replaced by estimates and cost accounts which give the necessary information promptly.

THE ADVANTAGES OF CO-ORDINATED ESTIMATES AND COST ACCOUNTS

A well developed account will cost both time and money. Most firms may find it necessary to collaborate with a book-keeping expert, in which case they should carefully consider beforehand what kind of information they want in the account and, also, when they want to see the results. The annual cost of book-keeping should be considered too. On the basis of this information a book-keeping expert will easily be able to find a suitable system for the firm concerned. The advantages represented by a cost account will, of course, vary in different companies but will generally include:

(1) The possibility of more orders, as the firm has a safer basis for estimating minimum prices.

(2) The company can avoid contracts involving a loss.

(3) Many jobs can be carried out on a cost basis if it has been agreed beforehand that the price should be the firm's own costs plus a specified profit. Many buyers have a latent suspicion of the calculations of boat builders who, on the date of payment, will be in a precarious state if they have not got accurately entered notes to go by. A reliable cost account will often save the builder much trouble and it should be realized that proof of the correctness of the bill undoubtedly rests with the supplier.

(4) To carry out the right policy when purchasing materials and equipment it is important to know the stock of the various kinds of goods and what has been ordered but not supplied. The average consumption of an article during a certain period of time should also be known. All this information should be given in a stock card-index. From the estimates it can then be determined what quantities are kept in stock and what items are to be purchased.

(5) Insurance premiums are high because boat-building yards are combustible. Large sums can therefore, be saved by a periodical adjustment of the insurable value. The cost account will indicate, at any time, the exact value of all stocks in hand.

(6) Labour will always constitute an essential item in the total cost of production. Even if the hourly wages are fixed in accordance with an agreed scale, the amount may be subject to great variations but clever management and rational production should achieve the best results. The demand for contract work must also be considered and a well developed calculation system giving all details of working hours will be extremely useful when organizing such work.

(7) It is important that the labour costs be divided into " productive " and " unproductive " work. If, for instance, the annual costs of transportation within the boatyard are known, one will no doubt find it appropriate to consider the whole transportation problem. First and foremost, buildings and machines should be placed in such a way as to facilitate smooth production so that,

during manufacture, materials are not needlessly moved about the premises. When the problem of minimizing transportation has been solved, the question of the cheapest way of handling movement of goods and materials will arise. For most yards, trucks with rubber wheels will be suitable for both indoor and outdoor transportation but in addition equipment must be installed to give adequate lifting power whenever needed in the boatyard. Such investment in suitable means of transportation will quickly pay for itself.

Clearing-up and cleaning will also be classified under " unproductive " labour and it is an essential cost item. A cyclone plant sucking away shavings from wood-planing machines as well as an industrial vacuum-cleaner for use in the works and on board hulls under construction, are good investments.

(8) In order that the boat builders can keep pace with developments, new working methods should be applied. Here, again, it is important that the account should show what can be gained when changing over to new methods.

Mass production is taking place in almost all industrial undertakings and might be applied successfully to boat building. In that case, shipyards will have to specialize in a few boat types. It ought to be possible to organize in each district some suitable distribution of production amongst the various builders.

It would be an advantage if several similar boats could be built at the same time. By using a system of measurement, most materials can be manufactured for a number of boats at the same time and various details of hull and fittings can be manufactured and assembled.

In this way a better distribution of the work is obtained with more specialized work for everybody. The machines are operated by special workers: the framework by one gang, the erection of the frame by a second gang, the planking by a third gang, etc. This procedure, however, requires several boats being built at the same time, so that whenever a job is completed on one boat, the gang can continue on another.

Such mass production demands detailed preliminary work. The time for the various operations must be calculated carefully and the right number of workers placed in each gang, so that production proceeds smoothly, i.e. one gang does not have to wait until another gang has finished its work. Without a good calculation system, it will be almost impossible to co-ordinate the various working operations.

(9) On the whole, the mechanical equipment of many boat yards is old-fashioned and in view of existing high labour costs, it is of vital importance to solve this problem satisfactorily. Here, also, the accounts—giving exact working hours for the various operations—will provide valuable guidance for an estimate of the savings to be gained by using improved mechanical equipment.

(10) Lamination of wooden materials will be advantageous in many cases. The cross section of a detail can be reduced without diminishing the strength, and a more economical use of the materials, for instance for frames, is thus obtained. The most important thing, perhaps, is that the standard saw-mill dimensions can be used, which makes access to dry materials easier. In districts where large keel and keelson dimensions are difficult to obtain, lamination is indispensable.

Lamination, however, generally means increased working-time. Here, also, the accounts should give the right comparison of the costs (materials+wages+overhead expenses) of the two alternatives.

(11) Other materials are being used more and more as a substitute for wood. Aluminium, for instance, is used for superstructures, fish rooms, etc. Many shipyards build the whole hull of aluminium; others use steel. Plastic also seems to have great possibilities as boat building material. But whatever the development, a comparison of the costs of the various construction methods will always be necessary, especially in a transition period.

CALCULATION SYSTEM

To calculate the cost of a fishing boat involves much work. The boat must be divided up into the smallest details. The more detailed the estimate is, the less will be the effect of a miscalculation of a single detail on the total sum. The calculation must be made according to a system which can be compared with the cost account as the construction of the boat proceeds.

Division of Details

The boat must be divided up into details according to a system which is suitable for the enterprise. The following division is adopted by the Norwegian Boat Builders' Association. In the case of large boats this complete division only should be used; for smaller boats groups 00–10–20 have to be used.

01 Keel—shoe
02 Stem—apron
03 Sterntimber—horntimbers
04 Dead woods
09 Miscellaneous
00 Total keel, stem and stern-post details

11 Frames—floortimbers
12 Frame—fastenings
19 Miscellaneous
10 Total framing

21 Planking
22 Keelson
23 Bilge stringer
24 Beam shelves
25 Ceiling—outside stringers
26 Shaft log
27 Engine foundation

29 Miscellaneous
20 Total planking

31 Deck
32 Deck beams
33 Covering board
34 Bulwark
35 Hatches
36 Deckhouses and superstructures
37 Bulkheads under deck
39 Miscellaneous
30 Total deck

41 Fastenings and nails
42 Knees and strapping
43 Stem
44 Rudder
49 Miscellaneous (galvanizing)
40 Total fastenings and nails

51 Carpenter work
52 Fittings (bull eyes)
53 Water and sanitary arrangements
54 Heating and Ventilation
55 Electrical arrangement and navigating lamps
59 Miscellaneous
50 Total accommodation

61 Caulking
62 Paying
63 Oiling and painting
69 Miscellaneous
60 Total finishing

71 Engine
72 Engine installation
73 Engine room (plates, sheets, asbestos)
74 Accommodation engine room
79 Miscellaneous
70 Total engine

81 Deck fittings
82 Steering gear
83 Spares, sail and rig
84 Loading gear
89 Miscellaneous
80 Total deck fittings

91 Interior equipment
92 Deck equipment
93 Engine equipment
94 Life saving equipment
95 Nautical instruments
96 Wires, chains and hawsers
99 Miscellaneous
90 Total equipment

13 Miscellaneous (slipways, templates, transports, etc.)
100 Total sum

Division of Expenses

The cost of production to be calculated can be divided into three main groups: material, productive wages, overhead expenses. The item materials should be divided into sub-groups, for instance: wooden materials, fastenings, other materials.

Under the various boat detail numbers there will thus be an estimate for each of these items:

(1A) Wooden materials;
(1B) Fastenings;
(1C) Other materials;
(2) Direct wages (productive wages);
(3) Overhead expenses.

(1A) *Wooden materials* should be classified separately according to each type of wood used. Quantity could be calculated in linear measure, grouped according to breadth and thickness; or square measure, grouped according to thickness. But the best thing is to use cubic measures this being the easiest comparable unit and, at the same time giving the necessary basis for estimating the weight of the wooden materials. The method of calculation chosen should, however, be used consistently for all later calculations. The next question is whether one will calculate the net cubic content of the finished detail or the cubic content of the wooden piece which, according to experience, will be used to provide the finished detail. In the latter case the off-cuts must be added, and for curved pieces, the breadth must be considered. Here, also, it will be necessary to carry out an absolutely uniform method of calculation for all estimates.

To this must be added a considerable percentage of loss which will vary for the different details and types of wood and will also differ according to quality and dimensions of available materials.

The cost account should indicate the exact quantity of wooden materials used for constructing the boat. If the " built-in " cubic is deducted from this, the loss in respect of the detail concerned or the whole boat is shown. Expressed in percentages it is:

$$\frac{\textit{Loss in cubic} \times 100}{\textit{cubic used}} = \textit{percentage of loss}$$

The most accurate result is obtained by calculating the percentage of loss for the various types of wood separately. The percentage of loss should be checked frequently to gain experience for future estimates of similar boats.

(1B) *Fastenings.* It is expedient to separate spikes, screws, bolts, nails, etc., into special groups.

(1C) *Other materials.* Under this group can be included all other than wooden materials and fastenings. This item may also be divided into different groups, but it should not be necessary as the groups of commodities specified under the various boat detail numbers will be uniform.

This splitting up of materials under the various boat detail numbers is of practical importance. When the prices change, each item effected, which on the whole, follows the same price level, can be corrected proportionately. Really accurate results are not obtained in this way but in any period with continual changes in prices it will be a hopeless job to correct all the calculations.

For firms with a saw mill of their own, the ready-cut wooden materials should be transferred to the rest of the production according to current prices.

For firms with, for instance, a forge, a plate shop, a carpenter's shop, etc., of their own, semi-manufactured parts, such as stem- and rudder-fittings, rudder, hanging and horizontal knees, bolts, doors, cupboards, tables, etc., should be calculated by the shop concerned so that only the price for the whole finished article is given in the boat estimates.

Freight expenses must be included in the price of the materials with such percentage added as experience has proved necessary.

(2) *Direct wages.* The splitting up of the total wages bill is taken up in the cost account section. (*Editor's note:* It is not included in this paper.). Only direct (productive) wages are mentioned here. The working hours are calculated in hours and decimals of hours. For instance 5 minutes=0.12 hours. The working hours are calculated for each detailed number in such a way that they permit a reasonable profit for contract work. The calculated working hours should form the basis of contract work and a very exact estimate must, therefore, be made. Time studies would be of very great importance.

(3) *Overhead expenses.* These are summarized in the overhead percentage. The current percentage is to be found in the cost account. The usual thing is to calculate the overhead percentage in its entirety, but larger enterprises may calculate different percentages for their different departments.

(4) *Special expenses.* These are expenses not included under the usual overhead expenses—for instance, guarantee-fees, provisions and expenses, for classifications, certificates, inspections, etc. They are added at the end of the estimates.

There should be special calculation forms for each of the three groups of materials and also for direct wages and, finally, a form where wooden materials, fastenings, other materials, number of hours, wages, overhead expenses and total costs are entered.

The first item to be entered in the estimates will then be 01 Keel—Shoe. It is compared with drawings and specifications:

Wooden materials are specified on one form, the percentage of loss is added, and the total is transferred to the final form.

Fastenings such as bolts and spikes are measured, added up and arranged according to dimensions on a second form. The total is transferred to the final form.

Other materials are entered on a third form in the same way.

All the data necessary to estimate the number of working hours are then given in the various forms and they can be specified on a form for " direct wages ", from where the total can be transferred to the final form.

The procedure is the same for each boat detail number. All materials are first entered on the respective forms and then the data necessary to estimate the number of working hours are taken from these forms.

The cost account should follow the same lines as the estimates so that for each boat detail number the exact cost is given. And here we arrive at the most important point, a comparison between the cost of the work carried out and the estimate of the same work.

The differences that may occur are:

For materials: Changes in prices, quantities, types of material or qualities.

For wages: Changes in working hours or wages. For instance, one can employ several assistant workers or apprentices and get a great number of hours but, nevertheless, get reduced wages—a fact which should be closely analysed.

To what extent these changes should be corrected in the estimates is an important question which must be carefully considered. On the whole, the estimates should express what the cost account has taught us after the construction of the boat. The estimates should, therefore, be corrected correspondingly so that they indicate what the next boat will cost. Such corrected estimates will be extremely valuable for future price quotations for similar boat types or when fixing contract work rates.

In the case of great differences one must try to find out whether these are permanent or caused by special circumstances during the construction of the boat concerned. For example, specially favourable or unfavourable purchases of material, or delayed supplies which have rendered rational working methods difficult, are deviations that should not be expressed in the estimates.

On the other hand, differences caused by the fact that the quantity of materials used is inconsistent with the estimates, and price differences, must be corrected. Wrongly calculated working hours and changes in hourly wages must, of course, be corrected.

(*Editor's note:* This paper has been translated from Norwegian. The original contribution also included a section on cost accounts but the translation was confined to the material here given. Readers are referred to the author's extensive article (Christensen, 1948) on the subject presented at the Scandinavian Fishing Boat Congress in 1947.)

BOAT TYPES—DISCUSSION

Mr. Jan-Olof Traung (FAO): Different fishing methods determine the various boat types. The main methods used are: trawling, purse-seining, gill netting, Danish seining, long lining, trolling, hand lining.

Trawling is done over the stern on the Pacific coast of North America, in the Mexican Gulf, in the Mediterranean and, to some extent, in New England, U.S.A. (especially at New Bedford, Mass.). On both side of the Atlantic—Canada, U.S.A., England, France, etc., trawling is carried out over one or both sides of vessels, while many fishermen of Portugal, Spain and Japan use the pareja method in which the boats work in pairs and drag one net between them. In the Mediterranean and in Sweden and Denmark, very light equipment is preferred, yet Dutch and English fisherman, working on the same fishing grounds as the Scandinavians, use quite heavy equipment. Floating trawls are normally used by two boats fishing in pairs.

Purse-seining is done over the stern on the U.S. Pacific coast and over the side in European waters. The handling varies in the Pacific boats, some using a turntable, some using a drum, and some using neither. Purse-seining is also carried out from small seine boats working with the " mother-ship ", such as those used in the menhaden fishery on the U.S. Atlantic coast, much in the same manner as it is done in Norway and Iceland, but in Sweden the purse-seine is used over the side on an ordinary fishing boat. Scottish fishermen also employ this method when using their type of purse-seine, the ring net. The gill net is sometimes used from the fore part of a vessel in Iceland and Scotland, but from the stern on the U.S. Pacific coast, where some boats are equipped with drums.

Danish seining is common in Europe and it is also practised in Japan and, to some extent, in Australia. It consists of a type of drag net with long ropes, which is laid out in a circle or triangle. The long ropes and the net are usually hauled in by the boat lying at anchor. A variation, in which the boat is steaming slowly while hauling in the ropes, called fly-dragging, is being used more and more, particularly in Ireland and the U.K. Danish seining is an efficient method but is not yet in use in North America, with the exception of some boats in Newfoundland. In long-lining, Norwegian fishermen use a long-line winch in the fore part of a vessel. Then there is the type of halibut long-lining used on U.S. and Canadian west coast boats and the Japanese type of tuna long-lining, which Mr. H. C. Hanson recommends as an additional method for American tuna vessels.

Hand-lining, considered to be a primitive method of fishing, can be carried out from almost any small boat, but the most complicated fishing boats in use to-day, the tuna clippers, are designed to use this simplest of all fishing methods.

The trolling boats on the U.S. west coast are very specialized and are fitted with much mechanical handling equipment so that fishing can be carried on by small crews. Some of them are designed as combination boats. This raises the point whether it is economically sound to design a boat to be used for two, three or, perhaps, all the various fishing methods. In a way every fishing boat is now a combination boat. Norwegian boats, for example, use long-lines, hand-lines, trolling lines, purse seines and gill nets. Yet it would be difficult to design a combination boat which would be most suitable for every area of the world. In some places the most important fishing method is trawling; in other places, purse-seining; so that a combination boat, to be the best boat for a particular area, would have to be designed to meet primarily the demands of the most important fishing methods used in that area.

The design of present-day fishing boats varies considerably from area to area. U.S. Pacific boats, for instance, have the steering house and engines ahead, but the boats used on both sides of the Atlantic have them aft. The Mediterranean type of boat has the steering house ahead, with the cargo room underneath it—and the engine room midships, a very awkward and unpractical arrangement—whereas the Irish type of fishing boat has the engine ahead, the steering house aft, and the cargo room midships.

The problem of deciding whether to have the engine and steering house forward or aft is difficult. Placed forward, as in the Pacific type of boat, it permits a large fishing platform aft, but it makes difficult the captain's task of watching fishing operations, especially when setting a trawl. Steering a straight course is also more difficult and pitching is felt more. From a naval architectural point of view, it cannot be considered right to have the main weights concentrated in the fore part of the ship and heavy tanks situated aft. There are extraordinarily large torsional stresses in the long propeller shafts while the hull is subject to hogging and sagging. For seaworthiness, the centre of gravity should be aft of the midship section and not before, as it is likely to be with such an arrangement. An engine forward raises the centre of gravity.

There appear to be differing opinions regarding the use of wood or steel and other materials in building fishing boats and, where preferences have been expressed, they seem to be based on familiarity with the material concerned. The exception is Mr. H. C. Hanson who, although he probably has the best knowledge of wood of any living naval architect, prefers steel. In Europe, sawn frames are used for almost every fishing boat down to 40 ft. (12.2 m.) in length, but on the Pacific coast of North America steam-bent frames are used in building boats up to 130 ft. (40 m.) in length. The advantages of using steam-bent frames are: they save space; they are cheaper; and, in bending them, a check is made on the quality of the timber.

There is considerable difference of opinion concerning the

merits of government regulations controlling the construction of fishing boats. Where, as in the United States, there are no rules, some builders construct boats that are too weak. In Europe there exist a number of classification societies, such as the Bureau Veritas, with flexible rules. If a boat builder can show that a certain design is safe, even if it does not conform to the printed regulations, the plans are approved. But many European governmental regulations tend to stifle development, although, of course, there are always exceptions.

There has been disagreement over the question of V-bottom design, some experts claiming it to be cheaper, some saying it is dearer. An important point concerning V-bottom design is that, if the cost of material is the same and the number of working hours is the same, a saving would still be made by using less skilled, cheaper labour.

A main obstacle to mechanizing fishing fleets in underdeveloped countries is the lack of harbours. Knowledge of design of beach-landing boats is, therefore, important. The Danes and the Portuguese, among others, have shown that mechanized boats can be landed on beaches. The problem is to design a boat which is simple and cheap enough to build for, say, India, and this is one of the acute difficulties faced by the Fisheries Division of the Food and Agricultural Organization of the United Nations.

A world standard fishing boat would be impractical not only because it could not be designed to be suitable for fishing everywhere but also because fishing boats are, in the main, built locally. The boat-building industry does not lend itself to the standardization and mass production that takes place, for example, in the motor car industry. Boats are more varied in size, design, purpose and so on and there will never be anything near the demand for them as there is for cars. In any case small private boat yards would still be in the position to build cheaper.

COMMENTS BY SOME OF THE AUTHORS

Mr. H. C. Hanson (U.S.A.): Small boat yards often tend to be slack about fastenings, etc., as there are no rules laid down concerning them. The timbering and fastening tables given may start a move towards setting up recognized standards.

V-bottom construction is cheaper than that of any other type, with the exception of barge building, when ordinary floor timbers and frames are used and built up and bolted together. The V-bottom type of construction will cost 25 per cent. less in labour than the round-bottom type, although the cost of materials will not vary much in either case. Steel is more often used for building trollers than for gill netters, and the vessels are usually developed into combination types, because it is of the utmost importance for a boat to be able to work the year round to enable the owners to recover their capital outlay. This is especially true for U.S. fishermen as they do not receive government subsidies.

The trend now in the U.S.A., in the case of combination vessels, is to use welded steel construction. Up to 1935 only about 10 per cent. of his designs were for steel vessels; to-day the proportions are 90 per cent. steel, 10 per cent. wood. The U.S.A. Register does not cover the construction of wooden boats under a certain size but it is hoped that a standard will be set up for them.

The 24-year-old tuna clipper, *Northwestern*, was shown in his paper because it was a fine example of sawn frame construction, built by Petrich of Tacoma, and would probably be good enough to remain in service for another 25 years. In an effort to offset the big labour costs to-day, aluminium was being used in the hulls of smaller vessels. It was hoped that more use could be made of such material, and of plastics, to reduce construction costs.

Instead of trawl winches being set up in the sides of a vessel, there should be a single combination winch in the centre of future vessels. Much time would be gained by using dynamometers or gauges with the trawl winches, as they would tell when it was time to haul the trawl. Hydraulic drives are used in U.S. vessels operating in near waters. Air is also used, but most drives are mechanical, because there are big distances to be sailed and mechanical drives are most reliable.

Some Europeans are critical of the light-weight engine but U.S. boat builders are grateful for its development. Many owners cannot afford heavier engines. That is not to say the light-weight diesel should be used for every fishing boat, but it has its place.

Costs are causing great concern in the U.S. boat building industry. A 57 ft. (17.4 m.) boat now costs about $70,000 (£25,000) and a 100 ft. (30.5 m.) one about $300,000 (£107,000)—prices that are far too high. In view of such costs it is prohibitive to build a one-purpose boat.

Laminates are being used in construction in an effort to reduce costs but in many cases the material is being overdried, which means that rot will set in within a few years and the boat's life will be finished in 10 years. Prices being what they are, an owner will be lucky to recover his capital in 20 years, which makes it economically unsound to build a boat that will last only 10 years. That is a basic objection to laminates and, if they are to be used, they should be used where they can be easily repaired.

Mr. Howard I. Chapelle (U.S.A.): In dealing with the effect of fishing methods and operations on the design and safety of boats, he referred again to the Jonesport or Cape Island boat. It is very easy to point out the deficiencies of this type, but it is less easy to produce a safe launch, even if changes are made, because there are certain requirements in a launch that occasionally put it in a dangerous condition. For example, it is necessary to have a low freeboard for handling certain types of gear; again, it is necessary to have a large cockpit for some types of fishing. No change in model, or in the basic design of the hull, will improve the safety of a boat if a large cockpit with low freeboard is a necessity.

There has been a trend towards steam-bent frames, simply because they save a great deal of labour, which is cost. Every sawn frame has to be bevelled and, even though part of the work can be done on the bandsaw, there is always the dressing off and hand fitting to be done, which amount to many hours of labour.

The use of the V-bottom design in the fishing industry is an economic one. It may cost either more or less than the round-bottom design, and this difference is not wholly controlled by the design. The size of the boat, where it is being built, the availability of materials and of skilled labour, are factors which influence the cost. Labour is one of the chief factors in the cost of construction of wooden craft, whether V-bottom or round-bottom.

V-bottom construction has been almost the only method used in building commercial boats in Chesapeake Bay (Virginia, U.S.A.) shipyards for more than 50 years. The method was first used for sailing craft and oyster dredgers, but now it is used also for cargo boats. The largest are about

70 ft. (21.3 m.) long. They have to be thoroughly seaworthy because they work in winter, when Chesapeake Bay can be very rough. The largest boat built by the Chesapeake Bay method was about 94 ft. (28.7 m.) long and she was later sold to the West Indies to operate as a fruit carrier.

He expressed the opinion that it is possible to build a V-bottom boat by omitting the standard framing system and by changing the planking system. By doing this, a considerable saving is effected in the cost of construction. But there are other factors involved. For instance, the construction shown in many of the U.S.A. Pacific Coast craft, either the round- or V-bottom, would be expensive on the Atlantic Coast because there are no timbers of large scantlings. If the boat yards on the Atlantic Coast adapted the Pacific Coast type of construction, particularly in small craft, it would increase costs, an instance of how locality affects the price.

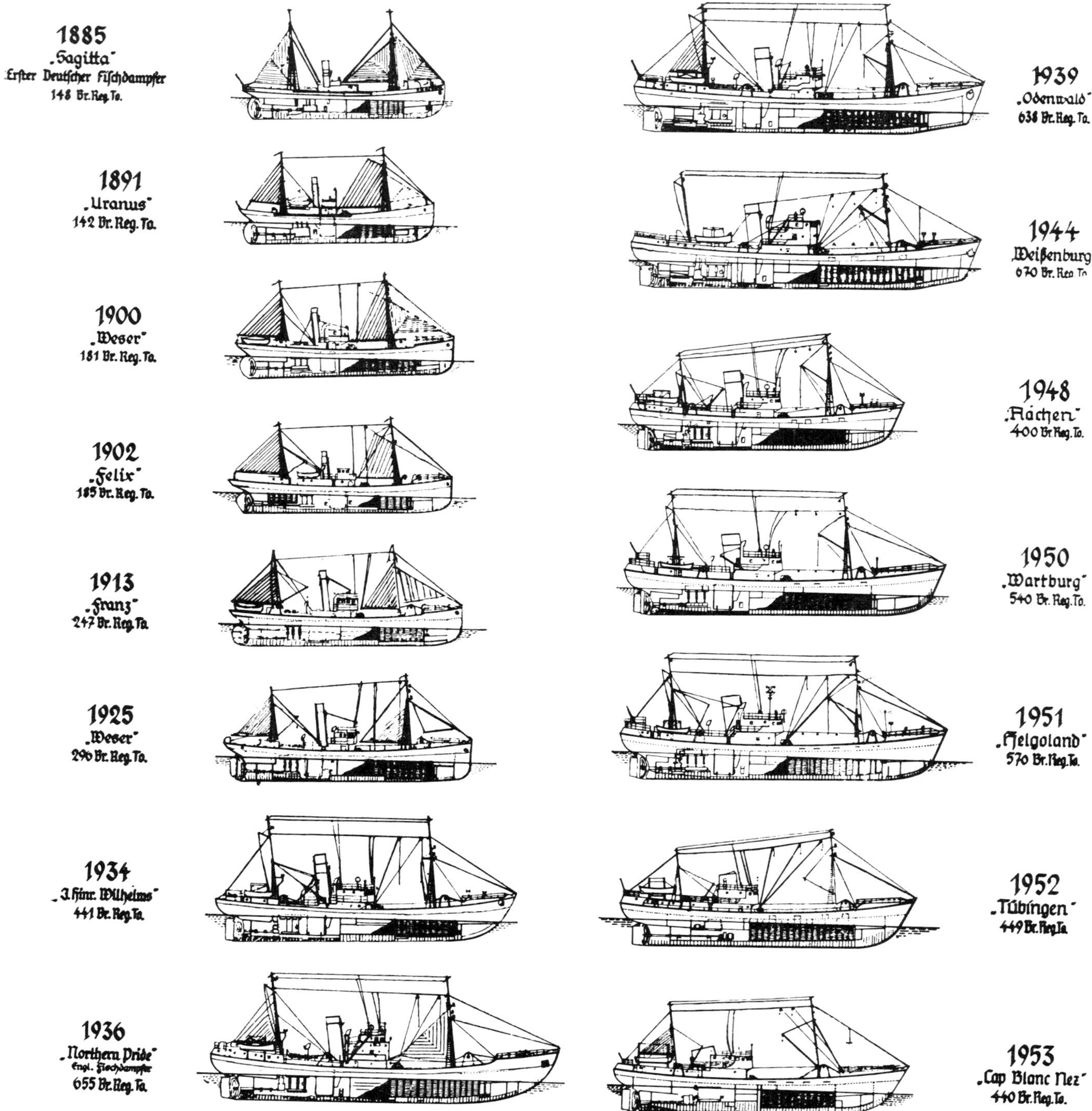

Fig. 310

Again, there are the different costs of materials and labour. In some places labour is of the greatest importance to cost but elsewhere materials may be the chief factor. Locality is, in fact, the chief consideration when determining round- and V-bottom designing cost. A great many boat yards cannot determine the true cost of a boat until it is completed, so it is not simple for a naval architect to determine what will be cheaper.

The speed-power characteristics of the V-bottom type have not been fully explored. A good deal of research has been done on V-bottom models for high speeds but they are unsuited for the relatively low-speed ratios of fishing boats. There can be a saving in constructing steel boats with V-bottoms but reasons of economy control the selection of steel or wood. In some areas it is cheaper to use steel but that is not applicable to all areas and there are many places where it is cheaper to build in wood.

Fig. 311

Mr. Fred Parkes (U.K.): As an owner he praised the part played by France in the post-war development of various types of trawlers, especially those with diesels designed by Mr. E. R. Gueroult. They proved to be excellent vessels. After more than 50 years in the fishing industry, Mr. Parkes found it paid to proceed cautiously with investments in fishing boats. His companies had built more than 60 fishing vessels of various types since the war—distant, middle and near water trawlers and drifters. All this experience has led him to believe that diesel-driven vessels are the boats of the future. The old coal-burning steam trawler is now too expensive to operate. Before World War II, coal used to be 21s. ($2.95) a ton, f.o.b., in England; it is now £4 ($11.20) per ton.

What owners chiefly demand is a constantly dependable diesel, otherwise operation of a diesel trawler may be very expensive. He illustrated this point by saying that if a trawler has to put into a port, say on the Norwegian coast, for engine repairs it costs a lot of money, sometimes as much as £2,000 or £3,000 ($5,600 or 8,400). Such a sum eats into the year's profit, and it can nullify the advantage of the lower fuel consumption of the diesel. Constant reliability is, therefore, a most important factor for trawler owners.

A great deal of thought for the welfare of the crew has gone into the design of the post-war vessels, and all crew are now sleeping aft. For example, a man can now pass to his quarters from the wheelhouse without going on deck, which was often dangerous in bad weather in the old type of trawler. More than one fisherman has been washed overboard in doing that walk.

He believed in going to an expert for the design of a new vessel. The best way is to give the naval architect the length, the breadth and the depth of the vessel, and the h.p. and speed required. The architect is then in a position to design the new trawler. It is always best to leave the matter entirely in the hands of the architect and not try to tell him how to make the design. Then, having obtained a first-class design, it is best to go to a first-class builder and keep a careful watch on the construction. He stressed the point that the technicalities of design and boatbuilding should be left in the competent hands of the experts.

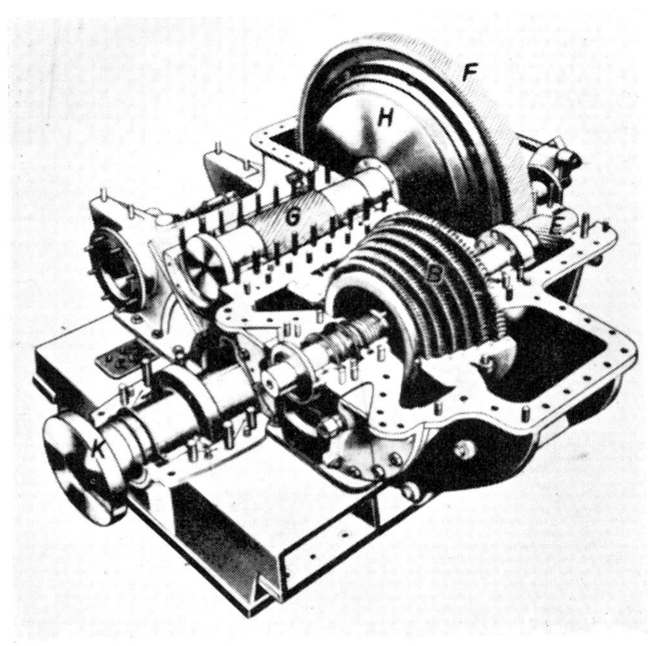

Fig. 312

Friends from Holland had asked him why he was building small ships when the North Sea is being over-fished. The reason is that the Government in Britain makes a grant of 25 per cent. of the cost of fishing vessels up to 130 ft. (39.6 m.) in length. For example, the Government pays the fishing boat owner £25,000 ($70,000) if the boat he is building costs £100,000 ($280,000). In addition to this money the Government also makes loans to owners at a low rate of interest to provide for at least part of the remaining £75,000 ($210,000). The Government gives owners further help by means of a subsidy if a vessel is losing money. The sum does not cover the difference between loss and actual operating expenses, but it may be £100 to £120 ($280 to 340) on a single voyage plus 4d. a stone ($0.0033 a lb.) on all fish landed for human consumption. These are the reasons why near- and middle-water fishing boats are being built in Britain, but construction of distant-water ships has almost ceased because they do not pay. They cost up to £200,000 ($560,000), their upkeep is very expensive, depreciation is heavy, and the price of fish is not high enough to make big fishing boats an economic success.

Mr. H. Kannt (Germany): A comparison between the first German steam trawler, *Sagitta*, and the latest motor trawler, *Cap Blanc Nez*, shows that there has been no significant change in the size of the vessels since 1885. On the contrary, during the past 10 years there has been a slight trend towards smaller trawlers (fig. 310).

Since 1945 an effort has been made to find out the most efficient method of propulsion of fishing boats of various countries—whether it should be coal or oil burning for steamers, with or without an exhaust steam turbine, or direct-drive diesel, diesel-electric or diesel "father and son" arrangements. German shipyards have also studied the problem of finding the most economic propulsion.

Bauer-Wach exhaust steam turbines have been fitted in steam trawlers and where, for financial reasons, such turbines cannot be fitted immediately, provision has been made for installing them at a later date, as in the case of the *Wartburg*. Exhaust turbines, either closed or open, have proved very efficient and trouble free (fig. 311 and 312).

There has been no fundamental change in catching methods but the conventional fishing gear has been strengthened and improved. For example, the fairlead has been arranged so that the crew can control the length of the trawl warps more easily and the starboard trawl can be used over the port side fairlead. A mechanical towing block for the warps has been developed to prevent accidents and to enable the warps to be handled better.

Another development is that of an oil-filled rigging screw (fig. 313). There is also a new type of safety hatch (fig. 314) which prevents water flowing into the coal bunkers when a vessel is swept by heavy seas.

A saving in top weight has been effected by reducing the weight of the davits, thus improving a vessel's seaworthiness and sea-keeping qualities. Fig. 315 illustrates a new one-legged davit, which is very simple to handle.

Efforts are being made to prefabricate equipment in trawlers and, in this respect, a new type of stanchion for fish-holds has been made (fig. 316), which is built in by merely welding two seams.

Professor H. E. Jaeger (Holland): The ideas expressed by Mr. Gueroult are similar to those contained in his paper dealing with large trawlers, but he did not feel so sure that very big trawlers with a fish-hold capacity of 53,000 cu. ft. (1,500 cu. m.) were suitable for general fishing. They are perhaps more suitable for long voyages and salting the fish which they catch. The problem of keeping fish in fresh condition while filling up such a big hold, or even a smaller one, has yet to be solved.

Mr. E. R. Gueroult (France): Between the 140 ft. (42.7 m.) French fresh fish trawler and the 225 ft. (68.6 m.) salt cod trawler there is a gap which is to be filled in the future. The fresh fish capacity of the 140 ft. boat will remain what it is (12,000 cu. ft. or 350 cu. m.), as this is considered to be the maximum quantity that can be brought to port in good condition. It represents about 12 days' fishing. The remainder of the catch will be salted, frozen or otherwise processed. The fish-hold capacity on any of the new French trawlers is the same as that of the 140 ft. boat, but on the large trawlers that fish off Newfoundland, the fish-hold is divided into two

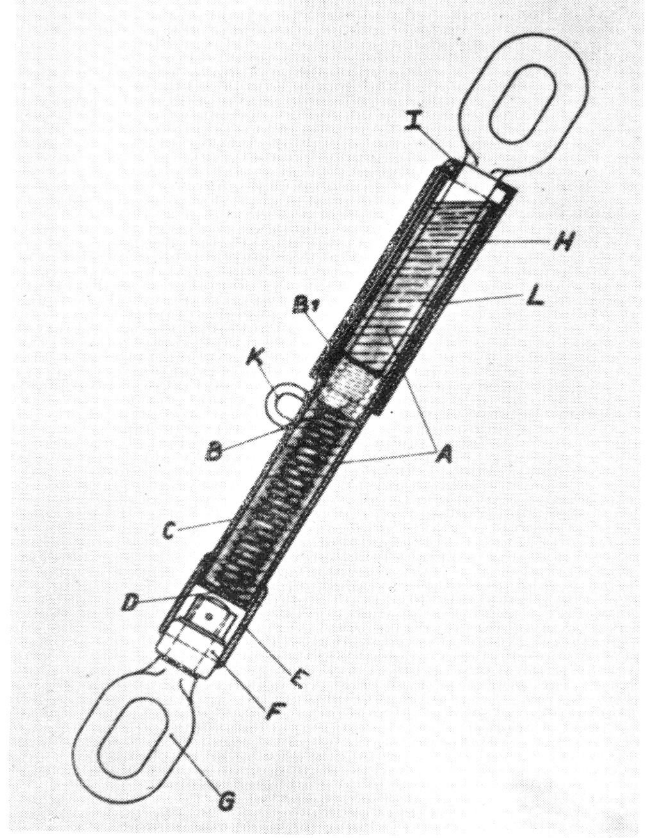

Fig. 313

Fig. 314

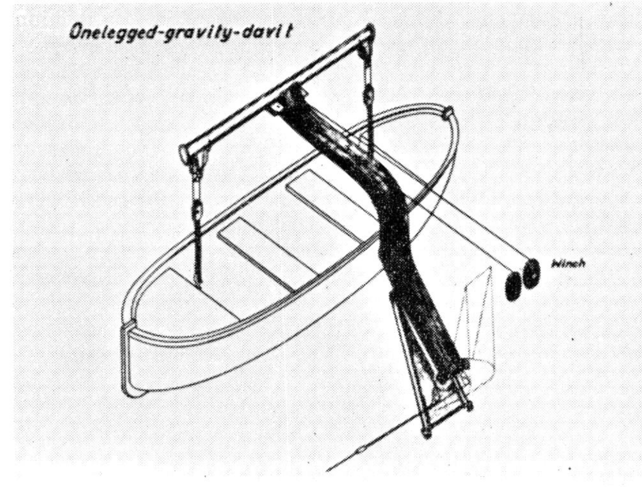

Fig. 315

watertight holds which, in turn, are each divided into two compartments by wooden partitions.

The problem of wood or steel construction was examined when it was decided to start the French reconstruction programme after the war. It was then decided that, due to the quality of labour and material available, all boats longer than 85 ft. (25 m.) would be built of steel but, because of the progress made in welding, the limit has now been lowered to 65 ft. (20 m.). He expressed the opinion that the choice between wood and steel construction depends on local conditions.

DIFFERENT TYPES

Mr. W. S. Hines (Canada): There are 5,000 Cape Island boats in use in the inshore fisheries of Nova Scotia and Chapelle's criticism of them, particularly about the loss of life, does not reflect on the hull form or the model but the use to which the boats are put. They are inshore boats but are being used for offshore fishing.

Mr. Howard I. Chapelle (U.S.A.): The point raised by Hines illustrated one of the problems of criticizing fishing boats. The standard Cape Island boat and the companion Jonesport boat of Maine (U.S.A.) were both excellent for inshore work, where shelter can be reached quickly, and for use in protected waters. Nobody could design and produce such a small boat, with relatively little power so cheaply or with better qualities, but both types of boat are being employed for offshore work, for which they are not suitable. On the other hand some of them are satisfactory for certain offshore operations because their model has been improved along the lines mentioned in his paper. That is, they have been deepened, and for certain offshore work it would be difficult to find better boats.

Mr. A. Scherrl (Austria): On inland lakes a fishing boat must not only be suitable for several methods of fishing but also usable for transport, for which reason the boat must be stable, seaworthy and spacious. These requirements are met by a design, fig. 317, for a 20 × 5.6 ft. (6.1 × 1.7 m.) boat which can accommodate a crew of two or three men, and provide plenty of space for fishing equipment. The boat has a 5 h.p. outboard motor and, in good weather, can sail at about 8 knots. The larchwood planking is ⅝ in. (15 mm.) thick. The weight of the hull is about 775 lb. (350 kg.) and, when taking into account the weight of the motor, equipment and crew, displacement is about 1,300 lb. (600 kg.). In this condition the freeboard is about 16 in. (40 cm.). When used for transport purposes the boat can carry eight people.

Mr. J. F. Petrich (U.S.A.): One of the problems arising from the papers and the discussions was to find suitable types of boats for the expansion of fisheries in the underdeveloped regions of the world. There are many areas, for example, in the South Pacific, which are now being developed and the ways in which they should be fished and the types of boats to be used had to be decided. It seems that small boats will have to be employed and it is thought that the long-lining method will be very suitable. This is a problem which must be common in other parts of the world, and it would be interesting to know if experts had in mind types of boats particularly suited for development in these new areas.

On the Pacific Coast of the U.S.A., for example, are many efficient fishing boat types. Here the greatest development of fishing took place when the combustion engine was being evolved, and, unhindered by former tradition of sail boat

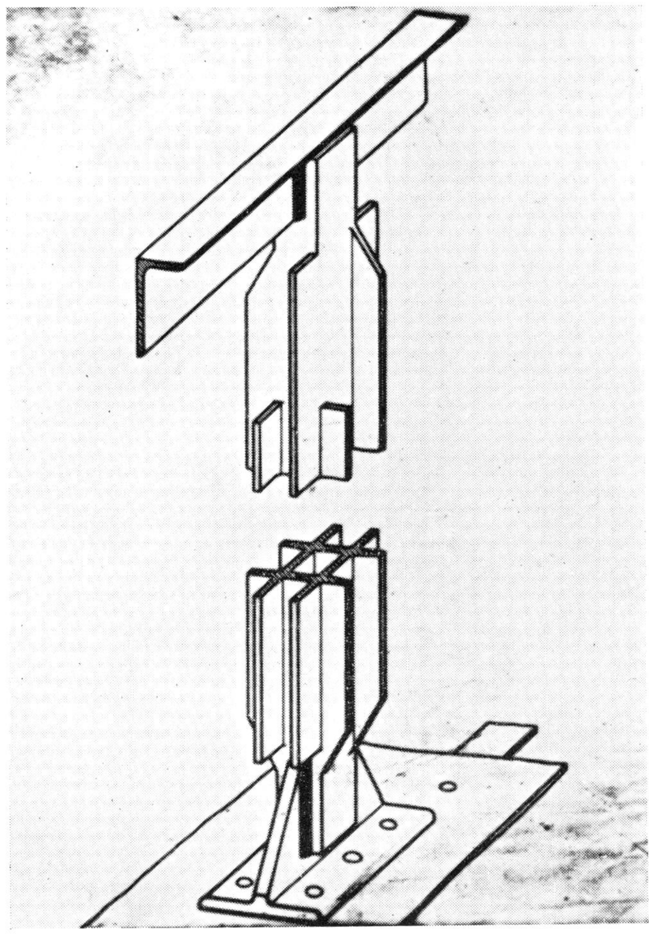

Fig. 316

designs, resulted in the development of very efficient types of powered fishing vessels. Adding an engine does not necessarily make the boat more efficient and might in fact make it more uneconomical by increasing costs. On the other hand, if the fisheries of the world are to be adequately developed and expanded, suitable types of power boats must be introduced.

Mr. L. C. Ringhaver (U.S.A.): The shrimp catching industry of the Mexican Gulf has grown rapidly. For example, a man who had $30,000 (£11,000) invested in boats in 1947 was considered to be a big operator; to-day a man who has $100,000 (£36,000) invested is thought to be a small operator. In 1947 his shipyard produced about three boats a month, usually about 50 ft. (15.2 m.) length. About 45 men were employed. In 1949 trade became bad and the company started to build a specially designed Florida type shrimp trawler, 60 ft. (18.3 m.) length, by using the idea of mass production. Since that time the yard has never been able to produce enough to meet the demand. It is now building 10 shrimp trawlers a month and has set up a programme to install three engines a week. The 2 × 4 in. (5 × 10 cm.) frames of oak are steamed and spaced at 12 in. (30 cm.) distance. The men start work at 7.30 in the morning and by 7 o'clock in the evening the boat is framed. They can plank a boat in three and a half days and complete a pilot house on a boat in four days. One of the reasons for the success of the methods employed is that Mr. Ringhaver was trained as a cost accountant and he knows exactly how much each and every operation costs. The result is a shrimp trawler, completely ready to fish, for less than $40,000 (£14,300).

One of the problems arising from such quick production is to get the right lumber, such as the oak frames, planking and the material to build the pilot houses. The supply must not only be of the right kind of wood but also delivered regularly so that there is no hold-up in the three production lines which the shipyard keeps going.

Mr. Jarl Lindblom (Finland): Finland had to tackle the task of building 90 300-ton fishing schooners for the Soviet Union which were to be delivered within four years as war reparations. She had no shipyards in which to do the work and these had to be built before construction of the schooners could be started. Up to 1953 his shipyard had built 45 of the 90 schooners and, as the result of a trade agreement between Russia and Finland, 30 more fishing and seal-hunting vessels have been ordered. All the boats have been built to the requirements of the Russian Marine Register. Measurements of the schooners are: 130 ft. (39.6 m.) over-all length, 30 ft (9.1 m.) beam, and 30 ft. maximum draught. The full load displacement is 730 tons and the speed is 9.3 knots. Cargo capacity is 6,900 cu. ft. (195 cu. m.), and they have very large fuel tanks which allow 900 hours of running at full speed. They have to carry a food and water supply sufficient for a three months' voyage. To save their fuel, they are equipped with exhaust-gas boilers. They have evaporators for producing fresh water and machines for making bread.

As the schooners have to operate in Arctic waters, their planking is very heavy. The outside planking is $4\frac{1}{2}$ in. (11 cm.) on double frames of 6 × 14 in. (15 × 36 cm.), spaced only $1\frac{1}{2}$ ft. (450 mm.) apart. Inside the frames a complete ceiling is built up to the deck. On top of the planking they have an extra thick ice-protection planking of $2\frac{1}{2}$ in. (6.3 cm.) of hardwood. Sterns are laminated from fir and Brazilian hardwood called perola.

Because of the adverse weather conditions in Finland the boats have to be built under cover. In the plant where he works, there is space for eight schooners to be erected inside the building at the same time.

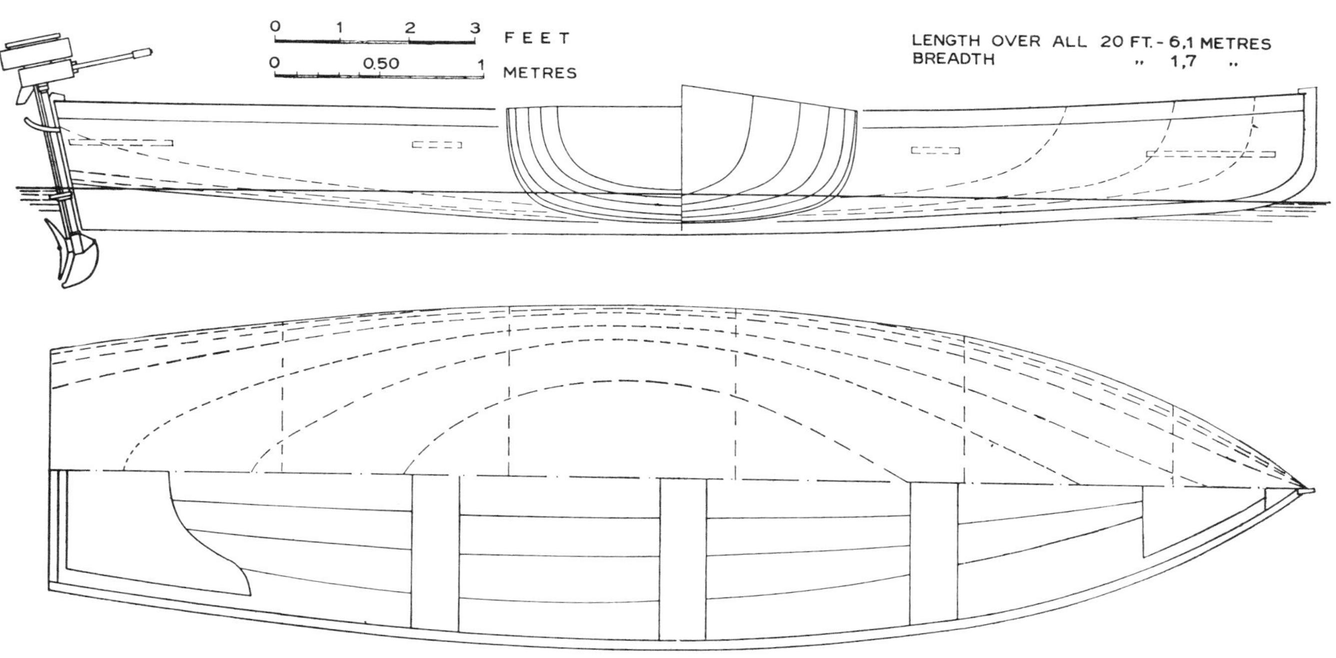

Fig. 317

The keel is laminated of hardwood and underneath there is another keel of solid wood. In accordance with Russian requirements all the frames in the foreship are placed together so that the foreships of the vessels are practically solid. The hydraulic deck machinery installed is of Norwegian design.

One schooner was built completely anti-magnetic, even the riggings, chains, anchors and everything being made of non-magnetic material. He added that he could give no information about the main engine of this anti-magnetic schooner as it was delivered engineless to the Soviet Union.

Mr. Knud E. Hansen (Denmark): Fishing has always been of the utmost importance to sea-girt Denmark, and the variety of the surrounding seas and the nature of the coast have demanded the development of different types and sizes of vessels. The most important and best known type is the North Sea cutter, usually from 30 to 50 tons gross, fig. 318. It has a reputation of being a most seaworthy boat and is used for fishing in distant waters, such as the Barents Sea and the areas off the west coast of Greenland. These cutters are engaged mainly in Danish seine fishing but they are often equipped with a combination type winch, making them suitable also for trawling, purse seining and long-lining. They also use the floating trawl.

The bigger type of these cutters, ranging from 50 to 100 gross tons, fig. 319, fish mainly in the Barents Sea and in the waters of the Faroe Islands, Iceland and Greenland. All the cutters are built of domestic hard oak, a wood unsurpassed for ship construction. The scantlings are determined by official rules. The bigger cutters are often built to a special Danish edition of Bureau Veritas rules, which give scantlings about 10 per cent. above the usual requirements. This is in keeping with the best Danish traditions, especially with regard to fastenings and transmission of stresses. Only galvanized fastenings are used, and both the equipment and the accommodation are of a high standard. Direction finder, echo sounder, wireless telephone, and decca (loran) are usually installed. The motors are mostly the Scandinavian semi-diesels, with controllable-pitch propellers, which have been in general use since motors were introduced to the fishing fleet. A number of cutters are now fitted with Danish built diesels of a very reliable type.

A smaller type of cutter—about 10 to 20 tons gross—is commonly used in the inshore fisheries. The cutters are similar to

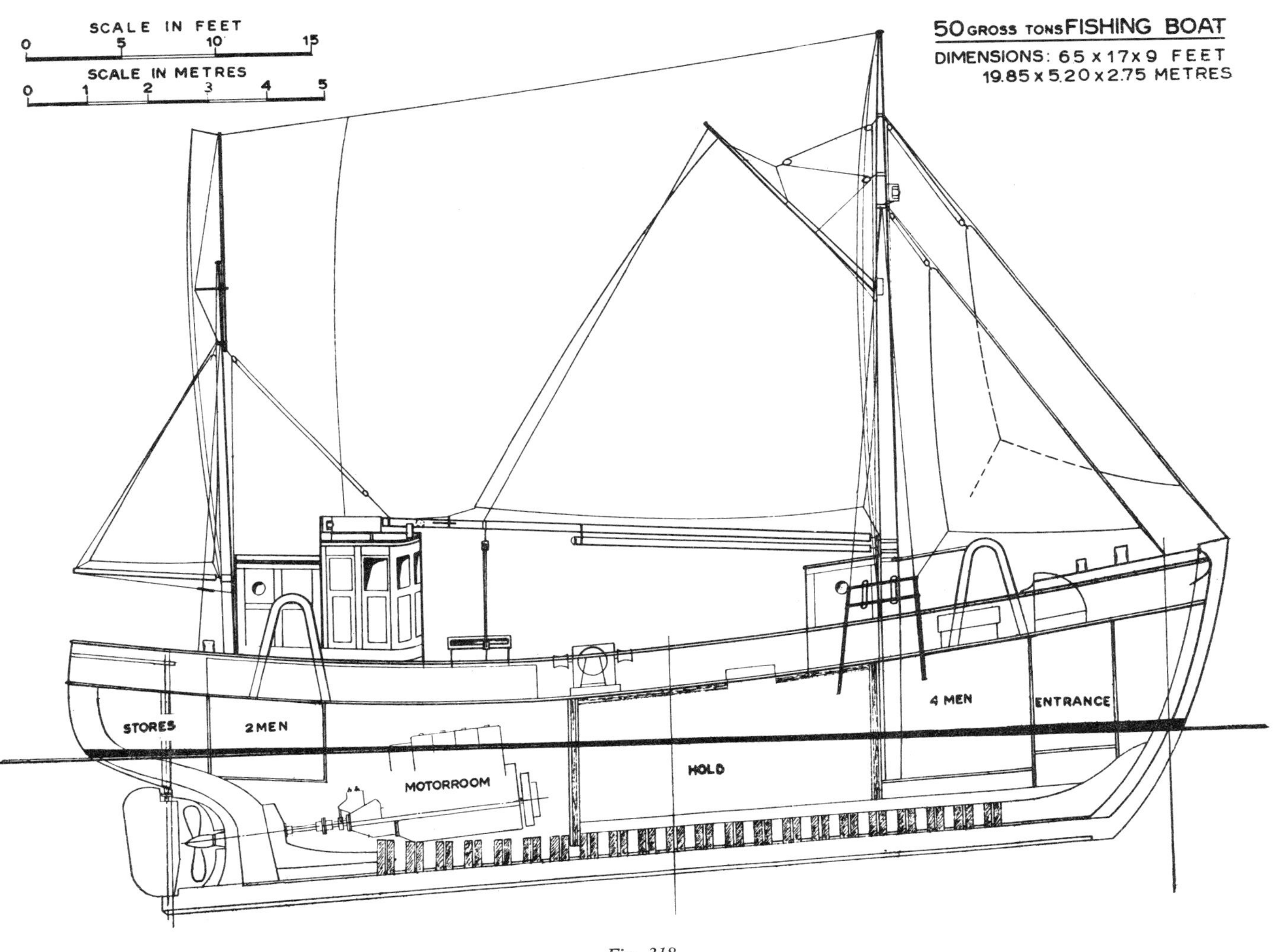

Fig. 318

the North Sea type but differ in being fitted with a well for carrying live fish. The well is located in the hold below the waterline, between two watertight bulkheads. A funnel to the main deck in the centre line gives access to the well from the deck, and leaves space for dry holds at both sides.

A typical boat for the coastal fishery is about 23 ft. (7 m.) in length but is seaworthy and reliable under all conditions, fig. 320. Such boats are generally clinker-built and are inexpensive. The motors are single-cylinder semi-diesels, very simple to operate. Many of these boats are shipped each year to Greenland for coastal fishery work there, for which they are excellently suited. Numbers are also exported to other countries.

The west coast of Jutland has few harbours and it is exposed to strong westerly winds and a special type of beach landing craft is used there. Such boats are generally from 30 to 35 ft. (9.1 to 10.7 m.) in length, are very seaworthy, and can be beached, stem first, even in rough weather. They are clinker-built of sturdy but not too heavy construction, and are usually powered with semi-diesels, although diesels are sometimes preferred. In addition to these typical boats, Danish shipyards also build steel motor trawlers of the bigger size. Many of these trawlers are built for French and Spanish owners.

Mr. Paul A. H. Lembke (Germany): Although the German fishing cutter fleet had suffered severe losses during the war, the shipyards could carry out only a limited programme of building replacement because of the shortage of raw materials. Among the new boats built there are two of remarkable design. The smaller type has a length of 69 ft. (21 m.), breadth of 18 ft. (5.50 m.), moulded depth of 7.8 ft. (2.37 m.), draught aft of 8.9 ft. (2.70 m.) and draught forward of 6.4 ft. (1.95 m.). The displacement is 105 tons (they measure 64 gross and 20 net register tons) and they are built completely of steel. The fish-hold has a capacity of 500 baskets of 110 lb. (50 kg.) of fish. The engine is 150 h.p., 6 cyl., 4 cycle, 375 r.p.m., with reverse gear, but each vessel is also equipped with auxiliary sails of 540 sq. ft. (50 sq. m.). Fuel space is provided for 7 tons. The crew consists of five or six men and the fishing gear and equipment are up to modern standards. These boats are designed for fishing in the North Sea but of the 15 that have been built some have been sold abroad.

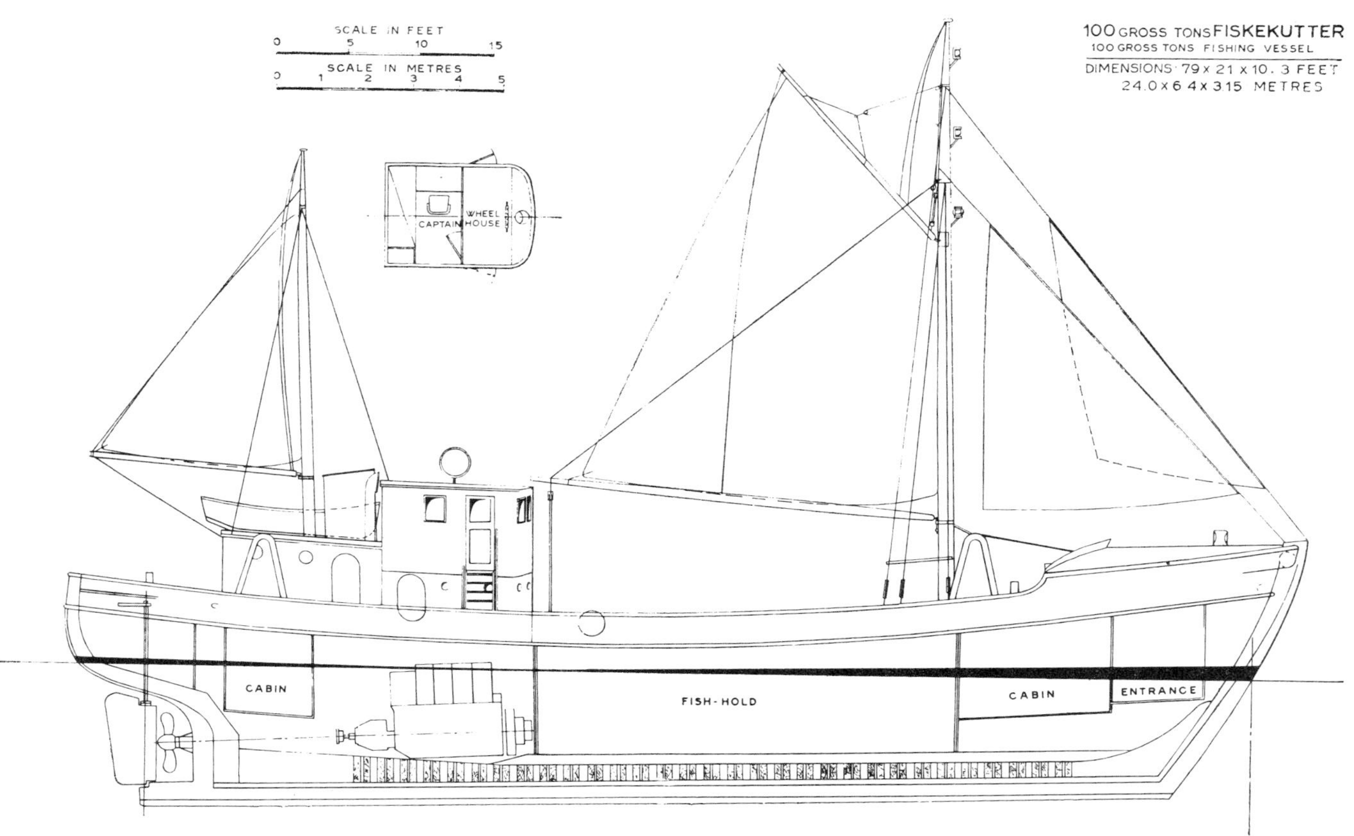

Fig. 319

The bigger type of cutter has a length of 84 ft. (25.5 m.), a breadth of 21 ft. (6.40 m.), a moulded depth of 10.6 ft. (3.23 m.) and a draught of 9.8 ft. (3 m.). The displacement is 158 tons and they measure 135 gross and 86 net register tons. The boats are built completely of steel. They are equipped with 180 to 200 h.p. 6 cyl. diesels with reverse gear, and carry 970 sq. ft. (90 sq. m.) auxiliary sails. The trawl winch, with two drums of 110 fm. (200 m.) of wire, $\frac{9}{16}$ in. (14 mm.) in diameter, is driven by the main engine, the trawl wire can be hauled at a speed of 72 ft. (22 m.) per minute.

The boats are equipped with echo sounders and radio

telephone. The fish-hold is 18 to 21 ft. (5.5 to 6.5 m.) in length and can store 800 to 1,000 baskets of 110 lb. (50 kg.) of fish. The crew consists of six or seven men. So far 10 boats have been built. They have proved very successful and should provide a suitable basis for further development of fishing cutters.

Commander A. C. Hardy (U.K.): It is impossible to produce a standard fishing vessel. On the other hand, it is interesting to know that the Pacific Coast type of tuna clipper and purse seiner is arousing much interest in other parts of the world. For example, such purse seiners are fishing at Walvis Bay, Union of South Africa, while in Australia another type of boat based on the tuna clipper is being used.

He recalled that the Pacific Coast type boat was fishing over the stern and said that the new British fish factory ship, the *Fairtry*, does the same. She uses a chute similar to that of a whale factory ship. She has an aft bridge from which she is controlled. The net is drawn up through the stern chute,

Fig. 320

and the fish are carried down through a hole in the deck into the factory deck below. The chute arrangement might be something for future Pacific boats.

Mr. Howard I. Chapelle (U.S.A.): Fishermen will resist for practical reasons any great change in arrangement of trawlers. For reasons of safety it is desirable to place the pilot house aft, so that the helmsman can watch the winch-man. It also has the advantage of simplifying and reducing the cost of the control equipment.

In Irish boats the engines have been placed forward. The designers in the United States would like to follow this example, but for several reasons it would not work very well. For instance, in most of the small U.S. trawlers a chain drive off the main engine is used to operate the winch, and to put the engine forward on the other side of the fish hold would call for the use of an hydraulic or electric drive. That would not only add considerably to the expense but would also increase maintenance problems.

Mr. John Tyrrell (Ireland): The best position of the engine depends on the use for which the vessel is intended. What may be ideal for trawling and Danish seining, will not suit a general purpose vessel, which engages in drift net fishing for herring.

In boats of 50 ft. (15.2 m.) there is much to be said for having the engine forward and the cabin aft, as the hold is then located about the longitudinal centre of buoyancy. Thus, when fully loaded, good trim is possible in a properly designed hold, while the cabin aft makes full use of the space up to the rudder post. While the longer propeller shaft is a disadvantage, it is acceptable because of the greater benefits found in other directions. In larger boats, of 60 ft. (18.3 m.) and over, there is more scope for placing engines. Even so, it may well be that for a specific purpose the forward installation is best.

His firm is building a standard 50 ft. hull (illustrated in his paper) and, of the 30 built in recent years, the engines have been installed in various places, such as right forward, less far forward, right aft and close forward of the cabin. The original hull design was prepared with these alternative positions in view. He added that it is usually the owners, who are extreme individualists, who insist on having the engine where they want it.

In answer to a question by Mr. Traung as to where he would personally place the engine in a boat to be used for Danish seining only, he replied that, in a boat of 50 ft. or over, he would prefer the engine aft.

Mr. William P. Miller (U.K.): In Scotland builders have long experience in positioning machinery, dating back to the conversion of the sailing boats of the herring fishing fleets to engines. At first it was thought that the crew must be accommodated aft and the fish-hold placed amidships, with the engine placed forward of amidships. But it was not long before builders were experimenting with the engine placed aft because a number of troubles arose from the engine forward arrangement. There were, in particular, difficulties with the propeller shaft and trim. As the result of this experience, the standard position of the engine to-day in Scottish boats is aft of amidships but forward of the crew's quarters. It may be said that the modern Scottish fishing boat is something of a compromise as it has evolved and developed from sailing boat types. The key to the problem is the trim of the vessel with a full hold of fish.

Commander A. C. Hardy (U.K.): It seemed to him logical in ship design to put the machinery in the part of the ship where it would not take up valuable space which could be used for other purposes. For that reason, apart from questions of stability and trim, the arrangement of machinery forward could be recommended because it leaves clear the best part of the amidships. It also allowed fish-holds, refrigerating rooms and bait tanks to be placed aft from the break of the fo'c'sle. If the machinery is placed forward there is the choice of having a high-speed diesel driving a generator, with small electrical propulsion motors at the stern, geared to the shaft, or the Pescara free piston generator could be used, with its generators forward and its turbine placed towards aft, thereby making it unnecessary to have a long shaft through the hold. Such a shaft is subject to torsional vibrations, and stresses due to possible sagging and hogging of the hull, disadvantages which are the main reasons why the machinery should not be placed forward.

Mr. H. C. Hanson (U.S.A.): He designed and built craft with engines both fore and aft, the decision depending upon the work the boat was expected to do. But practically no one on the Pacific Coast of U.S.A. would build a fishing vessel with the engine aft because it would then have to be a one-purpose boat, whereas, with the engine forward, it was possible to build a five-purpose boat, able to fish all the year round. In the larger wooden vessels, where the engine is placed forward, trouble is experienced with the shaft, but this is not true in the case of steel boats.

suggests thinking of the wheelhouse as a high-visibility control station, with remote rudder and engine controls available at the necessary vantage points. Aside from its other advantages in propulsion, the possibility of direct bridge control inherent in diesel-electric or controllable-pitch propeller drive lend further weight for its consideration on the merits of better manoeuvrability.

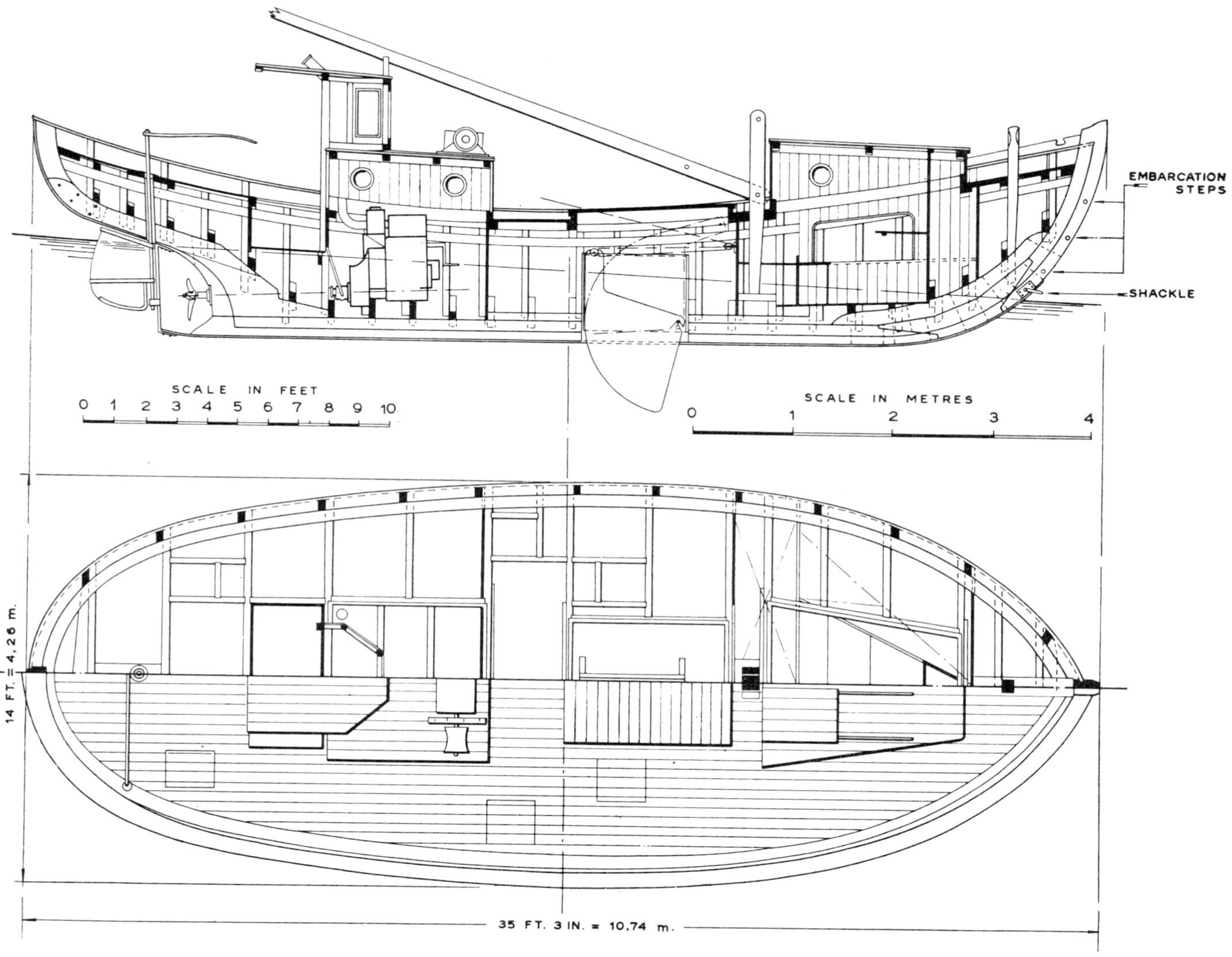

Fig. 321

He thought it would be a good idea to place electric propulsion motors aft, over the shaft log, and the diesel-generator forward. This would keep the hold clear and would dispense with the shaft alley, carrying conduits in the wings.

Mr. Philip Thiel (U.S.A.): Hauling and setting a trawl is a complicated and dangerous process. It is important that one man be in constant and complete control of the situation, with all means of control at his fingertips. His position is similar to that of a jockey riding a horse; and the engine and the helm should be manageable by him from a position of complete visibility of the gallows and fishing deck. This

The unsheltered unpowered steering station belongs to the past; it is recognized that better working conditions permit better work. In the trawler control station adequate heating and better window design can play their part. A continuous convector under the windows, with perhaps a perforated metal glove-drying rack above at sill height, should be considered. So should an existing type of sash windows incorporating a horizontal opening that can be quickly varied in extent and height.

Entrances to crew's quarters from deck should have some sort of vestibule to act as a wind trap. This provides a good place for hanging oilskins. Access to washrooms and toilets

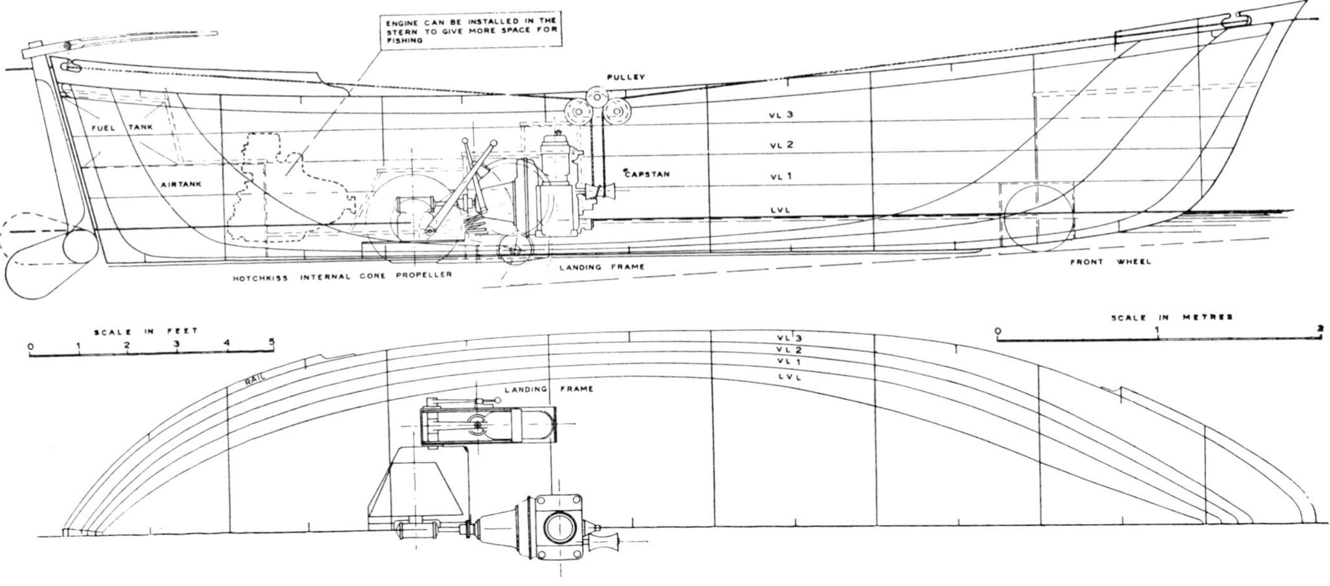

Fig. 322

can be from these vestibules, which are thus equally accessible from the deck and from berthing spaces. Washroom areas can then be made the centre of discharge of a mechanical exhaust ventilating system.

In the interest of sanitary berthing, the consideration of metal pipe berths is suggested with plywood or other wooden lining boards on the inside to prevent contact with the cold metal. Individual clothes lockers of open wire mesh would help the ventilation problem.

BEACH LANDING

Mr. Hans K. Zimmer (Norway): The Danish naval architect, Knud E. Hanson, recently designed a beach landing craft for use in the Mediterranean. One of the features of his design is the fitting of a centre board. This may be of importance as there is no possibility of beaching if the boat is caught in a storm. It must be able to ride out the bad weather under sail. The design appears in fig. 321.

Mr. Eric Estlander (Finland): Two disadvantages connected with beach landing to-day are: (1) the boats have to be heavily built to stand the wear and tear and strain of beach landing; (2) a big crew is needed for both launching and beaching. In an attempt to eliminate these disadvantages he had designed a very light fishing boat, fitted with wheels, and had devised a new system for launching and beaching. The boat, fig. 322, measures 24.6 × 8.2 ft. (7.5 × 2.5 m.), is open and has a cruiser stern. It has an enclosed section fore and aft, which makes it unsinkable, and has a rather flat bottom. The boat can be fitted with a normal propeller. This bottom shape may well be changed if experience shows that the landing frame gives good protection to the hull. In that case, it would be given better sea-going qualities.

The rudder is placed outboard and is pulled up when the boat goes into shallow water. It is propelled by an 8 to 10 h.p. 4-cycle diesel, with high r.p.m. and low weight. This operates a 24 in. (610 mm.) Hotchkiss internal cone propeller, which gives the boat a speed of about 6 knots. The engine is supplied with a simple winch device, consisting of a capstan

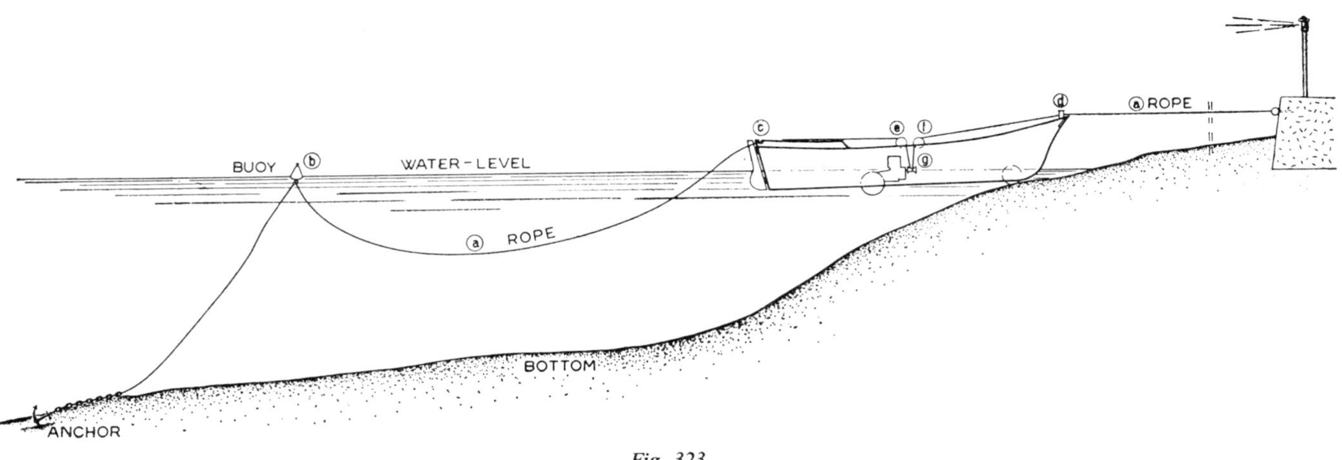

Fig. 323

which is fitted on the camshaft. It revolves with half the revolutions of the engine.

The landing frame fitted to the boat is operated by a lever and can easily be hoisted or lowered. The nose wheel is 16 × 4 in. (40 × 10 cm.) and the other two wheels are 18 × 7 in. (46 × 17.5 cm.). They are equipped with springs and have a total capacity to carry 7,000 lb. (3,200 kg.).

The landing method, fig. 323, consists of using the boat's own winch to pull it up the beach along a hawser, one end of which is anchored on the beach, and the other in the sea some 100 ft. (30 m.) offshore. The hawser is left slack and is supported by a buoy. In landing, the hawser is picked up from the buoy and is paid through the boat's hawser pipe fore and aft and run over two pulleys down to the capstan,

Fig. 324

around which it is turned. By this means, the boat is hauled up the beach.

In launching, the boat can easily be swung round on its wheels and then it is hauled out to sea by the same method with which it was beached.

He had built an experimental boat of 18 ft. (5.5 m.) length and 5 ft. (1.5 m.) width, and fitted it with a 3 h.p. one cyl. Stuart Turner, operating a Hotchkiss internal cone propeller, fig. 324. His experience showed that the Hotchkiss propeller has three advantages: (1) it is so well protected that it cannot be damaged by beaching; (2) it cannot get tangled with the fishing gear; (3) its situation on the hull enables the boat to ride safely and well in heavy seas. But it is heavier, less efficient and takes up more room than a normal propeller.

This experimental boat, fig. 325, operated excellently, landing with little trouble even in comparatively rough water, and launching, against quite heavy waves, was found to be much easier than had been expected.

Boats to his design could be built of aluminium or fibre glass. Using these materials would not only make the boat lighter and stronger than present beach fishing craft, but the costs of maintenance would be reduced. Such boats could be mass-produced and, for delivery to distant places, the hulls could be set into each other and all the machinery and equipment transported separately. The boat could then be fitted up with its engine, landing frame, and other machinery and gear, at the place from which it is to operate.

Mr. George C. Nickum (U.S.A.): He described an Army amphibious vessel, called the Barc, which is a development of the World War II DUKW. The Barc is more than 60 ft. (18 m.) in length, weighs 200,000 lb. (100 short tons, 90 tons) and has a capacity to carry 200,000 lb. (100 short tons, 90 metric tons) of cargo ashore, run up the beach and dis-

Fig. 325

charge it. Its four tyres are about 9½ ft. (3.1 m.) in diameter and when the amphibian beaches there is only a negligible impact on the tyres. In fact, tests with delicate recording instruments in surfs which ran up to 20 ft. (6.1 m.) gave negative results—very surprising but agreeably so because it meant a considerable reduction in required design criteria.

He endorsed Mr. Estlander's view that beaching with the help of wheels could be employed on fishing boats and the present round-bottom lifeboats. An interesting point is that the Barc has no springs on its wheels so that it would seem that a very simple set of wheels, on a retractable arm with no springs and no great supporting members, could be attached to a beach landing fishing craft. Such a method might well eliminate some of the difficulties now encountered in landing and launching fishing boats and it might well increase the number of days such boats could be at sea, and, therefore, increase their earning power.

He described also the development of a rubber-tyre beaching carriage. This was first thought of as a portable, marine railway which could be transported to any part of the world and used in the repair of vessels. It will have four rubber tyres of 9½ ft. diameter and 36 in. (0.9 m.) width and the carriage will have a capacity to take 150,000 lb. (68 ton).

Commander R. E. Pickett (U.S.A.): An interesting cargo handling system is used on the west coast of Africa. Cargo vessels sail via Freetown, Sierra Leone, to Matadi on the Congo river, making 25 to 30 stops on the way. Discharge of a vessel is completed at Matadi and loading starts for the homeward trip when much the same stops are made. Most of these stops are in the open sea, the ship anchoring two or three miles offshore where a heavy surf, varying from 14 to 60 ft. (4.3 to 18.3 m.) is almost always running. The loading and the discharging of cargo between beach and shore is handled chiefly by surf boats. The procedure starts at Freetown on the outward voyage. Here the ship takes on board a tug of about 28 ft. (8.5 m.) length and some 12 surf boats, with a crew of winch operators and cargo handlers, about 50 men in all. When the ship anchors off a surf port, the tug and surf boats are lowered overboard and take up positions where the cargo can be loaded into them. Each boat usually takes two sling loads or between 3,000 or 4,000 lb. (1,400 to 1,800 kg.). Then natives, who arrive from the shore, take up their stations in each boat and the tug tows them to the edge of the breakers, usually about ¾ mile offshore. At this point the surf boats cast off from each other and ride to the beach, each being steered by a sweep and propelled by 8 or 10 paddlers, who are ranged along and sit on the gunwhales as the cargo takes up most of the space in the boat. In this loaded condition the boat has 12 to 16 in. (0.3 to 0.4 m.) of freeboard.

As soon as the bow of a surf boat touches the beach, the paddlers jump out and, on the next wave, swing the stern up the beach and tilt the boat towards the shore. The cargo is then removed.

A surf boat might be described as follows: double-ender, single-chine V-bottom, 28 to 30 ft. (8.5 to 9.1 m.) overall, about 4 ft. (1.2 m.) in depth and having about 7 ft. (2.1 m.) beam. The sides seem to be flared about 2 in. (5 cm.) per foot amidships, and the stem and stern have a slight rake. The deadrise appears to be about 15 to 20 deg. from the horizontal. There are no thwarts and some bottom boards are used to keep the cargo out of the bilge water.

On the return voyage of the cargo vessel, when freight is loaded, the surf boats are launched into the heavy sea with, again, two sling loads of cargo on board. They are then paddled out past the edge of the breakers and are picked up by the tug. The usual cargo loaded consists of bags of cocoa, palm kernels or other baled or packed produce of the West African Coast.

He expressed the opinion that surf boats of this type might be fitted with two air-cooled petrol engines, the twin screws being protected by skegs and guards. The steering could still be done by sweeps.

He had seen nothing like these boats in other parts of the world. They were simply constructed and, in addition to being hoisted on and lowered off the ship several times a day, they make four or five passages through the surf in a day and are used to carry fresh water when lying on deck. They are very strong and seem to last for many years.

RESEARCH VESSELS

Captain H. Bertram (Germany): A German fisheries research vessel now under construction will have the shape and size of a modern trawler but will be built with a shelter deck. This will give the vessel a good range of stability and adequate seaworthiness. It will also provide a dry and completely protected working deck.

The fishing gear will be installed on the shelter deck to keep it clear of the working deck. The deck gear will consist of two pairs of gallows, two steam trawl winches and two deep-sea scientific winches. The fish-hold will have a capacity of 50 tons, and a freezing plant will be installed in the vessel.

A triple-expansion steam engine will propel the ship, which will also be equipped with an exhaust gas turbine, water tube boiler, and oil burner. The maximum speed will be 12 knots, and to increase manoeuvrability an "Active" rudder of 100 h.p. will be installed.

Mr. Hans K. Zimmer (Norway): The Norwegian Directorate of Fisheries took over a whale catcher hull, 170 ft. (51.9 m.) long, and adapted it to use as a fisheries research vessel. It is propelled by two two-cycle engines driving through friction couplings and reduction gear a single shaft which is fitted with a controllable-pitch propeller. Three diesel auxiliaries supply 70 kW. D.C. Deck gear includes two trawl winches, one big and one small, a windlass and two special winches for handling water specimens. All equipment is hydraulically driven.

The ship, *G. O. Sars*, is, of course, equipped with such instruments as echo sounder and radar. The work carried out through the vessel has been successful, particularly investigations of the herring fisheries which have proved to be of outstanding value.

A feature of special interest is the asdic (sonar) used for finding fish shoals on the sides of the vessels. The echo sounder, of course, registers shoals passing underneath the vessel, but the asdic can detect them at any angle. Mr. Zimmer mentioned also a recent Norwegian invention which, by the use of a metal mirror, makes it possible to employ an echo sounder as an asdic.

The *G. O. Sars* operates in cold and stormy waters and it is recognized that she is on the small side. She has, however, been so successful that a companion ship is being designed. It will be larger and have a different type of main engine installation.

Mr. T. C. Leach (U.K.): British research vessels are basically fishing boats modified to accommodate laboratories and seagoing scientific staff. They are manned by specially selected fishing crews so that all fishing gear is handled competently.

The research vessel *Ernest Holt* was launched in 1947 and is similar to the big British distant-water trawlers. She is 175 ft. (53.3 m.) in length and 30 ft. (9.1 m.) in beam, and has laboratories below deck and at deck level. She is fitted with all the latest navigational aids and is propelled by a steam engine with an oil-fired boiler. She is used for research within the Arctic Circle—Barents Sea, Faroe Islands, Spitzbergen, Iceland and Greenland. She works to carefully planned programmes with a different team of scientists for each voyage, depending on the nature of the work in hand. For example, one voyage may be devoted to plankton and echo trace observations and another to hydrological or fish tagging work.

Another British research vessel is the *Sir Lancelot*, a trawler of 126 ft. (38.4 m.) in length and 23 ft. 8 in. (7.3 m.) in beam. She is also propelled by a steam engine with an oil-fired boiler and her area of operation is the North Sea, the Channel and waters near the British Isles. She has a laboratory below deck and she is mainly engaged on echo surveys, plankton work and gear tests.

There are two other research vessels, the *Platessa*, an ex-Admiralty wooden motor fishing vessel of 88 ft. 6 in. (27 m.) length, 24 ft. 4 in. (7.4 m.) beam, diesel driven, and the *Onaway*, a small wooden drifter of 53 ft. 3 in. (16.3 m.) in length, and 16 ft. 3 in. (5 m.) beam. Both vessels work in near waters and their programmes include herring tagging, echo surveys and research on demersal and pelagic fish stocks. Both vessels have small laboratories on board.

In addition to these vessels there are those controlled by the Scottish Home Department. The largest is the *Explorer*, 135 ft. (41 m.) in length. She is being replaced by a 175 ft. (53.3 m.) distant-water trawler. Then there is the *Scotia*, a converted Admiralty trawler of 164 ft. (50 m.) and two wooden motor fishing vessels, the *Kathleen*, 55 ft. (16.8 m.), which is a seine net vessel, and the *Clupea*, 75 ft. (23 m.), which is a drifter.

The British Herring Industry Board also owns two motor fishing vessels, the *Silver Scout*, and the *Silver Searcher*, which are used on research work occasionally but sail mainly as reconnaissance vessels during the herring season. The Marine Biological Association at Plymouth recently launched a new research vessel of 115 ft. (35.1 m.) length, driven by diesel.

Mr. P. Mardesic (Yugoslavia): A new research vessel, owned by the Institute of Oceanography and Fishery in Split, operates in the Mediterranean, the Adriatic and the Red and Black Seas. The work done in her includes the study of the life of the sea, oceanography and experimental fishing of various kinds. The principal dimensions of the vessel are:

Length over all . . .	99 ft. (30.0 m.)
Breadth	22 ft. (6.7 m.)
Displacement . . .	250 tons
Main diesel . . .	380 h.p. at 375 r.p.m.
Maximal speed . . .	11 knots
Low-speed electric drive .	32 kW.
Giving a minimum speed of	0.25 knots

She is built of wood and has steel superstructures. Accommodation for scientists and the crew is located below deck and the laboratory is placed in the best position on the main deck. It measures 26 × 13 ft. (8 × 4 m.) which is large enough for 15 persons to work in.

The ship is equipped with a windlass driven by an 18 kW. motor. The windlass has two gypsies for ordinary chain, and one drum for deep-sea anchorage, carrying 1,100 fm. (2,000 m.) of ½ in. (12.4 mm.) wire. At the end of this wire there are 27 fm. (50 m.) of ⅜ in. (9.5 mm.) chain and a 100 lb. (50 kg.) anchor attached. Hand hydraulic steering gear is used. The main mast has two derricks of 3 tons capacity each. To avoid exhaust gases from the engines, all three exhausts are led through the mast. The cooling water from the auxiliaries is led to an insulated tank on the flying bridge. From there the water is piped to radiators and shower baths, the galley and the laboratory. Surplus water is led overboard. To prevent waste of drinking water, the tanks are connected to individual hand pumps wherever water is made available. Only the laboratory has running water. The main source of electricity is in two diesel generators, each 50 kW. of 230 volts D.C. A second supply comes from a 10 kW. dynamo driven by a belt off the main engine. In addition, both auxiliaries have 1½ kW. 24 volt generators for charging batteries. All instruments are connected to 220 volt. Those used in ports and on stations are of 24 volt, which is used for night work when the auxiliaries are stopped. A transformer is fitted for land attachment when the ship is lying in port.

There is accommodation for 22 people in 10 two-berth and two spare cabins.

The main winch, driven by two 32 kW. motors separately or in conjunction, is suitable for two purposes: (1) For hoisting the deep-sea net by a single wire of conical shape with a maximum diameter of ¾ in. (19 mm.) and a minimum diameter of ⅜ in. (9.5 mm.). There are 3,300 fm. (6,000 m.) of wire which is hauled by two central drums, 17.3 in. (440 mm.) and 23.7 in. (600 mm.) diameter respectively, with a speed of 100 to 165 ft. (30 to 50 m.) per minute. The maximum load taken by the wire is about 6.6 tons. (2) For trawling. In this case there are two drums each large enough to hold 1,400 fm. (2,600 m.) of ⅝ in. (16 mm.) diameter wire.

The wire, as it comes from the drums, is led under the deck to a winding up drum with a minimum tension of about 110 lb. (50 kg.). To overcome the difficulties of the constantly varying number of revolutions, the winding up drum is worked by an air-driven motor.

At the rear end of the deck there is a large roll to help haul the net, work which is normally done by hand in the final stages. The cylinder is turned by an air-driven motor of 14 h.p. A large platform, inclined against the stern, receives the contents of the net and keeps the deck dry and clean.

Two 3-ton derricks handle the trawl boards. They can swing in and outboard, and can be used to lift a boat across the stern.

On the port side are two serial winches for oceanographic work, each taking 3,300 fm. (6,000 m.) of 0.16 in. (4 mm.) diameter wire. They are each worked by a 5 h.p. air-driven motor. All told there are four air-driven motors on board and the compressed air for them comes from an electric-driven compressor in the engine room.

The main propelling machinery consists of a 380 h.p. 4 cycle direct reversible diesel of 375 r.p.m., giving a cruising speed of 10 knots and a maximum speed of 11 knots. A special electric drive is provided for the very low speed necessary for oceanographic work. This consists of a reduction gear and a motor of 32 kW., which enables the speed of the boat to be slowed down from 5 knots to about 0.25 knots. The propeller shaft is in two sections. There are two clutches, one each side of the gear, with a small servo motor acting upon each clutch. Both engines can be kept running, one ahead and the other astern. By pressing a button on the bridge the electric drive is engaged ahead. By pressing another button the diesel is engaged astern. This causes the boat to stop immediately.

Because of its relatively low running costs a vessel of this size has advantages over bigger ones. It can be kept running throughout the year and can undertake various types of work for which a big research ship would never be used.

Commander A. C. Hardy (U.K.): Two new research vessels have been launched recently in Great Britain. One, the *Africana II*, has been built for the South African fisheries. While she is a Newfoundland Banks size deep-sea trawler, she was designed at the start for research work. She has a double compound totally enclosed steam reciprocating engine. This takes steam at about 300 lb./sq. in. (21 kg./sq. cm.) pressure from a water tube boiler. It is reported that she is operating successfully. The second ship, built for the Torry Research Station, Aberdeen, is also a trawler type but she has a diesel electric propulsion. The high-speed diesels can be completely removed in one unit, a feature of that kind of installation.

Mr. H. C. Hanson (U.S.A.): The *Baird* was built for the U.S. Fish and Wildlife Service in 1946. She is a steel vessel, 138 ft. (42.1 m.) long, built as a tug with a raised deck forward and powered with two 900 h.p. diesels turning one propeller through a gear. The laboratory was placed where the towing winch had been, and a large trawl winch, with four reels, and a hydrographic winch were installed. She has a three-compartment, refrigerated bait well on the aft deck, and a deep-freeze compartment below deck. There was hardly any information available of the design of the equipment required and, because of this, he kept a record of the details concerning the vessel, together with plans and photographs, and he had read a paper about it before the Society of Naval Architects and Marine Engineers, New York.

Mr. Jan-Olof Traung (FAO): Research vessels are sometimes expected to do too many different types of work and are equipped with laboratories that are too elaborate. As the ships go out for relatively short periods, they could bring the samples they gather ashore and work on them there. There is also a tendency to provide too much accommodation on board for guests and government officials in addition to the scientists. This takes up space that should be provided for research work.

A research vessel for experimental fishing should preferably be built on fishing boat lines but, when most of the research is basic or oceanographic, there seemed to be no reason why a trawler hull should be used. In the Far East, where there is much need for basic research, boats of shallow draught, such as motor torpedo boat types, could be used in off-monsoon periods. He cited, as an example of unnecessary expense, a research vessel built on trawler lines for Indonesia. It does not have even a trawl winch, but only a few hydrographic winches which could be carried just as easily on a cheaper hull.

Mr. R. T. Whiteleather (U.S.A.): The U.S. Fish and Wildlife Service have had a lot of experience in converting all kinds of hulls into research vessels and in some cases very great difficulties have arisen. In his experience there is no question that to design and build a ship for a particular purpose proves economical in the long run, even if war surplus hulls could be obtained for only one dollar. The initial cost of the new ship is, of course, higher than that of the converted hull. But over a period of five or six years there is no doubt that a new ship is the better proposition from both the economic and operating point of view. If a research ship is wanted, he would strongly recommend building it, keel upwards, to the design of a good naval architect.

Dr. Milner B. Schaefer (U.S.A.): It seems perhaps impossible to build a research vessel to serve all oceanographic and biological purposes simultaneously. It will probably be necessary to design a research vessel not only for particular work but for operating in particular areas. For example, research vessels expected to work throughout the year in the North Atlantic must be different from research vessels designed to work in tropical waters.

Mr. Howard I. Chapelle (U.S.A.): The aim should be to define the purposes of a research vessel and try to build an ideal ship. In this case, all available information should be provided at the start and he thought that the catalogue of research vessels proposed by Mr. Kesteven should include the total cost of a research vessel.

Mr. Francis Minot (U.S.A.): The Geophysics Branch of the U.S. Office of Naval Research has compiled a catalogue of all research ships in America.

Dr. Waldo Schmitt (U.S.A.): He thought a definite effort should be made to find out what engineers, sub-officers and crew thought was wrong with each research vessel now afloat. He had worked on Navy vessels and, as a civilian, he got to know the crew better than their officers could. From the crew he had heard many complaints about vessels and engines and he thought that a compilation of such criticisms would be of value in designing an ideal vessel. The men who have to work a ship have very real practical knowledge of its faults and its virtues.

Mr. G. C. Eddie (U.K.): All research vessels have to do experimental fishing and they should, therefore, be basically fishing vessels. He spoke as one who has been associated with the design of two vessels for research purposes—one for fishing commercially and the other for exploratory fishing—and has watched the planning of three others. In more than one case the employment of consulting naval architects who, in Great Britain, do not normally deal with fishing craft, resulted in designs which were underpowered, had insufficient draught, and too much side area above the water line.

There is a tendency to make research vessels too versatile. Scientists tried to think of all the possible gear they would need and have it built into the ship. For instance, they have been known to ask for separate electric supplies of 12, 24, 110, 220 and 440 volts, and frequencies of 50, 60 and 1,000 cycles. It very often happens that, after one or two years' service, some of this excess gear has to be removed to make room for essential equipment not thought of at the time of planning. It is necessary, therefore, to distinguish between basic permanent equipment and portable temporary apparatus.

Mr. Francis Minot (U.S.A.): The purpose of his paper was to indicate what oceanographic ships should be, by determining their requirements and indicating how these requirements may be fulfilled. It also provided some standard of comparison between research ships in existence to-day and those which might be built in the future. Oceanographic and fisheries research are becoming so important and call for such specialized activities that the day of the converted ship is probably near its end and that, in future, vessels will be specifically designed and built for the jobs they have to do.

Noise control is of the greatest importance in oceanographic ships. The problem is so new, and the solution so obscure, that there may have to be many modifications of design before noise control is satisfactory. For example, the choice of machinery for the ship, and the arrangement and everything to do with the engine room, may be severely affected.

Mr. Minot referred to the oceanographic research vessel *Atlantis*, which he designed for the Woods Hole Oceanographic Institution 25 years ago. The only requirements mentioned at that time were that the ship should be able to carry a complement of scientists and scientific equipment, be able to go anywhere on the ocean and maintain herself, and also be able to move continuously at very slow speeds—about half a knot. As against those few specified requirements, the naval architect to-day is asked to fulfil a great many more.

The *Atlantis* has been repowered with a 400 h.p. diesel with fixed blade propeller as the power requirements have grown continuously since her launching. The controllable-pitch propeller, which was fitted to her originally, did 17 years service until it lost a blade during World War II. As it was a German propeller there were no means of getting replacements, so she had to be fitted with a fixed-blade propeller, and she cannot now produce the same slow-speed characteristics.

Replying to a question by **Mr. Nickum** regarding noise in research ships, Mr. Minot said it is the over-all noise level which so much affects underwater sound equipment. In other words, it is the engine room noise, the shaft noise, the propeller blade noise and the noise made by the waves outside the ship. He added that if underwater sound equipment is to be used by fishermen, then these problems concerning noise control will also have to be considered in designing and equipping fishing vessels. There has been some discussion as to whether many of these noise problems might not be solved by using a wooden ship, but there is not enough information available to settle that question.

One point concerning noise control is that, whatever the propulsion system used, the internal noise must not be transmitted externally by a completely rigid shaft. It must be broken by using some sort of flexible soft coupling. Another point is that acoustical experts are rigidly opposed to any form of reduction gear because all such gears have a noise frequency most difficult to suppress.

Mr. George C. Nickum (U.S.A.): A large steel tuna clipper was built on the west coast of the U.S.A. in 1937. The question of the noise level was then debated in great detail, as it was thought that it might reverberate through the steel hull and scare away the bait and the tuna. Experience showed that it had no effect on the fish.

Mr. W. C. Gould (U.S.A.): A steel vessel had been built in which absence of noise was required throughout the ship, except in the engine room. Much development work was done to effect noise control and the engine room itself was insulated. The engine room noise started at 91 decibels and had to be reduced to zero. The insulating consisted of 1½ in. (4 cm.) glass wool in the engine room itself and noise baffles placed in the air intakes. The air-conditioning equipment was placed in the engine room and, again, noise baffles and flexible couplings were used to suppress sound from it. The engines and the generator sets were installed on spring mountings to deal with the airborne and vibrating noises. Two Morse rubber couplings were used on the propeller shafts because a Morse coupling can take a thrust in both directions, therefore it was unnecessary to put additional thrust bearings on the shafts. The shafts were about 61 ft. (18.6 m.) long and carried intermediate bearings about every 12 ft. (3.7 m.) suspended in rubber in sheer. The owner's quarters were immediately aft of the after bulkhead of the engine room, 3 in. (7.5 cm.) from which a false bulkhead was installed, packed with two more inches of glass wool material. The walls and ceilings in the owner's quarters and the deckhouse were inserted in rubber channels instead of being secured to the steel hull directly by bolts and screws. The rubber channel method, incidentally, was cheaper than the normal method. As a result of this insulating there was never any noise noticeable at any point in the vessel except in the engine room itself. Such insulating might well be suitable for a research vessel as there was no noise externally or internally, other than normal propeller noises.

Mr. Francis Minot (U.S.A.): The present practice on oceanographic research vessels is to stop the engines when carrying out work which demands silence. This is not a practice which satisfies investigators; they only do it because they have to do it. In the future such work will have to be done with the ship under way and with machinery in operation.

Mr. D. E. Brownlow (U.K.): The underwater noise problem was probably more difficult than most others. In research ships the aim is to install the smallest and lightest engine with the greatest power, but the higher the speed, the bigger becomes the noise, especially high-frequency noise which is most difficult to deal with. Experiments in Great Britain show that combustion noises present one big problem, and high-frequency vibration in the framework of the engine, transmitted to the hull of the vessel, presents another difficult problem. Mountings and couplings and so on, are being insulated with rubber, and exhaust noises are being silenced, but there are certain high-frequency noises which do not seem to respond to insulation.

Commander R. E. Pickett (U.S.A.): A possible solution to the engine noise problem in the oceanographic research vessel might be to install a high-pressure steam turbo-electric plant similar to that installed in the T2 tankers. This installation consists of a turbine generator set driving a synchronous propulsion engine. The ratio of the number of poles on the engine to the number of poles on the generator is the reduction ratio. Speed is regulated by slowing down or speeding up the turbine. The main power plant has only three bearings and no gears, and only two bearings on the electric motor, which should ease the problem of eliminating engine noise. With a synchronous type of engine, with a squirrel cage starting line, it is possible, by using a controllable-pitch propeller, to obtain any speed control necessary. It is also possible to take off any power requirements of the ship from the main generator which, at normal speed, produces 60 cycle A.C., and is controllable down to about 15 cycles. Controllable-pitch propellers should be used with this machinery.

Another type of installation that might be suitable is the slow-to-medium speed direct reversing diesel, used in combination with an electro-magnetic coupling.

For fisheries research vessels, the uniflow steam engine with high-pressure boilers might be suitable. It could have a fixed-blade propeller as the lead and cut-off are adjustable on the engine. The fuel rate is between that of the turbo-electric plant and that of the modern quadruple expansion counterflow engine. Problems such as excessive ring and cylinder wall wear and fouling of boilers, and damage to tubes from excessive cylinder oil, have been found in practice to be capable of satisfactory solution. New cylinder lubricants and lubricating equipment, diatomaceous earth condensate filters and judicious surface blowdown seem to solve the problems.

Mr. R. T. Whiteleather (U.S.A.): One of the problems in research ships is to supply a constant voltage to the various delicate electronic instruments. When the voltage varies it upsets the accuracy of the instruments. On the *John N. Cobb*, the exploratory fishing vessel built for the U.S. Fish and Wildlife Service, the problem has been largely overcome by using hydraulic equipment for taking care of heavy loads. But in other U.S. vessels the problem is so pressing that the installation of a separate generator and switchboard panel has been considered.

Mr. W. C. Gould (U.S.A.): One way to solve the variable voltage problem is to have a 110 v. D.C. system on board the ship. A motor generator set is used, the D.C. motor being rated at 75 volt. Between the board and the motor generator set is a sensitive carbon pile voltage regulator which acts very quickly. The voltage seldom drops below 45 and it is possible to convert and run the current at 400–420 cycle A.C.

There are now small diesel generator sets available, equipped with icocronous governors, and they will keep the current within half a cycle with a voltage variation of about 35 per cent. load.

The icocronous governor increases the cost of the generator set by about $200 (£70).

Dr. B. Kullenberg (Sweden): In his valuable study of the requirements of the oceanographic research vessel, Mr. Francis Minot stated that long cores of bottom sediment are already accumulating more rapidly than they can be fully studied. A similar remark was made by Dr. E. C. Bullard at a meeting of the Society for Visiting Scientists in London in 1951. He said " It is now possible to get cores 80 ft. (24 m.) long which, in practice, represent a considerable embarrassment to their owners ". At the same time Dr. J. G. H. Wiseman stated that it is useless to take a core where there is interference from volcanic dust or glaciation or where the rate of sedimentation is small. Dr. Wiseman protested against the indiscriminate collection of miles of deep-sea cores.

Dr. Kullenberg said that, as he is responsible for the fact that cores of 80 ft. can now be gathered and as he had collected miles of the deep-sea cores complained about, he felt entitled to take part in the discussion about them. It is important to keep in mind that it is much cheaper to take a core than to make a detailed investigation of it. If a first-rate winch is available it is possible to take a core at a depth of 3,300 fm. (6,000 m.) in four hours. The Danish winch Mr. Minot referred to was developed for the Swedish Deep-Sea Expedition 1947/8 by Swedish engineers and was then transferred to the Danish research ship *Galatea*. It was most suitable for this work. The expense of a deep-sea expedition is little affected by the number of cores collected during the cruise and, of course, the cost of each core will decrease as more cores are taken. The difficulties arise when the time comes to make a thorough investigation of a long deep-sea core. Not only is an investigation expensive but it takes a lot of time, which is one reason why the cores are being accumulated. If the collection of deep-sea cores in great numbers is stopped, the alternative is to collect a few but good cores. This may sound very convincing but how is one to know that a core being taken is a good one? It is possible to select a suitable bottom topography but this may take so much time that several cores could be collected meanwhile, and one of them might be better than a core obtained from a selected site. It is difficult to choose a site where a good core may be expected. For instance, unless one is studying slumping, the site should be free from slumping, or, unless one is studying volcanology, the site must be free from volcanic ashes, and so on. Cores often display unexpected features, sometimes interesting, sometimes undesirable, and provide a valuable opportunity for research.

The difficulty in dealing with cores seems to be caused by an exaggerated idea of the value of a core and this leads to an unnecessarily thorough investigation which does not always pay. But it would be a serious error to cut down the number of cores to be collected. A new coring technique has opened a new field of research, and new methods are required to deal with the long cores now available. It is appropriate that extensive work is being carried out to obtain a general knowledge of deep-sea deposits within reach before starting research on special subjects. Until that general knowledge has been gained nobody can tell what astounding things a deep-sea core may reveal, and anybody who allows his present choice of coring stations to be limited by considerations of bottom topography, depth, the nearness of land, the influence of volcanism, etc., will miss many opportunities to make interesting discoveries.

When a great number of cores have been collected in an area it is advisable to make a preliminary examination, and this is not very expensive, which makes it possible to select cores which are best suited for a detailed investigation of physical, chemical, mineralogical and biological characteristics. The remaining cores will not be wasted. The geologist can use them in his attempts to evaluate the detailed analysis of the specially selected cores. They will give indications as to whether certain features are local or general, and they can be useful, without detailed investigation, in considering the stratification in distant cores which have been subjected to a full analysis. In this way some of the cores may reveal facts which, but for them, might have remained unknown for a long time to come.

Mr. E. R. Gueroult (France): It is important to have research vessels which are genuine fishing boats, preferably of modern design and at least as big as trawlers. They must also be able to be used as demonstration boats for fishermen. They need to be propelled by machinery which is very easy to control, so that there is no difficulty in working at very low speeds, such as from ½ knot to 3 knots. Such speeds are necessary in trawling the plankton nets and in using other oceanographic research gear. Controllable-pitch propellers seem to be necessary.

Scientific personnel, who may not be used to life at sea, must be provided with comfortable quarters as they may have to stay on board a long time and the success of their work depends a good deal on the provision of comfortable accommodation.

Instrument makers always complain that the recordings are not very good because the sensitive part of the equipment concerned is working in the wake belt of the ship. Attempts have been made to suspend the instruments outside the ship's path by means of a retractable device.

Mr. Francis Minot (U.S.A.): In the U.S.A. recording instruments are submerged and towed at a considerable distance from the hull of a research ship. This method might not be practical for a fishing boat searching for shoals of fish as it would call for the use of elaborate and delicate equipment.

Mr. R. T. Whiteleather (U.S.A.): Asked by Mr. Arthur de Fever about the sea scanner used in the *John N. Cobb*, Mr. Whiteleather replied that experiments indicate that it is a very satisfactory instrument of the asdic (sonar) type for locating not only shoals of fish but, in many cases, individual fish. It has a range of 1,600 ft. (490 m.) and is, presumably, effective throughout that range.

V-BOTTOM

Mr. H. C. Hanson (U.S.A.): The cost of labour is about 25 per cent. less in building V-bottom design boats than it is in building the ordinary round-bottom vessels. In his opinion, V-bottom design is to be recommended for use in the Far East.

Mr. G. O. Huet (U.S.A.): The question of the advantages or disadvantages of V-bottom and round-bilge boats depends on the use the vessel is to be put to. He cited his experiences in building 100 vessels of a certain type for the U.S. Army. The Army drawings showed a round-bottom hull but his employer, who is a practical man and had been an operator of boats, wanted to know why the boats should be built so and whether it was economical. Mr. Huet looked into the question and found that, instead of using the plain V-bottom design with a single chine, a better form could be obtained by using a double chine. A model was constructed of this design, together with a model of a round-bottom boat, and both were tested at the University of Michigan. It was found that for the purpose for which the boat was to be used and the speed it was to make, the double-chine type was the best. Efficiency had been increased by about 18 per cent., and there was no difference in the cost between the single- and double-chine hull because so many boats were being built to the one design: both types were much cheaper than the round-bottom hull.

In the case of building one boat there is no doubt that a V-bottom costs must less than a round-bottom design. He added that if the naval architect is given full information as to the speed of a proposed vessel, the type and cargo it is to carry, the waters it is to operate in, etc., then a V-bottom or modified straight section boat can be designed which will be more efficient than the round-bottom boat.

Mr. W. C. Gould (U.S.A.): There is growing interest in the fishing industry in the double-chine type of design which enables yards to build in steel without using bending rolls or furnacing the frames of the plates. Tank tests show that this design is as efficient in the water as the round-bottom hulls, certainly in the speed/length ratio of 1.25. He cited a case in which this firm had built an 86 ft. (26.2 m.) vessel, having a cruising speed of 18 knots. It had a speed-length ratio of almost 2.1, and several models were built and tank tested before it was decided to adapt the double-chine design. One of the advantages of the type of plating used in the double-chine vessel is that it is possible to weld the ribs to the plating before they are placed in the ship's hull. They can be placed on the forms made by the boat ends, sometimes with a few additional wooden frames, until they match the plates. They do not have to be exact entering the chines as they are welded at that point. This is often an advantage because of the waviness present in welded vessels of light plating.

Mr. Howard I. Chapelle (U.S.A.): The V-bottom design, or the chine hull, has been used for generations. The Japanese sampans, for example, have been designed that way for centuries. This is just another case of using a primitive idea in modern naval architecture and tank tests show that such an idea has advantages. Wooden V-bottom hulls are comparatively more expensive to build than the steel type.

Mr. Philip Thiel (U.S.A.): The greatest economies in building V-bottom steel hulls appear to result from the use of a longitudinal system of framing, where web frames only are fabricated from plate and bar stock by welding, and the longitudinals are flat bars sprung to shape cold. A 96 ft. (29.3 m.) vessel was lofted by one man, and entirely fabricated by one welder and one burner with no previous ship construction experience, under the supervision of a loftsman. Aside from one bow plate, no plate rolling was necessary. This vessel was built under survey of the American Bureau of Shipping and its scantlings appear in fig. 326.

MATERIAL

Commander E. C. Goldsworthy (U.K.): Hulls of vessels built of aluminium alloy to the correct specification show that after 15 years there has been little or no corrosion. The metal is in excellent condition. But a steel hull of the same age would have rusted, no matter how well it was kept painted. An important point concerning aluminium alloy hulls is that anti-fouling paint containing mercury should never be used. As a large number of anti-fouling paints do contain mercury, care should be taken to find out the actual composition of a paint before using it on a light alloy boat.

The radius of gyration of a light metal vessel is the same as for one built entirely of steel.

Mr. Wm. P. Miller (U.K.): For a long time past there has been talk in shipbuilding circles that the day of the wooden ship is almost over, but he felt confident that timber would always remain a shipbuilding material because almost every country in the world has natural home-grown stocks to draw from. It is true to say, however, that timber is not being used to the best advantage and few people recognize that it provides the basic food of a very large number of parasites. The life of the wood is measured by the number of parasites living in it. To put it in simple language, they live on the sugar content of the timber and the most common of these parasites are the death watch, toredo and gribble and those fungi causing dry and wet rot. If the life of timber is to be increased, it must be given treatment to resist the attack of these enemies. The sugar content must be reduced and, as far as possible, wood should be kept in its natural element—fresh air. When these two points are recognized in ship construction, the life of the wood increases enormously. But there have been a large number of cases in recent years where timber boats have been condemned. A study of the design shows the reason why: there were large areas of unventilated space and untreated timber.

His firm has made extensive use of the copper sulphate and a brand now being used, applied by a brush on one side only, can penetrate 1 in. (2.5 cm.) thickness of mahogany. To

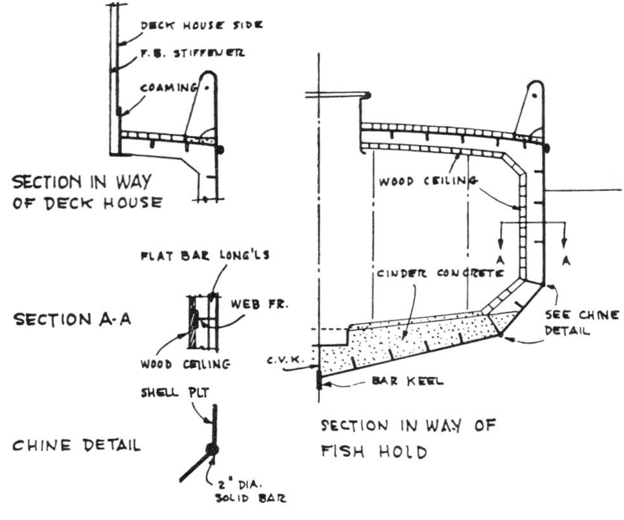

Fig. 326

eliminate stagnant air spaces, a special ventilator has been patented which forces fresh air into the bilges and other inaccessible places.

While he considered steel to be a very good material for shipbuilding, he pointed out that it has many drawbacks and is not superior to timber for building small craft. He added that there are and will be many more cases of steel being replaced by alloys and plastics than there will be of wood being replaced by steel.

Mr. S. A. Eaco (U.S.A.): The problem faced by owners in the U.S.A. was to get delivery of steel boats. It takes 18 months to get one. It seemed that few shipyards were building such vessels of good design and performance. On the coast stretching from New Orleans to Brownsville there are, perhaps, some 3,500 small or large shrimp trawlers in operation, and there is a strong desire among owners to replace or expand their fleets with all-steel boats. Would it not be possible with a system of prefabricated steel hulls? One of the problems is the high cost of construction, now about $1,000 (£360) a foot. Naval architects, builders and steel makers, should co-operate to bring down the cost of construction.

Mr. W. Zwolsman (Netherlands): The question of wood or steel belongs to the sphere of economics. The problem is: how can fish be produced cheaply as far as boats are concerned? The answer is, of course, that the life of steel-built fishing boats is longer than that of wooden boats and that, together with the low maintenance cost, makes the steel vessel a much more economical proposition.

It would be interesting to know why only the Netherlands have generally taken up construction of steel fishing boats. Countries such as Belgium, Germany, France and England, for the economical reasons stated, long ago gave up wooden constructions of harbour lighters and tugs, etc., which are used in great numbers, but their fishermen have not turned to steelbuilt boats, at least not in the smaller sizes.

Another advantage of the all-steel vessel is that deformation does not take place in it. It has been stated that wooden vessels built after the war have not shown deformation over a period of six to eight years and that it has not been necessary to correct the alignment of the engines, but an interesting fact is that many French marine engines are provided with a flexible coupling as the manufacturers consider this necessary to cope with the deformation of wooden vessels.

One of the first criticisms of steel boats is that they are subject to rusting. This objection was raised by the Dutch fishermen in the years 1914/1920 when small steel fishing boats were first introduced. But experience soon showed that these objections were not valid. The vessels, varying from 50 to 80 ft. (15 to 24 m.) and equipped with 80 to 300 h.p. engines, fish usually from Monday to Saturday. On arrival in harbour a member of the crew walks around the ship to see if there are any spots of rust. If there are, they are brushed with oil or coal tar. The latter is a cheap and effective material and is still much used on the underwater section of hulls of boats in the Netherlands.

Mr. Zwolsman then gave the following information regarding the price of steel vessels to facilitate comparison with wood. The price for the steel hull does not include woodwork or inventory, but does include built-in fuel and drinking water tanks, superstructure and steering gear. The types refer to Table XXXVIII of his paper.

TABLE XLVIII

Fig.		*L.o.a.*	*B.*	*D.*	Prices *Fls.*	U.S.$	£
221	ft.	49.2 ×	13.4 ×	6.4	20,000	5,291	1,878
	m.	15.00 ×	4.10 ×	1.95			
222/3	ft.	54 ×	15.4 ×	7.1	26,000	6,878	2,441
	m.	16.45 ×	4.70 ×	2.15			
224	ft.	59 ×	16.7 ×	7.5	34,000	8,995	3,192
	m.	18.00 ×	5.10 ×	2.30			
225	ft.	62.2 ×	17.4 ×	8	40,000	10,582	3,756
	m.	19.00 ×	5.30 ×	2.45			
226/7	ft.	69 ×	18.4 ×	9	54,000	11,640	5,070
	m.	21.00 ×	5.60 ×	2.75			
228	ft.	74 ×	19 ×	9	60,000	15,873	5,633
	m.	22.50 ×	5.80 ×	2.75			
229	ft.	79 ×	19.7 ×	9.8	72,000	19,047	6,760
	m.	24.00 ×	6.00 ×	3.00			

The cost of the upper part of the pilot house, the masts, panelling the crew's quarters and boarding the fish-hold, must be added to get the price of the complete hull. In the case of bigger vessels the cost of the wooden deck should also be added, but in smaller boats the steel deck is made of plate and is already included in the price.

Mr. Philip Thiel (U.S.A.): The deck of a trawler is often wet, and the lower part of the deck house sides is thus subject to a great deal of corrosion. A practical solution is to employ a 12 in. (30 cm.) coaming of a heavier plate at the deck, with the main deck house plating lapped inside of this along its upper edge. A light continuous weld on the outside and an intermittent weld on the inside are sufficient for attachment. Fig. 326 shows this detail.

Wooden vessels are better off with a steel deck house or, at least, a steel trunk over the engine room. In some cases this has proved cheaper than a wood house; and in any case it does reduce the problem of leaks into the engine room.

He regretted that in some wooden vessels the worthwhile practice of extending the stern post up to the deck is not followed. By pinning the head of this post into the adjacent deck beams, as shown in fig. 327, greater transverse stiffness of the after-end of the hull is assured. Check pieces extending past the post on both sides stiffen the counter framing. It is felt that if this practice was followed many of the mysterious " stern post " leaks of wooden ships would not occur.

Foundations for heavy diesels in wooden vessels are always a problem: here again steel permits a more satisfactory solution. It is important that a steel engine foundation be well bedded to the keel, as this will use the stiffest longitudinal member of the hull to advantage in preserving shaft alignment. In the case of some 132 ft. (40.2 m.) wooden trawlers the steel engine foundation girder plates were extended from the engine bulkhead well up into the horn timber. Aft of the engine the girders knuckled in to the sides of the shaft log, and were moulded to the bearding line. The thickness of the girder plate was increased where its depth was reduced. Through-bolting locked the entire stern assembly into one unit with the engine. Fig. 327 shows such an arrangement.

LAMINATION

Mr. John Tyrrell (Ireland): The plant and equipment necessary for laminated construction on any considerable scale would be far too costly for the resources of ordinary boat building yards. Orders normally received for new vessels do not warrant much expenditure on capital fixtures, especially as there is usually little standardization in size or shape of fishing boat hulls built at any one yard.

Mr. Howard I. Chapelle (U.S.A.): The U.S. Navy is making extensive use of laminated structure but it is not an economical method of construction so far because the specifications call for timber of superior quality. It is necessary to use low cost wood and depend upon preservatives to give it a long life, if laminated hulls are to be produced cheaply. The strength of lamination is very difficult to ascertain but, then, an analysis of the structure of wooden vessels is not easy.

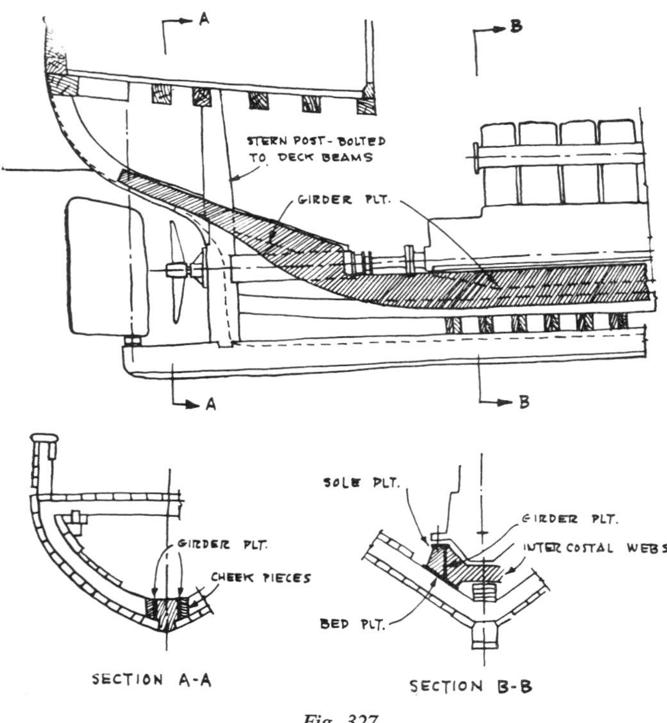

Fig. 327

Mr. G. A. Huet (U.S.A.): His firm are building about 30 minesweepers which have laminated construction of the frames, the stem and the keel. It was thought, at the start, that the cost was going to be very high but it is now found that it is actually less than it would be if these vessels were built by the sawn-frame method. One reason for this is that the cost for forms, clamps, bending slaps, heating equipment, moisture recording devices and so on, is distributed over the 30 boats. It is possible to use small sections of wood, that have no defects, for laminated construction. In sawn-frame construction, the strength depends on the fastenings, which are not needed in laminated construction. Again, with sawn-frame construction, it is impossible to avoid some cross-grain timber, but with laminated construction all the timber used can be straight-grained. Another advantage with laminated construction is that the risk of drying out the wood, as may well happen in steam bending, is eliminated. There are times in steam bending, when it is not possible to control the extent of the drying and the wood then becomes more subject to rot. Preservatives are, of course, being used in laminated construction to extend the life of the material.

The problem to-day in wooden ship construction is that of finding suitable timber. It is becoming scarcer and scarcer all over the world and that which is available is usually of mixed quality. The result is that, if two or more boats are built of the same grade of timber, one may be subject to rot in two or three years, while the other may remain in excellent condition for, say, twenty years. He cited the example of boats built in New Orleans 20 years ago that are still intact and without a trace of rot, but, on the other hand, there are boats, built at the same time, in which the keel or the chine or some other part has rotted. In view of this inequality of the timber available, laminated construction is a distinct advantage.

Mr. H. C. Hanson (U.S.A.): He designed laminated construction on high-speed vessels twenty-five years ago. But he was not impressed with laminations and would like to know how a laminated frame could be considered good when, if it is dropped from a height of 5 ft. (1.52 m.) it breaks into three separate pieces across the laminations. There is also the case of the laminated wooden keel which broke when lifted. In his view, the U.S. Government should not have spent untold millions of dollars on a plant for producing minesweepers constructed of laminated materials. They should have built an experimental vessel first. As it is, they have gone ahead with the building of a whole fleet of boats which, in his opinion, are likely to rot in a few years.

If laminations are not done properly the material becomes brittle and breaks just like glass. If there was a way to leave the moisture content in the timber when it is glued, it would be a different matter, but even then rot would set in.

Mr. Jarl Lindblom (Finland): He agreed with Mr. Hanson that, under certain circumstances the strength of lamination may be open to question. Oak, for instance, is not easy to laminate. Experiments carried on in Finland have shown that the Brazilian wood, peroba, can be laminated fairly easily but, of course, the best material for laminating is a softwood, such as pine. He agreed with Mr. Huet that for the bigger scantlings it is necessary to use laminations. In building the schooners for Russia natural wood was tried for the keels, which are 100 ft. (30 m.) long, but it was found to be impossible because, when a log had only a small defect, it was condemned. Tests showed that laminated frames with no glue between them were stronger than frames of ordinary construction. The laminated keel was found to be 30 per cent. stronger and about double the sitffness of a solid wood one.

Dr. C. A. Rishell (U.S.A.): His experience of laminating or glueing wood together for boatbuilding started 12 years ago when the U.S. Navy was in urgent need of laminated construction. In those days there were no suitable glues and they had to be developed. In the case of bending oak, in ordinary construction, only one out of 200 trees produced timber of the quality the Navy demanded, but now, due to laminating, it is possible to use wood from every oak tree.

During the Korean War, the U.S. Navy needed to build very large minesweepers. These could not be constructed of metal because of new types of mines that had been developed, and his firm was asked to carry out laboratory research with the object of producing a 70 ft. (21.3 m.) frame to withstand a pressure of about 19,000 lb. per sq. in. (1,350 kg./sq. cm.). One frame was laminated of fir and the other of fine oak, and they were both put into the testing machine. The oak frame finally broke at 56,000 lb. (3,900 kg.), roughly three times the strength required. To-day timber may be laminated to any size. It is also possible to treat timber, then glue it together, and re-bend it, so that it has a great deal more strength than could ever be given it by bolting.

Mr. George C. Nickum (U.S.A.): He asked Dr. Rishell two questions. The first concerned the elasticity of the wood after drying and before glueing. Mr. H. C. Hanson had seen laminations broken by being dropped and so had he. The fracture was very similar to that which occurs in cast iron and evidently the laminated frame was very brittle. He would like to have evidence or data on the modulus of elasticity in oak after drying or in the finished laminated frame. He would like to have test data of the relation between the elasticity of wood and/or a lamination of wood and the moisture content of it.

The second question was concerned with the life of laminated wood. Had Dr. Rishell any experience over a period of time of the life and resistance to rot of laminated fir and oak in actual service in ships?

Dr. C. A. Rishell (U.S.A.): Timber can be glued successfully up to a moisture content of 20 per cent., and it is thought possible to go further than that. Wood used for making furniture was once dried to about 4 or 5 per cent. in order to make effective the glue used in those days, but the modern type of glue in marine structures would not be effective on wood dried below 10 per cent. Another important point is that timber has greater strength at 10 to 15 per cent. moisture content than it has at 40 or 50 or 100 per cent. It may be slightly more brittle but that should give no trouble if the moisture content is kept up to about 10 or 12 per cent. and there is evidence to indicate that elesticity is actually higher if the wood is dried to 12 to 15 per cent.

The timber is normally kiln-dried or, preferably, air-dried. When the moisture content is brought below 20 per cent. all the organisms which cause the decay in the wood are killed, therefore, in laminated structures the boat builder is using sterilized material. New types of chemicals are now being used in treating laminated material and they should help to increase considerably the life of a boat. It is estimated that the laminated boats built for the U.S. Navy will have a life of 12 to 14 years. With proper treating of the frame and other sections, with the new preservatives, the life of a boat will probably be increased to 25 to 30 years.

Mr. Jarl Lindblom (Finland): Experiments in Finland on the question of moisture content led to the conclusion that 19 per cent. was very safe and very good and that the strength of the glue joint was actually higher at that figure than with humidity at 8 to 16 per cent. Later it was found possible to laminate wood with as much as 29 or even more than 30 per cent. of moisture content, and tests on the strength of these laminations gave splendid results.

Dr. C. A. Rishell (U.S.A.): He would not recommend a moisture content as high as 30 per cent. and he thought that the best adhesion takes place when the moisture content is between 12 and 20 per cent. In glueing, the critical spot is the surface of the wood and it is there that the moisture content is measured, not in the interior. If the surface is overdried, the water in the glue is absorbed by the wood and that does not allow the glue to have proper adhesion. As long as the surface of the wood has the right moisture content, then there is no need to worry about the interior moisture content.

Mr. Howard I. Chapelle (U.S.A.): One of the reasons why laminate frames are brittle is because they are very long and their cross section in proportion to the length becomes quite small, although, for structural purposes, it is more than ample. Speaking of his experience of the problem of laminated construction in the shipyard, he had found difficulty with laminated keels in small craft. They did not come through in exactly the same dimensions so that when the boat was planked many hours were lost in compensating for these differences. Such differences could probably be controlled by proper manufacture but, at the present time, it is a problem that some shipyards, building small craft, have not overcome.

With regard to fishing boat hull forms and fishing boat types, generally, laminated structure seems, theoretically, to offer important solutions to material problems. But, as things are at present, lamination adds to the labour cost, thereby raising the total cost of construction, unless a mass production system is used. A great deal of emphasis, perhaps undue emphasis, is being placed on the enormous strength obtainable through laminated structure. The question is: how much strength is needed? It is quite possible that these laminated structures are being made much too large. He referred to the U.S. Navy specifications which called for the use of very fine timber for laminate structure, another factor which adds to the costs and also increases supply difficulties. He thought that, perhaps, such stiff specifications are not necessary and that it might well be possible to use poorer grades of timber, a very important point if, in the future, laminated structure is to be used in building fishing vessels.

Many of the failures of laminated construction are not due entirely to the laminates or laminate technique. They may be due to the improper design of the structure. Too much is expected of it and, again, most boats have, or are always supposed to have, a certain amount of flexibility, but if a boat is built entirely of laminated material, it is a very rigid structure and, therefore, subject to damage by shock. Perhaps laminated materials should be used in shorter lengths.

HANDLING FISH AT SEA

Commander E. C. Goldsworthy (U.K.): Fish must be brought to the consumer in first-class condition, and it does not matter how good the fishing boats are or how skilful the fishermen are if, at the end of a voyage, the catch must go to the fish meal plant. That is wasteful, raises the price of good quality fish, impoverishes the owner and frustrates the effort to increase the quantity of food for human beings. For these reasons it is desirable to study experiment with and use new materials in constructing fishing boats, and in transporting the catch, as an aid to keeping fish in prime condition for the consumer. In Great Britain much thought has been given to the use of aluminium, plastics and other new materials, and many experiments have been made with them.

Mr. W. A. MacCallum (Canada): The fishing ports in Nova Scotia, Newfoundland and New England are two to two and a half days from the fishing grounds, therefore fishing trips may be limited to six to nine days. The industry concentrates on landing iced fresh fish which can reach consumers in good condition. With some improvement in handling aboard and ashore the product can be landed considerably cheaper than from freezer ships. It is felt that this method can be used as long as sufficient fish are available on the present fishing grounds. If this situation changes, factory ships may be called for.

Mr. Mogens Jul (FAO): When small fish, such as mackerel, herring, etc., are caught they are normally immediately iced in the holds. Larger fish, such as cod, salmon, halibut, etc., are usually gutted first. Government regulations in Norway require that every white fish must have the main artery cut the instant it is taken aboard; if the fish are to be sold fresh or frozen, they must be eviscerated, as soon as possible thereafter, preferably within one hour of capture. Then the fish must be washed, preferably in ice water or in running sea water, then placed in boxes and iced if the air temperature exceeds 46 deg. F. (8 deg. C.). Therefore, a Norwegian purse seiner fishing for cod must be provided with a grating for throat cutting, 26 in. (65 cm.) above the deck, and three or four bleeding ponds, where the fish are bled for at least 15 minutes.

On British trawlers, the gutted fish are nowadays often washed in a fish washing device, which consists of a trough filled by powerful water jets. An overflow takes the fish to a grated spillway over which they tumble through the hatch into the fish-holds.

In the holds, the fish may be boxed, with ice, or placed in pounds. The pound boards on British trawlers are normally 4 ft. × 8 in. × $1\frac{1}{2}$ in. (1.22 m. × 20 cm. × 3.8 cm.). A trawler may carry as many as 3,000 of them. The fish are placed on layers of ice. There must be a layer of at least 1 ft. (30 cm.) ice between the hull and the fish. The ice-fish ratio should be about 1:1 under tropical conditions. The pounds in Icelandic trawlers have a size of about 70 cu. ft. (2 cu. m.). During the first days of fishing two parts of ice are used to three parts of fish. There must be a free space below the bottom pound through which the ice water may drain away.

If fish are iced in thick layers weight losses may occur, while fish iced in thin layers, on single shelves or in boxes, may gain slightly in weight. It is for this reason that on Norwegian cod purse seiners the distance between the shelves in the fish-holds may not exceed 28 in. (70 cm.), except for the upper shelf where it may be 43 in. (110 cm.). The fish may be stowed in boxes instead of pounds, but bulk stowage is not allowed. Gratings on the floor of the hold to permit drainage of blood and water are also required.

Until recent years the interior of the fish holds was normally made of tongued and grooved wood, and the pound boards were of wood. This made cleaning difficult and caused excessive penetration of fish slime and ice water into the walls. This was particularly bad for insulated holds, as it resulted in rotting of wood and insulating material and loss of insulating properties.

Nowadays, the walls are often metal lined, frequently with an alumimium alloy. In construction the first plates are normally screwed to the wood, the following plates overlap the screws and are welded on. Transverse bulkheads are attached to T-sections of the lining. Similarly, pound boards, stanchions, etc., are now often made of aluminium alloy.

The pamphlet " Care of the Trawler's Fish " suggest that a heater be provided aboard trawlers to heat sea water, or town water when the trawler is in port, to nearly boiling point, for cleaning the deck and fish-holds. Steam may also be used for this purpose.

Prof. H. E. Jaeger (Netherlands): It is to be hoped that, in future, there will be more hygienic handling of fish and that no wood will be used in the construction of fish-holds.

Mr. G. Eddie (U.K.): The workers at the Torry Research Station have been getting a clearer idea of the relative importance of cooling, handling and hygiene in the preservation of fish in crushed ice. A series of observations on board trawlers have been supported by laboratory experiments. The work is not finished but it is clear that the most important factor is temperature. Whether anything more than washing the fish with ordinary care has a significant effect on stowage life has not yet been determined. But it seems quite practicable to keep a wooden fish-room in a satisfactory hygienic condition, although, naturally, more time and trouble must be taken to do this than in the case of metal fish-rooms. Considerations of durability, weight, convenience and saving of labour, rather than of preservation, should govern the choice between wood and metal for fish-rooms. The latest experience of the Torry Research Station in these matters is recorded in a new edition of the pamphlet " Care of the Trawler's Fish " (Cutting, Eddie, Reay, and Shewan, 1953). The main points brought out are: (1) The fish must not be allowed to accumulate on deck and must be gutted as soon as possible. Many British trawlers now use a washing machine to maintain a steady flow of fish into the hold. (2) The fish must be cooled quickly and kept cool, and crushed ice is the only tested method of doing this. Special precautions are necessary to prevent entry of heat through the sides and deck. Insulation or refrigeration or a combination of the two may be used for this purpose. It has been demonstrated that a catch can be satisfactorily cooled, even in a wood-lined fish-hold without insulation, by building a wall of ice against the linings, so that in Arctic trawlers insulation is only an extra precaution, and an expensive one. Some owners are now taking an interest in the method described by Mr. MacCallum, of circulating refrigerated air between double walls, a system which is coming into commercial use in Canada, and experimentally in the United Kingdom, for cold stores. An advantage is that, besides keeping the insulation dry, it allows the relative humidity inside the refrigerated space to be kept high. In this respect the jacketed cold store seems better than the Minikay system cited by Mr. Bain. It is also simpler. But the best use of mechanical refrigeration seems to be to pre-cool the holds and fittings and keep ice in an easily handled condition. Although many skippers do not start the refrigeration machine until the first fish is caught, it seems better to use mechanical refrigeration on the outward voyage and to rely mainly or entirely on ice during the rest of the trip. (3) Because of the pressure caused by the weight of the fish, losses in weight average about 8 per cent., with a maximum of about 16 per cent. on the trips of Arctic trawlers, which take about three weeks or so. These figures refer to shelf spacing in the vertical direction of 3 to 4 ft. (0.9 to 1.2 m.); closer spacing is advisable.

Mr. W. A. MacCallum (Canada): Experience on Canada's east coast, unlike that in Britain, reported by Mr. Eddie, is that wood hold linings are for the most part slime sodden and carry large populations of facultative anaerobes within a few weeks after the annual refit and thereafter throughout the ensuing months. The most careful washing is of no avail whatsoever in decontaminating these holds. Whenever fish lie in contact with porous, scarred, slimy wood surfaces for a few days (steak cod in this environment for as short a time as four days may be unfit to market) they develop strong bilgy odours which may sometimes be found throughout both fillets cut from the fish. The same cleaning procedure, however, has proved to be very effective in reducing the contamination in our metal-surfaced holds. It is an exception to find bilgy fish against a metal surface at time of discharge (MacCallum, 1954).

As long as wood-lined fish-rooms are employed, extreme care should be taken to ensure that crushed ice separates all fish from wooden surfaces. One of the methods to accomplish this which rules out the human factor is to use screens over contaminated surfaces (MacCallum, 1954).

The Canadian viewpoint is to consider ice as (1) a means of refrigerating the fish, and (2) a means of shielding the fish from contaminated surfaces and from one another. Moreover, the choice of metal or wood in the fish-hold must be considered from the standpoint of preservation of the catch, as explained above, as well as for savings in ice required for stowing and shielding the fish, for convenience, for saving in labour and efficiency in refrigerating the fish-hold by artificial means.

Mr. G. C. Nickum (U.S.A.): The jacketed refrigeration hold described by Mr. MacCallum is likely to be used for transportation of carcasses. The great advantage of the jacketed hold is that the circulating cold air can be kept away from the product, dehydration is minimized, and discolouration of the flesh of the fish is avoided. But installation of a jacketed hold in a vessel is very expensive and its value for iced products such as fish is doubtful.

Mr. W. A. MacCallum (Canada): The value of metal surfaces in Canada's east coast trawlers is not at all in doubt. The reasons for the use of metal have been given above. Granting this, and whether we use a closed jacket system of air circulation as in the *Cape Fourchu* and in the *Cape Scatari* or the closed jacket system integrated with unit pounds, or an open jacket system which also could be employed with the unit pounds or with metal boxes used to hold the iced fish, we have, in supplying and installing this metal, contributed to most of the cost of a *wholly refrigerated, metal surfaced fish-room.* Therefore, the actual cost of refrigeration is not expensive, as stated by Mr. Nickum.

Mr. Mogens Jul (FAO): Refrigerating coils in a fish-hold with a conventional refrigerating system might be 1 to 1¼ in. (2.5 to 3.1 cm.) in diameter and spaced with 5 to 6 in. (12.5 to 15.2 cm.) between centres under the deck. If coils are needed on the sides, they might be a little further apart. One or two coils may be run down the shaft alley and two or three around the hatch coaming.

Landmark (1950) estimates that a trawler with 15,900 cu. ft. (450 cu. m.) holds needs about 100 tons of ice, but only half that amount, if it is mechanically refrigerated. This gives a refrigeration need of about 16,000 B.Th.U. (4,000 Kcal) per hour. He also gives diagrams for refrigeration requirements under various conditions. Ofterdinger (1950) on the other hand feels that a trawler with a 7,060 cu. ft. (200 cu. m.) fish-room should have a 24,000 B.Th.U. (6,000 Kcal) refrigeration installation.

Lantz (1953) has experimented with systems whereby the dressed fish is placed in stainless steel sea water tanks. The sea water is chilled by mechanical refrigeration to temperatures close to freezing. He considered this method of stowage considerably superior to ordinary ice stowage, one advantage being that all bruises on the fish are avoided.

Mr. L. Fernandez Muñoz (Spain): Perhaps a fish-hold capacity of 3,500 cu. ft. (100 cu. m.) is most convenient for northern European trawlers, but in Spain, where the trawlers have to go longer distances from port to fishing grounds, a capacity of 6,600 to 7,700 cu. ft. (150 to 170 cu. m.) is thought to be the most economic.

Again, the problems connected with refrigeration of fish-holds are different in Spain from those of northern countries. For one thing, temperatures in the ports are different, and Spanish fishermen also go to African waters where the temperature of both air and water is much higher than in the northern seas, so that the problem of insulation of fish-holds needs special consideration. Deck insulation must be heavier than that on the sides because of air temperatures of about 104 to 113 deg. F. (40 to 45 deg. C.).

Another problem has been the stowage and carrying of ice. About 90 tons of ice are necessary to preserve 150 tons of fish, and as the ice melts the change of weights has a very adverse effect on the trim and stability of the trawler. But a solution to this problem was reached some years ago by using an auxiliary refrigeration plant solely to take care of transmission losses. This is most effective and made it unnecessary to carry more than 35 tons of ice for preserving 150 tons of fish.

ECONOMICS

Mr. Seamus O'Meallain (Ireland): The papers by Mr. C. Beever and by Mr. Christensen deal with two different sets of problems.

The problem presented by Mr. Beever is a fundamental one: how can capital be put into the catching side of the fishing industry to bring about a large increase in production within a relatively short time, especially in under-developed areas, without over-capitalization or over-specialization which could result in a fall in production, or tying up capital which could be better used elsewhere?

Urgent reasons are given against an attempt to effect a revolutionary departure from existing practice. A sudden expansion of fish supplies to a market, even in highly developed areas, usually causes a serious dislocation, sometimes with the effect of driving fishing vessels off the sea. Again, a sudden increase in catching power in an under-developed area might be followed by a decline to a lower level than had existed before the " improvement ". Many of the factors mentioned, and some not mentioned, apply equally to both highly-developed and under-developed areas. A great majority of vessels operating in European countries and in North America are owned by the men who work them. Such vessels seldom exceed 70 ft. (21.3 m.). This is a healthy condition. Boat-owning fishermen can derive some sort of livelihood from fishing even in a condition of depression in the industry, when large trawlers might well have to be laid up. Mr. O'Meallain expressed his belief that there is an advantage to be gained in having in any area a reasonable wide range of vessels, some perhaps too expensive to be owned by the ordinary working fisherman, providing ways can be found to prevent large vessels driving the small ones off the fishing grounds. Unfortunately there is often over-capitalization to be found among owner-fishermen. The big industrial fisheries, being more objective in their approach, do not usually tie up capital in unproductive assets. Owner-fishermen, compelled by a number of motives, including some which are not economic, tend to invest in craft, engines and equipment out of proportion to the return that might be expected. In addition to the provision of education and training and improvement of fishing craft and technique, as suggested by Mr. Beever, Mr. O'Meallain added two recommendations. The first is to encourage general purpose fishing so that boats can be employed profitably throughout the year, and the second is to discourage over-capitalization.

Dealing with Mr. Christensen's paper, Mr. O'Meallain pointed out that costing is not a means of providing a historical account of how the cost of a particular boat has worked out, but it is a part of the process of production. It is an investigation or analysis to establish the cost of the boat in advance of its production, and cost control is the process of verifying that the costs incurred during the construction are not excessive. Cost control must be applied at all stages of construction so that any falling off in production or increase in expenditure may be corrected without delay. Discussing the way in which Mr. Christensen divides materials into sub-groups, he disagreed with the method. It is most important that the cost control should proceed step by step with the construction of a boat. Materials should be subdivided, not as Mr. Christensen indicates, but in accordance with the actual boat sections. If this is done it is easy to check progress once each week or at even shorter intervals, and to do it without any complex calculation. For instance, if a particular section should be completed in four days but the actual time taken is longer, there is a loss. If it is completed in less than four days, there is a gain. This process is common in civil engineering works where operations are carried out on a time schedule along the lines mentioned.

An essential part of control is the ordering of materials. Ideally, they should be ordered in such sequence and at such time as to ensure delivery at the time they are to be used. This seldom can be done in practice because of the danger of a hold-up or non-delivery of any one item. On the other hand, ordering can and should be balanced. There is little point in having in stock three years supply of half-round galvanized iron and only three days supply of nails.

Treatment of overheads is another very important matter and should be applied to sections or departments of construction. Allocation on a percentage basis, although approximate, gives sufficiently accurate results.

Many boat builders think there is enough book-keeping in boatbuilding as it is, without adding to it. Mr. O'Meallain observed that, while a proper system of costing requires expert advice for its installation, it actually can have the effect of reducing clerical work. All that has to be done on the costing of the production side is to check progress by stages, a job that can be done on a simple numbered card.

Mr. Wm. P. Miller (U.K.): Economics in boat building is most important to the boatbuilder. The Scottish Fishing Boat Builders' Association have given this matter considerable thought and have taken action. Most small boatbuilders are practical men, highly efficient in their own trade, but with no office or business experience, and usually with very few office facilities. This being so, the Association decided that a simplified scheme of costing was essential and, after some experiment, a scheme was adopted and is now in general use. Standard hull specifications, much in excess of Lloyds and Bureau Veritas, were drawn up for the various lengths and types of boats. A costing experiment was tried on a tonnage basis but was unsatisfactory. A scheme for cubic measurement was then adopted and this has been in use for the past 15 years.

The boat-building trade was compelled by the trade unions to adopt a payment-by-results scheme, which meant having a fixed wage rate. This rate is not constant on all lengths of boats but varies and is controlled by graph. The smaller the boat, the higher the wage per cubic foot. Graph measurement is also used in dealing with the timber situation.

It is comparatively easy to calculate the cost of fastenings and other small items on the cubic foot basis. Standard outfit specifications have been drawn up and price calculation is fixed on a cubic foot basis. It is now a very simple and quick calculation to arrive at the price of a given size of hull.

Mr. John Tyrrell (Ireland): The lack of standard designs is one of the greatest obstacles to the reduction of building costs. The individualism of the ordinary fishing boat owner is to a large extent responsible for this. He suggested that there might be standardization of fishing boats of 50, 60 and 70 ft. (15.2, 18.3, 21.4 m.) lengths.

One serious problem faced by the boat builders is to keep in full employment a trained and skilled staff. The absence of a steady flow of orders creates the difficulty. The ideal staff consists of men who have started as apprentices in the yard and have continued in employment there because these men will have absorbed fully the established routine and will be thoroughly familiar with the plant and machinery. Such men are able to get the fullest benefit from their machines and tools, and from the management's point of view, all the men are known thoroughly and their individual capabilities are fully assessed. The most suitable man for a particular operation, therefore, can be picked without hesitation. If, through lack of work, any of the men have to be laid off, the whole balance of the team is upset. Furthermore, the men laid off sometimes leave the district and are permanently lost to the shipyard.

Mr. George C. Nickum (U.S.A.): There will be a decline, perhaps a very severe one, in the shipbuilding industry unless costs are reduced. Costs require a great deal of detailed study. The average fishing boat owner and the average small boat builder know, usually, only the total cost of the vessel concerned, but there should be a careful investigation into the cost of each item. It is very easy to add some gadget which, in itself, does not have much effect on the price, but it often happens that the gadget leads to an increase in other costs because, perhaps, a little larger power plant is now required or a bigger pump is made necessary by the presence of the gadget. One of the reasons why the building costs have gone so high is because nobody has considered all related costs and what are thought to be desirable improvements. If these so-called improvements were considered in such a way a lot of them would be left off and that would bring down the price of boatbuilding.

Mr. Howard I. Chapelle (U.S.A.): In many small boat yards, and even in some fair-sized shipyards, there is an old-fashioned approach to costing. The people concerned know the final cost of the boat but they do not know the cost of each operation, which makes the design of a low-cost boat very difficult to do. If a complicated set of drawings is presented to one of the smaller yards the cost of building the boat is likely to go up because the builder realizes that his supervisory personnel will have to stop work and study the plans. For this reason simplicity in design and in construction is a most desirable feature.

Mr. Jan-Olof Traung (FAO): Sometimes when a boat yard quotes for two equal boats, one drawn by a " cheap " naval architect, delivering a few sketchy plans, and the other by an " expensive " one, giving lots of detail drawings for easier work, it gives a higher price for the last design. Some boat yards do not yet appreciate the value of clear distinct plans and specifications.

Mr. G. O. Huet (U.S.A.): It is very hard to determine costs but the designer knows that the more details he puts into his plans the higher the cost is going to be. If the designs can be simplified then the cost must come down because cheaper labour can be employed to do the work.

He recounted his meeting with a fisherman in a shrimp boat. On talking to the man about his shaft log, Mr. Huet found he had a pipe and on the offside of the pipe, in the stern, he had a large oak bearing made out of wood. Inboard he had a live oak bearing and over the pipe he had a piece of raw hide, wrapped around and tied with string, to keep the water out. When the hide wore out he merely replaced it with another piece. For his throttle arrangement he had a piece of string and wire going from the throttle past the boathead. There was a loop in this wire into which he put his big toe to operate the engine. This fisherman had a boat that was more economical to build and operate than anybody could give him, although he might not have known it. The abilities of such fishermen should be recognized and their opinions should be respected.

Mr. John B. Bindloss (U.S.A.): In his opinion the cost factor has been over-stressed. He did not believe any person operating fishing boats is so cost-conscious as not to want improvements or new designs. Fishermen, like others, have to be shown and part of the work of architects, designers and builders should be to show the fishermen the way to improve their boats.

Mr. C. Beever (FAO): Credit schemes are usually designed in an attempt to meet the special economic problems which the fishing industry has to face. Mr. O'Meallain referred to the danger of over-capitalization. Mr. Beever agreed that his paper was much too brief and sketchy to contain more than a passing reference to the problem and in a very broad context. He was thinking of what has been called the " opportunity " cost of investing in a particular enterprise—for example deep sea trawling operations based on a hitherto unexploited and unpredictable resource. In other words, what alternative and possibly more attractive " opportunities " for investment could have been utilised with the same capital expenditure, e.g. small-craft mechanization.

However, Mr. O'Meallain also had in mind the temptation of the owner-fisherman to invest too heavily in his own particular type of fishing which is especially great when easy credit facilities are available. It is a very big step from employed fishermen to owner-fishermen or from owner of an old, long-depreciated craft to owner of a new craft of modern construction whose capital cost, depreciation, insurance and maintenance must be met out of current earnings. The fisherman may not be prepared for such a jump—his enthusiasm for a new boat and his confidence in his own fishing skill may warp his appraisal of the financial risks involved, the legal and managerial problems to be considered and in many cases he requires expert guidance. Where grants are awarded he may be apt to overlook that these, nevertheless, represent part of the capital cost of the boat; he may underestimate the cost of engine and hull maintenance; he may never have paid comprehensive insurance before; finally he may base all his calculations on the results of a successful trip or at best the earnings of an average trip. He is not always alone in this kind of fallacious reckoning, nor in his embarrassment when the " average " trip eludes him for a few weeks and he cannot keep up his repayments.

In the discussion on engines, Mr. T. S. Leach makes some cogent references to the disastrous costs of engine-breakdowns which are out of all proportion to the isolated cost of repairs. This is particularly true in the case of the craft newly-built on credit. Manufacturers' guarantees never cover the costly loss of fishing time, deterioration of catch, towage charges, agents' fees, labour, crews' pay, provisions, fuel and so on, that a mechanical breakdown incurs.

Mr. Beever did not know what could or should be done to reduce this risk, but it is one which should be carefully appraised both in supplying and taking financial aid. There are many other kinds of risk of course, but this one is selected chiefly because it is especially relevant to the question of fishing-boat efficiency and because it exposes so many of the problems which the owner/operator of a fishing boat has to face and which should therefore be considered when launching any kind of building programme by means of easy credit.

There is the question of ownership itself. There is so much more to successful ownership than the ability to catch fish and the fisherman often has little time, aptitude or inclination for the business side of his venture. A little guidance in legal and business matters can be very helpful indeed if applied with the necessary tact and understanding. It is equally important that the responsibility of ownership shall be fulfilled so that credit is not misused. " Easy come, easy go " is true in fisheries as elsewhere and may result in negligent maintenance and skimpy repair-jobs. In this connection the relative merits of mortgage and hire-purchase arrangements might well be considered where the loan is secured by a first mortgage on the boat, lengthy legal processes may be involved if foreclosure is applied and the boat may be laid up for a long period serving no useful purpose and probably deteriorating. Under an arrangement where possession only is given and legal ownership is contingent on repayment of the loan, it may be possible to exercise more effective supervision. Legal and other conditions vary but the aim should be to protect the investment adequately while recognizing the special circumstances of the fisherman.

This may dictate some flexibility in the timing and method of repayment. Fixed sums at fixed intervals may be out of line with fishing practice whereas repayment by share of catch earnings may reduce the incentive of some fishermen to get their boats paid for.

In connection with financial assistance schemes, it should be remembered that these, while they may have been designed to assist the fishermen directly, can also assist the boat-building industry. New construction, regular maintenance and repairs which are made possible by easy credit facilities, also promote more stability and efficiency in the boat-yards. This may be an optimistic view to take and no doubt it has not been confirmed by experience everywhere, especially where the increased demand for boats may have caused prices to rise. However, it is a consideration which should be taken into account. Financial assistance schemes should be designed with an eye to the boat-builder as well as the fisherman if the maximum benefits are to be secured. In practice, this would probably mean more joint consultations with boat-builders and fishermen, in regard to the flow of work deriving from credit facilities, the recognition of " standard " types evolved commercially, the cost of accommodating individual preferences among fishermen, the location of fishing centres and particular types of fishing in relation to building facilities and so forth. It may be true that in cases where financial assistance for new construction has failed to achieve its object, the boat-builder has been the only one to profit from assistance intended for the fisherman. This, however, is a reflection on the operation of the scheme and in the long run the builder, like the fisherman, has a strong interest in seeing that the investment in new craft and equipment proves economic in commercial practice.

It is interesting to note that quotations by boat-yards are often influenced by the quality of the drawings. Where two sets of drawings are prepared for identical boats, one set comprising a few sketchy plans and the other including a lot of detailed drawings, a boat-yard will often quote a higher price in respect of the latter. While it is true that the more precise detailed drawings will demand correspondingly careful supervision during the construction, they do provide a better guarantee that the finished boat will conform to the designer's requirements. If sketchy plans produce the same end-result, it simply means that the boat-yard is relying on " memory " and the builder's former experience of similar designs. Other things being equal, where a design incorporates certain innovations or departures from traditional practice, the sketchy plans would probably require longer to interpret than careful, clear and detailed drawings. The amount of details required may, therefore, be determined by the type of boat being built and the known ability of the builder to make the details "by heart". The main thing is that plans should be clear and distinct irrespective of the amount of detail, and there is a need for more emphasis to be placed on the value in terms of labour and accuracy of plans and specifications which can be readily interpreted by the builder

Insurance is perhaps too wide a subject to broach here but it is one that requires careful study in relation to fishing craft. Consideration might be given to some spreading of risk over as many fishing craft as possible and to mutual arrangements whereby claims are reduced, e.g. by charging the actual cost of towage (including lost fishing time) rather than taking into account the value of the boat under tow. Cases are known where the annual insurance premium has cost the fisherman as much as his annual loan-repayment charges. On the operational side the influence of breakdowns has already been mentioned and raises the question of trained mechanics. In many small craft the engine-man has to be a fisherman too and he is not always adequately trained in the maintenance and repair of modern diesels. Special training is required and perhaps manufacturers could help by encouraging courses for the purchasers of their engines. A few weeks spent in the factory might save a skipper/owner a great deal later on.

These are very superficial observations which are designed to indicate problems rather than to offer conclusions. It is believed, however, that there is room for more consultation between the respective authorities and interests when building programmes are being planned, in order that the technical, social and economic problems of the fishermen are understood by those who try to serve his needs, whether it is the designer, the builder, the engine-manufacturer or the Government.

TANK TESTING TECHNIQUE

by

W. A. P. VAN LAMMEREN

THE object of tank tests is to promote the efficient design of a ship. This is done by analysing the experiments to find the most economic hull shape and propeller design.

There has been a great advance in the theory of propulsion during the last two decades, but empiric investigations on small-scale models are still necessary to find concrete answers to specific questions concerning design and performance of fishing vessels. Naval architects generally make use of the results of systematic research and special model tests are usually made during the design of any ship of importance. It is important for the owner that his ship has the greatest possible speed for a certain dead-weight and engine power, or that she has the least possible power and fuel consumption for a certain speed. Such efficiency is also in the interests of the national economy.

Tank tests are usually carried out as follows: a provisional lines plan is submitted to the ship model basin and from this drawing a paraffin wax model is made. For sea-going vessels this model is usually 20 to 23 ft. (6 to 7 m.), because a smaller one would lead to flow-phenomena that would produce unreliable results. The model must also be large enough to contain the propelling machinery and various measuring appliances.

The model is first observed in the model basin, where it is towed in a straight line at the required displacement and draught reduced to scale. The resistance is measured for a range of speeds and is converted to the required effective or tow-rope horse-power (e.h.p.) for the full size ship. At the same time the wave pattern reveals possible defects in the shape of the hull.

Fig. 350. Ship model basin with adjustable floor

The model is then fitted with an electric motor to drive the propellers or paddles, etc., and with an inboard dynamometer to record the torque on the propeller shaft and the thrust of the propeller. The r.p.m. is also recorded. By conversion of these data the shaft horse-power (s.h.p.) is calculated and the quotient e.h.p./s.h.p. gives the quasi-propulsive coefficient (q.p.c.).

The results of e.h.p. and s.h.p. are then converted into non-dimensional coefficients and plotted in standard diagrams.

By comparing these non-dimensional figures with the results of previous tests it is possible to decide whether the results are satisfactory or whether they can be improved. If it is necessary to modify the design, the paraffin wax model is easily altered and testing is continued until satisfactory results are obtained.

By testing the propeller model separately (in open water) and by comparing the characteristics with those of the self-propelled ship model, all the factors affecting propulsion can be analysed and steps taken to improve them. Many other factors may have to be investigated, such as flow measurements, tests in shallow water, overload tests (for trawlers and tugs when towing), steering and turning tests, rolling tests, etc.

When the tests are concluded a report is made which contains all the necessary data and plans for the hull form and the propellers. From these details an accurate prediction can be made of the trial and service speeds.

Usually two or three weeks are required for the tests but, in urgent cases, they can be carried out in eight to ten days. The costs depend on the elaborateness of the tests, but generally they amount to a few thousand guilders (=a few hundred pounds.) Such an outlay is of minor significance when the ship is big and costly but in the case of small pleasure and fishing vessels the test programme has sometimes to be reduced to save money.

SHORT DESCRIPTION OF THE EQUIPMENT

A typical basin, such as that belonging to the Netherlands Ship Model Basin, is 827 ft. (252 m.) long, 34.4 ft. (10.5 m.) wide and 18 ft. (5.5 m.) deep. It contains about 495,000 cu. ft. (14,000 cu. m.) of water. At the bottom of the tank there is an adjustable floor, fig. 350, about 660 ft. (200 m.) long which consists of 25 heavy pontoons 26 ft. (8 m.) long. This floor can be fixed at any desired depth for shallow water tests with river and canal boats. The basin must be of large dimensions in order to simulate the motion in the open sea and eliminate the adverse influence which the walls and bottom might have on a travelling model. In fact, it is not practicable to make the model or the tank smaller.

The model is attached to a heavy tow carriage (fig. 351) which moves along rails on the top of both sides of the tank. This carriage is electrically driven at predetermined speeds, the limits of which are 1 to 2 in. (2 to 5 cm.) per second and 28 ft. ($8\frac{1}{2}$ m.) per second. Once fixed, the speed of the carriage must remain constant during the run, and the position of the machine-finished rails must, therefore, meet the most exacting requirements. The top of the rails must be parallel to the water surface to an accuracy of 0.004 in. ($\frac{1}{10}$ mm.). They must follow the curvature of the earth which, for a distance of 827 ft. (252 m.) is roughly $\frac{3}{32}$ in. (2.5 mm.). The speed of the carriage is controlled by the voltage transmitted to the driving motors. This voltage is kept constant to 1/1,000th by an ingenious regulating device equipped with high vacuum electrode valves.

The towing carriage has four motors of 20 h.p. each, and is controlled from a glass cabin at the end of the tank. The carriage weighs 15 tons, and is fitted with special devices to stop it quickly during high-speed tests.

The carriage is also equipped with various automatic recording and measuring appliances. The observers always stand on the carriage near the model to attend to and read the measuring instruments, to take photographs and to plot the results of the tests. A self-propelled model can travel independently under the carriage and is only kept on a straight course by this carriage. About six runs per hour can be made, depending on the speed and the wave-motion set up in the water.

Clients are encouraged to attend the important test

Fig. 351. Towing carriage

Fig. 352. Workshop for the manufacture of paraffin models

runs, as it is highly instructive to watch the model and it stimulates the exchange of ideas for improving the hull shape.

The model basin building contains offices for draughtsmen, workshops and rooms for the staff. The shop where the models are cast and shaped is the largest (fig. 352, and 353). Among the equipment in it is a model cutting machine which accurately copies the water-lines from the drawing.

In the instrument shop (fig. 354) measuring instruments are made, verified and calibrated. The main work however, consists of casting and finishing the propeller models. Here again are to be found ingenious copying and measuring devices to ensure the highest degree of accuracy (fig. 355, 356, 357). The propellers used are made of white metal and, as a rule, have a diameter of 8 to $10\frac{1}{2}$ in. (200 to 270 mm.). Propellers intended for cavitation tests are cast in bronze with a diameter of about 18 in. (450 mm.) .

When the pressure on a propeller blade is reduced

Fig. 353. Model workshop

Fig. 354. Instrument shop

below that of the vapour pressure in the surrounding liquid, vapour bubbles burst on the blade with great force. This phenomenon, known as cavitation, gains in intensity with increased speed and engine power and can badly damage a costly bronze propeller.

It is impossible to study cavitation in a model basin because the atmospheric pressure cannot be reduced to correspond to the scale of the model. It can only be done in a totally enclosed canal, the cavitation tunnel (fig. 358, 359, 360), through which water is pumped at high speed to pass the rotating propeller, which is kept in a longitudinal position.

Cavitation can be observed by means of stroboscopic lighting and the characteristics of the propeller can be determined by thrust and torque-measuring devices. A typical tunnel is 35 ft. (10.5 m.) long and 23 ft. (7 m.) high and in full operation requires 700 h.p.

MEASURING TECHNIQUE

Tests Carried Out in the Basin

The following tests are part of the normal routine work carried out in the basin.

1. Resistance tests

The paraffin wax ship model is attached to a resistance dynamometer on the towing carriage. The resistance of the model is measured at a series of speeds above and below the trial speed. According to Froude's model-law, the corresponding speed of the model is equal to the quotient of the ship's speed and the square root of the ratio of ship length to model length.

The first run of the model, before the water has been disturbed, is used to engrave the wave profile in the model. This is usually done at a previously-estimated

Fig. 355. Propeller cutting machine

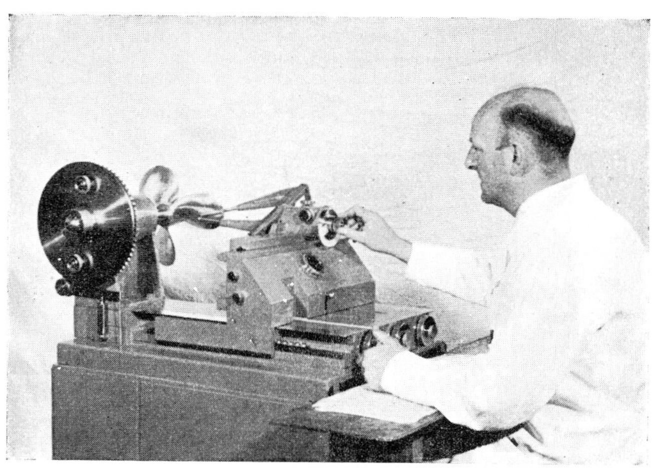

Fig. 356. Apparatus for measuring propeller models

Fig. 357. Finishing a bronze propeller model

service or trial speed. To assure a turbulent flow along the model in the boundary layer, a trip wire 0.04 in. (1 mm.) thick is fitted at 1/20 of the length of the bow. This causes an increase in resistance of 2 or 3 per cent. which is maintained in calculations as a safety margin.

The accuracy of these resistance tests is within $\frac{1}{2}$ to 1 per cent. for fine ships, 1 to 2 per cent. for full ships.

After a run during which measurements have been taken—runs are always made in the same direction—the towing carriage travels slowly back to enable the water to calm down again. This time is used for computing and plotting the recorded data. Four to six runs can be made per hour, depending on the speed of the model.

In fig. 361 the attachment of the model to the resistance dynamometer can be clearly seen. The motor and the propeller dynamometer have already been installed for the propulsion test, which usually immediately follows the resistance test.

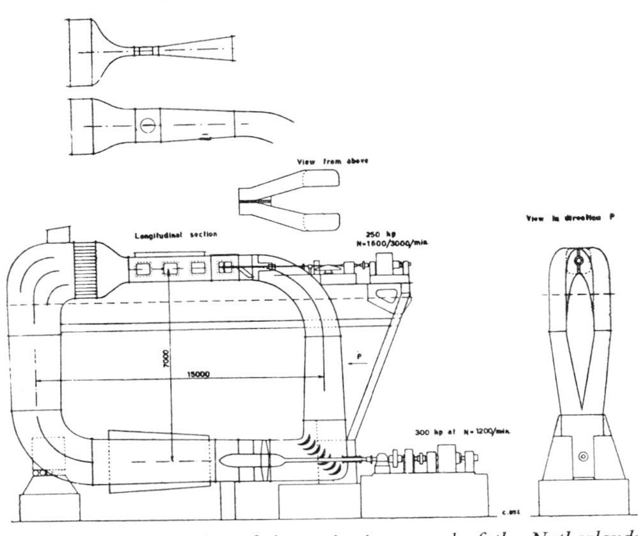

Fig. 358. General plan of the cavitation tunnel of the Netherlands Ship Model Basin

2. Propulsion tests (fig. 362)

The model is self-propelled by electric motors, which drive the propeller models.

The model is not connected to the towing carriage during the measuring period except through the trimming apparatus, which prevents it deviating from the course. This apparatus exerts, if necessary, small athwartships forces on the hull. The towing carriage travels with the model and at the same speed.

Thrust and torque of the propeller or propellers are recorded by dynamometers coupled between the electric motor and the propeller(s). The speed of the towing carriage and the r.p.m. of the propeller(s) are simultaneously recorded on the drum of the resistance dynamometer. In order to obtain the r.p.m. of the propeller model, the motor has an installation which enables a

Fig. 359. Lower part of a cavitation tunnel

current circuit to be closed every 25 revolutions, thereby causing a displacement of one of the recording pens.

A certain towing force is applied to the model to account for certain frictional differences between ship and model. The thrust and torques for the ship at corresponding speeds can be calculated by taking the respective model measurements and multiplying them by the third and fourth powers, respectively, of the model scale. The r.p.m. of the ship's propeller are found by dividing the model r.p.m. by the square root from the model scale, a method easily deduced from the model-laws. The accuracy of thrusts and torque measurements are within 1 to 2 per cent.

Propulsion tests are made over the same range of speeds as the resistance tests. The propulsive coefficient is found by dividing the tow rope e.h.p. by the corresponding s.h.p.

This coefficient, plotted in statistical form, shows the efficiency of propulsion.

The tests also show the engine power necessary to propel a given ship at a given speed. A certain allowance is given on the s.h.p. values to compensate the influence of greater roughness of the ship's hull, wind, rough seas, possible absence of appendages on the model and bearing losses of the propeller shaft. These allowances are different for trial and service conditions, and are based on experience. They depend very largely on the route to be followed by the ship.

3. Open-water propeller tests (fig. 363)

Resistance and propulsion tests are often followed by by an open water propeller test. The propeller is then operated in front of a slim, streamlined body because the object is to study the properties of the propeller alone. Thrust and torque are measured at constant r.p.m. over a range of speeds varying from zero to a value where thrust=0. The accuracy of these tests is equal to that of the propulsion tests. From these figures the efficiency of the propeller in open water can be determined. The test also provides the data for breaking down the propulsion coefficient into a number of propulsive components.

If these tests are carried out with series of propellers, having a systematic variation in the number of blades, blade area ratio, pitch ratio, etc., it is possible to develop open water propeller diagrams, which are of great assistance in designing efficient propellers.

4. Steering tests

A model basin can be used for testing steering properties of ships. A self-propelled model is kept on its course by two dynamometers placed athwartships fore and aft, and is able to move slightly on its natural revolving point, which lies at about one-third the length from the bow. It is possible to measure the torque exerted on the model caused by turning the rudder over at a certain angle. The rudder-head torque, or the forces exerted on the rudder in longitudinal and transverse directions, can also be measured.

These tests provide only a momentary reproduction of the conditions prevailing on the actual ship immediately after the rudder has been put over, and then only if it has been put over quickly. But the tests are very useful for comparing steering properties of various rudder shapes and the various rudder positions. River vessels, with one or more rudders and operating in restricted water, are also tested for steering properties.

Fig. 364 shows the stern of a twin-screw ship during a steering test. The stern dynamometer, fixed athwartships to the frame of the towing carriage, can be seen clearly. The centrifugal dynamometer is connected to a

Fig. 360. Upper part a cavitation tunnel

Fig. 361. Resistance tests

Fig. 362. Self propulsion tests

vertical lever which turns on the point of the frame, situated at approximately the same height as the clamp of the model. The bottom of the lever is connected to a vertical rod. The rod can slide up and down as the model trims during the run.

Zig-zag and turning-circle tests can only be carried out in specially constructed steering basins or ponds. The Netherlands Ship Model Basin uses a pond with a diameter of 180 ft. (55 m.) and a depth of 9 ft. (2.75 m.) Since it is not enclosed, it can only be used under favourable weather conditions and then only for taking approximate measurements of turning-circle diameter and deviation angle. Models used for this purpose are often made of wood and manned by a helmsman. They are propelled by electric motors run off a storage battery. The speed of the model is deduced as accurately as possible from the r.p.m.

The chief aim is to compare the efficiency of different rudders. Fig. 365 shows a turning-circle test being carried out with a destroyer model in a pond of the National Park, De Hoge Valuwe.

5. Rolling tests (fig. 366)

The aim of these tests is usually to determine the damping curve, when rolling, at rest, or at fixed translational speed.

From the formula for the rolling period of the ship

$$t=\frac{2\pi i}{\sqrt{g.GM}}$$

i = radius of inertia
g = acceleration due to gravity
GM = metacentric height.

it follows that when the rolling movement of ship and model are mechanically similar, the rolling period of the ship will be equal to that of the model, multiplied by the square root of the model scale.

The rolling period of the model can be adjusted to that of the ship by moving balance weights in a vertical and/or athwartships direction. The GM can be determined by an inclination test. It is not always possible to adjust the rolling period and the metacentric height to the correct values. In this case the GM is set at the correct value and the rolling period is adjusted as close as possible. Rolling movements are recorded on a revolving drum by a vertical pendulum suspended at the metacentre.

These tests are very useful for determining the influence of anti-rolling devices such as bilge keels.

During the tests, the model is attached to the resistance dynamometer and kept on its course by the trimming appliances. The points of attachment, about which the model rolls, are fixed at the same height as the metacentre. The towing carriage is not required when there is no ahead speed on the model.

6. Wake tests (fig. 367)

To make propeller calculations with the aid of systematic diagrams of propeller series, and to determine the propulsion components, it is necessary to know the mean wake figure:

$$w=\frac{v-v_e}{v}$$

where v = model speed and v_e = main speed of advance of fluid to the propeller. This figure can be determined

Fig. 363. Open water propeller tests

Fig. 364. Initial steering tests on a twin-screw model

according to Froude's method when the data resulting from the propulsion and open-water propeller tests are available.

Other methods must be used, however, if it is necessary to determine the wake distribution in order to calculate, by means of the vortex theory, the propeller adapted to a radical wake distribution. Fixed rings or revolving vane wheels are attached behind the model. They are calibrated in open-water condition on resistance and r.p.m. By measuring the resistance of the rings or the r.p.m. of the vane wheels the mean wake figure can be determined for any diameter of rings or vane wheels.

If it is necessary to find out the wake at every point of the speed field behind the model, Prandtl or Gebers pitot tubes must be used and the whole field systematically investigated.

The accuracy of these speed measurements are within $\frac{1}{2}$ to 1 per cent.

Tests in the Cavitation Tunnel

1. With propellers

The top horizontal part of the cavitation tunnel contains the bronze propeller which is to be tested. The r.p.m. of the propeller, driven by an electric motor, and the translational speed of the water in the measuring space, are chosen so that the slip of the model propeller and of the actual propeller are the same. This is the only essential condition for applying the law of Newton for converting the results on the model to the actual propeller. The static pressure at the centre line of the propeller shaft is regulated by means of a vacuum pump connected to the air space above the free fluid surface. This is done in such a way that the quotient of static and dynamic pressure is the same for the model and the full-size propeller.

The cavitation phenomena can be observed and photographed by stroboscopic lighting, synchronized with the r.p.m. of the propeller shaft (fig. 368). This is done in the vicinity of the propeller by an observer who also keeps water-speed and static-pressure constant. Another observer reads the r.p.m., the thrust and the torque.

The tests are carried out at a constant water speed in the tunnel by varying the number of revolutions of the propeller model. This is contrary to the method applied during the open-water propeller tests in the basin to ensure that the cavitation number remains constant as the slip varies. The first observer, therefore, has only to see that the static pressure, once established, remains constant.

Fig. 365. Steering tests in an open air pond

Fig. 366. Rolling tests

If the tip vortex and the cavitation remain within certain limits, so that there is no influence on thrust and torque, it is not necessary to determine the propeller characteristics and observation of the cavitation phenomena is sufficient. This offers the great advantage of allowing the propeller to be equipped with three or four blades of varying design. Each blade can be observed and photographed separately. It also eliminates the need to make two or three extra propellers which because of their large dimensions—up to 20 in. (50 cm.) diameter—are comparatively costly.

The tests are usually carried out in a homogeneous speed field. The horizontal portion in front of the measuring section is, however, so arranged that bodies can be introduced to excite a variable velocity field so that the influence of the irregularity of the speed-field on the cavitation characteristics can then be examined. The whole speed-field is measured in the usual manner by means of pitot tubes.

The accuracy of the equipment depends on the extent of the measuring range. For torque and thrust it varies in the lowest part of the range from 2 to 3 per cent. and in the higher from 1 to 2 per cent. thrust up to 2,800 lb. (1,270 kg.), torque up to 795 ft. lb. (110 kg.m.). The

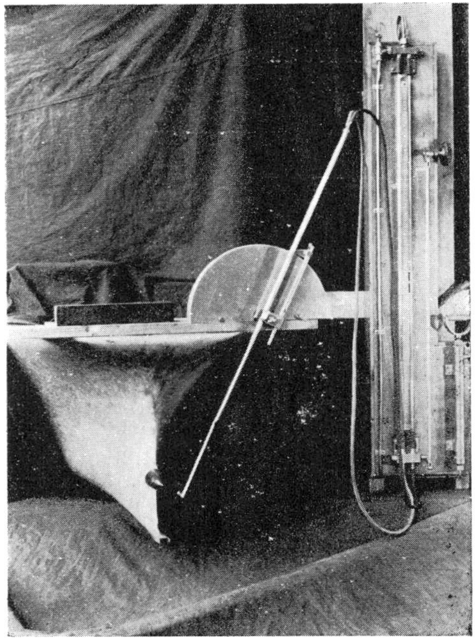

Fig. 367. Wake measurement device

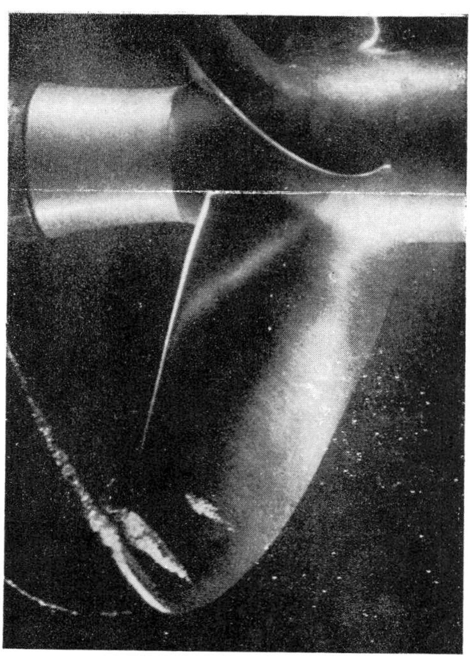

Fig. 368. Cavitating propeller

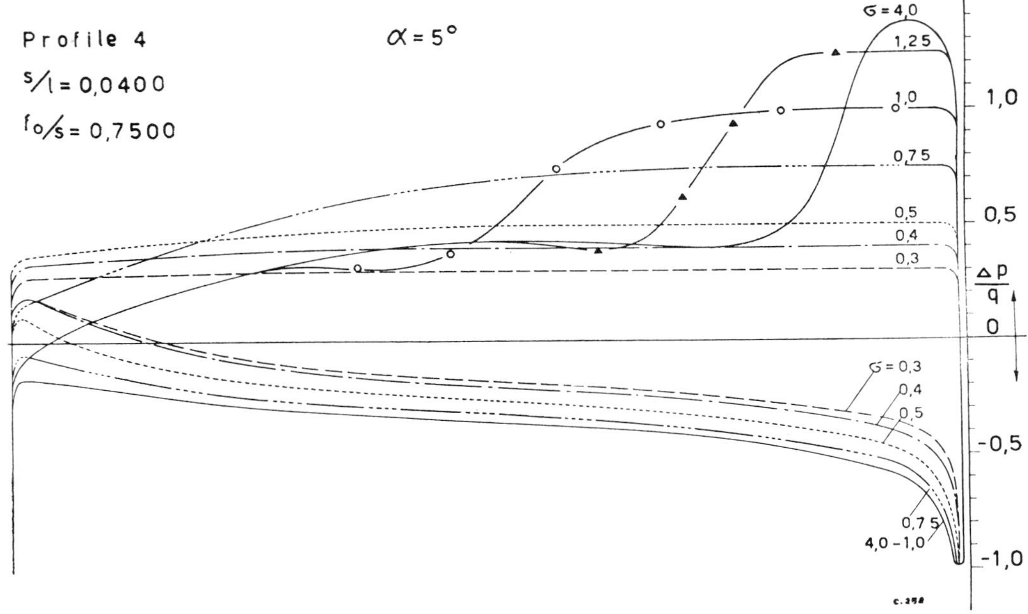

Fig. 369. Curves of pressure distribution at different cavitation numbers and a certain angle of attack

r.p.m. can be determined within ±3. As the r.p.m. varies from 300 to 3,000, accuracy is within 1 to 0.1 per cent. The speed measurements are accurate within 1 per cent.

2. With profiles

The Netherlands Ship Model Basin has also a measuring section for examining the cavitation properties of profiles. This is done by measuring pressure distribution and resistance in cavitating and non-cavitating conditions of the profiles. The data acquired form a valuable addition to the numerous measurements which have been carried out in air-tunnels. It is also possible to examine profile forms which are of special importance for propeller design.

The section consists of an entrance, a cylindrical middle where the measuring is done, and an exit. The profile to be measured is placed in the middle of the section together with the apparatus for measuring the speeds and pressures in front of and behind the profile.

The profile is fixed to two discs on either side. These can revolve in the tunnel wall, so that the angle of attack of the profile can be varied. The discs are fitted with windows to enable the flow phenomena around the profile to be observed. An upper and a lower window make it possible to observe the profile from other directions.

The determination of the lift and the drag, by means of a balance, is not sufficiently accurate because of the limited width of the profile in proportion to the length l ($b/l=0.75$). The circulation round the profile does not remain constant across the width because of the boundary layer along the vertical walls, which form the side limits of the profile. For this reason, lift and drag are determined from pressure and impulse measurements in the vertical symmetry area. The theory that

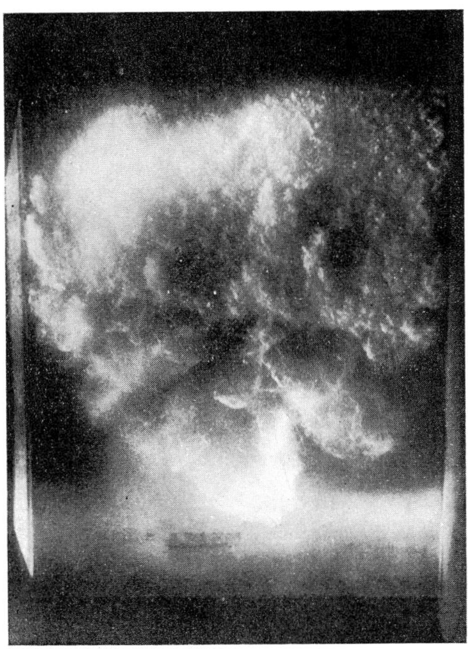

Fig. 370. Profile no. 4 ($s/l=0.0400$, $fo/s=0.7500$), $\alpha=5°$, $\sigma=1,0$, $v=10$ m./sec.

the resistance of the profile is equal to the change of impulse of the water flowing per unit of time along the profile is used in making the measurements.

To determine lift and drag as functions of the angle of

attack and the cavitation number, measurements are usually taken at angles of attack $= +8°, +5°, +3°, +2°, +1°, 0°, -1°, -2°, -3°, -5°$ and $-8°$. The cavitation numbers = are 4, 1.25, 1.0, 0.75, 0.50, 0.40 and 0.35. The last number is the lowest that can be reached with this apparatus.

With number =4, cavitation does not occur at any angle of attack in the range, so that in this condition lift and drag values of the non-cavitating flow can be determined. The intake speed at this cavitation number is 24 ft./sec. (8 m./sec.). At other cavitation numbers, the speed is 44 ft./sec. (13.5 m./sec.). The values of the Reynolds number corresponding to these speeds are 3.2×10^6 and 5.4×10^6, respectively. Fig. 369 shows the pressure distribution curves of a profile at these cavitation numbers and an angle of attack of $+5°$, while in fig. 370 the cavitation phenomena of the same profile at cavitation number=1.0 and angle of attack $= +5°$ can be seen.

Trial Measurements

Measurements taken on the trial and in service are of great importance for the evaluation of towing-tank results. They are indispensible when establishing proper allowances on the tank results, from which the trial prediction diagram (fig. 371) is made.

Measured mile runs are well known for establishing speed, power and r.p.m. Power is frequently determined by means of indicators, but the method using torsionmeters fixed to the shaft or shafts is more reliable. The Netherlands Ship Model Basin has a number of Maihak torsionmeters for this purpose, suitable for using with shafts of $7\frac{7}{8}$ to $19\frac{3}{4}$ in. (200 to 500 mm.) in diameter.

Trials can also be organized at sea if there is a reliable log available. The Netherlands Ship Model Basin has a self-recording log, designed by Kempf, which gives satisfactory results. The r.p.m. of the propeller shaft(s) and the wind speed and direction are measured at the same time as the speed of the ship. These measurements are necessary for making corrections if there is extra wind resistance. With trawlers, it is sometimes desirable to measure the bollard pull or the tow-rope force in the operating condition. To do this the Netherlands Ship Model Basin uses the Amsler tow-rope dynamometer which is connected between the boat and the trawl wire. It gives complete satisfaction.

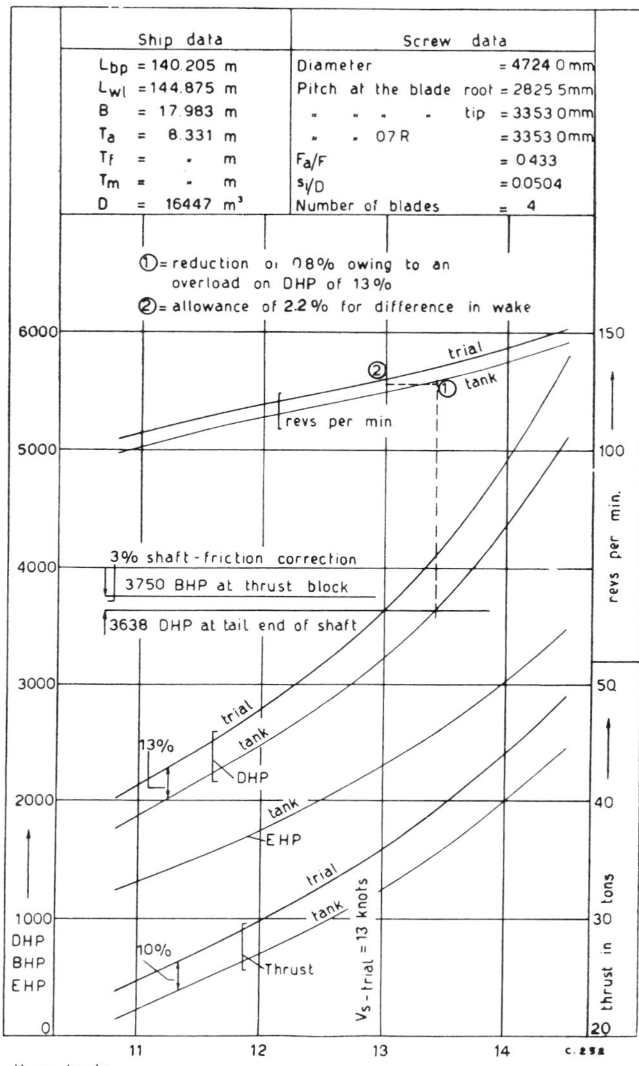

Fig. 371

SOME TANK TESTS

by

JAN-OLOF TRAUNG

TEXT-BOOKS on naval architecture and literature about the resistance and powering of ships are of limited value for estimating the resistance of a new fishing boat hull. It is obvious that knowledge of the resistance, before construction starts, would eliminate many mistakes and would help in the choice of engine and design of optimum propeller.

The author has tried to evolve a fairly simple system, based on tank tests, for estimating fishing boat resistance. It is similar to Ayre's (1948) system for large ships. The idea is that the naval architect himself should be able to study, with reasonable accuracy, the influence of length-displacement ratio, location of centre of buoyancy and the prismatic coefficient. He should be able to determine the difference in required engine power between an optimum hull shape and the hull determined by working conditions. The author has compiled many results of tank tests with fishing boats and similar types from published data and through personal contacts. He was given research grants to test several models and he also had opportunity to test models for both his previous employer and for FAO, so that he has collected the results from more than 120 different models tested on different displacements and trims and totalling more than 370 tests.

The efforts to design a simple, reasonably accurate system for estimating resistance, with the help of results of available tests, have been in vain so far. It is doubtful whether it will ever be possible to design an accurate and simple system covering all fishing boat types, as these vary to a large extent.

Much can be learned, however, from single or related tank tests. A uniform presentation of fishing boat tests should be valuable not only for practical design, but could also form the basis for further research.

FAO started, in 1952, to prepare data sheets of available tests. The sheets are not intended to be selective as both models with good and bad qualities are being covered. This paper contains 29 sheets, fig. 372–400, of model tests not published before. Work is in progress to issue a catalogue of fishing boat tank tests containing more than 150 sheets. (Copies can be had when ready, on application to Technology Branch, Fisheries Division, Food and Agriculture Organization (FAO) of United Nations, Rome, Italy).

DISCUSSION OF DATA SHEETS

Due to the limitation in staff and funds available to FAO, it has been necessary to develop a simple system of recording tank tests. The material is derived from original tank reports, technical papers and abridged magazine articles.

Calculations of the effective (tow rope) power by the different tanks have been considered adequate and they have been used in calculating the quasi-non-dimensional form value

$$C_1 = \frac{\nabla^{\frac{2}{3}} \cdot V^3}{P_e},$$

popularly called the " Admiralty Constant ".

This constant is based on the wrong assumption that the frictional resistance is two-thirds of the total resistance and that the frictional resistance per unit of surface area is the same for ships of different length. Although wrong, and although many other non-dimensional form values are being used, the use of the C_1 value has grown so popular with the years, that it has been selected for the non-dimensional presentation of the results.

Some calculations have been made, in order to determine the variation of C_1 with different L. For larger ships, there is a rule that one should use C_1 only for L $= \pm 10$ per cent. of the original length. In this way the results of a 65.5 ft. (20 m.) boat should only be compared with boats having L$=$59 to 72 ft. (18 to 22 m.). Now, the frictional resistance of large ships is often about 75 per cent. of the total resistance, but fishing boats have higher normal speeds and hull forms with proportionally less frictional resistance. An investigation of three models (1a—XXXIV, 2a—XXX, 5b—I) shows that, on Froude's number $=$0.30, the frictional resistance is about 34 to 48 per cent. and at 0.35, only 22 to 35 per cent. of the total resistance. Another calculation of the resistance of model No. 5b—I with 65.5, 100, 131 and 200 ft. (20, 30.5, 40 and 61 m.) waterline lengths showed that the increase of C_1 was rather small; the increase over the original 60.5 ft. (18.45 m.) waterline and Froude's number 0.30 was

Type	FISHING SCHOONER	Material, Finish	Paraffin	FAO No.	Ia
Tank	SSPA	Appendages	No	Roughness allowance	—
Test date	17/18 April 1953	Turbulence	Not induced	Reference	FAO files
Ship/Model Ratio	8.5	Friction Formulation	Froude	Notes	

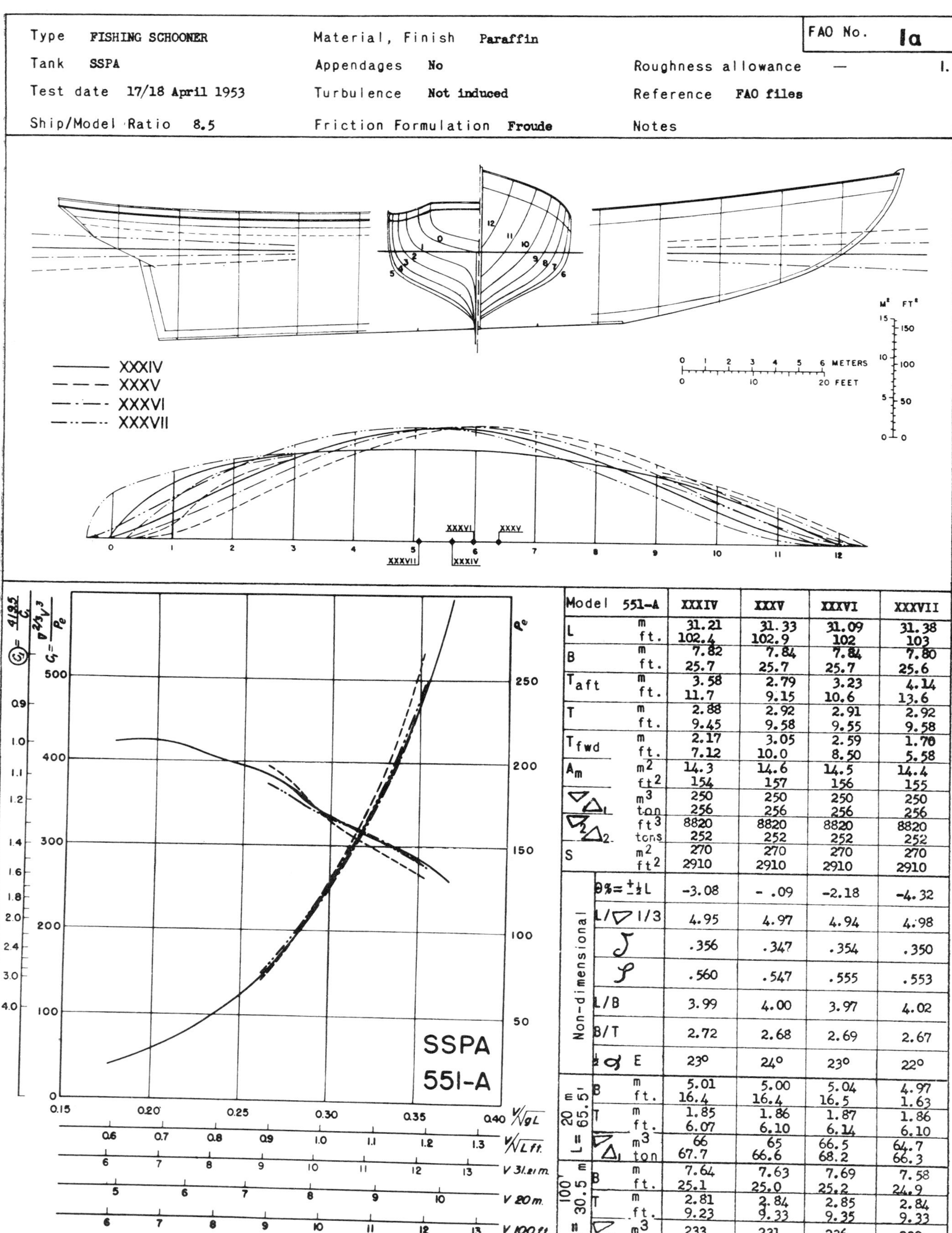

Model 551-A		XXXIV	XXXV	XXXVI	XXXVII
L	m	31.21	31.33	31.09	31.38
	ft.	102.4	102.9	102	103
B	m	7.82	7.84	7.84	7.80
	ft.	25.7	25.7	25.7	25.6
T_{aft}	m	3.58	2.79	3.23	4.14
	ft.	11.7	9.15	10.6	13.6
T	m	2.88	2.92	2.91	2.92
	ft.	9.45	9.58	9.55	9.58
T_{fwd}	m	2.17	3.05	2.59	1.70
	ft.	7.12	10.0	8.50	5.58
A_m	m^2	14.3	14.6	14.5	14.4
	ft^2	154	157	156	155
∇	m^3	250	250	250	250
Δ_1	ton	256	256	256	256
∇_2	ft^3	8820	8820	8820	8820
Δ_2	tons	252	252	252	252
S	m^2	270	270	270	270
	ft^2	2910	2910	2910	2910
Non-dimensional	$\theta\% = \pm\frac{1}{2}L$	-3.08	- .09	-2.18	-4.32
	$L/\nabla^{1/3}$	4.95	4.97	4.94	4.98
	δ	.356	.347	.354	.350
	φ	.560	.547	.555	.553
	L/B	3.99	4.00	3.97	4.02
	B/T	2.72	2.68	2.69	2.67
	$\frac{1}{2}\alpha$ E	23°	24°	23°	22°
L = 20 m 65.5'	B m	5.01	5.00	5.04	4.97
	ft.	16.4	16.4	16.5	1.63
	T m	1.85	1.86	1.87	1.86
	ft.	6.07	6.10	6.14	6.10
	∇ m^3	66	65	66.5	64.7
	Δ_1 ton	67.7	66.6	68.2	66.3
L = 30.5 m 100'	B m	7.64	7.63	7.69	7.58
	ft.	25.1	25.0	25.2	24.9
	T m	2.81	2.84	2.85	2.84
	ft.	9.23	9.33	9.35	9.33
	∇ m^3	233	231	236	230
	Δ_1 ton	239	237	242	236

Fig. 372

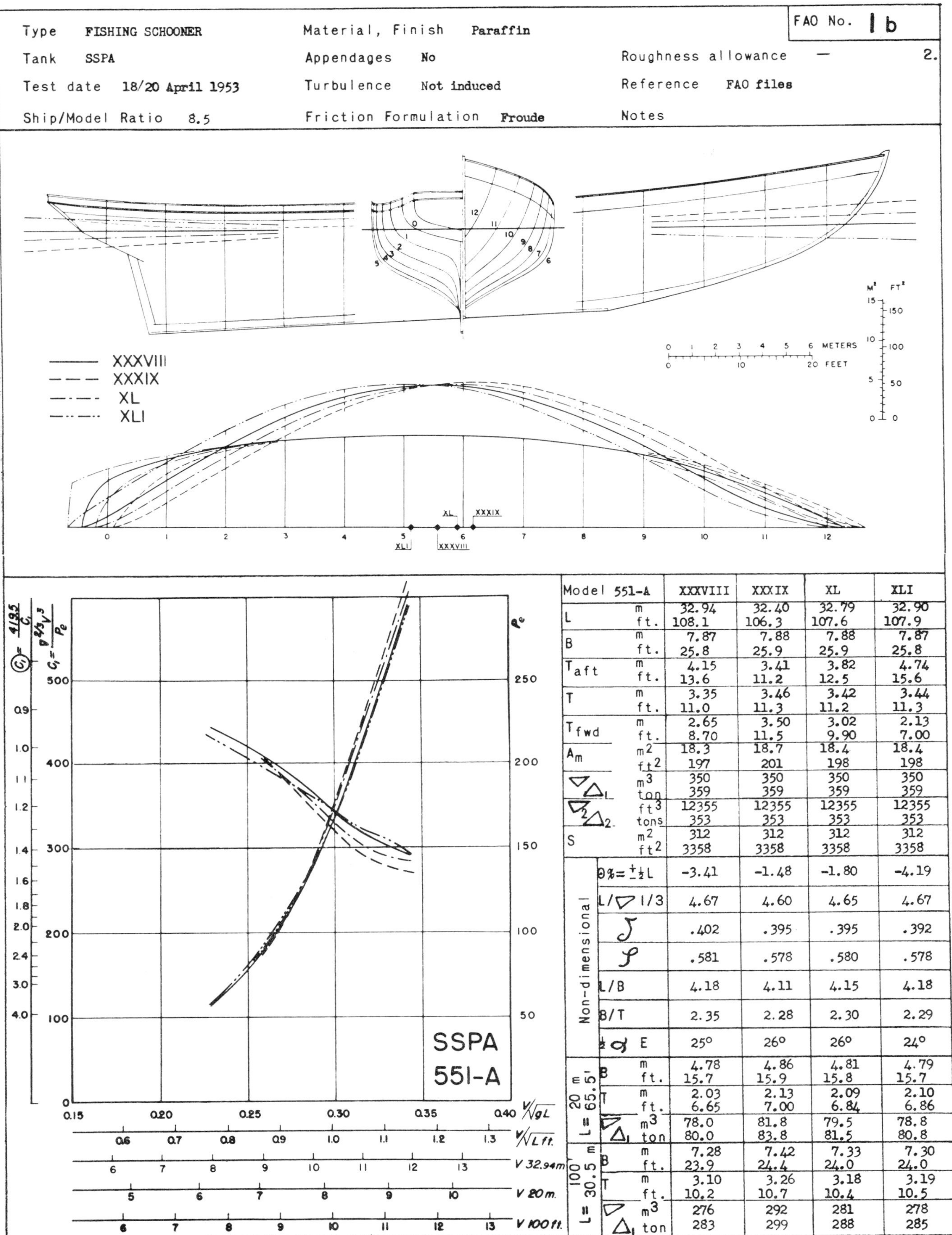

Type FISHING SCHOONER | Material, Finish Paraffin | FAO No. 1b

Tank SSPA | Appendages No | Roughness allowance — | 2.

Test date 18/20 April 1953 | Turbulence Not induced | Reference FAO files

Ship/Model Ratio 8.5 | Friction Formulation Froude | Notes

Model 551-A		XXXVIII	XXXIX	XL	XLI
L	m	32.94	32.40	32.79	32.90
	ft.	108.1	106.3	107.6	107.9
B	m	7.87	7.88	7.88	7.87
	ft.	25.8	25.9	25.9	25.8
T_{aft}	m	4.15	3.41	3.82	4.74
	ft.	13.6	11.2	12.5	15.6
T	m	3.35	3.46	3.42	3.44
	ft.	11.0	11.3	11.2	11.3
T_{fwd}	m	2.65	3.50	3.02	2.13
	ft.	8.70	11.5	9.90	7.00
A_m	m^2	18.3	18.7	18.4	18.4
	ft^2	197	201	198	198
∇	m^3	350	350	350	350
Δ_1	ton	359	359	359	359
∇_2	ft^3	12355	12355	12355	12355
Δ_2	tons	353	353	353	353
S	m^2	312	312	312	312
	ft^2	3358	3358	3358	3358
Non-dimensional	$\theta\% = \pm\frac{1}{2}L$	-3.41	-1.48	-1.80	-4.19
	$L/\nabla^{1/3}$	4.67	4.60	4.65	4.67
	δ	.402	.395	.395	.392
	φ	.581	.578	.580	.578
	L/B	4.18	4.11	4.15	4.18
	B/T	2.35	2.28	2.30	2.29
	$\frac{1}{2}\alpha$ E	25°	26°	26°	24°
L = 20 m 65.5'	B m	4.78	4.86	4.81	4.79
	ft.	15.7	15.9	15.8	15.7
	T m	2.03	2.13	2.09	2.10
	ft.	6.65	7.00	6.84	6.86
	∇ m^3	78.0	81.8	79.5	78.8
	Δ_1 ton	80.0	83.8	81.5	80.8
L = 100' 30.5 m	B m	7.28	7.42	7.33	7.30
	ft.	23.9	24.4	24.0	24.0
	T m	3.10	3.26	3.18	3.19
	ft.	10.2	10.7	10.4	10.5
	∇ m^3	276	292	281	278
	Δ_1 ton	283	299	288	285

Fig. 373

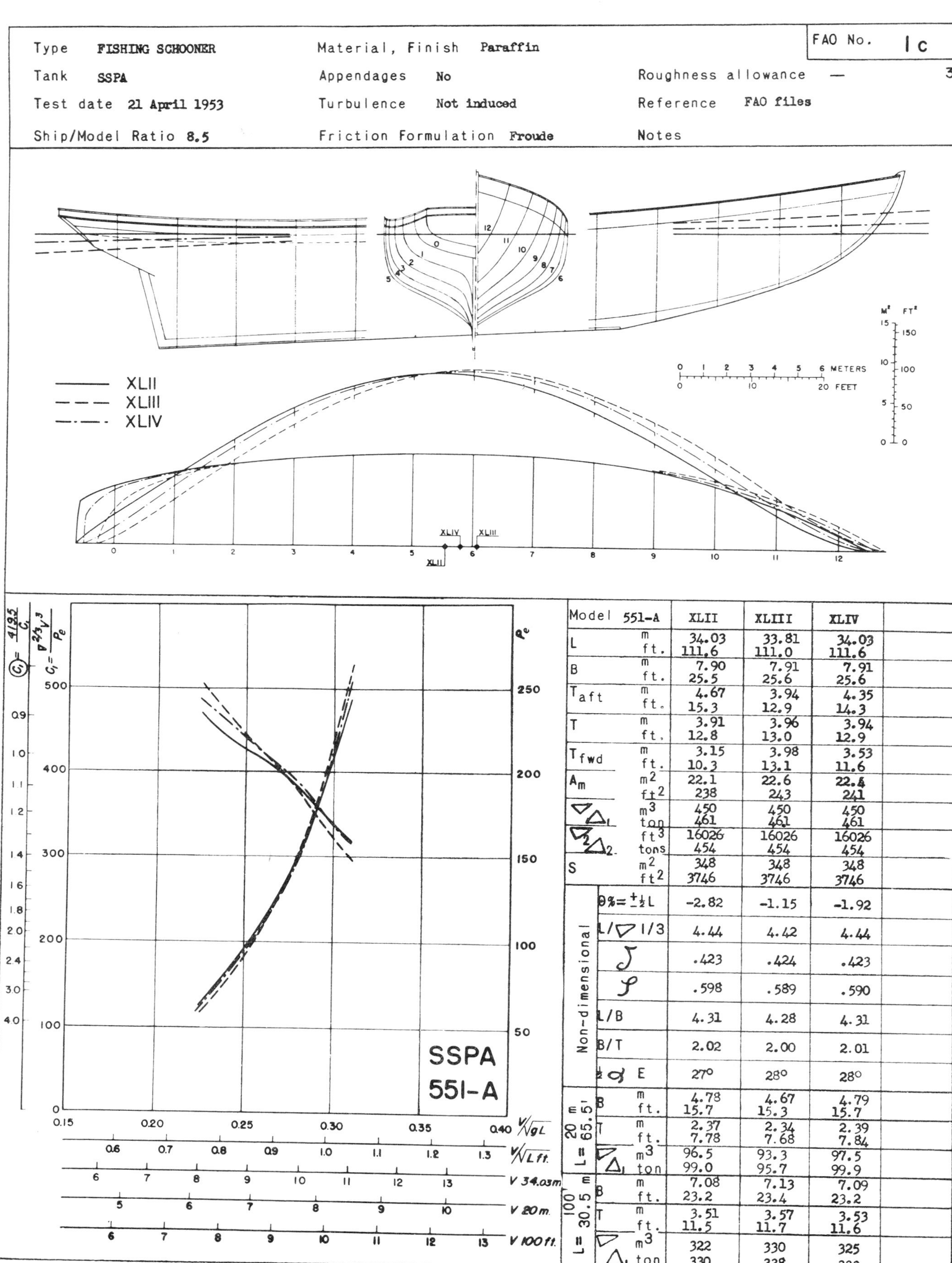

Type FISHING SCHOONER	Material, Finish Paraffin		FAO No. 1c
Tank SSPA	Appendages No	Roughness allowance —	3.
Test date 21 April 1953	Turbulence Not induced	Reference FAO files	
Ship/Model Ratio 8.5	Friction Formulation Froude	Notes	

Model 551-A		XLII	XLIII	XLIV
L	m	34.03	33.81	34.03
	ft.	111.6	111.0	111.6
B	m	7.90	7.91	7.91
	ft.	25.5	25.6	25.6
T_{aft}	m	4.67	3.94	4.35
	ft.	15.3	12.9	14.3
T	m	3.91	3.96	3.94
	ft.	12.8	13.0	12.9
T_{fwd}	m	3.15	3.98	3.53
	ft.	10.3	13.1	11.6
A_m	m^2	22.1	22.6	22.4
	ft^2	238	243	241
$\nabla_1 \Delta_1$	m^3	450	450	450
	ton	461	461	461
$\nabla_2 \Delta_2$	ft^3	16026	16026	16026
	tons	454	454	454
S	m^2	348	348	348
	ft^2	3746	3746	3746
Non-dimensional	$\theta\% = \pm\frac{1}{2}L$	-2.82	-1.15	-1.92
	$L/\nabla^{1/3}$	4.44	4.42	4.44
	δ	.423	.424	.423
	φ	.598	.589	.590
	L/B	4.31	4.28	4.31
	B/T	2.02	2.00	2.01
	$\frac{1}{2}\alpha$ E	27°	28°	28°
L = 20 m 65.5'	B m	4.78	4.67	4.79
	ft.	15.7	15.3	15.7
	T m	2.37	2.34	2.39
	ft.	7.78	7.68	7.84
	$\nabla_1 \Delta_1$ m^3	96.5	93.3	97.5
	ton	99.0	95.7	99.9
L = 100' 30.5 m	B m	7.08	7.13	7.09
	ft.	23.2	23.4	23.2
	T m	3.51	3.57	3.53
	ft.	11.5	11.7	11.6
	$\nabla_1 \Delta_1$ m^3	322	330	325
	ton	330	338	333

Fig. 374

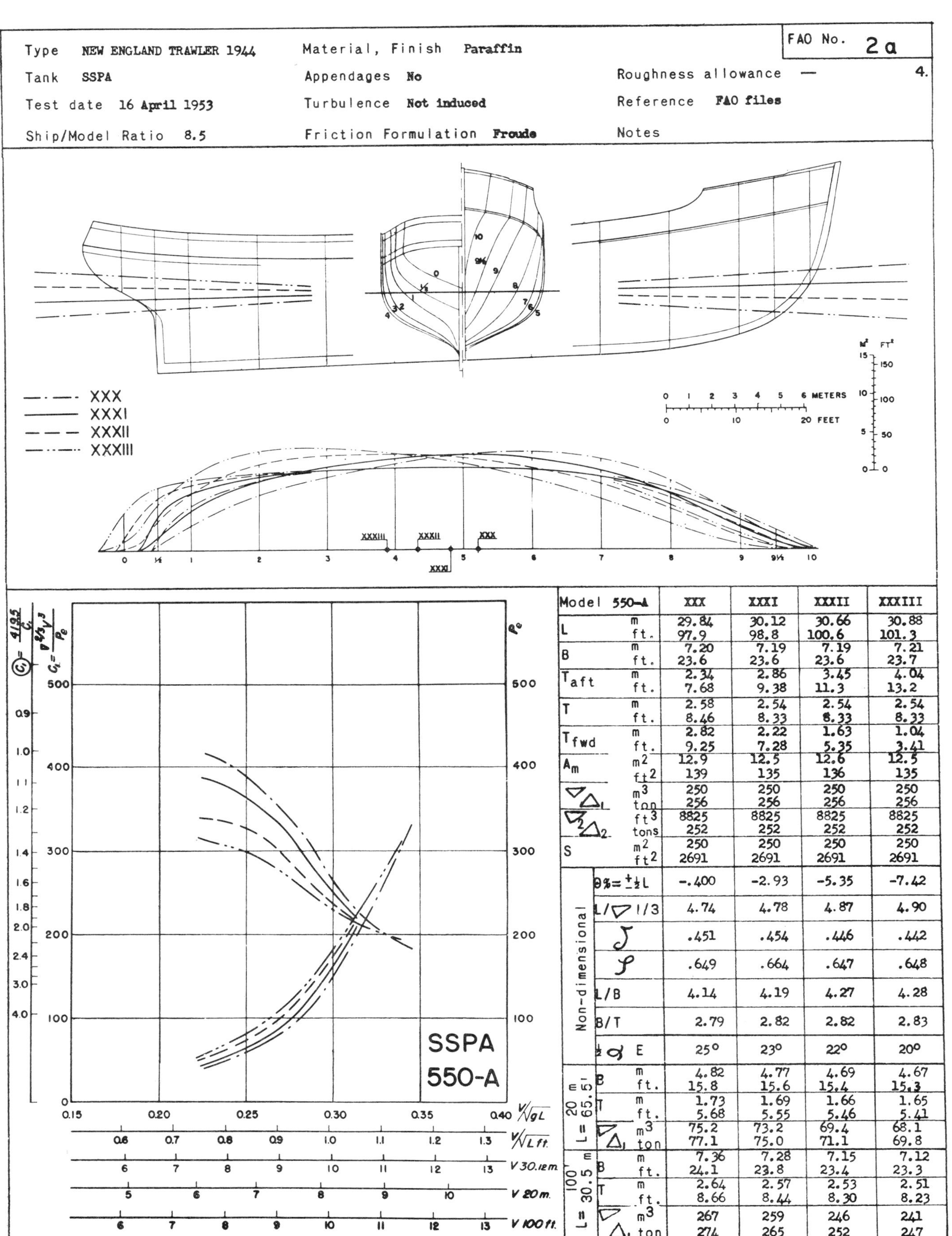

Type NEW ENGLAND TRAWLER 1944 | Material, Finish Paraffin | FAO No. 2a

Tank SSPA | Appendages No | Roughness allowance — | 4.

Test date 16 April 1953 | Turbulence Not induced | Reference FAO files

Ship/Model Ratio 8.5 | Friction Formulation Froude | Notes

Model 550-A		XXX	XXXI	XXXII	XXXIII
L	m	29.84	30.12	30.66	30.88
	ft.	97.9	98.8	100.6	101.3
B	m	7.20	7.19	7.19	7.21
	ft.	23.6	23.6	23.6	23.7
T_{aft}	m	2.34	2.86	3.45	4.04
	ft.	7.68	9.38	11.3	13.2
T	m	2.58	2.54	2.54	2.54
	ft.	8.46	8.33	8.33	8.33
T_{fwd}	m	2.82	2.22	1.63	1.04
	ft.	9.25	7.28	5.35	3.41
A_m	m^2	12.9	12.5	12.6	12.5
	ft^2	139	135	136	135
∇ / Δ_1	m^3	250	250	250	250
	ton	256	256	256	256
∇_2 / Δ_2	ft^3	8825	8825	8825	8825
	tons	252	252	252	252
S	m^2	250	250	250	250
	ft^2	2691	2691	2691	2691
Non-dimensional	$\theta\% = \pm \frac{1}{2} L$	-.400	-2.93	-5.35	-7.42
	$L/\nabla^{1/3}$	4.74	4.78	4.87	4.90
	δ	.451	.454	.446	.442
	φ	.649	.664	.647	.648
	L/B	4.14	4.19	4.27	4.28
	B/T	2.79	2.82	2.82	2.83
	$\frac{1}{2} \alpha$ E	25°	23°	22°	20°
L = 20 m 65.5'	B m	4.82	4.77	4.69	4.67
	B ft.	15.8	15.6	15.4	15.3
	T m	1.73	1.69	1.66	1.65
	T ft.	5.68	5.55	5.46	5.41
	∇ m^3	75.2	73.2	69.4	68.1
	Δ_1 ton	77.1	75.0	71.1	69.8
L = 30.5 m 100'	B m	7.36	7.28	7.15	7.12
	B ft.	24.1	23.8	23.4	23.3
	T m	2.64	2.57	2.53	2.51
	T ft.	8.66	8.44	8.30	8.23
	∇ m^3	267	259	246	241
	Δ_1 ton	274	265	252	247

Fig. 375

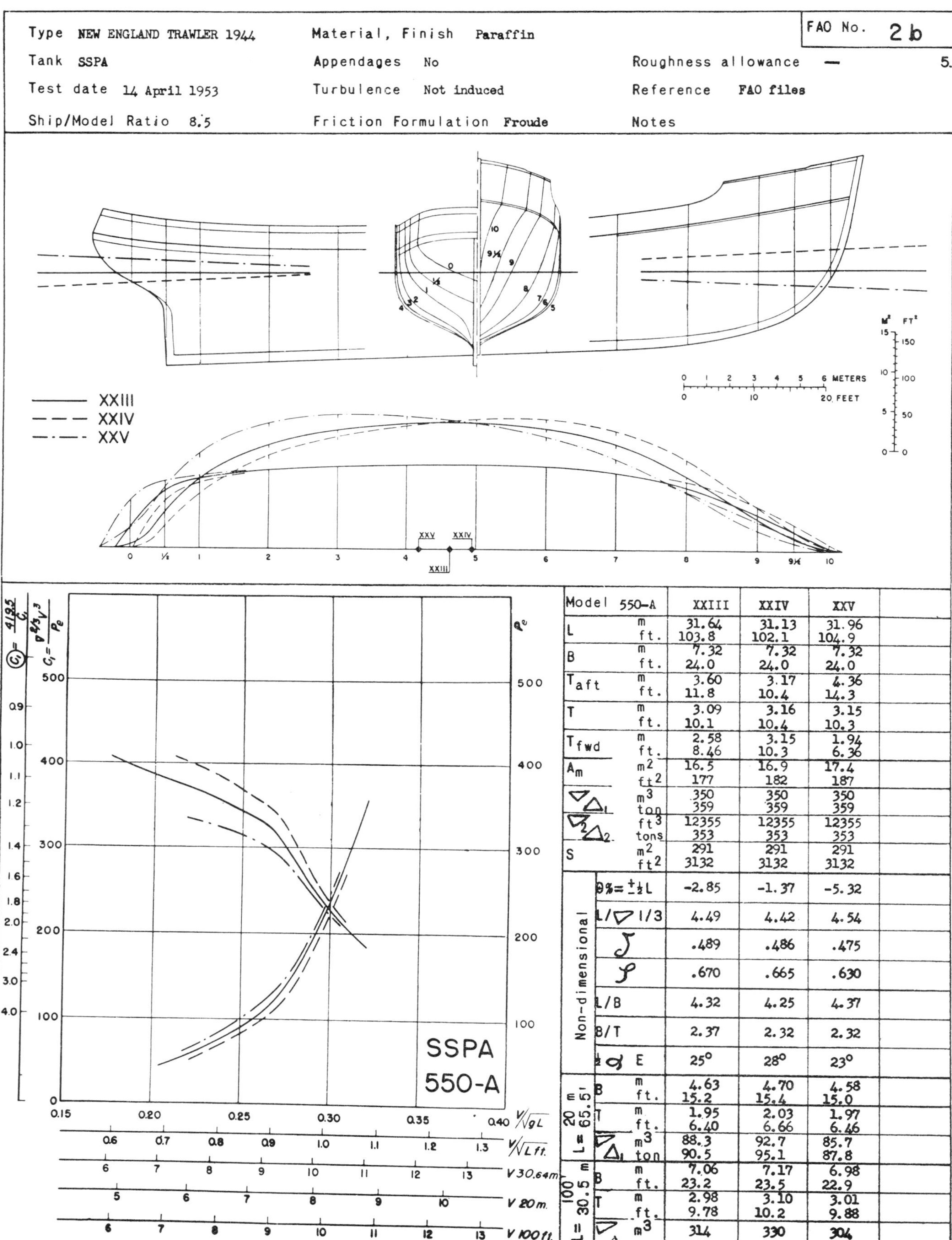

Model 550-A		XXIII	XXIV	XXV
L	m	31.64	31.13	31.96
	ft.	103.8	102.1	104.9
B	m	7.32	7.32	7.32
	ft.	24.0	24.0	24.0
T_{aft}	m	3.60	3.17	4.36
	ft.	11.8	10.4	14.3
T	m	3.09	3.16	3.15
	ft.	10.1	10.4	10.3
T_{fwd}	m	2.58	3.15	1.94
	ft.	8.46	10.3	6.36
A_m	m^2	16.5	16.9	17.4
	ft^2	177	182	187
$\nabla \Delta_1$	m^3	350	350	350
	ton	359	359	359
$\nabla_2 \Delta_2$	ft^3	12355	12355	12355
	tons	353	353	353
S	m^2	291	291	291
	ft^2	3132	3132	3132
Non-dimensional	$\theta\% = \frac{+}{-}\frac{1}{2}L$	-2.85	-1.37	-5.32
	$L/\nabla^{1/3}$	4.49	4.42	4.54
	δ	.489	.486	.475
	φ	.670	.665	.630
	L/B	4.32	4.25	4.37
	B/T	2.37	2.32	2.32
	½ α E	25°	28°	23°
L = 20 m = 65.5'	B m	4.63	4.70	4.58
	B ft.	15.2	15.4	15.0
	T m	1.95	2.03	1.97
	T ft.	6.40	6.66	6.46
	∇ m^3	88.3	92.7	85.7
	Δ_1 ton	90.5	95.1	87.8
L = 100' = 30.5 m	B m	7.06	7.17	6.98
	B ft.	23.2	23.5	22.9
	T m	2.98	3.10	3.01
	T ft.	9.78	10.2	9.88
	∇ m^3	314	330	304
	Δ_1 ton	322	338	312

Fig. 376

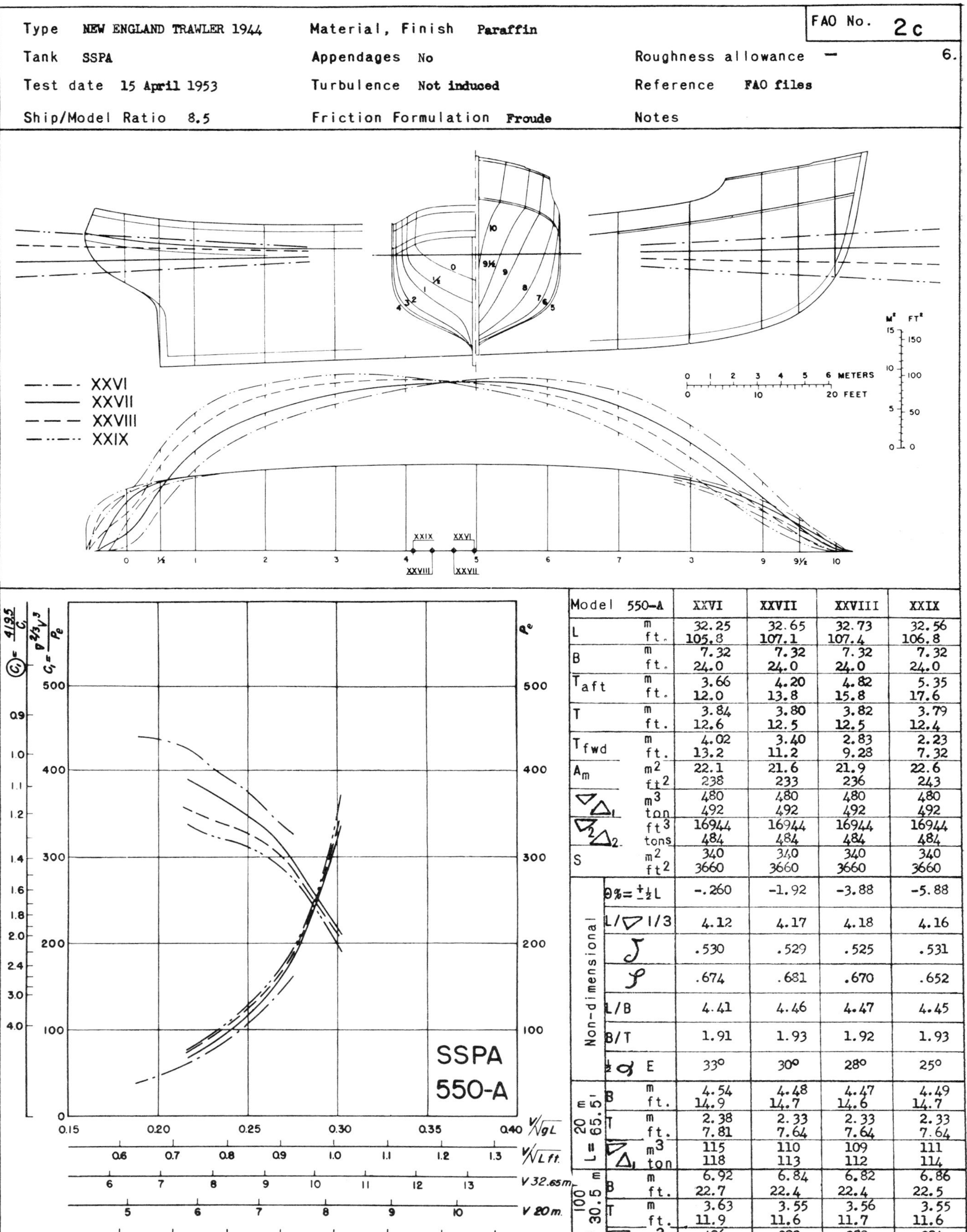

Type NEW ENGLAND TRAWLER 1944	Material, Finish Paraffin	FAO No. 2c
Tank SSPA	Appendages No	Roughness allowance — 6.
Test date 15 April 1953	Turbulence Not induced	Reference FAO files
Ship/Model Ratio 8.5	Friction Formulation Froude	Notes

Model 550-A		XXVI	XXVII	XXVIII	XXIX
L	m	32.25	32.65	32.73	32.56
	ft.	105.8	107.1	107.4	106.8
B	m	7.32	7.32	7.32	7.32
	ft.	24.0	24.0	24.0	24.0
T_{aft}	m	3.66	4.20	4.82	5.35
	ft.	12.0	13.8	15.8	17.6
T	m	3.84	3.80	3.82	3.79
	ft.	12.6	12.5	12.5	12.4
T_{fwd}	m	4.02	3.40	2.83	2.23
	ft.	13.2	11.2	9.28	7.32
A_m	m^2	22.1	21.6	21.9	22.6
	ft^2	238	233	236	243
$\nabla\Delta_1$	m^3	480	480	480	480
	ton	492	492	492	492
$\nabla_2\Delta_2$	ft^3	16944	16944	16944	16944
	tons	484	484	484	484
S	m^2	340	340	340	340
	ft^2	3660	3660	3660	3660
Non-dimensional	$\theta\% = \pm\frac{1}{2}L$	-.260	-1.92	-3.88	-5.88
	$L/\nabla^{1/3}$	4.12	4.17	4.18	4.16
	δ	.530	.529	.525	.531
	φ	.674	.681	.670	.652
	L/B	4.41	4.46	4.47	4.45
	B/T	1.91	1.93	1.92	1.93
	½α E	33°	30°	28°	25°
L = 20 m 65.5'	B m	4.54	4.48	4.47	4.49
	ft.	14.9	14.7	14.6	14.7
	T m	2.38	2.33	2.33	2.33
	ft.	7.81	7.64	7.64	7.64
	∇ m^3	115	110	109	111
	Δ_1 ton	118	113	112	114
L = 30.5 m 100'	B m	6.92	6.84	6.82	6.86
	ft.	22.7	22.4	22.4	22.5
	T m	3.63	3.55	3.56	3.55
	ft.	11.9	11.6	11.7	11.6
	∇ m^3	406	392	389	394
	Δ_1 ton	416	402	399	404

Fig. 377

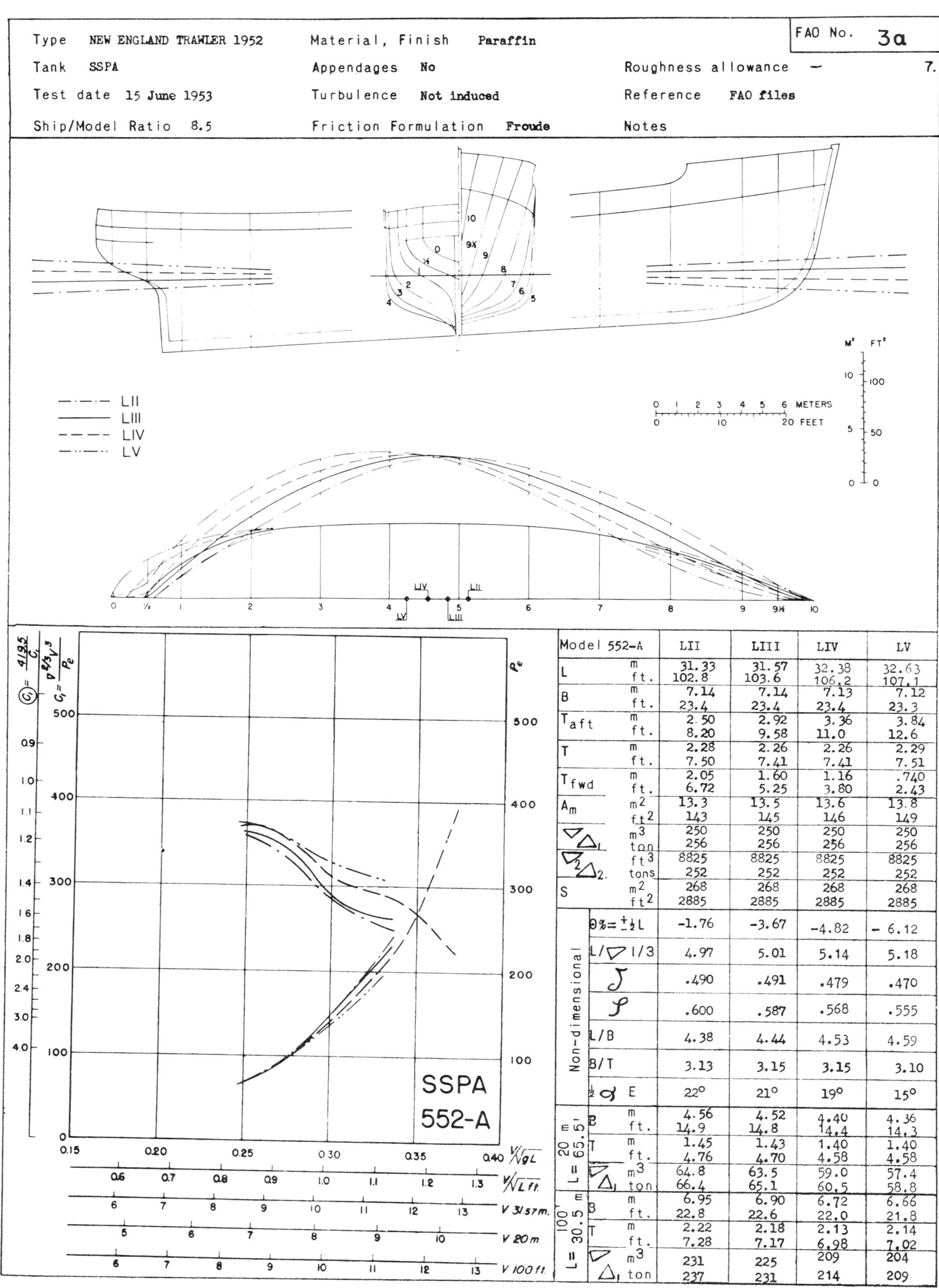

Type NEW ENGLAND TRAWLER 1952	Material, Finish Paraffin		FAO No. 3a
Tank SSPA	Appendages No	Roughness allowance —	7.
Test date 15 June 1953	Turbulence Not induced	Reference FAO files	
Ship/Model Ratio 8.5	Friction Formulation Froude	Notes	

Model 552-A		LII	LIII	LIV	LV
L	m	31.33	31.57	32.38	32.63
	ft.	102.8	103.6	106.2	107.1
B	m	7.14	7.14	7.13	7.12
	ft.	23.4	23.4	23.4	23.3
T_{aft}	m	2.50	2.92	3.36	3.84
	ft.	8.20	9.58	11.0	12.6
T	m	2.28	2.26	2.26	2.29
	ft.	7.50	7.41	7.41	7.51
T_{fwd}	m	2.05	1.60	1.16	.740
	ft.	6.72	5.25	3.80	2.43
A_m	m^2	13.3	13.5	13.6	13.8
	ft^2	143	145	146	149
∇_1 Δ_1	m^3	250	250	250	250
	ton	256	256	256	256
∇_2 Δ_2	ft^3	8825	8825	8825	8825
	tons	252	252	252	252
S	m^2	268	268	268	268
	ft^2	2885	2885	2885	2885
Non-dimensional	$\theta\% = \pm\frac{1}{2}L$	-1.76	-3.67	-4.82	- 6.12
	$L/\nabla^{1/3}$	4.97	5.01	5.14	5.18
	δ	.490	.491	.479	.470
	φ	.600	.587	.568	.555
	L/B	4.38	4.44	4.53	4.59
	B/T	3.13	3.15	3.15	3.10
	$\frac{1}{2}\alpha$ E	22°	21°	19°	15°
L = 20 m 65.5'	B m	4.56	4.52	4.40	4.36
	ft.	14.9	14.8	14.4	14.3
	T m	1.45	1.43	1.40	1.40
	ft.	4.76	4.70	4.58	4.58
	∇ m^3	64.8	63.5	59.0	57.4
	Δ_1 ton	66.4	65.1	60.5	58.8
L = 30.5 m 100'	B m	6.95	6.90	6.72	6.66
	ft.	22.8	22.6	22.0	21.8
	T m	2.22	2.18	2.13	2.14
	ft.	7.28	7.17	6.98	7.02
	∇ m^3	231	225	209	204
	Δ_1 ton	237	231	214	209

Fig. 378

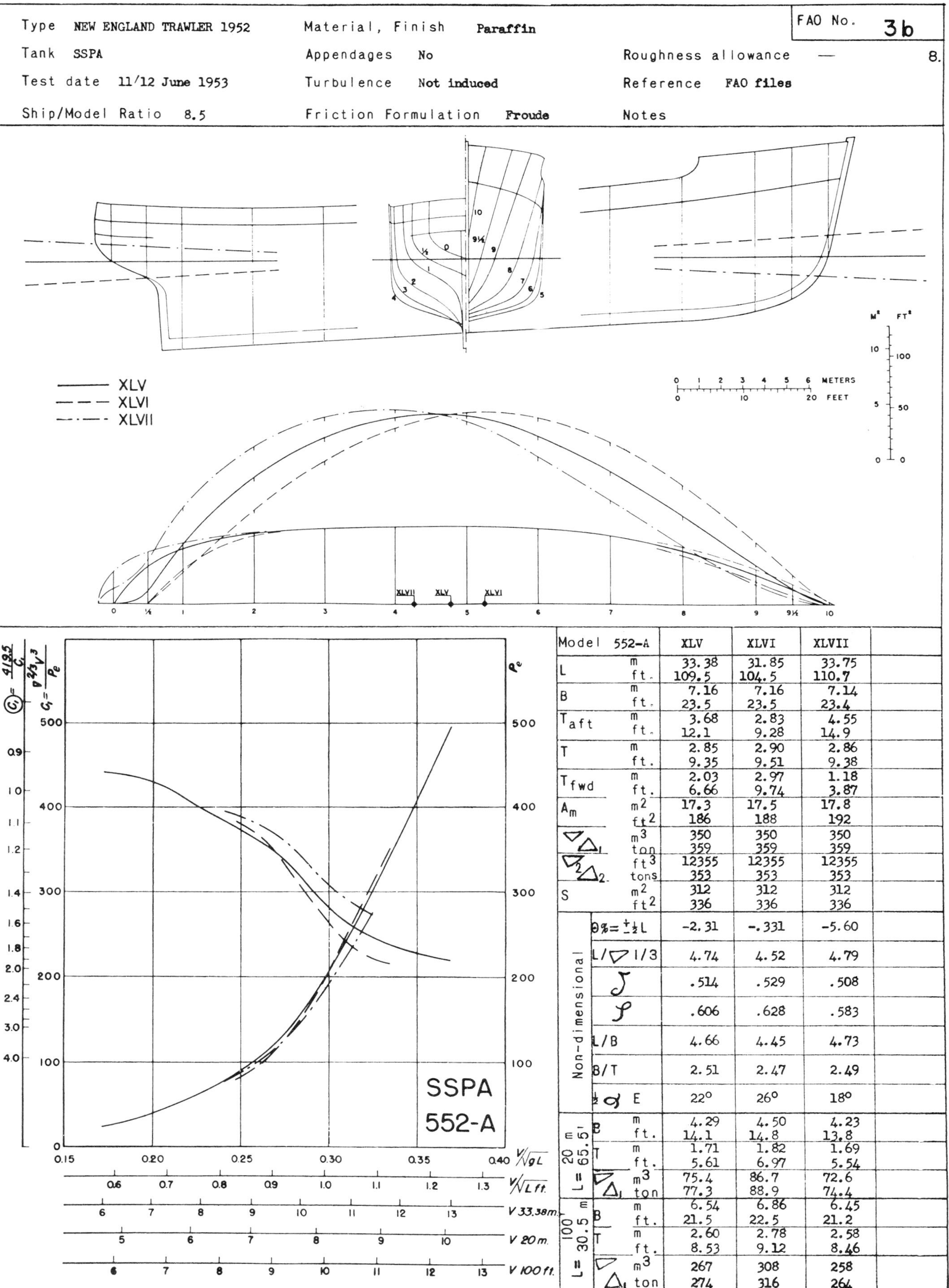

Type	NEW ENGLAND TRAWLER 1952	Material, Finish	Paraffin		FAO No. 3b
Tank	SSPA	Appendages	No	Roughness allowance	—
Test date	11/12 June 1953	Turbulence	Not induced	Reference	FAO files
Ship/Model Ratio	8.5	Friction Formulation	Froude	Notes	

Model 552-A		XLV	XLVI	XLVII
L	m	33.38	31.85	33.75
	ft.	109.5	104.5	110.7
B	m	7.16	7.16	7.14
	ft.	23.5	23.5	23.4
T_{aft}	m	3.68	2.83	4.55
	ft.	12.1	9.28	14.9
T	m	2.85	2.90	2.86
	ft.	9.35	9.51	9.38
T_{fwd}	m	2.03	2.97	1.18
	ft.	6.66	9.74	3.87
A_m	m^2	17.3	17.5	17.8
	ft^2	186	188	192
∇ Δ_1	m^3	350	350	350
	ton	359	359	359
∇_2 Δ_2	ft^3	12355	12355	12355
	tons	353	353	353
S	m^2	312	312	312
	ft^2	336	336	336
Non-dimensional	$\theta\% = \pm\frac{1}{2}L$	-2.31	-.331	-5.60
	$L/\nabla^{1/3}$	4.74	4.52	4.79
	δ	.514	.529	.508
	φ	.606	.628	.583
	L/B	4.66	4.45	4.73
	B/T	2.51	2.47	2.49
	$\frac{1}{2}\alpha$ E	22°	26°	18°
L = 20 m 65.5'	B m	4.29	4.50	4.23
	ft.	14.1	14.8	13.8
	T m	1.71	1.82	1.69
	ft.	5.61	6.97	5.54
	∇ m^3	75.4	86.7	72.6
	Δ_1 ton	77.3	88.9	74.4
L = 30.5 m 100'	B m	6.54	6.86	6.45
	ft.	21.5	22.5	21.2
	T m	2.60	2.78	2.58
	ft.	8.53	9.12	8.46
	∇ m^3	267	308	258
	Δ_1 ton	274	316	264

Fig. 379

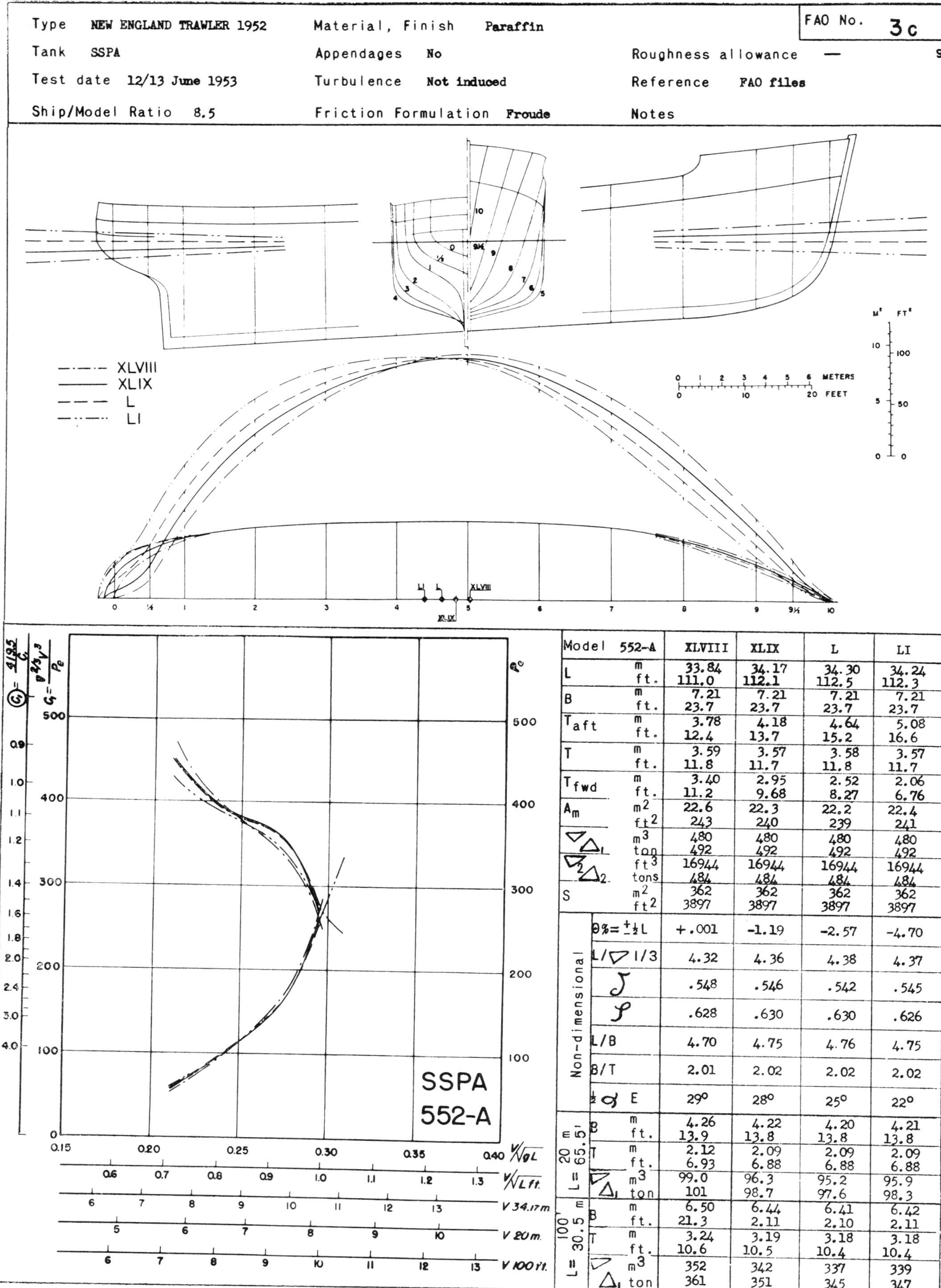

Type NEW ENGLAND TRAWLER 1952	Material, Finish Paraffin		FAO No. 3c
Tank SSPA	Appendages No	Roughness allowance —	9.
Test date 12/13 June 1953	Turbulence Not induced	Reference FAO files	
Ship/Model Ratio 8.5	Friction Formulation Froude	Notes	

Model 552-A		XLVIII	XLIX	L	LI
L	m	33.84	34.17	34.30	34.24
	ft.	111.0	112.1	112.5	112.3
B	m	7.21	7.21	7.21	7.21
	ft.	23.7	23.7	23.7	23.7
T_{aft}	m	3.78	4.18	4.64	5.08
	ft.	12.4	13.7	15.2	16.6
T	m	3.59	3.57	3.58	3.57
	ft.	11.8	11.7	11.8	11.7
T_{fwd}	m	3.40	2.95	2.52	2.06
	ft.	11.2	9.68	8.27	6.76
A_m	m^2	22.6	22.3	22.2	22.4
	ft^2	243	240	239	241
∇_1 Δ_1	m^3	480	480	480	480
	ton	492	492	492	492
∇_2 Δ_2	ft^3	16944	16944	16944	16944
	tons	484	484	484	484
S	m^2	362	362	362	362
	ft^2	3897	3897	3897	3897
Non-dimensional	θ% = ±½L	+.001	-1.19	-2.57	-4.70
	$L/\nabla^{1/3}$	4.32	4.36	4.38	4.37
	δ	.548	.546	.542	.545
	φ	.628	.630	.630	.626
	L/B	4.70	4.75	4.76	4.75
	B/T	2.01	2.02	2.02	2.02
	½α E	29°	28°	25°	22°
L = 20 m 65.5'	B m	4.26	4.22	4.20	4.21
	ft.	13.9	13.8	13.8	13.8
	T m	2.12	2.09	2.09	2.09
	ft.	6.93	6.88	6.88	6.88
	∇ m^3	99.0	96.3	95.2	95.9
	Δ_1 ton	101	98.7	97.6	98.3
L = 100' 30.5 m	B m	6.50	6.44	6.41	6.42
	ft.	21.3	2.11	2.10	2.11
	T m	3.24	3.19	3.18	3.18
	ft.	10.6	10.5	10.4	10.4
	∇ m^3	352	342	337	339
	Δ_1 ton	361	351	345	347

Fig. 380

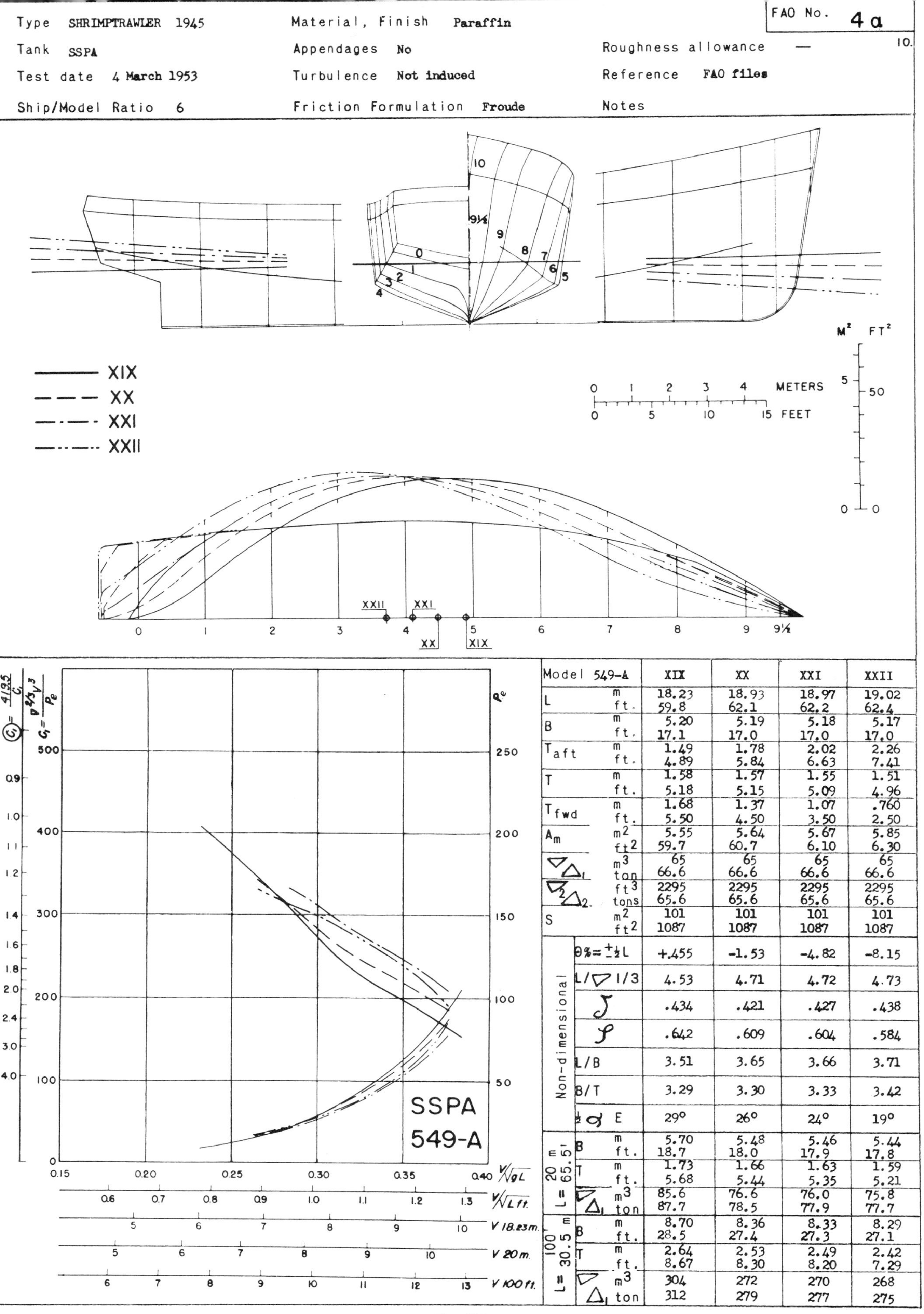

Type SHRIMPTRAWLER 1945	Material, Finish Paraffin		FAO No.	4 a
Tank SSPA	Appendages No	Roughness allowance —		10.
Test date 4 March 1953	Turbulence Not induced	Reference FAO files		
Ship/Model Ratio 6	Friction Formulation Froude	Notes		

Model 549-A		XIX	XX	XXI	XXII
L	m	18.23	18.93	18.97	19.02
	ft.	59.8	62.1	62.2	62.4
B	m	5.20	5.19	5.18	5.17
	ft.	17.1	17.0	17.0	17.0
T_{aft}	m	1.49	1.78	2.02	2.26
	ft.	4.89	5.84	6.63	7.41
T	m	1.58	1.57	1.55	1.51
	ft.	5.18	5.15	5.09	4.96
T_{fwd}	m	1.68	1.37	1.07	.760
	ft.	5.50	4.50	3.50	2.50
A_m	m^2	5.55	5.64	5.67	5.85
	ft^2	59.7	60.7	6.10	6.30
∇_1 Δ_1	m^3	65	65	65	65
	ton	66.6	66.6	66.6	66.6
∇_2 Δ_2	ft^3	2295	2295	2295	2295
	tons	65.6	65.6	65.6	65.6
S	m^2	101	101	101	101
	ft^2	1087	1087	1087	1087
Non-dimensional	$\theta\% = \pm\frac{1}{2}L$	+.455	-1.53	-4.82	-8.15
	$L/\nabla^{1/3}$	4.53	4.71	4.72	4.73
	δ	.434	.421	.427	.438
	φ	.642	.609	.604	.584
	L/B	3.51	3.65	3.66	3.71
	B/T	3.29	3.30	3.33	3.42
	$\frac{1}{2}\alpha$ E	29°	26°	24°	19°
L = 20 m 65.5'	B m	5.70	5.48	5.46	5.44
	B ft.	18.7	18.0	17.9	17.8
	T m	1.73	1.66	1.63	1.59
	T ft.	5.68	5.44	5.35	5.21
	∇ m^3	85.6	76.6	76.0	75.8
	Δ_1 ton	87.7	78.5	77.9	77.7
L = 30.5 m 100'	B m	8.70	8.36	8.33	8.29
	B ft.	28.5	27.4	27.3	27.1
	T m	2.64	2.53	2.49	2.42
	T ft.	8.67	8.30	8.20	7.29
	∇ m^3	304	272	270	268
	Δ_1 ton	312	279	277	275

Fig. 381

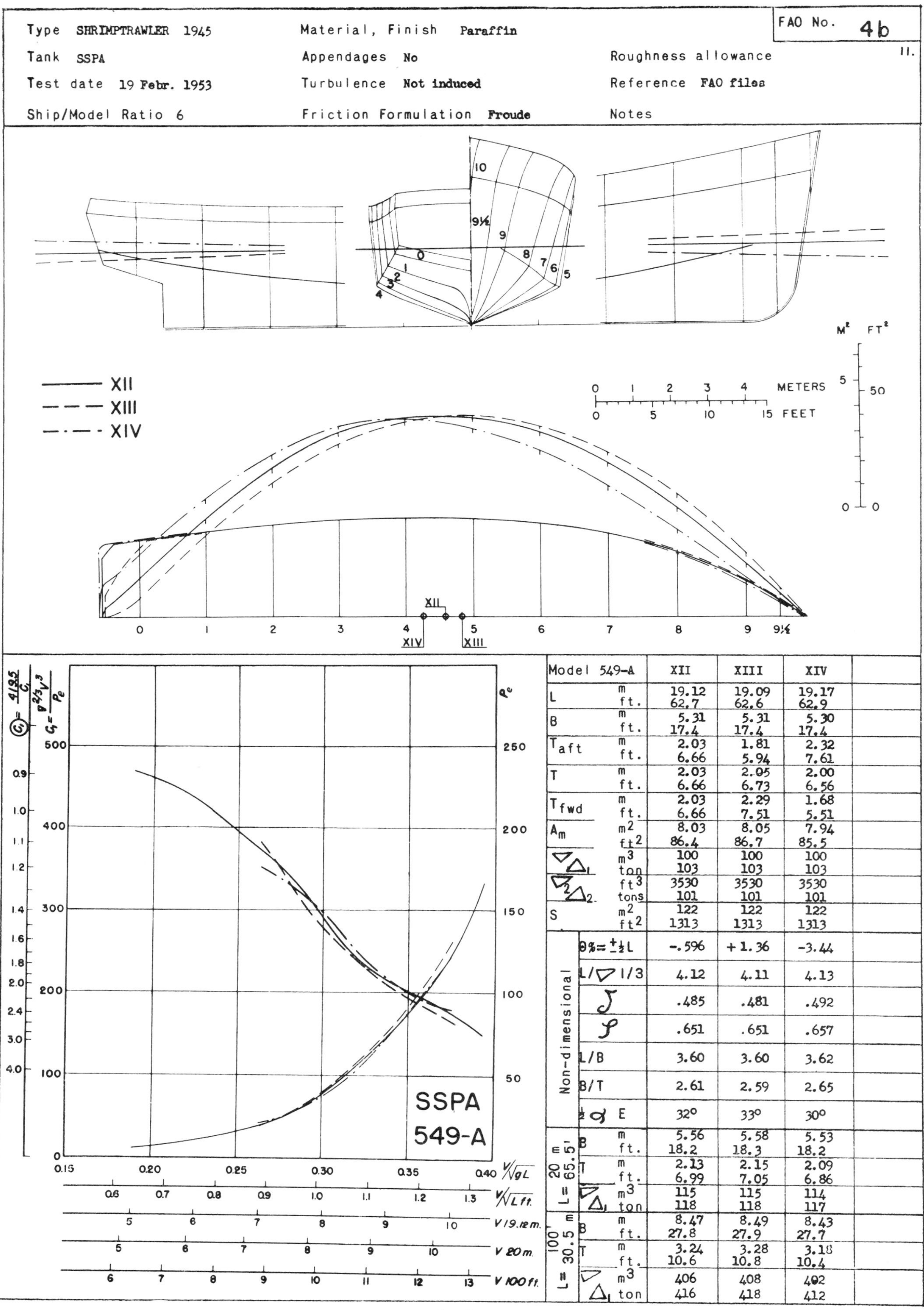

Type SHRIMPTRAWLER 1945
Material, Finish Paraffin
Tank SSPA
Appendages No
Roughness allowance
Test date 19 Febr. 1953
Turbulence Not induced
Reference FAO files
Ship/Model Ratio 6
Friction Formulation Froude
Notes

Model 549-A			XII	XIII	XIV
L		m	19.12	19.09	19.17
		ft.	62.7	62.6	62.9
B		m	5.31	5.31	5.30
		ft.	17.4	17.4	17.4
T_{aft}		m	2.03	1.81	2.32
		ft.	6.66	5.94	7.61
T		m	2.03	2.05	2.00
		ft.	6.66	6.73	6.56
T_{fwd}		m	2.03	2.29	1.68
		ft.	6.66	7.51	5.51
A_m		m^2	8.03	8.05	7.94
		ft^2	86.4	86.7	85.5
$\nabla_1 \Delta_1$		m^3	100	100	100
		ton	103	103	103
$\nabla_2 \Delta_2$		ft^3	3530	3530	3530
		tons	101	101	101
S		m^2	122	122	122
		ft^2	1313	1313	1313
Non-dimensional	$\otimes\% = \pm\frac{1}{2}L$		-.596	+1.36	-3.44
	$L/\nabla^{1/3}$		4.12	4.11	4.13
	δ		.485	.481	.492
	φ		.651	.651	.657
	L/B		3.60	3.60	3.62
	B/T		2.61	2.59	2.65
	$\frac{1}{2}\alpha$ E		32°	33°	30°
L = 20 m 65.5'	B	m	5.56	5.58	5.53
		ft.	18.2	18.3	18.2
	T	m	2.13	2.15	2.09
		ft.	6.99	7.05	6.86
	$\nabla_1 \Delta_1$	m^3	115	115	114
		ton	118	118	117
L = 30.5 m 100'	B	m	8.47	8.49	8.43
		ft.	27.8	27.9	27.7
	T	m	3.24	3.28	3.18
		ft.	10.6	10.8	10.4
	$\nabla_1 \Delta_1$	m^3	406	408	402
		ton	416	418	412

Fig. 382

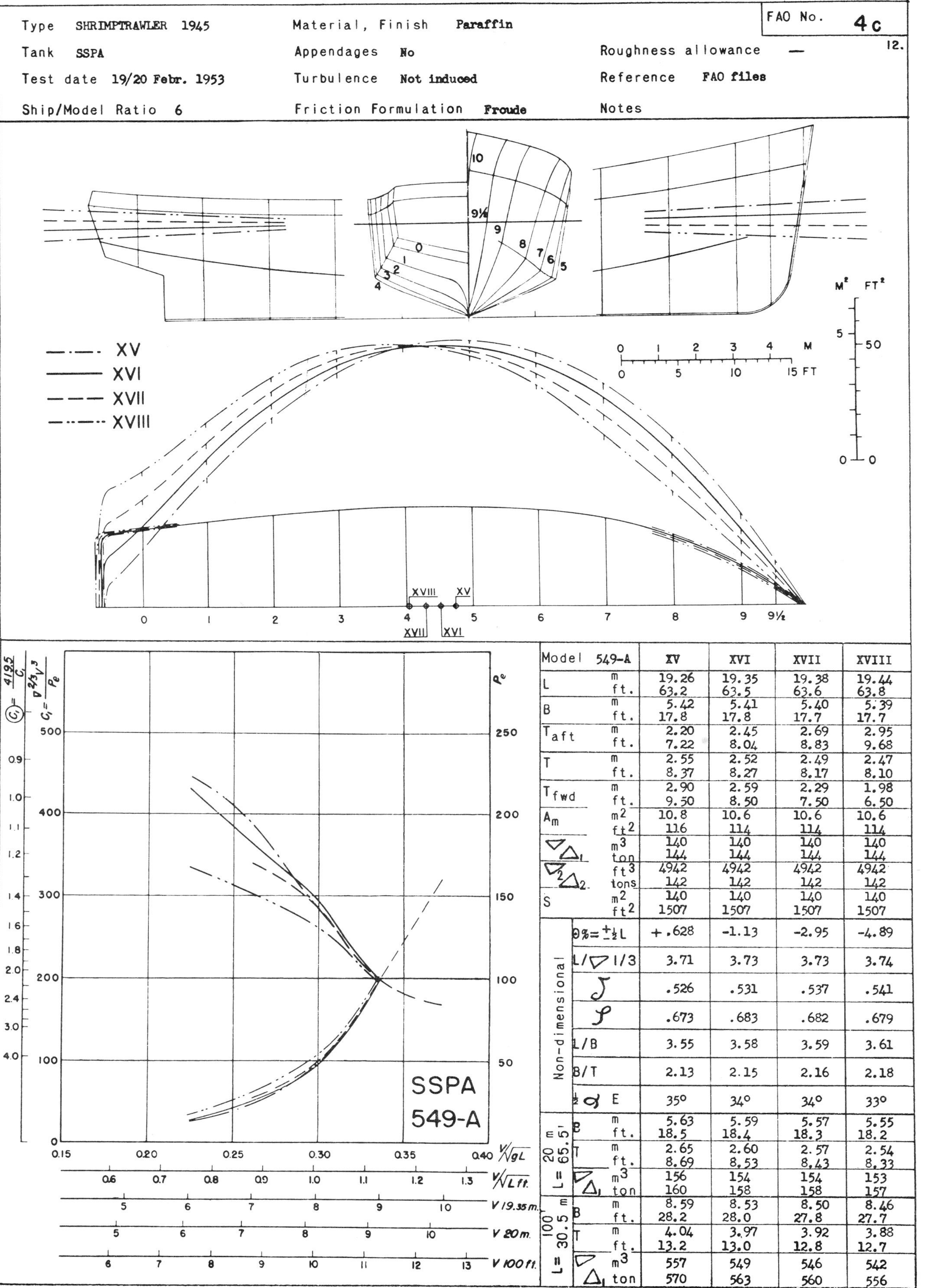

Type SHRIMPTRAWLER 1945	Material, Finish Paraffin		FAO No. 4c
Tank SSPA	Appendages No	Roughness allowance —	12.
Test date 19/20 Febr. 1953	Turbulence Not induced	Reference FAO files	
Ship/Model Ratio 6	Friction Formulation Froude	Notes	

Model 549-A		XV	XVI	XVII	XVIII
L	m	19.26	19.35	19.38	19.44
	ft.	63.2	63.5	63.6	63.8
B	m	5.42	5.41	5.40	5.39
	ft.	17.8	17.8	17.7	17.7
T_{aft}	m	2.20	2.45	2.69	2.95
	ft.	7.22	8.04	8.83	9.68
T	m	2.55	2.52	2.49	2.47
	ft.	8.37	8.27	8.17	8.10
T_{fwd}	m	2.90	2.59	2.29	1.98
	ft.	9.50	8.50	7.50	6.50
A_m	m^2	10.8	10.6	10.6	10.6
	ft^2	116	114	114	114
∇	m^3	140	140	140	140
Δ_1	ton	144	144	144	144
∇_2	ft^3	4942	4942	4942	4942
Δ_2	tons	142	142	142	142
S	m^2	140	140	140	140
	ft^2	1507	1507	1507	1507
Non-dimensional	$\theta\% = \pm\frac{1}{2}L$	+.628	-1.13	-2.95	-4.89
	$L/\nabla^{1/3}$	3.71	3.73	3.73	3.74
	δ	.526	.531	.537	.541
	φ	.673	.683	.682	.679
	L/B	3.55	3.58	3.59	3.61
	B/T	2.13	2.15	2.16	2.18
	$\frac{1}{2}\alpha$ E	35°	34°	34°	33°
L = 20 m 65.5'	B m	5.63	5.59	5.57	5.55
	ft.	18.5	18.4	18.3	18.2
	T m	2.65	2.60	2.57	2.54
	ft.	8.69	8.53	8.43	8.33
	∇ m^3	156	154	154	153
	Δ_1 ton	160	158	158	157
L = 100' 30.5 m	B m	8.59	8.53	8.50	8.46
	ft.	28.2	28.0	27.8	27.7
	T m	4.04	3.97	3.92	3.88
	ft.	13.2	13.0	12.8	12.7
	∇ m^3	557	549	546	542
	Δ_1 ton	570	563	560	556

Fig. 383

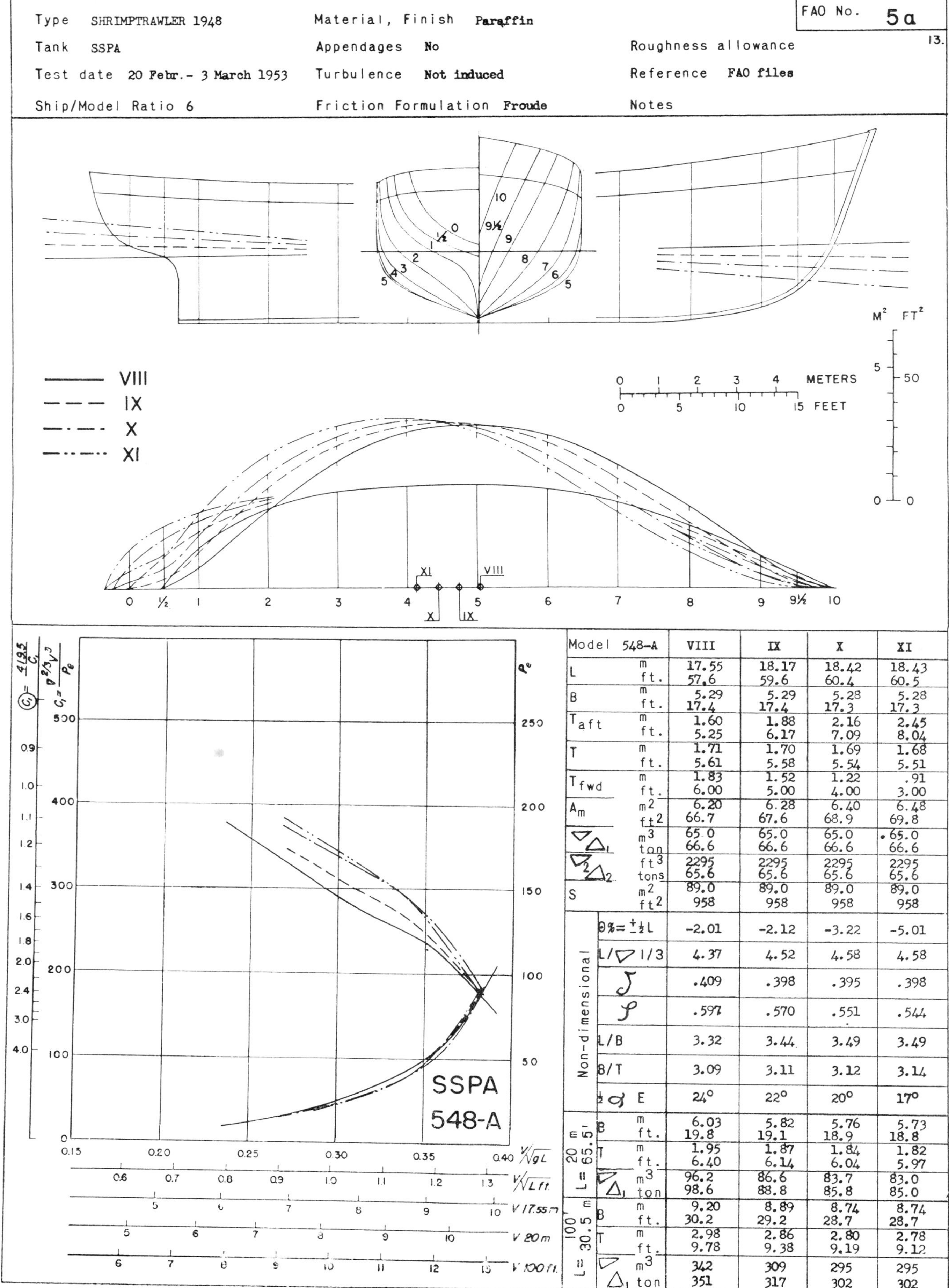

Model 548-A			VIII	IX	X	XI
L		m	17.55	18.17	18.42	18.43
		ft.	57.6	59.6	60.4	60.5
B		m	5.29	5.29	5.28	5.28
		ft.	17.4	17.4	17.3	17.3
T_{aft}		m	1.60	1.88	2.16	2.45
		ft.	5.25	6.17	7.09	8.04
T		m	1.71	1.70	1.69	1.68
		ft.	5.61	5.58	5.54	5.51
T_{fwd}		m	1.83	1.52	1.22	.91
		ft.	6.00	5.00	4.00	3.00
A_m		m^2	6.20	6.28	6.40	6.48
		ft^2	66.7	67.6	68.9	69.8
∇ / Δ_1		m^3	65.0	65.0	65.0	65.0
		ton	66.6	66.6	66.6	66.6
∇_2 / Δ_2		ft^3	2295	2295	2295	2295
		tons	65.6	65.6	65.6	65.6
S		m^2	89.0	89.0	89.0	89.0
		ft^2	958	958	958	958
Non-dimensional	$\theta\% = \pm\frac{1}{2}L$		-2.01	-2.12	-3.22	-5.01
	$L/\nabla^{1/3}$		4.37	4.52	4.58	4.58
	δ		.409	.398	.395	.398
	φ		.597	.570	.551	.544
	L/B		3.32	3.44	3.49	3.49
	B/T		3.09	3.11	3.12	3.14
	$\frac{1}{2}\alpha$ E		24°	22°	20°	17°
L = 20 m 65.5'	B	m	6.03	5.82	5.76	5.73
		ft.	19.8	19.1	18.9	18.8
	T	m	1.95	1.87	1.84	1.82
		ft.	6.40	6.14	6.04	5.97
	∇ / Δ_1	m^3	96.2	86.6	83.7	83.0
		ton	98.6	88.8	85.8	85.0
L = 100' 30.5 m	B	m	9.20	8.89	8.74	8.74
		ft.	30.2	29.2	28.7	28.7
	T	m	2.98	2.86	2.80	2.78
		ft.	9.78	9.38	9.19	9.12
	∇ / Δ_1	m^3	342	309	295	295
		ton	351	317	302	302

Fig. 384

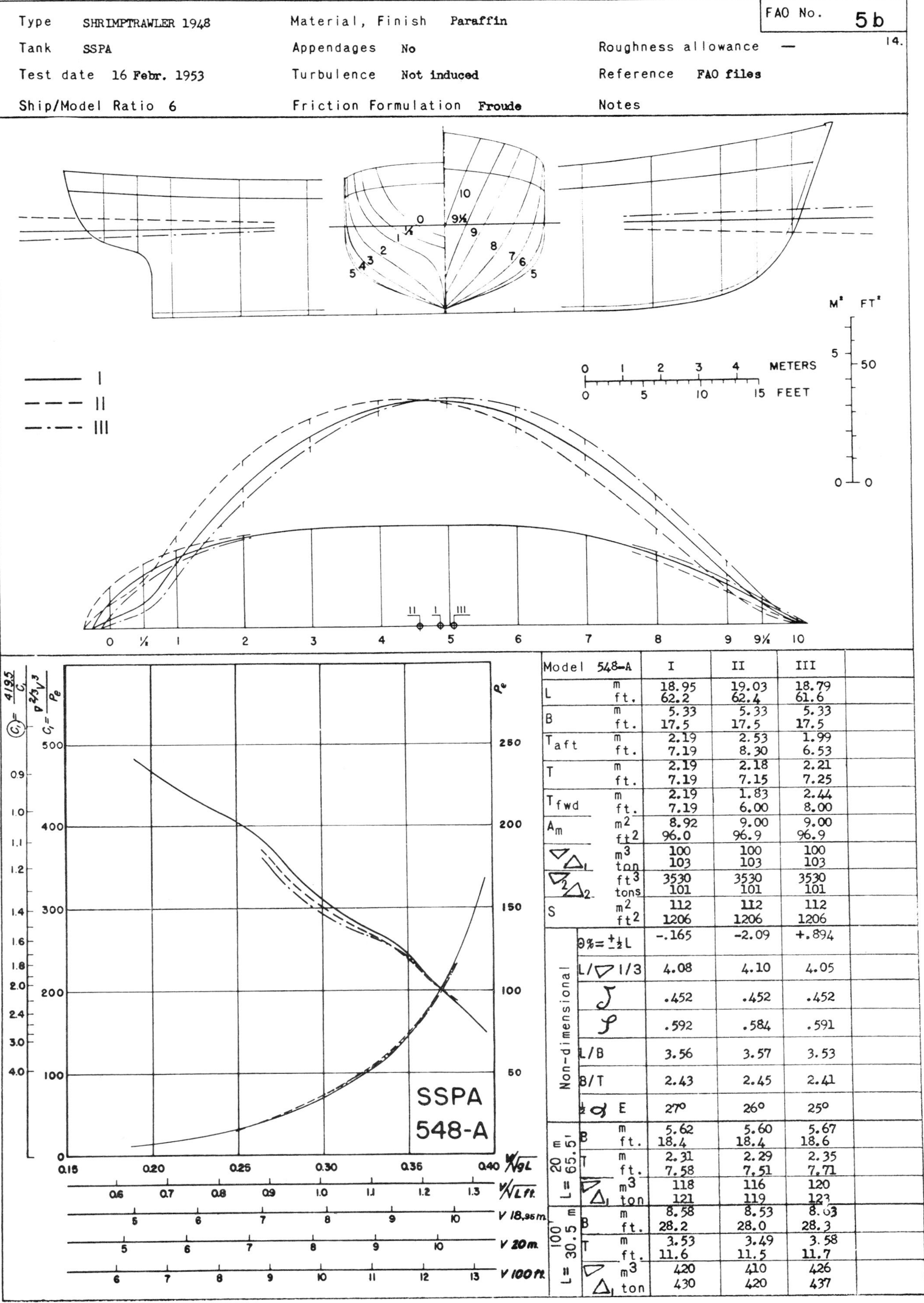

Model 548-A		I	II	III
L	m	18.95	19.03	18.79
	ft.	62.2	62.4	61.6
B	m	5.33	5.33	5.33
	ft.	17.5	17.5	17.5
T_{aft}	m	2.19	2.53	1.99
	ft.	7.19	8.30	6.53
T	m	2.19	2.18	2.21
	ft.	7.19	7.15	7.25
T_{fwd}	m	2.19	1.83	2.44
	ft.	7.19	6.00	8.00
A_m	m^2	8.92	9.00	9.00
	ft^2	96.0	96.9	96.9
∇_1	m^3	100	100	100
Δ_1	ton	103	103	103
∇_2	ft^3	3530	3530	3530
Δ_2	tons	101	101	101
S	m^2	112	112	112
	ft^2	1206	1206	1206
Non-dimensional	$\theta\% = \pm\frac{1}{2}L$	-.165	-2.09	+.894
	$L/\nabla^{1/3}$	4.08	4.10	4.05
	δ	.452	.452	.452
	φ	.592	.584	.591
	L/B	3.56	3.57	3.53
	B/T	2.43	2.45	2.41
	$\frac{1}{2}\alpha$ E	27°	26°	25°
L = 20 m 65.5'	B m	5.62	5.60	5.67
	ft.	18.4	18.4	18.6
	T m	2.31	2.29	2.35
	ft.	7.58	7.51	7.71
	∇_1 m^3	118	116	120
	Δ_1 ton	121	119	123
L = 100' 30.5 m	B m	8.58	8.53	8.63
	ft.	28.2	28.0	28.3
	T m	3.53	3.49	3.58
	ft.	11.6	11.5	11.7
	∇ m^3	420	410	426
	Δ_1 ton	430	420	437

Fig. 385

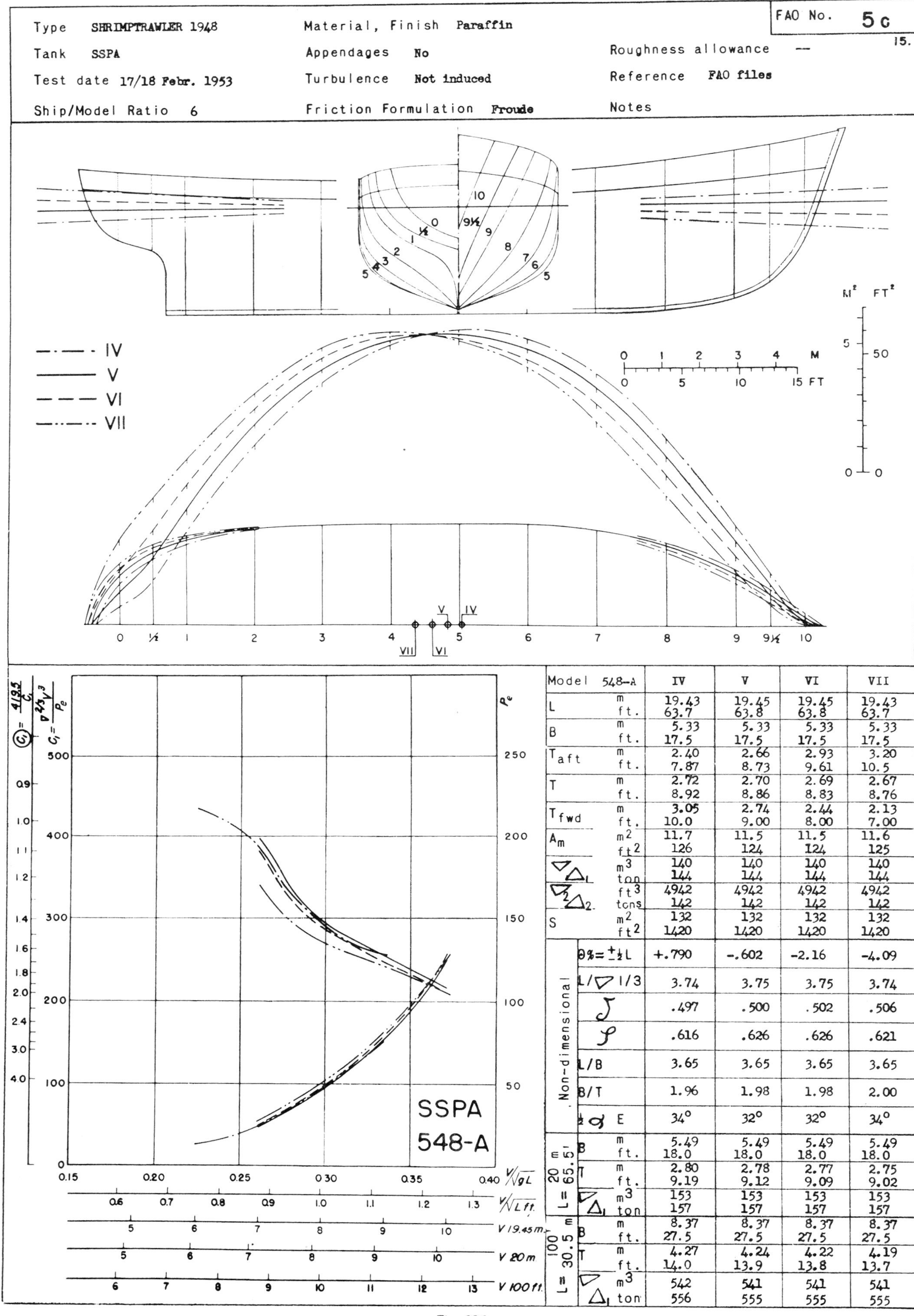

Type	SHRIMPTRAWLER 1948	Material, Finish	Paraffin	FAO No.	5c
Tank	SSPA	Appendages	No	Roughness allowance	—
Test date	17/18 Febr. 1953	Turbulence	Not induced	Reference	FAO files
Ship/Model Ratio	6	Friction Formulation	Froude	Notes	

Model 548-A		IV	V	VI	VII
L	m	19.43	19.45	19.45	19.43
	ft.	63.7	63.8	63.8	63.7
B	m	5.33	5.33	5.33	5.33
	ft.	17.5	17.5	17.5	17.5
T_{aft}	m	2.40	2.66	2.93	3.20
	ft.	7.87	8.73	9.61	10.5
T	m	2.72	2.70	2.69	2.67
	ft.	8.92	8.86	8.83	8.76
T_{fwd}	m	3.05	2.74	2.44	2.13
	ft.	10.0	9.00	8.00	7.00
A_m	m^2	11.7	11.5	11.5	11.6
	ft^2	126	124	124	125
[displacement] Δ_1	m^3	140	140	140	140
	ton	144	144	144	144
[displacement]$_2$ Δ_2	ft^3	4942	4942	4942	4942
	tons	142	142	142	142
S	m^2	132	132	132	132
	ft^2	1420	1420	1420	1420
Non-dimensional: θ% = ±½L		+.790	-.602	-2.16	-4.09
L/[displacement] 1/3		3.74	3.75	3.75	3.74
[symbol]		.497	.500	.502	.506
[symbol]		.616	.626	.626	.621
L/B		3.65	3.65	3.65	3.65
B/T		1.96	1.98	1.98	2.00
½ [symbol] E		34°	32°	32°	34°
L = 20 m 65.5' B	m	5.49	5.49	5.49	5.49
	ft.	18.0	18.0	18.0	18.0
T	m	2.80	2.78	2.77	2.75
	ft.	9.19	9.12	9.09	9.02
[displacement] Δ_1	m^3	153	153	153	153
	ton	157	157	157	157
L = 100' 30.5 m B	m	8.37	8.37	8.37	8.37
	ft.	27.5	27.5	27.5	27.5
T	m	4.27	4.24	4.22	4.19
	ft.	14.0	13.9	13.8	13.7
[displacement] Δ_1	m^3	542	541	541	541
	ton	556	555	555	555

Fig. 386

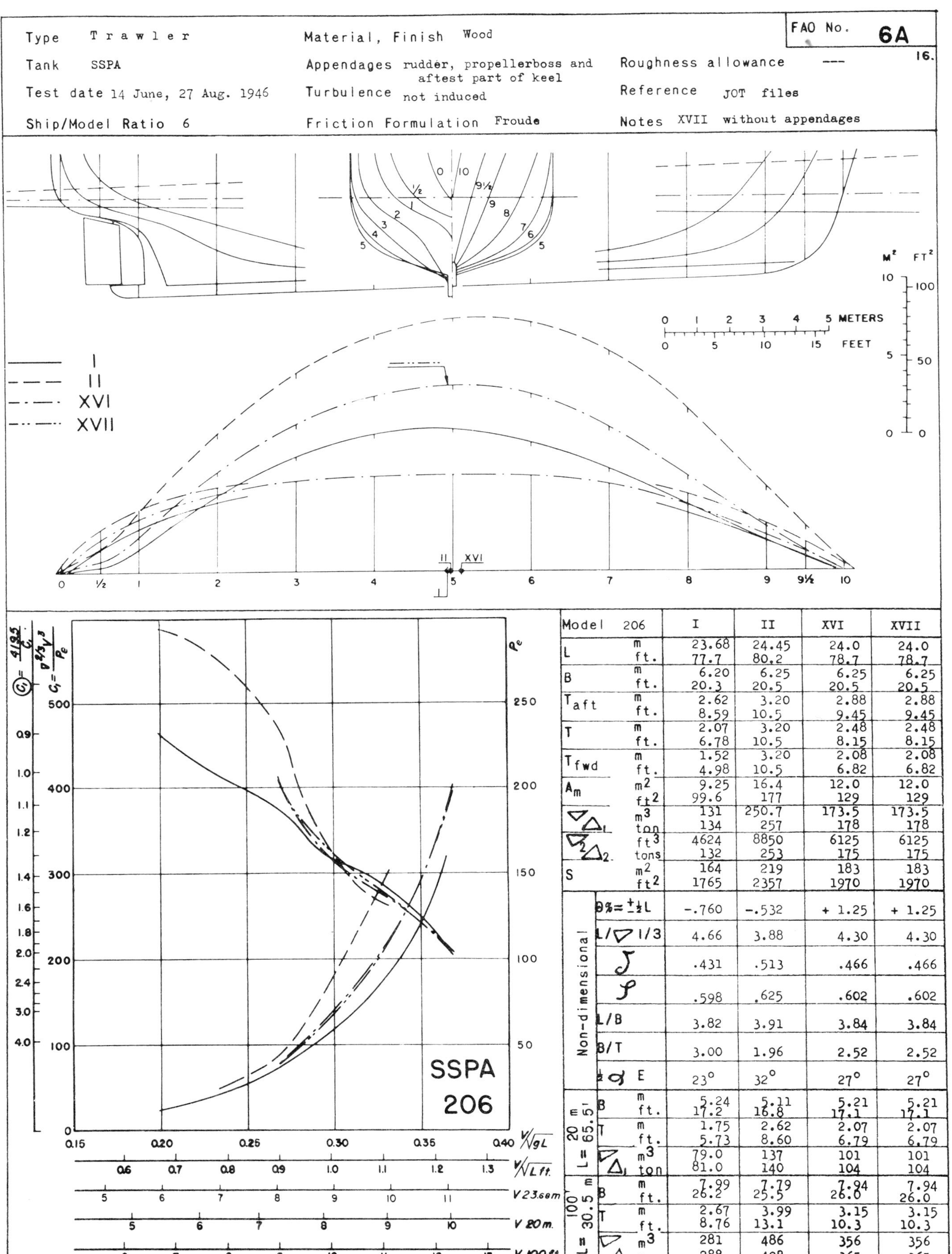

Type	Trawler	Material, Finish	Wood	FAO No.	6A
Tank	SSPA	Appendages	rudder, propellerboss and aftest part of keel	Roughness allowance	---
Test date	14 June, 27 Aug. 1946	Turbulence	not induced	Reference	JOT files
Ship/Model Ratio	6	Friction Formulation	Froude	Notes	XVII without appendages

16.

Model 206		I	II	XVI	XVII
L	m	23.68	24.45	24.0	24.0
	ft.	77.7	80.2	78.7	78.7
B	m	6.20	6.25	6.25	6.25
	ft.	20.3	20.5	20.5	20.5
T_{aft}	m	2.62	3.20	2.88	2.88
	ft.	8.59	10.5	9.45	9.45
T	m	2.07	3.20	2.48	2.48
	ft.	6.78	10.5	8.15	8.15
T_{fwd}	m	1.52	3.20	2.08	2.08
	ft.	4.98	10.5	6.82	6.82
A_m	m^2	9.25	16.4	12.0	12.0
	ft^2	99.6	177	129	129
∇_1	m^3	131	250.7	173.5	173.5
Δ_1	ton	134	257	178	178
∇_2	ft^3	4624	8850	6125	6125
Δ_2	tons	132	253	175	175
S	m^2	164	219	183	183
	ft^2	1765	2357	1970	1970
Non-dimensional	θ% = ± ½L	-.760	-.532	+ 1.25	+ 1.25
	$L/\nabla^{1/3}$	4.66	3.88	4.30	4.30
	δ	.431	.513	.466	.466
	φ	.598	.625	.602	.602
	L/B	3.82	3.91	3.84	3.84
	B/T	3.00	1.96	2.52	2.52
	½ α E	23°	32°	27°	27°
L = 20 m 65.5'	B m	5.24	5.11	5.21	5.21
	B ft.	17.2	16.8	17.1	17.1
	T m	1.75	2.62	2.07	2.07
	T ft.	5.73	8.60	6.79	6.79
	∇ m^3	79.0	137	101	101
	Δ_1 ton	81.0	140	104	104
L = 100' 30.5 m	B m	7.99	7.79	7.94	7.94
	B ft.	26.2	25.5	26.0	26.0
	T m	2.67	3.99	3.15	3.15
	T ft.	8.76	13.1	10.3	10.3
	∇ m^3	281	486	356	356
	Δ_1 ton	288	498	365	365

Fig. 387

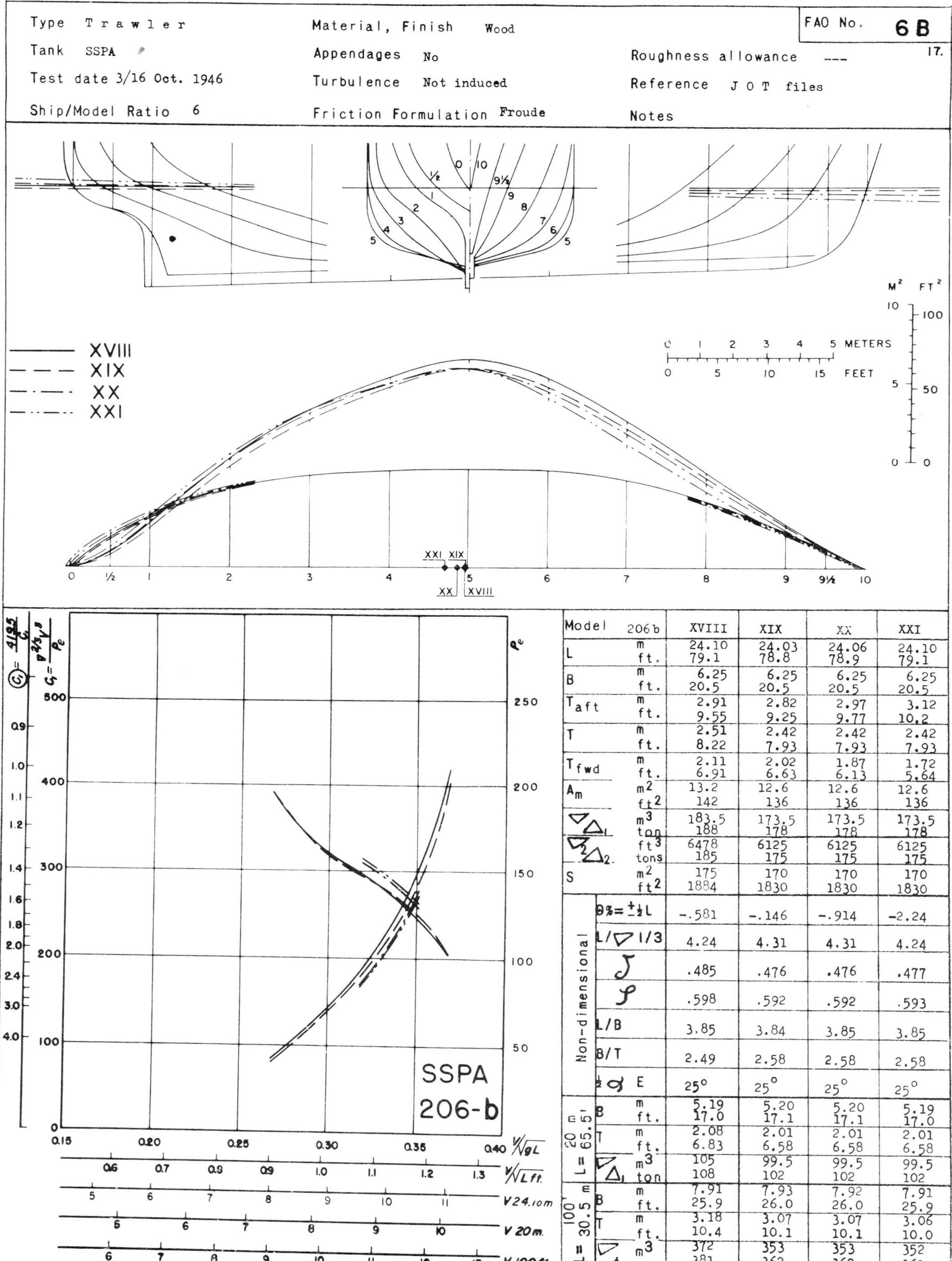

Type Trawler
Tank SSPA
Test date 3/16 Oct. 1946
Ship/Model Ratio 6
Material, Finish Wood
Appendages No
Turbulence Not induced
Friction Formulation Froude
FAO No. 6 B
17.
Roughness allowance ---
Reference J O T files
Notes

Model	206b	XVIII	XIX	XX	XXI
L	m	24.10	24.03	24.06	24.10
	ft.	79.1	78.8	78.9	79.1
B	m	6.25	6.25	6.25	6.25
	ft.	20.5	20.5	20.5	20.5
T_{aft}	m	2.91	2.82	2.97	3.12
	ft.	9.55	9.25	9.77	10.2
T	m	2.51	2.42	2.42	2.42
	ft.	8.22	7.93	7.93	7.93
T_{fwd}	m	2.11	2.02	1.87	1.72
	ft.	6.91	6.63	6.13	5.64
A_m	m^2	13.2	12.6	12.6	12.6
	ft^2	142	136	136	136
∇	m^3	183.5	173.5	173.5	173.5
Δ_1	ton	188	178	178	178
∇_2	ft^3	6478	6125	6125	6125
Δ_2	tons	185	175	175	175
S	m^2	175	170	170	170
	ft^2	1884	1830	1830	1830
Non-dimensional	$\theta\% = \pm\frac{1}{2}L$	-.581	-.146	-.914	-2.24
	$L/\nabla^{1/3}$	4.24	4.31	4.31	4.24
	δ	.485	.476	.476	.477
	φ	.598	.592	.592	.593
	L/B	3.85	3.84	3.85	3.85
	B/T	2.49	2.58	2.58	2.58
	$\frac{1}{2}\alpha$ E	25°	25°	25°	25°
L = 20 m 65.5'	B m	5.19	5.20	5.20	5.19
	B ft.	17.0	17.1	17.1	17.0
	T m	2.08	2.01	2.01	2.01
	T ft.	6.83	6.58	6.58	6.58
	∇ m^3	105	99.5	99.5	99.5
	Δ_1 ton	108	102	102	102
L = 100' 30.5 m	B m	7.91	7.93	7.92	7.91
	B ft.	25.9	26.0	26.0	25.9
	T m	3.18	3.07	3.07	3.06
	T ft.	10.4	10.1	10.1	10.0
	∇ m^3	372	353	353	352
	Δ_1 ton	381	362	362	361

Fig. 388

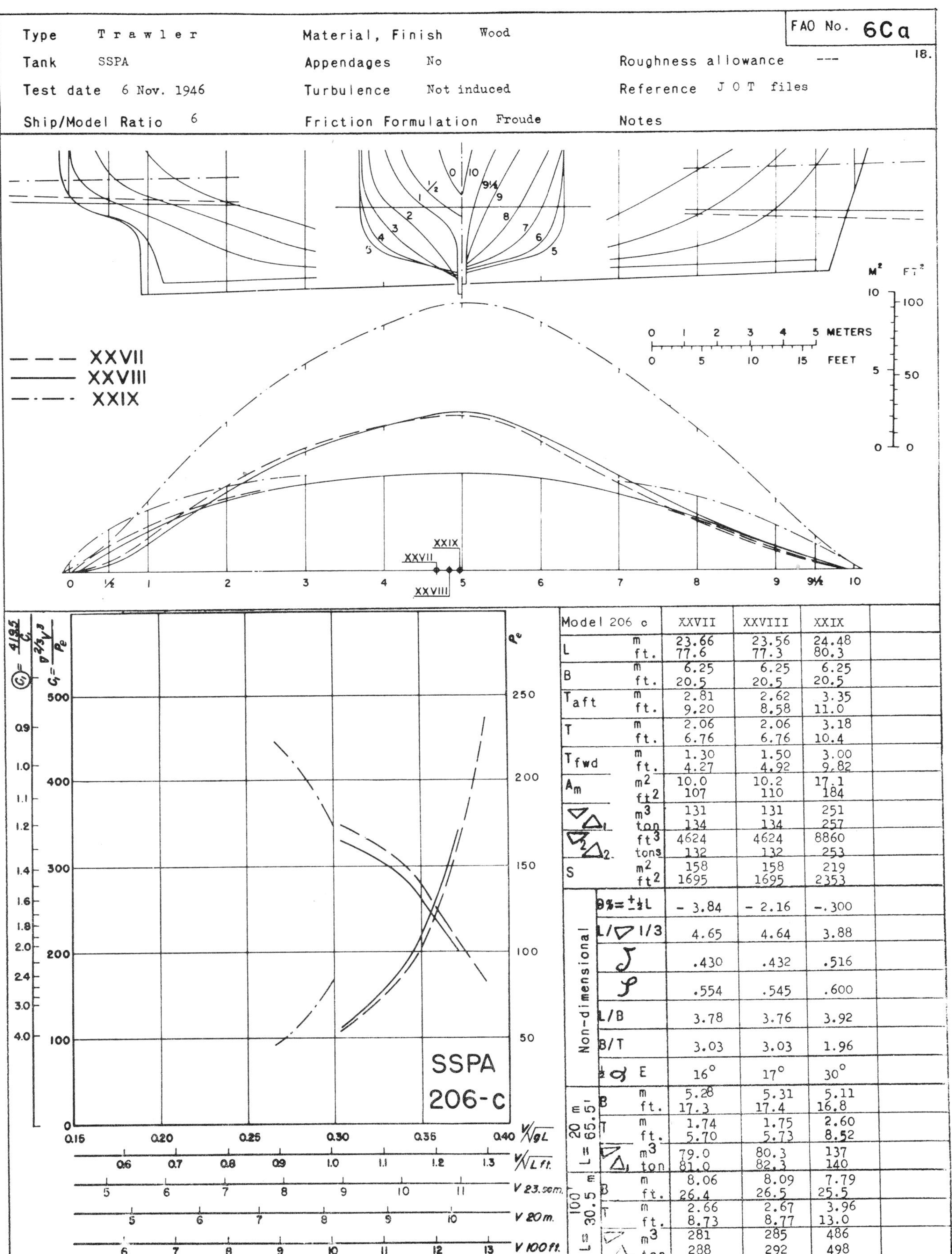

Type	Trawler	Material, Finish	Wood	FAO No.	6Ca
Tank	SSPA	Appendages	No	Roughness allowance	---
Test date	6 Nov. 1946	Turbulence	Not induced	Reference	J O T files
Ship/Model Ratio	6	Friction Formulation	Froude	Notes	

Model 206 c		XXVII	XXVIII	XXIX
L	m	23.66	23.56	24.48
	ft.	77.6	77.3	80.3
B	m	6.25	6.25	6.25
	ft.	20.5	20.5	20.5
T_{aft}	m	2.81	2.62	3.35
	ft.	9.20	8.58	11.0
T	m	2.06	2.06	3.18
	ft.	6.76	6.76	10.4
T_{fwd}	m	1.30	1.50	3.00
	ft.	4.27	4.92	9.82
A_m	m^2	10.0	10.2	17.1
	ft^2	107	110	184
∇ / Δ_1	m^3	131	131	251
	ton	134	134	257
∇_2 / Δ_2	ft^3	4624	4624	8860
	tons	132	132	253
S	m^2	158	158	219
	ft^2	1695	1695	2353
Non-dimensional	$\otimes\% = \pm\frac{1}{2}L$	- 3.84	- 2.16	-.300
	$L/\nabla^{1/3}$	4.65	4.64	3.88
	δ	.430	.432	.516
	φ	.554	.545	.600
	L/B	3.78	3.76	3.92
	B/T	3.03	3.03	1.96
	$\frac{1}{2}\alpha$ E	16°	17°	30°
L = 20 m 65.5'	B m	5.28	5.31	5.11
	B ft.	17.3	17.4	16.8
	T m	1.74	1.75	2.60
	T ft.	5.70	5.73	8.52
	∇ m^3	79.0	80.3	137
	Δ_1 ton	81.0	82.3	140
L = 100' 30.5 m	B m	8.06	8.09	7.79
	B ft.	26.4	26.5	25.5
	T m	2.66	2.67	3.96
	T ft.	8.73	8.77	13.0
	∇ m^3	281	285	486
	Δ_1 ton	288	292	498

Fig. 389

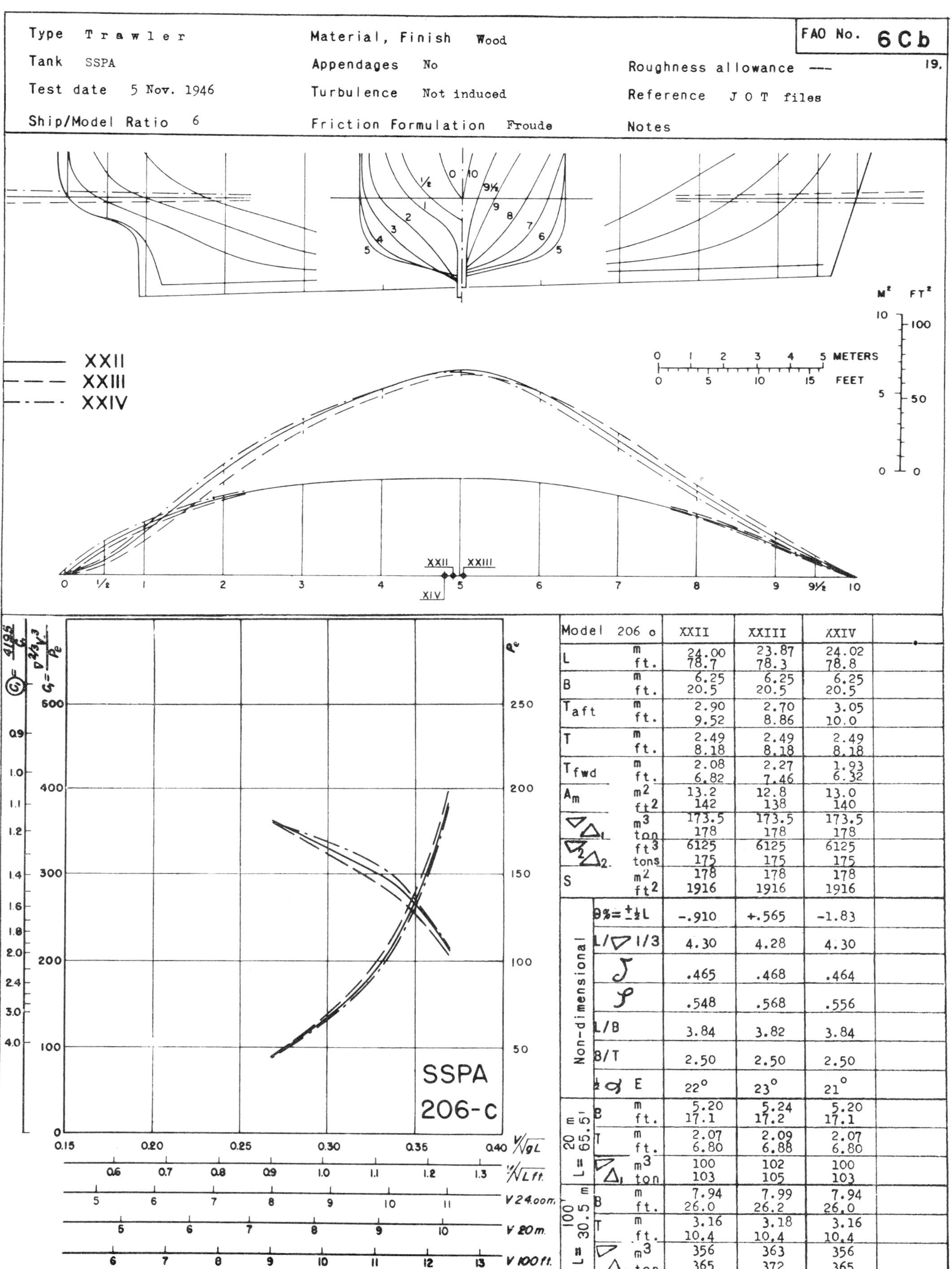

Type	Trawler	Material, Finish	Wood		FAO No. 6Cb
Tank	SSPA	Appendages	No	Roughness allowance	---
Test date	5 Nov. 1946	Turbulence	Not induced	Reference	J O T files
Ship/Model Ratio	6	Friction Formulation	Froude	Notes	

19.

Model 206 c			XXII	XXIII	XXIV
L		m	24.00	23.87	24.02
		ft.	78.7	78.3	78.8
B		m	6.25	6.25	6.25
		ft.	20.5	20.5	20.5
T_{aft}		m	2.90	2.70	3.05
		ft.	9.52	8.86	10.0
T		m	2.49	2.49	2.49
		ft.	8.18	8.18	8.18
T_{fwd}		m	2.08	2.27	1.93
		ft.	6.82	7.46	6.32
A_m		m^2	13.2	12.8	13.0
		ft^2	142	138	140
$\nabla_1 \Delta_1$		m^3	173.5	173.5	173.5
		ton	178	178	178
$\nabla_2 \Delta_2$		ft^3	6125	6125	6125
		tons	175	175	175
S		m^2	178	178	178
		ft^2	1916	1916	1916
Non-dimensional	$\theta\% = \pm\frac{1}{2}L$		-.910	+.565	-1.83
	$L/\nabla^{1/3}$		4.30	4.28	4.30
	δ		.465	.468	.464
	φ		.548	.568	.556
	L/B		3.84	3.82	3.84
	B/T		2.50	2.50	2.50
	$\frac{1}{2}\alpha$ E		22°	23°	21°
L = 20 m 65.5'	B	m	5.20	5.24	5.20
		ft.	17.1	17.2	17.1
	T	m	2.07	2.09	2.07
		ft.	6.80	6.88	6.80
	$\nabla \Delta_1$	m^3	100	102	100
		ton	103	105	103
L = 30.5 m 100'	B	m	7.94	7.99	7.94
		ft.	26.0	26.2	26.0
	T	m	3.16	3.18	3.16
		ft.	10.4	10.4	10.4
	$\nabla \Delta_1$	m^3	356	363	356
		ton	365	372	365

Fig. 390

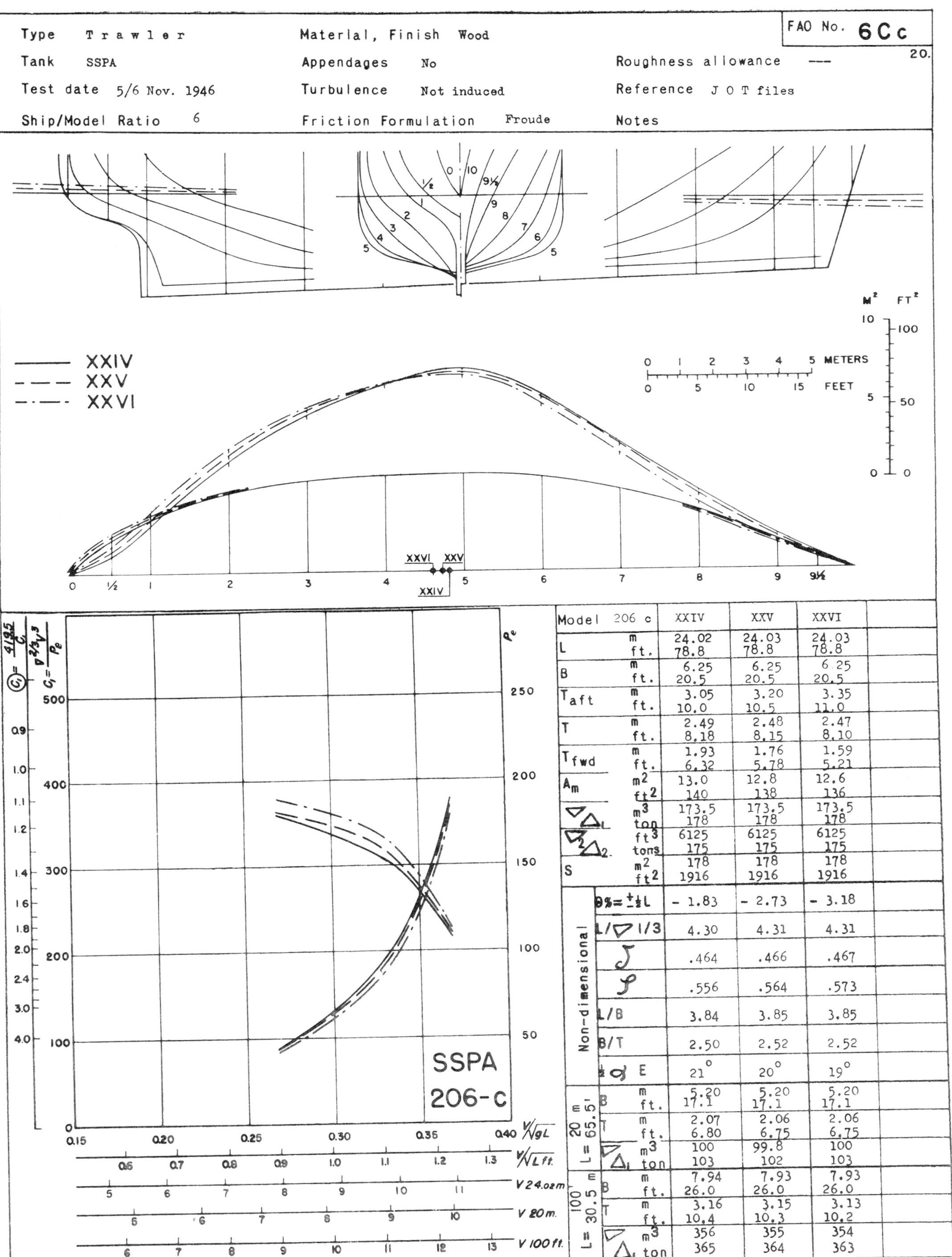

Type Trawler	Material, Finish Wood		FAO No. 6Cc
Tank SSPA	Appendages No	Roughness allowance ---	20.
Test date 5/6 Nov. 1946	Turbulence Not induced	Reference J O T files	
Ship/Model Ratio 6	Friction Formulation Froude	Notes	

Model 206 c		XXIV	XXV	XXVI
L	m	24.02	24.03	24.03
	ft.	78.8	78.8	78.8
B	m	6.25	6.25	6.25
	ft.	20.5	20.5	20.5
T_{aft}	m	3.05	3.20	3.35
	ft.	10.0	10.5	11.0
T	m	2.49	2.48	2.47
	ft.	8.18	8.15	8.10
T_{fwd}	m	1.93	1.76	1.59
	ft.	6.32	5.78	5.21
A_m	m^2	13.0	12.8	12.6
	ft^2	140	138	136
∇_1 Δ_1	m^3	173.5	173.5	173.5
	ton	178	178	178
∇_2 Δ_2	ft^3	6125	6125	6125
	tons	175	175	175
S	m^2	178	178	178
	ft^2	1916	1916	1916
Non-dimensional	lcb% = ± ½L	- 1.83	- 2.73	- 3.18
	$L/\nabla^{1/3}$	4.30	4.31	4.31
	δ	.464	.466	.467
	φ	.556	.564	.573
	L/B	3.84	3.85	3.85
	B/T	2.50	2.52	2.52
	½ α E	21°	20°	19°
L = 20 m 65.5'	B m	5.20	5.20	5.20
	B ft.	17.1	17.1	17.1
	T m	2.07	2.06	2.06
	T ft.	6.80	6.75	6.75
	∇ m^3	100	99.8	100
	Δ_1 ton	103	102	103
L = 100' 30.5 m	B m	7.94	7.93	7.93
	B ft.	26.0	26.0	26.0
	T m	3.16	3.15	3.13
	T ft.	10.4	10.3	10.2
	∇ m^3	356	355	354
	Δ_1 ton	365	364	363

Fig. 391

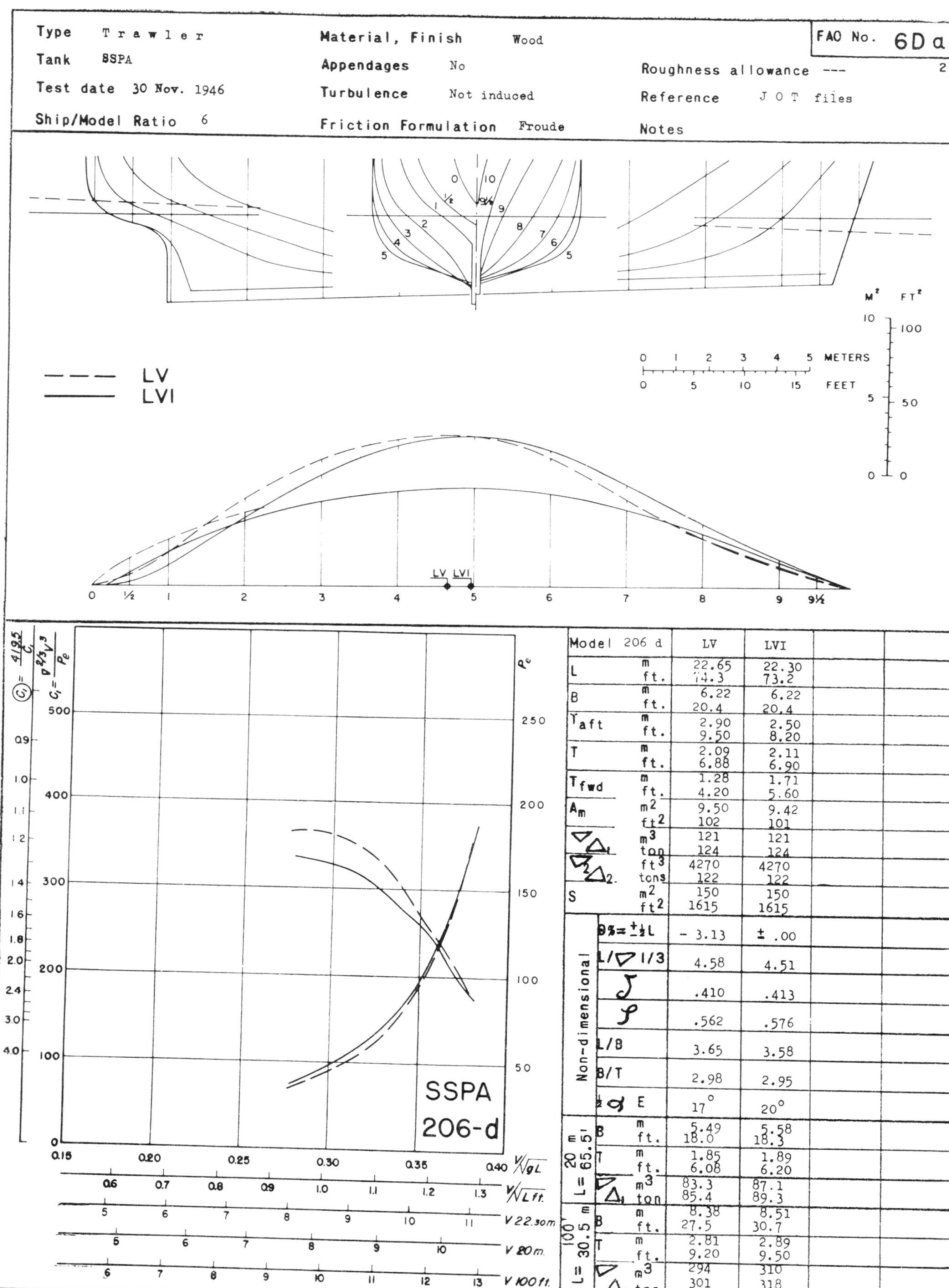

Type	Trawler	Material, Finish	Wood	FAO No.	6Da
Tank	SSPA	Appendages	No		21.
Test date	30 Nov. 1946	Turbulence	Not induced	Roughness allowance	---
Ship/Model Ratio	6	Friction Formulation	Froude	Reference	J O T files
				Notes	

Model 206 d		LV	LVI
L	m	22.65	22.30
	ft.	74.3	73.2
B	m	6.22	6.22
	ft.	20.4	20.4
T_{aft}	m	2.90	2.50
	ft.	9.50	8.20
T	m	2.09	2.11
	ft.	6.88	6.90
T_{fwd}	m	1.28	1.71
	ft.	4.20	5.60
A_m	m^2	9.50	9.42
	ft^2	102	101
$\nabla_1 \Delta_1$	m^3	121	121
	ton	124	124
$\nabla_2 \Delta_2$	ft^3	4270	4270
	tons	122	122
S	m^2	150	150
	ft^2	1615	1615
Non-dimensional	$\otimes\% = \pm\frac{1}{2}L$	- 3.13	± .00
	$L/\nabla^{1/3}$	4.58	4.51
	δ	.410	.413
	φ	.562	.576
	L/B	3.65	3.58
	B/T	2.98	2.95
	$\frac{1}{2}\alpha$ E	17°	20°
L = 20 m 65.5'	B m / ft.	5.49 / 18.0	5.58 / 18.3
	T m / ft.	1.85 / 6.08	1.89 / 6.20
	∇ m^3 / Δ_1 ton	83.3 / 85.4	87.1 / 89.3
L = 30.5 m 100'	B m / ft.	8.38 / 27.5	8.51 / 30.7
	T m / ft.	2.81 / 9.20	2.89 / 9.50
	∇ m^3 / Δ_1 ton	294 / 301	310 / 318

Fig. 392

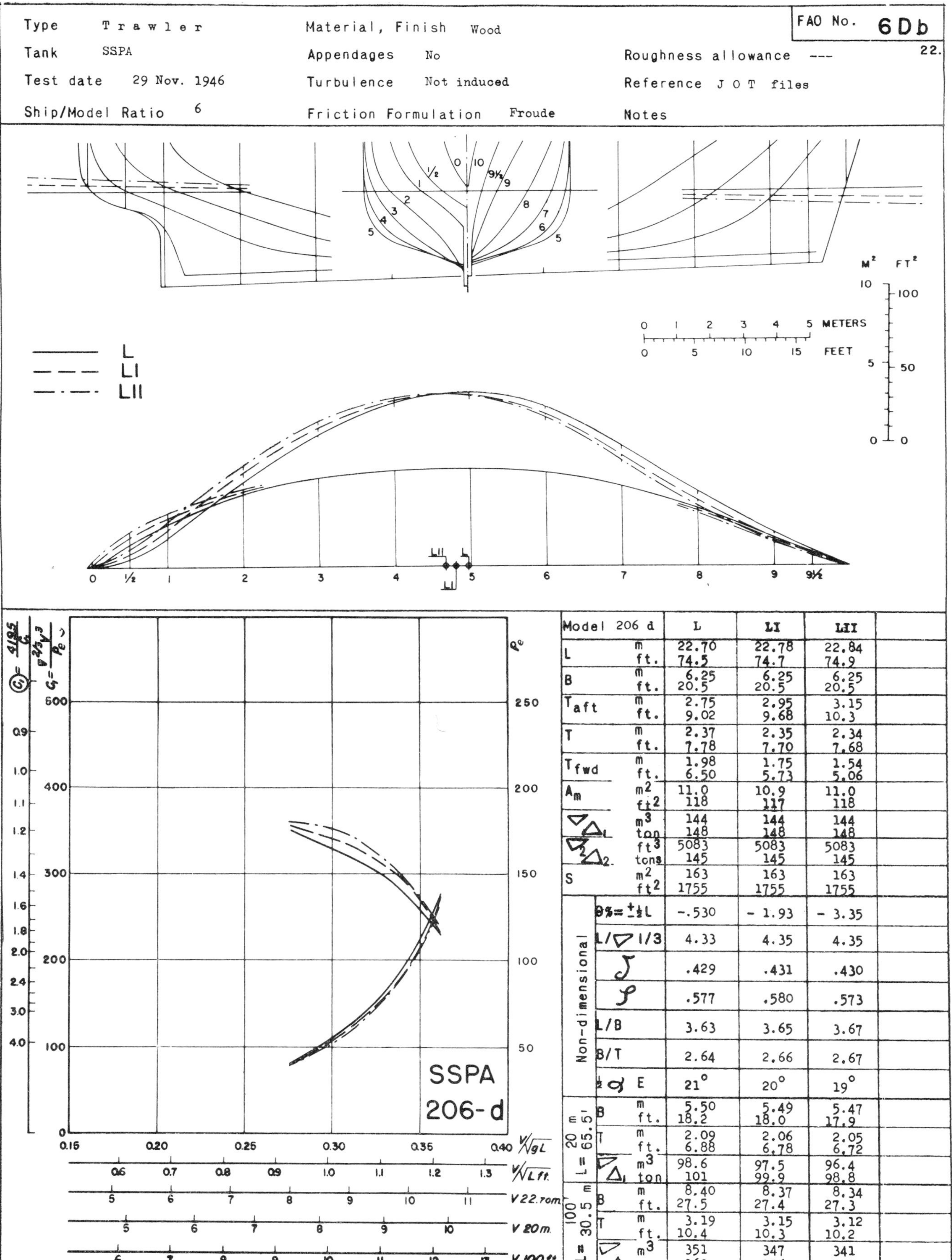

Model 206 d		L	LI	LII	
L	m ft.	22.70 74.5	22.78 74.7	22.84 74.9	
B	m ft.	6.25 20.5	6.25 20.5	6.25 20.5	
T_{aft}	m ft.	2.75 9.02	2.95 9.68	3.15 10.3	
T	m ft.	2.37 7.78	2.35 7.70	2.34 7.68	
T_{fwd}	m ft.	1.98 6.50	1.75 5.73	1.54 5.06	
A_m	m^2 ft^2	11.0 118	10.9 117	11.0 118	
∇_1 Δ_1	m^3 ton	144 148	144 148	144 148	
∇_2 Δ_2	ft^3 tons	5083 145	5083 145	5083 145	
S	m^2 ft^2	163 1755	163 1755	163 1755	
Non-dimensional	$\oplus\% = \pm \frac{1}{2}L$	-.530	- 1.93	- 3.35	
	$L/\nabla^{1/3}$	4.33	4.35	4.35	
	δ	.429	.431	.430	
	φ	.577	.580	.573	
	L/B	3.63	3.65	3.67	
	B/T	2.64	2.66	2.67	
	$\frac{1}{2}\alpha$ E	21°	20°	19°	
L = 20 m 65.5'	B m ft.	5.50 18.2	5.49 18.0	5.47 17.9	
	T m ft.	2.09 6.88	2.06 6.78	2.05 6.72	
	∇ m^3 Δ_1 ton	98.6 101	97.5 99.9	96.4 98.8	
L = 30.5 m 100'	B m ft.	8.40 27.5	8.37 27.4	8.34 27.3	
	T m ft.	3.19 10.4	3.15 10.3	3.12 10.2	
	∇ m^3 Δ_1 ton	351 360	347 356	341 350	

Fig. 393

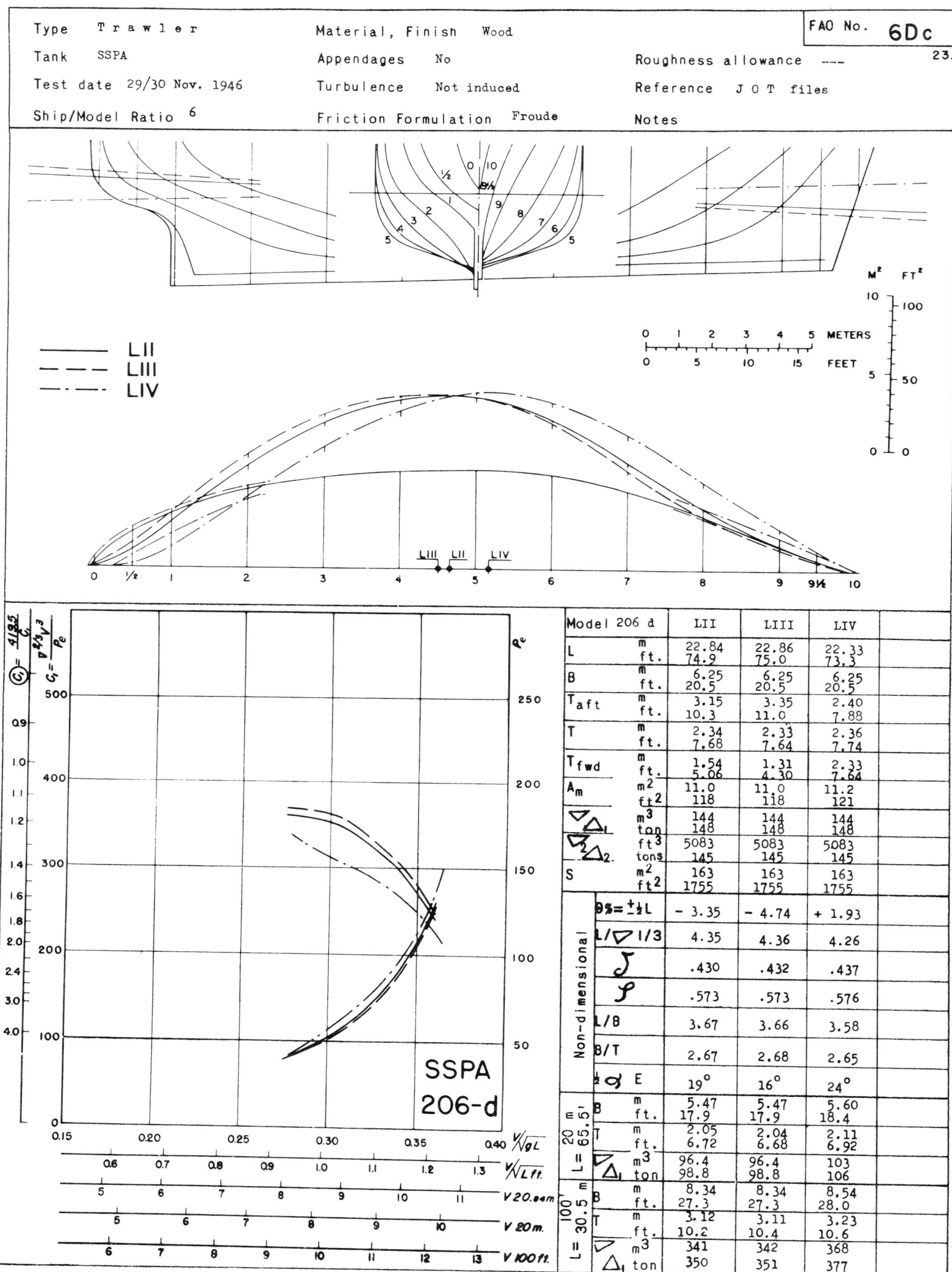

Type	Trawler	Material, Finish	Wood	FAO No.	6Dc
Tank	SSPA	Appendages	No	Roughness allowance	---
Test date	29/30 Nov. 1946	Turbulence	Not induced	Reference	J O T files
Ship/Model Ratio	6	Friction Formulation	Froude	Notes	

23.

Model 206 d		LII	LIII	LIV
L	m	22.84	22.86	22.33
	ft.	74.9	75.0	73.3
B	m	6.25	6.25	6.25
	ft.	20.5	20.5	20.5
T_{aft}	m	3.15	3.35	2.40
	ft.	10.3	11.0	7.88
T	m	2.34	2.33	2.36
	ft.	7.68	7.64	7.74
T_{fwd}	m	1.54	1.31	2.33
	ft.	5.06	4.30	7.64
A_m	m^2	11.0	11.0	11.2
	ft^2	118	118	121
∇	m^3	144	144	144
Δ_1	ton	148	148	148
∇_2	ft^3	5083	5083	5083
Δ_2	tons	145	145	145
S	m^2	163	163	163
	ft^2	1755	1755	1755
Non-dimensional	$\theta\% = \pm\frac{1}{2}L$	- 3.35	- 4.74	+ 1.93
	$L/\nabla^{1/3}$	4.35	4.36	4.26
	δ	.430	.432	.437
	φ	.573	.573	.576
	L/B	3.67	3.66	3.58
	B/T	2.67	2.68	2.65
	$\frac{1}{2}\alpha$ E	19°	16°	24°
L = 20 m 65.5'	B m	5.47	5.47	5.60
	ft.	17.9	17.9	18.4
	T m	2.05	2.04	2.11
	ft.	6.72	6.68	6.92
	∇ m^3	96.4	96.4	103
	Δ_1 ton	98.8	98.8	106
L = 100' 30.5 m	B m	8.34	8.34	8.54
	ft.	27.3	27.3	28.0
	T m	3.12	3.11	3.23
	ft.	10.2	10.4	10.6
	∇ m^3	341	342	368
	Δ_1 ton	350	351	377

Fig. 394

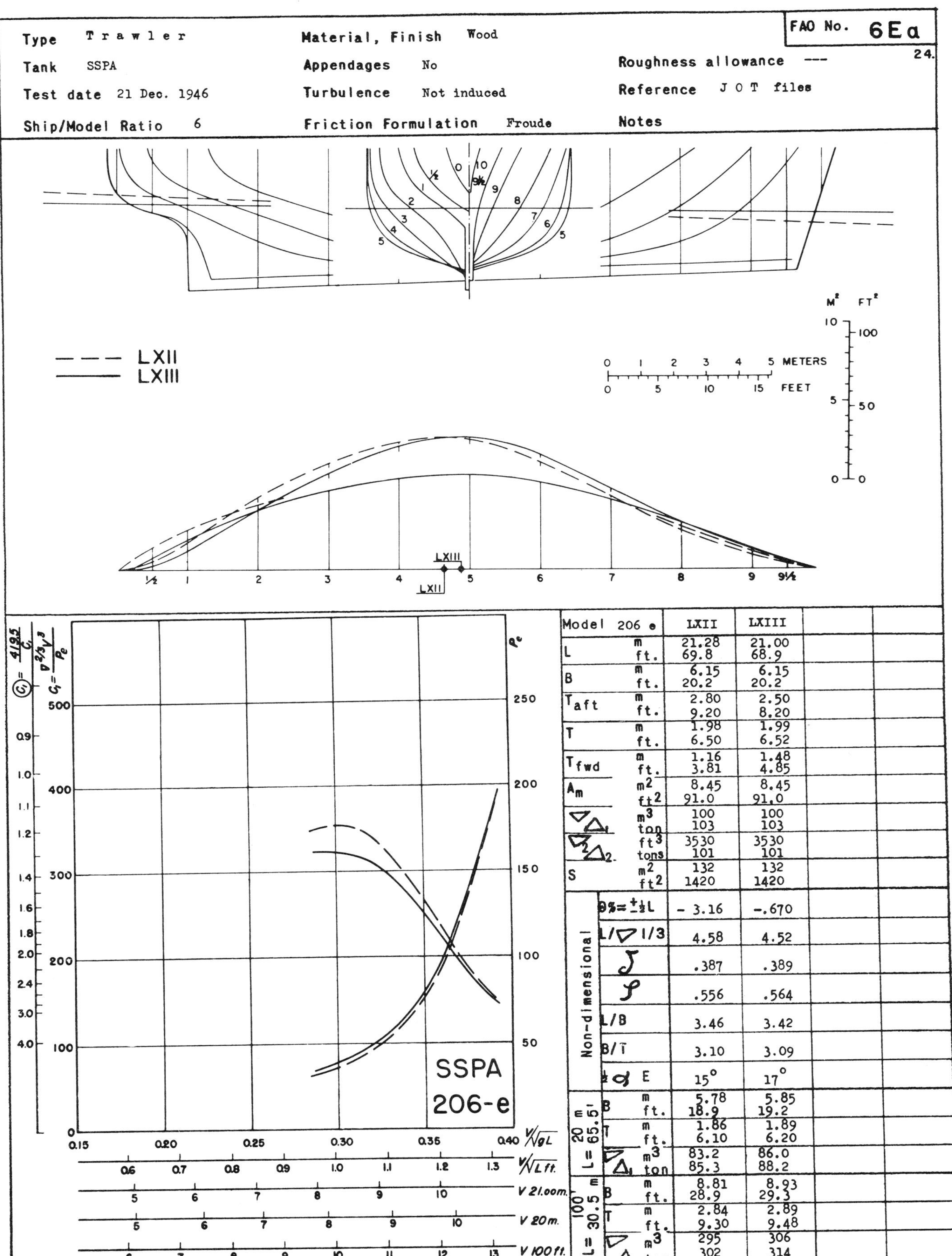

Model 206 e		LXII	LXIII
L	m	21.28	21.00
	ft.	69.8	68.9
B	m	6.15	6.15
	ft.	20.2	20.2
T_{aft}	m	2.80	2.50
	ft.	9.20	8.20
T	m	1.98	1.99
	ft.	6.50	6.52
T_{fwd}	m	1.16	1.48
	ft.	3.81	4.85
A_m	m^2	8.45	8.45
	ft^2	91.0	91.0
∇	m^3	100	100
Δ_1	ton	103	103
∇_2	ft^3	3530	3530
Δ_2	tons	101	101
S	m^2	132	132
	ft^2	1420	1420
Non-dimensional	θ% = ±½L	- 3.16	-.670
	$L/\nabla^{1/3}$	4.58	4.52
	δ	.387	.389
	φ	.556	.564
	L/B	3.46	3.42
	B/T	3.10	3.09
	½α E	15°	17°
L = 20 m 65.5'	B m	5.78	5.85
	ft.	18.9	19.2
	T m	1.86	1.89
	ft.	6.10	6.20
	∇ m^3	83.2	86.0
	Δ_1 ton	85.3	88.2
L = 30.5 m 100'	B m	8.81	8.93
	ft.	28.9	29.3
	T m	2.84	2.89
	ft.	9.30	9.48
	∇ m^3	295	306
	Δ_1 ton	302	314

Fig. 395

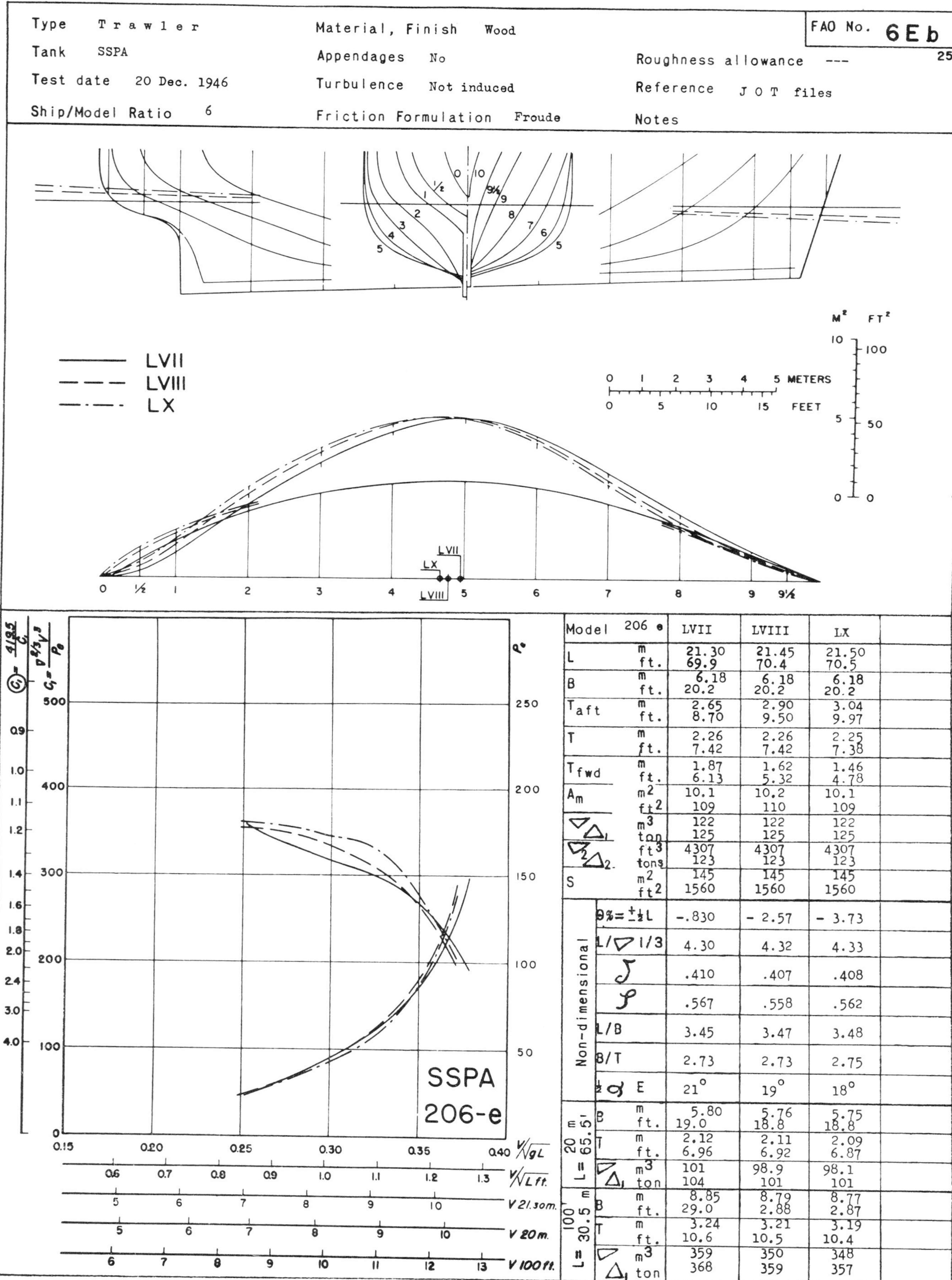

Type Trawler | Tank SSPA | Test date 20 Dec. 1946 | Ship/Model Ratio 6

Material, Finish Wood | Appendages No | Turbulence Not induced | Friction Formulation Froude

FAO No. 6Eb | 25.

Roughness allowance --- | Reference J O T files | Notes

Model 206 e		LVII	LVIII	LX
L	m	21.30	21.45	21.50
	ft.	69.9	70.4	70.5
B	m	6.18	6.18	6.18
	ft.	20.2	20.2	20.2
T_{aft}	m	2.65	2.90	3.04
	ft.	8.70	9.50	9.97
T	m	2.26	2.26	2.25
	ft.	7.42	7.42	7.38
T_{fwd}	m	1.87	1.62	1.46
	ft.	6.13	5.32	4.78
A_m	m^2	10.1	10.2	10.1
	ft^2	109	110	109
$\nabla_1 \Delta_1$	m^3	122	122	122
	ton	125	125	125
$\nabla_2 \Delta_2$	ft^3	4307	4307	4307
	tons	123	123	123
S	m^2	145	145	145
	ft^2	1560	1560	1560
Non-dimensional	$\theta\% = \pm\frac{1}{2}L$	-.830	- 2.57	- 3.73
	$L/\nabla^{1/3}$	4.30	4.32	4.33
	δ	.410	.407	.408
	φ	.567	.558	.562
	L/B	3.45	3.47	3.48
	B/T	2.73	2.73	2.75
	$\frac{1}{2}\alpha$ E	21°	19°	18°
L = 20 m 65.5'	B m	5.80	5.76	5.75
	B ft.	19.0	18.8	18.8
	T m	2.12	2.11	2.09
	T ft.	6.96	6.92	6.87
	$\nabla \Delta_1$ m^3	101	98.9	98.1
	ton	104	101	101
L = 30.5 m 100'	B m	8.85	8.79	8.77
	B ft.	29.0	2.88	2.87
	T m	3.24	3.21	3.19
	T ft.	10.6	10.5	10.4
	$\nabla \Delta_1$ m^3	359	350	348
	ton	368	359	357

Fig. 396

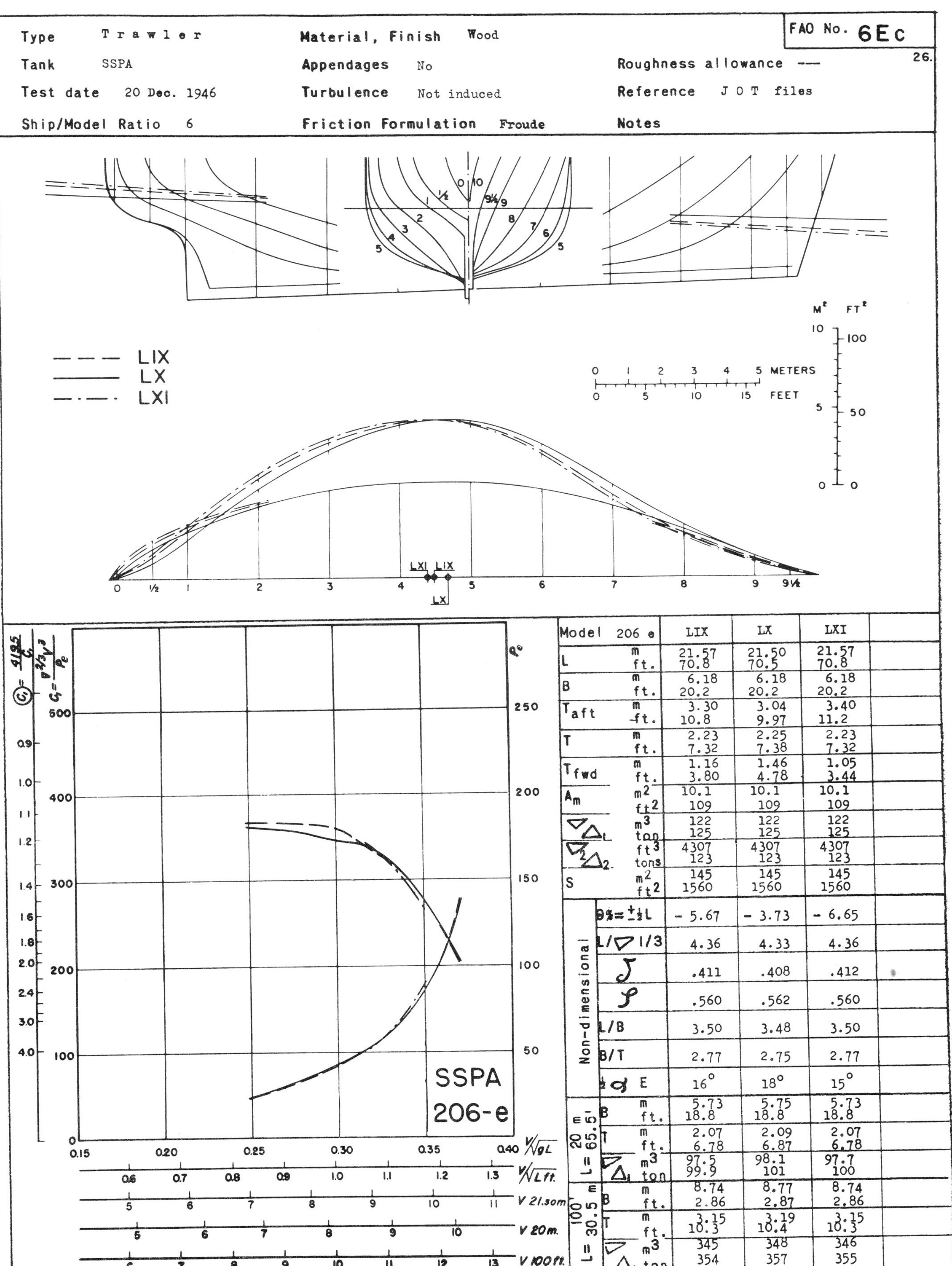

Type	Trawler	Material, Finish	Wood
Tank	SSPA	Appendages	No
Test date	20 Dec. 1946	Turbulence	Not induced
Ship/Model Ratio	6	Friction Formulation	Froude

FAO No. 6Ec

26.

Roughness allowance ---

Reference J O T files

Notes

Model 206 e		LIX	LX	LXI
L	m	21.57	21.50	21.57
	ft.	70.8	70.5	70.8
B	m	6.18	6.18	6.18
	ft.	20.2	20.2	20.2
T_{aft}	m	3.30	3.04	3.40
	ft.	10.8	9.97	11.2
T	m	2.23	2.25	2.23
	ft.	7.32	7.38	7.32
T_{fwd}	m	1.16	1.46	1.05
	ft.	3.80	4.78	3.44
A_m	m^2	10.1	10.1	10.1
	ft^2	109	109	109
∇_1	m^3	122	122	122
Δ_1	ton	125	125	125
∇_2	ft^3	4307	4307	4307
Δ_2	tons	123	123	123
S	m^2	145	145	145
	ft^2	1560	1560	1560
Non-dimensional	$\otimes\% = \pm \frac{1}{2}L$	- 5.67	- 3.73	- 6.65
	$L/\nabla^{1/3}$	4.36	4.33	4.36
	δ	.411	.408	.412
	φ	.560	.562	.560
	L/B	3.50	3.48	3.50
	B/T	2.77	2.75	2.77
	$\frac{1}{2}\alpha$ E	16°	18°	15°
L = 20 m 65.5'	B m	5.73	5.75	5.73
	B ft.	18.8	18.8	18.8
	T m	2.07	2.09	2.07
	T ft.	6.78	6.87	6.78
	∇ m^3	97.5	98.1	97.7
	Δ_1 ton	99.9	101	100
L = 30.5 m 100'	B m	8.74	8.77	8.74
	B ft.	2.86	2.87	2.86
	T m	3.15	3.19	3.15
	T ft.	10.3	10.4	10.3
	∇ m^3	345	348	346
	Δ_1 ton	354	357	355

Fig. 397

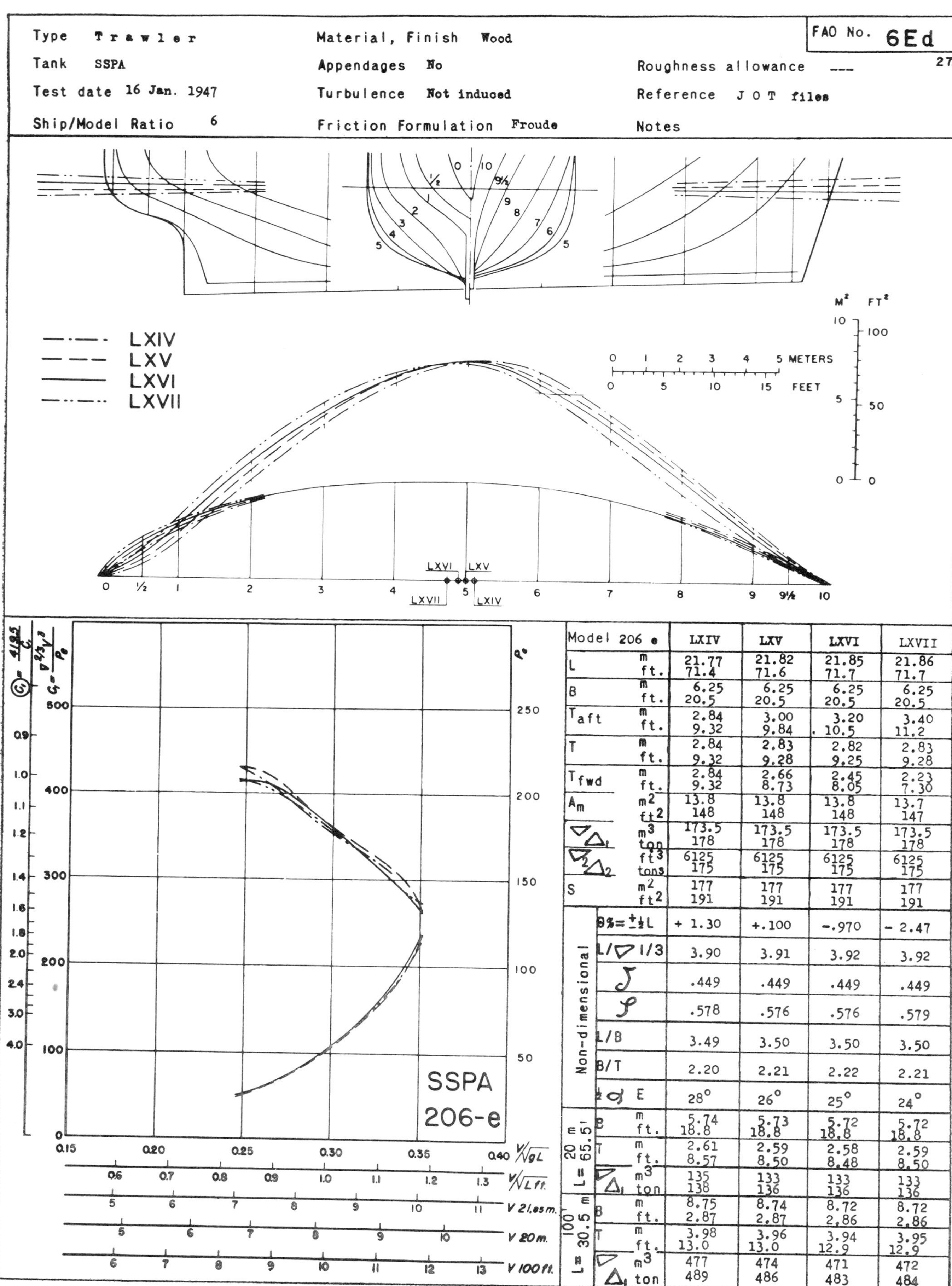

Type Trawler	Material, Finish Wood		FAO No. 6Ed
Tank SSPA	Appendages No	Roughness allowance ---	27.
Test date 16 Jan. 1947	Turbulence Not induced	Reference J O T files	
Ship/Model Ratio 6	Friction Formulation Froude	Notes	

Model 206 e		LXIV	LXV	LXVI	LXVII
L	m ft.	21.77 71.4	21.82 71.6	21.85 71.7	21.86 71.7
B	m ft.	6.25 20.5	6.25 20.5	6.25 20.5	6.25 20.5
T_{aft}	m ft.	2.84 9.32	3.00 9.84	3.20 10.5	3.40 11.2
T	m ft.	2.84 9.32	2.83 9.28	2.82 9.25	2.83 9.28
T_{fwd}	m ft.	2.84 9.32	2.66 8.73	2.45 8.05	2.23 7.30
A_m	m² ft²	13.8 148	13.8 148	13.8 148	13.7 147
∇_1 Δ_1	m³ ton	173.5 178	173.5 178	173.5 178	173.5 178
∇_2 Δ_2	ft³ tons	6125 175	6125 175	6125 175	6125 175
S	m² ft²	177 191	177 191	177 191	177 191
Non-dimensional	$\theta\% = \pm\frac{1}{2}L$	+ 1.30	+.100	-.970	- 2.47
	$L/\nabla^{1/3}$	3.90	3.91	3.92	3.92
	δ	.449	.449	.449	.449
	φ	.578	.576	.576	.579
	L/B	3.49	3.50	3.50	3.50
	B/T	2.20	2.21	2.22	2.21
	½ α E	28°	26°	25°	24°
L = 20 m 65.5'	B m ft.	5.74 18.8	5.73 18.8	5.72 18.8	5.72 18.8
	T m ft.	2.61 8.57	2.59 8.50	2.58 8.48	2.59 8.50
	∇ m³ Δ_1 ton	135 138	133 136	133 136	133 136
L = 100' 30.5 m	B m ft.	8.75 2.87	8.74 2.87	8.72 2.86	8.72 2.86
	T m ft.	3.98 13.0	3.96 13.0	3.94 12.9	3.95 12.9
	∇ m³ Δ_1 ton	477 489	474 486	471 483	472 484

Fig. 398

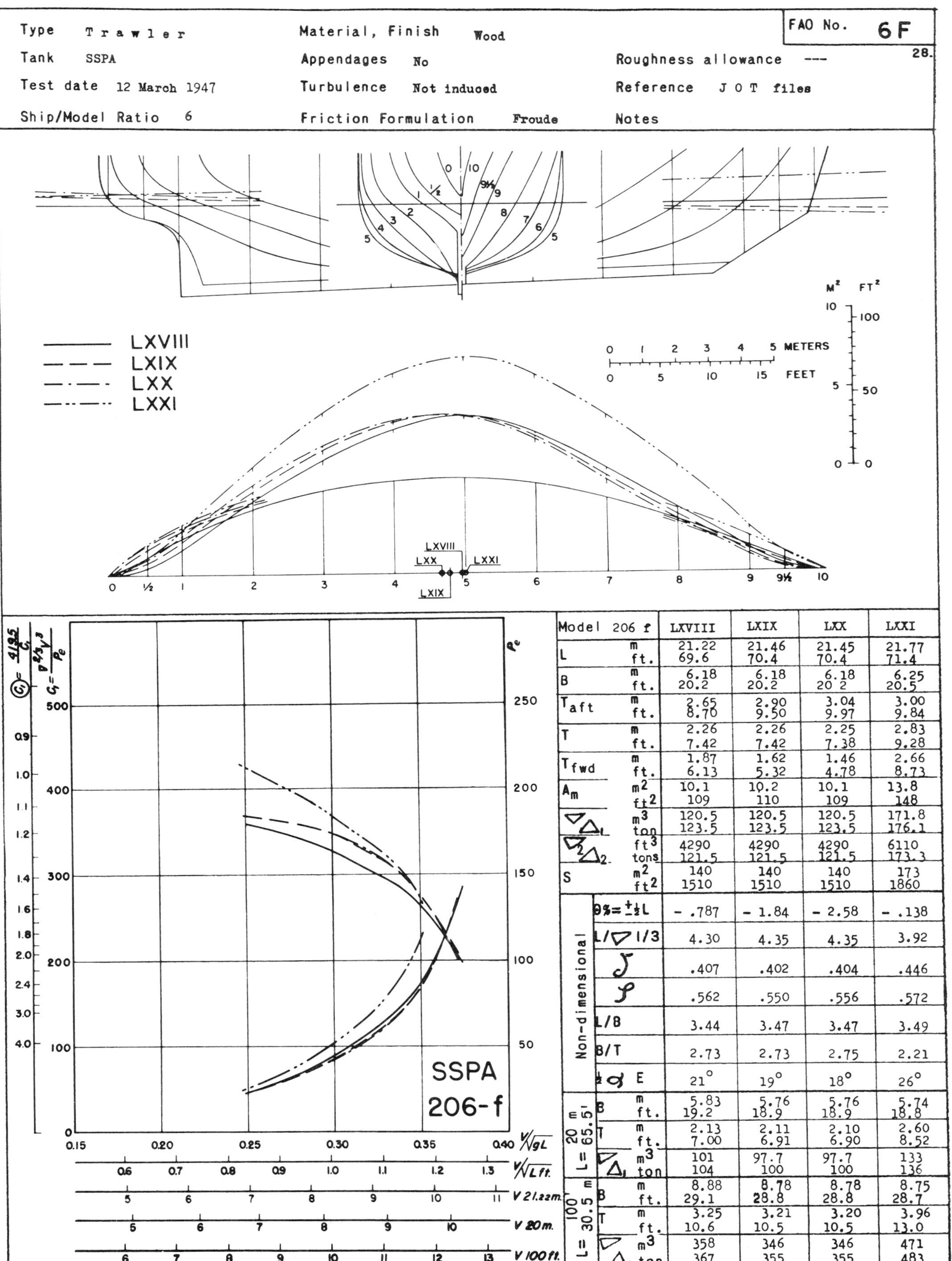

Model 206 f		LXVIII	LXIX	LXX	LXXI
L	m ft.	21.22 69.6	21.46 70.4	21.45 70.4	21.77 71.4
B	m ft.	6.18 20.2	6.18 20.2	6.18 20 2	6.25 20.5
T_{aft}	m ft.	2.65 8.70	2.90 9.50	3.04 9.97	3.00 9.84
T	m ft.	2.26 7.42	2.26 7.42	2.25 7.38	2.83 9.28
T_{fwd}	m ft.	1.87 6.13	1.62 5.32	1.46 4.78	2.66 8.73
A_m	m^2 ft^2	10.1 109	10.2 110	10.1 109	13.8 148
$\nabla_1 \Delta_1$	m^3 ton	120.5 123.5	120.5 123.5	120.5 123.5	171.8 176.1
$\nabla_2 \Delta_2$	ft^3 tons	4290 121.5	4290 121.5	4290 121.5	6110 173.3
S	m^2 ft^2	140 1510	140 1510	140 1510	173 1860
Non-dimensional	θ% = ±½L	- .787	- 1.84	- 2.58	- .138
	$L/\nabla^{1/3}$	4.30	4.35	4.35	3.92
	δ	.407	.402	.404	.446
	φ	.562	.550	.556	.572
	L/B	3.44	3.47	3.47	3.49
	B/T	2.73	2.73	2.75	2.21
	½α E	21°	19°	18°	26°
L = 20 m 65.5'	B m ft.	5.83 19.2	5.76 18.9	5.76 18.9	5.74 18.8
	T m ft.	2.13 7.00	2.11 6.91	2.10 6.90	2.60 8.52
	∇ m^3 Δ_1 ton	101 104	97.7 100	97.7 100	133 136
L = 30.5 m 100'	B m ft.	8.88 29.1	8.78 28.8	8.78 28.8	8.75 28.7
	T m ft.	3.25 10.6	3.21 10.5	3.20 10.5	3.96 13.0
	∇ m^3 Δ_1 ton	358 367	346 355	346 355	471 483

Fig. 399

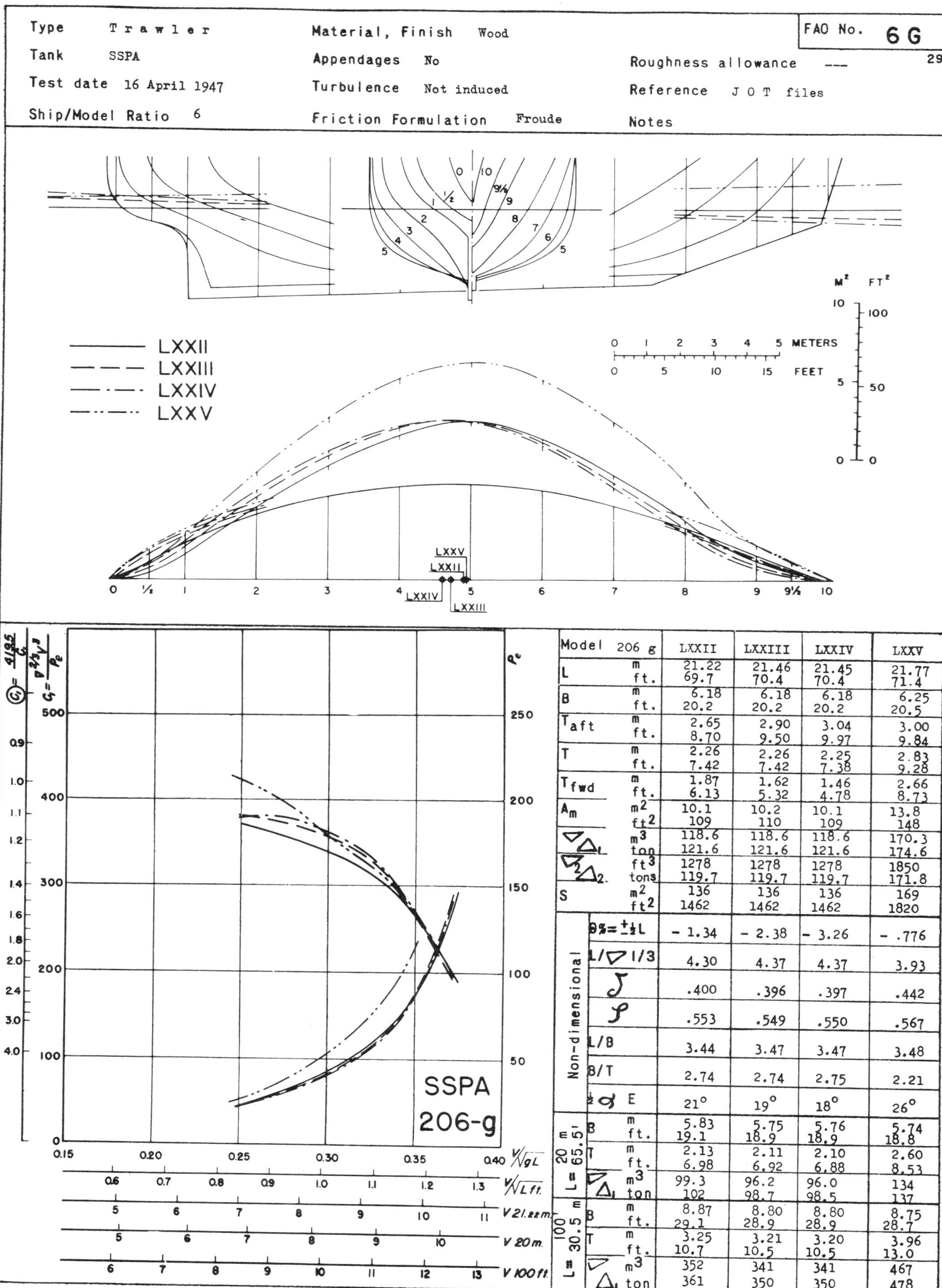

Type	Trawler	Material, Finish	Wood		FAO No. 6 G
Tank	SSPA	Appendages	No	Roughness allowance	---
Test date	16 April 1947	Turbulence	Not induced	Reference	J O T files
Ship/Model Ratio	6	Friction Formulation	Froude	Notes	

29.

Model 206 g			LXXII	LXXIII	LXXIV	LXXV
L		m ft.	21.22 69.7	21.46 70.4	21.45 70.4	21.77 71.4
B		m ft.	6.18 20.2	6.18 20.2	6.18 20.2	6.25 20.5
T_{aft}		m ft.	2.65 8.70	2.90 9.50	3.04 9.97	3.00 9.84
T		m ft.	2.26 7.42	2.26 7.42	2.25 7.38	2.83 9.28
T_{fwd}		m ft.	1.87 6.13	1.62 5.32	1.46 4.78	2.66 8.73
A_m		m^2 ft^2	10.1 109	10.2 110	10.1 109	13.8 148
∇ Δ_1		m^3 ton	118.6 121.6	118.6 121.6	118.6 121.6	170.3 174.6
∇_2 Δ_2		ft^3 tons	1278 119.7	1278 119.7	1278 119.7	1850 171.8
S		m^2 ft^2	136 1462	136 1462	136 1462	169 1820
Non-dimensional	$\theta\% = \pm\frac{1}{2}L$		- 1.34	- 2.38	- 3.26	- .776
	$L/\nabla^{1/3}$		4.30	4.37	4.37	3.93
	δ		.400	.396	.397	.442
	φ		.553	.549	.550	.567
	L/B		3.44	3.47	3.47	3.48
	B/T		2.74	2.74	2.75	2.21
	$\frac{1}{2}\alpha$ E		21°	19°	18°	26°
L = 20 m 65.5'	B	m ft.	5.83 19.1	5.75 18.9	5.76 18.9	5.74 18.8
	T	m ft.	2.13 6.98	2.11 6.92	2.10 6.88	2.60 8.53
	∇ Δ_1	m^3 ton	99.3 102	96.2 98.7	96.0 98.5	134 137
L = 30.5 m 100'	B	m ft.	8.87 29.1	8.80 28.9	8.80 28.9	8.75 28.7
	T	m ft.	3.25 10.7	3.21 10.5	3.20 10.5	3.96 13.0
	∇ Δ_1	m^3 ton	352 361	341 350	341 350	467 478

Fig. 400

2.5 and 3 per cent. for the 131 and 200 ft. (40 and 61 m.) waterline lengths respectively, and for 0.35 it was 1.6 and 2 per cent., respectively.

During recent years, the problem of laminar flow in tank testing has been discussed to considerable extent. Under certain conditions, the flow of water along the model is laminar, whereas it is always turbulent along the ship. When converting the model resistance figures to figures for the ship, one subtracts a calculated friction resistance based on turbulent conditions from the total model resistance, the rest being the residual resistance. If, actually, laminar flow occurred along the model, one would thus subtract a part of the residual resistance, with the result that the calculated total resistance for the ship would be too low. It is now fairly well determined that laminar flow exists when the models are small, when the model speed is low, when the block coefficient is high and when the angle of entrance is large. Small models are especially subject to the effect of laminar flow and different kinds of turbulence stimulating devices are used, such as trip wires, sandpaper, struts and water spray to ensure correct prediction of resistance. Turbulence is not always stimulated when larger models are used because fishing boat models are mostly tested at high speeds and their block coefficients are low. As the resistance normally decreases with smaller angles of entrance, there is not much danger that laminar flow on some tests should induce the selection of a wrong alternative. But it is recommended that naval architects should consider the problem and compare results from a large and a small model carefully. They must also remember that a turbulence stimulating device has resistance and that if of two similar models, the one with the trip wire has a somewhat higher resistance, this might be due to the trip wire and not to the quality of the lines.

It is rather unfortunate that there is no agreement to use a single friction formula to convert model results into those for the full-sized ships. The Froude formula is used in Europe and the Schoenherr in North America. Schoenherr gives somewhat higher figures than Froude. As the friction part of the total resistance of fishing boats is lower than for large ships, the difference is also lower. A comparative calculation of effective power for one model showed that the effective power, when calculated according to Schoenherr and including 12 per cent. allowance, was about 1 to 3 per cent. higher than when calculated after Froude. When judging the data sheets, one must therefore remember that results calculated after Schoenherr seem to be 1 to 3 per cent. worse.

A data sheet shows a model test as such and *does not at all* indicate that it is a report of a good hull shape. In many parts of the world, even in countries with well-developed boat-building industries, inefficient hull shapes are used. Since both good and bad models will be listed, the naval architect will be able to estimate the resistance of a variety of different designs and then, from the study of other sheets, be able to see if a better design could be developed. Many of the models shown in the data sheets were made from a purely exploratory point of view and therefore do not always correspond to the real displacement or trim of a ship.

Data sheets can be analysed in different ways. It is felt that when a naval architect is working on a specific design and has specific problems to solve, he wants to analyse only those factors which are relevant. In order to show how the sheets may be used only a few details will be discussed.

It is well known that a boat's performance is changed by trimming, and that skillful skippers get much more out of their boats by proper ballasting and stowing of the cargo. In trimming, the effect of different LCB (location centre of buoyancy) can be established along with more favourable entrance and run angles of the water lines. The author believes that fishing boat models should be tested on different trim angles as this is an economic way of studying what can be done with a specific design. It is also important that fishing boats are tested at different displacements, contrary to cargo boats where the load condition is of most interest.

DESCRIPTION OF MODELS

Models No. 1 to 5 have been tested by FAO in 1953. No. 1 is a typical Grand Bank fishing schooner which was built in Clarenville, Newfoundland. The author visited Newfoundland in 1951 to investigate the possibilities of improving locally-built fishing boats. He found that these boats generally had favourable coefficients from a modern naval architectural point of view and the advice was that one should continue building boats along the same lines, rather than to introduce new designs from abroad. The Department of Fisheries and Co-operatives in Newfoundland requested FAO to test a model of a typical fishing schooner so that it could be compared with modern fishing boats from other parts of the world. The type represented by Model No. 1 was selected, more because a drawing was available than because it was a so-called high-liner. The lines are considered typical for this class and a racing type of schooner, such as *Bluenose*, was deliberately avoided.

Nos. 2 and 3 represent typical New England trawlers, No. 2 being designed in 1944 and No. 3 in 1952. The tests were carried out at the request of the U.S. Fish and Wildlife Service. No. 2 is typical of the boats in use to-day, although naval architects are trying to convince owners that it would be better to change the design towards the lines of No. 3. The designer of No. 2 when submitting his plans, wrote the following:

> " This type might not be the best type of hull form for you to experiment with in a tank, even though she is as good as we could make her considering all of the factors involved. She is a very heavily displaced boat for her overall length and if you will check with the various dimensions of similar boats of her length, you will find that she carries more beam and a great deal more depth of hold than virtually anything you could put her up against. In 1951, I had occasion to

incline two North Sea trawlers of the same overall length in England and was amazed to find they were boats of relatively small body, low freeboard, and depth, when compared with such a vessel as this. They were, at the same time, 30 per cent. less in displacement. . . . Incidentally, this boat is fuller aft in the way of the machinery space than we like to see her, but these little vessels, on this side of the Atlantic, carry a tremendous amount of fuel and machinery weights concentrated in this spot."

The " ready to go to sea " displacement of No. 3 is estimated by the designer to be 367 tons on about 110 ft. (33.5 m.) waterline length, as compared with 332 tons on 102 ft. (31.1 m.) waterline length for No. 2. This gives length-displacement ratios of 4.7 and 4.5, respectively.

Models Nos. 4 and 5 are Gulf of Mexico shrimp trawlers. No. 4 is of a V-bottom type and these lines are used in a number of boats, but No. 5 is of quite a new design. There are other shrimp trawler types which should have been included in the tests, such as the round-bottom boats which have a fairly large transom stern. Models Nos. 4 and 5 were also tested at the request of the U.S. Fish and Wildlife Service, who submitted the plans.

No. 6 is a type tested by the author under a grant made by the Icelandic Legation in Stockholm. The first version, 6A, had been improved considerably over the original design but was again altered several times. Model 6B had a fuller midship section and the hull floated somewhat higher and had a lower prismatic coefficient. Model 6C was sharpened in the forebody so that the displacement was reduced to 6A's. This model was some 15 to 20 per cent. better than 6A. Models 6D and 6E were arrived at by cutting 6C in the middle and taking away 5 per cent. of the waterline length each time. This was an inexpensive way to get a model with the same entrance and run but a smaller L/B ratio. Models 6F and 6G had the forefoot cut away in order to study the influence of different types of forefoot.

DISCUSSION OF THE TESTS

Fishing schooner No. 1 was tested on 252, 353 and 454 tons (250, 350 and 450 cu. m.) displacement, each displacement with several trims. On the light displacement, the resistance increased somewhat when the boat was trimmed by the bow, the reason apparently being that the entrance was fuller and the good run made possible by the transom stern could not be used. The same is true of the other displacements. The tests indicate that a well-submerged transom stern helps to decrease resistance.

The tests with the 1944 New England trawler, No. 2, show quite a difference on the lower speeds for different trim angles, the stem trim having less resistance. However, with increased speeds, the resistance for all trims is practically the same. It looks as if, at least for the lower speeds, the sterns of these boats are too full.

The test results of the 1952 New England trawler, No. 3, are different. Here the resistance is almost the same at lower speeds, but stern trim gives considerably less resistance at increased speeds. The very heavy displacement of 484 tons (480 cu. m.) is an exception, on which the designed trim is more favourable. It seems as if No. 3 would benefit from a sharper entrance and a fuller aft body. The LCB would then be further aft, but as the engine could also be fitted further aft, one should be able to make up for the change in LCB.

The shrimp trawlers were tested at 65.6, 101 and 142 tons (65, 100 and 140 cu. m.) displacement. It is interesting to study the influence of a large transom stern in the tests with No. 4. On the light displacement, the well-submerged transom decreases the resistance up to 40 per cent. A contributory factor might be that the entrance angle is reduced from 29° to 19° and that the LCB moves aft. However, this type of boat is built with engine and steering-house forward and it is rather unfortunate that those heavy weights are not aft. The medium displacement tests show less influence of trimming. The same is true with the heavy displacement tests at higher speeds.

The tests with No. 5 show that the originally-designed trim is the best and that nothing can be gained by trimming or changing the LCB.

Model 6A was tested on three different displacements, but the form values were almost equal at high speeds. The traditional opinion that a lower length-displacement ratio gives lower form values is not confirmed by the experience gained in these tests. The tests with Model 6B were made with only slight trims, but the stern trim was more favourable. The same is true of tests with Model 6C. The tests with Models 6D and 6E showed that the shortening of the length or enlarging of the beam does not increase resistance.

COMPARISON OF THE TESTS

It is supposed that naval architects, when studying the tests, will pay greater attention to form value curves C_1 than to the curves for effective (tow rope) power. As the C_1 curves are non-dimensional and exclude the influence of length (if plotted to Froude's No.) and displacement, the tests can more easily be compared. Curves on each diagram, making it easy to determine if a model is bad, fair or excellent, would have been useful but that would have required the selection of a few representative tests or the calculation from some standard series such as Taylor's or Takagi's, with the resistance of models having the same main coefficients as the ones compared. In fig. 414 to 417, certain tests are compared with models No. 6C and 6E, but it is felt that it is premature to introduce in the sheets themselves such curves before all sheets are made up.

The simplest way of analysing the tests is to use a piece of transparent paper and transfer the C_1 curves from one sheet to another or to select a representative speed and compare how different models behave. Such a speed should be the average speed which a boat makes and not the trial speed.

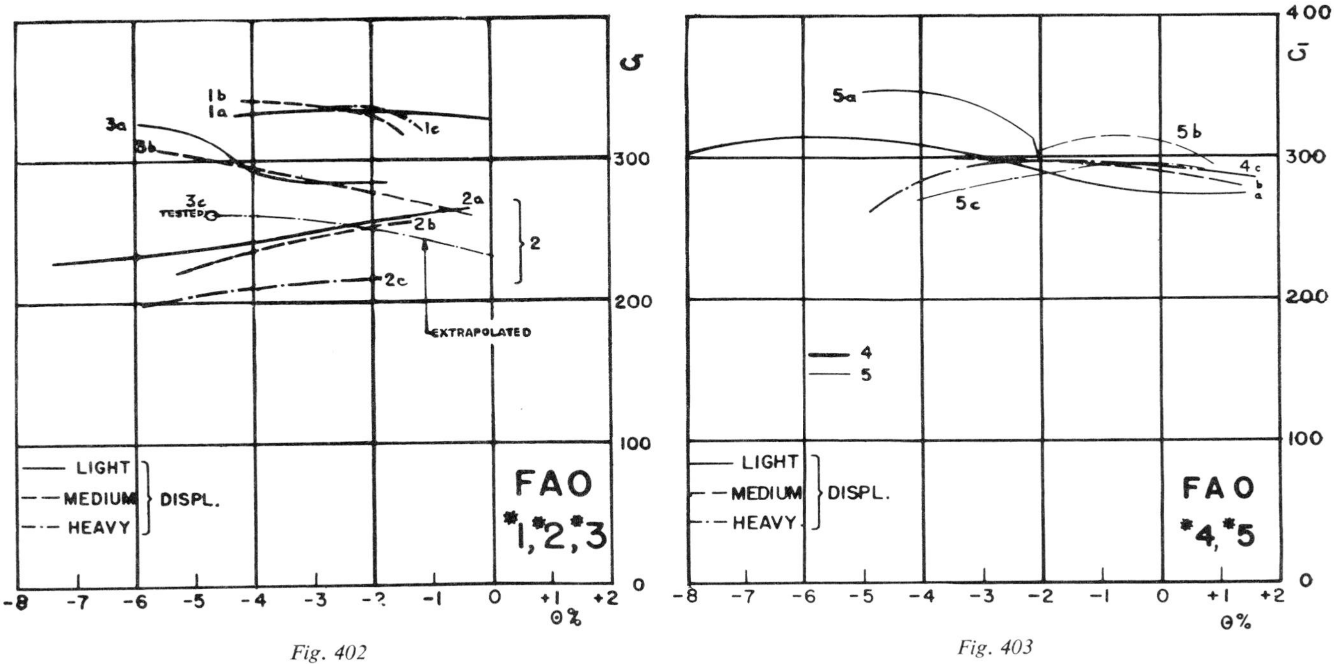

Fig. 402 *Fig. 403*

Models compared at $\frac{V}{\sqrt{gL}}=0.30$

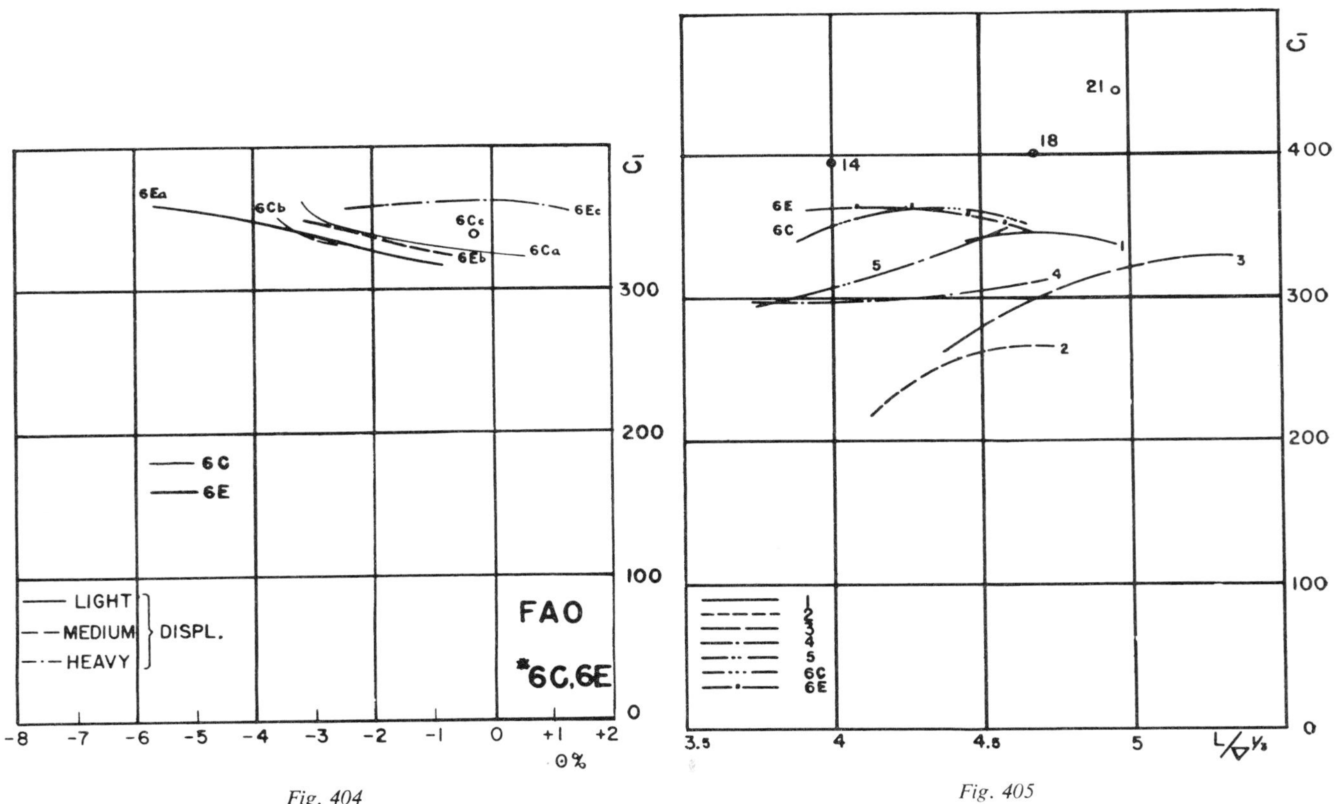

Fig. 404 *Fig. 405*

Models compared at $\frac{V}{\sqrt{gL}}=0.30$

The Froude's No. 0.30 corresponds to about 8 knots for a 65.5 ft. (20 m.) boat or 10 knots for a 100 ft. (30.5 m.) boat and represents fairly well such a general operating speed. Fig. 402 compares Models No. 1 to 3 at this speed. The form values C_1 are compared with the different LCB's and the diagram shows, what was already seen from the data sheets, that No. 1 does not change its resistance much with a different LCB, and that No. 2 does better with a forward LCB and No. 3 with an aft one.

Fig. 403 shows a similar comparison between Models No. 4 and 5 and fig. 404 shows a comparison between Models 6C and 6E, 6E being 10 per cent. beamier than 6C.

Fig. 402 to 404 show that on the selected Froude No. 0.30, Model 1 is better than No. 3 which, in its turn, is better than 2. The Newfoundland-built fishing schooner is therefore better than a modern New England trawler. Fig. 403 shows that the round-bottom shrimp trawler No. 5 is better than the V-bottom No. 4, but the differences are not as great as those between the trawlers Nos. 2 and 3. These shorter and beamier shrimp trawlers do also compare quite favourably with the longer trawlers.

Fig. 405 shows, for Models Nos. 1 to 6, the influence of length-displacement ratio at the most favourable LCB for each model, and on Froude's No. 0.30. In this way Model No. 2, which has more favourable values at the forward LCB, is compared with Model 3, which has its best C_1 values with the LCB aft. Fig. 405 shows that increased length-displacement ratio, i.e. less displacement per length, increases the form value C_1, for Models 2 to 5. Model No. 1, the fishing schooner, has little variation in the form values for the different length-displacement ratios. If compared on a length-displacement ratio of 4.5, the modern New England trawler, No. 3, is 7 per cent. better than the old one. Similarly, the V-bottom shrimp trawler is 16 per cent better and the round-bottom one 30 per cent. better. The fishing schooner, too, is 30 per cent better than the 1944 New England trawler and Model No. 6 is some 38 per cent. better. Fig. 402 to 405 only show the situation at the specific speed of Froude's No. 0.30 and do not take the whole speed range into consideration.

Models Nos. 1 to 3 and Nos. 4 to 5 were tested on the same displacements, but with a different length. Fig. 406 and 407 show the best horse power curves from each group. The values for 454 tons (450 cu. m.) in fig. 406, for Models 2 and 3, have been arrived at through interpolation. The difference between No. 1 and 3 on 353 tons (350 cu. m.) displacement and 10.5 knots speed is 9 per cent., between No. 1 and 2, 60 per cent. and between Nos. 2 and 3, 47 per cent. At 11 knots there is a difference between Nos. 1 and 3 of 8 per cent. and between Nos. 1 and 2 of about 70 per cent. to 80 per cent. (extrapolated). The heaviest displacement for No. 1 requires less horse-power than the lightest displacement of No. 2 on speeds higher than 10 knots. If both boats weigh the same, a boat built along the

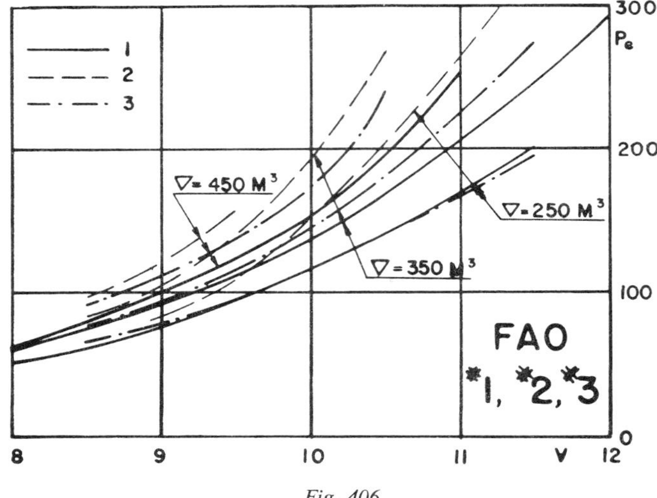

Fig. 406

lines of No. 1 would be able to carry 200 tons of cargo and still make the same speed.

The difference between the shrimp trawler tests is not so great in earlier examples, but at 9 knots and 101 tons (100 cu. m.) displacement, No. 4 requires 20 per cent. more power than No. 5.

HONEST COMPARISON

Fig. 402 to 405 illustrate a quick way of comparing tests non-dimensionally, by selecting a specific speed. They do not, however, show the influence of the whole speed range. Fig. 406 and 407 show a comparison on the same displacements, which is dishonest, because it does not take into account that the models are of different lengths and were therefore running at different relative speeds.

In the following, Models 1 to 5 will be compared on the same displacement and the same water line length with Model 6. The larger boats are compared on a length-displacement ratio of 4.65 and the shrimp trawlers on 4.1. The C_1 values are obtained by interpolation between tests with different length-displacement

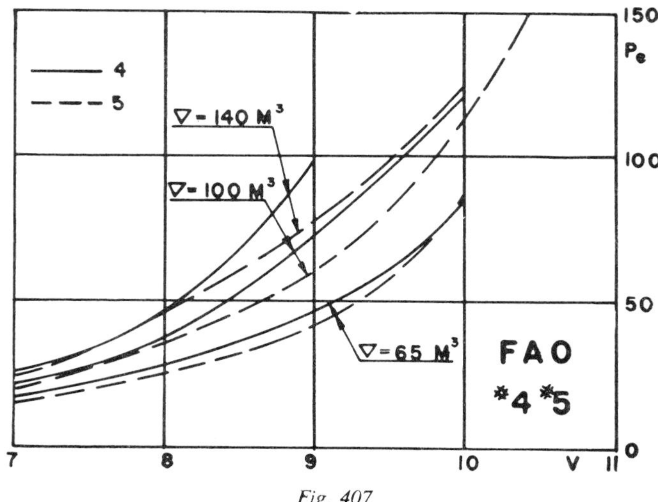

Fig. 407

ratios. To make the comparison as fair as possible, the best possible trim angle, not the designed trim, has been selected from each displacement. Fig. 408 shows the C_1-curves for the best series for the three different displacements of Model 3. Cross curves of these C_1 values are then plotted to the length-displacement ratio as the ordinata in fig. 409. The values for length-displacement ratio 4.65 have been selected and plotted in fig. 410. The dimensions for this interpolated test are shown in Table L and the profile, section area curve, shape of waterline and midship section are given in fig. 412. For Models Nos. 1, 2 and 6C, C_1 curves for the

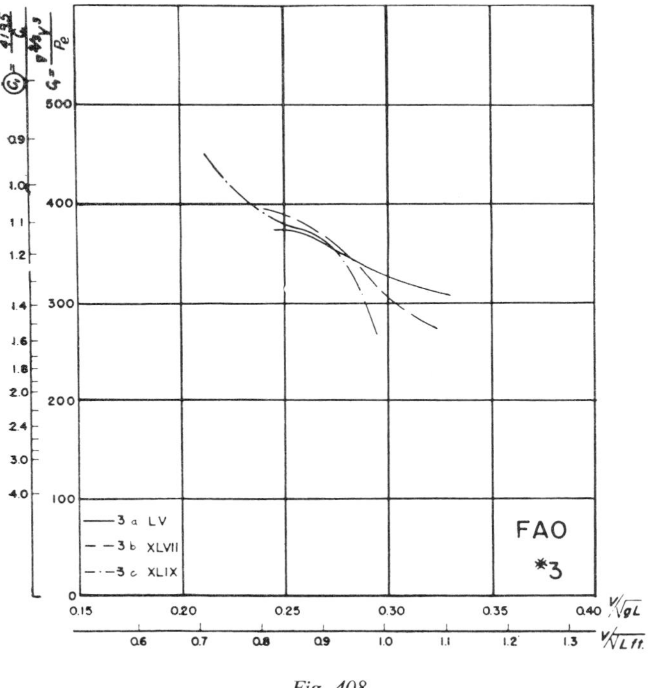

Fig. 408

length-displacement ratio of 4.65 have been interpolated in the same way and the 4 C_1 curves have then been converted to corresponding effective power and plotted on fig. 410. Models No. 1 and 6 C are almost equally good, the difference at 11 knots being only 6 per cent. Model No. 3, the modern New England trawler, requires 27 per cent. more h.p. than 6C and 20 per cent. more than No. 1, the Newfoundland schooner, and the 1944 New England trawler requires 54 per cent. more h.p. than No. 6C, 45 per cent. more than No. 3 and 25 per cent. more than the New England trawler (which, however, does not have such a favourable trim, under actual service conditions). Fig. 412 shows that Models Nos. 1 and 6C have a larger midship section and, therefore, less prismatic coefficients than Model No. 3. Also Model No. 6C has less entrance angle than the other models. Table L indicates that 6C has 10 per cent. more beam than the old New England trawler and Newfoundland fishing schooner and 25 per cent. more than the modern trawler. As all models are compared on the same waterline length and with the same displacement, they will have practically the same cargo capacity.

A similar comparison between the shrimp trawlers Nos. 4 and 5 have been made with Model 6E. All three

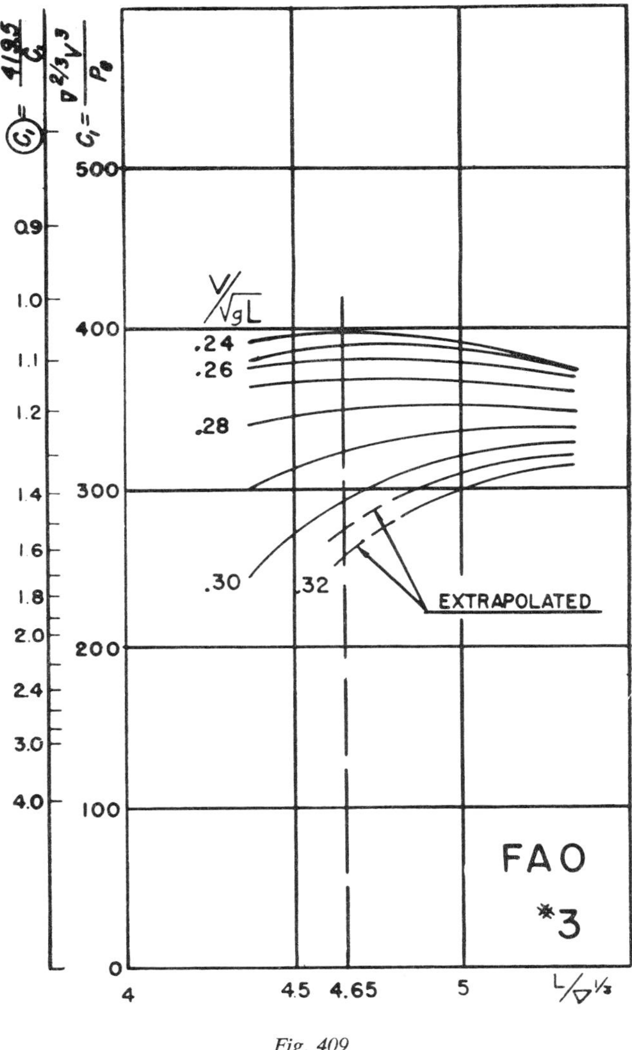

Fig. 409

tests were reduced to the same waterline length as Model No. 4 and the C_1 curves and h.p. curves can be seen in fig. 411 and the difference of shape in fig. 413, 6E having less prismatic coefficient and much finer entrance than both 4 and 5 and a slightly fuller run than 5. At 9 knots, the round-bottom type shrimp trawler required 16 per cent. more h.p. and the V-bottom 43 per cent. more h.p. than the 6E, the difference between the V-bottom and round-bottom shrimp trawler being 27 per cent. The dimensions for Models No. 4, 5 and 6E are shown in Table LI. Also these models will have practically the same cargo capacity.

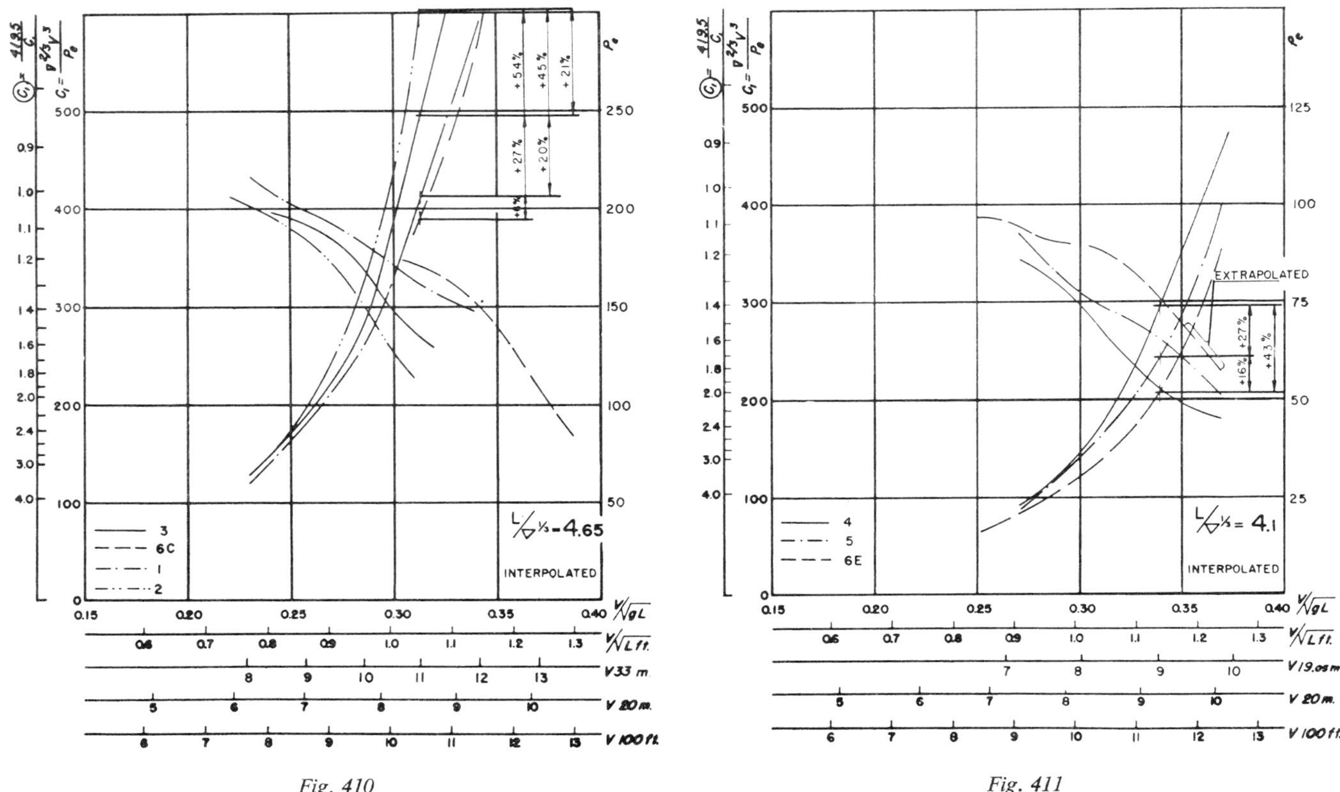

Fig. 410

Fig. 411

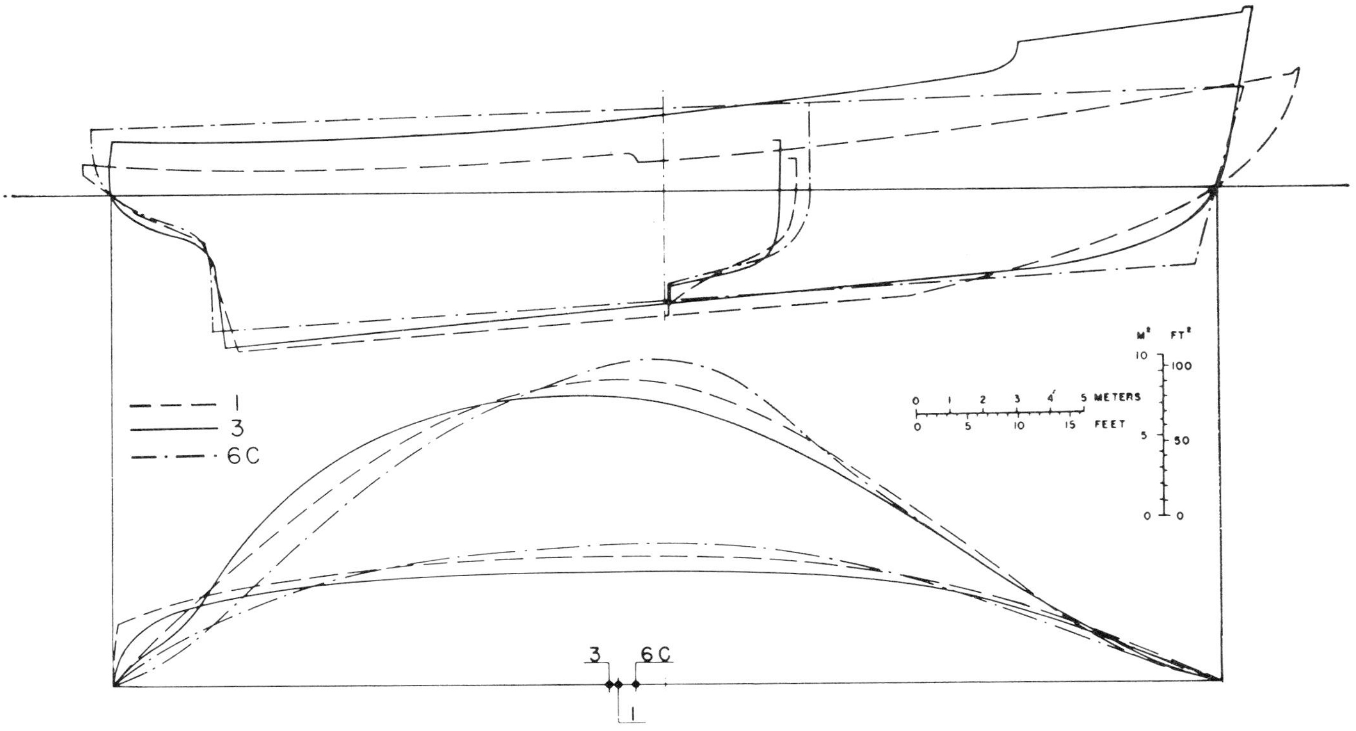

Fig. 412

TABLE L

Main dimensions for boats compared in Fig. 410 and 412

Model		*1*	*2*	*3*	*6C*
L	ft. m.	108 33	108 33	108 33	108 33
B	ft. m.	25.8 7.88	25.8 7.88	22.8 6.97	28.5 8.71
T	ft. m.	11.9 3.62	9.9 3.01	9.8 2.99	9.4 2.88
Am	sq. ft. sq. m.	203.5 18.9	177.5 16.5	193 17.9	217.5 20.8
∇ Δ_1	cu. m. tons	358 367	358 367	358 367	358 367
∇_2 Δ_2	cu. ft. tons	12,630 364	12,630 364	12,630 364	12,630 364
$\odot\% = \pm\frac{1}{2}L$		−4.22	−1.7	−4.97	−2.7
$L/\nabla^{\frac{1}{3}}$		4.65	4.65	4.65	4.65
δ		.396	.457	.52	.43
φ		.573	.657	.605	.54
L/B		4.19	4.19	4.73	3.78
B/T		2.18	2.52	2.33	3.03
$\frac{1}{2}\alpha E$		24°	26°	21°	16°

TABLE LI

Main dimensions for boats compared in Fig. 411 and 413

Model		*4*	*5*	*6E*
L	ft. m.	62.5 19.05	62.5 19.05	62.5 19.05
B	ft. m.	17.4 5.31	17.6 5.37	17.9 5.46
T	ft. m.	6.6 2.0	7.0 2.13	7.4 2.26
Am	sq. ft. sq. m.	86.5 8.04	96 8.92	101 9.37
∇ Δ	cu. m. tons	100 102.5	100 102.5	100 102.5
∇_2 Δ_2	cu. ft. tons	3,530 101.6	3,530 101.6	3,530 101.6
$\odot\% = \pm\frac{1}{2}L$		−3.06	−0.26	−1.63
$L/\nabla^{\frac{1}{3}}$		4.1	4.1	4.1
δ		.445	.446	.432
φ		.652	.588	.560
L/B		3.57	3.55	3.49
B/T		2.66	2.52	2.42
$\frac{1}{2}\alpha E$		30.3°	28°	21.5°

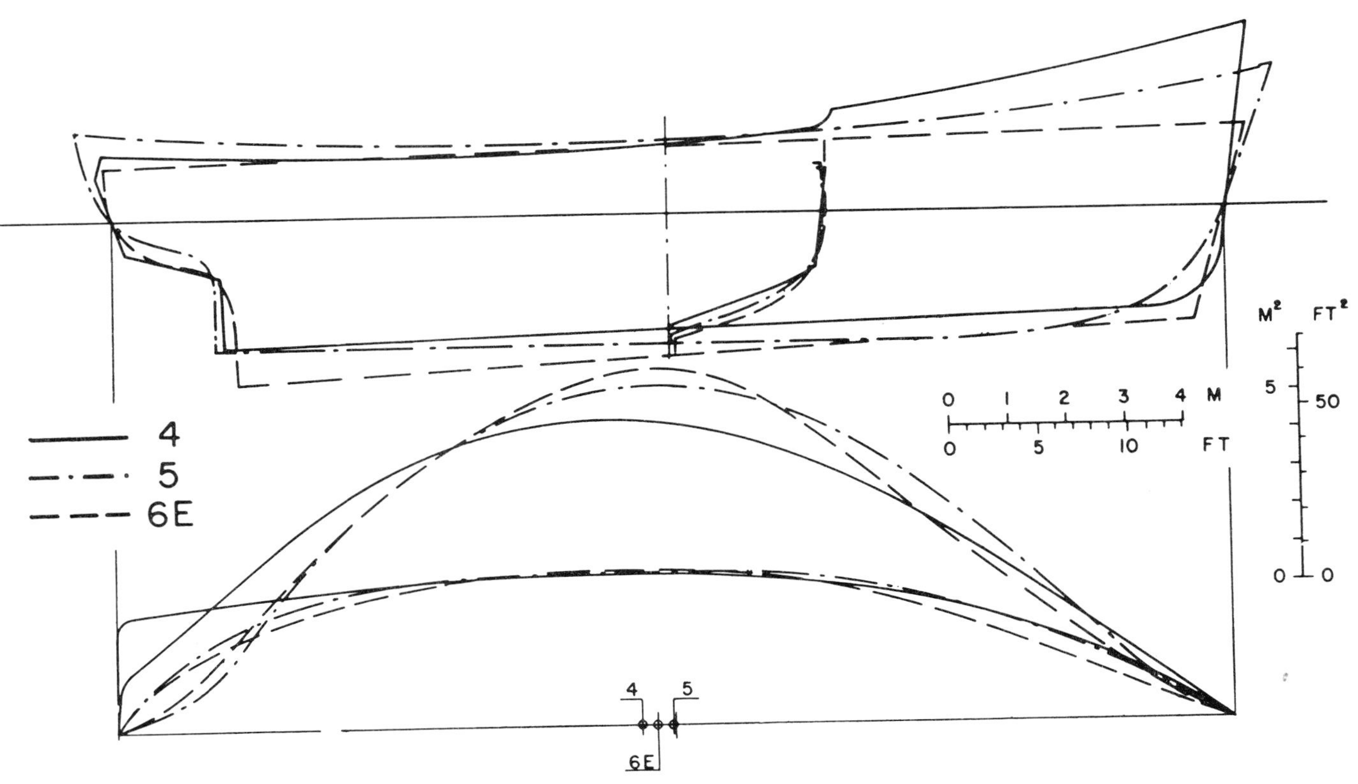

Fig. 413

TABLE LII

Main coefficients for models Fig. 414 to 415

Fig. No.	*Model No.*	$L/\nabla^{\frac{1}{3}}$	δ	φ	L/B	B/T	$\frac{1}{2}\alpha E$	⊙%
All	6C	4.65	.430	.540	3.78	3.03	16°	−2.7
	6E	4.1	.432	.560	3.49	2.42	21.5°	−1.63
	Norway							
414	85	3.68	.430	.583	3.02	2.35	32.5°	+1.15
	92	3.99	.438	.600	3.40	2.40	31°	+0.19
	191	4.78	.456	.610	5.04	2.00	20°	−0.79
	Norway							
415	165	3.7	.436	.593	3.13	2.25	30.6°	+0.3
	183	4.46	.436	.593	4.37	2.04	23°	+0.3
	Norway							
416	226	5.54	.423	.586	5.10	2.76	12°	−3.1
	217	5.46	.522	.575	5.78	2.55	13.5°	−1.9
	166	5.54	.499	.550	5.78	2.55	12.2°	−1.9
417	Allan			.600 .625 .650	5 to 5.5	2.5		
	NPL 2946C	4.96						

COMPARISON WITH OTHER FISHING BOAT TESTS

Fig. 414 to 417 show comparisons between Models Nos. 6C and 6E with some tests which have been published recently. Considerable work on fishing boats, whale catchers and Fjord buses has been done at the Norwegian tank in Trondheim and has yielded a number of interesting results. Fig. 414 shows the best tests for 40 ft. (13 m.), 65 ft. (21 m.) and 130 ft. (42.5 m.) boats (Astrup 1951) and fig. 415 shows C_1 curves for tests with 100 ft. (30.5 m.) boats (Sund 1951). The main coefficients are listed in Table LII. These models represent considerable improvements when compared with the original fishing boat types used in Norway.

Fig. 416 shows tests with whale catchers and Fjord buses (Voll and Walderhaug 1952) and smaller, fast passenger ships (Haaland 1951). In fig. 417 there are curves for some results of modern British trawlers

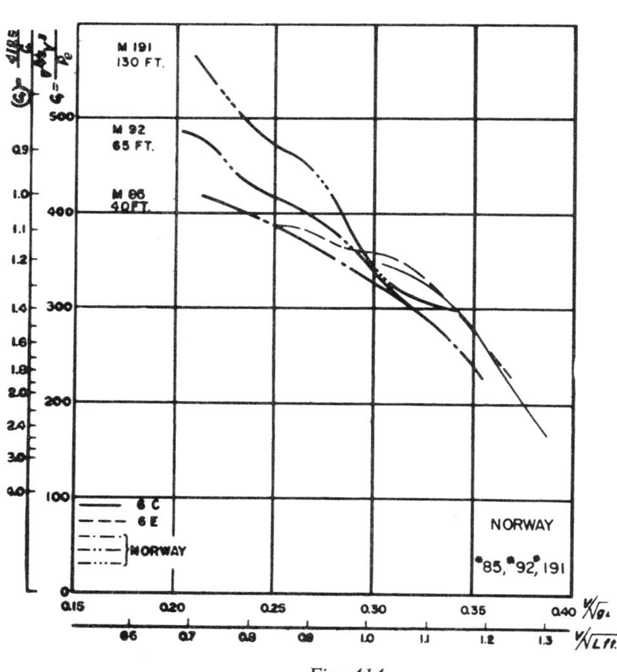

Fig. 414

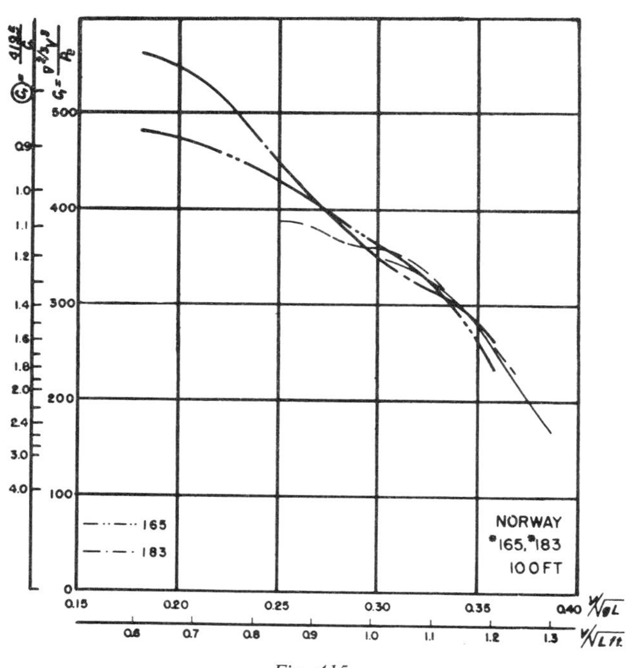

Fig. 415

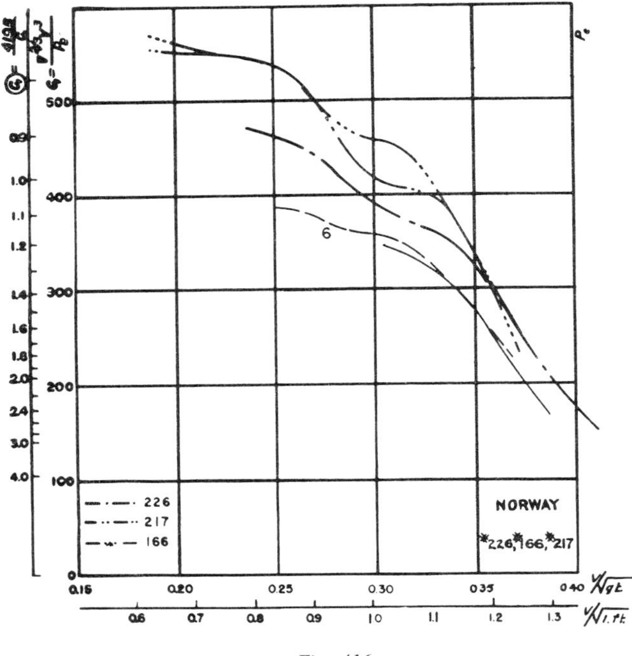

Fig. 416

(Allan 1953), showing the influence of different prismatic coefficients, fig. 417 also shows the curve for Model 2,946C, tested at Teddington (Hunter 1952).

This short review of tests is by no means comprehensive, but the differences are striking and it is evident that fishing boats can be improved. The author's experience is:

(1) The length-displacement ratio is of little importance;

(2) The beam does not increase resistance;

(3) The location of the centre of buoyancy should be far aft;

(4) Differences in the block coefficient, below 0.55, have no influence;

(5) The prismatic coefficient is very important and seems to be best at about 0.575;

(6) Transom sterns seem to give reduced resistance ;

(7) The entrance angle should be kept low;

(8) Parallel mid-body and sharp shoulders should be avoided.

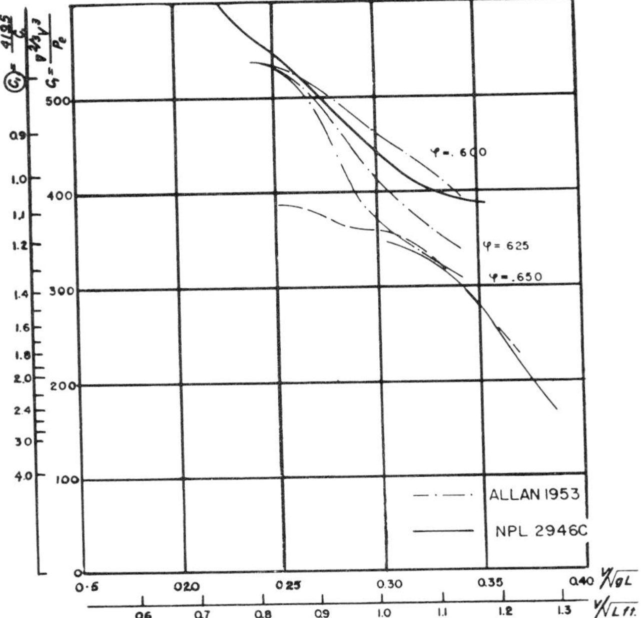

Fig. 417

PROPOSED STABILITY CRITERIA

by

GEORGE C. NICKUM

THE question of what constitutes a satisfactory margin of stability for a floating object is as old as the first floating structure and, up to comparatively recent times, it has been answered only empirically. Vessels which continued to go to sea and come back safely were judged to have a satisfactory margin of stability. Those which capsized and were lost had an inadequate margin.

In recent times man has applied his scientific tools to the problem of ship stability and has devised means of measuring the stability of any type of vessel. But these methods have generally been applied only to the larger ships and, as naval architecture has advanced, they have been forced into complications and elaborations which further limit, by economic reasons, application to the larger vessels. From simple calculations of statistical stability and cursory examination of the effect of free internal liquids, there has been progress to cross curves of stability and analysis of the dynamic forces, to analysis of the problems of rolling in synchronous seas, to the problems of combined rolling and pitching, and so on.

All of these features and all of the new methods are essential to stability studies, and all have their place. They have contributed immeasurably to the comfort and safety of marine transport. But, because of the complexities involved, the vast majority of floating structures in the world today are built without any analysis of stability other than the old empirical question: " Has it turned over yet? " This is particularly true of the world's fishing fleet.

Fishing vessels generally are evolutions of types which have been in use over a long period of time. A fisherman stepping into a new boat, which is usually only slightly modified from an existing type, has to decide by his senses whether the boat is safe or not. Does she heel too much when he puts the rudder hard over? Is she tender when she has a load of fish on deck? The fisherman uses his sense of feel to evaluate these points. To give him his due, he has generally been right in his decision and the number of casualties in fishing vessels due to an insufficient margin of stability is relatively low.

However, these are days of rapid change. Technological development in the methods of catching, handling, and storing of fish have brought about big changes in vessels used in certain industries. Snyder (1946) presents the stability problems of U.S.A. tuna clippers using the brine freezing method for the freezing and storing of tuna fish. In twenty years the average size of these vessels doubled, evolving from conventional small ice boats to large craft filled with tanks and large bait tanks on their decks. Voyages of 2,000 to 3,000 miles became common and the size of generators, pumping and freezing equipment grew proportionately. During this rapid evolution many lives and vessels were lost because there were no suitable standards to establish an adequate margin of stability and, to this date, no criterion has been generally accepted defining the margin of stability required for such vessels. Fishing boat designers and builders use different methods of measuring stability, and the underwriters depend on different designers and builders to examine and approve a vessel's stability.

If a simple stability criterion, easily understood and applied, had been available for tuna vessels, it is believed that many of the early casualties could have been prevented. In other fields of ship design, various criteria have been developed for particular types of vessels. For example, the International Conference for Safety of Life at Sea in 1948 accepted the criterion of wind heel and passenger heel which was in general use in most countries for determining margins of intact stability for freight and passenger vessels. The time should be ripe to establish such criteria for fishing vessels.

A criterion of stability for fishing vessels must meet definite requirements. First, it must be of such a nature that the designer may readily forecast a new vessel's compliance with the criterion without requiring elaborate and expensive calculations. Second, simple tests by relatively untrained personnel must be able to verify the criterion without requiring design plans.

The purpose of these two requirements is obvious. Fishing throughout the world is a highly competitive business and is generally operated by individuals who do not have resources to pay for elaborate designs and complicated calculations and tests. Any criterion which does not meet these requirements would never have a chance of being put into universal practice.

Here is the criterion proposed for this service:

(1) The vessel, in its most severe operating condition, must have the following characteristics:

(*a*) GM equal to or greater than either $\frac{B}{10}$ or 2 ft. (61 cm.)

(*b*) $\frac{F}{B} + \frac{FA}{L \times B}$ equal to or greater than ·15

Where: — GM = the distance between the centre of gravity of the vessel and the transverse metacentre;
B = maximum beam over planking or plating at the waterline;
F = freeboard from the waterline to the edge of the freeboard deck at the side measured amidships;
FA = freeboard area (projected on a vertical plane through the centre-line) between the waterline and the freeboard deck at the side;
L = the registered length.

(2) Compliance with the above characteristics must be proved by physical measurements with the vessel in its most severe operating condition. GM is to be calculated from actual measurements of period of roll in this condition, using the following formula:

$$GM_{feet} = \left(\frac{m_1.B}{t}\right)^2$$

Where: m_1 = .40, t = time, in seconds for one complete roll from maximum heel to one side to maximum heel to the other, and return.
(Editors note: m_1 includes here the 1.108 in the feet formula on p. 328. Hence $m_1 = 1.108m$).

The simplicity of the above parameters will be obvious. To the designer the estimation of GM, whether made by detailed and elaborate calculations or by approximations based on experience, is an absolute minimum requirement of any design. The determination of GM by the rolling period method is simple and requires neither a great amount of time nor elaborate equipment. Freeboard heights and areas can be easily obtained by physical measurement and by calculation. Most vessels, whether in the design stage or afloat, can be measured against this criterion with the minimum of time and expense.

The reasons for the parameters may not be as obvious. The use of and the need for inclusion of metacentric height in the criterion is undoubtedly apparent, since GM is an indication of the force available to put the vessel back on an even keel. The use of freeboard and freeboard area may not be so apparent. As the determination of GM gives us only the forces available to restore the ship to an even keel for small angles of inclination, some other means is required to evaluate the righting forces at larger angles of inclination. The conventional methods of determining these forces involve the calculation of righting arms at various degrees of inclination and at various drafts. These calculations are lengthy and tedious, and to be properly used require the accurate determination of the location of the centre of gravity by means of inclining experiments. In addition to the time and expense involved in preparing such calculations, they are of questionable value in small vessels, which are subject to large changes in trim due to heel. In one case, on a tuna vessel, a complicated calculation was gone through to determine the effect of trim on the actual righting arm. It was found that the arm at 10 degree inclination after the trim was 25 per cent. less than that given by a conventional calculation, which neglected the trim due to heeling. In this case, the variation was due to the fact that the deck edge at the stern was immersed at about 2 degrees of heel, and as the heel increased the deck immersion increased correspondingly with ever-increasing trim.

Because of the complication and expense of the use of cross curves and because of the need for careful analysis of the results to determine whether they are accurate or not, the use of freeboard and freeboard area is proposed as a substitute. The ability of a vessel to right itself, and the values of the righting arms, are primarily dependent on the location of the deck edge and how soon it is submerged when the vessel is heeled. Therefore, freeboard and righting arms are very closely related. Admittedly, freeboard as such does not give an accurate determination of the righting arm at varying angles of heel but after a careful check of large numbers of fishing vessels, it seems that it is a good indication of righting arm or reserve stability. Of several hundred vessels checked, those having freeboards equal to or in excess of the minimum freeboards set forth in the formula have had sufficient reserve stability to withstand the actions of the seas without capsizing.

Because of variations in the sheer of different types of fishing vessels, the minimum freeboard set forth is a combination of amidship freeboard and freeboard area. American tuna vessels, as an example, are so designed that in their worst operating conditions they have a zero freeboard at the stern. However, they are conventionally fitted with raised forecastles and a very high sheer forward. The loss of water plane at the stern due to heel can be accepted in these vessels; it represents only a small loss of the total water plane area because of the high sheer forward. The use of freeboard also gives proper credit to the raised poops and forecastles fitted on many vessels.

The use of freeboard and freeboard area provides a check of the most common reason for lack of stability in an operating condition—that of overloading. No fisherman ever wants to stop loading when he has a full net and it is probably a truism to say that more fishing vessels have foundered or capsized from overloading than from any action of the sea or other cause.

Many people will undoubtedly argue that the use of freeboard and freeboard area as an indication or evaluation of available righting arm is not correct, and that it certainly does not indicate the vessel's ability to right itself when the heel is, say, 45 degrees. But the fact is that large numbers of U.S.A. Pacific Coast fishing vessels, on which inclining experiment data and operational information are available, have been operated successfully for years with the minimum GM and minimum freeboard area set forth in the formula. The writer's opinion is that these fishing vessels do not roll 45 degrees. At least 50 vessels investigated have operated satisfactorily for ten years and would certainly go to the bottom should they roll 45 degrees, because of openings in their superstructure and in the freeboard deck, which would be exposed and flooded.

The term " freeboard deck " means the deck in which all openings are so protected as to prevent the ingress of water into the hull. This could be amplified at length and, as the international load line regulations show, is amplified in detail in the case of larger vessels. From a practical standpoint, however, the broad definition of the term seems adequate. The use of glass windows in an enclosure on the deck of a 400-ft. (122 m.) ship would be heresy, but the fact remains that thousands of small fishing vessels operate through storms without damage to the windows in their houses.

The use of period of roll to check the metacentric height is now a common practice throughout most of the world. But the period of roll has been used only as a check of the GM and never as a means of determining the GM. Variations in the " m_1 " factor between various types of vessels have made designers hesitate to depend on the rolling period formula. Fifteen years ago, the dependence of the rolling period in place of an inclining experiment would have been considered sheer lunacy. Since then, however, data on actual inclining experiments, in which the rolling period was checked on hundreds of fishing vessels of various types and sizes, shows that the " m_1 " factor ranges between ·385 and ·415 in 75 per cent. of the cases. Only in a few extreme instances has the factor been as low as ·37 or as high as ·43, and it is a considered opinion that this represents the range of inaccuracy of this factor on fishing vessels. If this is correct, such a margin of error can be accepted in setting up a stability criterion. This range gives a variation in GM of $\pm$15 per cent. With a 2 ft. (61 cm.) minimum GM, this means a possible error of $\frac{3}{10}$ ft. (18 cm.). Such an error can, it is believed, be accepted.

Earlier the statement was made that the period of roll could be readily obtained for most vessels. Many questions may be raised on this point, as the vessel must be in its most severe operating condition. Tuna vessels can, it is realized, be more easily put in such an operating condition than any other type of fishing vessel. The tuna vessel is nothing but a tanker. When loaded with fish and brine, the weight in the vessel is for all practical purposes the same as when loaded with water and by pumping water into the tanks almost any operating condition can be simulated. This, obviously, is not true of vessels which store their fish in the hold. However, on the U.S.A. Pacific Coast, net boats, when they have their full load of the nets on board and are light on fuel, are in a condition where they have less GM than when their holds are fully loaded. Thus an experiment made to determine the period of roll in the light operating condition, plus either a measurement of freeboard when the vessel is fully loaded, or a calculation of freeboard at load draft (which can be fairly readily made), gives an adequate check on the stability characteristics. Should conditions not be similar, obviously an inclining experiment and calculation of loaded condition is essential.

The proposed criterion of stability and the universal use of the period of roll to check stability on fishing boats, is based on experience in a limited type of fishing vessel because all the data used in arriving at the factors of GM, freeboard, and rolling period come from U.S.A. Pacific Coast type of vessels. Tuna clippers conventionally have a raised forecastle and a large, flat, low stern and the Pacific coast type of purse seiners and trawlers are of similar type. And there are indications that the American Atlantic coast trawlers and net boats are sufficiently similar for the stability criterion to be used, although the use of rolling period instead of the inclining experiment requires evaluation by experienced designers. The GM requirements also appear to be applicable to European trawlers. Taylor (1943) has stated that British trawlers should have not less than 2 ft. of GM in order to withstand the heeling moments of the trawl warps.

It would be surprising if the proposed criterion could be applied to every type of fishing vessel in the world but it is believed that the method of the formulas can be used, although there may have to be variations in the constants. The applicability of these constants can only be determined by designers familiar with the characteristics of the various types of fishing craft. After comparison of inclining experiments and rolling tests it may well be that boats of 30 and 50 ft. (9.1 and 15.2 m.) in certain trades may require a different value. The same thing is true of freeboard. On the rolling period it is probable that certain types of vessels with large amounts of concrete ballast may have a sufficiently different radius of gyration as to require a revised " m_1 " factor in the rolling period formula or, if the amount of ballast varies widely from boat to boat, the " m_1 " factor may be of no value. In particular, it is felt that considerable study and investigation should be made of determining the worst operating condition and that in which the vessel must be rolled.

Almost all naval architectural theories stem originally from assumptions, the validity of which have been proved empirically. The assembly and distribution of data on fishing vessel stability and the proving of a criterion by experience would be a real contribution to the art of naval architecture.

LOADING AND CHANGE OF TRIM ON SMALL TRAWLERS

by

WALTER J. McINNIS

THE term " small trawler " is used to describe fishing vessels less than 150 gross register tons. Approximately 92 per cent. of trawlers operating in the North Atlantic waters of the western hemisphere, fall into these categories:

Vessels under 50 tons	*Vessels 50–150 tons*	*Vessels over 150 tons*
66 per cent.	26 per cent.	8 per cent.

The figures given are compiled from the combined registers of United States and Canadian merchant vessels.

Virtually all of the vessels under 50 tons are constructed of wood, and 90 per cent. of those between 50 and 150 tons are also of wood, so it appears that the problem of securing and maintaining a proper loading trim is chiefly concerned with wooden vessels, a task more difficult than in a steel ship. It also goes without saying that the trim disturbances are generally in inverse proportion to the size of the vessel.

There are many underlying causes for the acute disturbances of trim due to loading. Some may be brought about by the actual mechanical or engineering problems that confront the architect, but these are in the minority and they could be easily remedied if due opportunity were given to the designer or builder to correct them at the source.

A very small percentage of the vessels are actually designed or engineered but are " just built ", either as a duplicate of some previous ship of the size wanted, with some modifications to suit the particular owner, or from a new " model " cut to the length required. At this stage, there is little thought given to the completed vessel as a unit loaded for sea, and small consideration of such important factors as propelling requirements, size of pay load, fuel and water capacities, crew and nature of waters to be fished in.

In the few cases where an architect is commissioned to prepare plans, or a competent building yard with a proper engineering staff is engaged to design and build a vessel, the problem is difficult because control of the situation is easily lost during construction and even after completion. This sounds like a harsh statement but, nevertheless, it is a true one which is recognized by those who are constantly seeking improvement.

The principal factors which contribute to most of the difficulty are as follows:

1. *Desire on the owner's part to secure (at the risk of overloading) the maximum pay load on the minimum overall length.*

The relationship of pay load to be carried is far out of balance with the " loaded for sea " port departure displacement of the vessel. In average cases, it ranges from as high as 45 per cent. in vessels of about 45 ft. (13.7 m.) waterline length and about 20 gross tons, down to 38 per cent. in vessels around 100 ft. (30.5 m.) waterline length and about 150 tons gross. This presents an obstacle at the start which is more difficult to overcome than in the sailing vessels where the centre of gravity of the pay load was located to correspond with the buoyancy centres under different conditions of displacement.

Effects. Extraordinary pay loads result in a trim moment that is virtually impossible to control at any time under conditions that generally exist.

Remedies. The following suggestions are offered as the first and most important steps to be taken in remedying these effects:

Attempts at cramming in the maximum pay load at the expense of other features should be checked. A reduction of at least 20 per cent. would make a good start. This would leave a maximum load of not over 36 per cent. of the vessel's " loaded for sea " port departure displacement for boats of about 45 ft. (13.7 m.) waterline length, and about 20 gross tons, down to not over 28 per cent. in vessels of 100 ft. (30.5 m.) waterline length, and about 150 tons. Naturally, this would result in shortening the fore and aft length of the hold. In almost all Atlantic trawlers, this means taking the weight out of the forward end of the hold, reducing the lever arm and disturbing moment. The few so-called mast head type trawlers with the steering house and engine forward and having the cargo room aft would be equally benefited by taking the excess weight out of the aft end of the hold and subsequently improving the trim.

Limiting the hold size to these amounts would work no hardship on the owner, as perusal of the reports of fish landings shows clearly that in more than 90 per cent. of the trips, vessels return to port with partial loads, some as low as 50 per cent. of capacity.

2. *Indecision at the start with respect to the size (particularly the length), weight, type, and horse-power of the main propelling unit.*

There seems to be no semblance of judgment used in determining in advance what the propelling plant shall be. And when the final decision is made, it is either based on the price of the unit or the alleged power to be delivered at the shaft, without due regard for the overall length, weight or suitability for the hull itself. Two schools of thought prevail, with the majority of operators selecting the biggest, heaviest engine at all times and completely forgetting all of the consequences of such a choice. Too little help is given to the architect or builder by the engine manufacturer when advising an owner, as they seem to be chiefly interested in selling an engine whether it best fits or not.

Effects. Failure to study the propelling requirements well in advance can only result in haste and improvising at a later date. There is a definite fixed relationship of engine to hull and fittings. If an engine selected is too long and too heavy for a vessel, the centre of gravity of the pay load automatically is moved out of position, and the trim moment increased. The emphasis for the most part appears to be on this kind of thinking.

When the total loaded displacement, overall vessel length, conditions for average fishing, etc., have been predetermined, it is poor judgment to select a motor which is not in complete harmony with these factors.

To secure adequate trim control, as well as for reasons of proper shaft revolutions and propeller thrust, an early decision is imperative. The effect of such a balanced set-up usually leads to satisfactory results; failure to plan far enough ahead means additional expenditure for " second guessing ", and a feeling that it is, after all, only a " makeshift ".

Remedies. All conditions must be studied early and the decision on the main engine must be made to meet them all, not just one or two. Let the engine be selected first, if necessary, and a proper vessel built around it, rather than otherwise.

The points to be remembered when selecting the motor are as follows:

Has it enough power at the shaft to give the best results both at free steaming and when trawling?

Is it the right type of engine to suit the hull for length, weight, and shaft revolutions?

If the motor has to be of the reduction gear type to best suit the hull for length, one should not arbitrarily insist on fitting a long, heavy, slow speed motor which upsets everything gained in other places. The reverse condition is also true in the larger vessels, where weight is more needed in the engine compartment to maintain trim.

3. *A general adherence to the arrangement of quarters, disposition of fuel and water locations found in vessels commonly used in the locality, and lack of desire to depart from such.*

The adherence to popularly accepted arrangements of quarters, hold, machinery space, fuel, water, and superstructure is easy to understand because of tradition. But it is desirable to depart from such age-old ideas when there is a chance to secure a better control over trim. So far, most of such approaches have been largely " cried down " as not being wanted, and often the only reason given has been " my grandfather had his vessel rigged this way and if it was good enough for him, it is good enough for me ".

It is not intended to overlook the needs in specific instances, where the fishing operation is peculiar to a locality, of staying as close as possible to the traditional arrangement, but to point out the importance of the ability to control trim, and the dangers when reasonable control is lost. And it is in no sense implied that traditional arrangements are not or cannot be considered suitable at times. But it is suggested that there is too much tendency to hold to these layouts regardless of the consequences.

It is difficult to prevail upon an owner or captain to split up or otherwise depart from the regular position of fuel and water tanks to ease the change of trim problem The location of living quarters is perhaps the most inflexible feature and perhaps rightly so, as their position in the fore part of the vessel uses the rapidly changing space form to the best advantage. All other arrangements, however, should be deemed adjustable. Where they are held inviolate, there is little hope for improvement in control, and the result is standstill or even back sliding type of thinking.

Remedies. The attempt should be made to locate the centre of gravity of the pay load over the average fore and aft common centre of buoyancy in various conditions of light and full load. This is not easy but can be accomplished. If necessary, a small trimming fish-hold, separate from the main hold, may be fitted. There will be considerable resistance to this on the part of the operator but the importance of it in trying sea conditions will more than offset the criticism levelled at the start.

It is also felt that now is the time to depart from the hide-bound ideas of fuel and water locations. These are successfully spread about in the larger steel trawlers, so why not do something similar in the smaller craft, particularly in the wooden vessels which predominate in the fishing industry? It would enable the engineer to exercise a very close control of the trim at all times, provided the pay load was properly located in the beginning. In other words, the pay cargo should come as near as possible to putting the vessel constantly down on even keel.

4. *The feeling that trim disturbance cannot be corrected even in part and it must be lived with and considered one of the evils of the operation.*

The mental attitude of fishermen on the subject of trim disturbance cannot be overcome easily. They think it cannot be avoided. But before long this feeling may be driven out by demonstrating its fallacy.

There is not much more to add to the defeatist attitude on the part of owners. Failure to recognize this fallacy can only result in more obstacles to progress.

Remedies. Sample vessels should be built in various localities, each having suitable trim control, to demonstrate that a condition which has existed for a long time can be corrected. Fishermen are hard headed, particularly owners of the numerous small vessels, and can only learn to accept something new by seeing it perform. They cannot generally visualize it on paper.

5. *Tendency on the owner's part to allow each captain complete freedom to re-arrange, re-rig, or otherwise alter the vessel, either during construction or after he takes command.*

A small trawler, built carefully from design in 1939, had little trim disturbance. In subsequent years, 44 more vessels were built from the same plans, and if these boats were placed side by side in the same basin, it would be difficult, even for the expert, to determine that they were presumed to be of the same class, so radical were the various conditions of trim.

The owner's relationship with his skipper's is again one which can be readily comprehended. The old adage that " no fishing vessel is any better than its skipper " is generally true, for there have been many cases of good skippers in inferior boats getting better results than poorer skippers in superior vessels. Due consideration should at all times be given to the captain in matters of rigging, deck and trawl gear, but it is dangerous to rely too much on his knowledge of matters which could possibly upset performance and balance of the vessel.

This comment is made with all due regard for the ability of the average fishing skipper, and is intended as a warning that the risk, if taken, can often become cause for regrets. The architect and builder are the authorities on hull form, disposition of fixed and expendable weights, type of machinery, etc. It is asking too much of a skipper to decide such matters and it is handicapping the engineer. Architects are not infallible and builders are usually careful and feel they have a reputation to sustain. For the most part, however, both give due attention to the weight and moment of trim calculation, and they see the picture of the vessel as a whole.

Remedies. The architect should work closely and early with the skipper, the engineer, engine manufacturers, and fishing gear makers. Their ideas should be solicited and the difficulties discussed. And then, after due study, the architect must insist on full control and must have the backing of the owner in all final decisions.

6. *A general trend towards over-ballasting (a throw-back from the days of sail).*

Over-ballasting is a sore subject. In the days when sail was the only means of propelling fishing vessels, removable ballast in quantities to suit summer or winter fishing, was used to gain stability. Despite the fact that many present-day vessel operators have had no immediate experience with fishing vessels of the sail type, tradition impels them to over-ballast, usually in the wrong place, and this is another cause of disturbance of trim.

SUMMARY AND CONCLUSIONS

Effects

The combination of all the factors discussed leads to a heavy fore and aft trim disturbance, ranging from as much as 5 per cent. of the load waterline length in the larger vessels, up to 7 per cent. in the smaller craft. Except for those few vessels which operate in Atlantic waters, with the so-called mast head type rig and the fish-hold located aft of the engine-room, the trim disturbance is forward. A heavy trim by the head when loaded is conceded by all to be bad, particularly in winter weather when chances of icing up on deck are prevalent.

The ill effects on comfort in a sea way, steering ability, excessive wetness on deck, increased sluggishness and lack of buoyant lift, increase at all times directly in proportion to the amount of trim change. In an effort to counteract this, many boat owners deliberately ballast their vessels and send them to sea badly out of trim in anticipation of straightening them up to even keel when loaded. Such a procedure is negative and seldom produces anything beyond normal trim with half to two-thirds of a pay load. And it still leaves the old problem of operating out of trim.

While it is conceded that a stern trim is to be preferred, if there has to be such a thing, it still remains axiomatic that any ship gives its best all-round performance when it floats and sails on its designed waterline or on a waterline closely parallel.

Remedies

All problems of engineering, construction, or otherwise, should be approached on the premise that there is always room for improving that which has been done before. Careful study of all conditions of operation, principal components, propulsion, gear, handling, etc., should be put into the hands of one individual and the responsibility should be his and his alone. He, naturally, is going to confer with and seek advice from the owner, captain, engine maker, fishing gear manufacturer, and others but the final decision and responsibility should be his only. No other approach will give such complete satisfaction.

Conclusions

Fishing vessels, regardless of size, undergo severe treatment from all sources when in service. Let it not be said that any of these failed to come home because they were not as well developed as possible to meet the usual conditions encountered in fishing at sea.

BEHAVIOUR OF TRAWLERS AT SEA

by

WALTER MÖCKEL

THE naval architect must know what stresses and strains a ship has to face. There are two ways of finding the necessary data:

(1) through systematic observations on board ship;
(2) through model tests.

Model tests can give practical results only if they are correlated to the data collected on board ship showing the behaviour of the ship in all conditions of service. Such research on large cargo boats has been carried out in several countries, but little information is available concerning fishing vessels.

The Hamburgische Schiffbau-Versuchsanstalt (Hamburg Shipbuilding Research Institute), during 1946 to 1949, had measurements taken by the author on board some German trawlers. The critical state of trawlers in heavy seas is known and has been proved by relatively frequent accidents.

The voyages were made in five trawlers sailing from the Elbe to the fishing grounds between Iceland and the White Sea. One or two trips were made in each vessel. The available results still possess a limited basis as yet so it is desirable to supplement the data with further measurements. But the results obtained so far embody some new knowledge that may be of use to the naval architect.

The characteristic data of the investigated ships (all coal burners) are summarized in Table LIII.

The ships have raked stems, V-sections and cruiser sterns with the exception of ship D which was built in 1913. The power of the steam engines was stated to be 750 i.h.p. for ship A, and 600 i.h.p. for ship B. According to a power measurement taken on Ship A, the performance was 580 s.h.p. at 110 r.p.m. at which the ship attained 11.1 knots. The speed of ship B was 10.8 knots.

The measurements covered the speed of the ships, the rolling and pitching angles, and the relevant periods. The periods of encounter between the ship and the waves (apparent period of the waves) were also measured. An observation report was answered four times per day (*see* Appendix). Special measurements were also made when noteworthy phenomena occurred.

Ships A and B were equipped with a pressure log in the bow for measuring their speed. As, in the case of very

TABLE LIII

	Ship A	*Ship B*	*Ship C*	*Ship D*	*Ship E*
Year of construction	1935	1948	1943	1913	1952
Length b.p.	157.4 ft. (48.00 m.)	141.0 ft. (42.98 m.)	164.9 ft. (50.27 m.)	132.1 ft. (40.25 m.)	170.51 ft. (51.97 m.)
Moulded breadth	26.17 ft. (7.99 m.)	26.2 ft. (8.00 m.)	30.1 ft. (9.16 m.)	23.23 ft. (7.08 m.)	28.68 ft. (8.74 m.)
Mean draught	13.85 ft. (4.23 m.)	13.7 ft. (4.18 m.)	14.1 ft. (4.31 m.)	11.83 ft. (3.62 m.)	13.75 ft. (4.20 m.)
Displacement	856 tons (870 cu. m.)	786 tons (799 cu. m.)	1041 tons (1058 cu.m.)	550 tons (559 cu. m.)	—
Block coefficient	0.537	0.556	0.533	0.541	—
Midship section coefficient	0.88	0.88	0.83	0.79	—
Prismatic coefficient	0.611	0.633	0.642	0.685	—
Length/breadth	6.00	5.37	5.49	5.69	5.95
Speed	11.1 knots	10.8 knots	11.2 knots	9.1 knots	12.0 knots

violent pitching, the log indication may be erroneous, astronomical and, when possible, terrestrial bearings were taken continually. By determining distances in this way, and by paying due regard to the current, speed measurements of maximum precision were obtained.

The rolling and pitching angles were determined, with the aid of a protractor fixed to the ship, by reading the reversion points of movement in the line of the horizon. Thus the angles are related to the horizon and not to the slope of the waves. The bi-section of these angles shows the approximate magnitude of the rolling angle to each side and of the pitching angle above and below the horizon. In fact, the rolling angle is somewhat greater to the lee side than to the windward. In the following description the total angles between the reversion points of movement have been retained.

The length and height of the waves were estimated by comparison with the known dimensions of the ship. The wave period was measured by timing the motion of a foam spot on the wave-crest and from this the wavelength in metres was calculated as follows :

$$\lambda = g/2\pi t^2 = 1.56t^2$$

where t is the period of the wave. The wavelength determined in this manner is consistent to a satisfactory degree with the published data of the Hydrographic Institutes.

Wind force (in Beaufort numbers) was estimated in accordance with nautical practice. The speed of wind was partly determined mathematically, after measuring the pressure by means of a column of water.

LOSS OF SPEED IN HEAVY WEATHER

Loss of speed in heavy weather is caused by increased air and rough water resistance. Rough water resistance cannot be calculated, owing to the multiplicity and impenetrability of the different factors of influence. It is only estimated on the basis of the measured speeds and engine data with the aid of propeller diagrams. On the other hand, air resistance can be calculated in a simple way.

In determining loss of speed of ships A and B in a head wind the speeds, measured at approximately identical torque, were plotted against the force of wind and from this a curve of the average speed loss in per cent. was determined. The speed loss curve has a parabolic course which, as shown in fig. 418, becomes straight or almost straight if the speed losses are entered as a function of the second power of wind or wave speed. This linear relation is valid as long as the wind and wave speeds are proportional within the observed range. The diagram shows that, under the same weather conditions, the ship A loses considerably less speed than the shorter and broader ship B.

Trim also exerts a considerable influence on the loss of speed. This is shown by a comparison of the two curves of ship A. On the outward voyage she was trimmed down by the stern and on the homeward voyage she was heavily down by the head because of the full fish hold. The speed loss in this condition, when sailing against seas, is considerably greater than in case of stern trim. For example, with the wind at Beaufort No. 6, the speed loss

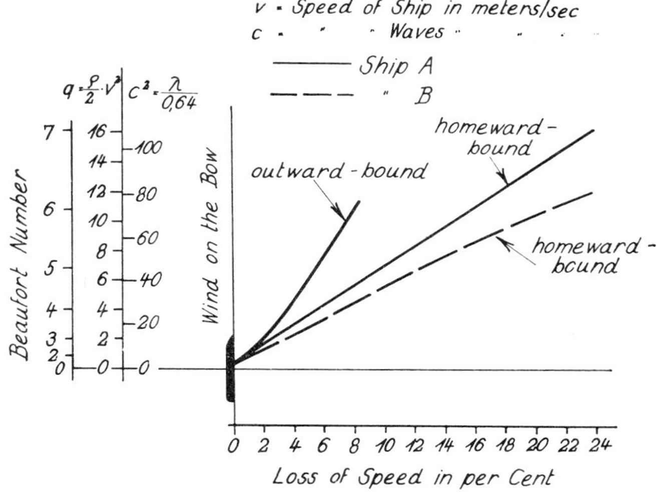

Fig. 418

on the outward trip amounted to 8 per cent., while that on the homeward voyage in similar weather was 15 per cent., i.e. nearly double.

On the homeward voyage of ship A, in fine weather, the Admiralty constant is:

$$C_w = \frac{D^{2/3} \times V^3}{\text{s.h.p.}} = 235$$

At Beaufort No. 5, this constant (with regard to the output loss through the reduction in propeller r.p.m. due to the increased load) is reduced to 179 and at Beaufort No. 7 to 109.

In fine weather and at cruising speed, the resistance of the ship is calculated to be 10,750 lb. (4,870 kg.). This approximately corresponds to the resistance as measured on board other similar fishing vessels. At Beaufort No. 7, the total resistance rises to 15,450 lb. (7,000 kg.) i.e. by 44 per cent. Of the absolute increase of resistance, 15,450—10,750=4,700 lb. (2,130 kg.), 1,500 lb. (680 kg.)=14 per cent. relate to air resistance and the balance of 3,200 lb. (1,450 kg.)=29.8 per cent. to rough water.

The situation is similar during trawling. The captain of ship A endeavoured to maintain a trawling speed of 3 knots. With increased head winds, power had to be increased. Fig. 419 shows the resistance and required power as a function of the wind. Trawling at 3 knots could be done up to 770 h.p. i.e. up to approximately Beaufort No. 7.5. But at higher wind speeds trawling had to be stopped, not because the handling of the trawls was too difficult but because speed could no longer be maintained. So if the design of ship and fishing gear are improved to the extent that trawling beyond Beaufort No. 7 is possible—this is already the case with some modern ships—it is also necessary to provide the engine power to maintain the required speed in rough water.

Fig. 419 also shows the calculated resistance curve for trawling. The total resistance in smooth water

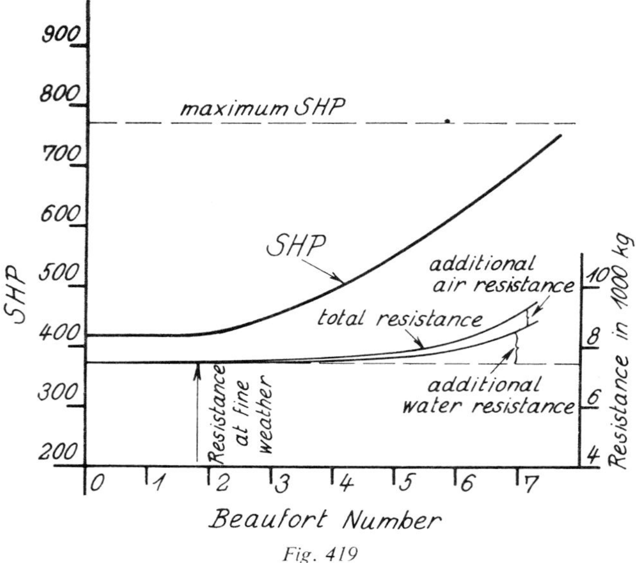

Fig. 419

amounts to 15,550 lb. (7,050 kg.). At a speed of 3 knots, without the trawl, the ship's resistance is about 880 lb. (400 kg.) so that 14,670 lb. (6,650 kg.) is accounted for by the trawl. At Beaufort No. 7, resistance increases to 20,060 lb. (9,100 kg.) i.e. by 4,510 lb. (2,050 kg.) = 29 per cent. Of this, 1,040 lb. (470 kg.) are calculated to be air resistance. The balance of 3,470 lb. (1,580 kg.) is additional resistance caused by the waves. According to these calculations, the additional resistance when trawling, and despite the lower speed, is only 4,700 − 4,510 = 190 lb. (2,130 − 2,050 = 80 kg.) less than when cruising at the same wind speed.

In order to know the required power to overcome air resistance during cruising, the mean power requirement was determined each day on the outward and homeward voyage. It was found that an average of 26 h.p. = 4.5 per cent. was required to overcome air resistance.

It should be added that these voyages took place in the spring, a season distinguished by bad weather.

STABILITY DURING THE TRIP

The behaviour of a ship in a seaway is determined by its shape and weight distribution. The period of roll depends, apart from the radius of gyration, on the metacentric height (GM), which can be determined either by an inclining test or a moment calculation. Because of the relation between the natural period of roll and GM, the latter can be estimated from measurements of the rolling period in accordance with the known formula:

$$GM = \left(\frac{2\pi i}{\sqrt{g}\, t}\right)^2$$

In this formula: i = radius of gyration; g = acceleration due to gravity; t = rolling periods, in seconds.

The radius of gyration depends largely on the breadth of the ship, and may be determined by multiplying the breadth of the ship, B, with a value $m = \frac{i}{B}$ known from experience. Hence i = m.B. In the feet system $\frac{\pi}{\sqrt{g}} = .544$ and in the metric $\frac{\pi}{\sqrt{g}} \approx 1$. Accordingly

$$GM_{feet} \approx \left(\frac{1.108.m.B}{t}\right)^2 \text{ or } GM_{metre} \approx \left(\frac{2.m.B}{t}\right)^2$$

The stability of the trawlers investigated was continuously determined during the test voyages by this method. Comperiod on fishing vessels have produced, with a fair degree of consistency, an m-value of 0.40 for the ship leaving the port, and of approximately 0.385 when returning from a fishing voyage. The radius of gyration of the outward bound ship is somewhat greater because the fish-hold contains only ice and reserve coal. A greater influence of the masses located on the periphery of the system is predominant.

The GM of the fishing vessels dealt with were calculated with the m-values. There were also available the stability calculations of the shipyards, relating to different loading conditions of the ships. The GM calculated from the rolling period agreed very well with this data, considering that the hypotheses of the shipyards sometimes differed from the actual load conditions. For example, according to the shipyard calculations, ship A should have a GM = 2.8 ft. (0.852 m.) on departure. The GM calculated from the rolling period was 2.76 ft. (0.84 m.). The calculation of the shipyard indicated a GM of 1.48 ft. (0.45 m.) for the fully-loaded ship on arrival while the rolling period gave a value of 1.41 ft. (0.43 m.). A similar consistency existed in the case of ship B, as shown in fig. 420, which gives curves of

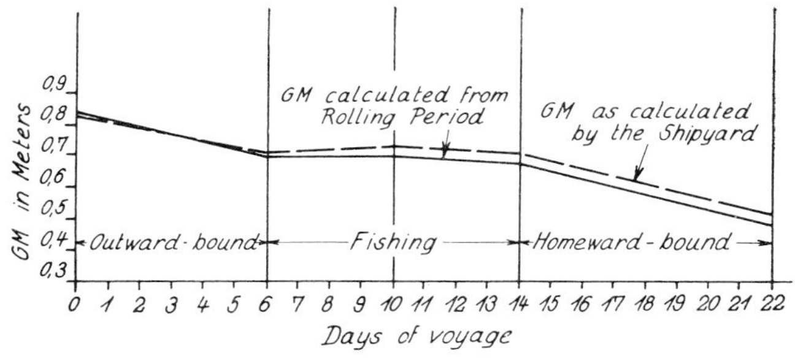

Fig. 420

the stability for an entire voyage, according to both shipyard calculations and measurements of the rolling period.

The coal and water consumption and the increase in the load by the catch were recorded daily, and the effect of these alterations of weight on the GM was calculated by measurements of the rolling period. The results are shown in fig. 421 which contains the curves of daily load increases and decreases, the total load and the GM. These data relate to both voyages of trawler A. Similar

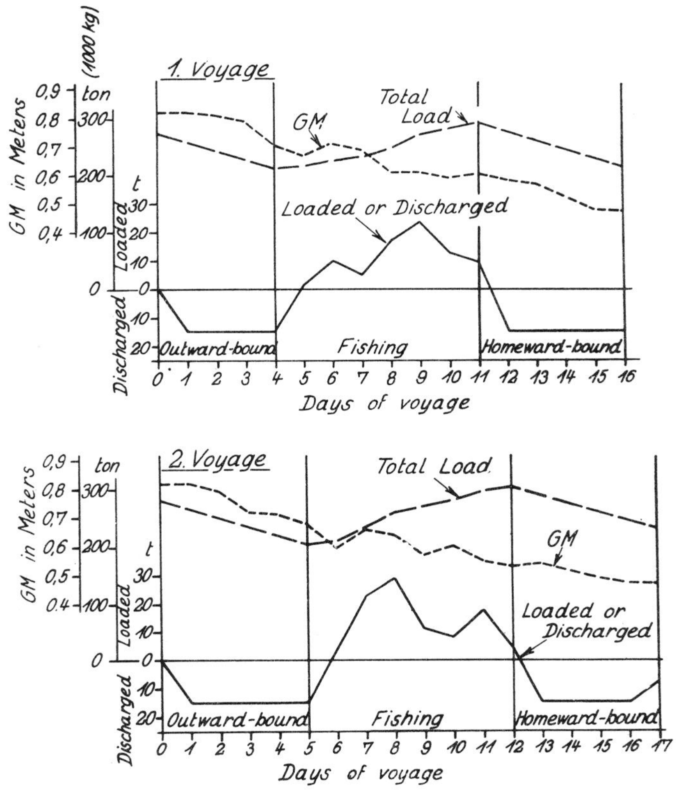

Fig. 421

curves were also established for other ships. In general, the trend of the total loading of fishing vessels is similar on all voyages. It is only on trawling days that deviations may occur, because of the contingency of the catch. For example, on the first voyage the big catches occurred

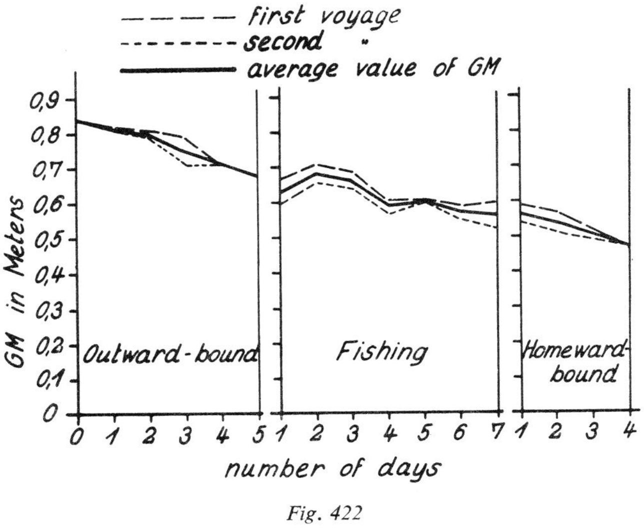

Fig. 422

during the last, and on the second voyage during the first, trawling days.

In fig. 422, the two stability curves of fig. 421 are entered together, with the mean stability curve resulting from the two voyages. The curves of the voyages show a strikingly parallel course, which is quite consistent with the tendency of the curves of ship B in fig. 420.

The development of the GM during trawling days is interesting. As shown in fig. 422, the GM increases from

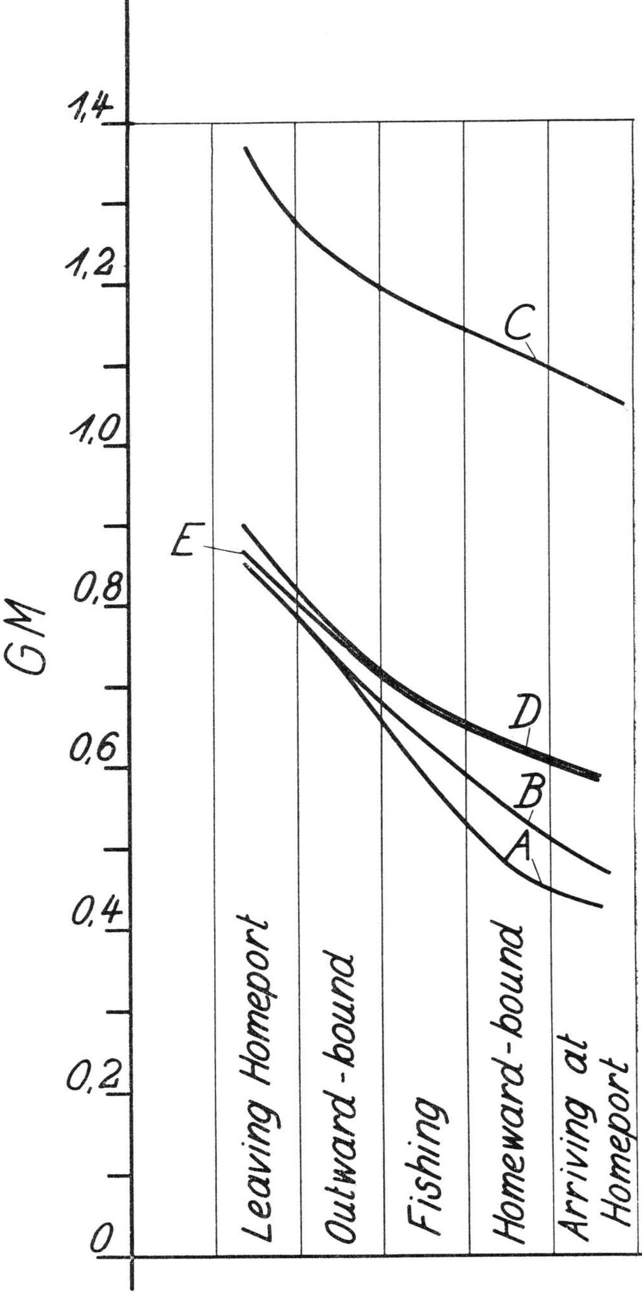

Fig. 423

the first to the second trawling day by 2 in. (5 cm.), then gradually falls again. The sudden rise of the metacentric height follows the stowing of the first catches on the bottom of the fish-hold. The centre of gravity shifts downwards and the ship becomes stiffer. During the succeeding days, when the fish are stowed higher and higher, the centre of gravity moves upward again, and the GM is reduced. The slight rise of the GM on the fifth day occurred when the loading of the partitioned small fish-hold was begun.

Fig. 423 shows the development of GM of the five fishing vessels investigated at typical stages of the voyages, i.e. departure, outward voyage, trawling, homeward voyage and arrival, all determined by measurements of the rolling period.

The stability of ship C differs from that of a normal fishing vessel because, while in the Naval Service, considerable top weights were fitted and compensated by heavy concrete ballast. Later, when the deck weights were removed, the ship became stiff and her rolling motions were so jerky that they raised serious complaints from the crew. The vessel also shipped so much water that she had to stop trawling when other vessels continued without much difficulty.

The rolling periods of the five vessels are compiled in Table LIV.

TABLE LIV

Ship		*A*	*B*	*C*	*D*	*E*
Breadth	ft.	26.2	26.2	30.1	23.2	28.7
	m.	7.99	8.00	9.16	7.08	8.74
		Rolling periods in seconds				
Departure		6.95	7.00	6.30	6.20	7.60
Outward bound		7.38	7.40	6.60	6.40	7.90
Trawling		8.14	7.80	6.60	6.70	8.00
Homeward bound		8.92	8.35	6.70	6.90	8.50
Arrival		9.34	8.90	6.80	7.10	8.80

In calculating the GM at departure and arrival the rolling periods were measured in the port or on the river Elbe, the ships being set in motion by use of the rudder. On the outward voyage, trawling and homeward voyage, the mean rolling period was calculated from daily measurements from which the GM was determined.

The figures indicate that ship C, owing to its abnormal weight distribution, has an unusually short rolling period for its breadth of 30.1 ft. (9.16 m.). In fact, the rolling period is exactly the same as that of trawler D, which is only 23.2 ft. (7.08 m.) broad.

Ships C and D caught few fish during the observation voyages so the holds were not full. Ships B and E also had some empty space.

If the GM values of the five ships are plotted against the rolling time, the influence of breadth, i.e. of the radius of gyration, becomes very clear, as shown in fig. 424. Stability being identical, the curves adapt themselves smoothly to the breadth of the ship, and the narrowest ship shows the shortest rolling periods and the broadest ship the longest. It is of special interest that the curves of ships A and B, whose breadths differ only by $\frac{3}{8}$ in. (1 cm.) almost coincide. Ship C cannot

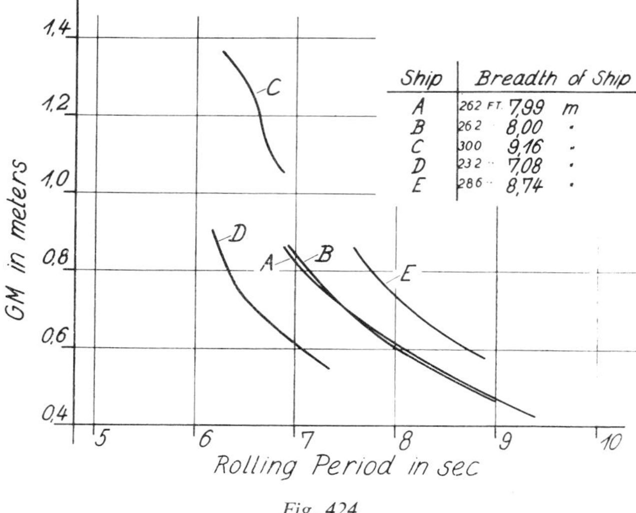

Fig. 424

be included in the comparison because of its abnormal weight distribution but the position of its curve, as compared with the curves for normal conditions, is of interest. The tendency of the curve suggests that, consistently with the breadth of the ship, it would run to the right of that of ship E if the GM were smaller.

In view of this result, it is obvious that GM as parameter should be plotted versus the rolling period and the breadth of the ship. This presentation has been chosen in fig. 425. From this diagram, which was made solely on the basis of measurements taken on ships, it is possible to find the GM for any rolling time of fishing vessels having normal weight distribution and breadths within the limits that exist here. It would be desirable to carry out similar observations on a large number of fishing vessels of various designs to obtain a clear picture of their characteristic stability. It would also be particularly desirable to measure rolling periods of ice-caked trawlers fishing in arctic waters, so that the designer could have data concerning the stability requirements for such conditions.

ROLLING AND PITCHING PERIODS

Consideration of stability also involves the question of a ship's behaviour in waves, because of the relation between stability and rolling period, rolling angle and acceleration. The comfort of the people on board depends largely on the acceleration.

The rolling periods of ships at sea can be measured easily but correct determination of the natural pitching period is difficult because they rarely pitch to the rhythm of their natural period. Instead, they adapt themselves to the motion of the waves passing under them.

In fig. 426, the rolling and pitching behaviour of ship A is plotted against the period of encounter. The influence of the waves is very apparent. The rolling periods are clearly shown in three groups, corresponding to the load

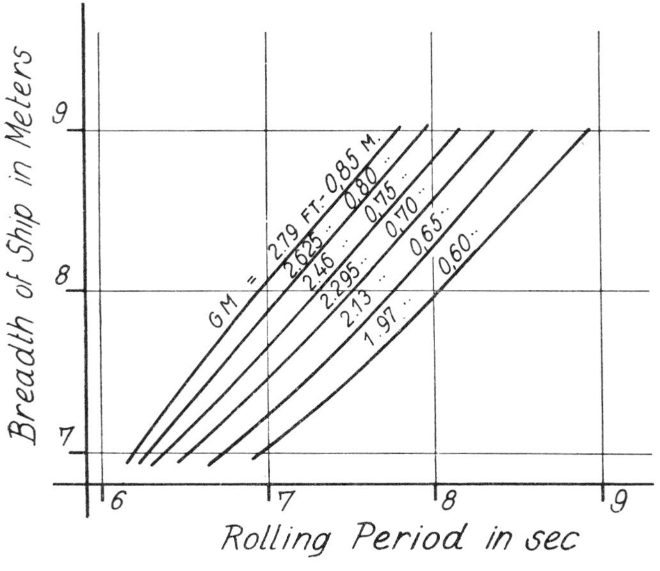

Fig. 425

conditions: outward bound, trawling and homeward bound. Those measured during the outward voyage constitute the lower, and those measured during the homeward voyage the upper limit of the rolling period range. The measurements on which this diagram is based were carried out in a light wind, but there was a pronounced swell, so that the ship was affected chiefly by one wave oscillation and not simultaneously by several wave surges.

The diagram clearly shows that the rolling period is little affected by the period of encounter. It is true that with an increasing period of encounter the rolling period also increases, but, within the range of these measurements, the increase amounts to a maximum of only 0.3 seconds. The situation is different with the measured pitching periods, which are grouped round the 45° line of the diagram. The pitching period adapts itself closely to the period of encounter i.e. to the motion of the wave passing under the ship. A similar oscillating behaviour has been observed on large cargo boats.

From a few measurements, the results of which may be regarded as natural pitching periods, there emerges for ship A a pitching period of 3.6 seconds on the outward voyage and 4.2 seconds on the homeward voyage. The pitching periods of the ships B, C and D are also of the order of 3.5 to 4.0 seconds. The differences between the pitching periods of these ships, despite the variation in length, cannot be very great, considering that the pitching periods of large cargo boats investigated are from 7.3 to 7.7 seconds. On the small ship D the heaving period t_h was several times measured in a transverse sea and found to be 3.5 seconds. Euler's formula,

$$t_h = 2\sqrt{\frac{D}{F.g}}$$

D = Displacement
F = Waterplane
g = Acceleration due to gravity

predicts a heaving period of this ship to 3.6 seconds.

The difference between the pitching periods on the outward and homeward voyages observed in ship A is explained by the load situation. Owing to the consumption of coal stored in the middle of the ship and to filling the fish-hold, the longitudinal radius of gyration is greater during the homeward voyage. This also causes an increase of the pitching period, provided the second root of the longitudinal stability of the head trimmed ship increases less than the radius of gyration.

If the measured natural pitching periods are assumed to be correct, then the value $m = i/L$, calculated with the aid of the longitudinal metacentric heights taken from the hydrostatical curves, will be 0.275 for the outward voyage and 0.32 for the homeward voyage. The m-value for loaded cargo ships, with the engine amidships, has been found to be approximately 0.36 to 0.38. The values established for the fishing vessels agree reasonably with the results calculated elsewhere.

ROLLING AND PITCHING ANGLES

The rolling and pitching angles were measured over the horizon, between the inversion points of the ships' motions. In the course of a long series of measurements, the relatively rare maximum values were separated and from these the average of the *maximum* angle was calculated. The *normal* angle was calculated from the remaining values in a similar manner. The angles thus determined, which present a comparative survey of the

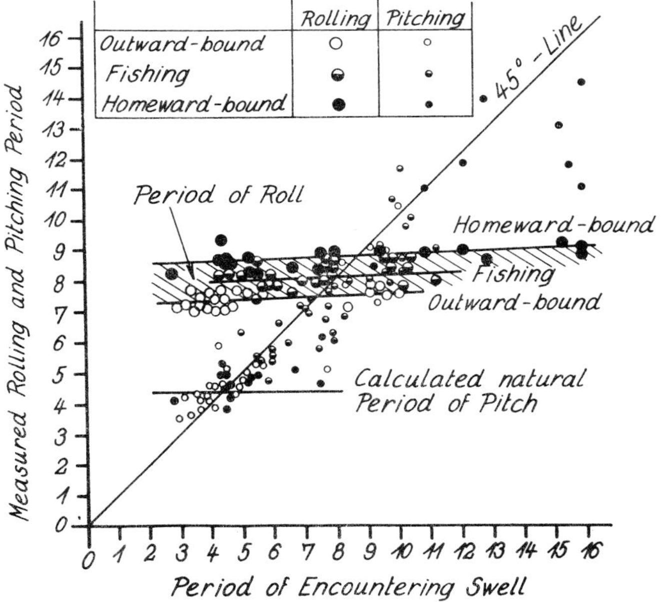

Fig. 426

oscillations of the ships A to D, are plotted against the Beaufort Numbers in fig. 427 and 428. Only such rolling angles were used when the angle of encounter was less than 15° from athwartship, and pitching angles when

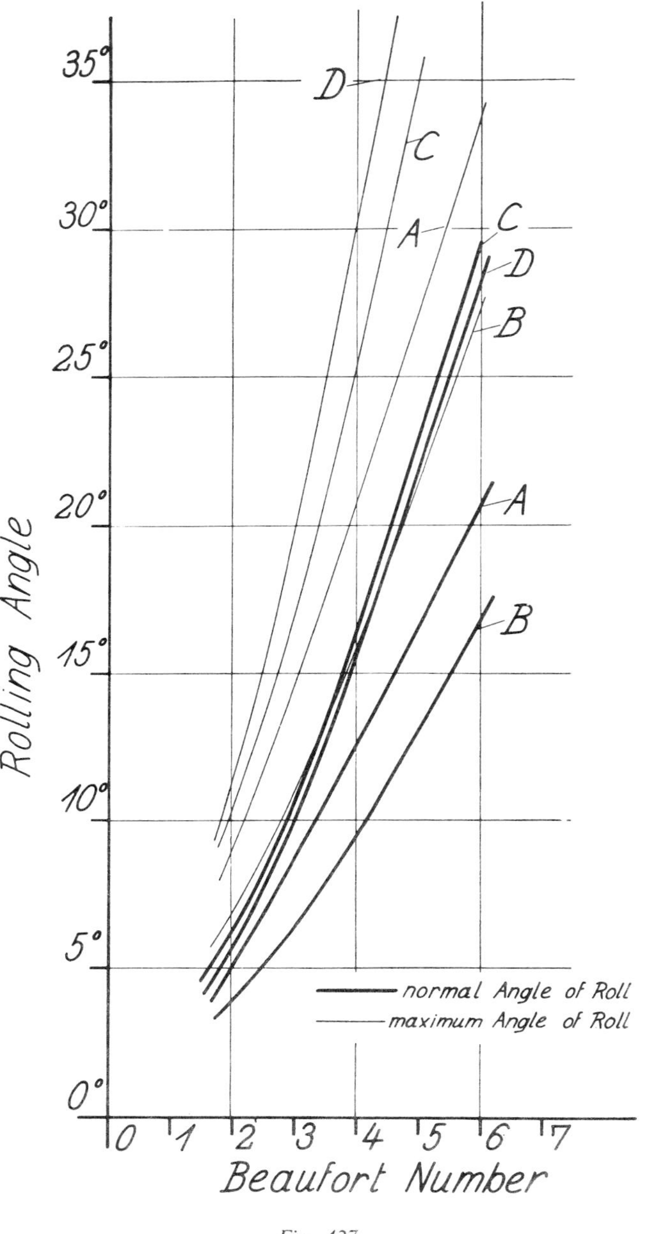

Fig. 427

the angle of encounter was less than 20° on each side of the bow. In the diagrams, the curves of the normal angles are drawn in heavy, and those of the maximum angles, in thin lines. In order to have, as near as possible, identical conditions with regard to the effect of the waves on the ship, the only measurements used were those taken when the swell accompanying the waves caused by the prevailing wind was slight.

Under identical wave conditions, ships C and D had the largest rolling angles. The reason for this in ship C must be in its great GM which gives a short natural rolling period corresponding to the most frequent periods

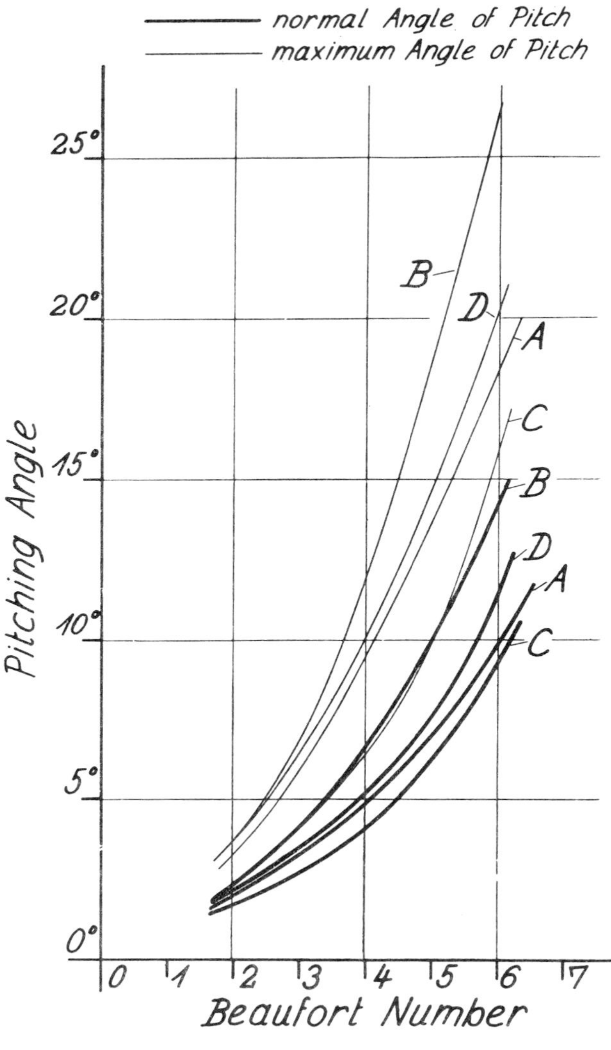

Fig. 428

of encounter. Ship D, owing to her smaller dimensions, particularly the breadth of only 23.2 ft. (7.08 m.), rolls more than the other ships. In addition, the midship section coefficient of ship D amounts to only 0.76, as compared with 0.88 in ships A and B, and 0.82 in ship C. The bilge radii and floor rise of ships C and D are relatively large. The resulting smaller damping effect explains to some extent the behaviour of the two ships. The rolling angles of ships A and B only differ by about 3°. Both vessels have the same midship section coefficient.

The curves of the angles of pitch in fig. 428 show that ship B makes by far the greatest amplitudes. The reason might be the full shape of the bow, which possesses a relatively great reserve buoyancy. When steaming against the waves the bow rises considerably. When dipping, the full bow sets into the water with short, violent jerks. Owing to these pitching motions, which the crew finds very unpleasant, the vessel rarely ships water so that handling the catch is seldom disturbed even in rough weather. And these motions, which are so violent during cruising, are considerably more gentle when trawling against the waves because of the longer periods of encounter.

Evidently, the large pitching motions and the short, jerky ploughing into the waves is one of the reasons why ship B loses more speed than ship A.

Experience with ships A to D in heavy seas teaches that, if fishing vessels are to be fast, the underwater part of the hull should be of sharp design, and for good cruising properties in heavy seas, the frames above water should be of slight convex shape. The transition from the underwater to the above water shape must be smooth, otherwise the ship will plough into the waves in a jerky manner.

The angles of pitch of ships A, C and D differs only slightly.

It is well known that pitching is influenced by the ratio between the length of the waves λ and the length of the ship L. If the four ships are compared, for example, on ratio $\lambda/L=1$, ship A pitches 6 to 12° normal-maximal, ship B 8 to 15° and ships C and D 5.5 to 10°.

During trawling in a slight seaway the rolling and pitching angles are almost the same as when cruising. It is only at wind speeds exceeding Beaufort 4 to 5 that the angles are smaller during trawling than during normal sailing.

To show what oscillations may occur on fishing vessels, the largest angles measured on the four ships are compiled in Table LV.

The periods of encounter show that there was synchronism, in nearly all cases, between the ship's natural periods and the periods of encounter. The great rolling angle of ship A at $\gamma=30°$ is also attributable to synchronism.

Synchronisms frequently occur in trawlers, as shown by the percentage of frequency of the period of encounter on ships A to D (*see fig. 429*). The short periods of encounter from about 2 to 4 seconds, due to the greater speed of the ship while cruising, occur more frequently than during trawling, where the maximum frequency is about 6 seconds. The longer periods measured during

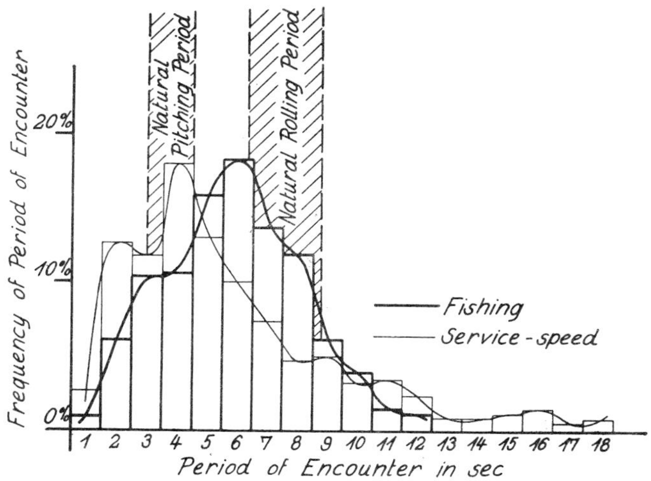

Fig. 429

cruising occur in stern seas. Under those conditions, the periods are shorter during trawling because of the low speed of the ship.

The diagram shows that in cruising the natural period of pitch of about 3 to 4 seconds coincides with the maximum frequency of the period of encounter, while the natural rolling period is less subject to synchronism, but during trawling the situation is reversed. There, the natural rolling periods come into synchronism with the waves more frequently than the natural pitching period. This applies particularly to ship C, with the short natural rolling period of about 6.6 seconds. As, however, there is a wide maximum range of periods of encounter during trawling, the pitching motions also often come into synchronism with the waves. Owing to the great pitching amplitudes that are to be expected for this reason, it is advisable to design the forepart of the ship so that the

TABLE LV

	ROLLING				PITCHING			
Ship	ϕ	γ	τ	*Beaufort Number*	Ψ	γ	τ	*Beaufort Number*
A	21°/40°	30°(T)	7.2 sec.	6-7	15°/27°	0°	4.4 sec.	10
B	28°/36°	90°	7.0 sec.	6	14°/25°	10°	4.5 sec.	6
C	29°/40°	85°	6.8 sec.	6	15°/22°	0°	6.8 sec.	7-8
D	25°/52°	75°(T)	6.6 sec.	6	13°/21°	20°	6.3 sec.	8

ϕ = Rolling angle, normal/maximum
Ψ = Pitching angle, normal/maximum
γ = angle of encounter
τ = period of encounter
(T) = measured during trawling

bow dips into the waves softly yet in such a way that little water comes on deck.

The maximum acceleration $^{b}max = \frac{4\pi^2}{t^2} \cdot h \cdot \sin \phi$, (h=lever arm) shown in Table LVI, may be regarded as typical of the crew's reactions to the ships' behaviour. The maximum relate to the motions of the ships at Beaufort No. 6.

If the ratio between the length of the wave and the length of the ship is higher than 1, then there is the additional danger that, owing to the contrary orbital motions of the water, she will veer across the crest and the trough, i.e. get out of control, and be rolled over by the sea, especially in breaking seas.

A further possibility of capsizing in seas running obliquely astern has been pointed out by Kempf (1938).

TABLE LVI

	ROLLING			PITCHING			
Ship	*Maximum angle*	*Maximum acceleration b max*		*Maximum angle*	*Maximum acceleration b max*		*Crew's view*
A	20.6°	2.25 ft./sec.²	0.687 m./sec.²	18°	12.6-ft./sec.²	3.86 m./sec.²	Good
B	17.0°	1.92 ,,	0.586 ,,	26°	14.4 ,,	4.38 ,,	Too hard pitch
C	29.0°	4.81 ,,	1.470 ,,	16°	12.1 ,,	3.68 ,,	Too violent roll
D	28.0°	2.83 ,,	0.862 ,,	20°	9.6 ,,	2.92 ,,	Very good

The lever arm h used for the calculation of maximum acceleration of the rolling motion was assumed to be from the centre of gravity to the bridge rail, and that for the pitching motion from amidships to half of the foredeck, where the catch is handled. The angles entered were half the angles mentioned in the table.

The biggest rolling acceleration is shown to be 4.8 ft./sec.² (1.47 m./sec.²) for ship C, whose crew also complained very much of the hard and restless rolling motions. On the other hand, ship B has the biggest pitching acceleration at b max=14.4 ft./sec.² (4.38 m./sec.²). On this ship the crew complained only of the unpleasant pitching. The crew of the small trawler D spoke highly of the sea-kindliness of their ship, which has the smallest acceleration of the four ships.

SAILING SMALL SHIPS IN A FOLLOWING SEA

Far more attention is paid to the behaviour of ships sailing against the waves than with them, although the latter are a bigger threat to safety. Grim (1951) demonstrated, both theoretically and by model tests, that even if a ship has adequate static stability she can be capsized if the waves strike her astern. There need be no synchronism between the ship's natural period and the period of encounter. In any case, a ship cannot be capsized through synchronism alone. As Grim has demonstrated by model tests, the greatest danger arises when the speed of the waves is equal to that of the ship, or when the waves overtake the ship so slowly that an almost static situation is created and the ship lies on the crest. In this position the lever arms and the degree of static stability can be reduced to such an extent that the ship capsizes even without the action of dynamic forces. The stability is reduced when the ship is travelling on the crest of a wave, as compared with calm water, but is increased when she is in the trough of a wave.

This danger arises when the ship rolls heavily, to which synchronism may be a contributory factor, and is struck by a big wave on the windward side when inclined far to the leeside. In this case the centre of buoyancy is moved to the windward side, so that a static capsizing moment is created. If, in this situation, the kinetic energy of the wave is added and the leeside rail dips under water, there is an immediate danger of capsizing.

It is well worth noting Grim's finding that synchronism is present not only in the case of synchronism between the ship's natural rolling period and the period of encounter. Grim found several points of synchronism where, on occasion, very large rolling angles were measured. These points arise when the ratio between the period of the change of the stability t_s and the rolling period t of the ship is 0.5; 1; 1.5; etc. The largest angles were measured at $t_s/t=0.5$. The rolling angles become smaller with increasing period ratios.

Another great danger arises when a big wave slowly overtakes the ship and floods the deck along its entire length. The danger is increased if the doors of the deck structures are open. Several small German vessels, including a trawler, were lost in this way in recent years.

Owing to the extra weight of the water above the centre of gravity and the free surfaces on deck, such a diminution of stability takes place that the resistance of the ship to the dynamic attacks of the sea is almost, if not completely, exhausted.

It must be expressly pointed out that the dangers referred to can become acute only if the waves pass the ship slowly, i.e. when they come from astern. The speed of the ship must be reduced to allow quicker passing of the waves so that the danger cannot develop.

The author experienced such a case on board trawler A. The ship, travelling with a full fish-hold and with the head down, was north of the Orkney Islands. The wind and the waves were about 70° astern in relation to the transverse direction. The wind force was Beaufort No. 9 and, in gusts, 10. The waves were not yet fully developed in relation to the wind speed. Their length was estimated at 220 to 230 ft. (65 to 70 m.) and their height at 16.4 to 19.7 ft. (5 to 6 m.) so they were very steep. The period of encounter, measured with a stop-watch, was 12.2 seconds, and the pitching period 11.8 seconds. The ship's natural period of roll, measured a few hours after the incident in calm water near land, was 9.01 seconds. Thus, there was no synchronism between the ship's natural rolling period and the period of encounter. From the rolling period, the GM was calculated to be 1.6 ft. (0.49 m.), which corresponds very well with a moment calculation on the basis of the shipyard data.

The rolling motions were extremely hard and jerky in the heavy seas, and there were rolling angles of 15.4° to 29.4°. Occasionally, some water dashed over the rail on the windward side. On the leeside, the vessel shipped green water over the bulkward when heeling. According to the trochoid formula, which of course applies only with reservations to undeveloped waves, the waves passed the ship with a speed of about 16 ft./sec. (5m./sec.). In fact, the wave-speed was probably somewhat smaller.

Suddenly the ship was struck with great violence by a breaking wave coming from astern, which flooded the low foredeck nearly to the level of the rail. The head trim of the ship increased, so that the following waves, coming in from both sides, flooded the foredeck to the top of the bulwarks and the rails were in the water. The situation of the ship, which was scarcely obeying the rudder, was very precarious.

As the breaker struck, the engine was immediately stopped and the ship was turned with her stern straight against the sea. She was held in this position by means of rudder and propeller manoeuvres.

According to the drawings of the ship, the weight of the water taken inboard was calculated to be about 124 tons. At the same time, the ship took a list of about 5° to 7°.

The increase of weight above the centre of gravity and the free surface extending over the entire breadth of the ship caused a negative GM, which was confirmed by the list of the ship and its heavy and irregular motions and the fact that she stayed in any position into which she was thrown by the sea. Her labile behaviour conveyed the feeling that stability and buoyancy were exhausted and that she might at any moment capsize and sink.

Other high waves flooded the foredeck and it took half-an-hour before the voyage, at slow speed, could be resumed. The ship is known to be seaworthy and she behaved very well, especially when she had resumed her normal trim. During the slow speed the overtaking wave passed the ship very rapidly, no further water was shipped and the deck remained entirely dry. This different behaviour of the ship before and after the accident, i.e. at full and at low speed, made it dramatically evident on what simple nautical measures the safety of a ship may depend.

From this incident, the following conclusions may be drawn:

1. Even a trawler known as seaworthy can incur the danger of capsizing by shipping large quantities of water.
2. The situation is most dangerous when travelling before the seas at a speed approximately, or actually corresponding to, that of the waves, as then great quantities of water may be shipped. The danger is particularly great if the water can flood the ship through openings in the deck superstructures.
3. Dangers of this kind are not the consequence of faulty design but chiefly of an unfavourable ratio between the speed of the waves and that of the ship. They can be largely or wholly eliminated through simple nautical measures, i.e. by slowing down.
4. In the case of small vessels with a head trim, additional danger arises when the head lies deep in the water, the light stern can be turned athwart the seas by the orbital motion of waves coming obliquely from astern, and the ship can be rolled over.

Hence, in designing trawlers, care must be taken to ensure:

(*a*) Sufficient reserve buoyancy, particularly forward. A good freeboard should be aimed at, though this is unpopular with many fishermen because they consider it makes the handling of the trawl difficult. Well flared sections above the waterline in the foreship, and as large a forecastle as possible, also contribute to extra buoyancy.

(*b*) In calculating stability, the shipping of considerable quantities of water should be taken into consideration. To keep rolling acceleration low, heavy equipment should be, if possible, arranged towards the ship's sides, as this will increase the radius of gyration without decreasing GM.

(*c*) Scuppers must be large enough to allow the water to run rapidly off the deck.

(*d*) The superstructures astern must be built over the entire breadth of the ship, as is sometimes done on modern trawlers, to protect them against overtaking waves.

(*e*) The stern above the waterline must be as full as possible, to provide extra displacement and to adapt itself easily to the waves coming from astern.

(*f*) According to experienced trawler skippers, a GM of about 2.8 ft. (0.85 m.) at departure, and one of 1.65 ft. (0.50 m.) at arrival, is considered to be very good.

(*g*) To obtain a satisfactory damping of rolling, it is not desirable that the floor rise and bilge radius should be too great.

In operating trawlers, care must be taken:

1. To keep the openings of superstructures closed in rough weather, and particularly in a high, overtaking sea.
2. To reduce the speed of the ship considerably in a high sea from astern.
3. Never to forget that there are natural limits to the safety of even the best-designed ships. These cannot be exceeded without exposing the vessel to danger.

The combination of experienced design and true seamanship only will ensure safety.

APPENDIX

Log

Year.............. Port of departure......... Draught at departure fore .... aft .... Mean draught .... Displacement..........

Month........... Port of return.............. Draught at return fore .... aft .... Mean draught .... Displacement..........

Fishing ground...

Name of ship.................... Owner.................... Name of master.................... Name of observer....................

							Wind		*Waves caused by prevailing wind*							*Swell*							*Pitching*			*Rolling*						*Remarks*
Date	Time	Position of ship	Course	Speed in knots	Trawling or cruising	r.p.m.	Direction	Beaufort No.	Direction	Force	Period of encounter	Period of waves	$\lambda = g/2\pi t^2$	λ estimated	Height	Direction	Force	Period of encounter	Period of waves	$\lambda = g/2\pi t^2$	λ estimated	Height	Period of pitch	*Angle* Average	*Angle* Maximum	Period of roll	*Angle* Average	*Angle* Maximum	Where water shipped ?	Cargo at observation	Fuel and water at observation	

SAFETY AT SEA

by

Wm. C. MILLER

TEN BILLION dollars are invested in the fisheries industry of the United States of America. Between 45 and 50 per cent. of this amount is in boats and gear, the remainder being devoted to operations ashore. The floating equipment is, therefore, the most important part of the industry, and no doubt this is also true of the fishing industries of other nations.

The financial loss on southern California fishing vessels and equipment has reached figures of such magnitude that there have been international discussions in the insurance world and a threat by underwriters to refuse fishing vessels as insurable risks.

Safety at sea results from a combination of wide experience and technical ability of naval architects and boat builders to design, lay down, construct and equip a good ship, but it is not effective unless the vessel is manned by competent officers and crew.

PERSONNEL

The serious financial losses in the fishing fleet along the south-western coast of the United States of America, are chiefly due to improperly trained and improperly equipped crews. For several years people who have realized this fact have been unsuccessful in impressing it upon others, such as underwriters, who are in a position to make certain demands and get results. The vessels' owners have been unable to take corrective measures because of the regulations of organized labour. There is no federal or independent agency in the United States of America empowered to establish and certify the qualifications for seagoing personnel on fishing vessels, other than some masters and engineers. The men are not compelled by labour unions to be seamen but are mostly fishermen, purely and simply. Everyone cognizant of the meaning of the word " seamanship " and aware of the big losses sustained by the southern California fishing fleet in recent years, will admit the incompetence of the crews and their inattention to duty. And each year recently the financial loss has been greater than in the preceding year.

The owner's position is like that of being in the centre of a devastating monsoon. It is possible to survive a storm with the existing ships, as they are adequately strong and physically safe, provided the crews do their duty as seamen rather than try to make their voyages pleasure trips. This problem of personnel holds the answer to safety at sea in all the fishing fleets of the world and, in view of this, it is worth considering those qualities of seamanship which crews should acquire for their own and their ship's safety.

SEAMANSHIP

Seamanship includes the art of handling a ship or boat under all conditions of weather, tide, or other influence affecting its movement or safety. Seamanship should not be confused with navigation, which is the art of determining the correct course of a ship when out of sight of land, nor with pilotage, which is the art of determining the correct course of a ship when working along a coast or up a buoyed channel. Seamanship is a companion art for without it practical effect cannot be given to good navigation or pilotage.

Good seamanship is essentially acquired by practice, and is a marked characteristic of certain races. It cannot be taught by precept or mastered by a study of books, although the basic practice can be learned that way. The capable seaman is made by long service in ships and boats where he can obtain the necessary practical experience. Seamanship has not died with the elimination of sail but it is not taught in certain fishing areas in a way to retain it from generation to generation.

The power-driven fishing vessel, despite all the mechanical aids to handling her, is ultimately dependent upon good seamanship. The officers and ship's company may not be efficient sailors in the old sense of the word but they should be proficient seamen. In sailing ships seamanship was, and still is, largely concerned with the rigging, the making and shortening of sails, the correct manipulation of ship's sails and rudder for such operations as putting about, wearing or club hauling, and for making the best passage whether " on a wind " or " running-free ". But the handling of all vessels, whether sailing or power-driven, requires a knowledge of the influence on their movement of wind, tide or current, their behaviour in a seaway, and various methods of securing them, by anchor in a roadstead or by moorings to a pier or dock. Seamanship also involves the proper loading of cargo, shifting of weights, battening down hatches and making the ship generally "snug" for heavy weather, together with lowering, handling and hoisting of ship's boats and equipment, proper attention to ship's fittings, and innumerable other incidentals to safety of life and equipment afloat. In power-driven ships there must be knowledge of the action of screws. While certain general laws govern the behaviour of nearly all such ships, individually they vary in their response to helm and engines. However handy or unhandy a ship may be, a good seaman will get the best out of her, but a captain who does not know or understand his ship, or who has a crew lacking in seamanship, will often find himself in difficulty. The same applies to the good order of the ship and her fittings. Proficient seamen will keep her clean and trim, pleasing to the eye and ready for any emergency. The ill-manned ship is slovenly in appearance, and some essential for her safety will be found wanting at a time of crisis.

If the crews of the world's fishing fleets were all seamen in this sense safety at sea would be largely achieved. For example, there are fishing vessels on the west coast of the United States whose crews are chosen and fairly treated to ensure they will remain with the ship. These vessels are no different from the average but their loss has been approximately 90 per cent. below that of the other vessels of the fleet. This fact bears out the need for first-class seamanship among fishing vessel crews.

SHIP REQUIREMENTS

A study of the safety and efficiency of fishing vessels has revealed that the average vessel frequently breaks down at sea and completes relatively few successful fishing trips each year. This points conclusively to the need for competent technical attention to design, construction, outfitting, operation and maintenance.

Potential fishing vessel owners do not always get the technical advice they need, and too frequently they accept it from unqualified sources. But when they have the advice of fully qualified experts they certainly profit from it.

Little interest is shown in small fishing craft design by the classification societies, and their regulations for construction of fishing vessels have never been established and there should be basic research to provide data on which such rules could be established. Data which have already been accepted by the societies might be useful, but, for instance, the minimum requirements for scantlings of fishing vessels should not necessarily be those demanded by the societies. They should be determined by common consideration of the problem by a qualified group of designers and engineers with world-wide experience in this class of construction.

Much information on scantlings, prepared by competent naval architects around the world, is already available. The economic condition of the fishing industry, however, would offer a major problem to international standardization. For example, scantlings for vessels operating in near waters would require a different treatment to those fishing in distant waters. Scantling sizes of fishing vessels have been determined to meet the need for a given type under local conditions and most of the vessels have been seaworthy. However, some disastrous results have occurred when vessels have ventured into areas for which they were not designed and, from this standpoint, standardization of design and assignment in a classification would be helpful to the fishing industry.

By way of example of variation in scantling and fastening sizes of similar size fishing vessels, it is interesting to compare those on the west coast of the United States of America with those on the east coast (*see tables* LVII and LVIII prepared by Simpson (1951) and Hanson (1951) of both localities). The difference in scantlings offers no special obstacle to the standardization of design or construction of fishing vessels in the United States, but presents a challenge to formulate data from which a set of standards might be developed.

Information pertinent to design and construction methods in the British Isles, northern European and Scandinavian countries would undoubtedly add priceless data to any such attempt at standardization.

Wood is the material most used in the construction of fishing vessels today and, because of the advance in design and in building equipment, it is no longer necessary to depend on natural timber shapes. Adequate strength to provide safety well beyond the required tolerances is derived from laminated, formed and scarfed sections of much finer fashion than ever previously used. Much attention is still paid to grades of timber, etc., in the design stage but it is impossible under today's conditions to find materials of the standard used throughout the history of wooden ships and there are fewer naturally grown crooks of timber available. Modern machinery, however, is able to duplicate strength through various ways even though using inferior materials.

Timber itself has no greater strength value than the fastenings which hold it to the structure, and, as every fastening reduces the strength of the timber to some extent, the timber sizes must be adequately computed to allow for this fact. All wood, regardless of its density or strength factor, is soft in relation to modern steel fastenings, and poorly mated structural parts will soon wear and loosen. On the other hand, closely fayed surfaces, securely mated with properly set fastenings, will allow a useful life of 30 to 50 years.

The quality of fastenings and methods of inserting them have been improved, and it is possible now to build a more seaworthy vessel in natural materials than it was in the past. In addition, modern wood preservatives, when properly used, add five or more years of useful life to a ship.

Approximately 98 per cent. of the vessels in use in the fishing fleets are built of wood, and many losses, particularly in the United States, have been caused by fire. These losses can be greatly reduced by using steel construction, and it is also possible to obtain better cubic capacity, better tankage, and stronger foundations. And if better foundations are provided, longer life can be expected of shafting, internal combustion engines and other moving parts, which should reduce the frequency of breakdowns at sea. Steel will deteriorate more quickly than wood if maintenance is neglected, safety is jeopardized and a high financial loss may result. The difficulties encountered in forming shell plates, bending of frames, etc., also present problems, especially as it is necessary to keep down costs by building the boats in small yards where only a minimum amount of equipment is available.

TABLE LVII A

Typical scantlings for wooden vessels of the east coast of U.S.A.

Length, over-all	53 ft. (16.2 m.)	62 ft. (18.9 m.)	71 ft. 4 in. (21.8 m.)	89 ft. (27.1 m.)	98 ft. (29.9 m.)	110 ft. (33.5 m.)	123 ft. 5 in. (37.6 m.)
Beam	17.0 ft. (5.2 m.)	16.87 ft. (5.1 m.)	17.92 ft. (5.5 m.)	20.9 ft. (6.4 m.)	23.0 ft. (7 m.)	23.0 ft. (7 m.)	23.5 ft. (7.2 m.)
Depth	7.75 ft. (2.4 m.)	8.66 ft. (2.6 m.)	10.0 ft. (3.1 m.)	12.5 ft.	12.0 ft. (3.7 m.)	12.0 ft. (3.7 m.)	12.8 ft. (3.9 m.)
Keel	8 in. × 11 in.	9 in. × 18 in.	10 in. × 18 in.	10 in. × 20 in.	11 in. × 22 in.	12 in. × 22 in.	12 in. × 24 in.
Keel shoe	3 in.	3 in.	3 in.	2½ in.	3 in.	3 in.	4 in.
Sternpost and log	14 in.	15 in.	15 in.	18 in.	20 in.	20in.	20 in.
Keelson	6 in. × 10 in.	8 in. × 10 in.	8 in. × 10 in.	10 in. × 10 in.	12 in. × 11 in.	12 in. × 12 in.	12 in. × 12. in.
Sister keelsons	None	None	None	8 × 8 in.	8 × 10 in.	8 × 10 in.	10 × 12 in
Frames, sided	3½ in.	3 in.	3 in.	4 in.	5 in.	5½ in.	6 in.
Frames, moulded*	1¾ in.	7–4½ in.	7½–4½ in.	8½–5½ in.	8½–5½ in.	10–5 in.	11–6 in.
Frames, spaced	10 in.	16 in.	16 in.	18 in.	20 in.	21 in.	21 in.
Beams	3½ × 5 in.	3¾ × 6 in.	6 × 6 in.	4¾ × 7 in.	5 × 7 in.	8 × 8 in.	7 × 8 in.
Clamps, No. of strakes	2	2	2	2	2	2	2
Clamps	2½ × 8 in.	2¼ × 8 in.	2½ × 10 in.	2¾ × 10 in.	3 × 10 in.	4 × 14 in.	4 × 12 in.
Shelf, No. of strakes	1	1	2	3	2	3	3
Shelf	3 × 4 in.	4 × 4 in.	4 × 4 in.	4 × 4 in.	5 × 5 in.	4 × 5 in.	3½ × 6 in.
Shelf lock strake	3 × 4¾ in.	4 × 4½ in.	4 × 5 in.	4 × 5 in.	5 × 6 in.	4½ × 6 in.	4½ × 6½ in.
Lodger	None	None	None	5 × 5 in.	None	6 × 6 in.	6 × 6 in.
1st garboard	2 in.	2 in.	2¼ in.	3½ in.	3⅜ in.	4 in.	5 in.
2nd garboard	None	None	None	3 in.	None	3½ in.	4 in.
Planking	1¾ in.	2 in.	2¼ in.	2½ in.	3 in.	3 in.	3 in.
Ceiling, bilge, No. of strakes	5	5	6	8	4	10	12
Ceiling bilge	2¼ × 4¾ in.	2¼ × 5 in.	2½ × 5 in.	2¾ × 5¾ in.	3 × 8 in.	4½ × 6 in.	5 × 6 in.
Ceiling normal	1¼ in.	1⅜ in.	1½ in.	1¾ in.	2⅜ in.	2¼ in.	2 in.
Decking	2 in.	2¼ in.	2¼ in.	2⅝ in.	3 in.	3 in.	3 in.
Rail stanchions	3½ in.	3¾ in.	4¼ in.	4¾ in.	5¼ in.	5½ in.	6 in.
Rail cap	2¼ × 6 in.	2½ × 6 in.	2¼ × 7½ in.	2¾ × 9 in.	3 × 9 in.	4 × 10 in.	4 × 11 in.

* The 53-footer has two bent frames, one inside the other with a floor timber 2 × 12 in. All other boats have double sawn frames.

SHIP EQUIPMENT

Equipment aboard fishing craft is generally of very good quality, although it is not always well maintained. Then it deteriorates rapidly and reduces the overall safety factor.

On the other hand, fishing gear is almost always the best to be found locally and is usually well maintained because of its importance to the livelihood of the fishermen, who apparently regard it with more concern than they do the vessel. This is a foolhardy attitude because their catch and even their lives may be lost if the vessel is not adequately maintained. The restrictions of labour organizations have been made with little concern for the safety of life at sea. For example, it is almost impossible to find a real seaman among the members of the fishermen's union who work the California tuna fishing fleet. They have been organized as fishermen primarily, and think that their major usefulness aboard is confined to catching fish. In their opinion the vessel's maintenance is the concern of the owners only. Yet the tuna clippers are at sea for three to six months on a single voyage, and the crews will only do minor maintenance merely to keep the vessel operating. Consequently, the clipper's condition deteriorates so much that proper maintenance cannot be paid for out of the share of the catch allowed for that purpose. Many losses have resulted, and much unnecessary expense to owners and crews has been experienced, because vessels again and again put to sea in the hope of making a fast, fruitful trip, the proceeds of which will allow owners to put their ships in good condition.

FIRE PROTECTION

In the past, safety equipment, meeting only the minimum standards, has been carried aboard the majority of fishing vessels. No authorized agency has ever established the proper requirements and even labour organizations have disregarded their members' safety in this respect.

For many years independent marine engineers, surveyors and equipment suppliers have urged the use of more positive safety equipment and methods. This would call for rigorous training programmes, and there has been little co-operation from owners because they can see no financial gain. Their attitude is strengthened by the ease with which they have been able to insure against loss.

Underwriters as a group are in a position to correct this situation by making demands which owners could not ignore. They are, however, divided in opinion and, on the west coast of the United States, one group of underwriters will not always agree to the requirements of another. In many instances such disagreement has been right because not all the requirements have reflected sound engineering principles.

For example, there is the existing underwriters' requirement for installation of carbon dioxide full flooding fire protection systems in the engine rooms in tuna class vessels. This type of vessel has always been fitted with full flooding inert gas systems, installed principally to protect the main engine room, and some total losses have recently resulted from fires gaining headway in the unprotected forepeak area ahead of the engine room. Underwriters immediately called for a revised system but, while the reasoning was sound, the engineering was poor. Sufficient gas was added to the engine room system to protect the cubic capacity of the forepeak. The discharge manifold and nozzles in the forepeak were piped directly into the main engine room system and, for no apparent sound reason, two release stations were installed for the single system—one on the main deck outside the engine room and one on the deck above, completely removed from visual or voice communication with the escape passage from the engine room. Such a system endangers the lives of men who might be unable to escape from the engine room before someone at the remote control station releases the gas, or the full benefit of the gas might be lost by release before those at the scene of action have been able to secure the compartment for proper flooding.

These are two of the deficiencies. Another is that, to control a fire in either the engine room or forepeak compartments, the entire volume of available gas must be released at one time. Should the fire be in the forepeak most of the gas is lost when released in the engine room and the fire hazard is increased because the gas deprives the internal combustion engines of needed oxygen and puts them out of operation. This, in turn cuts off the ship's electrical supply and puts the mechanical water pumps out of action. So, if the fire is not controlled by the single exposure to the carbon it must be fought by a " bucket brigade " and, possibly, the ship will have to be abandoned.

Unless the ships carry an adequate supply of spare inert gas tanks, they must return to port to recharge. They are not required by underwriters or any agency to carry spares. When proper rules are drawn up and adopted the owners will have to spend more money to convert the system which would not have been necessary had sound engineering principles been employed in the first place. Such lack of concern by underwriters for their own financial welfare encourages both owners and crews to disregard essential conditions of safety in vessels.

Fires can be repetitive—to some extent a more serious hazard exists after a fire than before—and mechanical pumping systems should be installed in all vessels for fighting fire with water, which is not only plentiful but is superior to anything else for that purpose. Breathing vents are needed for auxiliary machinery so that release of fire-fighting agents which deplete the supply of oxygen will not stop the generating units supplying power to the pumps. The use of solid water injection, now recommended by underwriters, is dangerous to life but this hazard can be eliminated by the use of controllable nozzles on approved type hoses. And, finally, there should be a rigorous training programme to make sure that the crews are able to use the equipment efficiently.

TABLE LVIII B

Typical fastenings for wooden vessels of the east coast of U.S.A.

Length	53 ft. (16.2 m.)	62 ft. (18.9 m.)	71 ft. 4 in. (21.8 m.)	89 ft. (27.1 m.)	98 ft. (29.9 m.)	110 ft. (33.5 m.)	123 ft. 5 in. (37.6 m.)
Keel members	—	1-$\frac{7}{8}$ in. B	1-1 in. B	1-1 in. B	1-1 in. B	1-$\frac{3}{4}$ in. B	1-1 in. B
Frames to keel	2-$\frac{1}{2}$ in. D	2-$\frac{1}{2}$ in. D	2-$\frac{1}{2}$ in. D	2-$\frac{5}{8}$ in. D	2-$\frac{3}{4}$ in. D	2-$\frac{7}{8}$ in. D	2-$\frac{7}{8}$ in. D
Keelson to keel	1-$\frac{5}{8}$ in. B	2-$\frac{3}{4}$ in. D	2-$\frac{7}{8}$ in. D	2-$\frac{7}{8}$ in. D	2-1 in. D	2-1$\frac{1}{8}$ in. B	2-1$\frac{1}{4}$ in. D
Sister keelsons	—	—	—	$\frac{3}{4}$ in. D	$\frac{7}{8}$ in. D	1-$\frac{3}{4}$ in. B 1-$\frac{7}{8}$ in. D	2-$\frac{7}{8}$ in. B
1st garboard	$\frac{5}{16}$ × 4 in.	$\frac{3}{8}$ × 4$\frac{1}{2}$ in.	$\frac{3}{8}$ × 5 in.	$\frac{5}{8}$ in. D	$\frac{9}{16}$ × 7 in.	$\frac{5}{8}$ in. D	$\frac{3}{4}$ in. D
2nd garboard	—	—	—	$\frac{1}{2}$ in. D	—	$\frac{9}{16}$ in. D	$\frac{5}{8}$ in. D
Clamps to frames	1-$\frac{1}{2}$ in. B 1-$\frac{1}{2}$ in. D	1-$\frac{1}{2}$ in. B 1-$\frac{1}{2}$ in. D	2-$\frac{1}{2}$ in. B —	2-$\frac{5}{8}$ in. B 2-$\frac{5}{8}$ in. D	1-$\frac{3}{4}$ in. B 1-$\frac{5}{8}$ in. D	2-$\frac{3}{4}$ in. B 2-$\frac{3}{4}$ in. D	2-$\frac{3}{4}$ in. B 2-$\frac{7}{8}$ in. D
Lodger	—	—	—	1-$\frac{7}{8}$ in. B	$\frac{3}{4}$ in. B*	1-$\frac{7}{8}$ in. B	1-1 in. B
Shelf	1-$\frac{1}{2}$ in. B	1-$\frac{5}{8}$ in. B	1-$\frac{5}{8}$ in. B	1-$\frac{7}{8}$ in. B	1-$\frac{3}{4}$ in. B	1-1 in. B	1-1$\frac{1}{8}$ in. B
Beams to clamp	1-$\frac{7}{16}$ in. D	1-$\frac{5}{8}$ in. D	1-$\frac{5}{8}$ in. D	1-$\frac{3}{4}$ in. D	1-$\frac{5}{8}$ in. D	1-$\frac{7}{8}$ in. D	1-1 in. D
Beams to shelf	1-$\frac{7}{16}$ in. B	1-$\frac{5}{8}$ in. B	2-$\frac{5}{8}$ in. B	2-$\frac{3}{4}$ in. B	2-$\frac{5}{8}$ in. B	2-$\frac{3}{4}$ in. B	2-$\frac{7}{8}$ in. B
Ceiling bilge	1-$\frac{3}{8}$ in. B —	1-$\frac{7}{16}$ in. B —	1-$\frac{7}{16}$ in. B —	1-$\frac{1}{2}$ in. B 1-$\frac{1}{2}$ in. D	1-$\frac{3}{8}$ in. B 1-$\frac{5}{8}$ in. D	1-$\frac{5}{8}$ in. B 1-$\frac{5}{8}$ in. D	1-$\frac{5}{8}$ in. B 1-$\frac{3}{4}$ in. D
Planking	$\frac{5}{16}$ × 4 in.	$\frac{3}{8}$ × 4$\frac{1}{2}$ in.	$\frac{3}{8}$ × 5 in.	$\frac{7}{16}$ × 6 in.	$\frac{7}{16}$ × 6$\frac{1}{2}$ in.	$\frac{9}{16}$ × 6$\frac{1}{2}$ in.	$\frac{5}{8}$ × 7 in.

B—through bolt; D—drift; 1—teco rings; *—knees.

RADIO AND RADAR

Radio equipment installed aboard the majority of fishing vessels is good, having long operating range and clarity, but few operators have been trained in proper radio procedure, although they do accomplish the desired results admirably.

Radar is not something to be dealt with by men who learn only the rudiments of its operation and who, as a result, fail to read properly its scanning screen. When so used the safety of vessel and crew is jeopardized and cases on record reveal total losses as a result of unskilled persons relying on radar in bad weather.

OTHER EQUIPMENT

Rescue apparatus, shorings, collision mats, etc. are not carried aboard the tuna vessels at all. This equipment should be, and there should also be proper instructions for its usage, supplemented by periodic drills.

With so much capital invested in the fishing industry, especially in boats and equipment, there should be a training programme for crews, and insurance underwriters, owners, suppliers, and others who get a substantial income from the fishing fleets, should willingly aid such a programme. The results would certainly yield high economic returns.

CONCLUSION

The following is a suggested four-point programme:

I. *Personnel, including fishermen as seamen, owners, and labour unions.*

(*a*) *Masters and officers of vessels.* They should be required to demonstrate proficiency when they take control of bigger ships and more valuable equipment.

(*b*) *Command of vessels.* The practice on large fishing vessels of carrying a federally-licensed master who, although responsible for the ship and crew, relinquishes his authority at sea to a so-called fishing captain, should be discontinued. Instead, licensed masters should be taught to operate as fishing captains. Similarly, capable fishermen should be helped to qualify for masters' papers.

(*c*) *Training of crews in seamanship.* They should be required to demonstrate their proficiency before being signed on.

(*d*) *Proper maintenance while at sea.* Crews should be trained to protect their own lives and the financial investment of owners and underwriters by proper attention to their duty in the care and maintenance of the boats. A requirement in the insurance policy for maintenance by the crew while at sea would relieve owners from certain labour union pressure on owners and would result in better physical maintenance and higher moral attitudes by owners and crews.

II. *Ship design, specifications, scantlings, and construction.* Owners should be made aware of the advantages of qualified technical attention to:

(*a*) *Design*, including use of proved modern developments and the merits of wood versus steel when contemplating new construction or conversion.

(*b*) *Construction*, including use of modern methods and materials for both wood and steel, such as laminated components, scarfed sections, improved fastenings, and preservatives.

(*c*) *Powering*, including use of modern engines.

(*d*) *Operation*, including use of qualified engineering and other personnel at sea.

(*e*) *Maintenance*, including application of modern improvements and methods.

III. *Ship's fixed equipment, including safety equipment.* The crews should be impressed with the need for proper maintenance of:

(*a*) *Ship's fixed equipment at sea.* Fishing equipment maintenance should be secondary.

(*b*) *Positive water pumping systems for fire-fighting* and other emergencies should be installed including:

1. *Auxiliary engines with breathing vents* to uncontaminated air supply.
2. *Proper fire hoses and controllable nozzles* to eliminate a definite existing hazard to life.
3. *Collision mats* with instruction in their use. Mattresses with hogging lines attached serve

TABLE LVIII

Typical scantlings for the west coast of U.S.A.

Length, over-all	57 ft. (17.4 m.)	74 ft. (22.6 m.)	88 ft. (26.8 m.)	65 ft. (19.8 m.)
Breadth	16 ft. (4.9 m.)	20 ft. (6.1 m.)	21 ft. (6.4 m.)	17 ft. 4 in. (5.3 m.)
Depth	8 ft. (2.7 m.)	10 ft. (3.1 m.)	11 ft. (3.4 m.)	8 ft. 6 in. (2.6 m.)
Keel	9½ × 9½ in.	11½ × 11½ in.	11½ × 11½ in.	11½ × 11½ in.
Keelson	13½ × 13½ in.	13½ × 17½ in.	13½ × 17½ in.	13½ × 13½ in.
Shoe	1¾ in.	1¾ in.	1¾ in.	1¾ in.
Sternpost	9½ × 15½ in.	13½ × 17½ in.	13½ × 17½ in.	11½ × 15½ in.
Sister keelson	7½ × 9½ in.	9½ × 19½ in.	9½ × 11½ in.	7½ × 7½ in.
Frames (white oak)	2¼ × 4 in.	3 × 4 in.	Dble 3½ in.	3 × 4 in.
Spaced	10 in.	12 in.	16–18 in.	12 in.
Beams	5–7½ in.	5½ – 8 in.	5½–9 in.	5¼–7¾ in.
Clamps	2¾ × 10 in.	3½ × 11½ in.	4½ × 13½ in.	2¾ × 11½ in.
Shelf	2¾ × 11½ in	2-3½ × 5½ in.	4½ × 11½ in.	2¾ × 11½ in.
Garboard	2 × 11½ in.	2½ × 11½ in.	3½ × 11½ in.	2 in. net
Planking	1¾ in.	2 in. net	2½ in.	1¾ in.
Bilge ceiling	6-2¾ × 5½ in.	6-3½ × 6 in.	7-5½ × 7½ in.	5-3½ × 6 in.
Decking	2 × 3½ in.	2¼ × 3½ in.	2½ × 3½ in.	2 × 3½ in.

the purpose and would have prevented many total losses in the past.

4. *Shoring timbers* and patching materials in minimum quantity.
5. *Rescue breathing apparatus* and minimum fire penetration equipment.

IV. *Underwriters should publish a pamphlet in all the necessary languages depicting the advantages of marine insurance, their own obligations to the assured, and the assured's obligations to underwriters.* It should contain the conditions always shown on the insurance policies, but should make them understandable.

Underwriters should periodically make available to fishing vessel owners a list of per cent. differences in insurance premium rates for varying classes of vessels, including certain commercial vessels outside the fishing industry, to arouse the owner's interest in paying lower rates through better maintenance and fewer losses.

SAFETY AT SEA REGULATIONS IN NETHERLANDS

by

J. G. DE WIT

IN the Netherlands the enforcement of various Acts and Regulations affecting fishing and other vessels and their crews is carried out by the Shipping Inspection Service of the Ministry of Traffic and Public Works. There are three districts to cover commercial shipping, and a fourth for the whole seagoing fishing fleet.

The Shipping Inspector in this fourth district is located at the Hague, and there are deputy inspectors and surveyors in the fishing towns of Iymuiden, Scheveningen, Den Helder and Flushing. There are 12 persons on the technical and nautical staff who supervise 740 fishing vessels of 72,000 gross register tons. Only 21 ships are under the supervision of classification societies, so that virtually all the fleet is controlled by the Shipping Inspection Service.

The Netherlands fishing fleet is composed mainly of trawlers, trawler-luggers and cutters, and some old types such as " botter ", " schokker ", " hoogaars " and " hengst ".

Ships of the first three types usually receive an unlimited Certificate of Seaworthiness. This entitles them to fish anywhere in the North Sea, and in more distant waters, provided the skipper or another member of the crew, possesses a supplement to the skipper certificate. In the case of trawler-luggers, the hatch covers must also satisfy certain requirements. The other boat types may receive a Certificate of Seaworthiness A, which entitles them to go up to 15 miles from the coast or if they satisfy certain additional standards of size and construction they receive a Certificate of Seaworthiness B, which entitles them to go 50 miles from the coast.

Docking must be reported to the Inspector and it is only occasionally that a shipowner or skipper is made to have his ship dry-docked because the compulsory limit of 12 months has elapsed. In particular cases the limit may be extended to 24 months. When a ship is grounded or has run into some other trouble where damage is presumed, dry-docking is compulsory.

Every year the life-saving appliances, safety devices, loading equipment and the engine installation are surveyed. Propeller shafts must be drawn every two years, but when they are protected by an oil seal between the propeller and the inside stern bearing and, consequently, rotate in an oil-bath, or when they are protected by a continuous bronze liner, they are only inspected every three years. The boilers are inspected every year and the ships and engines every four years. The inspection of boilers is the work of the Inspection Service of Steam Engines, which comes under the Ministry of Social Affairs. Air cylinders in motor installations are surveyed every five years.

When a ship is damaged, a provisional survey is made for the Shipping Council. The term " damage " has a very wide meaning within the Shipping Act, covering any damage to the ship, her cargo or crew or damage to another ship. One of the main reasons for such investigations is to find out whether something may be learned from the accident and whether new regulations are desirable.

Surveyors of the Shipping Inspection Service control the construction of ships. Drawings are subject to the approval of the central office of the Hague. The Service has issued no rules on the scantlings but uses those of Lloyd's Register, the Bureau Veritas or the Germanischer Lloyd. Where these rules do not cover the vessel concerned the Service is entitled to issue regulations, and it is important to note their significance in relation to the design of fishing boats.

STABILITY

The bigger trawlers and trawl luggers are now subject to stability tests since the Shipping Order 1952 reduced the exemption limit from 800 to 900 gross register tons. An inclining experiment must be made and sufficient data of stability for various angles of heel established before the Certificate of Seaworthiness is issued.

These provisions are of little use to designers. The question is whether the minimum stability should have been more definitely fixed, although the present rule leaves the designer free to express his views on stability. But he should know beforehand what yard-stick will be used when the Shipping Inspection Service considers his plans. On the other hand, it is difficult, if not impossible, to establish a standard for minimum stability applicable to both ocean liners and small fishing boats, but it would be highly desirable to reach agreement on a standard for fishing boats, especially as so much spade work has been done.

An example is the investigation made by the See-Berufsgenossenschaft (1904) in co-operation with the Germanischer Lloyd, into the losses due to lack of stability, of a number of fishing vessels in 1903. As the stablility of the lost ships was unknown, it was determined by the stability of a number of similar ships. The curves of statical stability of *August*, *Blexen*, *Breslau*, *Braunschweig*

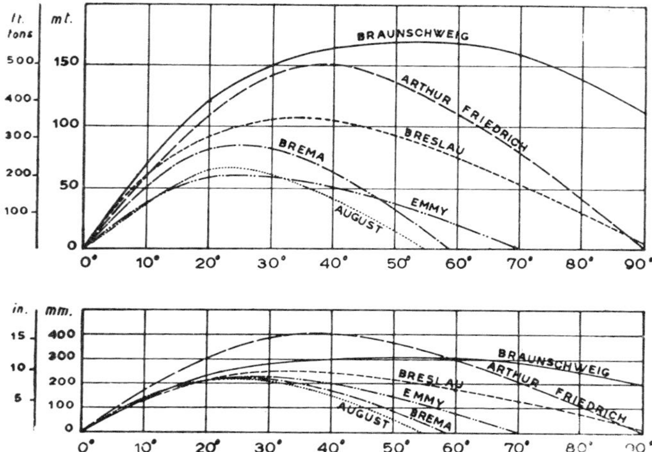

Fig. 430. Reproduction of Johon's data on the moments of statical stability and arms of statical stability of different types of fishing vessels (from Johow's Hilfsbuch fuer den Schiffbau, *5th edition)*

and *Arthur Friedrich* have since become known. They, or similar ones, have also been published by Johow (1928) and are reprinted in fig. 430 and Table LIX. It is only to be regretted that these curves are given without comment, especially as only the last three ships were judged in the See-Berufsgenossenschaft report to be safe against capsizing. These data, however, did not lead to a generally accepted definition of adequate stability.

A first step in this direction was taken by Benjamin (1913) who considered that capsizing is a dynamic process and is not determined by the stability moment, but by the work necessary to incline the ship. Benjamin used the dynamical lever and not the curve of arms of statical stability as a basis. He collected a great amount of comparative material on big and small ships and this led him to propose the following minima for the dynamical lever: 2 in. (50 mm.) at 30°, 8 in. (200 mm.) at 60°; with a capsizing angle of less than 60° the dynamical lever must be not less than 8 in. (200 mm.) at the capsizing angle.

This proposal was received with great reservation and, in retrospect, it seems that the values rather than the method were criticized. Too much importance was attached to the form of the curve of arms of statical stability beyond the point where the curve exceeded its maximum, at about 35° to 45°. Consequently, in 1927, Benjamin greatly reduced the values (*fig. 431*) but, nevertheless, they were not generally accepted mainly due to the way in which they were presented, although his method—the use of the dynamical lever—was theoretically well founded.

Pierrottet (1935) developed, along the same lines, a method for judging stability and laid down rules for determining the minimum stability. This proposal was based on two principles:

1. Even under the most unfavourable circumstances, in which all possible unbalancing forces act simultaneously and in the same direction, the heel of the ship may not exceed a certain " permitted angle ". In view of the risks of shifting cargo this angle was generally limited to 50 °, 25° for ferries.
2. In this condition the dynamical stability must be equal to or greater than the amount of work done by wind, waves, centrifugal forces and shifting of people on board, when acting between the 0° and the " permitted angle " of heel.

The discussion of these proposals brought no solution and, again, the theoretical side of the problem received

TABLE LIX

	Displacement	*GM*	*Freeboard*	*Superstructures*	*L*	*B*	*T*
Emmy	255	2.82 ft. .86 m.	2.13 ft. .65 m.	—	105.5 ft. 31.32 m.	20.7 ft. 6.31 m.	10.8 ft. 3.29 m.
August	275	2.56 ft. .78 m.	2.36 ft. .72 m.	—	110.0 ft. 33.53 m.	22.0 ft. 6.70 m.	11.8 ft. 3.6 m.
Brema	370	2.725 ft. .80 m.	2.2 ft. .67 m.	—	119.6 ft. 36.51 m.	22.5 ft. 6.85 m.	12.3 ft. 3.75 m.
Arthur Friedrich	373	2.905 ft. .885 m.	2.63 ft. .80 m.	—	128.0 ft. 39.00 m.	21.5 ft. 6.55 m.	11.9 ft. 3.63 m.
Breslau	440	2.49 ft. .76 m.	1.935 ft. .59 m.	Fo'c'sle and Q.D.	121.3 ft. 37.00 m.	22.95 ft. 7.0 m.	13.13 ft. 4.00 m.
Braunschweig	555	2.46 ft. .75 m.	1.97 ft. .60 m.	Fo'c'sle and Q.D.	134.5 ft. 41.00 m.	22.5 ft. 6.85 m.	13.6 ft. 4.15 m.

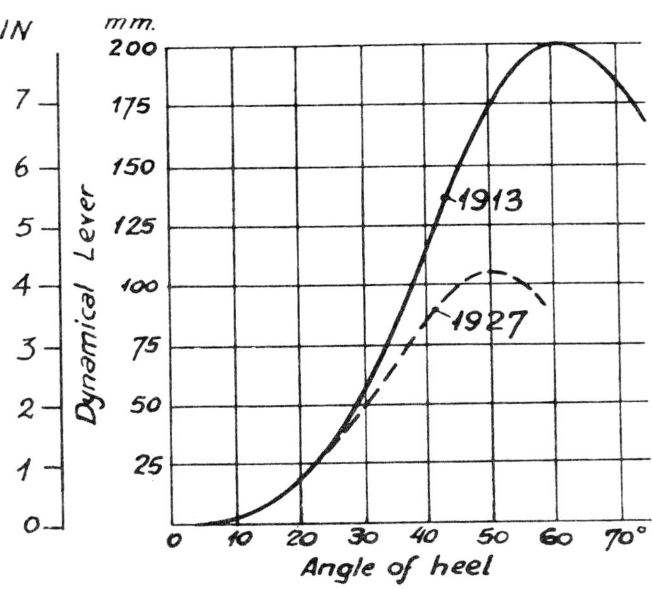

Fig. 431. Standards of dyamical levers, according to Benjamin

little attention. The criticism only concerned the proposed values and the " permitted angle " of heel.

Just as Benjamin some years after his proposal reduced to one half his second minimum, Pierrottet reduced during the discussion of his paper his " permitted angle " of heel. Based on these methods, Rahola (1939) made a new proposal. He collected data on the stability of capsized ships and ships that had been running successfully. He examined mainly small ships and his statistical investigations led to the following conclusions:

1. The values of the arms of statical stability must be: at least $5\frac{1}{2}$ in. (140 mm.) at an angle of 20° and at least $7\frac{7}{9}$ in. (200 mm.) at an angle of 30°;
2. The " critical angle " of heel of a ship must be more than 35°. By " critical angle " of heel is meant the angle of heel at which the curve of arms of statical stability reaches its maximum value.

For big ships, whose GM metacentric height may often be small—or proportionally smaller than that of small ships—it will generally be difficult to satisfy the minimum value for 20° as proposed and Rahola strikes a note of warning that even his standards for the stability of smaller ships might be too stiff. When the " critical angle " is reached before 35°, Rahola considers the stability to be sufficient, provided the dynamical stability offers sufficient security against exceeding the critical angle. So the designer is not limited by the standard of statical stability. Rahola uses the statical stability curve only as a starting point to find a more elastic determination of the minimum stability based on the dynamical levers.

The conclusion of Rahola's investigations is strikingly simple: The dynamical lever should at the " permitted angle " of heel be equal to or larger than $3\frac{1}{8}$ in. (80 mm.)

The " permitted angle " of heel is determined by the following conditions:

1. It should be equal to or smaller than the " critical angle " where the curve of statical stability reaches its maximum.
2. It should be equal to or less than 40°.
3. The non-watertight hatch-coamings and doorways, through which the water might flow into the ship, may not be submerged with the " permitted angle " of heel.
4. If the cargo is liable to shift, the dynamical angle of shift must be determined.

The proposal of Rahola deserves the greatest attention. Fig. 432 shows the stability data of a number of ships, known from literature or otherwise, which have been tested by the Rahola method. In the cases that have come to the knowledge of the Netherlands Shipping Inspection Service it has been found that the Rahola standard for dynamical stability is reliable in judging the stability of a fishing vessel.

The Shipping Act does not require the stability data to be on board, as it is felt that the crews are not capable of using them effectively. In judging stability the Shipping Inspection Service has to take into account nearly all loading conditions. Consequently, the calculation of the arms of statical stability should be made for the following conditions:

1. Departure;
2. Arrival at the fishing grounds;
3. Departure from the fishing grounds;
4. Arrival.

It should be assumed that on departure the ship is fully supplied with fuel, water, stores and ice. The voyage to the fishing grounds is assumed to last seven days and the consumption of fuel, water, stores and the melting of ice is calculated accordingly. On departure from the fishing grounds it should be assumed that the fish-hold is filled to its capacity. At this point the supply of fuel should be sufficient for nine days and, on arrival, it should be assumed that there is still enough fuel, etc., for two days. The loss in weight through ice water being pumped overboard should be calculated as follows:

Condition 1— 0 per cent.
Condition 2—10 ,,
Condition 3—30 ,,
Condition 4—45 ,,

These figures apply to ships with properly insulated holds and where insulation is bad or lacking the figures are higher. When ships carry salt instead of ice, it should be assumed that the salt is distributed over the whole fish-hold in conditions 3 and 4.

FREEBOARD

Netherlands freeboard regulations do not apply to vessels solely used for fishing purposes. Considerations which

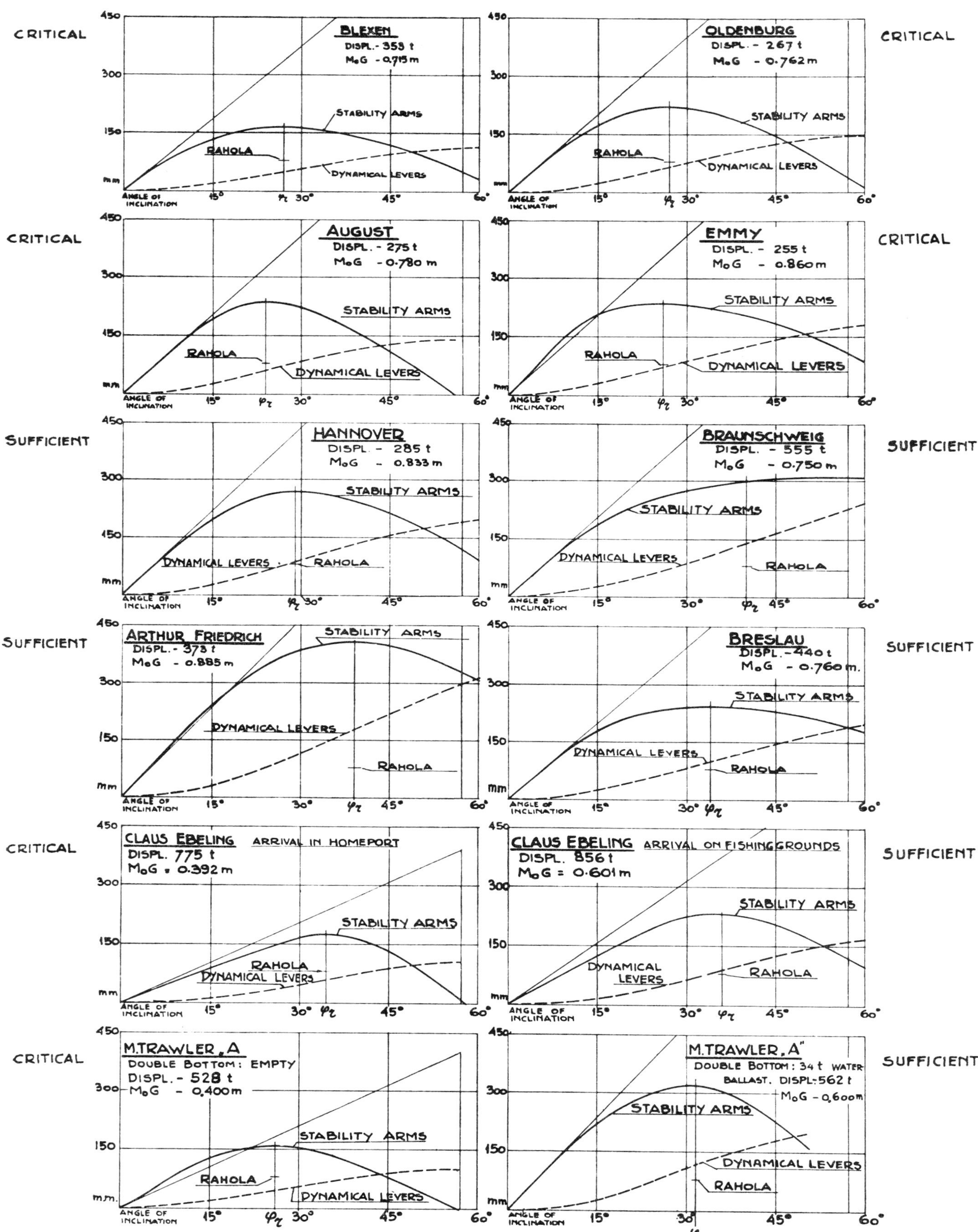

Fig. 432. Data of the stability of some fishing vessels

have led to the exemption of fishing vessels from these regulations are:

(*a*) Fishing vessels take their cargo at sea where it is difficult to ascertain whether a ship is already down on the mark.

(*b*) Fishing vessels with a full cargo hold generally have a fairly constant weight on board.

Nevertheless, the surveyors of the Netherlands Shipping Inspection Service are instructed to make sure that fishing vessels are not overloaded. In spite of the absence of a freeboard mark, any skipper who has overloaded his ship is summoned before the Shipping Council. Dangerous loading occurs oftener than might be expected and it is evident that a smaller freeboard than the minimum for which a ship is designed can nullify reasonable stability conditions. The question arises: Are the rules for the determination of the minimum freeboard for fishing vessels necessary?

Three causes of overloading, in the opinion of the Shipping Inspection Service, are:

1. Small steam trawlers which were originally built for fishing on the North Sea are now often fishing on more distant grounds. To make such long voyages they carry a larger supply of coal so that their freeboard is dangerously small on leaving port. The coal lies toward the stern of the ship, so that the vessel has a heavy trim aft.
2. In good seasons motor luggers, and motor trawler-luggers, fish inshore. They often make enormous herring catches which exceed their loading capacity and their fuel consumption is then very low. The catch is stored in the hold, and the barrels are filled and stowed on the after part of the deck. When, in addition, the deck-pounds also are full, it is clear that the freeboard may become dangerously small.
3. In bad weather big quantities of water may enter the hold through the hatches of luggers and trawler-luggers, which means a further reduction of freeboard.

These are reasons why regulations are needed to establish a minimum permitted freeboard for fishing vessels and why they should carry a mark. Measurements have been made of the freeboard of outgoing and incoming ships but a satisfactory freeboard regulation has not yet been drafted from the material collected.

Since World War II, many motor luggers have served as cargo vessels, mainly to transport barrels of herring to ports on the North Sea, the Channel and the Baltic. As these ships were used for purposes other than fishing they were no longer exempt from carrying a loadline mark and their freeboard was fixed according to international regulations. Generally, the freeboard of motor luggers of about 115 ft. (35 m.) length, varied from 20 in. to 2 ft. (0.5 to 0.6 m.). In these calculations it

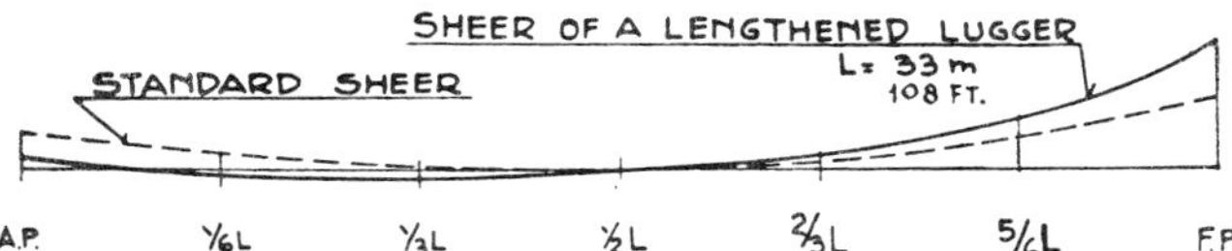

Fig. 433. Comparison of a standard sheer and the actual sheer of a motor lugger

appears that the freeboard largely depends on the sheer (*see fig. 433*). From the fact that the sheer has its lowest point at about $\frac{1}{3}$L from the after perpendicular, it follows that the sheer on this ordinate is negative in relation to the sheer at $\frac{1}{2}$L. and an increase of the table-freeboard is required. If the aft part of the sheer falls below the standard sheer, an " extra " should be given on the table-freeboard. These extras are generally not compensated by reduction on the table-freeboard because of the excess of the sheer in the fore-part of the ship.

HATCHES

The original lugger was a sailing vessel with a length of 75 to 82 ft. (23 to 25 m.) and had a reputation of being a seaworthy vessel which shipped little water. The hold was closed by a number of wooden hatches (*fig. 434*), on a coaming with a height of about 3$\frac{1}{8}$ in. (8 cm.) or even flush hatches, and they presented little danger. They were not perfectly watertight but, because the ship rode lightly, breaking seas were only dangerous when hatches were carried away and, owing to this, some sailing-luggers have been wrecked.

In 1926 some sailing-luggers were fitted with motors of 50 to 100 h.p. Because of the weight of motor and fuel, the ships increased in weight and draught and shipped water. To meet this situation, the hatch-coamings of the rope-hold and the store-hold were raised.

About 1930 a start was made with lengthening luggers and was continued until recently. At first these lengthenings were rather restricted but, in later years and especially after World War II, lengthenings by 26 to 30 ft. (8 to 9 m.) were not uncommon. It is obvious that such a lengthening must change the vessel's sea-going qualities. In addition, a number of luggers that had been requisitioned during the German occupation were returned with motors of 200 to 300 h.p. and since 1945 many motors of this capacity have been installed. This, of course, called for bigger and heavier superstructure and larger supplies of fuel and, because of the increased weight, length and motor power, the reconstructed

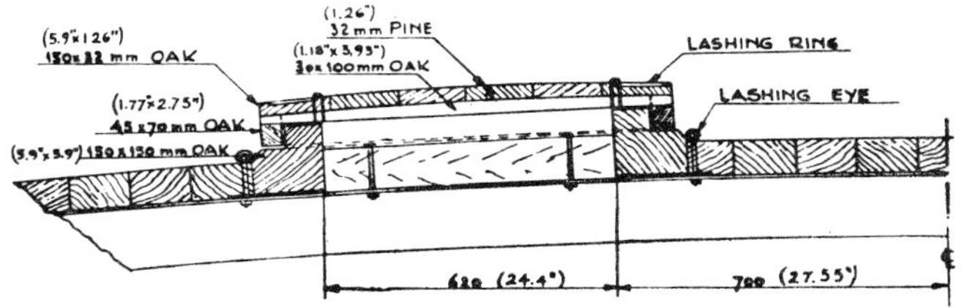

Fig. 434. Classical wooden hatchcovers on Dutch luggers

luggers no longer rode the waves but ploughed through them, especially when trawling. To meet this situation, in many cases an open forecastle was put on the stem and it was an improvement.

Though it may be true that the forecastle protects the crew from heavy seas, trawler-luggers ship most of the water further aft behind the forward gallows. This is not so bad when the hatches are closed but trawler-luggers have been frequently so flooded through open hatches that both ship and crew were in great danger and could only be saved by the utmost effort.

It was felt that the Shipping Inspection Service should pay more attention to this problem. At first little progress was made because, in the opinion of the crews, a change in the hatch system would interfere with the work on deck. This applied especially to herring drift-net fishing. The only thing the Service could do was to make it compulsory to have sufficient covers and battens on board to close the hatches with tarpaulins or steel covers in bad weather. Even this minimum requirement sometimes met with resistance.

In winter fishing at higher latitudes than 61° N., and in the Irish Sea, regulations now require hatches to remain closed during trawling and to have a fixed and closed inner hatch. A flush hatch must be caulked and nailed in place and a fixed raised steel coaming must be placed on hatches that have to remain open.

This method did not provide a solution for herring trawl fishing in which all hatches must be opened for lowering the barrels and, in 1947, some ship-owners began to change the existing system. Instead of two rows of wooden or flush hatches, raised coamings were built on the centre line of the ship. Though far from perfect, this system has worked well in practice. After some experimenting, the solution shown in fig. 435 gave the most satisfaction. With it the number of double tiers is not greater than usual and the danger for the stower in the hold is small.

When this hatch system had shown that it was practicable and that this arrangement of deck did not cause drastic changes in working methods, the Inspector-General for Shipping issued the following order on March 6th, 1952:

1. Steel coamings about 14 in. (350 mm.) in height shall be constructed round all hatchways on all future motor luggers.
2. The coamings shall also be constructed on all existing motor luggers which have a motor of more than 150 h.p., when:
   - (*a*) the ship is lengthened;
   - (*b*) the deck amidship is altered or renewed;
   - (*c*) the loaded ship has, in the opinion of the head of the Shipping Inspection Service, too little freeboard to be safe without raised coamings.
3. The coamings shall also be constructed on all existing luggers which are later provided with a motor of more than 150 h.p.

This order will ultimately lead to an improvement of the obsolete hatch system on lugger-trawlers. Already practice has shown:

1. That the coamings work rather well, although the arrangement below deck and the working method on deck have to be adapted to the system, and
2. That the luggers and lugger-trawlers are safe at sea, which cannot be said of lengthened boats with traditional hatch construction.

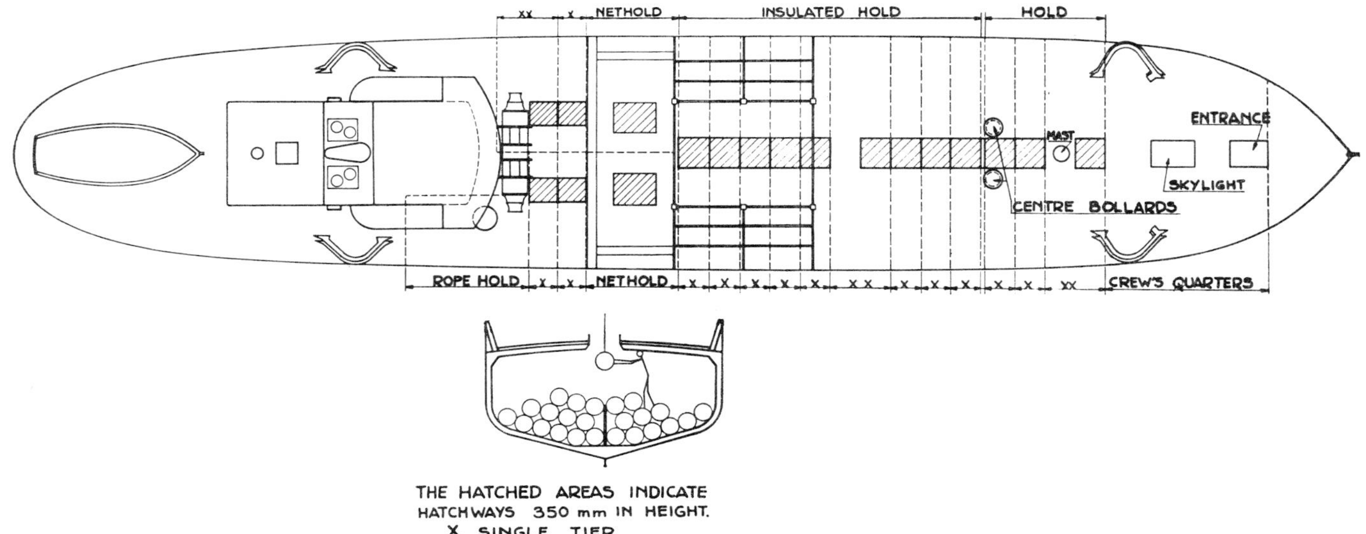

Fig. 435. Deck arrangement of a modern Dutch trawler-lugger

For safety with the existing system of wooden hatches, tarpaulins have to be used to run in one length, for and aft. They have to be nailed on the deck with battens. This solution, however, provides only a minimum cover of the hatches in stormy weather.

DRINKING WATER TANKS

Fishing vessels were allowed to use new herring barrels for storing drinking water. But since 1952 the water has had to be stored in suitable tanks, closed in such a way as to exclude any foreign matter. The tanks must be kept separated by cofferdams from fuel or spill water tanks and they have to be constructed so that they are left clean when pumped dry. The quantity of drinking water depends on the number of the crew, the duration of the voyage and the possible use of distillation which, in the case of fishing vessels, may be ignored. The duration of the voyage, especially for luggers in herring fishing with the drift net, is a rather uncertain factor. One solution is to determine the consumption per man per day, but there are no rules on this point. The Shipping Inspection Service calculates on the following data: the crew on herring luggers consists of 15 or 16, and on trawlers, 11 men. The duration of the voyage is assumed to be four weeks, and the consumption per man per day at 2.2 imp. gal. (2.6 U.S. gal., 10 l.). This means that luggers must carry 4½ or 5 tons and trawlers 3 or 3½ tons of drinking water. Fresh water for other purposes, such as washing, is not included in this calculation, and for ships having a showerbath the figures have to be raised by 50 per cent.

ACCOMMODATION

The Shipping Inspection Service is also in charge of the crew's quarters. They have to be arranged behind the collision bulkhead and the distance from the foremost berth to the fore perpendicular must be more than $\frac{1}{20}$ of the ship's length. On most Netherlands luggers and trawler-luggers 12 men are accommodated in the forward and three or four men in aft quarters. In some ships there is still a cabin on the bridge for the skipper but the arrangement is not much favoured to-day. In the latest vessels of the Netherlands fishing fleet the tendency is to accommodate the whole crew aft, which is far safer and more convenient.

A messroom is not yet required but the Service recommends one because it provides day quarters where the crew can eat and sit in their spare time. It also means that the separate night quarters are quieter, more restful and cleaner. The messroom is usually built on the upper deck where it can be kept clean easier.

In contrast to commercial ships, where the night quarters must have a cubic capacity of not less than 70 cu. ft. (2 cu.m.) per man, and the day quarters not less than 76 cu. ft. (2.15 cu. m.) and a combined accommodation for day and night not less than 124 cu. ft. (3.5 cu. m.), the cubic capacity of the combined day and night quarters of fishing vessels may not be less than 97 cu. ft. (2.75 cu. m.) per member of the crew. The berths may be included in this calculation and, after the furniture has been placed, there must be sufficient space for moving about. As a result, the cubic capacity per member of the crew is usually greater than 97 cu. ft. (2.75 cu. m.). A lining is desirable where the berths adjoin the hull, but no use is made of the regulation that a board of 16-in. (40 cm.) high is sufficient. Any space between the lining and the hull must be accessible for inspection and cleaning.

The berths must be at least 6 ft. 3 in. (1.90 m.) long and 2 ft. 2½ in. (0.68 m.) wide and not more than one above another. No berths are allowed under the openings of ventilation ducts. One berth must be available as a sick bay—which means that it must be so arranged as to allow the patient to be taken in and out easily.

The height of the crew's quarters of fishing boats may not be less than 6 ft. (1.83 m.) measured from the floor to the underside of the deck beams.

On ships less than 500 gross register tons ventilation may be natural but on those of more than 500 tons one of the systems of ventilation must be artificial. It must not cause troublesome draughts but it must ensure the removal of bad air and the supply of fresh air.

Generally, one lavatory is sufficient on fishing vessels and with a crew of more than four a lavatory is required. Proper washing facilities have to be arranged near the engine room for the engine room watch. These regulations on accommodation date from 1937, and reflect the social conditions of fishermen at that time. Conditions have improved since then and fishermen now show more interest in the lay-out of their quarters and everything connected with it. This means there is considerable support for all who make it their task to provide better accommodation.

LOADING AND UNLOADING EQUIPMENT

Loading and unloading seldom present great difficulties on trawlers. As a rule a steel wire is stretched between the main and the mizzen masts and is provided with eyes from which a block is suspended. The weight lifted seldom exceeds that of a basket of fish, about 110 lb. (50 kg.).

Greater weights have to be lifted on luggers, such as barrels of herring weighing about 265 lb. (120 kg.), and barrels of salt, weighing 330 lb. (150 kg.). Barrels will sometimes stick in the hold and it is difficult to assess the magnitude of the forces involved in hauling them out. From fig. 436 it is evident that the forces in the horizontal wire between the main and mizzen masts increase as this wire is tightened so that it is not desirable to fix the wire on the mizzen with a rigging screw, as is often done. A tackle with a manilla fall is preferable. The best is to fix the block in the mizzen mast with a closed eye and shackle. When the wire is stretched so tight that it sags less than 3.3 ft. (1 m.) in hoisting a barrel of herring, the block in the mizzen must be strong enough for a working load of 3 tons in the eye. With more sag, a working load of 1½ tons in the eye is sufficient.

In the fish tackle blocks in the main masts of trawlers and lugger-trawlers forces also act which are difficult to calculate and it is assumed that a working load of 6 tons in the eye is not exceeded. The whole loading and unloading equipment is surveyed annually by the Service.

LIFE-BOATS

Fishing vessels of less than 200 gross register tons and with an unlimited Certificate of Seaworthiness must have at least one life-boat. Such boats need not have air tanks or other means for increasing buoyancy but they must be large enough to take the whole crew. On ships exceeding 200 tons, but less than 400 gross register tons, one boat is sufficient but it must be provided with means to increase buoyancy by 10 per cent. of its cubic capacity. Fishing vessels of more than 400 tons must have two such boats, each capable of taking the whole crew.

The launching of life-boats present the least difficulty when the ship has two life-boats, because they are usually suspended from davits. When only one boat is required, it must be so placed that it can be lowered on either side, which means that it will have to be stowed on the centre line of the ship. In this case davits are of no use and usually the spankler boom is used for launching, which is no ideal solution for lowering the boat quickly because the mizzen sail must be furled and the boom topped. This disadvantage applies mainly when the boat has been placed on the superstructure. When it is placed on the upper deck, behind the superstructure, it may be lowered by hand over the bulwarks—again not an ideal solution. Despite these disadvantages, which are fully recognized, a better solution has not so far been found.

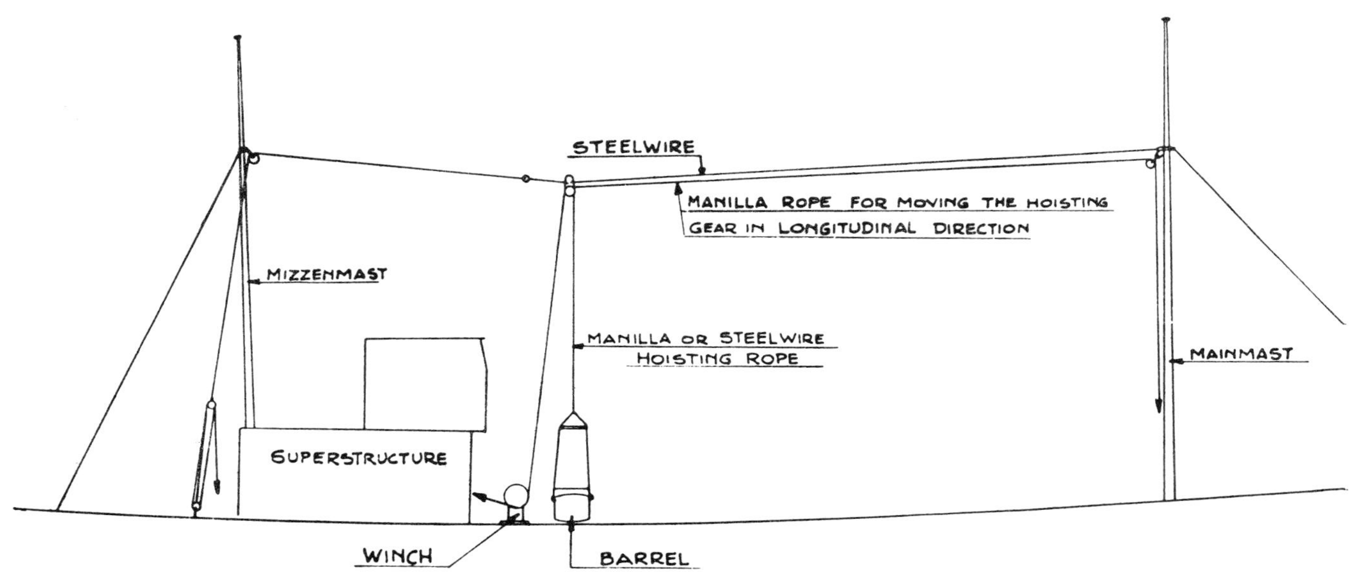

Fig. 436. Arrangement of loading and unloading equipment on Dutch luggers

PROBLEMS OF DESIGN AND CONSTRUCTION OF FISHING BOATS

by

LEANDRO FERNANDEZ MUNOZ

FREEBOARD

THE majority of craft in operation today have a very small freeboard, based on the regulations of the International Load-Line Convention, which is now considered to be too small. There is proof of this in the great number of fishing boats lost and men washed from the deck during fishing operations. These losses are not all due to foul weather.

A merchant ship can, during bad weather, head the sea and, by changing the speed, lay-to easily. Work on fishing boats often continues through bad weather and on the fishing grounds, because of the shallow waters, the waves lose their rhythm and break in heights disproportionate to the wave length. This means that decks are swept by water during normal fishing work.

The situation is aggravated by the fact that the hull of the majority of the fishing boats, during the return trip, are deeply loaded, even overloaded on leaving the fishing grounds in order to arrive at the home port with boat loaded to the load line. The fuel consumption in fishing vessels is very large in relation to the tonnage of the boat. At the same time, the ice melts and the water is bailed out by bilge pumps. In boats of special design, such as cod trawlers, where the haul is stowed in salt, the fish loses a considerable amount of water, thus decreasing its weight. The skipper always tries to land the greatest possible load and therefore he will overload the boat with the estimated consumption of fuel and ice, when leaving the grounds. There is no doubt that the form stability in loaded fishing boats is generally bad and, even if an excess of stability in the light condition compensates somewhat for this defect, sailing with the boat fully loaded is always disagreeable and dangerous.

There is a need to investigate this problem thoroughly and, in any case, the freeboard of fishing boats needs to be increased. A new table of freeboards, different from the one now in existence for merchant ships, should be drawn up.

For this purpose it is fundamental to make a detailed study of the waves on fishing grounds. Up to now very little attention has been given to this important factor in the design of fishing boats, although some governments have laws on the minimum dimensions of boats which can be used for fishing in specified places. Among them is the Spanish Government which regulates the fishing range according to the gross tonnage of the boats.

But this is not enough. Navigation on the fishing grounds is different from that in the open sea and the problem does not consist in calculating a boat length favourable to develop speed, a study that has already been made. The question is: how to build boats to ensure safety during dangerous work? Obviously, the circumstances are not the same on all the fishing grounds. There are places which are fished only during special seasons and under certain conditions; others, because of their proximity to a secure harbour, allow time for the boats to run to shelter. But in other places the speed of weather changes endangers the boat and the lives of her crew.

It is to be stressed that a number of fishing boats, as well as lives, are lost every year, due to lack of knowledge of the conditions under which the boats operate.

The nets could be hauled in, of course, and the boats go to port when the weather worsens, but this is not done, and even if it were recommended, it would not be done. Fishermen stay on the grounds as long as they can because they are far from port and must get a worthwhile catch. So the only thing to be done is to build fishing boats which assure a maximum amount of safety at sea.

Once the normal and maximum dimensions of the waves on every fishing ground are known, a study can determine the necessary dimensions for the ships, such as the freeboard.

LIFE BOATS AND RESCUE EQUIPMENT

Although much has been done, it is necessary to have compulsory adoption of complete rescue equipment. The majority of the men who are washed overboard cannot be rescued in spite of the available equipment. Lifeboats are often the first to be lost in a storm, and in the rough sea and in the height of the waves on the fishing grounds, the boats can seldom be used.

Regulations now require only one lifeboat in the

small fishing craft, and not only is it often smashed by the sea but is too small to carry all the crew. Life-rafts, with a greater capacity and buoyancy, should be used and, instead of one, the minimum carried should be two.

In addition, every man should carry, during his work, an intense colouring material which, should he fall into the sea, would indicate his position clearly. He should also wear a lifebelt which would automatically inflate on contact with the sea.

ANCHOR GEAR

The Classification Societies require that, in order to anchor a boat, a certain amount of chain, a windlass, and anchors of specified weight, should be carried.

It is well known in Spain that this equipment is on board only during the time the inspectors of the Classification Society and the marine authorities are on board to give their approval. The windlass disturbs the freedom of movement of the fisherman when using the pareja nets. The chains, in general, are not long enough for the boat to anchor where the fishing is done, so it is secured by attaching the anchor to a trawl warp operated by the trawl winch, which means that the boat is not safe nor are the regulations followed.

STANDARDIZATION OF AUXILIARY EQUIPMENT

Standardization of the auxiliary equipment (snatch blocks, sheaves, etc.) should be started as soon as possible after a detailed study of the manufacturing processes and materials. Experience shows that there are no identical or similar pieces of equipment, either in boats of different nationalities or of the same nationality, nor in boats built by the same builder. If one piece is broken, it is necessary to make new dies or to modify others, which adds to labour and increases the price considerably.

There is no constancy in the hardness of any one material, the foundries using all kinds of iron and steel. Because of this, at the end of every fishing voyage, it is usually necessary to refill the deep grooves in the sheaves or to replace them, but in other cases, the cables are worn by friction against sheaves which are too hard. Obviously here is a need to study the hardness of cast material in connection with the resistance of cables, and to determine the adequate diameters of the sheaf to reduce wear to a minimum. It should be kept in mind that skippers are against the installation of snatch sheaves of adequate diameter, as it is easier to manipulate lighter pieces, even if the results are not economical.

WOODEN DECKS

Wooden decks wear considerably in fishing boats and soon lose the required thickness, a problem that is acute today because of the bad quality of the wood used after the war.

Some ship owners specify that wood decks should be thicker than the dimension requested by the regulations, a demand that increases the cost of the work and only postpones the problem.

A solution suggested is to deck the boats according to the regulations and then use a second wearing deck that could be replaced periodically. This proposal is based on the results of different satisfactory tests.

USE OF RADAR

Radar should be compulsory in all the boats fishing on the International grounds because of the many ships that gather there at certain seasons. The high cost of the radar is an objection, but an increase in the number of ships that use it may reduce its price. Or, as it happened with the radio-telephone, companies might be formed to rent the apparatus for a fee so that even small ships could afford it.

NAVIGATION LIGHTS IN PAREJA TRAWLERS

Nowadays the pareja trawlers do not carry more lights than other trawlers because they only shoot nets by day, but it can be foreseen that in the very near future they will begin fishing by night. An additional navigation light should be required to indicate that the net is being trawled by the two boats so that no ship may pass between them.

GALLOWS

In the majority of the boats, gallows are installed on both starboard and port, but rarely, if ever, are both sides used. Some ship owners justify the installation of four gallows as a precaution in case of a breakdown of the fishing gear but it appears to be merely a question of routine and tradition. The elimination of two of them would save money and provide more space on board.

CREW ACCOMMODATIONS

The efficiency of the crew is naturally affected by the conditions of life aboard, a truism which, it appears, is not generally accepted, and it is a shame to see the conditions under which some crews sail. Great efforts have been made to improve conditions in Spanish vessels and modern fishing boats have wide, clean and well-ventilated crew accommodations.

An effort should be made even in small boats to protect the entire crew during bad weather.

NAVAL ARCHITECTURE — DISCUSSION

TANK TESTS AND RESISTANCE

Dr. R. Brard (France): There are few differences between various tanks because they are all mainly concerned with the measurement of resistance and prediction of power, but some establishments have, in addition, special tanks for channel and shallow water work, manoeuvrability, seaworthiness, and rough water testing.

The possibilities of using a special tank for investigating manoeuvrability and seaworthiness are particularly important. All fishing vessels, especially trawlers, need good manoeuvrability as they must be able to steer while trawling at one third or one quarter of their normal speed. And, of course, all fishing vessels must be able to keep a course in rough water. Measurements in the usual tank give only the instantaneous turning movement due to rudder action, and some information on the start of turning following that rudder action. A circular tank is needed to watch the whole evolution. It takes more rudder to stop than to start a turn. In a circular tank all the manoeuvres and behaviour of a ship at sea can be studied. Seaworthiness is not confined to rolling and pitching. Other factors influencing it are the real freeboard above wave crests, the shipping of water and the increased resistance due to rough weather. The final aim of seaworthiness is to design a ship to sail and keep its speed and course in average conditions.

The tests analysed by Mr. Traung show, in some instances, differences up to 50 per cent. between models, indicating that, after centuries of experience, naval architects are not yet in possession of practical rules for designing fishing boats, and much the same can be said with regard to propellers.

Fishing vessels have a high relative speed. The trawlers mentioned in Mr. Möckel's paper operate just under the last hump of the resistance curve, and from that point of view they belong to the same class as passenger liners. Those with $\frac{V}{\sqrt{gL}}$ of .31 are beyond the hump, while others are still faster and are in the range that Admiral Taylor considered suitable for cruisers. The resistance rules act for all ships having the same Froude's number. That explains why the transom stern, by increasing the apparent length of the hull, is advantageous. The main dimensions which have an influence on the resistance are the prismatic coefficient, the $\frac{B}{T}$ ratio, and the angle of entrance. A small angle of entrance always means reduced resistance and for this reason the bulb is often adopted on big ships. The $\frac{B}{T}$ producing the lowest resistance is always greater than 3.0. On fishing boats $\frac{B}{T}$ is often smaller than 2.5, so it is clear that B can be increased without increasing the resistance. The length-displacement ratio is without a marked influence on the resistance of fishing boats.

A research programme combining measurements at sea and tank testing in artificial waves would clarify many questions. It would show to what extent rules could be laid down without being unreasonable and making them impossible to apply.

In a sense, the hull of a fishing vessel is a big-scale model ($\frac{1}{4}$ to $\frac{1}{10}$) of a fast passenger ship, but with a greater metacentric height and more draught aft. What makes fishing boat design so difficult is that the hulls are not exactly similar to those of passenger or sailing ships, or destroyers or tugs, but have something of each.

Dr. W. A. P. van Lammeren (Netherlands): Every hull design should be tank tested and, as Mr. Traung has stated, it is important to collect and analyse systematically the data supplied by tests at various model basins. They should also be compared, if possible, with full-scale tests under trial and service conditions.

Mr. Francis Minot (U.S.A.): Mr. Traung has performed a very valuable service to the fishing industry and fishing boat designers and builders in compiling the data found in his paper.

The question next arises as to the application of this information. The two recently designed Boston trawlers mentioned by Mr. Traung provide a good example of what is happening all the time in the fishing industry. Improvements could have been made to them in the design stage. It is true that the designer has difficult problems to solve and he has got to take into consideration all sorts of requirements but, as Mr. Traung points out, if they can be satisfied and the design of the ship improved, it is certainly common-sense for the owner, builder and designer to consider these things while the ship is in the design stage.

Mr. G. S. Selman (U.K.): Nearly all the fishing boats in the United Kingdom below the length of 80 ft. (25 m.) are built without lines plans by craftsmen who have little technical knowledge. They certainly know nothing about tank tests and, in fact, very little interest has been shown in tank tests with fishing boats by members of the technical societies in the United Kingdom. Mr. Traung's paper records valuable data relative to the resistance of fishing boat forms and it is to be hoped that FAO will find means of bringing this information to fishing boat builders all over the world, and to educate them into the use of the data. With regard to technical details of presentation of tank tests, it seems unwise to follow the example of Taylor and Baker and use a universal maximum ordinate for both area and water-line curves. It is sounder to follow the example of Froude and express both area and water-line curves relative to a dimensional maximum ordinant. The reason is that with relatively high speed craft the curvature gradient is important above everything else and means must be found to express true shape as distinct from relative distribution of displacement.

With regard to the area curve of a boat with a large inclination of keel, plotting does not give a true indication of shape. Instead of this, the figures obtained by dividing each individual sectional area by its depth should be plotted as ordinates.

Dr. J. F. Allan (U.K.): It is unreasonable to expect that the design of fishing boats can be standardized in all countries because of the wide variation of local conditions. Co-operative research on an international scale is not likely to lead to useful results. Much more will be learned by consulting papers presented at international conferences and studying the methods adopted by other countries.

A fishing boat must provide a stable fishing platform, near the water-level and reasonably dry, even in severe weather conditions. It is a difficult problem to provide this on a comparatively small vessel and a compromise is required between the various factors involved.

Fishing boats generally have a maximum speed at or about the unity $V/\sqrt{L}$ $\left(\frac{v}{\sqrt{gL}}=0.30\right)$ wave hump. Some smaller boats are pushed towards the final wave hump around $1.5V/\sqrt{L}$ $\left(\frac{v}{\sqrt{gL}}=0.45\right)$. For minimum resistance in these conditions the entrance angle must be fine and the forward waterlines straight or only slightly rounded, with an easy curvature to the midship section which is as full as practicable. This leads to an LCB position as far abaft amidships as can be allowed by consideration of the internal arrangements and the trimming of the ship in the loaded condition.

Mr. Traung's paper gives a very useful collection of tank data on a wide range of craft of the type under consideration and shows in general the trend in form referred to above. The fining of the bow in the interests of maximum top speed causes some small increase in pitching, but a good compromise can be achieved by using a fairly straight waterline bow combined with good flare in the top body. A good design can also be achieved by adopting in the forebody the extreme " V " type of section favoured in Germany but this will show some increase in resistance at top speeds compared to fine entrance designs.

It is important to keep the after-body lines, especially the buttock lines, easy enough to ensure a good flow to the propeller, and the adoption of the modern type of cruiser stern helps in this direction. The stern profile must be raked well aft above the waterline for good seaworthiness in following sea conditions. The transom stern has been referred to, and if this is given a suitable rake aft it is not greatly different in behaviour from the cruiser stern and may be useful from a constructional point of view. With reference to the question of transom sterns it is interesting to note the extremely heavily-raked transoms indicated in the Pakistan vessels. The rake appears to be overdone.

In the paper by Mr. Simpson it is stated that designs have already been developed for a 140 ft. (42.7 m.) vessel to do 15 knots. This is just about the worst possible combination of speed and length that one can be faced with. It is known that the power curve in this region will be extremely steep, even if the vessel should have a reasonably fine coefficient.

The results given by Mr. Gueroult, when compared with recent experiments carried out by the National Physical Laboratory, indicate that, with the exception of the 0.62 block form, the other forms in his paper could be filled at least 5 per cent. in coefficient without loss of performance.

Mr. E. R. Gueroult (France): Mr. Traung has arranged the results of his tank tests of typical fishing boats in a way which allows easy reference and the paper gives an all-round view impossible for the ordinary naval architect to acquire in the normal course of his work. The U.S.A. tuna clippers, which have provoked widespread interest in Europe, are not included in the catalogue. The isolated tests on traditional hull forms are of great interest, but systematic trials on a series of models which allow for modification and improvement should also appear in the catalogue.

The latest systematic trials are those which Dr. Allan in 1953 presented at the Institute of Marine Engineers. The results published in this paper seem excellent and difficult to equal. But could Dr. Allan explain his statement on the longitudinal distribution of displacement, that is to give the centre of buoyancy more precisely? The impression is that a great part of the improvement obtained is due to fining the aft bodies and, despite what Dr. Allan has said, the tendency now is to restore the centre of buoyancy towards the middle.

Mr. Gueroult agreed with Mr. Traung's conclusions with the following exceptions: The optimum prismatic coefficient is from 0.55 to 0.60, but he would warn against a rigid adoption of 0.575 which would lead, together with other given factors, towards forced forms and more resistance. Excellent results could be obtained with a prismatic greater than 0.60 when the centre of buoyancy is correctly placed and the lines are faired well. Optimum prismatic should be related to the beam-draught ratio. As Sir Amos Ayre recommended, one should avoid being prismatic conscious despite the importance of the prismatic. The resistance results in a tank of smooth water, though interesting in studying lines, are insufficient for the prediction of power and speed. The resistances in the tank are generally increased to 25 per cent. or 30 per cent. to take into account the state of the sea, wind resistance, hull fouling, etc. Fig. 418 and 419 in Mr. Möckel's paper give for cruising and for trawling certain values of increased resistance as a function of the state of the sea. These are values for which naval architects have been waiting and much is owed to the work of Mr. Traung and Mr. Möckel.

The catalogues of tank tests should certainly be continued and perfected in the form which Mr. Traung has chosen, because the conventional system of reference is of little help. However, it is suggested that the non-dimensional comparison of the models should take into account the important differences of length from 50 to 230 ft. (15 to 70 m.).

In the reply to a question by Mr. Bordoli (Italy), Mr. Gueroult said they had not made tests on models shorter than 105 ft. (32 m.) before the French trawler building programme started. Since then a number of tests have been made on ships of similar dimensions but differing in proportion and fineness.

It is being said that still water resistance is not an important factor. He would like to put the point in a different way. Quite a lot is known about still water, which enables architects to devote more time to seaworthiness and manoeuvrability. As nearly all the operations in a fishing boat are carried out at slow speed or at no speed, it seems desirable to carry out future research work in these conditions. There is a big amount of money involved in the undue wear of the prime movers because of ignorance of the operation conditions at very low speed, such as trawling, and that should be made the main object of future research work.

Mr. N. Chr. Astrup (Norway): He accepted the general validity of the need for a low prismatic and a sharp bow, but for low resistance in still water a knife-sharp bow is not absolutely necessary unless an effort is being made to get an extraordinarily low resistance. He had tested a model, S.M.T. No. 92, of reasonably low resistance even though its angle of entrance was as high as 30°. When the load water lines are drawn with the same width and length it appears that this Norwegian model has a fuller waterline forward than any other vessel of equally low resistance.

In the values of prismatic coefficient and longitudinal position of the centre of buoyancy, it is possible to depart considerably from the data given in Mr. Traung's paper without suffering much of a penalty in increased resistance.

A boat must be able to carry a big load of fish to pay its cost and provide the fishermen with a living. It is not usually possible for a fisherman to get a motor powerful enough to double the speed of his boat so that he can do two trips in the time he previously took for one. But if he buys a boat of greater displacement, suitably powered, he may be able to return to port with a load of fish twice as big as he could carry in his smaller boat. The fuel consumption of the bigger boat will not be increased in the same ratio as the displacement of the boat has been increased. In fact, with a boat of good design the fisherman might find that fuel consumption per ton of displacement is much less, a saving which may help in paying for the new hull. Again, the bigger boat will need stronger construction and will cost more money in total, but it should not be double the price of the boat which has only half the carrying capacity.

From an economic point of view, an approximate measure of the quality of boats is the motor rating per ton of displacement, necessary to drive a boat of given length overall at a given speed. For example, when the Pakistan boats are compared with a Norwegian drifter the former are worse, compared in this way.

A speed-length ratio, containing the length of the waterline of the boat at rest, should not be used in this case as a base. This length is presumably less than the wetted length of the Pakistan Bedi at full speed, because of the sloping stern and stem, and is also much shorter than the overall length.

Mr. Traung states that the displacement-length ratio has little influence on Admiralty coefficients. The Swedish and Norwegian tests with drifter models show that it does not vary more than about 10 per cent. over a wide range of displacement variations. But for boats of very different design and for low displacement length-ratio, direct comparison of Admiralty co-efficients will not show whether the lines are good or bad. In the case of boats which do not depart too much in displacement-length ratio, in outline and in type of lines, the method of comparison can show if one of the lines is very good or very bad.

Considering all necessary corrections, there is little difference in resistance coefficients between the designs of able naval architects. Better resistance coefficients than those obtainable for a good multi-purpose boat may be got only by using less displacement-length ratios, less stability and seaworthiness or less available space forward. These conclusions are based on a comparison of the results of tank tests in Japan, Norway, Sweden, Scotland and England. French and American designers, according to their papers, also make excellent lines, and they and naval architects of other countries are urged to publish tank tests complete with detailed information. Tank tests of old boats designed by good craftsmen are interesting although, in under-developed areas, these boats do not have the high displacement of modern fishing vessels.

The remarks made in this discussion are mainly valid for European drifters and small trawlers.

Mr. Jarl Lindblom (Finland): He had been forced to try out Mr. Traung's theory about the influence of the prismatic coefficient and the increased fullness of the midships section of vessels with a displacement of 640 tons. The owners kept adding weight to the vessels after the contract stage. The speed was to be 9 knots minimum and the power of the main engines could not be increased because it would have affected the radius of action if the fuel capacity was kept the same. There were penalties for excess draught. The problem became one of how much displacement could be increased. He saw Mr. Traung in 1947 and got information about his tank tests, so he increased the midships section. Displacement was increased to 730 tons and when the first ship was put to her trials she had a speed of 9.3 knots with only 320 h.p. on a waterline of 120 ft. (36.6 m.).

Mr. Philip Thiel, Jr. (U.S.A.): The longitudinal prismatic coefficient is the key to easy propulsion of any ship There is an abundance of trial and test data available to indicate the optimum value of this coefficient for each condition. In the majority of fishing vessels the prismatic coefficient is too high, although the value depends on speed-length ratio. Complications in trim and stability because of the use of a lower prismatic coefficient than usual can be resolved by a little study on the drawing board. For example, detailed calculations on nine trawler hull forms, of identical dimensions (350 long tons displacement), 100 ft. (30.5 m.) waterline length, 24 ft. (7.3 m.) beam, 10 ft. (3.05 m.) mean draught and a prismatic coefficient of 0.605) showed that, for an assumed KG of 10.5 ft. (3.2 m.), the transverse GM could be varied from 2.19 to 0.60 ft. (0.665 to 0.183 m.) and the moment to trim 1 in. varied from 29.1 to 19.95 ft./tons (9 m. tons to 6.2 m. tons). The variations are due to different vertical distributions of the displacement in the sections at the ends.

Most fishing vessels require buoyant quarters and ample deck space aft, and this is an added incentive for the use of the transom stern with its superior resistance qualities. If some concern is felt over the use of the motor-boat type of flat transom, a conical form with an underwater knuckle, similar to that used for destroyer hulls, might be considered. With a sufficient rake aft and adequate deadrise in the sections of the run, this form would have the seaworthy qualities of a good cruiser stern. Some research and considerable practical experience are available to point the way to economy of construction of steel vessels. Unfortunately, this is a factor of design where conservatism and prejudice are most deeply embedded. Tank tests have indicated that double-chine simplified hull forms can be designed to equal the best conventional hull form, and to exceed the average in performance. At the same time this hull form is cheaper to construct. The design problem is little more involved than that of merely approximating the sections of a conventional hull form with straight lines. Consideration must be given to the trajectories of flow along the hull, to avoid their crossing the knuckle or chine lines. The above-water appearance can usually be

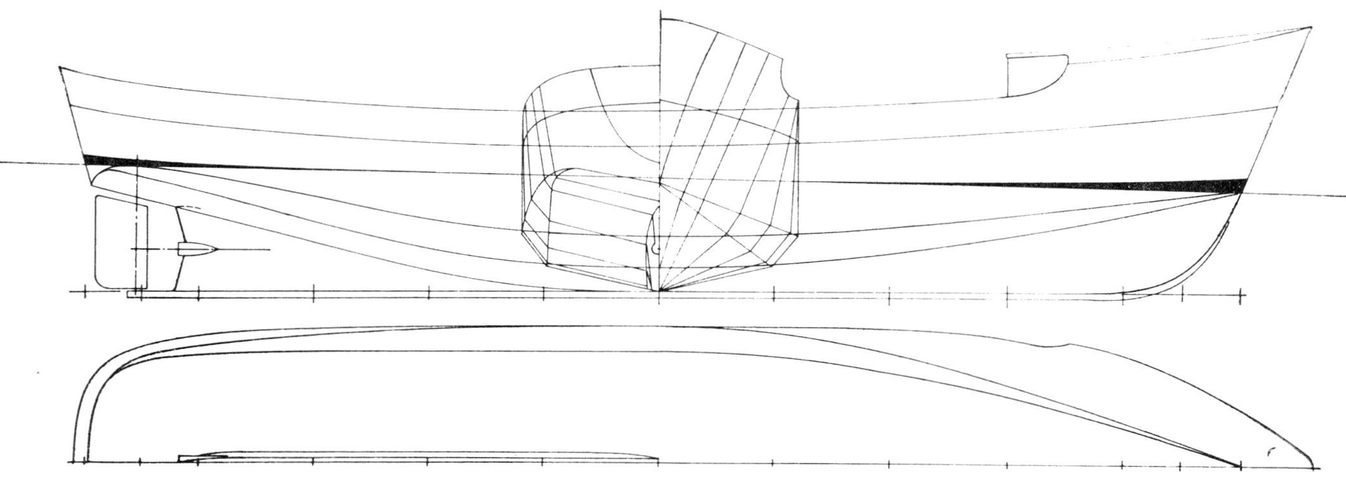

Fig. 438

designed to be indistinguishable from a conventional form. Fig. 438 is an example of the double-chine hull, one of the 350 long tons displacement trawlers with a 300 h.p. engine giving a speed of 10 knots. Fig. 326 gave a suitable scantling section.

Mr. Olin Stephens II (U.S.A.): The transom stern undoubtedly helps at the high speed-length ratios associated with vessels such as 60 to 65 ft. (18 to 20 m.) trawlers. These vessels are fairly fast from the standpoint of naval architecture and the transom stern is valuable for them, providing the rest of the

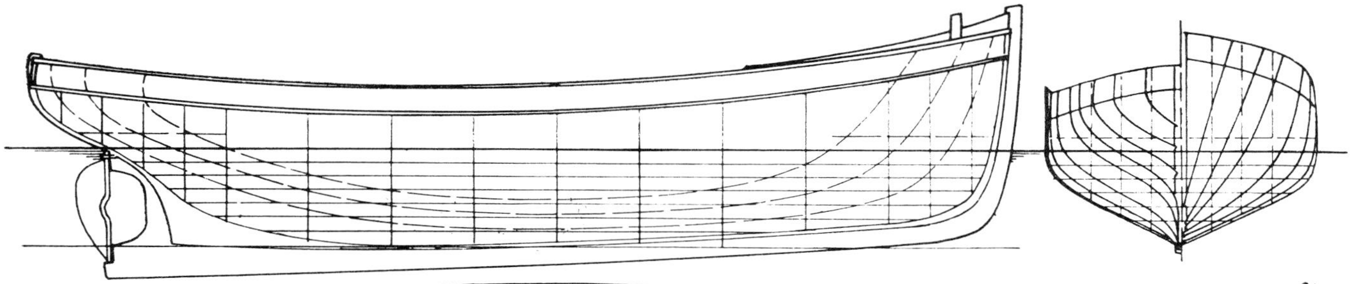

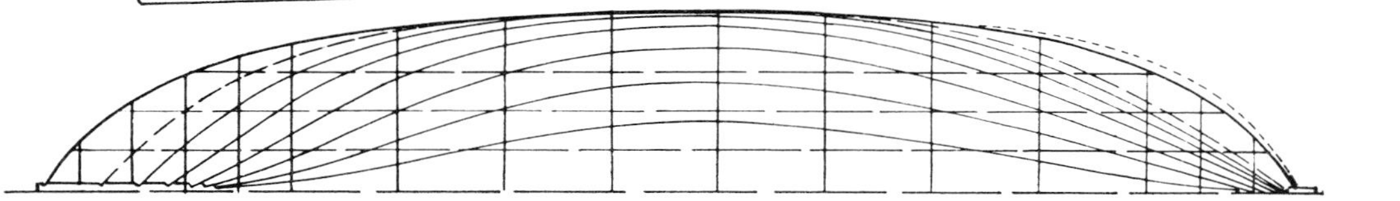

	m.	ft.
L b.p.	20.37	66.7
B mld.	6.25	20.50
D mld.	3.20	10.50
T	2.40	7.85
T max.	3.10	10.30

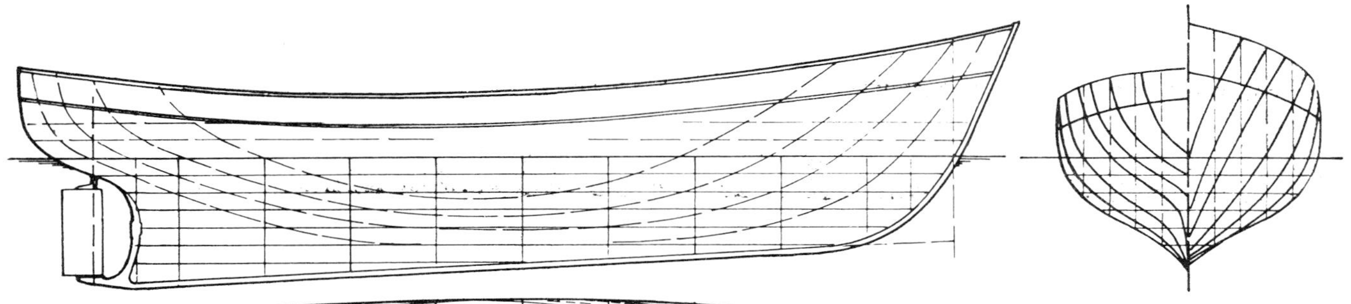

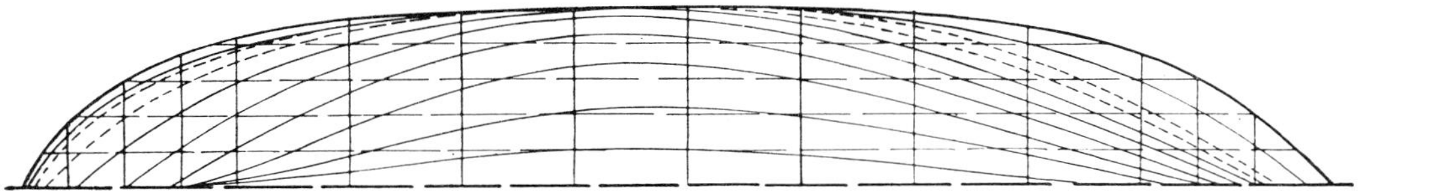

	m.	ft.
L b.p.	19.75	64.75
B mld.	6.25	20.50
D mld.	3.00	9.84
T	2.19	7.17
T max.	2.93	[illegible].61

Fig. 439 and 440. Original conventional lines above tested in comparison with new design below. According to the latter over 1,000 boats have been built.

design is carried out so that there is not too much change in the merging of the transom.

Mr. Stephens referred to the tabulation of smaller models of 10 ft. (3.05 m.) in Mr. Traung's paper and wondered, as they had not been tested with turbulence stimulators, whether the results given were truly reliable, particularly at a lower speed.

Mr. George C. Nickum (U.S.A.): A fine prismatic coefficient is essential for high speed of a vessel.

Mr. Frederick Parkes (U.K.): The assistance obtained from the National Physical Laboratory at Teddington is invaluable, particularly when one is building a new model. Although the owner or builder may have experience of the type of boat concerned, small differences in general dimensions can make a big difference to seaworthiness, speed and power required. The National Physical Laboratory has put forward many suggestions which have improved models materially at little or no additional expense. He now makes a practice of going to the Laboratory with any model that varies at all from boats he had previously built. In addition to tests of hull form the Laboratory's advice on engine power and propeller design is also of great assistance.

One of the biggest improvements in post-war trawlers is the development of a flared bow with a soft-nose, which gives the crew more protection on the foredeck where most of the work on a trawler is carried out. The design also helps a vessel when homeward bound in bad weather. It lessens the danger to other ships should there be a collision.

Dr. H. K. Kloess (Germany): The data collected by Mr. Traung will be important to naval architects in designing new boats, but as a comparative basis for trawlers of different form the data does not seem to go far. In judging the influence of ratio of main dimensions, of Froude's numbers and of displacement-length ratio, trawlers with affiliated lines can be studied in the way shown by Taylor (1933) and Takagi (1950). The qualities of different hull forms can only be determined by carrying out comparative tank tests with models of equal size and main dimensions.

The Taylor displacement-length ratio reaches values up to 250 ($L/\nabla^{\frac{1}{3}} = 4.9$) only but for trawlers this ratio is about 500 ($L/\nabla^{\frac{1}{3}} = 3.9$). The work of Takagi may be considered as a continuation of Taylor's standard series and it would have been enlightening to have carried out the comparison of fishing craft data collected by Mr. Traung on the Taylor or Takagi basis.

The data collected do not make it possible to separate the influence of the shape of lines from the influence of the ratios mentioned. A hull form should not be designed or altered on the basis of measuring resistance only. It is common knowledge that vessels with fine ends show low resistance but sometimes very poor hull efficiency, and vessels of low resistance may also lack stability. Seaworthiness is not directly dependent on resistance in calm conditions. Stability and seaworthiness are of the utmost importance to trawlers at all times. When trawling, the resistance in calm conditions has very little influence on speed because the pull far exceeds the resistance.

The answer to Mr. Traung's question as to whether boat model types should be developed in series is " Yes ". Similar work was carried out in Germany about 1940. Table LX gives the main dimensions (approved by the German Lloyd Classification Society) of German cutters built in series. The

TABLE LX

Data for cutter series

Type		*A*	*B*	*C*	*D*	*E*	*F*	*G*
Max. length overall	ft.	39.35	45.9	52.5	59.0	65.6	72.2	78.7
	m.	12.00	14.00	16.00	18.00	20.00	22.00	24.00
Length b.p.	ft.	29.85	37.7	43.6	49.2	55.1	60.3	67.5
	m.	9.10	11.50	13.30	15.00	16.80	18.40	20.57
Breadth	ft.	11.15	14.1	15.4	16.4	17.4	18.4	20.5
	m.	3.40	4.30	4.70	5.00	5.30	5.60	6.25
Depth	ft.	4.59	6.88	7.21	7.54	8.2	9.18	9.84
	m.	1.40	2.10	2.20	2.30	2.50	2.80	3.00
Draft, mean	ft.	4.53	4.69	4.86	5.05	5.51	6.1	6.92
	m.	1.38	1.43	1.48	1.54	1.68	1.86	2.11
B.h.p.		30	50	70	90	110	120	150
R.p.m.		550	500	425	375	330	310	300

graduation is made for every 6.5 ft. (2 m.) from 39 to 79 ft. (12 to 24 m.). The models have been tested in the Vienna experimental tank for resistance, propulsion and towing qualities, and have also been tested by measured mile trials. The type D comparison was made with the lines of a good Baltic Sea cutter, and type G with the lines of the best Finkenwerder cutter. Investigations included rudder, propeller and propulsion bulb. Fig. 439 and 440 show the respective lines and body plans of two of the 72 ft. (22 m.) type of boat.

The shaft horsepower and Admiralty coefficients of two of the 53 ft. (16 m.) vessels are plotted on the speed diagram in fig. 441. The table shows the respective dimensions. The Maierform trawler reached a speed of 9 knots with 33 per cent. less power than the best of Finkenwerder cutters. The 79 ft. (24 m.) type indicated 41.4 per cent. saving in power or .82 of a knot increase in speed when using the same power.

Power curves, pull and r.p.m. of the 53 ft. (16 m.) type are shown in fig. 442. They are the result of investigating different propellers.

The 39, 53 and 72 ft. (12, 16 and 22 m.) types have been tested in the experimental tank and the values for other types were determined from these tests. More than 1,000 vessels have been built on the basis of these standard models. During World War II many 72 ft. (22 m.) cutters were used as patrol boats.

These cutters were designed in 1940 and much experience has since been gained in them, but there does not seem to be much scope for improving their hull form. A better thrust and pull has been obtained by use of the propulsion bulb, which has been improved through experience. It was found that the efficiency of the bulb is increased with higher turbulence, which makes it suitable for use in trawlers and other fishing vessels (*see* fig. 443). The bulb improves weight distribution, corrects the vortex aft of the boss and eliminates the turbulence of water from the propeller. This improves propulsive efficiency and diminishes disturbing forces of the rudder. In making these improvements the performance of vessels before and after fitting the bulb was compared (*see* fig. 444). At a speed of 9.1 knots there is a saving of about 20 per cent. in fuel or an increase of .5 of a knot in speed.

Although the bulb device has been known for some years a good deal of time was required to test it out on full-size ships. The low Reynolds number obtained in the experimental tanks does not allow accurate tank testing of the bulb.

Mr. Arthur de Fever (U.S.A.): Tuna clippers have been given a fine hull shape to increase their speed. So far as hull design is concerned, the vessels can be divided into two categories: (1) those fishing the local banks; (2) those doing long-range fishing.

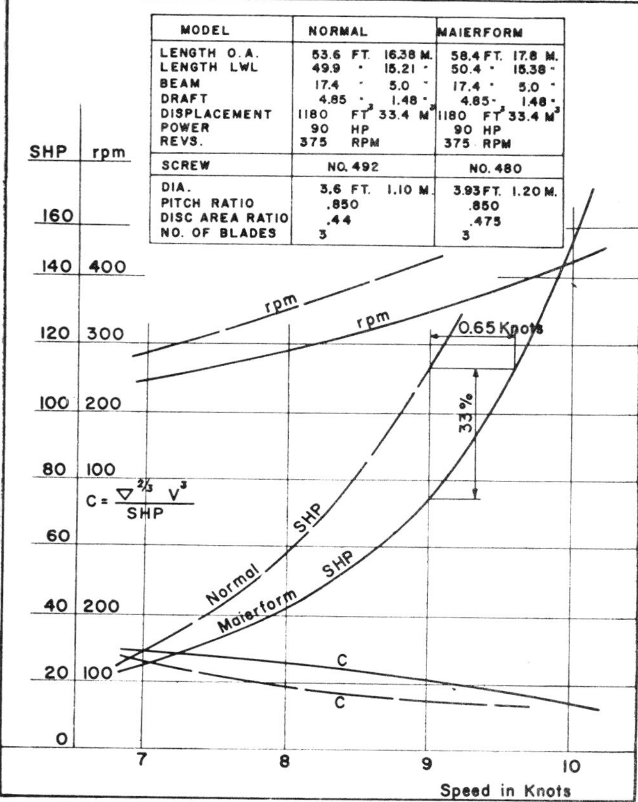

Fig. 441. 52.5 ft. (16 m.) Maierform trawler compared with trawler of conventional lines. Results based on tests at the model tank, Vienna.

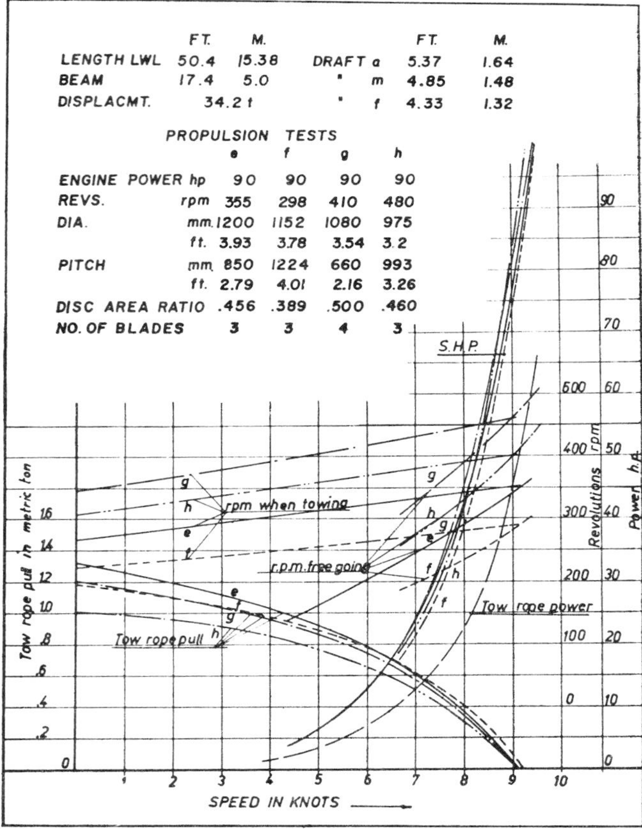

Fig. 442. 52.5 ft. (16 m.) trawler. Tow rope pull and free running power with various screws.

There is no great point in fining the hull of clippers fishing the local banks as there is little to be gained by an increase of speed, 9½ to 10 knots being sufficient for them. The longer, finer hull need only be designed for the long-range clippers.

Mr. J. G. de Wit (Netherlands): It seems from Mr. Chapelle's paper that U.S.A. east coast fishermen desire a high speed. Is there any reason for their wanting this? There seems to be a lot of waste in such high speeds.

Ballast in the form of sandbags is carried on deck level in the Bedi boats according to the paper on West Pakistan fishing craft. What is the aim of this? Although resistance seems to be fairly low, is it something to do with stability and seakindliness?

Mr. Howard I. Chapelle (U.S.A.): There has been a tendency to increase speed in fishing boats merely for the sake of it. Some craft on the U.S.A. east coast, such as the Chesapeake Bay launches, are capable of surprisingly high speed, and in some cases that speed does pay for itself. On the other hand, many boats are carrying a great deal too much power and can develop a much higher speed than they can use economically in their work.

When fishing craft develop high speed the cost of repairs and maintenance are also high and, of course, crews have to be better trained to operate high-powered engines than they do when low-powered engines are used.

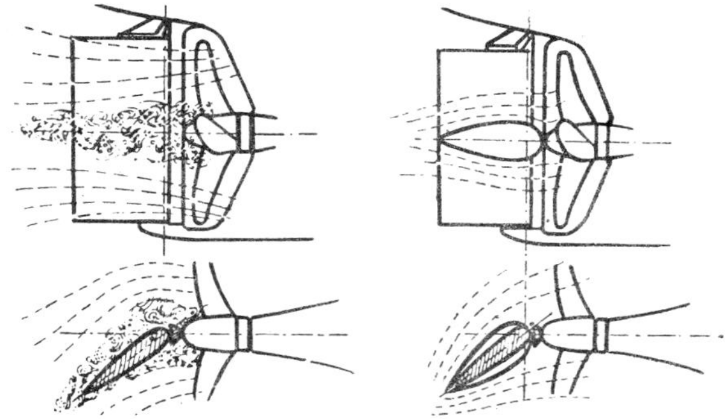

Fig. 443. The anti-turbulence effect of the Costa bulb is shown in these four drawings. At left are two views of a conventional rudder and propeller assembly with its vortex of turbulent water marked by heavy lines. At right, efficiency is increased as the bulb eliminates the vortex and guides the water aft without disturbance.

Experience shows that on the U.S.A. east coast a speed of 9 to 10 knots is ample for most purposes, although some small trawlers find it useful to make 10½ knots. There are some lobster boats in New England that can do 20 knots an hour but, of course, they do not pick up lobster pots at that speed. In fact, the speed is used on one day a year—for racing.

Mr. Jan-Olof Traung (FAO): Regarding the question about the ballast on Pakistan boats, it is understood that the sandbags are used on the windward side when tacking. Sandbags may also give the craft an easier motion in a rough sea.

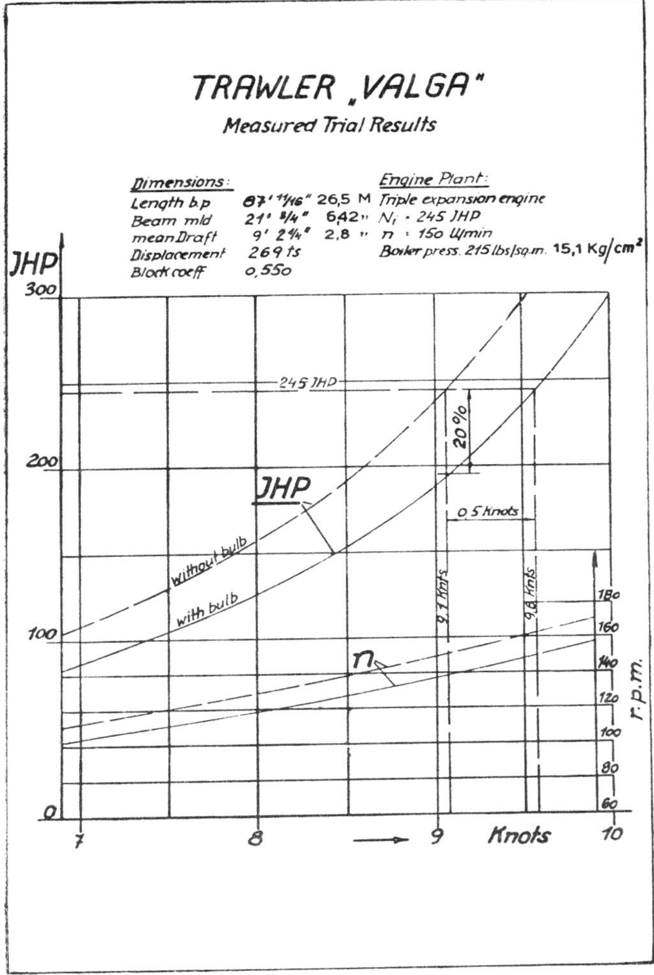

Fig. 444

Experience in making full-scale trials with a number of fishing boats shows that invariably the skipper over-estimates the speed his boat is doing. At the measured mile tests, the boat seldom makes the speed the skipper thought. Even 10½ knots by a 65 ft. (20 m.) boat having a 200 h.p. engine would be remarkably good if it was true.

Regarding a question by Mr. de Fever of the best angle of entrance, one could get reasonably good resistance with a half angle of entrance of about 20°. It might be possible to get better results at 15° in a low Taylor length displacement number, but about 20° is satisfactory.

Mr. G. O. Huet (U.S.A.): The average fisherman anywhere, who is brought up in the tradition of his father and grandfather, develops a very thorough knowledge of his craft whether or not he has had much formal education. It is important to realise that these fishermen are very ingenious in the way in which they have overcome obstacles in the design and working of their boats. It is true that the naval architect and technical man, with his knowledge of materials and design, and his experience of hull forms, has now reached a point where he has so much scientific information available that he should be able to find the proper hull form for any particular load condition and other circumstances. But this knowledge is not used as it should be. For instance, in the New Orleans area, the average boat owner will build a boat to have a speed of 8 or 9 knots an hour. Later, he finds that another owner has built a boat to do 12 knots an hour. That boat may have an engine twice as powerful but the owner of the slower boat does not consider anything but his wish to increase the speed of his boat to 12 knots an hour, so he installs a more powerful engine. He does not understand that the hull form of his boat could be modified in such a way that he could get the increased speed without any increase in power.

He once built 100 boats for a foreign Government, the original plans were for 50 ft. (15 m.) boats. One of the draughtsmen put on the drawing a speed of 9 knots instead of 9 miles an hour so that when the representative of the Government discussed the boats he insisted on 9 knots. It was impossible to drive that particular hull model at 9 knots with the power that was to be installed in it but, by lengthening the boat from 50 to 65 ft. (15 to 20 m.) it was possible to give additional cargo space and, with the same power, increase the speed to 10 knots. Incidentally, the cost of putting the extra 15 ft. (4.6 m.) on to the boat was small in relation to the advantages gained. So a larger vessel carrying a bigger payload and having more speed from the same power was built. This is a good instance of what a naval architect can do if he is acquainted with all the facts and has the co-operation of the owner and builder.

Mr. W. C. Gould (U.S.A.): There is a tendency to think much about the first cost of fishing vessels and little about the direct operating expenses. Building the hull does not always represent a major portion of the cost and there are instances where the hull can be lengthened to accommodate a bigger fish load and produce more economic advantages throughout the life of the boat. It seems invariable that the owner of each new boat claims to be able to carry another 10,000 or 15,000 lb. (4.5 to 6.8 ton) of fish and that creates problems of seaworthiness as well as cost. Yet an analysis of the average fish load carried by a boat over a year shows that it rarely takes up more than 50 to 60 per cent. of the vessel's capacity. If it were possible to design a boat to take the average fish load, a much better hull form would result. For example, by lengthening the hull slightly, it would probably be possible to arrange enough hatches in the deck to allow the equipment below decks to be easily and economically removed. If engines can be removed easily from the boat, the labour cost of overhauling can be cut by 50 per cent. and a much better job can be done. This is a point which naval architects should bear in mind in designing fishing boats. In addition, if a slightly longer hull is used, it should be possible to get far better speed from an engine of less power, and that in itself would be an economy in operating.

Mr. Howard I. Chapelle (U.S.A.) : It is very easy to suggest changes in design but even in lengthening a boat there may be difficulties. For example, in Canada there are certain laws which provide a bonus for vessels built under a certain length, so that fishermen want their boats built within that length. That is a definite limitation and boats must be designed to it. It would be simple enough, of course, to lengthen such a boat by 10 ft. (3.05 m.) and she would be a much better craft, but in that case the fisherman would not get his bonus and, consequently, he would not have the boat built. The naval architect must operate under such restrictions and do his best within the limits imposed on him.

Mr. Robert F. Allan (Canada): It is regrettable that the cruiser stern and fantail or " seine boat stern " designs according to fig. 278 in Mr. Hanson's combination boat paper are not compared on a basis of similar overall length. It seems that the cruiser stern model " B " is actually a much finer vessel—as indicated by an overall length of 95 ft. (29 m.) and midship draught of 7.75 ft. (2.36 m.)—as compared with the fantail stern model which has an overall length of 89 ft. (27 m.) and a draught of 8.70 ft. (2.65 m.). But there is no indication of any improvement coming from the adoption of the cruiser stern alone, nor is it clear that resistance would be reduced in the load condition.

Mr. Jan-Olof Traung (FAO): Speaking of his paper on tank tests, Mr. Traung said that it was not so much about tests and resistance as an outline to a catalogue of tank tests. So many tank tests on fishing boats are available to-day that it was felt they should be compiled in a form which could be used with advantage by all connected with the boat-building industry. From the data sheets published in the paper it is already possible to draw some practical conclusions. For example, the Boston trawler (model No. 2), which was built just after World War II, can be improved by 54 per cent. It may be said that such improvements in design would be at the expense of fish load capacity, but the comparisons have been done as objectively as possible, using exactly the same length, the same displacement, the same freeboard, and the same gross tonnage. Having done so, the fish-hold capacity should be the same.

Resistance tests have taken place in tanks in various parts of the world but there are very few naval architects who have been able to collect them all, yet it is of great importance to have available all data concerning resistance before tackling problems of propulsion and seaworthiness. Mr. Gueroult has asked if there will be reports on tuna clippers and other Pacific fishing boats in the tank-test catalogue when it is completed. The answer is yes. Mr. Hanson has already given two tests in his combination boat paper and has supplied his test data so that they can be worked up in sheets. The results given in Mr. Hanson's paper does not only show the influence of a cruiser stern. The new type tested with a cruiser stern at Ann Arbour was also longer and had a much finer entrance.

FAO will publish other tests of tuna clippers. For instance, Mr. Nickum has provided a series of four models in which he tested bulbous bows against V-bows and cruiser sterns against fantail sterns. Results of the tests of some Colombian boats, built in Sweden with transom sterns, will also be included and FAO will publish data sheets on tests being made by General Motors Corporation on V-bottom shrimp trawlers and on New England trawlers of double-chine design. The New England trawlers have been tested at Ann Arbour, Michigan, and have better C-values than the round-bottom 1952 New England trawler which appears in data sheets. The first edition of the FAO Fishing Boat Catalogue will be issued in 1955. It will be prepared in loose-leaf form and it will be easy to add sheets following later. FAO would appreciate it if naval architects, ship yards, ship experimental tanks etc. would contribute and submit tank results of fishing boats or related ship forms which could be worked up in data sheets. There are a number of such tests which have not been published and which would be of considerable interest to other workers.

The question of the round versus V-bottom design is not a simple one but if a round-bottom design is changed to a V-bottom, and the same centre of buoyancy, entrance angle and coefficients are retained, then the ship should have about 5 per cent. more resistance. But the data available are not conclusive. What usually happens, when a V-bottom design is tested, is that the coefficients are improved so that the hull shows better results.

The paper shows that the V-bottom shrimp trawlers tested are better than the New England trawlers. The tests have all been made in the same tank and with the same length of the models so that they are directly comparable. The same diagrams also show that the Newfoundland schooner design is better than the V-bottom design.

Regarding Mr. Gueroult's advice of not to think too much about prismatic coefficients, this may be true so far as it concerns naval architects who know all about it, but there are many who do not realize what it is. For that reason, this coefficient must be emphasized. It is true that the value of .575 is not the right coefficient for every occasion but it is useful as a guide for a normal boat with normal proportions if the design is to yield maximum speed combined with maximum seaworthiness.

As an example of how dangerous it is to take too easy a view on the prismatic coefficient, Mr. Traung referred to Mr. Gueroult's own fig. 215 indicating the economic speed of trawlers. The coefficient k given might well be correct for certain trawlers with a certain prismatic coefficient but it is inevitable that trawlers with a bigger coefficient will have a lower economic speed. Mr. Gueroult's formula says the contrary and designers should be very careful not to use his diagram for other boats than those having the same prismatic as the French standard trawlers.

While accepting the general validity of the need for a low prismatic coefficient and a moderate angle of entrance, Mr. Astrup says that it is possible to depart considerably from the optimum data given. This is in direct contradiction to the impressions one gains when reading Mr. Astrup's own report (1951) on the Norwegian tank tests and it would have been interesting to see proofs for this statement. If Mr. Astrup meant that one carefully designed boat with a prismatic of .600 can be as good as a badly designed one with .575, Mr. Traung was in full agreement—but why should not the careful designer take advantage of the possibilities to design a boat with lowest possible resistance? It seems sometimes to be forgotten that the largest influence of the prismatic coefficient actually is not at the top speeds of fishing boats, but at the service speeds under actual working conditions and they are often 2 to 3 knots lower.

Mr. Astrup's economic considerations are quite right for many types of fisheries and especially those in Norway. Conditions in the Far East are, however, not the same. There a large crew is used to living on open decks as they cannot stand the heat in cramped quarters. The Pakistan boats are sailing boats and a good length is essential to get a low

Froude's number. Mr. Astrup is right in saying that the Pakistan boats have a considerably different length-displacement ratio from normal fishing boats. Comparisons made between Pakistan and European boats on the basis of residual resistance per tons displacement shows a greater difference in favour of the Pakistan boats than the given comparison in fig. 61.

Mr. Traung did not agree with Mr. Astrup's remarks that less stability and seaworthiness should be the penalty for low resistance. The boats Mr. Astrup had developed for the Norwegian Board of Fisheries, which show considerably less resistance than the earlier boats, do not (so far as Mr. Traung understood) have less stability. The previous statement by Mr. Thiel clearly shows how the GM can be varied for lines with the same coefficients and approximately the same resistance. It is really up to the designer to ensure the necessary amount of stability and the skilled one can easily change his lines so that he can combine low resistance with sufficient stability. It is true, as Mr. Astrup states, that tank tests of fishing boats with low resistance are beginning to be published in different countries. It should however be remembered that boats being built to-day seldom are of that good design proposed in those tests. Very few boats have as an example been built after the fairly good lines represented by the models No. 206C and E in Mr. Traung's paper. Mr. Astrup's statement easily gives the reader to think that everything is all very good, but it is really not and operators of fishing boats in different places of the world should be aware of the fact that considerable improvements are possible.

Answering Dr. Kloess' question as to why the tests were not compared with Taylor and Takagi, Mr. Traung said that Taylor's standard series, as Dr. Kloess admitted, did not cover the length-displacement ratios of fishing boats. In Takagi's series, unfortunately, the basic model had a counterstern and the series therefore give a resistance which is higher than that obtainable with a transom or a cruiser stern. When all data sheets have been worked out it might be possible to find some representative tests to use as a kind of eye-mark.

Mr. Traung agreed with Dr. Kloess that the data published in his paper were not sufficient. More sheets will be published. On the other hand, it was difficult to understand how one could design some kind of Taylor series to cover all the different types of fishing boats. When publishing results of tank tests, Doubting Thomases, without knowledge of the tank-test technique, always asked: " What about the behaviour in bad weather, and what about the stability? " It is surprising to learn that Dr. Kloess, with his large experience of tank-testing, is not better informed on recent papers which definitely show that boats which were superior during calm-water tests were mostly superior in heavy weather too, and that boats having better resistance qualities often had better efficiency of propulsion. It is exactly this situation which is shown in Dr. Kloess's fig. 455, which demonstrates a model with 20 per cent. improvement during calm-water tests. This improvement was considerably greater according to Dr. Kloess, when the full size ship sailed in bad weather.

Dr. Kloess states that the German Classification Society has approved the main dimensions of his trawlers. It was news to Mr. Traung that it was difficult to get approval of dimensions by any classification society. If the statement should mean that the Classification Society also had approved the hull shape of the fishing boats, it was no proof of the boats being good. Classification societies never interested themselves in hull shapes, otherwise there would not be such a number of uneconomical boats sailing around. In fig. 441 and 445 Dr. Kloess gives curves of effective (tow-rope) horsepower of some Maierform designs. If the 53 ft. (16 m.) trawler is compared non-dimensionally with model 206 E on fig. 395 one will find that it has considerably higher resistance. The same is true of the larger trawlers shown in fig. 445 if compared with fig. 417.

Dr. Kloess' reports of tests with propulsion bulbs are striking. The question is, however, whether those tests have been done under absolutely equal conditions. Rumours are widespread that both weather and roughness of the ship's bottom have been different during some of the tests, otherwise it would be hard to explain the extraordinarily large differences.

Answering Mr. Stephens, Mr. Traung said laminar flow normally is present at lower speeds, high block coefficients, and at large angles of entrance. The published tests were done at relatively high speeds and the results show that a sharper angle of entrance gives less resistance. Although with laminar flow there should be less resistance with boats with a larger angle of entrance, results of tests indicate the contrary, and that is an indication of absence of laminar flow. And the tank mentioned in their reports that they had checked this problem.

Mr. Zwolsman explained in his paper that when the fishermen are asked which they prefer, a good sea boat or a fast boat, they choose the former. Everyone would probably choose likewise if the answer to the question was simply either/or. Probably Mr. Zwolsman believes that a fine forebody will produce a bad sea boat and when he advises fishermen so, they will naturally believe him, as a technician. But tests carried out in waves by men like Todd and Allan show the contrary, and all boats with fine forebodies, in Mr. Traung's experience, were excellent sea boats. Again and again it must be stressed that a tank-test in itself does not mean that the hull is good. Normally the tanks simply test a model according to the drawing submitted and give the results in hard figures. They do not improve it, neither do they pass judgment, but still too many people believe that because a special design has been tank-tested it is guaranteed that the form is good.

In redesigning a boat the usual arrangements must be considered. For instance, if the fish are stowed in the bow of the boat and the catch is big, as in the case of Scottish ring netters, then it is not possible to fine down the bow. In that kind of craft it is necessary to have a slower hull because of the working conditions. If it is not possible to alter the general arrangements of such a boat and if somebody fines down the forebody, the result will be a bad boat as it will trim too much by the head loaded. That is not the fault of tank-testing, but the wrong application of its lessons.

Dr. R. Brard (France): V-bottom design can be used in fishing boats because the beam-draft ratio is small as a fishing boat must be very manoeuvrable and sea-kindly. There is also the fact that a certain volume has to be carried fore and aft to ensure enough buoyancy to avoid the boat plunging into a wave or being flooded by it and, finally, the fining of the run is evidently important because of the flow of water to the propeller. Like all other boats, the fishing boat is a compromise because of the different and contradictory conditions it must satisfy. It has a higher resistance than other boats of the same coefficient and it must be designed so that the propeller has sufficient water to work it.

Much work has been done in testing fishing boat models

and comparing results with trials at sea and it seems that all the problems concerned with the boats can be studied through such tests. Even so, there is some doubt as to the validity of the conclusions reached in tank-testing, and for this reason it seems that boat owners wish to compare such data with the result of actual trials at sea.

Fishing boats can be regarded, with few exceptions, in the same way as bigger ships are regarded so far as tank-testing is concerned. But it is easier in the case of fishing boats to carry the tests from the tank to the actual boat at sea. The aim of a tank-test is not to study the model but to study the real ship or the real boat, and it is right that the tests should be extended to trials at sea.

STABILITY

Professor G. Schnadel (Germany): Under the new rules for the safety of fishing vessels in the Netherlands, stability curves must be calculated for ships of more than 200 tons register. Quite rightly, Mr. de Wit criticizes the failure to specify a minimum amount of stability. For 20 years curves of stability of fishing vessels have been collected in Germany and it has been found that statical curves give sufficient information. When the arm is 8 in. (20 cm.) at the maximum, the critical angle should be at least 30°. Dynamical lever does not seem to be so important. Being the integral curve of the statical stability curve, it is only a summary of the static stability. The metacentric height for special ships is found when the statical curve is determined. The freeboard determines the statical stability and it is possible to have ships with good seakindliness with a small GM if the freeboard is sufficient and very bad stability with a good GM if the freeboard is too low.

The difficulty is that freeboard regulations are not applied to fishing vessels. Nevertheless, the Netherlands shipping authorities forbid overloading which often endangers fishing vessels. This can be seen when ships load up for long voyages not anticipated when they were designed. This is quite a common danger as there is no agreement on a freeboard regulation for fishing boats. The skipper may over-estimate stability if he relies on his instinct. Unpleasant and quick motions are thought to indicate good stability. In some cases, the GM may be very great but the arms of stability insufficient. The skipper should have better information for judging stability. Mr. Möckel determines GM by the rolling period. Skippers should be told the admissible rolling period for every freeboard and every important loading condition. Such rolling period must be determined for each individual case. Insufficient seakindliness may be the consequence of an insufficient freeboard requiring high GM. Ships must primarily be safe from capsizing; making them seakindly is a secondary consideration.

Most losses of trawlers are caused by lowering the freeboard. German boats originally designed for fishing near the coast have had to go far to the north and have been overloaded. The skippers were satisfied by habit with the stability of the boats but when they met the big seas in the north the boats capsized and were lost. Möckel found that the radius of gyration of the trawlers was in the neighbourhood of 0.4 and gives the radius of 0.385 to 0.40. An American tuna clipper expert, Dickie, has given 0.36 to 0.39. This is a very small difference considering that the types are wholly different. The president of the Schiffbau Technische Gesellschaft (German Institution of Naval Architects) has formed a committee of designers, skippers and teachers of navigation to reach agreement on the formulae to be used, the admissible rolling periods and, particularly, to provide information that will enable fishermen to judge stability.

Small boats should have a greater amount of stability. It is a known fact that the chief dangers are ice, flooding, and wind pressure. In each case the small boat is in greater danger than the big boat, so it is necessary to give smaller vessels more stability. For instance, German fish cutters and luggers have greater stability than large trawlers.

Professor Jaeger demands in his paper higher stability for the steam trawler because it has a greater change of load. But the fact must be considered that motor trawlers have usually greater power and more open deck so that seas breaking over them have a greater influence on stability. If the static stability curve was used instead of only GM to judge stability, it should not be necessary to make a difference between steam and motor trawlers.

Mr. Gueroult suggests using permanent ballast in fishing vessels to minimize the big righting moment caused by the breadth of the ship. Many experienced fishermen prefer this and it is certainly a useful proposal for ships in which stability is not sufficient.

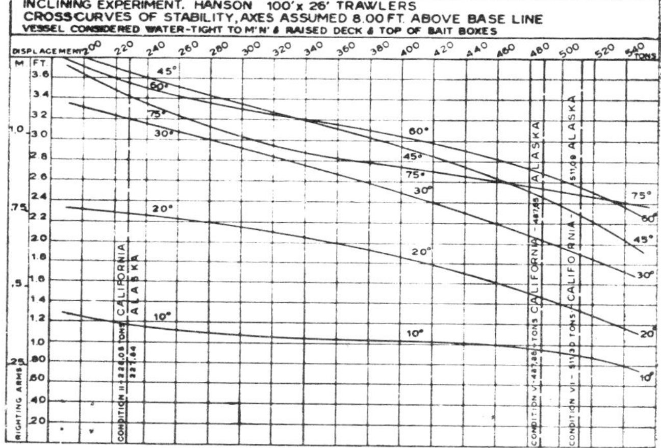

Fig. 445

Mr. H. C. Hanson (U.S.A.): Curves of statical stability for his 100 ft. (30.5 m.) combination boat are shown in fig. 445 and 446. They were calculated by conventional methods and the results are probably optimistic, but the fact that the ships have operated safely in a wide range of waters indicates that the curves are satisfactory. Nevertheless, it is folly to rely too much on statical stability curves as criteria for calculating the range of seaworthiness of tuna vessels, especially the raised deck clippers. There are four erroneous assumptions based on statical stability curves: (1) initial trim in some fishing conditions can be much more severe than in conditions for cruising trim for which the calculation is likely to be made; (2) increased trim aft while heeling is usually substantial and cannot be taken into account by the usual method of calculation; (3) free surface effect is generally ignored; (4) the combined effect of these factors may be so great as to make the curves unreliable at large angles of heel. This stresses the fact that theoretical calculations are not sufficient. The naval architect must have an inherent sense of stability and sound construction in designing a ship.

The arms of stability of the U.S.A. Pacific Coast combination fishing vessels are much higher than normally reached in European vessels which, because of their high freeboard, can have less arms.

Mr. James C. Aguinaldo (U.S.A.): From the standpoint of safety there are limitless advantages to be derived from statical stability curves. They provide data which lead to an extensive evaluation of overall stability when considering service heel and trim of a vessel.

For a vessel floating upright (fig. 447A) the centre of buoyancy B and the centre of gravity G are on the centreline. As weights are moved across the vessel (giving the effect of the crew crowding to one side as in fishing heel or trim) the centre of gravity shifts along a line parallel to the direction

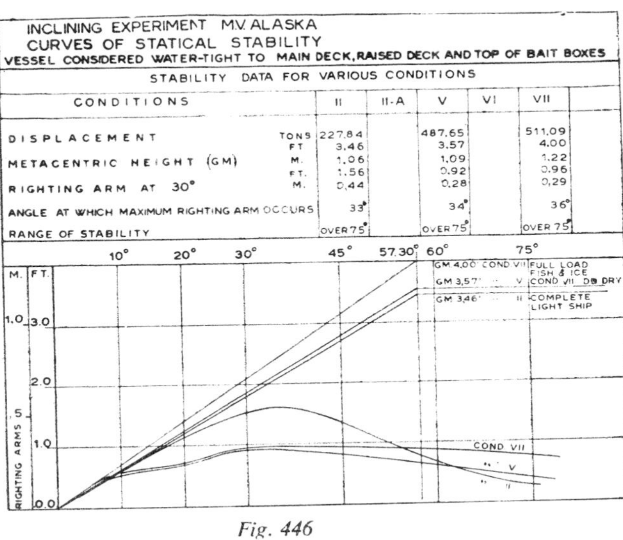

INCLINING EXPERIMENT M.V. ALASKA
CURVES OF STATICAL STABILITY
VESSEL CONSIDERED WATER-TIGHT TO MAIN DECK, RAISED DECK AND TOP OF BAIT BOXES

STABILITY DATA FOR VARIOUS CONDITIONS

CONDITIONS		II	II-A	V	VI	VII
DISPLACEMENT	TONS	227.84		487.65		511.09
METACENTRIC HEIGHT (GM)	FT	3.46		3.57		4.00
	M.	1.06		1.09		1.22
RIGHTING ARM AT 30°	FT.	1.56		0.92		0.96
	M.	0.44		0.28		0.29
ANGLE AT WHICH MAXIMUM RIGHTING ARM OCCURS		33°		34°		36°
RANGE OF STABILITY		OVER 75°		OVER 75°		OVER 75°

Fig. 446

of the weights. G moves to a new position from G to G_1, where

$GG_1 = w \times d/W$ w, the sigma of weights moved
GG_1 = the shift of weight moved
d = distance of weight moved
W = total displacement

After the weights have been moved athwartships G has moved to G_1 and the vessel heels over, changing the geometric shape of the underwater hull. This means that the position of the centre of buoyancy must shift until B coincides with G to restore equilibrium (447B). At this point, where buoyancy and weight are in the same straight line, a permanent angle of list is developed as long as the crew remain on one side of the vessel.

If then the fishermen cast their nets the centre of gravity will move further out from the centreline and a change takes place in the righting arm in all angles of heel (fig. 447C). At such an angle the remaining or " residual " arm, tending to restore the vessel to even keel, is G_1Z_2. If G remained at the centreline, the righting arm would be GZ_2 which is larger than G_1Z_2 by $GG_1 \cos \theta$. The latter is the loss of righting arm, commonly known as " upsetting " arm, of the boat. The effects of the horizontal shifting in the centre of gravity may be determined by plotting the value of upsetting arms at various angles of heel on the same diagram as the statical stability curves are plotted, and the effect on the overall stability is shown in fig. 447D. The area OABC shows what has been lost in the stability curve because of the movement of men and equipment to one side of the vessel, and the area ABD shows what remains of the curve. By evaluating the properties of the residual portion of the curve the stability of the vessel in any circumstances she may encounter in service may be exactly determined.

It is seen that statical stability curves directly contribute to the final estimate of service stability. They serve as a basis and/or criterion for judging the overall range and seaworthiness of any vessel regardless of type or service. The curves, naturally, have their limitations, but they should be established before any further study of stability is made. It is the duty of naval architects to instruct engineers and shipowners in the judicious use of the curves.

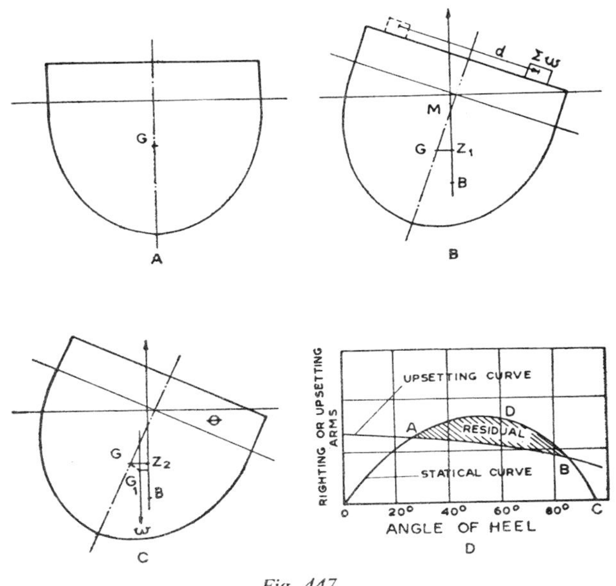

Fig. 447

Mr. L. E. Geary (U.S.A.): Mr. Hanson states that curves or righting arms (fig. 446) made in the conventional manner should not be heavily relied upon to represent all conditions of loading that may occur. The conventional method to-day is to make the cross curves (fig. 445) from a double body rotated about an assumed centre of gravity at various displacements and heeling angles, and then to correct the righting arms for fig. 446 when the true centres of gravity for various conditions have been computed.

In the United States Navy the practice is to incline vessels in light condition. The vessels usually have double bottoms and have to carry a very heavy weight of armaments on the upper decks. Some of the early aircraft carriers had negative metacentric height in light condition and had to carry liquid in double bottom tanks before they could be brought to Condition I for inclining. From this test it is usually easy to calculate the centre of gravity of Condition II, which is almost always the worst condition from the standpoint of initial stability and range of stability. From the known centre of gravity other conditions can be calculated and the curves of righting arms for various conditions of load can be made.

The raised deck tuna clipper is in her worst condition when ready for her first catch. She has a full load of bait water in all available compartments, nearly a full load of fuel and stores, and almost no freeboard. When inclined close to this condition the centre of gravity is accurately determined and from the cross curves the true stability curve may be calculated. Other less dangerous conditions may also be calculated although not so accurately.

A Navy ship is usually considered lost when damage causes unsymmetrical flooding or some other mishap overcomes her dynamic stability to about 50°. Because little attention is paid to making her watertight at the deck levels, a tuna clipper is usually lost if she takes a sustained list of 40°. To be sure of the direction of the curve of righting arms smaller angles at which her dynamic stability may have been overcome momentarily by the sea, it is desirable to make cross curves to 60°. The righting arm at any angle is the measure of the ability of a vessel to recover after heeling. The naval architect, shipowner or skipper with years of experience can be misled by the looks, feel or rolling period of a vessel, or even by the GM determined by tests.

INCLINING EXPERIMENT M.V. VIKING
CURVES OF STATICAL STABILITY
VESSEL CONSIDERED WATER-TIGHT TO MAIN DECK, RAISED DECK AND TOP OF BAIT BOXES

STABILITY DATA FOR VARIOUS CONDITIONS

CONDITIONS		II	II-A	IV	VI	VII
DISPLACEMENT	TONS	233.6		416,92		479,62
METACENTRIC HEIGHT (GM)	FT.	3.78		2,59		2,49
	M.	1.15		0,79		0,76
MAXIMUM RIGHTING ARM	FT.	1.34		0,93		0,85
	M.	0,41		0,28		0,26
ANGLE AT WHICH MAX. RIGHTING ARM OCCURS		29°		37°		43°
RANGE OF STABILITY		OVER 75°		OVER 75°		OVER 75°

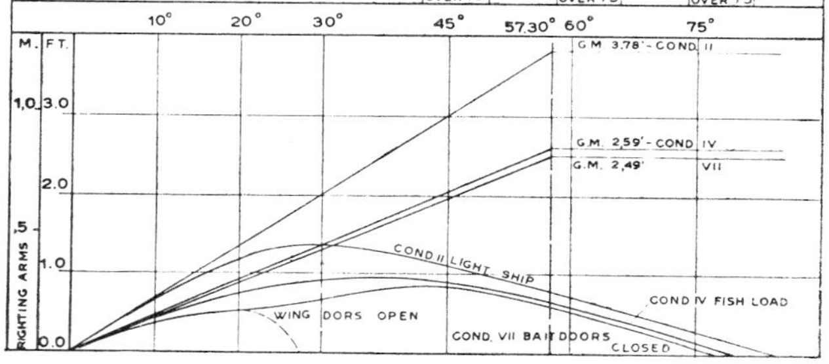

Fig. 448

INCLINING EXPERIMENT M.V. SANTA MARGARITA
CURVES OF STATICAL STABILITY
VESSEL CONSIDERED WATER TIGHT TO MAIN DECK AND TOP OF BAIT BOXES

STABILITY DATA FOR VARIOUS CONDITIONS

CONDITIONS		II	II A	IV	VI	VII	V
DISPLACEMENT	TONS	295,1		481,5		532,5	501.3
METACENTRIC HEIGHT GM	FT.	3,17		2,79		2.65	2,52
	M.	0,96		0,85		0,81	0,77
MAXIMUM RIGHTING ARM	FT.	1,05		0,50		0.30	0,43
	M.	0,32		0,15		0,09	0,13
ANGLE AT WHICH MAX. RIGHTING ARM OCCURS		28°		17°		11°	15°
RANGE OF STABILITY		63°		52°		31°	46°

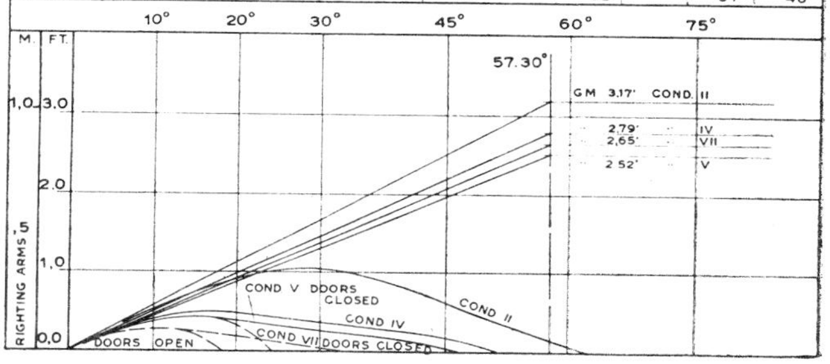

Fig. 449

A vessel's stability is determined by the righting moment and righting arms. They are calculated by using an integrator over a double body plan and not by using metacentric height, breadth of vessel, rolling period, freeboard and similar factors taken together to provide an empirical formula. It has been said that stability curves are helpful but too expensive, but hydrostatic curves must be made in any case. Anyone who can use the integrator can make the necessary integrations up to 60° and compute and plot the GZ arms of the cross curves. Even a novice, with some assistance, can do this in 3 days or 24 working hours. The expense involved in this calculation is relatively small, especially in relation to the cost of a tuna clipper which may vary from $350,000 to $500,000 (£125,000 to £180,000).

The story of some tuna clippers is told by the following examples: *Fig. 458:* From the standpoint of initial stability and range *Viking* is a better-than-average post-war tuna clipper. As she has righting arms of more than 30°, she can make a quick recovery from a wide heeling angle. These and succeeding curves show a typical dip to the righting arms at about 20° heel in full load condition. At this angle the low freeboard of the main deck is under water. The water would also flow over the 18 in. (460 mm.) coaming of the door to the upper engine-room. Vessels of this design become increasingly vulnerable to flooding through this door as the heel exceeds 20°, but they rarely capsize because the rush of water across the engine-room tends to bring them back on even keel before they founder.

Conditions II, IV and VII give figures respectively for light, homeward bound with full load of dry fish, and "worst condition", that is with a full bait load and ready to make the first catch.

Fig. 449: The *Santa Margarita* is one of the older live bait and ice boats. She was modernized in 1945 and now has brine freezing equipment, and bait and fuel wells below deck. She is a borderline vessel, having maximum righting arms up to 20° but declining rapidly after that figure. With a full bait load she has only 30° range, and the instructions to the master require her to be operated with 2 wells dry at Condition V. This is 31 tons lighter than at Condition VII and increases her maximum arm by more than 40 per cent. and gives an extra 15° range. By taking 31 tons of bottom weight off, her metacentric height has decreased from 2.65 to 2.52 ft. (0.81 to 0.77 m.).

Fig. 450: The *Rajo* is another of the older tuna clippers. One day when she was loaded with bait and ready to go fishing she was ordered to return to port to load with fuel. She went into San Diego fuel dock, removed the bait from the starboard well and started to pump the well dry to take on board diesel oil. She took a list against the dock and was flooded through the port side door and sank. She was later raised and repaired and put to an inclining test, the results of which are shown in fig. 451. As she had maximum arms before 20° the largest of three bait boxes on deck was sealed up so that it could not be flooded. After a year's fishing it was decided that all three bait boxes were needed, so they were reduced 11 in. (280 mm.) in height and the vessel was reinclined. It should be noted that this improved considerably her stability (fig. 451) and she had no further stability trouble but was subsequently lost by fire.

Fig. 452: The *Santa Barbara* is one of the well-known Courageous class tuna clippers of which five were built in 1945-46. The first plans for the class revealed a low stability and efforts were made to bring down the centre of gravity by eliminating a heavy foremast, steel fiddley casing, and other devices, and by using lighter stacks. The curves of the righting arms show an increasing righting moment between 30 and 60° for the load condition. But she might be in danger between 10 and 25° if the wells and boxes were to be flooded simultaneously and unsymetrically and she had too little righting arm up to the time her raised deck and

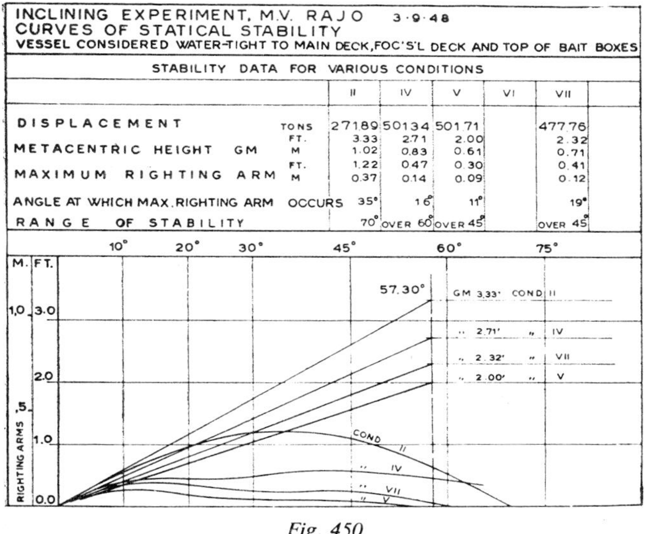

INCLINING EXPERIMENT, M.V. RAJO 3·9·48
CURVES OF STATICAL STABILITY
VESSEL CONSIDERED WATER-TIGHT TO MAIN DECK, FOC'S'L DECK AND TOP OF BAIT BOXES

STABILITY DATA FOR VARIOUS CONDITIONS

		II	IV	V	VI	VII
DISPLACEMENT	TONS	27189	50134	501.71		477.76
METACENTRIC HEIGHT GM	FT.	3.33	2.71	2.00		2.32
	M	1.02	0.83	0.61		0.71
MAXIMUM RIGHTING ARM	FT.	1.22	0.47	0.30		0.41
	M	0.37	0.14	0.09		0.12
ANGLE AT WHICH MAX. RIGHTING ARM OCCURS		35°	16°	11°		19°
RANGE OF STABILITY		70°	OVER 60°	OVER 45°		OVER 45°

Fig. 450

bait boxes lay in the water. This has happened to two of this class of vessel, which has induced the fishermen to be careful in handling them. The result is that the five boats have operated successfully despite their poor reputation for safety, and several of them have weathered tropical hurricanes off Central America. The main vessel of the class, *Courageous*

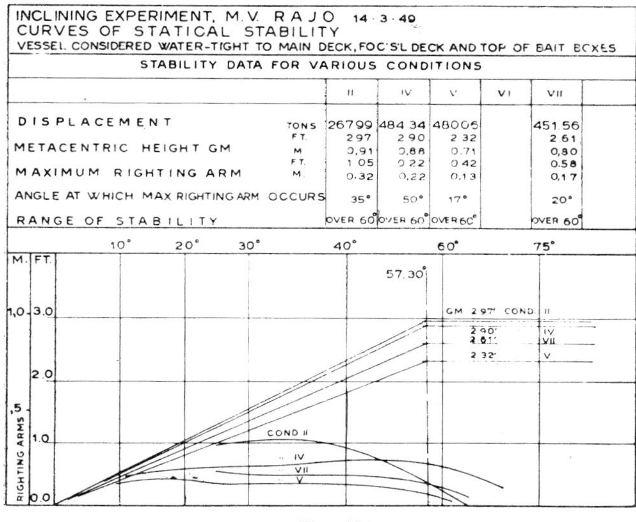

INCLINING EXPERIMENT, M.V. RAJO 14·3·49
CURVES OF STATICAL STABILITY
VESSEL CONSIDERED WATER-TIGHT TO MAIN DECK, FOC'S'L DECK AND TOP OF BAIT BOXES

STABILITY DATA FOR VARIOUS CONDITIONS

		II	IV	V	VI	VII
DISPLACEMENT	TONS	26799	484 34	48005		451.56
METACENTRIC HEIGHT GM	FT.	2.97	2.90	2.32		2.61
	M	0,91	0,88	0.71		0,80
MAXIMUM RIGHTING ARM	FT.	1.05	0.22	0.42		0.58
	M	0.32	0,22	0.13		0,17
ANGLE AT WHICH MAX RIGHTING ARM OCCURS		35°	50°	17°		20°
RANGE OF STABILITY		OVER 60°	OVER 60°	OVER 60°		OVER 60°

Fig. 451

is said to have been the first post-war clipper to pay for herself 100 per cent. and that was some years ago. Curve V on fig. 452 was plotted to determine if the Courageous class vessel would be safe without permanent ballast or seawater in the deck bait tanks when the fuel is burned out. The curve shows that she is.

But in 1953 the *Courageous* was at the pier in San Diego with a load of tuna valued at $100,000 (£35,700). The bait pumps were pumping water through the wells and boxes for thawing the frozen tuna prior to unloading, when somebody thought of saving time by putting oil into the empty tanks before reducing any top weight. The free surface created as the oil was pumped in caused her to list to port against the dock. She flooded through an open door and sank in 27 ft. (8.2 m.) of water. She lost her $100,000 catch but was raised and repaired at a cost of more than $100,000 and is now operating again.

Despite the disastrous losses sustained by the tuna fleet, the question of the stability of the ships is not the big problem. The usual tuna clipper has sufficient stability for safe operation if she is handled properly. The chief trouble is that the clippers are under no supervision or restriction from the Marine Inspection Services and this fact is chiefly responsible for the enormous losses arising from fires and strandings. It is up to the owners to take the necessary action to prevent such losses.

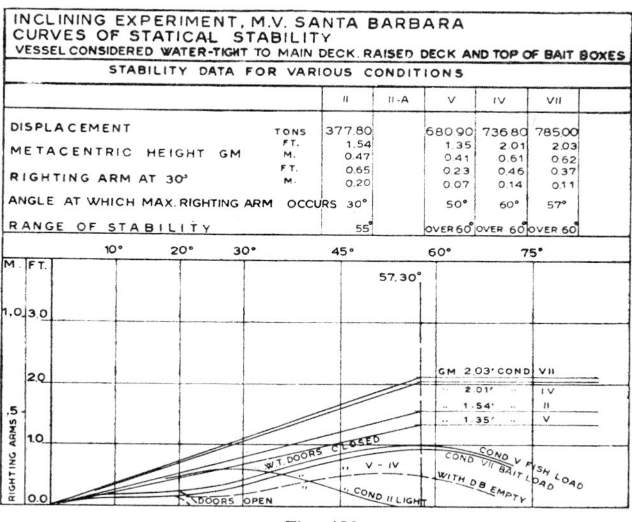

INCLINING EXPERIMENT, M.V. SANTA BARBARA
CURVES OF STATICAL STABILITY
VESSEL CONSIDERED WATER-TIGHT TO MAIN DECK. RAISED DECK AND TOP OF BAIT BOXES

STABILITY DATA FOR VARIOUS CONDITIONS

		II	II-A	V	IV	VII
DISPLACEMENT	TONS	377.80		680.90	736.80	785.00
METACENTRIC HEIGHT GM	FT.	1.54		1.35	2.01	2.03
	M.	0.47		0.41	0.61	0.62
RIGHTING ARM AT 30°	FT.	0.65		0.23	0.46	0.37
	M.	0.20		0.07	0.14	0.11
ANGLE AT WHICH MAX. RIGHTING ARM OCCURS		30°		50°	60°	57°
RANGE OF STABILITY		55°		OVER 60°	OVER 60°	OVER 60°

Fig. 452

The points raised by Mr. Hanson about stability curves are answered concisely by Charles L. Wright, Jr., naval architect, U.S. Bureau of Ships, in a paper entitled " Stability of Ships after Extensive Flooding ", read at the U.S. Naval Institute in 1948:

" Accordingly trim was considered to be a factor that could be neglected in all studies. This course of action was consistent with practically all studies ever made of stability before damage as few cross curves of stability had been based on trimmed waterlines. It is becoming quite generally agreed that righting arm values are essential in presenting a complete indication of the stability characteristics of a vessel. Factors such as change of trim (when heeled) must be neglected in studies of stability if the calculations are not to become too involved to be practicable. It has become evident that GM alone is not sufficient, that empirical and approximate methods are not dependable, and that righting arms determined by direct calculation are the only satisfactory indication of stability characteristics before or after extensive flooding; made in accordance with the methods described they are direct and not too involved to be practical ".

A steel tuna clipper will spill her first cupful at 27 to 30° heel and a wooden one at 35°. Most of the losses from instability occur to boats that take an initial list of 10° and at about

15° admit water to the low side double bottoms of fuel tanks long before spillage occurs. It would probably be all right if a listed boat would spill water sooner and relieve herself of load, but these accidents can happen to boats having no free water to spill.

Mr. Gordon C. Snyder (U.S.A.): It is recognized that statical curves are the standard means of determining the probable behaviour of a vessel at large angles of heel. In a practical sense, it is the best means available. However, it is also recognized that the method involves assumptions and approximations which are unavoidable but which must be evaluated for their unfavourable effects. Mr. Hanson is trying to emphasize that the effect of these assumptions is enlarged in a typical clipper, hence there is an increase of error in calculations applied to the clipper, as compared with the accuracy obtained in applying them to more conventional vessels.

It is customary in preparing statical curves to assume trim does not change when heeling. A glance at pictures of tuna clippers will show that these raised forecastle ships are very bulky forward in watertight enclosure but have only the bait boxes above the main deck aft. It is obvious that with severe heeling the stern will trim down rapidly, which is in contrast with the action of more conventional vessels which carry superstructure in a more central position.

The tuna clipper is essentially a tanker, with a large proportion of the tanks having insecure covers or none at all for most of the time when they are full of bait water. It is usually impracticable to calculate the very serious effect of spillage of bait water on the action of the vessel but it can obviously be great enough to affect seriously the accuracy of the statical curves.

Advantage can be taken of the large pumping plant on these vessels to incline in loaded conditions which closely approximate service load conditions. It is entirely practical to incline six or eight times a day, in six or eight progressively loaded conditions, from nearly light to near capacity, by gradually filling the wells with water. One set up of inclining weights and pendulums serves the whole operation. This procedure saves a great deal of calculation, eliminates a multitude of minor corrections, and gives a high degree of accuracy as far as initial metacentric height is concerned. It has the additional advantage of letting the crew actually see the water over the deck of a top-heavy boat. This generally makes a beneficial impression and encourages caution in handling.

Regardless of whether the results of inclining are used to produce statical curves or are interpreted by any other means, the use of multiple inclining experiments should be encouraged.

Mr. George C. Nickum (U.S.A.): There has been considerable discussion between those who are proponents of the use of the curve of the righting arm as the proper method of determining the stability of tuna vessels and those who feel that the curve is not the whole answer.

Even though accurately drawn, the curve of righting arms cannot be used as the sole guide in the investigation of the stability of tuna clippers. Mr. Snyder has mentioned that, in a vessel with so little freeboard at the stern, small angles of inclination immerse the deck edges which, in turn, changes the trim condition, thus seriously affecting the accuracy of the righting arm calculations. Mr. Geary also mentions the problem of spilling water from the hatches of the bait tanks as having considerable effect on the stability of the vessel, which is not taken in consideration in the curves of righting arms.

Mr. Nickum said that some years ago he took a typical tuna clipper and went through a laborious calculation to determine what the effect of trim would be on the righting arm figures. He took 5° and 10° angles of inclination and by trial and error methods determined the actual trim at each and then rechecked the righting arm at the correct trim. The vessel had a righting arm of about 2 ft. (0.61 m.) before trim. After trim the righting arm was reduced to 1 ft. 6 in. (0.46 m.), a 25 per cent. reduction due to trim alone. Such a percentage of error in small angles of inclination is enough to make a designer hesitate in stating that a vessel is safe when his decision is based solely on the curve of righting arms.

The most serious problem of the clipper is overloading. Mr. Geary mentions two vessels which had been inclined and in which the stability had been analysed, using the righting lever methods. Both these vessels sank. There is no doubt that the righting curves prepared were correct for the draft for which they were drawn but there is also no doubt that the vessels were loaded beyond that draft and because of that the deck edges were brought closer to, or under, the water. The residual stability was not adequate to withstand the poor operating judgment shown by the fishermen.

Two steps should be taken by naval architects to increase their knoweldge of stability. The first is to set up a standard of maximum operating conditions. There is general agreement that this is the first catch condition or the condition when the vessel reaches the fishing banks with a full load of bait and with the first catch on deck and with the brine wells filled with chilled water. Universal acceptance of this condition plus careful work on the part of the man in charge of stability investigation to ensure that the draft and freeboard under this condition are known, will go a long way towards eliminating the capsizing and foundering of tuna vessels.

The second step seems to be a matter of making multiple experiments to determine a constant factor for correction of curves of righting arms to allow for trim and for spillage of water from the bait tanks. Such model work would enable the designer to put his faith in the curves of righting arms and to make intelligent decisions based on the data supplied.

In these days of rapid advances in technology it is apt to be overlooked that good ship design still remains in large part an art. The designer still needs an experienced eye and the gift of using calculations with judgment, nowhere more so than in the field of small fishing boat construction. A stress calculation showing a particular fitting to be adequate in strength is of no help to a fisherman if that fitting tears loose at sea. Mr. Hanson's long record in design of successful fishing vessels speaks for his ability to decide where science leaves off and experience and judgment begin.

Mr. H. C. Hanson (U.S.A.): The comments on the use of the righting arm curves were intended to create discussion about the very important subject of stability, especially stability of tuna clippers because so many of these very expensive vessels have been lost in the past few years. Most of these losses are not the fault of the naval architect, but much of the responsibility is being put on him. That is why it is desirable to create constructive discussion.

There should be no unstable vessels to-day because to obtain a stable vessel is as simple as adding two and two. Leaders in the fishing industry, such as naval architects, builders, owners and, probably most important of all, underwriters need only to collaborate and most of the losses would disappear in a very short time. If the industry cannot collaborate and reach agreement on the voluntary basis, the inevitable consequence will be Government regulations, inspection, and more laws.

An example of a loss in the U.S. tuna fleet is that of an owner of a new 80 ft. (24.4 m.) craft who requested stability approval. A survey of the craft disclosed it was very unstable, and the owner was instructed to remove a large part of the bait tank on deck, place freeing ports in the aft bulwarks, and take on board 6 tons of permanent ballast. He and the underwriters were told that if this was not done he would lose his boat. He refused to do the work, and within a week the vessel was lost along with a member of the crew. In this case the underwriters were entirely to blame for not enforcing the advice as they were fully informed in writing of the facts, and they had to pay about $100,000 (£36,000). This is a comparatively small sum in relation to that involved in the loss of some 75 vessels in a three-year period. In some cases underwriters have had to pay as much as $600,000 (£215,000).

A combination fishing vessel correctly designed would need no ballast, but to ensure stability under the worst possible conditions and taking into account the human element, fixed ballast should be placed in sufficient quantities to offset the turntable and weights of the nets. Each owner has to adapt himself to his own vessel, and where one person is satisfied with a vessel, another may require more or fewer weights. This may be called " tuning the vessel and owner to each other ". It is particularly necessary where a boat goes to sea for extended periods.

It is important that the inexperienced designer understands what is being brought out here, because great numbers of these vessels were built and many of them capsized and were lost with members of their crews.

The suggestion that multiple inclinations should be made, merits consideration. Possibly, they should be carried to the next capsizing point. One must bear in mind that all stability tests are made in placid waters, turbulence not being considered in such tests, yet at the capsizing point in the open sea this is a very important factor.

Professor H. E. Jaeger (Netherlands): For big ships, Mr. Nickum's method of determining GM would give values that would be too great. Professor Jaeger perferred his own looser methods. Mr. de Wit seems to uphold Rahola's conclusions without critical observation. Fishing vessels work in special conditions and it is not possible to come up always to the Rahola standard.

Mr. J. G. de Wit (Netherlands): The freeboard of fishing-vessels presents a difficult problem. As Mr. Fernandez has stated in his paper, and Professor Schnadel has confirmed, there should be freeboard regulations for fishing vessels. Much progress would be made if the subject was studied thoroughly and Governments were advised by FAO Fisheries Division of the conclusions reached. Although the stability data for motor trawler " A " in fig. 432 meet Rahola's standard fairly well, this trawler, with empty double bottom, shows a lack of stability, according to Rahola. The trawler foundered in the North Sea, during a gale, while sailing with an empty double bottom. The crew was lost and so it is very difficult to determine the cause of this disaster with certainty. All indications, however, point to capsizing. This case confirms that a fishing vessel must also be judged as dangerous if it does not come up to Rahola's standard of dynamical stability.

The Netherlands authorities have studied the subject of the stability standards and have chosen Rahola's standard as the best one for the time being as it is mainly based on smaller ships, fishing vessels included.

Mr. S. A. Hodges (U.K.): Statistics show that British trawlers have been lost mainly by stranding and other navigational risks, and few have been lost solely because of insufficient stability and freeboard.

Adequate stability in any ship is of the highest importance, and changes in the design of trawlers may call for a very careful analysis of stability. For example, a comparison between diesel and coal burning steamer trawlers shows that a diesel trawler may still have a large amount of fuel at the fishing grounds and when a large catch is taken on board there may be little freeboard left and the curve of statical stability may be very poor.

Recently a small vessel carrying herrings in bulk took a serious list in bad weather because the cargo shifted. This case illustrates how changes in the design of fishing vessels and in the method of stowing fish may have an adverse effect on stability.

Owners of United Kingdom fishing vessels, under recent regulations issued by the Ministry of Transport, are not required to have stability information placed on board for the use of the skippers, but builders of fishing vessels should investigate both GM and curves of statical stability under different conditions of loading. Corrections will have to be made, of course, for the liquids in the ships. If this is done, it should be possible to indicate to skippers the conditions of loading which they should avoid. Elaborate stability information is not necessary but simple instructions should be prepared for every vessel and given to the skipper concerned.

Mr. E. R. Gueroult (France): It has been said that a fishing boat can be very rapidly overloaded when fishing, and if she is caught in bad weather and water floods the deck, the situation can quickly become serious. The safety rests on the water-tight superstructures, and this is normally not sufficiently taken into account in connection with the safety of the boat. For example, in the flooding incident described by Mr. Möckel in his paper, if there had been an open forecastle on the trawler concerned it would not have been possible to congratulate Mr. Möckel on a very fine paper. When the ship has a full deck-load of water, the first thing is to get rid of it, and the scuppers, which are normally fitted on French trawlers (complying with the merchant practice) are not big enough. In any case, on fishing boats they are usually obstructed by a board or something else. The classification societies should provide rules for the minimum area of scuppers.

Safety does not allow any compromise and it is from this point of view that stability must be examined. For many experts GM is not a sufficient criterion even if agreement has been reached on a GM of 2 ft. (0.6 m.). The value of 2.76 ft. (0.84 m.) given by Mr. Möckel for ship A is necessary for a steamship to ensure a sufficient GM on the return trip. Big steam trawlers catching cod have an even bigger GM when leaving port and on return have less than 1.64 ft. (0.50 m.) but they have a freeboard which ensures safety. French experience confirms the values given by Mr. Möckel concerning steam trawlers of 124 ft. (38 m.) length. The rolling period on the return trip with half a load is double the rolling period on departure, and there is a similar variation between the calculated stability on departure and return.

An increasing righting arm up to about 30° inclination seems to be essential for safety, and a GM of about 2 ft. (0.6 m.) is needed for stable working conditions on deck.

As Professor Jaeger has said, the GM varies appreciably during a voyage by a motor trawler. Small boats must have a considerable GM and big boats a low GM to behave well in a seaway. As this stability cannot be acquired by heavy ballast and small beam in smaller boats there will be excessive beam or rolling angle and that will make working conditions on deck impossible. This is confirmed by the data on the type C of the German trawlers described by Mr. Möckel. As sea-kindliness is a function of hull shape and distribution of weights, Mr. Möckel's paper could have been greatly improved by showing the lines and the exact situation of the centre of gravity, vertically and longitudinally. He has mentioned a margin of 8 in. (.2 m.) in his paper for the position of the CG. This assumes the fishing gear is in its normal place and not piled up on oil barrels. Ropes and other heavy weights are usually stowed as high as possible on the superstructure and because of this it is advisable to use light alloys for the superstructure.

In the different cases of stability reviewed, cargo is taken as homogeneous, but this is not always the case in fish-holds. Herring and sardines, for example, can be compared with a liquid cargo if the hold is not sufficiently sub-divided. There is insufficient sub-division in the holds of French trawlers. In this respect Spanish trawlers are much better.

He would like to ask Mr. Llanos whether his calculations have allowed for the floodable length to be adjusted so as to keep the ship floating with one compartment flooded. In dealing with stern seas in bad weather it is necessary to rely on careful navigation more than on naval architecture because sufficient protection cannot be given either by sheer or by superstructure aft. There are cases of ships of 92 and 157 ft. (28 and 48 m.) which have reached port in a heavily damaged condition after riding out a storm. They have owed their survival only to the engine stopping by accident.

Professor J. M. Gonzalez-Llanos (Spain): In answering Mr. Gueroult, Professor Gonzalez-Llanos emphasized that the fish-holds of Spanish long distance fishing vessels were divided principally to take better care of the catch. Security against flooding was only a secondary consideration.

Mr. George C. Nickum (U.S.A.): It is obvious, after reading other papers, that the criterion for the ratio of GM to beam in his paper is too high for European trawlers. Existing successful boats do not reach this standard and, from comments made, it seems that vessels designed with such a high ratio would be too stiff to be sea-kindly. Boats on the west coast of the U.S.A. can be stiffer than those on the east coast or elsewhere, probably because the Pacific waves are longer and not as steep or as quick. The main reason for using a $\frac{GM}{B}$ ratio of over .10 was primarily to give an adequate margin of safety, but he would be willing to accept a $\frac{GM}{B}$ ratio of .06 and probably .05 in vessels familiar to him. For years .05 has been used as a rule of thumb formula for determining the adequate stability of transport or passenger vessels when dynamic stability is not investigated. But to accept such a value, an inclining experiment to determine GM is essential rather than a simple check by means of the rolling period, which is subject to variations up to 30 per cent.

A standard of minimum righting arm established in relation to the definite angle of heel does have some value for making a quick check of the righting arm but such a rigid standard should not be established because the preparation of the statical stability curve and the analysis of it is the last and highest step in estimating stability. As such it should be the subject of the naval architect's analytical judgment. He should be allowed to take every factor into consideration—and there may be many factors after he has analysed stability from the righting arm—which would allow him to accept standards lower than those Rahola and others have set forth. Factors such as Mr. Gueroult has stated, that the righting arms are increasing again when the superstructures dip into the water, are normally not considered in connection with righting arm curves and still they have saved many ships. He was primarily interested in the GM by making multiple inclining experiments in the different loading conditions. This was easily made for tuna clippers which essentially are tankers.

A thorough investigation of a new type of vessel should include calculations of dynamic stability. Normally, it should not be necessary to prepare a curve of righting arms. The fishermen cannot afford to pay for a naval architect who has the equipment and experience necessary for the calculations. If a fisherman is pressed, he will probably resist to a point of having no stability check at all. Therefore the routine check should be done quickly and inexpensively on the average fishing boat. The detailed technical analysis of stability should be restricted in special or marginal cases where the standards cannot be met. For example, if every 50 ft. (15.3 m.) Gulf shrimp trawler was to be subjected to an inclining test and the preparation of a curve for righting arms and a curve of static stability, the cost would make the operation impractical.

Mr. James F. Petrich (U.S.A.): Mr. Nickum has the right ideas about stability, especially in relation to the tuna clipper, where there should be a check at different loadings of the vessel. This brings home to the skipper more than anything else—certainly more than a series of calculations—how safe his vessel is. The skipper is usually at the check tests because they are conducted shortly before the vessel is delivered. When she is loaded down to different displacements by filling her tanks with water and she is inclined, he can see exactly how she appears and behaves. Such multiple stability tests add weight to the recommendations of the naval architect.

Commander L. E. Penso (U.S.A.): Mr. de Wit was anxious that the Congress should establish some stability criterion. There is one which is used by the U.S. Coast Guard for passenger vessels. It is called the wind-heel criterion ($\Delta \times GM \times \tan \theta = A \times h \times wpc$). In this criterion, GM times the displacement times the tangent of the angle of heel is equal to the heeling moment which in turn, equals the projected area the wind can blow on, times an arm, times a wind pressure coefficient. The arm is from the centre of gravity of the area above the waterline to the mid-draught, which is called " h ". The criterion established in merchant vessels is that the heel shall not exceed 14° or half the freeboard, whichever is less. The theory is that the dynamic effects will roll the ship to a value just about double that under steady wind-heel conditions. It provides a very convenient way of analysing the various loading conditions. The standard may be too high for fishing vessels but it does provide a guide.

Mr. George C. Nickum (U.S.A.): He had investigated wind heel by this formula in working out the standard proposed in his paper. The values by the wind heel formulae were about half of what would be considered a minimum for fishing boats. This is primarily due to relatively low freeboard in relation to the length. A specific case was a wooden ATR of 165 ft. (50.3 m.) length, converted for processing and brine freezing in Alaska. The wind heel requirement on the vessel, speaking from memory, was in the neighbourhood of a GM of 0.6 ft. (0.183 m.). The minimum stability of the vessel occurred when she was anchored and was about to take fish aboard. She had a large freezing tank on her upper deck. A heavy gust of wind could come up and her $1\frac{1}{4}$ ft. (0.38 m.) GM was necessary and in excess of wind heel requirements. Out in the sea a 1 ft. (0.3 m.) GM would have been much too small to take the action of the sea.

Mr. Francis Minot (U.S.A.): Does the GM required by the wind heel formula take account of any damage requirements?

Commander L. E. Penso (U.S.A.): No. A series of six or seven curves for different conditions is used in the case of passenger vessels. There is a passenger heel requirement which usually does not affect the normal transatlantic liner but can affect an excursion vessel with 2,000 to 4,000 people on board. It is assumed in that case that all the people on one side of the vessel, in conjunction with the distance of their centre of gravity of the load is, perhaps, from the centre line, results in an upsetting moment. A curve is produced which looks similar to the wind heel curve. Then, of course, there are damage stability curves which are all plotted and have different slopes. Finally there is a series of loading lines and any theoretically dangerous loading condition is shown clearly. The deep tanks or bottom tanks are then ballasted to bring stability above the theoretical danger line.

Damage stability could be considered in fishing boats. If it is necessary to operate at a very low freeboard, made lower at times by working conditions, then it is necessary to consider that a hold may become half-full of water due to a sea coming on board. This is a condition for which damage stability calculations can provide data, but it is doubtful if a fishing vessel owner would pay for the information as the calculations take a long time and are costly.

Mr. Arthur de Fever (U.S.A.): Waterlogging increases weight of the wooden hulls and decreases the freeboard of tuna clippers. It can decrease the freeboard up to 3 or 4 in. (75 to 100 mm.) which means that the main decks are under water in heavy seas, and that affects the stability.

Fantails have been the most practical stern design for clippers. Vessels with cruiser-type sterns generally have less stability.

A tuna boat often changes her trim and/or conditions of load at sea because of fishing conditions, and the skipper should be instructed about the stability so that he is fully aware of the dangers involved. Many vessels have got into difficulty and some have been lost because the skipper was not sufficiently versed about stability and more particularly of his own vessel.

SEA-KINDLINESS

Professor G. Schnadel (Germany): Investigations into seasickness show that acceleration is the dominant cause. In a liner it is found that the passengers in the bow of the ship are the first affected by seasickness, then those in the stern and finally those in the first-class midships. In 90 per cent of the cases acceleration, and in 10 per cent. rolling, is the cause, but the crew and passengers always believe that it is rolling that makes them sick, which shows that feeling in the matter is not to be trusted. It is time that masters of ships were given some physical means of judging stability. As the difference in rolling and in acceleration may be very great in different ships, the master cannot judge by " feeling " the stability of the ship.

Dr. J. F. Allan (U.K.): The GM is one of the major factors concerned in rolling and certain minimum values of GM are considered necessary for safety reasons. The GM is the lever by which the sea rolls the ship, and the shorter this lever is, the better the ship from a comfort point of view. It is true that the linear acceleration in heave and pitch at many points on the ship is much greater than the acceleration in roll, but it is also true that the rolling of the ship makes moving about the deck much more difficult and inconvenient than does pitching and heaving. It is not possible to control the general shape of the ship from the point of view of increasing its resistance to rolling but it is important to provide adequate deadwood forward and aft and adequate bilge keels, all of which contribute powerfully to roll damping.

In the paper by Dr. van Lammeren there is an interesting statement of the variety of tests which can be carried out in a ship testing tank, and it is well known that similar facilities have been available at the National Physical Laboratory, Teddington, England, for many years. It is important to turn attention to rough water tests and to the study of steering and course stability. There are limitations to what can be done in this direction in the tank, but at Teddington methods have been developed to study both these aspects of the design of ships.

It has proved difficult to obtain satisfactory records of the performance of fishing vessels, either in calm water or in service conditions, and the paper by Möckel, which gives his observations on trips to the Arctic, is very valuable from this point of view. To obtain information of this sort is is important to find the correct personnel who can continue to function satisfactorily and make their observations under the difficult conditions on board a trawler at sea.

Prof. H. E. Jaeger (Netherlands): Mr. Möckel deals with behaviour of trawlers in a rough sea and he is right about the question of freeboard in this matter. Professor Jaeger expressed the hope that Mr. Möckel would accept the minimum freeboards he had given in his paper, which are higher than those prescribed by the International Load-Line Committee.

Möckel's value of metacentric height of 2.8 ft. (0.85 m.) is very high, certainly for motorships. Even if the rolling acceleration is limited, which means a diminution of the metacentric height, it is to be hoped there will be agreement that in each case the maximum metacentric height should be reconsidered. He believed that if Mr. Möckel did the same work for motorships the conclusions would be much the same as his own.

Captain W. Möckel (Germany): There is an impression that a metacentric height of 2.8 ft. (0.85 m.) is proposed for coal-burning trawlers. This is not so. It is merely a record of the fact that in all the trawlers investigated metacentric height was 2.8 ft. (0.85 m.) and that experienced skippers considered it good. He agreed with Professor Jaeger and pointed out that 2 ft. (0.6 m.) would be sufficient for a motor trawler. But for coal-burners the GM is 1.5 to 1.65 ft. (0.45 to 0.50 m.) when they arrive in port.

Commander C. Harcourt-Smith (U.K.): Tank-testing is undoubtedly an essential operation to determine the required power. Although the seaman is, to a certain extent, interested in the economy and lines angle, his primary consideration is seaworthiness and buoyancy. Any reasonably designed vessel will ride out a head sea if she is kept steaming gently into it, or even hove to, but this is only a small factor and covers a course of approximately 45° either side of the direction of the wind and sea. It is essential that a ship should be able to continue steaming without excessive roll if the wind is on the beam, and more particularly when the wind and sea come astern. In tank testing little or no attention is paid to this condition.

He referred to a ship he had commanded which had gone through exhaustive tank tests, but when she got into a following or quartering sea she became completely unmanageable. In fact, she was dangerous even at very reduced speeds, and if she had been allowed to yaw she would have ended up by broaching to.

This factor of seaworthiness, which is so often ignored in tank testing, not only concerns the crew of the ship but also the owner, because when she is hove to or must have her speed so reduced that she is literally hove to by the stern, she is costing time and money.

Mr. Frederick Parkes (U.K.): He expressed himself as being 100 per cent. for tank testing, and there was no point in blaming tank testing if things go wrong when the ship is at sea. For example, he had a vessel called the *Boston Hornet* which he transferred to some Newfoundland friends. The skipper put in command was not used to a vessel of such size but he had a tremendous success on his first voyage and would have done an equally quick trip on the second voyage if things had not gone wrong. He made a very large haul of fish, but instead of dividing it into two, three or four bags and bringing it inboard a bag at a time—as a practical fisherman used to that size of vessel would do—he had tried to bring the whole catch inboard at once. He rigged up his fish tackles at the top of the mast and left all the hatches open, and any landlubber would know what would happen from doing such a stupid thing. As he brought the heavy weight of fish inboard, the ship, of course, turned on her side and the water began to flood into the fishroom through the open hatches. Had the hatches been covered the situation would not have been so serious, but he should never have had the fish tackle on the upper part of his mast. Even so, had he cut away the net most likely he would have saved the ship. However, the water ran in and a brand-new ship was lost although, happily, another ship was close by and all the men were rescued. Nobody could blame the results of the tank tests for the sinking of the ship. It was sheer bad seamanship.

Another example is again of a brand-new ship, the *Boston Fury* (commanded by Skipper John Hobbs) a 185 ft. (56.5 m.) vessel fishing up in the Arctic Circle. A gale blew up and the skipper had to stop fishing. Then, suddently, came an S.O.S. from a ship in dire peril. She had sailed from a Russian port with a cargo of timber, some of it on deck. In the gale the cargo shifted and the ship turned turtle.

The *Boston Fury* was designed for a speed of 13½ knots, but in such a gale as was blowing the skipper would have been prudent to proceed at 5 or 6 knots. But he took a chance. He headed at 12 knots into the gale. Happily, she came through and he was able to save 27 of the crew of 29, picking them up from the sea where they were clinging to the timber floating around. In many cases men jumped overboard without a lifebelt, despite the skipper's orders, to rescue the shipwrecked crew. As a result, six of the crew, including the skipper, received Lloyd's Silver Medal, which is only given in cases where a man risks his own life to save others. Later the skipper was awarded the M.B.E.

The fact that the *Boston Fury* had been designed so carefully after tank testing at Teddington and was able to go through that gale at 12 knots, speaks well for the data which was used in designing her. It also shows, of course, the enormous value of seamanship. In the case of the *Boston Hornet* the ship was lost by an inexperienced skipper, but in the case of the *Boston Fury* the skilful seamanship of the skipper enabled him to take chances and make the fullest use of his ship.

Mr. Simpson has talked about a trawler doing 15 knots. It may be possible to get 15 knots, but at a most uneconomic cost. In his experience trawlers are not long enough to attain such a speed. A tank test would probably enlighten Mr. Simpson on this point.

Mr. J. G. de Wit (Netherlands): Mr. Frederick Parkes has said that tank testing pays dividends, so would it not pay to extend research more in the direction of sea-kindliness? Sea-kindliness is of great importance for all types of fishing vessels. For small ships in coastal waters it is more important than speed. For long distance trawlers it is equally important as speed, because wild movements of a ship may make fishing impossible in bad weather. The question arises as to whether the basins are able to conduct the model research in the direction of sea-kindliness and behaviour of the ships at sea.

Mr. Frederick Parkes (U.K.): A sensible suggestion has been made of having practical tests on sea-going vessels, similar to those made by Mr. Möckel on German trawlers; to compare data with tank testing results. As an owner of distant, middle and near water ships, he would be very happy for any organization to send scientists to sea in his ships to check on the results of tank tests.

Commander E. C. Goldsworthy (U.K.): Mr. Parkes' offer to place his fleet at the disposal of scientists to carry out tests on seaworthiness and seakindliness is of the greatest importance, provided the scientists who are sent to sea are capable of standing up to conditions in a trawler and to determine what their instruments are measuring and what they feel. The men to undertake this work should have had considerable experience at sea and knowledge of testing technique.

He had taken two of the *Boston Hornet* class of trawlers from England to Newfoundland. When one of them was within four hours of St. Johns, Newfoundland, a very heavy gale blew up. The ship was almost exhausted of fuel and water and had only about 20 tons of stores and spares in the fish room, but the vessel rode out the fierce gale satisfactorily and, according to the master, he had never been in such a good ship.

Dr. W. A. P. van Lammeren (Netherlands): Sea-kindliness is part of the correlation between tank tests and actual ship tests. He agreed with Dr. Allan that the major part of research in the future should be into sea-kindliness. At the Netherlands Ship Model Basin they intended to make a big steering pond in which to guide complex waves as well as winds in various directions so that sea-kindliness can be studied on a model scale.

He had given a paper before the Institution of Naval Architects in which he dealt with this particular problem and mentioned that fundamental research has been divided into two parts at the Netherlands Ship Model Basin. The first part deals with the problem of improvement of hull and propellers. That is the oldest part of research and it does not present a problem any more as every model basin can now design a hull and a propeller near to the optimum.

The second part of the research deals with the correlation problem which can be divided into three sections: (1) the scale effect; (2) the allowance for roughness; (3) sea-kindliness of the ship.

Several speakers have pointed to the necessity of carrying out extensive tests on actual ships. For the normal merchant ship this is done and owners give the opportunity to make extensive tests on the measured mile as well as to carry out tests while cruising. It is a good suggestion to make tests on fishing vessels during a voyage, but it will be difficult to find the right personnel for this work. In the Netherlands work of this kind is being carried out under the guidance of Professor Bonebakker of the Technical University of Delft. These people have been trained to carry out tests on actual ships and they have discovered important data.

Captain W. Möckel (Germany): Owners and builders who order resistance and propulsion tests in the Hamburg model tank often want to extend them to the behaviour of the ship in a seaway. This has been done on large vessels since 1936, with the help of officers and captains interested in such investigations and trained for the job. Since World War II similar investigations have been made in trawlers and it is hoped to make further investigations on a bigger scale. The chief difficulty is to find the right personnel who have nautical and technical training, being seamen as well as naval architects.

Experiments with models only in the tank cannot provide thorough and efficient results so that investigations must be made at sea if the quality of ships is to be improved. Such tests should be carried out on a very big scale, preferably by an international organization such as FAO. If FAO organized this work they could call on the personnel at various tank-testing establishments to co-operate in a programme for investigating ships at sea.

Mr. John Tyrrell (Ireland): In craft under 80 ft. (25 m.) in length the reduction of the resistance is not the most important factor to be considered in the design. Most fishing vessels, especially those which operate by towing, such as trawlers and seiners, are relatively high powered and commonly achieve a high speed-length ratio without running the engine at full power. The large engine is for towing power regardless of what the cruising speed may be and quite frequently small vessels obtain their maximum speed before the engine reaches its full revolutions.

Seaworthiness and trim are the first points to be considered and provided for in a satisfactory design. Trimming becomes of the utmost importance in vessels under 60 ft. (18 m.) in length, particularly if they have to carry large quantities of fish, such as herring in bulk.

Experience had led his firm to adopt a basic hull type which differs considerably from that in which reduction of resistance is sought as the first requirement. The type has a fairly full bow, moderate midsection, and a well drawn out midship section with the prismatic coefficient varying from .63 to .65. The fullness of the bow varies, being more emphasized in the smaller vessels. The half angle of entrance ranges from about 28° in a 50 ft. (15.2 m.) vessel to 23° in a 60 ft. (18.3 m.) vessel.

Over the past few years the bow has been made sharper below the L.W.L., while giving a correspondingly greater flare and reserve buoyancy above water. This alteration has been made chiefly to give easier motion in a short head sea.

Irish harbours and sea conditions make it essential that the fishing vessel should steer with certainty before a heavy following sea in shoaling water, so that a stern has been adopted in which the L.W.L. is fairly sharp. There is no pronounced fullness of the quarters which might drag a sternwave. For the same reason the rudder is placed as far aft as is practicable and its trailing edge is only slightly inside the stern overhang.

Mr. George C. Nickum (U.S.A.): Mr. Möckels data on angles of heel is very interesting. A short while ago he was asked for the degree of roll of the average fishing boat. He had found no published data, and, as far as he recollected, the first time he had ever seen a tabulation of maximum amplitude of roll of operating trawlers was in Mr. Möckel's paper. He was much surprised that the amplitude is so great but gratified that the information is now part of technical data available to the architect. Mr. Möckel's resolution of rolling and pitching in terms of acceleration is a new and interesting concept and he sets forth some minimum values for maximum rates of acceleration both in fore and aft transverse direction, and endeavours to relate these rates of acceleration to human comfort. This is a problem which should be further examined. Periods of roll, amplitude of roll, and periods and amplitude of pitch only cannot be used by the architect to make an accurate determination of whether a vessel will be comfortable or not, except, perhaps, in the particular class of ship he is familiar with. If it is possible to establish minimum rates of acceleration which will be comfortable, that rate should be related to all classes of boats.

Mr. Gueroult's method of establishing a sheer line by means of transposition of the vessel on a wave equal to the length of the vessel is again a new concept. It has considerable merits in the case of new types of vessels where there is no tradition of sheer. The method gives a guide as to what the sheer should be.

Mr. Olin J. Stephens II (U.S.A.): Asked by Mr. Gould as to whether anyone had experience of retractable hydrofoils to reduce rolling movements, Mr. Stephens said he had no actual experience to draw on but thought the device would be restricted to passenger ships where the comfort of the public was important relative to the expense. Such devices would be very expensive.

Mr. James F. Petrich (U.S.A.): The device mentioned for reducing rolling movements may relate to a type of paravane being used on some trollers on the west coast of the U.S.A.

This is a fairly successful type of paravane which hangs from each side of the vessel and cuts down the rolling movement considerably. As the troller heels one way, one door feathers and drops down deep in the water, while the other on the other side drags against the upward pull, so there is always a pull to the side away from which the boat is heeling.

Mr. R. T. Whiteleather (U.S.A.): The type of paravane mentioned by Mr. Petrich is a kind of otter door arrangement used on small trollers of 35 to 60 ft. (10.7 to 18 m.). The otter door is set on the horizontal plane and trolling poles are extended from each rail, port and starboard side of the vessel. From the trolling poles there are a series of lines which are principally used for catching albacore or salmon. In addition, the fishermen take a door-like arrangement and put it on the horizontal plane attached to the end of the trolling poles. It is allowed to ride at the proper angle of the horizontal plane as the boat moves along at very slow trolling speed. This gives the effect of some additional resistance to roll on each side of the ship. The fishermen have found that this device is quite effective in stabilizing very small vessels.

Mr. J. Laurent Giles (U.K.): In the case of fishing craft the vessels can be divided into large, medium, small, very small and minute. A division of this kind is necessary because of the very different incidence of numbers and also because of the possibilities of research which have existed for a considerable time for large, medium and even small vessels, if small is taken to mean 50 to 90 ft. (15 to 28 m.), but as far as fishing boats under this size are concerned, there has been hardly any research done at all.

The question of sea-kindliness, stability, safety, and movements at sea, are inextricably bound up with the size of the boat. Dr. Allan said the naval vessel provided a gun platform and the fishing vessel provided a fishing platform. In this connection, he had been concerned during World War II with the British " B " Fairmiles, round-bottom boats. He was responsible for their rearmament and other alterations. They were much liked by their crews because they were sea-kindly boats and they were built strictly to naval principles of minimum possible GM, which was 1.75 ft. (0.535 m.). Fairmiles produced then a " D " type designed with a high chine forward and a semi-displacement bow which had great stability. The B.M.L.'s, although so sea-kindly, were useless as fighting boats in any sort of sea because nobody could keep a gun anywhere near a target due to their rolling. But the " D " Fairmiles, which were entirely contrary to naval tradition in the matter of stability, were able to fight the whole time because they were so stiff and remained more or less upright.

Another example of what extreme conditions of stability are capable of producing was shown in a 48 ft. (14.6 m.) lighter built to carry aircraft from ship to shore or shore to ship. This boat had a beam of 16 ft. (4.9 m.) and had pontoons extending some 4 ft. (1.22 m.) outside the hull proper. The pontoons carried the deck on which the aircraft was placed. This boat seemed to be most unseakindly yet, when it went out on its first trials in reasonably bad conditions, it just ploughed its way, never more than 5° off the vertical, through considerable short seas in the English Channel.

These examples are worth keeping in mind in considering the stability of small craft. The relation between ship and sea seems to change radically when a boat is as small as 15 to 30 ft. (4.5 to 9 m.).

Pitching is another aspect of the same matter and it should always be borne in mind that very small fishing or other vessels are liable to find themselves in conditions of wave length and height which have never appeared in any textbooks.

The transom may be open to criticism for fishing boats—that it is liable to cause seas to break on board if the vessel is more or less stationary and/or held down by fishing gear—but if the vessel is free, the transom is as good a sea-going stern as any other. Mr. Giles went on to quote the example of a small boat of his design, 21½ ft. (6.5 m.), with a transom stern, which sailed from Singapore to England, 13,600 miles, in 140 days, sailing 100 miles a day at 4 knots. She never gave a moment's trouble. In fact, of the many boats designed, there has never been any suggestion that the transom stern, properly allied to the appropriate bow, caused anybody any trouble. The running—and it does come into the transom question too—is above all other points a question of balance of the form of the bow and of the stern. A geometrical system of matching the bow and the stern, incorporating a diagonal of exactly equal shape fore and aft, is used for sailing yachts. That involves a raking midship section, but it does seem to make a very big difference. These little things count. Sailing yachts are often very much the same in displacement-length ratio as fishing boats and it is usual to run two sets of lines of sailing yachts up to about $\Delta/.01L^3 = 450 (L/\nabla^{\frac{1}{3}} = 5.56)$. It is often a matter of very minute differences between the boat which is a charmer and the boat which is a pig. A flat transom form, such as that shown in the Bedi boat in the Pakistan paper and the Cape Island boat in Mr. Chapelle's paper, is likely to be disastrous because of broaching to. In the case of very small boats the tank tests must be treated on the understanding that these boats will never operate for more than a few hours at a time in smooth water.

Mr. Olin J. Stephens II (U.S.A.): There is a distinct division between large and small craft in the question of sea-keeping qualities and safety. With the small craft it is possible to have a very stiff body which periodically will have a quick roll but that roll will conform to the period of encounter of the sea when it is less than the period of encounter. When it becomes greater than the period of encounter, then it will establish itself on its own independent frequency. When that frequency comes close to the period of encounter there will be a condition where it will build up heavily, but where it is less than the period of encounter the vessel will conform to the surface of the water. For these reasons it is possible to have a very stiff small boat that will not be too uncomfortable and which gives a margin of safety to take advantage of. There is no need to cut down metacentric height to get reasonable comfort, but with larger trawlers, of course, it is essential. In this respect Mr. Möckel's paper is an extremely valuable contribution to knowledge. To some extent an analysis of that paper indicates that the pitching periods more or less conform to the periods of encounter, but the rolling periods are independent, having their frequency regardless of sea conditions. This bears out the point in connection with smaller vessels. He did not agree with Mr. Giles about the importance of minute differences. People became almost superstitious about these small differences, but, of course, these do become cumulative. An accumulation of good or bad differences in any combination affects a boat, but if that boat is good to start with, to make minute changes in it will not stop it from remaining a good boat.

Mr. Howard I. Chapelle (U.S.A.): Vessels of over 150 tons represent about 8 per cent. of the fishing fleet on the north-east coast of the United States. Changes in the fisheries in the north-east have required increased speed in the vessels and changes in location of the fish have increased the distances the boats have to travel. As Mr. McInnis has pointed out the holds on these vessels are larger than they need to be and fishermen do not bring in full loads usually.

Most of these vessels are subject to overcrowding and have too much power in proportion to the size of the boat, and there is an inherent difficulty in the design of a wooden vessel in that it is almost impossible to calculate the hull weight. Most of the trawlers on the New England coast and in Nova Scotia and East Canadian waters have their engine-rooms aft, their crew's quarters forward and fish-holds about amidships. The cabins and engineers' quarters are aft. The popular size of the small trawler seems to be in the range of 60 to 65 ft. (18.3 to 20 m.), a size that was imposed by economic factors such as the cost of the vessel, licensing laws in Canada and financial requirements. Trawlers of this size need 175 to 200 h.p. engines to be really efficient. As a result of the power on board, fuel consumption becomes an important factor and it is normal to find that a small trawler has a very large capacity. The water tank must be placed forward to help balance here. She carries the main engine, generating plant, batteries and fuel and air tanks in her engine-room. As a result almost every small trawler has excessive trim by the stern, which has a marked effect on running speed. The general practice is to use a very full run to float this weight aft, as shown in Mr. Simpson's paper. Actually, on these vessels the general movement of the centre of gravity aft is to carry much weight in light condition. When these vessels load their great fish-holds, many of them trim by the head and in that condition become unseaworthy. They lose speed and steer very badly. Most of the owners and skippers of small trawlers in North America seem therefore to prefer the round stern, and in wood construction it is very heavy. The stern has been changed in profile in recent vessels in order to lower the centre line of the stern at the waterline. In some vessels it approaches a rather full sharp stern, called a " canoe stern ". It is a practical stern which gives room aft and is good in a following sea. It has an effect upon the trim of the vessel because there is not enough bearing aft, and when the weight is concentrated there almost every vessel with a round stern shows excessive stern trim. Another type of stern which has come into use, particularly in Nova Scotia, is a wooden adaptation of the cruiser stern. This is a more desirable structure and is lighter and stronger than a round stern. The transom stern has not been particularly popular except in the very small inshore trawlers. Mr. McInnis has a number of 60 ft. (18.3 m.) trawlers with this type of stern building in Canada. Their trim is superior in light condition to any of the other types of small trawlers. What would happen if additional weight were put in those vessels is an open question, but it is apparent that there is a definite advantage in using a transom stern. This stern is likely to be necessary if an effort is to be made to maintain the present trend in power in these boats. Those who design small trawlers have approached the question of " drag " or trimmed keel with a remarkable amount of caution. The larger European trawlers, such as those designed by Mr. Gueroult, have a great amount of drag. If the small American trawlers had this drag they would have more bearing aft, without having to fill up the run excessively. The run is very important in these small vessels. They are obviously no more than sea-going tug-boats and it is important that the run is very well designed and that the water reaches the propeller with a minimum amount of disturbance.

There has been much trouble because the centre of buoyancy has moved gradually aft and the bow has tended to sharpen in an effort to reduce the trim of the stern. McInnis' suggestions for shortening the fish-hold are desirable, although it is doubtful whether this can be accomplished because of the desire to keep the fish-hold large. The solution of the trim problem is to increase the drag and obtain the maximum amount of displacement without causing excessive disturbance in the run.

These trawlers are now being designed for a speed in the neighbourhood of 10½ knots. This seems to be as good a speed as can be reached with the power put in the vessels, and with the poor form accepted. It should be noted that some of the best vessels for cruising and towing efficiency were not designed by naval architects. For example, Mr. Simpson shows in his paper a trawler called the *Golden Eagle* which was built from a model. She happens to be a very efficient vessel and is marked not only by a good run for a boat of her capacity but also by a very marked drag to the keel. It seems that a combination of her run form with drag is one of the reasons why her efficiency has been so high.

In recent years icing up has not been a serious problem on these trawlers because of the warming up of the North Atlantic, but it is a problem that will occur from time to time and in the past it has been a very serious matter, as small and large trawlers have been lost at sea, with all hands, through icing.

A great many designers have cut away the forefoot of fishing craft, approaching the Maierform hull. A vessel so designed has a tendency to " tramp ". When she lifts out of the sea her forefoot comes up, and when she goes down she does so with a slam that throws up spray. Anyone who has been to sea in winter in the North Atlantic will know that spray will ice up more rapidly than solid water. To cut away the forefoot on the schooner type of bow, driven at the speeds obtained in these small trawlers, is to invite rapid icing. It seems that the trawlers would benefit from a very angular forefoot which is deep enough and fine enough to prevent slamming in a head sea. There have been various suggestions about using heat to reduce icing, but it would take an enormous amount of heat to be of any practical use. The best solution seems to be to design the fore end of the vessel so that she throws the minimum amount of spray.

It has been the experience of many fishermen that excessive flare forward can be harmful. He had designed vessels of the small trawler class with very little flare. The sides and shoulders of the bow come out straight, V-shaped. With a deep forefoot and a rather angular turn it is possible to get very fine lines forward, and the lower water-lines may be even slightly hollow at the forefoot. But if the lower water-lines become almost straight for some distance aft it will move, of course, the buoyancy centre aft, and it has the disadvantage of making the forecastle rather confined.

Mr. McInnis dealt at some length with ballast. Because of trim, particularly excessive trim aft, a great number of small trawlers carry 4 to 10 tons ballast forward to keep the stern high and make them safe. If they do not have much drag when they fill their fish-holds they become down by the head and steer very badly. For this reason it seems rather remarkable that there has been so much caution in the United States in putting drag into vessels. He had moved slowly in that direction but found that in Europe they had moved much more rapidly in the large trawlers. It had many advantages because, when the vessel is trimmed by the head she still steers well, regardless of an angular forefoot, because she does not draw more water forward than aft. In some areas of the United States the old problem of the combination fishing boat has come up. So far very little has been done to produce a combination trawler and long-liner. A few boats have been built and tested but in each case difficulties have arisen which affected their economic desirability. It is a problem that designers of small trawlers must face. His own idea is to fit one side of the vessel for trawling and the other side for long-lining. This will add to the initial cost as, for instance, compared with a long-liner, but will make the vessel cheaper than a trawler of equal size. It is doubtful that money-making small trawlers can be built under 60 ft. (18.3 m.) length because if such a trawler is built, it will be so low-powered as to be unable to trawl in deep water, and deep-water trawling is a necessity.

Speed is controlled in these vessels by the hull form. There is ample power available to drive such a vessel at a higher speed, but the hull forms are not suitable. The capacity of a trawler is an important consideration and it is not possible to reduce the depth of the small trawlers very much. In fact, the trend is towards deeper vessels in order to shorten the fish-hold and carry extra weight. The solution does not lie entirely in using a very full mid-section. The general experience seems to be that designs show a good deal of deadrise, beam and some longitudinal distribution of the displacement. As a matter of fact, the hull lines of the small trawlers are really distorted. In some way a hull form has developed which approaches that objected to in an inshore vessel, but because the trawlers are deeper they do not have the same failure and are more seaworthy. The centre of buoyancy often approaches the after bulkhead of the fish-hold, about 57 per cent. of the length of the hull.

If towing requirements are allowd to control the design, the result will be a vessel that will sail faster, which is very easy to obtain in running light condition or even in load condition, in a vessel from 65 to 75 ft. (20 to 23 m.) length, without adding extra power. It is possible to obtain 10½ knots without excessive disturbance and without too much danger of the vessel losing seaworthiness.

Mr. H. C. Hanson (U.S.A.): Referring to the statement that a deep forefoot makes a vessel that handled better, Mr. Hanson said he could not agree to this, however he did not realize how bad the ice conditions were on the U.S.A. East Coast. The deep forefoot section he believed made for bad handling from the steering point of view but he could see that the spray cast by a long-running forefoot could be harmful if it iced up the ship.

Mr. Arthur de Fever (U.S.A.): He was in agreement with Hanson on the forefoot design, particularly as applied to U.S.A. West Coast vessels. He had found better success with the rounded forefoot than with the deep forefoot, again due to the trimming of the vessel. In some cases clippers had been designed on an even keel without drag and on their return voyage from the fishing grounds they had become head down because their brine wells were dried and they had used the fuel in the double bottoms. The steering then became a problem, but this situation had not arisen when the rounded forefoot design had been used.

Mr. Jan-Olof Traung (FAO): The problem in a country such as Sweden is that fishing boats have too much stability, so that they have a short period of roll and are uncomfortable. A GM of 4.5 ft. (1.37 m.) and a rolling period of 3½ seconds are common in boats of 80 ft. (24.4 m.) length and 20 to 22 ft. (6 to 6.7 m.) breadth. Such a boat will have more than 20 tons of concrete ballast and the more the boats rolls and rocks, the more ballast the fishermen put into it.

He had come upon a similar problem of ballast when visiting New Bedford, Mass., U.S.A. Two boats of exactly the same type, built from the same drawing and in the same yard, behaved very differently. One was a " charmer " and the other a " pig ". He inspected the boats and saw that one was rolling much faster than the other. The one with the shorter rolling period had 16 tons of ballast on board but the other boat had none.

In another instance, at St. Johns, Newfoundland, a Norwegian type of boat had been built but it rocked badly. It had 10 tons of ballast on board and the rolling period was about 4 seconds. He advised the Director of Fisheries, who was present, to remove the ballast. The Director said " You are crazy " but agreed to take it out. When 5 tons had been removed the rolling period increased to 4.75.

In certain areas there are boats with too little stability, as is the case of many boats on the U.S.A. West Coast, but there are also many areas where boats have too much stability, such as 3 to 4 ft. (0.9 to 1.2 m.) GM and nobody knows about the relation between stability and sea-kindliness, so they put ballast on board and increased the discomfort of the boat. To get a sea-kindly boat it is much more important to look into the ballast and the position of the GM than to look at the lines.

The first essential is to get the GM giving good motions, then a start can be made in the refining or altering of the hull shape. Dr. Allan is right in saying it is necessary to get a good flare in the foreship. The plans in Mr. Gueroult's paper show a good flare and also that the bulwark is faired along the ship's sides. It might be better if those bulwarks were given a knuckle at the deck to provide some kind of compromise flare. That would give the same buoyancy which is obtainable with a good flare, and, at the same time, give a much more practical bulwark, not likely to be damaged at a quayside. It would also provide a waterbreak, which would enable the water to leave at the deck and not at the rail from where it can sweep more easily into the boat.

Mr. I. Bromfield (U.S.A.): Talk about ballast reminded him of the attitude of fishermen. If they come in with a good catch everything on board ship, including the machinery, is excellent, but if they come in with a poor catch, then the owner must be ready to face a long list of repairs stretching from stem to stern.

In 1938 he had a vessel of about 128½ ft. (39 m.) length, 24 ft. (7.3 m.) beam and about 12½ ft. (3.8 m.) draft. She was the first all-welded ship in the Boston, U.S.A., fishing fleet. She was supposed to carry 30 tons of ballast but, by oversight, it was left out. It was decided, however, to see how she behaved without ballast and she turned out to be one of the best fishing vessels in the fleet.

Mr. H. C. Hanson (U.S.A.): He liked to keep every bit of excess weight out of the vessel, so much so that he used only very lightweight filling instead of ballast between the floors for hold drainage only. Provision must be made for the fact that a wooden vessel will absorb moisture and in a few years she may have 3 or 4 in. (0.075 to 0.1 m.) more draft than when she was built.

Mr. Olin J. Stephens II (U.S.A.): As one with little experience of fishing vessels he had never been able to understand the need for ballast except, perhaps, where many small trawlers are constructed from models designed by builders under the erroneous impression that by keeping down the beam they can improve the speed. Ballast is then put into the vessel to make her stable but stability should come from the beam which could probably help the resistance characteristics rather than harm them.

Mr. E. R. Gueroult (France): The answer to Mr. Traung's question about continuous flare in the bulwarks of the new French trawlers is that French designers and owners do not like the look of knuckles and do not like the idea of losing any buoyancy in the ends. The problem has been discussed with the owners and they all decided that it was better to have the boat damaged against a quayside at rail level rather than at deck level.

Stability in light condition is very difficult to achieve because of the big rise of floor and small inertia of the water lines down to the keel. If good stability is attained in that condition, it is likely to be lost in a loaded condition. That is one of the reasons why some ballast has been put into the new French trawlers. The storing of ballast has followed the English practice. British steam trawlers used to feed the boiler on salt water. They filled the bottom of the ship up to the floors with cement or sand and the bottom of the fish-hold was simply a layer of cement over this ballast. In designing the new French steam trawlers it was decided to stop the habit of feeding the boiler on salt water and to carry fresh water in double-bottom tanks, so that there was no more room for permanent ballast.

Although acceleration is very interesting to naval architects, the fisherman is more concerned with amplitude. Acceleration is related to comfort and is important for passengers, but fishermen are concerned with the angle of roll. When it is too big they must stop fishing. The stability value obtained from an inclining experiment of a ship in service which is not really dry, is not of much use. A lot of ships have been so inclined and the experiments have yielded very unreliable results. Data must be worked out from a dry ship. Measurements at sea of rolling periods have been made but the men in charge, unlike Mr. Möckel, were not specialists. It is very interesting to know that Möckel's measurements coincide with stability data calculated by the builders of the trawlers in which he made his measurements.

Dr. H. K. Kloess (Germany): The only way to examine the qualities of a vessel in a seaway is to carry out observations and measurements on board the ship. This was done in 1928 when a comparison of two different performances of hull was made in actual service condition. Fig. 453 and 454 show the two different forms which are named " conventional lines " and " old Maierform lines ".

Fig. 455 shows the tank results and main dimensions of both ships. A maximum improvement of 20 per cent. in calm condition was shown in the model in the tests but this advantage was considerably increased in the full sized ship when she sailed in bad weather conditions.

In assessing the differences in behaviour of two different designs of a ship, the following conditions are essential: (1) equal main dimensions; (2) equal ratios of dimensions; (3) equal block coefficients and centres of buoyancy; (4) equal loading as to draft, displacement and trim condition;

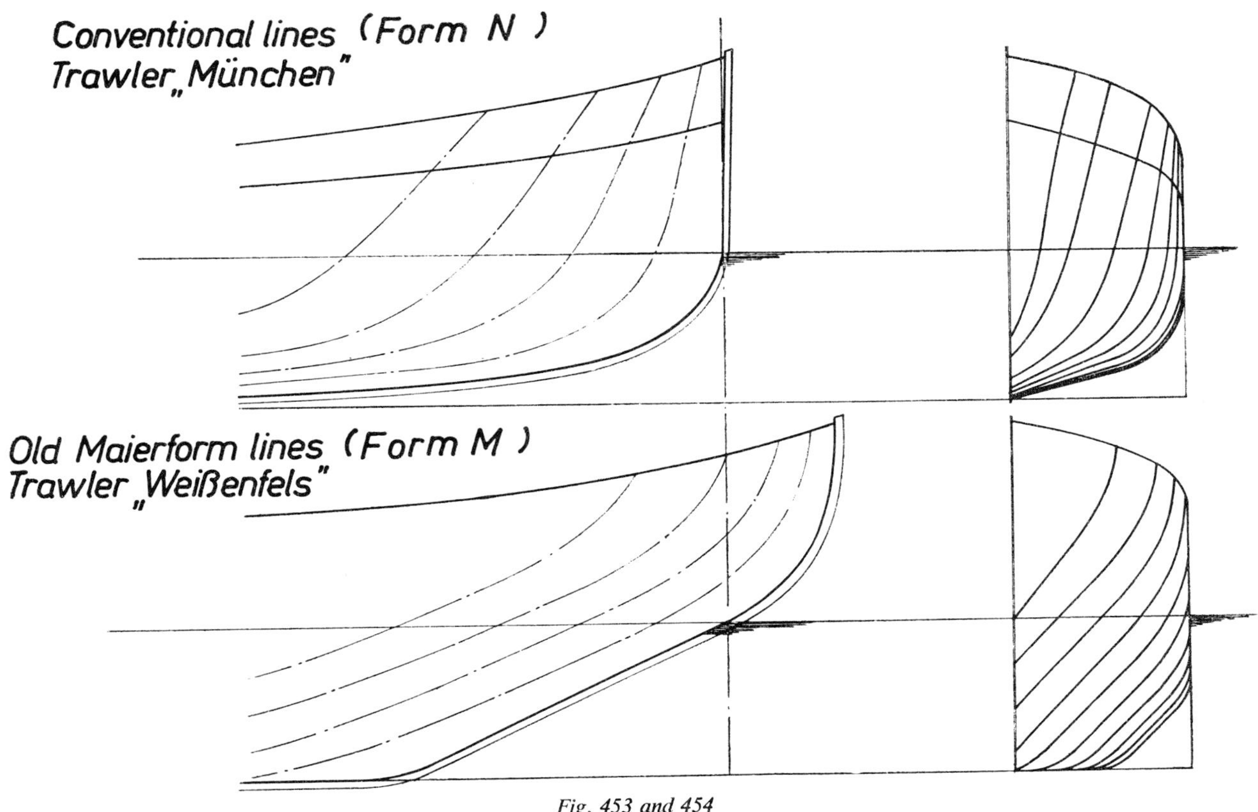

Fig. 453 and 454

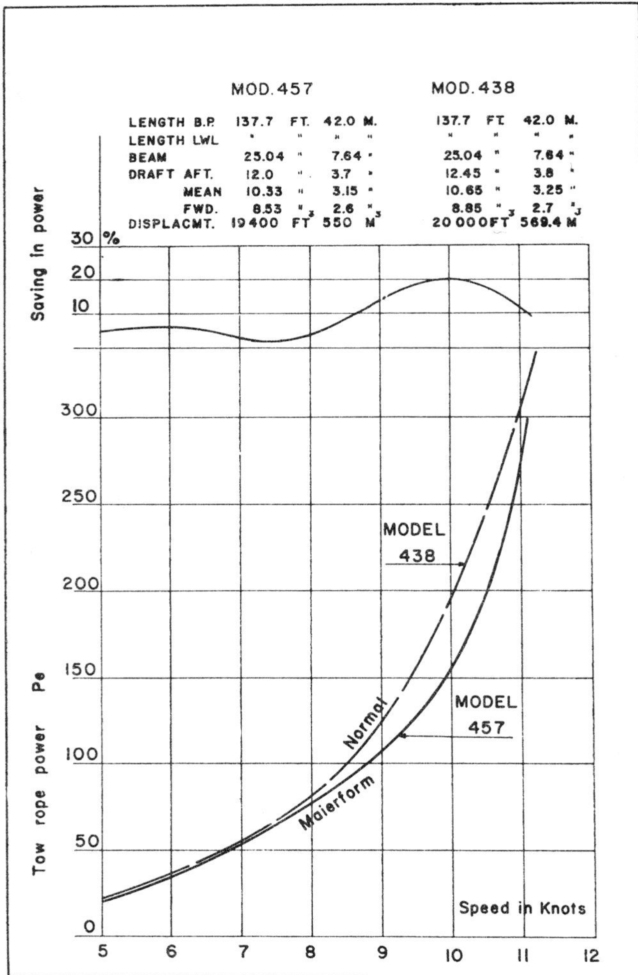

Fig. 455. Maierform trawler comparison with a trawler of conventional lines. Results based on tests at the model tank, Hamburg.

(5) equal engines as to type of machinery, output and revolutions; (6) equal fuel and fuel coefficients; (7) equal type of rudder (stream lined or not); (8) equal condition of hull as to shellplating, riveting, welding and last docking; (9) equal routes in similar seasons, and (10) equal conditions of wind and sea.

The two ships concerned fulfilled these conditions. Fig. 456

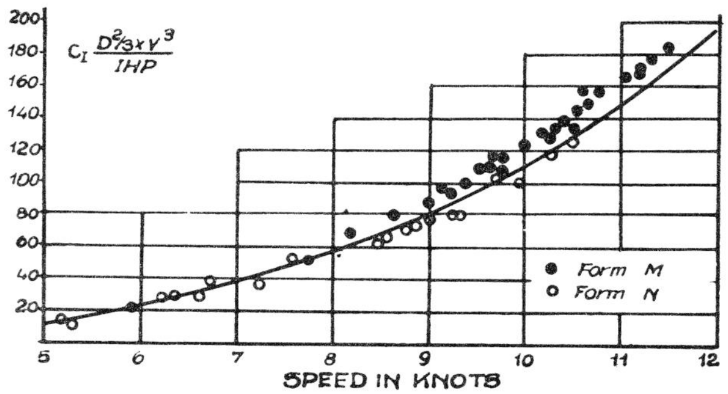

Fig. 456. Service results of two trawlers of different form of hull

shows the Admiralty coefficients for both, plotted to knots. The Maierform hull gave more speed for the same power than her sister trawler designed on the conventional lines.

Fig. 457 shows the difference in Admiralty coefficients of the two trawlers plotted on windforce to Beaufort's scale.

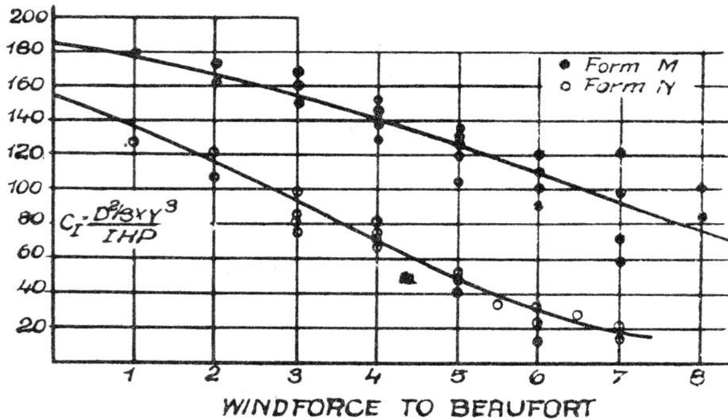

Fig. 457. Difference in Admiralty coefficients of two trawlers of different form of hull plotted on windforce

The diagram shows the increasing efficiency of the wedge-shaped Maierform as the weather becomes worse.

Fig. 458 is plotted on windforce at the highest speed the trawlers could attain. In calm condition the difference between them is about .75 of a knot but this is increased to nearly 3.5 knots when the wind reaches force No. 6—a moderate gale.

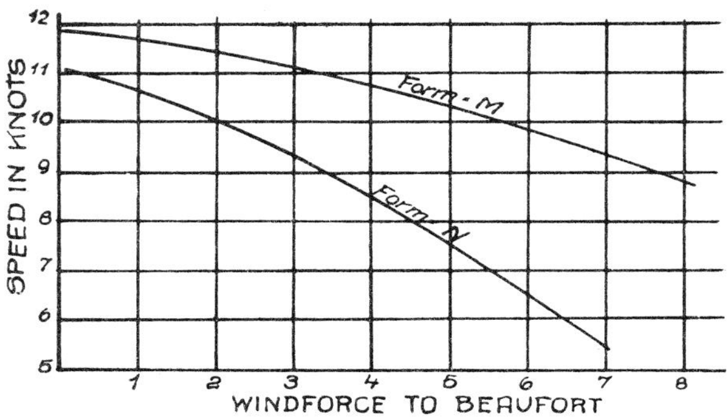

Fig. 458. Highest speed reached by two trawlers of different form of hull plotted on windforce

Fig. 459 shows the loss of speed and the increase of power. At windforce 6 the Maierform trawler required 95 per cent. more power but the trawler with conventional lines needed nearly 500 per cent. more power.

Similar measurements were repeated in 1935 with another trawler designed on similar lines but improved. Fig. 460 shows in the dotted lines the results obtained.

Captain W. Möckel (Germany): Dr. Kloess' contribution is of principal interest as it shows the magnitude of additional resistance of the two extreme shapes of hull in a seaway. However, in consequence of the development of both shapes of hull since 1928, the so-called "conventional lines" and "Maierform" as well as their service results differ considerably less as shown in the example cited by Dr. Kloess.

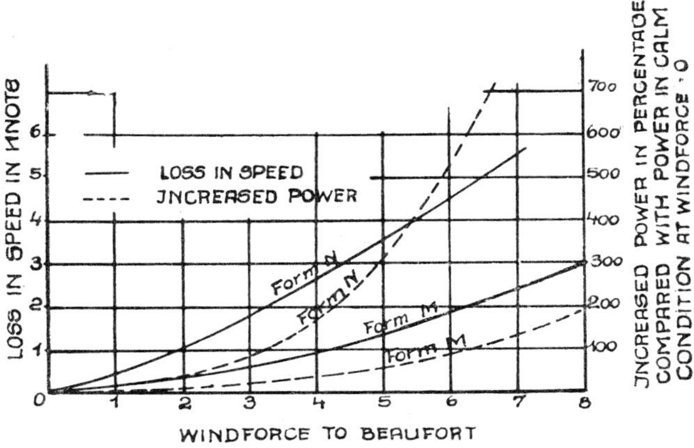

Fig. 459. Loss in speed and increased power of two trawlers of different form of hull in comparison to calm weather condition (windforce = 0)

Because of the superior seagoing qualities of V-shaped section ships this type of frames is commonly employed for many years already in the construction of trawlers. To this development in design the Maierform undoubtedly contributed to a remarkable degree. In consequence of the features of design the sea-going qualities of Maierform-ships also differ as those of trawlers designed by other naval architects. Fig. 418 e.g. shows the loss of speed of two Maierform-ships. The difference in loss of speed of the two ships may be explained by the difference of the block-coefficients. As proved by many service data, the loss of speed in a seaway increases with the block coefficient.

If the aim of service investigation is to find out the effect of the design features such as block coefficient, distribution of displacement and weights, stability, horse-power, etc., on the behaviour of ships in a seaway, the uniform conditions cited by Dr. Kloess are not wanted in all cases. In fact they are scarcely to be met on account of variations in the design of the ships.

It is still to be said that the measurements in ship A mentioned in Mr. Möckel's paper were executed in 1946 and not as Dr. Kloess assumes in 1935.

MIDSHIP SECTION

Mr. H. C. Hanson (U.S.A.): There is a tendency in Europe and on the U.S.A. east coast to make full deep mid-sections

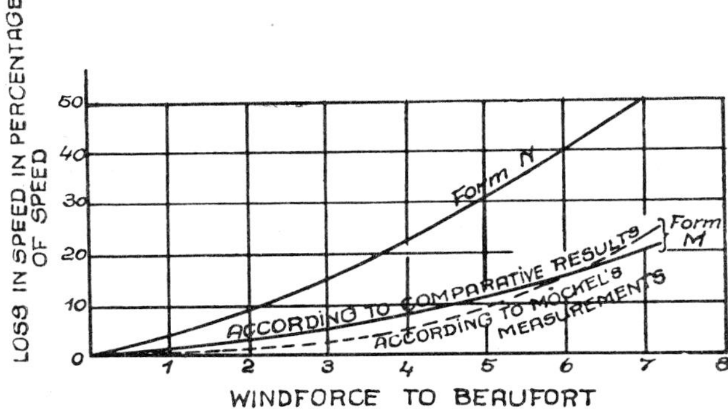

Fig. 460. Loss in speed for trawlers form M and N

but, said Mr. Hanson, he used a deadrise by means of which he obtained a faster vessel. The only way to get speed is to have long body section lines which were produced by a good deadrise. Wooden vessels have a tendency to hog and a good deadrise gives a stronger vessel which eliminates the tendency to hog even if the boat is small. He could not agree to make hard bilge sections and then having to place ballast to compensate.

Mr. Robert F. Allan (Canada): Mr. Hanson has contributed a valuable paper on the combination fishing vessel of the Pacific and as a pioneer effort, it is largely descriptive and there are some points of technical interest which merit further discussion.

In figs. 149 and 261 he draws attention to a trend away from narrow boats of full mid-section form towards the wider type with increased deadrise. This has been dictated largely by the tendency to raise the centre of gravity and increase the total weight by adding equipment and tankage.

There is no intrinsic merit in the finer section form nor has it appreciable bearing on resistance. The type having slack bilges and deep draft, as illustrated by dotted lines on the sketch in fig. 261, has several faults. It compels a full prismatic, has poor range of stability because of the reduction in the width of the immersed wedge and, because of flatness of the righting lever curve the vessel has a long and unpleasant rolling motion.

Mr. Arthur de Fever (U.S.A.): Mr. Hanson is right about a vessel slacking if the deadrise is too small. It causes an undue strain on the boat. But this can be offset in the case of the Pacific coast clippers by using shaft alley bulkheads for stiffening.

Mr. Olin J. Stephens II (U.S.A.): In the case of the midship section he was not altogether in agreement with Mr. Hanson that the shape of the section has very much to do with speed. It is a very important consideration but, assuming satisfactory stability for the fore and aft lines and the right prismatic coefficient, it is doubtful whether the shape of the section has anything to do with the resistance of the hull.

Mr. Jan-Olof Traung (FAO): In the Norwegian tank they have tested models with exactly the same dimensions and coefficients but with different types of midship sections and they confirm what Mr. Stephens said, that the type of midship section has no influence on the resistance whatsoever. The area influencing the prismatic coefficient has, however, great importance.

So far as sea-kindliness and stability are concerned, it is clear that if the midship section is made fuller, the centre of buoyancy of the hull is lowered. It is also clear that a fine water-line fore and aft means less moment of inerita of the waterplane and the position of the metacentre is thereby lowered. If the gravity remains at the same position there will be a smaller GM and a longer period of roll. Hanson has said that with a beamier midship section he gets more sea-kindliness and more comfortable (slower) roll or, in fact, less stability. On the contrary, Hanson with his section gets less sea-kindliness but more stability.

Mr. James F. Petrich (U.S.A.): The shape of the midship section does not have anything particularly to do with the speed

It is possible to get a good speed with a full midship section as with a good deadrise. There are other factors that influence the design of smaller boats, especially with the restrictions imposed on them. Mr. Hanson said he can get better speed with a boat of greater deadrise by adding more beam and that changes the lines fore and aft considerably. In larger tuna clippers it is possible to use a full midship section and get a better design throughout. He was in favour of that design as the draught must be restricted because the clippers have to go into shallow waters to catch their bait. He had designed a clipper with a great beam and increased stability, but the beam had not reduced the speed. In fact the vessel concerned, the *Mary E. Petrich*, has turned out to be one of the fastest in the tuna fleet and has very good resistance qualities. In such a full midship section vessel with such a large beam it is possible to distribute weights so that there is no need to make full lines fore and aft. The vessel has fine ends and can travel at full speed in weather that slows down a tuna clipper of normal hull design.

The dimensions of the *Mary E. Petrich* are:

Wood construction	
L.O.A. . . .	150 ft. (43.7 m.)
L.B.P. . . .	140 ft. (42.7 m.)
Beam . . .	34 ft. (10.4 m.)
Draft (full load, dry fish) . . .	14 ft. (4.27 m.)
Displacement .	946 tons
Prismatic coefficient	0.5975
L.C.B. . . .	0.0081 L aft amidships
Trial speed . .	13¾ knots at 1,600 h.p.

Mr. Howard I. Chapelle (U.S.A.): It is not safe to look at the midship section only as it is an isolated factor of design. In discussing a specific type, whether tuna clipper or North East trawler, there are certain conditions to be faced. For example, a shallow draught is considered an advantage in the tuna clipper. The exact contrary is the case in the North East coast trawlers where for working reasons a sharp-bottom and great draught, relatively speaking, are desired because the trawlers need to " hold on " (not make leeway) when they are sailing across the wind.

More power than can ever be used economically has been installed in small fishing vessels in the U.S.A. But as it has been installed some attempt should be made to use it, which suggests refinement in hull form. This might be made by modifying the mid-section, but the scope is limited by the type of boat.

In discussing a full midship section, Mr. Chapelle said he was confining himself to the New England and Nova Scotia trawlers of small size. Due to the unusual and unfortunate distribution of weight in these vessels, an attempt to lower the bilges rapidly creates a run problem, and there is a very short length in the vessel to form the run. But there is no stability problem of any great consequence as far as these trawlers are concerned. In the main the dimensions of the trawlers are about 65 ft. (20 m.) of deck length and 17½ to 19½ ft. (5.3 to 6 m.) beam. With the high bilges, the straight rise of floor, the marked deadrise, and the low weights placed in the vessels, the heavy engine, tanks and auxiliaries, the stability of the trawlers is generally excellent. They can operate at sharp angles of heel, as, for example, when they occasionally hook their trawl and run in circles until they break it free. At these times the vessels are often run at full speed with one rail under water. The general form of the trawlers is such that there is very limited application of the full midship section because of the combination of size, weight and depth involved. The trawlers are very deep in proportion to their length, which can be seen if the drawings are compared with those of the U.S.A. Pacific coast vessels, but it is really not feasible to compare the vessels because the tuna clippers are of great size in length and beam and have relatively small depth.

BOATS

Professor G. Schnadel (Germany): The new Dutch regulations call for one lifeboat for a ship not exceeding 200 gross tons, a lifeboat with air tanks for ships of 200 to 400 tons, and 2 lifeboats with air tanks, for ships exceeding 400 tons. In Germany two lifeboats are required for all ships over 200 tons because experience has shown that, in some cases, it has been too difficult to launch a lifeboat stowed amidship. When two lifeboats are carried, one will have to be stowed on the side of the ship. Small lifeboats have been capsized by launching and therefore Mr. Fernandez proposed the use of life rafts instead of boats. A short time ago most of the crew of a German trawler were able to take to the lifeboat but it capsized three times and only two men were saved. Had a life raft been used probably all the men would have been saved. Another proposal by Mr. Fernandez is that fishing boats should be equipped with radar, but this is not likely to be possible at present because men will have to be trained to operate radar efficiently.

Mr. J. G. de Wit (Netherlands): Foam plastic may take the place of present-day air tanks in lifeboats.

Lifeboat launching equipment needs to be improved. Davits should not have faulty gear. They should be able to make a greater swing to launch lifeboats stowed in the superstructure. The launching of boats placed in the centre line of the ship is unsatisfactory and needs further improvement.

The Netherlands Research Centre for shipbuilding and navigation recently carried out stability experiments with Dutch and French standardized lifeboats. One result of the tests showed that the 10 per cent. air tank capacity of the standardized Dutch boat No. 1, with a complement of 13 persons, is insufficient. This means that bigger tanks must be used, taking up passenger space.

Mr. Philip Thiel, Jr. (U.S.A.): A solution to the lifeboat storage and launching problem is to suspend the boat from the apex of an A-frame of welded pipe: the lower ends of the frame are pivoted on brackets stepped on the deck beside the deck house (or on the side of the deck-house itself). Half-brackets support the boat at a suitable height above the deck, and it is lowered by means of a tackle stretched from the apex of the frame to the opposite side of the deck-house top (*see* fig. 461).

Commander E. C. Goldsworthy (U.K.): It has always been a problem to launch lifeboats. He had the experience of being in a wrecked ship when lifeboats were launched on one side to act as fenders to the destroyer which came alongside to take off the crew. Those lifeboats could not have been lowered to get anyone away, but in this case they did make excellent fenders. More concentrated thought should be given to the problem of launching lifeboats, especially in small fishing vessels which carry only one boat.

Commander A. C. Hardy (U.K.): In view of the possibilities of air-sea rescue and the development of inflatable rubber dinghies, the retention of lifeboats and the complicated heavy gear which goes with them seems to be merely a carry-over from the old sailing ship days, when lifeboats were used not only for life-saving but also for communication with the shore. Like so many things in ship design, they have gone on from generation to generation without anyone really questioning why they should still be there and why they should still retain the same design.

It is undeniable that lifeboats have done magnificent work in saving life, but the facts show them to be most dangerous and more people are killed because of them than are saved

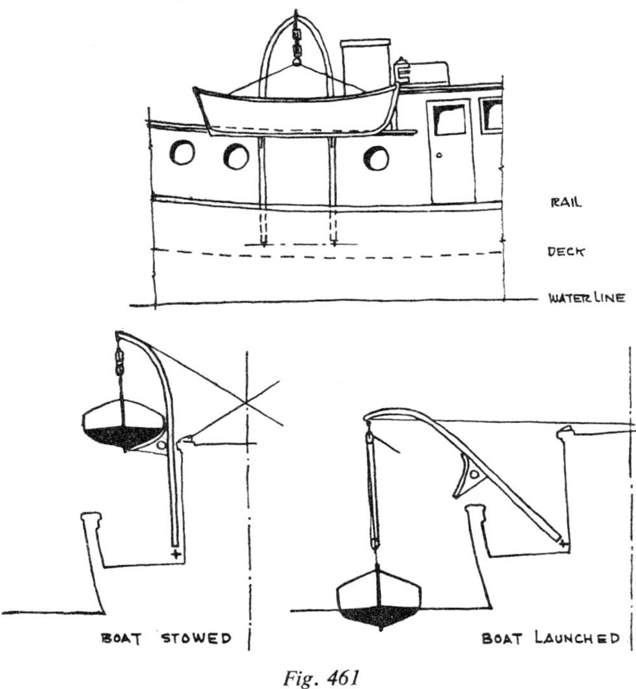

Fig. 461

by them. The time has come to scrap the conventional lifeboat and use the inflatable rubber dinghy which can be thrown overboard—where it automatically inflates—complete with food, radio equipment, etc. Rubber dinghies can be moored in groups so that, by use of the radio, rescuing aircraft can be guided to the spot where they are. Another advantage in using rubber dinghies is that valuable deck space is saved and tons of top weight is eliminated.

Captain H. Bertram (Germany): The question of replacing wooden lifeboats by rubber dinghies is being studied in Germany, and fishery protection vessels are all equipped with rubber boats. The skippers of the vessels think, however, that wooden lifeboats should still be carried because the crew needs special training to handle rubber boats. If these boats are not properly handled they cannot, at the critical moment, be inflated and then they are useless.

Mr. S. A. Hodges (U.K.): The Ministry of Transport in the United Kingdom has been conducting tests with rubber dinghies and is fully aware of their advantages. But there is difficulty in adopting them because life-saving appliances which must be inflated before they can be used are prohibited by the 1948 Safety Convention. The proper maintenance of them is also an important consideration.

Commander C. Harcourt-Smith (U.K.): As Commander Hardy has said, the old-fashioned lifeboat is virtually obsolete and has come down from the old sailing ship days when one of its functions was to tow the sailing ship out of danger of shoals or rocks in a calm, or the last mile or so into harbour.

It has been frequently found impossible to launch lifeboats from a sinking ship. For example, in the *Princess Victoria* disaster, not a single lifeboat could be launched by the designed methods, nor was it possible for the rescuing ships to launch their lifeboats. If the *Princess Victoria* and the other ships had been equipped with rubber dinghies, designed and adapted for life-saving at sea, most of the passengers and crew would have been saved.

The modern rubber dinghy is very difficult to capsize and has an inflatable canopy which completely protects its passengers. It can be handled by one man and put over the side. It has been said that a rubber dinghy is very difficult to get into. This is incorrect. Every rubber dinghy has a rope ladder hanging from the side which makes it easy to board, whereas with lifeboats the assistance of people in the boat is needed to get on board.

Rubber dinghies can be stowed in boxes on the main deck and it is not a difficult design problem to make the dinghies strong enough to be inflated on board and, if necessary, lowered complete with personnel and emergency equipment. Many lifeboats launched in heavy weather are smashed against the side of the ship, but the inflatable rubber dinghy can be bashed in this way without being damaged. The dinghies could be stowed in boxes on both sides of the ship and, if it is convenient only to launch from one side, they can easily be carried to that side.

Rubber dinghies could be launched by attaching to the deck head a light girder carrying a sliding member, from which can be suspended a frame platform containing the inflated raft with its crew and equipment. Launching would require only one man on a light winch, using a brake, to lower the raft. The same man could wind up the frame again ready for the next lift.

Rubber rafts or dinghies are not capable of being navigated in a conventional sense, but almost every vessel in distress is able to send out an SOS, giving its position, before the crew leave the ship.

Each raft is fitted with a small hand transmitter to send out an SOS. Rescuing vessels could locate them even in thick weather by using a direction finder. Meanwhile the occupants would be safely protected from the weather by the canopy.

There is little difficulty so far as maintenance is concerned and the responsible authority in each country could very quickly provide the necessary instructions to skippers on this point.

Mr. A. P. Schat (France): Great care is taken of the safety factor in the construction of a ship, but so far as gear for launching lifeboats is concerned, the law requires it to operate only at a 15° list on merchant ships. This puts the master of the ship in a difficult position. He should abandon ship before it takes a list of 15° but, of course, no master will ever issue the order until he is forced to. So far as fishing boats are concerned there is no compulsion to launch lifeboats when the vessel is listing 15°. That is only a moderate list and, in the case of fishing boats, the lifeboats should be able to be launched at a 30° list. Very often they cannot be safely launched and when they are they have very little freeboard so that water breaks over them. Furthermore, the occupants are exposed to the wind which acts like a fan. The normal temperature of the human body is 98.6 deg. F. (37 deg. C.) and if that temperature drops to 95 deg. F. (35 deg. C.) for several hours the person is bound to die. One way to protect the occupants of a lifeboat would be to place the air tanks in a different position so that all the people could sit down and remain under the boat cover. That would dispense with the need for a weather hood and give the same sort of protection which rubber dinghies give.

Mr. Schat referred to a crossing of the Channel by two young men in a 10 ft. (3 m.) rubber dinghy in a 50 miles/hr. (26 m./sec.) gale. The dinghy was equipped with a 4-cycle outboard motor.

One of the young men concerned was Dr. Nui Bombard who recently crossed the Atlantic on his own. He carried no food but lived from the sea. He proved that there is no reason why people in lifeboats should perish. There is plenty of fish and water to live on. The Dutch research organization T.N.O. has made a report on the possibilities of catching fish from lifeboats by using electricity. People in a lifeboat need to have exercise to keep them in good condition and they can get that by generating a 1 h.p. hand-operated motor, which not only can provide the electricity for fishing but can make drinking water from seawater. One man can produce 1.5 imp. gal. (7 l.) of drinking water per hour by the electricity he generates himself, so that, provided the lifeboat is equipped with a generator, there is no longer a probem of carrying fresh water in special tanks. The electricity generated can also be used for sending out strong wireless signals.

The cost of life-saving appliances is heavy and the authorities concerned with safety at sea should recognize that fact. Many fishing boat owners cannot afford to fit expensive equipment.

Dr. Gian Guido Bordoli (Italy): Much has been said about rubber dinghies but they are not always efficient because it is necessary to inflate them before putting them into the water. Little has been said of small rafts. They cannot, of course, be used in the Arctic Ocean because of the cold and wind, but they can be used in the Mediterranean and along the European and American coasts. He thought the rafts would be most useful if carried on vessels of 100 to 150 tons, and recollected that during World War I torpedo boats found the rafts were very satisfactory. During World War II many sailors were saved by the rafts.

The London Convention asks for rafts in addition to lifeboats. He would suggest that fishing vessels should carry, in addition to regulation lifeboats, small wooden boats built of cork or some similar substance.

FIRE

Professor G. Schnadel (Germany): The use of carbon dioxide for fire protection, proposed by Mr. Miller, is not thought necessary in Europe for fishing boats but only for big vessels. Water pumps and chemical apparatus are considered adequate. The reason may be that in Europe only boats of 100 tons or so are built of wood but in the United States big boats are built of wood so that the danger of fire in them is greater.

Mr. Jan-Olof Traung (FAO): In recent years many tuna clippers have been lost by fire. The reason seems to be found in the large quantities of ammonia carried in the clippers for their refrigeration plants. When there is a certain mixture of air with the ammonia it reaches an explosive condition. The danger is increased because of the extensive use of electrical equipment in tuna clippers (they have 30/40 pumps driven by electric motors). This is the reason why Miller has stressed the need for good fire-fighting equipment and not so much because the boats are built of wood.

Mr. E. R. Gueroult (France): France has made it a rule to include carbon dioxide (both foam and gas) fire protection in the engine room. This protection seems to have been very successful.

Commander L. E. Penso (U.S.A.): The improperly installed CO_2 system described by Miller does not seem possible in this day and age. But it has happened and it illustrates the need for proper technical supervision in all such matters.

Mr. Arthur de Fever (U.S.A.): Although Miller's idea for a separate fire protection system could be adapted to some boats in the fishing fleet, it would not be right for the majority of them because, particularly in small vessels, the forepeak is not fully enclosed. Many of the smaller vessels have been designed with an opening in the forepeak bulkheads while others have no bulkheads at all or only a passage way between ships' stores boxes and dry stores. This opens the area to the upper engine room and a separate fire protection system provided in that area would not work satisfactorily because the gas would be dispersed rapidly. This is the reason why CO_2 protection has failed in the clippers because only the lower engine-room has been protected. There is a large hatch over the main engine varying from 4×16 ft. (1.2×4.9 m.) to 6×20 ft. (1.8×6.1 m.). The gas in the lower engine-room is dispersed through these openings so that the rest of the deck area and the upper engine-room are added to the protection. About 50 per cent. or more of the clippers do not have a fully closed bulkhead in the forepeak and of the remainder only a part have middle doors providing ventilation up to the forepeak. In these cases the dispersion of gas would not be affected. In the remaining ships, if a fire occurred in the forepeak where paints and other inflammables are stored, the crew could close the door but in many cases the doors are secured on hooks and are not quick closing. In the turmoil the door is likely to be left open so that, again, the system would not be very effective.

Mr. A. M. Doxey (U.S.A.): It is difficult to believe that safety and fire protection standards have been kept because there has been no trouble in obtaining insurance, but underwriters are not, of course, versed in the technical details of fire protection equipment. Marine engineers and surveyors have pressed for more positive safety equipment and the chief engineer in a vessel seems to be the logical person to deal with the equipment. Contrary to Miller's proposal, there seems to be no necessity for a rigorous training programme. It is probable that the reasonableness of underwriters has made owners feel secure as regards safety but, in the end, a balance is established by experience and whoever is losing will, in due course, stiffen his demands. The fact that underwriters have not been strong in pressing for better safety explains to some extent the use of poorly engineered systems. The statements made by Miller, which are derogatory to the CO_2 systems, would not be possible if there were a set of good regulations which had to be observed when installing them. The U.S. Coast Guard have adequate rules and regulations to assure the owner and underwriter of safety, particularly against engine-room fires. One control near the main exit from the engine-room appears to be adequate. If the flooding system is to be used the presumption must be that portable extinguishers have failed to control the fire and in such circumstances nobody will be in the engine-room. If the system is in accordance with Coast Guard specifications there should be no need to secure the compartment.

Regarding the danger to lives of men in the engine-room, an experience on a large diesel vessel is illuminating. A crankcase explosion led quickly to an extensive fire going up the engine-room casing and three men came out badly burned. The chief engineer was at the controls of the total flooding CO_2 system. A third engineer was still in the engine-room. What was to be done? It was obvious that the engineer could not get out alive so the gas was discharged and volunteers, fitted with suitable breathing apparatus, went into the engine-room as soon as the fire was extinguished. They brought out three men, all unconscious. Two of these men are alive to-day.

The protection of engine-room and/or forepeak space together or separately depends upon the tightness of the bulkhead between the spaces and the contents of the forepeak. Experience shows that simplicity in operation is important and a system protecting both spaces simultaneously is the safest. The loss of gas in an area not on fire is compensated by possible success in extinguishing the fire.

If a fire in the engine-room is not extinguished quickly there will be no chance to use water pumps or other equipment. In this respect the value of the CO_2 flooding system was illustrated some time ago on a large diesel vessel. A fire starting from a donkey boiler caused a short circuit in the main electric cables so that all the water supply was cut off and the fire spread rapidly until it was put out by CO_2. The electric power was then re-established and water hoses were used to extinguish the accommodation fires.

It is possible, of course, to carry a spare set of CO_2 cylinders, but experience and economics are against it on all sizes of vessels. The possibility of a second fire on the same voyage is remote.

The need to change fire protection systems because of unsound engineering is unfortunate and is due to lack of enforceable regulations and specifications. This has led to the use of unreliable war surplus equipment, installed by persons without adequate technical skill. Although such an installation is cheap in the first cost, it is ultimately very expensive because of its unreliability. It would seem desirable to foster practical regulations to ensure installation of adequate fire protection systems.

Mr. William C. Miller (U.S.A.): Inert gas systems should be primarily designed to protect a compartment in which they are installed, and a single system is good in a compartment which can be completely closed. In tuna boats the main deck divides the engine room into an upper and lower area but the cubic atmospheric content is undivided due to a large open hatchway centrally located in the deck generally above the vessel's propulsion engine. There is a bulkhead across the area between the engine-room and the forepeak. Mr. de Fever mentioned that the bulkhead is not tight in all cases and that is correct. In some instances there is nothing more than a framed opening through the bulkhead or an opening with an expanded metal lathe-type door fitted between the forepeak and engine room. Rather than maintain this condition it would be better to alter the bulkhead or fit a tight closing door, to make separate compartments of the forepeak and engine room. Two separate inert gas systems could then be used, giving greater efficiency and greater fire protection than that of a single system. Actually portable fire extinguishers are considered best if the crews would stand a continuous watch thereby being on hand at the start of any fire.

In the southern California fleet double acting manually operated portable hand pumps are being installed and represented to produce adequate pressure to furnish fog to cool the internal atmosphere of compartments in which a fire exists. By trial, the pressure produced has proved inadequate to furnish other than a heavy spray. Mechanically-driven pumps capable of developing constant pressure at the nozzles will produce a good heat-quenching fog.

TRAINING

Mr. William C. Miller (U.S.A.): Naval architects can contribute towards safety at sea by designing strong ships with good stability, plenty of freeboard, and with engines of the right power for the hull. The quality of " good design " includes crew accommodation as well as the actual hull design. In the southern California fishing fleet there has been relatively little loss of life as the result of fishing operations. Even in launching lifeboats and using small boats the crews seem to be accomplished. The factors involving safety at sea are both physical and human, and they are important in the mechanical operation of fishing boats.

Many people have expressed the opinion that the crews of fishing boats cannot be taught the proper use of their ships and equipment through a safety training programme. Such is the attitude of a defeatist. A training programme can be formulated and carried out if the people behind it have faith, will, and the determination to work. In devising such a programme FAO might contribute much.

The need to instruct fishermen in safety at sea seems to be an international problem, although it is more acute in some areas than in others. The huge financial investment in the southern California fleet—where the boats range in size from 45 to 165 ft. (14 to 50 m.) and where one boat may cost upwards of $750,000 (£270,000)—is in itself an incentive to owners to promote safety at sea. Many vessels have been lost because the crew failed to pay attention to a particular item at the proper time. They have sometimes abandoned ship too soon without taking available measures to forestall the ship sinking or, in the case of fire, have not fought it with the means at their disposal.

About 32 per cent. of the losses in the southern California fleet arises from stranding, and the major cause of stranding has been lack of applied seamanship.

Mechanical failures have little to do with these losses. Mostly the reason has been navigational error. For instance, in catching bait tuna clippers work in shallow coastal waters where, because of their great draft, the danger of stranding is always present and can only be offset by the crew taking accurate depth soundings and maintaining the vessel in condition to operate clear of obstructions at all times. Strandings often occur when skippers, devoting their attention to catching bait, take their vessels beyond the limits of their draft.

About 29 per cent. of the losses in the tuna fleet have been caused by fire, probably as the result of poor seamanship or mechanical failure, or a combination of both. Clipper crews are notoriously lax in their watches during voyages, consequently fires gain headway before they are detected, and then it is often too late to extinguish them. From reports by the crews and from general experience with these boats, it is estimated that 90 per cent. of all total losses by fire could have been prevented had the man on duty watch in the particular area been there and used available fire extinguishing equipment at the start of the fire.

About 25 per cent. of losses are due to the clippers sinking. Here again lack of seamanship and mechanical failures both contribute to the disasters. The equipment which should be used in some circumstances to prevent sinking is often lacking, or if it is available on board, the fishermen lack the knowledge of how to use it. For example, it has been said that as much water runs through a clipper as around her when she is at sea and loaded with bait, because the bait wells are supplied with water through two through-hull fittings of 10 in. to 14 in. (250 to 350 mm.) diameter each and this water is always under pressure from electrically driven centrifugal pumps. If the pipes are allowed to deteriorate, a break may occur and, unless it is quickly detected, the vessel may be very rapidly flooded.

Due to the construction of the boat the water runs forward and she begins to go down by the head. Because of the size of the through-hull openings and the potential volume of inrushing water, it is dangerous for the crew to go into the engine-room area or the shaft alley to close the hull valves when a break occurs, or to combat the situation inside the ship. This condition again stresses the need for attention-to-duty watches. However, if crash mats and hogging lines were provided and used it would often be possible to save the boat. There have been a number of occasions when a crash mat placed over the entrance suction would have saved the ship.

Collisions cause about 7 per cent. of losses, and while there is reluctance to ascribe them to wilful neglect there is no doubt that on many occasions they could be avoided. For example, reports from skippers on such collisions frequently ascribe the cause of the accident to the fact that two or more boats head for the same school of fish and a collision occurs because neither skipper will give way until the last moment. The skippers seem to ignore the International Rules.

Capsizing counts for another 7 per cent. of losses, and that is concerned with stability. Skippers should have proper knowledge of loading and the use of the boat as well as the use of bait systems, piping systems and brine systems. Tuna boats have been lost, possibly because a bulkhead, subject to pressure of water or brine, had given way, rapidly flooding adjacent compartments and causing the vessel to roll over. There is no definite proof that this has happened as these vessels have usually been lost in deep water, but the crew's reports on the facts tend to indicate that a shifting of internal weights due to structural failure has been responsible.

In all these cases of loss of ships the lack of ability on the part of the crews had some influence, so that it seems a very sound proposal to draw up a training programme regardless of how difficult and how long it may take to obtain results from it. The programme is certainly worth while because when a ship is lost everybody concerned economically loses in one way or another apart from the possible disaster of loss of life.

The whole point of training crews to do their duties properly and take the necessary safety precautions is excellently illustrated by the record of clippers operated by Mr. Petrich and his family. The ships are kept in first-class condition and the crews are trained to carry out their duties properly. As a result the loss ratio of their clippers is about 80 per cent. lower than that of the average.

Mr. Frederick Parkes has asked for complete dependability and reliability of vessel, engines and equipment. Generally speaking, these factors exist as he wants them. For instance, engineers have developed machinery over long years of experience and usually the engine itself will stand up to the service demanded of it, but often it is in the care of inadequately trained personnel. If all the crews of fishing vessels throughout the world were properly trained, particularly those in very expensive vessels in which is installed complicated machinery, the cost of maintenance would be very much lower and so would be the expense of losses from various causes. If a good boat is not properly operated it can become an unsound vessel in a very few years.

Commander A. C. Hardy (U.K.): Mr. Miller clearly makes one point, that the machine is tending to become the master of the man. Scientists, naval architects and marine engineers produce devices which make a ship work almost automatically, but that calls for human beings with trained mental and physical qualities to operate such a ship successfully.

Mr. Francis Minot (U.S.A.): There is always some question regarding who are to be trained and what they are to be trained in. The New England trawling industry provides an example of the problems involved. Most of the New England trawlers are individually owned or owned by small groups. In most of these bodies the crews get good wages and the skipper is well paid but the owner, who has made the initial investment and who meets the cost for maintenance and repairs, is lucky to make any money. As ships become more and more expensive to build and operate it has become increasingly difficult for an owner to make a profit. If a trawler's engine does not develop the power the skipper thinks necessary, the owner will be asked to put in a new engine, probably with 50 per cent. more horsepower. If such an engine is put into a vessel which is, perhaps, 10 years old and never intended to have that amount of power, there are going to be difficulties with the hull. That would cost a lot of money so the owner, because his profit is small, skimps on maintenance and repairs. As a result, the trawler probably will not operate efficiently during the following year so the skipper, if he is a good one, will leave to take over a better ship.

The whole tendency in these circumstances is a downward spiral of performance which may be aggravated if a trawler has a bad catching season or the market price of fish drops. In any case, owners are unlikely to have money available to start training schemes. Of all the trawlers sailing from New England docks there are, probably, only about 400, commanded by the best skippers and fitted with the best equipment, which yield reasonable profits for their owners.

Mr. A. J. Wegmann (U.S.A.): As an operator of small vessels in the Gulf of Mexico he had built a carefully planned boat and had installed modern equipment such as an automatic light plant, automatic steering gear and expensive shore-to-shore radio. The skipper of this vessel was an honest but inefficient operator and, in an effort to extend the night range of the very expensive radio equipment, he used a screwdriver to tinker with it. Then the automatic equipment on the light plant went wrong. He removed some of it, and then found the equipment would not work efficiently, so he decided to operate it manually. He dealt in a similar fashion with the other equipment and the result was that not only did the efficiency of the trawler suffer but a very expensive bill for repairs had to be met. It would, of course, have been easy to dismiss the skipper, but the chances were that the next man would have been even more inefficient.

This case provides an example of the need to instruct skippers and crews in the use of modern equipment. It would surely be common sense to spend money on such an educational programme rather than constantly to be faced with heavy bills for repairs—as much as $5,000 or $6,000 (£1,800 or £2,100) at a time. In addition, there is the fact that each time the vessel is brought into port she must be laid up for repairs, so that not only must a lot of money be spent unnecessarily but operating time is lost. Further, misuse of the equipment may mean a breakdown at sea so that no catch is made, which leads to a complete loss of income.

Would it not be an advantage if crews were trained primarily as to operation duties and the necessity to perform these duties?

Mr. G. O. Huet (U.S.A.): Training of personnel can be very difficult. For example a new 50 h.p. engine was installed in a 50 ft. (15.2 m.) tug-boat, and a propeller designed for it to give a towing speed of 5 or 6 knots an hour. After the first trip the skipper reported " That propeller is all wrong ". The skipper thought that he should use the full power of the engine but, if he did that while towing, he would overload it and the exhaust temperatures would go up. That would lead to trouble with valves and burned pistons and so on. This was explained to the skipper, but for about 4 months he was continually urging that the propeller should be replaced until one day experience during a heavy tow, while working with a bigger vessel, having a larger but improperly designed propeller, suddenly convinced him of the efficiency of his own engine and propeller.

Mr. William C. Miller (U.S.A.): A training programme of the type seconded by Mr. Wegmann should be put into operation among fishermen of all countries. Naval architects, engineers, boat builders and other people seem generally to favour such a programme. Motion pictures could be used ashore and afloat to educate fishermen. Tuna clippers on the U.S.A. West Coast carry some motion picture equipment with them which they use to entertain crews during their leisure hours. Such equipment would lend itself naturally to educating crews, especially as they have very little diversion at sea other than watching motion pictures. The value of such a method was proved in World War II when the military services used films for instruction, and it would certainly pay the fishing industry to make this use of motion pictures. To this end he has studied the needs to produce such films and has developed plots and characters to create interest. He has discussed this idea with the United States motion picture industry and also with certain vessel owners, insurance and underwriters and suppliers of vessel equipment and commodities, but has not found any willingness to provide the money for such a programme although all of these people recognize the need for and the merits of the plan.

Commander A. C. Hardy (U.K.): If the idea of using motion pictures to instruct the fishermen in seamanship and safety measures was carried to an international basis, it is something FAO could take up. A film certainly could be used to instruct fishermen in the use of new machinery and new methods.

Mr. Howard I. Chapelle (U.S.A.): The question of training crews has long been a problem in the fishing industry on the U.S.A. North East Coast. It is not a simple matter to solve. For example, in certain parts of New England the unions require that an engineer be a full-fledged fisherman and only a part-time engineer. It must be kept in mind that a fishing vessel goes to sea to do one thing as long as it can, and that is to fish. The fishing industry does not exist for the purpose of creating boats. Boats are developed for the practical purpose of fishing. It is unreasonable to expect to change the whole industry to provide a better boat, although it is quite right to look for a change. But how many people or firms in the fishing industry have the time or money to spend in helping to set up training programmes in North America? There is room, of course, for training programmes in fisheries, sponsored by government agencies. But that is very different from organizing a privately-sponsored programme and while such a programme may be desirable it is not likely to be developed by the North American fishing industry.

Mr. Olin J. Stephens II (U.S.A.): Fishing boats are very much like other small craft; they are getting more and more complicated all the time with the introduction of high-speed engines, electronic equipment, etc. With all this specialization going on there is less and less contact between designers, builders and fishermen when in fact there should be an increase in co-operation. An effort should be made to get naval architects to go out in fishing boats so that they will produce designs and specify equipment in the light of working knowledge. It is no use putting in equipment which crews cannot understand or work. Development must be gradual. Crews must be introduced to new equipment and machinery and trained in its use, step by step, and that calls for co-operation between all concerned.

Mr. William S. Hines (Canada): Ownership seems to have some bearing on how much training and education can be promoted. Fishermen who do not own their boats tend to be careless in looking after and operating them. The Department of Fisheries in Halifax, Nova Scotia, has therefore promoted a programme for fishermen to own their own boats and be more interested in taking care of them.

Mr. R. T. Whiteleather (U.S.A.): Although the U.S.A. fishermen are generally good seamen they are not sufficiently trained to cope with all emergencies. Efforts to instruct and educate the fishermen have not produced encouraging results. Demonstrations and exhibitions are not well attended. There seems to be no real incentive among fishermen to seek the information which is available and some better procedure needs to be established.

Mr. George C. Nickum (U.S.A.): There is no doubt about the incompetence of some of the crews of fishing vessels so far as seamanship is concerned. For example, about 1938 he took a delivery trip on a tuna vessel which, at present-day prices, would probably cost $400,000 (£140,000). He woke up at 2 o'clock in the morning as the vessel was sailing down the Pacific Coast and went to the bridge. The vessel was off the Columbia River and he chatted with the captain. In the course of the conversation he asked what light was shining from the coast. The captain told him but he felt the information was inaccurate, so he pointed out several other lights and then said to the captain " You are heading right smack for Peacock Spit. That is the lightship over there and we should head that way ". The captain replied quite casually " Oh yes, maybe you are right. Let's go that way ". So the vessel was turned about and fortunately his advice to the captain had been right and they headed towards the ocean. Such was the navigation skill of the captain of a $400,000 vessel.

He did not like the Government regulations but if such regulations for licensing a crew would ensure the employment of people who could navigate and keep the vessel off the shore, then it seems there is a lot to be said for them.

Mr. Poul A. Christensen (U.S.A.): Safety at sea is not only a question of life and death for the individual but also for the industry. If, for example, a man owns an inshore trawler in the New England area, he must pay from $5,000 to $8,000 (£1,700 to £2,900) a year for insurance premiums. This is a lot of money to be taken out of the boat's share of income every year, and if these insurance rates could be lowered by having better-trained crews, the saving effected could be used by owners to keep their boats in a better state of repair.

The New York *Journal of Commerce* gives lists of marine casualties and it seems that almost every day an East Coast trawler is disabled and is towed in by the U.S. Coast Guard. There are about 2,000 trawlers actively engaged in fishing in the area, which means that one out of six is towed in every year. This is a very poor record.

Certain Coast Guard regulations do apply to skippers of inshore trawlers, but none apply to the engineers. While it is true that people generally want as little interference from the Government as possible, nobody objects to the fact that every person must face a driving test before he is licensed to take out a car. It seems reasonable that engineers in fishing boats should pass some qualification tests. The common procedure at present is that, when a vessel is ready to sail, the skipper goes to the nearest bar and employs the first man who says he is an engineer.

Similarly, it seems reasonable that members of the crew should possess some qualification as seamen. He, personally, would prefer to go fishing with an excellent crew and a bad boat instead of an excellent boat and a bad crew.

Mr. William C. Miller (U.S.A.): Manufacturers of engines and other equipment ought to give more help than they have in the past in teaching fishermen the correct way to operate and handle engines and equipment.

INSURANCE COMPANY'S FAULT

Mr. H. C. Hanson (U.S.A.): The insurance companies are largely to blame for the bad situation in the fishing fleets on the U.S.A. West Coast. They insure vessels without insisting on an agreed standard of safety. For example, he had surveyed a 103 ft. (31.4 m.) wooden vessel and in his opinion she was not fit to go to sea. She was built of very light sawn frames, and the floor timbers and keelson were made for small boat instead of large boat construction. Frames were spiked into the keelson and there was only a 4 in. (10 cm.) ceiling. The only bolts in the vessel were used in the shelf. The boat was badly constructed but, despite the adverse report, the insurance company issued a cover for it.

Another instance of bad construction was that of a brand new boat whose owner asked him to test it for stability. Intuitive sense is needed in stability assessment and a man who has all the technical qualifications is still not the right person to assess stability if he lacks this intuitive sense. In this case, when he stepped on the boat the water came half way up the bulwarks. There was no need for any stability test. He told the owner that it would be necessary to take off some of the deck tanks and put ballast in the hold. The owner refused to do so although warned that he might lose his boat and probably drown himself and his crew. The facts concerning the boat were sent to the insurance company but they still insured the vessel which a few weeks later was lost at sea. The mate was drowned.

These two examples show very clearly why insurance companies are to be criticized. They could and should insist on adequate standards. It is a matter of importance and urgency that the American Bureau of Shipping or some other classification society should set up standards of construction for wooden fishing vessels.

Mr. William C. Miller (U.S.A.): What Mr. Hanson has said about insurance underwriters is true. They have not paid attention to anything other than the dollar premiums and, consequently, they have lost much financially.

He had done many general condition, value and damage surveys for underwriters and his experience has been the same as that of Mr. Hanson, particularly when a risk is not recommended. The undesired information is shelved and the underwriters then get a report from another surveyor which enables them to insure a vessel. In fact, there are many instances of underwriters sending out a surveyor who knows nothing about the boat and is not qualified to survey it.

Mr. Arthur de Fever (U.S.A.): The facts stated by Mr. Hanson and Mr. Miller are true. The 103 ft. (31.4 m.) boat mentioned by Mr. Hanson no longer exists. There seems to be an unlimited number of underwriters so that if one refuses to insure a bad boat, there is always another who will. He spoke from experience as he did much survey work for underwriters.

Many owners do not want to listen to surveyors mostly because the recommendations made involve extra cost in construction. There is also the fact that underwriters in some cases do not insist on such recommendations being complied with by owners and many unsound vessels are covered by insurance despite the most strenuous objections of surveyors. One of the problems involved is that some agencies handle 30 boats or more which represents, perhaps $500,000 (£180,000) in premiums. The agencies in some cases can bring pressure on underwriters to insure poor risks by threatening to take away the entire account. In fact, these poor risks may be considered by some underwriters as an overhead factor in their business but if enforceable regulations were issued to cover standards of construction it would put an end to this practice.

Mr. George C. Nickum (U.S.A.): There is a great number of design defects in many of the larger wooden vessels. For instance, there was a case of a tuna clipper in which the difference between loading conditions caused the vessel to hog about 4 in. (100 mm.) between emptying and loading tanks. Obviously she was weak and this was reported to the underwriters. They should not have let her go to sea but they issued a cover for her.

However, he was opposed to Government regulations for

the construction of vessels because they would tend to be inflexible, add to the cost of construction and would be difficult to enforce. Private enterprises should be able to agree on suitable standards and it is to be hoped that the losses sustained by the underwriters will make them keep a close watch on construction and other features of fishing boat design.

Mr. R. T. Whiteleather (U.S.A.): Underwriters should set up a system whereby the owner who operates his boats with a high degree of safety gets cheaper insurance. This would provide a dollar and cents incentive to observe good standards of construction and to train crews to guard the safety of vessels.

Mr. G. O. Huet (U.S.A.): When a concern is placing a great deal of insurance on boats and they have a vessel which a surveyor will not pass, they bring pressure to bear on underwriters to cover the unsound ship. Naturally, the underwriters are not going to lose the total business for the sake of one unseaworthy ship so they insure it. If regulations are to be set up and enforced, there must be an enforcement body. Merely to study the problem and make recommendations is not going to do any good.

Mr. H. C. Hanson (U.S.A.): If underwriters are happy to take bad risks in order to get business, it must be evident that the serious operators have to subsidize the careless ones. It penalizes the good operator or increases the price of fish unnecessarily.

Mr. William C. Miller (U.S.A.): As Mr. de Fever pointed out in agreeing with several speakers, some of the general agents acting between the underwriters, the brokers and the owners, have not only allowed a situation detrimental to the financial interests of underwriters and owners to develop, but have actually aided the development of the situation. The desire on the part of some of the general agents to collect their commission seems to override their moral obligation to the underwriters who have placed confidence in them.

RULES

Professor G. Schnadel (Germany): Mr. Miller, in discussing the losses of tuna clippers on the U.S.A. Pacific Coast pointed out that there are no rules laid down by classification societies for the construction of wooden vessels in U.S.A. This was a state that existed in Germany and it was ended by Germanische Lloyd laying down rules for fishing boat construction in 1939. There was some opposition to the rules at the start but now they are used by boat builders and are accepted by fishermen themselves.

Mr. S. A. Hodges (U.K.): In the United Kingdom the Ministry of Transport and Civil Aviation is responsible for regulations regarding the safety of life at sea. It has safety regulations for fishing vessels in respect of lifesaving appliances, fire extinguishing appliances, navigation lights and distress signals. In addition, of course, there are the general powers which the Ministry's surveyors have for detaining any ship they consider to be unseaworthy and the powers of the Minister to order a formal inquiry in open court into the loss of any ship. All reports of casualties to British ships, including fishing vessels, are also carefully examined and if the Ministry considers there is useful information in reports, it is published in the form of Notices to the Industry. For instance, when difficulties were found in clearing bilge water from certain fishing vessels the Ministry issued a Notice pointing out the great value of the bilge injection in dealing with such leakage and of the necessity for keeping the bilge injection in good condition. The same Notice also pointed out the advantage of carrying a flexible suction hose in the engine room and provision for its attachment to one of the pumps. The use of such a hose would make it possible to keep the bilges, in the machinery space, clear of water if the permanent bilge suctions were choked with dirt or ashes.

Dr. Gian Guido Bordoli (Italy): In Italy all boats, including fishing boats, must be constructed according to the regulations of Rina (Registro Italiano Navali), but there is no special section for wood or steel fishing boats. The result is that the boats are often too heavy in construction, as a comparison between Italian wooden fishing boats and those of other countries will show.

Mr. H. C. Hanson (U.S.A.): It is a constructive move to set up regulations or standards covering fishing vessels. In Canada the Steamship Inspection Service have rules concerning all types of vessels, including fishing craft, and many of these rules are good and practical and have led to improvements in construction.

On the U.S.A. Pacific Coast there are practically no regulations so far as fishing vessels are concerned, except in the case of lights and lifesaving appliances. Small wooden vessels in the U.S.A., particularly tuna clippers, are mostly constructed to builders' own specifications and in many cases that construction is not sound. It is for this reason that he had tried to set up a good standard of construction for such vessels. Steel vessels are generally built to the American Bureau of Shipping Rules so that no great difficulty is encountered with them. In the case of tuna clippers the freeboard is usually non-existent aft and, unfortunately, owners will do all they can to eliminate whatever freeboard is provided by the designer. If there were regulations governing such a factor, then designers would have the right sort of backing to produce better vessels. Marine underwriters form the only force that could back up designers in the U.S.A. at the present time but they in many cases will insure a vessel in the full knowledge that its construction and stability are unsound.

Mr. George C. Nickum (U.S.A.): Government regulations stem from the desire of people to see that no individual can hurt the public, economically or physically. In his opinion no harm has been done in the U.S.A. by lack of Government regulations, but he felt that improvements in standards of construction should be made by the U.S.A. fishing industry if it is to avoid a situation where the Government and people believe regulations should be set up. The underwriters, the technical societies, and the fishing vessel owners, should take an active interest in establishing standards of construction so that the need should not arise for Government regulations.

Mr. L. Fernandez Muñoz (Spain): There should be regulations making it necessary to have two watertight bulkheads in wooden boats, one in the forward end of the engine room and another as a collision bulkhead. It may be said that in the ordinary way boats have these bulkheads, but in practice they have two holes in the bottom of the two bulkheads, so that the water comes into the engine-room and must be dealt with by a Lewis pump. When fish are put into the hold the boat trims forward and the water cannot get into the engine-room. There is then no means of removing the water and in the past year he had seen five boats lost because of this fact. There should be a regulation making it necessary to install a small hand pump for emptying the water.

Commander L. E. Penso (U.S.A.): The U.S. Coast Guard does not wish to regulate the fishing industry. The Service is faced with cuts in its budget and will find difficulty enough in coping with its present responsibilities without adding fishing vessels to its list. The Service would prefer the fishing industry to regulate itself. If regulations were introduced by the Government for the fishing industry the situation might be parallel to that existing in small pleasure boat building. Coast Guard motor boat regulations are so loose that it is possible to operate a boat which is actually unsafe and there is not much the Coast Guard can do about that fact without recommending that the U.S. Congress promulgate a new law. Rather than do that the Coast Guard has been working closely with the Yacht Safety Bureau, the National Association of Engine and Boat Manufacturers and the underwriters. An organization, known as the American Boat and Yacht Council, is now being planned by these bodies. Its purpose will be to formulate a set of standards for safe construction and equipage of boats. The fishing industry could join in that organization or set up a similar body on its own. Once such an organization is functioning, it will have a technical board which will be divided into a number of committees, one on equipage, one on construction, and so on.

It took a tremendous amount of work to start this new organization. It will have a president, a vice president and a board of directors, but the technical board, composed of 24 technicians, will be the real working group. It will deal with the practical problems of yacht design and construction.

Mr. Olin J. Stephens II (U.S.A.): It might be a good thing for the U.S.A. fishing industry if it became associated with the American Bureau of Shipping. At present the Bureau's rules do not cover wooden boat construction in a practical way but there is evidence that the Bureau would be interested in the problem if they were invited to consider it. With the help and co-operation of fishing vessel designers and builders they would be able to set up standards and rules as a guide for boat builders and underwriters.

Mr. William C. Miller (U.S.A.): The problem has been discussed with the American Bureau of Shipping but they do not want to become involved in the fishing boat business if it would mean they would have to argue all the time with owners about the cost of construction at specified standards. If a fishing boat is to be built according to the specification of any classification society there will be extra expense so that the first need is to convince owners that the standards involving the extra expense are essential.

He, personally, did not want Government regulations for the fishing boat industry. It would be far better for the industry to set up its own standards for construction and safety.

Mr. H. C. Hanson (U.S.A.): There are many practices current in boat building and in boat handling which call for some sort of regulation either by an authority set up by the industry or by the Government. For example, he had provided on certain vessels low hatches for use in handling salmon. The hatches were secured by cleats and covered with canvas, which was to be sealed. But, on inspection, he noticed that the crew had all the hatch cleats off, and there was loose canvas laying around. They had taken the cleats off to make work easier, so he had informed the insurance company that he would accept no responsibility if the vessels were submerged for lack of proper hatch covers.

In another instance he had seen a 65 ft. (18.3 m.) boat being built with 1 in. (2.5 cm.) planking. He would never have passed her as fit to go to sea.

These are two more examples of the reason why there should be an authority to issue and enforce regulations governing construction and safety at sea.

Mr. Arthur de Fever (U.S.A.): The American Bureau of Shipping will co-operate in the design and construction of a wooden vessel if they are requested to do so.

Mr. Jan-Olof Traung (FAO): They were invited to contribute a paper on safety at sea but they declined as they were apparently not interested in fishing boats.

Mr. G. O. Huet (U.S.A.): Who wants rules to be established in the fishing industry? Is it the owner or the builder or the naval architect? An owner who wants to feel reasonably sure that his boat will be properly constructed goes to a competent naval architect who has had experience with the type of vessel concerned. The owner will get a good vessel, and no rules adopted by any society, unless they could be enforced, would be of any value.

His experience with insurance surveyors in the south of the U.S.A. shows that they had been severe in their appraisals. Most of them are competent men who have had many years of experience in marine work. He had seen cases where surveyors, not thoroughly familiar with small boats, have called in somebody with such experience to give an opinion and make an appraisal.

The Bureau Veritas have rules for small wooden boat construction and at one time Lloyds also had rules for building small boats and yachts. Rules have also been made in the U.S.A. and they can be used if anyone wishes to use them. A technical section can be established in an organization to study rules, whether it is an association of naval architects, or of people interested in boating. The function of technicians in the group concerned would be to find out where failures occur and try to avoid them in future boats. It is all a matter of organization. Establishing codes in any industry calls for the co-operation of the various members of that industry. That is the only way in which progress is made and it is continually being done in every profession or industry. That applies to progressive naval architects or to mechanical engineers.

Mr. Huet asked what was being attempted in the discussion, what was the object of setting up the rules and what were they trying to protect?

Mr. H. C. Hanson (U.S.A.): Boats are defective.

Mr. Huet: Who builds these defective boats? The boat builder or the naval architect?

Mr. Hanson: The insurance companies, the boat builders and owners are all involved. The attempt to set up rules and regulations is an effort to make progress. Apparently the American Bureau and many naval architects and boat builders think it is logical to set up standards. Mr. Huet was the first person who had said there was something wrong with forming a better set of regulations. Mr. Hanson added, that, as a boat builder, he would not be afraid of any regulation.

Mr. R. L. Eddy (U.S.A.): Mr. Huet brought up a pertinent point when he asked who wants regulations, the builder, the owner or the Coast Guard? Nobody suggested the fisherman and the fisherman is certainly the man mostly in need of regulations.

The fisherman depends on people with knowledge to build him a vessel with good sea-keeping qualities. He depends on engineers for power plants that will require a minimum of technical know-how to prevent breakdowns and blow-ups. Regulations would be excellent for the fisherman if he was helped economically. For example, the State of Louisiana have some 6,500 vessels that often fish at a distance of more than 15 miles at sea. Four attempts to introduce regulations, allegedly sponsored by unions, failed to consider that in effect those regulations would have prevented the fishing population in Louisiana in participating in any off-shore industry that might later be developed. A fishing vessel was defined as any vessel over 15 gross tons operating offshore, in excess of 15 miles at sea, engaged in commercial fishing. In effect that definition would have prevented some of the 6,500 vessels from operating in Louisiana. Other Bills introduced had the same effect and no consideration was given to the way in which a shallow draft vessel operated. They would have been required to have licensed mates and pilots. Who can get lost in the Gulf of Mexico? There are only two ways of sailing there—in and out. It is only necessary to drift along to land on a safe shore.

Regulations would also have required all kinds of engineering equipment which was not necessary and would have increased the draft of a boat to a point where it would not have been able to pass round marsh and tide lands. Due consideration has not been given to the fact that the fishermen vary around the coast. They are not all deep water sailors and they do not all sail big ships which have to catch 90 tons of fish on each trip to pay their way. Some of the vessels need only 5 tons of fish and they make money on that small catch.

Mr. H. C. Hanson (U.S.A.): Regulations of some kind will be established, perhaps not by the Government but by the industry setting up a standard as a guide. The discussion has been relative to larger vessels and they certainly need regulations. There are boats being built which he certainly would not pass as seaworthy. For that reason some regulations should be set up. This does not mean that laws have to be passed. What is needed is a set of regulations and a standard of construction that can act as a guide.

Mr. R. L. Eddy (U.S.A.): There is no question but that is true. But a letter of recommendation to owners signed by important naval architects, must have weight and some day might have the effect of a law. Consideration must be given to all phases and consequences of setting up standards or a greater injury may result to the fishing industry.

Mr. H. C. Hanson (U.S.A.): Referring to the examples he had given of loss of life through bad construction, he said that he would be glad to be instrumental in setting up standards which insisted on vessels being so built that the loss of life could be avoided. All naval architects certainly wish to design boats that are seaworthy, and no boat builders wanted to construct a boat to go to sea, knowing that they are building a wreck.

Mr. Eddy said he appreciated the point.

Mr. Howard I. Chapelle (U.S.A.): Regulations and safety at sea concerned all connected with the fishing industry. There are two distinct points of view involved, and naval architects and those interested in safety want fundamental and sensible rules to be adhered to. In many places the livelihood of a number of people are involved in small boat operation. Much of that operation is close to the shore where safety regulations are not really required, and where, if accidents occur, the fishermen can usually walk ashore.

There is one group of people involved in both safety at sea and the economic problem—the owners. If they are not interested enough to put pressure on the unions, the underwriters, and the classification societies, then there seems little hope that much can be accomplished by proposing rules or standards. It is in their interest to do so although, of course, there are many difficulties in the way of establishing a set of regulations. As pointed out by Mr. Huet, there is no restriction on the owner and designer co-operating with the builder to construct a vessel according to the standards of the American Bureau, the Bureau Veritas or Lloyds rules. There is no rule that prevents architects and builders from adopting the U.S. Coast Guard regulations so far as they are applicable to small vessels in such matters as the insulation of fuel tanks, fuel lines, fire extinguishing equipment, etc. All the information is available, ready to be taken. If the owners will not take it and do not want to take it, or if they would rather build cheap and take risks, what can be done about it? There seems to be little more possible than to express an opinion.

Mr. Hanson has mentioned cases of bad construction such as occur all through the U.S.A.

There are two things that to some extent nullify the statement made. In Canada, for example, a great number of smaller boats are built under the rules of their Steamship Inspection Service. The Service must approve the plans before long voyage vessels are constructed and this is a very healthy situation in many ways. It does add to the cost of the vessels, and it is very difficult to depart from certain standard forms laid down, but the scheme has worked well. The men concerned with the regulations are sensible and competent. There are differences of opinion between designers and the Inspection Service, but this happens also in dealing with the U.S. Coast Guard. He had dealt with the Coast Guard while in the Army and they were reasonable, but their requirements added to the cost. The matter goes back consistently and constantly to cost. If the owner is not willing to spend the money to have a safe vessel there is nothing the designer can do, or the boat builder or others connected with the vessel. Designers and other people can only enforce standards of construction when those standards are supported by law. So the problem resolves itself into a choice of two conditions: (1) that in which owners of vessels want sound construction and will support naval architects and boat builders in building safe craft; (2) enforcement of standards by Government regulations. One or the other will come about in the U.S.A. and when this happens the standards will have to be enforced either by the Government or by an organized group within the fishing industry.

Mr. H. C. Hanson (U.S.A.): Mr. Chapelle seems to think that no more than a letter to the American Bureau and to the underwriters is necessary. That is not a question of law but something for the betterment of the fishing industry. It provides a guide and that is what he (Hanson) is striving for, not to make laws but to introduce standards of good construction throughout the fishing industry.

Mr. George C. Nickum (U.S.A.): There has been some confusion about what is proposed and what is not proposed. He had suggested means of meeting these problems by forming a committee, an international committee. Hanson had suggested writing a letter and that action be taken immediately.

Mr. William C. Miller (U.S.A.): Wherever one goes there is certain to be some degree of laxness in building, surveying and engineering. He recalled an incident concerning several tuna clippers sailing from New Orleans to San Diego, where they were given a rigid inspection because they had encountered great mechanical difficulties on the way to San Diego. The boats had been given a clean bill by a surveyor at New Orleans although they had been fitted with war surplus gear. In fact the surveyor's report went into detail about how the gear had been taken out and examined and passed as being perfectly all right. But when the clippers got to San Diego the bearings on the shafts were revolving in the case bore and the gear teeth had been galled and broken. The machinery broke down so badly that one of the boats had to be towed the last several hundred miles. In all probability the gear installed had been used during the war and had been, presumably, repaired on the battlefield. Shim brass had been used and the shims had been placed round the outer circumference of the bearing race and between the case. Once in operation they began to rattle and loosen and fall into the gear. Putting that ship back into operation was a very expensive job because of the laxness and inefficiency of a surveyor in New Orleans.

Mr. G. O. Huet (U.S.A.): The Bureau Veritas and Lloyds have rules and regulations for the construction of fishing vessels and small craft, but the American Bureau have no such rules. The Society of Naval Architects and Marine Engineers are concerned naturally with larger vessels but they have taken an interest in the last few years in trying to establish information regarding small vessels. Such boats, and the construction of them, had been discussed at local sections of the Society and several papers have been written on the subject.

Mr. Huet said that, as an architect, he is very much interested in progress towards establishing correct methods, and is a member of the Society of Small Boat Designers, formed to study and correct failures in small boat construction.

Commander L. E. Penso (U.S.A.): People talk about laws and regulations and recommended standards but there is a great difference between the three. The organization previously suggested for the fishing industry, composed of operators, builders, naval architects, underwriters and other interested parties, would be concerned with producing a set of minimum standards and an owner would not have to build to these standards if he did not wish to do so. However, insurance companies would use it as a basis for insuring boats, and surveyors would use it as a standard for recommending insurance to underwriters. In that way a man who did not meet at least the minimum standards would have to pay more for his insurance, which would be a fair thing.

With respect to Mr. Eddy's remarks, when the Lane Bill was introduced, the people behind that Bill were the widows of men lost at sea and the wives of fishermen who did not want their husbands to be drowned. They did not have a union or pressure group, so they did not get very far. If more pressure were exercised through such groups, some day a law might be passed. This is the sort of situation which the fishing industry might wish to avoid, therefore it must set its own house in order.

Mr. George C. Nickum (U.S.A.): The failure of ships can be classified under four headings: (1) stability and freeboard; (2) life-saving equipment; (3) fire fighting and bilge pumping equipment; (4) structural strength.

There have been international conferences dealing with all four of these categories, and international regulations have been issued for larger ships. It would be a good plan if the U.S. Fish and Wildlife Service sponsored four committees, with members drawn from the underwriters, fishing vessel owners, skippers, architects and the Coast Guards, to draft recommendations in these four categories. If this was done the data presented in Mr. Hanson's paper would be immensely valuable in providing a framework. On this data strength standards could be set up. The four committees could make tentative recommendations and then FAO could do an excellent job in sponsoring international co-operation in developing such standards. Such a scheme as this would take years to become effective but, if properly carried out, it would eventually ensure two things: (i) it would show that the industry did not need Government regulations; (ii) if public opinion insisted on Government regulations, the recommendations of the committees would provide acceptable standards.

Mr. Arthur de Fever (U.S.A.): The proposal to set up an industrial or safety committee is excellent and is in line with proposals he had made to underwriters. He had suggested a committee of naval architects, and surveyors from various areas, to get more knowledge of how the vessels in these areas were built and operated. So far there had been little response from the underwriters and no doubt the rivalry in the insurance business stands in the way of forming such a committee.

Surveying a vessel should start from the time when the keel is laid. After the vessel is completed and fitted out a multitude of faults can be hidden. For example, he had seen vessels under construction where the ceiling was not doubled, particularly in thicker dimensions of 3 to 4½ in. (75 to 115 mm.). In some vessels edge fastening and shoe boltings of bilge strakes are very important for stiffness but they would be of no value without doubling the ceiling.

Recently it was discovered in a 96 ft. (29.3 m.) vessel that the bait tank was made of $\frac{1}{8}$ in. (3.2 mm.) metal. That is criminal. The boat was built on speculation and no inspection services, no engineers and no naval architects were concerned in its construction. It is a type of vessel that is dangerous.

Mr. H. C. Hanson (U.S.A.): Referring to the question of standards he said his object was to introduce better construction methods into the U.S.A. fishing industry.

Mr. William C. Miller (U.S.A.): In the original draft of his paper, "Safety at Sea", was included a full set of actual specifications for the construction of a west coast tuna clipper as prepared by the naval architect and owners concerned. The specifications stated that a 600 h.p. 6-cylinder diesel engine, two six-cylinder auxiliary engines, two 10 in. (254 m.) vertical bait pumps, one 3 in. (75 mm.) fire pump, one 3 in. bilge pump, and many other items of necessary equipment were to be installed, but actually neglected to specify whether these things were to be marine or stationary equipment, etc. The only reference to the construction of the hull was dimensions to be 104 to 105 ft. × 25 ft. × 12 ft. 8 in. (31.7 to 32 m. × 7.64 × 3.86 m.), raised deck type tuna clipper, and it was also stated that the deck-house was to be of tear-drop design. All in all, the specifications for the construction of this clipper at a cost of $320,000 (£115,000) appeared highly inadequate to lay down the details from which to build any satisfactory vessel. The naval architect's drawings were also incomplete and unsatisfactory from a detail point of view. From his survey and examination of the vessel after her construction and the reports he issued to interested underwriters it could be the same vessel Mr. Hanson spoke of when he referred to the lightness of her construction and the inadequacy of her fastenings. Regardless of these facts and conditions this vessel was insured on the strength of other survey reports.

Professor G. Schnadel (Germany): There are great differences concerning safety regulations governing ships of different countries, but the aim must be to get acceptable rules for all countries and to guard the safety of life at sea. This can best be done through an international organization, such as the United Nations, although it is difficult to change the laws and the rules and regulations in the various countries. It is apparent, however, that naval architects and others competent to judge the safety of ships are of much the same opinion on the subject, and so it should be possible to improve the safety of fishing craft in a manner similar to that brought about in merchant shipping by the International Convention for the Safety of Life at Sea.

OIL ENGINES FOR TRAWLERS

by

G. S. HEPTON

ONLY a small percentage of home and middle water, and none of the long distance, trawlers in the British white fishing fleet were propelled by diesels before 1939. This was mainly due to adequate supplies of good quality cheap coal and the uneconomic initial costs of new construction incorporating diesels.

Since the war, the high cost of running steam engines and boilers with inferior coal, and the unstable price of boiler oil, has caused the owner to look to the diesel as a more economic form of power. There are considerations in selecting the best type of machinery for arduous fishing duties. Manning, economy, easy maintenance and initial cost all have bearings on the ultimate decision.

It can be appreciated, therefore, that the change from steam to diesel has created great difficulties for the engine-room personnel, and training ashore has had to be organized for the various types of machinery which have been installed.

Diesel manufacturers sell their products on the brake horse power output. The post-war designs submitted by the majority of them have been, for production reasons, engines which are principally used in industry, with only a small percentage going into vessels. As land engines are not successfully adaptable for marine use, difficulties have often arisen.

All diesel engines selected for use in fishing vessels must be derated by 15 per cent. to 20 per cent. for continuous running to insure against overloading and the inevitable breakdowns which can occur at sea.

The home and near water fleets are now gradually being replaced with diesel-driven vessels and the steam engine will no longer have a place in new construction.

DRIFTER TRAWLERS

Types of machinery fitted for the drifter trawler class of vessel, with a length of between 80 to 90 ft. (24.4 to 27.4 m.), are listed in Table LXV.

TABLE LXV

No. of cylinders	*Cycle*	*Brake horse power*	*Engine r.p.m.*	*Propeller r.p.m.*
6	4	220	475	215
5	2	300	300	300
6	2	300	300	300
4	4	300	300	150

No major defects have arisen with this varied class of power drive, but the chief cause of mechanical troubles has been as follows:

Two-cycle engines: Piston and liner wear is excessive, due mostly to overheating caused by engines running overloaded.

Four-cycle engines: The lighter and fast running engine has proved very successful and gives a high standard of performance, both in the two- and four-cycle types. The future unit for the drifter trawler may well be a high-speed engine, driven through a reduction gear box, having an isolating clutch for the engine to operate the winch drive unit at the forward end.

Winch: In this class of vessel, the winch, belt-driven from an extension of the forward end of the main engine shaft, has drawbacks, namely: (1) slipping of the belt under load; (2) limited power; (3) non-reversibility.

To overcome the slipping of the belt, an Airflex clutch has been tried with a chain and sprocket drive and, although this is a great improvement, the use of a hydraulic drive from the main engine is really the efficient method. This principle is now being adapted in new construction. The power required is 100 b.h.p.

For 100 ft. (30.5 m.) vessels, where drifting is not usually part of their duties, similar propelling machinery has been adapted and no major difficulties have been encountered. With good lines, these vessels can obtain 10 knots steaming at 350 b.h.p.

MIDDLE WATER AND FAROE CRAFT

More powerful vessels have always been used for hake fishing, as well as for Faroe and Iceland fishing in the finer seasons.

In selecting the machinery, it must always be remembered that it is common to use more power when trawling than when steaming. This is because of the heavy iron bobbins and the extra large trawl doors that have to be towed in 200 to 250 fm. (366 to 457 m.), usually at full speed. In these conditions, propellers should be designed to absorb 25 per cent. less power when steaming. By doing this, exhaust temperatures will not be excessive when trawling.

Table LXVI shows a typical example of a four-cycle turbo-charged engine:

TABLE LXVI

	Brake horse-power	*r.p.m.*	*Exhaust temperature*
Trawling	600	200	650°F. (343°C.)
Steaming	500	230	500°F. (260°C.)
Test bed	723	250	700°F. (371°C.)

If the propeller has been designed to absorb all the power when steaming, the engine would be overloaded when towing the gear, with the inevitable result of trouble.

The engine should not be less than 700 h.p. for this class of fishing and the length of the vessels is usually 120 to 130 ft. (36.6 to 39.6 m.).

Best results have been obtained without reduction gears for this higher power, and slower-running machinery, about 230 r.p.m., is recommended.

Due to the deep-water work and the need for manoeuvrability of the vessel, electric winches are recommended. The "Ward-Leonard" system is the usual type in service and is very reliable and efficient. A diesel of 175 b.h.p., to drive the electric generator, is required to give a winch output of 150 h.p. and it should not be any less. Extraordinary loads and overloads often occur but diesel-driven trawlers in this class have been remarkably successful from all points of view.

DEEP SEA TRAWLERS

Diesel machinery recommended for large trawlers should again be well derated: it must stand up to continuous running without trouble. Initial costs are increased by this method of selection but they are repaid by the engine being adequate for the job.

TABLE LXVII

Length of vessel between perpendiculars	*Derated b.h.p. recommended*
150 ft. (45.7 m.)	850
160 ft. (48.8 m.)	950
170 ft. (51.8 m.)	1,000
180 ft. (54.9 m.)	1,200

Reduction gears have not generally been used, and the straight drive is to be preferred.

The well-known opposed piston engine is very suitable and has been used with great success and only the present initial cost prevents much more liberal use of it. In one instance, such an engine using marine fuel and running for three years on arduous duties, showed little cylinder wear and required only piston ring renewals.

For these larger vessels, the electric winch power should be at least 250 h.p., to have sufficient reserve for all conditions.

GENERAL

FRESH WATER COOLING

It is imperative that all diesel machinery should be fresh-water cooled so that temperatures can be fully controlled at all times. Much trouble has arisen through the crude method of using sea water, which has led to cracking and corrosion of liners and cylinder heads, and to other breakdowns.

AUXILIARIES

Auxiliary machinery must be derated to reasonable revolutions and adequate reserve of electric power supplied, which is usually on the low side. When the demands of radar, echo sounder, winches, direction finders, searchlights, electric pumps, heating, electric windlass, etc., are considered, it is clear that a reserve of power is a necessary asset.

ELECTRICAL POWER

In diesel trawlers the electrical power is supplied by the auxiliary engine generators. Some are also fitted with a belt-driven generator off the main engine, having a voltage control regulator. This is a great value on long runs and is strongly recommended as a standard fitting.

Where an electric winch generator is fitted it is good practice to increase the field generator exciter so that in an emergency, or a breakdown of the supply from the auxiliary machines, fishing can be continued by using current from this source, the winch diesel becoming the power drive.

LUBRICATION

Correct lubrication through filters and separators is the keynote to efficiency in the engine room and to the life of the engine.

FUELS

The grades of fuel, generally made specifically for diesel use, available in Britain are gas oil, which is a distillated fuel, and a marine diesel fuel, which may be either a distillate or a blended fuel. The fuels may be said to be covered by British Standard Specification No. 209/1947 which is subdivided into Class "A" and Class "B", embracing in both cases limits for tests such as ash, Conradson carbon and sulphur. In marine circles, gas oils would approximate to Class "A" stipulation, although not necessarily so in respect of ignition quality. Those within Class "B" are more familiarly known as marine diesel fuels. All fuels within Class "A" in British Standard Specification No. 209/1947 also meet

Class " B " stipulation. The specification does not, therefore, attempt to describe two completely different grades, but lays down outer limits for Class " B " grades with closer ones for Class " A " fuels, so that users may have a choice according to the type of equipment.

In high speed main and auxiliary machinery, with revolutions from 300 to 1,000, gas oil should always be used, because of the trumpeting of injectors and stoppages for cleaning that happen when marine diesel oil is used.

Marine diesel oil is quite a satisfactory and economical fuel for slow-running main engines, providing it is pre-heated and centrifuged. Little liner wear has been found under these ideal conditions but care should be taken to avoid water and sulphur deposits, which leave a small percentage of sulphuric acid in the crank case and lead to shaft corrosion.

Injection Equipment

It is of the utmost importance that the injection equipment is properly maintained, to ensure that nozzle deposits, which are also influenced by sprayer performance, are kept to a minimum under all working conditions. Faulty combustion arising from these deposits can result in troublesome operating difficulties, such as excessive combustion chamber deposits, ring sticking and port blocking, increased contamination of the lubricating oil and quicker cylinder and piston ring wear.

Propellers

The common four-bladed cast-iron propeller has been generally used in the smaller vessels, but this practice is changing gradually. Many diesel vessels are still fitted with inefficient propellers, mostly due to high initial cost and the fear of possible damage from trawl doors and warps.

The controllable pitch propeller, tested out in one British trawler, has shown its advantages over all other types, but it is expensive and there is danger of a long delay in repairs after damage, causing loss in days at sea.

Correctly designed bronze propellers should be a standard fitting on all diesel trawlers, to ensure perfect balancing of the machinery and again it must be stressed that the present rating of diesels should be taken into careful consideration when the propeller is designed. Experience has been very costly to the trawler owner in this field, and instances can be shown where propellers designed to absorb the power output of diesels have caused overloading of the engine.

A very bold step was taken by B. A. Parkes in introducing the diesel-engined trawler into all classes of fishing, and the diesel builders have gained a lot of experience at his expense. But the ideal marine diesel engine is still to be produced for the fishing industry.

MOTORIZATION IN CHILE

by

PAUL ZIENER

THERE are about 20,000 small fishing craft in use in South America. Motorization presents different problems in the different countries. One problem is common: the present fleet is made up largely of small sail and rowboats, generally so small and primitive that their motorization is not at all advisable from an economical point of view. Proper mechanization of the boats demands modified, larger hulls.

More than 2,000 motorized fishing boats of different types and sizes are in use in Latin American countries, some of them having been introduced as early as 20 years ago. In certain zones of Argentine, Peru and Northern Chile, fishing boats have been developed which are admirably suited to the work they have to do. But, generally speaking, these boats—including modern imported vessels—have not reached the same standard of efficiency in South America as in other parts of the world. This is due to several factors, some of which are not purely technical and, therefore, change from country to country.

Although small steamships and motorboats have been used in Chile since 1912 (nine steamships and three motor launches in 1915), the real motorization of the fishing fleet started about 1935, and it has progressed very slowly. Of a total of approximately 4,500 fishing boats (1952), about 1,200 have motors, some 330 being small outboard engines. The number of motorized deep-sea fishing boats, not counting foreign vessels actually fishing on the coast, may be about 70, and there may be about 400 powered coastal fishing boats of more than 3 tons displacement. The rest is made up of small craft.

The slow progress is due to economic factors, and the poor results of the motorization is due to difficulties the origin of which nobody has bothered to discover. In general—and mistakenly—they are ascribed to the individuality of Chilean fishermen and to their inadequate technical training, but a closer study reveals that they are caused by factors far beyond the fishermen's influence.

Complexities are found in the Chilean marine engine and fuel oil trades, in which commercial interests sometimes prevail in such a way that they hamper the satisfactory development of the mechanization. Other reasons are strictly technical and, therefore, controllable.

THE COMMERCIAL ASPECTS

The dominating factor of Chilean fisheries is its low economic level. Importers therefore import the cheaper motors such as lightweight gasoline and diesel motors, generally called " high-speed " engines (1,800 to 3,600 r.p.m.). Conditions in Chile are unfavourable for their functioning and durability.

Importers very often do not have the technical knowledge to judge the suitability of an engine for fishing, and fishermen generally have only vague ideas about engines and rely on the salesman's arguments more than a buyer would in technically advanced countries. Another reason for the extensive use of such engines in the Chilean fishing fleet is that during and after World War II they were the only ones available.

The unsuitability of the motors for Chilean fishing is becoming more and more obvious, especially with the gradual increase of the tonnage of the boats. The principal reasons are economic: in the case of the gasoline engine, the high cost of fuel consumption, and for the high-speed diesel, its high maintenance cost.

Canadian motors of heavier design and burning gasoline/kerosene were introduced during the war years but, due to the characteristics of the Chilean motor fuels and other factors, these motors did not give the result hoped for.

After the war, when the European motor manufacturers again could offer their products to Chile, there was a rising demand for engines of solid and simple construction and of lower cost of operation. Light semi-diesels* and heavier gasoline and kerosene engines have been introduced, the great majority of them small, of 5 to 16 h.p. and 800 to 1,000 r.p.m. They offer great economic advantages to the fishery, despite their higher purchase price, and are used increasingly in coastal fishing.

The heavy, slow-running semi-diesels, the real " fishery

* The term " semi-diesel " is used for low-compression diesels (up to 260-290 lb./sq. in. ; 18-20 cg./cm^2 compression).

motors " of other countries, have been known in Chile for years, because the coastal freight companies quickly recognized their excellent qualities. They have not been installed generally in fishing boats because of their high price per h.p. Custom duties, up to 1950, were about 110 per cent. higher for these motors than for high-speed diesels, and about 300 per cent. higher than for gasoline engines, partly because of the higher weight per h.p.

A new law from 1950 excludes all marine engines intended for professional fishing activities from import duties. This measure favours the use of heavy motors and there is a trend towards them.

Full diesels with low revolution numbers are not used due to their high initial cost and the maintenance problem.

The four types of motors available (with the exception of the outboard engine which cover a transitory necessity) are as follows:

- the gasoline motor;
- the high-speed diesel;
- the small, light semi-diesel;
- the heavy-duty semi-diesel.

The price of these engines per h.p. is an almost constant ratio of approximately 1 : 2.2 : 2.6 : 3.4. This is to say a heavy-duty semi-diesel costs 3.4 times as much as a gasoline motor. These figures are free from import duties.

An important difference in quotations ought to be pointed out: American engines are generally quoted without (or with very little) installation material; European ones with most of the necessary elements, and Scandinavian ones generally with complete installation, equipment and some spare parts.

COST OF OPERATION

Based on the characteristics of the motor fuels obtainable on the Chilean market and the fuel consumption of the common types of motors along the coast, the fuel cost for one brake horsepower (b.h.p.) per hour has been calculated as follows (prices April, 1953)*:

		Relation
Slow-running semi-diesel (20–100 h.p.)	CH$ 1,14 per b.h.p./hr.	1.00
Lightweight semi-diesel (8–20 h.p.)	„ 1,43 „ „	1.26
High-speed diesel (20–100 h.p.)	„ 1,32 „ „	1.16
High-speed gasoline engine (8–60 h.p.)	„ 3,70 „ „	3.24
Slow-running gasoline engine (8–20 h.p.)	„ 3,86 „ „	3.40

While the above figures vary little and can be considered as constant (with a constant price), the efficiency of the propeller varies considerably from boat to boat and is an important factor in establishing the true economy.

There are boats that work with an extremely low propulsive efficiency. The average for the whole fleet may be estimated at 30 per cent., but several boats work with no more than 15 to 18 per cent. Considering 50 to 55 per cent. efficiency as obtainable, it is evident that such boats consume fuel in a very unprofitable way, wasting up to three-quarters of it.

The highest efficiency is obtained with a slow-running propeller on a normal fishing boat. The motors with high revolutions, directly coupled, or with insufficient reduction ratio, give the lowest efficiency, and generally burn the most expensive fuel, gasoline. Amazing results are often obtained when these motors are replaced by engines working on heavy fuel oil, because the advantage of the cheaper fuel is added to that of the more efficient propeller.

For example, a boat with a 25 h.p. gasoline motor, 3,600 r.p.m., which obtained a speed of 7½ knots at a fuel cost of CH$76.— per hour, changed to a small semi-diesel of 10 h.p., 800 r.p.m. and adequate propeller. It obtained the same speed at a fuel cost of CH$14.40 per hour, a reduction to one-fifth.

The consumption of lubricating oil is generally lower with slower revolutions of the motors and a fleet composed of larger units will use both fuel and lubricant more efficiently.

The maintenance cost of the motors used in Chile is high, the high-speed diesels being the most expensive. The frequency of serious breakdowns in these motors in Chile is really remarkable. The loss of a vital part such as a crankshaft is, in itself, an economic disaster. A replacement shaft for a 60 h.p. engine cost (1952) approximately CH$40,000 (£130; U.S.$365) and up to CH$150,000 (£485; U.S.$1,360), for the larger engines in use.

Such expensive spare parts are rarely kept in stock and delivery takes normally three to four months. If this occurs in the best fishing or lobster season, economic losses may be great, and there have been cases in which such replacements broke again a short time after installation.

It is very difficult to arrive at even approximate figures to show the maintenance cost of the Chilean fishing fleet. In some parts of the country, where facilities for efficient repairs are scarce, there is a tendency among the fishermen to leave the motors unrepaired and out of use when severe breakdowns occur. Motors that have become unserviceable shortly after purchase are sometimes found removed from the vessels and abandoned, even when repair is wholly justified. If such losses are considered within the general maintenance cost of the fleet—as they should—the resulting figures must be high.

About 330 outboard motors are used in the Chilean deep sea fishery. Although there are well-founded reasons for avoiding the use of such motors under severe conditions, the best makes of them have, in certain circumstances, given good service. But they are high consumption engines and they suffer rapid wear and, in tropical waters, extreme corrosion. The result is a high maintenance cost.

* All the prices in this paper are given in Chilean Pesos (£1 = CH$ 308 ; US$ 1 = CH$ 110)

The use of an outboard motor, generally of 10 to 12h.p., in the small sea skiffs in Central Chile, is found to increase the catch by two or three times. To the individual fisherman this represents a slight economic advantage and to the Chilean fishery as a whole the advantage is more apparent than real. The fuel and maintenance cost of the skiff and gear is approximately CH$1,800 (£5.85; U.S.$16.50) per ton of fish caught, which is very expensive compared to that of the modern motor-trawlers operating on the same grounds, for which it is only $260 (£0.85; U.S.$2.35) per ton of fish (a reduction of one seventh). The trawlers obtain the same price for their catch as the skiffs and it is felt that the outboard motor fishery will soon disappear.

In the lobster fishery, in the Juan Fernandez Islands, outboard motors have been used for 30 years. There are 57 specially designed dories operating around the islands with their outboard engines placed inboard. In this case the high consumption cost is of less importance because of the high value of the catch. Certain technical features and facilities for sending the motors to the continent for periodical overhaul, justify their use in this case.

Common repairs on both outboard and inboard motors are sometimes made difficult by commercial rigorism. One trouble caused by the American commercial pattern is that certain spares cannot be bought from the same factory as the engines (parts manufactured by contractors). There have been instances of foreign contractors not delivering orders because they were too small or a needed spare part cannot be obtained without buying a whole assembly, of which the spare is only a minute part.

Experience in Chile has shown that motor designs based on cheap replacement parts of short lifetime should be avoided. The expenses and inconveniences of importing such parts into Chile completely annul the advantages of the system. Marine motors often arrive damaged. While European motors arrive in boxes so carefully packed that damage is held below 1 per cent., the North American motors arrive frequently in partially broken boxes and damage runs to 5 to 20 per cent.

THE TECHNICAL ASPECTS

Some technical problems are common to all parts of the long Chilean coast. Perhaps the most important are the breakdowns of vital motor parts, which should not occur on engines working under favourable conditions. No thorough investigation has been made, but the causes are known in part. Breaks on crankshafts are the most serious. They occur mostly on high-speed diesels and they are more frequent in vessels constructed locally. Breakage of crankshafts has also occurred on the heavy semi-diesels, because of faulty installation on curved foundation beds in locally built vessels.

The causes of breakages in high-speed diesels are not so easily discovered. The main bearings of the motor can hardly be brought to misalignment; they are perfectly aligned even with a very careless installation. Torsional vibrations in the shaft itself may be the cause as breaks are invariably located close to the flywheel. Torsional vibrations may be accentuated, for instance, by unsatisfactorily selected diameter of propeller shafts, choice of the number of propeller blades, or a combination of both. Deficient reverse gears, bad timing of injection with severe fuel knocks and, on wooden boats, over-heavy propeller posts, may bring vibrations up to a dangerous point. Starting and combustion troubles, accompanied by severe fuel knocks, have been noted during periods prior to the breakdown. Fractures of the reverse gear and of the crankshaft have occurred simultaneously, which seems to indicate the direct influence of badly-selected characteristics of propeller shafts and propellers or a severe misalignment of these parts. Misalignment of the propeller shaft (which may develop in wooden boats after launching) is a common fault in Chilean boats. Faulty bolting of engines is still more frequent.

Fractures of connecting rods, camshafts, etc., on the diesels are also troublesome where repair facilities are scarce, but they do not occur with unusual frequency.

Fractures of vital parts are rarely seen in gasoline motors and troubles of importance are limited to the reverse gears. The general cause is incorrect lubrication. The fishermen do not distinguish between the reverse gears working in an oil bath and those working dry with grease lubrication.

Outboard motors are very little affected by fractures or unforeseen stops. They are, however, subject to very rapid wear, nearly all on replaceable bushings. They are exposed to intense corrosion in Chilean waters which, with the extreme wear, has successively eliminated all makes of inferior quality. Indeed, one Swedish factory now supplies 93 per cent. of all outboards used and this has led to a sort of standardization. Bronze propellers and bronze underwater parts are used and maintenance is limited to frequent change of shaft bushings and cylinder heads. With this upkeep and minor repairs, several outboard engines of this make have been used for more than 20 years and are still working.

The life of cylinder heads is three months, even if they are flushed with fresh water after each trip. Bushings last two to four months and experience indicates that they must be of white metal for protection of the steel shafts, bronze bushings being liable to attack the shafts. Reversing mechanisms are rejected. On a few such engines the mechanisms are purposely set out of function, as they cause excessive wear and trouble. The gasoline and lubricating oil obtainable in Chile are excellent for outboard motors and there is no record of combustion and lubrication troubles.

The semi-diesels which work to perfection in other countries show, in Chile, abnormalities. A carbon residue accumulates in the combustion chambers and on the pistons to such an extent that motors have to be dismantled and cleaned every 24 to 48 hours of running. The motors are delivered from the factory adjusted for a different fuel to that found in Chile, but an adjustment to the proper injection point does not alter the combustion process to any extent. It is evident that these abnormalities in combustion, found on diesels and semi-diesels and observed all along the coast, must hold some relation to the fuel oil used.

THE MOTOR FUELS

Diesel oil is imported from Aruba in the West Indies, or from Talara, Peru. The oil from the Aruba refinery (called 115) shows a slight advantage with regard to flash point and cleanness over that from the refinery of Talara (called 116), but one can not select the first one as both oils are sold under the same label.

Analyses show that both fuel oils are quite satisfactory for large diesel power plants with slow running motors used extensively in Chile for mining, manufacturing, electric power, etc. But for engines of the type and size common in fishing, the oil is inadequate. Its deficiency is not in heating value, nor in excessive water content and impurities, as is often believed. It is in the poor ignition and burning qualities (not completely defined by physical characteristics only), and most of all, in its tendency to form residues in the smaller motors.

An American manufacturer of diesels used extensively in Chilean fishing boats, states that the maximum content of residue (gum) tolerated by their engines is 75 milligrams per litre of fuel. After that there is excessive sticking of the rings. Tests made with oil available in Chile show a residue average of 188 milligrams per litre.

Another firm supplying semi-diesels states that with Chilean fuel the injection of their motors must be advanced 18 to 20 deg. from normal on smaller sizes (under 10 h.p.) and 6 to 10 deg. from normal on the bigger engines. Semi-diesels with hand-operated regulation of the injection for different loads, should obviously be the easiest to work on different fuels. In Chile, however, there is considerable trouble, because the fishermen are not " engine minded " enough to care about the regulation.

Although the analyses of Chilean fuel give characteristics close to the qualities of A.S.T.M. grade No. 1-D, generally recommended for the engines, some high-speed diesels, designed for 2,600 to 3,000 r.p.m., frequently show defective combustion with abundant smoke when working at revolutions above 2,200. This has mistakenly been blamed on water and excessive impurities and a few boats have installed strainers in their fuel lines, without convincing results. When, on occasions, a Mexican fuel has been obtained (all other conditions being unaltered) the motors have run smoothly up to their maximum revolutions.

Peruvian diesel oil (gravity 41.0 A.P.I., Flash 165 deg.F. (74 deg. C.), End point 580 deg. F. (304 deg. C.), density 0·818) shows the same effect as the Mexican oil.

Mixing the fuel oil with 20 to 25 per cent. of kerosene, to improve the starting, has met with some success. A certain number of motors are provided with double carburettor for starting on gasoline and working on kerosene when the motor is heated up. Difficulties have been encountered in the use of kerosene because of abundant deposit of carbon residue on the spark plugs, pistons and cylinder walls. This may occur if the motors are worked too cold (they should be kept hot), but even on engines with automatic temperature control, there is trouble. Research shows that available kerosene is not as suitable as that used in the countries where the motors come from. In Norway, a mixture of such low grade kerosene with 25 per cent. of gasoline gives excellent results.

The facts are that Chile is importing diesels and semi-diesels in increasing numbers but is not importing the right fuel for their operation. But it is worth mentioning that Chile possesses potential oil fields and is planning a petroleum fuel refinery.

GALVANIC ACTION AND ELECTROLYSIS

Galvanic action is exceptionally rapid in Chilean waters. It is believed, but not confirmed, that the action is more intense in ports where the Chilean nitrate is shipped. A European motor factory is actually testing new designs with regard to corrosion on the Chilean coast, before including them in continuous production.

Most motor factories have given little thought to the importance of this problem in tropical and sub-tropical waters. There are, however, some engines, mostly North American post-war designs, which are profusely provided with zinc plugs and bars for protection. The bars are sometimes provided with special threads not used in Chile (or in South America) which is impractical.

The usual metallic packings (copper-asbestos) are found to accelerate corrosion under certain circumstances and have to be replaced by non-metallic ones. Rubber spacers and ebonite incrustations prevent corrosion on certain parts.

Motor installations made without due consideration to galvanic action are sometimes found. For instance, there are records of iron fittings, in brass cooling water tubes, being eaten away in less than two weeks. The motors and boats were flooded. And there are instances of bilge water pumps delivering the water inboard in emergency cases. Imported boats are liable to have such accidents because generally they are not adequately protected against corrosion.

PROPELLER EFFICIENCY

Mistakes in establishing the correct relation between propeller and boat are increasingly frequent in Chile. When the motorization of the fleet had just begun there was no such problem. Marine motors were almost identical in design and standard propellers gave a reasonably good result in most cases.

Now the motors are offered in special designs for a great variety of hull shapes and speed ranges. And each motor is commonly offered with different propellers corresponding to direct drive or different reduction ratios. Fixed or controllable pitch propellers are available.

The selection of the right motor and propeller for a fishing boat requires knowledge and experience in naval architecture and fishing. It is not enough that the propeller corresponds to the engine; it must also correspond to the shape, speed, range and use of the boat. The same motor in three different types of boats may require three different types of propellers. In fact, propellers should be selected in each special case, a standard propeller will not suit every boat.

The Chilean fishing fleet operates with an estimated average propeller efficiency of 30 per cent., while 50 to 55 per cent. should be obtainable. Trial and error during several years have established a fairly good propeller efficiency in boats up to 3 tons, especially in the northern zone. It is on the medium size boats, from 5 to 15 tons, that the trouble is alarming, and these boats are being used in increasing numbers.

Six semi-diesels of a well-known European make, with their standard propellers, were installed in tuna launches of the Northern Chile type and gave excellent results. Six more motors were ordered. In the meantime, the factory had changed their propeller type, which was of an old design, for one incorporating all modern scientific improvements. The new propellers were fitted to the second lot of motors and there was a considerable decrease in speed. The fishermen put the old propellers on the new launches and obtained results as good as with the first lot.

An investigation gave the following result: These boats of fine lines and narrow beam, of light construction necessary for the speed required by tuna fishing, should have propellers of approximately 800 to 1,000 r.p.m., narrow blade width and positive blade rake (because of the stern lines) and approximately 11 in. (28 cm.) pitch. The propellers of the old design fulfilled these requirements, while the new propellers, with wide blades without rake and with 9 in. (23 cm.) pitch, were less efficient.

Based on this experience, there is a reason to question the propulsive efficiency of the other tuna launches (about 200) in Chile and Peru. They are equipped with similar motors of other makes, and with modern standard propellers. No investigation has been made, but another propeller design may improve their efficiency.

Controllable pitch propellers are not used in Chile, being rejected without reason. This is astonishing, because several Chilean fishing methods require controllable pitch propellers for best performance.

Cases of wrong motorization are seen. For example a new-built boat of 40 ft. (12.20 m.) length, 10 gross register tons, should, for a speed of 9 knots, have had a motor of about 60 h.p. at 600 r.p.m. and a propeller with 28 in. (71 cm.) diameter and 24 in. (61 cm.) pitch. Instead, on recommendation of the motor-importing firm, a motor of 90 h.p., 3,600 r.p.m., direct drive, with a propeller of 13 in. (33 cm.) diameter and 13 in. (33 cm.) pitch, was installed. In sheltered water the boat did not exceed 5 knots and outside the breakwater, under average weather conditions, less than 4 knots. The manoeuvrability was dangerously bad even in sheltered water. The small propeller of 13 in. (33 cm.) hidden behind a sternpost of 7 in. (17.6 cm.) width, worked with an efficiency estimated at 13 per cent. The lack of slipstream impaired the rudder action. The boat was useless for its purpose and the motorization proved a complete failure. Generally, such cases can be remedied by the installation of a reduction gear and a bigger propeller, but it was not possible in this instance because the stern was built too small to accommodate a normal propeller.

GENERAL ASPECTS

The Chilean fishermen do not resist the idea of mechanization. Apart from the logical desire of the small boat fishermen to keep trawlers away from their special fishing grounds, no serious antagonism between new and ancient methods is found and the fishermen are quick to take advantage of technical improvements within the limit of their resources. In fact, it can be said that whatever technical progress is being made is due to the individual fishermen and to private fishing companies, Government institutions taking care principally of social problems.

Motor representatives are doing very little to promote sound mechanization. A certain lack of technical emphasis in the nation's fishery development programme may be the cause. The technical improvements in Chile may be characterized as isolated experiments, and errors committed and corrected in one locality are repeated in another, which indicates a need to co-ordinate plans and efforts to promote mechanization of Chilean fishing vessels.

REMARKS ON FRENCH TRAWLER DIESELS

by

JEAN FAURE

THE following comments apply to four-cycle, single-acting, slow-running, reversible, six-cylinder diesels, fitted in the French standard post-war trawlers.

ENGINES FOR 42 m. (138 ft.) TRAWLERS

Power 800 h.p. at 200 r.p.m. After five years in service, or 30,000 hours of running and continuous operation during 11 months a year, these motors are in good condition. Periodical checks have shown that the cylinder liners have worn, not inside, in the region of piston ring friction, as expected, but on the external side in contact with the circulating cooling water. Among the factors which probably caused this wear are corrosion due to electrolytic action, chemical and biological corrosion, and friction by the circulating water containing residues of zinc, sediments and sand. One can take effective action against these corrosions by cleaning the circulation chambers at regular intervals, giving the liners and the frame a coat of protective paint, cleaning the filters, renewing the zinc plates and, best of all, by replacing sea water cooling by fresh water cooling.

ENGINES FOR 38 m. (125 ft.) TRAWLERS

Power 600 h.p. at 200 r.p.m. The motors are supercharged with turbo chargers running on exhaust gas. Some criticism has been directed against supercharging, which is not generally accepted in the French fishing industry. The author is of the opinion, however, that supercharging cannot be questioned for modern diesels when it increases the power by 30 or even 50 per cent. It seems that the trouble experienced so far is due to a bad correlation between propeller and motor torque. Supercharging should be designed for trawling conditions with a by-pass for excess air when the boat is steaming. The turbo blowers are built for the overload condition imposed by the classification societies, but they are run normally at speeds which are very far from this overload test and even from what is considered the normal running condition, they have therefore a safety margin.

ENGINES FOR 32 m. (105 ft.) TRAWLERS

Power 450 h.p. at 300 r.p.m. These motors have been working satisfactorily and do not call for special remarks.

ENGINES FOR 28 m. (92 ft.) TRAWLERS

Power 400 h.p. at 375 r.p.m. The winch is driven off the main engine as in most small fishing boats and the engine has therefore to be non-reversible. The owners very often use the power take-off for driving pumps, dynamos and heavy pulleys without any consideration for the main engine. This should not be done without approval by the engine builder, who can advise about critical speeds and the means to avoid them. The motor foundations are very often lacking in strength and rigidity, and strong vibrations can then occur which are very detrimental to bedplates and cast iron frames. Some trouble is also experienced by owners who are using fuels with a high sulphur content. This leads to undue wear of the cylinder liners and fuel injectors.

DEVELOPMENTS

Future developments anticipated are: (*a*) Motors with interchangeable elements either in line or V. (*b*) Change-over from slow-running to medium-speed engines, 400 to 500 r.p.m., with reduction gear in order to gain weight and space. (*c*) Cast steel or welded steel bedplates and frames instead of cast iron (lightweight, safety and reduced cost). (*d*) Improvements of supercharging in order to achieve more power per unit weight. (*e*) Fresh water cooling. (*f*) Injectors and piston rings suitable for burning heavy fuels of lower price.

CONSIDERATIONS ON DIESELS

by

A. DUSSARDIER

THE problem faced by marine engineers is how to adapt the diesel with a maximum efficiency to two widely different but equally important running conditions:

(*a*) cruising;
(*b*) trawling, which requires the motor to develop high torque at reduced revolutions.

The two conditions were solved by the steam engine through its flexibility. Since the diesel has been adopted, the propeller has been designed in most of the cases for the cruising condition, and this has resulted in the motor being highly overloaded for long periods when trawling.

SMALL TRAWLERS

Trawlers of 60 to 90 ft. (18 to 27 m.) are mostly built of wood and are normally at sea from 10 to 15 days at a time. They trawl at a depth of 27 to 55 fm. (50 to 100 m.) in the English Channel or the North Sea or from 110 to 140 fm. (200 to 250 m.) along the Atlantic coast. The main engine drives all auxiliaries by belt and pulleys from the fore end. This suggests the choice of a non-reversible engine with reverse gear. During the short stay in port, and owing to a small crew, the maintenance of the engine is usually neglected and the following conclusions have been drawn from the study of boats in service since 1945, which are equipped with both two- and four-cycle engines.

(1) To obtain a good service it is recommended to install a power equal to three and three-and-a-half times the gross tonnage of the vessel.
(2) For this operation, and especially if, as often happens, the power is on the small side, the four-cycle has shown definite superiority for powers up to 400 b.h.p.

With a multiple reduction gear, the motor is always used nearer its normal load and the advantage of the four-cycle over the two-cycle is correspondingly less. The controllable-pitch propeller, with its progressive change of speed, appears to be the ideal solution although there are objections to its use.

Another solution, which is expected to meet with great success in the near future, is to have the diesel supercharged by means of a blower working on exhaust gas. By such supercharging the air supply to the engine can be adjusted to the load required. Additional advantages of great importance for small boats are:

less weight and less floor space are required for a total given power, which all other things being equal increases fish-hold capacity;
maintenance costs are lower due to the smaller number of cylinders in the main engine, or to the smaller bore of the cylinders, in which case all parts are smaller, and more easily handled and cheaper.

At the present time, motors with 50 per cent. or 60 per cent. supercharging are running continuously with great safety and less thermal stresses.

Supercharging meets to-day with the fisherman's distrust of comparatively new devices. Trawling with small boats in fishing grounds hundreds of miles away is hard and dangerous work, especially in bad weather. It requires the absolute confidence of the crew in the boat's equipment, particularly in the engine, on which the safety of the ship and the size of the catch depend. So, fishermen have good reason to trust only engines that have stood the test of long service under very special conditions.

But this has not prevented fishermen from adopting technical advances as soon as their value is proved. For example, trawlers now have echo-sounders, radar, radio and radio telephone. And endurance tests that have been made with some boats fitted with supercharged diesels will certainly convince them of the value of the turbo-blower run by exhaust gases. This automatic apparatus is not connected with the engine mechanically and it is less delicate than the rudder or even a simple generator. It does not require more upkeep than a wrist watch and does not expose the fishermen to the risks of engine stalling. On the other hand, it greatly improves the ship's manoeuvrability, increases the trawling power and lowers operation costs. Even a total breakdown of the turbo-blower would not prevent the engine from running satisfactorily, at a lower power, but adequately for safe navigation. Two small Breton wooden trawlers have tried out four-cycle 250 h.p. engines with supercharging and have experienced no trouble. The *Joli 2* from Concarneau has used the engine for four years, and the *Kercadic* from Hennebont has used the engine for two years.

To get better performance, both in cruising and trawling, the machinery of small trawlers should be selected with the following rules in mind:

(1) Installed power should be about 3 h.p. per gross ton.
(2) The engine should be a non-reversible, preferably four-cycle, diesel with reverse and reducing gear, and a propeller running 200 to 250 r.p.m.
(3) The reduction gear could very well have multiple gears, allowing the installed power to be reduced by at least 10 per cent.
(4) With or without such gear, a good solution would be to have an engine supercharged by a turbo-blower on exhaust gases.

DEEP SEA TRAWLERS

In this class are included steel trawlers of 90 ft. (27 m.) or more in length between perpendiculars operating on fishing grounds as deep as 270 to 330 fm. (500 or 600 m.).

The 114 to 147 ft. (35 to 45 m.) trawlers have an auxiliary motor for driving the powerful trawl winch and other auxiliary equipment. The main engine is usually a diesel of 600 to 1,000 h.p.

The weight and power of these ships make them less liable than the smaller boats to sudden stalling while trawling. As systematic tank tests have been made on the pull when trawling, and of the necessary torque depending on speed and depth of trawling, it is possible to know in advance what power has to be installed without leaving too wide a margin for safety. However, it is still true that continual shifting in clutch resistance during trawling, requires a combination of engine and propeller line with enough inertia to ensure smooth operation despite such shifting. This is a consideration that favours comparatively heavy, slow-speed engines.

Experience seems to prove that in this power range the two-cycle diesel is the more advantageous because of its smaller weight and floor space and the extreme simplicity of its mechanical construction.

The shipowners, the only judges of actual results, agree that, within the range of the various types of boats built in France since 1945, the 138 ft. (42 m.) trawlers are among the best, with excellent speed, seaworthiness and trawling qualities. They have directly reversible, 750 h.p., 210 r.p.m. diesels, directly connected to the propeller shaft. While most of them have two-cycle engines, six have a four-cycle engine of the same power and working at the same r.p.m.

The identical features provide a rather rare opportunity to make a comparison of the respective advantages and disadvantages of the two- and four-cycle and settle the controversy, at least temporarily, that has been

TABLE LXVIII

42 M. (138 FT.) TYPE TRAWLER

COMPARISON OF OPERATING COST ACCOUNT FIGURES

Power equipment: directly reversible 750 h.p., 210 r.p.m. diesel

	SHIPOWNER A *Averages 1951/52/53*				SHIPOWNER B *Averages for two years* **for four ships** (*figures for 1953*)	SHIPOWNER C *Averages 1950/51/52* C	SHIPOWNER D *Averages 1951/52*
Ship	A_1	A_2	A_3	A_4	B_1 to B_4	C_1	D_1
Cycle of operation	4	2	2	2	2 2	4	2
(*a*) Gross catch per day at sea	1,135	1,105	973	1,040	1,325	1,085	1,290
(*b*) Direct expenses for fuel, oil, repairs maintenance, per day at sea	224	205	179	194	227.5	170	213
(*c*) Earnings in per cent. $\frac{a-b}{a}$	0.802	0.814	0.815	0.812	0.830	0.842	0.836
(*d*) Corrected earnings in per cent.	*0.822*	*0.832*	*0.835*	*0.830*	*0.830*	*0.842*	*0.836*
(*e*) In action factor	0.775	0.735	0.705	0.775	0.778	0.732	0.806
(*f*) Practical earnings d×e	*0.623*	*0.606*	*0.575*	*0.628*	*0.646*	*0.615*	*0.675*

The above figures are averages for two, three or four years of service. Because of price fluctuations, these figures which pertain to different times and different shipowners are not directly comparable. Only the figures in italics, which are for the same periods for each boat, are truly comparable.

raging for years between the advocates of each cycle. What counts in the final analysis is the practical operating results, and only the users know them. Owners of this type of boat have kindly given figures from their book-keeping accounts, but, unfortunately, a strict statistical analysis is hard to make because of the different break-downs used by the shipowners for various expense items. Furthermore, comparable data could only be obtained for a few ships.

In Table LXVIII, figures for a few of the boats have been assembled and can be compared because they deal with comparable factors. They cover, in the main, statistical averages for three or four consecutive financial years of operating costs limited to consumable materials (fuel, oil and grease), and maintenance and repair costs, as compared with the gross return from the catch on these trips. All figures have, of course, been reduced to the same basic unit.

Actually, some of the figures supplied include maintenance and repair costs for the entire ship and not for the propulsion machinery alone. If a fixed annual sum deducted from those figures for each boat, representing a reasonable estimate of the cost of maintenance of the hull and auxiliary equipment, the factor d is arrived at, called " corrected earnings."

Another possible criticism is that the figures do not apply to the same periods for all shipowners. For that reason, the figures in italics are the only ones that are of real value for comparative purposes.

Lastly, an interesting appraisal item called the " in action factor " was added to the table. This is the ratio, for each ship, of the number of days at sea to its total operating time. The product of d and e gives a " practical earnings " which shows whether or not the ship in use is a paying proposition.

There are factors which are independent of either the ship or its engine, such as the skipper's competence and ability, the proficiency of the crew, perils at sea, accidents and, above all, the impossibility of laying down a law for analyzing certain isolated cases. In view of this, the conclusion from the inquiry is that the practical returns, called " earnings ", are very close for all these ships, with the two-cycle engine having a slight advantage

In fact, analyzing the significant factors it is found:

TABLE LXIX

	Two-cycle	*Four-cycle*
1. Coefficient d (corrected earnings) average 8 two-cycle engine ships and 2 four-cycle engine ships . . .	0.8316	0.832
2. Coefficient (practical earnings), ditto .	0.633	0.619

If this sampling covering only two four-cycle engines is of rather low statistical value, it is nevertheless compensated by the fact that these were two very good boats with excellent skippers. The conclusions drawn from these figures agree with the appraisals of most users. It is surprising to learn that the two-cycle engine uses less fuel in service per day at sea than the four-cycle type. It has the additional advantage of being very simple to assemble or to take apart and is easy to overhaul. This considerably reduces labour costs.

These facts show why, after half a century of experience, the greatest specialized engine builders are tending to use the two-cycle beyond a certain power limit. The limit is still not absolute but is apparently around 100 to 150 h.p. per cylinder, depending on the cylinder bore and the number of r.p.m.

The case chosen for this study, the 42 m. (138 ft.) trawler is in the zone of uncertainty where both types of engines ought to give nearly equivalent results. For larger sizes it seems logical that preference should be given to the two-cycle main engine. Improvements can still increase its performance, with possible economies on parts that wear out. Maintenance costs can be reduced by various processes such as nitriding, chromium plating, and sulphuration. Experiments are being made with these processes and should lead to a great reduction in the wear and tear on the parts subjected to friction. Such developments should give the two-cycle engine top place for economy in service.

The growing practice of supercharging may, of course, alter to a certain extent the conclusions reached in this paper.

HEAVY DUTY FISHING ENGINES

by

R. G. ANDERSEN

THE so-called semi-diesels (sometimes also named crude oil engines) are overwhelmingly preferred in Scandinavia. But neither semi-diesels nor crude-oil engines are satisfactory names and they often provoke misunderstanding. The name "crude oil" is directly misleading as the modern Scandinavian heavy-duty engine is constructed for the use of gas oil of ordinary commercial quality with a thermal value of about 40,000 B.Th.U. (10,000 kcal.)—the same fuel oil as is employed in diesels of corresponding sizes.

The engine had its origin in the old hot-bulb engine and has been developed independently of the diesel. The designers of diesels primarily concentrated their efforts upon creating an engine which, as closely as possible, approached the theoretical ideal of thermal efficiency, but the designers of the Scandinavian heavy-duty engine concentrated primarily upon attaining simplicity in construction and a minimum of working parts. They have also striven to attain the greatest possible fuel oil economy, within the limits principally imposed.

It is this principal variation from the diesel which makes the name semi-diesel misleading, but it is internationally the most common name for these engines.

PRINCIPAL CHARACTERISTICS

The principal characteristics of the semi-diesel may be expressed in this way: it is a combustion engine in which the end pressure in the combustion chamber is not sufficiently high to ignite the fuel oil by the compression temperature. When starting, the injected fuel must hit a pre-heated surface which vapourizes and ignites it. This pre-heated surface may be part of the walls of the combustion chamber or a cartridge or glowing plug inserted into the chamber.

The engines are made from about 5 h.p. to about 125 h.p. per cylinder with one or more cylinders. They are used in freighters, tugboats, etc., as well as in fishing boats. Owing to the demand for simplicity and dependability (as few working parts as possible) the 1-cylinder engine is usually preferred, and the factories have for many years steadily increased the size of the cylinder. For example, Tuxham A/S, Copenhagen, produces a 1-cylinder engine at 275 r.p.m. developing 115/130 h.p. The bore is 16.35 in. (415 mm.) and the stroke 18.5 in. (470 mm.).

HISTORY

The first step towards the development of the semi-diesel was Capitaine's paraffin engine constructed about 1884. This was a four-cycle engine which followed the usual Otto system. On the cylinder head was fixed a small vaporizer which, while the engine was running, was kept red-hot by means of a blow-torch. The paraffin was injected into this vaporizer and ignition took place when the mixture of air and paraffin met the glowing wall of the vaporizer.

At about the same time another four-cycle paraffin engine, based on the Otto system, was developed by Priestman. Like Capitaine's engine, it was fitted with a vaporizer. The injection was continuous and the vaporizer was kept at the right temperature by the exhaust gases. Ignition was obtained by means of a glowing plug. The use of two blow torches was necessary to start the engine.

In 1892 the Hornsby-Akroyd engine made its appearance and it differed from the other four-cycle engines in that almost the whole of the combustion chamber was not cooled and was only connected with the cylinder through a narrow passage. The fuel oil was injected into this separate combustion chamber by a fuel pump, and the walls of the chamber were red-hot, while the engine was running, the whole thus acting as vaporizer. The intake of air was direct into the cylinder through a guided valve and did not first pass the vaporizer. During compression it was forced through the narrow passage into the vaporizer where it mingled with the vaporized oil. Combustion commenced in the vaporizer during compression, but owing to the narrowness of the passage between the vaporizer and cylinder, the velocity of the air which was pressed into the vaporizer was greater than the velocity of the combustion products in the vaporizer, and therefore the gases from the vaporizer did not press down into the cylinder until the piston had reached its top dead centre. The vaporizer was heated by a blow-torch which was put out once the engine started, and while the engine was running ignition was attained entirely through contact with the hot walls of the vaporizer. The engine could be run on paraffin as well as on heavier and cheaper oil products.

In 1893 Dr. Rudolf Diesel presented his well-known treatise "Theorie und Konstruktion eines rationellen Wärmemotors zum Ersatz der Dampfmaschine und der heute bekannten Wärmemotoren" (Theory and Design of a rational Heat Motor as a substitute for the Steam Engine and the Heat Motors of to-day), and Krupp and Maschinenfabrik Augsburg undertook the practical testing of the theories. This resulted in 1897 in the first serviceable diesel. In practice it deviated considerably from the theoretical ideal but it worked. Its fuel oil consumption was only about 0.497 lb./h.p./hr. (0.225 kg.) and it could be run on inferior oils, which was a great step forward. Its major disadvantages were the high pressure and the necessity for compressed air of about 1,150 lb./sq. in. (80 kg./sq. cm.) for the fuel oil injection. It was, therefore, necessary to fit the engine with a compressor to supply the air demanded. This was an undesirable complication and necessitated constant supervision of the engine when it was subjected to greater variations in load. In such cases the compressor had to be regulated by hand, otherwise the injection pressure would be too low at minor loads of the engine. It was 25 years until compressorless diesels were constructed.

Despite these drawbacks of the diesel, its advantages, especially in fuel oil consumption, were such that the paraffin engines then on the market could hardly compete with it and they were faced with the necessity of improving their fuel oil consumption considerably.

A major cause of the poor economy of the paraffin engines was that the fuel oil had to be injected well before ignition took place. This made the exact moment of ignition uncertain and an optimum blend of air and fuel impossible. Low compression had to be accepted to avoid pre-ignition. By changing to the two-cycle system and introducing a delayed fuel ignition it would be possible to increase the compression without the risk of pre-ignition, and several firms worked along these lines.

About the year 1900 there appeared almost simultaneously the Swedish engine Avance and the American engine Mietz and Weiss, both of them two-cycle, using the underside of the piston and a closed crankhouse casing as scavenging pump, a design which was introduced by Daimler in 1883. These two-cycle engines immediately proved to be more economical than the four-cycle and, what was of great importance, they could be run on inferior oils. They were also considerably simpler in construction and they met with great success in Scandinavia, especially in fishing vessels and smaller commercial craft, the mechanization of which had just begun.

Even though the semi-diesel then produced resembled in principle, and partly also in construction, the semi-diesels produced to-day, essential developments have taken place. The first engines had the disadvantage that it was necessary to inject water into the combustion chamber. The aim of water injection was partly to keep the ignition bulb of the engine at a suitably low temperature (a too high temperature of the ignition bulb caused carbonization) and partly to add vaporized water to the combustible mixture. The vaporized water acted as a catalyst and improved combustion. Water injection was undesirable, especially in the case of marine engines, which were compelled to keep a storage of fresh water because sea water could not be used. The injected water also had a tendency to wash away the lubricant from the cylinder, which resulted in increased wear, and it called for increased supervision of the engine, because the quantity of water had to be regulated according to the load.

All the leading firms interested themselves in constructing an engine which could be run without water injection but, until 1922, it could not be dispensed with at loads higher than 65 to 76 per cent. of the theoretically possible. Tuxham solved the problem in 1922, after years of systematic experimenting and research, by adjusting the proportions of the combustion chamber and bulb and by correct forming and dimensioning of the elements deciding and guiding the air streams in the combustion chamber, and the quantity of fuel injected.

All important makes of semi-diesels now work without water injection. Their mean effective pressure is 40 to 43 lb./sq. in. (2.8 to 3.0 kg./sq. cm.) at normal load and their consumption of gas oil is, for instance, for engines of about 75 h.p./cylinder, about 0.485 lb./h.p./hour (0.22 kg.), corresponding to a thermal efficiency of 28.5 per cent. in relation to b.h.p.

It is possible, through exact tuning of the length of the exhaust pipe, to obtain greater mean efficiency pressure and lower oil consumption, (less than 0.44 lb./h.p./hour (or 0.20 kg.) has been measured on testing benches) but such tuning is difficult to obtain in vessels.

DESIGN

Fig. 475 shows a section of a typical modern one-cylinder semi-diesel. It is characteristic of Scandinavian semi-diesels of all recognized makes, that they appear, from a technical point of view, over-proportioned. Dimensions of working parts, walls, etc., are, in many instances, determined by the practical fact that engines are specially built for fishing craft and other small-work boats and will be subject to careless treatment by people with no special knowledge of machinery.

Fine mechanism of any kind must be banned and the dimensions of minor parts, even screws and nuts, must be sturdy enough to resist rough treatment and to be suitable for dismantling or assembly by unskilled persons even under the most difficult conditions. The engines are designed with these considerations in mind and the result is great durability.

Owing to the specially trying conditions to which the fishing boat engine is subjected, Scandinavians place considerations of dependability and durability above everything else; all others are of secondary importance. This line of reasoning makes itself evident in many ways in the apparent over-proportioning of the engines.

The engine works in two cycles and uses the crank-case as scavenging pump. The upward stroke of the piston

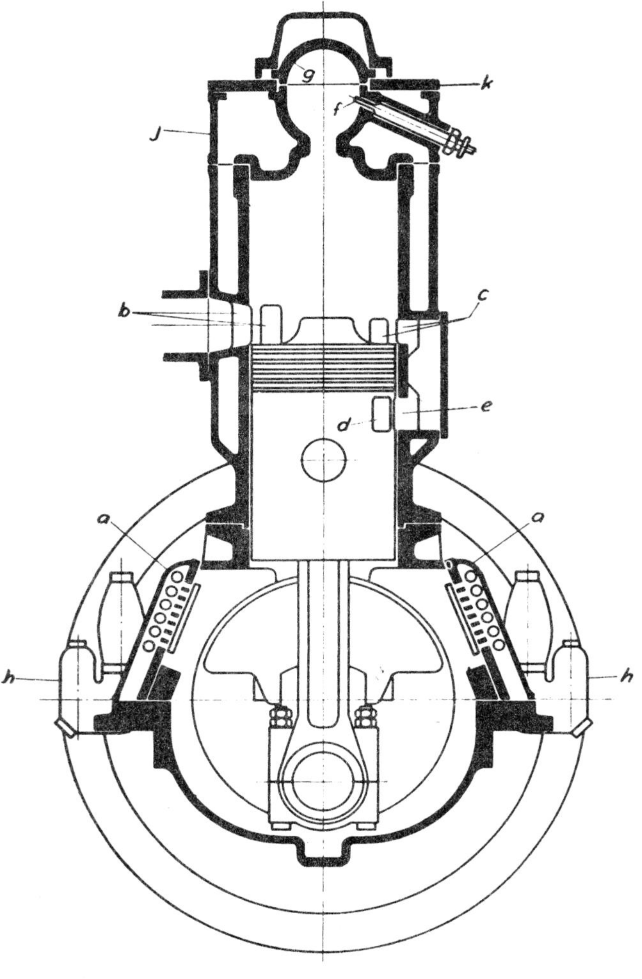

Fig. 475

sucks fresh air into the crank-case through the non-return clack valves " a " placed in the side doors. This design is advantageous because the ordinary inlet and exhaust valves, which traditionally result in interruptions of work in ordinary combustion engines, are entirely done away with. When the piston is at top dead centre, the crank-case will be filled with fresh air, and as the piston is pressed downwards, the air in the crank-case is compressed. When the piston has almost reached bottom dead centre it first frees the exhaust exit " b " and immediately afterwards the entrance " c ". At the same time the aperture in the piston " d " corresponds with the aperture " e " in the cylinder. In this position the exhaust gases escape through " b ", the fresh air from the crank-case being blown at the same time into the cylinder through " d ", " e ", and " c ", where it is used for scavenging by helping to expel the exhaust gases. When the piston, during its following upward movement again closes " b ", the next compression starts.

The form of the piston top shown is such that it forces the fresh air in a curve of half a circle so that it is compelled to scavenge the entire combustion chamber. The technical problems of combustion with this type of engine are but little explored theoretically and, instead of being based upon investigations in laboratories, the solutions are almost entirely derived from experience accumulated by the individual factories.

When injected through the nozzle " f ", part of the fuel will hit the wall " g ". This upper part of the combustion chamber, the bulb, is not water-cooled, and its surface of contact with the cylinder head is carefully designed to a size which permits transmission of heat sufficient for the bulb to ignite the fuel injected under all working conditions from idling to full load. Before the engine is started, the bulb is heated to a suitable temperature. A modern semi-diesel requires two to five minutes to heat, a matter of no practical importance when a ship is getting ready to sail.

There are semi-diesels in which the ignition proper, or the spray-angle of the injected fuel, must be regulated to maintain a suitable temperature of the bulb under varying loads. Either solution makes the machinery more complicated and demands increased supervision. With regard to the type of engine illustrated in fig. 475, this problem has been solved effectively by means of an automatic regulation of the ignition. The governor is driven through skew-gears and a vertical shaft directly from the crank-shaft, and activates the cam dial which, in turn, sets in motion the piston of the fuel pump. The eccentricity of the cam dial increases in a vertical direction and the quantity of fuel injected is decided by this vertical position. Further, the cam is formed as a screw line, which causes the quantity of fuel, as well as the time of injection, to be automatically regulated with varying load, and no extra adjustment of ignition or of time of injection is required.

Cooling and bilge pumps are shown at " h ". They are identical and interchangeable piston pumps directly driven from the crank-shaft through an eccentric. In this way the cooling pump can, in an emergency, be replaced by the bilge pump.

As a result of the comparatively low compression and the principle of combustion which does not require specially fine atomizing of the fuel, the nozzle pressure is only about 710 lb./sq. in. (50 kg./sq.cm.) as compared with an approximate 5,700 lb./sq. in. (400 kg./sq.cm.) for a modern diesel with solid injection.

The piston in the crank-case acts as a scavenging pump, and the main bearings (roller type of ample dimensions) of the crank-shaft are fitted with sealing rings, held in place by springs. The use of roller-bearings was made possible by the lower maximum combustion pressure and it has great advantages in lubrication and maintenance. Under normal conditions it is 20 to 30 years before the bearings have to be replaced.

To aid dismantling and inspection the crank-case is usually divided in half. The cylinders are, even where there are several, separate units, which saves time when it is necessary to replace a cylinder. The cylinder head "j" is often open at the top and closed by a cover "k". This design is used in order to avoid internal stresses in the material and it allows easy access for cleaning the cooling jacket, a point of special importance in marine engines which are cooled by sea water.

In spite of its ample dimensions the weight per b.h.p. of a semi-diesel is less than the weight of a diesel of the same r.p.m., because of the lower maximum combustion pressure as compared to the mean effective pressure.

A semi-diesel with a mean effective pressure of approximately 43 lb./sq. in. (3 kg./sq. cm.) will have a maximum combustion pressure of approximately 340 lb./sq. in. (24 kg./sq. cm.), whereas a four-cycle diesel, not supercharged, with a mean pressure of $\frac{87}{2}$=43.5 lb./sq. in. ($\frac{6}{2}$=3 kg./sq. cm.) will have a maximum combustion pressure of approximately 710 lb./sq. in. (50 kg./sq. cm.). Thus, the maximum combustion pressure is about eight times the brake mean effective pressure for the semi-diesel against a maximum combustion pressure for the diesel of approximately 17 times the brake mean effective pressure, more than twice as high as the ratio for the semi-diesel.

It is characteristic of the semi-diesel that all manoeuvring components, etc., are designed according to the simplest mechanical principles, and the more vulnerable electric or hydraulic systems are not used. Fig. 477 and 478 show a one-cylinder and a two-cylinder semi-diesel.

THE CONTROLLABLE PITCH PROPELLER

More than 95 per cent. of all engines in Scandinavian fishing craft from 10 h.p. to 200 h.p. are to-day fitted with mechanical controllable pitch propellers. Fig. 476 shows a typical design. The propeller blades "l" are rotated in the propeller boss "m" by the double rack-and-pinion mechanism "n" when the propeller shaft "o" is moved ahead or astern. This system replaces the ordinary reverse gear. Rotation of the propeller blades from position "p" to position "q" involves the propeller being switched from full speed ahead to full speed astern, but neither engine nor propeller shaft changes direction of rotation. Movement of the propeller shaft is controlled by a hand-wheel from the pilot house.

Fixed propellers have to be selected individually for vessels, the pitch being decided by consideration of the lines, the h.p. of the engine, and the estimated speed of the vessel. Often, however, it is difficult, if not impossible, to obtain exact advance information of the lines of the vessel, especially when changing the engine in an existing vessel or when exporting the engines through agents. A fixed propeller must, therefore, in most cases be chosen from a pre-determined standard set of propellers. If it is specially constructed, design must often be based upon certain presumptions which cannot beforehand be verified. If the pitch is too high, the engine will be overloaded and must be run at reduced speed; if it is too low, the engine will not be fully loaded even at the maximum speed. In both cases it will prove impossible to obtain a speed which the engine power and propeller speed should justify.

All difficulties originating from incorrect ratios between vessel, engine power, engine speed and propeller are eliminated by the controllable pitch system because the pitch of the propeller may be adjusted until the optimum pitch has been reached. This adjustment may be made without disengagement of the engine, simply by turning the wheel in the pilot house, and it enables the engine to be run with maximum torque in any vessel. In this way the propeller blades may be adjusted to any intermediate position, including neutral, in which position the screw propels the vessel neither ahead nor astern. By adjusting the propeller to slow ahead or slow astern, the vessel can be kept at a standstill, in spite of wind or sea, while the engine works at full or any other speed on the winch during the handling of fishing gear.

Experience in Scandinavia shows that the controllable pitch system, with the possibilities of variation which it presents, is indispensable, especially for Danish seining or for any kind of fishing with lines or net, because it makes available the maximum hauling power at any

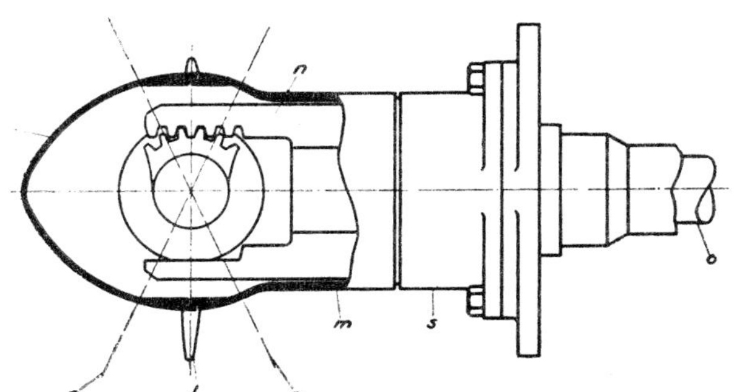

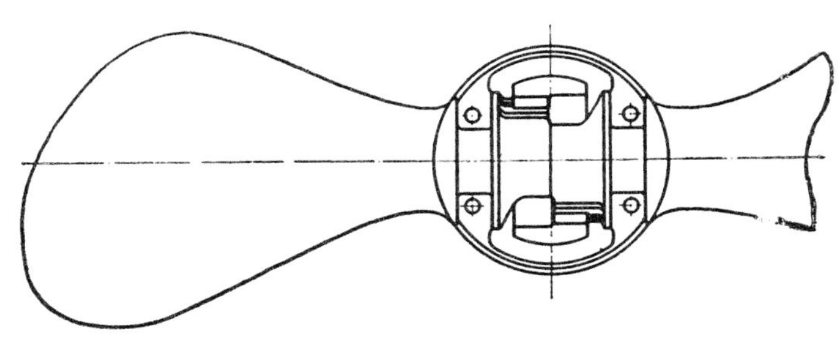

Fig. 476

speed of the vessel, whether high, low or stationary. When fishing is finished, the pitch of the propeller is increased to make use of the full engine power to obtain maximum speed of the vessel. It is the only system to provide full engine power for trawling as well as for speed.

Another great advantage of the controllable pitch system is to make it possible to combine wind power and engine power on all occasions. When a vessel carries sails, the pitch of propeller is increased to obtain a higher speed by the interaction of the two forces. In other words, the controllable pitch propeller acts as an over-drive in a motor vehicle.

This is not the case when a fixed propeller is used. If the sails alone give the vessel a certain speed lower than that obtained by the engine, the added engine power is only partially used. If wind power alone gives the vessel a speed considerably higher than that given by the engine power, the propeller acts as a brake. Even for a vessel not carrying sails it is, for the same reason, advantageous to be able to adjust the propeller to a lower pitch when sailing against the wind and to a higher pitch when sailing with the wind.

The alternative to the controllable pitch propeller is either a fixed propeller combined with a reverse gear or a fixed propeller combined with a reversible engine. Either system requires a considerable number of working parts, even for the least-complicated design, compared with the controllable pitch system, which does not reverse the propeller shaft but simply reverses the position of the propeller blades. It is, therefore, justifiable to state that the controllable pitch system, under comparable circumstances, will give less trouble in use.

When a controllable pitch propeller is damaged, repair or replacement of the blade is quickly and easily carried out.

The controllable pitch system as well as the engine shown in the figures may be considered prototypes. Various firms employ varying designs, and for bigger engines the controllable pitch propeller may be supplied

Fig. 477

with three or four blades. There are types in which the solid shaft, as illustrated, is not moved when manoeuvring the propeller blades. The rack-and-pinion mechanism is fixed to a special rod, which is moved ahead or astern inside the hollow propeller shaft. In this case the thrust-bearing will usually be placed in the astern part of the engine proper, but the thrust-bearing of the illustrated design, " s ", is placed on the aft side of the stern post, so communicating the pressure of propulsion directly to the hull from this point.

than 90 per cent. of the vessels above 5 tons in the Danish fishing fleet are supplied with semi-diesels. And it is probable that the fishing fleets of Norway, Sweden and Iceland are powered in much the same way except that there may be more diesels in Iceland because the country was cut off from its traditional suppliers in Scandinavia during World War II.

The picture in non-Scandinavian countries is essentially the opposite. It is true that Scandinavian builders of semi-diesels have, in the course of the years, created

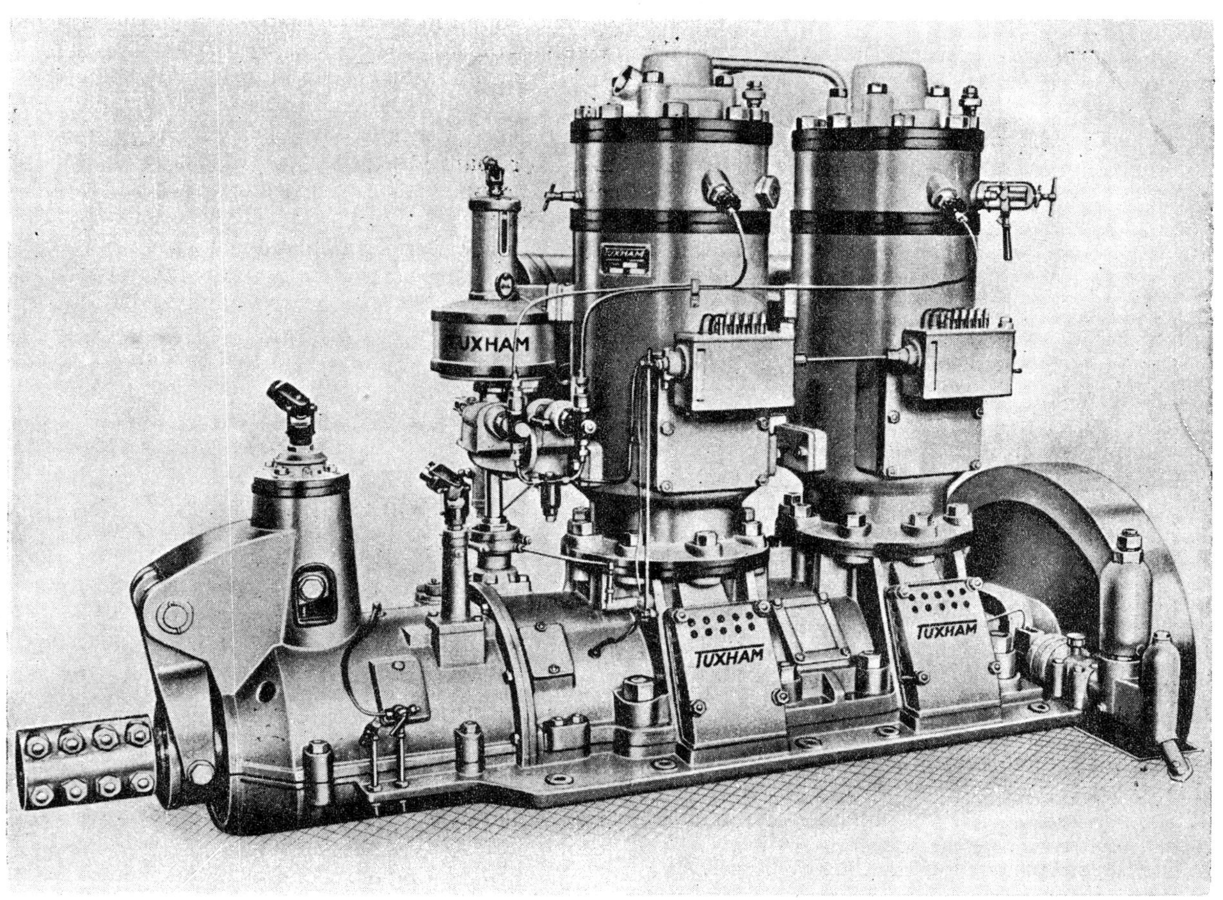

Fig. 478

satisfactory export markets in several countries, especially in the Mediterranean area, Poland, South America, Thailand, and a few other parts of the world, but the trade has not grown in relation to the size of the world market.

NUMERICAL SURVEY

The Danish fishing fleet consisted in 1950 of 7,635 motorized vessels—4,024 below 5 tons, 2,899 from 5 to 20 tons, and 712 above 20 tons. It is unlikely that there are diesels in Danish fishing vessels under 5 tons. They use petrol, paraffin and semi-diesel engines only. There are no exact figures for vessels between 5 and 20 tons, but it is certain that the number of diesels installed in these vessels is so small as to be of no importance. Perhaps 5 per cent. have diesels, the rest being equipped with semi-diesels. For the vessels above 20 tons the numerical survey shows that 189 vessels are fitted with diesels and 523 with semi-diesels. This means that more

SEMI-DIESELS VERSUS DIESELS

Scandinavian fishing methods have, in several instances, set the pattern for the development of fisheries in other countries and an endeavour to analyse the reasons why the semi-diesels in the Scandinavian fishing fleets are so predominant should presumably be of interest.

Semi-diesels of good quality have extremely long life

and low costs of maintenance. It is reckoned that an average life for pistons is eight to ten years and 20 years for the cylinders. It is common to find engines 25 to 30 years old still giving satisfactory service. And, because of the way a semi-diesel is designed and built, it may be kept in service even with a considerable cylinder wear. For example, two two-cylinder engines were installed in a freighter and a cylinder wear of about $\frac{1}{4}$ in. (6.35 mm.) in all four cylinders was found. The diameter was 14.2 in. (360 mm.). The reason for this fantastic wear, which appeared after only two crossings of the North Sea, proved to be very bad fuel oil, with a high content of sulphur, on which the engines had been forced to run because of the war. Of course, the pistons made quite a lot of noise, but the engines still worked.

The low combustion pressure and the working principle of the semi-diesels enable repairs to be done in small local workshops.

Without giving the vessel too much weight to carry, the semi-diesels may be built to a suitably low speed to ensure good propeller efficiency, and the low number of revolutions helps to increase the longevity of the engine. In making a practical comparison between various engines, the number of revolutions plays an important but, unfortunately, often disregarded part. The higher number of revolutions is a cause of inferior propeller efficiency and requires a propeller diameter which is often too small in comparison to the heavy stern timber common in fishing vessels.

A single instance will illustrate this. In a Danish fishing vessel of approximately 35 tons a diesel was installed developing 180 b.h.p. at 1,000 r.p.m., through gear reduced to 500 r.p.m. on the propeller shaft. It was re-engined with a semi-diesel of 120 b.h.p. at 275 r.p.m. The efficiency of the two propellers theoretically obtainable is 48.5 per cent. and 57 per cent. respectively, in other words, of the diesel engine's 180 b.h.p., $0.485 \times 180 = 87$ b.h.p. were made effective and of the semi-diesel's 120 b.h.p., $0.57 \times 120 = 69$ b.h.p. were made effective. The speed proportion is thus $\sqrt[3]{\frac{87}{69}} = 1.08$, meaning that the speed obtained from the diesels should have been 8 per cent. higher than that from the semi-diesel. The result, however, was exactly the same, namely 8.25 knots in each instance. Undoubtedly the reason for this must be found in a poorer hull efficiency for the propeller of the diesel which, being selected for 500 r.p.m. maximum, naturally had a smaller diameter.

The care and interest with which the owner or user daily treats the engine has a most decisive effect on its dependability and life. The fisherman usually commences without any special qualifications in operating and repairing his engine, and the efforts of engine builders in Scandinavia have, for the past 50 years, been concentrated on designing an engine which can meet the most exacting demands for dependability even under the most trying and—from a technical point of view—unsatisfactory conditions of servicing.

In countries such as the U.S.A., replacement of parts is relied upon in preference to re-conditioning the parts through repairs. The U.S.A. have a highly-developed industry with wide distribution, and it is possible, in these circumstances, that it may be advantageous to use, in fishing vessels, high-speed engines of types also used for many other purposes, because spares can be quickly supplied from local stores. But this does not alter the fact that a fishing cutter, often at sea for a month or more and perhaps hundreds of miles from the nearest harbour, cannot apply to the nearest repair-shop when trouble occurs. The crew and vessel are entirely at the mercy of the engine, and the foremost task of the engine builder is, therefore, to concentrate upon the dependability of the engine, all other considerations being secondary.

The less exacting demands on local repair shops made by semi-diesels should be a deciding factor in planning and carrying out mechanization of the fishing fleets in undeveloped areas. The simple operation and maintenance of the semi-diesel, which justify its predominant position in countries as highly-developed technologically as those of Scandinavia, must be of still greater importance in countries such as India and those of South America.

Scandinavian firms make slow-running and comparatively fast-running marine diesels with reduction gear, so they have experience of all types of engines But as more than 90 per cent. of the Danish fishing fleet use semi-diesels, it is evident that they are the most economical.

When evaluating the economy of an engine, the figures for fuel and lubricant consumption are not enough The total expense of an engine comprises the cost of oil, repairs and maintenance, and depreciation and interest. Practice has shown that the lower costs of repairs and maintenance of the semi-diesel more than offset any saving in fuel by the diesels. And, in addition, with the semi-diesel less time is lost in carrying out repairs, while its longer life is, itself, another hidden saving.

MEDIUM SPEED DIESELS

by

D. E. BROWNLOW

THE selection of the most suitable type of diesel, for the duty required, is very important and there are a variety of engines from which a choice can be made, e.g. four-cycle normally-aspirated, four-cycle turbo-charged and two-cycle. They may be low, medium or high-speed engines.

Consider a fishing vessel having approximate dimensions, $120 \times 25 \times 12\frac{1}{2}$ ft. ($36.6 \times 7.6 \times 3.8$ m.), and a tonnage of about 320 gross and 115 nett. Such a vessel has a fish-hold capacity of 7,000 cu. ft. (200 cu. m.), bunkers 70 tons of fuel, and requires an engine having 600 to 700 s.h.p. For many years, such a vessel was powered by a steam engine, coupled directly to the propeller shafting and running at a slow speed. Such engines gave reliable service, but the modern diesel, which is more economical to operate, can claim to give equal reliability.

A fishing vessel is required to be continually in service and the engine is only shut down for a day or two between trips. While at sea heavy weather and storms are often encountered, which demand the utmost reliability from engine and vessel. The following points, therefore, must be considered in choosing an engine:

1. *Reliability:* a foremost requirement.
2. *Simplicity*: the men employed to run the machinery are not highly skilled diesel engineers and they want an engine needing the minimum of attention.
3. *Maintenance:* the engine must be able to run for long periods between overhauls and, when maintenance work is done, the design should allow pistons and cylinder heads to be removed easily and there should be easy access to all parts for inspection.
4. *Size of engine:* space taken up by the engine should be kept to a minimum.
5. *Economy:* the fuel and lubricating-oil consumption must be considered when assessing the merits of diesels for installation in fishing vessels.

DESCRIPTION

To meet the requirements of the fishing industry, the Mirrlees KS type engine was designed and developed for speeds from 200 to 450 r.p.m. Up to the present time, the preference has been largely for engines running at speeds of 230 to 300 r.p.m. directly coupled to the propeller shafting. When the engines are required to operate at higher speeds and powers, it is necessary to fit a reduction gear between the engine and propeller, to maintain propeller efficiency.

The Mirrlees KS type is a four-cycle, turbo-charged engine, having a bore of 15 in. (381 mm.) and a stroke of 18 in. (457 mm.). It is fitted with a Buchi-type of turbo-charger and built with six, seven and eight cylinders. It is direct-reversing, or may be uni-directional, when coupled to a reverse-reduction gear, for the higher engine speeds of 300 to 450 r.p.m.

All engines are arranged for air starting, each fuel pump tappet being designed as a control valve which allows starting air to pass to the starting air valve fitted in the cylinder head.

The cylinder block, engine column and bed-plate are secured by through bolts, four of which can comfortably take the maximum firing pressure of one cylinder. It will be appreciated that, with this construction, the combustion load is taken by the through bolts and the main framing is relieved of tensile stresses.

The crank-shaft has crankwebs of generous proportions and large diameter overlapping pins and journals. This design gives an exceedingly stiff crankshaft and results in the engine having a high natural frequency of torsional vibration and helps to provide a running range clear of troublesome critical speeds. At the same time, the large diameter journals and crank pins provide ample bearing areas, with consequent low-bearing pressures, which ensure long periods of trouble-free operation.

Fig. 479 shows the torsional vibration spectrum for a six-cylinder engine coupled directly to the propeller shafting. Up to 300 r.p.m., there are no critical speeds of any importance and the engine can be run at the most convenient speed for trawling or any other duty required.

For an installation having an engine speed above 300 r.p.m., where a reduction is required to give a low propeller speed, a small damper is fitted to eliminate critical engine speeds up to 450 r.p.m. Some form of elastic coupling between the engine and gearbox is necessary to prevent gear hammer and to eliminate critical speeds in the propeller shafting.

The connecting rod is machined from an H-section steel stamping, having a palm end with four bolts for the white-metal-lined marine-type big end bearing. The small end of the rod is fitted with a phosphor bronze bush.

The piston is of cast iron and oil cooled. Oil is delivered to the big end bearing from the crankshaft and

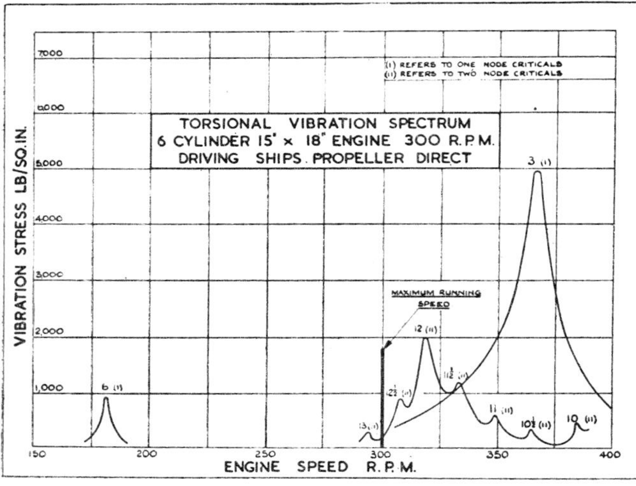

Fig. 479

flows through a drilled hole in the connecting rod, passing up to the piston pin and into an oil chamber formed inside the pin. It is then fed through the pin to lubricate the small end bearing, and to the crown of the piston, at two points, for cooling it. The return oil from the piston flows down through a pipe to a collecting tray at the bottom of the liner and is piped away to a box, outside the engine. The box has sight glasses at each side and from it the oil drains back into the bed-plate. Thus the hot oil from the pistons is not discharged on to the big end bearings and crank webs and is, therefore, prevented from being thrown up into the liner.

Five piston rings are each fitted with three pressure rings and two scraper rings. The bottom scraper ring is of deeper section than the gas rings and, at the bottom of its stroke, reaches the chamber at the bottom of the liner. The face of the ring is parallel with the liner for half its width and is then tapered 2 deg. to a $\frac{1}{16}$ in. (1.6 mm.) radius at the top edge. This radius and taper allows the ring to ride over the oil on the upward stroke and the bottom edge gives an effective scraping action on the downward stroke. The bottom of the piston is turned to a smaller diameter than the skirt, allowing the ring to be passed into position without undue strain. This smaller diameter also leaves a space for the oil which is scraped away. Many experiments and tests have been carried out to find the best type of scraper ring and, for a number of years, this form has proved to be the most effective. It ensures very economical consumption of lubricating oil.

The piston pin is fully floating, of large diameter and well supported. No oil pressure is allowed to reach the piston end plates so there is no danger of oil escaping at these points. The camshaft is situated in the engine column. In this position, it is well lubricated with oil mist and splash and there is no possibility of oil leakages to the outside of the engine.

The cast iron cylinder liners have a continuous uninterrupted bore and are free to expand downwards through water-tight joints. There are no cut-aways for valve heads or connecting-rod clearance.

There are two air and two exhaust valves, in addition to the starting air valve, relief valve and fuel injector, and a fitting for the maximum pressure indicator. There are many advantages with the four-valve head—40 per cent. greater valve area can be obtained with four small valves than by fitting two large ones. Higher volumetric efficiency and improved scavenging are achieved, resulting in a greater output per unit swept volume. The four-valve design also has the advantage of making a much better structure, as the combustion plate, at the bottom, is connected to the top plate at four places instead of two.

Attention is particularly directed to the water spaces of the cylinder head. There is a horizontal division plate cast inside the head, which forms two chambers. The water enters the lower chamber at the front of the head, and circulates round all the valves and injector, before being allowed to enter the upper chamber. This method ensures that the hottest part of the cylinder head, which is the combustion plate—particularly around the fuel nozzle—is adequately cooled. The exhaust valves are provided with cages, while the air valves are seated directly in the cylinder head, and there are large cleaning doors for access to the water spaces.

The engine operates normally on the dry-sump system and has two gear-type lubricating oil pumps, one being the scavenge pump which takes the oil from the bedplate and delivers it to a tank, and the other, the pressure pump, which takes the oil from the tank and delivers it to the engine system under pressure of approximately 25 lb./sq. in. (1.76 kg./sq. cm.). These two pumps are located in an accessible position at the forward end of the engine. They are interconnected with cocks so that, in the event of the failure of either pump, the engine can be run on a wet sump system, which requires only one pump. Such an arrangement meets with Lloyds and other classification societies' requirements for a stand-by pump.

Two double-acting plunger water pumps are driven by eccentrics at the forward end of the engine. A closed-circuit fresh-water cooling system, incorporating a heat exchanger, is employed for engine cooling and the sea water, which is used as a coolant, is also used for cooling the oil.

The engine is under the control of the governor from full to idling speed. The governor is of the centrifugal type, fitted with an oil operated servo-piston, and the speed of the engine is varied by movement of the engine control wheel.

Controls are at the forward end of the engine, a single

control wheel being provided for stopping, starting and running of the engine, and a small lever for reversing. This lever cannot be moved until the control wheel is in the stop position and requires no manual effort to operate as the manoeuvring shaft is worked by air. The camshaft is provided with a double set of cams, ahead and astern, which operate the valve push rods and fuel pumps. The reversing gear consists of a manoeuvring shaft, with cam followers, which is operated by air cylinders. On moving the reversing lever, the cam followers are withdrawn from the cams and are moved, with the manoeuvring shaft, to the appropriate ahead or astern cams and brought into the running position. The engine is then ready for working in the opposite direction.

These engines are designed and built as pressure-charged units. A Buchi-type of pressure charger is incorporated, without increasing the length of the engine. There are great advantages in a turbo-charged engine for fishing duty and the following are of particular importance:

1. The h.p. is increased by 50 per cent. without any increase in the size of the engine, and recent developments in turbo-charging are making available increased powers up to 100 per cent. The space taken for a given power is used economically, which allows a maximum space for the catch.
2. The turbo-charger is driven by the exhaust gases, thus eliminating gears or chain drives and couplings.
3. As no fuel is required to drive the turbo-charger, the fuel consumption of a modern turbo-charged engine is economical.

CONCLUSION

How does such an engine measure up to the five main points laid down?

1. *Reliability*. During development, the prototype engine was run satisfactorily for many thousands of hours, at much higher powers than offered to users to ensure that components had factors of safety to meet any emergency, that there were margins in the capacities of lubricating oil and water pumps, and reserve power in the governor.
2. *Simplicity*. All control operations are carried out by the movement of a single wheel and small reversing lever. There is simplicity in the design of all engine parts. For example, although the piston is oil-cooled, it is in one piece and there are no joints which could give rise to leakage, no joint faces to fret and no bolts to break. The oil chamber is free from ribs and can be thoroughly cleaned out. Such a simple piston can be relied upon for years of satisfactory service.
3. *Maintenance*. The engine can be run continually in service for a year without having to withdraw the pistons. The turbo-charger supplies air for scavenging and cooling down the top of the piston and valves, and the generous oil cooling of the piston ensures that the piston rings are kept free in their grooves.

   The exhaust valves are provided with cages and can be removed for cleaning without disturbing the cylinder head. Large inspection doors are fitted to the crankcase, through which the big ends and main bearings can be inspected and through which the pistons can be removed during the annual overhaul.

   As the water and lubricating oil pumps are accessible and can be removed from the engine without difficulty, their inspection and maintenance is easily undertaken.

   All engine parts are jigged and machined to close limits and are carefully inspected. There is, therefore, complete interchangeability and, if it is necessary to replace a part, it can be done quickly.
4. *Size of engine*. It is to be expected that a medium-speed engine, for a given power, may take a little more space than a high-speed engine but, bearing in mind that there is less wear, tear and maintenance, and that the turbo-charged engine give att least 50 per cent. more power than the normally aspirated engine, the space is economically used.

   Mirrlees are also producing a higher speed turbo-charged engine, running up to 900 r.p.m. with powers up to 2,000 h.p. It takes up less space for a given power but fishermen show a marked preference for the medium-speed engine.
5. *Economy*. The fuel consumption of these turbo-charged engines is of the order of .345 lb. (0.156 kg.) per s.h.p. per hour, as compared with the fuel consumption of .37 to .4 lb. (0.168 to 0.181 kg.) per s.h.p. per hour of normally aspirated and two-cycle engines. The engines can run satisfactorily on marine diesel fuel and even heavier grades when necessary heating and centrifuging equipment is provided. The oil scraping arrangements already referred to under piston design, combined with a cylinder liner having an unbroken surface without ports, ensure a lubricating oil consumption of .003 lb./s.h.p./hour (0.0014 kg.).

HIGH-SPEED DIESELS

by

WILLIAM C. GOULD

WHEN diesels were first accepted, they had to be built in the slow speed range (75 to 500 r.p.m.). The injection systems then available, combustion chamber design and metallurgical and engineering knowledge, were not advanced enough to permit the consideration of higher r.p.m. It was also felt that a long stroke was essential to gain highest efficiency, and that required low r.p.m. to keep piston speeds down to limits then thought satisfactory.

The engines were very large and heavy, necessitating air starting, direct reversing, and drip-type lubrication that used much oil. They were expensive in initial cost, but still more compact and much more economical to run than the steam engines.

For use in small fishing craft, smaller versions were developed by reducing the number of cylinders to one or two. While these engines were reliable, they vibrated violently to the detriment of the hull structure. Their advantages were: (*a*) reliability; (*b*) economy of fuel, about .38 lb. (0.173 kg.)/b.h.p. per hour; (*c*) reduction in engine room force. Their disadvantages were: (*a*) high initial cost; (*b*) need for an oiler or engineer to be on duty at all times; (*c*) high cost of overhaul because all parts had to be hand fitted; (*d*) high lubricating oil consumption; (*e*) heavy vibration.

Four advances permitted the introduction of medium speed diesels in the 1920s. They were: (*a*) better fuel injection systems (pioneered by Bosch); (*b*) better materials, particularly for wearing surfaces; (*c*) better lubricating oils that could stand higher temperature; and (*d*) better reduction gears, which permitted the medium speed diesels to turn the propellers at low r.p.m.

The manufacturers had a difficult time introducing their engines to the fishing industry. It was true then, as it is to-day, that this industry is most reluctant to adopt any change without having its value proved to them over several years of operation.

With the introduction of medium speed diesels came other notable changes: the use of forced feed lubrication to the bearing surfaces, reverse gears, and mechanical fuel injection. The increased use of motor trucks, and the constantly increasing cost of gasoline, inevitably forced automotive engineers seriously to consider the diesel cycle. During the '20s the motor truck industry grew by leaps and bounds, and it brought to the diesel engine industry vast new engineering skills, much larger testing and development facilities and, most important, huge financial resources. Ricardo of England and Mercedes-Benz of Germany were pioneers and were rapidly followed by the major automotive engine manufacturers throughout the world.

At this time the diesels developed from 50 to 100 b.h.p. at 1,400 to 1,800 r.p.m. Some were adapted to marine propulsion and auxiliary units, and were used in smaller fishing vessels because no other diesel would fit them.

World War II probably did more than anything to change the minds of fishermen about high-speed diesels. They were widely used on small landing craft and many a fisherman's son operated them. His experience convinced him of their reliability and he was most influential in selling the idea to his father. Then, at the end of the war, thousands of such engines came on the market at ridiculously low prices so that fishermen could not afford to ignore them. To their surprise, they found them generally to be as reliable as the older types of engines they had used.

Most of the surplus engines were of the high-speed two-cycle type which had been adopted by several manufacturers to obtain still more horsepower from smaller engines. This cycle was successfully used by introducing a positive displacement type of scavenging blower.

An analysis of fishing craft of the United States shows that their main propulsion engines are used from 2,500 to 5,000 hours a year, with an average probably around 3,500 hours. There are 8,760 hours in a year so that the engines are actually operating about 40 per cent. of the total available time. For the vast majority of American fishing vessels, trips last from 4 to 15 days, with a good average of around 10 days or 240 hours of continuous operation per trip. These hours provide some guidance for designers trying to produce an engine to meet the fishing industry's requirements. High-speed diesels, properly installed and with good propellers, to-day do meet these requirements, as proved by the thousands of fishing vessels using them.

PISTON SPEEDS

One old engineering term has been propagated throughout the world: that engines should not have piston speeds in excess of 1,500 ft. (454 m.) per minute in order to give long, reliable service. Such a limitation completely disregards development and engineering work on the design, finish and materials used in cylinder sleeves, pistons and piston rings. It is now known that piston speeds have very little to do with the actual wear pattern on the cylinder walls and piston rings. Despite this, most high-speed diesels, at the ratings given to them for fishing boat use, still do not exceed 1,500 ft.

In connection with piston speed, it is interesting to note that service records of later models of high-speed diesels, running at an average piston speed of 1,750 ft. (534 m.) a minute, are giving approximately 32 per cent. longer life between overhauls than the older, slower models. It seems to be forgotten that smaller cylinder diameters mean shorter strokes, consequently the engine can turn faster and still have the same piston speed as a slower turning engine with a much longer stroke. In many cases the actual piston speed of high-speed diesels is even less than those of the medium- and slow-speed engines. It is also interesting to note that present-day high-speed diesels have a frictional horsepower loss, in relation to the maximum b.h.p. rating of the engine, less than the percentage of frictional horsepower loss on the medium- and slower-speed engines.

INJECTION SYSTEM

Engineers have been able to develop high-speed diesels because of the improvements made in the past twenty years in the design and construction of mechanically-operated fuel injection systems. Not only have the injection and cut-off characteristics been improved, but the injector now permits smaller amounts of fuel to be injected at higher pressures. Furthermore, the hydraulics involved have been studied most carefully and are much more thoroughly understood to-day than they were at that time. The National Advisory Committee for Aeronautics in 1932 published a report on some tests which they had made concerning fuel injection. This indicated that pressures beyond certain limits did not increase the fineness of the spray. But practical considerations have since indicated that higher pressures are necessary for high-speed diesels because the time available for injecting the fuels is much smaller. Most modern high-speed diesels now inject at 10,000 to 17,000 lb. per sq. in. (700 to 1,200 kg./sq. cm.) pressure.

COOLING SYSTEMS

The diesel is essentially a heat engine but certain temperatures cannot be exceeded without deteriorating the material very rapidly. For this reason it is necessary to cool the engines. The hottest spot, of course, is in the combustion chamber of which one side is formed by the piston itself. The greater the distance the heat has to travel before being removed by the cooling medium, the more severe the problem is from a metallurgical and operating point of view. High-speed diesels, with their smaller diameter cylinders, therefore have an advantage over the medium- and slow-speed diesels because of the smaller distance the heat has to travel before reaching the cooling medium.

While it is true that this is materially assisted on all diesels, regardless of their speed, by oil coiling of the pistons, oil is not nearly as efficient as water as a cooling medium. Virtually all high-speed diesels to-day are fresh water cooled, either through the use of a heat exchanger or through so-called " keel cooling ", which permits the engines to be run at much higher temperatures than if sea water were used, as salt would be precipitated. It is of real advantage to run the engines at these higher temperatures as all internal working surfaces are closer to the same temperature and the wear is much reduced. And when operating in climates that reach freezing or below, an anti-freeze solution can be added to the fresh water system which reduces the chance of serious damage by freezing. Furthermore, engineers are able to put in a better system of thermostatic control on the cooling medium to ensure the engine running at a more even temperature regardless of outside temperatures and most high-speed diesels now are run at 165 to 195 deg. F. (73.5 to 90.5 deg. C.).

Fresh water also permits use of a so-called pressurized cooling system of about 12 lb./sq. in. (0.84 kg./sq. cm.) which means that the water will not boil until it has reached 225 deg. F. (107 deg. C.). There is every indication that in the future this temperature will be increased. In addition to reducing the wear on the rotating parts, the increased temperature will allow use of fuels of higher sulphur content. At the present time there are indications that if engines were run at 212 deg. F. (100 deg. C.) or higher, the sulphur in the fuel would not form sulphuric acid as a product of combustion. This, however, has not been definitely established.

VIBRATIONS

The smaller means of rotating weight in a high-speed diesel are, of course, much more easily balanced than they are on the medium- or slower-speed engines, and the power impulses are of smaller force and much more frequent. All of this results in a virtually vibration-free engine.

REVERSE GEARS

The development of reliable reverse gears was an essential step towards building high-speed engines. The older gears required constant adjustment on the clutch plates, as well as frequent greasing. Modern engines generally have hydraulically operated gears that require no adjustments to be made on them between overhauls. They have their own complete and independent oiling system with forced feed lubrication to all bearings and gears, and are practically trouble-free. They do away with all the mechanism that once was necessary to reverse the engine, and operate much more quickly than the old directly reversible motors ever did.

Due to advanced techniques in gear design and building, they are now compact units instead of being almost the same size as the engine itself. The horsepower loss through these gears—including the reduction gears—in order to get the slow propeller speeds, is very small, seldom exceeding 2 per cent. of the h.p. put through it. But this advance has not been made without some additional cost to high-speed diesels. The gears, built in relatively small quantities, are still expensive and represent a good percentage of the total cost.

BEARING SURFACES

Another adage that still seems to be accepted, is that the hand-made article is better than one made by machinery. While it is true that expert machinists could build a hand-made engine with as close tolerances as are required on present-day high-speed diesels, it is doubtful if they could build one diesel after another on this basis and still maintain the same tolerances. Because of their many uses, high-speed diesels are built in large numbers, which permits the introduction of specialized machines and heavy duty tooling. These tools are capable of turning out pieces with very little variation. For example, all of the main and connecting rod bearings on high-speed diesels are of the pre-fitted type. No hand scraping is required in order to fit them even for replacement purposes. The advantage of this can be appreciated by any fisherman who has recently tried to find someone to scrape in bearings for him. Furthermore, because of the large quantities involved, a better basic type of bearing can be fabricated than is generally found on the slow-speed diesels. All of these things contribute to reducing the frictional horsepower in the engine and to making it very easy to repair.

LUBRICATION

No one can overestimate the importance of properly lubricating any internal combustion engine. On high-speed engines this is taken care of by a forced feed lubrication system to every part on the engine, which assures the supply of the correct amount of oil at all times to every bearing surface. High-speed diesels, as a group, generally use less lubricating oil than the other types of diesels. Most of the high-speed class have a guaranteed lubricating oil consumption of .0025 lb. (.0016 kg.) per b.h.p. per hour. This saving in lubricating oil consumption is considerable over a 3,500-hour period.

FUELS

It is quite true that the high-speed diesels cannot properly burn the lower grade fuels that are used in the slow-speed and medium-speed diesels. They will, however, run satisfactorily on ASTM No. 1 or No. 2 diesel fuel. While some of the very slow-speed large diesels can burn No. 6 fuel oil, this requires a lot of very expensive auxiliary equipment, such as centrifuges and pre-heaters, to prepare the fuel for injection purposes. On a fishing boat of 150 ft. (45.7 m.) or less, the use of such fuel, even in this type of engine, is therefore not practicable. And because of the large use of No. 1 and No. 2 fuel oil throughout the world it is the most readily available of all diesel fuels. Its higher price is more than offset in high-speed diesels by the reduction in cost of parts at overhauls.

High-speed diesels, with their small diameter pistons, make it more difficult to get complete atomization of the fuel injected. For this reason they generally burn slightly more fuel than engines having larger diameter pistons, although all of the fuel is burnt cleanly. The specific fuel consumption for high-speed diesels is in the neighbourhood of 0.43 lb. (0.195 kg.) per b.h.p. per hour, but research which is being carried on indicates that this specific fuel consumption will be reduced.

INTERCHANGEABILITY OF PARTS

High-speed diesels are manufactured on a large production basis as they sell in many markets. The marine industry, incidentally, is one of the smallest users in numbers and seldom takes more than 10 per cent. of the production of the factories.

The engine parts are so nearly alike that they are completely interchangeable without hand fitting. This means that a part can be bought anywhere in the world and will fit the engine concerned. When improvements are made in the parts they are still interchangeable on the older models so that, when overhauls are made, the engines are not merely repaired but are also modernized.

Mass production brings about many other advantages. Not only are parts readily available in different areas of the world, but they are much cheaper than the comparable parts on medium- and slow-speed engines. And it is not unusual for a complete overhaul to be carried out in four or five days. Such an overhaul will require 100 to 190 man-hours of labour and from $300 to $800 worth of parts. In contrast to this, there are instances where owners of medium- and slow-speed diesels have had to wait 30 days to obtain piston rings.

The fact that high-speed diesels are generally overhauled every one or two years means that the fisherman knows his engine is in first-class shape and even though the overhauls are more frequent than in the case of slow- and medium-speed engines, the total cost of doing them, and the time consumed, is considerably less. The best operators using medium- and slow-speed diesels still spend time going over their engines annually when the boat is being refitted. This does not approach a complete overhaul but to do the job well still requires about the same number of man-hours and, quite frequently, more expense in parts.

SIZES OF HIGH-SPEED DIESELS

At the present time, for marine propulsion purposes, the heavy duty ratings on high-speed diesels are limited to approximately 250 s.h.p. for individual units. This means that all sizes of fishing boats up to 65 ft. (19.8 m.) can be supplied with engines as single units. The high-speed diesel manufacturers have introduced, however, multiple units where two or more engines are geared together to turn a single propeller. This is one of the more recent innovations in the application of these engines to fishing craft and the advantages to be gained are being more and more recognized by fishermen as it gives them the reliability of a twin screw boat with the advantages, from a fishing point of view, of still having only one propeller. Such units are now available with continuous duty ratings up to 520 s.h.p. In these cases individual diesels are clutched into heavy duty herringbone gear boxes and can operate with one or more of the engines disengaged. When the engines are clutched in together, they are locked into the gear box and all turn the same r.p.m. By means of a simple pyrometer set-up, recording the temperature of the exhaust gasses from each engine, it is a simple matter to adjust the governors so that each engine is carrying its fair share of the load. Equalizing mechanism is also incorporated to assist in this operation.

The thing that surprises most new owners is the small drop in speed, when running free, they experience when cutting out one engine. It seems to be forgotten that a propeller is nothing but a pump and the horsepower required to turn it at 75 per cent. of its normal r.p.m. requires less than 50 per cent. of the s.h.p. As a result, without overloading the individual engine, it is possible to cut off one engine in a multiple unit and still obtain 70 to 75 per cent. of normal speed.

This is a most important economic factor in the operation of fishing craft; and it is felt that the tendency to higher and higher horsepowers, particularly in the trawlers, is a very expensive move and not entirely necessary. Since trawling speeds are slow and the trawler becomes, in effect, a towboat, material increases in thrust at this time can be obtained by the use of Kort nozzles and contra-guide rudders. This efficiency can be further improved by the use of controllable pitch propellers. At 4 knots, these combinations can increase the thrust by as much as 30 per cent. That means a saving in the initial cost and in the maintenance and fuel bills, and the ability to get the same results as with an engine of 30 per cent. more horsepower.

The multiple engine units come with much larger reduction gear ratios than on the single units and are available with up to 6:1 reduction. At an average cruising r.p.m. of, say, 1,600 r.p.m., this means that the propeller shaft is only turning 267 r.p.m. In the 400 to 550 h.p. class of multiple units, it is not unusual to have propellers with 72 in. (1,830 mm.) diameter and 38 in. to 52 in. (965 to 1,320 mm.) pitch.

AUXILIARY EQUIPMENT

High-speed diesels have long been used on larger fishing vessels as auxiliaries for generating and pumping purposes. It is not unusual to find marine generating sets running as long as 15,000 hours between major overhauls. On many boats they drive the winch generator when an electrically driven trawl winch is employed. Their great flexibility is of material assistance in operating this equipment. With the further advancement of torque converters and fluid couplings, the wide operating range of high-speed diesels may be used for direct drive of larger winches. By means of a proper oil cooler on the torque converter, it is possible to adjust the speed of the diesel so that it, with the torque converter, becomes a brake and the hand-brake does not have to be set. This is a great advantage in heavy seas where surging on the wire occurs and also when the trawl is caught on the bottom. When this equipment or a separate diesel-driven electric winch drive is installed, it is possible to incorporate an A.C. electrical system throughout the vessel. The result is a material reduction in the initial cost of such equipment, and in maintenance of the electrical system. And it also eliminates the need for expensive, large, heavy duty storage batteries, which must be replaced every four to six years, a saving in space, time, money and labour.

SUMMARY

Compared with slow- and medium-speed diesels, the high-speed engine has the following disadvantages:

(*a*) Slightly higher fuel consumption;
(*b*) Noisier;
(*c*) Must be overhauled more frequently.

Against these, however, from the fisherman's point of view, there are the following advantages:

(*a*) Lower initial cost, running from 50 to 75 per cent. of the cost of the larger engines installed;
(*b*) Lower lubricating oil consumption;
(*c*) More compact, requiring much smaller engine rooms;
(*d*) Lighter weight, permitting better balance of the ship when the engines are either in the stern or in the bow;
(*e*) Less down time;
(*f*) Much lower overhaul cost, both in man-hours and in cost of parts;
(*g*) Much more flexible in its power application;
(*h*) Spare parts easily obtainable;
(*i*) Larger number of experienced mechanics available

These advantages show why, in the small and medium sized fishing vessels of the world, more and more high-speed diesels are being installed.

SEMI-DIESELS VERSUS DIESELS

by

IVAR STOKKE

THE first hot-bulb engines for fishing boats were four-cycle and a great improvement was the change into a two-cycle engine with crankcase scavenging. This gave simpler, stronger and more reliable engines, having up to four-cylinders with 6 to 100 b.h.p. output each. The lower outputs were generally used for single-cylinder engines, and the higher for multi-cylinder ones.

Fig. 480 shows weights per b.h.p. and overall lengths of modern semi-diesels. A one-cylinder 100 b.h.p. semi-diesel weighs about 176 lb. (80 kg.) per h.p., while a two-cylinder 100 b.h.p. engine weighs about 110 lb. (50 kg.) per h.p. One cylinder semi-diesels of a higher output than about 75 b.h.p. should not be built. The length of a two-cylinder engine is not much more than that of a 100 b.h.p. one-cylinder engine.

Modern semi-diesels have compression pressure of 213 lb./sq. in. (15 kg./sq. cm.) and combustion pressures of 355 to 426 lb./sq. in. (25 to 30 kg./sq. cm.). Contrary to the old hot-bulb engines, modern semi-diesels use an internal hot-bulb for ignition on starting and lower loads, while ignition at full load usually is achieved through compression ignition (variable fuel-spray angle), similar to diesels.

Diesels with open combustion chambers have compression ignition at all loads (apart from certain types of precombustion chamber and swirl chamber diesels, which are seldom used for propulsion of fishing craft).

The fuel consumption of modern one-cylinder 20 b.h.p. semi-diesels is about 0.485 lb. (0.22 kg.) per b.h.p./hr. at normal load. Modern semi-diesels with higher compression and injection pressure, and internal hot-bulbs allowing higher temperatures than the old external hot-bulbs, have a flatter fuel consumption curve. The specific fuel oil consumption curve for different loads looks more like that of the diesel's. The idling consumption is considerably lower than for the old hot-bulb engines, which had relatively high consumption compared with the diesels. Fishing boat engines are very often idling for long intervals and this consumption is of great importance for fuel economy.

Crankcase scavenging is especially well suited to semi-diesels of one- and two-cylinders, with cylinder outputs up to 75 b.h.p. The scavenging air, being produced on the under side of the piston, requires little power and the mechanical efficiency is, therefore, high—up to 90 per cent.—which ensures low specific fuel consumption at all loads. Semi-diesels work at lower combustion pressures than diesels and the wear on bearings and cylinders, under equal conditions, is less. The fuel injection pressures from ca. 1,140 to 2,130 lb./sq. in. (80 to 150 kg./sq. cm.) allow very robust fuel injection systems, which means minimum wear of fuel pumps and fuel nozzles. The result is that their guaranteed fuel oil consumption is retained long after the running-in period. The maintenance cost of a good semi-diesel is

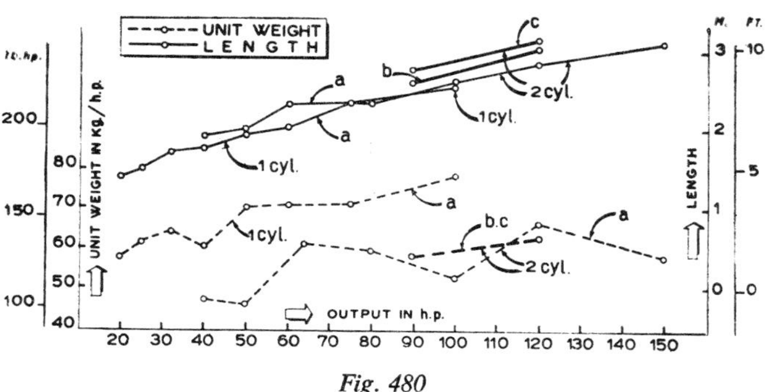

Fig. 480

lower than for a diesel, and the total operating cost is at least as favourable. Three cylinders and about 250 b.h.p. is usually the maximum size for semi-diesels. The mean effective pressure at the most economical load is about 45 to 50 lb./sq. in. (3.2 to 3.5 kg./sq. cm.) with a specific fuel consumption of about 0.43 to 0.41 lb./b.h.p./hr. (0.195 to 0.185 kg./b.h.p./hr.).

Several firms have begun to build three- to four-cylinder engines with a separate scavenging pump, which has a larger scavenging capacity than can be obtained by crankcase scavenging. The separate scavenging pump volume is about 1.3 to 1.5 times the cylinder volume and a better scavenging of the cylinder is obtained, as well as a higher mean effective pressure at normal load of 57 to about 60 lb./sq. in. (4 to about 4.2 kg./sq. cm.). The specific fuel consumption of a fully-scavenged semi-diesel engine is hardly any better than for a crank-case scavenged one because the special pump causes mechanical losses which decrease the mechanical efficiency. The top-volume cannot be scavenged thoroughly on a semi-diesel even if a special pump is used. This is a draw-back compared with fully-scavenged diesels with open combustion chambers. One- and two-cylinder semi-diesel engines, with a special scavenging pump, are uneconomical to build.

Crankcase scavenged semi-diesels are usually built with separate cylinders, but the overall lengths are not much longer than for fully-scavenged two-cycle diesels. Fig. 480, curves b and c, show the overall lengths of two-cylinder, fully-scavenged 2-cycle diesels with controllable pitch propellers or reverse gears. In addition, the weight per b.h.p. is higher for 2-cylinder, fully-scavenged, two-cycle diesels, as is shown by the dotted line marked b, c.

When fully-scavenged two-cycle diesels are built with three or more cylinders, with at least 50 b.h.p. cylinder output, they have advantage over semi-diesels in weight. Such diesels, with high pressure fuel injection, have a consumption of 0.375 to 0.397 lb./b.h.p./hr. (0.17 to 0.18 kg.). The compression pressures are about 470 to 510 lb./sq. in. (33 to 36 kg./sq. cm.) and the maximum combustion pressures about 710 to 850 lb./sq. in. (50 to 60 kg./sq. cm.), i.e. twice the semi-diesel's. Bearings, piston lengths, etc., must accordingly be bigger. A diesel, with the same reliability and lifetime as a semi-diesel, must be of first-class construction and materials. The lubrication system and the lubricating oil quality must also be top grade.

Maintenance and operation must generally be more careful than in the case of semi-diesels otherwise it will have shorter life and be less reliable. It may, therefore, be said that the semi-diesels are easier to manufacture, are more robust and simple, are easy to operate and do not require much attention, as long as they get cooling water, lubricating oil and fuel.

Fully-scavenged two-cycle diesels for larger fishing craft are usually built with three to eight cylinders and 50 to 130 b.h.p. cylinder outputs. For deep sea trawler, still higher cylinder outputs are used.

Crankcase scavenged diesels are seldom used in the fishing fleet, except as small, high-speed auxiliary engines, generally having a pre-combustion or a swirl combustion chamber with a fuel injection pressure of about 2,130 lb./sq. in. (150 kg./sq. cm.). Their optimum specific fuel consumption is about 0.44 to 0.46 lb./b.h.p. (0.20 to 0.21 kg.).

TABLE LXX

DEVELOPMENT OF SEMI-DIESEL

Characteristics		*External Hot-bulb*				*Internal Hot-bulb*
Test year		1931	1937	1938	1939	1951
Normal output in b.h.p.		33.55	51.6	17.5	32	60
Normal r.p.m.		401	390	545	430	375
Fuel consumption	kg./b.h.p./hr.	0.252	0.222	0.241	0.239	0.195
	lb./b.h.p./hr.	0.554	0.490	0.531	0.527	0.430
Effective mean pressure	kg./sq. cm.	2.49	2.89	2.33	2.10	2.80
	lb./sq. in.	35.4	41.1	33.2	29.8	39.8
Idling r.p.m.		163	108	177	170	107
Idling consumption	kg./hr.	1.218	1.286	0.642	1.24	0.792
	lb./hr.	2.68	2.83	1.415	2.74	1.75

Four-cycle diesels are also used for propulsion of fishing boats but they have only about 60 to 65 per cent. of the output of two-cycle diesels of corresponding weight and r.p.m. If they are not to be heavier and bigger than two-cycle types of corresponding power the r.p.m. of four-cycle engines must be increased. Another solution is supercharging. Higher r.p.m. can be used, because the four-cycle engine has one whole piston stroke of forced air scavenging, while the scavenging and charging period of the two-cycle engine takes place during a short part of a revolution at the lower deadpoint of the crank. Good propeller efficiency requires low propeller r.p.m. so four-cycle engines are frequently equipped with reduction gears.

Engine frames and cylinder blocks have been built of welded steelplates during recent years. With rational production and careful welding, they are as cheap as cast iron parts because miscastings are avoided. The engines are stiffer and the weight can be decreased by at least 15 per cent. More welding will require larger workshops with annealing furnaces to treat the welded parts to relieve welding stresses.

Table LXX shows characteristic test results of Norwegian semi-diesels from 1931 to 1951. The idling consumption per hour has decreased from 2.68 lb. (1.218 kg.) for a 35 b.h.p. semi-diesel in 1931, to 1.75 lb. (0.792 kg.) for a 60 b.h.p. engine built in 1951. Results obtained in 1953 are still better.

The maximum output for a Norwegian semi-diesel 60 h.p. one-cylinder, Wichmann built in 1951, has been recorded in fig. 481 with the mean effective pressures relative to the r.p.m. At maximum output the r.p.m. should not be higher than 400, otherwise the time cross-section for the exhaust and scavenging ducts will be too small. Fig. 482 shows mechanical, thermal fuel and effective fuel efficiencies, as well as specific fuel consumption at maximum output. Fig. 483 shows efficiencies and mean effective pressure at 375 r.p.m. (constant). Fig. 484 shows the efficiencies relative to the propeller load curve. As with the diesel, fuel efficiency is fairly constant from 30 to 70 b.h.p. and fuel consumption is also almost constant within a very large load area.

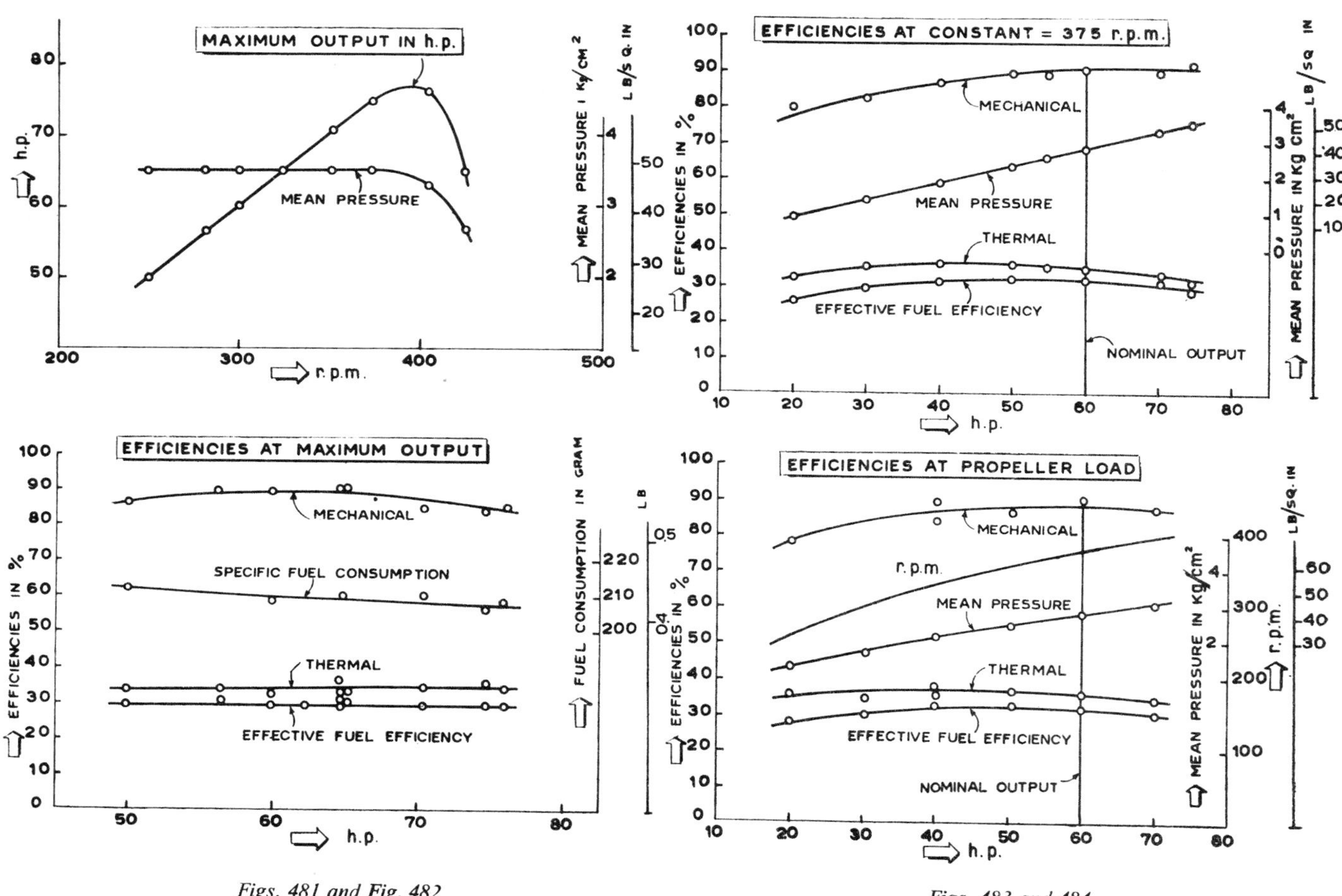

Figs. 481 and Fig. 482

Figs. 483 and 484

MODERN PROPULSION PLANTS

by

K. SCHMIDT AND TH. SCHUMACHER

IN 1927 the internal combustion engine was almost the exclusive means of propulsion in German fishing cutters, and sails began to lose their importance. The fishermen had to be warned against any hasty decision to install bigger engines in order to maintain the economy of the operation. On the basis of the wide experience gained, the Deutscher Seefischereiverein (German Sea-fisheries Association) and the Reichsanstalt für Fischerei (Federal Fisheries Institute) decided to study systematically the economic power for different sizes of cutters. Until then, the construction of wooden cutters had been, with little exception, a matter of the craftsman's experience. Extensive research was done in the years 1940–1942 with the financial aid of German engine manufacturers. Messrs. Maierform, naval architects of Bremen, Germany, analysed the line drawings of the vessels and designed new plans. The Shipbuilding Experimental station in Vienna carried out tank tests. These showed that the traditional empirical shapes could be improved considerably. Propeller tests were also carried out. Unfortunately, the valuable results did not get the deserved attention in the reconstruction of the cutter fleet of Western Germany after World War II, but they were widely used in the Eastern Zone. For this reason there are excellent fleets of modern cutters in the Baltic Sea to-day.

For cutters with directly driven propellers the tests indicated the optimum power and revolutions, given in Table LXXI.

The additional investment and the higher fuel consumption of higher power than recommended would not result in a correspondingly larger catch of fish because the higher speed and towing performance, achieved with stronger engines, is not proportional to the output. In certain circumstances, however, more engine power could be advantageous, i.e. if a reduced propeller speed was provided. It is essential to judge engine and propeller separately; each element has its optimum speed range, and a reduction gear has to be arranged between engine and propeller.

A summary of the optimum values is to be seen on Table LXXII.

TABLE LXXI

Overall length	ft.	32.8	39.3	45.8	52.4	59	65.5	72	78.6
	m.	10	12	14	16	18	20	22	24
Propeller power	h.p.	30	50	70	90	110	130	150	180
Maximum propeller speed	r.p.m.	550	500	440	380	350	330	320	315

TABLE LXXII

Overall length	ft.	32.8	39.3	45.8	52.4	59	65.5	72	78.6
	m.	10	12	14	16	18	20	22	24
Propelling power	h.p.	50	70	90	110	130	150	180	210
Maximum optimum propeller speed	r.p.m.	390	375	350	320	290	260	240	220
Maximum propeller diameter suitable for the Maierform design	in.	39.4	41.7	47.3	51.2	57.1	61	67	72.8
	m.	1.0	1.06	1.20	1.30	1.45	1.55	1.70	1.85

CONSUMPTION CURVES AND POWER DIAGRAM

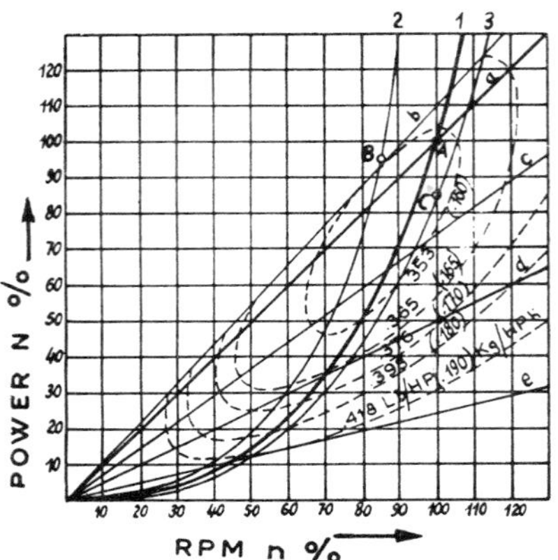

1) POWER ABSORBED BY CONTROLLABLE PITCH PROPELLER
2) " " BY CONSTANT PITCH PROPELLER TOWING
3) " " BY PROPELLER FREERUN
A) 100 % ENGINE POWER
B) 110 % " "
C) 75 % " "
D) 50 % " "
E) 25 % " "

Fig. 485. Consumption diagram of a two-cycle diesel, of 400 h.p. at 625 r.p.m.

The authors are concerned with the Deutz diesels, which have low weight as compared with the output. If one, therefore, selects an engine having a higher maximum output than necessary for continuous cruising, one would be able to use this engine with reduced output and still not get a too heavy engine. Then one will have a power reserve for trawling, which also will be valuable when the ship encounters wind and waves. Using such an engine with different revolutions is by no means disadvantageous, as the specific fuel consumption is about the same between half and full load, as is evident from fig. 485.

With two-speed reduction gear it is possible to use the engine power in the best way for both trawling and cruising, and recently the first three-speed gear was installed. The entire control of the engine and gearing is concentrated in a single operator's stand, by air or oil hydraulic pressure, which relieves the operator from heavy work. The operator's stand can be transferred from the engine room to the bridge. Some captains still prefer the engine room telegraph. Both systems can also be connected.

Two engines of equal output can be combined by a multi-state gear which can give different reductions and reversing. Furthermore, one engine can be cut off from the propeller shaft for driving the electric generator or the trawl winch.

Deutz designed and supplied " father and son " power plants for three trawlers, making both engines directly reversible so that all manoeuvres could be performed with the " son " engine alone. This was smaller in order to save weight and the system has proved very satisfactory in fishing. Each of the two engines operates by way of a clutch on to the common reduction gear. The electric generator is placed between the " son " engine and its clutch, because then the rotating masses are on one side of the crankshaft and the total system is accessible. This arrangement is favourable as regards its oscillations. " Disengaging " of the electric generator is effected by switching off the field excitation.

V-BUILT TWO-CYCLE ENGINES

Short engine length is especially advantageous for fishing craft in order to make room for the fish-hold and

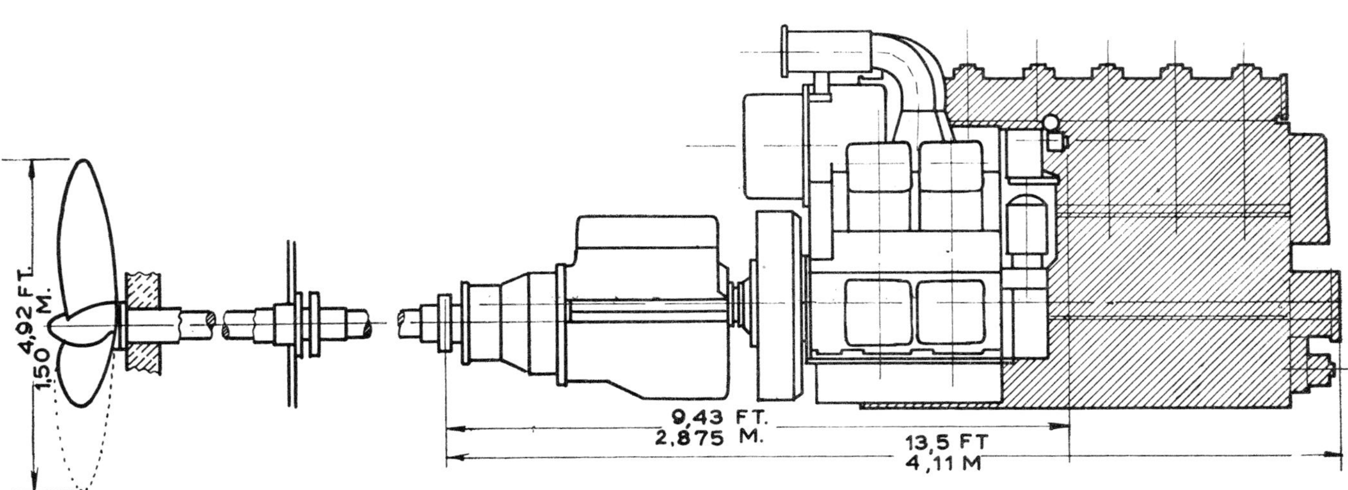

Fig. 486. Comparison of a 6 cyl., four-cycle diesel (shaded), with a 4 cyl., two-cycle, V-built diesel, both driving a reverse reduction gear of the same size

comfortable quarters for the crew. V-built, two-cycle engines with four cylinders in the length of a two-cylinder unit, have turned out to be very favourable. Such important elements as the injection pump and the fuel valves can be placed so that they are easily accessible. The dismantling of the pistons, at general overhauls, is also facilitated by the V-form and requires less height.

water cooling is recommended in order to combine steady cooling with minimum attention. The temperature is controlled by an automatic thermostat and the temperature of the lubrication oil is kept constant by arrangement of a by-pass piping at the cooler, fed with sea water. The heating of the vessel can be combined with the cooling circuit in a simple way.

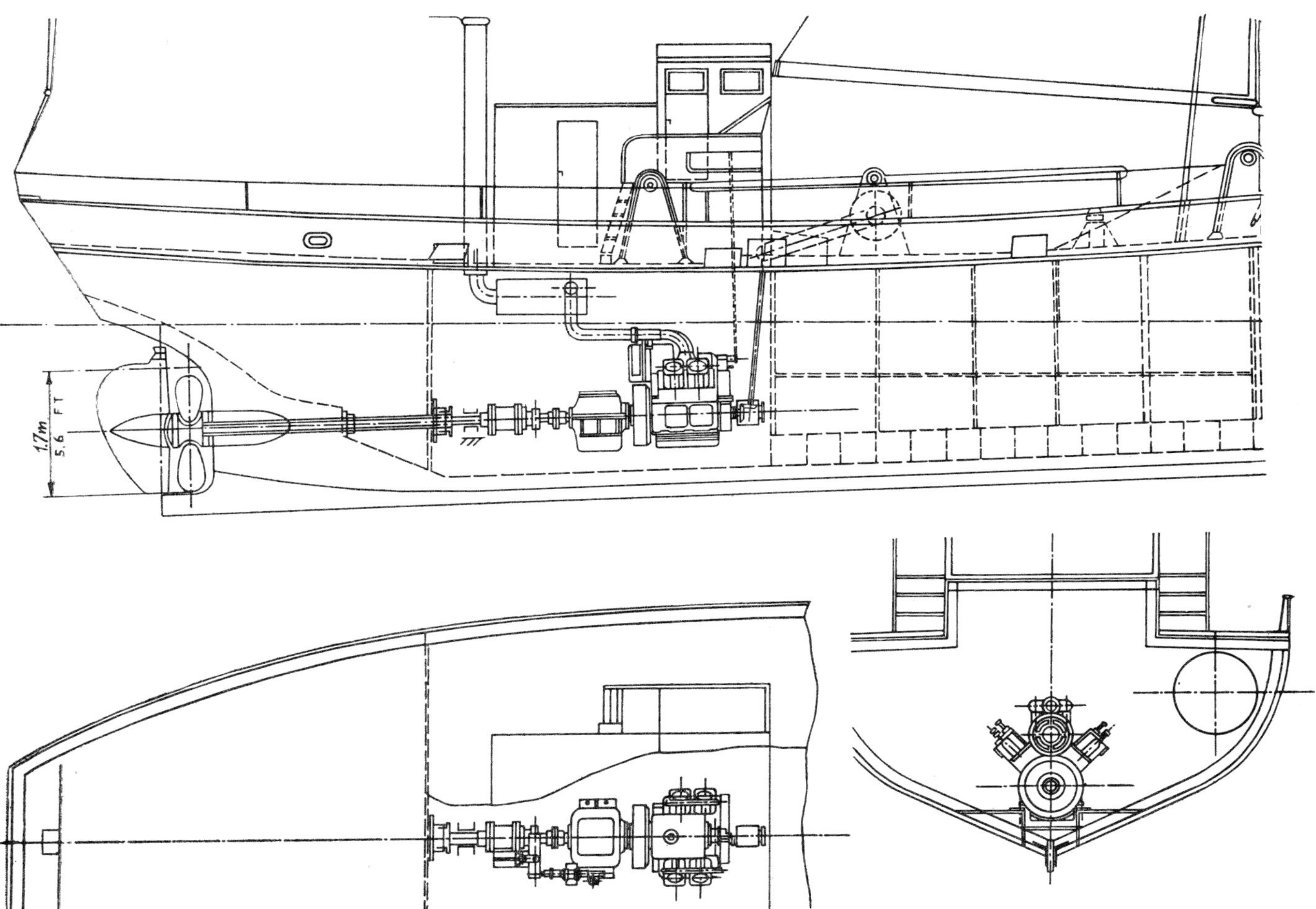

Fig. 487. Installation of a V-built 190 h.p. diesel with Reintje's controllable pitch propeller in a 78.7 ft. (24 m.) cutter

Fig. 486 shows a comparison of a V-built engine and a straight engine of about the same effect and number of revolutions. Fig. 487 shows a V-built engine installed in a fishing boat.

Table LXXIII lists V-built engines suitable for fishing cutters. The older series 425 has a Roots-type scavenging blower but the new types 525 and 320 are provided with a simple centrifugal blower, the air delivery of which is better related to the propeller output. The new types have low specific weight and short overall length.

For trouble-free operation, cooling of the cylinders and cylinder heads, the circulating lubrication oil, and the air compressors, are of particular importance. Fresh

AIR COOLED ENGINES

Water-cooling can, however, be a source of trouble, especially in shallow coastal waters, in estuaries, and on shallow lakes. Since 1942, Deutz has developed a series of air-cooled diesels, thousands of which have already proved satisfactory in land traffic, in severest cold as well as in great heat. Conditions of service are simpler on board ship than on land and the same air-cooled diesels have proved satisfactory in ships for the last three years. The working temperatures are reached in one-fifth of the time needed by a water-cooled one and it cannot be damaged by frost. Wear and corrosion are exceptionally low, allowing higher piston speeds. Air-cooled diesels have been installed in great numbers in fishing

TABLE LXXIII

WATER-COOLED IN-LINE AND V-BUILT TWO-CYCLE DIESELS FOR FISHING CUTTERS

Size of the cutter, overall length	ft. m.	32.8 10	39.3 12	45.8 14	52.4 16	59 18	65.5 20	72 22	78.6 24
Motor type		ST2M 425	ST3M 425	ST4M 320	ST4M 320	ST4M 320	ST4M 525	ST4M 525	ST4M 525
Number of cylinders		2	3	4	4	4	4	4	4
Bore	in. mm.	5.9 150	5.9 150	6.3 160	6.3 160	6.3 160	7.86 200	7.86 200	7.86 200
Stroke	in. mm.	9.84 250	9.84 250	7.86 200	7.86 200	7.86 200	9.84 250	9.84 250	9.84 250
Medium effective brake pressure	lb./sq. in. kg./sq. cm.	64 4.5	64 4.5	64 4.5	64 4.5	64 4.5	64 4.5	64 4.5	64 4.5
Output	h.p.	30/50	50/70	70/90	90/110	110/130	130/150	150/180	180/200
Motor speed	r.p.m.	450/600	450/600	420/540	540/660	660/780	450/470	470/560	560/625
Propeller speed	r.p.m.	450/600	450/600	350	320	280	260	240	220
Weight	lb. kg.	4,530 2,100	5,800 2,630	3,310 1,500	3,310 1,500	3,310 1,500	6,700 3,050	6,700 3,050	6,700 3,050
Weight with reduction gear	lb. kg.	5,850 2,650	7,485 3,380	5,080 2,300	5,080 2,300	5,080 2,300	9,000 4,100	9,000 4,100	9,000 4,100
Length	ft. m.	3.97 1.20	4.75 1.45	4.5 1.37	4.5 1.37	4.5 1.37	4.63 1.41	4.63 1.41	4.63 1.41
Length with reduction gear	ft. m.	6.85 2.09	8.17 2.49	7.87 2.4	7.87 2.4	7.87 2.4	9.41 2.87	9.41 2.87	9.41 2.87

TABLE LXXIV

INSTALLATION PROPOSALS FOR AIR-COOLED FOUR-CYCLE DIESELS

Length of boat	ft. m.	16.4 5	19.7 6	23 7	26.2 8	29.5 9	32.8 10	39.3 12
Motor type		SA1L 612	SA1L 612	SA2L 612	SA2L 514	SA3L 514	SA4L 514	SA6l 514
Number of cylinders		1	1	2	2	3	4	6
Bore	in. mm.	3.54 90	3.54 90	3.54 90	4.34 110	4.34 110	4.34 110	4.34 110
Stroke	in. mm.	4.73 120	4.73 120	4.73 120	5.51 140	5.51 140	5.51 140	5.51 140
Speed	r.p.m.	1,350	2,000	2,000	1,500	1,500	1,500	1,500
Mean piston speed	ft./sec. m./sec.	17.4 5.4	23.6 7.2	23.6 7.2	23 7.0	23 7.0	23 7.0	23 7.0
Mean effective brake pressure	lb./sq. in. kg./ sq. cm.	72.5 5.1	72.5 5.1	71.0 5.0	79.6 5.6	79.6 5.6	79.6 5.6	79.6 5.6
Output	h.p.	6	10	20	25	37.5	50	75
Propeller	r.p.m.	650	600	600	550	500	500	500

vessels and have operated quite successfully, for instance, in the warm waters of Indonesia and Malaya and off the African coast. Fig. 488 shows a typical installation. As air has a lower heat capacity than water (referring to the volume), comparatively large air quantities must be used for cooling. The hot air evacuation has to be arranged, therefore, to avoid resistance and inconvenience to personnel. The simplest solution is to lead the hot air to the funnel, together with the engine exhaust. The hot air can also be used for heating if required.

Air-cooled diesels are now delivered with outputs from 8 to 150 h.p. Table LXXIV lists some types for installations in fishing vessels of 16 to 40 ft. (5 to 12 m.) lengths.

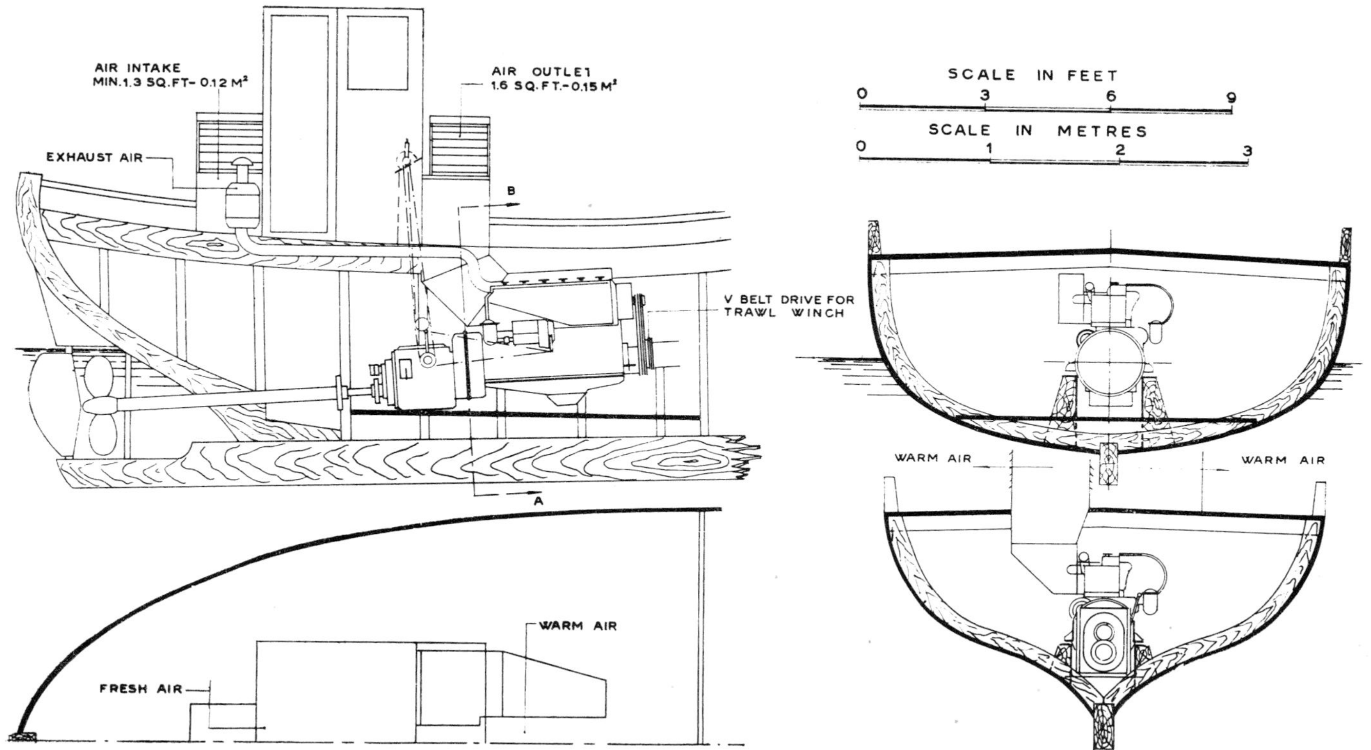

Fig. 488. Installation of an air-cooled diesel in a 39.4 ft. (12m.) cutter

JAPANESE DIESELS

by

I. TAKAHASHI, M. AKASAKA and K. TANAKA

THERE were 129,048 powered fishing boats in Japan in 1952, of which 45 were steam driven, 9,301 had diesels, 47,426 hot-bulb engines and 72,276 electric ignition engines. The average size for the steam engines was 915 h.p., for the diesels 66 h.p., for the hot-bulb engines 27.2 h.p. and for the electric ignition engines 5.4 h.p. The number of steam-powered vessels—mainly large whale catchers, medium trawlers and refrigerated fish carriers—is decreasing gradually.

In 1947, diesel boats numbered 2,311, with a total horse-power of 223,038. In the next five years the number increased by 300 per cent., but the average horse power per vessel decreased from 96 in 1947 to 66 in 1952, because between those years small diesels of less than 25 h.p. became increasingly popular.

Between 1947 and 1950, semi-diesels or hot-bulb engines increased in number by 49 per cent. and in total horse power by 45 per cent., indicating that many small fishing vessels had been equipped with hot-bulb engines, but after 1950 the number decreased because diesels became more popular.

Electric ignition engines are the most common fishing boat engines. There were 52,960 in use at the end of 1947 totalling 281,646 h.p. and by 1952 they numbered 72,276 (392,112 h.p.)—an increase of 30 per cent. in number and 39 per cent. in horse power. Their percentage of the total powered fleet is almost constant, namely 36 to 41 per cent. of the number and 7 to 8 per cent. of the horse power.

The first Japanese diesel for fishing was installed in 1922 in the *Hakuho Maru*, a fisheries patrol boat (2 by 320 h.p. twin screw), owned by the Japanese Fisheries Agency. Since then the modernization of fishing boats has accelerated. At first the manufacturers lacked knowledge of the requirements of the fishing industry, and the engineers knew little about the function of diesels, but many difficulties have been overcome.

The primary requirement for a fishing boat engine is reliability because a fishing boat:

1. Has to be used as much as possible through the year.
2. Frequently navigates alone on the rough, off-shore ocean, and
3. Has very often a home port where transportation facilities on land and repair shops are scarcely available.

Even though medium- and high-speed engines are now of fairly good construction, engines of heavy-duty type are still widely used.

WHALE CATCHERS

The first Japanese catchers used steam because it was believed that diesels were so noisy that they would frighten whales away. But to meet the general demand to use diesels for whale catchers, continuous research was made to produce a noiseless diesel. In 1936, a diesel catcher, *Seki Maru No. 1*, 320 gross tons, 1,050 h.p., was built. Although she was not noiseless at slow speeds, she made good catches in the Antarctic and her fuel consumption was only one-third of a steam catcher. It was clear that more speed and better manoeuvring would mean more efficient whaling, and two 1,200 h.p. diesel catchers, of 360 gross tons each, were built.

When Japan returned to Antarctic whaling after World War II, it was necessary to reconstruct the whaling fleet in a short time. The 1,800 h.p. medium-speed, four-cycle, engines, which had been used by the submarine chasers, were installed in several new catchers of 300 gross tons. These boats worked fairly well but it was found that heavy-duty main engines were needed.

A Whalers' Equipment Improvement Committee was established with members from government agencies and private companies. The Committee concluded that an engine of 1,800 b.h.p., 200 r.p.m., weighing about 80 tons, was desirable. A seven-cylinder 2,300 h.p., 200 r.p.m. diesel was installed in a 400 gross tons catcher in 1952. She has made bigger catches than any other boat in the northern Pacific since the spring of 1953.

There is not much difference in construction between the two-cycle engine for the whale catcher and the trawler diesel as far as scavenging or other fundamental functions are concerned. But as a whale catcher has two different working speeds—chasing and towing—a two-cycle engine with automatic scavenging valves has the widest flexibility.

When the first diesel catchers were built, it was found that an engine brake was essential. When a harpoon hits a whale and the engine is stopped with the fuel supply shut off, the propeller shaft continues to revolve because of the ship's remaining speed. So, the harpoon rope often gets tangled with the propeller shaft and is cut off. The harpooned whale is sunk and lost. If the engine is reversed then the torque is too big and may cause damage to the shaft or engine. Accordingly, an engine brake method was devised in which the main cylinders work like a compressor when the fuel supply is shut down. The torque produced by the propeller revolving, due to the ship's speed in the water, is absorbed by the cylinders, and within several seconds the propeller stops.

Remarkable improvements have been made in whalers during the past few years. Catchers have, contrary to expectation, become larger. Norway has now a 2,700 h.p. diesel catcher of 900 gross tons. England has completed a 192 ft. (58.5 m.) diesel catcher with 3,200 b.h.p. max. output of the two-geared diesels. She is fitted with a controllable pitch propeller.

OTTER TRAWLERS

Trawlers navigate alone in deep seas. The time to discuss whether diesel or steam should be used for large trawlers has passed: the question is, what kind of diesel should be used?

A six-cylinder 750 h.p. engine was built for the *Kushiro Maru* in 1927, on licence from the Nobel Co., Ltd., Switzerland. Many of the engines were later installed in large trawlers, and a seven-cylinder 1,050 h.p. engine for the *Suruga Maru*, a 1,000 gross tons trawler, was manufactured in 1930.

After World War II, a standard type five-cylinder 750 h.p. two-cycle diesel was manufactured. This was followed by a seven-cylinder engine, and now an eight-cylinder 1,200 to 1,300 h.p. engine is being planned. Four-cycle diesels, of course, are used for trawlers of about 300 gross tons.

Scavenging methods have not been changed from those introduced by the licence of the Nobel Co., Ltd. Some experiments have been made on an automatic scavenging valve which is very effective because it is capable of over-load running.

Although the mean piston speed and the mean effective pressure at rated output are 15 ft./sec. (4.64 m./sec.) and 68 lb./sq. in. (4.77 kg./sq. cm.) respectively, continuous operation is possible up to 75 lb./sq. in. (5.3 kg./sq. cm.) of mean effective pressure. The scavenging pump is double acting and fitted on the fore end of the engine. The weight of the engine, including the thrust bearing, is about 104 lb./b.h.p. (47 kg./b.h.p.).

TWO-BOAT TRAWLERS

Many trawlers have their ports in the southern part of Japan, and their fishing grounds stretch out to the south, in the East China Sea. They are wooden or steel boats of about 70 to 130 gross tons with diesels from 210 h.p. to 300 h.p. and they operate in pairs.

Slow speed four-cycle diesels of 380 to 400 r.p.m. with reverse gear are mainly used. Direct reversible engines are not suitable for trawlers because the trawl winch is driven through a pulley on the fore end of the crank shaft.

The fly wheel is fitted on the fore end of the crank shaft, and the engine bed, the thrust bed and the reversing bed are cast in one piece to make up a strong unit. The fore end of the crank shaft has one pulley to transmit a torque of about 580 to 650 ft.lb. (80 to 90 kgm.) to the trawl winch, and another to drive an auxiliary electric generator. The strength of the crank shaft and the bearing is sufficient for such a load. Auxiliary pumps and the operating handle of the engine are placed near the reverse gear.

The Mietz and Weiss type clutch is mainly used because of its simple structure and smooth operation for diesels from 300 to 350 h.p. transferring a torque of about 4,700 ft.lb. (650 kgm.).

SMALL TRAWLERS

Most of the fishing boats operating off the northern part of Japan are wooden trawlers of about 30 to 80 gross tons with main engines of 120 to 210 h.p. with reverse gear. A thrust of about 3 tons is needed when trawling, and the boats generally have a speed of 9 knots or more while cruising under loaded conditions.

Medium-speed engines with reduction gears have been made in Japan but they are not yet very popular for trawlers. On the contrary, many hot-bulb engines are being used because they are not damaged by overloading, and they can be repaired at small iron works. The cruising radius of small trawlers is very limited, but the number of diesels being installed is gradually increasing and they reduce running expenses and extend the navigation range.

LARGE TUNA CLIPPERS

Large-type tuna clippers were not built pre-war because base ports in the southern islands were available in those days. Comparatively small boats with diesels from 320 to 400 h.p. were used.

The size of the tuna clipper has increased gradually following the expansion of the fishing grounds outside the MacArthur Line. The engines have increased from 500 to 850 h.p.

Most engines now used are of the four-cycle, single-acting, solid-injection and trunk-piston type. Two-cycle is seldom used. A few manufacturers have begun to produce superchargers for the four-cycle engine. Typical types are listed in Table LXXV.

Large tuna clipper engines operate continuously for about 1,200 to 1,300 hours during each voyage with only five days in the harbour between trips. Time for repairs,

TABLE LXXV

Type	Four-cycle I		II		III		IV		Two-cycle V	
Rated b.h.p.	500		650		750		850		750	
R.p.m.	330		320		295		280		240	
Overload b.h.p.	600		780		900		1,020		900	
Number of cylinders	6		6		6		6		5	
Diameter of cylinder	13.1 in.	335 mm.	14.6 in.	370 mm.	15.8 in.	400 mm.	16.5 in.	420 mm.	14.2 in.	360 mm.
Stroke	18.5 in.	470 mm.	20.5 in.	520 mm.	22.5 in.	570 mm.	23.5 in.	600 mm.	23.0 in.	580 mm.
Mean effective pressure	78 lb./sq. in.	5.47 kg./sq. cm.	77 lb./sq. in.	5.42 kg./sq. cm.	80 lb./sq. in.	5.61 kg./sq. cm.	80 lb./sq. in.	5.60 kg./sq. cm.	72 lb./sq. in.	5.03 kg./sq. cm.
Mean piston speed	16.9 ft./s.	5.16 m./s.	18.2 ft./s.	5.54 m./s.	17.4 ft./s.	5.32 m./s.	17.9 ft./s.	5.46 m./s.	15.6 ft./s.	4.76 m./s.
Fuel consumption	0.38 lb./h.p./hr.	172 gr./h.p./hr.	0.36 lb./h.p./hr.	164 gr./h.p./hr.	0.36 lb./h.p./hr.	162 gr./h.p./hr.	0.35 lb./h.p./hr.	160 gr./h.p./hr.	0.40 lb./h.p./hr.	180 gr./h.p./hr.
Lubricating oil consumption	0.003 lb./h.p./hr.	1.2 gr./h.p./hr.	0.002 lb./h.p./hr.	1.0 gr./h.p./hr.	0.002 lb./h.p./hr.	1.0 gr./h.p./hr.	0.002 lb./h.p./hr.	1.0 gr./h.p./hr.	0.008 lb./h.p./hr.	3.5 gr./h.p./hr
Weight, tons	20.1		22.8		27.0		30.5		34.0	

therefore, is limited. Although fishing boats usually dock twice a year, the docking time is only two days. The Japanese law requires owners of fishing boats to have a survey every two years.

The four-cycle engines mostly used have wet liners, except for Type I, which has a block cylinder with a water jacket in it. The engine bed, frame and cylinder are made of cast iron and are connected together by tension bolts. The crank shaft is solid forged. The fuel pump is a spill valve type. The starting valve is an air control type and a pilot valve is fitted on every cylinder. The friction clutch is fitted between the crank and thrust shafts, and is operated by oil pressure. The clutch of a Japanese tuna clipper is operated 800 to 1,000 times a day when fishing. Each trip lasts 25 to 30 days.

An all-speed governor is used, the handle being fitted near the operating seat. The main engine is operated at about one-fourth of its maximum output when fishing and smooth governor running is needed in this condition. The engine has a gear type lubricating oil pump, a piston or plunger type cooling water pump, and a bilge pump. An air charge valve is fitted on one of the cylinders and compressed air for starting purposes is loaded in an air reservoir. The pulley for the auxiliary machinery is driven through a friction clutch fitted on the fore end of the crank shaft.

Some tuna clippers have a remote control device to operate engines from the bridge, but it is not very popular yet in Japan. In future, the trend will be to supercharge four-cycle engines, and develop reliable two-cycle, high speed engines.

SMALL OR MEDIUM TUNA CLIPPERS

The small- or medium-size tuna clippers of 55 to 200 gross tons have in general 160 to 400 h.p. diesels. Japanese tuna clippers in the past were mostly below 100 gross tons but recently bigger types have been built, and now a 300 gross tons type is becoming popular.

The main engines are, with a few exceptions, four-cycle, single-acting trunk-piston type with four to six cylinders. Those under 300 h.p. are equipped with Mietz and Weiss type reverse gears. Engines over 300 h.p. are directly reversible.

Table LXXVI gives particulars of a typical 400 h.p. main diesel engine.

Particular attention is given to easy operation, reliability, durability and economical driving in low speed. A friction clutch is used for the frequent manoeuvres. A pulley is attached to the fore end of the crank shaft. The generators, bilge pumps and line haulers are driven by belts off the main engines.

The weight and overall length are comparatively small, thereby effectively using the small space of the engine room. One man, without moving his position, can start, reverse, adjust fuel oil consumption (speed adjustment) and use the friction clutch. If especially required, a servomotor is attached to manoeuvre the friction clutch.

The speed regulator is a centrifugal type and operates very sensitively during all stages of output. Even if the

TABLE LXXVI

Number of cylinders	6	
R.p.m.	350	
Output	400 h.p.	
Maximum output	480 h.p.	
Diameter of cylinder	12.2 in.	310 mm.
Stroke	17.3 in.	440 mm.
Piston speed	17 ft./sec.	5.13 m./sec.
Mean effective pressure	74 lb./sq in.	5.16 kg./sq. cm.
Length, overall	209 in.	5,301 mm.
Breadth of engine bed	45 in.	1,144 mm.
Maximum breadth of engine	73 in.	1,844 mm.
Height above centre of shaft	74.5 in.	1,892 mm.
Height below centre of crank shaft	22 in.	560 mm.
Height, overall	97 in.	2,452 mm.
Weight	about 16 tons	

propeller is disengaged in low speed, the revolutions are not changed, and no harm is done in manoeuvring.

The engine can be controlled from the wheel house or the look-out platform on top of the wheel house. The advantages of the remote control system were demonstrated by an American type purse seiner of 120 gross tons in 1949, and it was installed with the same good results in a steel tuna clipper of 300 gross tons in 1953.

The engine bed and crank case are made of special cast iron and a lubricating oil tank is arranged at the bottom of the bed. The cylinders are made from cast iron and they are equipped with liners. The engine bed, crank case and cylinders are connected by tension bolts. The crank shaft is made of single-block wrought steel.

A spill valve type fuel oil pump is fixed to each cylinder. The fuel valves have many holes, and the discharging pressure is 4,000 lb./sq. in. (280 kg./sq. cm.).

Compressed air for starting comes from the two air tanks through pilot and starting valves on the cylinder covers. The capacity of each air tank is 13.5 cu. ft. (380 l.) of 430 lb./sq. in. (30 kg./sq. cm.). The air is delivered to the air tanks by an independent compressor, or by one of the cylinders of the main engine and, when charging is done with the cylinder, fuel oil is not sprayed into it.

The self-reversing type of reverse gear used is operated by a handle which moves a lever with two rollers.

The fuel oil consumption is less than .375 lb./b.h.p./hr. (170 gr.) using heavy oil of 0.92 specific gravity. The exhaust temperature is low, as shown in fig. 489, and the engine can be overloaded 20 per cent. for an hour without danger.

Fishing vessels will reach higher speeds by the further development of engines and/or the installation of superchargers. The performance of the engines is directly connected with safety of life, and they must be reliable and durable. High revolutions are not necessarily suitable and the low speed engines, which are now in the majority, may be developed to give medium speed.

The remote control system will certainly gain popularity and by its help the engines will be operated directly from the wheel house or look-out platform.

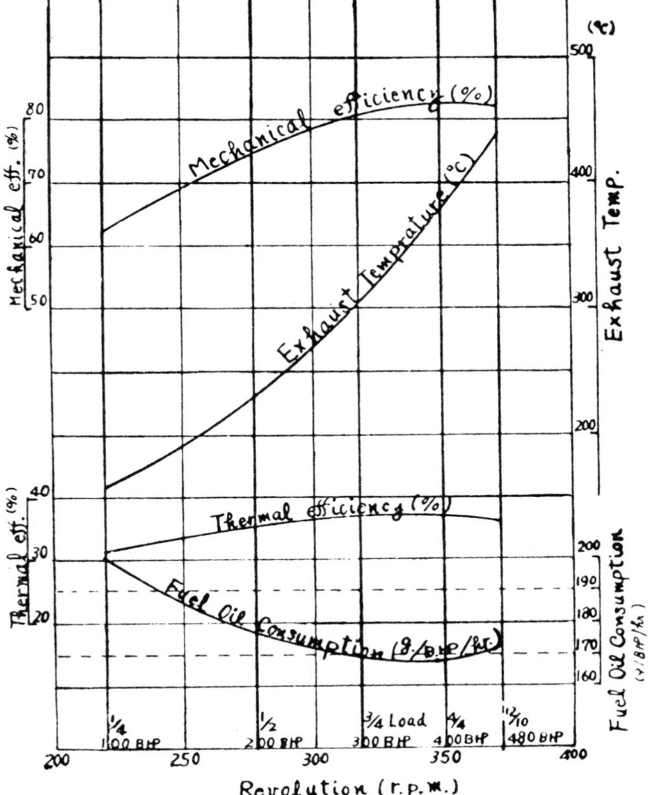

Fig. 489. Characteristic Curves of Hanshin Diesel Engine (400 B.H.P.).

GAS TURBINE PROPULSION

by

A. AUGUSTIN NORMAND Jr.

DESCRIPTION AND PRINCIPLE OF OPERATION

A gas turbine propulsion plant may consist of:

(*a*) one or more free piston gas generators;
(*b*) one or more gas turbines with several forward wheels and, if required, one or two astern wheels;
(*c*) one reduction gear per shaft with or without reverse gear;
(*d*) oil and fresh water pumps for the lubricating and cooling of the generators, turbines and reduction gears;
(*e*) sea water pumps and heat exchangers for the circulating fresh water and oil;
(*f*) air compressors and air bottles for starting the generators;
(*g*) other auxiliary equipment, such as pumps for oil and fuel transfer, filters, etc.

A free piston gas generator, fig. 490, consists of: a casing, closed at both ends with back-flow valves, containing a cylinder liner with injectors, scavenging and exhaust ports, and casings for the two synchronised connecting rods; two compression chambers at the end of the generator casing; two free pistons, each consisting of a working part towards the centre and with a small. high-pressure area and a compression part with larger diameter on the outside, the latter working on an air cushion on the outside and as a compressor on the inside; two connecting-rods for synchronization of the pistons.

When the pistons have reached the inside dead point and the injected fuel is ignited, the pistons are pushed outward again and part of the developed power is accumulated in the air cushion.

During this outward movement, air is sucked into the compressor chambers and the exhaust ports and later the scavenging ports (these are communicating with the engine casing filled with air under pressure) open up, enabling the exhaust gases to escape and scavenging air to enter, similar to a two-cycle engine with opposite pistons. When the pistons have reached the outside dead point, the air cushions restore the energy that they have accumulated and the pistons return to the inside dead point.

During this return movement the compression pistons push the air contained in the compression chambers into

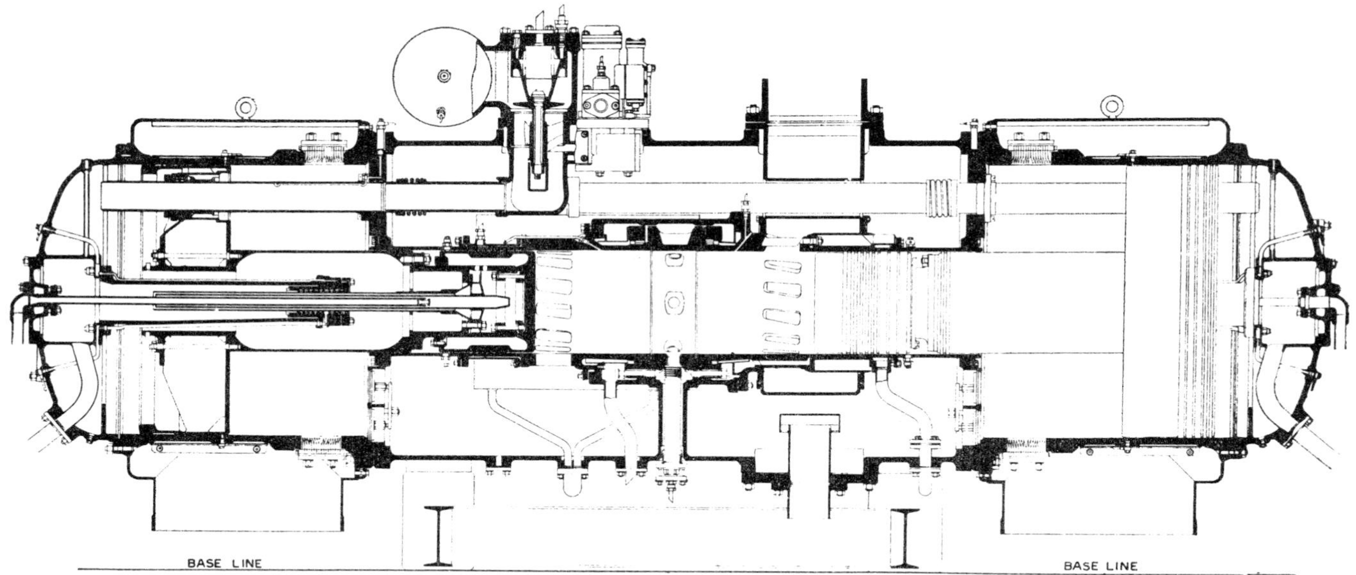

Fig. 490 Section through a free piston gas generator

the inside of the cylinder casing through the valves, and the working pistons close the scavenging and exhaust ports and compress the air to be used for the combustion in the next cycle. The mixture of air and gas that escapes is led to the turbine. Its temperature is between 840 and 930 deg. F. (450 and 500 deg. C.) and it is under a pressure of 50 lb./sq. in. (3.5 kg./sq. cm.). Metallurgically, the problems raised by these temperatures and pressures can be readily solved. More details on the operation and output of free piston generators are given by Eichelberg (1948) and Peillon (1949).

Reversing can be done by one or several reverse wheels in the turbine, a reverse gear incorporated in the reduction gear or a propeller with controllable pitch that, in the case of a trawler, has the further advantage of permitting both cruising and trawling efficiently.

ADAPTATION TO A TRAWLER

A study has been made on the adaptation of free piston gas generators and a gas turbine to a trawler with the following main dimensions:

Overall length	233 ft. (71.1 m.)
Length between perpendiculars	217 ft. (66 m.)
Overall breadth	37 ft. (10.6 m.)
Depth to the main deck	20 ft. (6.1 m.)
Volume of fish-hold	46,200 cu. ft. (1,310 cu. m.)

The plant has two free piston gas generators that together develop 1,700 h.p. in continuous operation and 2,100 h.p. when supercharged. These generators feed one turbine which drives the shafting through a reduction gear.

The following auxiliary equipment is provided for:

Fresh water

Two electric 92 imp. gal./min. (110 gal./min.; 25 cu. m./hr.) pumps

Two fresh water coolers.

Piston cooling oil

Two electric 66 imp. gal./min. (80 gal./min.; 18 cu. m./hr.) pumps

Two oil coolers

One electric 7.3 imp. gal./min. (6.8 gal./min.; 2 cu. m./hr.) pump

One hand pump.

Turbine and reduction gear lubricating oil

One electric 92 imp. gal./min. (110 gal./min.; 25 cu. m./hr.) pump

One emergency 46 imp. gal./min. (55 gal./min.; 12.5 cu. m./hr.) electric pump

One oil filter

One oil cooler.

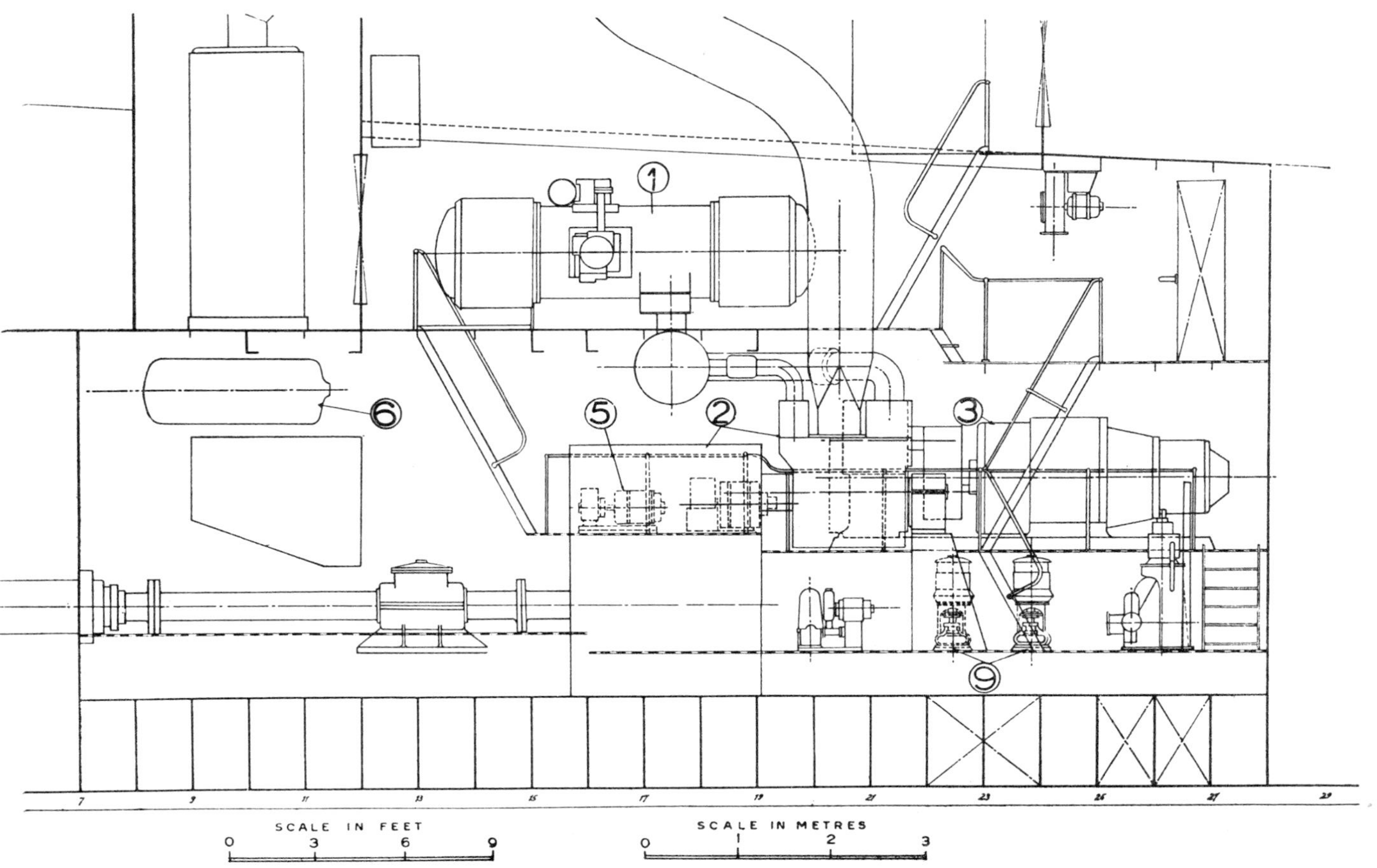

Fig. 491 Profile of engine room of a trawler with gas turbine propulsion

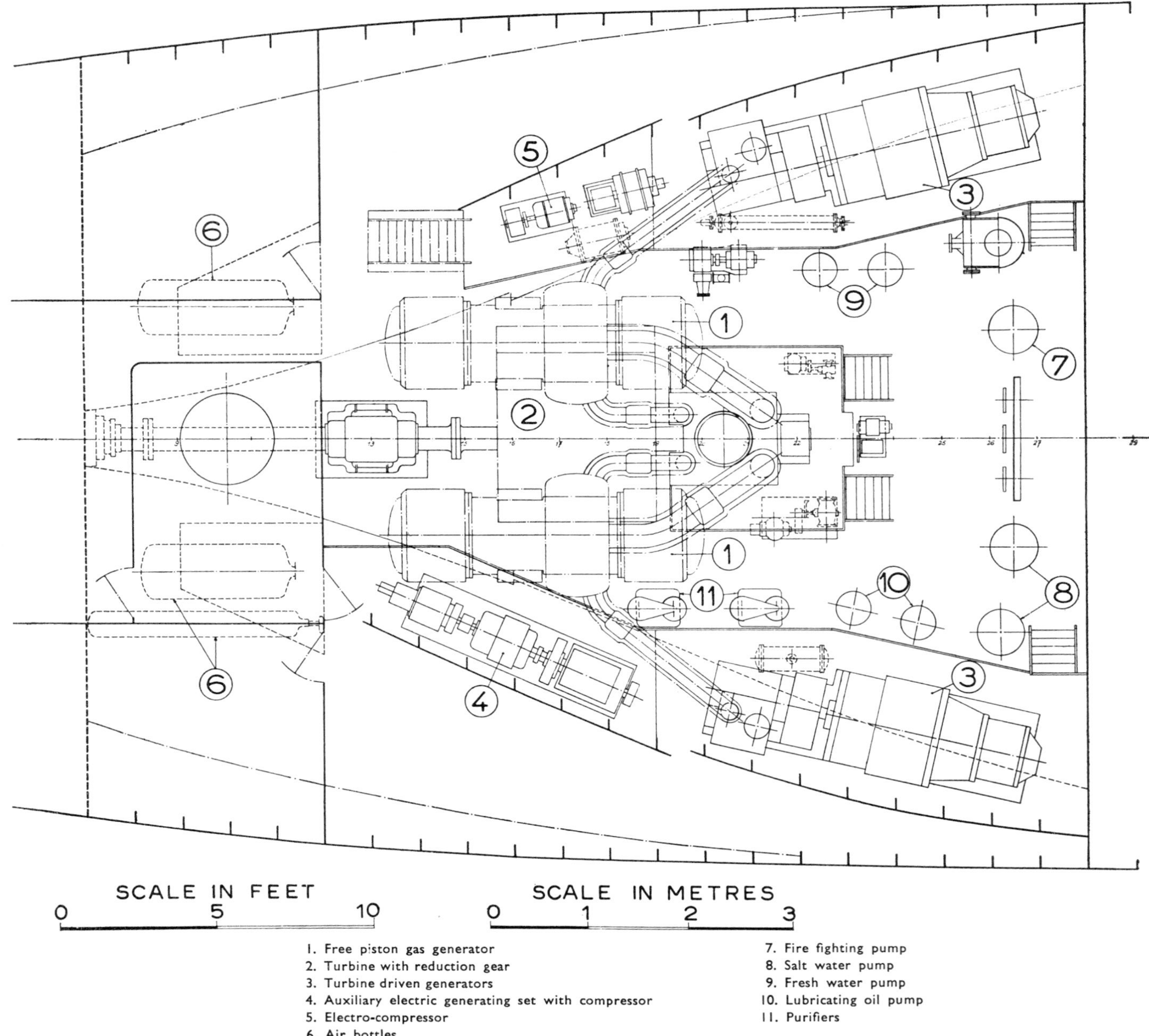

Fig. 492 Plan of engine room of a trawler with gas turbine propulsion

Salt Water

Two electric 290 imp. gal./min. (350 gal./min.; 80 cu. m./hr.) pumps for the refrigerating compressor.

Fuel

One transfer 81 imp. gal./min. (97 gal./min.; 22 cu. m./hr.) electric pump

One service 18 imp. gal./min. (22 gal./min.; 5 cu. m./hr.) electric pump

One hand pump

One fuel filter.

Compressed air

One diesel 95 imp. gal./min. (115 gal./min.; 26 cu. m./hr.) compressor

One emergency 95 imp. gal./min. (115 gal./min.; 26 cu. m./hr.) electric compressor.

Electric power

Two gas turbine driven generator units of 250 h.p.

Two 45 kw. exciters for the electric generators and other auxiliary equipment.

One emergency diesel 40 kw. generator-set that also drives the above-mentioned 95 imp. gal. min. (26 cu. m./hr.) air compressor.

Holds—ballast and service

One 92 imp. gal./min. (110 gal./min.; 25 cu. m./hr.) electric pump for the holds and ballast

One 280 imp. gal./min. (360 gal./min., 75 cu. m./hr.) electric pump for washing and fire fighting.

Ventilation

Two 14,700 cu. ft./min. (2,500 cu. m./hr.) ventilators.

The arrangement of the details of the engine room of this trawler is shown in fig. 491 and 492.

The arrangement of the two generators and the turbine makes it possible to reduce considerably the length of the engine room to allow for larger fuel bunkers or tanks or fish holds while allowing ready access to all apparatus or auxiliary equipment.

A study has also been made of an engine room fitted with a standard, direct-driven, four-cycle supercharged diesel developing 1,100 h.p. at 195 r.p.m. The arrangement of the details is shown in fig. 493, 494 and 495.

A comparison of the design of these two possible arrangements shows that the gain in length when free piston gas generators and gas turbine are used is four frames or 7.9 ft. (2.40 m.). The weights in either case are shown in table LXXVII.

The diesel alternative, therefore, weighs 122 tons, and the free piston generator one 112 tons. The figures should not be compared directly but through the weight per horsepower developed. In practice, the use of a gas turbine makes it possible to produce 1,700 h.p. in continuous operation while, with the use of the diesel, output is limited to 1,100 h.p. The total weight per h.p. is 245 lb. (111 kg.) and 145 lb. (65.7 kg.) respectively, which shows a distinct advantage in the case of the gas turbine.

There is an obvious advantage with generators and

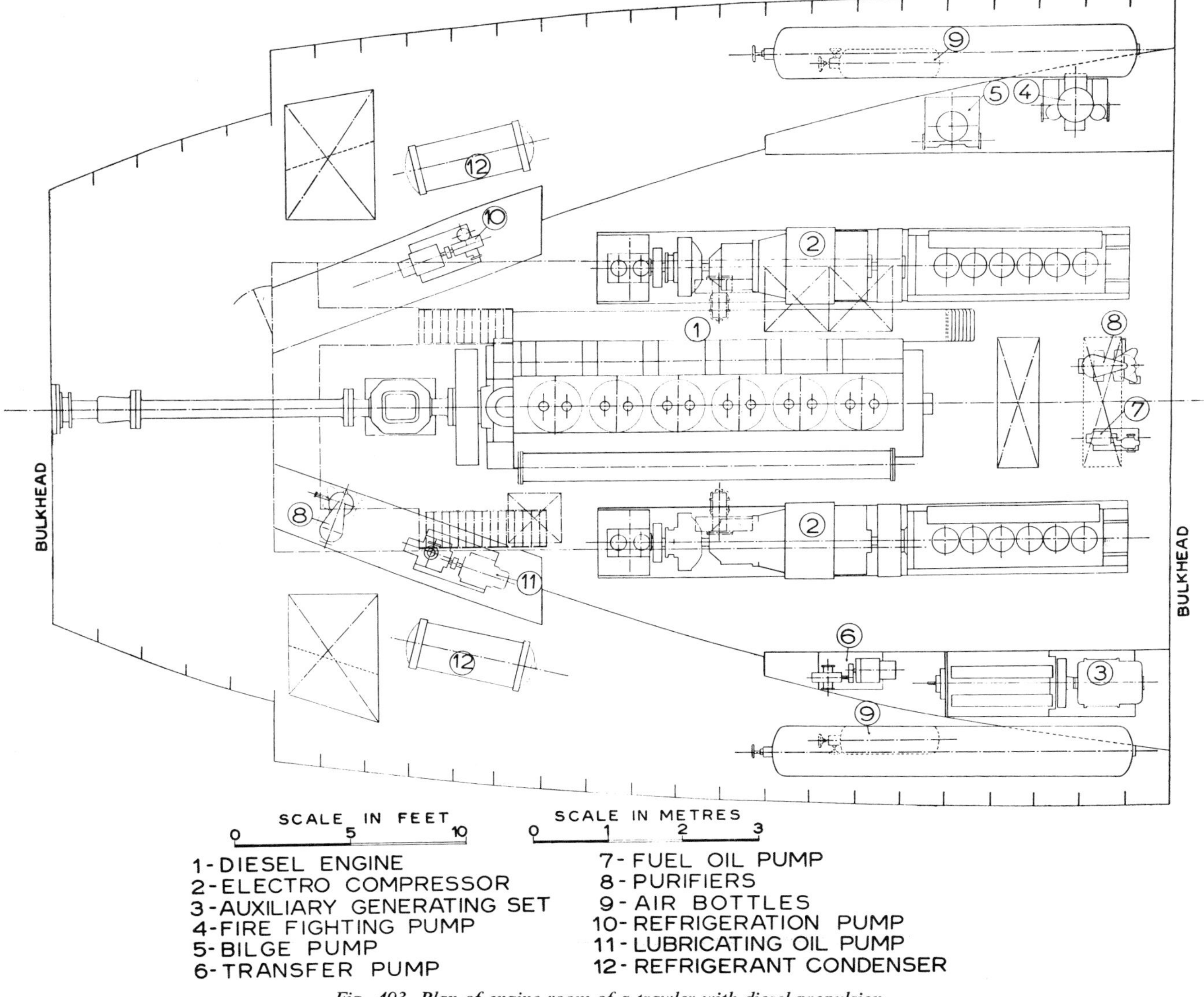

Fig. 493 Plan of engine room of a trawler with diesel propulsion

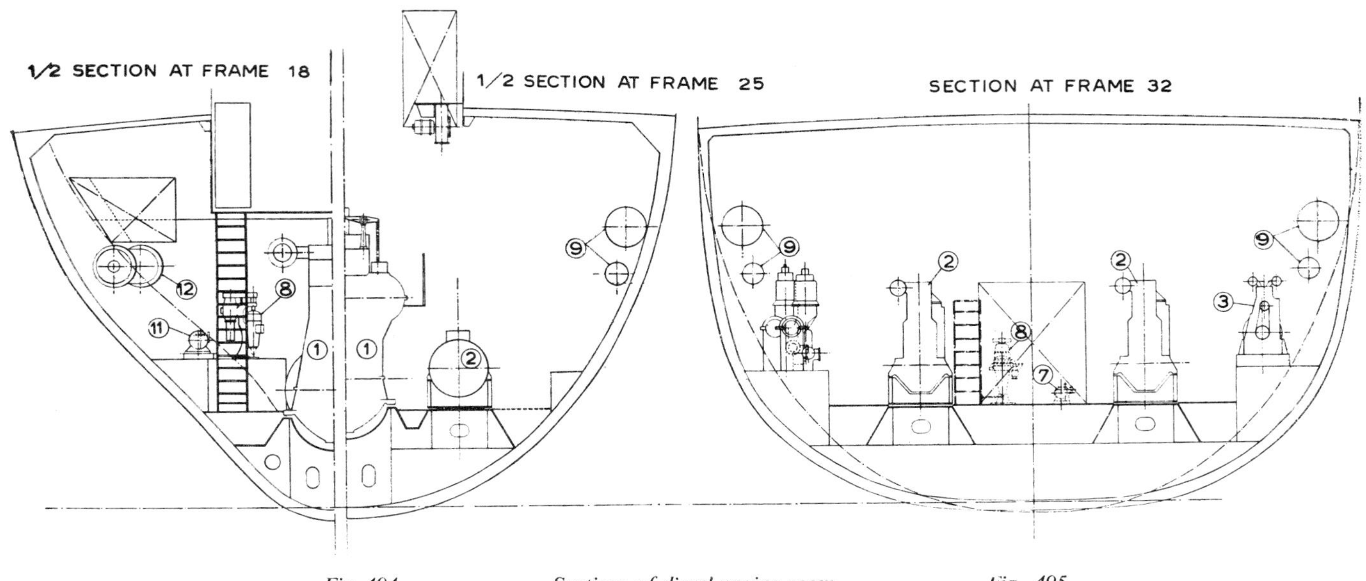

Fig 494 — Sections of diesel engine room — Fig. 495

TABLE LXXVII

	Diesel (metric tons)	Generator and gas turbine (metric tons)
The main machinery	50.000	33.700
Shafting	10,150	10.150
Cooling equipment	1.690	3.680
Air compressor	3.025	2.660
Fuel	2.760	2.680
General piping	11.800	14.780
Storeroom and workshop	0.950	0.950
Ladders and flooring	5.000	5.000
Generators	30.000	30.600
Auxiliary boiler	4.210	4.210
TOTAL	122.145	111.800

gas turbine as regards required space lengthwise and power-to-weight ratio. Further advantages are:

1. Flexibility of operation due to the fact that the generators adjust themselves instantaneously to changes in power output required of them;
2. Simplicity of reversing in the case of a turbine with a reversing wheel, as the generator continues in operation while the change is made and the reversing gear wheel can instantaneously receive the full impact of the power load as it is in permanent operation at a temperature of about 570 deg. F. (300 deg. C.);
3. The added safety of dividing power between two generators;
4. Ease of dismantling (about half-an-hour is needed to remove movable equipment) and replacing of a generator, because of its comparative light weight;
5. Elimination of vibration because of perfect balancing of the parts in motion;
6. The ease of finding spares because the generators have no large forged or moulded steel parts;
7. The possibility of feeding the generators with heavy fuel without difficulty because of the excess of air available.

VIBRATION IN SMALL SHIPS

by

JAMES WHITAKER

HULL VIBRATIONS

The Hull as an Elastic Structure

A SHIP'S hull is a complicated structure but from the point of view of vibration it is, nevertheless, just as susceptible as, for example, a reed or a steel bar, and it can vibrate in all the different ways common to such a member.

Hogging and sagging describe the bending of a ship in a vertical plane, due to it being alternately supported, first by one wave in the middle, then by two waves, one at each end. When the ship is vibrating in this way, there are two points along its length which are stationary (one at about a quarter ship's length from the stern and another about a third ship's length from the fore end); the two ends are vibrating up and down together in phase, and a point approximately midships is vibrating up and down out of phase, with the ends. This is a vertical vibration. The same thing can happen in a horizontal plane.

Exact mathematical treatment is impossible. The different parts of the hull are loaded differently at different times and these loads must be included even in an approximate solution to the problem. There are certain general rules, however, which determine the probable natural frequency of vibration of a ship. The longer the hull and the weaker its vertical section, the lower will be its vertical frequency.

Similarly, the narrower its beam, and consequently its sideways stiffness, the lower will be its horizontal frequency of vibration. The converse is true in each case and it is clear that a trawler will, for example, have a higher natural frequency of vibration than a shallow draught barge.

Possible Forms of Vibration

Vibration is rarely of one particular form but is a combination of several components, one of which will be more predominant. The principal types of vibration are: (*a*) in a vertical plane, (*b*) in a horizontal plane, (*c*) longitudinal (fore and aft movement), (*d*) twisting.

Of these, (*c*) and (*d*) are much closer to the modes of vibration of a simple steel bar than are (*a*) and (*b*). Their natural frequencies are normally so much higher than the frequency of engine impulses that synchronism is very unlikely and their effect can be ignored for practical purposes.

The severest mode of vertical and horizontal vibration has already been mentioned, i.e. where there are two nodes (or stationary points) in the vibrating hull. There are other forms of vibration, one in which there are three nodes—one amidships and one near each end—with four sections from stem to stern vibrating up and down alternately. Another in which there are four nodes—one forward and one aft of amidships and one close to each end—with five sections from stem to stern vibrating up and down alternately.

All modes, except the principal one, are of relatively high frequency and are not likely to be excited, unless very high-speed machinery is installed in the ship.

Symptoms

Rippling in still water at the sides of the vessel is an indication of vibration. This usually shows quite clearly the frequency at which maximum amplitude occurs (i.e. the natural frequency) and the position of the nodes. The points at which there is little or no rippling of water are the nodes and the points at which there is the most violent disturbance are the anti-nodes.

At the nodes we have high structural stresses with no apparent movement. At points of anti-node there is excessive movement and little stressing. In dangerous cases an indication of vibration is sometimes given by damage to the structure—typified by loosening of rivets and other fixings.

" Working " of the structure is a form of damping and it contributes to a reduction in the amplitude of vibration. As progressive weakening takes place, due to the damage caused, so the vibrations increase in severity because of a corresponding reduction in damping. At the same time the natural frequency is lowered and the vibrations begin to appear at lower engine speeds.

In addition to the vibrations due to synchronism between engine impulses and hull frequency, there are others which are generally more local in character and easier to deal with. Examples of these can be found in engine seatings which are too weak, large areas of unsupported deck plating and long, slender handrails.

Forces Causing Vibration

There are many and diverse forces attempting to excite vibrations in the hull. These are periodic in character and the principal ones are:

(*a*) The unbalanced force and couple due to the motion of the reciprocating parts of the engine. These force oscillations in a vertical plane.

(*b*) The unbalanced force and couple from the revolving parts of the engine. The horizontal and vertical components of these force vibrations in a horizontal and vertical plane respectively.

(*c*) The variation of turning moment on the propeller shaft. This tends to force twisting oscillations. Following from this there is a corresponding variation of thrust on the propeller, which tends to excite longitudinal vibrations.

(*d*) Inaccuracy of pitch or shape of propeller blades.

All these forces and couples acting simultaneously seek to vibrate the ship. Any of them which coincides with a natural frequency of the hull will intensify vibration at that frequency and make the ship uncomfortable to live in.

This agreement between frequency of vibration force and hull frequency may take place at relatively low speeds, or at operating speeds, or, as is desired by the designer, it may not appear until beyond the maximum operating speed. The limits of uncomfortable speed vary with individual cases but they are usually in the region of ± 5 per cent. to ± 10 per cent. For instance, a ship may be in violent vibration at three-quarter speed and quite comfortable at full speed.

The unbalanced forces and couples in the engine are the most important factors in producing vibrations and, general speaking, vertical vibrations are the ones most usually encountered.

The severity of the vibration depends on the magnitude of the forces or couples and on the position of the engine in the ship. As a rule, forces are more dangerous than couples; if the engine is situated at anti-node its out-of-balance forces are more dangerous than couples, but an out-of-balance couple takes full effect when the engine is situated at a node and is relatively unimportant at an anti-node. In a trawler where the engine is about $\frac{L}{4}$ from the stern, couples must be taken into account.

Transmission of Forces

Attempts have been made to prevent the transmission of these forces to the hull by using flexible mountings. These have been quite successful on small lightweight engines, especially on some auxiliary sets running at constant speed, but they are not yet favoured as a practicable proposition on heavier propulsion engines. This is due to difficulties encountered at low r.p.m., and high amplitude. Cork and rubber will not stand up to excessive fatigue deformation and there is too much " give " in a spring system. Suitable spring mountings can be designed, theoretically, but the difficulties of lining up to the thrust block, and making satisfactory pipe connections, have so far proved insurmountable. Further, there are many critical speeds inherent in a spring system and these make it less satisfactory for a propulsion unit than for a constant speed auxiliary set.

Propulsion engines are usually bolted firmly to their seatings, and it is important to have a seating as rigid as possible and well tied to the hull. The frequency of natural vibration of the ship depends upon its stiffness. The stiffer it is, the higher its natural frequency and the better chance there is of keeping it above normal running speeds. The question of forced vibrations must also be remembered. If the seating is not adequately tied to the hull, the forces are not evenly distributed, and high localized pressures can cause trouble.

The engine framing itself, or even the seating, can have a natural frequency of its own to coincide with the applied forces. Tall engines with light framing (e.g. welded frames) can have a big amplitude of movement at the top and a low natural frequency within themselves. If the engine is mounted on a light seating, such as a tank top, the amplitude is further increased and the natural frequency is still lower. When this frequency coincides with engine-exciting forces the engine vibrates badly. A method of overcoming this trouble is to tie the engine to the ship's structure from its top end by means of tie bars or steel cables.

It is regular practice in twin screw installations to tie the two engines together by stay bars, in a similar way, so that each lends stiffness to the other and prevents the kind of vibration just mentioned.

There is a special kind of vibration which occurs in twin screw installations. The two engines cannot be run at precisely the same speed, owing to difficulties of exact governing, so the speed of one is continually gaining on that of the other. Each forces a vibration on the hull and these two vibrations, which are very nearly equal in period, combine to give a vibration of varying amplitude. At one instant it is equal to the sum of the components and at another it is equal to the difference of the components. The number of surges between maximum and minimum amplitude per minute is equal to the difference in r.p.m. of the two engines. This is termed " hunting " or " beating " and it produces a shuddering vibration which comes and goes several times every minute.

This type of vibration is encountered wherever two engines are installed in a ship and are independently governed for speed, and it is the same when two engines are geared together, through hydraulic or magnetic slip couplings, to a common output shaft.

Estimation of Hull Frequency

Formulae have been devised (e.g. by Schlick and Todd) to give estimates of first order vertical and horizontal frequencies. These operate on the basis that for a given type of ship—liner, tanker or trawler—the frequency depends principally upon length, breadth, moulded depth and displacement. It also depends upon a coefficient which varies according to the type of ship and according to its particular design and scantlings.

These formulae have to be used with some discretion because the choice of coefficient is made arbitrarily from experience, but it is wise to use them in the early stages of design. It is then possible to check approximately whether the engine forces are likely to synchronise with hull frequency at important operating speeds (e.g. steaming or trawling).

Experimental Determination of Hull Frequency

When the ship has been launched it is advisable to check the natural frequency by experiment. This is not practised to any great extent at the moment but is a fairly simple precaution. The method is to bolt a vibrating machine in the machinery space and run it at varying speeds. By this means it is possible to confirm, before the engines are installed, whether there is likely to be any trouble.

Lloyds have devised and used such a machine for investigating troublesome vibrations.

Methods of Stiffening to Raise Natural Frequency

It is possible to get differences in natural frequency of the order of 20 per cent. between apparently similar ships, due to differences in design and scantlings. The main factors in determining these differences are:

For vertical vibrations—the resistance to bending in a vertical plane, e.g. depth of fore and aft girders.

For horizontal vibrations—the resistance to bending in a horizontal plane, e.g. width and stiffness of deck-plating and side girders.

If any adjustment is needed to raise the natural frequency it can be effected by additional stiffening on these lines. General or local vibration, due to weakness at isolated points, can be cured fairly easily by local stiffening, e.g. wheelhouse floors, large unsupported panels, handrails, etc.

Comparison of High- and Low-Speed Engines

High powers, piston speeds and maximum cylinder pressures are all features of modern diesel machinery. From the design point of view, considerable improvements have been made. Cylinder centre distances have been reduced and present-day engines are much more rigid than formerly. The disturbing forces and moments causing vibration are correspondingly less and a good modern engine can now be treated as a rigid whole, without fear of vibrations being set up in its own framing.

Higher maximum pressures mean smaller pistons for a given power, hence the masses of the working parts are reduced and out-of-balance forces and couples are also reduced. Internal moments within the engine framing, caused by balancing, are reduced, the stresses in the framing are comparatively low and the deflection of the engine frame is negligible.

Torque variation is important from the point of view of stresses in the crankshaft and of causing possible disturbance to the hull structure. This has improved on modern engines. The higher maximum pressures occur before the piston has moved far on the working stroke: the pressures of the expansion curve are proportionally less and consequently result in lower maximum torques.

Other things being equal, the slow-speed engine has a distinct advantage over one of high speed. For a given power output, imagine two engines working on the same cycle, at the same brake mean effective pressure with the same number of cylinders, both having the same bore/stroke ratio, one being $12\frac{1}{2}$ in. (317 mm.) bore and the other 10 in. (254 mm.) bore, the speed of the former 300 r.p.m. and the latter 585 r.p.m. The slower speed engine shows obvious advantages from the point of view of piston speeds and, in the matter of balancing and vibration, it is very much better. Assuming the same materials for piston, connecting rod, etc., the comparative weights of reciprocating parts are $(12\frac{1}{2}/10)^3 = 1.95$. The comparative out-of-balance forces are, therefore, $1.95 \times \left(\frac{300}{585}\right)^2 \times \left(\frac{12\frac{1}{2}}{10}\right) = .64$, thus showing that the high-speed engine has vibrating forces 56 per cent. greater than the slow-speed engine.

The above considerations apply to engines working on the same cycle (i.e. both two-cycle, or both four-cycle). There are some further noteworthy differences on this point.

Comparison of Two- and Four-Cycle Engines

The two-cycle engine has smaller cylinder dimensions than the four-cycle engine for the same power, number of cylinders and piston speed. Average figures for sers vice brake mean effective pressures on modern engine-are: two-cycle 65 lb./sq. in. (4.6 kg./sq. cm.), normally aspirated four-cycle 80 lb./sq. in. (5.6 kg./sq. cm.), and supercharged four-cycle about 110 lb./sq. in. (7.7 kg./sq. cm.). For the same power, the *ratio* of cylinder bore dimensions is—two-cycle 1.0, normal four-cycle 1.275, supercharged four-cycle 1.09.

The two-cycle engine, therefore, has an advantage in lighter running gear and, as already explained, if the four-cycle cylinder dimensions are reduced by increasing the r.p.m., the difference in disturbing forces widens rapidly in favour of the two-cycle engine.

As regards torque variation, the two-cycle engine is inherently better than the four-cycle one. With a power stroke every revolution, compared with the four-cycle engine's every alternate revolution, and smaller cylinder dimensions with consequently lower peak torques, the two-cycle torque variation is only about half that of a corresponding four-cycle engine.

SHAFTING VIBRATIONS

Whirling

When long and comparatively slender shafts are rotated at high speeds they sometimes suffer from a phenomenon termed " whirling " in which the shaft executes the motion of a skipping rope. If the speed at which this occurs is maintained, the deflection becomes large and the shaft will be fractured, but if this speed is quickly run through, the shaft will become straight again and run true until, at another higher speed, the same thing happens again.

This trouble is met with in turbines but very rarely with reciprocating engines because the shafts are relatively stiff and well supported and the speeds comparatively low.

Torsional Vibrations

Torsional vibrations are very common with reciprocating engine installations and they constitute a real danger to the shafting if they are allowed to occur in the normal range of operating speeds.

Torsional vibrations are superimposed on the normal rotation of the shaft. Relative to some fixed point on the shaft, certain masses (such as flywheel), are oscillating in one direction, while other masses (such as propeller) are oscillating in another direction. In these circumstances, there is an extra periodic twisting in the shaft over and above that due to the mean torque, which sets up excessive stresses.

The speeds at which this comes about are when engine-firing impulses per minute coincide with, or are a prime factor of, a natural frequency of the shafting system in vibrations per minute. The mose severe vibrations occur when they coincide exactly.

One- and Two-Node Frequencies

A shafting system in torsional oscillation is similar in conception to a hull in vertical or horizontal vibration. The one-node (or principal mode of) vibration occurs in a normal installation when cylinder masses and flywheel are vibrating together in phase at one end of the shafting system and the propeller is vibrating out of phase at the other end. At some point between engine and propeller there is a section of shaft which is not vibrating at all. This point is the node.

The two-node (or second mode of) vibration occurs usually when some, or all, of the cylinder masses and the propeller are vibrating together in phase and the flywheel is vibrating in the opposite direction out of phase. In this case there are two nodes, one aft of the flywheel and one just forward of it.

The frequency of one node vibration depends largely on the stiffness or flexibility of the shafting and the moment of inertia of the propeller. A long flexible shaft and heavy propeller have a low natural frequency.

Two-node vibration frequency is practically unaffected by propeller and shafting in a normal marine installation. It depends almost entirely on crankshaft stiffness and moment of inertia of cylinder masses and flywheel. Two-node frequency can, therefore, be fixed by the engine designer and remain constant.

Modes of vibration with more than two nodes can occur but their effect is usually of no importance.

Symptoms

When a shaft is operating at a critical speed of vibration it oscillates with a periodic swing and at regular intervals of swing it receives an exciting impulse from the firing pressure on the pistons. If there were no damping, or resistance to vibration, the amplitude of swing would increase very rapidly and continue increasing until the shaft fractured. This would happen very quickly.

In practice there are many factors which offer resistance to vibration. These are called damping factors. In a marine installation the propeller itself has a big damping influence on one-node vibrations because it has a relatively high amplitude of vibration and a capacity for absorbing torque. In the case of two-node vibrations it does not have much influence because its amplitude is generally low. Friction of bearings and pistons constitutes a damping factor. So does hysteresis loss in the shafting: this is the term for energy used in twisting a shaft backwards and forwards without doing any useful work.

The damping factors provide the symptoms of a torsional vibration. Hysteresis damping causes warming up of the shaft at, or near, a node; vibration friction at bearings causes heating and lubricating difficulties. Oscillation of running gear at points of maximum amplitude causes noise, and if there is a gear drive at the forward end of the engine, this usually causes a great deal of noise.

The Danger Points

The danger zones from the point of view of shafting stresses are the nodal positions and shafting failure will occur at, or near, a node. In one-node vibrations, this is usually somewhere in the thrust shaft or intermediate shafting. In two-node vibrations, there are two likely regions for trouble: one in the crankshaft and one near the propeller.

Bearing troubles may be looked for at places of maximum amplitude (i.e. the forward cylinders in one- and two-node vibrations and the stern bearing in one-node vibrations).

From a consideration of the cause and effect of damping, it will be appreciated that the aim of torsional tuning in a marine installation is not merely to prevent fracture of shafting. That an installation has never suffered a shafting failure does not imply that the system is satisfactory from the torsional aspect. It can suffer many kinds of excessive wear and tear without showing obvious fatigue. This emphasizes the fact that careful torsional tuning should be carried out for every installation.

Influence of Engine Design and Layout

Engine design and layout plays an important part in determining the two-node frequency of vibration, the major orders of vibration (i.e. those which will be serious) and their amplitudes. They also determine, to a considerable extent, the effect which minor orders of vibration will have on engine components.

Close cylinder centres and a stiff crankshaft produce a high natural frequency and the designer should ensure that this is always high enough to prevent the principal order of two-node vibration from appearing in the operating speed range of the engines. The number of cylinders plays a big part in this, engines with a small number of cylinders being better able to avoid trouble from two-node vibrations.

The number of cylinders has an influence also on other forms of vibration and experience suggests that the best number is from three to eight.

Accessories, such as camshafts and fuel pump shafts, are best if driven from the after end of the engine, near the flywheel, so that they cannot experience high amplitudes of vibration.

The same applies to such things as the scavenge pump on a two-cycle engine, which does not need to have very strong scantlings for the work it has to do. But when driven from the forward end of the engine its position renders it liable to extra stresses from vibration swing.

Importance of Propeller and Shafting

For any given engine, the propeller and its shafting are the vital factors in fixing one-node vibrations. They must be controlled so that the combination of engine and propeller produces torsional characteristics which are satisfactory in shafting stresses and engine running gear.

There are two general methods of approach to this problem. One is to keep the major order vibration speed well above all operating speeds of the engine. The other is to get the natural frequency so low that the major critical speed is below the normal operating range. With engines placed amidships, the second course is the usual one. The shafting is long and reasonably small in diameter, consistent with the power it has to transmit, and there is generally no difficulty in tuning to a low frequency. When engines are placed well aft, on the other hand, attempts are usually made to get the natural frequency above operating speeds. This is often the case in trawlers. The application of each method depends on individual cases and each has to be treated on its merits.

Desirability of Isolation of Forward End Drives

There are variations and exceptions to " normal marine installations " such as geared drives and occasional extra duty drives on certain classes of ship. Trawler installations come into this category when they include an extension shaft to the forward end of the engine crankshaft for driving a winch.

When this is done it can alter the two-node torsional characteristics considerably and bring about serious vibration troubles in the engine. For this reason such drives should be thoroughly investigated for their effect and, once the details of design have been fixed, they should be strictly adhered to.

The question of a forward end drive to a winch can be a difficult one from a torsional point of view and it can be taken as a maxim that such a drive should preferably be isolated from the engine system, if that is possible. A slip coupling of the hydraulic or magnetic type will provide such isolation and, to a great extent, so will a belt drive.

Correct Tuning

Every portion of shaft which rotates and every mass which is attached to the shafting contributes in some degree to the final characteristics of the system. It follows that the stiffness of every piece of shafting and the moment of inertia of every revolving mass, has to be computed with considerable accuracy.

Troubles sometimes occur because sufficient care is not taken in working exactly to the approved design. For instance, propellers are sometimes changed, and occasionally bronze shafts substituted for steel ones, without any thought being given to their effect on the torsional characteristics. To lend emphasis to this point it might be mentioned that an average of 50 hours work is required by a skilled investigator to complete the full calculations required for a normal marine installation. It will be seen, therefore, that accurate and reliable data is an absolute necessity.

Comparison of High- and Low-Speed Engines

There are so many other variables combining to fix the torsional characteristics of an engine and propeller system, that the question of engine speed is not, in itself, a governing factor. Other things being equal, however, it is easier to keep the major one-node criticals above the operating range, in after end installations, such as is usual in fishing vessels when low-speed engines are used.

One of the main differences between high- and low-speed engines arises because the former are usually connected to the propeller through a gearbox to get a reasonably effective propeller speed. This introduces complications into the dynamic system. The mass of the gears affects the two-node, as well as the one node, characteristics of the system if the gear drive is rigidly connected to the engine, and for this reason it is customary to fit a slip coupling between engine and gearbox.

The presence of backlash in gearing has the effect of spreading the objectionable speed range of a critical. Further, if reversal of tooth contact occurs, the gears are noisy and liable to damage through impact.

Comparison of Two- and Four-Cycle Engines

The task of investigating the torsional vibration characteristics of a two-cycle engine is somewhat easier than in the case of a four-cycle because the former has no half order components and the firing order is the same as the crank sequence.

A four-cycle engine has a greater number of possible critical speeds in a given range of r.p.m., because of its half order components. These criticals are not all serious, from the point of view of stressing or wear and tear, but they are objectionable and can cause unpleasant noise in the gearing.

FIXED BLADE AND CONTROLLABLE-PITCH PROPELLERS

by

J. A. VAN AKEN

FIXED BLADE PROPELLERS

It is seldom possible to design a propeller that is exactly right and the usual custom is to proceed by the trial and error method, fitting various propellers until a satisfactory one is found.

The three most important factors in determining the design of a propeller are:

1. The speed of the ship;
2. The wake fraction;
3. The engine power and the r.p.m. of the propeller.

In this discussion it is assumed that the speed is known but the power and r.p.m. will have to be determined. The normal practice is for the engine manufacturer to tell the propeller designer the power of the machinery and the r.p.m. The relation between power and r.p.m. is obtained from the test bed and is determined by the formula:

$$\text{s.h.p.} = \frac{2\pi M.N}{4{,}500} \quad \ldots \quad (1),$$

where s.h.p. = the output at the shafting, shaft horsepower, (1 s.h.p. metric 75 kg.m./sec.)
M = the torque in kg.m. and
N = the number of revolutions per minute

The engine power is directly proportional to the torque and the r.p.m. Power may be determined by a brake applied to the engine, but generally the indicated horsepower (i.h.p.) is calculated from readings of the pressure in the cylinders. The formula is:

$$\text{i.h.p.} = \frac{\frac{1}{4}\pi D^2 \cdot S \cdot N \cdot 2 \cdot Pmi \cdot Z}{4{,}500} \quad \ldots \quad (2),$$

where i.h.p. = indicated horsepower,
D = cylinder diameter in cm.,
N = number of revolutions per min.,
S = piston stroke in m.,
Z = number of cylinders, and
Pmi = mean pressure in the cylinders in kg./sq. cm.

assuming

$$\frac{\frac{1}{4}\pi \cdot 2}{4{,}500} = 0.000349 = f_1,$$

$$\text{i.h.p.} = f_1 \cdot D^2 \cdot S \cdot N \cdot Pmi \cdot Z \quad \ldots \quad (3)$$

For a given engine,

$$f_1 \cdot D^2 \cdot S \cdot Z = f_2 \text{ (a constant);}$$

hence

$$\text{i.h.p.} = f_2 \cdot N \cdot Pmi \quad \ldots \quad (4)$$

Formula (4) is valid for a double-acting two-cycle engine, while for a single-acting two-cycle engine,

$$\text{i.h.p.} = \tfrac{1}{2} f_2 \cdot N \cdot Pmi \quad \ldots \quad (5)$$

and for a four-cycle engine,

$$\text{i.h.p.} = \tfrac{1}{4} f_2 \cdot N \cdot Pmi \quad \ldots \quad (6)$$

Assuming

$$\frac{\text{s.h.p.}}{\text{i.h.p.}} = \eta_m$$

to be the mechanical efficiency, formula (2) is translated into

$$\text{s.h.p.} = \frac{\eta_m \cdot \frac{1}{4}\pi D^2 \cdot S \cdot N \cdot 2\ Pmi \cdot Z}{4{,}500}$$

while, according to formula (1)

$$\text{s.h.p.} = \frac{2\pi \cdot M \cdot N}{4{,}500}$$

so that

$$\frac{\eta_m \frac{1}{4}\pi \cdot D^2 \cdot S \cdot N \cdot 2 \cdot Pmi \cdot Z}{4{,}500} = \frac{2\pi M \cdot N}{4{,}500}$$

or

$$\eta_m \tfrac{1}{4} D^2 \cdot S \cdot Pmi \cdot Z = M \quad \ldots \quad (7)$$

Supposing

$$\tfrac{1}{4} D^2 \cdot S \cdot Z = f_3$$

$$f_3 \eta_m Pmi = M \quad \ldots \quad (8)$$

Formula (8) shows that the torque is directly proportional to the mean pressure and the mechanical efficiency, so that the torque may be replaced by the mean pressure. It is assumed that mechanical efficiency is kept constant. This is not exactly true but is accepted to avoid complications.

If the power, r.p.m. and ship's speed are given, the propeller can be designed for these conditions:

I: power 100%; revolutions 100%; torque 100%
II: power 90%; revolutions 90%; torque 100%
III: power 90%; revolutions 100%; torque 90%

For both conditions II and III, 90 per cent. of power is assumed, while 90 per cent. of the revolutions and 90 per cent. of the torque are assumed for the conditions, alternatively.

Condition I: power 100 per cent., revolutions 100 per cent., torque 100 per cent. (*fig. 496*). If the propeller is designed for this condition, the ship's speed will be 100 per cent. If the revolutions are reduced to 90 per cent. the power, torque and speed will be reduced, respectively, to 73, 81, and 90 per cent., and if the revolutions are reduced to 96.5 per cent., the respective figures will be 90, 93.3 and 96.5 per cent. This shows that a reduction in r.p.m. decreases regularly engine power, torque and speed, so that condition I is, therefore, normally used for large engines in merchant ships.

If the number of revolutions is 103 per cent., power becomes 109 per cent., torque 106 per cent., and speed 103 per cent., in which case the torque is overloaded by 6 per cent.

Condition II: power=90 per cent., revolutions= 90 per cent., torque=100 per cent. (*fig. 497*). If the propeller is designed in accordance with these data the ship's speed will be 96.5 per cent., and if the r.p.m. are increased to 93 per cent., power will be 100 per cent., torque 107.5 per cent. and speed 100 per cent., a condition rarely found with internal combustion engines. The pitch of the propeller is too large. But in the case of steam engines, particularly the old triple-expansion which can no longer attain its full power at 100 per cent. r.p.m. because pressure in the boilers is too low, this condition is often found. The number of revolutions never exceeds 90 per cent.

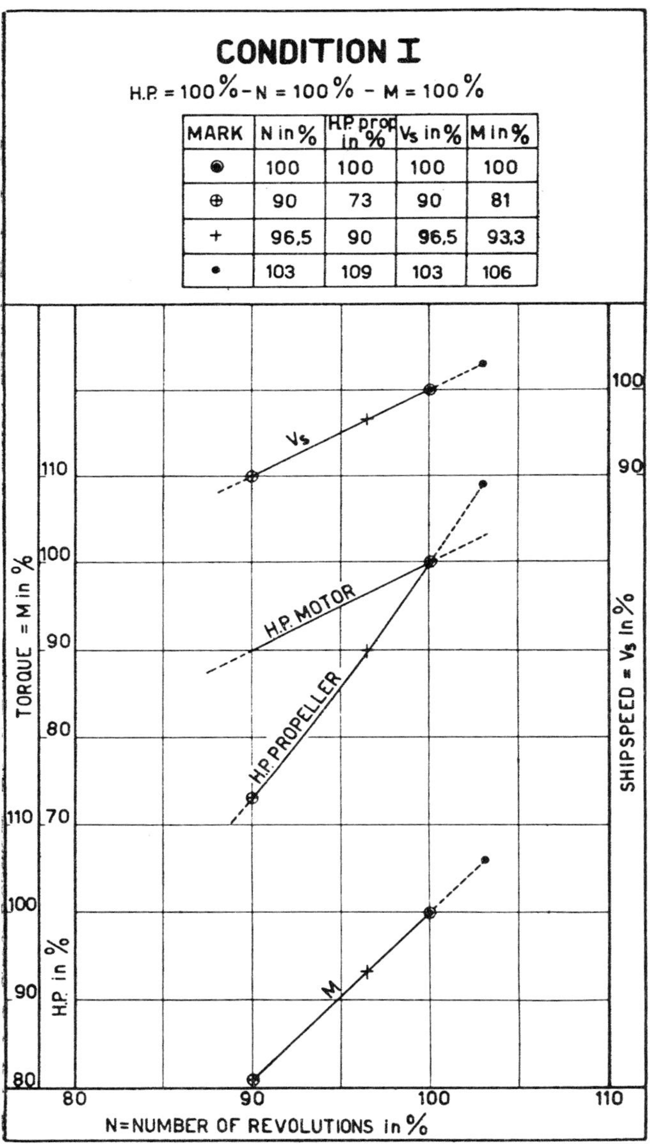

MARK	N in %	H.P. prop in %	Vs in %	M in %
◉	100	100	100	100
⊕	90	73	90	81
+	96,5	90	96,5	93,3
•	103	109	103	106

Fig. 496. Relation between horsepower, revolutions and speed for condition

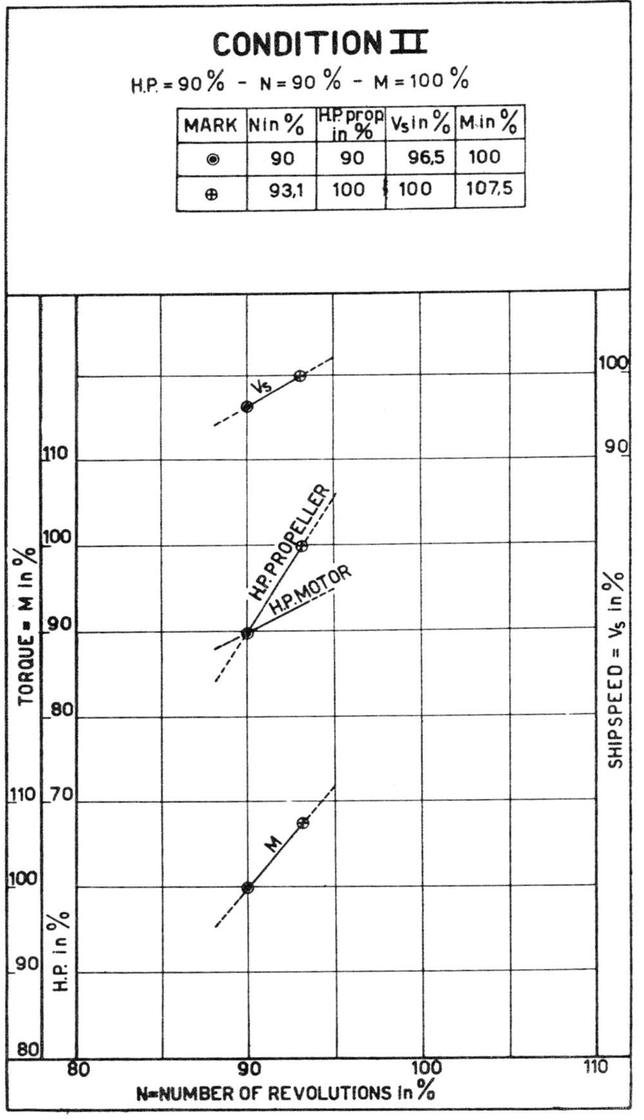

MARK	N in %	H.P. prop in %	Vs in %	M in %
◉	90	90	96,5	100
⊕	93,1	100	100	107,5

Fig. 497. Relation between horsepower, revolutions and speed for condition II

Condition III: power=90 per cent., revolutions=100 per cent., torque=90 per cent. (*fig. 498*). If the propeller is designed for this condition, the ship's speed will be 96.5 per cent. If the number of revolutions is increased to 103.5 per cent. the power will be 100 per cent., the torque 96.7 per cent. and the speed 100 per cent., a highly recommendable condition, especially for

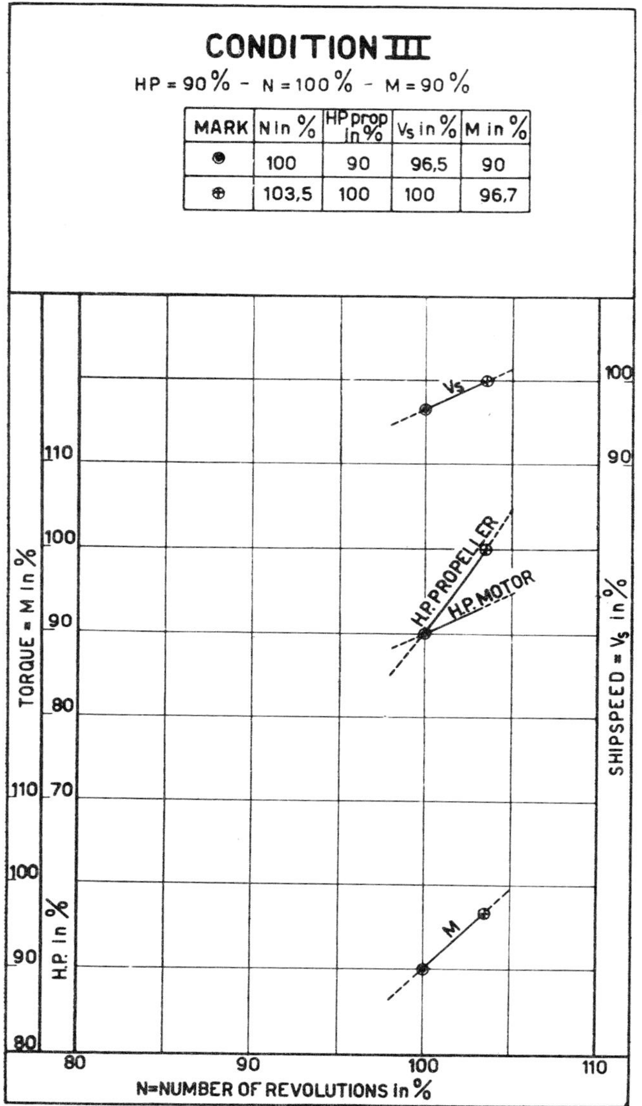

Fig. 498. *Relation between horsepower, revolutions and speed for condition III*

light engines. When, in the course of service, the roughness of a ship's bottom increases and her machinery can no longer produce full power, a propeller designed to meet Condition I would become too " heavy ", that is to say, would have too large pitch. But if the propeller is designed in accordance with Condition III, the engine may be kept running at full r.p.m. and the torque will not increase beyond 100 per cent. For this reason a propeller should be designed to meet Condition III.

Condition III is also recommended for tugs because the torque during towing will not be excessive while, without tow, the r.p.m. may be raised to 103.5 per cent. to obtain a power of 100 per cent. This is also the reason why Condition III is recommended in the case of fishing vessels. A compromise type of propeller has been adopted for trawlers in free-running condition and with the r.p.m. at 100 per cent., it does not yield the maximum power and, therefore, does not give the maximum speed. Further, in the tow condition, the tow-rope pull is not raised to its maximum. But postwar conditions demand speed for a vessel going to and from the fishing grounds. This calls for the maximum power to be attained in the free-running condition at full r.p.m. but is detrimental

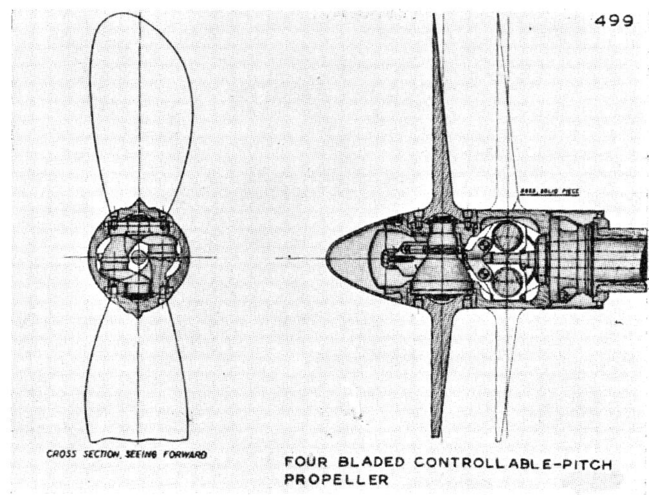

Fig. 499. *Four-bladed controllable-pitch propeller*

to the tow-rope pull in the tow condition. The problem has been solved by the adoption of the controllable-pitch propeller.

CONTROLLABLE-PITCH PROPELLERS

To-day a large number of controllable-pitch propellers have been patented. Among the principal types in use on the European continent are:

1. Bretagne, Nantes, France;
2. Voith-Schneider, Heidenheim, Germany;
3. Lips-Schelde, Drunen, Netherlands;
4. Kamewa, Karlstad, Sweden;
5. Escher-Wyss, Zurich, Switzerland.

The Voith-Schneider propeller has three to six blades rotating around a vertical axis and it requires a flat-bottom to the ship. All the other types are similar to fixed-blade propellers but, of course, the blades are controllable. A feature of the Lips-Schelde propeller is that the four blades are arranged in pairs on the propeller boss, fig. 499. This gives it the advantage of a smaller boss than that of the other types.

The advantages in using a controllable-pitch propeller on fishing vessels are:

1. Non-reversible propulsion machinery;
2. Constant r.p.m. in any condition of loading;
3. Maximum speed of the ship;
4. Maximum pull for trawling;
5. One direction of running the engine ahead and astern;
6. Rapid manoeuvrability as the engine need not be stopped to be reversed and the propeller can be controlled from the bridge;
7. Low speeds for entering the harbour or dock, and so on;
8. Reduced over-all cost, and smaller size and weight of engine;
9. Economy in fuel consumption and engine maintenance;
10. Practical means for avoiding critical torsional-vibration conditions.

The gear in the propeller boss, fig. 499, is the most important part of all controllable-pitch propellers. Fig. 500 shows the arrangement of a controllable-pitch propeller manoeuvred from the bridge.

The advantage of using a controllable-pitch propeller was demonstrated when a ship fitted with one hit a sunken landing craft. She was going at full speed and one of the blades of the propeller was damaged. But the reversing mechanism and the propeller shaft were unaffected. A spare blade was fitted in place of the damaged one and the ship went straight into service again.

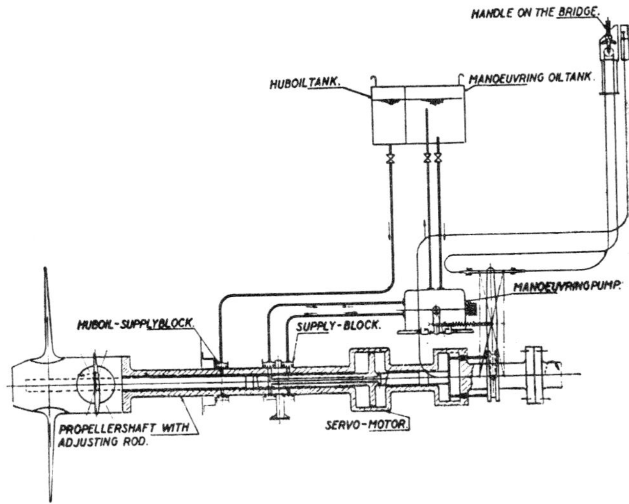

Fig. 500. Schematic arrangement of a controllable-pitch propeller

CONTROLLABLE-PITCH PROPELLERS FOR TRAWLERS

by

M. ROUCHET

CONTROLLABLE PITCH propellers have been used for a long time past but it is only about 20 years ago that the first was used with a big engine. The reason for this is that the flexibility and the manoeuvrability of the reciprocating steam engine enabled good use to be made of the fixed blade propeller. But with diesels the fixed blade propeller is much less satisfactory, as an analysis of the working conditions and maintenance costs shows. The main characteristics of the propeller are determined by speed, power and r.p.m. An estimate must also be made of the ship's maximum speed and the thrust in the two extreme conditions: standing pull and maximum speed. If the fixed blade propeller has been designed for maximum speed and if the governor is brought into full speed position when the boat is towing, the engine will not be able to reach full r.p.m. Any attempt to restore the full r.p.m. by increasing the fuel supply will overload the motor. Again, if the propeller has been designed for the standing pull, when in free route condition, the fuel supply will have to be reduced to avoid racing the motor. In both these cases only part of the total power of the motor will be absorbed by the propeller, therefore the propeller must be designed to effect a compromise between the running-free and trawling conditions.

When a fixed blade propeller is used, reversing the boat implies reversing the main engine or using a reverse gear. If the main engine is of the reversible type large quantities of compressed air are needed during manoeuvres. The engine is subjected to quick changes of temperature which impose high thermal stresses detrimental to the upkeep of cylinder covers and liners. Reverse gears are limited to 500/600 h.p. and have relatively a short life.

A controllable pitch propeller provides a better solution to these problems. By choosing the right pitch, the total power of the engine can be used at normal r.p.m. when running-free, and the thrust and the speed of the boat are increased. An economical speed can be found at an intermediate pitch corresponding to the optimum revolutions recommended by the engine maker. Fuel consumption will be lower than it is with the fixed blade propeller. In the case of the standing pull, which is almost the same as when trawling, the most important factor is the maximum pull. The pitch must be reduced so that the engine can run up its maximum r.p.m. and deliver its maximum power and thrust.

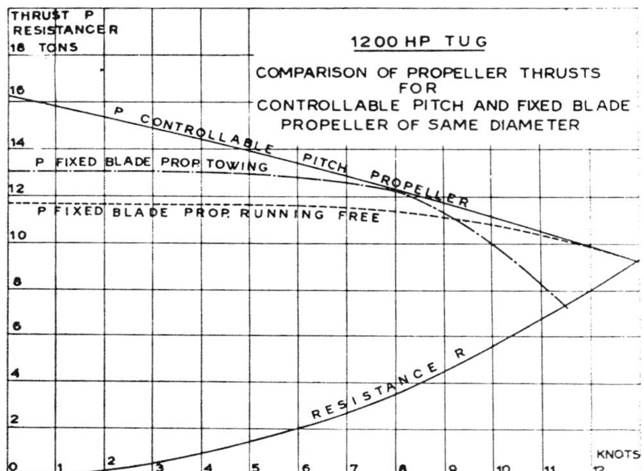

Fig. 501 Note:— The fixed blade propeller —·—·— is designed especially for trawling, the ------ one for steaming.

Fig. 501 illustrates the differences in speed and thrust between a controllable pitch and two fixed blade propellers. The thrust figures are:

	Free-running		Standing	
	lb.	*kg.*	*lb.*	*kg.*
Fixed blades	16,000	7,300	28,600	13,000
Controllable pitch	20,500	9,300	35,400	16,000
Gain	+27 per cent.		+23 per cent.	

A comparison of fig. 502 and 503 for the controllable pitch propeller in running-free and standing conditions will also show these results. If the pull attainable with a fixed blade propeller is considered to be sufficient the use of a controllable pitch propeller would enable a less powerful engine to be installed.

The advantage of the controllable pitch propeller for reversing does not need to be stressed. The engine is

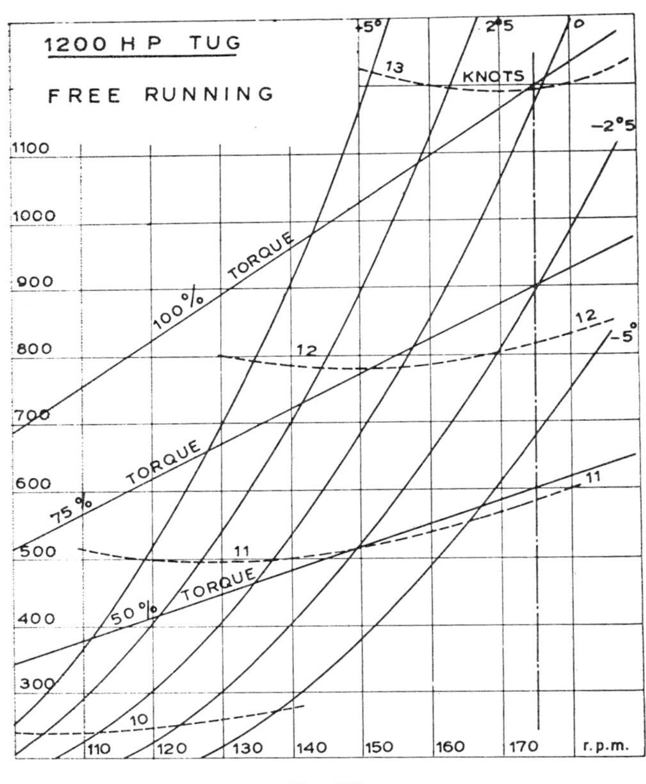

Fig. 502

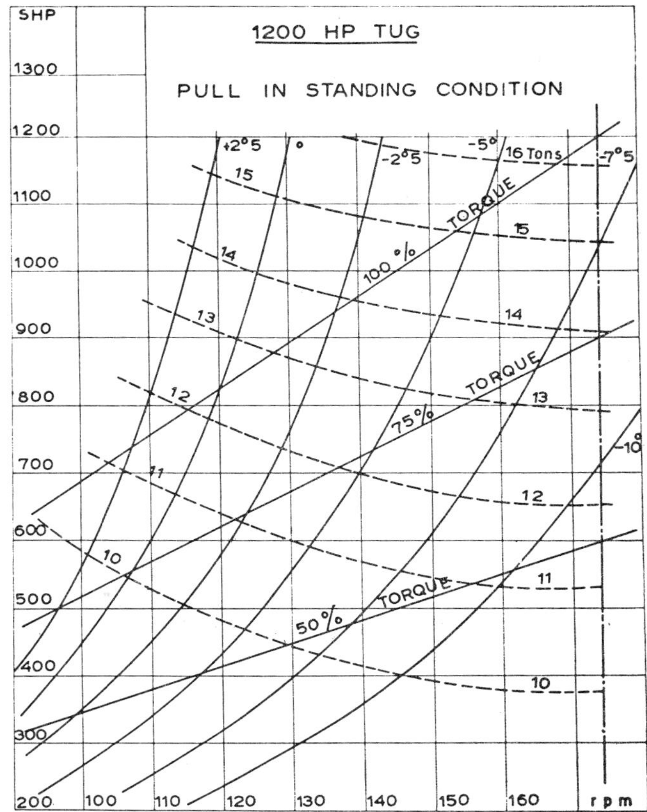

Fig. 503

simpler and less expensive and the compressed air installation is reduced to a minimum. Manoeuvring is entirely in the hands of the skipper, being controlled from the bridge. Because of this, many accidents are avoided.

Another advantage is that damaged blades can be removed without work on the hub, and such replacements can be done quickly and carried out without dry docking. The expenses are also less than those incurred by the renewal of a fixed blade propeller.

The extra initial cost of fitting a controllable pitch propeller is quickly offset by lower running costs and there is no doubt that it makes a trawler a better fishing vessel.

"FATHER AND SON" AND DIESEL ELECTRIC PROPULSION

by

ROBERT KOLBECK

"FATHER AND SON"

RECENTLY in Germany some large trawlers have been built with diesels arranged in a gear system, the so-called "father and son" method. This is the latest development to solve the dilemma of using a diesel to drive a fixed propeller efficiently at both steaming and trawling taking into consideration the divergent requirements related to the operation of trawlers, as:

high engine power for cruising;
high propeller torque for trawling;
good manoeuvrability;
high power for the trawl winch.

The first trawlers so equipped with the "father and son" system were *Bahrenfeld* and *Barmbeck*, built in 1950.

Fig. 504 shows two diesels working through a common propeller gear. The "father" is a reversible, super-

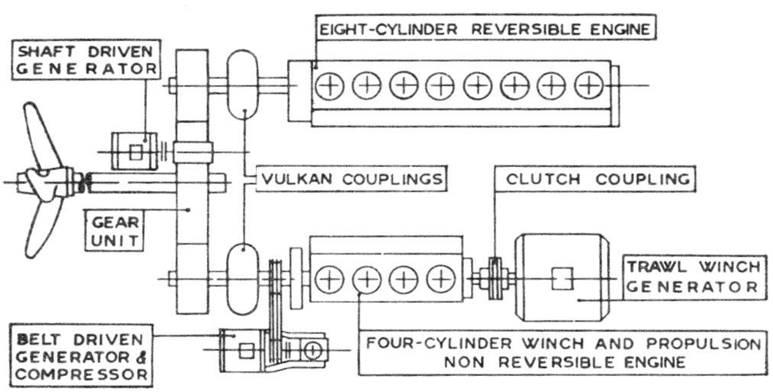

Fig. 504

charged eight-cylinder, four-cycle engine, developing 720 h.p. at 375 r.p.m., and the "son" is a non-reversible, four-cycle engine of 250 h.p. at the same r.p.m. The engines are connected to the gear through Vulkan couplings and both of them can be used to drive the propeller. Both engines are of the same type to make operation easy and to simplify the problem of carrying spare parts.

The smaller engine is also connected at the forward end, through a clutch-coupling, to a 162 kw. Leonard generator which supplies current to the trawl winch motor, which develops 190 h.p. at 600 r.p.m.

A second generator is driven by a belt off the shaft of the smaller engine, which also drives an air-compressor through a clutch.

Manoeuvring the boat ahead or astern is done solely with the reversible eight-cylinder "father" engine. During this operation the four-cylinder "son" engine is disengaged from the gear and may be used to drive the winch generator only. But if the demand for electricity is small, then the "son" engine can remain connected to the propeller gear. In any case, engaging and disengaging of the small "son" engine is done without interrupting regular propulsion.

When both "father" and "son" are working together there is an output of 970 h.p. available for propulsion but when, for the purpose of manoeuvring, the "son" engine is disengaged, the output is, of course, 720 h.p.

The energy balance shows that, after subtracting the losses of 50 h.p. in the gear, and of 10 h.p. in the shafting and allowing for a power demand of the trawl winch generator of 40 h.p., there is 870 h.p. available at the propeller. This gives the ship 11.5 to 12 knots at a propeller speed of 115 r.p.m.

For trawling at 3.5 to 4 knots, with an assumed propeller speed of 86 r.p.m., the propeller will absorb 640 h.p., supplied, of course, by both engines. The "father" engine will, at trawling speed, deliver 560 h.p. and the "son" engine 185 h.p. or in total 745 h.p. to the gear. The losses in gear, shafting and shaft driven generator is 100 h.p. A total of 645 h.p. or slightly more than the power absorbed, is available at the propeller, therefore the engine will not be overloaded.

The Vulkan coupling of the "son" engine is equipped with an automatic shut-off device which is set in motion by the reversing mechanism of the large engine. This device protects the non-reversible "son" engine from being rotated the wrong way when the propeller is reversed. If the propeller only runs ahead it allows the trawl winch to be connected to the system. This automatic interlocking and shut-off device is shown in fig. 505.

An individually-driven pump forces oil into the lubrication system of the two engines. Oil flows also to the servo-motor " c " and the control piston " d " through a cock " A " in a side branch. Piston " d " controls the engaging and disengaging of the Vulkan coupling of the small engine. The servo-motor and the control piston work hydraulically through valves " a " and " b ". The position of valve " a " depends on the position of the camshaft and that of valve " b " depends on the position of the control wheel. According to the position of these valves the servo-motor " c " and the control piston " d " are energized. The hydraulic couplings of both engines can be manually controlled. In the case of the " son " engine, it is always possible to disengage it manually but it can be manually engaged only if the position of the automatic device permits. Should the big engine break down, cock " A " is then closed and the " son " engine can take over for " ahead " propulsion.

The " father and son " system can be designed for bigger power requirements and it is, of course, feasible to use a more powerful " son " engine if more power is required for the trawl winch. It should always be kept in mind:

1. That the same engine type is chosen for both " father " and " son " to simplify maintenance and carrying of spare parts.
2. That the " son " engine should be non-reversible to avoid complicated interlocking devices.

The over-all arrangement of the engine room is shown in fig. 506. Both engines are arranged to permit observation and maintenance from the centre aisle. The water and lubricating oil pumps are separately driven by electric motors. The engines are fresh-water cooled, and there is a separator for purifying the lubricating oil, and a turbulo oil filter to clean the oil.

An oil-heated auxiliary boiler is provided for heating purposes and for cooking liver oil. Additional heat is recovered from the " father " engine in an exhaust boiler.

DIESEL-ELECTRIC

Two diesel-electric trawlers, *Freiburg* and *Tübingen*, were built in 1950, and in both vessels the propelling power of 700 h.p. is supplied by three diesel-generator sets. Each set is powered by a supercharged six-cylinder four-cycle diesel delivering 330 h.p. at 500 r.p.m. The direct current Leonard system of electric equipment is installed in both ships.

Fig. 507 shows the arrangement of machinery. Diesel-generator sets are located on the intermediate deck which allows excellent use of the space. Engine and generator are mounted on a common frame which includes the circulating lubricating oil tank. Each set is placed on elastic mounts to reduce noise. Both engines have lubricating oil and fresh water pumps attached. An oil-heated auxiliary boiler and three La Mont boilers heated by the exhaust of the three diesels supply the required steam. Two electric motors with an output of 350 h.p. each of 1,000 r.p.m. drive the propeller at 110 r.p.m. through a gear. Depending on the operation conditions, three generators, or two or one, can be connected to the two propelling motors and any generator can be disconnected from the engines and connected to the trawl winch, which can be done without interrupting propulsion. The engines can be controlled from the engine room as well as from the bridge.

COMPARISONS

Table LXXVIII shows data for steam and motor trawlers with identical fish-hold capacity, namely the steam trawlers *Buxta*, *Sonne* and *Hans Kunkel*, and the diesel trawlers *Barmbeck* and *Bahrenfeld*. With the same fish-hold capacity, the motor trawlers have 10 to 13 per cent. less gross tonnage and 18 per cent. less displacement, and they carry two less in crew. The liver oil tanks have 90 cu. ft. (2.5 cu. m.) more capacity.

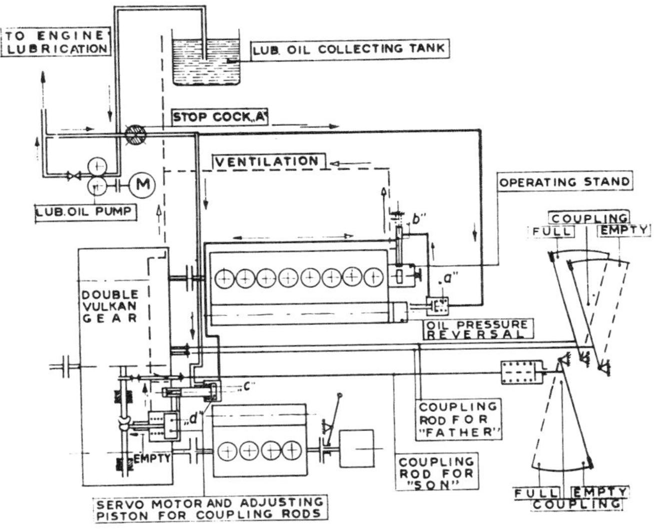

Fig. 505

The difference in displacement is caused by the lighter weight of machinery and the lower fuel consumption of the diesels. This saves bunker capacity. The length of the motor trawler is much less than that of the steam trawler. It should be emphasized, however, that the length of ships deserves particular consideration because the longer ship runs better and needs less power for a certain speed. In comparing the power installed in steam and motor trawlers it is remarkable to note that the steam trawler has a lower rated output measured in i.h.p. Reduction of the propeller speed during trawling is probably smaller, therefore the steam engine can supply the required towing power even with smaller rated output if the boiler is large enough. In Table LXXVIII also the steam trawlers *Alteland* and *Heinrich Coleman* are similarly compared with the diesel electric ships *Freiburg* and *Tübingen*.

A comparison of fuel consumption and operating range of steam and motor vessels is shown in Table LXXIX. It can be seen that, with the given bunker capacity the motor ship (equipped with "father and son" system) can make four herring trips or two fresh fish trips without refuelling, compared with the steam trawler's three herring trips or one fresh fish trip. The figures are based on experience covering several years of fishing. Because of the simplicity with which liquid fuels can be stored, the trim problem in the motor trawler is easier solved, a fact which appeals particularly to the naval architect. For the shipowner, of course, it means a saving in manpower because the ship can be brought to the best trim in minimum time.

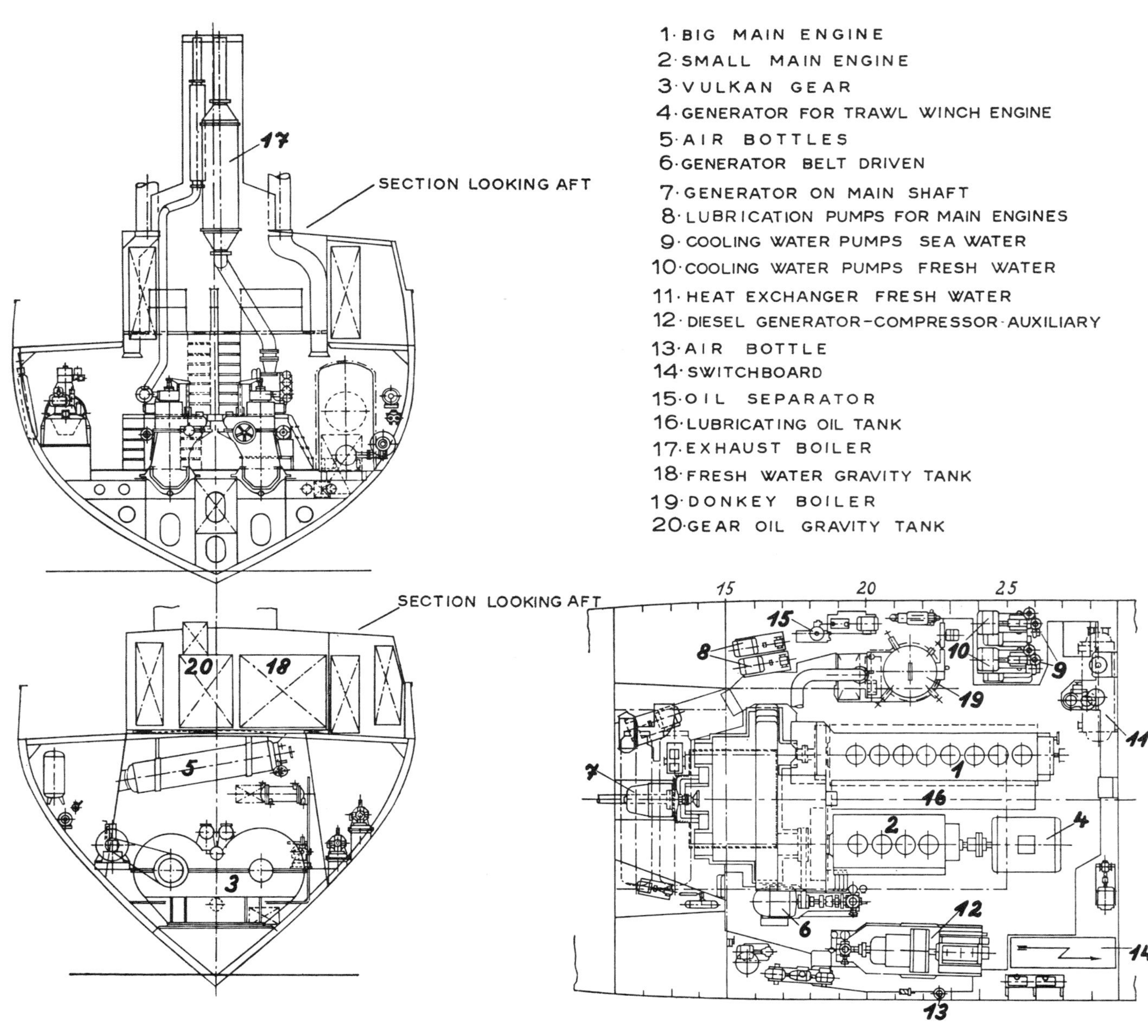

Fig. 506

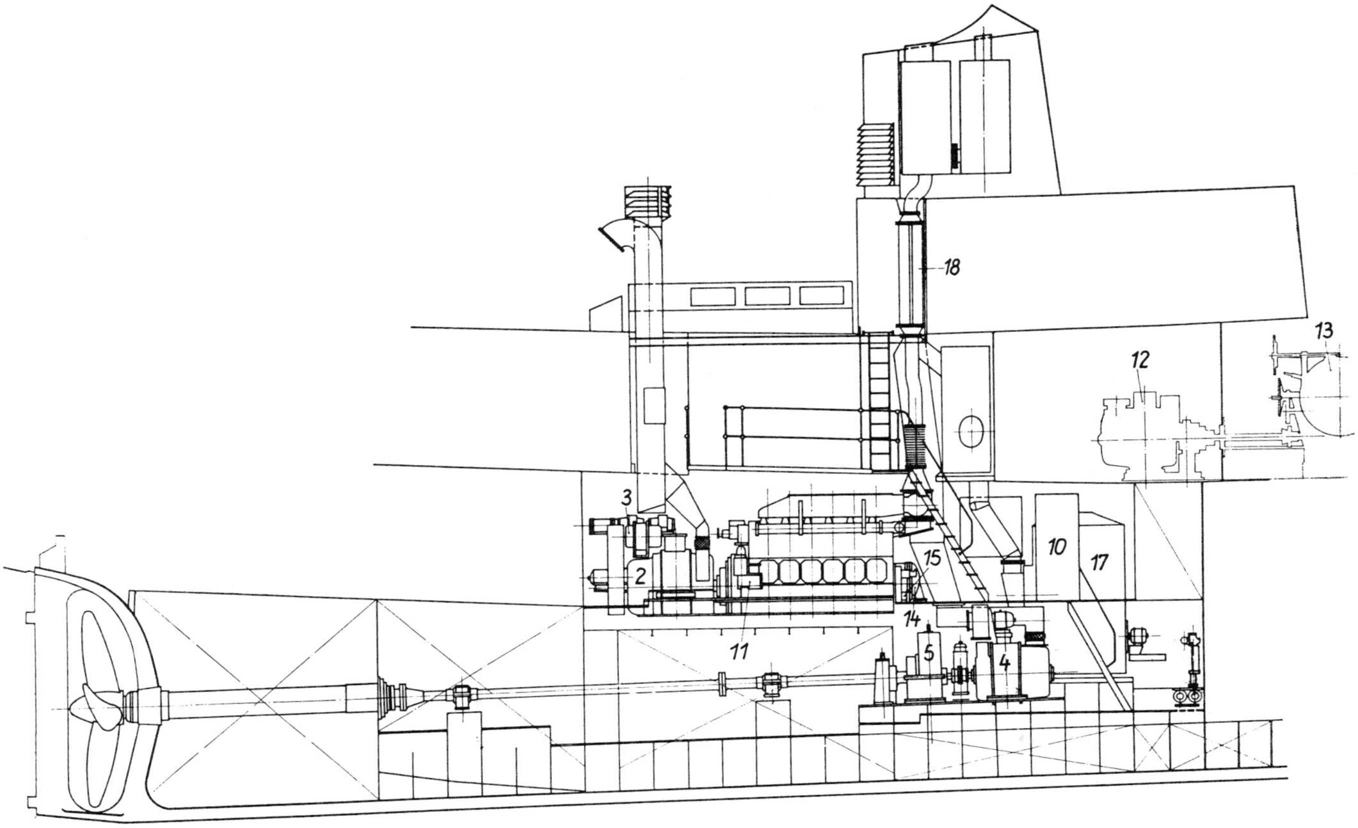

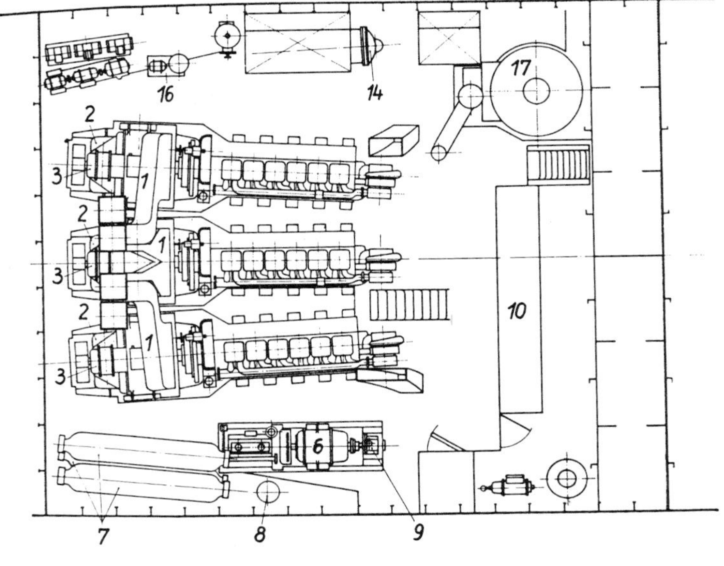

1 - DIESEL ENGINE
2 - MAIN GENERATOR
3 - BOARD NET GENERATOR
4 - PROPELLING ENGINE
5 - HELICAL GEAR WITH PRESSURE BEARING
6 - DIESEL GROUP FOR HARBOUR USE
7 - AIR BOTTLES
8 - AIR BOTTLE FOR ⑥
9 - AUXILIARY COMPRESSOR
10 - MAIN SWITCHBOARD
11 - COOLING WATER PUMP (FRESH WATER)
12 - TRAWL WINCH ENGINE
13 - TRAWL WINCH
14 - HEAT EXCHANGER (F W)
15 - LUBRICATION PUMP FOR DIESEL
16 - OIL SEPARATOR
17 - DONKEY BOILER
18 - LA-MONT EXHAUST BOILER

Fig. 507

TABLE LXXVIII

Steam trawler compared with motor trawler
(for same fish storage capacity)

	Steam trawler Buxta	*Steam trawlers Sonne Hans Kunkel*	*Diesel trawlers Barmbeck Bahrenfeld*	*Steam trawlers Alteland Heinrich Coleman*	*Diesel electric trawlers Freiburg Tübingen*
Fish room capacity, baskets	4,300	4,500	4,500	5,400	5,357
metric tons	215	225	225	270	268
Gross register, tons	522	513	454	551	449
Length ft.	170	162	144	168	147
m.	52.00	49.56	43.80	51.27	45.00
Beam ft.	27.5	28	26.8	28.5	27.9
m.	8.40	8.52	8.20	8.70	8.50
Depth ft.	16	16.3	15.6	16.2	15.3
m.	4.90	4.98	4.75	4.95	4.65
Displacement, metric tons	874	1,170	960	1,230	—
Fuel capacity (main and auxiliary tanks, tons)	295	325	135	320	100
Liver oil tanks, cu. ft.		460	550	565	635
cu. m.		13	15.5	16	18
Crew	22	22	20	26	22
Power, i.h.p. or b.h.p.	1,000	860	970	850	700
R.p.m.	120	120	375/115	115	110
Engine	Reciprocating steam engine (triple expansion) Exhaust steam turbine	Reciprocating steam engine (triple expansion) Exhaust steam turbine	G8V42 supercharged, G4V42	Reciprocating steam engine (triple expansion)	3 × G6V33 supercharged with generators
Speed, knots	13	12	12	11.3	12

Advantages of motor trawler:
- For same fish storage capacity (4,500 baskets, 225 tons)
- Gross tonnage, 10 to 13 per cent. less
- 2 crew members less
- Displacement, 18 per cent. less (made possible by smaller weight of propulsion plant and much smaller fuel tank capacity)
- Liver oil tank, 90 cu. ft. (2.5 cu. m.) larger

Advantages of motor trawler:
- For same fish storage capacity (5,400 baskets, 270 tons)
- Gross tonnage, 18 per cent. less
- 4 crew members less
- Fuel tank capacity, *220 tons less*
- Liver oil tanks, 70 cu. ft. (2 cu. m.) larger

TABLE LXXIX

Steam trawler compared with motor trawler

Fuel consumption — Operating range

(all tons metric)

Steam (triple expansion engine with exhaust turbine)	All values are measured	*Diesel engine* (" father and son " arrangement)
Coal 325 tons	Bunker capacity	Fuel oil 135 tons
9 to 9.5 tons/day	Fuel consumption per day	*ca* 3 ton/day during steaming *ca* 2.5 ton/day during trawling
	Fuel consumption per trip	
108 tons	Herring trip . . 12 days	31.5 tons
215 tons	Fresh fish trip . 24 days	65.0 tons
Bunker capacity suffices for 3 herring trips, or 1 fresh fish trip		Bunker capacity suffices for 4 herring trips, or 2 fresh fish trips
	Lubricating oil consumption per trip	
	Herring trip . . 12 days	48 imp. gal. (220 litres)
	Fresh fish trip . 24 days	100 imp. gal. (450 litres)

MULTIPLE REDUCTION GEAR PROPULSION

by

ALEXANDRE CHARDOME

NOW that diesels are replacing steam engines in trawlers, two problems have become more acute:

1. How to cope with the characteristics of an engine which develops a practically constant torque at all r.p.m., while driving a propeller requiring different torques at different speeds of advance;
2. How to utilize, while cruising, the source of energy which drives the trawl winch on the fishing grounds.

Different systems have been tried in seeking a satisfactory solution. While the " father and son " system solves the second problem, the controllable pitch propeller and the diesel electric drive are able to solve both problems. The advantages claimed for each system are neither convincing for the shipyard nor for the shipowner, as they are mostly sales talk and seldom give comparative figures. In technical matters it should be possible to check accurately any claim and to estimate the relative importance and influence of different factors.

A new system, the multiple reduction gear, has recently been installed in the *Belgian Skipper*, a trawler owned by the S. A. Zeevisserij & Handelsmaatschappij " Zeehandel ", Ostend, Belgium, and built by Béliard, Crighton and Co.

Fig. 508. The Belgian Skipper *making 14 knots during trials*

Main particulars of the trawler are:

Length, over-all . . .	158 ft.	(48.25 m.)
Length, between perpendiculars	141 ft.	(43.00 m.)
Breadth, moulded . .	26.5 ft.	(8.10 m.)
Depth	14.7 ft.	(4.50 m.)
Fish-hold . . .	7,000 cu. ft.	(197 cu. m.)
Fuel oil	115 tons	
Fresh water . . .	11 tons	
Crew	14 men	

It was proposed at the start to install a " father and son " system in which a power of 1,000 h.p. would be delivered by 2 diesels of approximately 750 h.p. and 250 h.p., but, when the problem had been fully studied, it was decided to use two diesels of 500 h.p. each, driving a single-propeller through the new type multiple reduction gear.

Fig. 508 shows the trawler during the trials. Loaded with 90 tons of fuel and fresh water, she made a speed of 14.1 knots, having both motors of 500 h.p. connected to the propeller. Having made her delivery trials, the trawler is now in full and successful fishing activity.

Fig. 509 indicates the propulsion arrangement. The two diesels A and B are two-cycle eight-cylinder V-type and develop 500 continuous h.p. at 500 r.p.m. The multiple reduction gear C was built to the patents of Dipl. Ing. Frantz Sueberkrueb, of Hamburg. Each motor can be connected to the propeller through three different ahead gears and one astern. The ratios are 3.45, 4.24 and 4.78. Each gear is manoeuvred by oil pressure from a central control stand.

The electric generators D and E are connected straight to the diesels through the multiple reduction gear and their ratio is always 3.00 to the r.p.m. of the diesels. One generator delivers current to the electrical trawl winch motor according to the " Kramer " system, the other,

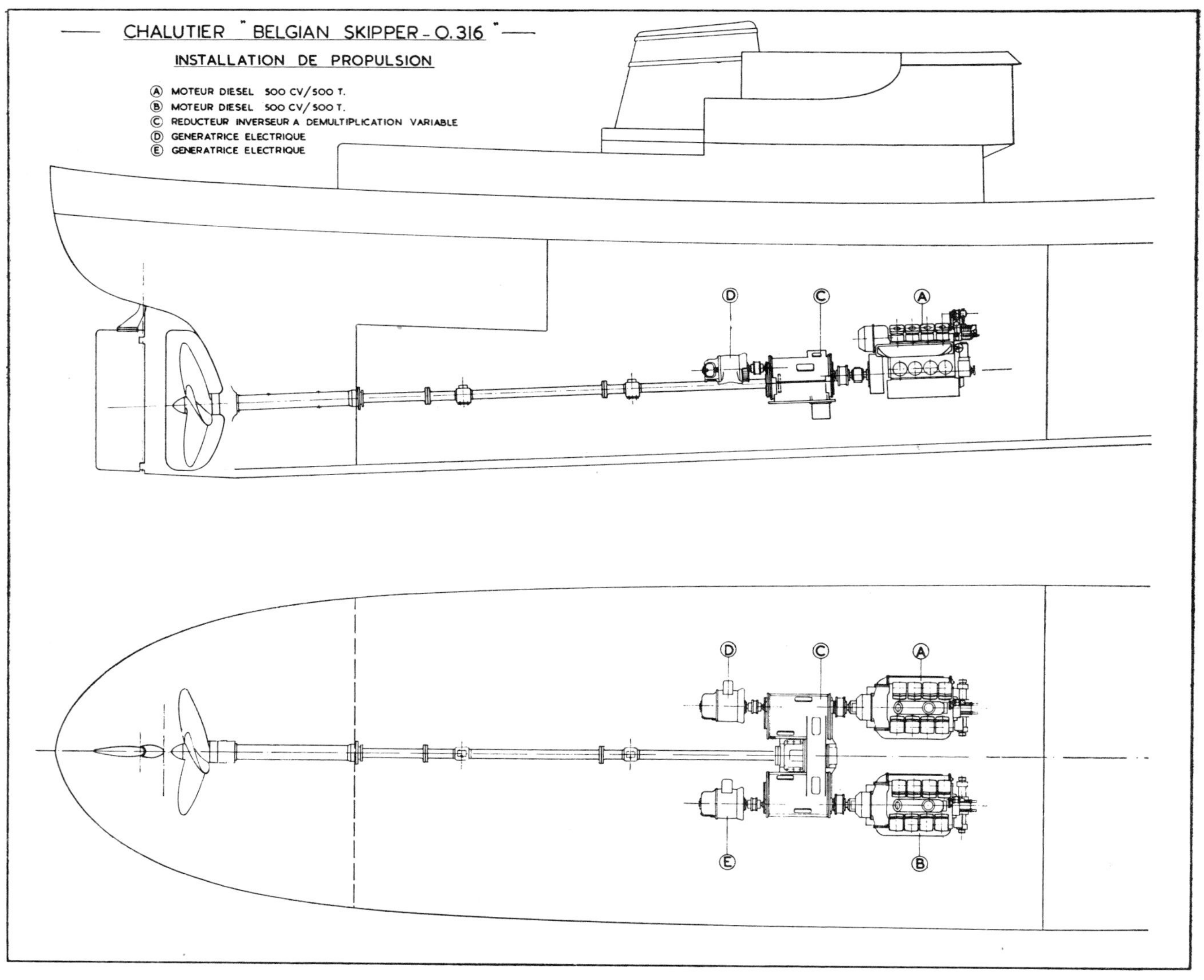

Fig. 509

giving constant voltage, is intended for lighting and small power purposes. Both generators are easily interchangeable. A better technical arrangement, but more expensive, would have been to provide the generators with double windings, so as to enable each of them to serve both purposes.

Fig. 510 is a view taken from the top of the multiple reduction gear C when looking forward between the two diesels. Fig. 511 is a view of the central control stand situated in the centreline of the vessel in front of the diesels. Fig. 512 is a view of the back of the multiple reduction gear C taken from above the intermediate shaft and looking forward. Fig. 513 is a view of the multiple reduction gear C open during assembly in the workshop.

Fig. 510. View of the engine-room

The speed trials confirmed the calculations and their results can be found in the trust-speed diagram (*fig. 514*). Special attention is drawn to two of the results as they were decisive factors in selecting the system:

(*a*) With the two motors developing 500 h.p. at 500 r.p.m., connected through the reduction gear to the propeller turning 145.5 r.p.m., the trawler made a speed of 14.10 knots. This speed was checked with the greatest care during four runs at about high-tide level. The tide speed curve obtained was perfectly regular;

(*b*) With a single motor of 500 h.p. and the propeller turning 118 r.p.m., the trawler made a speed of 12.05 knots, which was similarly checked.

Trials were also made to check the calculated static pulls, but certain difficulties could not be overcome. Trials were carried out in a dock with different positions of the trawler, with the propeller at a maximum distance from the dock walls of about 260 ft. (80 m.) at the rear and 130 ft. (40 m.) at the sides. The draught aft of the trawler was 14.1 ft. (4.30 m.) and the dock water depth about 26 ft. (8 m.). Although the dynamometer had been carefully checked, the readings were always about 10 per cent. higher than expected, and varied with the time, until the moment when the rate, direction and power of the water flow in the dock reached an established maximum. At the Ostend Congress in 1951, Dr. Dieudonné mentioned tests of the French Navy showing similar difficulties.

It seems one must be sceptical about commercial claims for static pulls. The common shipyard tests are apparently not sufficient to check the results of tank tests on this point.

Fig. 511. The central control

THE THRUST-SPEED DIAGRAM

Fig. 512. The back of the multiple reduction gear looking forward

The data given by the tank for the propeller, tested in open water, can be utilized in the following manner. A thrust P, easily calculable according to the usual presentation of the test results, corresponds to a certain speed of advance V and a certain number of revolutions N. It is possible to infer N=f (V,P). The speed V being plotted in the diagram along the x axis and the thrust P along the y axis, one gets a set of interrelated curves of the revolutions N. There is no difficulty to add a set of curves of efficiency, torque and power. The thrusts are shown along the y axis but to simplify the presentation, the advance speeds have not been drawn along the x axis. In fact, the diagram should not be used for the propeller in open water, but for the propeller behind a specific hull. If the speeds and corresponding powers are accurately measured during the trials, a power-wake coefficient can be found which makes it possible to show the curve of hull resistance on the diagram in terms of propeller thrust and water speed through the propeller. This is the L.M.S. curve of the diagram. Therefore it was possible to replace the speed of advance of the propeller by the ship's speed, in knots, on the x axis. The thrust-wake coefficient might be slightly different from the power-wake coefficient but, as already stated, it seems that shipyards are unable to check this point with customary means. The possible resulting error must be very small and is, in any case, the same for all the different systems considered.

What would the " father and son " system have given according to this diagram? Sea conditions, losses in the shafting, reduction gear and Vulcan couplings taken into consideration, a maximum power of 900 h.p. is assumed to be available at the propeller. From this comes the representative point S and the torque of 32,000 lb./ft. (4,440 kg./m.). The " father " motor of 750 h.p. alone would have operated with a torque of 24,000 lb./ft. (3,330 kg./m.), and the " son " motor of 250 h.p. with a torque of 8,000 lb./ft. (1,110 kg./m.). Assuming a trawling speed of 3.5 knots, the results would be: Line DS—both " father and son ": cruising speed 14.1 knots (900 h.p.); propeller thrust when trawling 21,000 lb. (9,525 kg.) corresponding to 640 h.p.; Line GQ " father " alone: cruising speed 13.2 knots (610 h.p.); propeller thrust when trawling 15,650 lb. (7,100 kg.) corresponding to 420 h.p.: Line KL—" son " alone: cruising speed 8.55 knots (125 h.p.); propeller thrust when trawling 5,100 lb. (2,300 kg.) corresponding to 90 h.p. From these data emerge two important points:

Fig. 513. The multiple reduction gear during assembly

1. The " father " motor of 750 h.p. is only capable of developing $420 \times \frac{10}{9} = 466$ h.p. when trawling, because of the increased torque of the propeller;

2. Because of the fineness of the hull, the cruising speed with the " son " motor alone, although being 8.55 knots (125 h.p.) provides a water flow to the rudder which would be insufficient for steering in normal weather conditions.

What are the results of a multiple reduction gear with two motors of 500 h.p. each, or a total output of 1,000 h.p.? The Vulcan couplings are eliminated, but several disconnected gears are permanently rotating in the reduction gear. It may be assumed that one factor compensates the other and that 900 h.p. are available for the propeller when two motors operate together, and 450 h.p. with one motor alone. The results are: Line BS—two motors of 500 h.p.: cruising speed 14.1 knots (900 h.p.); propeller thrust when trawling 25,500 lb. (11,565 kg.) corresponding to 845 h.p. Line EP—one motor of 500 h.p.: cruising speed 12.35 knots (450 h.p.);

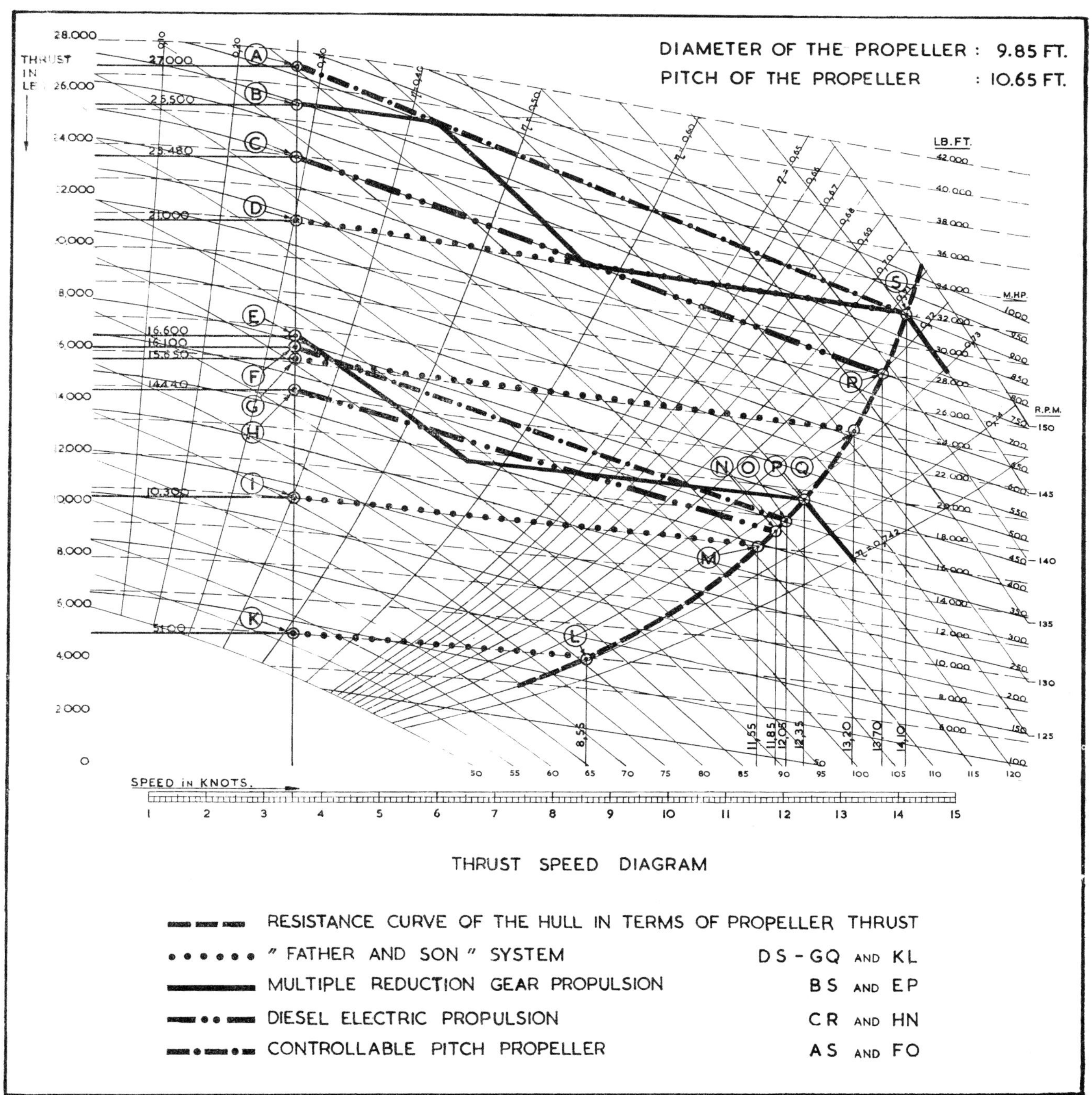

Fig. 514

propeller thrust when trawling 16,600 lb. (7,500 kg.) corresponding to 450 h.p.

The propeller thrust when trawling is thus 20 per cent. higher than in the " father and son " system when both motors are operating together, and each of the two 500 h.p. motors can provide alone a propeller thrust 5 per cent. higher than the " father " motor of 750 h.p. By choosing the three propeller speeds 145, 120 and 93 r.p.m., 12.35 knots cruising speed can be obtained by one motor only, instead of 13.20 with the " father " and 8.55 with the " son ". By using propeller speeds of 145, 115 and 93 r.p.m., the propeller thrust can be increased from 25,500 lb. (11,565 kg.) to 27,000 lb. (12,225 kg.) an increase of 28 per cent. compared to the " father and son " system. In this case the cruising speed with one motor decreases from 12.35 to 12 knots. The compromise in reduction ratio, which must be made when the gears are limited to three, does not have to be considered when a multiple reduction gear is applied to a single propulsion motor.

Diesel electric propulsion can also be studied in fig. 514. With the same propeller and diesels it is necessary, in order to consider comparable installations, to decrease the power disposable at the propeller by 15 per cent., to account for the electrical loss of efficiency through generators and electric motors. With two motors, the propeller can thus absorb 750 h.p., or 375 h.p. with one motor. The representing lines of the operation are CR and HN. In comparison with the multiple reduction gear system, each of the propeller thrusts during trawling is about 2,200 lb. (1,000 kg.) less, and each of the cruising speeds is about half a knot slower. These less favourable results would require a much more expensive installation.

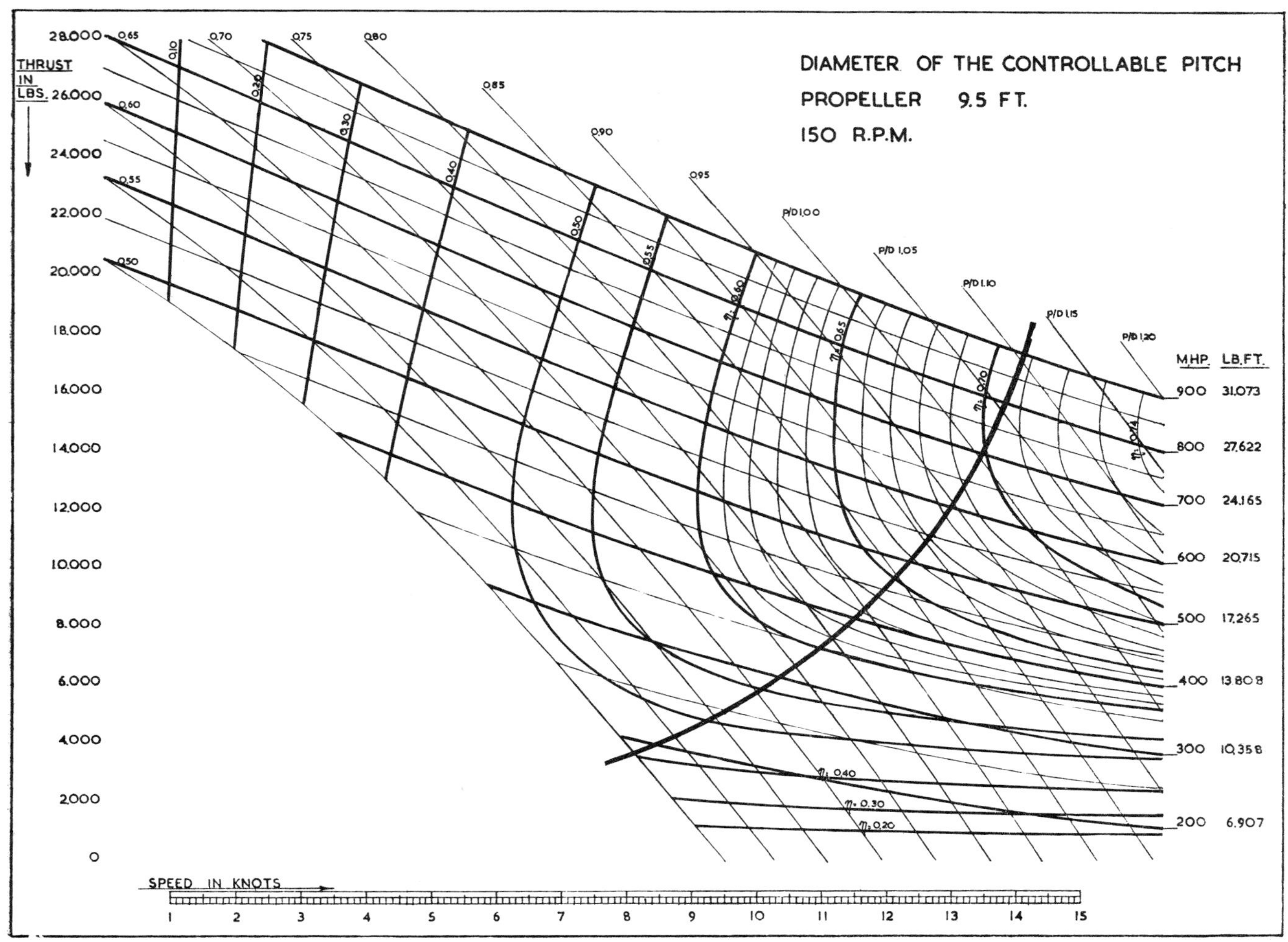

Fig. 515

With a controllable pitch propeller (*fig. 514*), the corresponding lines are AS with two motors and FO with one motor. These two are constant power and constant revolution lines. Here, also, 900 h.p. were assumed to the propeller with two motors, and 450 h.p. with one motor, the imperfect radial pitch distribution being supposed to be equivalent to mechanical losses, either of the Vulcan couplings or of the disconnected gears. As the risk of cavitation during trawling is smaller with a controllable pitch propeller, a diameter of 9.5 ft. (2.90 m.) instead of 9.85 ft. (3 m.) was adopted. It must be well understood that the lines AS and FO are really thrust-speed lines, but that the lines of power, torque, revolutions, and efficiency of fig. 514 are not valid in the presentation of this case. The revolutions were fixed at 150. The pitch/diameter proportion varies from 0.53 at point F to 1.1 at point S. Each thrust is calculated assuming the radial pitch distribution to be the usual distribution at that moment and not the perturbed distribution resulting from adjusting the blades in order to change pitch.

The diagram indicates that the results of a controllable pitch propeller are practically equivalent to those of an ordinary propeller connected to a multiple reduction gear. As a trawler operates in very distinct speed zones, the progressivity of the curves AS and FO are not, apparently, more advantageous than the broken lines BS and EP.

New lines of power, torque and efficiency for a controllable pitch propeller with a diameter of 9.5 ft. (2.90 m.) and 150 r.p.m. are shown in fig. 515. When comparing fig. 514 and 515, it is obvious that for a propeller with fixed blades the resistance curve of the hull (in terms of propeller thrust and speed of the ship) is running similar to the efficiency curves of the propeller. For the controllable pitch propeller, however, the resistance curve of the hull cuts the efficiency curve of the propeller nearly at a right angle. The result is that the fixed blade propeller has an efficiency of 0.715 at 14 knots and 0.733 at 12 knots, while the controllable pitch propeller has an efficiency of 0.705 at 14 knots and 0.64 at 12 knots.

In all ships where the maximum speed is obtained by two engines driving a single propeller, and cruising speed by the use of one engine only, the system having a fixed blade propeller will be hydrodynamically better than that using the controllable pitch propeller. On the other hand, in ships using a single engine with a controllable pitch propeller, the cruising speed will best be obtained if the blades are left at their position of maximum speed and the revolutions of the engine reduced. In other words, the controllable pitch propeller should be treated as a fixed blade propeller.

Regarding economical cruising speed, the controllable pitch propeller brings disadvantage in some cases, and no advantage in the other cases.

The real advantages of the controllable pitch propeller are to make the thrust perfectly progressive, to absorb the complete power when trawling and to simplify the problem of torsional vibrations.

These advantages are nevertheless limited:

1. In the *Belgian Skipper*, although there are several vibration systems, running is perfectly silent and there is no critical speed between 130 and 500 r.p.m.
2. The ability of the diesels to work, at revolutions down to one-third of their normal r.p.m., combined with three different reduction ratios, will meet the requirements of a fishing vessel in a very satisfactory and much less expensive manner than the controllable pitch propeller.

CONCLUSION

From the comparisons made for the *Belgian Skipper*, it can be concluded that the controllable pitch propeller with ordinary reduction gear, on the one hand, and the ordinary fixed blade propeller with multiple reduction gear, on the other, give similar results. The diesel electric drive and " father and son " system are definitely less efficient.

The " father and son " system and the multiple reduction gear system would have cost approximately the same, but the controllable pitch propeller and the diesel electric drive would both have increased the total cost of the trawler by 10 per cent.

ENGINES — DISCUSSION

Mr. J. Messiez-Poche (France): Although most fishing boat experts believe that the steam engine is becoming an obsolete form of propulsion for trawlers, it is coming into favour with the builders of tankers. There is always the possibility that some new development in the steam engine will bring it back into favour with fishing boat owners.

There seems to be agreement among experts that the controllable pitch propeller provides the best way of adjusting engine and propeller to a vessel. It is to be hoped that great care will be taken to make sure that efficiently-designed controllable-pitch propellers are used.

Too many engine designers think that their work is completed when an engine leaves the test bed. This is a mistake. An engine designer should make a close study of the vessel in which he is installing an engine so that he can avoid critical speeds, make sure that the engine beds are properly constructed, avoid unnecessary, and sometimes dangerous, auxiliary controls, and see that the propellers are properly designed. He should also be concerned with safe operation and maintenance of the engine. He should ensure that spare parts are available at reasonable prices and that there is no delay in delivering them.

It seems that the supercharged four-cycle engine is most suitable for fishing boats, but before it is installed a study should be made of the particular conditions in which the trawler concerned will operate.

Mr. Gould was most informative about the use of high-speed diesels in fishing boats, but while they may be excellent for the U.S.A. fishing industry, they may not be so useful if installed in European fishing boats. Certainly the initial cost of such an engine and the expense involved in maintenance and repairs tend to make it uneconomic from a European point of view.

Mr. J. A. van Aken has given some details about a controllable-pitch four-bladed propeller with the blades arranged in two different planes. It would have been helpful if he could have given figures to show the efficiency of this uncommon propeller arrangement, especially as other experts have said that the results of such an arrangement are disappointing.

Mr. Kolbeck dealt with modern diesel engine trawlers built in Germany and it would have been interesting if he could have shown the costs of the different engine arrangements he has studied.

In the paper on the free piston generator and gas turbine, presented by Mr. A. A. Normand, jr., the advantages of this system are shown. It would have been instructive, however, if the author had compared fuel consumption under three headings: (1) cruising; (2) trawling, and (3) homeward voyage with a full load.

Commander A. C. Hardy (U.K.): The engine papers may be grouped under four headings: (1) cycle of operations—should two-cycle or four-cycle engines be used? (2) speed of operations—should fast or slow engines be used? (3) method of operation—should diesels or semi-diesels be used? (4) method of drive—should the direct drive with a direct-coupled, slow-turn, direct reversible engine be used? Or a medium-speed engine driving through a reduction reverse gear? Or should two or more engines be geared to a single shaft? Should various combinations of this method be employed, using a controllable-pitch propeller? Or should the free piston gas generator, in combination with a reciprocating engine or turbine, be used?

It was not the business of the Congress to reach an " omnibus " conclusion but to exchange views and ideas so that the problem may be seen from its many angles.

Mr. Hepton's paper deserved careful study, particularly where he accuses diesel builders of overrating their engines. He suggested that many engines sold for a certain horse-power have, in practice, not been able to develop that power in arduous sea conditions.

The paper by Mr. Normand dealt with a matter of importance in the future—the free piston gas generator. A prominent British firm of marine engineers, which has only built steam engines in the past, has entered into agreement to build free piston gas generators. This system seems to have some of the advantages of diesel electric drive and the great advantage of flexibility. It can also be used for trawl winch propulsion.

The papers by Mr. Ziener and Dr. Setna showed some of the difficulties experienced by users in Chile and India, where there are special problems in the use of diesels in fishing boats. Mr. Ziener stressed the point that good servicing is essential.

The paper by Mr. Gould on the high-speed engine is again a pointer to the future and to the differences of opinion which exist. In Europe there is a tendency still to regard the high-speed engine with suspicion but in America it is regarded with great favour.

The " father and son " and the diesel electric drive methods have much to recommend them for trawlers provided that: (*a*) they are not too complicated, and (*b*) that the ordinary marine engineer is made to appreciate them.

Mr. Dussardier, in his paper, expressed the opinion that a good diesel manufacturer should not stick to a rigid form but should be willing always to adapt his design to changing engine requirements.

The papers by Messrs. Stokke and Anderson, dealing with semi-diesels, indicated that the idea of a hot-bulb engine being an anachronism is not supported by facts. Numerous fishermen in Scandinavian countries find it reliable and that is most important. Both papers dealing with semi-diesels supported the belief that there is still need for this type.

How far the controllable-pitch propeller is to be adopted for fishing vessels is bound up with questions of manoeuvrability. Many people say the controllable-pitch propeller should not be regarded from the point of view of manoeuvrability in trawlers but as an instrument in which pitch can be adapted to the duty which the vessel is carrying out.

It is significant that no papers dealing with steam engines were submitted. Although there are still steam trawlers being built there is no doubt that the diesel has taken the place of the steam engine in the fishing fleets. Indeed, Mr. F. Parkes, a trawler owner of great ability and distinction, sounded the death knell of steam in the Boat Type discussion by pointing out the efficiency and economic operation of diesel as compared with steam trawlers.

Mr. Frederick Parkes (U.K.): Because of the investment required, the wages and salaries to be paid, and the increase in price of fuel, engines, fishing gear, technical equipment, etc., the costs are too high to risk mechanical breakdowns on fishing trips. So the trawler owner puts dependability first and foremost because without it he cannot operate profitably. A saving of 10 per cent. on fuel may seem very important to the engine builder but, in one year, that would only represent in value about one-quarter of the net loss caused by a mechanical breakdown which results in a broken voyage or a false call at a port. Three days' delay in port would also largely absorb a year's saving of the 10 per cent. economy in fuel consumption. The ship owner wants reliability before anything else, even before the most modern equipment. In studying a ship's balance sheet over several years, he will pay greater attention to the number of days lost in carrying out repairs, or in making false calls in foreign or home ports, than to fuel or oil consumption. If, for example, after five years he can see that operation of a trawler has consisted of mechanical breakdowns and sailing has been delayed because of repairs, he will be all the more ready to spend money on maintenance and spare parts. This is why the ship's owner requires the manufacturer to give him engines made to ample and reliable calculations so that there is a very large security margin latent in all parts. This is the strength of the steam engine and it should be the strength of the modern diesel.

Mr. Parkes referred to an experience nearly 40 years ago when he installed diesels in two new vessels. One was a single cylinder engine and the vibration it caused in the ship was so violent that it nearly shook people off their feet. That ship was an utter failure. In the second vessel he installed a four-cylinder engine. It sometimes took a day to get the engine started. The vessel did actually get to sea once or twice but, after his experience with these two ships, Mr. Parkes decided to have no more to do with diesels. A few years ago, however, his son pointed out that diesels were being accepted in the fishing industry so he agreed to experiment again. The first vessel fitted with diesels spent about six months of her first year in port, but it was decided to persevere and about twelve or fourteen different types of diesels from the U.K., U.S.A., Switzerland, France, Holland, Belgium and Germany were tested. As a result, it now appears that diesel trawlers pay and that the steam trawler will be left in the background before long. The economy in running a large type trawler equipped with a diesel is considerable when compared with the same type of trawler equipped with a steam engine.

The National Physical Laboratory at Teddington, London, has been of great assistance in advising on the power of the engine needed for vessels and on propeller design. New ideas are always being tried in connection with diesels. Recently, for instance, an American invention, an air controlled clutch, has been used. The experiment has already cost several thousand pounds because of broken trips but it is going on.

British trawler owners are handicapped by having to train their engineers who usually start as greasers and work up to the position of chief engineer. An experiment which has been successful is to send the engineers to the manufacturer's works for a month or two so that they may see the engine in the course of construction. They come back with a good working knowledge of the machine and they make better engineers because of their experience.

Mr. E. G. Bergius (U.K.): The development of the Scottish fishing fleet provided a good example of how to help the under-developed fishing fleets of the world. The policy of that development could be summed up as helping the fishermen to help themselves.

The mechanization of the Scottish fleet has taken place during the last 50 years and medium-speed engines of 600 to 1,000 r.p.m. have been used. Reliability, simplicity and accessibility are the main considerations in the choice of engine because a fishing boat is of no use while lying idle. That the engines selected by the Scottish fishermen are simple and reliable is emphasized by the fact that the inshore fleets of Scotland, composed of boats up to 80 ft. (24.4 m.) in length, go to sea without engineers. Simplicity and accessibility of an engine contribute to reliability. For instance, it should always be possible for fishermen to carry out minor repairs. If a piston seizes on a multiple cylinder engine a fisherman should be able to remove that piston within five minutes and run on the remaining cylinders. That is the essence of simplicity, accessibility, reliability and service.

Mr. T. S. Leach (U.K.): The United Kingdom Government instituted in 1946 a scheme of financial assistance for fishermen. It is operated by means of grants and loans for new boats, insurance and gear, for vessels up to 70 ft. (21.3 m.). These are inshore fishing craft and are usually operated by skipper-owners whose capital resources are small, so that they cannot take the risk of buying an untested engine. For these fishermen the important considerations in the selection of an engine are: (1) the first essential is that the engine is absolutely reliable and easily accessible; (2) it should be simple and basically designed for use in a fishing boat; (3) it should be a slow-speed engine, capable of operating a small winch or line-hauler. Experience of high-speed engines, involving use of reduction gears, has not been good; (4) the lubrication and fuel systems must function under conditions of slow-running for many hours, particularly in the case of line-fishing vessels; (5) most inshore fishermen operate from small ports where there are no large engineering facilities. Perhaps the nearest source of help is the local garage. This is another reason why the engine must be absolutely reliable and very simple in operation, so that it can be serviced either by the fisherman himself or by local mechanics; (6) neither skipper-owners nor, indeed, the fishing industry generally can afford to be used as guinea-pigs for experiments with untried engines; (7) engine manufacturers should give a guarantee that the type of engine they are offering has been subject to tests at sea under normal fishing conditions.

Mr. William P. Miller (U.K.): The Scottish fishing fleet has been largely reconstructed since the end of World War II. The general result has been good, but some fishermen have suffered because they have installed unsuitable engines. This is unfortunate, not only from the fisherman's point of view, but also from that of the independent manufacturer. It does not make very pleasant reading to see that some million-pound engine manufacturing concern is paying a big dividend when fishermen are in distress because they bought unsuitable engines from that company. This is a vital matter which should receive the careful and serious attention of all engine manufacturers. It is most unfair that fishermen should be penalised for carrying out experiments for large engineering concerns, and it would be a good move if independent manufacturers issued some form of insurance policy against selling unsuitable engines to fishermen.

Mr. Jan Olof Traung (FAO): It is interesting to note that, so far as guarantees to fishermen are concerned, FAO has recommended one member nation to stipulate in a fisheries loan regulation that engine manufacturers should give a specific guarantee of engine performance. The usual sort of guarantee given with an engine is more in the nature of an exemption from guarantee. It is so written that normally it is not in accordance with the business laws of the country.

Mr. John Tyrrell (Ireland): Engine manufacturers should pay more attention to the external design of the engine. For stern installation, the engine base should be narrow to fit in the run of the ship and, if possible, it should be tapered. The flywheel is better on the forward end of the engine, which should extend the least possible distance below the shaft centreline. The foundation should be as high as possible in order to allow a reasonable size of bearer aft.

These remarks apply only to engines installed aft. Where engines are installed forward a reverse arrangement is more suitable.

Engine manufacturers should supply all the installation material with the engine. To a large extent successful running will depend on the quality and suitability of this. Tanks may be provided by the shipyard to suit the hull, but the engine manufacturers should give instructions to ensure that they are properly fitted with sumps, etc. The stern tube outer flanges should be of substantial size and have enough fastenings to resist propeller vibration. Simple and clear installation drawings, showing the necessary foundation details, should be supplied with the engine.

Mr. George C. Nickum (U.S.A.): An analysis and comparison was made recently of small high-speed light-weight engines ranging from 200 to 300 h.p. and it was discovered that, up to a cruising range of 25 hours, the high-speed diesel, the high-speed petrol engine and the modern gas turbine were within about 15 per cent. in total weight of plant and fuel. Over the 25 hours the diesel rapidly drew ahead and the gas turbine rapidly dropped behind. But the comparison indicated that the time is not far distant when the gas turbine will have to be considered for installation in fishing boats.

Many of the papers on engines are excellent, but some are poor. This is particularly true of papers presented by the manufacturers themselves, chiefly because they try to sell engines in the same way as other manufacturers sell bath tubs and automobiles. This may be the best way to sell diesels, but it is not the best way to treat the naval architect. He is an intelligent human being and can digest a little heavier fare than is usually sent to him by engine manufacturers. Naval architects need more data on engines, and there is need for diesel manufacturers and naval architects to work more closely together.

Mr. Andersen's paper on semi-diesels is particularly interesting when read in connection with Mr. Ziener's dealing with engine problems encountered in Chile. American manufacturers evidently have not supplied Chileans with properly-crated engines nor with sufficient spare parts.

Mr. Hepton has discussed candidly and honestly the problems of trawler owners and Mr. Gould has provided a good paper on high speed diesels although he could have included more technical information and comments on, for example, the b.m.e.p. for two-cycle high-speed engines or whether there is a standard limiting exhaust temperature.

Mr. Normand's paper would have been improved if he had given factual data on the results the French are getting in their test programmes for gas turbines, particularly the free piston generator.

Mr. E. R. Gueroult (France): Naval architects usually have to work out the pressures, piston speeds, and the quantities of heat going through a diesel, to find out for themselves the best characteristics of the engine. Manufacturers should provide complete and accurate information.

Much of the expense incurred in operating fishing boats has to do with the diesel. This is usually the result of overloading the engine when running at low speed of advance. Diesel manufacturers usually blame the propeller, especially the bad habit of designing it on a basis of compromise between the trawling and free route speeds. Most of the difficulty arises from ignorance of the real torque which can be accepted at reduced revolutions. Diesel manufacturers never provide this vital information, sometimes because they have not taken the trouble to find out what the torque should be.

As far back as 1927, when the first diesel trawlers were being built for fishing off Newfoundland, owners asked for the same reliability in diesels as they obtained from steam engines. An effort was made to have diesels running at revolutions near to those of the normal steam engine. Tank tests to establish the best optimum revolutions for towing showed there was no necessity to go as low as, say, 100 r.p.m., which was usual with steam engines. It was found possible to use 150 to 200 r.p.m. without great loss of efficiency, and since that time diesels have been running at these revolutions. The direct drive was also adopted. Trawlers with such slow-running main diesels, and with medium-speed auxiliary engines, gave satisfactory results. It is probable that in future the medium-speed engine, in conjunction with an electric drive, will be used in the big trawlers. For small powers it is impossible to avoid using the high-speed diesel. It lacks flexibility and, in addition, French naval architects are disinclined to drive the propeller direct from high-speed diesels. Owners are always asking for very low revolutions to be able to reduce the speed of the boat in bad weather, and even 200 r.p.m. engines are not satisfactory in this respect. In the more recently designed trawlers, owners have asked for a small electric motor to be coupled to the shaft, to give a very slow speed. This is related to the safety of the ship and is, of course, a first step towards the diesel electric.

The French owners are fortunate in having good engineers trained by the Navy and in commercial ships. These men do much to maintain the engines and, consequently, help to make the boats profitable in operation.

Mr. Howard I. Chapelle (U.S.A.): Engine manufacturers should provide naval architects with information in a compact form. Some engine makers provide only the h.p. and r.p.m. curves but not the torque, while others show h.p. curves, fuel consumption and torque. No two manufacturers seem to provide fuel consumption curves on the same basis. An instance of the uncertainty of the information is found in the case of a small trawler where the expected fuel consumption was calculated on the information provided. The owner had second thoughts, however, and as a margin of safety he added 500 gal. (420 imp. gal., 1900 l.). The vessel trimmed very badly at the stern because of the extra weight. Then, at the end of the first year, it was discovered there was an excess of tanks. Not only was the extra 500 gal. not required but the original calculations provided for an excess of fuel. Here was an engine consuming less fuel than had been calculated, but because no accurate information on fuel consumption had been provided, a lot of useless weight had been placed on board the ship and useful space taken up by unnecessary tanks.

There are no trained engineers on board small trawlers, which in itself creates an engineering problem, and it is unlikely that a training programme for engineers would function very well. The small trawler owner generally wishes to see the engine sealed up so that nobody can tamper with it. The time to make an overhaul is when the vessel comes into port and the work can be handled by men who have long experience of maintenance. It is not a job for a part-time mechanic. That was all right in the days of the old make-and-break petrol engine when a complete overhaul job could be done with a monkey wrench and a screwdriver. These are facts which must be accepted in dealing with engines to-day. The small diesels in U.S.A. trawlers have, by European standards, a very high r.p.m., a very high h.p. in relation to that used abroad, and they are of very light weight. The engines have been installed for economic reasons. What now is needed is a more extensive study of maintenance and of maintenance methods to be applied especially to the trawling industry.

Mr. Fredrik Dahllöf (Sweden): (1) Eighty per cent. of all fresh fish landed are brought in by small vessels; (2) big fishing vessels have turned out to be less profitable than anticipated and fishing companies are now building smaller fishing boats. The problem, then, is to find the engine most suitable for small- or medium-size fishing vessels.

Dependability is the factor most stressed by fishermen and, of course, the most dependable engine in the world is the steam engine, but it is being abandoned because of its poor fuel economy. The dependability of the steam engine lies in the simplicity of its design, and to achieve a similar dependability oil engine manufacturers must also rely on simplicity of design. In this respect the highest degree of simplicity is achieved in the two-cycle engine with crank-case scavenging and low combustion pressure. The low pressure makes it possible to employ simple and sturdy equipment for the injection of fuel into the combustion chamber. It has been said that the crank-case scavenging is poor but such a statement refers to older types of engines. In the modern engine this method of scavenging is now so efficient that the fuel economy is comparable with that of any engine of more complicated design. The only advantage to be obtained from a more complicated scavenging system is a slight reduction in the size of the engine. This might be important in some vessels but in small fishing boats the reduction is too small to be of any importance.

The two-cycle low-pressure oil engine with crank-case scavenging costs little to maintain and repairs can be made very quickly as its parts can be readily replaced. The engine is, therefore, well suited for installing in vessels up to 150 ft. (45.8 m.) in length.

Commander R. E. Pickett (U.S.A.): The small fishing vessel engine up to about 300 h.p. should be an internal combustion oil engine, preferably a port scavenging two-cycle type. The requirements are: (1) on long voyages the fishing vessel must not lose its propulsion or winch power although both may be reduced in extreme emergencies. The engine should be so constructed that a defective cylinder can easily be cut out without the engine ceasing to operate; or there should be a multi-engine installation; (2) the starting system should be so simple that it is proof against breakdown. This is one reason for the popularity among Gulf of Mexico fishermen of starting diesels with air-cooled petrol motors, an undesirable practice. A manually-operated hydraulic starting system for small and medium-size engines has recently been developed, and might be of general interest to fishermen; (3) a simple and reliable cooling system is necessary. A double bottom tank as a cooling water reservoir and heat exchanger combined might be used in steel constructed vessels; (4) cooling water and lubricating oil pumps should be installed separately and in duplicate to avoid failure of this important auxiliary equipment.

If a ship can retain her winch and propulsion power although reduced she can stay safely at sea for fishing and get safely back to port.

Mr. C. H. Bradbury (U.K.): The reference by Mr. Hepton to the derating of engines by 15 to 20 per cent. for continuous operation must not pass unchallenged. Derating from what power? The reputable manufacturer designs his engine to a specification which he knows he can fulfil and, for continuous running, the rating offered is well below that on which the engine operates on a shore installation. Because less experienced manufacturers do not follow this practice, that is no reason for condemning the reputable. Exhaust temperature quoted by itself means nothing at all. Some engines are perfectly safe at 800 deg. F. (430 deg. C.), while others, particularly certain blown two-cycles, are unsafe at 750 deg. F. (400 deg. C.). Exhaust temperature must, therefore, relate to test conditions and nothing more.

Mr. H. C. Hanson (U.S.A.): It is unfortunate that manufacturers tend to oversell their engines as to horsepower. They will sell a customer a 300 h.p. engine, and will then insist on the use of a propeller absorbing only 200 h.p. For example, in 1952 two vessels were built to the same design by the same shipyard and were installed with engines of the same power. A propeller with, say, a 45 in. (1,140 mm.) pitch was specified by the naval architect, but the engine manufacturer gave an order to the propeller manufacturer for a 42 in. (1,070 mm.) propeller. When it came to the trials the manufacturer's representative refused to go at full speed because he said the normal speed was the high speed.

The boats made only 9½ knots whereas they had been designed to make 10½. Later, when the two vessels were fishing together, both started for home at the same time when one of them hit a log. From that moment the vessel headed rapidly away from her companion because the log had increased the pitch, which proved the point that a 45 in. and not a 42 in. pitch was needed. The propellers of the two boats were then adjusted to the right pitch and the desired speed was achieved. This example should prove to engine manufacturers that they should allow those with experience of vessel design to specify the propellers. But if it is felt that they do not dare run their engines at full output, diesel manufacturers should specify lower h.p.

He believed one revolutionary improvement would be to place diesels in the forepart and an electric propulsion motor in the stern. This would do away with shaft-alley, intermediate bearings and shafts, and all the wiring would be placed in the wings. It would be a very practical and useful design in his opinion.

Mr. W. Zwolsman (Netherlands): Mr. Ziener's account of repairing and maintaining engines in Chile shows that manufacturers have not properly studied the market and seem indifferent to the considerable sales possibilities there. They should establish a service to supply spares and maintain engines in running condition, otherwise they will create an atmosphere of hostility and suspicion towards internal combustion engines.

In the fuel oil specification mentioned by Mr. Ziener the Conradson carbon value is missing. Contrary to Mr. Ziener's opinion, this fuel is perfectly suitable for high-speed diesels but it might be advisable to use a lubricating oil conforming to the American Petroleum Institute specification, such as Army spec. Mil-0-2104 or 2-104B, supplements 1 and 2. If the gum content is translated into Conradson carbon value, 188 mgs. per litre gives a figure of roughly 0.22 per cent. This is acceptable although the British standard specification No. 209 (1947) gives 0.1 per cent. as maximum.

Mr. C. E. Dietle (U.S.A.): There is undoubtedly a need for better collaboration between engine manufacturers and naval architects. But many of the troubles that have arisen cannot be blamed entirely on the engines. The human element often plays a very important part. For example, his firm at one time had made an engine which was greatly favoured by fishermen. It was the soul of reliability but it was not very efficient. That engine is not being built to-day, and will not be, because it has no place in the market due to its great weight and high fuel consumption. Probably the reason why some countries still find semi-diesels successful is because they have not felt the impact of certain economic factors which have been effective in the United States. When they do feel that impact the fishermen will no longer use the semi-diesel.

Mr. Georg Bruce (Sweden): There should be better co-operation between the boat-builder and the engine manufacturer to improve the design and construction of seatings for engines in small- and medium-size fishing boats. A sturdily-built seating is essential for: (*a*) the motor, to prevent damage to the crankshaft and the bearings; (*b*) for the boat, to prevent vibrations opening up the seams of the hull; (*c*) for the fishermen's comfort, as they have to spend about half their lives aboard boats and do not want to be constantly shaken by the vibration.

Mr. W. C. Gould (U.S.A.): It would be very useful if the naval architects would give engine manufacturers a definitive list of information wanted because there seems to be a divergence of opinion among architects as to what is needed. Unfortunately, 75 per cent. of the engines are sold directly to customers who are not technically qualified to make an analysis. Manufacturers are forced to explain engines in a manner that can be understood by fishermen, but it does not seem so certain that this effort is 100 per cent. successful. It would be desirable to have one set of literature for the fishermen and one set of power curves for the naval architects. If, in the case of U.S.A. manufacturers, the engines were being installed only in trawlers on the north-east coast or in tug boats on the Hudson, the matter would be simple. But the engines are used in various types of boats and areas in which power requirements are different.

A comparatively simple method has been developed of testing whether or not a propeller is right for the engine. It is a rule of thumb method, but it does give a quick check. The trawler or tug-boat is secured to a wharf to resemble the worst possible towing condition and the engine is run to its governed speed at maximum injection. If it runs 75 to 100 more r.p.m. than the normal working speed, it is considered free from risks of overloading. In other words, if it is to run normally at 1,600 r.p.m., in the test it should run at 1,675 to 1,700 r.p.m. and as the trawler or tug-boat skipper generally uses only two throttle positions, shut down or wide open, this means that trawling or towing operations can be carried out without overloading the engine. In the case of other vessels, where higher powers are used, the test is to run about 200 r.p.m. more than the average cruising r.p.m. would be. Tomalin (1953) suggested that engine manufacturers should use, as a continuous duty rating, the horse-power obtained at the minimum specific fuel consumption. Generally speaking, it will be achieved in those conditions where engines are putting out maximum torque.

His firm had an amusing experience in producing fuel consumption curves based on propeller load curves, and carefully worked out with regard to r.p.m. and speeds. Fuel consumption curves crossed the propeller load curves and gave the answer in total consumption per hour. But this diagram had to be scrapped because the customers thought that the engine used too much fuel. The reason was that most customers do not run their engines at the h.p. they were sold at. For example, they are used to a so-called 130 h.p. engine consuming 5 to 6 gal./hr. (4 to 5 imp. gal.; 19 to 23 l.). Then when they replace the engine with one of 200 or 225 h.p., consumption is shown at 10 or 11 gal./hr. (8 to 9 imp. gal.; 38 to 42 l.). In this case the consumption only went up to about 8 gal./hr. (6.7 imp. gal.; 30 l.) with the bigger engine indicating that it was not used at full output. But as the information meant full output and was misunderstood, the firm had to revert to specific fuel curves based upon consumption per horsepower per hour.

In re-powering a boat it is often necessary to use a reduction gear of a ratio which is not suitable from a propeller efficiency point of view. This gear has to be put in because of the size of the shafting already in the vessel. If a higher reduction gear ratio was used to give a better propeller efficiency, it would be necessary to put in a new shafting block, new stern bearings and new shafting, which is far too costly. In the case of high-speed diesels the average cost of the engine installed, complete with controls and exhaust system, is from $42 to $55 (£15 to £20) per horsepower.

Preventative maintenance is an important factor. Owners of large fleets can, by careful plotting, find the optimum period for overhauling the engines. That period is not necessarily the maximum the engines will run between overhauls. The cost of overhauling and the use of spare parts over a long period will be less when the optimum period is used. His firm had started a preventative maintenance programme in which engineers meet a boat when she comes in and, for a very small charge, check over the engines.

High-speed engines can now be completely enclosed so as to sound-proof them, and also prevent the crews from interfering with them. Probably a better measure for determining intervals for overhauls is the quantity of fuel oil to be burned rather than the number of hours an engine should run. For example, there is an engine that burns between 15,000 and 20,000 gal. (12,500 to 17,000 imp. gal.; 57 to 76 cu. m.) of fuel before overhaul. How soon that overhaul is needed can be determined by how fast the fuel is burned. It is not so much the r.p.m. that wear out modern engines, but the heat put through them which is indicated by the fuel consumption.

His company set up training programmes long ago, including mobile schools, but they have not been used extensively by the fishing industry although any owner could send his men to a school free of charge. The only cost to him would be the man's time. The mobile schools had met with better success by moving into a locality for three days to teach the fishermen maintenance on engines. But this is a very costly service to operate and it is not possible to visit each territory as often as is desirable.

Mr. C. E. Dietle (U.S.A.): A high-speed diesel of 100 to 175 h.p. that is not perfectly adjusted and maintained will burn as much as 10 gal. (8.3 imp. gal., 38 l.) of fuel an hour while the boat is trawling. The average consumption may be about 8 gal. (6.7 imp. gal., 30 l.). Taking Mr. Gould's figure of burning 15,000 gal. (12,500 imp. gal., 57 cu. m.) of fuel as a measure of when the engine should be overhauled that would give, roughly, a running time of 1,500 to 2,000 hours. That suggests very frequent overhauls. Is that right?

Mr. W. C. Gould (U.S.A.): No. He was not talking about a 175 h.p. engine but one having a maximum rating of 130 h.p. burning about 5 or 6 gal. (4.1 to 5 imp. gal., 19 to 23 l.) per hour.

Mr. C. E. Dietle (U.S.A.): Many would find that this method would call for too frequent overhauls for many engines, and he did not want anyone to conclude that it established the maximum period for all engines. Operation and maintenance have a great influence on the maximum periods between overhauls.

Dr. Gian Guido Bordoli (Italy): The naval architect is lucky to find an intelligent owner like Mr. F. Parkes who has said that he puts his trust in the experts. While this may be true in deciding the dimensions and form of the hull, Mr. Parkes has his own ideas about engines and these impose limits on the naval architect.

The naval architect of experience can quickly decide the overall length of the vessel he is designing because he knows the weight and size of the equipment which must be installed. He can also decide on the length in relation to speed and choose the hull form, while from the known results of tank tests he can calculate the engine power needed. But when he leaves his own field of activity he meets with difficulties. Even if he is a mechanical engineer he will find trouble in selecting a suitable engine and may be carried away by his wish to install the most modern design. Weight is an important factor so he will probably choose a medium-speed supercharged diesel because supercharging increases an engine's power 1½ times. It also means that he can install a more powerful engine in less space, which makes it possible to reduce the size of the engine-room yet still retain sufficient space to allow easy access to the engine and the auxiliaries, an important aid to servicing and maintenance.

He next runs into difficulties over the choice of propeller. He will probably look for ways and means to increase the propeller efficiency having selected an engine, say, of 500 r.p.m. This means he must use a reduction gear. And he must also decide whether to have a fixed blade or controllable-pitch propeller which, again, will confront him with many problems, not only of efficiency but of cost and the owner's preferences.

When it comes to steering equipment, the naval architect may prefer the Kort nozzle. It is interesting to note that Mr. Zwolsman has reported in his paper on the excellent results obtained by fitting Kort nozzles in some trawlers built in Holland since World War II.

Given a free hand, the naval architect might be tempted to take the initiative and install a turbine fed by free-piston generators. But in all his work and decisions the architect must face the preconceived ideas and the usual disinclination of the owner to install new types of engines or systems which may cost more than conventional, established types, and have yet to be proved in actual operation. And even if the architect is lucky enough to work for an owner who will allow him to install the most modern engines and equipment, he is faced with the all-important problem of service and maintenance on board ship. There seem to be few trained engineers and mechanics among fishermen and the architect must bear this fact firmly in mind when deciding on any of the machinery to be installed.

Mr. L. Varriale (Italy): Experience gained over 15 years, during which more than 5,000 cylinders have been installed in fishing vessels, has proved that the supercharged engine is the most economical type for a trawler.

In Italy, as in the rest of the world, supercharging is still considered by many to be a bold innovation, although it has been known for more than 20 years.

This type of engine certainly works better if supercharged than it does with a natural air flow. A maximum combustion pressure of 870 lb./sq. in. (61 kg./sq. cm.) is maintained and the exhaust temperature is kept constant at 700 deg. F. (370 deg. C.). The 50 per cent. gain in power is due to an exhaust turbo-blower. Italian fishermen, despite their traditional conservatism, have been convinced by the performance of these supercharged engines.

The big factory ship could be much-improved functionally and economically by the use of a supercharged engine because any saving in space and weight per h.p. is of special importance in such a vessel. For example, the Portuguese factory ship *Allan Villiers*, which operates in the Newfoundland waters, is equipped with a six-cylinder, 14.5 in. (370 mm.) bore engine, direct reversible and supercharged by a turbo-blower. With a length of only 19 ft. (5.80 m.) and a weight of 35 tons, it develops 900 h.p. at a normal speed of 275 r.p.m. and 1,100 h.p. at maximum power.

In view of the excellent results obtained, it seems reasonable to install four-cycle supercharged engines of 1,000 h.p. in fishing vessels.

Supercharging the two-cycle engine is not advisable but the supercharged four-cycle engine can replace it up to the power limits at which the engines meet.

Mr. D. Coste (France): The gas turbine resolves the problem of adapting the engine to the propeller. In fact, there is no problem. The generator is independent of the speed of the propeller, and the turbine has an extremely flat curve of efficiency which adapts itself very easily to the needs of the propeller. In a system with free pistons and gas turbine, with direct transmission, installed in a train running between Paris and Valenciennes, the turbine is connected and relayed directly to the axle trees. It has now run more than 30,000 miles (50,000 km.). The efficiency of the installation is, of course, the product of the efficiency of the generator, the pipe system and the turbine.

The efficiency of the generator—the gas efficiency—is a minimum of 43 per cent. In the near future this should reach 44 and even 45 per cent.

The efficiency of the turbine and the reduction gear must be taken into account to find the overall efficiency. The efficiency of the turbine depends, of course, on the unit power. For a turbine of 1,000 h.p. there is a different efficiency than with a turbine of 8,000 or 10,000 h.p. Various makes of turbines give different performances. One of the best results obtained was that of the Rateau turbine—85 or 86 per cent. output—which is remarkable because the average figure for engines of this power is 84 per cent. For the 8,000 h.p. group the figures of 86 or 87 per cent. are quite normal. The reduction gear has an efficiency of 98 per cent. for a simple reduction and 96 per cent. in the case of double reduction.

Engines installed in Cuba and in Detroit and Chicago have a consumption of 0.404 lb./h.p./hr. (0.183 kg.) with a turbine of only 80.5 per cent. efficiency. One can conclude that on taking 43 per cent. efficiency of the generator, 84 per cent. of the gas turbine-reduction gear from 1,000 to 2,000 h.p. and an efficiency of 98 to 98.5 per cent. for the pipe system, the specific consumption is 0.36 lb./h.p./hr. (0.175 kg.). This is comparable to that of a relatively fast diesel, with this difference: the generator is a machine which is extremely well adapted for combustion of heavy fuel. The Renault engine, which is in service, works normally with light fuel, which does not require reheating and is fluid at an ordinary temperature. The 1,000 h.p. engines installed by the Gafsa Co. in Tunisia has been in service for one year, using fuel No. 1.

In big installations the consumption would fall below 0.375 lb./h.p./hr. (0.170 kg.).

Mr. W. C. Gould (U.S.A.): There are two major problems connected with gas turbines: (1) the gas temperatures are limited at present to about 1,450 deg. F. (790 deg. C.); (2) the compression ratios of the blowers and superchargers are now about 8 to 1. When they are raised to about 10 to 1 the blades will be able to run at about 2,100 deg. F. (1,150 deg. C.) and it will then be possible to get specific fuel consumption of less than .4 lb./h.p./hr. (.18 kg.).

The temperature problem is the most difficult to overcome. One way of doing it is to use an internal combustion engine as a burner and then employ the exhaust gases from that engine in a gas turbine geared to the propeller shaft. He had particular experience of an engine with a volume of 1,710 cu. in. (28 l.), a compression ratio of 7 to 1, and a two-stage mechanically-driven air blower. The mercury pressure was raised to 110 in. (2.8 m.). Working at b.m.e.p. in the neighbourhood of 37.5 lb./sq. in. (2.6 kg./sq. cm.), the exhaust ran at 50 in. (1.27 m.) of mercury back pressure into a small turbine, which then put out more than 700 h.p. at 2,000 r.p.m. On 90 in. (2.29 m.) of mercury back pressure the turbine put out 750 h.p. at 2,800 r.p.m., and on a specific fuel consumption of .35 lb./h.p./hr. (0.159 kg.). On the test bed the actual consumption was .33 lb. (0.150 kg.). It seems that, to make progress with a gas turbine, an internal combustion engine, such as a free piston generator, should be used as a burner.

Mr. P. H. Hylton (U.K.): The gas turbine is not ready yet for the fishing industry but Mr. Normand's paper on a "partial" system is a pointer to the future. The system is "partial" because it involves reciprocating parts working under similar thermal, lubrication and fuel injection conditions as in conventional diesel cylinders. Presumably it needs the same maintenance. If the gas turbine advantages are to be fully exploited all parts must be rotative. To compensate for high fuel consumption there must be longer periods between overhauls, and the costs must be low. This can only be achieved by a simple machine, preferably having not more than two shafts, and having no reciprocating parts. There is a gas turbine of 1,000/1,250 h.p. in quantity production in Britain. It is eminently suitable for a trawler in every respect except for its fuel consumption. Related to the power available at the output coupling, fuel consumption is equal to or slightly better than that of the steam turbine of similar power. Large quantities of steam can be raised from the residual heat and exhaust gases and more of this must be used to bring the overall plant efficiency up to 70 per cent. (for a diesel with an exhaust gas boiler it is 49 per cent.). The most highly stressed blades have a maximum life of 100,0C0 hours. The engine needs to be inspected every 2,500 hours and dismantled and overhauled every 10,000 hours. The turbines, compressor, and reduction gear can be inspected in two hours without disturbing any ducting. A simple single fuel burner is used and the engine can be started instantly so that no warming or attendant standby charges are incurred.

Mr. Ivar Stokke (Norway): Commander A. C. Hardy is right in stating that high speed two-cycle diesels may not be economical engines to install in European fishing boats. No doubt the engines Mr. William P. Miller refers to were over-rated (too high b.m.e.p.) and it is a fact that if two-cycle engines run too fast and with too high mean effective pressure, they wear out too quickly. Mr. Stokke disagreed with Mr. Chapelle about the need for torque curves of the engines when designing a fishing boat. Such curves are only necessary for tractors and automobile engines. Instead, he would like the manufacturers to give curves of b.m.e.p. He agreed with Mr. C. H. Bradbury that a blown two-cycle diesel with 750 deg. F. (400 deg. C.) exhaust gas temperature was overloaded.

Mr. Stokke placed no trust in free piston gas generators with gas turbines. Such machinery is too complicated, has too little reliability and it is too difficult to regulate to make it suitable for the propulsion of deep sea trawlers. He was of the opinion that the fully-scavenged two-cycle moderately slow-running diesel, working with either controllable pitch propellers or with fixed blade propellers, was the most reliable and economical engine for larger fishing boats and deep sea trawlers. The super-charged four-cycle diesel directly coupled to the propeller shaft is also a very useful engine for such vessels.

COOLING

Mr. W. Zwolsman (Netherlands): The difficulty, reported by Dr. Setna, concerning the maintenance of engines in the Bombay fishing fleet seems to arise chiefly because they are cooled by salt water. This means that there is salt precipitation and severe corrosion must occur if the engines are at a standstill for a considerable time. To overcome this problem a Dutch firm has been using a simple, fresh-water cooling system specially suitable for small fishing vessels. This system (fig. 516) has been used successfully in Indonesia, and on African rivers, where vessels operate in dirty water which contains much sand. The equipment costs little more than the normal type because the mudbox, water inlet, and weed strainer have been omitted.

Mr. C. H. Bradbury (U.K.): The presence of sand in the water leads to considerable trouble with water pumps and cylinder jackets in water-cooled engines used for inshore fishing in England, but the skipper-owners wish to keep their capital charges to a minimum and are not willing to pay for heat exchangers, etc. They are now turning to air-cooled engines and a number have already been installed in boats. What has impressed the skipper-owner is that the air-cooled engine in the smaller sizes is not only more reliable than the water-cooled type, but is cheaper in the first cost and just as cheap to operate.

a certain extent influenced by the installation of air-cooled diesels. Comparatively large quantities of hot cooling air must be abducted; if it is short-circuited the engine will be overheated.

Air-cooling can only be applied to cylinders of from 92 to 120 cu. in. (1.5 to 2 l.). The engines reach their normal running temperature far quicker than the water-cooled type and they also have high cylinder wall temperatures at partial loads. This means that the wear caused by chemical action, the so-called inner corrosion, is substantially reduced and the overall wear is less than that of a fresh-water cooled engine.

A firm in Cologne has produced about 250,000 cylinders since World War II. They have been used satisfactorily for marine purposes. The engines have a combustion system—swirl chamber and single-hole nozzle—which is not influenced

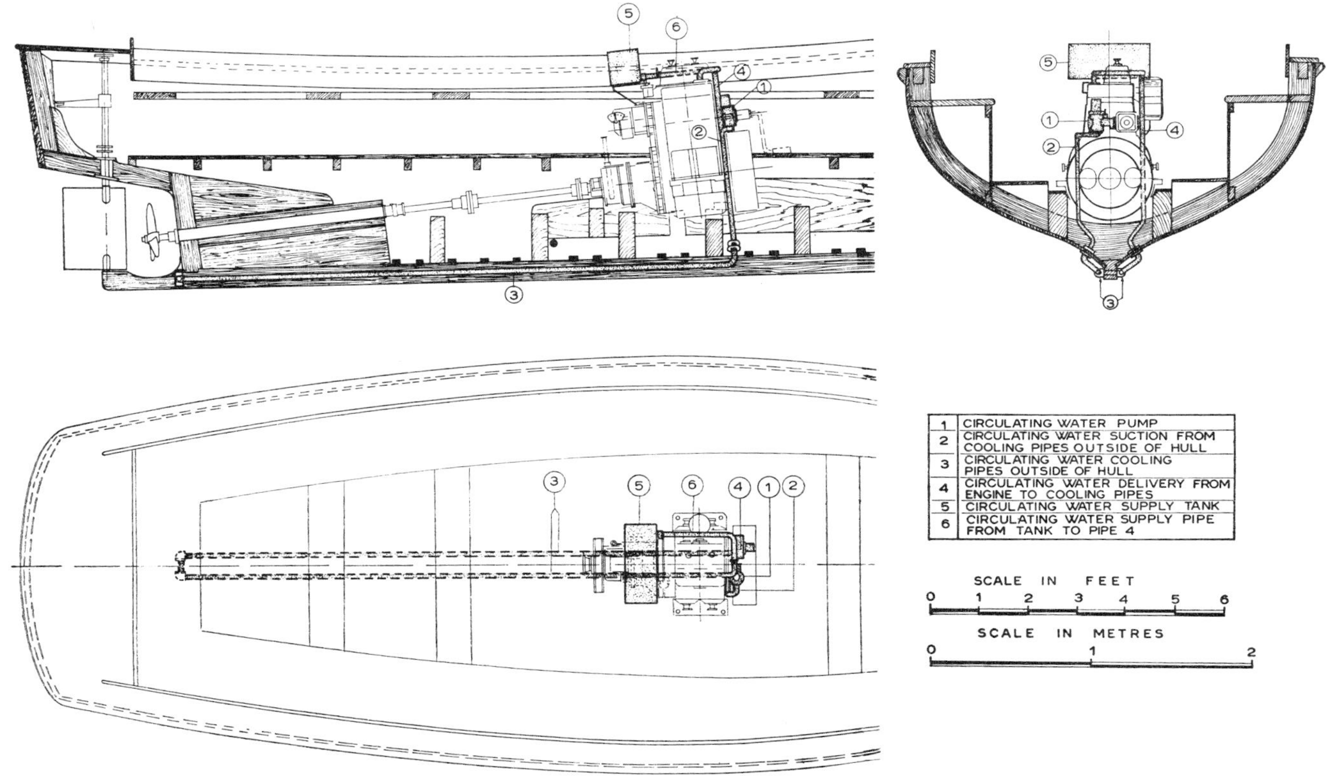

Fig. 516

Mr. Pierre Bochet (France): Some of the factors that affect marine engines are: (1) the galvanic effects of sea-water used for cooling; (2) the effects of condensation of steam in the crank cases; (3) the effects of corrosion by sprays or even by salt-laden air; (4) the variations of the temperature of the cooling water; (5) the absence of air-cooling around the lower crank cases; (6) the need to cool all elements of the engine to avoid excessive heating of the atmosphere nearby and injuries to the crew by burning.

Mr. Johann Ertl (Germany): Air-cooled engines are being built by several European factories. They range from 6 to about 150 h.p. Their installation is simple; there are no difficulties from cooling water pumps or from breakage of cooling water pipes because of vibration, etc. There is no danger from frost or corrosion. The design of the boat is to

by differences in quality of fuel. For example, they can be run on a mixture of petrol and 5 to 10 per cent. lubricating oil, as well as on lignite, tar oil, crude oil and paraffin. With any of these fuels the rated output is maintained on clean exhaust and without altering the injection timing.

The engines are made with an output up to 150 h.p. In engines of higher power, corrosion by cooling water can be largely reduced by using fresh water cooling in a closed circuit.

TWO-CYCLE VERSUS FOUR-CYCLE

Mr. E. R. Gueroult (France): French experience with two-cycle and four-cycle engines has revealed very little trouble with the four-cycle type, but few of the two-cycle diesels have been satisfactory in big trawlers. However, the slow-running four-cycle will not be built in future because it is very expensive and there is little demand for it. This means that the two-cycle diesel will have to be adapted to the needs of trawlers. One trouble is that, when trawling, the two-cycle diesel does not get enough scavenging air. Diesel manufacturers should provide a special device, such as a pump or blower, to give sufficient air during trawling. So far, manufacturers have treated the suggestion politely but have done nothing about it.

Mr. I. Bromfield (U.S.A.): Naval architects are basically familiar with the theory of diesels but they do not know the practical side. For example, it is important to know when, talking about two-cycle diesels, that there are two types. There is the old crankcase scavenging type which has its limits. It depends on the pressure in the crankcase for the volume of air delivered to the combustion chamber. It is important that the pressure in the crankcase is kept at its maximum, but due to the wear of the piston rings and faulty replacement of gaskets after overhauling, much pressure is lost. This results in poor performance.

On the other hand, in the two-cycle scavenging pump type of engine a separate pump is used for supplying air to the combustion chamber. This results in a considerably greater output which is not affected by loose bearings and worn gaskets.

Mr. L. Varriale (Italy): As M. Dussardier has shown, two- and four-cycle engines are very suitable for trawlers and Italian experience suggests more use could be made of the four-cycle engine. The following engines are in use on many Italian and foreign trawlers of all sizes: trunk piston, four-cycle, bore 8.5 in. (215 mm.), 10.5 in. (265 mm.) and 14.5 in. (370 mm.) respectively, available as three-, four-, six-, seven- and nine-cylinder engines from 100 to 1,500 h.p.

About 80 per cent. of the trawlers operating in the Tyrrhenian and the Adriatic Sea are driven by these engines with reverse gear and direct drive. They usually have three to four cylinders and are of the 100 to 200 h.p. and 375 r.p.m. group.

Mr. C. E. Dietle (U.S.A.): Some 25 years ago his firm built only four-cycle engines but, after a survey, they had realized that both two- and four-cycle diesels would have a place in the world. They decided to concentrate on the two-cycle type because it was felt that this engine would be the most reliable and would be lowest in cost. He was glad to know that engineers in England were following the U.S.A. lead and developing a two-cycle opposed piston engine. It seemed to him that the two-cycle opposed piston diesel was the engine that the fishing industry needed—reliable, low in cost, and taking up minimum space in the boat.

It is noted in connection with four-cycle engines that the exhaust valve cage is beoming a thing of the past now that triple charging has been introduced. Apparently the valve area no longer permits the use of cages. They were good from a maintenance point of view because they could be so easily removed. The valves could be ground and the new cage put in with the valves. It seems that in the interests of getting higher power into smaller space the designers of four-cycle engines have gone a step backward.

Mr. Brownlow's comment about four-cycle engines shows them to be a formidable competitor but he still felt the two-cycle engine is best.

Mr. J. Messiez-Poche (France): Members of the Congress seem to agree generally that four-cycle engines are able to cover the whole range of power needed in trawler engines. The simplicity, easier drive and upkeep are in the favour of two-cycle engines, but for low power it seems necessary to develop fast two-cycle engines, which means that scavenging would be restricted. Consequently, there is a minimum power below which the two-cycle engine does not seem to operate efficiently.

HIGH, MEDIUM, LOW SPEEDS

Mr. A. L. Gravenor (U.K.): His firm has been manufacturing internal combustion marine engines up to 30 h.p. for more than 50 years and it has been a cardinal principle that the engines should be slow-running, fitted with the minimum of accessories, and that all parts should be more than adequate for the maximum load they may be called upon to carry. The object has been to reduce maintenance to a minimum. A slow-running engine is less likely to develop trouble or break down and is more likely to keep running even if defects do develop in it. As a marine engine in operation is continuously delivering its full output, robust construction is essential if it is to give satisfactory service for many years. The engines are frequently in the hands of unskilled engineers and if a high standard of technical knowledge is necessary to maintain them, then trouble is likely to result. Many fishermen run their engines in conditions which appal the manufacturer, and that again is a reason why all working parts should be very strong. Such experience shows that a hand-built robust engine which develops its power at low speeds (650 to 1,100 r.p.m.) may be a little more costly at the start, but is incomparably more economical as years go by. For example, it is not uncommon for his firm to have engines returned for overhaul which are 30 years old, and after the overhaul they will last another 15 years.

Mr. Poul A. Christensen (U.S.A.): If it is assumed that a fisherman expects his boat to have a life of about 20 years, he is faced with the problem of deciding whether to have a high-speed engine that will need to be replaced two or three times during those 20 years, or to have a slow running heavy-duty engine that has the same life as the boat. The answer would seem to be to install an engine of the simplest design possible. If this is so then the two-cycle valveless engine is the simplest design available and has no unnecessary moving parts. It also takes up minimum space.

It has been stated that per horsepower a high-speed engine is very much cheaper than a low-speed engine as well as being very much lighter. But before determining the actual relationship it should be made certain that the same ratings are being discussed. One company quotes one rating that is the amount of horsepower available from the engine to-day, in continuous operation, as well as the horsepower available from it in 20 years' time. On the other hand, a competitor recently issued a diagram in which were listed ratings in such a manner that on top of the chart is given the maximum performance of the stripped engine in the laboratory. This is followed by ratings for pleasure boats, light work boats, heavy work boats, fishing boats, and tug boats, with an asterisk calling attention to a note at the bottom of the page which mentions that the ratings are dependent upon intermittent heavy duty, or continuous heavy duty.

For maintenance and replacement it is always an advantage

to have an engine so simple that a person with the most limited technical knowledge can look after it quickly and easily. In the case of the two-cycle heavy-duty slow-running engine there is no need to send it to a repair shop for overhauling as the routine maintenance done on board makes this unnecessary.

It is not uncommon to find that the expense of the upkeep of reduction gear, which is necessary for high-speed engines, exceeds the cost of maintaining the engine itself. In the case of slow-running direct-connected engines such gear does not add to the expense of upkeep, and so far as the initial price is concerned, the two-cycle valveless engine, complete with controllable-pitch propeller, is cheaper than the high-speed engine with reverse gear.

Mr. D. L. Brownlow (U.K.): He had been concerned with the design and development of high-speed engines for more than 20 years. He knew their behaviour and he thought very highly of them but for fishing purposes the medium-speed engine is to be preferred.

In making comparisons of high-speed and medium-speed engines there is a very important point to be appreciated. The inertia forces of the connecting rod and piston and the rotating forces of the crankpin, large end bearing and unbalanced portion of the crankwebs, which have to be taken by the main bearings, vary as the square of the speed. It is a simple matter to deal with these forces in a medium-speed engine but they become an increasingly difficult problem as speed is increased.

The high-speed engine designer is tempted to reduce reciprocating and rotating parts to a minimum to keep bearing loadings from becoming excessive. This is not the case with the medium-speed engine designer. He has more scope; there is no need to cut things too fine. He can allow greater margins for safety and these extra margins mean increased reliability for the medium-speed engine and reliability is of utmost importance to the fishing industry.

Referring to Mr. Gould's paper, Mr. Brownlow said it seemed as though Mr. Gould thought the finish of parts in high-speed diesels was better than that of parts in medium-speed engines. This was misleading. The standard of precision, finish and interchangeability of parts is up to that of high-speed engines. Cylinder heads, pistons, liners and bearings are all interchangeable and can be taken from stock and put into engines without any fitting. It has been suggested that high piston speeds are as good as low piston speeds. He had no objection to high piston speeds as long as they were not too high but, like most engineers, he preferred the lower. Medium-speed engines can be run 6,000 hours without having to remove cylinder heads or pistons—two or three times longer than possible with high-speed diesels. All parts are accessible and none is delicate and intricate. As the medium speed engine is simple and does not require a skilled engineer to run it, less maintenance is required.

Discussions with trawler owners and skippers on diesel requirements for trawlers from 100 to 190 ft. (30 to 58 m.) in length and having engines from 300 h.p. to 1,400 h.p. turning 230 to 650 r.p.m., made it quite clear that they wanted: (1) reliability; (2) simplicity; (3) low maintenance and running costs. They believe that these requirements are best met by using a medium-speed engine. It would have been quite easy to install diesels running at 800 to 1,000 r.p.m., fitted with suitable reduction gears, but trawler owners are not yet ready for really high-speed engines. They have had little experience of diesels but much experience in operating trawlers. It is necessary to blend their ideas with those of the progressive engine designer.

A comparison of the fuel consumption given in Mr. Gould's paper on high-speed engines with that given in his own paper for turbo-charge medium-speed engines, shows a saving of 25 per cent., not 5 per cent., in favour of medium-speed diesels. This lower fuel consumption is another factor leading to greater reliability. Perfect combustion means freedom from overheating, from piston ring sticking, from cracked pistons and from cracked cylinder heads. There is also the important fact that for the same h.p. a trawler powered with a medium-speed engine has to carry some 10 to 20 tons less weight of fuel.

The medium-speed engine is also just the right weight for trawlers and it meets the requirements of the ship builder for the trim of the vessel. The modern turbo-charged engine is neither excessive in weight nor is it bulky. Before turbo-charging was introduced a nine-cylinder engine was required to do the work a six-cylinder engine can do today.

Mr. W. C. Gould (U.S.A.): Naval architects have stressed the difficulties in trimming fishing boats, especially when the engines are in the stern, and of designing the hull so that there is a good run of water to the propeller. High-speed engines assist architects in coping with these problems because of their light-weight and compact design. Multiple engine units make it possible to get twin-propeller reliability in a single-propeller vessel.

The horsepower absorbed by a given propeller is a cube curve. A slight reduction in r.p.m. materially reduces the horsepower absorbed by a propeller. One of two engines connected to a single shaft will turn the propeller, without overloading, at 70 to 75 per cent. of the normal r.p.m. for both engines.

It is not necessarily true that the hardest service for an engine is to be found in the fishing industry. Tugboats, often working against current, load their engines as much as any other craft. In an experiment with one tug using a controllable pitch propeller, it was found that to get the same exhaust gas temperature (*e.g.* the same horsepower) there had to be up to 9 in. (225 mm.) change in pitch going up or down stream, or in running from shallow into deep water. The maximum pitch was 52 in. (1,320 mm.)

The operations of two identical boats, one with three quad engines and one with two medium-speed engines of 700 r.p.m., have been compared over two years. The quad engine boat has been in operation 91.2 per cent. of the available hours and she has never been held up because of engine trouble. The figures showed at the end of the period of comparison that the high-speed engines were slightly cheaper to operate, despite the fact that maintenance had to be done by a shore crew which had to travel 150 miles. All their travel time, and often the time for waiting up to 12 hours for the boat to arrive, had to be charged against the high-speed quad diesels.

It is difficult to understand the continued reluctance to accept reduction gears because it is now possible to build such gears to stand up to any wear and tear. An outstanding example is the steam turbine on big ships. There is nothing to prevent installing reduction gears on any engine to give required propeller speed. The present gear boxes are designed to stand up, between overhauls, to at least 25,000 hours. It is a mistake to make equipment to run for more than certain periods because it is better to inspect it occasionally to ensure that it is in first-class condition. Even in the case of very large engines, it has been found desirable to dismantle them once a year to make sure that they give continuous service in the next twelve months.

Mr. W. Vanaarsen (Netherlands): The development in design and engineering, and in materials used for such items as liners, piston rings, etc., the improvement in lubricating oils, injector systems, and the development of gear manufacturing, are all factors which have made the high-speed diesel a reliable power plant. This continuous all-round development has also made it possible to produce a two-cycle valveless diesel for marine use. This new engine runs with a maximum of 3,000 r.p.m. or 2,200 r.p.m. for continuous operation. Its total weight is only 1,500 lb. (680 kg.) and it has a 2½:1 reverse reduction gear. It develops a maximum of 87 h.p. and about 54 h.p. for continuous operation, which gives at the continuous h.p. rating a weight of 28 lb. (12.5 kg.) per h.p. The engine is suitable for small fishing boats.

Mr. H. C. Hanson (U.S.A.): Properly used, the light-weight, medium, medium-heavy and heavy-weight diesels have all done good work. During World War I the first large heavy diesel was installed in a 300 ft. (91.4 m.) freighter. Ten years later it was installed in a tug and is still being used to-day, so that in the past 35 years the engine has probably been in operation for 30 years. Nobody can criticize that performance.

In 1935 and 1936 high-speed diesels were designed to operate for five years. Two were installed, with reduction gears, in 1936 and they are being used to-day 18 years later. One of the advantages arising from the development of the light high-speed diesel is that it has forced manufacturers of heavy engines to make those engines lighter.

He had built some hundreds of small steel bulldozer tugs, 16 to 22 ft. (4.9 to 6.7 m.) long and from 8 to 9 ft. (2.4 to 2.7 m.) in beam, for use on the logging rivers of the Pacific North-West of the United States and in Canada. These vessels are powered with 100 to 200 h.p. diesels. Each tug represents a direct saving in costs of seven man-days of labour every day. In such small craft the light diesels are particularly good because they save space which can be used effectively for other purposes.

A 65 ft. (19.8 m.) tug, in which four light-weight diesels were installed to one shaft through a reduction gear, has 660 h.p. No auxiliaries are needed as two of the engines are fitted with power take-offs. In the case of a 100 ft. (30.5 m.) vessel, the use of two 600 or 800 h.p. engines, fitted on gear reduction to a single shaft, has given super performance. One engine can be shut down and repaired while the other is in operation.

In his opinion, the introduction of the light diesel and welded construction represent the greatest step forward in boat building in the past 20 years. Engine requirements are probably different in the U.S.A. from those in Europe. In any case, he was grateful to manufacturers who continued to refine their engines so as to reduce costs. If it were not for this, much design work he had done in recent years would not have been done at all.

Mr. C. E. Dietle (U.S.A.): There had been an attempt to put engines into categories of high-, medium and low-speed: they could all be classed as light-duty, medium-duty and heavy-duty. It does not necessarily follow that the high-speed diesel is less reliable than other types.

Any attempt to classify by speeds, and to say that a diesel of 1,000 r.p.m. is sure to have poor fuel consumption, is not accurate. While certain engines of 1,200 to 1,400 r.p.m. might have a fuel consumption of 0.46 lb./h.p./hr. (0.21 kg.) his company has an engine of a similar speed with a fuel consumption of 0.40 lb. (0.18 kg.). Running at the speeds mentioned up to its 750 h.p. rating tests have shown its fuel consumption to be 0.38 lb. (0.172 kg.). It is of the opposed piston design. Differences in design as well as speed must be considered.

Mr. J. S. Robas (U.S.A.): He operated a 90 ft. (27.4 m.) menhaden vessel equipped with a 400 h.p. diesel turning 400 r.p.m. and he had to face a repair bill of upwards of $10,000 (£3,600) because of a scored crank shaft, but during the five years he had been working with the engine he had not spent $400 (£140) on repairs or for maintenance. The engine had given excellent service and dependability but, having now to overhaul the engine, he had decided to replace it with a high-speed diesel. His present engine is over 17 ft. (5.2 m.) long, direct reversible, and take up space where he could carry fish. For this reason he has chosen a compact high-speed engine with reduction gear, for which spare parts are readily available. This is a very important point. In the case of the old engine, the 17 ft. (5.2 m.) crank shaft had to be loaded on a semi-trailer and sent far away to be metallised and turned down.

In addition to the cost of having an engineer fly in to inspect the engine, and the transport and cost of sending it away, such a breakdown would have been more serious if it had happened during the fishing season. It might well have caused him to lose the whole season because his crew of 22 men would not be content to wait the five weeks it took to repair the crankshaft. In that time they would have joined other vessels or taken other jobs and it would have been difficult to replace them. For these reasons he wanted a diesel with a crank shaft that can be replaced in three days. That meant a high-speed diesel.

There is another very important reason for this decision. The compact high-speed diesel enables the engine room bulkhead to be put back about 9 ft. (2.7 m.) and that provides accommodation for another 35 tons of fish. It also meant extra buoyancy as the vessel rose about 18 in. (460 mm.) out of the water. The old engine weighed 44,000 lb. (20,000 kg.) and it paid no fisherman to carry unnecessary weight.

He wanted to see the use of the electricity eliminated on fishing boats. He did not know anyone in the menhaden fishing industry who could operate a diesel electric drive. In his own vessel every piece of auxiliary equipment runs off an engine and if it is too much trouble to put up a chain drive, or some such arrangement, a 10 h.p. engine is installed instead. If such an engine broke down he wanted to be in a position to get another one immediately and resume fishing the next morning. An hydraulic system is best for running auxiliary equipment and the use of hydraulic equipment generally is a step towards the ideal of simplicity and dependability of machinery and equipment in fishing vessels.

Mr. I. Bromfield (U.S.A.): Vessel owners at the Boston Fish Pier have followed the old custom of overhauling their boats once a year, preferably during summer when the price of fish is low. Up to the time of that overhaul, the maintenance of the engine and the fishing gear is kept to a minimum, and when an engineer submits a list of minor repairs which should be made, it is generally ignored, and what might have been only minor repairs develop into major repairs and even lead to breakdowns at sea. If repairs were made after each trip, much expense and time would be saved.

One cause of serious problems has been the method of selecting engineers. There have been no standard requirements for licensed engineers and the practice has been to appoint as chief engineer a member of the crew who might be mechanically inclined. The second engineer has been chosen by the same procedure. Such " stop and start " engineers have not usually been capable of making minor adjustments, which have often led to major trouble by the time the vessel reaches port. However, the Steamboat Inspection (a United States Government agency) now requires engineers to take an examination before they receive a licence although, because of the type of people who usually apply for posts as engineers in fishing vessels, only an oral examination is given. This is, however, a step in the right direction as it provides an incentive for applicants to study diesels and acquire enough knowledge to make minor repairs at sea.

Mr. Howard I. Chapelle (U.S.A.): The owners, builders and designers of small trawlers have accepted light-weight diesels with the greatest reluctance. When they were first used there were many breakdowns which cost money and time. Many of the early trawlers were converted schooners and they carried a great deal of ballast, and the effect of the engine weight was not excessive on trim. As a design problem the selection of engines was not particularly difficult. Since then, however, fishing methods and trawling gear have been improved and there has been a demand for increased horsepower. The new small trawler is still in the process of development and there is a long way to go before a wholly satisfactory vessel is produced. Weight has become an extraordinarily important factor, particularly in the popular 60 to 65 ft. (18.3 to 20 m.) class. The problem is to deal with the effect of weight upon the trim of the vessel in light condition, and the handling of the excessive change in trim between the light and loaded conditions. The only answer to the problem has been the light-weight diesel. He had complained about it loudly but there seems to be no escape from using it. Many of the engines give surprisingly good performances when they are properly maintained. It might be possible to improve maintenance records of light-weight diesels by some planning. Unfortunately, there is a good deal of uncertainty in a year's fishing, and to plan ahead on a specific time basis, and make the necessary contract arrangements for maintenance, is not economically practical.

There is a tendency to overdo the deckhouses in U.S.A. boats, and when the necessary winch is added, deck space is limited. In the older trawlers it was possible to have a hatch so that the engine could be removed without seriously disturbing the deck structure of the vessel. The general trend in recent years, however, has made that almost impossible, and it is leading to a serious problem of maintenance.

Certain shortcomings have to be accepted in all engines. The problem is partly financial. A better engine could be produced if there were unlimited funds available but, in the fishing industry particularly, the money is restricted and there is a limit to how much may be paid for engines. The first cost of the high-speed diesel is favourable and it has weight factors required by the naval architect.

It is useless to compare fishing industry problems of engine maintenance with those in other industries. For example, the tugboat mentioned by Mr. Gould makes severe demands on the light diesel, but the problem of maintenance is not comparable with that of fishing boats. The river boat has relatively regular runs and closely spaced repair facilities. It also has skilled diesel engineers available. The small trawler, working in the North Atlantic, goes great distances to sea in proportion to her size and she works under all kinds of weather conditions. Very often she breaks down and has to be towed to port by one of the U.S.A. Coast Guard cutters. That is an expensive item.

There is a tendency on the part of U.S.A. trawler owners to require more horsepower than they can actually use, and also to install gear that is heavier than necessary. For example, the same gear is carried in a 65 ft. (20 m.) trawler as in a 110 ft. (33.5 m.) trawler. This is probably the result of salesmanship. Most likely the skipper goes to the supplier of winches, gallows frames and lead blocks and asks for gear for a 65 ft. (20 m.) boat. The proper gear is shown in the catalogue but most likely the salesman then points out the heavier gear and persuades the skipper that it is much superior and is really what he ought to have. The result is that the 65 ft. (20 m.) trawler goes to sea with gear for a 110 ft. (33.5 m.) trawler. It is carrying unnecessary weight, so the boat is overloaded and the light-weight diesel is probably subjected to undue strain.

The sensible thing to do is to realize the inherent restrictions and limitations in the use of light-weight diesels. They are usually well-designed and seem to be as reliable as can be expected under the trying and extreme conditions in which they operate.

Mr. J. Messiez-Poche (France): There are good slow-, medium- and high-speed diesels but the slow engine is safest. The breakdown of an engine on land or in a train or other vehicle is of minor importance but it is a catastrophe if it happens in a trawler operating 200 or 300 miles off the coast in a storm. People who want to install fast engines in trawlers should think first of the safety of the crew and of the maintenance needed by high-speed engines. This is a serious problem for fishermen with little experience of diesels—much more serious than in the case of a slow engine. As every engineer knows, a fast motor will break down if the connecting rod end is tightened too much or if the connecting rod has not been made of specially treated steel. Fast engines have already been installed in fishing boats, of course, but there is a need to be particularly careful to ensure safety in operation and maintenance and there is no doubt that the slow engine is to be preferred

SEMI-DIESEL

Mr. John Tyrrell (Ireland): Single cylinder engines create vibration which cannot be eliminated even by the most expensive construction. On the other hand, the single-cylinder Scandinavian semi-diesels are unsurpassed for reliability and general economy.

Mr. Alan Glanville (U.K.): As one who had used both semi-diesels and full diesels in various parts of the world, he had found both types of engines satisfactory but the choice of engine, especially for use in underdeveloped countries, was as difficult as important. First and foremost a fisherman wants reliability; the engine that runs every day is the one that makes money; and there is nothing more harmful to a programme of mechanizing a fishing fleet than to see engines constantly out of order. In such places as, for example, India and the Far East, where fishermen are unfamiliar with machines, engines will be frequently overworked. Fuel will often be supplied in dirty barrels containing a few gallons of water so that every engine installed should have a separate fuel filter and sediment bowl. If the fisherman considers that the engine " drank " too much lubricating oil one day he will probably try to teach it better manners the next day by not giving it any oil. It is only to be expected, therefore, that breakdowns occur and how long an engine may remain out of order often depends on the availability of spare parts. That may make a difference of a day or two or a month or more.

If large numbers of diesels of a certain make have been operated locally on plantations, as generating units, or for motive power in agriculture and industry, then it is a good plan to select the same type for marine installation because there will be persons in the neighbourhood with experience of operating the engines. And there will be spare parts available. Mr. Glanville cited the case of two villages in Bombay State, India, where 170 diesels had been installed in fishing boats during the past five years. These engines had all been kept running because diesels of the same make had been widely used ashore in that area, and there were stocks of spares immediately available at Bombay.

He also referred to his experience in Ceylon. Two years ago there were no engines in local fishing boats; now there are 20 and another 30 are on order. At the start he had tried to introduce both semi-diesels and diesels but he found that the fishermen preferred the diesels because they were lighter, quieter and started immediately without blowlamps. They were also acquainted with the engine as thousands of small diesels are used in the plantations on the island, and a good organization exists for servicing and providing spare parts. For these reasons it was soon decided to concentrate entirely on diesels. This did not mean opposition to semi-diesels which he considered to be more fool-proof, more robust, and better able to stand up to rough treatment. If, in some isolated area of an underdeveloped country there are no engines operating and no servicing or spare parts available, then he would advocate the use of semi-diesels. Choice of engine should be governed by local conditions rather than by the particular merits of any one type of engine.

Mr. Jan-Olof Traung (FAO): When the first engine is put into a fishing boat in an underdeveloped country the installation is closely watched, not only by the fishermen concerned but by the whole fishing community, indeed, by all the chief fisheries officials. If the engine does not operate well, then a good deal of harm can be done to the programme of mechanization. The choice is not necessarily between semi-diesel or diesel. It might well be that kerosene or petrol engines should be introduced. Almost everywhere in the world there are cars and lorries and there is more knowledge of, and facilities to repair and maintain, engines of the type used in them. Petrol engines are much easier to operate and maintain, although there is naturally a greater fire hazard. It is true that such engines, especially when there is a tax on petrol, are more expensive to run, but they cost so much less in the first instance. This means that operators can run them for a considerable number of hours before the effect of higher running costs is felt. And there is always the possibility that the Government of the country concerned may be persuaded to exempt from tax any petrol used in fishing boats because usually the tax on petrol is imposed to provide funds for building and maintaining roads.

Mr. Traung had been in favour of diesels, thinking the hot-bulb engine to be somewhat out-of-date, but experience with Swedish fishermen had changed his opinion. He had seen them start the engine and then leave the engine-room and the engine alone, not for a day but for several days. This was the reliability that fishermen expected of their engines and is one of the reasons why semi-diesels have also often been recommended for the fishing fleets of underdeveloped countries by the FAO experts.

A semi-diesel is normally more difficult to start than a diesel or petrol engine because it has to be preheated, and that takes time even if done by using cartridges or electricity. The smaller, hand-started, semi-diesels sometimes start in the wrong direction but, with a little skill, it is easy to reverse them. There may be also some difficulty in getting operators to understand immediately the right adjustment of the fuel sprayer for the idling and load conditions although this skill is usually acquired very quickly. A mechanically-minded fisherman would have no difficulty in handling and running a semi-diesel but it is a fact that in many underdeveloped countries the fishermen have little or no mechanical skill. In such cases it might be necessary to recommend an engine which requires no preheating, no adjustments during running and minimum maintenance while new. A diesel or petrol engine might then give the best results if the operators were prohibited from touching the engine and all repairs were done by trained mechanics. To avoid breakdowns at sea, such engines should be subject to preventive maintenance.

On the other hand, if the fishermen have acquired sufficient mechanical skill, are of average intelligence, and wish to maintain the engines themselves to save expense, then Mr. Traung felt that semi-diesels would be, in the long run, the most economical.

Although there was a trend recently in Scandinavia to switch over to diesels for fishing boats, this seems to be changing and boats being built to-day are often fitted with semi-diesels, principally because they are much cheaper and more dependable, even if fuel consumption is about 10 per cent. higher. It is a mistake to think that semi-diesels are only used in Scandinavia. Many are used in the Far East and South America. In Singapore most of the semi-diesels are of Japanese manufacture and when, during World War II, new engines were not available, parts of the old engines were used for replacements and the fishing boats were kept in operation. Engines were also made in local foundries, using the parts of old engines as the patterns. In Bangkok there is a harbour department with 66 launches equipped with 33 different makes of engines. An inquiry showed that the semi-diesels were the most popular.

As an example of FAO's approach to the problem of mechanizing fishing fleets in underdeveloped countries, Mr. Traung cited the case where 200 high-speed American-made diesels were used in tractors and other agricultural machinery. They were regularly overhauled in a Government workshop. In this instance, the problem was to select the best engines for a research vessel and it was suggested that high-speed diesels should be installed and that an extra engine should be bought so that there was always one of the ship's engines being overhauled by the Government's Engine Maintenance Department. In this way it was not necessary to have a skilled engineer on the research vessel. Even if a semi-diesel had been used, the arrangement could not have been simpler.

Mr. Georg Bruce (Sweden): From the manufacturer's point of view the high-speed engine is best for mass production but that does not mean to say a semi-diesel is an antiquated source of power if the fisherman, who is the customer, wants semi-diesels.

Manufacturers must be prepared to deliver spare parts over a period of many years. For instance, a firm had recently to make spares for engines delivered in 1910 and to produce them at a price comparable to the cost of modern parts. One of the reasons why parts should be simply designed is to help fishermen to do makeshift repairs in remote places. This is especially desirable in engines up to, perhaps, 300 h.p.

Mr. Olai Mollekleiv (Norway): It is not necessary that semi-diesels should take more space than high-speed diesels. There is a special two-cylinder type of semi-diesel (150 h.p.) which takes up less space than a high-speed diesel of four to six cylinders with a reduction gear.

Mr. Johann Ertl (Germany): The modern trend in engineering is to develop higher speeds in lighter and smaller engines. For smaller cylinders—up to about 183 cu. in. (3 l.) of cylinder volume—different types of precombustion chambers and swirl chambers have been developed satisfactorily. The injection pressures are comparatively low and do not exceed 2.850 lb./sq. in. (200 kg./sq. cm.) at the rated output. The engine needs only single-hole nozzles, so that differences in the quality of fuel do not matter. Combustion systems of this type have specific consumption figures of 0.41 lb./h.p./hr. (.185 kg.). In the case of air-cooled engines up to 91 cu. in. (1.5 l.) volume the consumption is even lower. This means considerable economy in fuel consumption, as compared with the lowest consumption figures of semi-diesels of 0.46 to 0.49 lb./h.p./hr. (.210 to .220 kg.). Greater economy can be obtained if direct injection is used, which is possible with present technical skill without impairing the reliability of engines with, say, a cylinder volume of more than 180 to 240 cu. in. (3 to 4 l.). In such a case, the specific consumption is less than 0.39 lb./h.p./hr. (.175 kg.). In modern two-cycle engines, with centrifugal blowers, fuel consumption may be reduced to 0.33 lb./h.p./hr. (.150 kg.). This means a saving in fuel of 25 per cent., as compared with the semi-diesel consumption. The specific weight of a two-cycle engine is about 64 lb./h.p. (29 kg.) including the reduction gear. A semi-diesel is about 155 lb. (70 kg.) at the rated output. This means a saving in space. In addition, a semi-diesel with direct drive gives, at a propeller speed of 310 r.p.m., a much lower efficiency than the geared engine with a propeller speed of 240 r.p.m. The specific fuel consumption of a modern diesel for cruising and trawling is below 0.36 lb./h.p./hr. (.165 kg.) from 10 per cent. overload down to 70 per cent. load on cruising or 60 per cent. load in trawling. In view of these figures, it seems reasonable to question whether the high fuel consumption and weight of the semi-diesel makes it too costly to-day.

Mr. F. C. Vibrans, Jr. (U.S.A.): The economic factor has largely influenced the popularity of semi-diesels in Scandinavia and high-speed diesels in the U.S.A. Scandinavian fishing grounds are within short distance of fishing ports, but U.S.A. fishermen have to go considerable distances to reach theirs. For example, the Georges Bank off Gloucester is about 250 miles away and for a fisherman to get there quickly the boat's engine must have high power. The semi-diesel weighs from 100 to 200 lb. (45 to 90 kg.) per h.p., as against the 22 to 25 lb. (10 to 11 kg.) per h.p. of the high-speed, light-weight diesels. This means that a lot more power and fuel can be put into the space available. An example of this is found in a boat recently repowered with a high-speed diesel. The new engine is half the weight of the old and it has double the power. The saving in space allowed the fish-hold to be lengthened about 6 ft. (1.8 m.).

Mr. James Whitaker (U.K.): Reliability, simplicity, slow speed, direct drive, direct reversing and maximum accessibility, are expected from engines of 200 to 1,000 h.p. by fishing boat owners. The question, therefore, is what type and cycle of engine gives the best answer to these demands? The answer is: the two-cycle engine with positive pump scavenging.

There is no doubt that semi-diesels have proved themselves in Scandinavia as most reliable engines. They are simple, robust and work with low cylinder pressures, which are the secrets of their success, but they have poor fuel consumption. That is their weakness. It happens because the scavenging of air in the cylinder cannot be made perfect, and combustion of fuel is, therefore, incomplete. The unburnt products of combustion affect cylinder lubrication and cause wear.

The four-cycle engine can obtain perfect combustion because it devotes one complete stroke in every four to drawing a full charge of fresh air into the cylinder. The disadvantages of the engine are its complicated cylinder head, with attendant valve gear, and the fact that it cannot be made direct-reversing if it has less than six cylinders.

The engine which combines the advantages of both semi-diesel and four-cycle is the two-cycle pump scavenging engine, with inlet and exhaust ports. This engine is a development from the semi-diesel and has the same simple cylinderhead, a minimum number of working parts, and maximum accessibility. But it also has 100 per cent. air supply, with good scavenging, and low fuel and lubricating oil consumption. It is an engine that meets the requirements of fishermen for extreme reliability.

Mr. Bent G. Andersen (Denmark): In Denmark there are no big trawlers such as are found in Holland, France and Great Britain, and the reason why more than 90 per cent. of the Danish fishing cutters are fitted with semi-diesels, and less than 10 per cent. with full diesel, petrol or paraffin engines, is that Danish fishermen have found that it pays to use simple, robust and reliable engines. This preference for semi-diesels is found throughout Scandinavia where fishermen will not be bothered with the intricate problems of the internal combustion technique nor with complicated gearings. Experience has taught them which is the best engine for making their boats pay. This is an important economic fact and on it depends whether or not a fisherman is to survive.

It had been shown that, although working conditions may differ considerably from country to country and from coast to coast, like problems may often be solved in like ways. He believed that the mechanization of fishing craft, as carried out in Scandinavia, provided experience and example for other parts of the world. This is especially true of the mechanization of fishing craft in under-developed areas where, as a rule, only small engines are needed. The problem differs considerably between engines for fishing cutters and for bigger trawlers. Semi-diesels are built up to about 400 b.h.p. but the majority are not above 250 b.h.p.

Referring to Mr. Brownlow's paper on medium-speed diesels, Mr. Andersen endorsed the description of what the ideal fishing boat engine should be. But he would like to add two points: easy manoeuvrability and long life. A direct comparison of the descriptions of the medium-speed diesels and the semi-diesels showed that semi-diesels fulfil the requirements stated by Mr. Brownlow but, of course, the engines described by Mr. Brownlow are meant for bigger vessels.

Referring to Mr. Gould's statement with regard to high-speed diesels that a complete overhaul can be done in four or five days by 100 to 190 man-hours of labour, at a cost of £110 to £290 ($300 to $800) for spare parts, Mr. Andersen said that if Scandinavian fishermen were asked to pay such expenses once or twice a year they would discard the engine. The usual cost of maintenance for a semi-diesel of good Scandinavian make is about £5 to £8 ($14 to $22) a year for the first eight or ten years. This refers to engines of 50 to 150 h.p. and overhauls of such engines are not regular. They are first made when the first piston has to be shifted and that usually only happens after 10 years' hard service. Mr. Andersen cited a recent instance in which a semi-diesel had been in use 26 years and during all that time the cylinder cover had never been lifted.

Mr. Stokke's opinion, expressed in his paper on semi-diesels, is that cylinder sizes above 75 h.p. are not recommendable because they are too heavy. But the weight of a Danish semi-diesel is only 145 to 175 lb./h.p. (65 to 80 kg.).

Mr. Chapelle (see chapter on Boat Types) warns against over-mechanization in under-developed areas where the skill and facilities to maintain engines may be lacking. In this case he suggests local sailing craft should be improved. In this connection it should be pointed out that in Denmark skilled mechanics are not usually employed in fishing boats and often the boy in the crew is put in charge of the engine. Nor is it necessary to choose between sails and engine because a combination of sail and engine may be used with a controllable-pitch propeller. Mr. Chapelle is right in warning against too much emphasis on speed in fishing boats and gives, as the best range, 9 to 10 knots. Nearly all Danish fishing boats operate between these limits.

Mr. Hans K. Zimmer (Norway): Mr. Leach pointed out that usually skipper-owners of small fishing boats have little capital and would only buy an engine which had proved itself to be reliable. Skipper-owners, however, have a tendency to buy second-hand engines—the most risky experiment of all.

Mr. Andersen's statement concerning maintenance of semi-diesels gives a figure that may be possible in insolated cases, but he wished to stress it is far from average. The maintenance cost for a 50 h.p. semi-diesel is more than double the figure given and somewhere about £15 ($42) a year would be a better average. He also confessed himself sceptical about the engine in which the cylinder head had not been touched for 26 years and suggested that the engine probably had not been running for the last 20 years. A further investigation of the vessel's log book might be desirable.

There are obvious reasons for the popularity of the semi-diesel, but it has its limitations and drawbacks such as heavy weight, large dimensions, and fuel and lubricating oil consumption. Above all, its vibrations put an unnecessary strain on the hull and loosen the fastenings. In this connection, the usual old type of bearing, which can be remetalled and adjusted by handscraping, is to-day both expensive and uncertain to maintain. The precision type of steel-backed white metal lining, which can be renewed when worn, is the best and safest solution in fishing boat engines.

Mr. Bent G. Andersen (Denmark): If Mr. Zimmer's figure of £15 ($42) a year for maintenance of a 50 h.p. semi-diesel was accepted it would still compare very favourably with the maintenance figure for high-speed diesels, quoted by Mr. Gould. But, so far as his own semi-diesels are concerned, the figure of £5 to £8 ($14 to $22) a year was on the safe side. For other makes of semi-diesels the cost may be higher.

Regarding Mr. Glanville's remarks, it seems that he is of the opinion that semi-diesels are better suited for operating in primitive conditions, give less trouble and require fewer spare parts to be stocked. If this is a correct interpretation of Mr. Glanville's remarks, then semi-diesels should be preferred to diesels in under-developed countries where repair facilities and stocks of spares are scarce.

Mr. Karl Hugo Larsson (Sweden): The name " semi-diesel " is not correct. He would prefer to call the type " low-pressure oil engines ". They are used in Swedish fishing boats of 65 to 90 ft. (20 to 27.5 m.) length. The engines develop 120 to 300 h.p. Nearly all the boats are owned by the fishermen themselves and with these small craft they fish in the Skagerrak, in the North Sea and in the Atlantic, as far up as the northern part of Norway. Controllable-pitch propellers are standard and the controls are so arranged that the skipper is able to handle the engine from the wheel-house.

These Swedish vessels normally carry a crew of six or seven men, all fishermen. Nobody is employed specially to look after the engine. They handle their trawling gear and take care of the fish and the engine takes care of itself. It is this great confidence which the fishermen have in the reliability of their engines which seems to be their main reason for preferring semi-diesels. Another important fact is that they themselves can do much of the maintenance and repair work because of the simple design of the engines. As the pressures are low there is not much wear on cylinders and, if a bearing is damaged, it can be repaired at any little workshop. No special alloys are needed. A skipper with a single-cylinder engine found that the crank-shaft bearing had broken down and there was not a spare one available on board. He was in the North Sea, in bad weather. He took the bearing out of the engine, lined it with pieces of the trawl winch belt, and was able to work the engine and bring his boat safely to home port.

Semi-diesels can be run in any direction. In one vessel the shaft controlling the pitch of the propeller broke and the fishermen were able to fasten this shaft by a line put into the stern tube. This meant the propeller was fixed for running astern, so the engine was started in the reverse direction and the boat was sailed safely to harbour.

During World War II, when fuel was very scarce, semi-diesels were run on ordinary black tar made out of wood. No diesel could have been operated on such fuel. Swedish manufacturers have improved design and construction considerably during the past 20 years, resulting in much more economic fuel consumption.

He did not think semi-diesels would always stay dominant in Scandinavian countries as demand for bigger engines would make it necessary to use diesels. It is probable that semi-diesels will never be built for more than, say, 400 b.h.p. But such a simple, robust, long-life engine should prove a boon in under-developed areas where fishermen are learning to use engines for the first time in their lives.

Mr. J. Messiez-Poche (France): The selection of a diesel or semi-diesel depends largely on the conditions in which the engine is to be used. In circumstances where fishermen have no knowledge of engines and repair facilities are limited, the slightly higher fuel consumption of the semi-diesel is of no consequence because it is more than compensated by economy in maintenance and operation.

DRIVE

Mr. P. H. Hylton (U.K.): However good an engine may be as a power producer at the flywheel it may still be inefficient unless this power, or as big a proportion as possible, is transferred to the water. Surprisingly, there had been little direct reference to this point. Propeller speed and diameter are important factors and, as Mr. Paul Ziener had illustrated, immense wastage may take place when this point is not understood or is ignored. Boat designers and users would welcome more co-operation and information from engine manufacturers.

Propeller speed is too often governed by engine speed. Designers would prefer to make the propeller to suit the vessel rather than the engine. This is particularly true of the powerful machinery used in trawlers where low propeller speed is so important for towing efficiency. To obtain this low speed with direct drive the engine must be large, heavy, expensive and "tailor made". The reasons given for preferring low crank shaft speed are usually long life and longer time between maintenance work, but how much of this is tied to the necessity for low propeller speeds?

Piston speed is a better criterion than crankshaft speed and experience has shown that the best of both worlds can be had by the use of medium-speed 400/600 r.p.m. engines with reduction gears. Up to 350 h.p., these can be had at standard ratios which, combined with slight variations of crankshaft speed, enable the designer to approach close to his ideal propeller speed. Above 350 h.p., the gears are usually made to order and the ratio can be chosen to match exactly the engine and propeller.

These reduction gears can be made to reverse, and the oil operated type is robust, light and has a long life. Usually a uni-directional medium-speed engine of the four-cycle type is used with such reverse reduction gear. But it is essential that the engine and gearbox be designed as a whole. When compared with direct drive direct reversing machinery such a combined unit has the following advantages for a trawler:

(1) Propeller speed to choice.

(2) Uni-directional constant running engine: (*a*) a drive suitable for mechanical, hydraulic or electric trawl winch operation can be provided at the forward end; (*b*) such an engine is not dependent on compressed air for manoeuvring, large compressors, air tanks, and complicated reversing mechanism are eliminated; (*c*) the refrigerating blast of manoeuvring air in hot direct reversing cylinders is not present; (*d*) the engine can be started, run up, and warmed up before manoeuvres.

(3) The combined unit is: (*a*) simple and positive of operation in semi-skilled hands; (*b*) reasonable in size and weight and, lending itself admirably to exhaust gas turbo pressure charging, increasing power; (*c*) easy to install, the difference in shaft levels permitting a deep and stiff seating; (*d*) obtainable from manufacturers who, having outlets for similar engines in other fields, build in large quantities on precision machines, jigs, and fittings which cannot be found in smaller, more specialized plants. This guarantees continuity of supply of interchangeable spare parts.

The system has proved a success in practice in a number of vessels.

Referring to Mr. Basil Parkes' comments on the variation in power between direct and gear box drives, Mr. Hylton asked if this was because the gear box enables more power to be efficiently utilized. There is a small mechanical loss through a reduction gear but this is more than counterbalanced by the larger slow propeller, a fact illustrated by lower fuel consumption for given vessel speed, better performance in head seas and when trawling.

The propeller speed may be chosen when the diesel electric drive is used (motor size can be reduced by use of reduction gears) while maximum power is available at reduced trawling revolutions and simple uni-directional generating units can be in numbers best to suit the load factor, such division also increasing the safety factor. These advantages may be said to compensate for electrical transmission losses. However, first cost remains high although the price differential compared with mechanical drive is lessened when auxiliary loading is high as in a quick-freeze factory ship.

The additional safety factor and the added flexibility that goes with division of units are advantages of the "father and son" method. To gain full available benefits from this idea both engines can be uni-directional and coupled through a reverse reduction gear.

Simplicity and other advantages of the uni-directional engine may be exploited with the use of controllable-pitch propellers. British manufacturers of this type of engine would welcome more experience with controllable-pitch propellers but their high cost probably means that they would be used at speeds higher than desirable for maximum efficiency, so they would suffer by indirect comparison with the performance of fixed-blade propellers. All reputable manufacturers try to design and produce robust simple engines which can operate efficiently for long periods between overhauls; and spare parts, machined to fit, are always available. Co-operation between manufacturers and trawler owners must eventually provide machinery better suited to the industry's requirements, but correct use of existing engines and equipment can, however, effect immediately an acceptable compromise between the ideals of the user and designer.

Mr. I. Bromfield (U.S.A.): The impression that it is complicated and expensive to install electric drive in a vessel is untrue. He had installed such equipment on boats twenty years ago and no overhaul had been needed since. In 1928 when the first electric drive was used on some large trawlers, it had so-called armature control. When a vessel was trawling the net might sometimes be fouled, with the result that the motor, generator and engines would be overloaded. In most cases there would be a circuit-breaker which would then act. After it had operated several times it was quite likely that the engineer would take a bucket of sand and hang it on the lever. The net would then be hauled in, but the motor or grids would burn out, and that would lead to another broken trip. This problem had interested him and he developed the so-called torque control. Instead of using the armature control of probably 300 or 400 amperes, he used the control of the fields, which was down to approximately 10 amperes of control. Since then there has been no trouble because when the load reaches its maximum the motor stalls momentarily but picks up its speed directly the load is relieved.

Mr. C. E. Dietle (U.S.A.): The diesel electric drive is high in first cost and high in fuel cost and, in diesel locomotives, more than 75 per cent. of the failures are electrical rather than mechanical. If a highly complicated diesel electric drive is to be installed in a fishing boat, then there should be a skilled engineer on board to look after it.

There are certain advantages which derive from using multiple engines on single shafts, but there are also disadvantages. For example, the quad engine, two-cycle design, rates about 550 h.p. and competes with a certain seven-cylinder engine. The length of the quad is 14½ ft. (4.4 m.), the length of its competitor is 6 ft. ⅝ in. (1.85 m.). The weight of the quad is 12,065 lb. (5,450 kg.); the weight of its competitor is 9,800 lb. (4,450 kg.). The speed of the quad is 1,600 r.p.m. against the 1,200 r.p.m. of its competitor. It is obvious that by arranging engines in a quad, space and weight are sacrificed.

Professor H. E. Jaeger (Netherlands): He could not agree with Mr. Kannt that the triple expansion engine with coal burning Scotch boilers is generally to be preferred. In Germany they had again come back to the ordinary coal burner, although elsewhere in the world the diesel is being more and more used for trawlers. There are great advantages in using diesels.

Mr. Kannt refers to the German "father-and-son" system. Although he liked the solution it provided, he thought an even better solution was to be found in the system invented by Mr. Barbiot of A.C.E.C., Charleroi, but the simplicity of the big non-reversible motor with controllable-pitch propeller is a better and more economic proposition. He liked the "father-and-son" system better than diesel-electric, which is heavy and expensive. In view of the name invented for the system, "father-and-son", he liked to think the time would come when shipowners would be able to order and get the Holy Ghost.

Mr. P. S. Stoffel (Germany): The first controllable-pitch propeller was put in service in 1934 on a Lake of Zurich vessel. The safety of operation is due to the great amount of experience obtained with the Kaplan water turbine. The efficiency of the controllable-pitch propeller, when it is well designed, is the same as for the fixed blade propeller. Two German trawlers of 400 b.h.p. are equipped with the controllable-pitch propellers.

Mr. K. Rosenthal (Denmark): In Scandinavia 95 per cent. of the fishing boats are fitted with controllable-pitch propellers, and his firm is making such propellers up to 1,000 b.h.p. The rack and pinion solution is suitable for small power within the range of semi-diesels. For high power, his firm uses a crank mechanism with sliding blocks for controlling the propeller blades, thus avoiding back-slash and obtaining well-lubricated parts transmitting the main forces. Such a mechanism can be made so strong that when the blades are bent or damaged the inside mechanism remains intact. Up to 240 h.p. the propellers are hand-controlled. From 180 h.p. they can be hydraulically operated. It was mentioned that the price of an engine with clutch and controllable-pitch propeller was lower than that of the same engine with reversing gear and propeller with fixed blades.

Mr. J. Hojsgaard (Denmark): Controllable-pitch propellers made by his firm are quite as strong as fixed propellers. They have never experienced a blade break at the root and never had a case of a broken shaft or any damage to the internal mechanism (push rod, bearings, etc.). The boss is made quite large to contain a very robust inside mechanism. He confirmed Mr. Stoffel's opinion that even with a large propeller the efficiency is not reduced if the boss is well streamlined. The upper limit for manual control is 200 b.h.p. Boats operating in pack ice for a part of the year should be fitted with a thick-bladed propeller which could be replaced by a thin-bladed type in summer time.

Mr. M. Tusseau (France): It has been said that one of the advantages of the controllable-pitch propeller was that it permitted an extra 20 to 25 per cent. pulling power compared with a fixed-blade propeller. It would be interesting to see how this extra thrust could be put to practical use. The first idea that comes to the mind is to increase the speed of the trawler. With an increase of thrust of 20 per cent. the trawler's speed would be increased by about 10 per cent. This would obviously allow a much larger area to be fished and would improve the total catch; or the duration of present fishing trips could be shortened. But the resistance of the trawl and preservation of the fish may be factors which set a limit to the speed of the trawling operation.

Mr. E. R. Gueroult (France): French naval architects have not had much experience of controllable-pitch propellers in big trawlers. It seems, however, that the propeller is excellent for all fishing boats except the trawler, because there is the danger of the trawl boards getting into it. If this happens, say, twice a year, the value to be gained by using the propeller is greatly reduced.

The controllable-pitch propeller must be brought to the lowest pitch when towing and would have to be brought to the lowest pitch for good efficiency in bad weather. As a trawler is towing or going through bad weather about three-quarters of the time she is at sea, it can be seen that the controllable-pitch propeller has not a very great advantage over a fixed-blade propeller with a low pitch. Another point is that when the skipper thinks he should have more pull while towing he will increase the pitch of the propeller, and so overload the motor. These are important factors to consider as they affect safety and reliability.

Mr. I. Bromfield (U.S.A.): A controllable-pitch propeller was used on the U.S.A. Fish and Wild Life Service vessel, *Albatross*, and while it was found to have many advantages, the maintenance cost was high, and after a year it was replaced with a conventional propeller.

Mr. H. C. Hanson (U.S.A.): Controllable-pitch propellers are not new. He had used them since the start of the century in small fishing boats to reduce the speed of the vessel to 2½ knots. They had been of great value in trawlers, and also in vessels sailing in Arctic waters.

Mr. F. C. Vibrans, Jr. (U.S.A.): Controllable-pitch propellers have, admittedly, many advantages, especially in the matter of control, but the argument of full power availability at all speeds seems to be exaggerated in the trawler industry. The general belief is that where trawling speed is only one-third of steaming speed the engine should slow down similarly with the constant torque maintained. This is not true. The reduction in engine speed is only 10 to 15 per cent. This is shown in Professor Troost's new propeller charts, together with the fact that wake fraction varies with speed. With higher speed, when all power is devoted to propelling the boat, there is a higher wake fraction than when the boat is held back by the net. Thus, propeller speed of advance is reduced less than in proportion to the change in vessel speed when going from steaming to trawling.

Mr. Poul A. Christensen (U.S.A.): The question of cost of upkeep of the controllable-pitch propeller is often raised but, for example, an hydraulically-operated unit installed in a New England boat 26 months ago has so far not cost a penny in repair or upkeep. Again, in connection with the danger of damaging the propeller so that it cannot be operated, there is the instance of another New England boat in which a trawl door damaged one of the propeller blades. About ⅜ in. (10 mm.) of the blade was broken off from the leading edge, over a length of about 7 to 8 in. (175 to 200 mm.), but the skipper of the boat had no difficulty in steering nor with handling the propeller unit, and on his return to harbour a new blade was installed with small loss of time or money.

Commander R. E. Pickett (U.S.A.): As Mr. Frederick Parkes has said, a vessel must be free of vibration and noise. This is not generally the case with small trawlers in the U.S.A. The controllable pitch propeller has not yet been accepted in the U.S.A., and if it can operate successfully despite debris in U.S.A. harbours and rivers it would be of great value to small fishing boats.

Mr. Jan-Olof Traung (FAO): There is nothing new about controllable-pitch propellers and 99.9 per cent. of fishing boats in Scandinavia are fitted with them. There they are much cheaper than reverse gear and fixed-blade propellers and more dependable. It is a mistake to think that the controllable-pitch propeller calls for fancy designs. In Scandinavia it is of two types: one is mechanical and is used up to 250 h.p. at 300 r.p.m. and the other is hydraulically controlled. The mechanical type is inexpensive but the hydraulically-controlled type is very costly, probably as expensive as the engine itself. It also costs a lot in maintenance. It is a wonder, however, that an effort is not made to copy the cheap mechanically-controlled type. For example, in the United Kingdom they have tried to produce controllable-pitch propellers for the mechanical system but have made something more complicated than is necessary. There is a fear that a controllable-pitch propeller will be damaged by trawl doors and trawl warps. This does not happen during trawling in Scandinavia and if, as can sometimes happen with any propeller, some damage is done, all the fisherman has to do is to change the blade. This is a very simple procedure, simpler than to change the whole propeller. Incidentally, the blade is always designed weaker than the hub.

Mr. Bent G. Andersen (Denmark): Most experts agree that slow-speed engines are to be preferred in small vessels, and the general interest expressed in controllable pitch propellers is surprising to a Scandinavian as nearly all Scandinavian vessels are fitted with such propellers. But Scandinavian boat builders often find that their export customers do not want controllable-pitch propellers, probably because they have had some unfortunate experience with propellers of bad design and performance. Scandinavians have used controllable pitch propellers from the beginning of mechanizing their fishing fleets and have developed a propeller of simplicity and great reliability. They have been used for nearly 50 years and the Scandinavians long ago developed a very dependable simple mechanical manoeuvring of the propellers on engines up to about 400 h.p. For higher power, hydraulic manoeuvring or some other power operation is necessary. It is true that badly-made propellers have been put on the market. In Scandinavia the invariable practice is to use controllable-pitch propellers but when engines are exported difficulties arise if the propellers are not properly introduced into a market where they are novel or where unreliable controllable-pitch propellers have previously been used.

It has been pointed out that in some under-developed countries it is desirable to combine the use of sails and engines, a problem which is faultlessly solved by the use of a controllable-pitch propeller. This is the answer to Mr. Setna's problem. And the propeller, when a boat is under sail, allows the use of a smaller engine.

Mr. Andersen knew only one instance where fixed propellers held any advantage over controllable-pitch propellers and that is waters filled with seaweed as, for example, on the Amazon river. A propeller will catch the weeds until it can work no longer. A fixed propeller can free itself by reversing but this is normally impossible for a controllable-pitch propeller because an engine fitted with one is seldom made reversible.

Dr. J. F. Allan (U.K.): The controllable-pitch propeller has been adopted to an appreciable extent on fishing vessels in some European countries but not much in the United Kingdom.

The controllable-pitch propeller is ideal from the operational point of view because it enables full power to be absorbed at any speed of advance without material loss of efficiency. It can also be arranged to give reverse thrust, which eliminates the need to reverse the propeller by reversing the main engine or by providing a suitable gear box.

A change of about 5 deg. in propeller blade angle is required to absorb full power at the same r.p.m. running free and when trawling at 3 to 4 knots, but to provide reverse thrust the blades must be rotated through a large angle. The main problem of the controllable pitch propeller is to mount the blades on the boss so that they are adequately supported but can be rotated accurately in their mountings. It is not of very great importance whether the blades have to be rotated 5 deg. or 90 deg.

A controllable-pitch propeller is little superior to a fixed pitch propeller designed to compromise between the free running and the trawling condition. The advantage will be unimportant in the case of a steam reciprocating engine, and of no major importance even in the case of a less flexible diesel. The widespread use of steam engines in trawlers in Great Britain accounts to some extent for the comparative lack of interest in controllable-pitch propellers. The advantage gained with a controllable-pitch propeller providing both ahead and astern thrust coupled to a diesel running at constant revolutions is that it eliminates the reversing mechanism on the engine and reduces wear and tear.

The danger of warps fouling the propeller and causing damage to the controllable-pitch mechanism is also important, and an over-riding consideration is the extremely high cost of the propeller compared with one of solid bronze or the much cheaper cast iron propeller.

The case for a controllable-pitch propeller on fishing vessels is not strong but there is more advantage in using one with a diesel than with a steam-reciprocating engine.

VESSELS AND GEAR

by

A. VON BRANDT

FOUR factors were decisive in the development of fishing technique:

(1) The transfer of fishing from shallow to deep waters, which happened in the large lakes as well as in the littoral areas. The use of fishing vessels encouraged the development of pelagic fishing in deep sea areas and bottom fishing in large depths.

(2) The development of bulk fishing which was greatly influenced by the production of powerful engines because they gave the necessary speed. For example, trawling speed has increased from 1 knot in 1870 to 5 knots at the present time.

(3) The development of partly or wholly mechanized fishing methods. The use of larger gear for bulk fishing required the installation of auxiliary machinery.

(4) The detection of fish shoals by means of electronic devices. These are of great practical importance to commercial fishing.

A compromise had to be made in the shape of trawlers which are so very important for North European fisheries. German trawlers, for instance, must be capable of high speeds to reach distant fishing grounds, but they must also be able to lie-to safely at the grounds even in bad weather. High speeds call for sharp bow and stern but when the fishermen have to work on deck, the trawlers should have a fuller form. The correct distribution of the displacement and the position of the centre of buoyancy are particularly important and there must, of course, be suitable proportions of length to breadth, and breadth to draught. Naval architects and fishing gear technologists have not reached full agreement in their efforts to solve this problem and they have had to compromise.

Fishing technique itself calls for special attention by naval architects, raising questions which can be summarized:

(*a*) Mechanization of ancient fishing methods and development of new and very specialized methods, as, for instance, electric fishing;
(*b*) Observation and control of fishing gear in action, particularly from the bridge;
(*c*) Detecting fish in vertical and horizontal directions.

Deep sea fishing, which is most essential to North European fisheries, made it necessary to consider the fishing gear when designing new vessels. The first trawlers (sailing from North Shields in 1877, from Boulogne in 1881 and from Geestemünde in 1885) had no remarkable fishing equipment. But when the beam trawl was replaced by the otter trawl (about 1892) it was necessary to invent the gallows. At first these consisted of two slightly tilted wooden poles with a beam operating a roll. Then, as trawling was done at greater depths, the old capstans were replaced (about 1885) by the first large winches for steel wires. The otter trawl, with its greater resistance, could not be towed by sailing vessels, and the larger size of nets required stronger gallows, more powerful winches and, as already mentioned, greater speed.

These few facts reveal the trend of development and show how comparatively new are the ideas of mechanization and the knowledge of the correlation between vessel and fishing gear. And development is still going on. Larger gear and higher trawling speed will be used and greater depths will be exploited. It is doubtful whether the present equipment can be improved to meet such new demands. Perhaps new types of auxiliary machinery will have to be designed.

Most of the gallows are criticized by trawler fishermen because the trawl lines get untwisted by badly operating rolls and are left in unequal lengths. This brings the fishing gear into a wrong position and risks the catch.

The design of winches will have to be considered when trawling is carried on at greater depths. Most of the present winches will not take the strain of hauling much longer trawl wires and, in any case, mechanical hauling of the trawls is still unsatisfactory. Driftnet fishermen have invented electrically driven rolls for hauling the nets but trawling always includes plenty of manual labour when hauling.

If electric fishing develops, it will be necessary to instal generators on deck and more auxiliary machinery below deck. There is only one vessel, so far, engaged in electric trawling. For the purpose of experiment, a generator of very large dimensions, producing 10.000 amperes, has been installed and it takes up a space of 6.5 by 13 by 6.5 ft. (2 by 4 by 2 m.). It must be large to overcome the 500 fold conductivity of seawater and to produce an adequate catching area. Even though the size of the generator might be reduced some day, the safety for the people on board must be taken into account. Moreover, electric fishing needs special winches for the electric cables, and a suitable type has not yet been developed. For reasons of safety the stern of the vessel should be covered.

It is of great importance to observe and control a trawl in action, particularly when the net is being filled. This is indicated by a change in speed and, simultaneously, increased thrust and decreased revolutions because of increase of resistance. The difference can be recognized particularly if the trawl is overloaded or torn. Instruments have been invented in Germany which inform the skipper when the trawl is filled.

The equal length of the two trawl wires is essential in trawling. The tension of each wire can be tested by a comparatively simple device and can be adjusted by slackening one wire or the other as required, so long as the warps are not exposed to friction after the measured spot. Although there are numerous types of slip hooks, most of them do not ensure smooth working of the wires, each of which should run over a separate roll. Moreover, the trawl wires are pressed against the hull even with the slightest turn of the vessel and this adds to the differences in the pull. A few vessels are so designed that the wires can run out without trouble. Then an exact control of the net is impossible because any difference of pull cannot immediately be felt by the wires.

It should be possible to record continuously on the bridge the height and breadth of a trawl, and the draught if it is a floating trawl. Extra cable winches would have to be installed to connect the bridge with electronic depth recorders on the trawl. But electrical cables are difficult to handle in rough seas, so it would be a better idea to watch the behaviour of the trawl by means of optic or acoustic devices, except that the race of the propeller and air bubbles under the vessel would prevent this. Here again is a problem which naval architects and fishing gear technologists should study together.

The firms making echo sounders have varying opinions about attaching the transmitter and receiver to the ship's bottom. The race of the propeller and air bubbles from the bow produce most of the trouble in both vertical and horizontal sounding. It would be desirable to have a well in the ship, through which the sounder could be lowered, but, at present, this is only practical for research vessels and it remains for future study to show how the sounder can best be attached to the bottom of a fishing vessel.

Improvements of fishing boats are not confined to nautical questions and existing fishing gear problems should be considered by naval architects. Up to the present, all fishing equipment has been attached to the exterior hull without altering the vessel herself, but progress in fishing gear technology will compel the naval architect to pay more attention to new fishing methods when designing vessels in the future.

RECENT DEVELOPMENTS IN DECK GEAR

by

CARL B. CARLSON

THE fisheries of the United States are faced with increasing shortages of skilled manpower and this is encouraging more mechanization of fishing operations and improvement of deck gear.

PUMPS FOR UNLOADING MENHADEN

Pumps have been used for many years in the California sardine industry for transferring fish from a hopper alongside the vessel to the shore plant or processing ship. More recently pumps were introduced into the Maine sardine and the Atlantic and Gulf menhaden fisheries to transfer fish from the vessel's hold to the plant. Previously, in the menhaden fishery, the dry fish were shovelled by hand into a vertical conveyor for delivery to the measuring hopper. This was hard and undesirable work, and the labour shortage often slowed operations. On days when fish were plentiful, a vessel might wait 12 hours or more to be unloaded. With centrifugal or reciprocating pumps, a menhaden vessel can be unloaded in a matter of minutes.

Menhaden vessels have a rectangular pipe with removable manhole plates built fore and aft into the bottom for the entire length of the hold. This terminates in a vertical pipe through the deck to which is attached an 8 to 10 in. (20.3 to 25.4 cm.) diameter suction hose. The fish in the hold are wetted and sluiced to the manhole plates where they and the water are removed. When shovelling by hand some fish were usually left in the hold and the wash down was inadequate. The use of pumps and large quantities of water have greatly reduced obnoxious odours and flies and has contributed to general cleanliness.

PUMPS FOR BRAILING MENHADEN

Drying the menhaden in the fish bag and brailing the fish aboard the vessel is also a time consuming and laborious task. The success of pumps for unloading led to testing them on the brailing operation. Menhaden are caught by two boat purse seining and at least 10 men per boat are required to " harden " or dry the fish for brailing. When the catches are large, or the fish are " hard worker ", it is almost impossible to harden the fish by hand, and it may even be necessary to use a powered fall for assistance. Under these conditions, the seine is likely to tear and most of the catch lost. The use of pumps greatly reduces the effort required to raise the fish because they need not be dried to the degree required for conventional brailing. The intake end of the suction hose can be lowered considerably below the surface and beyond depths that can be reached by a brailer. The time saved can be used for making additional sets.

The operators at Fernandina, Florida, U.S.A., believe the centrifugal pump (*fig. 525*) is more practical than the reciprocating type because it requires less space. A very compact unit, consisting of an 8 or 10 in. (20.3 to 25.4 cm.) pump and a 2 in. (5.1 cm.) diameter priming pump mounted on a single frame, is now being manufactured. This same unit, with the addition of a 5 in. (12.7 cm.) diameter pump for sluicing the fish in the hold, can be used for a shore installation to unload the vessels. Power is supplied to the main pump shaft by high-speed gasoline or diesel engines rated at 65 to 100 h.p., and the auxiliary pumps are driven from the main shaft by sprockets and chain. Reinforced suction type hose 8 or 10 in. in diameter, and from 20 to 30 ft. (6.1. to 9.1 m.) in length, is used to direct the fish from the net to the pump. The hose is too heavy to manipulate easily so a hand tackle, or a powered single fall, is used to govern its position. The intake end is fitted with ½ in. (13 mm.) diameter iron bars in the form of a hemisphere to prevent the net from clogging.

After leaving the pump, the fish and water are discharged on a screen consisting of ½ in. diameter bar irons at an incline of about 30° to drain most of the water and to direct the fish to the hold. The separated water may either be discharged to the deck or piped overboard on the idle side of the vessel. Surplus water carried with the fish is removed by the vessel's bilge pumps.

POWERED SEINE REELS IN THE SALMON FISHERY

The purse seine fishery for salmon has long been considered highly efficient from a mechanized viewpoint because a crew of eight or nine men can haul a seine, over 300 fathoms (550 m.) in length and 25 fm. (46 m.) in depth, six or seven times per day. In the Alaska fishery, twelve or thirteen hauls per day with a seine 200 fm. (366 m.) in length and over 15 fm. (27.4 m.) in depth, are not uncommon for a crew of seven or eight men. But the recently introduced power reels, or drums, indicate that five men can make up to 20 sets per day with a 300 fm. seine.

Salmon are caught by the " one boat " method with the aid of a powered skiff. In established practice (Carlson 1945) the seine is set and hauled from a turntable mounted on the stern, which can be swung through a complete circle. The turntable is fitted with a roller which is powered to assist in hauling by friction between the net and roller, and a clutch for free-wheeling while setting. The seine is set over the stern and hauled over the side. Upon completing the circle, the purse lines, one from forward and the other from aft, are led through blocks on a davit amidships to the winch for power pursing. When pursing is completed the purse rings and lead line are lifted aboard. Meanwhile the cork line has been bunched on the working side of the vessel by men on the stern and in the skiff attached to the bow. When the turntable is in the hauling position, the cork line is piled on one of the tables over the stern; the lead, purse and ring lines are piled on the forward side; and the netting laid between them. This arrangement enables the crew to haul various portions of the seine according to pre-assignment and keeps the lead and cork lines well separated while setting.

By the new drum system (Drum seining 1953), the net is set and hauled from a fixed reel set in a tank in the stern. The net is set in a fairly straight line, or an elongated hook, and the circle may be completed by towing on one or both ends by either or both the fishing vessel and the powered skiff. The procedure depends on the type of set most advantageous for the locality and the behaviour of the fish. When " pursing ", or closing the bottom, only the forward purse line is taken to the winch, while the after purse line and a portion of the seine are wound on the drum. The purse line finally becomes taut between the davit and the reel. The remaining purse rings are picked up on an elongated " U " iron, or " clothespin ", suspended in a near horizontal position by a bridle from the boom. As the net continues to be reeled on, successive rings slip off the clothespin as the lead and ring lines become tight. A proportionate length of purse line is paid through the rings to be reeled on the drum with the seine. No effort is made to keep the cork and lead lines separated as in conventional practice, but one of the problems is to reel the seine evenly and tightly so that it will not bind while being set.

Fig. 525

Before being wound on the drum, the seine passes over a horizontal roller 8 to 10 in. (20.3 to 25.4 cm.) in diameter and then between a pair of vertical rollers mounted on a traveller controlled by hand through a level wind gear (*fig. 526*). The vertical fair leads are about 8 in. in diameter and 30 in. (76 cm.) high with a free space of about 10 in. between them. They are mounted on a hinge so they can be tipped inboard to avoid tearing the net while setting.

The power skiff plays an important part in drum seining because one of its functions is to tow the vessel from the non-working side to maintain proper fair lead of the seine for hauling. The drum system requires close co-ordination of the entire crew because the chances of fouling the gear on the rudder or propeller are increased.

The drums are made of steel and have ample capacity if the seines are tightly wound. The dimensions vary from 5 to 9 ft. (1.5 to 2.7 m.) in diameter and are of a width to suit the space on the stern of the vessel and the volume of the seine. Reels now in use have a capacity from 225 to 400 cu. ft. (6.37 to 11.3 cu. m.) but one reel, 9 ft. (2.7 m.) diam. by 10 ft. (3 m.) in width, with a capacity of 625 cu. ft. (17.7 cu. m.), is to be installed on a Canadian purse seiner. The reels are set in a tank in the stern, made watertight from the rest of the vessel but fitted with a self-bailing device for drainage. It would not be surprising if some enterprising fisherman devised a tight top for these tanks to provide extra fuel capacity for use in other fisheries where longer cruising range is required. Such tanks might also be made suitable for carrying live bait in the albacore tuna fishery.

Fig. 526

The reels must be adequately powered with variable speeds, forward or in reverse. When hauling, it may be necessary to reverse the drum if it becomes evident that a "bind" is being made, which will interfere with a subsequent set, or to help start the seine when a set is being made. Furthermore, the pursing strain may exceed 5,000 lb. (2,270 kg.) line pull and variable speeds are required, depending on the strain and the need to remove gilled fish or those caught in a "pocket". The drum must be free wheeling so that the seine can be set easily, and a good brake must be provided to stop and hold the net in the water.

The reels are powered by a hydraulic motor or a mechanical system. In the latter, power is taken from the main engine and transmitted through a system of sprockets, roller chain, and shafting, to a truck transmission short-coupled to a truck differential and brake drum on the reel. Speed reduction up to 84:1 and reversing the reel, are possible by this method. The hydraulic drive (*fig. 527*) functions by oil under pressure from a pump, through a reversible and free-wheeling hydraulic motor coupled by a chain and sprocket drive to the drum. Fishermen who favour the hydraulic drive claim smoother and more positive control of the drum speed in either direction, while those who prefer the mechanical drive claim greater pulling power in hauling the seine.

The present system of locating the drum on a fixed horizontal axis presents certain limitations which did not exist with the turntable system. The turntable roller can be moved from the stern to a fixed position over either side, depending on how the set is made, and better fair lead can be obtained while hauling. Fishermen are considering how to mount the horizontal reel on a pivot to suit best the angle of the seine for hauling. If this is done, the hydraulic drive system may well gain added popularity because of the flexibility of the oil supply hoses compared with the fixed limitations of a mechanical drive. Some thought is being given to mounting a drum on a vertical axis but level winding above and below the deck line then becomes a serious problem.

Fig. 527

A different method of hanging the seine is also essential for the drum system. Customary practice is to hang salmon seines in flights of 10 fm. (18.3 m.) consisting of 10 fm. of cork line, 11 fm. (20.1 m.) of stretched measure netting and 9 fm. (16.5 m.) of lead line. This is known as 10 per cent. hanging, i.e. 10 per cent. more netting than cork line, and 10 per cent. less lead line than cork line. This type of hanging has proved effective both in shallow water, where it tends to keep the purse and lead lines clear of the netting, and in deep water, where it tends to form a bottom while towing the seine. When hauling by the drum method, this system is unsuitable because the lead line is too short for proper spooling. Distortion of the meshes and loops in the cork line would take place and they would bind when setting. However, the lead line can be hung about 3 ft. (0.9 m.) shorter per flight, or 5 per cent. instead of 10 per cent. shorter than the cork line, and be spooled evenly because of the greater bulk of the cork line. Experience indicates that 5 per cent. hanging of the lead line will not hold fish in deep water, but fishermen who favour the drum system maintain this can be overcome by hanging in an extra 2 or 2½ fm. (3.7 or 4.6 m.) or 30 to 35 per cent. of extra netting per flight, thereby creating a great bag in the seine.

The use of drums for hauling nets is not a novel idea. It has been used in the salmon gill-net fishery for a number of years but, in this fishery, the problems are simpler; the cork and the lead lines are of equal length, the net is shallow, and only one or two hauls are made per night. The drum seining method was developed by Canadian fishermen several years ago and has been used by one boat on Puget Sound, Wash., U.S.A., for some time. In 1953 11 boats were fitted for drum seining.

Fig. 528

Further development and modifications of drum seining may well reduce manpower requirements in the menhaden, herring and mackerel fisheries.

Menhaden seines, for example, are always set in a circle and the working of the fish toward the stern of the purse boats while the seine is being hauled suggests that the reels could advantageously be set at an angle to the boats. The use of pumps has lowered the effort required to brail fish and power reels may carry this a step further.

APPLICATION OF CHILL TANKS IN THE SHRIMP FISHERY

Recently several shrimp trawlers, using equipment designed by a commercial engineering firm, have been fitted with tanks of refrigerated sea water as a medium for cooling and holding shrimp. The University of Miami Marine Laboratory is doing considerable research work on holding shrimp in this medium and on means of surmounting some of the technological problems involved, (Higman and Idyll 1952). The results to date have demonstrated that the formation of " black spot ", a discoloration caused by enzymatic action under the shell, can be retarded by holding shrimp in refrigerated sea water. An oxidase enzyme seems to be involved. At temperatures of 30 deg. F. (—1.1 deg. C.) or lower, spoilage of shrimp can be retarded for a longer period in sea water than in regular ice. After 18 days in ice, shrimp have deteriorated to the point of being unmarketable but, held in refrigerated sea water, they have been palatable after 24 days. Unfortunately, off-odours develop in the raw uncooked shrimp after 14 days and it is causing some buyer resistance.

A practical use for chill tanks is found aboard freezer vessels. Several advantages occur because the headed shrimp can be chilled more rapidly and can be packaged for freezing as time permits. If large catches are made, the whole shrimp can be chilled pending disposition. Furthermore, the temperature of the shrimp is reduced from the ranges of 85 to 90 deg. F. (29 to 32 deg. C.) to 34 to 40 deg. F. (1 to 4.5 deg. C.), thereby retarding deterioration and easing the load on the freezer. The tanks are about 10×3×3 ft. (3×0.9×0.9 m.) in size, well insulated with 4 in. (10 cm.) of cork, and are refrigerated by the main system for freezing and holding the catch. With the freon system, the use of ⅝ in. (16 mm.) diameter copper tubing on approximately 8 in. (20 cm.) centres is a common practice. One exploratory vessel now operating from remote Central and South American ports has a similar tank with a capacity of about 5,000 lb. (2,270 kg.) headed shrimp. The refrigeration system is driven by a compressor requiring about 2 h.p., and temperatures of 30 to 35 deg. F. (—1 to 1.6 deg. C.)

are maintained, depending on the quantity of shrimp put in the tanks. Before departure, and when *en route* to the fishing grounds, the operators try to reduce the temperature of the sea water to 28 deg. F. (−2.2 deg. C.), or lower, to form slush ice as a reserve of cold. To reduce contamination, the shrimp are thoroughly washed with sea water before immersion in the tank. It is reported that shrimp can be held in excellent condition by this method for at least five days. But operators of freezer vessels maintain that shrimp should not be held for more than 48 hours in chill tanks because a difference in quality is detectable, through loss of flavour, compared with those frozen soon after catching.

AIR CONTROLS FOR WINCHES

Mechanical controls for the brake and friction of trawling and other winches are standard practice, but these have inherent hazards which can be minimized by air controls. When controlling the brake and friction of trawl winches, the use of two hands, or one hand and a foot, is required where two men operate the winch. In the southern shrimp fishery, where one man operates the double-drum winch, the use of both hands and both feet are required. Operators of shrimp trawling fleets are faced with increasing difficulties in finding skilled crews and have generally avoided winches with superior brakes and frictions of a mechanical control type. They fear that full brake may be applied against a full friction, causing undue strain on, and damage to, the drive mechanism. When the vessel is rolling in a heavy sea, the difficulties in manipulating mechanical controls are increased.

By using a two-way valve with a neutral position, full brake or full friction and intermediate degrees of each can almost immediately be applied by a simple movement of one hand. In the neutral position the drum is in free-wheeling. Air from the vessel's tanks is delivered through a reducing valve which maintains a working pressure of 125 lb./sq. in. (8.8 kg./sq. cm.) at the control valve. In one extreme position full air pressure is delivered to a piston attached to the brake hand. An unloading spring on the non-pressure side, together with a return release line to the valve, completely releases the brake as the lever is moved to neutral or free-wheeling position. At the opposite extreme position, full air pressure is delivered to an oval-shaped, air-flex clutch tube fixed to the driving side. Air causes the tube to expand and engage the drum to provide full friction for hauling. A return relief line to the valve, and the desire of the air-flex tube to resume its normal moulded form provides free-wheeling of the drum as the control valve reaches the neutral position. By this system one operator can easily control two drums, one with each hand.

Air control systems for trawl and purse seine winches (*fig.* 528) have been used for several years on the Pacific coast and have proved satisfactory. Operators believe them to be safer than mechanical controls because of more rapid response and greater freedom in rough weather. The disadvantages are the need for an air supply and the greater cost. The air, after compression, is reasonably dry but if the tanks are not periodically bled of water, in accordance with good marine practice, quantities of water could be drawn into the system and freeze in cold weather.

ELECTRIC DRIVE FOR TRAWL WINCHES

by

MAURICE GRAFTIAUX

THE power at which a trawl winch may operate is limited by the normal strength of the warps and the average hauling speed. The best control system is to have the winch operate with constant power between the extremes of maximum pull and maximum speed. The control must also permit flexibility and quick manoeuvring, safety and easy maintenance.

The main operations of the winch are as follows:

1. *Paying out.* During this operation the drums are running loose on the shaft as the trawl, immersed, is run out by the boat moving forward. No power is required from the winch. The warps are kept tight all the time by the brakes with which the lengths are also kept equal.
2. *Trawling.* During this operation the drums are declutched and the brakes are tight. No power is absorbed.
3. *Hauling in.* Both drums are in clutch and the winch motor is running in full power. This is the real working period for the winch. The power required varies because of the friction of the trawl on the sea bottom and of the behaviour of the boat in rough sea.

FLEXIBILITY AND MANOEUVRING

The principal manoeuvres are:

Low speed when paying out the trawl, the bobbins and doors, and high-speed when paying out the warps.

High speed when hauling in the warps, medium speed when the trawl approaches the boat, and very slow speed when taking the trawl doors on board.

The flexibility of the winch does not depend solely on the progression of the speed range but also on quick response. A fisherman will think that the winch is slow when it is slower than his own reactions.

SAFETY AND EASY MAINTENANCE

The trawl winch is the most important equipment on board a trawler and if it is damaged or stopped, the vessel is put out of action. Dependability is, then, more important than any other factor. The control system must be such as to prevent overloading or misuse of the circuit breaker. A special danger is overloading, which is always apt to happen at a critical moment and may disrupt the boat's electrical installation. Conditions of fishing are such that maintenance of the winch is difficult and only the simplest and most robust type is suitable.

ELECTRIC WINCHES

Electric winches usually get their power from independent direct-current generators but some get it from the general circuit of the boat. The winch motor always operates on direct current with separate excitation. The speed variation is obtained by variating the voltage in the armature and sometimes by a variation of the excitating current. In case of independent drive, the motor receives a variable voltage current from a diesel generator set. In the other cases the motor draws its current from the general circuit through an auxiliary generator operating as a reversing booster. The first system has the advantage of complete independence, and the disadvantage of a relatively powerful diesel remaining idle when the ship is not fishing. In practice, it is the best solution when the power of the winch generating set is greater than the average power supplied for the ship's use. The second system is justified when the winch power is less than the power of the other auxiliaries.

Independent generating sets have been installed in the French trawlers built after World War II under the reconstruction programme. The winch motor receives its current from a generator with a " falling characteristic ", i.e. with a voltage variable with the load, the curve being determined by three points:

(*a*) maximum voltage when running idle;
(*b*) maximum power for normal current;
(*c*) no voltage when stalling.

This automatic regulation is obtained by three different excitation coils:

(*a*) one separate excitation coil for the control of speed;
(*b*) one series anti-compound excitation coil;
(*c*) one shunt excitation coil for amplification.

Inside this curve different intermediate values can be obtained through controller connected with the winch control lever and acting on the separate excitation coil.

All the auxiliary circuits are fed by an exciter driven by the generating set, which also supplies direct current for the ship's use.

Control instruments are reduced to a minimum and, in some of the ships, the controller is connected to the exciter through a battery of contactors. These have never given any trouble.

The characteristics of the winch have been improved by a few devices of which the most important is the third position, so that a satisfactory stalling torque can be kept on the first position without modifying the characteristics of the curve at the last position.

Another useful addition is the *electro magnetic brake* on the shaft of the motor. When the winch lever is brought to zero the brake stops the motor immediately and corrects, to a certain extent, the lag of the generator, especially at slow speed.

The magnetic brake is also slow, which is a hindrance when the winch is being started. It increases the inertia of the motor shaft and in some ships it has been replaced by a friction brake worked by a foot pedal. Winches would be improved if given more flexibility and if some partial recuperation of energy by the diesel was admitted, as already achieved in diesel electric locomotives.

Table LXXX

Standardization of French trawl winches

Particulars	*Trawlers size (length)*		
	102 ft. (32 m.)	*138 ft. (42 m.)*	*213 ft. (65 m.)*
Nominal power	80 h.p.	110 h.p.	150 h.p.
Nominal hauling speed	3.3 ft/sec. (1 m./sec.)	3.3 ft./sec.	3.3 ft./sec.
Nominal load (mean radius)	4.500 tons	6.000 tons	8.500 tons
Stalling load	6.500 tons	8.000 tons	12.000 tons
Uncoiling load (trawl stalled, winch motor energized, boat moving forward)	9.300 tons	11.400 tons	17.000 tons
Speed at no load	6.6 ft./sec. (2 m./sec.)	6.6 ft./sec.	6.6 ft./sec.

torque adjuster. The speed and the pull are determined for a variable ratio which depends on the coiling radius on the drum. A maximum stress on the warp being imperative for safety, the maximum torque of the motor must correspond to the smaller coiling radius, with inevitable reduction of the pull when the warp is coiled on the drum at the greatest radius. This happens precisely when the trawl is near the surface and the heaving effect is most felt. The torque adjuster must automatically modify the separate and shunt excitation proportionately to the coiling radius and maintain an almost constant effort. Unfortunately this is difficult to reach and the present system needs some adjustments.

Another addition would be to control the speed by combining the separated and shunt excitation from the

Trawl winches running on current supplied from the ship's general installation have a much quicker response in manoeuvring. In this case, the installation must be more powerful than the winch, otherwise the exchanges of energy by the intermediate variable voltage generators will start the circuit breakers and stop the winch.

The variable voltage generator is of the three-coil excitation type but with a much shorter time constant (the power is smaller and the speed greater). It is easier to adjust the characteristics curve but a reversing relay must be installed for reversing the motor.

HYDRAULIC TRAWL WINCHES

by

PAUL GUINARD

A TRAWL WINCH should fulfil the following requirements:

Great variation of speed in both rotating directions.
Increasing torque when the hauling speed decreases.
Stalling of chosen torque and paying out if the pull exceeds the limited torque.
Easy and flexible operation.
Great sturdiness.
A minimum maintenance—repairs to be done on board without a specialist.

All these requirements are fulfilled by the hydraulic winch.

GENERAL

Basically an hydraulic installation includes:

1. A pump, generally driven by an internal combustion engine.
2. An hydraulic motor (fed by the pump).
3. A regulator for speed and reversing, very often incorporated in the hydraulic motor.
4. A valve controlling the pressure (i.e. the torque).

The valve is automatic and adjusts the hoisting speed without manual assistance. The speed depends only on the pull of the trawl.

The oil circuit can be open or closed, open if the pump operates from an oil tank, closed if it is directly cônnected by piping, without any tank, to the hydraulic motor.

Pressure

285 to 425 lb./sq. in. (20 to 30 kg./sq. cm.) pressure allows use of commercially-made pipes and valves and avoids gaskets and high-pressure piping, which are difficult to repair on a trawler. With a high pressure system there would be a small saving in weight, but the oil cooler would be bigger.

Driving Fluid

The usual driving fluid is oil, which should be pure and have a viscosity around 6 deg. Engler.

Choice of Equipment

The most commonly used devices are the " volumetric " type, for pumps as well as for motors. Typical equipment, taking into consideration the simplicity of installation and operation and the best efficiency, is:

(1) *A pump with a fixed cylinder capacity* driven by a constant speed diesel and placed in the engine-room. This can be run without any control and almost without maintenance. With a constant flow at the maximum pressure, controlled by a by-pass valve, it will deliver constant power and not overload the diesel.

(2) *An hydraulic motor with variable cylinder volume* driving the winch. Variation in the cylinder volume is easily made by turning a wheel. The operator is able to adjust the winch speed, including acceleration, slowing down, stopping and reversing. According to the required speed, the motor will transform the power supplied by the pump into variable torque. If the resistance is small, the hoisting of the trawl will be possible at a great speed but if it is great then the speed of the hydraulic motor is reduced. In both cases the total power of the pump will be used.

(3) *A by-pass valve limiting the torque*, placed either on the hydraulic motor or cônnected to the piping, is an important item. Calibrated to a slightly higher than normal working pressure, it switches off the circuit when the limit is reached, and keeps the motor charged at its maximum torque (stalling torque).

If the resistance increases over the pre-set maximum torque, the winch motor reverses because the pressure in the oil pipe remains constant, so that neither the pump nor the diesel is overloaded.

When the resistance of the trawl becomes normal the hydraulic motor automatically starts again, progressively and smoothly.

Such a by-pass requires the most careful design and construction.

AN EXAMPLE OF AN INSTALLATION FOR A 75 h.p. WINCH

Pump

Constant flow pump, (*fig. 529*) directly coupled to a 120-h.p. diesel, 575 r.p.m. working capacity 4,450 cu. ft./hr. (126 cu. m./hr.); working pressure 290 lb./sq. in.

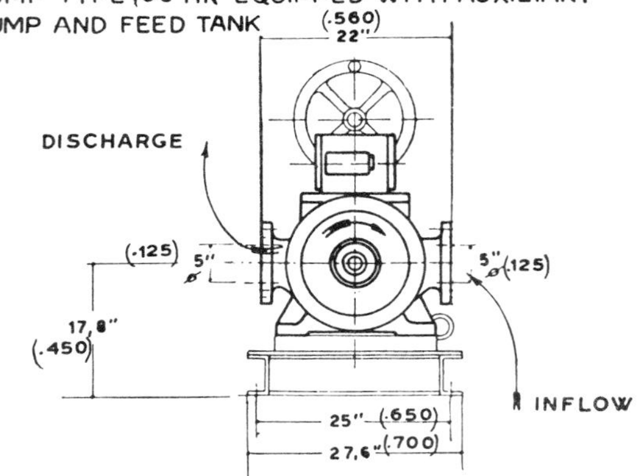

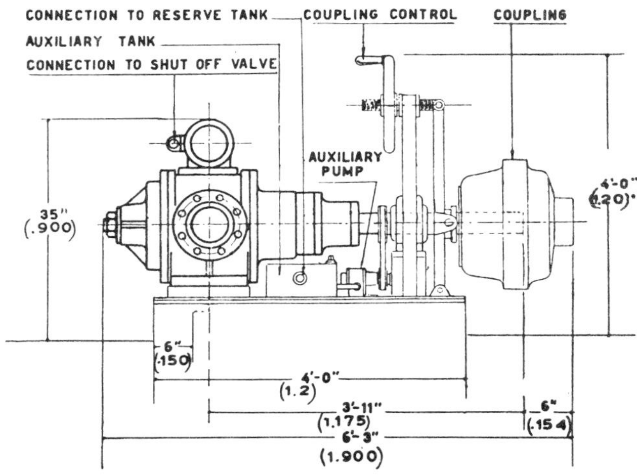

Fig. 529

(20.5 kg./sq. cm.); at this pressure the output is 85 per cent. and the absorbed power is 112 h.p. The by-pass valve limiting the torque is calibrated to 310 lb./sq. in. (22 kg./sq. cm.).

Hydraulic Motor

Variable cylinder volume and reversing arrangement type, fig. 530, variation of speed by a hand-operated wheel, maximum speed 500 r.p.m., 74 working h.p. at a normal pressure.

Speed 250 r.p.m.—76 working h.p. at a normal pressure. Decreasing pressure from 291.5 lb./sq. in. (25 kg./sq. cm.) to nil. At this by-pass pressure the hydraulic motor develops a power of 81 h.p., corresponding to a maximum stalling torque of 1,663 ft. lb. (230 kg.m.).

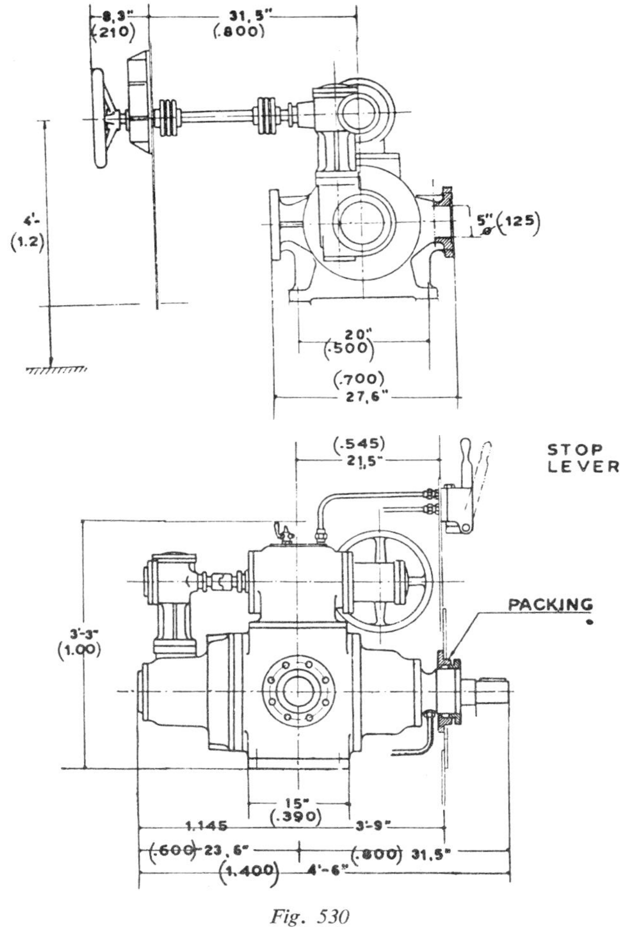

Fig. 530

Same range of power and speed in reverse direction.
By-pass equipped with anti-hammering device.

Closed circuit with an auxiliary pump, feeding the return circuit (pressure below 14.7 lb./sq. in. (1 kg./sq. cm.).

Fig. 531 shows the characteristics for this installation.

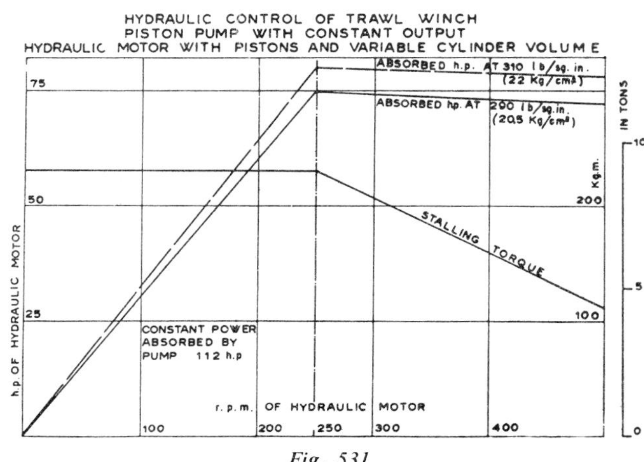

Fig. 531

CONSIDERATIONS OF THE HYDRAULIC TRANSMISSION APPLIED TO TRAWL WINCHES

When hauling, it is necessary to adjust the speed according to the resistance but, at the same time, it is necessary to take account of the condition of the sea. A transmission which automatically adjusts the speed according to the resistance torque is desirable.

" Volumetric " devices, notably the piston type, need a particular care in their manufacture. Great accuracy of machining is required for increased pressure. The following system is not so difficult to manufacture and is simpler in operation:

(1) *Pump.* A centrifugal-type pump is chosen because it can give high pressure with a low output and, inversely, a low pressure with a high output, according to requirements.

Fig. 532 and 533 show that the curve with the best pump efficiency corresponds to the most commonly-used speeds of the hydraulic motor. The speed of the hydraulic motor varies automatically from 0 to the maximum, according to the pull of the trawl.

The principle of the centrifugal pump permits the flow to be reduced, for a given pressure, while still maintaining the same pull. The reduction of the flow is obtained by the simple operation of the slide-valve and the power absorbed by the centrifugal pump decreases.

This type of pump has no fragile parts. It is composed of impellers in series. The periphery

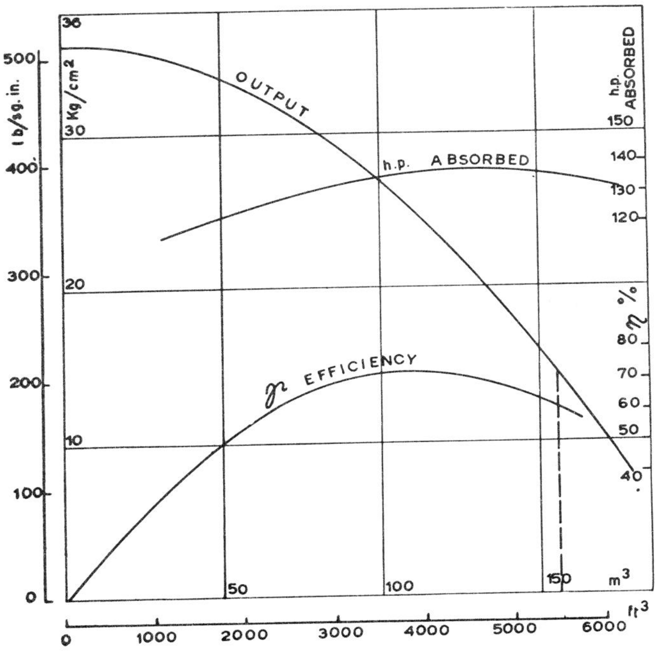

Fig. 532

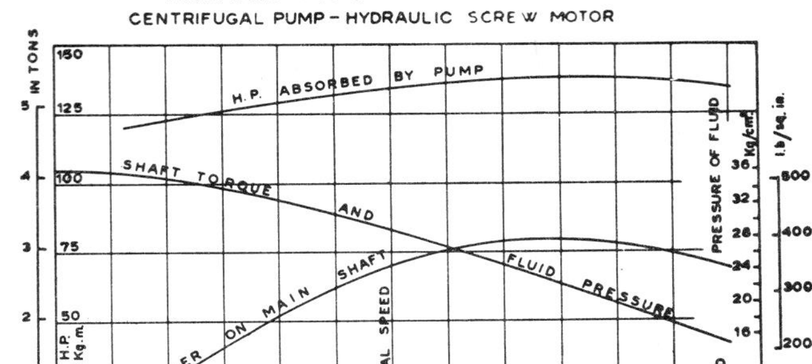

Fig. 533

clearance between impellers and leakage ring is only 0.012 to 0.016 in. (0.3 to 0.4 mm.).

The shaft rotates on ball-bearings which take hardly any strain in a well-constructed pump where the impeller is axially and radially balanced.

The centrifugal pump eliminates the troubles usually associated with the volumetric pump, such as undue friction leading to wear and tear and lack of mechanical precision. It does not need a by-pass relief valve.

(2) *Hydraulic motor.* Another simple type of hydraulic pump, not so well known because its kinetic characteristics are less evident than those of the gear pumps, is remarkable for its sturdiness. It can operate in both directions and experience with it has led to the development of some excellent hydraulic motors based on the same system. Numerous types are in service. A suitable motor to drive a trawl winch, fig. 534, has two long screws on parallel axes which rotate in corresponding bores. The screws have a large square-cut thread and they work together, i.e. they are engaging along their whole length, therefore they must have opposite threads. They rotate in opposite directions through gears keyed on their shaft, so there is no friction between the two screws. The opposite pitch of the two pinions assures an absolute axial balance of rotating parts. The thrust bearings have, therefore, no action load to support. The ends of the pinions are sunk in chambers which are connected by the outlet. The " Guinard " motors have no mechanical friction either between the threads of the screws or between the screws and the bores because of appropriate clearances. No wear can occur.

(3) *Slide-valve for reversing direction and adjusting speed.* To reverse the rotation of the hydraulic motor, the inlet and outlet of the driving fluid must also be reversed.

The statically and hydraulically balanced, hollow, slide valve operates in a cylinder fitted with

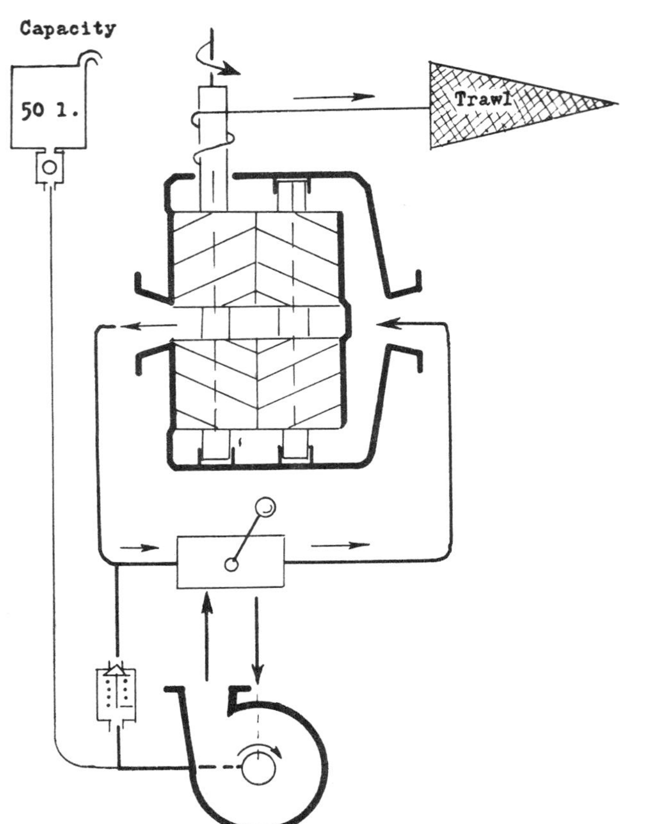

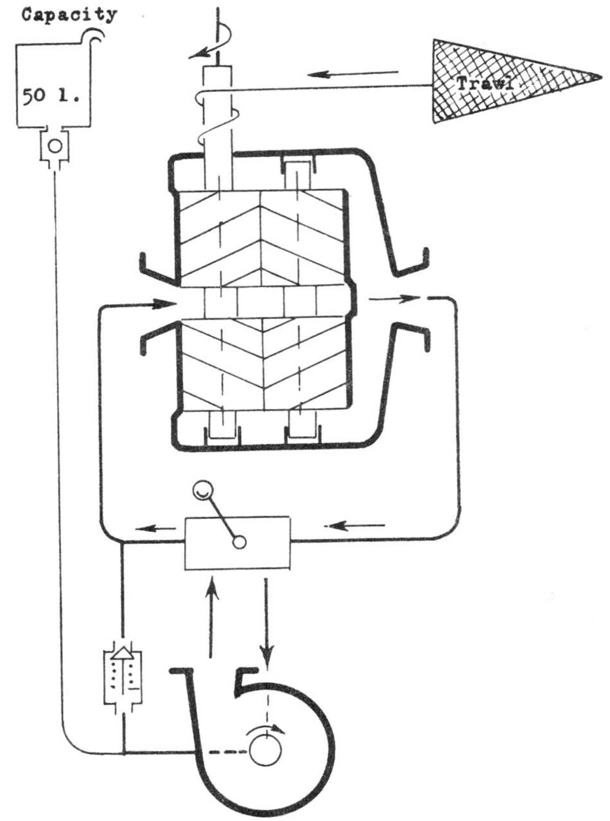

PAYING OUT OF TRAWL

Reversing by Slide-Valve.

The hydraulic motor gives only low power at low pressure and turns with max. speed.

The pump is fed under pressure, and if the trawl pays out too fast, there is no risk of cavitation, as it is easy to brake the motor by means of the slide valve and so control the paying out speed.

HAULING IN TRAWL

(A) The hauling is done at the most favourable speed which depends on the resistance of the trawl.

(B) As soon as the stalling torque is reached, the motor stops. The motor maintains a max. permissable tension on the cable.

(C) If the stalling torque is surpassed, the motor pays out. A damping valve prevents overcharging the diesel.

(D) Slow manoeuvres are made by using the slide valve.

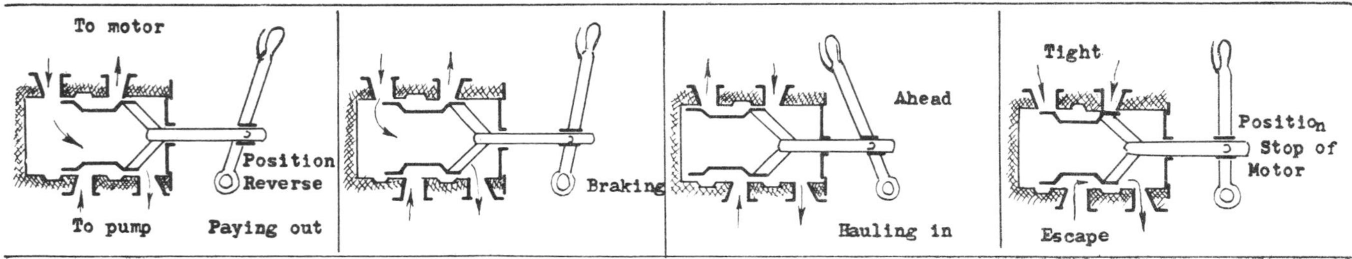

Fig. 534

four vents (*fig. 534*). It controls the motor fluid circuit and regulates the pump for slowing the hydraulic motor. It is used as an hydraulic brake when there is rapid paying out and makes slow manoeuvrings possible in both directions. When stationary it closes the vents of the motor to allow the pump to rotate at a low but sufficient flow to avoid abnormal heating of the circuit. The valve is easily operated by the control lever (*fig. 534*).

CONSIDERATIONS OF THE WHOLE INSTALLATION

The centrifugal pump is usually placed in the engine room and does not require supervision. A slide valve is mounted on the motor which can be mechanically or hydraulically controlled by foot pedal, if necessary, to free the hands of the winch operator. A valve prevents overloading of the diesel when hauling at full load.

A small tank (50.1 imp. gal.) keeps the system charged to prevent air entering and to replace fluid that has leaked from the stuffing-boxes. The driving fluid used

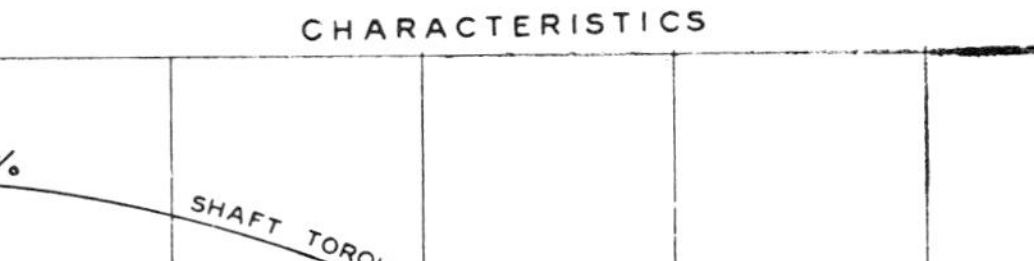

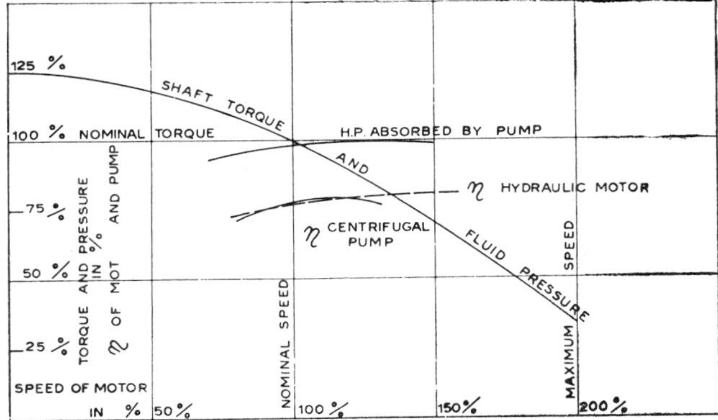

Fig. 535

is a low viscosity oil (3 to 4 Engler). Fig. 535 shows that when the pull increases the speed decreases and, at the same time, the power absorbed by the pump decreases.

The overall efficiency of the drive is 0.62 to 0.63 at normal speeds of the winch. The installation is flexible, simple and sturdy, and it is easily operated by fishermen.

HYDRAULIC DECK EQUIPMENT

by

HANS VESTRE HUSE

THERE are various ways of transmitting power from an internal combustion engine to the winch, In smaller ships the arrangement often consists of small auxiliary engines on deck directly connected to the winches by couplings or gears. Fishing vessels often use a belt or chain transmission from the main engine.

Electric winches are so far used only on the bigger trawlers and similar ships, and it is apparent that they can compete very well with steam winches. Ordinary electric winches, without Ward-Leonard couplings and, to a still higher degree, purely mechanical transmissions, are too rigid, and the regulating possibilities are often limited. The hoisting speed depends mostly on the r.p.m. of the engine and to achieve variations in this speed the r.p.m. must be changed. This is especially inconvenient when the main engine is used for driving the winch, especially if it is connected to a fixed-blade propeller, but if a controllable pitch propeller is used, the pitch can be altered to maintain the same speed of the ship. This dependence between speed of the ship and drive of the winch is naturally very inconvenient.

The hydraulic system of power transmission is more often used where the demand is for good regulating and adaptability. Another great advantage is the hydraulic motor's excellent acceleration. Up to 250,000 radians per second^2 are possible, whereas the maximum acceleration for fractional horse-power electric motors is only 50,000 radians per second^2. An electric winch has to lift a load 26 to 30 ft. (8 to 9 m.) before the full hoisting speed is reached; an hydraulic winch reaches full speed almost immediately after the start. Experience shows that, because of its quicker acceleration, the maximum hoisting speed for an hydraulic winch may be only 60 per cent. of that of an electric winch to handle a given quantity of cargo in the same time.

The high torque, compared with the weight and space required, is another feature of the hydraulic system. Fig. 536 shows a comparison between an hydraulic and an electric motor of the same h.p. and r.p.m. The principle of hydraulic drive of winches on board a fishing vessel is shown in fig. 537. An hydraulic pump is connected to the main engine through a claw coupling. This pump forces an oil flow through a system of pipes to the winch. There the oil at first goes through a control valve which regulates the quantity of oil to the hydraulic motor. The engine rotates at a speed corresponding to

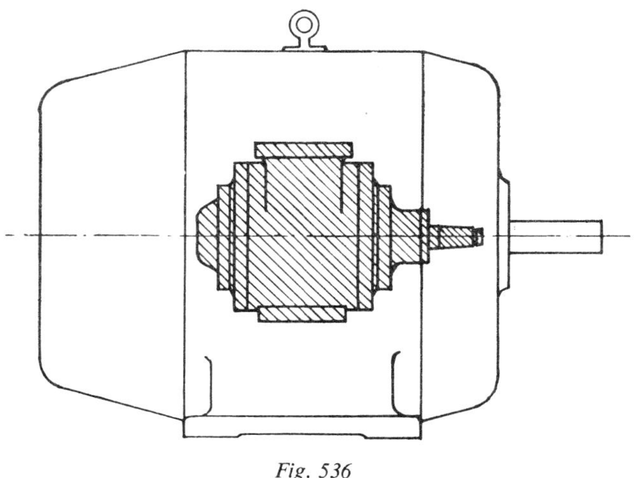

Fig. 536

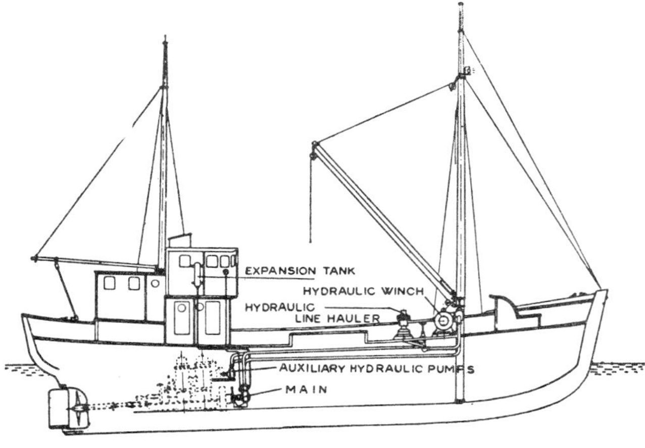

Fig. 537

the oil flow. Then the oil returns through the suction pipe to the pump. Consequently, the same quantity of oil is circulating all the time in the closed system.

To eliminate variations of the oil volume at different temperatures, there is an expansion tank connected to the suction side of the pump, placed above the highest part of the system. The air which happens to be in the oil will partly ascend to the tank and be separated there. The static pressure from the tank will also prevent air leakage into the pump through the packings. A manometer is connected to the pressure side of the pump. This may be placed in the engine room, wheel house or on the winch and it enables the operator or the skipper to control the oil pressure and the load on the hoisting wire

Fig. 538

or fishing gear. The oil pressure is always proportional to the load of the winch and a safety valve on the pump prevents overloading.

The combination indicated in fig. 537 is much used in Norway. Besides the main engine-driven pump, another pump is connected to an auxiliary engine, and the pipes from the two pumps are coupled to the main pipes to the winch. From fig. 537 it may be seen that in addition to the winch there is a vertical long-line hauler connected to the same system of pipes. This is operated by a special control valve inserted in the return pipe from the winch, so that the long-line hauler and the winch are connected in series. The pipes from the pumps, on the other hand, are connected in parallel, making it possible to obtain double hoisting speed by running both pumps simultaneously.

Fig. 538 shows an hydraulic pump and its connection to the main engine. The fixed part of the claw coupling is connected with a flange to the belt pulley of the flywheel and inside this is a seat for the guide bearing of the head shaft. The slide claw runs on sliding keys on the shaft. A universal coupling between the pump shaft and the fixed claw eliminates strains caused by misalignment of the engine and pump shafts. This arrangement is only used for slow running main engines.

The pump rests on a cast-iron bed plate which, in turn, is anchored to cross-planks bolted to the keelson and frame through the ceiling. On steel ships the pump rests on a steel frame.

Fig. 539 shows a cut-away drawing of the pump, which is of the positive displacement vane type, delivering oil at a pressure of 350 to 400 lb./sq. in. (25 to 28 kg./sq. cm.) at full load of the winch. Compared with the high pressure type, the advantages of this low-pressure system are the economies in manufacturing and maintenance, in spite of the higher weights and increased pipe diameters. Everyone familiar with hydraulic transmissions will know that presence of air in the system is the main cause of trouble and improper operation, and the ability of oil to absorb air is heavily increased by higher pressure. In the low pressure system this trouble is reduced.

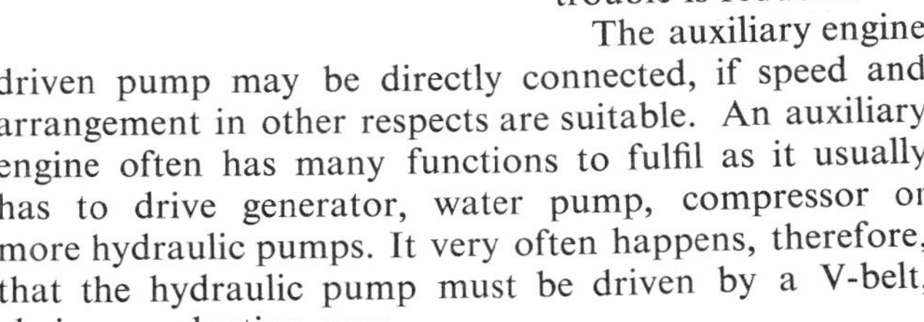

The auxiliary engine driven pump may be directly connected, if speed and arrangement in other respects are suitable. An auxiliary engine often has many functions to fulfil as it usually has to drive generator, water pump, compressor or more hydraulic pumps. It very often happens, therefore, that the hydraulic pump must be driven by a V-belt, chain or reduction gear.

A usual type of cargo winch and its parts are shown in fig. 540 and 541. The hydraulic motor is a rotating vane type and in the casing there is a rotor wheel with radially movable vanes that follow the eccentric guide curve of the casing. This guide curve is designed in such a way that, under load, the vanes make no radial movement.

The control handle works on the operating valve slide. The position of the slide determines the quantity and the direction of the oil to the hydraulic motor, and

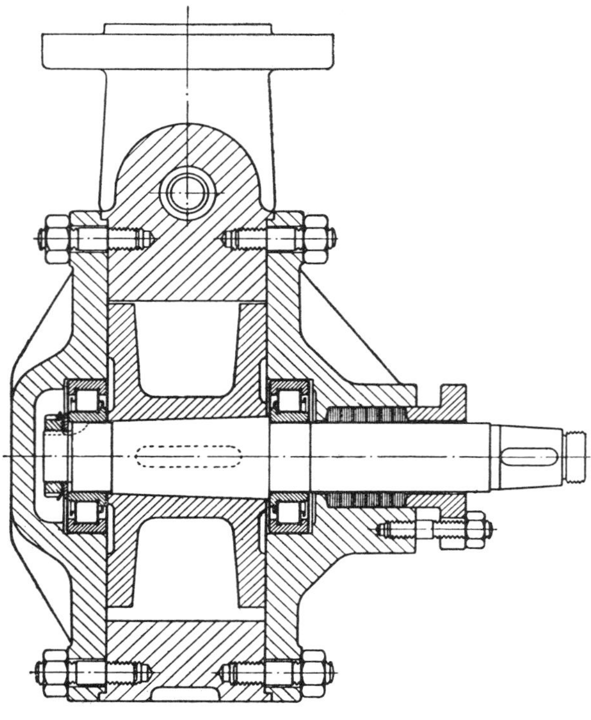

Fig. 539

Fig. 542

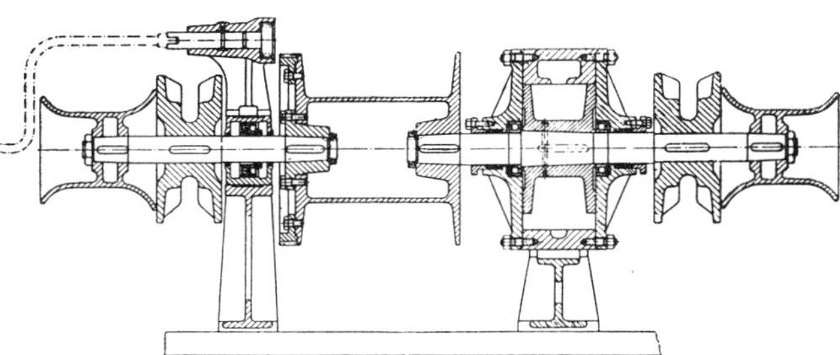

Fig. 540

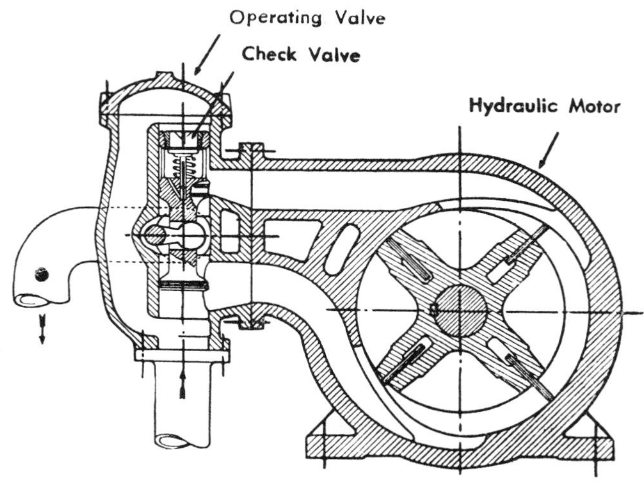

Fig. 541

Fig. 543

Fig. 544

thereby the speed and rotation direction of the winch. The slide is equipped with a check valve that prevents return of the oil. If, whilst lifting, the oil pressure in the pipe is interrupted, the load will remain suspended without slipping because of the oil pressure built up in the winch motor between the vanes and the check valve. There is no need for mechanical brakes as braking and holding, as well as manoeuvring, is hydraulically carried out.

The hydraulic motor is directly connected to the shaft of the drums and gypsies. This eliminates power losses and noise from gear and chain transmissions. The winch may also be used as a windlass, in which case it is equipped with gypsies for the chains. This winch type is normally delivered for 2, 3 and 4 tons test load.

Fig. 542 shows a combined net and long-line winch. At the bulwark, in the background, is the manoeuvring wheel. Its valve is placed under the deck. During the hauling of the long-line the control valve is adjusted for a certain maximum pressure corresponding to a pull somewhat lower than the strength of the line. When the vessel rolls or pitches, causing an increased strain on the line, the winch automatically stops hauling. When the vessel goes down the next wave and the strain is reduced, the winch automatically begins to haul again. Nothing like this can be achieved so simply by mechanical or other winches.

The first skipper to use this winch for long-line fishing (off Iceland and Greenland in 1938) found that he reduced his gear losses almost 50 per cent. compared with those sustained when using mechanical winches. How many fish are pulled off the hooks and lost by use of a direct mechanically-driven winch, because of the jerks on the long-line, is not easy to say. Furthermore, shocks directly transferred from the combustion engine must adversely influence the result, while the steady haul of an hydraulic winch must be favourable.

The hydraulic winch allows more independence of weather conditions in, for example, halibut fishing, because the long-line may be hauled in considerably higher seas than is possible with mechanical winches, The maximum hoisting speed of the long-line winch. by normal r.p.m. on the motor, is usually fixed to 230 ft. (70 m.) per minute. The winch is equipped with two gypsies for handling nets, one above and one below the line sheaves.

A number of fishing vessels use wire instead of chain for the anchor. The wire drum, which is usually fitted in the cabin hatch (placed amidship in front of the mast on Norwegian fishing boats), is often driven by a heavy chain from the cargo winch, placed behind the mast. A more convenient solution, however, is to let a separate hydraulic motor, mounted in the cabin hatch (*fig. 543*), drive the drum directly or through a roller chain in an oil tight case.

During the war ordinary Norwegian fishing craft took up whale catching to help the food situation. These vessels were equipped with a whaling winch type, as shown in fig. 544.

In 1947, the Norwegian Fishery Department started experiments in trawling with larger vessels. One of them, the *Uran*, was equipped with a 12-ton hydraulic trawl winch, and another with a smaller hydraulically-driven trawl winch. Based on experiences obtained, the Fishery Department ordered complete hydraulic deck machinery for their new deep-sea research ship, *G.O. Sars*, built in 1950. This machinery includes a 16-ton trawl winch (*fig. 545*), a 5-ton trawl winch (*fig. 546*), a long-line winch, two special winches for plankton nets and water specimens, and a windlass. All these items are driven from two pump aggregates, each consisting of two duplex pumps driven from the main engines (*fig. 547*). A similar installation is also ordered for a new research ship being built for the Department.

Fig. 545

A 4-ton hydraulic trawl winch is installed on board the research ship *Eystrasalt* of the Royal Swedish Fisheries Board, and fig. 548 shows a winch for water specimen and plankton nets on this vessel.

The hydraulic trawl winch arrangement is

very suitable for diesel trawlers. Another advantage is that the hydraulic motors are connected directly to the drum shaft and, therefore, no space is required in the deck house or under deck.

Purse seines are important in the Norwegian herring, cod, coalfish and tuna fishing. Approximately 90 per cent. of the total landings of one million tons of herring per annum are caught by purse seines. By this method, the winch gets a strain which is unusually hard. The herring *must* be taken out of the net in less than an hour or they will die, sink and drag down the net. The quantity loaded into the big purse seine boats may amount to 400 tons of herring and great quantities of water. As a comparison, the average figure for general cargo loading for a cargo liner is 14 tons per *pair* of winches an hour, whereas *one* winch on a purse seine net fishing boat has to take 400 to 500 tons an hour.

As a measure of dependability, it may be mentioned that 300 to 400 vessels with hydraulic winches took part in the " big herring " season last winter (1953) without a single mishap during the whole time. The popularity of the winches among the fishermen is proved by the fact that the best net-bosses, for whom the skippers and owners strongly compete, will not take the job unless the boat is equipped with hydraulic winches.

Besides the simplex type of cargo winch (*fig. 540*), duplex or multiple speed winches are now used to an increased extent in both fishing vessels and cargo ships. They are equipped with two hydraulic motors and have two ranges of speeds, enabling a light hook and a half load to be hoisted or lowered at higher speeds than for full loads. This is made by distributing the oil to one or to both of the motors simultaneously In this case the manoeuvring of the winch is also carried out with one handle or handwheel. The control is equipped with a built-in valve that automatically distributes the pressure oil to both of the motors when the load is increased. This prevents the winch being overloaded, irrespective of the position of the control handle.

With multi-speed winches in 3 and 5 ton sizes, light hook speeds of six times the full load speed can be obtained, and the winches are fully " dynamic " in action as the speed is automatically controlled by the load. The overall efficiency, reckoned from the pump shaft to the load, is 65 to 77 per cent. at full load, according to the conditions.

The demand for dependability, simplicity, increased efficiency and lower running costs is increasing in the fishing industry, and the world-wide interest in hydraulic winches suggests that this equipment is meeting the demand.

Fig. 546

Fig. 547

Fig. 548

LIVE-BAIT EQUIPMENT

by

CARL B. CARLSON

THE pole-and-line tuna fishery is dependent on supplies of live bait, and the range of the tuna vessel depends in part on her ability to carry large quantities of bait. The equipment used to capture bait and keep it alive varies from the simple net and row-boat with a tank of sea water to the very elaborate system on the Pacific tuna clippers.

CUBA

The method and equipment for catching and holding live bait in the Cuban fishery, as described by Rawlings (1953), consists of a short, shallow fine-mesh net set from a row boat, a receiver for transporting the bait from the shoals to the fishing vessel, scoops for transferring the bait and live wells. The live wells are fitted with plugs for draining after the bait has been expended but no provisions are made for the circulation of water. Since the fishing grounds are near the bait grounds this system is practical, for bait caught in the morning is generally expended the same afternoon. But a circulation system would enable the fishermen to hold larger quantities of bait—mostly the majua (*Jenkinsia lamprotaenia*)—for a longer time. Experiments in 1953 using tanks with circulating sea water, showed that majua can be kept alive for prolonged periods.

HAWAII

More elaborate equipment, as described by June (1951), is used in the Hawaiian fishery. The bait is small and is caught in fine-mesh nets, at night by using lights to attract the bait to the nets, or in the day by using large encircling nets. The net boats are powered by outboard motors and the transfer of bait to the vessel is made with buckets to avoid injury through loss of scales. Live bait wells are built in the hull and are fitted with screened holes below the water line to permit the entrance and exit of seawater. Bait may be kept alive for from one to several weeks in wells of this nature. By plugging the holes and draining the water into the bilge, the wells may be converted to carry tuna. When underway or rolling in a sea, the circulation is adequate to keep the bait alive, but when lying in calm waters the fishermen must periodically rock the boat to encourage circulation. There is a trend toward the installation of power-driven circulation systems. Both the Hawaiian and the Cuban tuna fisheries were pioneered by Japanese immigrants, and their influence predominates.

PACIFIC ALBACORE FISHERY

The system of capturing and holding bait alive in the Pacific albacore fishery was derived from the tuna clippers, but it is less elaborate and the vessels are smaller. Most of the vessels were designed for other kinds of fishing but they have adequate stability to carry live-bait tanks on deck. Circulating water is supplied by pumps and the bait is caught in large lampara nets set either from a powered boat or from the fishing vessel.

TUNA CLIPPERS

The most elaborate system for capturing and holding bait alive is found on the Pacific tuna clippers. The equipment required consists of tanks for a large quantity of bait, a water circulation system to sustain life, boats and gear to catch the bait, and facilities for transferring the bait from the nets to the tanks.

BAIT TANK DESIGN

The structure of the tanks must be adequate for the weight of water and safety factors for service at sea. Hatches at least 18 in. (46 cm.) high are required to maintain a head of water to eliminate air pockets and to prevent sloshing. Waterproof lights of at least 100 watts should be installed in the accessible sides of the tanks to allow the bait to mill properly at night. Portable or fixed lights over the hatches are also desirable, and it is essential to have smooth surface screens with adequate openings to reduce the flow rate to limits which will not injure the bait.

Controversy exists among fishermen as to the amount of live bait which can be carried in a tank of a given volume. The amount of bait is probably dependent on a number of factors, including oxygen content, temperature, and salinity of the waters; the abuse to which the bait is exposed between capture and being put in the tanks;

variation in the hardiness of the bait because of species, area or season of capture, and the proper design and manipulation of the tanks and circulation system. General practice is to place more bait in the tanks than is expected to survive.

When the U.S. Fish and Wildlife Service vessel *Oregon* was operated in the commercial tuna fishery off central America in the Pacific Ocean, 300 and 400 scoops of bait (each containing about 12 lb. (5.4 kg.)) were placed in deck tanks having a capacity of 3,400 and 5,100 imp. gal. (4,100 and 6,130 gal., 15,520 and 23,200 l.) respectively, exclusive of baffle and screen areas. When doing investigation work in the Japanese tuna fishery, Cleaver and Shimada (1950) reported that a sardine $2\frac{3}{4}$ in. (7 cm.) long required 0.07 cu. ft. (2 l.) of water in tanks having natural circulation, i.e. screened holes in the hull to admit sea water. The amount of water per fish, under forced circulation, could be reduced to 0.05 cu. ft. (1.4 l.) with a water temperature of 64 to 68 deg. F. (17.7 to 19.9 deg. C.). In another instance, approximately 1,000 lb. (454 kg.) of bait were held in a tank $8 \times 8 \times 7$ ft. ($2.4 \times 2.4 \times 2.1$ m.) without mechanical circulation. The same authors also report 50 buckets, containing from 15 to 20 lb. (6.8 to 9.1 kg.) of bait per bucket, in a tank $6\frac{3}{4} \times 6\frac{3}{4} \times 9\frac{5}{6}$ ft. ($2.36 \times 2.36 \times 2.9$ m.) without mechanical circulation.

The design of the screens in the bait tanks aboard a clipper is very important. They must have a relatively smooth surface and openings small enough to hold in the bait but big enough to admit and discharge the water at flow rates that will not injure it. Too rapid flow on the inlet side will affect the milling of the bait while an excessive rate through the discharge side will draw the live bait or cause dead bait to collect and clog the screen. Typical screens are made of $\frac{3}{4}$ in. (19 mm.) waterproof marine plywood with $\frac{1}{8}$ in. (3 mm.) slots cut by a circular saw on 1 in. (2.5 cm.) centres. Very little, if any, reliable published data are available on the optimum area of the openings, and fishermen may

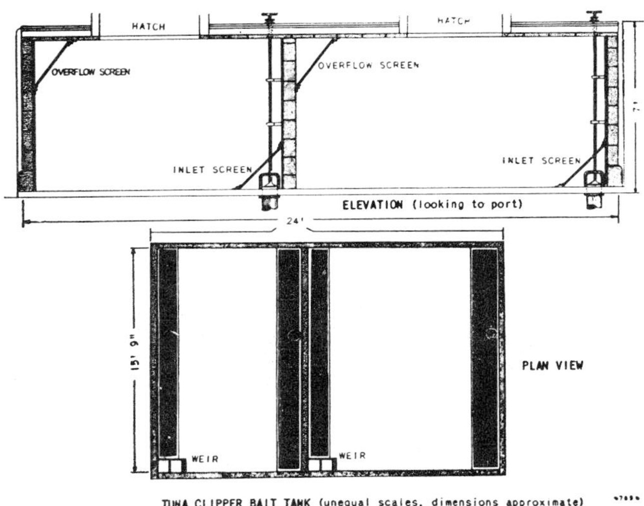

TUNA CLIPPER BAIT TANK (unequal scales, dimensions approximate)

Fig. 549

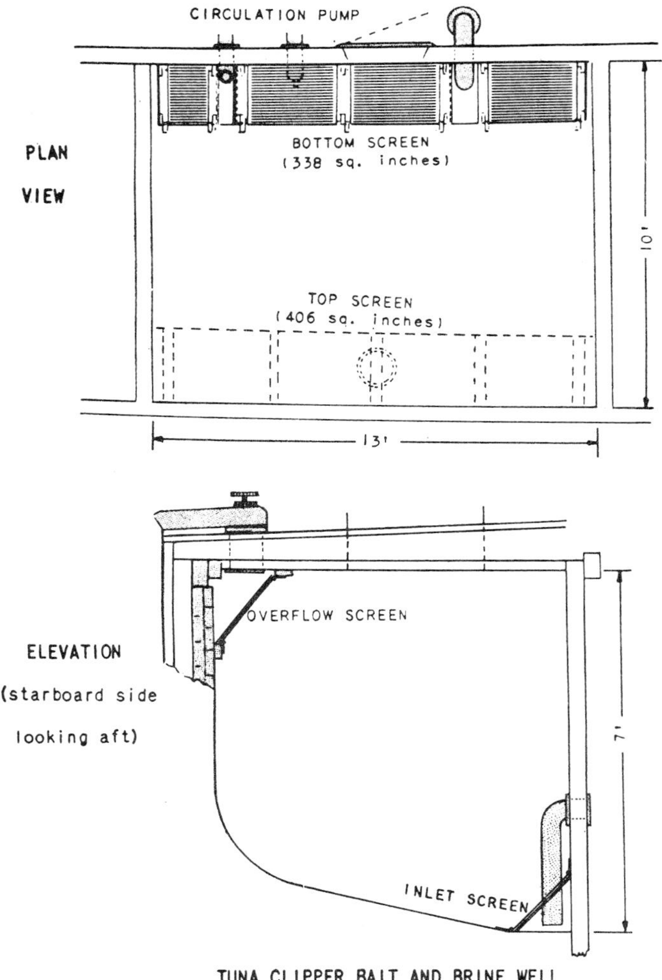

TUNA CLIPPER BAIT AND BRINE WELL
(unequal scales, dimensions approximate)

Fig. 550

provide additional slots if the bait behaves improperly. One tank, having a volume of 910 cu. ft. (25.75 cu. m.) was fitted with inlet and outlet screens having areas of 338 and 494 sq. in. (2,180 and 3,186 sq. cm.), respectively.

The deck bait tanks, fig. 549, are rectangular to suit the available space, but the below deck tanks (or wells), fig. 550, are vertical on three sides while the fourth side and bottom conform to the shape of the hull. Water for the deck tanks is introduced at one corner behind the screen and allowed to overflow through a screen and baffle in an opposite or adjacent corner, so tending to create circulation in a horizontal plane. Water for the tanks below is admitted at the bottom on the shaft alley side and allowed to overflow in the upper outboard side, which tends to create circulation in a vertical plane. The flow of water to the tanks is throttled by a valve. In the deck tanks the overflow is governed by the height of baffle boards, which are set to maintain a constant level of water in the hatch. The rate of discharge and the consequent level of water in the hatch in the wells below is controlled by a valve behind the discharge screen.

PUMPS AND PIPING

The pumps for supplying seawater to the tanks are of the vertical impeller type designed to deliver large volumes at low head pressures. High head pressure types are undesirable because they may churn and separate dissolved gases from the water. The sea chests for the pumps are located near the keel and close to amidship to allow the least disturbing suction of water when cruising. The pumps change the tank water five to seven times per hour. To achieve this performance a safety factor of at least two must be considered because the water can be throttled by the tank supply line valves. And live bait is so important that dual pumps should be installed, each able to carry the load for most of the tanks. The pumps discharge to a header on each side with a cross-over connection so that one or both pumps may be used for the system. Each take-off to a tank is fitted with a valve to govern the flow or to isolate the tank.

BAIT FISHING BOATS AND GEAR

Most of the bait is taken in sheltered and shallow areas along the coasts of Mexico and central America. Small boats are necessary in the shallow water because if the bait pump intakes on the clipper get too close to the bottom, mud may be drawn into the tanks and kill a lot of the fish.

Three boats are used to find and capture bait: a power boat capable of speeds up to 30 or more knots per hour, a net skiff, and a " dry boat ". The power boat may be fitted with an echo sounder to aid in locating the bait, and may vary from 16 to 18 ft. (4.9 to 5.5 m.) in length and 4 to 5 ft. (1.22 to 1.52 m.) in beam. The engine is usually high-speed gasoline type of 100 to 150 h.p. Some engines are fitted with reduction gears which allow the use of larger diameter propellers to reduce slippage when towing the net and dry boat.

The net boats are from 16 to 18 ft. (4.9 to 5.5 m.) in length, from 6 to 7 ft. (1.83 to 2.3 m.) in beam, and from 20 to 30 in. (0.5 to 0.76 m.) in depth. The dry boats are flat-bottom skiffs about 12 ft. (3.66 m.) in length and are used to hold the end of the net while setting, and to support the fish bag.

Two men are required in the speed boat; one serves as operator while the other searches for bait and directs the fishing operation. From six to eight men are required to set and haul the net. When not in use, the boats are carried in nests aboard the clipper.

Most of the bait is caught in hand-operated lampara nets varying from 130 to 160 fm. (239 to 293 m.) in length and from 50 to 75 ft. (15.2 to 22.9 m.) in depth at the bag. The end meshes in the wings vary from 5 to 8 in. (13 to 20 cm.), stretched measure, and gradually diminish in size to ½ in. (1.3 cm.) stretched measure mesh in the bag. The lampara nets are lightly floated and leaded in the wings, about one cork and one lead every 12 to 18 in. (30 to 40 cm.), but they are heavily buoyed and leaded in the bags, about three corks alternating with a 6-in. (15-cm.) space. The bottom line is almost solidly leaded at the throat. The lampara nets are used along the mainland for day fishing in shallow water or at night in deep water when the bait can be located by phosphorescence.

A smaller fine-mesh net from 125 to 150 ft. (38.1 to 45.7 m.) in length and 20 to 30 ft. (6.1 to 9.1 m.) in depth is used for catching bait in the clear waters near the rocky shores of the Galapagos Island. The nets are moderately buoyed and weighted, with about two corks and leads per foot (30 cm.), and a mesh size of from ¾ to 1 in. (1.9 to 2.5 cm.) by stretched measure, depending on the preference of the fishermen. These are used as a surround net and extend from the surface to the bottom. Because of obstructions it is necessary to send divers to the bottom to free the net and to keep the lead-line together while hauling. The diving equipment may be either a helmet or a face mask supplied by compressed air from the surface or a face mask and air tank. The face mask is preferred because of the possible loss of air from a helmet when bending to clear the net.

If the bait is caught in shallow or hazardous areas it is transferred to a collapsible receiver, fig. 551, which can be towed. The receiver consists of a transom stern and midships section, about 10 ft. (3.0 m.) long, 7½ ft. (2.28 m.) wide and 3 ft. (0.9 m.) deep, and a pointed bow section about 6 ft. (1.83 m.) long. The bow section is solidly planked but the after section is covered on the sides and bottom with netting to provide circulation.

In making the set, the dry boat holds one end of the net and the circle is completed by the speed boat towing the net skiff. Dumping the bag is a difficult operation. It must be specially piled so that it can be set as a unit to avoid tearing and if the length of the circle exceeds the length of the net, a running line is paid out so that both ends may be pulled to the net skiff. It is retrieved

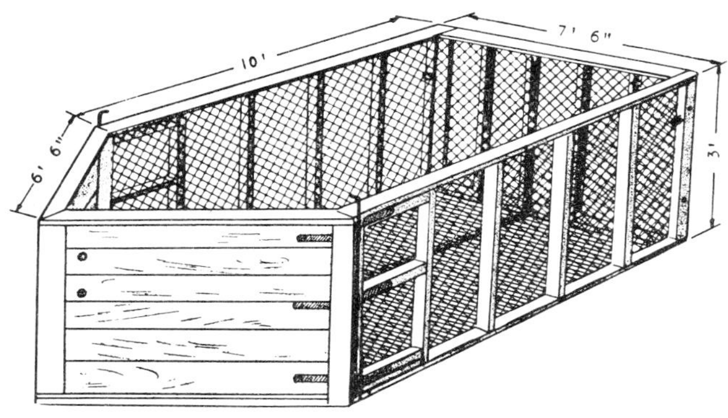

FLOATING BAIT RECEIVER (dimensions approximate)

Fig. 551

by hauling from both ends. Until the lead-line is aboard, one man throws a weighted line into the opening to scare the fish into and keep them in the bag while the dry boat picks up the cork line at the bag for additional support. After drying up, the fish may be transferred directly to the clipper, if the water is deep enough, or transferred to a receiver and towed to the clipper. Transferring is done with scoop nets of ½ in. (1.3 cm.) mesh holding from 10 to 15 lb. (4.5 to 6.8 kg.) of bait. Then follows a " rest period ", the clipper remaining from one to several days in quiet water to acclimatize the fish to life in the tanks.

A new method of using pumps to transfer the live bait is described in *Pacific Fisherman* (" Pumping Live Fish . . . " 1953).

Recently a new type trap lift net for catching tuna bait fish has been developed by the technical staff of the exploratory fishing vessel *Oregon* (Siebenaler, 1953). This net is operated from the fishing vessel, using lights to attract the schools of bait at night. To date, it has been very successful in catching anchovies and other small species in the Gulf of Mexico. It eliminates the need for special bait boats and large bait seines. It differs from ordinary lift nets because it has the advantage of quick lift of the sides, independent of lift of the bottom and frame of the net.

HATCH COVERS AND MAST WITHOUT SHROUDS

by

P. BAIN

WATERTIGHT METAL COVERS FOR FISH-HOLD HATCHES

TRAWLERS usually have four or five small hatches to the fish-holds to preserve the low temperatures in them but when the hatches are opened air enters and starts to melt the ice. The hatches are usually made with insulated inner covers, wood covers and tarpaulins or steel outer-covers, a design which has not been improved for a long time.

About one hundred English trawlers have been equipped with watertight metal hatch covers which have been found satisfactory after being in service for several years. Among the trawlers using these new hatches is the *Kingston Garnet*, 184 ft. (56.1 m.) long between perpendiculars, built by Cook, Welton and Gemmel, Ltd., Hull. Fig. 552 shows the longitudinal section of the vessel and fig. 553 shows one of the MacGregor hatch covers specially adapted for quick manoeuvrability. They are also designed so that the hatches need only be partly opened, which reduces the volume of air entering the holds. The design makes it possible to have hatches of 13 to 14 ft. (4 m. to 4.3 m.) long for unloading but to reduce them to very small openings when loading. Working on the same principle, hatches of smaller size can also be fitted with covers by using one or two sections instead of three. Both loading and unloading are easy, whether done by hand or winch.

The British factory ship, *Fairtry*, is equipped with MacGregor hatch covers. The experience gained by use of them on British trawlers shows that they retain all the qualities of the conventional hatch cover.

NEW TYPE OF " BIPOD " MAST

About 100 ships have now been fitted with the new " bipod " mast developed by the Swedish naval architect, Eskil Hallen. The mast is 15 to 20 per cent. lighter than the conventional type and is not held by stays. It is made of tubular girders which form an " A " by means

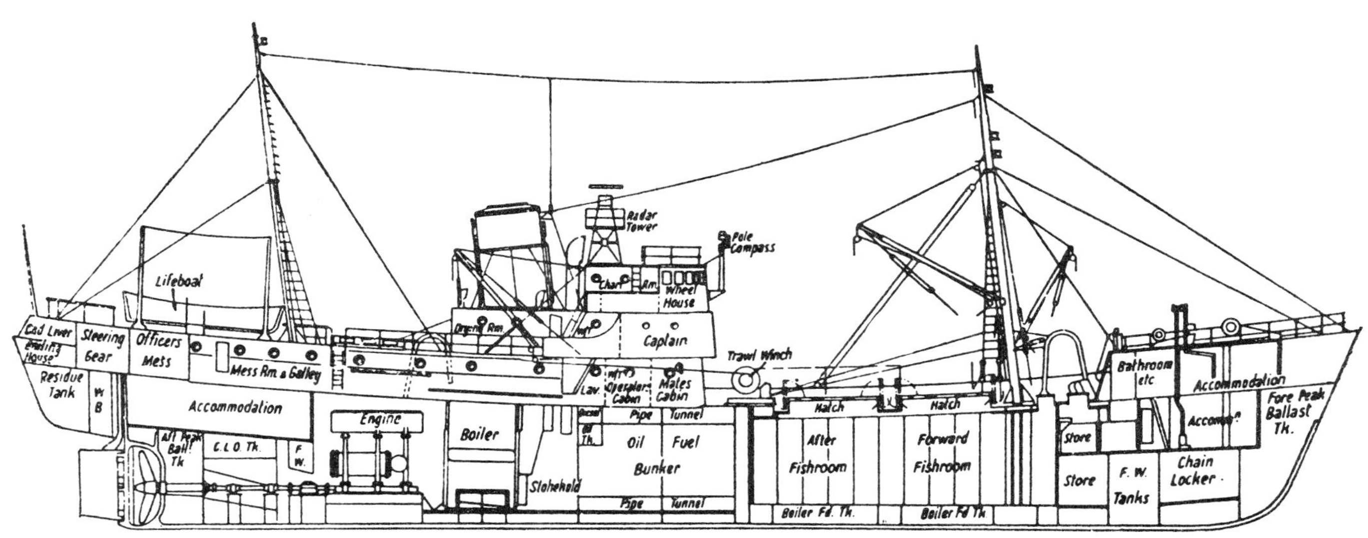

Profile of the British oil-burning steam trawler 'Kingston Garnet'

Fig. 552

of a bracket at the top. The angle is about 20 deg. At the top is a crosstree and a small mast. The tubular girders have a special section, with high inertia in the fore and aft direction. This makes it possible to do away with stays and standing rigging. The mast can be fitted with derricks and fig. 554 and 555 show two methods of installing it on fishing boats. In fig. 554 the mast is placed well forward, leaving more room for the fish ponds. In fig. 555 a horizontal girder is supported on the mast, which does away with the gallows.

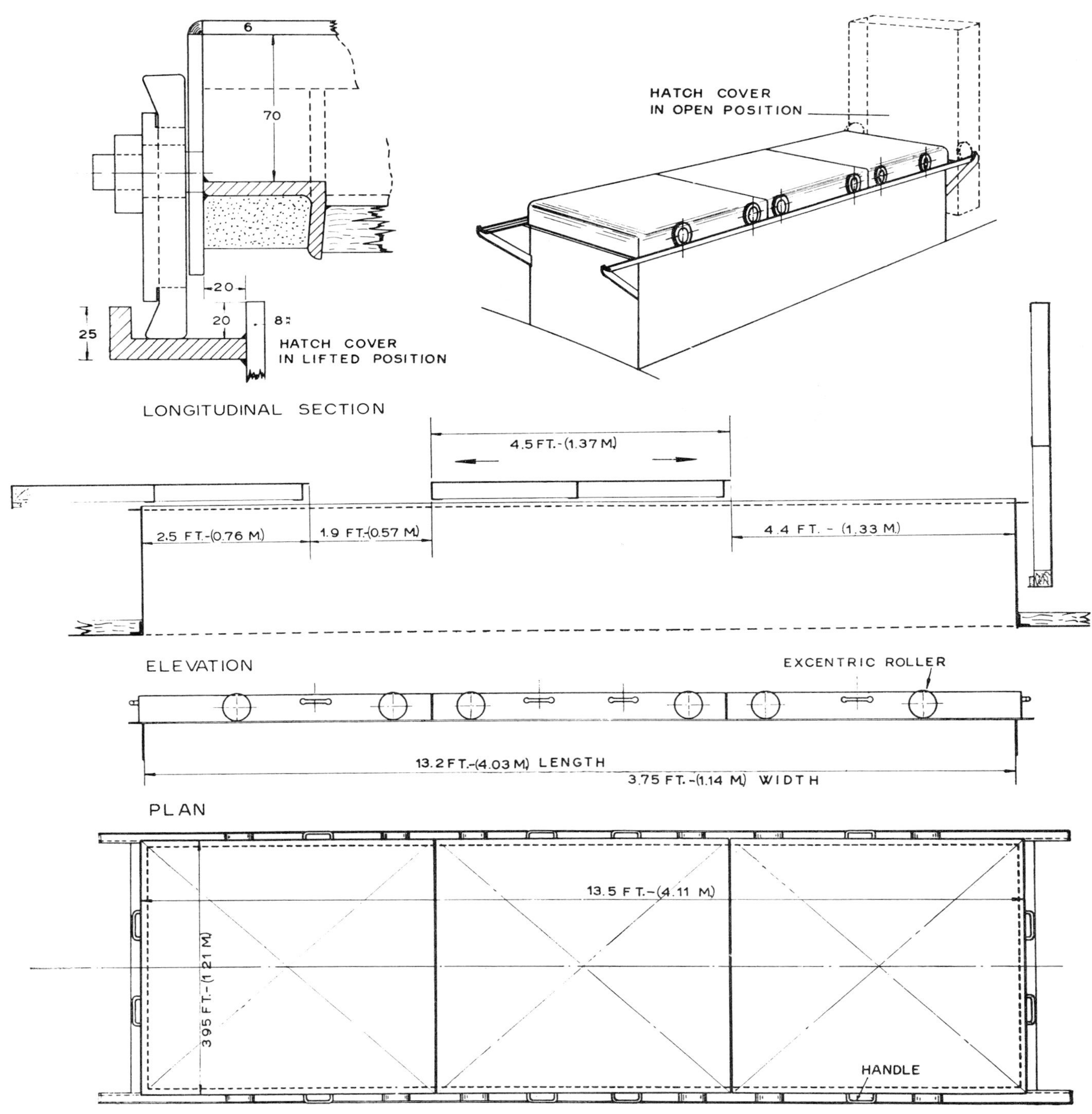

Fig. 553

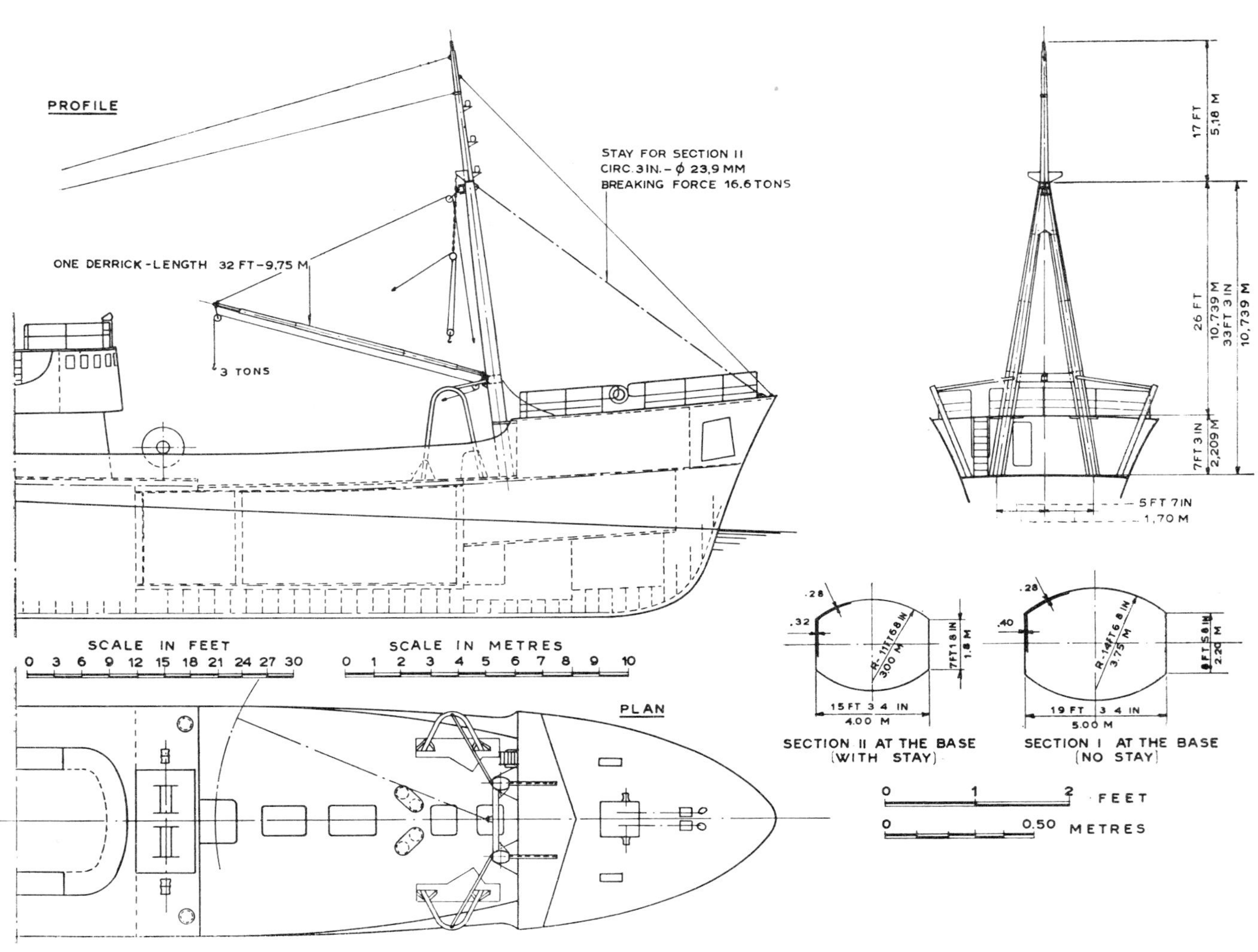
PROFILE
STAY FOR SECTION II
CIRC. 3 IN. – φ 23,9 MM
BREAKING FORCE 16.6 TONS
ONE DERRICK-LENGTH 32 FT – 9,75 M
3 TONS
SCALE IN FEET
0 3 6 9 12 15 18 21 24 27 30
SCALE IN METRES
0 1 2 3 4 5 6 7 8 9 10
PLAN
17 FT
5,18 M
26 FT
10,739 M
33 FT 3 IN
10,739 M
7 FT 3 IN
2,209 M
5 FT 7 IN
1,70 M
SECTION II AT THE BASE
(WITH STAY)
SECTION I AT THE BASE
(NO STAY)
15 FT 3 4 IN
4.00 M
19 FT 3 4 IN
5.00 M
FEET
METRES

Fig. 554

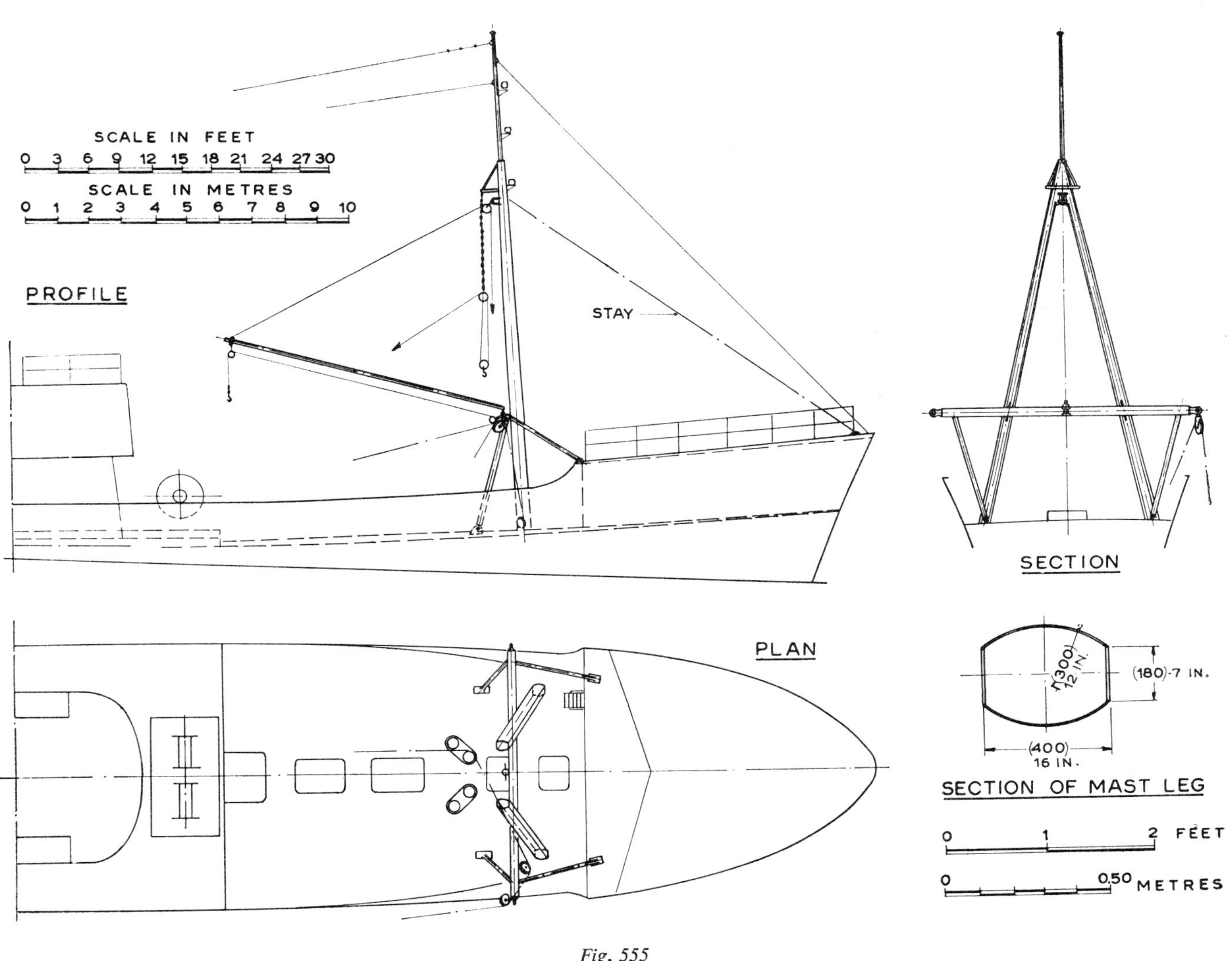
SCALE IN FEET
0 3 6 9 12 15 18 21 24 27 30
SCALE IN METRES
0 1 2 3 4 5 6 7 8 9 10
PROFILE
STAY
SECTION
PLAN
(300) 12 IN.
(180) 7 IN.
(400) 16 IN.
SECTION OF MAST LEG
0 1 2 FEET
0 0.50 METRES

Fig. 555

DERRICK ARRANGEMENT OF A TRAWLER

by

MARIO COSTANTINI

WHEN the Italian fishery company, Genepesca of Leghorn, had two deep sea motor trawlers built in 1949 they would not have conventional derricks for hoisting the cod end of the trawl net on board. They suggested fitting twin derrick posts, cross-braced together, and provided with an outrigger extended to the sides of the ship. The system had plenty of transverse stability but seriously lacked longitudinal stiffness. This could be obtained by extending the base of the derrick posts into the fish-hold or fitting large brackets on deck, both unsatisfactory makeshifts.

The problem was finally solved by fitting two sheer-legs to the foremast to form a rigid tripod and installing a tubular steel double outrigger flush with the bulwark on the starboard side and somewhat shorter on the port side (*see* fig. 556).

The system is rigid and the only obstruction below deck consists of a solid pillar fitted under the foremast.

The starboard outrigger is used for handling the trawl net, while the port side supports the upper block of the topping lift of a small ($1\frac{1}{2}$ ton) derrick which hoists the net and provision stores forward.

A portside gangway, connecting the midship deck-house to the forecastle, bridges the obstruction formed across the deck by the ponds and fish cleaning installation.

The arrangement has proved in every respect satisfactory. (For further particulars about this type of trawler see paper " Liver conveyor system of a trawler " on page 545.)

Fig. 556

JAPANESE LONG LINING

by

K. YAGI

BEFORE line haulers were used in tuna long-lining, much labour and time were spent hauling in the gear. Fishermen would cut their hands doing this and continuous operation was often impossible. The length of line that could be used was restricted, which limited the catch. When development of tuna fishing was accelerated through the use of the hot-bulb engine, it was necessary to improve the primitive method of hand hauling. A mechanical line hauler was conceived and in 1923, at a base port on Shikoku Island, a prototype was completed and used successfully.

Fishermen were at first reluctant to try the line hauler, but were finally persuaded to use it. The demand for the device continually increased and to date no less than 8,230 have been built and used, not only in Japan, but also in Taiwan, the United States, Hawaii, Brazil and other countries.

For the purpose of description, the line hauler shown in fig. 557 is divided into three parts. In the lower part is the motor, the gears to change the speed, and the clutch handle. In the middle is the governor, controlled by a stop handle, which reacts, even if the motor is at full power, to the stresses caused by wave resistance or big catches. This enables the operator to adjust the stress of the long-line which minimizes wearing, prevents

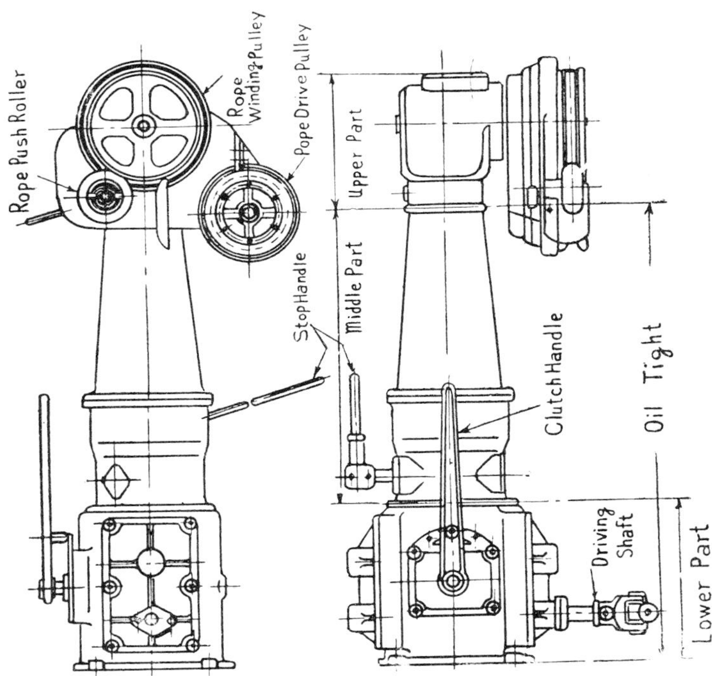

Fig. 557. Japanese long-line hauler

TABLE LXXXI

Model *Size of boat*	*Height*		*Weight*		*R.p.m. of shaft*		*Winding speed*		
	in.	*mm.*	*lb.*	*kg.*	*normal*	*maximum*		*ft./min.*	*m./min.*
Special size Over 100-ton	58	1,480	790	356	220	300	High Low	600 530	184 161
Large size (standard) Over 30-ton	54	1,380	620	280	200	300	High Low	470 310	144 96
Large size (lower) Over 20-ton	49	1,240	570	260	200	300	High	470 310	144 96
Medium size: Over 10-ton	45	1,150	410	185	230	280		250	75
Small size: Less than 10-ton	33	840	240	110	170	200		220	68

the lines being cut and keeps the fish on the hooks while hauling. In the upper part is the gurdy proper consisting of three sheaves by which, through friction on the pulleys, the lines are hauled mechanically. Rubber sheave liners are used on the pulleys to give greater friction on the lines and also to minimize wear.

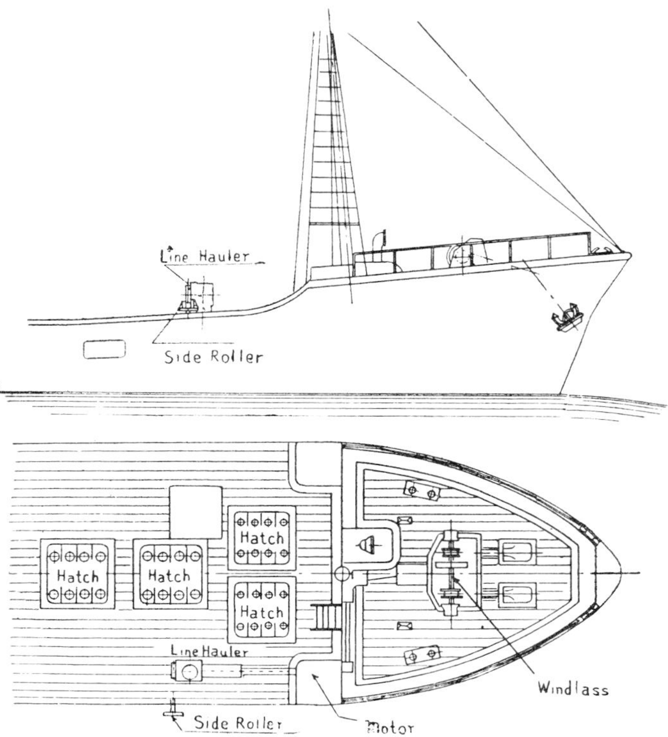

Fig. 558. Arrangement of Japanese long-line hauler, driven by separate motor

The line hauler is driven by an electric motor, fig. 558, or from the main or auxiliary engine through an intermediary shaft leading to the deck, fig. 559. The line hauler is made of cast steel and the pulleys of gunmetal.

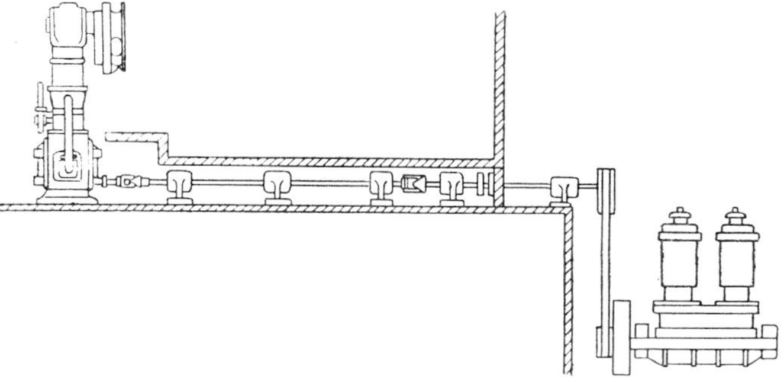

Fig. 559. Long-line hauler, driven by main engine

Usually 3 h.p. is needed to drive the line hauler at about 220 to 250 r.p.m. The hauling speed of the lines is 500 to 600 ft./min. (150 to 180 m./min.). Under normal fishing conditions it takes about 12 hours for a vessel using 300 units or baskets of long lines to haul in the 60,000 fm. (110 km.) of line.

THE FISHING TECHNIQUE

One long line consists of many short complete main lines which are coiled and stacked in baskets when not in use. For fishing, the hook at the end of each branch line is baited and is kept at a suitable depth from the surface of the water by varying lengths of buoy lines attached at intermediate sections of the main line.

On fishing vessels operating in equatorial waters a 200 fm. (370 m.) main line is usually treated as a unit. A branch line is usually 15 fm. (27 m.) long and a buoy line varies in length up to 15 fm. Vessels of 200 gross tons and over usually operate 300 to 350 units of long line, the total length of which reaches 60,000 to 70,000 fm. (110 to 127 km.).

One branch line with a big-size tuna hook is used between two buoys, 100 fm. (180 m.) apart, when fishing for blue fin tuna. In fishing for albacore, 12 or 13 branch lines with small tuna hooks are used and the diameter of the main line is a little smaller. A few big hooks are used along some sections of the line.

Fig. 560. Hauling a tuna with long-line hauler

The length of buoy line varies according to the fishing ground and the main species of tuna it is desired to catch. In fishing for blue fin tuna, 6 to 20 fm. (11 to 37 m.) are used and for albacore about 30 fm. (54 m.).

A tuna fishing vessel usually finds a good fishing ground by observation and begins to throw out the lines before sunrise. They are set at a rate of about 100 units per hour, the hooks being baited as the lines are paid out. When the setting is finished, the vessel comes around to the end of the line which was thrown out first and starts to haul it in. The time needed to haul the lines depends upon the catch. Normally it takes about 12 hours to haul in 300 units. Fig. 560 shows a tuna being hauled in.

DECK GEAR — DISCUSSION

Hilmar Kristjonsson (FAO): Most news about electric salt water fishing has come from Germany and it is therefore interesting that Dr. A. von Brandt is not optimistic about its prospects in the immediate future. Conventional methods of catching fish will still be used for some years to come, and the use of electrical current to drive or lead fish into the trawls will not yet present design problems to naval architects.

Devices for detecting fish, and observing and controlling the behaviour of nets while fishing, are more urgently needed to-day than ever before, especially by fishermen using off-the-bottom and mid-water trawls. These will certainly come into more general use in the next few years. But devices are also wanted by fishermen who use submerged purse seines for catching fish in very deep water, as is done, for example, in the Lofoten cod fishery.

Asdic (sonar) and similar echo-ranging devices will soon be considered essential equipment in medium and large size fishing boats. Their introduction will not involve serious design problems as hull attachments for most echo-ranging equipment are now made in a size that can be housed in a cylinder of about 3.3 ft. (1 m.) diameter. Such a device can be installed in the fish-hold or some other suitable place and can be operated through a retractable transducer extending through the bottom of the boat. Equipment of this kind is produced in Norway and the U.S.A.

C. B. Carlson describes the use of pumps for unloading fishing boats in the U.S.A. Some experience has been gained in loading fish from the seines into the boats and it seems to be a promising method. It certainly should be tried in the herring fisheries of northern Europe. For example, in the winter fishing off Norway the rough sea often makes the conventional methods of brailing very difficult and pumps might be effectively used in this case. When pumps are used, the fish do not have to be so closely confined and there is less strain on the net from the roll and heave of the vessel. On the other hand, the use of pumps has not been altogether successful in unloading the fat summer herring caught off Iceland. The fish are very sensitive to rough handling because they are very fat, and as they are caught during feeding runs their bellies are quickly weakened by the enzymes they contain.

The power requirements quoted by Carlson are high and the suggested use of auxiliary petrol engines poses a problem of finding the necessary space on board. It also creates a fire hazard. Some better solution should be found, such as driving the pumps off the main engine, which is otherwise idle during brailing. The air pressure controls for winches, described by Carlson, invite remote control from the pilot house or some other desirable location.

Graftiaux analyses the work done by the trawl winch. In this connection, mention should be made of the recent developments in British-built diesel trawlers, in some of which the winch generator is driven by the main engine through a clutch.

The hydraulic winches described by Huse can be operated at suitable speeds irrespective of the speed of the engine. This is clearly of major importance in long-line and cod-net fishing where the boat must constantly manoeuvre into a position for hauling. It involves a great deal of speed variation and even reversing if a fixed-blade propeller is used. Guinard described a different type of drive for hydraulic trawl winches, but in both types there is the problem of speed regulation. Some sort of speed governor is needed to regulate the maximum speed. When a long line, or net, is fouled and then is suddenly released, the winch inevitably increases its speed considerably until the regulating valve has been readjusted for a lower torque. While that is happening the line or net comes in faster than it can be hauled off the sheave. This disrupts the hauling operation and it seems that some simple limiting speed governor of a fly-ball type might provide a practical solution to this problem.

Costantini describes a fixed derrick system for power handling of the trawl. Fishermen have been much too slow in eliminating the heavy manual labour of hauling the trawls or getting them into the sea. This is a task which lends itself easily to power handling. An arrangement similar to the one proposed by Costantini may be more practical than the present method of swinging the booms from the mast. But such fixed derricks should give the least possible windage and problems of icing in northern waters should also be considered.

H. C. Hanson records the increased use of labour-saving equipment and methods on board U.S.A. fishing boats, a development that has been hastened through the demand by American fishermen for more wages and a higher standard of living. They were the first to realize the need to employ mechanical power for handling gear wherever possible and a similar development is now taking place in other parts of the world. If fishermen are to have higher standards of living, and the fishing industry is to be profitable, then much more use must be made of mechanical power for saving manual labour on board the boats.

TRAWL WINCHES

Dr. A. von Brandt (Germany): A big stride forward in fishing after World War II was the introduction of echo sounders, and the most recent development in European fishing is the study of the behaviour of fishing gear in action. This development was started by three British underwater films made by the Lowestoft and Aberdeen research stations. The films dealt with various trawls and seines in action.

The skipper on the bridge should be able to check the behaviour of the gear while fishing. He wants especially to know when the net is full, and the angle between the two warps cannot tell him that with certainty. A better method is to make use of the fact that there is a correlation between the r.p.m. and the speed. The resistance of the net is indicated by the change in speed and r.p.m. of the engine relative to changing thrust. The French and German fishing industries have gathered much experience in using this method to judge when the net is full.

In other countries a dynamometer has been used. This may be applicable to small or floating trawls but it does not give reliable figures when trawling on rough ground or against a current. One advantage is that when the resistance is equal on the two warps the dynamometers indicate that the fishing gear is in good working order. If the resistance is different, the length of one of the trawl-wires must be changed.

The skipper also needs to know the height of the opening of the trawl, the distance between the otter boards, the depth of pelagic gear, etc., and instruments to permit such observations are being developed.

The difficulty is to install them so that the skipper can read the recordings on the bridge. It is not possible as yet to link the recording apparatus with the underwater gear by electric cables to the bridge, but scientists of the Netherlands and Germany are working together to find suitable methods to transmit the measured values.

Mr. Philip Thiel, Jr. (U.S.A.): In a wooden ship a serious problem is to distribute over a sufficient area the stresses and shock load to the trawl gallows and bollards, and to reduce the loosening and leakage of through-fastenings. The difficulties may be ameliorated by two means: (1) by combining adjacent fittings on a common foundation plate of steel, and (2) by bedding this steel plate to the deck structure by studs, welded water-tight at the upper plate, and extending through inter-beam blocking to a lighter steel plate under the deck (see fig. 561). Fittings that may be combined (by welding) on one plate are the two double bollards and the foremast step, the forward gallows frame and single bollard. Admittedly this is expensive, but it makes for true economy by insuring greater safety on deck and less depreciation of the hull.

Anyone who has been to sea in a trawler, or has noticed in port the effect of gear chafe on the hull, will realize the importance of using stiffer, corrosion-resisting plate for the gallows frames, towing block and quarters, and of using only solid, not tubular, chafing strips on the side of the hull.

Mr. Maurice Graftiaux (France): The electric drive of trawl winches has taken a long time to develop, chiefly because the diesel has taken a long time to replace the steam-engine. Reliability is of prime importance in a winch. If damage to the main engine is a catastrophe for a trawler, so is damage to the trawl winch, because if it stops working the trawler cannot fish. It is also necessary to see that the warps do not break. It is preferable to work with constant pull, which can be done with a diesel-electric drive. The electric system allows an automatic and progressive selection of speed, according to the pull.

The simplest and the safest solution seems to be to use D.C., which has been widely adopted in nearly all countries. A French firm has developed the Ward Leonard system with special falling characteristics. The speed of the winch motor is controlled by the voltage of the generator, which is in turn determined by the current it supplies. By varying the field of the generator, the speed of the winch is varied as a function of the torque, which it develops. This represents an entirely self-governing system.

There is another solution for driving winches from the general circuit of the boat. This solution has been adopted specially in England on big trawlers where the power available is greater than the power needed for the winch. In this instance it would be less economical to have an independent electro-power system for the winch, which is idle during cruising periods.

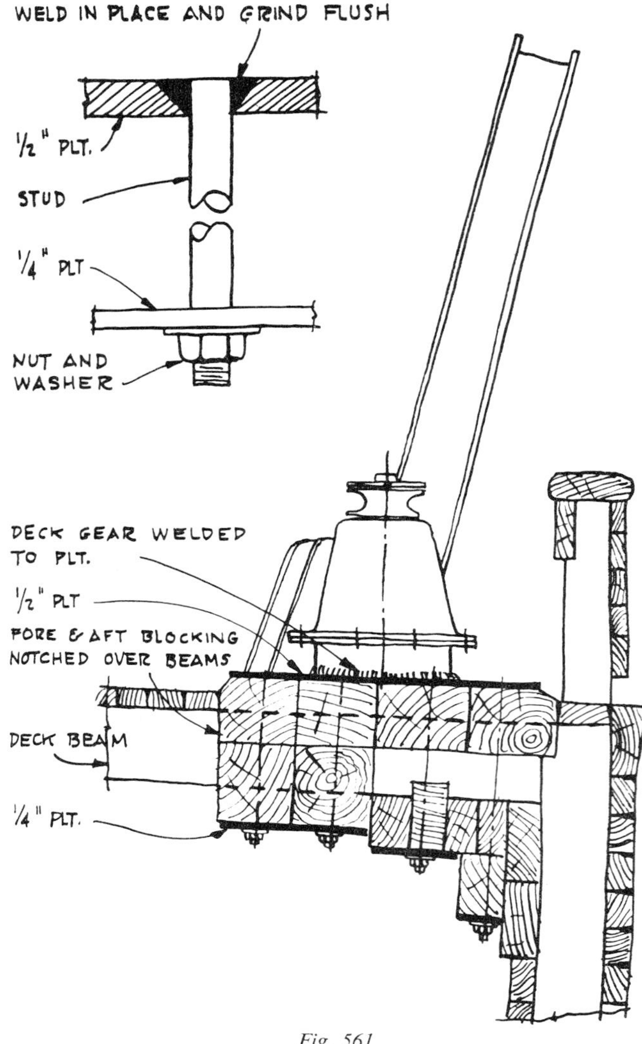

Fig. 561

Many developments of the electric drive for trawl winches are likely to take place. For instance, electric propulsion can very well be combined with the drive of trawl winches because most of the work of the winch is done when the boat is stopped or moving slowly. The winch uses a maximum force when hauling the net, and that is when the general electrical installation is not fully used for other purposes.

Mr. George C. Nickum (U.S.A.): The value of electrical deck machinery is immediately apparent, and so also is the effect of the climate in which it operates. On the U.S.A. south-west coast, for example, a great deal of difficulty has been experienced with electrical machinery because of the very high humidity of the atmosphere. On the other hand, a vessel can operate in Alaska during the summer season, where temperatures are moderate and the humidity is low, and no difficulty is experienced with the electrical machinery. Again, vessels can operate the year round in Puget Sound, the Oregon and the Northern Californian waters, and there is no trouble. But if they are sent south, particularly to Central America, where temperatures go up to 100 deg. F. (38 deg. C.) and humidity ranges from 85 to 95 per cent., the maintenance problem of electrical gear is exacting and very troublesome. For that reason the machinery chosen for a ship is usually determined by the area in which she is to operate.

Mr. Howard I. Chapelle (U.S.A.): During World War II, intensive study was made of the effect of climate on mechanical and electrical equipment, and a great deal of testing and research was done on the effect of tropical conditions on insulated surfaces. This information should be available to the fishing industry on application to the U.S.A. Department of Defence, Washington, D.C.

Mr. W. C. Gould (U.S.A.): The advantages of A.C. over D.C. current in lower cost and greater safety at sea, even in the case of small vessels, are such that efforts have been made to develop suitable A.C. equipment. The latest is a 12½ kw. generator set with a permanent magnet bolted directly to the flywheel of the auxiliary diesel so that there are no bearings of any kind in the generator, nor any slip-rings and brushes to maintain. The equipment gives a perfectly regulated three-phase current. According to the Bureau of Standards in the United States, the permanent magnet loses only 1 per cent. of its magnetism every 100 years.

Mr. Hans Vestre Huse (Norway): A type of governor may be used to control the increased speed of the hydraulic winch when the load is decreased. Mr. Kristjonsson is the first person, however, to object to the increased speed. One firm has installed the hydraulic winch on 900 long-liners, and the fishermen have usually found the automatic increase in speed to be advantageous. The maximum hoisting speed for long-line haulers is usually fixed at 230 ft. (70 m.) per minute, and as the pump and hydraulic motor are both of the positive displacement type, the speed cannot exceed this figure even when the light line is hauled in. At that speed, experience shows the line may easily be hauled off the sheave.

A hydraulic motor now available is still of the positive displacement vane type, but instead of having one radial pressure chamber it has three. This makes it possible to have three different speeds according to the load, and by using two motors coupled to the same shaft, six different speeds are obtained. The hand-operated control valve is equipped with an automatic valve system to limit the hoisting speed of smaller loads even if the handle is put in the full hoisting position. The new motor is especially suitable for trawl winches and a number of Norwegian trawlers are installing it.

The three pressure chambers are arranged in such a way that the motor is pressure balanced at full load.

Mr. Hilmar Kristjonsson (FAO): Perhaps Mr. Vestre Huse missed the point about the need for speed control on hydraulic net and line haulers. The automatic increase in hauling speed with decreased torque or resistance is, of course, an extremely valuable characteristic of the hydraulic drive, but what Mr. Kristjonsson had in mind was the need for setting the maximum speed not only at 230 ft. (70 m.) but at other lower values as well, if fishing conditions so required.

Mr. G. O. Huet (U.S.A.): The efficiency of hydraulic systems for the transmission of power is very low compared with electricity, especially when the transmission is carried out over long distances. High-pressure hydraulic systems are very sensitive and the installation must be perfect. If dirt gets into them it gives considerable trouble. It is probably better for small vessels to have mechanical equipment, but with larger vessels hydraulic equipment may be installed because it is less troublesome.

Most people agree that the electrical equipment is the most efficient for the transmission of power and for cost of operation. A lot of development work has been done to make cables and other electrical equipment suitable for use in the tropics.

He referred to two ships, one of which was installed with A.C., which was used for the cargo winches. The ship has been very successful. About the same time an L.S.T. was converted and fitted with D.C. but it was unsatisfactory and every time it came into port there was work for an electrician to do on board.

Mr. Olai Mollekleiv (Norway): Even owners of small vessels can afford to install Norwegian hydraulic deck equipment because it is inexpensive. It is also simple in design and construction so if anything goes wrong, which is seldom, the trouble is easily remedied.

This equipment was first developed in Norway and many improvements have since been made in it by Norwegian manufacturers. For example, it is now possible to have two changes of speed with one hydraulic motor without affecting design or cost. This type of motor works at a much higher speed in handling the load, such as hoisting catches. It is especially useful in the small boats where two hydraulic motors would be too heavy and expensive. The motor has two working chambers which distribute the oil pressure to opposite sides of the shaft. This practically eliminates bearing pressure. It also has duplex pumps double-acting in one rotor house, which is an advantage over duplex pumps in two houses with two rotors. When a lower speed is wanted, one of the chambers may be cut out. The winch retains full lifting capacity with only half the engine power. This is especially advantageous when, for instance, the auxiliary engine is small. In this case, however, smaller loads may be lifted at full speed.

Another Norwegian development is the magnetic filter or mudbox which can be cleaned without taking the oil out of the system. In the case of pumps on both main and auxiliary engines it is possible to drive two hydraulic winch motors, when coupled in serie, at full torque by a single pump but, of course, at a lower speed. Such an arrangement provides an element of safety when one auxiliary engine is out of action.

Nearly all the Norwegian fishing boats now use hydraulic machinery. Even the small seine boats have winches of 1,100 lb. (500 kg.) capacity. Purse seiners in particular find the hydraulically operated derrick very useful, especially when the sea is rolling and the catch has to be hoisted tons at a time into the listing vessel.

Commander C. Harcourt-Smith (U.K.): One of the difficulties of the closed hydraulic system is that, owing to the very high pressure, there is always the risk of a pipe bursting or of a leakage at one of the unions. Even a slight leak of oil might make the decks very slippery and dangerous, and there is, of course, the need to carry special fluid for topping up the system.

An hydraulic system employing water instead of oil uses only one pump in the engine-room. This pump can be used to operate an ejector system for clearing the bilges and for hydraulic gear—winches, line haulers or windlasses. It can also supply pressure to connections for deck washing, fish washing, etc. The system of drive to the winches and line haulers consists of a form of water turbine geared to the machines. One great advantage is that the control wheel can be adjusted to regulate the hauling power of the winch to pay out at any predetermined strain so that there is no need for broken wires or warps. This function is also useful in rough weather as the winch can be set to pay out again when wave action causes a sudden snatch. Similarly, when a trawl catches on the bottom the paying out automatically operates.

Mr. I. Bromfield (U.S.A.): A high-pressure hydraulic winch developed in association with M.I.T. and Vickers, has been very satisfactory in replacing a mechanical drive in a trawl winch. A special feature of the winch is the means of limiting the overload condition. In the previous installation with mechanical drive from the main engine maintenance costs were very high because of the wear and overload. There was no way of controlling the volume of horsepower transmitted through the winch. The main engine developed about 560 h.p. and when it was operating at trawling speed (about 300 h.p.) the clutches and gears wore and the bushings were actually squeezed out because of the strain on the cable. The high-pressure hydraulic system which replaced this mechanical drive has worked almost five years free of trouble. The high pressures were troublesome at the start as leaks occurred in the solid pipes because of vibration, but this difficulty was overcome by using flexible pipes.

Mr. Bromfield wanted to know what was the difference in size between the pump and motor in a low-pressure hydraulic system and those in the high-pressure system.

Mr. F. R. Stobinski (U.S.A.): The pump and motor in the low-pressure system are both bigger.

Mr. Bromfield: How much bigger? Let us take for example a 50 h.p. motor.

Mr. Stobinski: To get 50 h.p. out of a pump in a low-pressure system would call for the use of 3½ in. (89 mm.) diameter tubing. On a high-pressure circuit the tubing would be about 1 to 1¼ in. (25 to 32 mm.) in diameter.

Mr. Bromfield: How much larger would the motor be, physically?

Mr. Stobinski: Definitely larger but I can give you no exact idea. There is a great variety of hydraulic motors. The piston type, for example, is very compact, while the vane type is bigger, so it is very difficult to provide a comparison. Again, sometimes it is necessary to gear down the drum speed in a high-pressure system and the gearing, of course, takes up some space.

Mr. Bromfield: That is true of the piston type only, but not of the vane type of pump. If a 1,000 r.p.m. piston type motor was used, the size of the hydraulic unit would be about 1/5, and the hydraulic motor would drive the winch direct through a worm gear with a ratio of about 40 to 1.

Mr. Stobinski: The Norwegian system is completely different. It has no worm gear and the motors are connected direct to the shaft.

Mr. Bromfield: What you have in mind would not be adaptable for the U.S.A. fishing fleet. To put a large hydraulic motor between the drums would sacrifice a considerable amount of cable and the length of cable is very important because the fish are caught in deep water.

Mr. Stobinski: That would depend on the size of the ship.

Mr. Bromfield: Yes, but we are talking about the type of ship we are using to-day in fishing.

Mr. R. T. Whiteleather (U.S.A.): Referring to the hydraulic winches on the Norwegian vessel, the *G. O. Sars*, he said that a big hydraulic trawl winch had been installed and operated very effectively.

Mr. Bromfield: The ship has a 30 to 35 ft. (9 to 10 m.) beam and U.S.A. trawlers have, at the most, only a 25 ft. (7.6 m.) beam. The electric drive seems to be the most foolproof and least troublesome drive available to-day, although most people seem to be under the impression that electricity means complications, continual repairs and maintenance. This is untrue. The type of drive he installed on vessels during the past 30 years has given no trouble. In some cases not even the brushes in the motor or generator have been changed. There are no contactors and no overload relays. The generator is connected direct to the auxiliary diesel. There are two leads from the armature of the generator to the armature of the motor. One of the leads, instead of going direct to the motor, has an exciter in the circuit, whereby one shunt coil of the exciter passes through a shunt field attached to the exciter which is mounted on the generator whereby the current passes through this shunt field, bucks the field of the exciter, and thereby decreases the voltage of this exciter which in turn excites the field coils of the generator. This so-called Bromfield torque control system eliminates overload to the winch motor, generator and diesel engine. The amount of overload on such a winch motor may be predetermined at, say, 10 or 15 or 20 per cent. The moment the current increases over the point established, it is automatically controlled as above.

In the system previously used in ships, the winch operator would not pull his lever back when there was overloading, or the winch would reach momentarily a stalling condition. The result was that, in the case of an armature control, the grids were burned out, or the motor or the diesel was overloaded and would begin to slow down. But in the system described above it is impossible to overload the motor, the generator or the diesel. For example, when a vessel is on the crest of a wave an extreme load condition at once develops and the motor automatically slows down. As the vessel comes down from the crest of the wave the motor speeds up, taking in the slack, which eliminates the chance of the cable snapping as sometimes happens when a mechanical drive is used. The control is exercised through the current from the battery. Two or three amperes go into the fields of the exciter of the main generator and therefore the current is very small passing through the contractor. No contactor has had to be replaced on any installation during the past 20 years. The resistors are very small and carry a maximum of 2 or 3 amperes. Experience shows that the performance and flexibility of the electric drive is 100 per cent.

The reason for installing the hydraulic system on the smaller vessels in place of the electric drive is that it is not satisfactory to run a generator from the main engine because that engine is not running at constant speed. This causes a problem of voltage variation. So, in the case of small vessels, the hydraulic system can replace the mechanical drive and eliminate problems of overload.

Mr. Bromfield concluded by asking if Mr. Huse was correct in stating in his paper that a hydraulic winch could hoist 400 tons an hour.

Mr. Hans Vestre Huse (Norway): The brailing rate for herring will seem very high compared, for instance, with the U.S.A. west coast purse seiners, but the quantities in each catch, and, therefore, the special practice in Norway, is entirely different. The biggest Norwegian purse seiners have a full load capacity of 400 tons of herring and, as this may be taken in a single catch, it is necessary to brail the seine in about one hour, or the herring will die, sink and drag down the net and the small seine dories. During the brailing great quantities of water follow the herring and have to be lifted by the winch. About 90 per cent. of the vessels are fitted with hydraulic winches.

Mr. H. C. Hanson (U.S.A.): On the U.S.A. Pacific Coast mechanical drive has been used because of the great distances the fishing boats have to sail. Some of them will go 1,500 miles or more from one fishing ground to another. Manufacturers of hydraulic equipment are partly to blame for the continued preference of the mechanical drive because they have not developed their equipment as they should have done. But in British Columbia, Canada, in recent years almost exclusively hydraulic systems have been used for trawl winches and steering gears, even in very small boats. It has been successful, but these vessels do not range great distances.

Twin trawl winches are now becoming out of date, and in future the combination winch will be installed in smaller vessels built on the U.S.A. Pacific Coast.

Much time is lost by not using dynamometers when trawling.

Mr. R. T. Whiteleather (U.S.A.): The U.S.A. Fish and Wildlife Service research vessel, the *John Cobb*, which is engaged in exploratory fishing in the Pacific North-West and Alaska, is equipped with a good deal of hydraulic machinery. It was chosen because of its flexibility and to conserve space. The variety of operations carried out demand flexibility in equipment. For example, equipment is sometimes shifted on the stern deck. Sometimes it is desirable that the trolling gurdies be replaced with a net hauler. With equipment being moved about in this manner, it is undesirable to have shafts or chain-drive or gear of that kind on the deck. A low-pressure hydraulic system is used in the *John Cobb* for the windlass, the trolling gurdies, and the drive for the refrigeration compressors. The equipment has been in operation for four years and has given no trouble.

The hydraulic system in the main winch for vessels operating on the west coast of the United States is under scrutiny by the fishermen because of the heavy conditions met in deep trawling. Winch equipment must face a severe shock problem and for that reason a torque convertor arrangement was chosen for the *John Cobb*. This has worked very satisfactorily.

Mr. James F. Petrich (U.S.A.): With regard to the auxiliary engine driving the torque convertor, the drive chains and shaft went through the hold to the deck winch. Another sprocket was put on one of the shafts and also one on the main drive shaft of the propeller, so that if the main engine broke down it was still possible to operate by using the general drive through an hydraulic coupling from the engine. What is there to say regarding this arrangement?

Mr. Whiteleather: The drive from the auxiliary engine to the main shaft of the propeller on *John Cobb* was installed for two reasons: (1) to be able to proceed at very low trawling speeds of, say, $\frac{1}{2}$ to $\frac{3}{4}$ knots, for long periods; (2) it was thought that the easy way of coping with the problem was to use a fixed-blade propeller, which would ensure reliability. It did not work out well.

Commander A. C. Hardy (U.K.): A tug, installed with hydraulic machinery, has been built for use on the River Thames. It has four high-speed engines athwartships, each driving an hydraulic pump and supplying power to a small hydraulic motor on the propeller. The vessel has been successful.

Mr. A. Conhagen (U.S.A.): Hydraulics have been used on board ship for many years. All steering gear is hydraulically operated, so crews are familiar with it, and it must have been reliable, otherwise they would have objected to further installation of such equipment. A variable capacity type pump is generally used and it is doubtful if there is any piece of machinery made which can be more easily controlled or which gives better results. This installation is different from the fixed displacement, the fixed vane type of hydraulic motor and pump, which is usually discussed.

The cleanliness of the oil in the hydraulic system is very important. Use of a filter until the system is run in normally ensures cleanliness. But before the system is assembled, the pipes and the bends are treated and the insides of the tubes are wire-brushed or sand-blasted. After that, rags are run through and then the filter is put on.

Mr. I. Bromfield (U.S.A.): Pickling the pipes of the hydraulic system after they have been formed to a shape is very important because, no matter how many filters are used, the scale remains on the pipes whether they are made of copper for use in low-pressure, or steel for use in high-pressure, systems.

Mr. George C. Nickum (U.S.A.): More than pickling is needed for the pipes in an hydraulic system. For example, one shipyard set up an elaborate sand-blasting and pickling equipment, but was surprised some months later when a test revealed there was still sand in the pipe. It is very difficult to be sure that all the sand has been taken out after pickling, but the internal parts of the hydraulic system can be partly cleansed by flushing with hot oil.

It is not customary to put in heat exchangers, but there are installations where they are necessary. When there is any doubt in the matter it is certainly worth while including a heat exchanger.

One of the problems to be faced on the U.S.A. Pacific

Coast is that of finding shipyards capable of properly installing hydraulic equipment. To keep the lines clean and ensure that dirt does not get into the system requires highly skilled labour. Hydraulic equipment manufacturers should help the shipbuilding industry to train personnel for this important work. There is nothing worse than an improperly installed hydraulic system as it can wear out very quickly.

In the past ten years manufacturers of hydraulic equipment have made the same mistakes as all other equipment manufacturers make—they have over-rated their equipment and have ignored the problem of heat. It is essentially true that the oil lines are too small in the vane type of hydraulic prime movers.

Mr. F. R. Stobinski (U.S.A.): A heat exchanger is not necessary in a low-pressure hydraulic system where operations are intermittent, but if operations are continuous it is worth installing one. That will eliminate throttling losses and the heat generated by throttling. The exchanger is generally placed on the return line and cools about 15 per cent. of the flow.

One of the problems concerning hydraulic equipment is that it requires good installation. The reason for the success of the Norwegian type of equipment is that it is built for average use and is easily understood.

Mr. J. S. Robas (U.S.A.): In the menhaden fisheries each vessel carries two 33 ft. (10 m.) launches with small gasoline or diesel engines. The deck equipment of the mother ship is mostly petrol-driven because fishing boat crews understand petrol engines better than they do diesels. In recent years the trend has been to get more h.p. into the launches, rising from 40 h.p. to more than 90. On the 40 h.p. standard launch, a crank starting system is used, but the 60 and 80 h.p. engines cannot be hand-started so an electric starter is used. This has led to a host of troublesome problems. Sometimes on Monday morning the batteries are dead and the starter will not work, and on Saturday the crew, with their minds on the pay-day, spray brine on the engine instead of on the net. The cost of electrical equipment has risen steeply and is very heavy. The result of this experience is to reverse the trend. Once again 40 h.p. engines are being installed solely to get away from electrical starting equipment. There should be a promising market for hydraulic starters for such small engines. One American manufacturer is now equipping tractors with hydraulic starters.

Mr. W. C. Gould (U.S.A.): A nurser starter is available to-day for starting diesels of 20 to 300 h.p. It is hand-operated and completely enclosed. One machine that has been tested has done more than 10,000 starts without a single failure. The machine is being installed in some menhaden fishing boats powered with diesels. At present it is made only in one size and costs about $520 (£190) in the U.S.A.

Mr. I. Bromfield (U.S.A.): Much of the damage and losses caused to fishing vessels come from the use of mechanical drives, particularly by the winch. With a mechanical drive on the winch it is only possible to haul in, not pay out. To pay out, two men have to handle the cables, although, once the doors are dropped, the weight tends to pull the cables out. In hauling the net the doors go up first and are placed between the gallows and the rail. In most cases the doors cannot fall into place because the weight is not sufficient to haul the cables off the winch, so they have to be slackened to let the doors drop a little while still on the outboard side. They are then moved by the hanging block and made to drop into position. Many men have lost a hand or arm or foot—in some cases a leg—because of this awkward manoeuvring. The speed of the winch is regulated by slipping the clutch because, as the engine runs at a constant speed, there is no other way of varying the winch speed. Sometimes the clutch sticks, and when that happens the men at the doors are in danger of serious injury. The advantage of the hydraulic drive is that it eliminates much of the hazard in this work because it is possible to go into reverse, a most important safety factor. The mechanical drive is not reversible.

Mr. W. C. Gould (U.S.A.): There have been two or three types of direct mechanical drives installed, operating through torque convertors between the main or auxiliary engine and the winch. Fluid couplings have been used to reduce the shock loads on the drive from the main engine and, in one sense, to limit the load to be put on the auxiliary drive. The oil in the system was cooled in a heat exchanger, so that it is possible to operate the winch without breaks. By throttling, the winch can be made to pull as conditions demand. When the vessel surges the winch will pay out the cable and bring it back in the same way as the electrical drive does. This system has been used on a U.S.A. Pacific Coast vessel for about four years with great success. The additional cost is about $250 to $300 (£90 to £110) for 65 to 70 h.p. torque converters, and $700 to $800 (£250 to £290) for 100 to 150 h.p.

Mr. William C. Miller (U.S.A.): The consensus of opinion on the U.S.A. west coast is that most fires in vessels are started from electrical sources. Investigation showed that many were the result of over-heating certain resistances and some lines below deck. This is one reason why there should be more use made of hydraulic equipment.

Mr. Arthur de Fever (U.S.A.): Generally speaking, all auxiliaries on tuna clippers are A.C. This is especially true of installations since 1944, and nearly all vessels that still use D.C. were built before that time.

On one purse seiner an hydraulic system was installed. Unfortunately it gave some trouble at the start and the first trip resulted in very poor catches. This has made owners and fishermen on the Pacific Coast of the U.S.A. cautious about using hydraulic machinery. On any complete new and different installation there is usually a certain amount of trial and error and it may be that ample trial was not given by fishermen in this case.

Commander R. E. Pickett (U.S.A.): Mr. Huse stated that an hydraulic motor has about one-third the volume of an electric motor, rated at the same r.p.m., and is more powerful, but to compare the two motors fairly it is necessary to have the following information: (1) the effective pressure on the hydraulic motor; (2) the voltage impressed on the electric motor. In either case an increase in hydraulic pressure or in electrical pressure respectively will cause the machine to increase its power rating. A high-voltage motor of the same size as a low-voltage motor will be of greater power.

If an engine developing 100 h.p. is driving the hydraulic motor and is connected to a dynamometer, what is the efficiency of the system? The answer is from 68 to 73 per cent. This is interesting because it indicates a safe, convenient, dependable and low-cost means of supplying power requirements for the trawl winch and anchor winch, and even main engine starting motors on small vessels. On many vessels this equipment is operated less than 200 hours per year.

A cheap petrol engine could be placed in a safe and convenient location above deck, sheltered from the weather and directly connected to the hydraulic pump. The pump, pipe line and motors could be a fixed installation. There are now available hydraulic systems of this type that are for all practical purposes foolproof and repair free.

Mr. I. Bromfield (U.S.A.): The efficiency of an electric drive is about 80 to 85 per cent.—that is from the generator to the motor of the winch.

Mr. Howard I. Chapelle (U.S.A.): In small trawlers the mechanical drive system is installed as a matter of economics and of space and weight. To a large extent it controls the position of the trawl winch in relation to the main engine. This is unfortunate because it has prevented the rearrangement of deck gear to cope with surging and shocks on the winch and cable, which have caused a good deal of trouble. For this reason it is desirable to have a torque convertor or hydraulic coupling on the power take-off.

There seems to be no chance of putting an electric drive in small boats. Mr. Bromfield has explained the very sound reason for this. There is, however, an opportunity to use some kind of hydraulic drive, providing a design can be produced which will enable the deck gear to be arranged to advantage, without becoming involved in a mass of troublesome details. But, of course, the use of such hydraulic equipment would be controlled by its cost.

BAIT

Dr. S. Shapiro (U.S.A.): In Japan, fisheries for bait and tuna are conducted separately. The bait is caught and held in containers in protected harbours; tuna vessels collect the bait just before leaving for the fishing grounds. Live bait is used in much the same way as American tuna fishermen use it for pole and line fishing. The bait is thrown into the water and hides under the boat, and then the tuna come in for it.

Fishing for bait is conducted with many types of gear. One method of fishing in use in southern Japan is done at night, four vessels operating together as follows: four small fishing craft—one at each corner of the lift net—lower the net in the water and a light is placed above it. As the sardines are caught they are placed in floating containers made of bamboo or, sometimes webbing. The water circulates through the containers and the bait is kept in them for a week. Nothing further is done except to remove the weak fish so that the survivors when taken on board the tuna vessels, are hardy, and can be kept in good condition merely by circulation of sea water through holes in the hull of the vessel. This simple method of catching and keeping bait is most effective.

Mr. William C. Miller (U.S.A.): The Japanese method of keeping bait is the same as that used by the commercial sport fishing organizations on the U.S.A. west coast. The only difference is that in the U.S.A. the bait is not held for any time to condition it.

One of the problems of the U.S.A. tuna clipper fleets is to find sufficient quantities of bait. They usually have to go down to Central American waters to catch it. The system of keeping bait on board the clippers is very effective and not much of it is lost.

Some chemical manufacturers are experimenting with the use of chemicals to attract schools of tuna and there are other companies experimenting with electronics, etc., in the hope of replacing live bait. But so far the most economical way for tuna clippers to operate is still to catch and use live bait.

Mr. Mogens Jul (FAO): Some of the Latin American countries feel that U.S.A. tuna clippers are taking too much bait from their waters. It might save time and money for the tuna clippers if local fishermen were encouraged to start bait fishing for sale to U.S.A. tuna fishermen. The clippers could then, for example, sail to Puntarenas and pick up the bait without having the trouble of fishing for it and endangering the boat by operating in shallow water.

Mr. H. C. Hanson (U.S.A.): The fishermen in the Hawaiian Islands have a very difficult time in catching bait. They use vessels in which the hulls are divided into six compartments, each with water inlets. As the bait is caught it is dropped into the compartments. When they go tuna fishing they empty the bait from each compartment and replace it with the tuna they catch.

Mr. James F. Petrich (U.S.A.): If there was a source that could guarantee bait it would be an excellent arrangement for tuna clippers, but probably the chief objection to the scheme proposed by Mr. Jul is that the clipper has been developed as a self-supporting boat. The fishermen like to have control of every operation, and that includes catching bait.

The arrangement for keeping live bait in tuna clippers is well developed. Water is pumped into the fish wells and deck boxes, passing through screens which disperse its flow down to the bottom of the well. From there it moves to the top and out through another screen, which is of considerable area to ensure that the speed of the flow is not so strong that it traps the bait against the screen. The flow is enough to change the water once every 12 minutes. Weir boards are used to control the height of the water in the deck boxes.

Mr. Arthur de Fever (U.S.A.): The U.S. Fish and Wildlife Service in Honolulu have been experimenting with fluids to attract tuna and some commercial fishing vessels from San Diego have used it, but so far without much success.

Much depends on the behaviour of the tuna. For example, sometimes only a small amount of bait needs to be used for catching a great quantity of fish. At other times a lot of bait must be thrown into the sea to attract the tuna and probably only a very few of them are caught.

The tuna occasionally get into a frenzy and anything thrown on the surface will excite them. For example fishermen have used spaghetti or rice, when out of bait at the end of a voyage and have made big catches proportionately. It all seems to depend on whether the tuna are hungry and excited or are feeling torpid.

The suggestion of building up bait reservoirs in Central and South American countries, made by Mr. Jul, has often been considered. Four years ago in Panama a man tried to obtain a concession to set up a bait farm but because of the political situation the concession was never granted. If a bait farm was organized it would probably be successful provided the tuna clippers agreed to buy the bait regularly, otherwise they would only buy it when they were unable to catch it themselves.

Replying to a question by Mr. Traung about the use of water-spray as complement to bait in fishing for tuna, Mr. de Fever said that it was quite true that sprays were used, and he had designed an independent water-spray system for three or four boats. It was not a new method, as fishermen from Honolulu and Japan use it regularly. All the vessels he had seen in Honolulu were equipped with sprays. The main purpose of the spray system is to ruffle the water when it is very calm so that the fish are not frightened off by the reflection of the poles and fishermen.

Having experimented with aeration of water in the bait wells, he had found that the principle behind changing the water every 10 or 12 minutes is not merely to clarify or keep it fresh but also to aerate it. The mere process of changing the water is sufficient to cause aeration. If, however, a device was used to increase the aeration of water in the bait wells it should be possible to operate them with smaller pumps.

Mr. H. C. Hanson (U.S.A.): He described a tuna-fishing trip at Hawaii in a 75 ft. (22.9 m.) boat with a speed of about 6 knots an hour. When the tuna were sighted the bait was thrown out and the water spray was used. The tuna seemed to be travelling at about 9 knots an hour, and appeared to leap aboard the boat. The seven or eight fishermen caught 115 tunas (about 30 to 35 lb., 14 to 16 kg.) in about nine minutes.

The exploration vessel *Baird*, operated by the United States Bureau of Fisheries, was designed so that a study could be made of bait-keeping. Aeration has been tried but it did not seem to make much difference. The tanks, with the continuous change of water, kept the bait in good condition.

Mr. R. T. Whiteleather (U.S.A.): Some thought has been given by the U.S. Fish and Wildlife Service to aeration in the bait tanks during experiments in finding bait in the Gulf of Mexico. But if there is the right amount of circulation of clean water in the tanks and enough use and replenishment of bait, there seems to be no need to spend extra money installing aeration devices.

Mr. E. R. Gueroult (France): Replying to questions asked by Commander Hardy, Mr. Traung and Mr. Hanson, Mr. Gueroult said that French fishermen used live bait—sardines—both for rod and net fishing for tuna. French trawlers in in the North Sea had caught tuna by net, but that was accidental. The live bait tanks are usually made of aluminium alloy and are painted white. Some have water and air circulation as well as electric light, which is kept on all the time.

He added that modern French tuna boats are motorized and equipment is up-to-date. The old smack type of sailing boat is disappearing rapidly because it is not now profitable to operate.

MISCELLANEOUS

Mr. R. T. Whiteleather (U.S.A.): In reply to a question about the use of drum seines in fisheries other than for salmon, Mr. Whiteleather replied that not much could be said about the effectiveness in the Californian sardine fishery, because of the size of the nets. It might be more effective if used in the menhaden fishery on the U.S.A. east coast, but it probably would be limited in the Pacific Coast sardine fishery because the seines have to be heavier and deeper. The first trouble met with the drum seine equipment was the bulkiness of the deep net. Fishermen had to re-hang the nets and re-design them to some extent. They found that the five-strip net, which is quite shallow, was most satisfactory.

The menhaden fishery, on the other hand, might lend itself to the drum seiner type of fishing. The fish are taken in shallow water and it is the largest fishery in the U.S.A. Small boats could be used, and they were fastest in seining operations.

Commander A. C. Hardy (U.K.): There has been deck gear ever since there have been fishing vessels, but two important developments seem to have been overlooked. They concern hatches and masts. If the masts described by Bain and Costantini are as revolutionary as is suggested, why have they not been fitted to more ships? They ought to be more closely investigated as both seem to be developments away from the traditional type of mast.

Mr. Jan-Olof Traung (FAO): In Boston, Massachusetts, U.S.A., about half the trawlers of more than 100 ft. (30.5 m.) length are equipped with the tripod masts with no shrouds and stays. The masts are heavier than the common rigged type and in winter they accumulate an enormous weight of ice. Perhaps it is because of this that the trend in Boston is no longer to install tripod masts.

Mr. P. Bain (France): About 100 modern British trawlers have been equipped with a new type of watertight metal hatch covers that can be opened or closed or adjusted rapidly and make for safety during loading operations. He did not suggest that the new covers were suitable for all ships but he thought they could be adapted to meet the particular working conditions of fishing vessels.

There are two big advantages in using a bipod mast: (1) it abolished the use of shrouds which encumber deck space and require maintenance; (2) the mast has relatively a low weight.

Until recently this type of mast has not been used on trawlers. Mr. Kristjonsson had asked about the absence of back stays for steadying the cod-end when hoisting the fish on board. This is a problem which must be studied but it does not seem to be insoluble.

Mr. Philip Thiel, Jr. (U.S.A.): The problem of staying the foremast has received a good deal of ingenious attention in the U.S.A. One successful solution is to consider the mast as one leg of a tripod, with the other two legs as struts extending from the outboard after end of the forecastle to the hounds. This leaves the space about the forward gallows free from stays, and makes a simple sturdy arrangement. Another approach is to replace the two struts with wire stays and add two other stays leading from the hounds aft to the deck, inboard of the after leg of the forward gallows frame. This arrangement is lighter. Incidentally, to avoid the nuisance of seagull droppings a 12 in. (30 cm.) metal spike at the top of the mast is useful.

The weight and expense of both an anchor windlass and chain cable may be avoided by using the trawl winch and trawl wires in conjunction with the anchors shackled to a suitable length of chain. Usually it is necessary to fit some fairlead rollers for the trawl wire at the break of the forecastle. This is an arrangement which has been approved by the American Bureau of Shipping.

PACIFIC COAST PROCESSING VESSELS

by

GEORGE C. NICKUM

THE use of vessels on the Pacific Coast of North America which are large enough to allow the fish to be actually processed on board is a development which has come about within the past 30 years. The first fish-processing vessel was probably the *Santa Flavia*, a World War I wooden freighter which was converted to a floating salmon cannery in the spring of 1922. She operated up to 1934 when she was burned and abandoned. In 1923 the *Mazama* entered the same service. She was also a World War I wood freighter, powered by steam engines, and she operated until the early thirties.

The first steel processing vessel appeared in 1929. She was the *International*, a 250 ft. (76 m.) freighter. She was used exclusively for the canning of salmon and operated up to the start of World War II. She was eventually sold to foreign interests.

The *Memnon*, one of the standard 5,000 ton World War I steel freighters, was converted in 1934 to a salmon cannery and operated until the start of World War II. After reconversion to a freighter, she became a war casualty. The *La Merced*, a three-masted sailing schooner with auxiliary diesel power, was also converted to a salmon cannery in 1934. She is the only one of this pre-war class of processing vessels still operating. In 1939 her owners purchased the *Ogontz*, a Hog Island type freighter of World War I, and converted it to a salmon cannery. She was requisitioned for war service and was eventually torpedoed in the Mediterranean.

All these vessels were used in the salmon canning industry in Alaska. They were generally similar in their arrangement but, as the *Ogontz* was the latest of her class, she will be used to illustrate salmon canning vessels.

Fig. 570 gives her deck arrangement and the inboard profile. The salmon cannery on the *Ogontz* was typical of the early class, but was designed to allow the ship to be used as a freighter when not serving as a cannery. The cannery was almost entirely enclosed in the bridge house structure. The iron chinks, sliming tables, fish bins and conveyors, located fore and aft abreast of No. 2 and 4 hatches, were made portable to be readily removed. The only other cargo space used for anything than general stowage was the No. 1 'tween deck, converted to accommodate the cannery crew. This was simply a mass berthing and messing space. The hatch trunk proper was made portable as were the berths and tables, and the 'tween deck was used for general cargo when in freight service. The fishermen's berthing was on the poop deck and the cannery crew toilets and galley facilities were in the forecastle head. The main work of conversion on the *Ogontz* was the installation of the canning machinery, provision of berthing and messing facilities, revision of the boat deck quarters and galley to take care of the increased complement, increasing the boatage, and installing turbine generators and an auxiliary condenser to cope with the increased electrical load.

In the mid-thirties there was a very short period during which a number of large floating herring reduction plants were operated off the California coast. These vessels were later legislated out of business.

The end of World War II brought a resumption of interest in fish processing vessels and a number of interesting craft were converted for use in the salmon industry and in other fishing. The first was the *Pacific Explorer*. An experimental vessel, she was designed to develop the unexplored bottom fish grounds in the Bering Sea, and to operate as a mothership in the tuna trade. Completed in 1946, she operated for several seasons in the Bering Sea and South American waters.

Fig. 571 gives the arrangement of the upper decks and the outboard profile, and fig. 572 gives the arrangement of the lower decks. The conversion of this vessel from a standard three-island type World War I freighter was an extensive operation. In effect the main machinery spaces were retained without change and the basic steel hull was used, but almost everything else was installed new. The wells between the poop bridge and the forecastle enclosures were covered to make a complete shelter deck. Orlop decks were installed fore and aft of the machinery spaces. An auxiliary machinery space was installed aft of the present engine room and the space abreast of the shaft alley in the lower hold was used for fuel tanks.

Next came the trawler *Deep Sea*, fig. 573. She was built to fish the vast shelf underlying the Bering Sea.

A conventional trawler from outward appearance, she was designed to process fish on board. Fig. 574 gives the deck arrangement. The distinctive features are the processing house aft of the trawl winch and the blast freezer located below it.

Two L.S.T.'s, the *Saipan* and *Tinian*, were converted for use as fish-freezing vessels in the tuna trade. They were the first of several conversions which included two World War II net tenders and another L.S.T., the *Oceanic 5*.

Fig. 575 is a profile drawing of the M/V *Tinian*. The *Saipan* is basically a sister ship. Conversion work on these vessels consisted of refrigerating the holds, installing brine tanks, quarters, refrigeration machinery and cargo handling facilities.

In 1948 a salmon cannery firm made plans for the conversion of the *L.S.T. 662* into a combination salmon canning and freezing vessel. This project was completed in 1950, and it represents the latest evolution of fish processing vessels. Its name is *Neva*. Fig. 576 and 577 show the inboard profile and arrangement of decks. The entire hold area is refrigerated and a salmon cannery is put in a new house extending almost full length of the main deck. Boilers and generating equipment, cargo handling facilities and a pilot house are installed.

CONVERSIONS

All but one of the processing vessels that have operated in the past were originally built for some other purpose. The *Deep Sea* is the exception to the rule and is the only vessel actually operating at sea for more than a few months of the year. Tuna motherships have in the past made a maximum of only four trips a year and the large bulk of their time is spent at anchor, receiving or discharging tuna. This is a basic reason why the fishing industry cannot afford new vessels, other than the *Deep Sea* type, designed specifically for fish processing work. Fuel and other operating economics are of real advantage only during full operation of a vessel. As fish processing vessels operate at full power only one-quarter to one-sixth of the time of a regular freighter, these savings would have to be four to six times as large to make the capital investment of a modern vessel worthwhile. Fish processing vessels can be justified only if the net cost per ton of product is less than the same product processed in shore establishments.

The first design problem in developing a fish processing vessel (other than the *Deep Sea* type) is the selection of the vessel which is to be converted. The importance of this from the standpoint of overall cost and operating efficiency cannot be over-emphasized. The naval architect can be of more value in the selection of the vessel than he can in the remainder of the programme. To have a vessel with adequate space for the required quarters, handling equipment, services, storage spaces, etc., makes a tremendous difference in the overall cost.

The next most important consideration is labour cost in the processing operation. Processing vessels are at a fundamental disadvantage to shore establishments because the operating crew does not walk out the front door at night and provide their own food, accommodation, and amusements. A large part of the total cost is in providing facilities and taking care of the men's needs after working hours so that every man hour saved by the proper arrangement of processing and handling saves on a man's wages and the cost of keeping him on board.

NEED FOR SPACE

Adequate space is essential, a most difficult thing for an inexperienced owner or architect to appreciate. They should keep a picture of the average food processing plant constantly in mind and remember it is not possible to add extensions to a ship after the initial conversion has been made. Piling nets or spare gear on " the vacant lot next door " is not possible on a ship.

The vast majority of fish-processing vessels are essentially motherships and are responsible for a fleet of small boats that look to their parent ship for the same type of services they get at a dock. The fishing gear, in particular, is bulky and cannot be stowed neatly on shelves like spare parts for machinery. The *Ogontz* appeared to have ample room for any conceivable operation yet, when canning at full capacity, both forward and after weather decks were completely covered with salmon cans being cooled. During the canning operations at least one of the 'tween deck spaces had to be assigned to the storage of gear and similar materials. Had the *Ogontz* been of smaller size and of smaller hold capacity, this would have restricted her potential capacity.

The *Pacific Explorer* when operating as a tuna mothership in the South Pacific had no use for her fillet line or crab line on the shelter deck so this whole area could be used for storage. Reports from the skipper showed it was so used and he thought that the ship was in no way over-supplied with general storage and working space. In the Bering Sea, when both the crab cannery and the fillet line were being operated, the crew were limited to some open deck space and to the 'tween decks for storage of spare gear. This did seriously handicap them as the full potential processing capacity could not be used and some cargo space was appropriated for gear storage.

The *Deep Sea* is again the exception because she does not act as a mothership and does her own fishing. The *Saipan* and the *Tinian* are good examples of how space is really needed and used. The entire wing wall spaces, with the exception of the machine shop and the refrigerating machinery areas, are available for storage. In addition, the *Tinian* has a large open deck on which gear can be piled. It would seem impossible to fill all of this space, but it can and has been done. The wing walls are filled with sacks of salt and spare provisions for the fishing vessels, plus spare parts, ropes and gear of all nature.

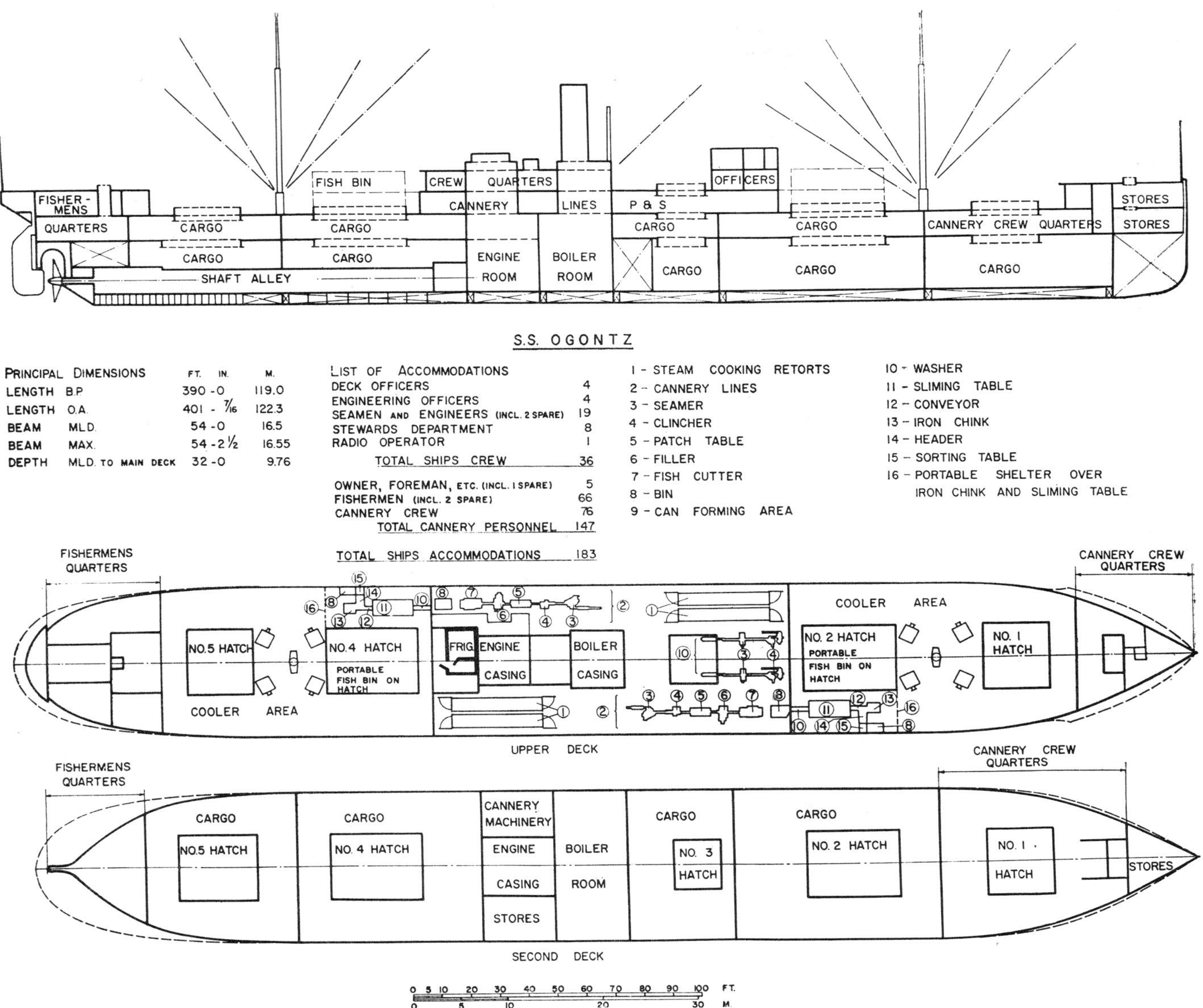

Fig. 570. Profile and deck arrangement of Ogontz, *a Hog Island type freighter converted into salmon cannery. Naval architects: W. C. Nickum and Sons, Seattle, Washington, U.S.A.*

The deck is usually cluttered with all sorts of gear from the small vessels plus large numbers of oil and gasoline drums.

QUARTERS

Quarters present the same problem. The standard of accommodation for ship operating personnel in fish-processing vessels is the same as that of merchant ships, although large open berthing rooms are still common for the cannery personnel and fishermen. On the *Ogontz* all of the ship's personnel and some of the cannery personnel were located on the boat and bridge decks. The practice of berthing more than one officer in a stateroom is no longer followed. However, fishermen on a cannery ship spend a relatively short time on board, so their quarters are of a minimum standard. The rest of the time they are fishing and have only an occasional meal on the mothership. Their quarters, of course, are also used by the fishermen on the trip to and from the fishing grounds.

On the *Ogontz* the No. 1 'tween deck was designated as a mass berthing space, following the common practice in transporting cannery workers to the fishing grounds. The berths, however, were two-high instead of three-high, which was usual in the Alaska steerage trade. All berths were easily portable. It was relatively simple to remove the berths and the hatch trunk bulkhead and open the whole 'tween deck for stowage. Items such as the sparring on the frames, protection for pipe covers, etc., were left undisturbed. Neither berth lights nor lockers were provided for this class of personnel, although minimum locker storage is now usually provided for all personnel. On the *Pacific Explorer*, the cannery crew berthing spaces in the No. 1 'tween deck area had special foot lockers designed to fit underneath the lower berths.

Accommodation on the *Saipan* and *Tinian* is very similar, in number and size of rooms, to that on first-class freighters. The heat conditions encountered by tuna motherships anchoring in tropic waters are severe and personnel required to live under these conditions for two or four months at a time, appreciate the value of room. Naturally, the provision of awnings and folding camp cots on the tuna mothership is an essential as a good many of the crew sleep on deck.

Neva follows the arrangement of the *Saipan* with additional quarters to take care of a slightly larger crew. The wing wall compartments have been rearranged to accommodate cannery personnel.

Steam vessels, such as the *Pacific Explorer*, are subject to U.S. Coast Guard inspection and the crew, other than fishermen and cannery men, come under the jurisdiction of the sea-faring unions. Their quarters must meet the union standards for merchant ships. Motor vessels, however, since they are not inspected, generally come under the jurisdiction of the fishermen's union. Perhaps because of the share basis on which they are paid, or perhaps because they are accustomed to small boats, the fishermen generally have no rigid standards for quarters and the owner's opinion as to what is required to keep a good crew contented is the governing factor.

Ships' stores facilities present no great problem. In all cases the vessels carry a considerable quantity of consumable stores but they make a fairly even rate of catch throughout the operating period and it is usual to stow stores in the cargo holds on the outgoing voyage. As they are consumed, their place is taken by the fish.

STABILITY

Stability of fish-processing vessels has, of course, to be investigated, but in the types afloat to-day no stability problem has been introduced by the special character of the work. The vessels usually follow freighters in stability requirements. They leave with large tankage and cargo space filled with a fair amount of stores, provisions, supplies, etc. As they load up the catch, the oil and water are used up in the bottom tanks. The worst condition usually occurs on arrival home with full cargo and no fuel. Generally, however, the maximum load is well under the cargo deadweight that would be carried by a freighter of similar size, and consequently the average processing vessel can handle without difficulty the higher centre of gravity due to equipment being placed high up. In the case of the L.S.T.s, of course excess stability and GMs, which range from 15 ft. (4.6 m.) upwards, cause the designer some thought because of the stiffness of the vessel but again the special nature of the service prevents this from becoming a problem. Alaskan operations take place only in the summer when the weather and sea conditions are favourable. Operations on the tuna grounds in South America can generally depend on good weather and trips can be planned to avoid severe conditions.

The case of the *Pacific Explorer* was somewhat of an exception. Because of the large number of crew aboard and also because the vessel would be operating in Alaskan waters for longer than was customary, with the consequent possibility of bad weather, the sub-division was investigated and the vessel met one-compartment sub-diivsion standards. Most of the vessels in the trade can meet this requirement as their draft, because their deadweight, is considerably less than a merchant vessel of the same size.

BUNKER AND WATER PROBLEMS

The provision of an adequate supply of fuel oil and fresh water is always a problem. The mother-ship type of vessel has to provide large quantities of diesel oil for the attendant fleet and fresh water for the cannery service. The fishing type of vessel, such as the *Deep Sea*, has an unusual oil problem due to the long distance between Seattle and the Bering Sea in which she operates. She needs to carry a supply of oil to take her up there and back and fish for several months. Fresh water is not much of a problem as it can be replenished at Dutch Harbour, but the ideal situation, of course, would be to be self-contained in oil and water.

Vessels other than the *Deep Sea* have no particular problem in providing tank capacity for the main engines and generators, as the trips to and from the grounds are no longer than usually expected on an average commercial vessel. Their problem, however, is to have sufficient tank capacity to supply their fleet of boats. In the *Pacific Explorer* the space abreast of the shaft alley under No. 4 and 5 holds was fitted with tanks for diesel and fish oil. By means of spectable flanges in the pipe lines, the tanks can be changed from fish to diesel oil service. This was not found to be necessary in operation.

In the *Saipan* and *Tinian*, the total capacity of diesel oil available for use on board and for supplying the attendant fishing boats is 285,000 gal. (238,000 imp. gal., 1,080 cu.m.), adequate for southern operation.

The problem of providing adequate fresh water for salmon canning is a difficult one. The large crew requires a considerable amount for normal services, but some is used for the salmon canning process, even though the amount is kept to a minimum and is considerably less than on shore installations. A small can washer is always installed on canneries at the after end of the canning line. This was omitted on the *Ogontz* to save water, but is installed on *Neva*. The water which

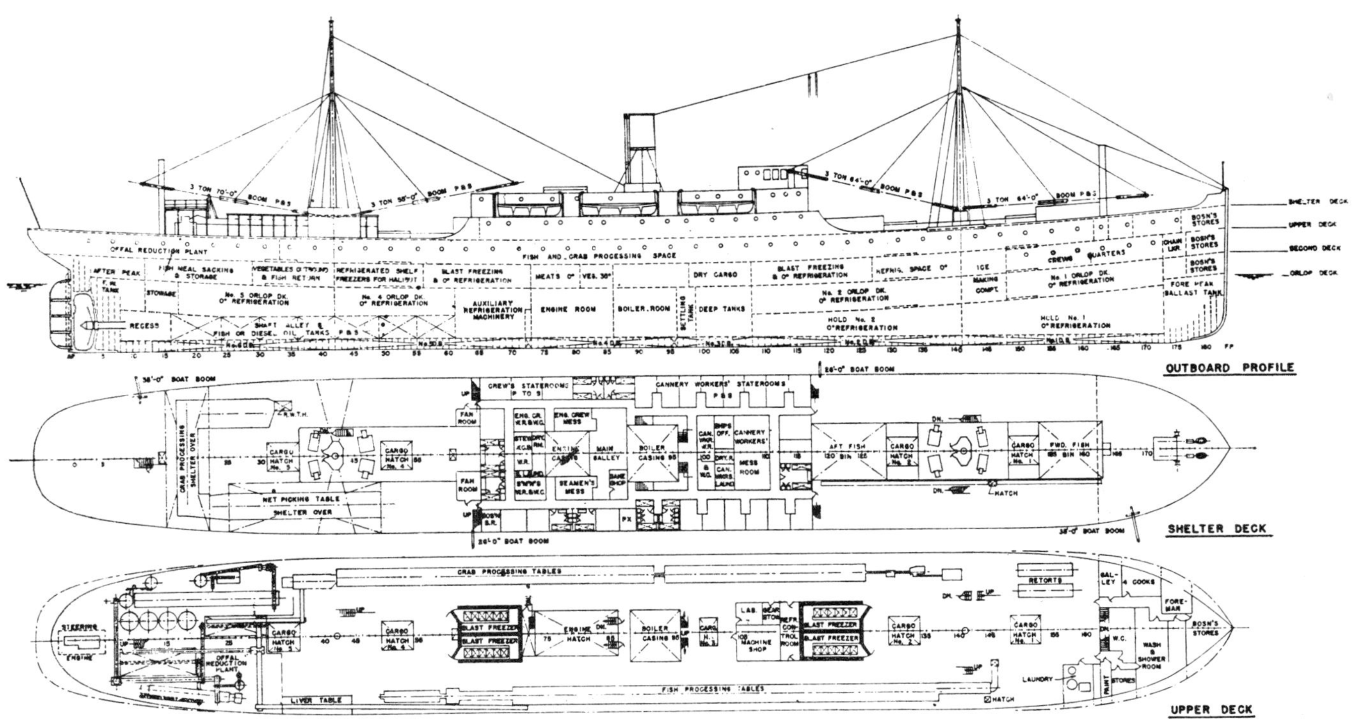

Fig. 571. Profile and upper decks of 410 ft. (125 m.) Pacific Explorer, *experimental factory ship for tuna and crab freezing and canning. Naval architects: W. C. Nickum and Sons, Seattle, Washington, U.S.A.*

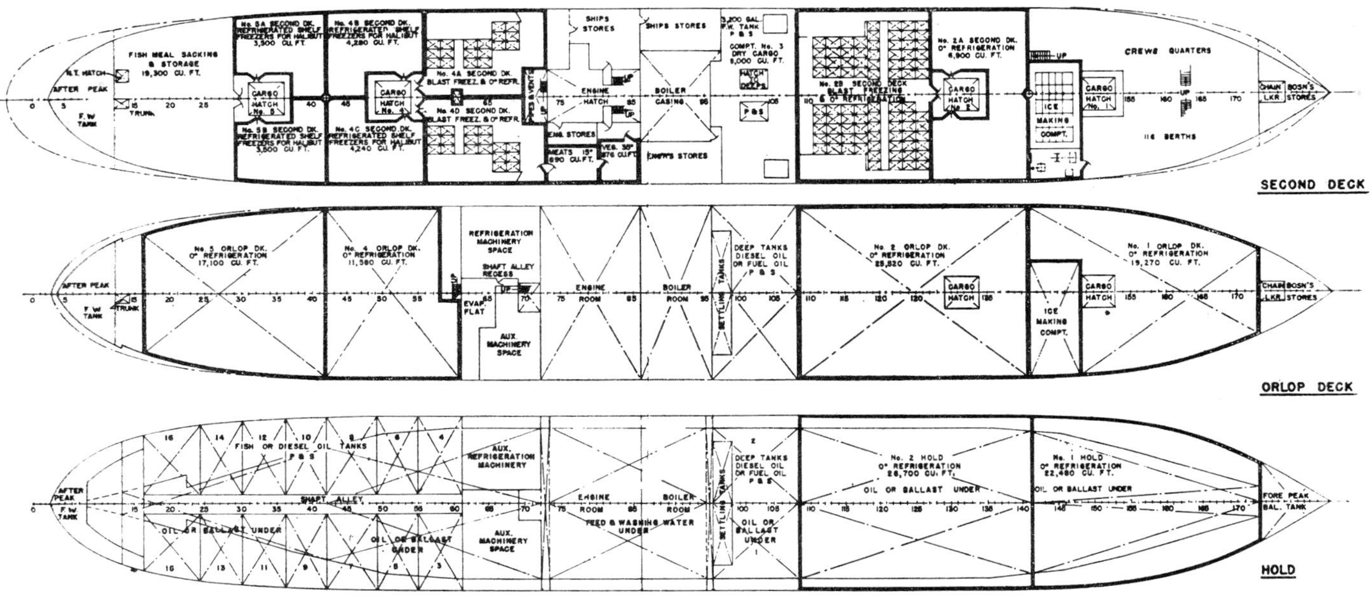

Fig. 572. Lower decks of 410 ft. (125 m.) Pacific Explorer, *experimental factory ship for tuna and crab freezing and canning.*

condenses from the steam in the cooking retorts cannot be recaptured and used again in the boiler because it is full of impurities. Various methods have been tried to solve the problem. On the *International* the condensate from the retorts was led through a single effect evaporator and condensed in a regular surface condenser and then returned to the feed system. On the *Ogontz* the deep tank was filled with fresh water from barges after the vessel reached the fishing grounds. It was planned to put a coffer dam around the settling tank in this operation, but because of the cost involved it was not done. When the vessel was at anchor the oil was drained into the double bottoms and the oil service pump drew directly from them. This, of course, is not recommended practice for a long-term operation, but it was done successfully. The fresh water in the deep tanks was connected so that it would be circulated by a special pump through the retorts and the retort drains were led directly into the deep tanks. To minimize the amount of cooling required, spray headers were fitted in the retorts. When cooking was completed the spray headers were turned on to give an initial quick cooling to the salmon. This, of course, reduced the time the salmon had to be piled on deck before being loaded into the holds. The scum was taken off every morning by suction connections, keeping the water clean enough to be used for cooling.

Another scheme was used on the *Pacific Explorer* which allowed spray cooling in the retorts yet restricted the amount of fresh water required. A fresh water tank was placed in the 'tween deck below the retorts and fitted with pipe cooling coils connected to the salt water system. A circulating pump drew from the tank and discharged to the spray headers in the retorts. When cooking was completed the pump was turned on and the water circulated through the retorts until the can temperature was reduced to an acceptable level. It is not felt that the *Pacific Explorer* operated for a sufficient length of time to determine whether this was of particular value in reducing the time and space required for cooling.

On *Neva* the cans are cooled by air circulation after they leave the retorts. The retort water is sent through a heat exchanger to give part of its heat to the make-up feed water. The water is then passed into the suction side of the cannery salt water service pump to take part of the chill of this water when it is being used for sliming.

In the southern operation, the *Explorer's* water problem was aggravated because she had to supply water to the small vessels and to the ice-making tank. This required 10 short tons (9 metric ton) of water per day. In the northern operation the ice-making plant was not operated, but a large amount of fresh water was required for washing the fillets and for washing the crab. The reduction plant steam cooker also required a considerable amount of direct contact steam. The evaporators on the *Pacific Explorer* had a total capacity of 12,000 gal. (10,000 imp. gal., 45 cu. m.) per day between the two of them, sufficient, but none too much.

GOVERNMENT REGULATIONS

Fish-processing vessels are free from the majority of Government regulations. The motor ships, even though over 300 register tons, are exempt from any regulation by the Coast Guard. Steam vessels are subject to the regulations for inspection of freighters but do not have to meet passenger ship requirements. They do not have to carry lifeboats for a full complement on both sides of the vessel.

The *Pacific Explorer* was a special case. Because she was owned by the Government, carried a very large crew and was engaged in experimental work which might require her to operate in all weather conditions, it was decided to comply as nearly as was practical with Coast Guard passenger ship regulations. The installation of a submersible bilge pump was the major exception. Quarters were constructed of non-inflammable material throughout; provisions for fire detecting, fire extinguishing and fire control were carried out in complete accordance with Coast Guard regulations. All the electrical work and the piping installation followed Coast Guard rules. In the case of the *Ogontz*, because of having to comply only with freighter standards, the existing wood joiner bulkheads were left in.

The fish-processing vessel is also out of the jurisdiction of the U.S. Public Health Service because it does not make a voyage from port to port. While most installations are made in accordance with Public Health Service rules as a matter of course, full compliance would be onerous.

The products canned or processed are, of course, subject to the Pure Food and Drug Act but this is simply a matter of inspection of the final products.

United States fish-processing vessels are not required to comply with the Loadline Regulations as they are not engaged in commercial trade from port to port but when the *Saipan* and *Tinian* were changed to Honduras registry they had to comply with the regulations.

SANITATION

Sanitation is given serious consideration on all fish-processing vessels although those exclusively engaged in tuna freezing have no particular need for worry. Tuna are caught and frozen in the round and consequently there is no great danger of contamination of the flesh from insanitary surroundings. In salmon canneries sanitation is very important to ensure the purity of the product.

On the *Pacific Explorer* special care was taken to ensure the purity of the large quantities of salt water used in washing the fish and crab. A sea chest was installed forward of the forepeak bulkhead and a special salt water cannery service pump was fitted over the forepeak tank. It was reasoned that, with the vessel lying at anchor, water taken from close to the bow would be uncontaminated by the vessel's sanitary outlets.

On the *Ogontz* the main emphasis on sanitation was on the all-steel character of the canning equipment and surrounding structure, and the fact that it was installed partly in the bridge deck space and partly in the open air where everything could be cleaned with steam and water hoses.

The *Pacific Explorer* went a step further. Care was taken to ensure that the fish ponds had no hidden nooks or cranies in which matter could collect, and the decks, particularly around the sides of the cannery deck, were cemented to provide a slope to all the drains. Sanitation is even better on *Neva* where there are no blind corners or dead spaces in the tanks and the steel fish ponds are either smooth or the stiffeners are of pipe sections which can be easily washed down. The usual steam and salt water hoses are provided. The cement deck covering has been built up to a thickness of 6 in. (15 cm.) at some points to allow the installation of proper gutters and a slope towards the drains. All drains are oversized and can be easily cleaned. The entire deck area of the cannery can be flushed out continuously.

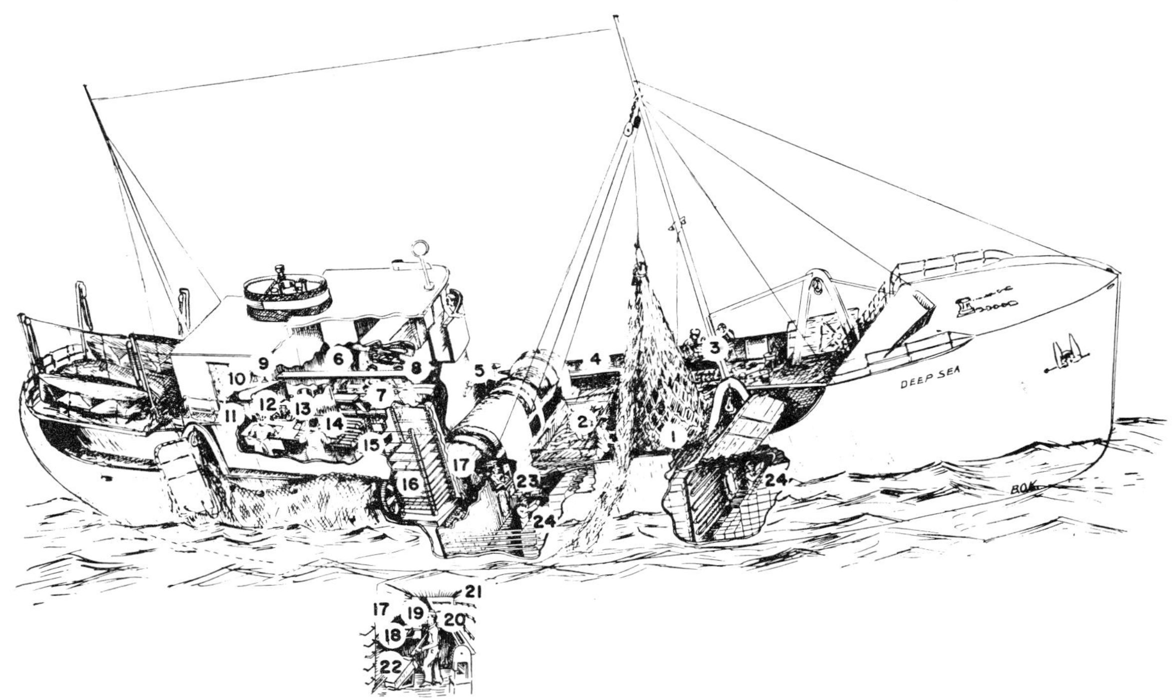

Fig. 573. Deep Sea, *freezer trawler for king crab.* (*1*) *landing the catch;* (*2*) *sorting species of fish;* (*3*) *butchering, washing and cleaning king crabs prior to cooking;* (*4*) *sluice conveyor to cooker;* (*5*) *cooker conveyor leading to processing room;* (*6*) *chute to processing conveyor;* (*7*) *processing conveyor to filleters;* (*8*) *fish conveyed to spray washer;* (*9*) *spray washer;* (*10*) *conveyor to hopper;* (*11*) *hopper;* (*12*) *inspection table;* (*13*) *weighing;* (*14*) *packing in freezer trays;* (*15*) *chute for clean tray supply;* (*16*) *freezer conveyors;* (*17*) *frozen trays;* (*18*) *defrost stage;* (*19*) *defroster tank;* (*20*) *glazing tank;* (*21*) *racks for glaze to set;* (*22*) *packing cartons;* (*23*) *full carton conveyor;* (*24*) *carton storage forward in ship. Naval architects: Coolidge, Hart and Brink, Seattle, Washington, U.S.A.*

SALMON CANNING

The starting point of the salmon canning operation is the fish bin (ponds). On the *Ogontz* the bins were on top of No. 2 and No. 4 hatches. They were portable and could be readily removed at the end of the canning season. Divisions are not indicated in the drawing, but there were internal transverse bulkheads in each bin which divided each into four sections. The bottoms of the bins had a slope of about $1\frac{1}{2}$ in. per ft. (125 mm. per m.) and vertical sliding watertight doors were provided at the lower end. The fish were dumped directly into the bins from the dump boxes or brailing nets.

On *Neva*, the fish bins are permanently installed, located under cover in the forward part of the main deck. Each bin has a hatch directly over the centre so that the fish can be dumped straight into them. A sorting belt, served by hatches and chutes, port and starboard, is located forward of the bins. Unsorted fish are carried along the belt and an attendant fills each bin with the proper species of fish. An indexor, by means of which the fish are mechanically fed into the iron chink, is placed between the supply bin and the chink.

The " iron chink " is the machine that cuts the head off the salmon and removes the fins and the entrails. This work was done in the old days by Chinese, and the automatic machines which replaced them were called " iron chinks ". From the chink the fish are taken by an elevating conveyor to a sliming table where operators, standing on each side, scrub them clean and place them on another conveyor for delivery to bins ahead of the filling and canning machines.

The can bodies are loaded on the vessel in a flattened condition. The heads and bottoms are packed separate from the bodies. In the can reform line the bodies are placed in the machine where they are put into a round shape and the bottom is put on. The cans then go by cable conveyors to the filling machine. The lids are loaded in stacks into the clinching machine.

The filling machine cuts the salmon into even lengths and fills the empty cans. The cans then go over a patching table where, if there is doubt about their weight, they are put on a scale. Modern shore canneries use automatic weighing machines which operate on the balanced beam principle but they require a steady floor and will not operate when a ship is rolling or pitching. With a cannery line operating at 250 cans per minute, the patching table is a busy place during full production.

The cans next go through the clincher and seamer where the lids are put on, the cans clinched and evacuated under vacuum, and the final closing seam made. The cans then generally go through a fresh water wash and on to what is termed a "cooler loader", an inclined chute at the base of which is the cooler tray. This is a heavy shallow wire basket made of flat bar approximately 36 in. (90 cm.) square. As the cooler trays are filled they are stacked one on top of the other on four-wheeled dollies. Flat bars welded to the deck are generally used for the tracks. The dollies are wheeled into the retorts and the cooking process is started. Steam at 15 lb./sq. in. (1.05 kg./sq. cm.) pressure is generally used. On *Neva* steam is produced at 250 lb./sq. in. (17.6 kg./sq. cm.) pressure at the boilers and is used to drive certain pumps, the cannery line turbines and steam generators. Steam is exhausted from the turbines at 15 lb. (1.05 kg.) back pressure and put from there into the retorts. A surface condenser condenses the steam when the retorts are not in operation and a pressure reducing valve station automatically feeds extra steam into the retorts if the available back pressure steam is not sufficient.

After being cooked in the retorts the packed cans are wheeled out of the retorts into a position for cooling. The early canneries cooled their salmon on the open deck. Cooling is necessary to prevent excessive "burning" of canned salmon because if the cans are put in the hold while hot, the cooking continues. Cooling on the open deck has disadvantages, however, because of the large amount of rain encountered in Alaska. This means that it is necessary to rig tarpaulins and to keep a careful watch to see that the cans are not touched by the rain which will spot and rust the cans. On *Neva* the entire cooling area is placed inside of the main deck house. The cans cool within 10 hours without the use of mechanical ventilation but fans can be added to increase the cooling rate. On *Neva* the large side ports installed in the main deck house can be opened to allow air to circulate over the salmon and then up through the hatch and ventilator openings.

On the *Ogontz* the cooled salmon was lowered into holds by sheet metal spiral chutes set into the hatchways after being put through a simple casing machine. On *Neva* the product is labelled and cased on board. There is a cooler inverter at the forward end of the cooling area. This machine turns over the cooler trays and feeds them into a chute leading to the labelling and packaging machines. From the packaging machine they are led into the hold for stowage.

Every emphasis in a salmon canning operation is placed on speed of packing and dependability because, in many of the areas in Alaska, the season averages three weeks or 15 days packing and the loss of one day means a loss of 7 per cent. of the season's operations.

CRAB CANNING

The crab cannery installed on the *Pacific Explorer* differs from a salmon cannery in that it is essentially a manual operation. Automatic machinery has not yet been developed for processing crabs. A crab cannery requires twice the personnel of a salmon cannery, yet on the *Pacific Explorer* output never reached more than 40 cases per hour as against the 400 cases of salmon per hour output of the two lines on the *Ogontz* and the 300 cases of one line on *Neva*. The crab cannery on the *Pacific Explorer* was the first and only American cannery designed to handle Alaskan King crab for an output much greater than the average small crab cannery in the U.S.A. home water. The Russian crab fishing vessels, equipped by the U.S. War Shipping Administration, have never equalled the output of the *Explorer* because their installations were very crude and laid out with little thought to saving manpower. They require approximately twice the personnel for the production obtained on the *Pacific Explorer*. The data from preliminary tests by the U.S. Fish and Wildlife Service on a small vessel in Alaska were used in designing the crab cannery. An estimate was made of the number of people required per weight of product and of the number of raw crabs which had to be handled to obtain the finished pack. That the cannery produced close to its designed rate speaks well for these preliminary man-hour figures of the Fish and Wildlife Service. But assumptions had to be made. For example, in laying out the continuous cooker, it had to be assumed that with a given weight of broken crab legs coming into the cooker, the legs would stack approximately 3 in. (7.5 cm.) high and that in the given length of the tank and a given conveyor speed it would be possible to get a sufficient cooking period. This is hardly precise engineering and as the nearest King crab, at the time the design work was being done, was some 2,000 miles away on the floor of the Bering Sea, there was little possibility of making tests to confirm the assumptions.

King crab cooked by the *Pacific Explorer* was caught by conventional trawlers and by tangle nets. King crab is all leg with a relatively small body. It is not uncommon to find crabs which measure 72 in. (1.8 m.) from tip of one claw to the other. Basic design assumptions called for an average crab weight of 6.2 lb. (2.8 kg.) It was presumed that a large percentage of the crabs would be caught in tangle nets and a house with a large work table was installed on the after end of the weather deck on the starboard side. The tangle nets came in the forward end of the weather deck and were pulled over this table. Men picked the crabs out of the nets and placed them on a slatted wood conveyor running below the table. The space aft of this house was arranged to take the trawled crabs which went on to the same conveyor belt. The belt extended into another house built the tank, then dumped down an inclined chute which fitted into a small raised watertight hatch. A conveyor carried the crabs through salt water which reduced the temperature to allow them to be handled. The line extended the full length of the port side of the cannery deck. A 200 ft. (61 m.) row of tables serviced by three belts was installed, with stations for various parts of the crab process. The top belt fed in the raw material, the bottom took away the offal which fell from chutes, and the middle belt carried the finished product to other

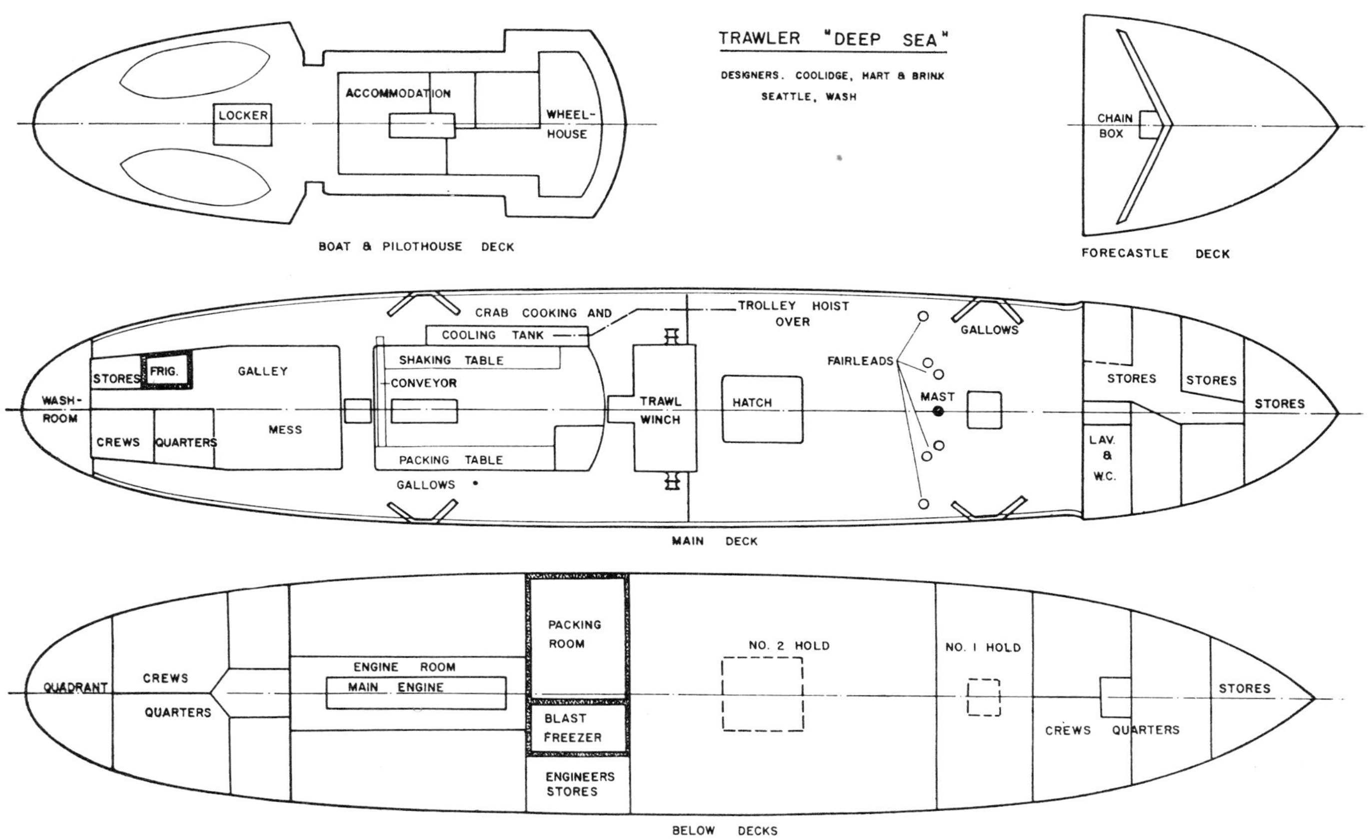

Fig. 574. Deck plans of Deep Sea.

across the ship aft of No. 5 hatch and dumped on to the distribution belt which feeds the crabs past the main butchering tables. A special hooked knife was permanently fixed in front of each operator. The men ripped off the backs of the crabs and dropped them into a chute which discharged overboard. They next broke the crabs into halves and placed them on another feed belt which dumped on to a conveyor running forward on the port side through the cooking tank. This conveyor, the tank, including the shell, steam heating coils, and all parts in contact with water, were made of stainless steel because of the peculiar chemical characteristics of the crab. The meat is subject to excessive discolouration when brought in contact with copper or conventional mild steel. The crabs were given their initial cook in

stations. The first tables consisted of crab picking and butchering stations. There were about twelve stations provided with small rotating saws to enable the joints to be cut neatly. At the picking stations the meat was shaken out of the shells into pans which were taken through a brine and acid dip to tables where the crab was packed by hand into cans. From here it went past checking and weighing stations, then into the clincher and seamer and then into the retorts on cooler trays. The *Explorer* was unconventional in that a lift truck was used to pick up the load in the cooler trays and place them in the retorts, and to remove them and place them in the cooling area. No can reform line was installed because the packing capacity was small. The empty cans were carried in the round.

FILLETING

As the *Pacific Explorer* was also designed to handle Bering Sea bottom fish, a large fillet line was installed. The fish were brought aboard from the trawlers and were placed in ponds forward of No. 1 and aft No. 2 hatches. The fish went on to a conveyor belt which dumped them into a watertight hatch leading through to the shelter deck. A chute conveyed them into a rotary fish washer from which they were fed to the top belt of the fillet line with spaces provided for 60 men, 30 on each side. Like the crab table, it had three belts. The top belt brought the raw material, the middle took away the finished product, and the bottom belt removed the offal. At the end of the line the finished fillets were discharged through a continuous dip tank (which was subsequently found to be unnecessary) to packing tables, then fed to packaging machines which labelled and wrapped them. Next the fillets were placed by hand on the trays of large dollies which fitted the blast freezers.

BY-PRODUCTS

The offal belt on the fillet line discharged on another belt which ran past a packing table where personnel removed the livers. They were frozen in drums for processing ashore. After the carcasses left the liver picking table they were taken through a hog in which they were chopped into small pieces and discharged into a pond. Both crab and fish offal ponds were provided with screw conveyors in the centre line at the bottom which discharged through a series of screw conveyors into the reduction plant cooker, from which it was discharged into a press of the screw type. The oil and water squeezed out was taken through a " fletcher " machine, a basket type centrifuge which effectively separates all remaining solid ingredients from the liquid. From the " fletcher " several alternate procedures could be followed. A series of four settling tanks, all provided with steam jets and heating coils and arranged at different levels so that the oil could gravitate over the top of one settler into the other, were provided. A final finishing centrifuge could be used or could be used in combination, depending on the species and character of the fish or offal from which the oil was obtained. The oil was discharged into tanks. The solid material was taken by a screw conveyor through a disintegrator, which reduced the particles into smaller bits, and then into a dryer, the most imposing piece of equipment in the reduction plant. On the *Explorer* it was 50 ft. (15 m.) long by 72 in. (1.8 m.) outside diameter. It rotated slowly and was arranged at an angle so that the meal being fed into one end would flow by gravity, tumbling and in contact with steam pipes, until it reached the other end in a dry condition. The meal was discharged by air into a receptacle where a screw conveyor carried it into a finishing grinder which further refined the meal. From there the meal went through a fan and cyclone and was dropped into sacking bins which were provided with a screw conveyor leading to an elevator. This discharged the meal into the sacking machine. After being sacked the meal was stored in the open for 48 hours and then piled in a hold or in the sacking area.

It is nearly impossible to convince an equipment manufacturer or owner's representative, who is used to a 12 to 14 ft. (3.6 to 4.3 m.) head room in a shore cannery, that there is any necessity to know within an inch the exact height of a piece of machinery. The reduction plant probably requires the greatest amount of attention to detail because of the large number of conveyors. During the design stages of the reduction plant on the *Explorer* the amount of slope for the dryer caused much discussion. It was placed at the stern and the natural sheer of the vessel allowed $\frac{1}{4}$-in. slope per ft. (21 mm. per m.) when she was on an even keel. It was considered that no difficulty would be encountered in maintaining an even keel due to the large tankage and the fact that the cargo could be stored in any hold. Shore installations are generally made $\frac{3}{8}$ to $\frac{1}{2}$ in. per ft. (33 to 42 mm. per m.). The *Explorer* only operated her reduction plant for about two months in the Bering Sea. While the operation was successful, the short length of time does not give conclusive evidence that the slope would be satisfactory for all types of processing vessels.

FREEZING

The *Pacific Explorer* froze some crab, processing it in the same way through the cooker and cooling tank. The first sections of the tables were used in trimming the legs to proper lengths which were put on trays on the dollies and moved into the blast freezers.

The *Deep Sea* planned to freeze both crab and fillets but soon gave up filleting because of the high labour cost and low price of the product. Crab is caught by the *Deep Sea* by trawl and taken from the trawl to a table, protected from the very inclement weather found in the Bering Sea. The crab backs are removed and a conveyor carries the legs under a high pressure water jet to remove the bulk of the sand and other foreign matter. The legs are discharged into containers which have wood sides and ends, portable wood tops and heavy expanded metal bottoms. A small electric trolley hoist, running on a rail, takes the containers back to cooking and cooling tanks which are made of wood and fitted with steam coils and live steam jets. There are three sections, two of which are used for cooking and the third for cooling. Sea water is circulated through the cooling tank to take much of the cooking heat out of the crabs. A transverse mesh conveyor takes the crabs to the final processing table. On their way, they pass behind the engine casing where they are given another cleaning spray. Crab legs may be taken off on the starboard side of the final processing table where the joints are cut to size for freezing whole, or they may be taken off on the port side for shaking when the product is loose crab meat. The meat is put into pans on the same conveyor and again transported across the ship to the final processing table. The blast freezer extends from the after end of No. 2 hold up to a trunk extending into the fish-processing room where a door in the after end allows the product to be fed directly into the blast freezer. Legs and meat are placed is removed. In the case of loose crab meat, the meat is packed on the final processing table into trays of such size that the product becomes a flat bar weighing 5 lb. (2.3 kg.). Each tray takes three bars. The trays have locking covers which are not quite watertight but allow for the expansion of the product. When the trays are loaded, water is added to the crab meat to glaze it and

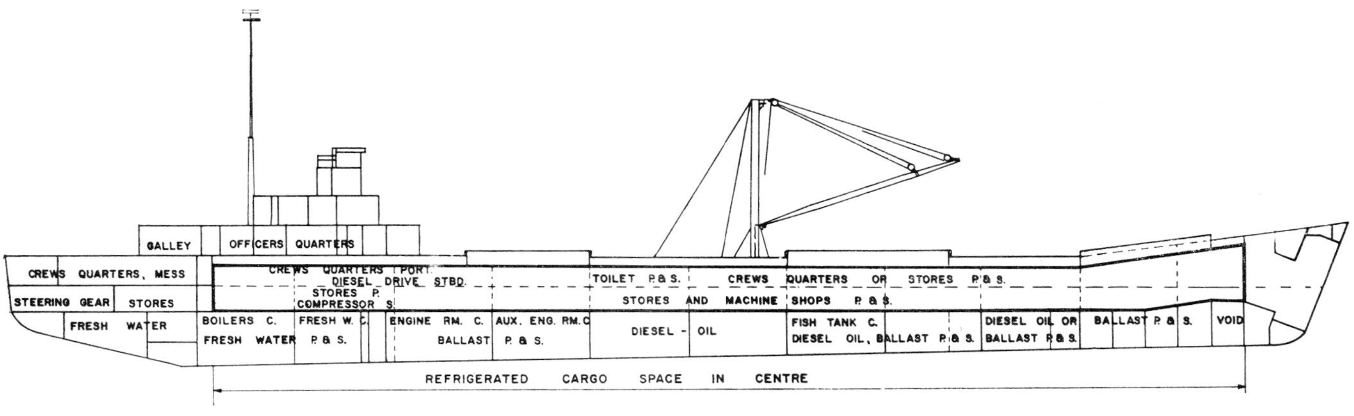

Fig. 575. Profile of Tinian, *L.S.T. converted to tuna freezing ship. A similar ship was* Saipan. *Naval architects: W. C. Nickum and Sons, Seattle, Washington, U.S.A.*

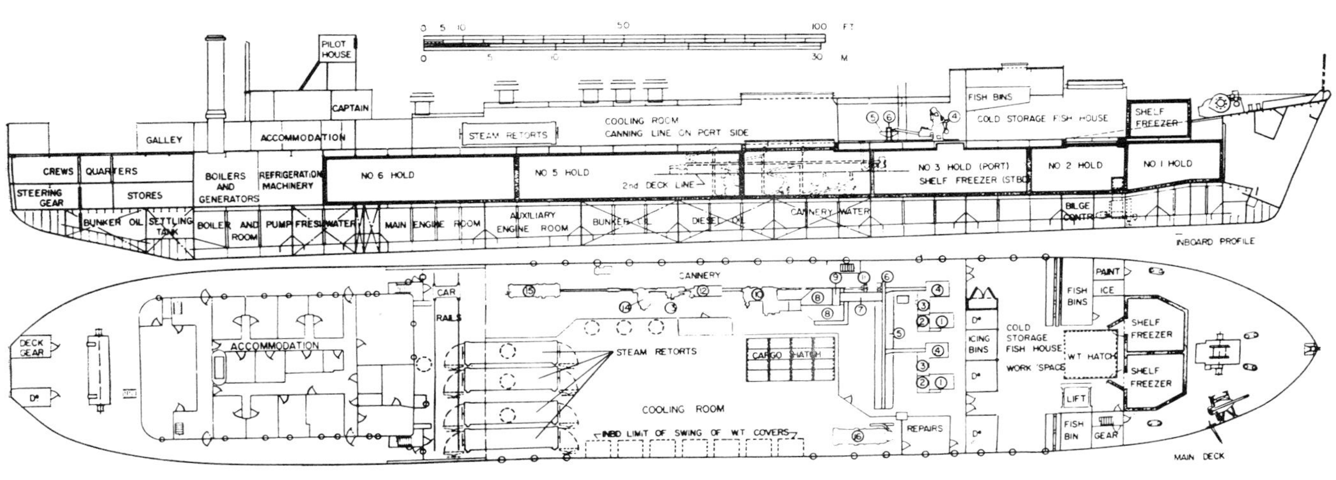

Fig. 576. Profile and main deck of Neva, *an L.S.T. converted for both salmon canning and freezing. Length o.a., 328 ft. (100 m.); length p.p., 316 ft. (96 m.); beam, 50 ft. 1½ in. (15.3 m.); depth, 24 ft. 10 in. (7.5 m.). (1) hopper; (2) feed table; (3) indexor; (4) iron chink; (5) sliming table; (6) conveyor belt; (7) fish elevator; (8) fish pond; (9) chute; (10) filling machine; (11) can elevator from below; (12) patching table; (13) curler clincher; (14) vacuum closing machine; (15) cooler loader; (16) cooler inverter. Naval architects: W. C. Nickum and Sons, Seattle, Washington, U.S.A.*

on plywood shelves in the blast freezer on angles secured to a vertical continuous conveyor which moves down from the top of the trunk. Fig. 578 shows the freezer arrangement. The crabs may be frozen continuously by adjusting the conveyor speed and adding shelves as the conveyor moves down, or in a batch process. Experience has proved the batch method best and it is now usually followed. In this process all of the shelves on the conveyor are filled with the product. The conveyor remains still while the fan is on. After the batch is frozen the conveyor is started again and the product after freezing, the trays fall out by gravity into the final processing room on the port side in No. 2 aft hold, where they are immersed in a defrost tank to allow removal of the frozen crab bars. Glazing tanks and the packaging area where the bars or crab legs are put into cartons, are also in this room from which the finished cartons are moved forward on a roller conveyor to the hold. The *Deep Sea* has successfully operated several seasons on crabs. It is believed that the handling costs are close to the minimum and the consumer demand shows that the process yields a very satisfactory product.

TRANSFERRING AND HANDLING OF CATCHES

Handling facilities are a problem on all fish-processing vessels and the more products handled and the more varied the processing work, the more difficult the problem becomes. On salmon canneries conventional gear is used to transfer the fish from the small vessels into the bins. No method has yet been devised which works better than this. Portable elevators have been tried. One operator has tried a portable elevator every year for some five years and has always ended up with one elevator completely broken each year. The problem of adjusting the depth of the elevator to small fishing vessels and the control of the relative motion of the two vessels has so far not been overcome. In the opinion of the writer the answer to the problem is the use of a small steel receiving barge with very little freeboard, moored tightly against the hull of the mothership, into which fish can be sluiced from barges or pitched from small boats. An elevator can be led to the top deck from the moored barge. Nobody has yet tried this method.

In bringing tuna aboard ships such as the *Saipan* and *Tinian* conventional burtoning type cargo gear is used. The *Saipan*, when purchased from the Navy, was equipped with a diesel electric rotating crane mounted on the upper deck. This was left aboard but it was found that even in calm water when tied up to a dock the movement of the vessel was sufficient to make control of a load suspended from a single whip inadequate. The *Saipan* and *Tinian* used a bottom dump type basket for bringing the fish aboard, with a hook type weighing scale inserted between the regular cargo hook and the dump basket. This enabled the load to be weighed as it came aboard. This is not necessary in a salmon operation as payment is on the basis of a unit price per fish.

Two portable elevators were provided on the *Pacific Explorer* and were very useful in lifting products from one deck to another, such as bringing cans up from the 'tween decks to the canning and processing area. Two lift trucks were also provided and were used extensively for transporting materials along the cannery deck for storage in the holds or for delivering materials to the various processing stations.

The handling of frozen tuna between the brine tanks and the holds represents a large labour cost. The ideal location for a tank on a tuna mothership is on the upper decks, as high as possible to allow discharge by gravity chutes.

STORAGE

The storage of the tuna after it was frozen was given considerable thought on the *Pacific Explorer* because no one had previous experience of how to handle it on a large scale. It was felt that it would be necessary to provide storage bins to stack the fish and to prevent them from falling off the stacks. This was right for the wrong reason. There was no difficulty in maintaining the stacks of tuna. The addition of a very light fresh water spray, which froze immediately upon striking the tuna, effectively prevented any movement of the fish. It was, however, found to be necessary to segregate fish received from each tender because they were paid for on the basis of being in good shape and any spoilage after delivery was deducted from the boat's share.

During the design stage of the *Pacific Explorer* the estimates of stowage factors varied from 50 to 75 cu. ft. per short ton (0.64 to 0.43 metric ton/cu. m.) of tuna. The actual figure worked out to be, on average, 59 cu. ft. per short ton (0.54 ton/cu. m.). In some cases holds have been stowed to a factor of 57 cu. ft. per short ton (0.56 ton/cu. m.), but this has been with very small tuna averaging not over 10 to 15 lb. (4.5 to 6.8 kg.) apiece. Large tuna running up to 150 lb. (68 kg.) apiece will stow at the rate of 65 cu. ft. per ton (0.49 metric ton/cu. m.).

MOORINGS

No special provisions were made for mooring the small vessels alongside the motherships in any of the earlier salmon canneries, as they anchored in quiet water.

The *Pacific Explorer*, however, was expected to operate in the fairly open Bering Sea and perhaps in open South Pacific waters where running seas would be a problem. For this reason special mooring arrangements were provided, based on the experience of the reduction plants operated in the mid-thirties, 15 miles at sea off Monterey on the California coast. The key to this mooring arrangement is a counterweight secured to the inboard end of a mooring line which can move up and down as a small craft surges against the line. In detail it consists of a tower which is an open frame structure with the counterweight on the inside. A cable leads from the end of the boat boom over fair lead sheaves to the top of the tower and is then secured to the counterweight. The other end of the cable is attached to a 4 in. (10 cm.) manila line which is secured to the small boat and acts as a safety link because it will break from heavy strains without damaging the boat boom or the mooring fittings. This line is secured to the bow of the small boats. At the other end it is secured to a similar line which, in turn, is secured to a heavy spring on the mothership's deck. The counterweight tower takes care of the extensive movements of the small boat and the spring takes up any heavy shock load. This type of mooring worked very successfully on the reduction plants when they were anchored in mid-ocean for long periods of time but it is not believed that any salmon cannery or tuna mothership operation in the future will require such mooring equipment.

ELECTRIC POWER

As most processing machinery is commonly driven by electric motors, the installation of fish-processing equipment almost invariably requires additional generating capacity. On the *Ogontz*, two turbine driven A.C. generators, each developing 100 kw., 220 v., three-phase current, were installed in the old bunker space on the 'tween decks abreast of the engine room. The cannery installation was connected to an A.C. system and was made separate from the ship's regular D.C. installation. This was one of the first commercial installations of A.C. equipment on board ship in the Pacific north-west. The use of A.C. was influenced by the second hand generators available and the fact that all canning equipment for shore installations was universally equipped with A.C. Even with new generators the first cost worked out to slightly less than the cost of D.C. equipment. No difficulty was experienced with the operation of the subject to continual wash-down and surface water and steam. Transformers were installed to provide 220 v. current for these motors. A few conveyors were driven by D.C. where speed regulation was important. Transformers also delivered 110 v. single phase power for the lighting and fractional horse power motors. To increase the power factor of the installation the two 175 h.p., one 225 h.p and one 50 h.p. motors driving the refrigeration compressors were of the synchronous type, and a motor generator set was installed to provide D.C. motor excitation.

No difficulty at all was experienced with the electrical installation while operating in Alaskan waters. In southern waters, however, it was a constant source of trouble.

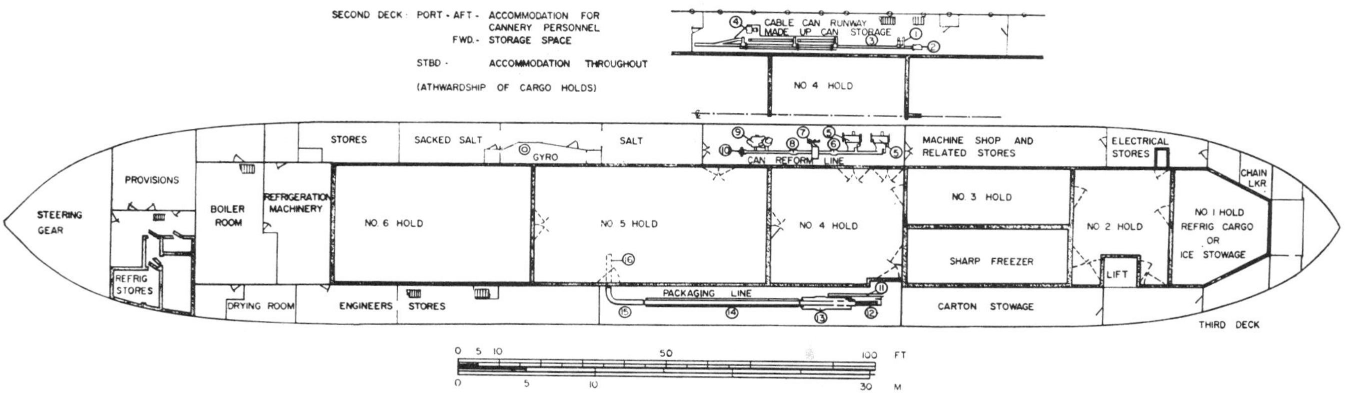

Fig. 577. Second and third deck of Neva, *an L.S.T. converted for both salmon canning and freezing. (1) can elevator; (2) cable drive stand and motor; (3) cable runway; (4) motor for seamer; (5) can reformer; (6) can elevator to flanger; (7) flanger; (8) can elevator to seamer; (9) seamer; (10) can elevator to second deck; (11) can packer; (12) discharge conveyor; (13) gluer; (14) 36 ft. (11 m.) compression unit; (15) roller conveyor; (16) portable roller conveyor. The holds occupy the whole space on both second and third decks in the centre.*

equipment. This rather startled the writer when he found out that at the end of each season they had to take off about 25 per cent. of the motors in the cannery and send them into a shop for baking out and rewinding. Even though totally enclosed motors were used, the service in the cannery was particularly arduous as there was the large amount of water and steam around and at frequent intervals the entire area was hosed down with water and steam. The owners accepted the necessity for this repair work as a matter of course, as they were used to doing it in Alaskan shore canneries.

The *Pacific Explorer*, because of the size of the refrigeration plant and the number of motors required, needed a large new electrical installation. The generating plant as installed consisted of three 300 kw. 440 v., three-phase diesel driven generators. The existing two 15 kw. D.C. generators in the main engine room were left installed and by means of double throw switches the lighting circuits were arranged so that should the *Explorer* ever sail as a freighter, the diesels could be shut down and the auxiliary machinery space left unmanned.

All the motors of the *Explorer* received 440 v., except those which drove conveyors and other equipment in the crab canning and fish-filleting areas where they would be The generators laboured continuously under southern conditions. One of the diesel engines broke down and the trip had to be completed with two generators in operation. After six months' service the two remaining generators had to be completely overhauled. Practically every electrical motor which had done any kind of service during this six months stay in the tropics had to be overhauled on its return. The motors that gave the most difficulty were those which were in contact with the refrigerated air. The blast freezer fans were axial flow type with the motor directly in the air stream. Under these conditions the motors had an expected life, before rewinding, of approximately three months. In addition, it was found that contactors and other delicate control parts of the electric system were a source of continual trouble due to the high humidity and the temperature conditions.

On the *Saipan* and *Tinian*, because of the relatively large available generating capacity already in the vessels, no additional equipment was installed. However, it was necessary in both cases to drive two of the four refrigeration compressors by diesels. The original 230/115 v. D.C. electrical installation was left unchanged. No serious breakdowns have occurred in this equipment but it requires constant attention and maintenance. Experience has shown that electrical equipment is the largest item of maintenance when operating in extremely hot and humid waters.

On *Neva* the existing installation was 230/115 v. D.C. The generators did not have sufficient capacity for the necessary new electrical equipment. The problem was resolved by the use of surplus equipment. Surplus boilers, with a steam capacity far in excess of that required for just salmon cooking and washing service, were available at a very low cost. A 200 k.w. turbine-driven D.C. generator was also available from surplus. The operators were familiar with the upkeep cost of motor-driven canning equipment and were glad to have a steam turbine installation using line shafting and belts for the cannery line so the installation had steam drives for the canning machinery, steam winches, and a steam generator large enough to carry the balance of the load without operating the diesels. When the vessel is at anchor, the operation is almost entirely by steam. When she is under way and no processing is going on, the operation is by diesel. Economic studies showed that the first cost of this type of installation was the lowest and operating costs about the same. It was felt that the few dollars additional cost in oil required to generate electricity and drive the canning equipment by steam would be offset by the reduced maintenance of the steam drive canning line.

The *Deep Sea* has diesel driven 220 v. A.C. generators. No particular electrical problems were encountered in her operation. Fortunately, the blast freezer fan motor could be installed in the engine room with an extended shaft into the blast freezer so that it gives no trouble.

VENTILATION AND HEATING

Ventilation and heating presented no problem on the *Ogontz* and the *Deep Sea* and *Neva* as in Alaskan waters conventional commercial ventilation and heating installations are practicable.

On the *Saipan* and *Tinian* some special precautions were taken to ensure adequate air distribution around the refrigeration machinery but otherwise the original Navy installation and heating system was not disturbed and it proved to be adequate for tropic conditions.

On the *Pacific Explorer*, as the vessel was required to operate in both tropic and in northern waters, a combined heating and cooling system was installed. The heating was done by a conventional hot air system, the air being warmed by coils in the inlet ducts and distributed by fans through ducts into the living quarters. Approximately 70 per cent. of the air passing through the quarters was continually being re-circulated with 30 per cent. being continually removed. The heating coils in the ducts were supplied with hot water from a water tank heated by steam coils. A conventional continuous circulating pump was provided. In order to cool the air when in the tropics a duplicate water tank cooled by ammonia coils was installed and the two tanks were so connected that either hot or cold water could be circulated through the duct coils. A few radiators were installed in the pilot house, passageways, etc. The system worked satisfactorily for heating but, unfortunately, it never had a real test as a cooling system. On the way to southern waters, on its first and only trip under tropic conditions, the chilled water tank froze, due undoubtedly to failure or mis-adjustment of the ammonia back pressure valve. The only thing proved was that simple ventilation alone was not adequate for quarters similar to those in the *Pacific Explorer*. The experience of the writer has been that unless the air is actually cooled, simple ventilation does not give any sense of relief from oppressive heat, no matter how many times the air is changed. The air must be given velocity by a bracket fan or discharge jet from a duct, so that a person can feel an impact on the skin. This was proved by the fact that there were no complaints on the *Saipan* and *Tinian*, although the air changes were less than those on the *Pacific Explorer*. They had bracket fans but *Pacific Explorer* had not.

The other special ventilation problem on the *Pacific Explorer* was the reduction plant, located in the stern. The general odour around a reduction plant, no matter how sanitary it is and how often it is washed down, is hardly conducive to a good appetite. Besides the smell, however, there is a considerable amount of steam discharged from the press, and over the basket type centrifuge. The ventilation was entirely exhaust, so as to discharge it as far over the stern as possible. While the rate of change in the whole reduction plant area was made once every four minutes and 2,000 cu. ft. (57 cu. m.) per minute of local exhaust ventilation was provided in the press area, it was not adequate to keep the area free for condensation. When the plant was operating there was a continual fog of steam within a radius of about 15 ft. (4.5 m.) from the basket centrifuge. Future installations of this nature should provide separate enclosures for the press and basket centrifuge and that not less than 10,000 to 15,000 cu. ft. (280 to 400 cu. m.) per minute exhaust fans should be used to change the air as rapidly as possible.

Another problem was proper local ventilation in the refrigerated holds. Deep floor gratings and sparring around the shell and bulkheads are essential to allow air circulation around the refrigerated cargo. On the *Tinian*, portable blowers were necessary to provide some air movement. A small blower of only 500 to 1,000 cu. ft. (14 to 28 cu. m.) per minute per hold was adequate and in any coil type of refrigeration storage this type of ventilation should be permanently installed.

REFRIGERATION

The *Ogontz* was the only vessel of the group discussed which did not have facilities for freezing and frozen storage of fish.

The refrigeration installation on the *Pacific Explorer* was designed to be able in southern waters to: freeze tuna at the rate of 120 short tons (109 metric ton) per day; make 10 short tons (9 metric ton) of ice per day; and to hold the entire refrigerated storage space at 0 deg. F. (—18 deg. C.).

In northern waters the plant had to freeze: fillets at the rate of 5,000 lb. (2.3 metric ton) per hour or 60 short tons (55 metric ton) per day; to freeze halibut on shelf type freezers at the rate of 25 tons (23 metric ton) per day; and keep the refrigerated holds at 0 deg. F. (−18 deg. C.).

Requirements on the *Deep Sea* were to freeze 1,000 lb. (450 kg.) of crab or fillets per hour and to hold the cargo at 0 deg. F. (−18 deg. C.).

Requirements on the *Saipan* and *Tinian* were to freeze tuna from +90 deg. to 0 deg. F. (+32 deg. to −18 deg.C.) at the rate of 60 short tons (55 metric ton) in 24 hours and to keep the holds at 0 deg. F. (−18 deg. C.).

Requirements on *Neva* were to freeze 70,000 lb. (32 metric ton) of salmon or halibut from +48 deg. to −10 deg. F. (+9 deg. to −23 deg. C.) in 24 hours and to keep the hold at 0 deg. F. (−18 deg. C.).

The design assumptions on ambient temperature varied slightly but basically they were as follows:

(*a*) Tropic waters: water temperature, +88 deg. F. (+31 deg. C.); air temperature, +110 deg. F. (+43 deg. C.), with 25 per cent. to be added for decks and sides where the insulated surfaces were in direct contact with the sun.

(*b*) Alaska and Pacific northwest waters: water temperature, +60 deg. F. (+16 deg. C.); air temperature, +80 deg. F. (+27 deg. C.) with 25 per cent. added for sun effect on exposed decks and sides.

These design assumptions proved to be only just adequate in the southern waters and slightly in excess of requirements in northern waters. Water temperatures of +88 deg. F. (+31 deg. C.) were recorded on various occasions on the *Saipan*, *Tinian* and *Pacific Explorer*, and the average for months at a time was +84 deg. F. (+29 deg. C.). Air temperatures exceeded +110 deg. F. (+ 43 deg. C.) but it was felt that the sun effect took care of this. Air temperatures in excess of +60 deg. F. (+16 deg. C.) were never found in the Bering Sea and water temperatures averaged +42 deg. F. (+5.6 deg. C.). On very rare occasions in inlets and bays the temperature went to +50 deg. F. (+10 deg. C.).

When vessels bring their catch back to northwest ports where +60 deg. F. (+16 deg. C.) water and +80 deg. F. (+27 deg. C.) air temperatures are found, it is believed that only the holding coils need be designed to cope with these higher temperatures. Capacities of compressors and condensers can be based on the lower temperatures and the freezing compressors used to make up any deficiencies in holding compressor capacity when in Puget Sound waters.

The *Pacific Explorer* and the *Deep Sea* used the air blast method, the *Pacific Explorer* and *Neva* used the contact method, and the *Saipan* used the brine immersion method of freezing fish. All of these methods have their advantages and disadvantages and the selection of one of them depends on the product to be frozen. The air blast method has the advantage of a fairly quick rate of freezing and using a small amount of space when the product is of a uniform shape and size. Its disadvantage is that the flesh of raw fish brought in direct contact with the air blast will get a " freezer-burn " which spoils the quality of the product. Another disadvantage is that products not uniform in size cannot be loaded symmetrically on to uniform dollies or on uniform freezing trays designed to fit the air passage in the tunnel.

The contact method, which is commonly used for halibut or other large fish which must be dressed before freezing, has the advantage of handling products which lack uniformity in size or are too large for packaging without discolouration or burning. The term " direct contact " is a misnomer, as the shelf type of coil used on the *Pacific Explorer* and *Neva* is not solely direct contact. Thermal convection of air current in the compartment undoubtedly plays an important part in the freezing action. As yet no freezer of the double contact type have been used on ship board, although it has advantages for freezing fillet or crab.

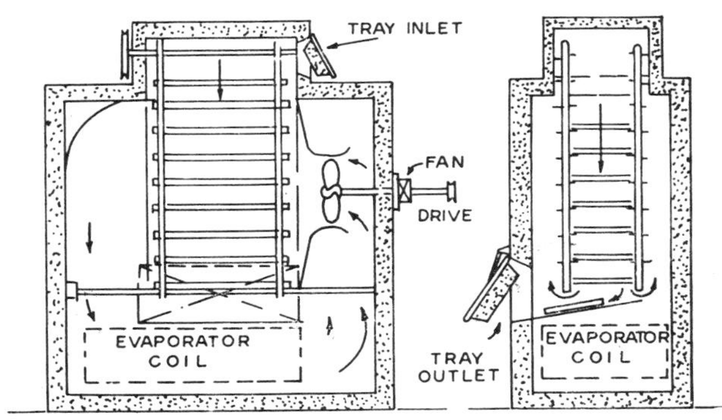

Fig. 578. Deep Sea. *Arrangement of continuous tray blast freezer*

The brine immersion method used on the *Saipan* and *Tinian* for tuna has many advantages. Stowage of the product during freezing is no problem. It is only a matter of dumping the tuna into the brine. Due to the high heat transfer values obtained between the fish and brine the tuna can be frozen in a shorter period of time than under the contact or air method. If a large tank of brine is already down to a low temperature, it has a potential refrigerating effect which is available immediately. If the fish introduced into the tank area are over 65 deg. F. (+18 deg. C.) and are one-third the weight of the initial brine, raising the brine temperature from zero to +10 deg. F. (−18 deg. to − 12 deg. C.) will lower the temperature of the fish more than 30 deg. F. (17 deg. C.) without any additional refrigerating effect from the compressors. Also, the capacity of the compressor increases as suction pressures go up, and this initial rise in brine temperature allows an increase in suction pressure, with a consequent increase in capacity. With properly designed evaporator coils in the tanks, the maximum capacity of the compressor can be used during all stages of the freezing action. The brine tank has the disadvantage, of course, of not being useable on packaged products, and salt penetration is judged to be a disadvantage when the raw flesh of the fish is subject to immersion. This is questioned by some experts and there are many people who claim that brine freezing will work well on dressed salmon, at least for salmon that will ultimately be canned. The present use of brine freezing is limited almost exclusively to the tuna industry where the fish are never sold to the fresh fish market.

All the vessels described use the direct expansion ammonia system for refrigeration. Many people have questioned why freon is not used extensively in the fish processing field and many manufacturers of freon equipment have spent considerable time and money investigating the industry, but direct expansion ammonia continues to remain a favourite for several reasons. The first cost is in favour of the ammonia system and because of low temperatures the power required generally is less. Thus the cost of fuel is less and the operating maintenance costs are certainly equal if not lower than that of freon units. The use of freon has certain disadvantages. The majority of fish processing vessels store their products unpackaged in coil type, rather than circulating air type, holding rooms. Copper coils would be exorbitantly expensive and the use of steel coils has well-known disadvantages. Also there are no rules and regulations similar to the U.S. Coast Guard rules prohibiting ammonia on passenger ships, which affect the installation of the machinery, nor are there any hazards involved, providing common sense precautions are taken.

The investigation made at the time the *Pacific Explorer* was designed shows the economic advantage of the ammonia system. A check indicated that an ammonia plant with 575 h.p. compressors was sufficient but a freon system required about 900 h.p. It was estimated that CO_2 or other refrigerant would also require greater power than ammonia, and the use of any other system would require the installation of large brine heat exchangers and pumps. The space required for the machinery would be about double that for an ammonia plant.

The refrigeration installation on the *Pacific Explorer* is believed to be not only the largest but the only two-stage system installed on ship board in the U.S.A. fleet. Five ammonia compressors were installed; two were boosters and three were secondary compressors. The boosters are one four-cylinder 15 × 10 in. (382 × 254 mm.) unit and one four-cylinder 13 × 9 in. (330 × 229 mm.) unit driven by 125 and 50 h.p. synchronous motors. They were designed for operation at zero lb./sq. in. (0.07 kg./sq. cm.) suction pressure and a discharge pressure of 24 lb./sq. in. (1.7 kg./sq. cm.). The boosters were operated at anywhere from 5 to 9 in. (127 to 228 mm.) of vacuum and at discharge pressure between 15 and 25. lb./sq. in. (1.05 and 1.75 kg./sq. cm.). The large high-stage compressors were two-cylinder $11\frac{1}{2}$ × 10 in. (292 × 254 mm.) machines, each driven by 175 h.p. motors. The third high-stage compressor was a small $6\frac{1}{2} \times 6\frac{1}{2}$ in. (165 × 165 mm.) machine driven by a 40 h.p. motor used for the icemaking plant and for pump out. The evaporating coils, for the shelf and blast freezer spaces, and the holding rooms, are all of the flooded type. Booster suctions were connected to seven low-pressure receivers, one for the storage rooms and six, two forward and four aft, for the blast- and shelf-freezers. The liquid from the receivers was pumped through the evaporator coils and back into the receivers. Fig. 579 shows the arrangement in, and fig. 580 shows the arrangement forward of, the machinery space. The installation was the conventional two-stage type. The discharge gas from the boosters was put through a water-cooled gas cooler, and a direct expansion liquid and gas cooler, before entering the high-stage compressors. These discharge directly to shell and tube condensers and into a high pressure receiver from which the liquid is returned to the low-pressure receivers.

Float controls were used for maintaining the levels in the receivers. The ice plant was fed from the high-pressure liquid receiver. A combined hot gas defrost and pump out line was installed to all of the evaporator controls. Piping was so arranged that boosters could be by-passed and the whole plant operated on a single stage.

The plant on the *Pacific Explorer* was operated for six months under tropical conditions which were as severe a test as has ever been given any refrigerating plant. The equipment was first class, but personnel inexperienced in two-stage compressor operation were incapable of handling it properly. Heavy fluctuations in loads threw the system out of balance and careful and continuous attention was required to prevent suction pressure going up, ammonia being robbed from one part of the system by another part, etc. Certain modifications could and should be made to the basic layout which would simplify operations. For example, the boosters suctions were common from all the freezers and storage spaces. This means that load fluctuations in one freezer affected the other freezers and the holding rooms. Separate suction lines and more compressors of a smaller size would possibly eliminate some of the operational difficulties encountered on the *Pacific Explorer*. However, it is the writer's considered opinion that two-stage operation is beyond the capacity of marine personnel.

The refrigerating machinery on the *Deep Sea* was single-stage and operated very satisfactorily. All refrigeration compressors were motor-driven and consisted of two four-cylinder $3\frac{1}{2} \times 3\frac{1}{2}$ in. (89 × 89 mm.) compressors and one two-cylinder $5\frac{1}{2} \times 5\frac{1}{2}$ in. (140 × 140 mm.) compressors. The latter compressor was generally used in the holding load and the quads for the freezing load. On her first trip the *Deep Sea* suffered from lack of flexibility. Later, the compressors were connected up so that any compressor could be used on either freezing or storage suctions. The holding compressors were operated at 5 to 10 lb./sq. in. (0.35 to

0.7 kg./sq. cm.) suction pressure, depending on the load conditions. The compressors working on the blast freezers were operated around zero lb./sq. in. (0.07 kg./sq. cm.) suction pressure. With average condensing water temperature in Bering Sea of +42 deg. F. (+5.6 deg. C.), compression ratios at zero lb. (0.07 kg.) were well within acceptable and standard levels.

speeds were kept lower than this. The installation proved very satisfactory.

Refrigeration on the *Saipan* and *Tinian* illustrates the difficulty in selecting the size of a plant for a vessel using the brine freezing system. In the preliminary design stage, the operators felt that they would like to have the capacity of 50 to 75 short tons (45 to 68 metric ton) per

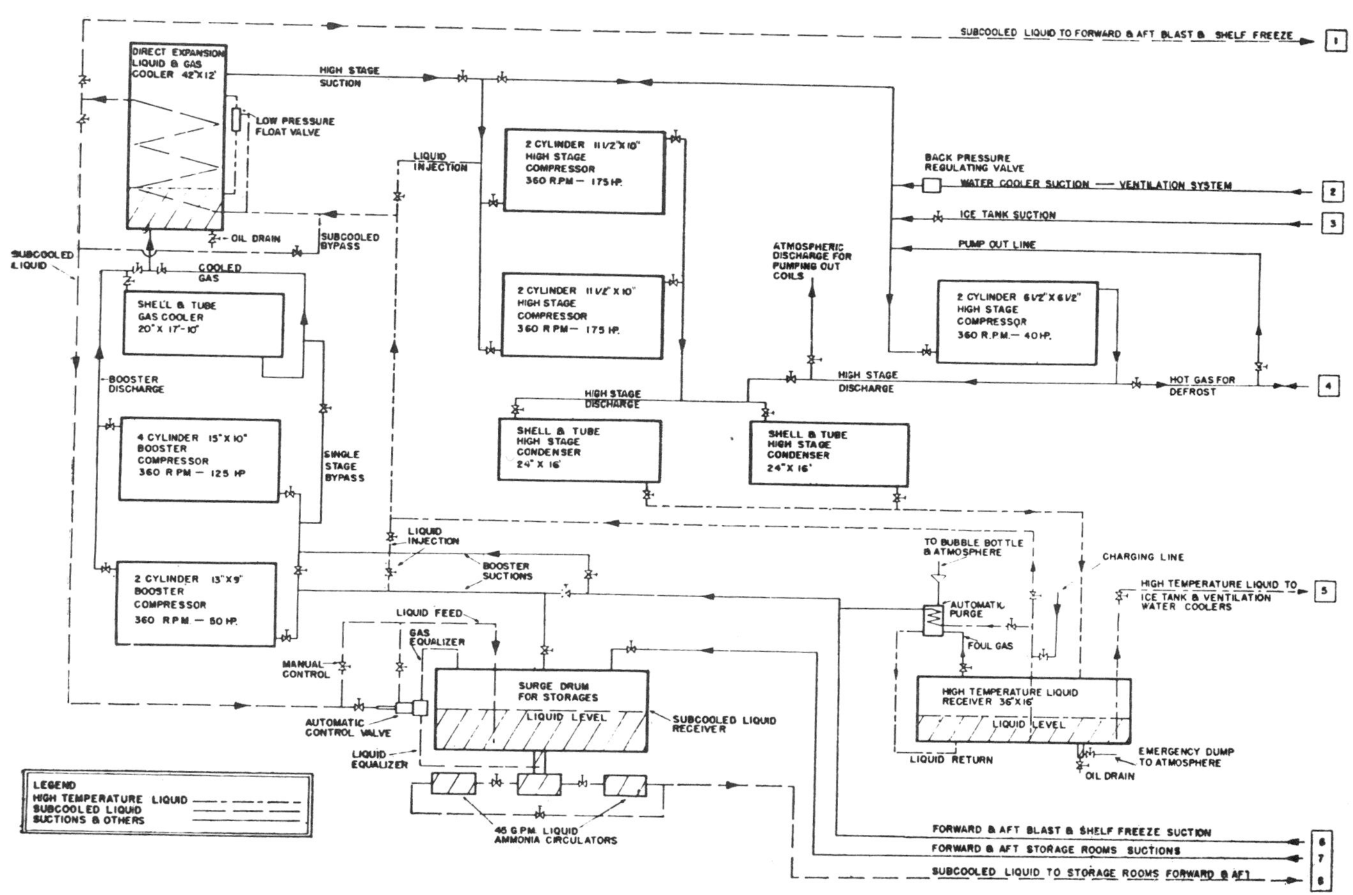

Fig. 579. Pacific Explorer. *Ammonia flow diagram in the compressor room*

The *Saipan* and *Tinian* were each equipped with four 2-cylinder ammonia compressors. Two of the compressors were driven by 75 h.p. electric motors and two were driven by diesels. Separate suction lines were provided to the storage holds and to the brine freezing tanks and the compressors were arranged so that each can be thrown on either load. Holding compressors operated at 5 to 10 lb./sq. in. (0.35 to 0.7 kg./sq. cm.) suction pressure and the brine tank compressors at 10 to 15 lb./sq. in. (0.7 to 1.05 kg./sq. cm.). With +88 deg. F. (+31 deg. C.) cooling water, head pressure of necessity had to rise to 185 lb./sq. in. (13 kg./sq. cm.). With compressors operating at 5 lb. (0.35 kg.) suction pressure, compression ratios were above accepted standards. However, the compressors were never operated at more than 350 r.p.m. and 95 per cent. of the

day. It was decided that the basic requirements would be 60 tons (55 metric ton) every 24 hours with the fish brought on board at +80 deg. F. (+32 deg. C.), the final fish temperatures leaving the brine tank to be 0 deg. F. (−18 deg. C.). Refrigeration size was, therefore, based on the suction pressures remaining at 5 lb. (0.35 kg.) unloading fish semi-continuously from the tanks at 0 deg. F. (−18 deg. C.). It was realized that the plant would actually freeze a good deal more than 60 short tons (55 metric ton) of tuna, as the brine temperature and suction pressures could rise with greatly increased compressor capacity, and because of the stored up potential in the brine. It was felt that it was advisable to provide excess capacity which might be needed for surge loads. It was not realized, however, just how much excess capacity was built into the plant.

On one trip the *Saipan* froze a total of 175 short tons (160 metric ton) of tuna in 24 hours. But the tuna came aboard at +32 deg. F. (0 deg. C.) not at +90 deg. F. (+32 deg. C.), as it had been in the brine tanks of the small boats for at least 24 hours. It was brought down to +10 deg. F. (−12 deg. C.). Just raising the brine temperature from 0 deg. to +10 deg. F. (−18 deg. to 12 deg. C.) was the equivalent of 25 American tons of refrigeration per 24 hours (23 British tons of refrigeration, 300,000 B.Th.U.; 76,000 kcal./hr.). The fish were put into the storage holds and the holding coils brought them down to 0 deg. F. (−18 deg. C.) in 24 to 36 hours. This potential capacity for peak loads is one of the great advantages of the brine freezing system.

Refrigeration on *Neva* was designed for freezing halibut or salmon on shelf freezers and keeping the fish at zero in the holds. The refrigeration machinery is of two types, conventional compressor equipment and an ammonia absorption plant. The compression equipment consists of a 6 × 6 in. (153 × 153 mm.) two-cylinder vertical compressor to take the holding load in northern waters and two two-cylinder 9×9 in. (229×229 mm.) vertical compressors to handle the lower deck shelf freezer. There is one large shelf freezer on the lower deck and two smaller units on the main deck. The capacity of the lower and main deck shelf freezers are about equal. The ammonia absorption plant is for the main deck freezers. The suction lines are cross-connected so that either the absorption plant or the compression equipment may be put on the storage coils or on either of the shelf freezers. The capacity of the plant is 30 American tons of refrigeration per 24 hr. (27 British tons of refrigeration, 360,000 B.Th.U.; 91,000 kcal./hr.) at −50 deg. F. (−46 deg. C.) suction temperature. There are several advantages of the system for marine service. Low evaporator temperatures, i.e. −50 deg. to −60 deg. F. (−46 deg. to 51 deg. C), can be obtained without the necessity of going to the complicated two-stage compression equipment. The elimination of lubricating oil from the ammonia system means continuously effective heat transfer surfaces and no oil drains, etc., in the ammonia lines. The ability of the absorption system to take liquid ammonia back into its absorber without danger does away with surge drums, knock-out drums, etc., and the operator does not have the problem of constantly watching his suction lines to prevent surges.

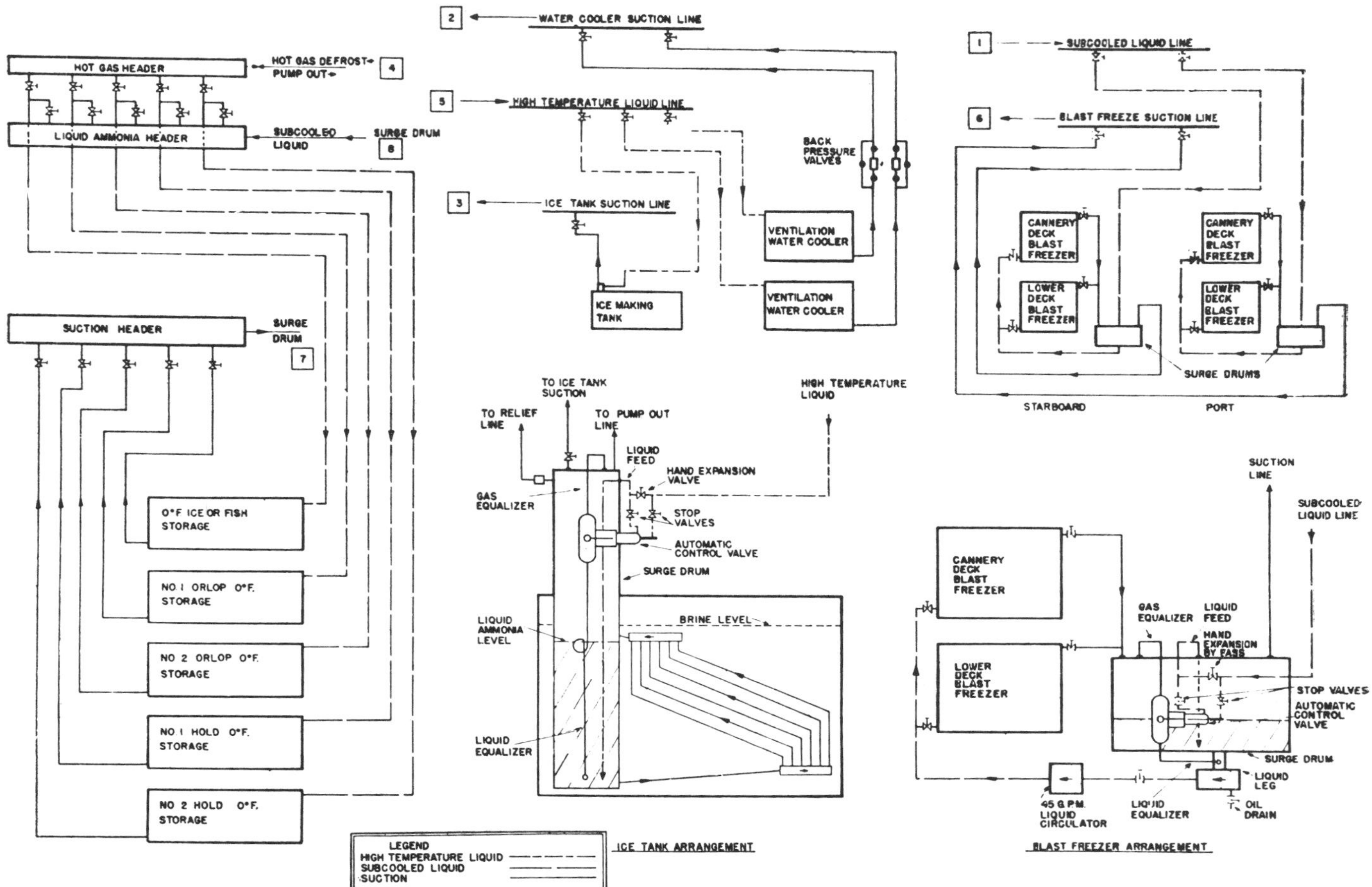

Fig. 580. Pacific Explorer. *Ammonia flow diagram in the forward part of the ship*

Many vessels, particularly cannery vessels, have steam readily available, whereas electric power for compressor drives often is expensive and difficult to obtain.

The difference between a good and a barely acceptable design is the provision of flexibility and proper balance. Flexibility enables the operator to divide the load and to use any and all of the equipment as the circumstances arise. Balance means the careful design so that the maximum capacity of the equipment may be used under varying conditions. For example, the compressors on the *Saipan* and *Tinian* each require approximately 50 h.p. when operating at 10 lb./sq. in. (0.7 kg./sq. cm.) suction and 185 lb./sq. in. (13 kg./sq. cm.) discharge pressure. However, when a load of warm fish is dropped in a brine tank, the temperatures may rise to where the suction pressure can be raised to 35 lb./sq. in. (2.5 kg./sq. cm.) when the horsepower required is approximately 65. By installing a motor developing 75 h.p., the maximum capacity of the compressor can be used under extreme load conditions. Similarly, a condenser with the capacity based on the low suction pressures only will be considerably under capacity when the compressor gets up to a high suction pressure. The *Saipan* and *Tinian* compressors had a capacity of 21.8 American tons of refrigeration (19.8 British tons; 263,000 B.Th.U.; 66,000 kcal./hr.) at 10 lb. (0.7 kg.) suction and 185 lb. (13 kg.) discharge pressure. At 35 lb. (2.5 kg.) their rating was 35.8 American tons of refrigeration (32.5 British tons; 433,000 B.Th.U.; 108,500 kcal./hr.). Obviously it was no use putting large motors on the compressors for peak loads unless the condensers were able to handle the load. This provision of adequate condenser capacity is of particular importance in southern waters. The average refrigeration engineer can scarcely believe that +88 deg. F. (+31 deg. C.) sea water actually exists, nor can anyone who has not seen the fouling action that takes place in southern waters realize just how severe this can be. On the *Pacific Explorer* the head pressure of the circulating water pump went from 30 to 80 lb./sq. in. (2.1 to 5.6 kg./sq. cm.) in three months simply due to fouling. Divers cleared the sea chest and the pressure dropped to the neighbourhood of 45 lb. (3.2 kg.). An examination of the condenser tubes when the *Explorer* returned revealed an unbelievable amount of growth and shells which, of course, easily explains why the head pressure on the *Pacific Explorer* went to 250 lb./sq. in. (17.5 kg./sq. cm.) during certain periods. The tentative standards for tuna boat refrigeration put out by the

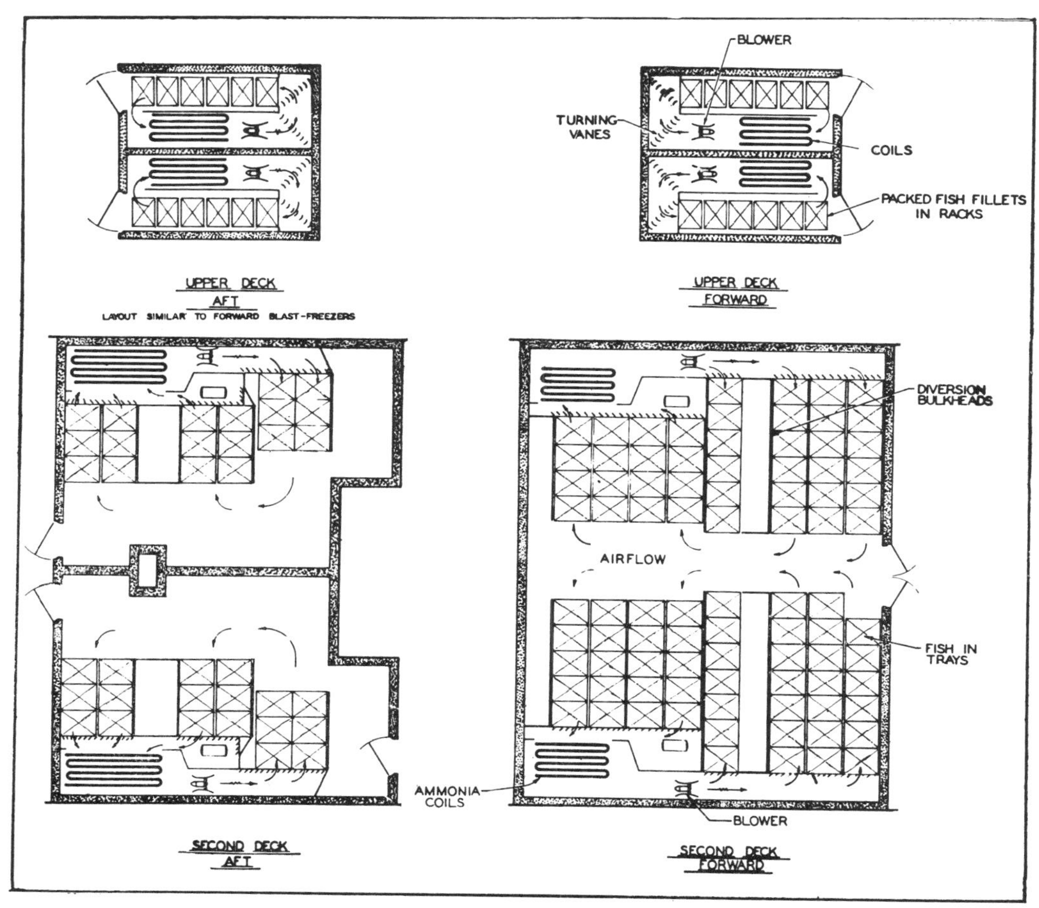

Fig. 581. Pacific Explorer. *Arrangement of blast freezers*

Southern California Refrigeration Installers in 1938 recommends a condenser capacity of 15 sq. ft. (1.4 sq. m.) per American tons of refrigeration per 24 hr. (16.5 sq. ft. or 1.53 sq. m. per British tons of refrigeration per 24 hr. or 13,300 B.Th.U. or 3,340 kcal./hr.) at 185 lb. (13 kg.) discharge pressure, and 15 lb. (1.05 kg.) suction pressure. The errors that can be introduced by the use of rule of thumb formulas, such as this, can be understood by comparing this rule to the condensers on the *Saipan* and *Tinian*. These condensers were designed for an 80 per cent. clean tube factor and the ratio was 23.4 sq. ft. (2.18 sq. m.) of surface per American ton of refrigeration per 24 hr. (25.8 sq. ft. or 2.4 sq. m. per British ton or 13,300 B.Th.U. or 3,340 kcal./hr.).

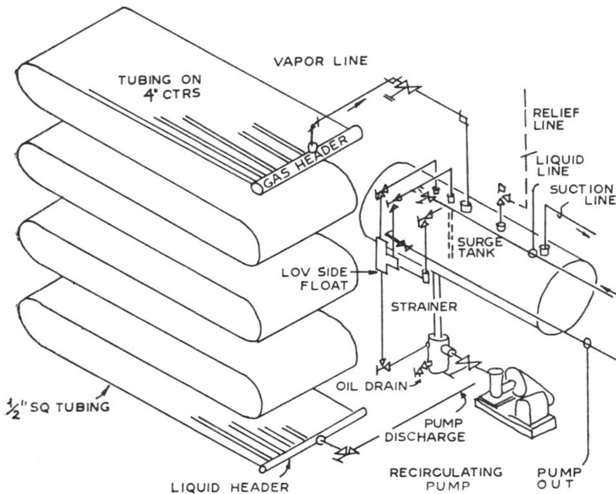

Fig. 582. Pacific Explorer. *Flooded circulation type shelf freezer*

FREEZERS

The blast freezers units installed on the *Pacific Explorer* followed conventional land practice. They were essentially a tunnel in which the product could be loaded and through which air could be circulated by means of a high capacity fan and the evaporator coils. They proved effective and met their rated capacity. However, certain difficulties were experienced with their operation. It is recommended that the motor proper be not placed in the blast freezer chamber. Defrosting on the *Pacific Explorer* was accomplished by a water spray introduced over the blast coils. The headers were effective and defrosting could be accomplished but the problem, which was never solved, was how to make the operating personnel defrost the coils as often as was required and to do a complete job of defrosting. Unless the coils were defrosted at least every eight hours, their capacity was reduced practically to zero. The operator of the *Deep Sea* had actually to submerge the coils for a short period to ensure a complete defrosting.

Fig. 581 shows the arrangement of the blast freezers on the *Pacific Explorer*. Fig. 578 showed the arrangement of the blast freezers on the *Deep Sea*. Both freezers operate on the same principle, a continuous circulation of air over the product and the coils which effects a rapid transfer of heat. The freezer on the *Deep Sea* has proved to be very satisfactory. The small freezers on the cannery deck of the *Explorer*, which were designed and used for package products, were also effective. The large blast freezers on the second deck of the *Pacific Explorer*, which were designed to take tuna or any other big fish, were not too satisfactory. It took from 24 to 36 hours to freeze big fish, and the freezers had to be defrosted three or four times in the process. The fish were loaded manually on portable shelves in the freezer, an expensive and tedious job. The low pressure receivers and pumps were located in the second deck freezers. The pumps were subject to shaft seal troubles due to suction pressures which varied between vacuums and positive pressures, and ammonia leaks around the seals, which occurred when the freezers were full of tuna, were naturally troublesome. The oil drains had to be checked and cleared continuously to prevent the float controls from sticking. This became such a problem on the *Explorer's* southern trip that the direct expansion liquid cooler was by-passed. On any future marine installations, the float valves and pumps should be placed where they are easily accessible for adjustment. The use of a packless type pump would also be very desirable.

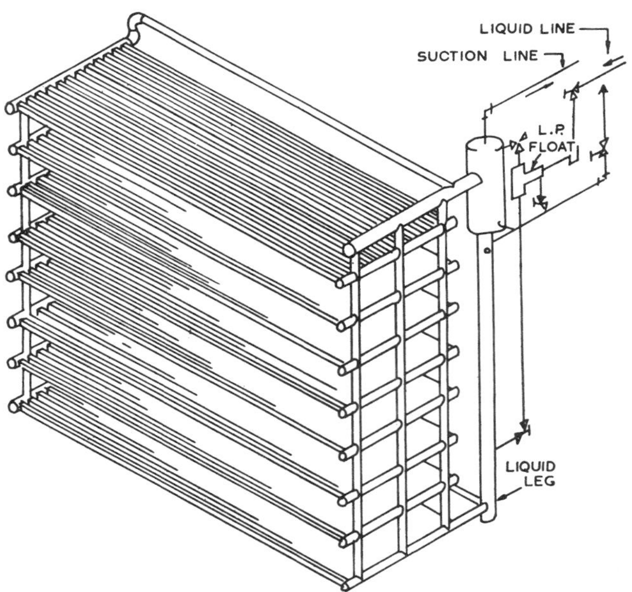

Fig. 583. Neva. *Flooded gravity recirculation type shelf freezer*

The coils on the *Deep Sea* differed from the *Explorer*, in that they were designed for natural rather than forced circulation by a pump. A float control was used to maintain the coils in a flooded condition. No difficulties were experienced with the float, possibly because the liquid was not sub-cooled.

The shelf coil freezers on the *Pacific Explorer* were located on the second deck aft. They were designed primarily for halibut, although they were used effectively on tuna. Fig. 582 shows a typical shelf coil, low pressure receiver, pump and control. The coils were arranged in groups and two pumps and two receivers fed all the coils. Other than some pump and float difficulties, they gave no trouble. They had a very high capacity and did a good job of freezing.

On *Neva* it was decided to use a gravity circulation type coil. A special coil, shown in fig. 583, was designed. Level control is maintained by the usual float and the liquid feeds are oversized and arranged to feed in any of the vertical legs. The headers and vertical legs were made sufficiently large that the gas velocities would be well below the point where they could pick up any liquid. Tests of small scale models showed the coil surfaces to be effective under extreme conditions of concentrated local heat loads, and under high angles of heel and trim.

Brine tanks are arranged on the ordinary tuna clipper as shown in fig. 584. On some of the smaller vessels the outside brine coolers are omitted, tank coils alone being used for refrigerating.

The *Saipan* and *Tinian* arrived with large loads which would take a cannery weeks to process. This meant that the tuna had to be in dry storage and the brine tank was just a production freezing tank.

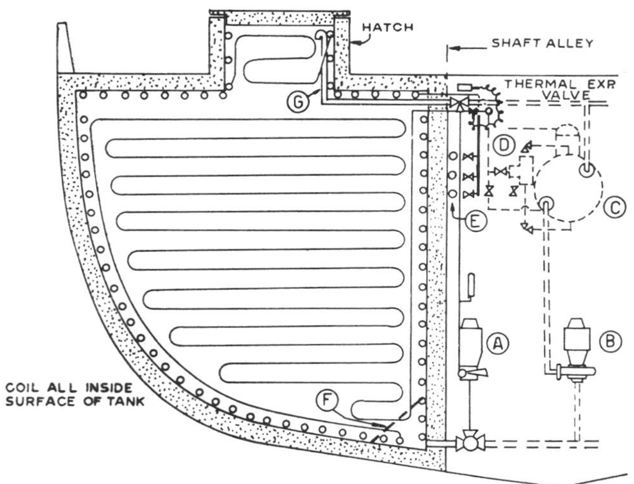

Fig. 584. Typical arrangement of tuna clipper brine freezing tank. (A) tank circulating brine pump; (B) brine transfer pump; (C) brine cooler (optional) (suction to each compressor); (D) brine cooler feed (L.P. float); (E) compressor suction mains (through suction pressure regulators); (F) strainer plate brine suction; (G) brine discharge

The arrangement of brine tanks on the *Saipan* and *Tinian* is shown in fig. 585. In place of external brine coolers, regular raceway coils similar to an ice tank installation were provided and impellers circulated the brine through the coils, out through the tanks and back through the coils again. By using high brine velocities through the raceway (150 ft. or 46 m. per minute) a very high heat transfer rate between the brine and ammonia was obtained and the large volume of the tanks ensured that the fish would give up their heat to the brine as fast as the coils would receive it. Impellers mounted on vertical shafts driven by belts from motors in the wing wall compartments were used for circulation. The tanks were very successful in operation.

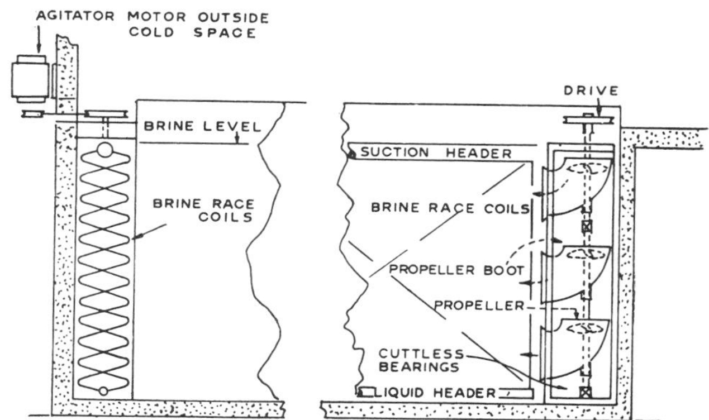

Fig. 585. Tinian *and* Saipan. *Arrangement of brine freezing tank*

The total volume of the tanks was about 10,000 cu. ft. (280 cu.m.). Approximately 80 short tons (73 metric ton) can be placed in the tanks and frozen at one time without the fish sticking together in the freezing operation. The writer believes that the sticking is due to the buoyancy of the fish in the dense brine. As successive layers of fish are added, the pressure exerted by the fish on the bottom layer reaches a point where the fish actually freeze together. Thus, depth is the controlling factor and it is recommended that future tanks be shallow and broad. Experiments are going on now with the use of horizontal partitions which may increase the effective capacity of the *Tinian* tanks.

Brine temperatures on the *Saipan* and *Tinian* are maintained between 0 deg. and +5 deg. F. (−18 deg. and −15 deg. C.). As the fish have always been either in ice or brine on the small boats, no " chilling " is done before they are loaded into the brine. Tuna clippers conventionally use cold sea water to chill the fish when first caught, and the water also cleans off the blood and gurry.

The *Oceanic V*, another converted L.S.T. uses brine tanks for both freezing and storage and the entire hold is simply a series of large brine tanks. On the first trip they depended on shell and tube brine coolers and pumps which meant that they had to keep the fish in the brine for the entire voyage. The rolling and motion of the ship damaged the fish so ammonia coils were put in the tanks. This now allows the fish to be frozen with the brine coolers. The brine is then pumped out and the fish held dry with the coils. This method of storing the tuna increases the capacity of a given vessel by about one-third as the fish will stow at the ratio of at least 45 cu. ft. per short ton (0.71 metric ton/cu. m.). However, the conversion cost per ton of capacity is about the same and the fish must be thawed before they are removed, so they must be canned immediately they are brought ashore.

The holding coils on the *Pacific Explorer* were the flooded ammonia recirculating type. Holding coils on the other vessels are all the " dry " type controlled by thermal and hand expansion valves. Flooded coils are undoubtedly more effective, and have a higher heat transfer valve, and on large installations the saving in pipe justifies the increased cost of the low pressure receiver and pump and the additional ammonia. An overall coefficient of heat transfer of 1.6 B.Th.U. per sq. ft. of surface per deg. F. per hour (7.8 kcal./sq. m., hr., deg. C.) can be used for the flooded coils, whereas the dry coils should not be rated at more than 1.25 B.Th.U. per sq. ft. per deg. F. (6.1 kcal./sq. m., hr., deg. C.).

Provisions are generally made for defrosting the hold coils by hot gas from the compressor discharge. Defrosting over the tuna load is done wherever necessary. The drip from the coils adds a glaze to the stacks of fish. On the *Deep Sea* hot gas defrost was not installed, as the vessel carried a package product and could not therefore defrost with a load. A quick turn around was of no importance at the unloading terminal and there was ample time to defrost by natural air circulation.

ICEMAKING

The *Pacific Explorer* had the only ice tank ever installed to the writer's knowledge on a sea-going ship. It was conventional, as far as the tank, agitator, ice cans, ammonia coils, thaw tank, can filler, can dump, and can hoist were concerned. It was rated at 10 short tons (9 metric ton) per day at a 36-hour freezing rate using 300 lb. (135 kg.) cans. It met its rated capacity in calm waters, but had difficulties whenever a sea was running.

The unusual features of the tank were the insertion of swash bulkheads in an attempt to minimize the effect of the movement of the ship, and the addition of a brine dump tank under the main deck. The dump also acted as an overflow collection tank. Its capacity was sufficient to allow the brine to be lowered to half the normal level in heavy weather. The greatest difficulty encountered in operation with any roll or pitch was the spilling of fresh water from the ice cans, which diluted the brine, and the spilling of brine into the cans, which made them harder to freeze.

INSULATIONS

Insulation in the refrigerated spaces of the *Deep Sea* was cork throughout. The other vessels used a combination of cork and fibre-glass. The experience of the writer is that there is no better material for cold insulation than cork. Where weight and fire-retarding qualities are no consideration, the only reason for using other materials is cost. Because the difference in cork and fibreglass is so great, every effort was made on the *Pacific Explorer* and *Saipan* and *Tinian* and *Neva* to use fibreglass. To be practical, the installation must provide a perfect vapour seal on the warm side of the insulation. This is essential in tropic service where high humidities and temperature increase the vapour pressure differential between the cold and warm sides of the insulation. The only good vapour seal in the opinion of the writer is a steel surface against which the insulation can be placed with the warm area on the opposite side of the steel. Wherever this condition existed on the vessels named, fibreglass was used, with the exception of the blast freezers on the *Pacific Explorer*, the brine tanks on the *Saipan* and *Tinian*, and the decks of all three vessels. On the blast freezers cork was used because of the high temperature differences. The brine tanks had a steel lining on the inner cork surface because leaks would ruin the fibreglass. Cork was used on the decks for the same reason.

The insulation on all vessels gave a minimum coverage of 2 in. (5 cm.) over all structural members. With low cost materials such as fibreglass the space between frames and beams can be filled in solid at less cost than is required for boxing. The result was 8 to 15 in. (20 to 38 cm.) of insulation, depending on the size of frames and beams. Through experience, the writer has adopted the practice of using an overall coefficient of heat transfer of .1 B.Th.U. per sq. ft. per deg. F. per hour (0.5 kcal./sq. m., hr., deg. C.) for ship side, deck head and deck insulation, regardless of whether the thickness is 8 or 14 in. (20 or 35 cm.).

The mobility which allows fish-processing vessels to move to the most productive fishing areas is likely to result in the industry making more use of them.

THE EXPERIMENTAL FREEZER TRAWLER *DELAWARE*

by

C. G. P. OLDERSHAW

THE objectives of the project to freeze fish at sea being carried out during 1952–1954 at the Boston Technological Laboratory of the U.S. Fish and Wildlife Service's Branch of Commercial Fisheries are: (*a*) the development of handling, freezing, and storage facilities which can be installed and used in existing vessels of the New England fleet, and (*b*) the establishment of the technical and economic feasibility of freezing fish " in the round " at sea for later processing (i.e. thawing, filleting, and re-freezing) ashore. Methods that would require extensive conversions or the designing and building of new vessels are not being considered.

The trawler *Delaware* used in these studies is a typical, present day, large New England trawler with the following dimensions:

- Length, over-all: 147 ft. 6 in. (44.8 m.);
- Beam: 25 ft. (7.62 m.);
- Depth: 14 ft. 8 in. (4.47 m.);
- Gross register tonnage: 303 tons;
- Net register tonnage: 172 tons;
- Main engine: 7 cylinder, two-cycle, 735 h.p.;
- Auxiliaries: two units supplying the 115 v. D.C. electrical system; one 40 kw. diesel generator set, one 25 kw. diesel generator standby set;
- Trawl winch power: one 80 kw. diesel generator set connected to the 100 h.p. electric winch motor;
- Fresh water tank capacity: 11.0 short tons (10 ton);
- Fuel-oil tank capacity: 63.2 short tons (57 metric ton);
- Lubricating oil tank capacity: 335 imp. gal. (400 gal., 1514 l.);
- Cruising range: 8,000 nautical miles;
- Speed: approximately 10 knots;
- Crew accommodation: 20.

ALTERATIONS TO FISH-HOLD

The fish-hold of the *Delaware* originally had a volume of about 8,000 cu. ft. (226 cu. m.), with its 36½ ft. (11.12 m.) length divided into seven pound sections. It has now been partitioned by cork-insulated bulkheads as follows (*see* fig. 586): one pound section forward for storing iced, gutted fish; two pound sections insulated and refrigerated for storing frozen whole fish; three pound sections insulated and refrigerated for storing frozen whole fish and housing the fish freezing tank; and one pound aft, housing the refrigeration machinery.

The original hold insulation of 4 in. (10 cm.) of cork, laid in hot asphalt and covered with 1½ in. (3.8 cm.) tongue and grooved sheeting, was thought sufficient for the experimental work to be carried out. However, there may be more economical methods of insulation in fitting a commercial vessel for cold storage of frozen fish.

REFRIGERATION EQUIPMENT

The refrigeration equipment installed on the *Delaware* consists of the following:

(1) BRINE TANK AND FISH-FREEZING MECHANISM

The method of freezing the whole fish by immersion in refrigerated liquids was chosen in preference to refrigerated coils or plates, refrigerated air blast, or refrigerated moulds, because it appears to offer advantages in economy of space and manpower, short freezing time, and simplicity of machinery.

The freezing apparatus consists of a tank of chilled brine through which the fish are carried in baskets mounted on an endless chain conveyor (fig. 587). The rectangular steel tank, approximately 8 ft. (2.4 m.) long by 5 ft. (1.5 m.) wide by 14 ft. (4.3 m.) deep, is situated lengthwise in the centre of the fish-hold approximately 5 ft. (1.5 m.) forward of the refrigeration machine room bulkhead. It extends upward through the main deck about 2½ ft. (0.75 m.). Chilled brine is pumped from the brine cooler in the refrigeration machinery room to four inlets located in the tank line about 1 ft. (0.3 m.) below the deck on the starboard side. The brine flows out of the tank and returns to the circulating pump by way of an overflow trunk at the forward end of the port side, having been made to circulate through the depth of the tank by a central baffle. The overflow trunk is fitted with a removable strainer, which is accessible from the deck for cleaning.

The mechanism for carrying the fish through the brine consists of 11 cylindrical steel-mesh baskets, 7 ft. (2.1 m.) long by about 2 ft. (0.6 m.) in diameter, mounted centroidally at each end on continuous travelling chains. The chains run on 30 in. (76 cm.) diameter sprockets mounted on a supporting framework. This framework can be hoisted as a unit out of the tank to facilitate maintenance or repairs. The path of the chains brings each basket, in turn, up beyond the top rim of the tank, where the fish can be loaded or discharged by personnel on the deck. After being removed from the baskets, the frozen fish are lowered through the hatches and fed into the pounds by light metal chutes. The top of the tank is covered with a hood fitted with hinged doors. A 2 h.p. motor and reduction gear, located above deck in a watertight enclosure, drives the conveyor at the rate of one cycle per minute.

Each basket is fitted with a full-length door about 20 in. (51 cm.) wide, which can be opened from either side to make loading and discharging easy. The baskets, 20 cu. ft. (0.57 cu. m.) in volume, hold about 300 lb. (136 kg.) of round fish, giving the freezer a total capacity of about 4,400 lb. (2,000 kg.). The loading density of about 20 lb./cu. ft. (320 kg./cu. m.) provides sufficient space for agitation of the fish in the baskets.

(2) Hold-Cooling Equipment

The frozen fish storage holds, where the temperature is 0 deg. F. (—18 deg. C.), are refrigerated by banks of 1¼ in. (3.1 cm.) iron pipe coils mounted on bulkheads, deckhead, and port and starboard surfaces. Wooden gratings are laid on the floor to aid in circulation of air under the piles of frozen fish. The refrigerant, a solution of ethanol (ethyl alcohol) and water, is pumped through the coils of a shell-and-tube cooler in the refrigeration machinery room.

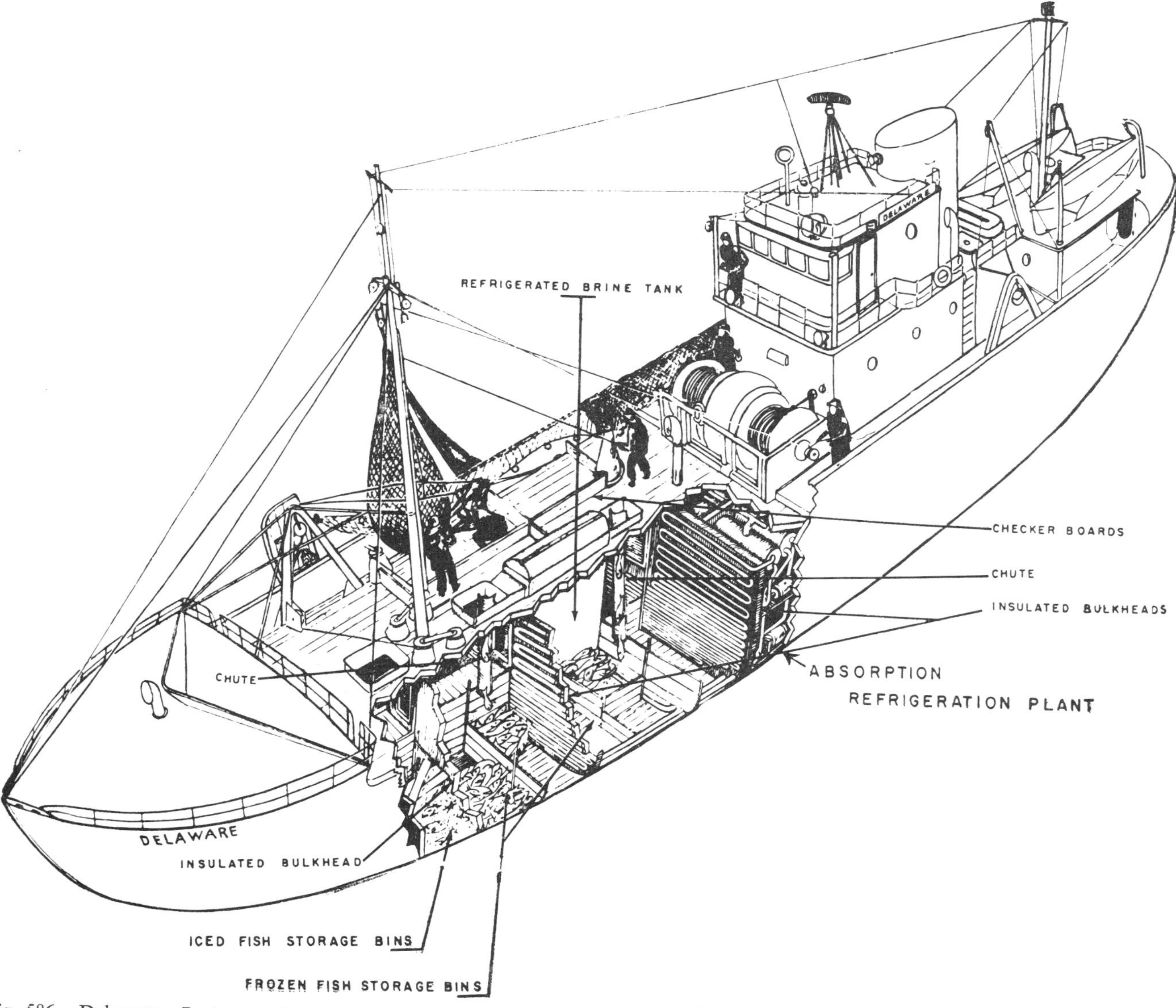

Fig. 586. Delaware. *Boston trawler converted for experimental brine freezing of whole white fish. Owned and operated by U.S. Fish and Wildlife Service*

(3) Refrigeration Plant

A 25 U.S. ton (23 Br. tons, 300,000 B.Th.U., 75,000 kcal./h.) ammonia absorption refrigeration machine is used for freezing and cold storage. A number of these plants have been used on the U.S.A. West Coast salmon freezer ships for several seasons. They appear to offer a number of advantages over the more conventional compression system: low initial cost; small electrical

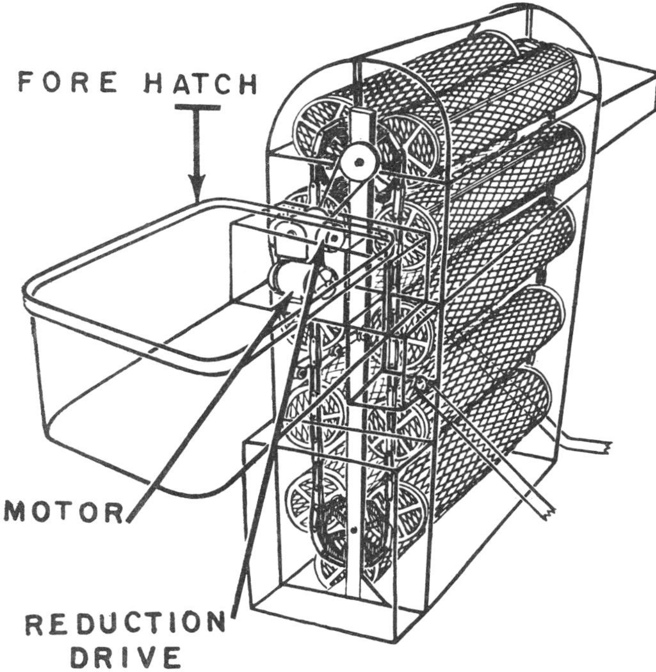

Fig. 587. Delaware. *Arrangement of perforated containers of freezing mechanism*

load, as low pressure steam energizes the plant ; a few moving parts, simplifying maintenance; ability to adjust to widely fluctuating loads without complicated controlling devices; little change in efficiency with reduced loads; little reduction in capacity with lowering of load temperatures, thus eliminating the need for " staging "; and comparatively small floor space requirement. One disadvantage is the necessity of using ammonia in the plant. To overcome this hazard, all of the refrigeration plant on the *Delaware* is enclosed in a water-and-gas-tight compartment which, in case of dire emergency, could be flooded.

Two shell-and-tube liquid coolers are used, one rated at 20 U.S. tons of refrigeration (18 Br. tons; 240,000 B.Th.U., 60,000 kcal./hr.) for maintaining the freezing medium at +2 deg. to 5 deg. F. (—17 to —15 deg. C.), and one rated at 5 U.S. tons (4.5 Br. tons, 60,000 B.Th.U., 15,000 kcal./h) for maintaining the holds at 0 deg. F. (—18 deg. C.). A 30 h.p. oil-fired vertical boiler, located in the engine room, supplies steam to the refrigeration machine and heat to the vessel.

Electric motors, totalling 9 h.p., are used in the refrigeration equipment. This includes power for circulating the brine and ethanol solutions, circulating aqua-ammonia solution and cooling water in the refrigeration machine, and driving the fish-freezing mechanism.

COLD STORAGE CAPACITY AND FREEZING RATES

The refrigeration machinery room, originally the aft pound-section of the hold, occupies approximately 1,300 cu. ft. (37 cu. m.), while the brine tank and associated pipes occupy about 700 cu. ft. (20 cu. m.) in the centre or " slaughter-house " area of the present after hold. Using an observed load factor of 33 lb./cu. ft. (525 kg./cu. m.) for frozen whole fish, the vessel, as now fitted for experimental work, has an estimated total capacity of about 130,000 lb. (59 tons) of whole frozen and 15,000 lb. (6.8 tons) of iced gutted fish.

The design of the freezing plant was based on an average catch rate of 1,000 lb. (450 kg.) per hour, with an appreciable overload capacity. The fish, depending on size, are frozen in one to three hours in the 5 deg. F. (—15 deg. C.) brine, having been sorted into species and size groups. After the correct freezing time has elapsed for each size group, the machine is stopped for discharging the appropriate numbered baskets, then run again until the next size group is frozen. The larger fish, requiring about three hours to freeze, remain in the freezer during the course of two one-and-a-half-hour trawling hauls but there are sufficient baskets to accommodate the ensuing catch during normal fishing.

FREEZER TRAWLERS

by

DAVID B. CUNNINGHAM

THE whale factory ship was developed by necessity and perseverance and it is now a highly scientific and meticulous operating unit. Fish, as distinct from whales, are more plentiful and can be found in almost every sea and ocean of the world so that a wider scope should exist for the fish factory ship.

Fishing boats vary from country to country and the diversity of size, design, and opinion in planning them, arises from the need to meet the actual operating conditions in each locality. Bearing in mind that these vessels are all designed and built for fresh fish, one may consider the distinct advantages that the freezer factory ship can offer. The factory ship has no need to rush home for the market so she does not need to be as fast as a

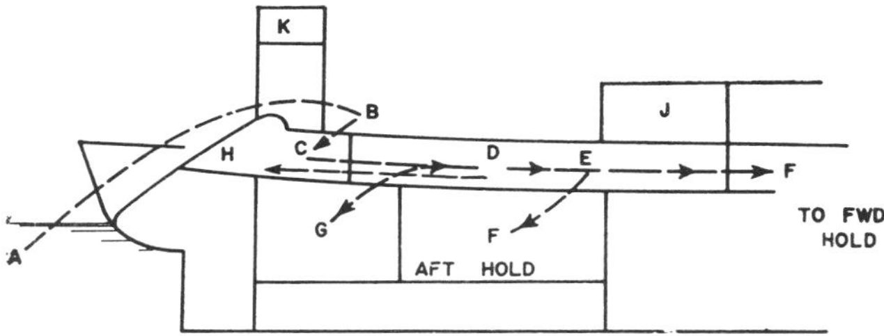

Fig. 588. Arrangement of a freezer trawler illustrating passage of fish from time caught until processed. (A) fish in cod end; (B) on trawl deck; (C) in cooled fish ponds; (D) factory deck (heading, gutting, filleting); (E) freezers; (F) low temperature holds; (G) fish meal plant; (H) liver processing; (J) trawl winch; (K) fishing bridge

" wet fish " vessel. She can stay longer at sea, and fish longer. She also makes a considerable saving in fuel because of her slower speed and fewer trips to the fishing grounds.

In warm climates the freezer factory ship may be very important, as she can operate at sea for many weeks, catching her fish in the cool of the morning or evening, and processing the catch inside the factory deck, which is enclosed and cooled so that work may go on irrespective of outside temperature.

BASIC DESIGN

In planning a freezer factory ship, such as the *Fairtry*, there are many salient points which affect the design. The handling of the nets over the side of the vessel, as in a normal trawler, would require low freeboard and this would detract from the seagoing qualities in a ship which has to remain at sea for many weeks. This problem has been overcome through experiments carried out in the steam yacht *Oriana* and later in the *Fairfree*, both described by Lochridge (1950). A stern ramp or chute was designed to enable the nets to be shot from the stern. This reduced the handling of nets to a minimum and ensured maximum recovery of the catch instead of losing a portion, as often happens, when bringing the net alongside. The arrangement also enables the sides of a vessel to be built up, or another deck to be carried out to the side to provide cover and accommodation for factory work.

As speed is not of prime importance, beam can be introduced to give greater transverse stability and a wider working platform for shooting and recovering the nets, and a steadier 'tween deck for the factory workers.

Factory ships can be built to any size to suit the requirements of owners and may vary in length from 100 ft. (30.5 m.) to 500 ft. (152.5 m.) depending on the work the ship is to do and how much may be spent on it. In designing, two salient questions must be answered:

1. What quantity of frozen fish does the owner require to be processed during the voyage?
2. In what time, as regard the number of days at sea, has this to be carried out?

If, for instance, the owner wishes to process 100 tons of frozen special type fish and this is to be brought back in 20 days, the ship will be quite different in design from that planned to handle, say, 600 tons of a common type fish, brought back in 60 days. The length, breadth, depth and draft of the vessel, her fuel and fresh water capacity, stores, refrigerating machinery, generators, and number of crew, are all affected.

The size of a vessel is determined also by the area in which she will operate. In northern waters she must be robust to stand up to heavy storms, but in southern waters the weather may be generally placid and the ship will not be subject to such arduous strain. And, again the catch may be smaller than that normally taken in Arctic waters.

As the beam can be increased, will the vessel be best powered by single- or twin-screws? When the nets are streaming astern, the warps are pivoting from the centre of the top of the ramp and not on a quarter of the ship, as is normal. If a single propeller is used, the trawling depth must be watched to avoid traversing " crab-wise ", because the rotation of the screw does not help to keep the trawl warps as taut as they should be. Use of twin propellers partly overcomes this problem, provided the engine room staff are experienced in operating the engine revolutions to steady the ship on her trawling course. But the propellers should be protected to avoid the net catching on them.

Sea-worthiness is essential. Stability is a contributory factor toward this and the ideal is to create a vessel which will provide a steady platform for the crew and the factory workers.

The entrance waterlines of the hull can be straight or slightly full. This gives a little extra buoyancy forward, and some extra capacity to the forward fish-hold.

The run of the lines to merge into the stern is worthy of careful attention as a following sea could be a nuisance if it arrived under the overhang at the crucial moment of taking the cod end up the slope.

OPERATING CREW

Each member of the crew of a small ship has several duties to perform, but the larger the vessel, the more clearly defined are the specific duties. These can be roughly divided into three categories:

(*a*) the navigating crew and operating personnel who run the vessel as a sea-going ship;
(*b*) the fishing crew who shoot and recover the nets during fishing operations;
(*c*) the factory workers who operate below decks to process and freeze the fish.

Assuming, for a medium-sized factory ship that there is a crew of about 90 people, it is wise to appoint a head for each department, all under the command of the navigating captain.

The suggested distribution of the crew would then be:

Navigating captain	1
Fishing skipper	1
Factory deck manager	1
Chief engineer	1
Navigating officers	6
Fishing officers	2
Engineering officers	6
Refrigeration engineers	2
Seamen	8
Catering staff	6
Fishermen	6
Engineroom hands	10
Factory workers	40
Total	90

The navigating captain and his crew take the vessel to the fishing grounds where the fishing skipper and his crew take over and maintain control from the fishing bridge during fishing operations. Should any emergency occur which might affect the safety of the vessel, the navigating captain at once assumes command.

When the fish are caught the nets are hauled up the ramp to the trawl deck and shot through the hatches into the fish ponds below, where the factory deck manager is in charge. The fishing crew are then free to shoot their nets again. On completion of trawling, with all gear safely on board, the ship reverts to the command of the navigating captain.

Meanwhile, below decks the processing of the catch has been taking place. When these operations are completed, the frozen fish is transferred to the refrigerated holds for storage, the livers are canned or processed, and the offal is processed into fish meal.

These products are stored in their respective compartments and it is the responsibility of the navigating captain to ensure safe carriage and correct storage temperature.

It can be assumed that the navigating captain is in command of the vessel at all times but that he delegates his authority to the specialized branches which he has to control. They can be divided into the following groups:

(*a*) Navigation and machinery;
(*b*) Fishing operations;
(*c*) Fish processing and by-products;
(*d*) Refrigerated cargo holds, temperatures and storage;
(*e*) Safe transit of processed fish and all by-products;
(*f*) Catering and administration personnel.

FISHING OPERATIONS

The trawl deck of the *Fairtry* is situated well above the sea and, being wide and clear, gives adequate space to spread the trawl net for examination and repair.

About 8 ft. (2.44 m.) from the ship's side on the trawl deck is fitted a small bulwark, about 2 ft. (0.61 m.) high, running fore and aft. This enables the net and the wet fish and slime to be confined in the centre of the ship leading aft to the trawl ramp. The space on the sides can house numerous articles such as spare nets or bobbins, etc., and all the ancillary gear for trawling.

To shoot the net or nets if they are run in pairs (this is possible with stern operation, and nets are normally run parallel to each other) the lines are led from the outer ends of the rear rope aft into the slipway and round the vertical rollers or pillars at the stern and then back to the warping ends of the trawl winch.

When the trawl winch is started, warp ends, port and starboard, operate simultaneously. The mouth of the net is pulled aft until it is suspended over the slipway, with the belly and cod end lying on the trawl deck. These are pushed into the chute and, sliding down into the water, they stream aft. Then the remainder of the net is lowered and the trawl warp is payed out until the whole net streams astern of the vessel. Next, the trawl doors are attached from the outer after-quarters and the trawl winch pays out the length of warp required. This method of using the stern slipway (for which patents are already in existence) enables the trawl winch operator full and complete vision of the net and warps during the entire operation of shooting, trawling and recovery. Both the winch and operator are fully protected from the weather, and the ship is usually so "dry" while fishing that operators have been known to wear house-slippers instead of the usual sea boots.

During the shooting or hauling operations the fishing bridge maintains control of the ship, including instructions to the engine room and trawl winch. During the actual trawling the ship is steered from the wheel inside the fishing bridge.

To recover the net the ship is slowed to about 2 knots and the trawl warps heave in until the trawl doors come up. They are then unshackled and the net is hauled in, with the head rope and mouth first, the belly next and, finally, the cod end which comes straight up the stern slipway. It continues over the safety hump at the top of the ramp and on to the trawl deck, finishing just forward of the hump on top of the flush deck hatches. The rip cords are pulled and the hatches opened, mechanically or hydraulically.

The catch is shot aft into the cooled 'tween deck fish-ponds and the hatches closed. The rip cords on the cod end are re-tied and the net is checked over and repaired immediately as it is spread put on the trawl deck. No man-handling over the bulwark is required. The deck is well above the water line and the danger of a man being washed overboard is remote. The catch is immediately cleared which enables the crew to concentrate on fishing without having to wade about waist deep in fish, as normally happens in a side-operated trawler where a heavy catch has to be stowed before another trawl can be brought on board.

TRAWL WINCH

The trawl winch is the heart of trawling operations and fishing operations cannot be carried on if anything happens to it. In a factory ship it is even more important for the winch to function efficiently, with complete freedom from breakdown, otherwise all the factory deck machinery, freezers, processing equipment, etc., would come to a standstill. As the trawl winch of the *Fairtry* is placed slightly abaft amidships at the fore end of the trawl deck, and is fully protected, there is little likelihood of a breakdown.

A factory ship is of greater displacement and bulk than the normal trawler, consequently the horsepower of the main engines is greater. Should a net become fast—bearing in mind that the nets are recovered with the vessel moving slowly ahead—the trawl winch must be able to take the sudden strain. The size of the electric motor on the trawl winch in a factory ship, such as the *Fairtry*, would be in the region of 300 h.p., but if there is any doubt in assessing the size it is a safe policy to over-power to ensure reliability.

Some owners have considered the advisability of carrying a spare motor on board, and there is an instance where one has been placed beside the main motor to form a pair. They are so linked by clutches that either can be brought into use in a matter of a few seconds if the other fails.

FACTORY DECK PONDS

The fish are shot into the 'tween deck factory ponds for sorting. These ponds are similar to those in the normal trawler except that they are in the stern of the vessel and are under cover, between decks.

The rate of catching must be governed by the amount of fish which the freezers can handle in 24 hours. But as catching may be uncertain because of bad weather or some other factor, it is necessary to store enough fish to keep the factory machines at work. For this purpose it is advisable to separate the deck ponds from the factory deck by a bulkhead because the temperature in the pond must be maintained at 32 deg. F. (0 deg. C.) to keep the fish fresh. On the factory deck a temperature of, say, 50 to 60 deg. F. (10 to 16 deg. C.) is required for the comfort of the workers.

The bulkheads in the pond should be fitted with refrigerated cooling grids to maintain an even temperature. Draughts must be excluded so the compartment and the door between it and the factory space may be insulated.

FISH WASHING MACHINES

After sorting the fish pass along the trough or conveyor belt into the factory. The first machine washes them, removing all slime, blood and dirt. There are many types of fish washing machines on the market and selection is governed by the size and type of fish the vessel is most likely to catch. A simple type is a rotating, circular tank, with continuous flowing water, which swills the fish round its periphery. The slime and scum overflow to a discharge pipe led overboard. Such tanks have been used for several years and one machine can adequately wash all the fish the freezers can handle in 24 hours, which may be in the region of 20 to 30 tons.

The fish are not damaged by the mechanical washing machine as they often are on a normal trawler because fishermen walk about in the ponds while washing the catch.

HEADING AND GUTTING

The test of efficiency of the heading and gutting machinery is its ability to do the job in one operation, cleanly and with no loss of edible flesh. As the freezers govern the amount of fish to be processed, the most economical method is to deal with the heading and gutting in batches, processing the tonnage required to provide sufficient fillets to fill the freezers.

FILLETING

Automatic filleting machines are now reasonably efficient and one or two of the more prominent makes have been fitted on board ships. They can safely handle most of the fish to be filleted. The fish are hand-fed to the machines. To guard against breakdowns, it is advisable to install tables for hand filleting. As these will not be in constant use, they may be built of aluminium alloy so they are easy to store, handle and clean.

Adjacent to the machines and tables should be placed a conveyor leading into the fish meal compartment where the offal may be stored in hoppers before processing. If it is allowed to accumulate in bins or in deep troughs on the factory deck, it becomes congested and unsanitary.

PACKING IN TRAYS

From the filleting machine the fish are conveyed on a table, or a slow-moving belt, to the packing benches. They are placed in packs or trays and weighed to ensure that the average weight is consistent.

The trays are usually selected to hold the following sizes: personal or small domestic pack, or multiples of this pack, making a large domestic or commercial pack. Multiples of the larger packs determine, of course, the size of trays which, in turn, decide the breadth of the freezers. The number of shelves are determined by the total capacity required.

Calculations should be made on all multiples to ensure the trays fit into the freezers. If trays were to get loose inside the freezer, the motion of the vessel would amplify the movement and damage the freezer and the fish.

FREEZING AND REFRIGERATION HOLDS

In view of the restricted space available, a blast freezer with finned refrigerating tubes as the shelves is a satisfactory type. The cold blast is circulated around the trays. The freezer is about 10 to 12 ft. (3.1 to 3.7 m.) in length, 6 ft. (1.8 m.) high and 3 ft. (0.9 m.) wide and has a door at each end. The machine is compact and the shelves are spaced to provide for the maximum number of trays. These are pushed in one after the other, travelling the full length of the 12 ft. (3.7 m.) shelves. The shelves are loaded at one end and unloaded at the other. It may take about two hours for one tray to travel the full length of the freezer. That is sufficient time to freeze fillets or small whole fish.

Many other types of freezers are available and the owner must select one that is suitable for handling the fish he intends catching. Bulky freezers are costly to carry and maintain. The unit must be compact and require the minimum amount of refrigerant.

The amount of pipes in refrigerated fish holds must be sufficient to ensure a steady storage temperature of –5 deg. F. (–21 deg. C.). The holds must be free of draughts and be thermostatically controlled, and so planned as to eliminate any blind spots. Protective batten should be provided to separate the fish from the pipes.

Of available refrigerants, freon has many advantages compared with brine, carbon dioxide or ammonia. The refrigerant used may, in the event of leakage or other accident, have the disastrous effect of giving the fish an off-smell. In some cases it is not possible to detect a leakage until the damage is done.

PACKING IN CARTONS

After freezing, the fillets or whole fish are tipped onto the packing tables for making up into boxes.

Assuming, for example, commercial packs are being dealt with, they would probably weigh about 14 lb. (6.4 kg.) each and measure about 18×12×2 in. (457×305×51 mm.). Four of these 14 lb. packs are then placed into a cardboard carton, giving a total weight of 56 lb. (25.4 kg. or ½ cwt. British measure).

The cartons are sealed and stored in low temperature holds. They can be transported to the holds by electrically operated conveyor belts and gravity chutes. The after-hold is situated below the factory deck and is best supplied by means of a chute from the packing tables. The chute should be of smooth polished, stainless metal and should start steeply to give the carton initial impetus. It should level out at the bottom and run the full length of the hold. This enables stops to be set anywhere along its length, adjacent to the pound being filled.

The use of a conveyor belt is a decided asset for the fish-hold forward. Accommodation is normally admidships and the cartons have to be taken through the crew's alleyway. A conveyor offers a very compact method of transport and does not interfere with the normal use of the alleyways. It starts from the packing benches and is covered to protect the crew because a vessel at sea is not always steady. It carries the cartons to the top of a chute where they travel down through insulating felt curtains. and are stowed.

INSULATED FISH-HOLDS

Particular care and attention is needed in arranging lay-out, accessability, storage, cooling and insulating to maintain a steady temperature of –5 deg. F. (–21 deg. C.) in the fish-holds. Modern weight-saving thermal insulating materials, such as glass fibres, aerated rubber or spun stranded asbestos, should be used. The floor, bulkheads, and ceiling of the holds should all be insulated to a thickness of about 12 in. (30.5 cm.) .

Opinions vary on whether the inside of the fish-hold should be covered in wood suitably varnished or sheathed in aluminium. In a normal " wet " fish-hold aluminium has considerable advantages. It is easy to clean and does not harbour bacteria. But in a factory ship the fish are already processed and in cartons. In view of this, some owners have preferred wood facing covered with insulating varnish.

All hatches to the fish-hold should be of the plug type and insulated. Size of the hatches is governed by the number of cartons to be lifted from the hold and the need to maintain the low temperature while unloading.

The entrance door into each hold should be covered with two curtains about 12 in. (30.5 cm.) apart, made of 2 in. (5.1 cm.) thick wadded felt. Many experiments have proved felt to be the best for this purpose. The openings used are about 2 ft. (0.6 m.) square. When the carton comes down the chute it pushes its way through these flaps which automatically fall back into place as it moves into the hold.

Floor, deck head, bulkheads, and ship sides of the hold are covered with refrigerating pipes, in some cases placed between sections or between the cargo.

The cartons have to be carefully stowed to avoid crushing and damage by movement of the ship. The standard size of the pound in the normal fresh fish trawler is about 3 ft. (91 cm.) in length and forms an approximate cube of this dimension. In a freezer ship the cartons are all cubical so that their mobility is reduced. Because of this the pound space can be increased to approximately 6 ft. (1.83 m.) cube or the nearest multiple of the size of the cartons. Tolerances must be allowed for easy accessibility for stowing, but not too much or the motion of the ship may cause damage. The distance between the shelves is about 6 ft. (1.83 m.) but it should, in some instances, be reduced to about 4 ft. (1.22 m.), depending on the type of fish processed. Too much weight may crush the bottom cartons. This actually happened in a freezer trawler a few years ago.

LIVER PROCESSING

One of the essential by-products is liver oil. Livers are collected in ponds and moved aft to the liver oil room, normally situated in the stern because of the obnoxious gases given off in the boiling.

In some vessels an electric pumping system has been used for transporting livers from the collecting hopper, placed near the factory deck where the livers are extracted. This is a very clean, efficient method. A pump of slow revolutions pushes the livers through. The system gives satisfactory results and does not damage the livers. The storage tank is placed above the boilers.

The livers can be cooked in a standard steam liver boiler. The installation in a ship, such as the *Fairtry*, comprises five or six boilers, each about 2 ft. (61 cm.) in diameter and 3 ft. (91 cm.) high and fed with a perforated steam coil at the bottom of each boiler. Boiling merely opens the liver cells, releasing the oil which floats to the surface. The boiler is allowed to settle and then the oil is drained off to the storage tanks.

Another method coming into use is electric centrifugal separation. The livers are placed in a shredding machine and, with the addition of hot water, reduced to a thick porridge which is put through strainers. Then it passes into the centrifuges which extract the clear liver oil. This method gives a slightly higher yield of oil than the steam extraction process, but the initial cost of equipment is higher and maintenance calls for more skilful operators.

Some vessels have been very successful in canning livers and it seems likely that the canning process will be more widely adopted, especially, perhaps, for canning small, choice portions of fish.

FISH MEAL PLANT

The fish meal plant is best sited below the factory deck, preferably aft, so that the offal can be sent down to it by chutes. Opinion differs on whether the batch or the continuous type of processing plant is the better. The decision must rest with the owner and depend on the type of fish to be caught and processed.

For normal white fishing—bearing in mind the limitations of space—the batch type seems to be most suitable because it occupies less space. If a number of hoppers are installed an almost continuous process can be organized.

The final grinding and packing of the fish meal can be done by the operator attending the plant. It is stowed in a dry wood-lined compartment adjacent to the plant.

CONCLUSION

The freezer factory ship has been developed by a very long process of painstaking advancement. Its use commercially widens the gap between big and little fishing vessels, but there is a market for the catches landed by all of them. The demand for wet fresh fish will always exist, but the factory ship makes it possible to store first class fish in big quantities.

NOTES ON FACTORY SHIPS

by

H. E. JAEGER

BECAUSE of overfishing the boats have to go to more and more distant grounds. This is the reason why freezer trawlers (Jaeger 1949) are contemplated. Such a vessel, which has been built as an experimental ship in Great Britain, is based on the consideration that the consumable part of the fish is only one third of the weight of the whole fish and if it was frozen immediately after the catch, the ship could stay longer on the fishing grounds. The time spent in fishing could be 75 per cent. or more of the voyage instead of 25 per cent. or 30 per cent. Fig. 589 shows a design of such a freezer trawler or factory ship, capable of dealing with 25 tons of fish per day during 60 days.

These 25 tons of fish yield 9 to 10 tons of fillets, 1 to 2 tons of liver oil, and 14 tons of refuse, to which can be added 6 tons of small size fish for fish meal production. From 25 tons of fish, 4 to 5 tons of fish meal can be obtained. After 60 days the ship returns to port with 500 tons of fillets, 300 tons of fish meal and 35 tons of liver oil. The displacement is 2,400 tons and the power 1,400 h.p. for a speed of 11½ knots. The temperature in the freezer is —40 deg. F. (—40 deg. C.) and the refrigerated store should be kept 0 deg. F. (—18 deg. C.).

The various equipment for filleting, for extracting liver oil and fish meal are installed under the deck. The refrigeration machinery is not very voluminous and it is placed in a chamber next to the fish-hold. The electrical installation is important as the compressors require large power to produce the low temperatures. The working deck does not extend to the full breadth of the ship and consequently a good space is available on the main deck between the gallows. It is estimated that such a factory ship could replace six or seven normal trawlers, but the large sums which must be invested, have slowed down considerably the interest in this kind of ship.

(See fig. 589 on next page.)

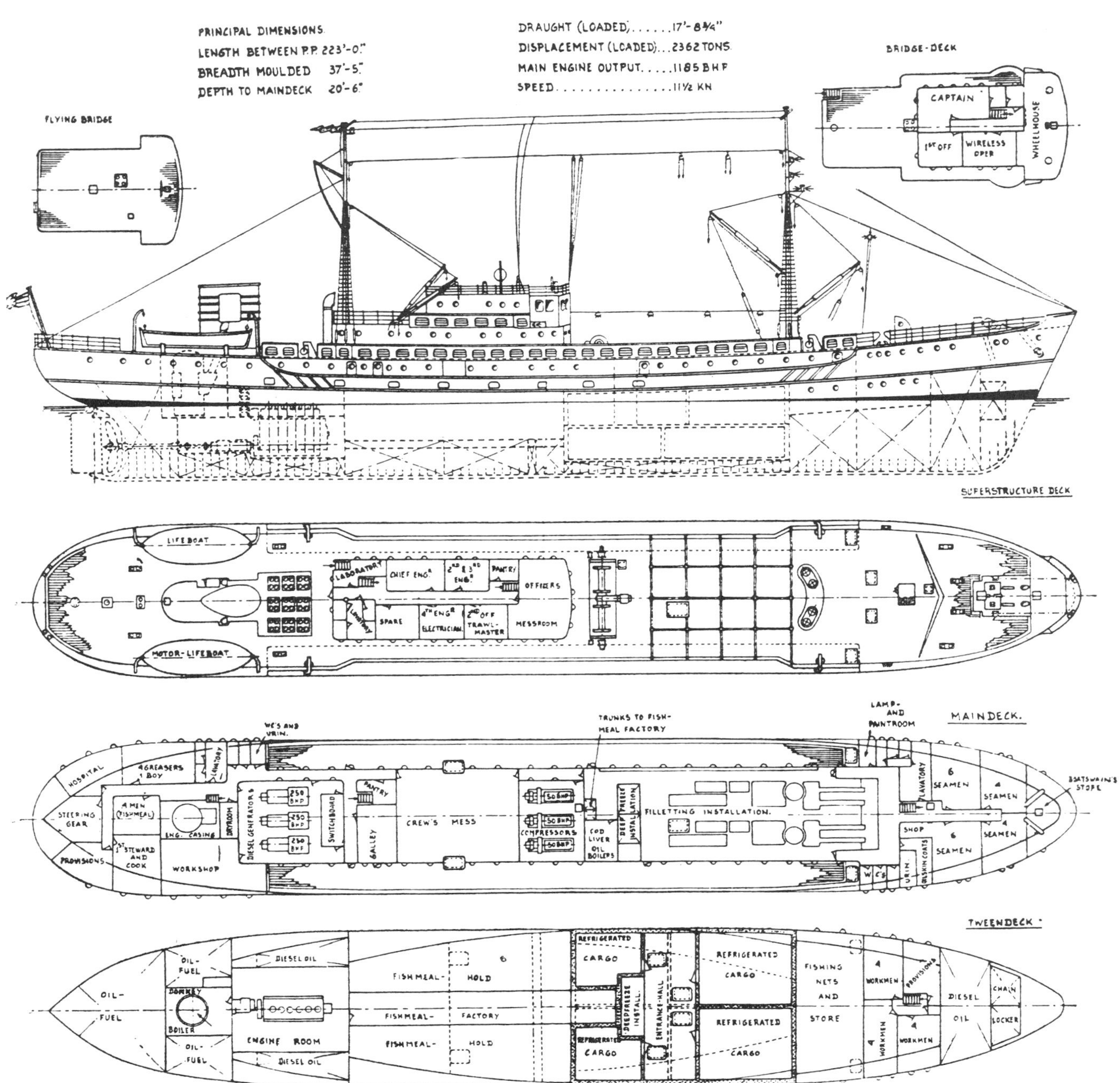

Fig. 589. General arrangement of a proposed freezer trawler

LIVER CONVEYOR SYSTEM OF A TRAWLER

by

MARIO COSTANTINI

IN 1949 the Italian fishery company, Genepesca of Leghorn, placed an order with the shipbuilders, Cantieri Riuniti dell' Adriatico, for two deep sea motor trawlers (fig. 590 and 591). They were to fish for cod on the Newfoundland banks and the catch was to be cleaned and salted on board, ready for marketing. This meant that the processing system adopted had to be capable of dealing with the fish as fast as they were caught. It was decided that, to do this the fish must always move in one direction only to avoid double handling and that labour saving devices must be used wherever possible.

ARRANGEMENT OF DECK PONDS AND GUTTING BENCHES

In the system finally adopted the cod move across the deck from starboard to port. The fish ponds and the trawl gallows are on the starboard side. The system for sorting, gutting and cleaning the fish is shown on the deck view of the general arrangement (fig. 592). When the trawl is hauled in the fish are dropped into the ponds (A) to be sorted and from there are moved to the gutting benches (3) where heads, guts and bones are removed and thrown into a trough, while the fish, after being washed in tanks (C), are returned to the ponds (A) for stacking in salt in the holds. The offal is washed aft by a stream of water.

LIVER CONVEYOR SYSTEM

Meanwhile the livers are collected in another trough and from there they have to be carried aft. In some trawlers this is done by hand which calls for a journey over a considerable distance on an exposed and slippery deck. In the trawlers under discussion this was, in any case, an impossible method because, with the limited crew available, production would outpace disposal of the livers.

This is not a new problem and attempts have been made to solve it by using conveyor belts, mono pumps and water pressure but in each case there were disadvantages, such as scalding, chaffing or breaking the livers in transit. In this instance, however, the shipbuilders, working in collaboration with Messrs. Gallieni, Vigano and Marazza of Milan, devised an hydraulic conveyor system based on the ejector principle. Fig. 593 shows the arrangement of the plant. A wooden trough (1) with running water (2) is placed at one end of the gutting benches (3). The livers are thrown into the trough and from there a stream of water conveys them to the ejector (4) where they are drawn into the jet chamber by suction. They are then carried by water through a conveyor pipe (6) to the processing room (7). The mass of water in the pipe is sufficient to float the livers so that they reach the processing room undamaged.

A swivel spout is fixed at the end of the conveyor pipe to enable the water and the livers to be discharged into either of two wire baskets in a steel tank (9) which drains overboard (10). The livers collect in the baskets and while one is being filled the other is emptied into the disintegrator.

The ejector unit (fig. 594) consists of a funnel-shaped receiver (12) and the ejector (13). The lower end of the receiver fits into a circular chamber, being in communication with it through a series of vertical slots. Air may be drawn into the chamber through a spring-loaded non-return valve. If there is a stoppage in the system the livers soon pile up in the chamber to a level higher than the vertical slots in the receiver. When this happens the suction created by the ejector draws air through the non-return valve and the vertical slots, which sets the livers moving again by stirring up the bottom layer.

Sometimes cod heads and bones, etc., get into the ejector (fig. 595) and clog it and have to be forcibly cleared. To do this the jet nozzle (15) is forced forward by means of a wheel (16) which works, through the bevel gears (17). The jet nozzle cuts and forces the obstruction into the tapering discharge nozzle (18) and there the full pressure of water from the jet clears it.

Fig. 596 and 597 show the finished installation. The capacity of the plant is about 1 ton of livers an hour. Some 30 tons of water are used at 40 lb./sq. in. (2.8 kg./sq. cm.) pressure. The conveyor pipe has an internal diameter of 4 in. (10.2 cm.) and a length of 135 ft. (41 m.).

The plant has been working since 1950 and has given satisfactory results.

Fig. 590. Photograph of the salt cod trawler Genepesca II

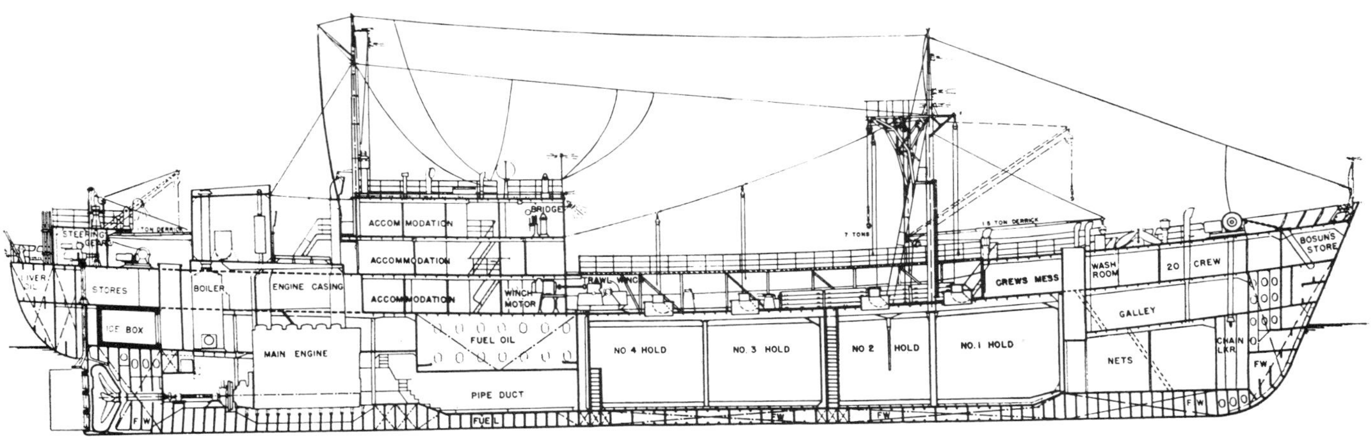

Fig. 591. Profile of Genepesca II

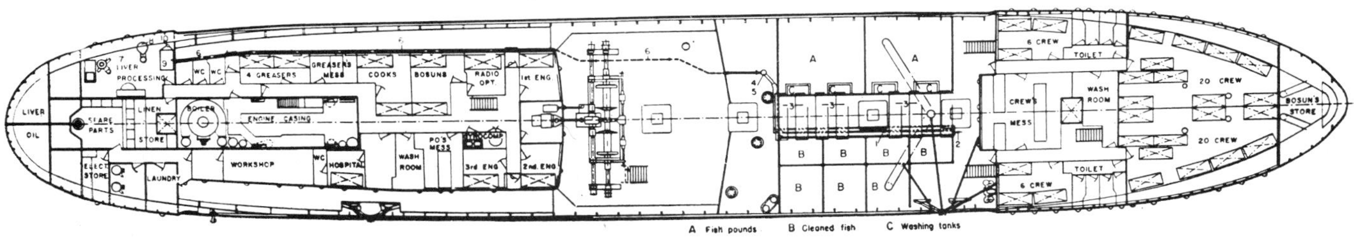

Fig. 592. Deck plan of Genepesca II. *(A) fish ponds; (B) cleaned fish; (C) washing tanks. (1) liver trough; (2) sluicing line; (3) gutting benches; (4) propelling unit; (5) ejector; (6) conveying pipe line; (7) liver processing room; (8) disintegrator; (9) receiving baskets and tank (10) discharge overboard*

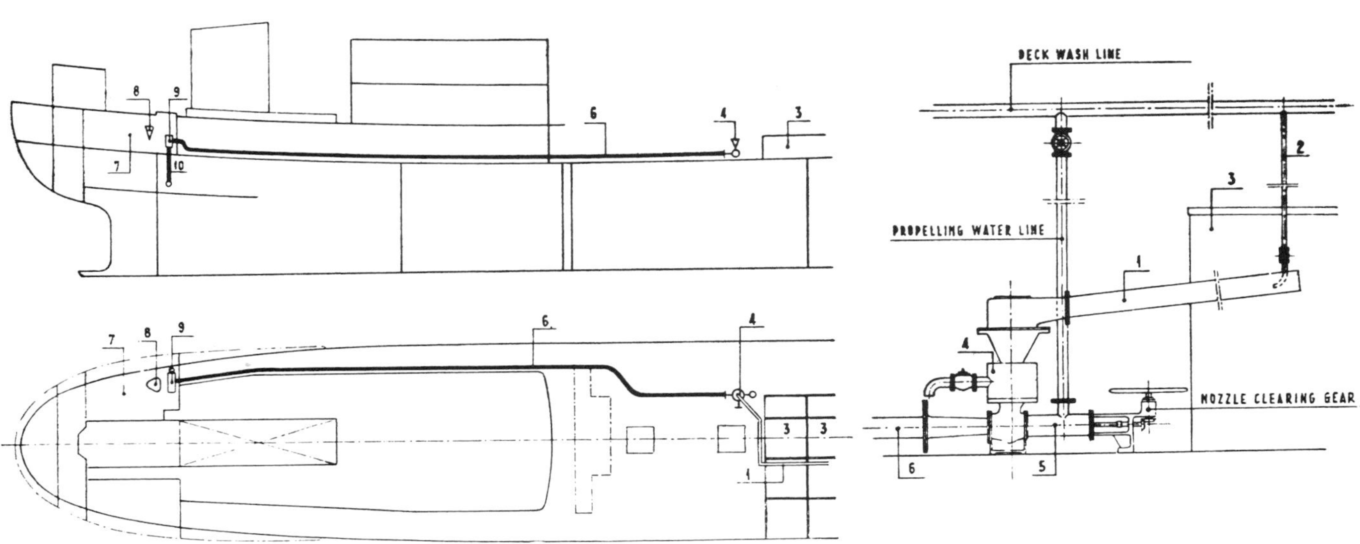

Fig. 593. Genepesca II. *General arrangement of hydraulic liver conveyor.* (1) *liver trough;* (2) *sluicing line;* (3) *gutting benches;* (4) *propelling unit;* (5) *ejector;* (6) *conveying pipe line;* (7) *liver processing room;* (8) *disintegrator;* (9) *receiving baskets and tank;* (10) *discharge overboard*

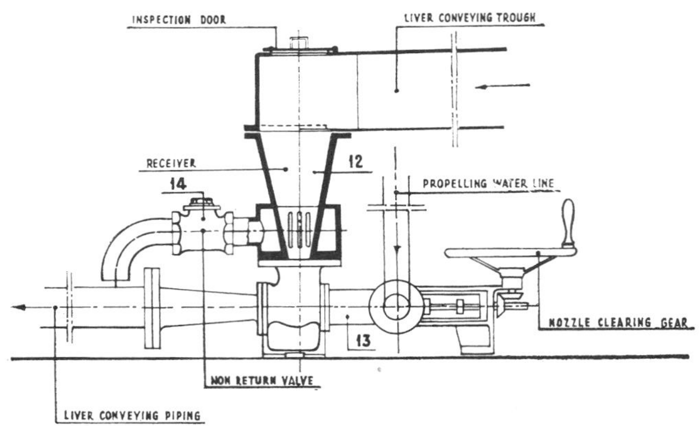

Fig. 594. Genepesca II. *Ejector unit of the hydraulic liver conveyor*

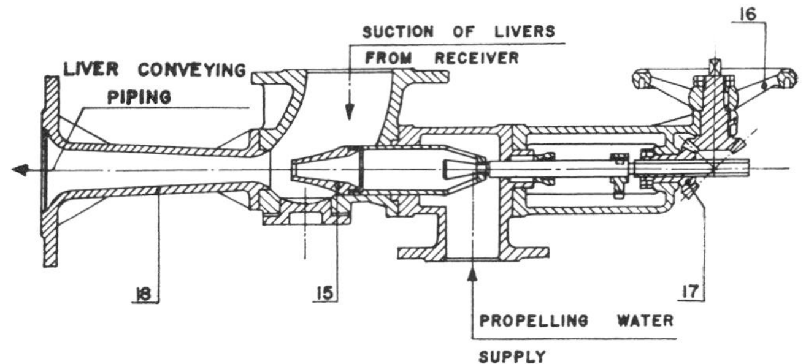

Fig. 595. Genepesca II. *Ejector with jet cleaning nozzle*

Fig. 596. Genepesca II. *Photograph of the ejector unit*

Fig. 597. Genepesca II. *Photograph of the steel tank in the liver processing room*

FREEZING AT SEA

by

G. C. EDDIE

It has been possible to put some kind of freezing plant on board fishing vessels for at least 50 years, and during the last 20 years enough has been known about the quick-freezing and low-temperature storage of fish to produce a satisfactory refrigerator fishing vessel. The real problem is matching the installation to the conditions of the fishery—and it has to be solved for each fishery. What is satisfactory for tuna, which are to be dissected and cooked twice, is not satisfactory for white fish, which may not be processed beyond freezing. Even where projects are concerned with more or less the same species, as are the *Delaware* and the current British projects, the designs differ a great deal because of the different operational conditions on the fishing grounds, the different final forms of the product, and the different processing and distribution problems on land. But there are points common to all such schemes—namely, choice of refrigeration plant, layout, process planning and so forth.

The final product from the *Delaware*, after processing on shore, will be frozen skinned fillets. It is apparently possible to produce satisfactory frozen fillets by brine freezing whole fish and storing at 0 deg. F. (—18 deg. C.) even if they are subsequently thawed and refrozen on shore. In Britain, on the other hand, much of the cod landings is distributed either whole or as steaks or long fillets, or is split and smoked, or filleted and smoked, according to a demand which varies daily. In contrast to U.S.A., British inland transport is of short enough duration for distribution against daily orders in the unfrozen state, even after non-sterile filleting. Indeed, there are not yet enough cold stores for anything else. The processors, therefore, require whole fish, and in this respect the *Fairtry* project is a new departure. But it will be many years before British public taste is converted to frozen fillets to an extent to justify an entire fishing fleet producing only or mainly such food.

In Britain, therefore, external appearance of the whole fish when thawed is important; and cold smoking (which much of the catch undergoes) is a most delicate test of the quality of freezing and cold storage technique. On the criteria of external appearance and smoke curing, and as a result of severe tests, air blast and plate contact freezing has been adopted in place of brine freezing. It is believed that brines being developed in U.S.A. and Denmark do not produce such undesirable effects on the surface of the fish and in the flesh as ordinary sodium chloride, but there are other objections to them. Again, tests have shown that fish must be gutted before freezing at sea, otherwise the blood gives them an unfamiliar appearance and they rapidly develop off-flavours. It takes as long to bleed a fish as to gut it.

Cold store temperatures of —22 deg. F. (—30 deg. C.) are currently recommended and used in the United Kingdom as compared with 0 deg. F. (—18 deg. C.) in U.S.A. Freezing always has some effect on external appearance, and it is not considered worth while to freeze fish which will be less than 10 days in ice at landing, whereas the New England, U.S.A. fleet makes entire trips of not more than about ten days. Some authorities believe the American fish, although of similar species, spoil faster at ice temperature than N.W. European fish but, again, some stowage temperatures quoted for New England and Canadian vessels show that the icing technique could perhaps be improved.

The size of fish caught by *Delaware* gives freezing times of one to three hours; one British project is concerned almost entirely with Arctic cod, which takes about four hours, so that the proportion of hold space occupied by the freezer would alter radically. Finally, the catching rate quoted of less than half a ton per hour requires, in Arctic fishing, that fish are at times left in ice for five or six days before freezing. The British are not sure if this is acceptable: two to three days is satisfactory, seven is not, and four to six are as yet doubtful, pending larger scale tests.

It can be seen how very difficult it is to say whether any particular project is a good solution to the problems it was designed to overcome, but very easy to say why it will not suit another fishery.

All schemes for freezing at sea have two objectives, namely, improvement in quality and steadying the markets. Another is to build ships which can spend more of their life on the fishing grounds, and sail at more economical speeds. This latter may in some cases require the building of an entirely new fleet. Oldershaw's remark that every ship is a potential factory ship, must be considered with reserve in the light of the following facts: *Delaware*, a trawler of 147½ ft. (45 m.) and 8,000 cu. ft. (226 cu. m.) capacity, can be fitted for the New England fishery to freeze almost 90 per cent. of the catch, whereas a 175 ft. (53.3 m.) British trawler of 13,000 cu. ft. (370 cu. m.), for the reasons given, cannot be fitted to freeze more than 25 per cent. of the catch without reducing the total catch to below the present average of 100 tons. Oldershaw unfortunately does not mention whether the *Delaware's* hold capacity is now less or greater than the average landing. There is only one British Arctic trawler on which it would be at all possible to freeze more than 50 per cent. of the catch. The best example at present of the size of ship and crew economically necessary to freeze the whole catch is the *Fairtry*.

Reduction in size and cost of factory vessels can only come through better design of freezing and auxiliary plant and better planning and mechanization of the process. The size of the freezing plant can be substantially reduced by arranging for temporary storage of fish at ice temperatures. This can be done, as mentioned, in the case of cod, for two to three days provided the frozen fish is left at —20 deg. F. (—29 deg. C.) after freezing. For example, a prolonged study of catching rates on British Arctic trawlers gives roughly the following requirements in freezing capacity expressed as weight of fish processed per hour:

Freezing all fish at once	6 tons/hr.
,, ,, ,, within 12 hours	2–3 ,,
,, ,, ,, within one day	1–1½ ,,
,, ,, ,, within two days	0.8 ,,

This raises problems of double handling, calling for novel systems of storage, unless freezing by brine in wells in the hold, as on tuna boats, is proved acceptable.

Filleting or cutting into steaks reduces the size and cost of the freezer and of the hold. Not only is the volume reduced by more than 50 per cent but, as the freezing time varies (at least for plate freezing) nearly as the square of the thickness, the space occupied by freezers may be as little as one quarter or one-eighth of that required for whole fish. In this connection the successful development in Germany of mechanical filleters for certain species is a big step forward, otherwise an Arctic trawler would require between ten and twenty filleters per shift. Conventional designs of freezers require too much labour and the use of vertical plate types is being investigated.

A saving in weight in cooling grids for the hold would be achieved by adopting the jacket system of cold store cooling, as developed in Canada. This also has the advantage that no labour is required for glazing fish or for removing frost from the pipes.

The biochemical problems of freezing at sea are by no means solved. Besides the question of buffer storage, already mentioned, there are effects associated with *rigor mortis* which require more study. The results of tests are so far conflicting.

Freezing at sea is a very complicated subject and success does not come by merely putting a freezing plant of a certain capacity on board. A great deal of thought and experiment lies behind such projects as *Delaware* and *Fairtry*—a work which has very little to do with refrigerating engineering and naval architecture in the narrow sense. Success requires not only the co-operation of engineers and naval architects but also, and above all, of biochemists, bacteriologists and fisheries experts. Very few schemes have so far had complete success because some part of this team was missing. It is necessary for the technical people to go out on the fishing grounds and record what the operational conditions are. There have been too many schemes merely drawn up in design offices. *Delaware* and *Fairtry* are not among them.

The factory ship producing domestic packs eliminates the shore processor, and this may go far to offset the present necessity to carry (relative to normal trawlers) a very much bigger crew who will get higher wages at sea than on shore for similar work. Also one or two ships may be able to catch a higher proportion of expensive fish than the fleet as a whole, and this makes it difficult to set up a firm criterion of economic success.

ASPECTS OF FACTORY SHIP OPERATIONS

by

MOGENS JUL

MUCH progress has been made in the techniques of handling and stowing fish in fishing boats, but fishing trips may be so long that the catch cannot be sufficiently preserved by icing and chilling. British trawlers make trips lasting up to 21 days, although 17 days is considered the maximum for landing fresh fish. Some maintain that fish from such trips could still be distributed and retailed in a satisfactory state of freshness if it were handled with better care ashore. While this is undoubtedly true in some cases, it is equally true that there is often room for improvement in the handling at sea. The real difficulty is that, with fishing trips up to 21 days, one is so close to the upper limit of the keeping time for iced fish that they cannot always be in prime condition when put on sale in the fishmonger's shop.

For such long fishing trips, the use of factory ships, i.e. vessels equipped for processing the catch into a non-perishable product, seems to be the answer. The idea of factory ships has long appealed to designers of fishing boats and many have been proposed and some built. Much has also been written about different ways of fitting out such boats.

All modern plans for factory ships are based on freezing, canning, or the manufacture of meal and oil. When considering factory ships, one should distinguish between four different types:

1. The mobile plant, i.e. plants installed on barges or boats but depending on shore facilities for labour, water, etc. Several salmon cannery scows fall into this group.
2. The mother ship, i.e. a completely self-contained plant aboard a ship, but which is dependent on catcher boats for raw material. Mother ships generally also act as supply boats for the catchers. Whale factory ships are in this category.
3. The factory ship proper, i.e. boats equipped for both fishing and processing, e.g. the *Deep Sea* and the *Fairtry*.
4. The fishing boat equipped for processing part of the catch, e.g. trawlers with fish meal or freezer installations.

TECHNICAL CONSIDERATIONS

Cargo Space

Nickum has dealt in his paper with the problem of the large space needed in a fish-processing vessel. This seems to be one of the most difficult things to appreciate, and Nickum suggests that one should bear in mind the appearance of the ordinary fish-processing plant, noting how much storage space is used and how much extra space has been added since it was built. All such space must be provided for in the original design of a factory ship.

One of the most frequently mentioned reasons for factory ship operations is the possibility of better use of the cargo capacity. In canning or filleting about 30 to 40 per cent. of the volume of the round fish is used, figures differing according to the kind of fish. This means, it is argued, that a factory ship should be able to land about three times more edible fish flesh.

A study has been made of figures from some factory ships. The large Boston trawler, *Cormorant*, with a 15,000 cu. ft. (425 cu.m.) cargo hold, was converted into a freezer trawler, *Oceanlife* (fig. 598) by constructing a 'tween deck with a 1,500 cu. ft. (42.5 cu. m.) fresh fish storage hold, a 5,300 cu. ft. (150 cu. m.) factory department, and a freezer cargo space of 5,720 cu. ft. (162 cu. m.). About 2,476 cu. ft. (70 cu. m.) of the fish holds were used for insulation, 'tween deck, etc. The cargo room could store 200,000 lb. (90 metric ton) of frozen packaged fillets. This corresponds to 600,000 lb. (270 ton) round fish compared with some 500,000 lb. (226 ton) or almost as much, which could have been stored in ice in the fish holds of the ship before conversion.

On the *Fairfree*, space was used roughly as follows:

	cu. ft.	*cu. m.*
Fish ponds	1,400	40.0
Fish working deck	5,600	158.5
Freezer	935	26.0
Freezer storage	13,200	374.0
Refrigerating machinery	4,160	118.0
Insulation, 'tween deck	4,930	140.0
	30,225	856.5

The total space, 30,225 cu. ft. (856 cu. m.) could store about 1,030,000 lb. (470 metric ton) of iced round fish, as compared with 1,400,000 lb. (630 ton) of round fish needed to produce the 210 tons of fillets which was the storage capacity on the *Fairfree*.

As seen from this, the *Fairfree* used about 44 per cent. of the original space for cargo while the *Oceanlife* used about 38 per cent. Not included in the figures is the additional space required to lodge the factory workers. On the *Oceanlife* this space was 1,560 cu. ft. (44 cu. m.). If fish meal plants had been provided for, still less cargo space would have been available, so it appears that this type of conversion does not materially increase the catch with which a fishing boat can deal. It may well be, however, that a ship which is carefully designed from the beginning as a factory ship, e.g. the *Fairtry*, may show more favourable figures.

Another, and much more important consideration is, that a fishing boat may often have to return to port before the holds are filled in order to land the catch while it is still fresh. Actually, in 1948/49, i.e. before her conversion, the *Cormorant* (later *Oceanlife*) averaged 76,500 lb. (35 ton) of fish per trip, compared with an actual storage capacity of 500,000 lb. (226 ton). After her conversion the *Oceanlife* (ex *Cormorant*) returned from the first voyages with fillets from 15,000 to 30,000 lb. (7 to 14 tons) of round fish. These catches were very much lower than when she operated as a trawler. This was due to labour difficulties which were so serious that the venture was eventually given up and the ship reconverted to a conventional trawler. This shows how important non-technical factors may be in factory ship operation.

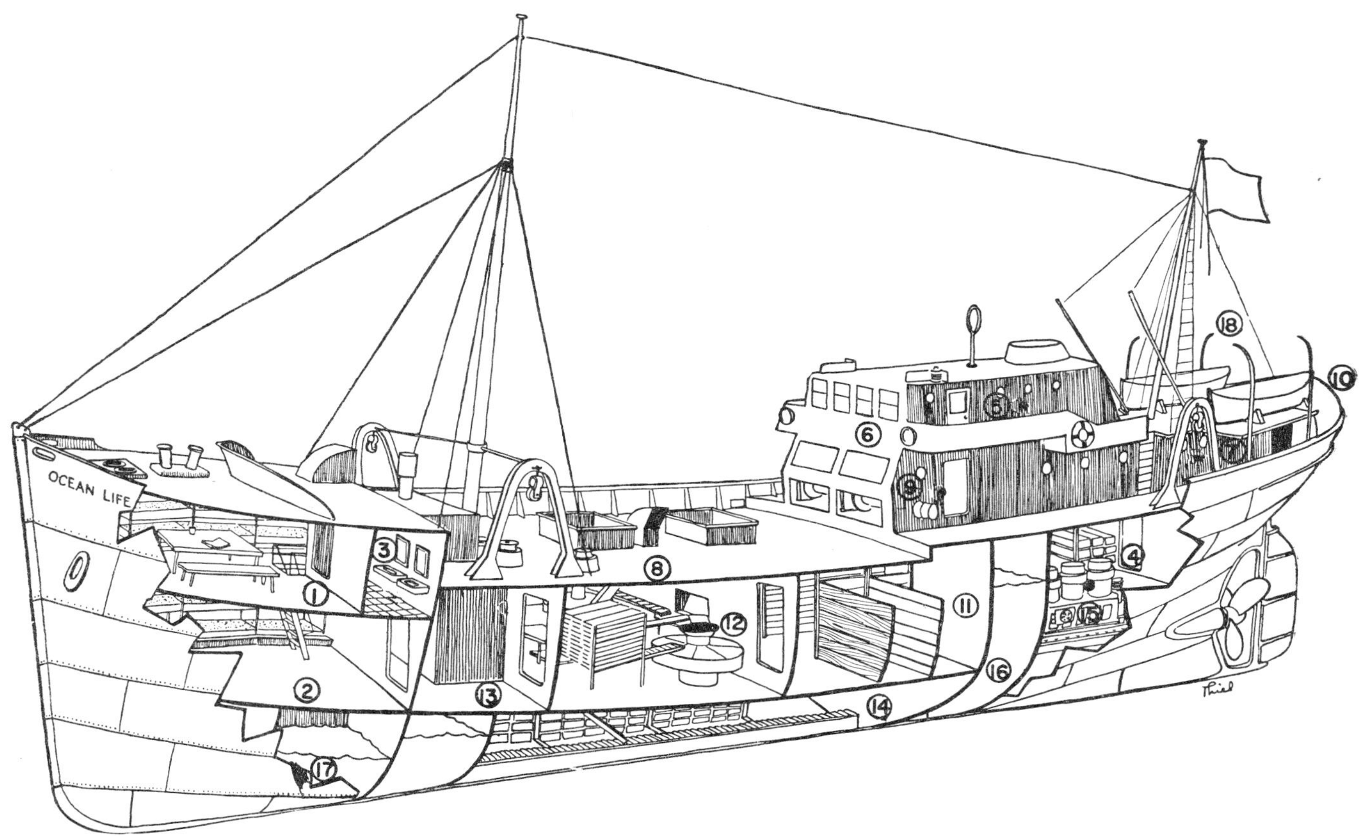

Fig. 598. Oceanlife *ex* Cormorant. *Boston trawler converted for freezing fillets. Length o.a., 150 ft. (46 m.); length p.p., 139 ft. (42 m.); beam, 25 ft. (7.6 m.); depth, 14.5 ft. (4.4 m.). (1) upper crew's quarters; (2) lower crews quarters; (3) washroom and shower; (4) officers' quarters entrance; (5) captain's quarters and chartroom; radio direction finder, echo sounder, loran, radio telephone; (6) wheelhouse; (7) galley and mess-room; (8) fishing deck; (9) trawl winch and controls; (10) fishing gear stowage and steering engine; (11) fish-hold; (12) processing room; chute from deck and fish hold; scaling table; filleting table; weighing and packing table; freezing tray racks; conveyor belts; (13) freezers; (14) frozen fillet holding room; (15) engine room; 690 h.p. diesel, refrigeration plant, generators, compressors, pumps, batteries; (16) diesel oil tanks; (17) fresh water tanks; (18) lifesaving equipment*

On the other hand, schemes for freezing fish in the round at sea are promising because the space requirements for processing equipment and extra crew are modest.

Processing Space

Factory ships are often very ingeniously designed to give just enough room for every machine and every process, but they frequently lack space for unforseen storage, stops due to mechanical breakdowns in the processing line, etc. Difficulties relating to space requirements are particularly common for processing vessels which operate as mother ships, as these have to serve a fleet of small boats with fuel, fishing gear, etc. Often their processing capacity is limited, not by the processing machinery, but by the storage space available. Reports on the operations of *Pacific Explorer* indicate that she had about enough storage space but only because she was fitted with both a canning and a freezing line. When one line was being used, the other was completely taken up for storage. The ship could not have carried on full-scale filleting and canning at the same time as was anticipated in the original design.

Lund and Kramhøft (1953) give a detailed description of the construction of the *Refrigerator* series built in Denmark for Russia. It is interesting to note how considerable space on these ships is allowed for storage of fish before processing, storage after washing and before packing, and for the storage of empty boxes, etc. The wooden boxes used for packaging the fish are manufactured aboard. On the whole, it is obvious that much experience forms the basis of the design.

A British trawler can catch up to five tons per hour. Processing installations would use excessive space and labour if adjusted to the maximum catching rate, therefore, processing vessels must generally have buffer storage of iced fish.

The freezer capacity was 0.25 ton per hour on *Oceanlife*. On the aft part of the 'tween deck there was space for about 80,000 lb. (36 tons) iced fish, equal to about 12 tons of fillets, i.e., about two to three days' supply for the freezer. From these and similar figures for other factory ships, it is seen that it is normally not possible to process the fish in a state as fresh as one might expect. Here, again, ships freezing fish in the round are the exception.

Water Supply

A major problem for the factory ship is to provide fresh water for processing and to meet the needs of the increased crew during prolonged stays at sea.

One floating salmon cannery was supplied with fresh water from barges after the vessel reached the fishing grounds. The Icelandic floating reduction plant *Haeringur* has had to operate most of the time at a pier in Reykjavik, to take fresh water aboard from the city supply.

Water supply problems are particularly important on a fish canning vessel because so much water is required for washing and cooking, for the steam retorts and for cooling the processed cans. Most floating canneries cool the cans in air. This easily leads to overcooking the can contents, and it requires large space. It has been abandoned in most shore canneries. It causes even greater difficulties at sea because frequent heavy rains, especially in Alaskan regions, make it difficult to protect the cooled cans from humidity. If the cans are even slightly humid when put in storage they rust. According to " Pacific Fisherman " (Made in America . . . 1947), the *Alma Ata*, a Russian floating cannery, cools cooked crabs in sea water and also uses sea water for initial cooling of the processed cans. The cans are then rinsed in fresh water and finally dried in air.

Operation in Rough Sea

Many processing machines, i.e. fish cutting machines, automatic scales, ice plants, etc., may cause problems because they are designed for a shore-plant having a motionless floor. On a ship there are difficulties, not only with balances, but also with inclinations, the drainage of water, and extraordinary bearing stresses, especially in the longitudinal direction of the shafts, etc. A normal fishing crew is used to work on a boat in a seaway, and can cut, wash, ice-pack and split fish in difficult weather conditions without much trouble, but not so in the case of a factory crew used to stable shore conditions. Even if they remain free from seasickness, work in cramped quarters on board ship is much more fatiguing for them than work in a shore plant.

Fishing from Factory Ships

Factory ships are, as a rule, designed to be larger than deep sea trawlers. To pay their way they must catch more fish per day, and this necessitates the use of larger gear. Fishing gear normally consists of trawls but, because of the larger size of the ships, it is necessary to devise new fittings so that the gear can be properly handled. The *Fairfree* had paravanes to open the net and these paravanes are said to have been so successful that the *Fairfree* continually outfished the other trawlers. Catches of 10 tons for half an hour's trawling were not uncommon.

Because of the necessity for a large amount of space on a factory ship, Lehmann (1953) suggested having a half shelter deck on the port side of the ship, fig. 599. This is to gain more space, and keep the trawling deck down at its normal level.

The *Fairtry* has a full shelter deck and uses a stern set trawl because her freeboard is too high for trawls to be handled over the side as they are generally on British distant water trawlers. The stern set arrangement is similar to that on a whale factory ship. Anderson (1951) mentions that a stern set trawl was used on the 150 ft. (45.7 m.) trawler *Alaska*. This trawl had about the same capacity as the conventionally set trawl on the *Deep Sea*, but *Alaska's* could not be handled in as rough weather. However, it was not exactly the same as *Fairtry's* and corresponded to those commonly used on U.S.A. Pacific trawlers.

Mooring and Transfer of Catches

A difficulty in mother ship operations is that of transferring the catch on the high seas. The success of the whale factories is in no small part due to the ease with which whales may be transferred.

Roscher (1953) mentions that the 10,000 ton German factory ship *Hamburg*, built shortly before World War II, it was intended to operate as a mother ship for a number of normal trawlers, but she was tested in this way only a few times before she was sunk. During these trials it was found that the difficulties of transfer at sea were so considerable that she had to stay inside sheltered waters on the Norwegian coast. The catchers had to go there to deliver their fish, which was not what had been intended.

Wigutoff and Carlson (1950) describe how the *Pacific Explorer*, in a crab fishing venture where she served as mother ship for ten fishing boats, found that mooring the boats was difficult.

Japanese trials in 1948–49 with mother ships in tuna fishing are reported to have had only a limited success, principally because the weather delayed transfers and caused the fish to spoil as the catchers were not equipped with refrigerating plants.

Fig. 599. Model of a proposed semi-shelter deck on a freezer trawler as suggested by Dr. G. W. Lehmann, naval architect of New York

Conventional ship's gear is sometimes considered sufficient for transferring the catch. This is the case of the ships in the *Refrigerator* series.

"Pacific Fisherman" (Cannery, 1950) describes how salmon is transferred from the scows to the *Neva* by the aid of a two compartment metal outrigger frame attached to the side of the scow. A perforated metal bottom dump box is placed by the ship's tackle in each of the compartments. Fish is sluiced into one box while the other is being hoisted and unloaded.

Roscher (1953) suggests a novel method in which trawlers use net bags 33 ft. (10 m.) long and 5 ft. (1.5 m.) in diameter. The bags are kept afloat by large air cushions and are filled with cleaned and gutted fish. When the bags are full, they are left with a buoy to be picked up by the factory ship and hauled aboard through the stern chute.

Fick (1953) feels that this method, which is inspired by whale factories, would be impractical, because it would be difficult to have a net bag lying alongside a trawler in heavy seas without it being torn and the fish being seriously damaged or even lost. He also feels that, in anything but calm weather, it would be time-wasting for the mother ship to locate the net bags, and it might prove difficult to haul them aboard.

It is very interesting to note that Roscher thinks that a ship depending on its own gear alone runs a greater financial risk in case of poor catches. On the other hand, careful experiments and calculations made by Lochridge (1950) conclude that the cost of operating a factory ship and trawlers together leads to a much higher cost per ton of fish, and that the system would be less flexible than that using the combined freezer-trawler.

Quality of Raw Material

It is generally assumed that the quality of fish processed at sea is much higher than that processed ashore. This is not always the case. While the shore processor can select on the market the type, size and quality of fish suited for its production, the sea operator has to process whatever catch he happens to make, and he has to process practically all of it, regardless of its quality or composition.

ECONOMIC AND SOCIAL CONSIDERATIONS

The cost of paying, lodging and feeding the processing crew on a factory ship is probably the most important consideration of all. Lehmann (1953) arrives at a mathematical formula for the relation between capacity of fish holds and fuel tanks, distance from fishing grounds, number of fishing trips per year, etc.; but he does not consider mechanical breakdowns, labour, or other factors, such as ordinary fishing luck; yet their influence is likely to upset the calculations.

So far as the crew is concerned, one should keep in mind the conditions under which a shore plant employs staff. Labour is hired whenever there are fish to process, and paid by the hour until the catch is treated; then the workers go home. But when the workers are taken to sea on a factory ship, they have to be paid, fed and lodged all the time, whether or not there is anything to process. A calculation on the operation of the *Oceanlife* may be of interest. Before her conversion, the *Oceanlife*, when still a fresh fish trawler, in 1948/9 seldom landed more than 135,000 lb. (61 metric ton) a trip. Average landings for this vessel during the years 1948/9 were actually little more than 76,500 lb. (35 metric ton) and she made only 2.23 trips per month. The average landings for other larger trawlers at Boston, U.S.A., during 1949 were 41 tons per trip and each boat made 2.6 trips per month.

On Boston trawlers, the crew is paid on a share basis, and investigations have shown that the deck-hands got on an average $20-22 (£7 10s. 0d.) per day at sea in the 1948/9 period. A normal trip takes about 10 days. No pay is made for the days between trips.

If a boat takes about 10 days to bring ashore 40 tons of fish and five days are used for fishing, the average catch is 8 tons per day on the fishing ground. A factory ship such as *Oceanlife* needs about 270 tons of fish to produce a full cargo of fillets (200,000 lb. or 90 metric ton), so it would take about 34 fishing days to produce them. With five days for steaming to and from the ground, the total trip would take almost six weeks. As the ship is expected to stay in port at least one week, that would mean about seven weeks for a complete round trip.

The *Oceanlife* (the *Cormorant*), as a fresh fish trawler, would have been able to make 3.9 trips in about seven weeks and catch about 300,000 lb. (136 ton) at an estimated value of \$18,000 (£6,400). If the boat had caught as much as an average Boston trawler, i.e. 89,800 lb. (41 ton) per trip and had made 2.6 trips per month, the corresponding catch in seven weeks would have been over 400,000 lb. (185 ton), valued at \$24,000 (£8,600).

The cost of the crew would be about

4.5 trips × 10 days × \$21 × 14 men = \$13,200 (£4,700).

This means that 55 per cent. of the gross income goes to direct labour. The rest, or \$10,800 (£3,900), is used for bonuses to officers, for food, ice, fuel, fishing gear and boat repairs, insurance, depreciation, etc., and what is left is profit.

Now the same boat working as a factory ship, could *theoretically* catch and process 200,000 lb. (90 ton) of fillets in 45 days, and if it were possible to get a crew for the usual pay per day, in spite of the fact that they must stay at sea during a longer continuous period and catch and handle 50 per cent. more fish, the labour cost would be:

Fishing crew:		
45 days × \$21 × 14 men	= \$13,200	(£4,700)
Factory crew:		
45 days × \$21 × 14 men	= \$13,200	(£4,700)
	\$26,400	(£9,400)

If the fillets could be sold for \$0.25 (1s. 9½d.) per lb. or \$50,000 (£18,000), the difference would be \$23,600 (£8,600). The "trawler cost" was \$10,800 (£3,900), which means that \$12,800 (£4,700) would be left to pay for depreciation of the factory, the extra food and bonuses, insurance, wrapping material, etc. This is 6 c. (5d.) per lb. fillet. If the conversion cost \$250,000 (£89,000) and we reckoned on 10 years depreciation and 5 per cent. interest, the annual cost would be \$37,500 (£13,400). The maximum processing capacity would be 1,380,000 lb. (630 ton) per annum, and accordingly the cost of depreciation 2.7 c. (2.3d.). per lb. The extra insurance would perhaps take another cent. and if, then, repairs, upkeep and other costs took as little as 2 cents (1.7d.) per lb., little would be left for the owner.

In the above calculation, it was assumed that the boat would operate at maximum capacity. If production dropped 25 per cent., the factory ship would land 50,000 lb. (23 ton) at 25 c. or \$12,500 (£4,500) less. The same would happen if the fillets should sell for 20 c. (1s. 5d.) instead of 25 c. (1s. 9½d.) per lb.

Savings or earnings, compared with the operation of the ordinary trawler, are made because as no ice is required aboard, the fish are landed filleted, and not round. Further, total fuel consumption might be less. These factors may possibly offset the extra cost items, not included in the calculation, such as provisions for the processing crew and maintenance of processing equipment.

This is only a *very rough* calculation, but it does indicate that every single cost factor must be watched very closely if such a venture is to be an economic success. In fact, it tends to confirm the opinion that factory ship operations can only be economic successes if they are doing something which cannot be done economically from a shore plant, or if very cheap labour can be employed.

When the latter is the case, conditions are much more favourable. This is assumed to apply to Russian and Japanese factory ship operations. The Russian crab processing vessels which were built in the U.S.A. during World War II were fitted with processing lines which required a very large crew. Wages paid to cannery workers on Japanese factory ships are said to be less than 3s. 6d. (U.S.\$ 0.50) per day. In addition, this type of labourer requires only modest accommodation and food, and accept long working hours.

The indirect cost of the factory crew is also considerable. Quarters take up much space and are expensive to install.

Stanley Hiller, who was engaged in the operation of floating reduction plants off California in 1935-37, states in a letter that: "There is no reason why a floating fish factory cannot be successful under the register of governments who have more liberal labour laws. But years of experience have taught me that a shore plant is more satisfactory than a floating factory because it uses local labour for both fishing and operating, and is more economical in its operations. A ship must have a large capacity to overcome the great difference in cost of operation, as there are many days in the year when it is not possible to make a catch because of the seasonal habits of most schooling fish. A ship must be used a large part of the year to cover the extra costs of its operation, whereas on shore the men can do some other kind of work during the six months the plants are idle." It is thus mainly the cost of labour which make a factory ship more difficult to operate economically than shore plants.

"Pacific Fisherman" (*Deep Sea*, 1947) describes how a special arrangement was worked out with the fishermen's unions while the *Deep Sea* was being planned and gives examples of production bonus schemes and other conditions agreed with the unions.

Similar efforts to meet the increasing requirements of the crew as to accommodation, etc., are indicated in a brief reference to the Russian floating cannery, the *Vsevolov Sibertsev*, which is said to be equipped with stores, dining rooms, libraries, barber shop, tailor shop, and numerous recreation facilities.

On the other hand, personnel considerations may ultimately make factory ships a necessity in some fishing operations. It is evident that fishermen on conventional trawlers, long liners, etc., often have accepted inferior working conditions with excessively long working hours, great risk, exposure to very rough climate, and so on, simply through force of habit. It is not likely that modern workers will continue to accept these conditions. Superior accommodation and working conditions may be one of the most important justifications for such ventures as the *Fairtry*.

EXPERIENCES WITH FACTORY SHIP OPERATION

The cost is naturally the determining factor in factory ship operation. Therefore when considering the experience gained from such ventures in various parts of the world, one should examine their balance sheets. As these are not readily available one may, for a moment, consider what might be termed the world's balance sheet of factory ship operations. Fiedler (1937), Anderson (1951) and Hardy (1947 and 1953) review such operations throughout the years. It is evident from these papers that many, probably most, operations have resulted in economic failure.

Notable exceptions have been cod and herring salting in the Atlantic, whale factories in the Antarctic, tuna freezing in the Equatorial Pacific, salmon and crab freezing in the North Pacific and shrimp freezing in the Gulf of Mexico. In making this observation it is at once evident that none of these operations could have been carried out from shore bases without great difficulties, and that this seems to be the main criterion for successful factory ship operation.

Much information on problems of this nature can be obtained from a review of the methods used in the successful operations mentioned, but whale factories will not be considered as they are beyond the scope of the present subject.

Salting Aboard

Salting aboard cod fishing vessels has been done in the Grand Bank fisheries for centuries. The U.S.A. and Canada discontinued such operations when refrigeration came into common use, and landed the fish in ice because this was preferred by the consumers, But countries further away from the Grand Banks, especially France, Spain, Portugal and, in later years, Italy, have continued the trade.

Salting of herring aboard is carried out by Russia, the Scandinavian countries and the Netherlands. In the summer season of 1950, Russia is said to have had four mother ships producing salted herring, working with 50-60 catching vessels off Iceland. In 1954, the Russian fleet operated almost the year around in the area between Jan Mayen and the Faroes. Some modern trawlers, e.g. the *Jacques Coeur*, are equipped for salting the major part of the catch and freezing the remaining part.

One of the reasons for the economic success of salting at sea is that it is a very old tradition. Fishermen are used to salting the catch and accept it as something for which they cannot expect any extra pay. But, fishermen not used to this work may refuse to do it, as happened in Denmark when it was proposed to have herrings salted aboard. The fishermen were not used to it, and were not willing to do it without such extra pay that the operation was economically impossible.

Floating Canneries

Floating canneries operate mainly on the North Pacific where they are used for canning salmon and crabs. According to Anderson (1951) several American floating salmon canneries have operated in Alaska every season and, with two exceptions, they are all dependent on adjacent shore facilities.

The most successful canning-at-sea operations appear to be carried out by Russia. It is possible that the secret of their success lies in the relatively low pay of the workers, or in state support of the operations.

In general, the tendency is to change from canning to freezing at sea. Frozen fish, e.g. salmon, is canned ashore; frozen crab meat is marketed frozen. This development is due to the fact that freezing at sea can be carried out with less personnel than is used in canning, which offsets the cost of double processing—freezing aboard, canning ashore—as in the case of salmon.

Freezer Ships

A French trawler, *Zazpiakbat*, formerly engaged in the salt fish trade, was equipped with freezing installations in 1928, and a little later an Italian company also fitted out a vessel for freezing aboard. In 1931 and succeeding years, several other French and some Greek freezer trawlers followed. The fish were in most cases frozen in brine in a revolving wire mesh drum.

A number of fish-freezing ships were built in the thirties and after World War II, but most of them have been unsuccessful. A sad example is the Norwegian filleting and freezing ship *Thorland* (Norway, 1947). She showed a deficit of about a half million Norwegian kroner (*ca.* £36,000/U.S.$100,000) in her first year of operation. This ship had an annual fillet production of 863 tons instead of the anticipated 4,500 tons, and she was later put to other uses.

It is believed that one reason for these failures may have been that the designers followed the trend in freezing fish ahore, i.e. towards the freezing of filleted, packaged fish, a process requiring a great deal of labour and special handling, which is difficult to provide aboard.

Schemes based on freezing fish in brine or other liquid have, on the whole, been much more successful. Particularly well known is the method used on tuna clippers. Putting the fish in brine is actually less time-wasting and more convenient than icing them. In this respect, the tuna clipper is different from most factory ships, and it is easy to see why this type of vessel had gained a popularity which had never been approached by any other type of fish-processing vessel. The fact that the fish freeze together in the brine tank is of no consequence, as defrosting is begun one or two days before the boat reaches port, so they are thawed out when unloaded at the cannery.

As soon as one changes the system—for instance, by introducing mother ships, such as several companies have done—complications arise. It is not convenient to have a big ship tied up while the cargo is being defrosted, one part at a time according to the processing capacity of the cannery. On one tuna mother ship, the *Oceanic V*, the catch was first kept in brine during the entire voyage but the rolling of the ship caused so much damage to the fish that she had to be converted so that the holds could be emptied and cooled with conventional coils.

It appears that the salmon industry in Alaska is giving up floating canneries and relying more and more on floating freezers in which the designers are using to a considerable extent the experience gained with tuna clippers. Some of this experience is described by " Pacific Fisherman " (Freezership Revolution, 1951). These boats have processing capacities as large as four to eight tons per hour because they all serve as mother ships.

The success of the tuna clipper has led to several experiments in freezing white fish at sea, particularly trawler-caught fish for refreezing or smoking ashore. In most of the experiments, labour-saving immersion freezing (mostly salt brine) has been used. By this method the freezing medium can store up refrigeration for quick use when a catch is made. A difficulty encountered in brine freezing is that the fish freeze together, but experience has shown that this can be avoided if fewer fish are put into the tank at one time and if the direction in which the brine circulates is changed. Another difficulty is brine penetration. This can, to some extent, be prevented by using a carefully controlled temperature close to the eutectic point of the salt brine (Holston and Pottinger, 1954). Also attempts are being made to find other freezing liquids, i.e. ethanol or sugar solutions, to avoid the undesirable taste of brine and its tendency to cause rancidity. Methods based on the use of ethyl alcohol solutions run into difficulties with regard to tax on alcohol and control of its proper use.

There is now a tendency for designers of freezer ships to turn to compact designs of contact or air blast freezers. A specially designed contact freezer was used on the *Oceanlife*. Clement (1951) states that as filleting machines, etc., would reduce the carrying capacity too much, the *Marbrouk* was fitted for freezing the fish after gutting and washing. Pans of fish are frozen on flat finned brine coils at —31 deg. F. (—34.6 deg. C.). Air is circulated above the pans at about 1,500 ft. (460 m.) per minute.

With the *Fairtry* the British have taken the lead in experimenting with freezer boats not equipped for brine freezing. The *Fairtry* uses contact freezers and combination contact-airblast freezers of the type sometimes referred to as " Murphy " freezers.

According to Lund and Kramhøft (1953) the *Refrigerator* series has air blast tunnels with a total capacity of 50 tons of fish per 24 hours, at an evaporator temperature of —35.5 deg. F. (—37.5 deg. C.). This is expected to give complete freezing in four to five hours, reducing the temperature in the interior of the fish to 0 deg. F. (—18 deg. C.). The coils are placed below, and can be completely shut off from the tunnels. They are defrosted with hot refrigerant and heating through a steam pipe placed in the water trays below the coils. The connection between the freezer tunnel and the coils is closed every time the tunnel doors are opened to prevent the entrance of large amounts of warm air. The fish are packed in trays. The trucks, which take 18 trays each, are hung on overhead rails attached with four wheels so as to prevent longitudinal swinging. They are moved by a chain. Whenever a truck has to be moved, a bar on it can be shifted up and engaged on the chain. The fish are loosened from the trays in the glazing process, after which they are placed in boxes and taken to the storage rooms. The latter are maintained at 0 deg. F. (—18 deg. C.), and are cooled by brine in $1\frac{1}{2}$ in. (3.8 cm.) coils. The capacity of the holds is 46,400 cu. ft. (1,315 cu. m.).

The Torry Research Station, Aberdeen, Scotland, has developed a plate freezer which will take big Arctic cod and is suitable for installing in any large trawler. Present conditions in the fishing industry of the U.K. were kept in mind in developing this equipment. There are a large number of well-built trawlers representing a huge investment, and every effort is therefore made to use that fleet to the best advantage as, for instance, by developing freezer equipment capable of being installed in these ships. Similarly, the large demand in the British market is at the moment, and will be for a long time to come, for fresh or smoked whole fish, therefore, the freezer in demand is one which is adapted to such a production.

Some of the most promising fields of freezing at sea have been in the crab, shrimp and crawfish fisheries. Crustaceans spoil quickly and, therefore, should be frozen very soon after capture; and, as they are highly priced products, they can well carry the added expense.

Freezing shrimp at sea takes place in the Gulf of Mexico, some on independent catching units and some on mother ships. Some of the latter are converted naval salvage tugs, and they also supply ice and fuel to the catchers. This activity has been taken up partly because operators from the U.S.A. find considerable legal and technical difficulties in setting up processing plants in the other countries with shores on the Gulf. Recent experiments carried out by the U.S. Fish and Wildlife Service have, according to Dassow (1954), demonstrated that it may be feasible to freeze shrimp at sea in much the same type of equipment as is used on the *Delaware*.

The king crab of the North Pacific is frozen at sea by Russian and American ships. The fishing grounds are rich, but are so far from any shore base that factory ship operation is the only suitable method of exploitation. The *Deep Sea* and other ships are engaged in this trade.

A third type of activity is carried out in the Southern Hemisphere. Boats from France are venturing down to the small groups of islands southeast of Madagascar, and returning after five or six months with frozen crawfish tails. The *Cape Horn*, based in Valparaiso, is carrying out similar operations around the Juan Fernandez and other island groups west of Chile.

Processing Part of the Catch

Many Icelandic, German or British trawlers freeze the prime fish and take the rest, generally the larger part of the catch, home in ice; French, Italian or Greek trawlers often salt the remaining fish. Operations of this kind require only a small processing crew and very limited factory installations but, of course, production is low.

Fish Meal Installations

The last type of processing aboard is that of producing fish meal and oil. The ships manufacturing whale meal and oil in the Antarctic confirm the general principle that factory ships can be economically successful where shore-based plants are not feasible. Whale factory ships have seldom been operated in waters other than the Antarctic. Even so, shore-based plants in Chile, on the South Georgia Islands, and in Australia, seem to compete successfully with them, although they can exploit whales from only a very limited area.

The whale factory ship also confirms the principle that, in order to be really efficient, a factory ship should be designed specifically for the particular operation she is to carry out, especially if she is to operate on the high seas and not in sheltered waters. An advantage in the operation of whale factory ships is that they have no mooring problems.

Some nine fish reduction boats, serving as mother ships for fleets of purse seiners, operated off the coast of California in 1935-37. Their appearance was due, in part at least, to State laws which put a tax of about U.S.$1.00 (7s.) per ton on pilchards used in the manufacture of meal in the State, i.e. ashore, and also limited the quantity of pilchards that could be processed into meal.

Mr. Stanley Hiller (in a letter) states that their operations were prohibited by law, but what really destroyed the industry was the labour regulations which finally forced the operator to have a separate crew to operate the processing plant. Hiller recalls that the first plant was operated with 36 men in three eight-hour shifts. When the ship anchored and started to receive the fish, the sailors became fish handlers, the engineers became operators, foremen became superintendents, and the captain plant manager. However, in 1937 it took about 96 men to do the same job. The payroll became so high that it was impossible to continue the operation, but Hiller thinks that such boats could be used in countries having different labour union regulations.

Hiller took one of the boats to Newfoundland and Labrador, where it operated very successfully for a season but the outbreak of World War II prevented continuation of such activities in these waters. In any case, the U.S.A. had such need for cargo vessels at the time that all the floating reduction ships were reconverted.

Many of the floating reduction plants operated in Alaska are in effect mobile plants, i.e. dependent of shore facilities.

The Icelandic herring reduction ship *Haeringur*, was a rather unsuccessful experience, but better results have been obtained with the *Clupea*, fig. 600, an L.S.T. converted into a floating reduction plant in Norway. She operated very successfully off Esbjerg, Denmark, during 1952-53 when the catches of herring exceeded what could be processed by the shore plants in Esbjerg, but by the beginning of the 1954 season new shore plants were in operation and she was no longer used in this fishery. The *Clupea* did not operate on the fishing grounds as originally visualized, but was anchored in sheltered waters. She must have been profitable, as the owner converted another smaller ship, the *Bras*, into a herring reduction boat.

In recent years a great many trawlers have installed fish meal plants for processing offal and trash fish, and there may even have been some over-enthusiasm for such installations. Meyer (1952) has calculated the percentage of offal in relation to the weight of round fish to be about 12 to 15 per cent. An average German trawler producing 2,500 tons of fish annually would produce about 50 tons of fish meal valued at DM.30,000 (£2,700; U.S.$7,500) which would only pay running expenses and depreciation of the fish meal plant. This indicates that fish meal plants should be installed only on trawlers more than 175 ft. (53.3 m.) in length. The installation of fish meal plants aboard trawlers has been described in "Allgemeine Fischwirtschaftzeitung" (Fischmehlfabrik, 1952) and in "Ship and Boat Builder" (Fish Meal, 1952).

Some people think that factory ships should use all the various parts of the fish processing them into foods, pharmaceuticals and other products. However, the quantities available of some parts are usually too small to make processing profitable; in any case, to produce several different products is contradictory to the need for extreme simplicity in factory ship operation. The problem of the manufacture of fish meal aboard trawlers is an example of this.

GENERAL OBSERVATIONS

Many consider factory ships an efficient tool in the development of fisheries in technically under-developed areas. After a considerable number of studies of this problem, FAO has concluded that it is very difficult to find persons in under-developed areas with the necessary experience in fishing, maintenance of ships and machines, and management. Yet it is not inconceivable that a mother ship, carrying a large number of dories or similar small boats, might be used in such cases. Russian ships, e.g. the *Refrigerator* series, have frequently used this principle; a similar example is the *Silver Lord*, fishing with long lines off Mauritius.

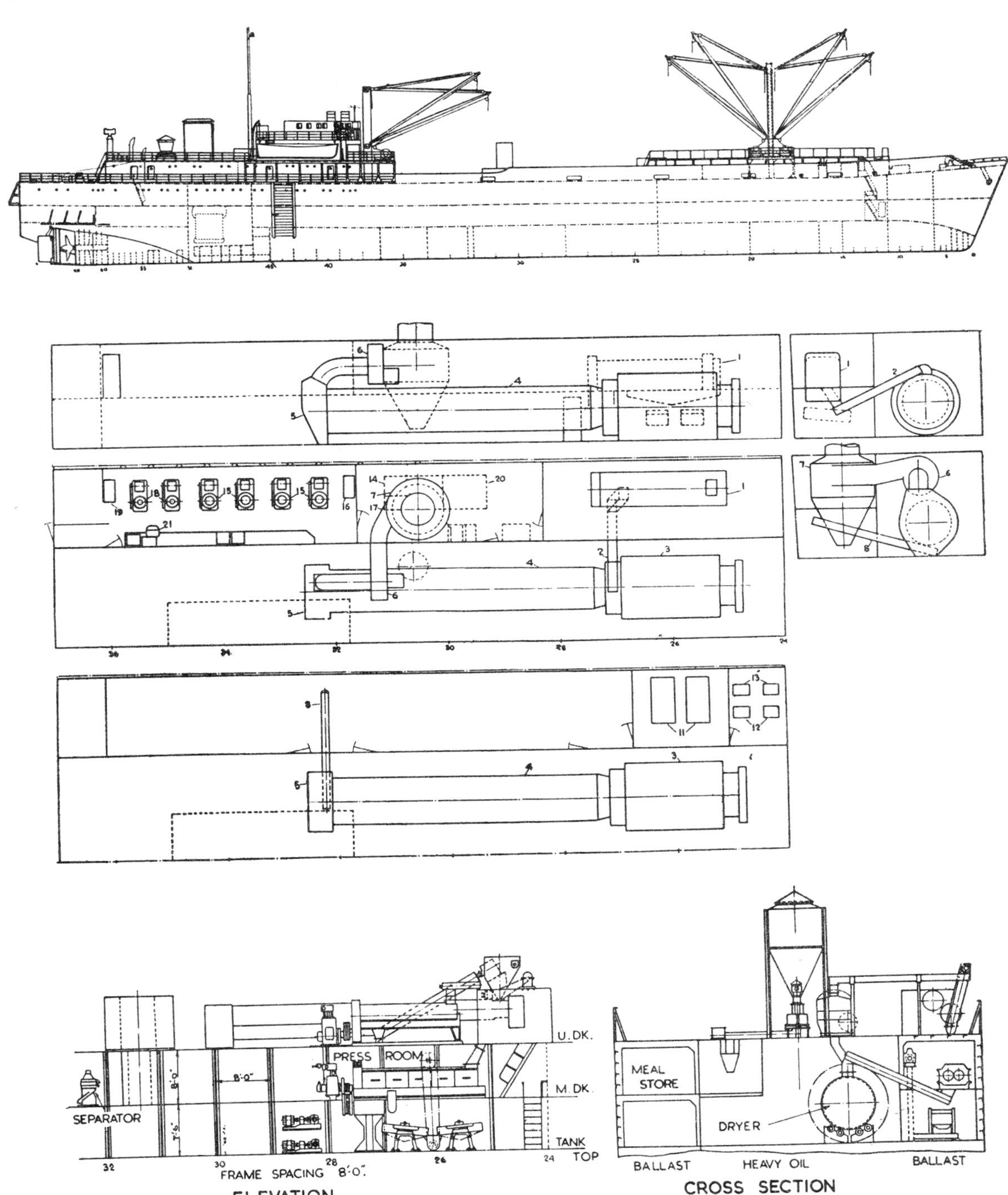

Fig. 600. Arrangement and details of processing equipment in the Norwegian fish meal factory ship Clupea, *a converted L.S.T.*

In such cases, however, one would probably have to export the product or limit the processing aboard to salting. The reason for this is that a system for frozen fish distribution is not likely to exist in technically under-developed areas, and the purchasing power of the population is not likely to justify the production of either frozen or canned fish.

It has sometimes been suggested that freezing aboard is particularly desirable in the tropics, because high temperatures make the fish spoil quickly. However, the amount of refrigeration required for freezing would always greatly exceed that required for keeping the fish chilled with ice, and in a country such as Venezuela, for example, where the climate is very hot on the coast, excellent results are obtained by icing the fish in insulated holds.

A factory ship of the mobile plant type may be useful where large landings of fish occur irregularly and over a short period. Examples are, the use of the *Clupea* in Esbjerg, the *La Merced*, or several of the salmon freezers in Alaska.

Most references to the *La Merced*, the *Deep Sea*, the *Clupea*, and the *Fairtree-Fairtry* ventures refer repeatedly to the usually efficient management under which these boats have been operated, an essential factor in successful factory ship operations.

Factory ship operations are most likely to be successful when they exploit a resource which cannot be brought to a shore processing plant. Only small parts of the oceans are now being exploited, and one must hope that factory ships may eventually find and use fish stocks not at present available to man.

FACTORY SHIPS — DISCUSSION

U.S. PACIFIC COAST PROCESSING VESSELS

Mr. James F. Petrich (U.S.A.): On the tuna clippers, after the tuna has been initially frozen, the brine is pumped overboard or into another well and the catch is maintained dry frozen at 0 deg. to 10 deg. F. (−18 deg. to −12 deg. C.) until a few days before the clipper arrives in port. At this point the refrigeration is switched off and the well is filled with circulating sea-water which gently thaws out the fish so that they are ready for immediate despatch to the cannery upon landing.

A similar method is used in the Alaskan salmon fishery. Salmon is more sensitive to salt penetration and salt is more easily tasted in them because they are canned complete with skin and bones. The problem has been to devise quicker freezing to avoid excessive salt penetration. It was found, for example, that if a well was filled with salmon, the fish, because of their smaller size, packed tightly, the brine could not circulate effectively and the fish were spoiled. This difficulty has been overcome by partly filling the well with fish so that the brine can circulate effectively and freeze the fish faster. The brine is then pumped out and the salmon are transferred to a dry well. This involves extra handling of the catch but the method is successful.

Salmon freezing ships also use tanks on deck in which brine race coils are fitted. The fish freeze satisfactorily in 6 to 12 hours, then the brine is pumped out and the fish are stowed in refrigerated holds.

These successful methods of handling salmon have been developed in the past few years and tests have shown that the fish delivered frozen by the ships to the canneries in the Puget Sound area often yield a superior product than the fish canned in shore factories in Alaska, especially when there is delay in landing and canning.

Mr. George C. Nickum (U.S.A.): Mr. Petrich has described the salmon freezers which are now being built. The reason for this development is an economic one, because the ships will operate in Bristol Bay where the catching season only lasts about 20 days. The Bristol Bay salmon have traditionally been canned in large shore-based canneries and the fish have always come in great gluts. The expense of running the shore canneries has increased enormously and owners are now combining their factories in an effort to cut down costs. A lot of this cost arises from the fact that the companies have to fly in equipment and men, a journey of 3,000 miles from Seattle. Mechanics have to be flown to the canneries months before the season opens to ensure that once the fish start coming in, there is no hold-up. It is obviously uneconomic to-day to fly upwards of 600 men to the site and to maintain a complete establishment of this kind. It is much cheaper to have a ship which will anchor in Bristol Bay, freeze the salmon as it is caught, and take it down to Seattle. The salmon can be held in the ship in prime condition until the cannery there is ready to take them.

Once there were a number of floating salmon canneries but now there are only two in service. Most of the floating canneries were built and operated in the past by salmon operators who had not arrived in time to get good sites near fishing grounds. Of the two cannery ships now operating, one is very successful and the other unsuccessful. This difference in results has nothing to do with the installations on board the ships. The successful ship is run by a very shrewd and intelligent man who moves his boat from place to place to get full benefit of the seasons and the various qualities of fish available. In other words, he makes excellent use of a ship's great advantage—mobility. Salmon is a very valuable product and if, by being on the scene with a floating cannery at the right moment, the ship is enabled to can 30,000 cases of fish, then upwards of $700,000 (£250,000) are earned. The ship is an old three-masted schooner, built probably 50 years ago, with a 600 h.p. diesel. She has probably made more money for her owner per foot of length or per ton of displacement than any other ship afloat.

The unsuccessful salmon cannery-vessel is owned by a salmon fishing company. It was designed to be based at one site where there was a good regular catch of salmon, but this site had to be shared with another cannery, with the result that there has not been enough fish to keep the ship's cannery busy.

The capacity of the refrigeration plants in modern salmon freezing ships is from four to eight short tons an hour. Means should be developed to even out the flow of fish to the freezing or canning plants. This has been neglected because of the concentrated effort to speed up processing. For example, in the Alaska salmon fishery the climatic conditions are such that fish can be held for at least twelve hours in ponds with circulating chilled seawater or ice. One reason why brine freezing is so popular among American fishermen is that it enables them to store fish for later processing. However, this can also be done by using ice. A tank full of sea-water at 28 deg. F. (−2 deg. C.) or brine at 10 deg. F. (−12 deg. C.) has a large capacity of immediately available refrigeration.

The flow of fish for processing has been regulated by one Alaska salmon fishing company through the use of a 202 ft. (62 m.) ex-naval landing craft, fitted with two ice-making machines capable of producing between them 20 short tons (18 tons) of ice a day. The machines are located over a large insulated compartment built on the deck aft. Forward of the compartment is the iron chink, the machine for heading and gutting salmon. The ex-landing craft operates in conjunction with a ship fitted out as a floating cannery. The fishing boats bring their catches to the ice-making vessel to be boxed and iced in the conventional manner. They can be kept for

72 hours in prime condition for delivery to the canning ship according to an agreed schedule. As a result of this " regular flow " scheme, the output of the canning ship has been substantially increased.

There were once several tuna freezers operating from the Columbia River area. The *Saipan* and *Tinian* were two of the best known of them. One has been sold to Japan but the other is still in operation. They were economic for their operator when they were built but the time came when the industry could buy tuna from the Japanese at less cost than they could fish it off South America, so the vessels were converted into tuna packers and cold storage ships. The vessel bought by the Japanese will, however, be used for a different purpose.

powered vessels or barges and they seem to have operated with economic success for years where larger vessels used for the same purpose have failed.

OCEANLIFE

Mr. I. Bromfield (U.S.A.): The *Oceanlife* was converted to a filleting freezer trawler because the New England fishing boats were having to go further and further afield to make catches. Going to the Grand Banks or Western Banks meant that they had to steam about six days, and when they got there they

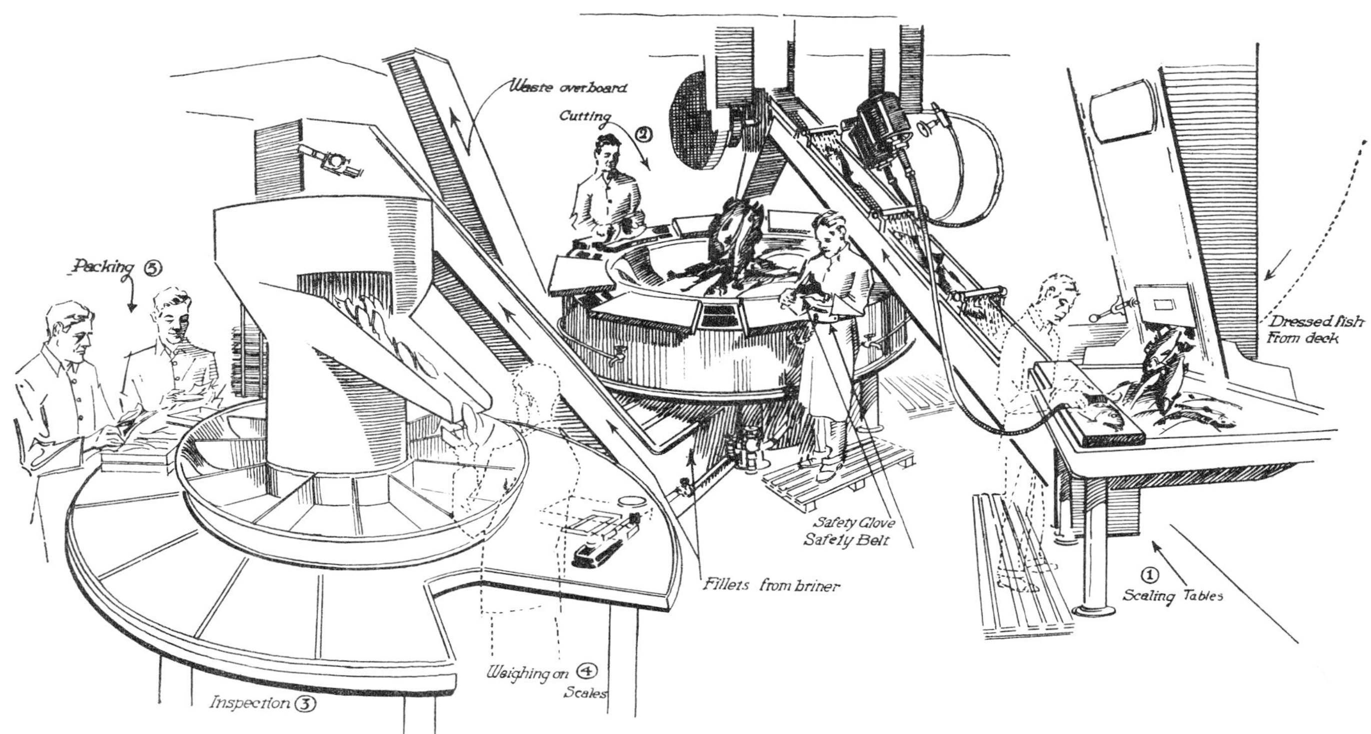

Fig. 601. Arrangement of the processing room of Oceanlife

Freon seems to be the popular refrigeration agent in Europe. In the United States, especially on the Pacific Coast, ammonia is generally used, probably because the big shore-based refrigerating industry uses ammonia. The danger arising from the use of ammonia on board ship has been greatly overrated; the fact that ammonia is soluble in water means that the best way to get rid of an ammonia leak is to use water. Another reason why ammonia is popular among United States fishermen is the general use of an evaporator in contact with water in brine freezing. A water leak is less damaging to the equipment in an ammonia plant than it is in a freon plant. Further, big quantities of refrigerants are used. For example, about 60 short tons (54 metric ton) were used in the system on the *Pacific Explorer*. If that had been freon at $1.00 (7s.) per lb. and had been lost, the cost would have been enormous (U.S.$120,000; £43,000).

Mr. H. C. Hanson (U.S.A.): On the West Coast of the United States several operators use mobile plants installed on small

had to catch fish quickly because their ice was depleted. They were able to fish only for about two to four days and then had to return, otherwise the catch would be spoiled.

From the point of view of design, the main problem in the *Oceanlife* was to find the right proportion between the catching and processing capacity. She has a length of 150 ft. (46 m.), a beam of 25 ft. (7.6 m.) and a depth of 14.5 ft. (4.4 m.). As an ordinary trawler she carried about 500,000 lb. (226 metric ton) of iced fish. In the conversion an upper deck was used for the processing machinery and the lower deck for storing the product. It took more than a year to design machinery to fit the space available. For example, it was necessary to design a completely new plate freezer because there was no suitable machine available on the market. An absorption type plant was selected for refrigeration because of (1) limited space, (2) the light weight of the plant, and (3) the small amount of maintenance it required. The plant needed only two small circulating pumps and a steam boiler.

The filleting process (*see* fig. 601), started with the scaler, and then the fish, having been washed on the way, were carried by a conveyor to a rotating filleting table. After processing, the fillets were washed through a brine tank and carried by another conveyor to a rack which held about 15 lb. (6.8 kg.) of fish, the capacity of one shelf in the plate freezers. It took almost three hours to freeze approximately 1,500 lb. (680 kg.) of fish in 5 lb. (2.27 kg.) packages, 2½ in. (63 mm.) in thickness. After freezing, the fish were packed into cartons of 50; the cartons were sealed and slid down to the lower hold where they were securely packed. The hold had a capacity of about 200,000 lb. (90 metric ton) of processed fish. The temperature of the plate freezer was about −30 deg. F. (−34 deg. C.) and the temperature in the hold was maintained at −1 deg. F. (−18 deg. C.).

In the aft part of the processing deck there was space for about 80,000 lb. (36 ton) of fresh fish. This was provided so that if the catch was too great for the cutters to handle, part of it could be stored until they could fillet the fish.

When *Oceanlife* was being converted there were many problems to be solved. The fishermen, for example, considered it impossible to cut fish at sea because of the motion of the vessel so that belts, similar to those used by window cleaners, were fitted to the filleting table. A stainless steel mesh glove, leaving two fingers free and the rest of the hand protected, was designed. But experience showed that no matter how rough the weather and how much the boat pitched and rolled, if the men could fish, then the cutters were able to carry on with their filleting and they didn't need to use any of these devices.

The machinery and equipment on the *Oceanlife* gave no trouble at all, but there were many difficulties with the crew. The fishermen were always willing to work on deck, but the cutters—experienced men from the shore installations—were supposed to work six hours on and six off. Fishing is always uncertain and sometimes, perhaps, three or four hours would pass before fish were located and caught; then the cutters might refuse to work the next six hours on the pretext that they had been ready to cut fish but there had been none available. So the fishermen might bring the catch on board and it might have to lie there six hours before the cutters would touch it, by which time it had lost quality.

These circumstances made it impossible to operate the vessel. Later two red fish filleting machines were installed on deck in an effort to overcome the problem; they did an excellent job. On one trip a catch of 30,000 lb. (14 ton) of fish was dealt with in less than 10 hours. Unfortunately, on that trip difficulties arose with the processing crew and the vessel returned to port without shooting her nets again. That was the end of the experiment.

The fillets processed in the *Oceanlife* were excellent. The U.S. Fish and Wildlife Service compared a 5 lb. (2.27 kg.) package of them with a similar package of frozen fillets processed on shore. The report of the analysis showed that the *Oceanlife* fillets were rated at 90 per cent. compared with 60 per cent. for the fillets which had been conventionally processed on shore. Not only was the condition of the *Oceanlife* fillets better, but the colour was pure white. The Atlantic and Pacific Tea Company, after making tests in their laboratory, were so impressed by the good colour and the excellent cooking of the fillets that they were ready to pay a special price and launch an advertising campaign to support a marketing scheme if sufficient and regular supplies could be guaranteed. They felt that the *Oceanlife* fillets could be sold in considerable quantity in the Mid-West of the United States where fresh fish generally arrive in poor condition.

DELAWARE

Prof. H. E. Jaeger (Netherlands): It is not economic to make a trawler so big that her fish-hold is larger than 17,700 cu. ft. (500 cu. m.). Above this size, only a deep-freeze trawler has a chance of success. The freezing trawler *Delaware* described by Mr. Oldershaw is too small and does not take full advantage of the deep-freezing possibilities.

Mr. G. C. Eddie (U.K.): It is difficult to understand, especially on a motor trawler, the choice of an absorption refrigeration system, in which the horse-power demand ranges only from 50 to 100, because efficiency is not of prime importance but space is. It is difficult to believe that a boiler and its fuel and an absorption plant are smaller than a diesel generator and compressor plant. Experience of similar lay-outs suggests that freon compressors of similar capacity could go into the same place and be more accessible for servicing and removing, as well as less dangerous. The overload capacity is illusory because the evaporator must be big enough to take the maximum overload without the tube metal temperature falling below the freezing point of the brine. The capacity is in the end fixed by the boiler and the compressor plant can be made flexible in capacity.

Mr. J. F. Puncochar (U.S.A.): The *Delaware* is not strictly a factory ship; she is an experimental vessel for developing methods of handling and processing fish at sea. Mr. Eddie raised the question as to why an absorption unit was selected for the *Delaware*. The answer is simple. It was the cheapest unit available at the time. The second reason is that fishermen do not like complicated machinery. Mr. Eddie also raised the question as to space requirements for the boiler which provides the energy used to run the refrigeration machine. The boiler was already in the ship. It was enlarged from 15 h.p. to 25 h.p. and used to supply heat to the vessel as well as to run the refrigeration plant. A diesel generator and refrigeration compressor would have taken more space than the boiler.

Fish frozen in the freezer designed for the *Delaware* come out very straight so they pack neatly and closely into the ship's hold. The *Delaware*, as she is now fitted out, can bring in 50 per cent. more fish than normally done by a New England trawler of her class. The average load of a large New England trawler is about 100,000 lb. (45 ton). The present capacity of the *Delaware* is 130,000 lb. (59 ton) of frozen fish and 15,000 lb. (6.8 ton) of iced fish. The old hold had a volume of 8,000 cu. ft. (226 cu. m.) of which 1,300 cu. ft. (37 cu. m.) were taken for the refrigerating machinery and about 700 cu. ft. (20 cu. m.) for the freezing equipment. If the refrigerating machinery had been placed in some other part of the vessel, the capacity of the *Delaware* would have been about 185,000 lb. (84 ton). This was not done because it was not essential for an experimental vessel.

Mr. G. C. Eddie (U.K.): Oldershaw's stowage rate of 33 lb./cu. ft. (525 kg./cu. m.) if it includes shelves, stanchions and boards, disagrees with British figures for almost rectangular, solid slabs of frozen cod, which, with the shelves, gives a figure of 28 lb./cu. ft. (448 kg./cu. m.). The figure for fish stowed in ice, including shelves, is 32 lb./cu. ft. (512 kg./cu. m.). There is also the point that fish frozen in brine come out in awkward shapes. Brine also makes it difficult to glaze the fish for cold storage if they must be held for any time on shore, and brine also produces corrosion problems. It is, however, the fastest form of freezing apart from the very low temperature air blasts of −50 deg. or −60 deg. F. (−46 deg. or −51 deg. C.) now being used in British Columbia.

The Torry Research Station is engaged in experiments regarding bulk freezing to be used aboard trawlers or ashore for pelagic fish. The Station feels that, with the present marketing system in the U.K., there is a greater demand for frozen fish from the fish processors than from the consumers, therefore the freezing method should be adapted to fit the processors' needs. Contact freezers are used with slabs 15 to 100 lb. (7 to 45 kg.) in weight so as to require very little hand labour.

Mr. Mogens Jul (FAO): Mr. Eddie expressed surprise in his paper to see that in the U.S.A. the storage temperatures used are around 0 deg. F. (−18 deg. C.). In the U.K. −10 deg. to −15 deg. F. (−23 to −26 deg. C.) is common. There appears to be some confusion over this matter as an American refrigeration expert recently commented that he was very surprised to see that in Europe they used temperatures of around 0 deg. F., while in the United States freezer storage temperatures far lower than that were in use. It seems that there is a variation in freezer storage temperatures from plant to plant both in Europe and in the United States.

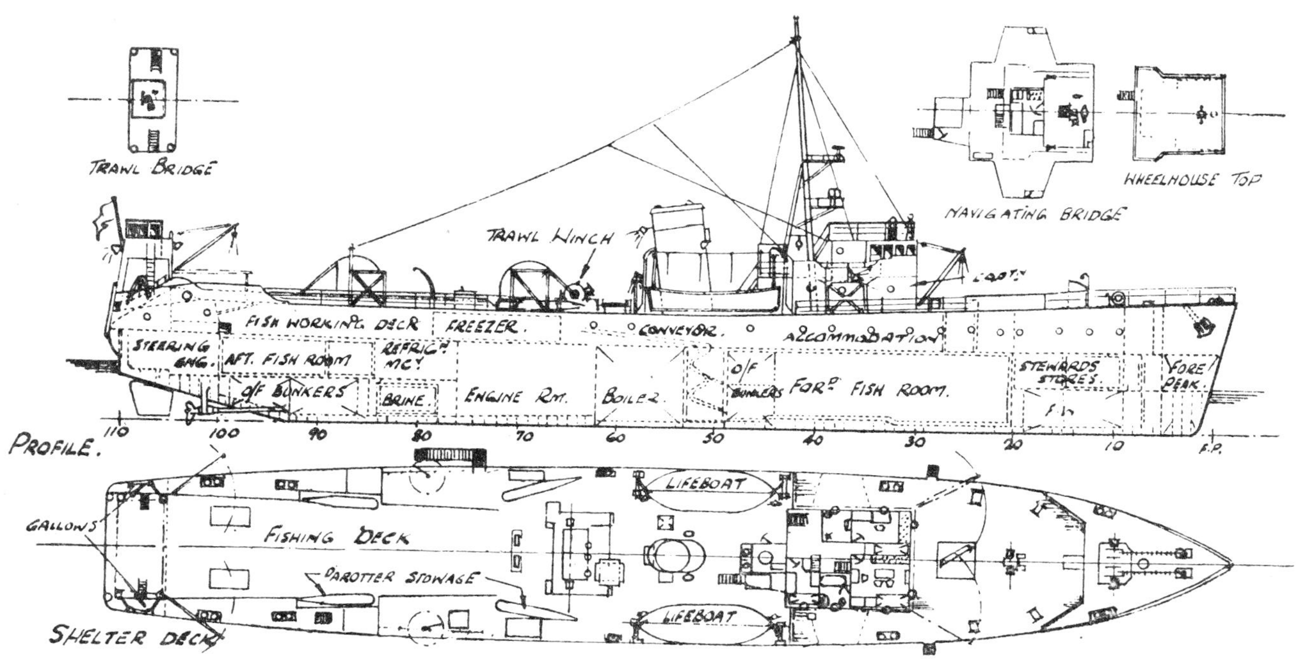

Fig. 602. Profile and shelter deck of Fairfree

FAIRFREE AND FAIRTRY

Mr. L. M. Harper Gow (U.K.): Before Messrs. Chr. Salvesen & Co. decided to risk substantial finance on building the *Fairtry*, they bought and converted an experimental vessel called the *Fairfree*, fig. 602 and 603. They thought that she might pay her way and they knew that they would learn a lot about factory ships through her. After three years of operating her they felt that they had reached the stage when they could plan a new factory ship which would meet most of the requirements. What they have thought right in a fish factory vessel is not necessarily right for everyone, because they have had to think of their specific conditions, i.e. market, location of fishing grounds, type of product wanted and many other things. Experience with the *Fairfree* convinced them that the fish factory vessel has certain advantages over the deep-sea trawler. They look upon it as a vessel which will compete with the deep-sea trawler, not with the inshore fishermen or even the middle water fishermen. One advantage is that the ship can be designed and built to have long endurance and the productive days on the fishing grounds can be extended far beyond those of the normal trawler. For example, the *Fairtry* has an endurance of 80 days, which allows, theoretically, four voyages a year with more than 60 days per voyage on the fishing grounds. To that must be added the possibility of getting a big haul of fish on the last few fishing days so that processing can be carried on during the homeward voyage.

It was important to find out if the fish frozen on the fishing grounds was worth having. That was not for the Company to say. They had to catch, process and sell the fish to find out what the wholesalers, retailers and the consumers thought. The answer, they discovered, is that, if a high quality is maintained, the fish is superior in texture, taste and smell, to almost any fish brought ashore in a wet state and then frozen. But that fact does not solve all problems because production at sea can go wrong very easily. It is not like buying a lot of standard quality fish in the market and freezing it. Fish have to be taken as they are caught and a constant effort must be made to keep up the standard of production. It is necessary to keep a very careful watch throughout all operations and never to lower quality for the sake of quantity; otherwise difficulties may be created and perhaps months afterwards sales will suffer.

Another advantage is that however long the factory ship remains out, the catch can always be brought home in good condition. There is no risk of having part of it condemned and consigned to the fish meal plant. That does happen on deep-sea trawlers, resulting in heavy losses. There is also the fact that the factory ship can visit distant fishing grounds more economically than can the deep-sea trawler because the range of the vessel is greater and fuel consumption is relatively lower than in the big steam trawler.

Finally, the factory ship makes it possible to escape the periods when fish are plentiful and when vessels come back to port and find it impossible to sell magnificent fish. That problem is avoided, but, on the other hand, the factory ship operator does not enjoy the very high, often ridiculously high prices, fetched by individual catches when there is a shortage.

These were the points that Messrs. Chr. Salvesen & Co. saw as advantages, but they felt that the disadvantages, although fewer, were just as serious. The first was the very high financial investment at the outset. It is far more expensive to build a factory vessel, with all its equipment, than a trawler. Then there is the high operating cost and, thirdly, in Great Britain, it is necessary to overcome the prejudice in the minds of many members of the public with regard to frozen fish.

Making the most of the advantages, and offsetting the disadvantages, is a matter of common sense applied specifically to the *Fairtry*, fig. 604. The vessel has been designed to give as much range as possible and to keep fishing all through the year. Time for steaming back and forth to the fishing grounds is cut to the minimum. The point concerning the quality of the catch is a matter of management, of careful checking of processes from start to finish.

Every owner of a factory ship will have to watch operating costs very carefully. Flexibility is necessary. There must be reserves to take and freeze all the fish caught. There must be no hold-up because of a bottle-neck somewhere. One must be able to deal with many types of catches and put them up in different ways. For example, a full catch should not be frozen in consumer size packages as that would require too large a crew. The variety of production sought in the *Fairtry* consists of large packs for processors, individually wrapped fillets in boxes, and consumer packs. The decision has to be made in accordance with the size and composition of the catch. Thus, large catches of cod should mainly be put up in 7 or 14 lb. (3.2 or 6.4 kg.) packages. All flat fish can be kept on ice and processed in consumer size packages when convenient, as for instance, when the weather is unfavourable for fishing. The policy should be to try to produce the most valuable product from the catch.

Problems of insufficient space apply particularly to a converted vessel. Nearly all previous factory ships have been converted vessels and converted vessels are never satisfactory, and that one fact tips the economic scales against them. The heavy cost of crews is an important point. In the U.K. when something new of this kind is being tried, the procedure is to discuss the problem with the labour unions. If they feel that the employers are being fair, they will agree to reasonable terms. Incidentally, Messrs. Chr. Salvesen & Co. feels that the rates of pay on a factory ship should be lower than on the ordinary trawlers because superior working conditions are provided indoors and the work is done in shifts. Men cannot work 16 hours a day doing detailed sorting, packing, freezing, and filleting. The work has to be carried out with care and the workers are required to work only so many hours a day, therefore there is scope to negotiate reasonable wage agreements.

With regard to high operating costs, the object is to try to keep the crews working every day. There is work for them on the way out, preparing everything, and work for them the whole way back, finishing off production and getting ready to discharge the cargo.

The *Fairtry* will have greater endurance (about 80 days at sea) than had the *Fairfree* but lower fuel consumption, 8¼ tons. Her carrying capacity is much increased (600 tons) and by the installation of filleting and freezing machinery sufficient to deal with very large catches (up to 40 tons a day), her catching capacity in periods of good fishing will not be limited by her processing capacity, as was the case on the *Fairfree*.

Mr. G. C. Eddie (U.K.): Referring to Cunningham's paper, the circulation of chilled sea water in factory deck ponds rather than refrigerating coils is a better method of cooling fish rapidly. Another method is to mix crushed ice with the fish. In handling the fish on a factory ship, the ideal process is continuous flow, not by batch.

Mr. W. Lochridge (U.K.): The *Fairfree* was a converted Navy vessel which had a stern chute for its fishing operations and an aft bridge from which fishing was directed. It fished satisfactorily except that there were difficulties in handling large catches. The limited processing facilities also created difficulties and some of the catch often had to be thrown overboard after 24 hours. Fishing was frequently stopped so that processing could catch up. The slipway of the *Fairtry* is in the water so that no lifting of the net is required and the trawl is dragged straight on board. A heavy catch cannot be taken on board in several bags as in an ordinary trawler so the trawl is equipped with two, three or even four cod ends, to distribute the load and thereby the catch can be taken on board in one haul. The fish are dropped directly through hatches into the fish ponds on the factory deck, generally in the side ponds, from which they are taken, bled and gutted, and transferred to the centre ponds. When the fish are stored they are springled with an ample amount of flake ice.

Air blast freezers are used for bulk freezing the fish, while plate freezers are used for the fish fillets packed in catering or consumer size containers.

Accommodation for the crew on the *Fairtry* is similar to that on a merchant ship. She has surplus accommodation for extra crew if experience proves that necessary.

The *Fairtry* is equipped with a Doxford opposed piston engine developing 1950 h.p. at 135 r.p.m. All accessories are electrically driven, powered from four 240 kw. diesel-driven generators placed forward on the main engine. The refrigerating machinery, using freon as its refrigerant, is of ample capacity and is placed aft of the main engine. Evaporators and brine pumps are located in a special insulated room and brine is used as the cooling medium for all refrigerated low temperature storage spaces.

The vessel carries drinking water for the full voyage as well as about half of the domestic and boiler feed water supply. An evaporator is installed so that the balance of fresh water required can be produced during the voyage.

OTHER FISH PROCESSING ABOARD SHIPS

Mr. E. R. Gueroult (France): In the 1930s the Greeks had ships in which the fish was frozen in brine. The process was continuous and the capacity was 15 to 20 tons a day. After that the fish was kept in a hold at a temperature of —6 deg. F.

(−21 deg. C.). At present, the French industry has devoted a few salt cod trawlers to freezing part of the catch. Up to a quarter of the hold is used for fillets. Maintenance of filleting machines is very expensive and as the crew can fillet up to 1,400 tons of cod by hand per voyage, the necessity of machines is questioned. The main problem from a design point of view is to find the right proportion between the catching and the processing capacity. For instance, a self-contained factory ship should not be bigger than 250 ft. (76 m.) in length because beyond that size trawling becomes very difficult. A distinction should also be made between the type of fish caught and processed—between high quality filleted fish, frozen and packaged and sold at a high price, and whole fish frozen in bulk. The quality fish finds a ready market in Europe and North America, but the bulk frozen cheaper fish are probably more suitable for the markets of underdeveloped countries. A 200 ft. (61 m.) French trawler is catching and freezing fish south of Madagascar. After a voyage of 5½ months this ship has brough back crayfish tails and whole frozen fish in excellent condition. The main features of present-day French designs are: trawling over the stern, diesel electric drive with all machinery forward, bunkers in the double bottom and cranes for handling nets and fish on the deck. A mother ship and catcher is a probable solution and may be devised by using the Norwegian whaling practice as an example. The difficulty, however, is to transfer the fish from the catcher to the mother ship in a moderate sea.

Mr. Hilmar Kristjonsson (FAO): Several of the newest Icelandic trawlers have freezing installations for processing part of the catch. In some instances, the plant is installed in an extension of the raised forecastle on the port side of the foredeck. Many of the trawlers also have fish meal plants in which they can process a great deal of the trash fish and offal which would otherwise have to be discarded.

Mr. I. Bromfield (U.S.A.): In the larger factory ship there is the advantage of using by-products. For instance, the head and bones, which form a base for chowder after being cooked could be put into gallon cans and sold to hotels and restaurants. In addition, of course, the offal when fresh can be used for pharmaceutical purposes and also processed into fish meal or fertilizer. There is much of value in the entrails of fish and no doubt they could be used if some thought were given to the problem of handling them.

Mr. L. M. Harper Gow (U.K.): Mr. Jul in his paper put his finger on the point when he mentioned fish meal. It is very doubtful if it pays to make fish meal on a freezer trawler. The capital cost is quite considerable, and installation of reduction equipment introduces the difficulty of using steam on board the ship which may not otherwise be needed with a diesel but, considering all factors, it probably does pay to make fish meal in a large factory ship producing fillets, although it is not certain that in every part of the world it would be advisable to fillet the catch. Where the catch is to be frozen after cleaning and gutting only, or frozen in the round, an owner would need to consider carefully before deciding to install a fish meal plant, particularly in a diesel vessel.

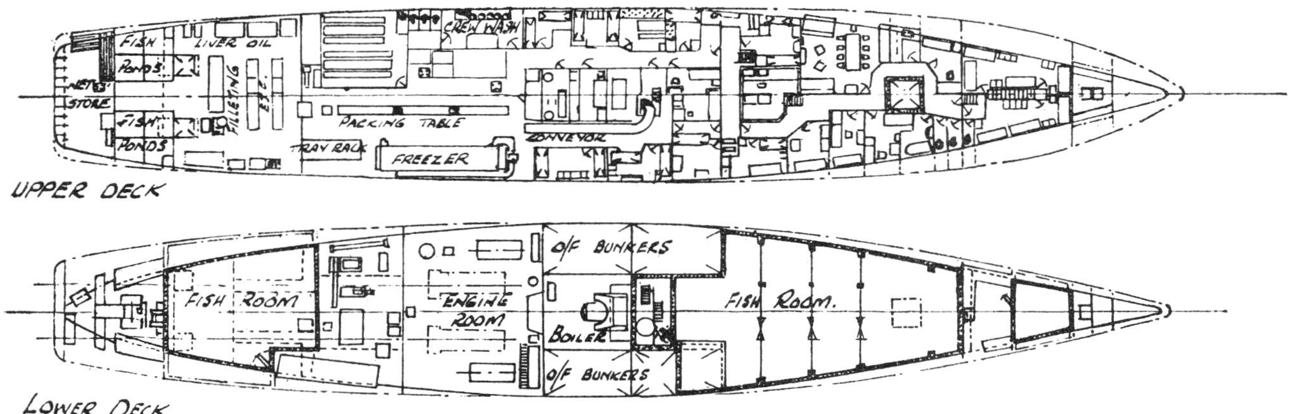

Fig. 603. Upper and lower deck of Fairfree

Mr. Hilmar Kristjonsson (FAO): An Icelandic floating reduction plant is installed on a 7,000 ton converted Great Lakes steamer called the *Haeringur*. She is equipped with California press reduction machinery. The main difficulty in operating her as a factory ship without connection with land is that water consumption is too high because direct steam is used for cooking the fish. The plant is said to have a capacity of 500 tons of fish per day but has only been operated at capacity for a total of a few weeks in its six years of existence. Almost all of this time the *Haeringur* was tied up to a pier during operations and obtained her water supply from the shore.

CONCLUSIONS

Mr. George C. Nickum (U.S.A.): A factory ship must be seaworthy and hull and machinery must be in good condition and acceptable to ship regulations. There must be adequate space for the processing plant and the extra factory crew, and considerable deck area for fishing or handling the fish. As such ships are intended to stay at sea for long periods, their endurance must be longer than common for fishing boats or freighters, i.e., a considerably larger proportion of the dead weight capacity must be devoted to fuel tanks, stores for provisions, etc. The type of engine must be chosen with a view to low fuel consumption; some consider that steam engines have already proved unfit for factory ship operations because of their high fuel consumption. Most of the equipment, processing machinery, etc., must be of special design to meet the difficulties of continuous sea movements, different trim according to different loading, corrosion by salt water and extreme limitation of space. A mother ship must, in addition, have stores, fuel supply, and repair and unloading facilities for the catchers.

Labour costs in a factory ship can be disastrous. This was the case of the *Pacific Explorer*. It is impossible to pay a man $700 (£250) a month to shake crabs on a trot-line, house him and take care of him, and still sell the catch at a low price. Similarly, one cannot pay $1,000 (£360) a month for a greaser when the ship is anchored off-shore, packing tuna, and expect to make a profit.

When approaching a factory ship scheme one should start with the premise that whenever one can operate a shore plant, where labour goes home at night and takes care of housing and feeding itself, it is more economical than processing on a ship where the owner has to provide quarters and food for the entire crew and pay them throughout the trip.

Mr. H. C. Hanson (U.S.A.): One of the reasons why the factory ship has not been very successful on the Pacific Coast

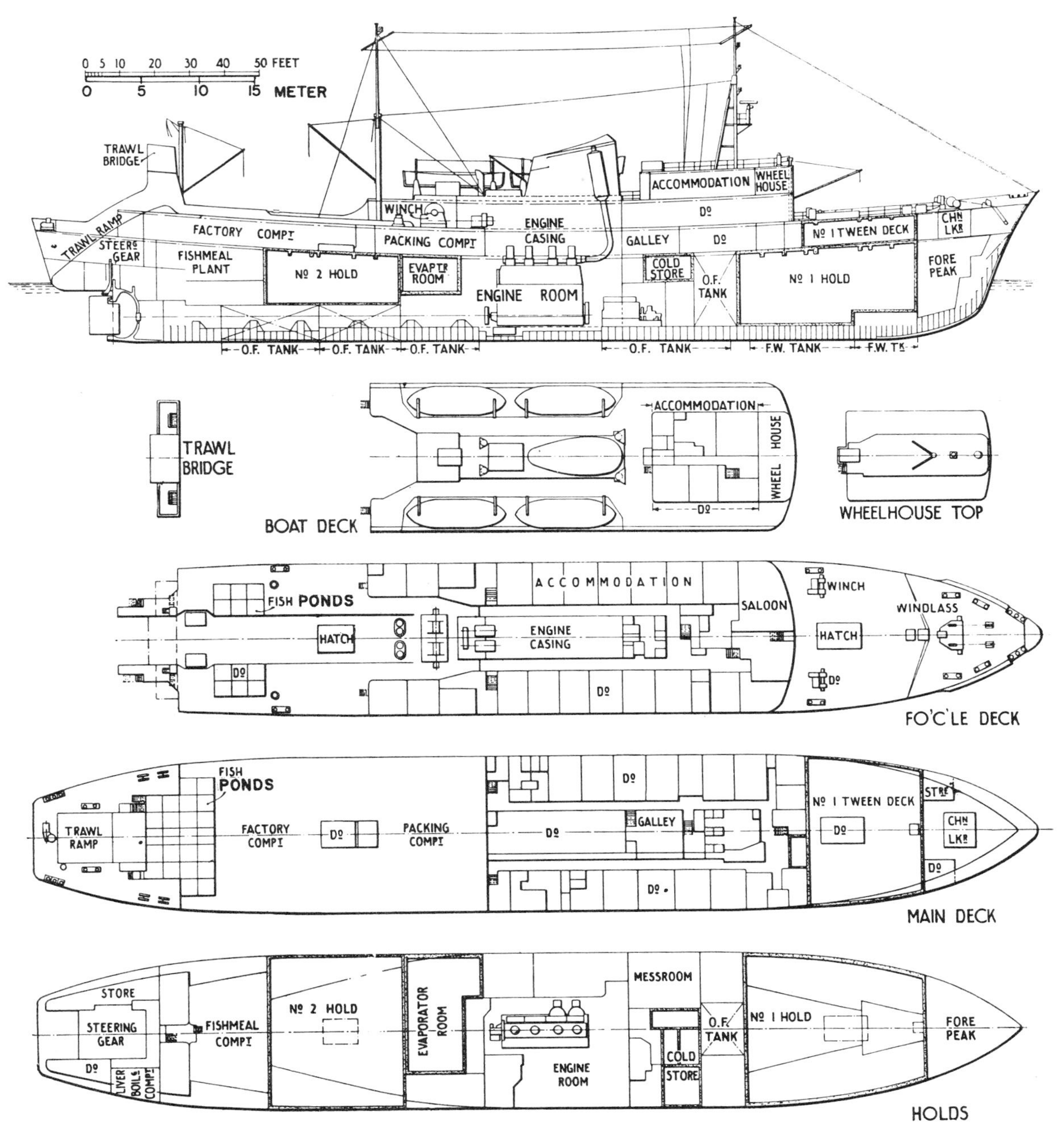

Fig. 604. Arrangement of Fairtry

of the United States is that some operators use too large a vessel with the consequent increase in labour required and a tremendous increase in labour costs. The very high pay, overtime and fringe benefits make the operation uneconomic.

Elsewhere in the world the factory ship could have a great part to play. For example, in the Gulf of Mexico the shrimp industry is wasting as much as 90 per cent. of its catch, retaining only the shrimp. The retention of the larger part of the catch for other purposes would be valuable. Again, in the Indonesian region and like areas the factory ship might prove most valuable, particularly where labour is plentiful.

Mr. A. Thurmer (Netherlands): As the operation of freezing factory ships is extremely costly, it is much more economical for Dutch long distance trawlers to salt the fish, and it is also easier for the Netherlands to find an export market for salted fish than for frozen fillets.

Mr. Ronald Kendell (U.K.): Attention is called to the main development in food technology, especially to the possibility of dehydrating fishery products. It is felt that a product has been developed which is so satisfactory that it may considerably change processing on board ship and would do away with the need for refrigerated holds. The system is also promising because the product is much easier to distribute than frozen fish.

Mr. Mogens Jul (FAO): Mr. Kendell has mentioned the possibility that new fish preservation processes, such as vacuum dehydration, might make freezing or canning ships obsolete, but it seems that no process so radically new is yet in sight.

Commander A. C. Hardy (U.K.): It is of no use landing first quality fish if the fishmonger does not know how to deal with it and British fishmongers have so far not shown sufficient care in the handling of fish. It seems necessary to build up a complete distribution system and to educate both distributors and consumers in dealing with the product.

Thirty years ago whale factory ships were converted cargo liners or tankers. From them gradually evolved the whale factory ship. It is probable that the day of the converted factory ship is over. Many of them have failed in the past chiefly because they were not designed first and foremost for the job. When conversions are indicated, the L.S.T. is a good basic hull design for mother ship operations.

The owners of the *Fairtry* do not necessarily have in mind exploitation of fishing grounds as known today. They have had great experience in whaling and it is quite possible that they intend to use the *Fairtry* in fishing in Antarctic waters. If Messrs. Chr. Salvesen & Co. are right and the operation of the *Fairtry* proves economically successful, will that lead to a marked change in trawling operations? Fish factory ships are being designed to deal with fish in the same way as the oil extraction plant deals with the oil. From the first crude refining operations, where only the essential parts of the oil were used, processes have been improved so that now every single fraction of the oil is utilized. It appears that the factory ship can take care of all parts of the fish in the same manner.

There will always be different types of factory ships. Some will be completely self-contained, others will fish with dories. It does seem that there is a place and a function for factory ships, but they need to be designed and built as such, not merely converted from former trawlers or other types of vessels.

Mr. I. Bromfield (U.S.A.): There are many plans and designs for proposed factory ships, some of them misleading. The idea that such a ship can go out and catch perhaps 300,000 lb. (135 ton) of fish in a week or so and fillet them is all wrong. There is no guarantee that fish will be found in the right quantity or of the right quality or species. In a day's fishing it might be possible to catch 100,000 lb. (45 ton) or maybe catch less than 5,000 lb. (2.25 ton). It is impossible to make a sound plan for a factory ship on the basis that a certain quantity of fish will be caught in a stated time.

Mr. E. R. Gueroult (France): There are more factory ships on paper than at present afloat. Experiments in design and in coping with the various problems connected with the ships must go on and will be solved.

Mr. Mogens Jul (FAO): Some have said that the factory ship is the fishing boat of the future. People have also said that the aeroplane is the means of transport of the future. Both may be true, but only in this sense: for a long time to come the world will probably have cargo liners, river barges, bicycles and mules as well as aeroplanes. In the same way it will have factory ships along with trawlers, purse seiners, long-liners, catamarans and other fishing craft.

REFERENCES

Allan, J. F. Research on Design of Drifters. Transactions
1951 of the Institution of Naval Architects, London, S.W.1, Vol. 93, No. 1, p. 56, 16 pp., illus. (104, 361)*

——— Improvements in Ship Performance. Transactions of
1953 the Institute of Marine Engineers, London, E.C.3, Vol. 65, No. 5, p. 117, 16 pp., illus. (143, 319, 354)

Anderson, A. W. Technological Development in Fisheries
1951 with Special Reference to the Factory Ship in the United States. Proceedings of the United Nations Scientific Conference on the Conservation and Utilization of Resources, New York, Vol. 7. (553, 556)

A New Type of Lifeboat. The Motorboat and Yachting,
1946–1950 London, E.C.1, Feb., 1946, p. 52, 2 pp. The New Mumbles Lifeboat, same publication, Aug. 1947, p. 294, 3 pp. For Life-saving and General Duties, Nov. 1949, p. 454. R.N.L.L. Lifeboats of To-Day, Mar. 1950, p. 82, 3 pp. (68)

A Stainless Steel Boat. The Welder, Waltham Cross, Herts,
1938 U.K., p. 178, 2 pp. (68)

Astrup, N. C. Modellforsøk med Fiskefartøyer I (Model
1951 Tests with Fishing Boats). Skipsmodelltanken, Trondheim, Norway, No. 6, 39 pp. In Norwegian with English Summary. (318, 360)

——— Modellforsøk med Fiskefartøyer III (Propeller
1952 Tests with Fishing Boats). Skipsmodelltanken, Trondheim, Norway, No. 14, 82 pp., illus. In Norwegian, with English summary. (281)

Ayre, A. L. Approximating E.H.P.; Revision of Data Given
1948 in Papers of 1927 and 1937. Transactions of the North East Coast Institution of Engineers and Shipbuilders, Newcastle-upon-Tyne, U.K., 13 pp., illus. (281)

Benjamin, L. Über das Mass der Stabilität der Schiffe.
1913 (On the Measure of the Stability of Ships). Jahrbuch der Schiffbautechnischen Gesellschaft 1914, Hamburg 56, Germany, p. 594, 20 pp., illus. In German. (334)

Bjursten, G. Normer och forskning i USA rörande spik-
1947 förband (Standard USA Specifications and Research about Nail Fastenings). Statens Kommitte för Byggnadsforskning, Stockholm, Sweden, 41 pp. In Swedish. (71)

Butler, C., Punchochar, J. F., and Knake, B. O. Freezing
1952 Fish at Sea—New England. Part 3—The Experimental Trawler *Delaware* and Shore Facilities. Commercial Fisheries Review, Washington 25, D.C., U.S.A., Vol. 14, No. 2, p. 16, 10 pp., illus. (535)

* The figures in brackets are the page numbers of this book where the references have been quoted.

Butler, C., and Magnusson, H. W. Freezing Fish at Sea—
1952 New England. Part 4—Commercial Processing of Brine-Frozen Fish. Commercial Fisheries Review, Washington 25, D.C., U.S.A., Vol. 14, No. 2, p. 26. (535)

Cannery–Neva–Freezer. Pacific Fisherman, Seattle, Wash.,
1950 U.S.A., Vol. 48, No. 11, p. 26, 8 pp., illus. (554)

Carlson, C. B. Experimental Purse Seining for Menhaden
1945 with the *Jeff Davis*. Fishery Market News, Washington 25, D.C., U.S.A., Vol. 7, No. 5a, supplement, 48 pp. (477)

Change in Stern Design. The Motorboat and Yachting
1948 Manual, London, E.C., p. 20, illus. (57)

Carr, F. G. G. Types of Coastal Craft of the British Isles.
1934 University Press, Cambridge, U.K., p. 5, 15 pp. (57)

Christensen, A. N. Bokholdert og Kalkyle (Accounting and
1948 Estimation). Unda Maris 1947–48, Sjöfartsmuseet, Göteborg, Sweden, Vol. 6/7, p. 257, 23 pp., illus. In Norwegian. (241)

Cleaver, F., and Shimada, B. Japanese Skipjack (Katsu-
1950 wonus Pelamis) Fishing Methods. Commercial Fisheries Review, Washington 25, D.C., U.S.A., Vol. 2, No. 11. (495)

Clement, P. Le Chalutier congélateur *Marbrouk* (The
1951 Freezing Trawler *Marbrouk*). La Revue générale du Froid, Paris, France, No. 12, p. 1163, 11 pp., illus. In French with English and Spanish summaries. (557)

Cunningham, D. B. Notes on Trawl Fishing. The Institu-
1949 tion of Engineers and Shipbuilders in Scotland, Glasgow, Scotland, 70 pp., illus. (151)

Cutting, C. L., Eddie, G. C., Reay, G. A., and Shewan,
1953 J. M. The Care of the Trawler's Fish. H.M.S.O., London, 14 pp. (264)

Dassow, J. A. Freezing Gulf-of-Mexico Shrimp at Sea.
1954 Commerical Fisheries Review, Washington 25, D.C., U.S.A., Vol. 16, No. 7, p. 1. (557)

Deep Sea—a Pioneering Venture. Pacific Fisherman, Seattle,
1947 Washington, U.S.A., Vol. 45, No. 7. (555)

De Reddingboot (The Lifeboat). Mededelingen van de
1952 Koniklijke Noord- en Zuid-Hollandsche Redding-Maatschappij, Amsterdam, Netherlands, No. 73, 20 pp. In Dutch. (68)

Dieudonne, M. Discussion on Paper by V. Albiach—
1951 Determination rapide de caracteristiques de helice de remorqueurs (Rapid Determination of Propeller Characteristics in Tugs)—given at 4th Congrès International de la Mer, Ostend, Belgium, p. 446. In French. (451)

Drum Seining. Pacific Fisherman, Seattle, Washington, 1953 U.S.A., Vol. 51, No. 8, p. 25, 4 pp. (477)

EDWARD, J. and TODD, F. H. Steam Drifters: Tank and Sea 1938 Tests. Transactions of the Institution of Engineers and Shipbuilders in Scotland, Glasgow, U.K., p. 51, 57 pp., illus. (281).

EICHELBERG, G. Les générateurs de gaz à pistons libres (Free 1948 Piston Gas Generators). Le Génie Civil, Paris, France, 15 Nov., 1 Dec., and 16 Dec. 1948. In French. (429)

EICHLER, H. Deutsche Fischereifahrzeuge IV (German 1942 Fishing Vessels IV). Schiffbau, Berlin S.W. 68, Germany, Vol. 43, p. 543, 7 pp., illus. In German. (153)

Fairtry Goes to Sea. World Fishing, London, E.C.4, Vol. 3, 1954 No. 6, p. 209, 5 pp., illus. (564)

Fairtry—World's Largest Trawler. World Fishing, London, 1953 E.C.4, Vol. 2, No. 9, p. 353, 6 pp., illus. (564)

FICK, H. Zum Problem des Fischerei-Fabrikschiffes (About 1953 the Problem of Factory Ships). Die Fischwirtschaft, Bremerhaven-F., Germany, Vol. 5, No. 12, p. 281. In German. (554)

FIEDLER, R. H. The Factory Ship—Its Significance to our 1937 World Trade and Commerce. Transactions of the American Fisheries Society, Washington, D.C., U.S.A., Vol. 66, p. 427, 15 pp. (556)

Fischmehlfabrik an Bord (Fish Meal Processing on Board). 1952 Allgemeine Fischwirtschaftszeitung, Bremerhaven-F, Germany, Vol. 4, No. 5, p. 1. In German. (558)

Fish-Meal Plant Afloat. Ship and Boat Builder, London, 1952 E.C.4, Vol. 5, No. 8, p. 277 (558)

Floating Processing Ship—*Oceanlife* Makes Boston Début. 1950 Fishing Gazette, New York, Vol. 67, No. 8, p. 28, 4 pp., illus. (562)

Freezership Revolution. Pacific Fisherman, Seattle, Washing-1951 ton, U.S.A., Vol. 49, No. 8, p. 15, 6 pp. (557)

GRAFF, W., and HECKSCHER, E. Widerstands- und Stabilitäts-1941 versuche mit drei Fischdampfermodellen (Resistance and Stability Test with Three Trawler Models). Werft, Reederei, Hafen, Hamburg, Germany, Vol. 22, No. 8, p. 115, 6 pp., illus. In German. (145)

GRIM, O. Das Schiff in von achtern auflaufender See (The 1951 Ship in High Following Seas). Yearbook of the Schiffbautechnische Gesellschaft, Hamburg,Germany. In German. (334)

——— Rollschwingungen, Stabilität und Sicherheit in 1952 Seegang (Rolling, Stability and Safety in a Seaway). Research Studies in Shipbuilding Technique, Hamburg, Germany, No. 1. In German. (334)

HAALAND, A. Noen systematiske Formvariasjoner med 1951 Modeller av Lokalskip og deres Innflytelse på Motstands- og Framdriftsforholdene (Some Systematic Form Variations on Local Ship Models and their Effect upon Resistance and Propulsion). Skipsmodelltanken, Trondheim, Norway, No. 4, 40 pp., illus. In Norwegian, with English Summary. (318)

HANSON, H. C. Small Craft Construction and Design. Trans-1951 actions of the Society of Naval Architects and Marine Engineers, New York 18, Vol. 59, p. 598. (194, 338)

Hanstholm Udvalget 1946: Hanstholm Havn, hvorfor den 1946 bør fuldføres (Hanstholm Harbour, Why It Should Be Finished), Thisted, Denmark. 31 pp., illus. In Danish. (55)

HARDY, A. C. The Floating Fish Factory—an Important 1953 New Ship Type. World Fishing, London, E.C.4. Vol. 2, Nos. 1–2, pp. 5, 60. (556)

——— Seafood Ships. Lockwood and Son Ltd., London, 1947 248 pp., illus. (556)

HARTSHORNE, J. C., and PUNCOCHAR, J. F. Studies on Round 1951 Fish Frozen at Sea, New England Species. Food Technology, Champaign, Ill., U.S.A., Vol. 5, No. 12, p. 492. (535)

——— Studies on Round Fish Frozen at Sea, New England 1951 Species. Food Technology, Vol. 5, No. 12, p. 492, 4 pp. Also appeared under the title " Freezing fish at sea—New England. Part 1—Preliminary Experiments " in Commercial Fisheries Review, Washington 25, D.C., U.S.A., Vol. 14, No. 2, p. 1, 7 pp., 1952. (535)

HEARMON, R. F. S. The Elasticity of Wood and Plywood. 1948 H.M.S.O., London, S.E.1, 87 pp. (71)

HECKSCHER, E. Form- und Antriebsgestaltung des 22 m. 1941 Motor-Fischkutters (Shape and Propulsion of 22 m. Motor Fishing Boat). Werft, Reederei, Hafen, Hamburg, Germany, Vol. 22, No. 21, p. 315, 5 pp., illus. In German. (281)

HENDERSON, F. Y. Timber. Crosly Lockwood and Son, 1948 Ltd., London, E.C.4, 148 pp. (71)

HIGHMAN, J. B., and IDYLL, C. P. Holding Fresh Shrimp in 1952 Refrigerated Sea Water. Proceedings of the Gulf and Caribbean Fisheries Institute, Fifth Annual Session, Marine Laboratory, University of Miami, U.S.A. (479)

HOLST, H. Spikade och limmade träkonstruktioner (Nailed 1952 and Glued Wood Structures). Teknisk Tidsskrift, Stockholm, Sweden, p. 939, Dec. 1952. In Swedish. (71)

HOLSTON, J., and POTTINGER, S. R. Freezing Fish at Sea—1954 New England, Part 8. Some Factors affecting the Salt (Sodium Chloride) Content of Haddock during Brine-Freezing and Water Thawing. Commercial Fisheries Review, Washington 25, D.C., U.S.A., Vol. 16, No. 8, p. 1. (535)

HOLT, W. J. Admiralty Type Motor Fishing Vessels (M.F.V.s). 1946 Transactions of the Institution of Naval Architects, London, S.W.1, 14 pp., illus. (281)

HUNTER, A. The Art of Trawler Planning. Ship and Boat 1953 Builder, London, E.C.4, No. 2, p. 257, 6 pp., illus. (319)

JACKSON, H. E. Puget Sound Naval Shipyard Developes 1952 New 36-Foot Plastic Landing Craft. Marine Engineering and Shipping Review, New York 7, Oct. 1952, p. 82, 4 pp., illus. (69, 72)

JAEGER, H. E. Idées sur la flotte de pêche future (Ideas for
1946 the Future Fishing Fleet). Transactions du Congrès de la Mer, Ostend, Belgium. In French. (153)

——— Moderne Fartøjer til Saltvandsfiskeri (Modern Sea
1949 Fishing Boats). Ingeniøren, Copenhagen V, Denmark Vol. 58, No. 29, p. 591, 9 pp., illus. In Danish. (153, 543)

JAHN, W., KÜHL, H., and OSTERNDORF, A. Über die Bohr-
1951 muschel Teredo in der Elbemündung (On the Borer Teredo in the Elbe Estuary). MS. privately published, 36 pp. In German. (228)

JAPAN, FISHERIES AGENCY. Illustrations of Japanese Fishing
1952 Boats, Tokyo, Japan, 311 pp., illus. (80, 137, 424, 503)

JOHOW–FORSTER. Hilfsbuch für den Schiffbau (Shipbuilding
1928 Handbook). V Edition. Springer Verlag, Berlin, Germany. In German. (344)

JUNE, F. C. Preliminary Fisheries Survey of the Hawaiian-
1951 Line Islands. Part III—" The Live Bait Skipjack Fishery of the Hawaiian Islands ". Commercial Fisheries Review, Washington 25, D.C., U.S.A., Vol. 13, No. 2. (494)

KEMPF, G. Die Stabilitätsbeanspruchung der Schiffe durch
1938 Wellen und Schwingungen. (The Effect on Stability in Ships of Waves and Vibration). Werft, Reederei, Hafen, Hamburg, Germany, No. 13. In German. (334)

KENT, J. The Design of Seakindly Ships. Transactions of the
1950 North-East Coast Institution of Engineers and Shipbuilders, Newcastle upon Tyne, U.K., Vol. 66, p. 417, D. 159, 40 pp., illus. (216)

KÜHL, H. Über die Wirkung von Kontaktinsektiziden in
1948 Bewuchsschutzanstrichen (On the Efficiency of Contact Insecticides in Anti-Fouling). Akademische Verlagsgesellschaft, Leipzig, Germany, p. 183. In German. (229)

——— Über die Besiedlung Metallischer Oberflächen durch
1950 Balanus Improvisus Darwin (On the Attachment of Balanus Improvisus to Metallic Surfaces). Neue Ergebnisse u. Probleme der Zoologie, Akademische Verlagsgesellschaft, Leipzig, Germany, p. 461, 17 pp. In German. (229)

LANDMARK, J. Om Kjøling av Fisk i. Traalere (On Chilling of
1950 Fish in Trawlers). Kulde, Copenhagen, Denmark, Vol. 4, No. 4, p. 48. In Norwegian. (265)

LANTZ, A. W. Experiments with the Use of Chilled Sea
1953 Water in Transporting Fish. Western Fisheries, Vancouver, B.C., Vol. 46, No. 6, p. 16. (265)

La pêche aux îles St. Pierre et Miquelon (Fisheries on the
1953 Islands St. Pierre and Miquelon). La pêche maritime, la pêche fluvialè et la pisciculture, Paris 8è, France, p. 15, 3 pp., illus. In French. (70)

LEHMANN, G. Technisch-Wirtschaftliche Fragen zum Tief-
1953 kühl-Fabriktrawler (Technical-Economic Questions of the Freezer Factory Trawler). Die Fischwirtschaft, Bremerhaven F., Germany, Vol. 5, Nos. 1–2, pp. 21, 51. In German. (553, 554)

LOCHRIDGE, W. *Fairfree*. Paper No. 1130 presented to the
1950 Institution of Engineers and Shipbuilders in Scotland, Glasgow, U.K. (538)

LUND, S. O., and KRAMHØFT, O. Fiskefryseskibe. Ingeniøren,
1953 Copenhagen, Denmark, Vol. 62, No. 33, p. 599. In Danish. (553, 557)

MACCALLUM, W. A. Refrigeration and Storage of Fish in
1953 Trawlers. The Engineering Journal, Montreal, Canada, Vol. 36, No. 6, p. 707. (231)

——— Pen Construction for Refrigerated Fish-Holds.
1954 Progress Reports of the Atlantic Coast Stations, Halifax, N.S., Canada, No. 55, p. 27 (231, 265)

Made in America—*Alma Ata*—Russia's Great Floating
1946 Cannery. Pacific Fisherman, Seattle, Washington, U.S.A., Vol. 44, No. 4, p. 35. (553)

MAGNUSSON, H. W. Equipment and Procedures for Thawing
1952 Fish Frozen at Sea. Commercial Fisheries Review, Washington 25, D.C., U.S.A., Vol. 14, No. 7, p. 18, 2 pp. (535)

MAGNUSSON, H. W., and HARTSHORNE, J. C. Freezing Fish
1952 at Sea—New England. Part 5—Freezing and Thawing Studies and Suggestions for Commercial Equipment. Commercial Fisheries Review, Washington 25, D.C., U.S.A., Vol. 14, No. 12a, p. 8, 15 pp. (535)

MAGNUSSON, H. W., POTTINGER, S. R., and HARTSHORNE, J. C.
1952 Freezing Fish at Sea—New England, Part 2—Experimental Procedures and Equipment. Commercial Fisheries Review, Washington 25, D.C., U.S.A., Vol. 14, No. 2, p. 8, 7 pp. (535)

Marine Fouling and Its Prevention, United States Naval
1952 Institute, Annapolis, Maryland, U.S.A., 398 pp., illus. (228)

MEYER, A. Fischmehlfabrik an Bord? (Fish Meal Processing
1952 on Board?). Allgemeine Fischwirtschaftszeitung, Bremerhaven F., Germany, Vol. 4, No. 12, 1 p. In German. (558)

MÖCKEL, W. Über die Stabilitätsgefährdung von Fischdampf-
1949 ern bei hohem Seegang von achternlichen See (Danger to the Ship Through Reduced Stability in High Following Seas). Schiff und Hafen, Hamburg, Germany, No. 1. In German. (326)

NORDSTRØM, H. F. Försök med fiskebåtsmodeller (Tests
1943 with Fishing Boat Models). Meddelanden från Statens Skeppsprovningsanstalt, Göteborg, Sweden, No. 2, 32 pp., illus. In Swedish, with English summary. (281)

NORÉN, B. Bultförband i trä (Bolted Wood Fastenings).
1948 Teknisk Tidsskrift, Stockholm, Sweden, p. 535, 4 pp. In Swedish. (71)

Norway Fiasco. The Fishing News, London, Vol. 35, No.
1947 1792, p. 1, Short notice. (556)

PEILLON, M. L. Le rendement des générateurs de gaz à pistons
1949 libres. (The Output of Free Piston Gas Generators). Paper presented at 1949 Session of A.T.M.A. (429)

PIERROTTET, E. A Standard of Stability for Ships. Transactions of the Institution of Naval Architects, London, S.W.1, pp. 208, 14 pp., illus. (344)
1935

Plastics in Boatbuilding. The Motor Boat and Yachting, London, E.C.1, p. 42, Jan. 1951 and p. 171, April 1953 (72)
1951–1953

POTTINGER, S. R., HOLSTON, J., and MCCORMACK, G. Fish Frozen in Brine at Sea: Preliminary Laboratory and Taste-Panel Tests. Commercial Fisheries Review, Washington 25, D.C., U.S.A., Vol. 14, No. 7, p. 20, 4 pp. (535)
1952

POTTINGER, S. R., KERR, R. G., and LANHAM, W. B., Jr. Effect of Refreezing on Quality of Sea Trout Fillets. Commercial Fisheries Review, Washington 25, D.C., U.S.A., Vol. 11, No. 1, p. 14, 3 pp. (535)
1949

Pumping Live Fish Simplifies Problem of Bait Handling Aboard Clipper. Pacific Fisherman, Seattle, Wash., U.S.A., Vol. 51, No. 8, p. 55, 3 pp. (497)
1953

PUNCOCHAR, J. F. Freezing Fish at Sea for Later Thawing, Processing and Refreezing Ashore. Fishing Gazette, New York 1, Vol. 65, No. 12, p. 206, 210. (535)
1949

——— Freezing Round Fish Aboard Boats. Atlantic Fishermen, Golfstown, New Hampshire, U.S.A., Vol. 30, No. 5. (535)
1949

PUNCOCHAR, J. F., and POTTINGER, S. R. Freezing Fish at Sea. Food Technology, Champaign, Illinois, U.S.A., Vol. 7, No. 10, p. 408. (535)
1953

RAHOLA, J. The Judging of the Stability of Ships and the Determination of the Minimum Amount of Stability. Yhteiskirjapaino Osakeyhtiö, Helsinki, Finland, 232 pp., 81 illus. In English. (151, 344, 367)
1939

RAWLINGS, J. E. A Report on the Cuban Tuna Fishery. Commercial Fisheries Review, Washington 25, D.C., U.S.A., Vol. 15, No. 1 (494)
1953

Report of the 4th International Lifeboat Conference. Swedish Lifeboat Association, Gothenburg, Sweden, 1936. In Swedish. (68)
1936

Report of the 5th International Lifeboat Conference. Norwegian Lifeboat Institute, Oslo, Norway. In English. (68, 69)
1947

Report of the 6th International Lifeboat Conference. Belgian Marine Administration, Brussels, Belgium. 246 pp. In English. (68, 72)
1951

ROORDA, A. Beoordeling van de metacenterhoogte MG en keuze van een passende MG bij het ontwerpen van zeeschepen (Calculation of GM and Device for a Suitable GM in the Design of Ships). Schip en Werf, Rotterdam, Netherlands, Vol. 11, p. 39, 3 pp., illus. In Dutch. (151)
1944

ROSCHER, E. K. Zum Problem des Fischerei-Fabrikschiffes (The Factory Ship Problem). Die Fischwirtschaft, Bremerhaven-F, Germany, Vol. 5, No. 10, p. 238. In German. (553, 554)
1953

RUPP, L. A. Controllable Pitch Propellers. Transactions of the Society of Naval Architects and Marine Engineers, New York 18, Vol. 56, p. 272, 87 pp., illus. (218)
1948

SAUNDERS, H. E. The Multiple Skeg Stern for Ship. Transactions of the Society of Naval Architects and Marine Engineers, New York 18, Vol. 55, p. 97, 73 pp., illus. (217)
1947

SCHAFFRAN, K. Modellversuche für Fischereifahrzeuge (Model Tests with Fishing Vessels). Schiffbau, Vol. 22, Nos. 22–23, p. 507; same Vol., No. 24, p. 541; same Vol., No. 25, pp. 578–84; same Vol., No. 27, pp. 633–641; same Vol., No. 42, pp. 1028–37; same Vol., No. 43, pp. 1071–76; same publication, Vol. 23, No. 1, pp. 6–11; same Vol. as above, No. 2, pp. 43–50. Illus. In German. (281)
1921–1922

SCHLEUFE, F. Der Fischdampferbau (Construction of Steam Trawlers). Clasing und Co., Bielefeld, Germany, 90 pp., illus. In German. (151)
1948

SIEBENALER, J. B. Trap Lift Net for Catching Tuna Bait Fishes. Commercial Fisheries Review, Washington 25, D.C., U.S.A., Vol. 15, No. 8. (497)
1953

SIMPSON, D. S. Small Craft, Construction and Design. Transactions, Society of Naval Architects and Marine Engineers, New York 18, Vol. 59, p. 554, 58 pp., illus. (338)
1951

SMITH, H. M. The Design and Construction of Small Craft. Association of Engineering and Shipbuilding Draughtsmen, London, 299 pp., illus. (127)
1924

SNYDER, G. C. Stability of Tuna Clippers. Paper read 3 May, 1946 before Pacific North-west Section of Society of Naval Architects and Marine Engineers, New York 18, Photostat. (320)
1946

Society of Naval Architects and Marine Engineers. Model Resistance and Expanded Resistance Data Sheets 1 to 150. New York 18, illus. (281)
1950–1953

SPANNER, W. F. Some Notes on the Design of Trawlers and Drifters, With Particular Reference to Sea-worthiness and Stability. Transactions of the Institution of Naval Architects, London S.W.1, p. 32, 17 pp., illus. (151, 212)
1946

STEVENS, W. C., and TURNER, N. Solid and Laminated Wood Bending. H.M.S.O., London, S.E.1, 71 pp. (71)
1948

SUND, E. Modellforsøk med Fiskefartøyer II (Model Test with Fishing Vessels II). Skipsmodelltanken, Trondheim, Norway, No. 7, 31 pp., illus. In Norwegian, with English summary (318)
1951

Svetsaren (Welding). ESAB, Göteborg, Sweden, No. 38, p. 1067, 5 pp. In Swedish. (68)
1943

SYMONDS, R. P., and TROWBRIDGE, H. O. The Development of Beam Trawling in the North Atlantic. Transactions, Society of Naval Architects and Marine Engineers, New York 18, Vol. 55, p. 359, 26 pp., illus. (127)
1947

TAKAGI, A., INUI, T., and NAKAMURA, S. Graphical Methods for Power Estimation of Fishing Boats. Fisheries Agency, Tokyo, Japan, 47 pp., illus. In English, with Japanese summary. (357)
1950

TAYLOR, A. R. Fishing Vessel Design. Transactions of the Institution of Naval Architects, London, S.W.1, Vol. 85, p. 95, 8 pp., illus. (152, 322)
1943

REFERENCES

TAYLOR, D. W. The Speed and Power of Ships. U.S. Government Printing Office, Washington, D.C., U.S.A., 3rd edition, 2nd revision, 310 pp. (72, 357)
1943

TRAUNG, J.-O. Några erfarenheter från tankförsök med fiskebåtar (Some Experiences from Fishing Boat Tank Tests). Unda Maris 1947–48, Sjöfartsmuseet, Göteborg, Sweden, p. 83, 70 pp., illus. In Swedish. (281)
1948

——— Improving the Design of Fishing Boats. FAO Fisheries Bulletin, Rome, Italy, Vol. 4, No. 1–2, 26 pp., illus. In English, French and Spanish. (105, 281)
1951

VAN LAMMEREN, W. P. A., TROOST, L., and KONING, J. G.
1948 Resistance, propulsion and Steering of Ships. The Technical Publishing Company H. Stam, Haarlem, Netherlands, 366 pp., illus. In Dutch or English. (212)

VENUS. Marine Developments in Aluminium. Transactions
1951 of Engineers and Shipbuilders in Scotland, Glasgow, U.K. (223)

VOLL, A., and WALDERHAUG, H. A. A. Slepe- og propulsjons-
1952 forsøk med modeller av fjordbusser og hvalbåter (Towing and Resistance Tests with Fjord Bus and Whale Catcher Models). Skipsmodelltanken, Trondheim, Norway, No. 13, 57 pp., illus. In Norwegian, with English summary. (318)

WIGUTOFF, N. B., and CARLSON, C. B. *Pacific Explorer.*
1950 Part 5—1948. Operation in the North Pacific and Bering Sea. Fishery Leaflet, U.S. Fish and Wildlife Service, Washington 25, D.C., U.S.A. No. 361. (513, 554)

WHITE, E. W. British Fishing Boats and Coastal Craft,
1950 Part I (54 pp.) and Part II (48 pp.). Science Museum, H.M.S.O., London, S.E.1, (57)

WINDOLF. Der Bohrwurmbefall an Holzbauten in den
1936 Cuxhavener Häfen. (The Attack of Teredos on Timber Work in Cuxhaven Docks). Werft, Reederei, Hafen, Hamburg, Germany, Vol. 17, No. 24, p. 404. In German. (227)

Wood Handbook. U.S. Department of Agriculture, Washing-
1940 ton, 25, D.C., U.S.A., 325 pp. (71)

INDEX

Other books published by Fishing News Books Limited Farnham, Surrey, England

Free catalogue available on request

A living from lobsters
Advances in aquaculture
Aquaculture practices in Taiwan
Better angling with simple science
British freshwater fishes
Coastal aquaculture in the Indo-Pacific region
Commercial fishing methods
Control of fish quality
Culture of bivalve molluscs
Eel capture, culture, processing and marketing
Eel culture
European inland water fish: a multilingual catalogue
FAO catalogue of fishing gear designs
FAO catalogue of small scale fishing gear
FAO investigates ferro-cement fishing craft
Farming the edge of the sea
Fish and shellfish farming in coastal waters
Fish catching methods of the world
Fish farming international No. 2
Fish inspection and quality control
Fisheries oceanography
Fishery products
Fishing boats of the world 2
Fishing boats of the world 3
Fishing ports and markets
Fishing with electricity
Fishing with light
Freezing and irradiation of fish
Handbook of trout and salmon diseases
Handy medical guide for seafarers
How to make and set nets
Inshore fishing: its skills, risks, rewards
International regulation of marine fisheries: a study of regional fisheries organizations
Marine pollution and sea life
Mechanization of small fishing craft
Mending of fishing nets
Modern deep sea trawling gear
Modern fishing gear of the world 1
Modern fishing gear of the world 2
Modern fishing gear of the world 3
Modern inshore fishing gear
More Scottish fishing craft and their work
Multilingual dictionary of fish and fish products
Navigation primer for fishermen
Netting materials for fishing gear
Pair trawling and pair seining – the technology of two boat fishing
Pelagic and semi-pelagic trawling gear
Planning of aquaculture development – an introductory guide
Power transmission and automation for ships and submersibles
Refrigeration on fishing vessels
Salmon and trout farming in Norway
Salmon fisheries of Scotland
Seafood fishing for amateur and professional
Ships' gear 66
Sonar in fisheries: a forward look
Stability and trim of fishing vessels
Testing the freshness of frozen fish
Textbook of fish culture; breeding and cultivation of fish
The fertile sea
The fish resources of the ocean
The fishing cadet's handbook
The lemon sole
The marketing of shellfish
The seine net: its origin, evolution and use
The stern trawler
The stocks of whales
Training fishermen at sea
Trawlermen's handbook
Tuna: distribution and migration
Underwater observation using sonar